HISTORY OF BIOLOGY

Year	Name	Country	Contribution
1628	William Harvey	Britain	Demonstrates that the blood circulates and the heart is a pump.
1648	Jan B. van Helmont	Belgium	Shows that plants derive little substance from the soil.
1665	Robert Hooke	Britain	Uses the word *cell* to describe compartments he sees in cork under the microscope.
1668	Francesco Redi	Italy	Shows that decaying meat protected from flies does not spontaneously produce maggots.
1672	Marcello Malpighi	Italy	Microscopic studies enable him to discover that capillaries link arteries to veins.
1673	Antonie van Leeuwenhoek	Holland	Uses microscope to view living microorganisms.
1700	John Ray	Britain	Describes many plants, and classifies flowering plants as either monocots or dicots.
1735	Carolus Linnaeus	Sweden	Initiates the binomial system of naming organisms.
1772	Joseph Priestley	Britain	Demonstrates that plants give off a gas required by animals.
1779	Jan Ingenhousz	Holland	Declares that sunlight is required for the green parts of plants to photosynthesize and purify the air.
1786	Luigi Galvani	Italy	Discovers that nerves can be electrically stimulated, leading to muscle contraction.
1796	Edward Jenner	Britain	Shows that vaccination with cowpox protects individuals from smallpox.
1809	Jean B. Lamarck	France	Supports the idea of evolution but thinks there is inheritance of acquired characteristics.
1825	Georges Cuvier	France	Founds the science of paleontology and shows that fossils are related to living forms.
1828	Karl E. von Baer	Germany	Establishes the germ layer theory of development.
1833	William Beaumont	United States	Presents evidence that digestion is a chemical process.
1837	René Dutrochet	France	Realizes that green pigment, chlorophyll, is necessary to photosynthesis.
1838	Matthias Schleiden	Germany	States that plants are multicellular organisms.
1839	Theodor Schwann	Germany	States that animals are multicellular organisms.
1851	Claude Bernard	France	Concludes that a relatively constant internal environment allows organisms to survive under varying conditions.
1858	Rudolf Virchow	Germany	States that cells come only from preexisting cells.
1858	Charles Darwin	Britain	Presents evidence that natural selection guides the evolutionary process.
1858	Alfred R. Wallace	Britain	Independently comes to same conclusions as Darwin.
1865	Louis Pasteur	France	Disproves the theory of spontaneous generation for bacteria; shows that infections are caused by bacteria, and develops vaccines against rabies and anthrax.
1866	Gregor Mendel	Austria	Proposes basic laws of genetics based on his experiments with garden peas.
1869	J. Friedrich Miescher	Switzerland	Discovers that the nucleus contains a chemical he called nuclein, now termed DNA.
1878	Joseph Lister	Britain	Devises a method of sterilizing the operating room to prevent infection in surgical patients.

Anton van Leeuwenhoek

Charles Darwin

Louis Pasteur

Inquiry into Life

Inquiry into Life

Sylvia S. Mader

Seventh Edition

WCB Wm. C. Brown Publishers
Dubuque, Iowa • Melbourne, Australia • Oxford, England

Book Team

Editor *Kevin Kane*
Production Editor *Renée Menne*
Design Assistant *Kathleen F. Theis*
Art Editor *Miriam J. Hoffman*
Photo Editor *Lori Gockel*
Permissions Coordinator *Karen L. Storlie*
Art Processor *Amy L. Ley*

Wm. C. Brown Publishers
A Division of Wm. C. Brown Communications, Inc.

Vice President and General Manager *Beverly Kolz*
Vice President, Publisher *Kevin Kane*
Vice President, Director of Sales and Marketing *Virginia S. Moffat*
National Sales Manager *Douglas J. DiNardo*
Marketing Manager *Carol Mills*
Advertising Manager *Janelle Keeffer*
Director of Production *Colleen A. Yonda*
Publishing Services Manager *Karen J. Slaght*
Permissions /Records Manager *Connie Allendorf*

Wm. C. Brown Communications, Inc.

President and Chief Executive Officer *G. Franklin Lewis*
Corporate Senior Vice President, President of WCB Manufacturing *Roger Meyer*
Corporate Senior Vice President and Chief Financial Officer *Robert Chesterman*

Copy edited by Mary Jean Gregory

Cover and interior design by Mark Christianson

Photographs researched by Michelle Oberhoffer

Cover photo: Canadian polar bear. © Art Wolfe

The credits section for this book begins on page 721 and is considered an extension of the copyright page.

A Times Mirror Company

Library of Congress Catalog Card Number: 92–75412

ISBN 0–697–13680–9 (paper)
0–697–13679–5 (case)

Printed in the United States of America by Wm. C. Brown Communications, Inc.,
2460 Kerper Boulevard, Dubuque, IA 52001

10 9 8 7 6 5 4 3 2 1

Welcome baby Sylvia!

BRIEF CONTENTS

CONTENTS

Part One

Life and Its Chemistry 1

1

The Study of Life 3

2

The Chemistry of Life 17

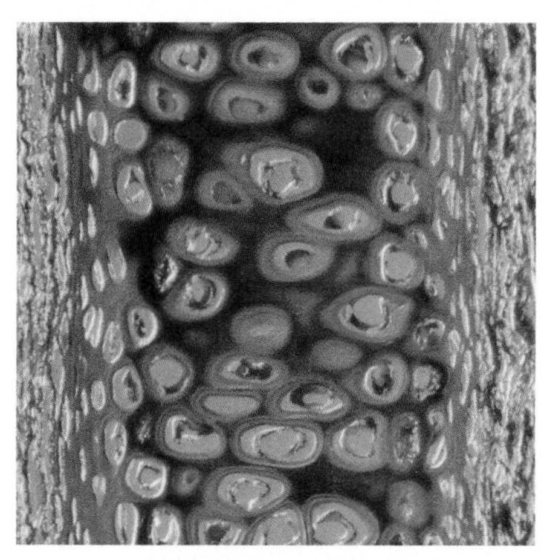

Part Two

The Cell, The Smallest Unit of Life 41

3

Cell Structure and Function 43

9 Plant Physiology and Reproduction 138

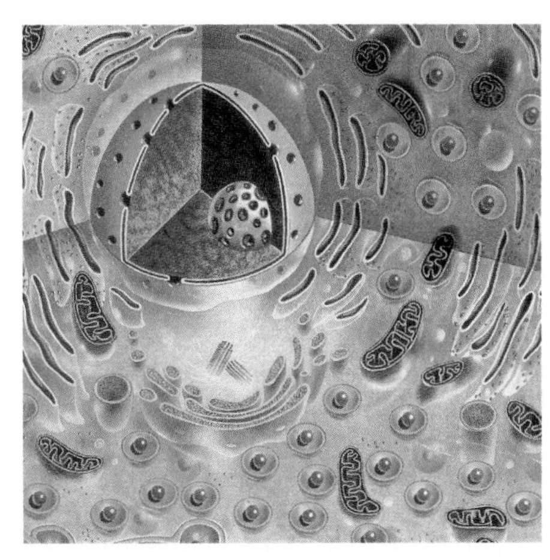

Part Four Human Anatomy and Physiology 153

10 Human Organization 155

11 Digestion 171

12 Circulation 197

13 The Lymphatic System and Immunity 220

Hormones 343

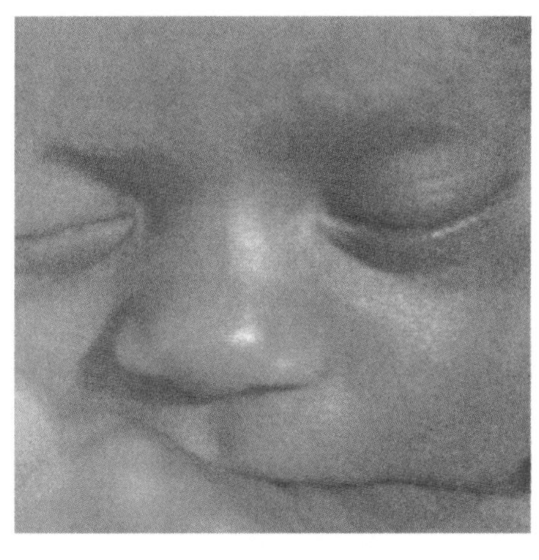

Part Five

Human Reproduction, Development, and Inheritance 367

20

The Reproductive System 369

21

Development 393

22

Patterns of Gene Inheritance 421

Part Seven

PREFACE

INQUIRY into Life is written for the introductory-level student who would like to develop a working knowledge of biology. Educational theory tells us that students are most interested in knowledge of immediate practical application. This text attempts to remain true to this idea. Its basic theme is knowledge about and understanding of human biology. Plants and other animals are included, however, because humans cannot understand themselves unless they understand other living things. All organisms share an evolutionary history, but through the evolutionary process they are adapted to particular ways of life. Humans can better understand themselves when they appreciate this unity of life, while at the same time seeing its diversity. Concerned citizens need to realize that humans are not the pivot point, nor even the culmination, of life. They are part of a great, interrelated network.

While this text covers the whole field of basic biology, it emphasizes the application of this knowledge to human concerns. The student will then be able to see that biology is truly relevant. Together with this approach, there is an emphasis on concepts and principles, rather than on detailed, high-level scientific data and terminology. The latter can always be added by more advanced study after acquisition of a core of knowledge. Each chapter presents the topic clearly, simply, and distinctly so the students can achieve a basic understanding of biology.

Usually, biology texts get larger with each edition, but this is not the case with *Inquiry into Life*. The use of a two-column rather than a one-column format has enabled us to save paper and to present a book that is shorter and lighter, but that is just as comprehensive as before. The seventh edition of *Inquiry into Life*, however, has a slightly different organization than the previous edition. A general chapter on bioenergetics precedes one that takes up specific pathways found in photosynthesis and cellular respiration. The general chapter covers the principles; therefore, some instructors may wish to include only this chapter in their curriculums.

The illustration program continues to improve, and in this edition there is even greater coordination between photographs, drawings, and tables. In several instances, all 3 elements are now part of one illustration. You have told us how to improve the illustration to meet your needs, and your suggestions have been acted upon. For example, outlines have been added or darkened so that better overhead transparencies can be produced for lecture.

Coordination between text elements has been strengthened in yet another way. The chapter concepts are more closely based on the chapter outline, and they are page referenced to their discussion in the chapter and in the summary. In this edition, the summary utilizes art pieces to tie concepts together, and some of the objective questions are based on diagrams taken from the chapter.

Inquiry into Life is even more user friendly than it was before. Every effort has been made to meet the students where they are and to bring them along. The heads and the readings have been revised, as witnessed by their snappy new and interesting titles.

Organization of Text

Inquiry into Life is available in either cloth or paper cover. The chapters can stand alone and so can be assigned in any order. For example, some instructors may wish to begin their course with the chapters in part 7, "Behavior and Ecology."

Part One: Life and Its Chemistry

The first chapter reviews the characteristics of life and discusses the process of science. Biology, the study of life, is a science, and the information presented in this book was gathered by using the scientific method. The first part of this chapter was rewritten for this edition, and the number of life's characteristics was expanded for better clarity.

A knowledge of chemistry is useful before our study of life begins because all living things are composed of chemicals.

The properties and characteristics of chemicals determine the anatomy and physiology of living things.

Part Two: The Cell, the Smallest Unit of Life

In order to understand how the multicellular organism functions, it is necessary to understand how the cell functions. Our study of cells includes their structure, how they acquire and utilize energy, and how they grow and reproduce.

The presentation of bioenergetics has been reorganized. A general chapter now precedes one that discusses photosynthesis and cellular respiration in detail. Some instructors may feel that the general chapter is sufficient to meet the needs of their students.

Part Three: Plant Biology

The flowering plants serve as the basis for the study of plant anatomy and physiology. As with the sixth edition, there are 2 chapters in this part. The first chapter concerns plant organization and growth, and the second chapter discusses plant physiology and reproduction. The chapter on plant organization was rewritten to clarify the topics transport and photoperiodism.

Part Four: Human Anatomy and Physiology

As before, humans serve as the basis for discussing animal anatomy and physiology. However, each chapter in this part now contains a reading that gives comparative information and discusses the system in other animals. Instructors who prefer the comparative approach will be especially pleased by the inclusion of these readings.

This part in the seventh edition contains one less chapter—the circulation chapter now includes the topic blood, which previously was an independent chapter. All information pertinent to immunity is now included in the next chapter, "The Lymphatic System and Immunity," which was rewritten and simplified according to adopter's suggestions.

Part Five: Human Reproduction, Development, and Inheritance

Humans, like all living things, reproduce and pass on their organization to their offspring. The human reproduction chapter gives updated information on common birth-control methods and AIDS. All information on AIDS is now included in this chapter. The development chapter was rewritten to clarify the presentation on differentiation and morphogenesis.

As before, there are 4 chapters on genetics. The first 2 chapters cover Mendelian genetics but also include information regarding human genetic disorders and mapping the human chromosomes. The other chapters concern biochemical genetics and discuss the latest recombinant DNA techniques. Cancer-causing oncogenes and tumor suppressor genes are covered, and discussion of therapy in humans is now more comprehensive.

Part Six: Evolution and Diversity

Life first evolved and then diversified. "The Origin of Life," chapter 26, was revised to include various hypotheses about this topic. The five-kingdom system of classification is also discussed, and there is a summarizing table.

The evolution chapter was completely rewritten with input from several experts in the field. The plant life cycles are simplified in chapter 29, "The Plant Kingdom," and there is an expanded section on the importance of angiosperms. The section on vertebrates was reorganized for the animal kingdom chapter, which now includes an updated section on human evolution.

Part Seven: Behavior and Ecology

In this part, a single chapter devoted to behavior occurs before biomes, ecosystems, and human population concerns are discussed. The behavior chapter was updated and rewritten for this edition. Adopters will now find that this chapter completely meets their needs concerning this important topic.

The future existence of human beings is dependent on our preserving the natural world, and the goal of this part is to make students aware of this dependence and what should be done to protect the balance of nature. The last chapter considers modern ecological concerns, such as global warming, tropical rain forest destruction, and the growing ozone hole.

Aids to the Reader

Inquiry into Life includes a number of aids that have helped students study biology successfully and enjoyably.

History of Biology Endpapers

The front and back inside covers list major contributions to the field of biology in a concise, chronological manner. Students may refer to these whenever it is appropriate.

Part Introduction

An introduction for each part highlights the central ideas of that part and specifically tells the student how the topics within each part contribute to biological knowledge.

Chapter Concepts

Each chapter begins with a list of concepts that correlate with the chapter outline. This listing introduces the student to the chapter by organizing its content into a few meaningful sentences. The concepts provide a framework for the content of each chapter.

Chapter Outlines

In addition to the chapter concepts, each chapter begins with a chapter outline. These allow the student to tell at a glance how the chapter concepts are organized and how they have been included in the chapter. The chapter outline includes the first- and second-level heads for the chapter.

Readings

Two types of readings are included in the text. Readings chosen from popular magazines illustrate the applications of concepts to modern concerns. These spark interest by illustrating that biology is an important part of everyday life. The second type of reading, usually written by the author, is designed to expand in an interesting way on the core information presented in each chapter. New to this edition are the comparative readings, which are found in the human systems chapters. These compare the anatomy and physiology of other animals in an evolutionary context.

Tables and Illustrations

Numerous tables and illustrations appear in each chapter and are placed near their related text discussion. The tables clarify complex ideas and summarize sections of the narrative. The photographs and drawings have been chosen and designed to help the student visualize structures and processes. New to this edition are integrative illustrations: both a photograph and a drawing are integral parts of the illustration. Other illustrations include information formerly found in tables.

Boldfaced Words

New terms appear in boldface print as they are introduced in the text. They are also immediately defined in context. Key terms are defined in the end-of-chapter glossary, and most boldfaced terms are in the text glossary. Phonetic pronunciations are provided for unfamiliar terms.

Internal Summary Statements

Summary statements are strategically placed throughout the chapter. These immediately reinforce the concept that has just been discussed. The summary statements will aid student retention of the chapter's main points.

Chapter Summaries

Chapter summaries offer a concise review of material in each chapter. Students may read the summary before beginning the chapter to preview the topics of importance, and they may also use it to refresh their memory after they have a firm grasp of the concepts presented in each chapter. New to this edition are art pieces that help summarize key concepts in the chapter.

Chapter Questions

Three types of questions—study questions, critical thinking questions, and objective questions—appear at the close of each chapter. *Writing across the curriculum* recognizes that students need an opportunity to practice writing in all courses. The study questions review the chapter, and their sequence follows that of the chapter. When students write out the answers to the study questions, they are writing while studying biology. The critical thinking questions require the student to form a hypothesis, come to a conclusion, or apply information in a new or different way. Writing out the answers to the critical thinking questions also fulfills any *writing across the curriculum* requirement. The objective questions allow students to quiz themselves with short fill-in-the-blank questions. In this edition, there is even more frequent use of labeling exercises in the objective question section. Answers to the objective questions and the critical thinking questions appear in appendix E.

Selected Key Term Lists

Major boldfaced terms within the chapter are defined at the end of each chapter for more convenient review. Selected key terms are carefully defined and page referenced. Most boldfaced terms are still listed alphabetically with their definitions and page references in the text glossary at the end of the book. Phonetic pronunciations are provided for unfamiliar terms.

Further Readings

For students who would like more information about a particular topic or who are seeking references for a research paper, each part ends with a listing of related articles and books to help them get started. Usually the entries are *Scientific American* articles and specialty books that expand on the topics covered in the chapter.

Appendix and Glossary

The appendix contains optional information for student referral. It includes the periodic table of the elements and a review of the metric system. An important part of the appendix is the classification system of organisms used in the text.

The text glossary defines the terms most necessary for making the study of biology successful. By using this tool, students can review the definitions of the terms most frequently used.

Index

An index is also included in the back matter of the book. By consulting the index, it is possible to determine on what page or pages various topics are discussed.

Additional Aids

Instructor's Manual/Test Item File

The Instructor's Manual/Test Item File, revised by Les Wiemerslage, is designed to assist instructors as they plan and prepare for classes using *Inquiry into Life*. Possible course organizations for semester and quarter systems are suggested, along with alternate suggestions for sequencing the chapters. A general discussion and an extended lecture outline are provided for each chapter; together, these give a brief overview and a complete set of lecture notes for each chapter. For previous users of the text, seventh-edition changes are noted for convenient comparison to the sixth edition. Approximately 50 objective test questions and several essay questions are provided for each chapter. A list of suggested audiovisuals for the various topics and a list of suppliers are included at the end of the Instructor's Manual.

Student Study Guide

The Student Study Guide that accompanies the text was also revised by Les Wiemerslage. For each text chapter, there is a corresponding Student Study Guide chapter, which includes a chapter outline and numerous study questions and objective questions. The study questions are designed to help students carefully digest the content of the chapter, while the objective questions allow students to practice for an examination. Answers are listed at the end of each chapter.

Extended Lecture Outline Software

The highly detailed outlines for each text chapter that appear in the Instructor's Manual are also available on IBM, Apple, and Macintosh diskettes. These offer instructors flexibility and convenience.

Laboratory Manual

The Laboratory Manual that accompanies *Inquiry into Life* has been thoroughly revised. Its 33 exercises provide enough variety to meet the needs of a broad spectrum of class designs. They help students appreciate the scientific method and learn the fundamental concepts of biology and the specific content of each chapter. All exercises have been tested for student interest, preparation time, and feasibility.

Customized Laboratory Manual

The Laboratory Manual's 33 exercises are now available as individual "lab separates" so instructors can tailor the manual to their particular course needs. The separates, which are published in one color at a greatly reduced price, will be collated and bound by WCB on request.

Laboratory Resource Guide

Helpful and thorough information regarding each lab preparation is found in the Laboratory Resource Guide. The guide is designed to help instructors make the laboratory experience more meaningful for the student. Each chapter in the resource guide has 2 parts: "Materials" gives preparation information, and "Exercises" gives expected results and answers to the questions.

Lecture Enrichment Kit and Transparencies

The Lecture Enrichment Kit is a series of optional lecture notes to accompany 250 transparencies. The transparencies are grouped according to topic, and for each group there are 3 or 4 "extensions," topics not discussed in the text. Extensions are drawn from popular periodicals of general interest, scientific periodicals, or more advanced texts. They vary in detail and degree of difficulty.

Slides

Often instructors prefer to use slides rather than transparencies. Slides of all transparency art pieces are available upon request.

Slides of Photomicrographs and Electron Micrographs

This addition to the ancillary program features 50 slides of high-interest photomicrographs and electron micrographs, most of which are scanning electron micrographs.

Visuals Testbank

A set of 50 transparency masters are available for use by instructors. These feature line art from the text with labels deleted for student quizzing or practice.

Critical Thinking Case Study Workbook

Written by Robert Allen, this ancillary includes 30 critical thinking case studies. These case studies are designed to immerse students in the "process of science" and to challenge them to solve problems in the same way biologists do. The case studies are divided into 3 levels of difficulty (introductory, intermediate, and advanced) to afford instructors greater choice and flexibility.

Critical Thinking Case Study Workbook Answer Key

The answers to each critical thinking case study are presented in this key.

Testing Software

WCB provides a computerized test generator for use with this text. It allows you to quickly create tests based on questions provided by WCB and requires no programming experience to use. The questions are provided on diskette in a test item file. WCB also provides support services, via mail or phone, to assist in the use of the test generator software, as well as in the creation and printing of tests.

A computerized grade management system is also available for instructors. This allows you to track student performance on exams and assignments. Reports based on this information can be generated for your review.

Software to generate quizzes can also be provided. These quizzes can be used to allow students to prepare for exams on their own.

Bio Sci II Videodisc

This critically acclaimed laser disk, produced by Videodiscovery, features more than 12,000 still and moving images with a complete, bar-coded directory. The disk can be used for lecture support, individual student study, or student group activity. The instructor's manual that accompanies *Inquiry into Life* contains a bar code directory that correlates specific frames on the laser disk to topics in the text.

Bio Sci II Stacks Software

This software, using Macintosh HyperCard or IBM Linkway, puts the entire textual database for Bio Sci II at the instructor's or student's fingertips while simultaneously controlling the videodisc.

Classroom Testing Software

A complete test item file is on diskette for use with IBM, Apple, or Macintosh computers.

Animated Illustrations of Physiological Processes

Thirteen illustrations taken from WCB's library of anatomy and physiology texts are brought to life through animation. The animations have been placed on a 35-minute videotape and illustrate such physiological processes as conduction of an action potential, synaptic transmission, electron transport and oxidative phosphorylation, and viral replication of HIV. These full color animations enable the student to more fully and easily grasp complex physiological processes. Free to qualified adopters. (ISBN 0–697–21512–1)

Other Titles of Related Interest from Wm. C. Brown Publishing

You Can Make a Difference

by Judith Getis

This short, inexpensive supplement offers students practical guidelines for recycling, conserving energy, disposing of hazardous wastes, and other pollution controls. It can be shrink wrapped with the text at minimal additional cost. (ISBN 0–697–13923–9)

How to Study Science

by Fred Drewes, Suffolk County Community College

This excellent new workbook offers students helpful suggestions for meeting the considerable challenges of a college science course. It offers tips on how to take notes, how to get the most out of laboratories, and how to overcome science anxiety. The book's unique design helps students develop critical thinking skills while facilitating careful note taking. (ISBN 0–697–14474–7)

The Life Science Lexicon

by William N. Marchuk, Red Deer College

This portable, inexpensive reference helps introductory-level students quickly master the vocabulary of the life sciences. Not a dictionary, it carefully explains the rules of word construction and derivation, in addition to giving complete definitions of all important terms. (ISBN 0–697–12133–X)

Biology Study Cards

by Kent Van De Graaff, R. Ward Rhees, and Christopher H. Creek, Brigham Young University

This boxed set of 300 two-sided study cards provides a quick yet thorough visual synopsis of all key biological terms and concepts in the general biology curriculum. Each card features a masterful illustration, pronunciation guide, definition, and description in context. (ISBN 0–697–03069–5)

The Gundy-Weber Knowledge Map of the Human Body

by G. Craig Gundy, Weber State University

This 13-disk Mac-Hypercard program is for use by instructors and students alike. It features carefully prepared computer graphics, animations, labeling exercises, self-tests, and practice questions to help students examine the systems of the human body. Contact your local Wm. C. Brown representative or call 1–800–351–7671.

The Knowledge Map Diagrams

1. Introduction, Tissues, Integument System (ISBN 0–697–13255–2)
2. Viruses, Bacteria, Eukaryotic Cells (ISBN 0–697–13257–9)
3. Skeletal System (ISBN 0–697–13258–7)
4. Muscle System (ISBN 0–697–13259–5)
5. Nervous System (ISBN 0–697–13260–9)
6. Special Senses (ISBN 0–697–13261–7)
7. Endocrine System (ISBN 0–697–13262–5)
8. Blood and the Lymphatic System (ISBN 0–697–13263–3)
9. Cardiovascular System (ISBN 0–697–13264–1)
10. Respiratory System (ISBN 0–697–13265–X)
11. Digestive System (ISBN 0–697–13266–8)
12. Urinary System (ISBN 0–697–13267–6)
13. Reproductive System (ISBN 0–697–13268–4)

Demo—(ISBN 0–697–13256–0)
Complete Package—(ISBN 0–697–13269–2)

GenPak: A Computer Assisted Guide to Genetics

by Tully Turney, Hampden-Sydney College

This Mac-Hypercard introductory-level program features numerous interactive/tutorial (problem-solving) exercises in Mendelian, molecular, and population genetics. (ISBN 0–697–13760–0)

Acknowledgments

The personnel at Wm. C. Brown Publishers have always lent their talents to the success of *Inquiry into Life*. My editor, Kevin Kane, directed the efforts of all. Jim Daggett, my developmental editor, served as a liaison between the editor, those in production, and me. Renee Menne was the production editor, Miriam Hoffman the art editor, Michelle Oberhoffer the photo researcher, Mark Christianson and Elise Lansdon the designers, Vicki Krug the permissions editor, and Amy Ley the visuals processor. My thanks to each of them for a job well done.

A special word of thanks goes to Carlyn Iverson, who created many fine drawings and also coordinated the work of the other artists for the book.

The Reviewers

Many instructors have contributed not only to this edition of *Inquiry into Life* but also to previous editions. I am extremely thankful to each one, for we have all worked diligently to remain true to our calling and to provide a product that will be the most useful to our students.

In particular, it is appropriate to acknowledge the help of the following individuals. For the seventh edition:

Jane E. Aloi
Saddleback College
James Averett
Nassau Community College
Joyce Azevedo
Southern College
Carl W. Candiloro
Pace University
Jerry A. Clonts
Anderson College
Stanley Cohn
DePaul University
Shirley A. Crawford
SUNY–Morrisville
Menter H. David
Barton County Community College
Rebecca McBride DiLiddo
Suffolk University
Lee C. Drickamer
Southern Illinois University
Thomas C. Emmel
University of Florida
Carl Fiore
Central Connecticut State University
Colleen T. Fogarty
St. John's University
Leon J. Gorski
Central Connecticut State University
Michael F. Gross
Georgian Court College
James R. Jackson
Missouri Southern State College
Sheila A. Johnson
Massasoit Community College
Arnold J. Karpoff
University of Louisville
Kerry S. Kilburn
West Virginia State College
Peter A. Kish
Harrisburg Area Community College
Ron W. Leavitt
Brigham Young University
Delores McCright
Texarkana College
Pamela Monaco
St. John's University and Molloy College
Keith Morrill
South Dakota State University
Sanford A. Moss
University of Massachusetts–Dartmouth
Ava Nickerson
Cooke County College
Ezequiel R. Rivera
University of Massachusetts–Lowell
Kelvin F. Rogers
Kent State University
Patricia Rugaber
Brunswick College
Stephen M. Shimmel
Chipola Junior College
John Sowell
Western State College
John F. Utley
University of New Orleans
Ann B. Vernon
St. Charles County Community College
Robert P. West
Lee College
Steven W. Woeste
Dr. William M. Scholl College of Podiatric Medicine
Jeffrey S. Wooters
Pensacola Junior College
Shanna D. Yonenaka
San Francisco State University

THE *INQUIRY INTO LIFE* LEARNING SYSTEM

Chapter Concepts

Each chapter begins with a list of concepts stressed in the chapter. This listing introduces the student to the chapter by organizing its content into a few meaningful sentences. The concepts provide a framework for the content of each chapter and are page referenced for your convenience.

1

THE STUDY OF LIFE

Chapter Concepts

1.
Although life is difficult to define, it can be recognized by certain common characteristics. 5, 15

2.
There are levels of biological organization proceeding from the simplest level to the most complex level. 5, 15

3.
Living things are classified into categories according to their evolutionary relationships. 9, 15

4.
Biologists often use the scientific method to gather information and to come to conclusions. 12, 15

5.
All persons have the responsibility to decide how scientific information can best be used to make ethical or moral decisions. 15

Leaves of a lily (*Worsleya rayneri*)

Chapter Outline

3

Chapter Outlines

In addition to the chapter concepts, each chapter has an outline. These will allow students to tell at a glance how the chapter is organized and what major topics have been included in the chapter. The outlines include the first and second level heads for the chapter.

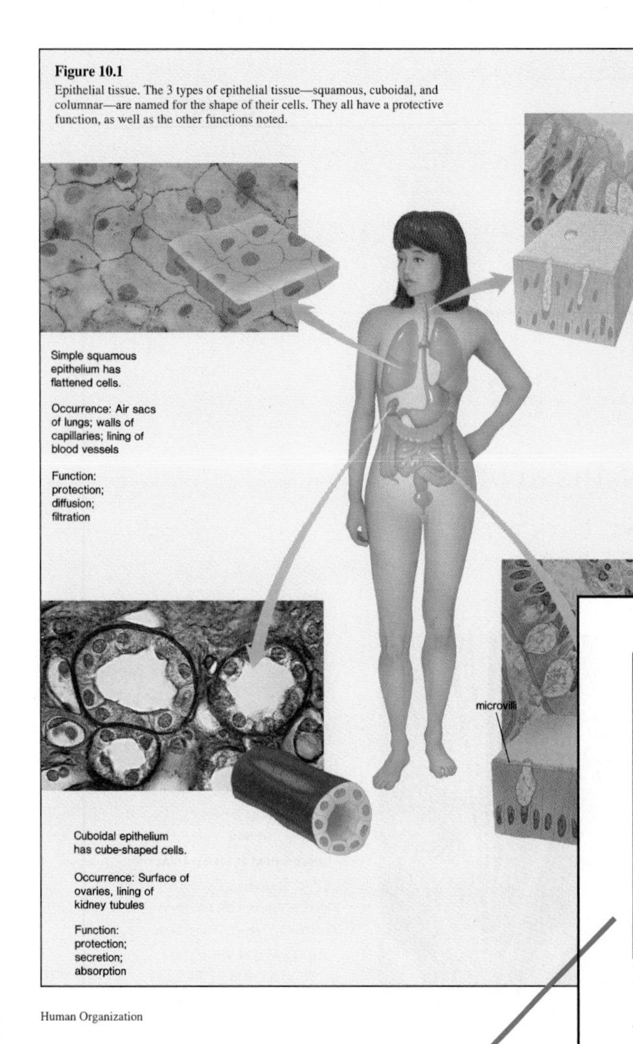

Dramatic Visuals Program

Colorful, informative photographs and illustrations enhance the learning program of the text as well as spark interest and discussion of important topics. **New** to this edition are illustrations that combine photographs and drawings to more clearly present important concepts.

Text Line Art

Graphic diagrams placed immediately after or within textual passages help clarify difficult concepts and enhance learning.

Table 16.3
Path of a Simple Reflex

1. Receptor (formulates message)[*]	Generates nerve impulses
2. Sensory neuron (takes message to CNS)	Impulses move along dendrite (spinal nerve)[†] and proceed to cell body (dorsal-root ganglion) and then go from cell body to axon (spinal cord)
3. Interneuron (passes message to motor neuron)	Impulses picked up by dendrites and pass through cell body to axon (spinal cord)
4. Motor neuron (takes message away from CNS)	Impulses travel through short dendrites and cell body (spinal cord) to axon (spinal nerve)
5. Effector (receives message)	Receives nerve impulses and reacts: glands secrete and muscles contract

[*]Phrases within parentheses state overall function.
[†]Words within parentheses indicate location of structure.

The reflex arc is the main functional unit of the nervous system. It allows us to react to internal and external stimuli. Do 16.2 Critical Thinking, found at the end of the chapter.

Autonomic Nervous System: For Internal Organs
The autonomic nervous system (fig. 16.12), a part of the PNS, is made up of motor neurons that control the internal organs automatically and usually without need for conscious intervention. The sensory neurons that come from the internal organs allow us to feel internal pain. The cell bodies for these sensory neurons are in dorsal root ganglia along with the cell bodies of somatic sensory neurons.

There are 2 divisions of the autonomic nervous system: the sympathetic system and the parasympathetic system. Both of these (1) function automatically and usually subconsciously in an involuntary manner; (2) innervate all internal organs; and (3) utilize 2 motor neurons and one ganglion for each impulse. The first of these 2 neurons has a cell body within the CNS and a **preganglionic axon.** The second neuron has a cell body within the ganglion and a **postganglionic axon.**

The autonomic nervous system subconsciously controls the function of internal organs.

Sympathetic System: Fight or Flight
The preganglionic fibers of the **sympathetic nervous system** arise from the *thoraco-lumbar* (middle) portion of the spinal cord and almost immediately terminate in ganglia that lie near the cord. Therefore, in this system, the preganglionic fiber is short, but the postganglionic fiber that makes contact with an organ is long.

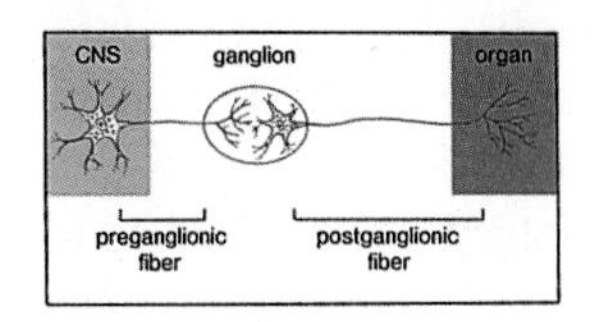

The sympathetic system is especially important during emergency situations and is associated with "fight or flight." For example, it inhibits the digestive tract, but it dilates the pupil, accelerates the heartbeat, and increases the breathing rate. The neurotransmitter released by postganglionic axons is primarily norepinephrine, a chemical close in structure to epinephrine (adrenalin), a medicine used as a heart stimulant.

The sympathetic system brings about those responses we associate with "fight or flight."

Parasympathetic System: Housekeeper
A few cranial nerves, including the vagus nerve, together with fibers that arise from the sacral (bottom) portion of the spinal cord, form the **parasympathetic nervous system** (fig. 16.12). Therefore, this system is often referred to as the *craniosacral portion* of the autonomic nervous system. In the parasympathetic nervous system, the preganglionic fiber is long and the postganglionic fiber is short because the ganglia lie near or within the organ.

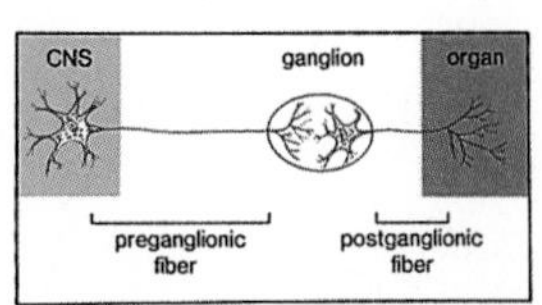

The parasympathetic system, sometimes called the "housekeeper system," promotes all the internal responses we associate with a relaxed state; for example, it causes the pupil of the eye to contract, promotes digestion of food, and retards the heartbeat. The neurotransmitter utilized by the parasympathetic system is primarily acetylcholine.

The parasympathetic system brings about the responses we associate with a relaxed state.

The Nervous System 289

Tables

Numerous strategically placed tables list and summarize important information, making it readily accessible for efficient study.

Internal Summaries

Summary statements are placed at strategic locations throughout the chapter. These immediately reinforce the concept that has just been discussed, helping students to retain the chapter's main points.

Boldfaced Words

New terms appear in boldface print as they are introduced within the text and are immediately defined in context. Most key terms are defined in the end-of-chapter glossary and in the text glossary, where a phonetic pronunciation is given along with an appropriate page reference.

Readings

Throughout *Inquiry into Life*, selected readings reinforce major concepts in the book. Readings chosen from popular magazines spark interest in biology. Readings by the author expand on the core information in the chapter. **New** to this edition are comparative readings. These compare the anatomy and physiology of other animals in the evolutionary context. These readings are found in the human systems chapters.

A N I M A L • B I O L O G Y

ANIMAL RESPIRATORY ORGANS

Animals do not have a storage area for gases; therefore, they must continually acquire oxygen and rid the body of carbon dioxide. Some animals, like hydras and planarians, are small and shaped in a way that allows their body cells to carry out gas-exchange. Among larger animals, the earthworm's shape provides an extensive outer surface for respiration. Glands keep the surface moist, and the worm is behaviorally adapted to remain in damp soil during the day.

In most complex animals, vascularization—close association with an extensive capillary system—enhances the effectiveness of the respiratory organ. Aquatic animals take oxygen from the water. Quite often they have gills, which are finely divided and vascularized outgrowths of either an outer or inner body surface. In the crayfish and similar crustaceans, the gills are located on the thorax, just beneath the exoskeleton. In many fishes, the gills are outward extensions of the pharynx. When the mouth opens, water is drawn in. When the mouth closes, water flows through the gill slits located between the gill arches.

Terrestrial animals take oxygen from the air. Among invertebrates, insects have a tubular respiratory system. Air enters the system at valvelike openings, and then the tubes called trachea branch and rebranch, until finally tracheoles are in direct contact with the body cells. Terrestrial vertebrates, in particular, have evolved lungs, which are outgrowths from the lower pharyngeal region. The lungs of amphibians are simple sacs, and most amphibians also make use of the skin as a respiratory surface. The lungs are more finely divided in reptiles and are especially divided in birds and mammals. It has even been estimated that human lungs have a total surface area that is at least 40 times the surface area of skin.

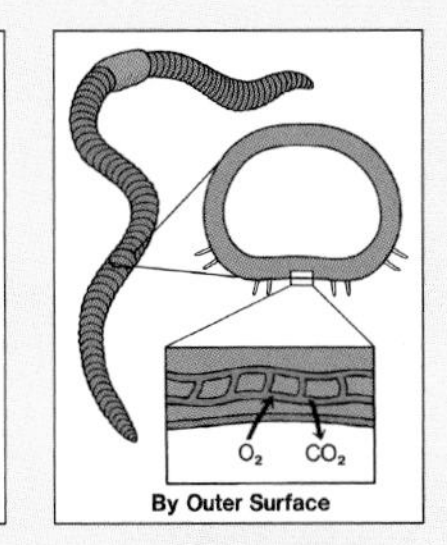

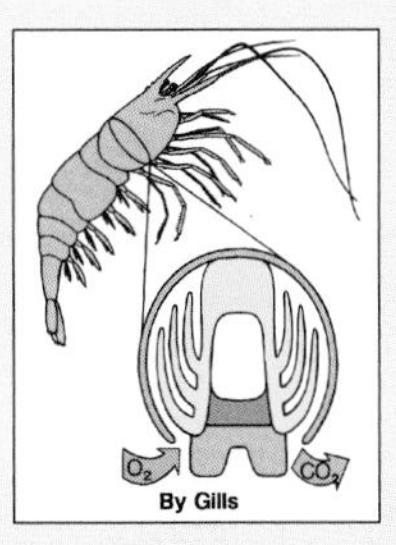

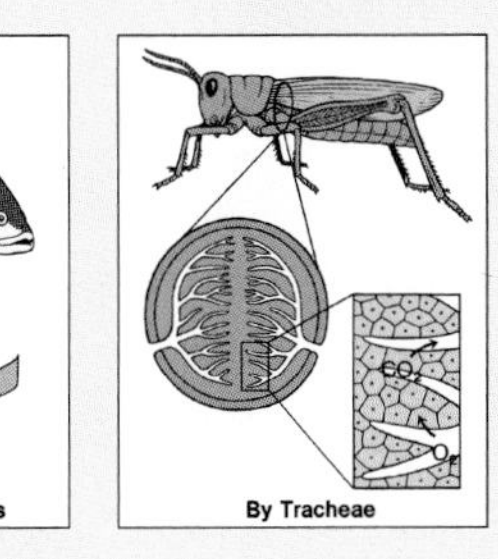

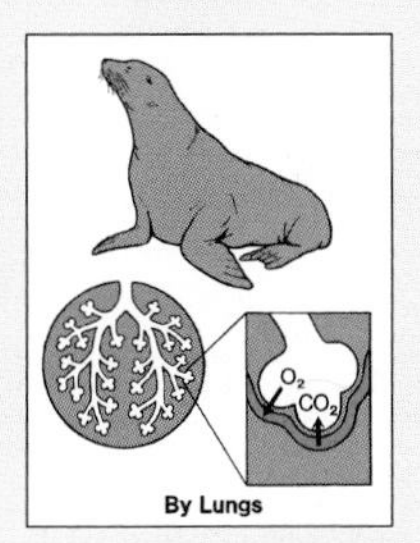

Human Anatomy and Physiology

SUMMARY

Water transport in plants occurs within xylem. The cohesion-tension model of xylem transport states that transpiration creates tension, which pulls water upward in xylem. This transport works only because water molecules are cohesive. Most of the water taken in by a plant is lost through stomata by transpiration.

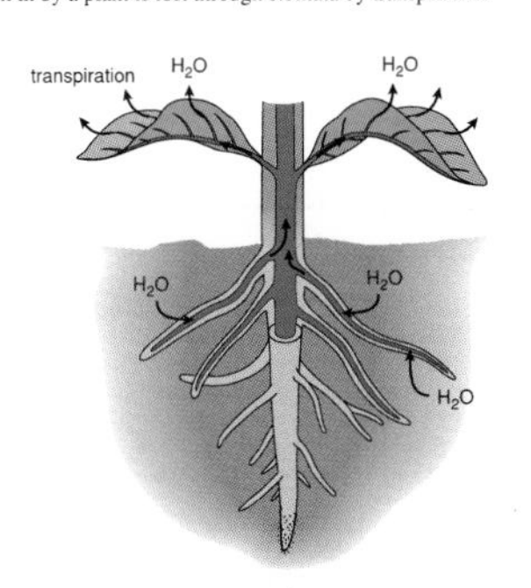

Stomata open when guard cells take up water (H_2O), stretching their thin side walls. Water follows potassium ions (K^+) into the guard cells.

Transport of organic nutrients in plants occurs within phloem. The pressure-flow theory of phloem transport states that sugar is actively transported into phloem at a source and water follows by osmosis. The resulting increase in pressure creates a flow, which moves water and sucrose to a sink.

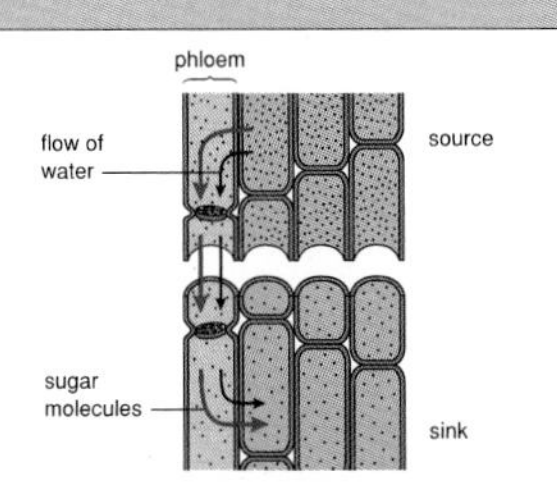

Both stimulatory and inhibitory hormones help to control certain plant growth patterns. Some hormones stimulate growth (auxins, gibberellins, and cytokinins), while others inhibit growth (ethylene and abscisic acid). Plant hormones most likely control photoperiodism. Short-day plants flower when the days are shorter (nights are longer) than a critical length, and long-day plants flower when the days are longer (nights are shorter) than a critical length. Some plants are day-length neutral. Phytochrome, a plant pigment that responds to daylight, is believed to be a part of a biological clock system that in some unknown way brings about flowering.

Flowering plants have an alternation of generations life cycle, which includes separate male and female gametophytes. The pollen grain, the male gametophyte, is produced within the stamens of a flower. The female gametophyte is produced within the ovule of a flower. Following pollination and fertilization, the ovule matures to become the seed and the ovary becomes the fruit. The enclosed seeds contain the embryo (hypocotyl, epicotyl, plumule, radicle) and stored food (endosperm and/or cotyledons). When a seed germinates, the root appears below and the shoot appears above.

STUDY QUESTIONS

In order to practice **writing across the curriculum,** students should write out the answers to any or all of the study questions. The study questions are sequenced in the same order as the text.

1. Explain the cohesion-tension theory of water transport. (pp. 139–140)
2. What events precede the opening and closing of stomata by guard cells? (pp. 140–141)
3. Explain the pressure-flow theory of phloem transport. (pp. 142–143)
4. Name 5 plant hormones, and state their functions. (p. 143)
5. Define photoperiodism, and discuss its relationship to flowering in certain plants. (p. 143)
6. What is phytochrome, and what are some possible functions of phytochrome in plants? (p. 144)
7. How do plants reproduce asexually? sexually? (p. 146)
8. Describe how a female gametophyte forms in flowering plants. (p. 147)
9. Describe how a male gametophyte forms in flowering plants. (p. 147)
10. Contrast the monocot seed and seedling with the dicot seed and seedling. (pp. 148–149)

150 Plant Biology

Chapter Summaries

Chapter summaries offer a concise review of material in each chapter. Students may read them before beginning the chapter to preview the topics of importance, and they may also use them to refresh their memories after they have a firm grasp of the concepts presented in each chapter. **New to this edition are art pieces that help summarize key concepts in the chapter.**

Study Questions

Writing across the curriculum recognizes that students need an opportunity to practice writing in all courses. These page-referenced, review questions provide the student the opportunity to write while learning biology.

Objective Questions

The objective questions allow the students to quiz themselves with short fill-in-the-blank questions. In this edition, there is frequent use of labeling exercises. Answers to the objective questions are listed in appendix E.

OBJECTIVE QUESTIONS

1. The transport of water is dependent upon _______, which occurs whenever the stomata are open.
2. Stomata open when _______, followed by _______, enters guard cells.
3. The _______ theory explains the transport of sugar in sieve-tube cells.
4. Short-day plants _______ (will, will not) flower when a longer-than-critical-length night is interrupted by a flash of light.
5. _______ is the pigment that is believed to signal a biological clock in plants that exhibit photoperiodism.
6. Plants have a life cycle called _______.
7. The female gametophyte develops within the _______ of a flower, and the male gametophyte develops within the _______.
8. Monocots have seeds with one _______, while dicots have seeds with 2.
9. Label this diagram of the alternation of generations life cycle and the reproductive parts of a flower.

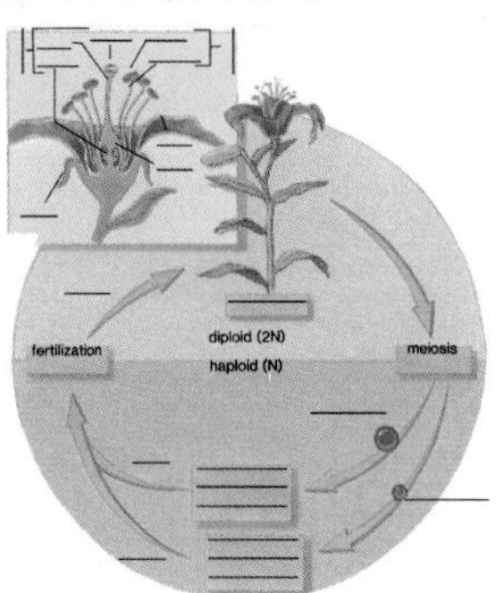

Critical Thinking Questions

These questions not only offer the student further opportunity to fulfill any *writing across the curriculum* requirement, but also require the student to form a hypothesis, come to a conclusion, or apply information in a new or different way. Answers to the critical thinking questions are listed in appendix E.

CRITICAL THINKING

In order to practice **writing across the curriculum,** students should write out the answers to any or all of the critical thinking questions. Suggested answers to the critical thinking questions are in appendix E.

9.1

A twig with leaves is placed in the top of an open tube. The tube contains water above mercury:

1. Atmospheric pressure alone is sufficient to raise mercury (Hg) only 760 mm (760 mm Hg = 10.4 m water). What is atmospheric pressure, and of what significance is this finding for a tree that is 120 m high?
2. Why does the mercury rise higher than 760 mm Hg when a twig with leaves is placed in the top of the tube?
3. What does the experiment suggest about the ability of transpiration to raise water to the top of tall trees?

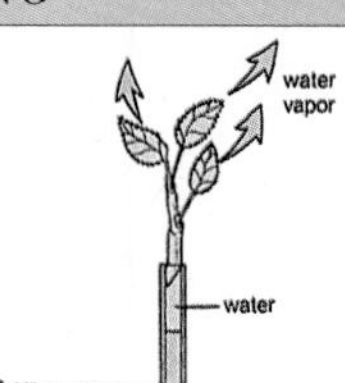

Selected Key Terms

A selected list of boldfaced key terms from the chapter appears at the end of each chapter. Each term is accompanied by its definition.

SELECTED KEY TERMS

alternation of generations a life cycle typical of plants in which a diploid sporophyte alternates with a haploid gametophyte.
cohesion-tension theory explanation for upward transportation of water in xylem based upon transpiration-created tension and the cohesive properties of water molecules.
companion cell a small nucleated cell that lies adjacent to and helps with the activities of a sieve-tube cell.
embryo sac the female gam
contains an egg cell.
epicotyl (ep˘ĭ-kot´il) the pla
cotyledons; contributes
hypocotyl (hi˝po-kot´il) the
cotyledons; contributes

photoperiodism a response to light and darkness; particularly in reference to flowering in plants.
phytochrome (fi´to-krōm) a plant pigment that enables a photoperiodic response in plants.
pollen grain the male gametophyte of flowering plants that, upon maturity, contains sperm.
pollination the delivery of pollen by wind or animals to the stigma of a pistil in flowering plants.
pressure-flow theory explanation for phloem transport; osmotic pressure following active transport of sugar into phloem brings about a flow of sap from a source to a sink.
radicle the embryonic root of a plant.
sieve-tube cell a phloem cell that functions in transport of organic nutrients. During development, sieve-tube cells align vertically and form a continuous pathway for transport.
spore a haploid reproductive cell produced by the diploid sporophyte of a plant; asexually gives rise to the haploid gametophyte.
tracheid (tra´ke-id) a component of xylem made of long, tapered nonliving cells.
transpiration the evaporation of water from a leaf; pulls water from the roots through a stem to leaves.
vessel element a conducting cell in xylem. During development, vessel elements lose their contents and end walls so that they form a continuous vertical pipeline for transport of water.

Further Readings

A list of readings at the end of each part suggests references that can be used for further study of topics covered in the chapters of that part. The items listed in this section were carefully chosen for readability and accessibility.

FURTHER READINGS FOR PART THREE

Alberts, B., et al. 1989. *Molecular biology of the cell.* 2d ed. New York: Garland Publishing.
Barrett, S. C. H. September 1987. Mimicry in plants. *Scientific American.*
Bazzar, F. A., and E. D. Fajer. January 1992. Plant life in a CO_2 rich world. *Scientific American.*
Bold, H. C. 1980. *Morphology of plants and fungi.* 4th ed. New York: Harper & Row, Publishers, Inc.
Brill, W. J. March 1977. Biological nitrogen fixation. *Scientific American.*
Cronquist, A. 1982. *Basic botany.* 2d ed. New York: Harper & Row, Publishers, Inc.
Epel, D. November 1977. The program of fertilization. *Scientific American.*
Hesslop-Harrison, Y. February 1978. Carnivorous plants. *Scientific American.*
Jansen, W., and F. B. Salisbury. 1971. *Botany: An ecological approach.* Belmont, Calif.: Wadsworth.
Niklas, K. J. July 1987. Aerodynamics of wind pollination. *Scientific American.*
Raven, H., et al. 1986. *Biology of plants.* 4th ed. New York: Worth Publishers, Inc.
Rayle, D., and H. L. Wedberg. 1980. *Botany: A human concern.* Boston: Houghton Mifflin.
Rost, R., et al. 1984. *Botany: A brief introduction to plant biology.* 2d ed. New York: John Wiley and Sons.
Salisbury, F. B., and C. W. Ross. 1985. *Plant physiology.* 3d ed. Belmont, Calif.: Wadsworth.
Shepard, J. F. May 1982. The regeneration of potato plants from leaf-cell protoplasts. *Scientific American.*
Stern, K. 1991. *Introductory plant biology.* 5th ed. Dubuque, Iowa: Wm. C. Brown Publishers.
Zimmerman, M. H. March 1963. How sap moves in trees. *Scientific American.*

Part One

LIFE AND ITS CHEMISTRY

Parnassian butterfly on a thistle

LIFE AND ITS CHEMISTRY

Biology is the study of life, a phenomenon that is not easily defined but one that can be recognized by certain characteristics. Biology is also a science, and the process of science allows us to come to conclusions concerning the natural world. The information in this text has been gathered using the scientific method.

All of the sciences are interrelated, and a knowledge of chemistry is particularly useful before we begin our study of biology. All living things are composed of chemicals, and their properties determine the structure and function of the human body.

1

THE STUDY OF LIFE

Chapter Concepts

1.
Although life is difficult to define, it can be recognized by certain common characteristics. 5, 15

2.
There are levels of biological organization proceeding from the simplest level to the most complex level. 5, 15

3.
Living things are classified into categories according to their evolutionary relationships. 9, 15

4.
Biologists often use the scientific method to gather information and to come to conclusions. 12, 15

5.
All persons have the responsibility to decide how scientific information can best be used to make ethical or moral decisions. 15

Leaves of a lily (*Worsleya rayneri*)

Chapter Outline

Figure 1.1
Some of the animals that live in the rain forests of Sumatra. These are all mammals, organisms that have hair and nurse their young. Human population growth, accompanied by increased agriculture and intense logging, threatens their continued existence. Similar pressures threaten the diversity in all the rain forests of the world.

THIS book is about living things (fig. 1.1); therefore, it is appropriate to first define life. Unfortunately, this is not done so easily—life cannot be given a simple, one-line definition. Life can be recognized, however, by certain properties.

The Characteristics of Life

The properties that characterize all living things are as follows:

1. **Living things are organized;** their parts are specialized for specific functions.
2. **Living things take materials and energy from the environment;** they need an outside source of nutrients.
3. **Living things are homeostatic;** they stay just about the same internally despite changes in the external environment.
4. **Living things respond to stimuli;** they react to internal and external events.
5. **Living things reproduce;** they produce offspring that resemble themselves.
6. **Living things grow and develop;** during their lives they change, sometimes undergoing various stages from fertilization to death.
7. **Living things are adapted;** they have modifications that make them suited to a particular way of life.

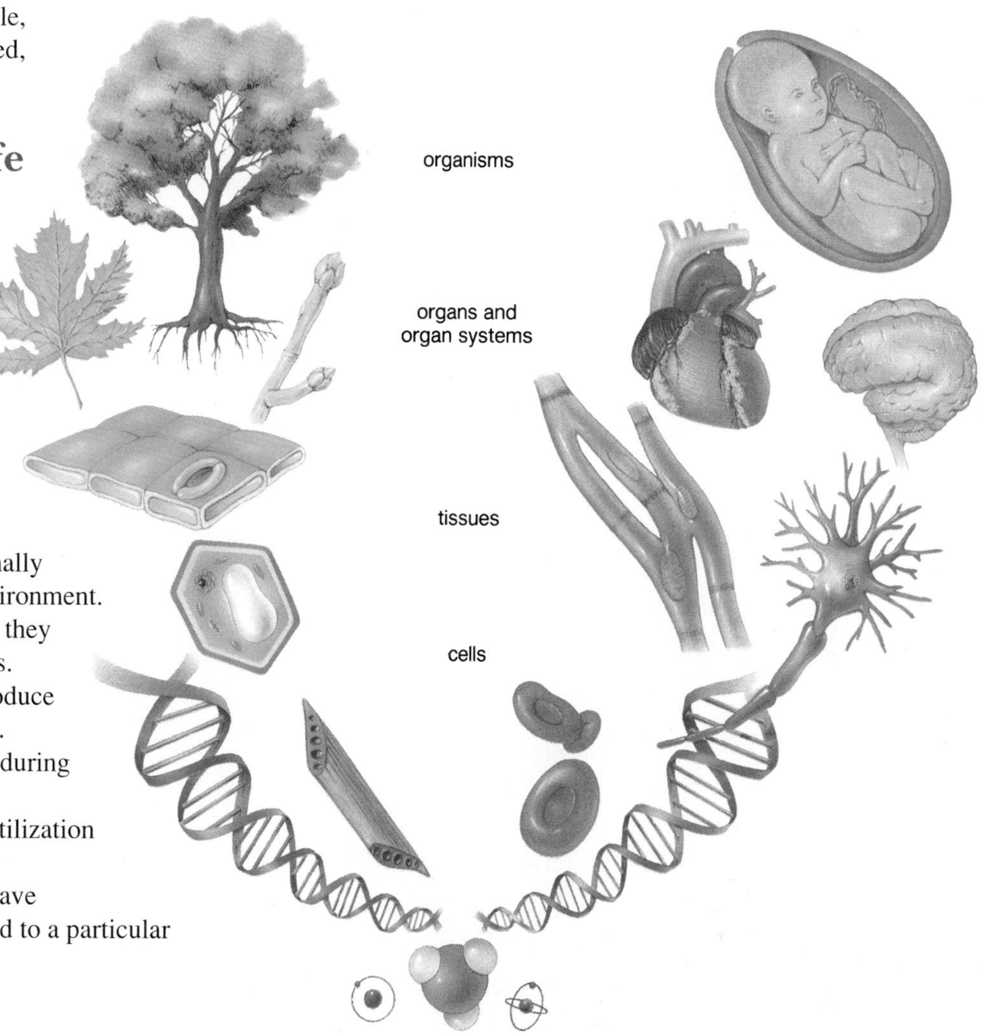

Figure 1.2
Levels of biological organization. Cells are composed of molecules; tissues are made up of cells; organs are composed of tissues; and organisms, such as a tree or a human, contain organs.

Life Has Organization

Living things are highly organized. In humans, the heart pumps blood and its structure normally suits this function. If a heart attack occurs, a disorganization in structure causes a portion of the heart to no longer function as it should. Death brings total disorganization of living things.

Living things have various *levels of biological organization*. In trees and humans (fig. 1.2), for example, each organ has a specific structure and function. Because the heart is muscular and contains cavities it can collect and pump blood. The stomach wall, on the other hand, has deep folds, allowing it to expand and to store food. Organs are composed of tissues, and each tissue contains groups of similar cells. A **cell** is the lowest level of biological organization to have the characteristics of life. Multicellular organisms contain many, many cells; humans have 60 trillion–100 trillion cells. A cell is the smallest unit of life—there are even unicellular organisms.

Cells are made up of molecules. Just as food contains the molecules we call proteins, carbohydrates, and fats, so do cells. Molecules are made up of atoms. It is of interest to note that water, a molecule that contains the atoms hydrogen and oxygen, accounts for over 50% of a cell's weight.

Acquires Materials and Energy

Living things cannot maintain their organization nor carry on life's other activities without an outside source of materials and energy. Human beings and other animals, like ospreys (fig. 1.3), acquire materials and energy when they eat food. (Only photosynthesizers, like trees, are capable of using carbon dioxide, water, and solar energy to make their own food.)

Food provides nutrient molecules, which are used as building blocks or for energy. **Energy** is the capacity to do work, and it takes work to maintain the organization of the cell and of the organism. When nutrient molecules are used to make

Figure 1.3

An osprey's way of life. **a.** An osprey is a large bird that preys on fishes. **b.** An osprey nests on tops of trees, rock pinnacles, and even telephone poles. From 2 to 4 eggs are laid in the large nest made of sticks. It is 40–50 days before the fledglings leave the nest.

Figure 1.4
A plant's response to sunlight. The stem of a plant bends toward the light as it grows. Plants can capture solar energy and carry on photosynthesis, a process that allows them to make their own food.

their parts and products, cells carry out a sequence of synthetic chemical reactions. Some nutrient molecules are broken down completely to provide the necessary energy to carry out synthetic reactions. **Metabolism** is all these chemical reactions that occur in a cell.

Stays Just about the Same

Homeostasis means "staying the same." Actually, the internal environment stays *relatively* constant; for example, human body temperature fluctuates slightly during the day. Homeostasis can be overcome. The body's ability to maintain a normal internal temperature is somewhat dependent on the external temperature—we will die if the external temperature becomes overly hot or cold.

All systems of the body contribute to homeostasis. The digestive system provides nutrient molecules; the circulatory system transports them about the body; and the excretory system rids blood of metabolic wastes. The nervous and hormonal systems coordinate the activities of the other systems. One of the major purposes of this text is to show how all the systems of the human body help to maintain homeostasis.

Responds to Stimuli

Living things respond to external stimuli. All of us notice when a living thing moves in response to a stimulus. Animals move because they have a nervous system. Other living things use a variety of mechanisms in order to respond to both the physical and biological environments. Leaves of plants track the movements of the sun during the day, and when a houseplant is placed near a window, its stem bends to face the sun (fig. 1.4).

The responses of an organism to the external environment constitute the organism's **behavior.** Behavior is directed toward minimizing injury, acquiring food, and reproducing.

Reproduces

All living things **reproduce**—they make more of themselves (fig. 1.3*b*). Like begets like, because each new organism acquires hereditary information in the form of genes. Unicellular organisms reproduce asexually simply by dividing. The new cells have the same genes and structure as the single parent. Multicellular organisms usually reproduce sexually. Each parent, male and female, contributes one-half the total number of genes to the offspring which then has characteristics of both parents and does not resemble either one exactly.

Grows and Develops

Growth, recognized by an increase in size and often the number of cells, is a part of development. In humans, **development** includes all the changes that take place between conception and death. First, the fertilized egg develops into a newborn, and then a human goes through the stages of childhood, adolescence, and adulthood. Even aging is now considered a phase of development. Development also includes the repair that takes place following an injury.

All organisms undergo development. An oak tree progresses from an acorn to a seedling before it becomes an oak tree, and a spotted salamander goes through an aquatic stage before becoming a land-dwelling adult (fig. 1.5).

Adapts

Adaptations are modifications that make an organism suited to its way of life. Consider, for example, a bird like an osprey (fig. 1.3), which catches and eats fish. An osprey can fly because it has hollow bones to reduce its weight, massive flight muscles to depress and elevate the wings, and asymmetrical flight feathers, which give the wings a shape suitable for flying. As an osprey cruises above the surface of the water, its white breast and belly camouflage its presence; its strong feet take the first shock of the water as it dives; and then its long and sharp claws hold onto the slippery prey. These few examples tell how an osprey is adapted to its way of life.

Adaptations come about through evolution. **Evolution** is the process by which characteristics of species (a group of similarly constructed organisms that share common genes) change through time. When new variations arise that allow certain members of the species to capture more resources, these members tend to survive and to have more offspring than the other, unchanged members. Therefore, each successive generation will include more members with the new variation. In the end, most members of a species have the same adaptations to their environment.

Figure 1.5

Growth and development. **a.** Stages in the development of an oak tree, from an acorn to a seedling to an adult. **b.** Stages in the development of a spotted salamander, from an egg deposited by a female to an aquatic stage with gills to a land-dwelling adult.

a.

b.

Adaptations to various ways of life explain why life is so diverse. Evolution, which has been going on since the origin of life, explains both the unity and the diversity of life. All organisms share the same characteristics of life because they can trace their ancestry to the first cell or cells. Organisms are diverse because they are adapted to different ways of life.

All living things share the same characteristics of life: living things are organized; they take materials and energy from the environment; they are homeostatic; they respond to stimuli; they reproduce; they grow and develop; and they are adapted to their way of life.

The Classification of Living Things

Taxonomy is that part of biology dedicated to naming, describing, and classifying organisms. Taxonomists give each type of organism a scientific name in Latin. The scientific name is a binomial (*bi* means *two*; *nomen* means *name*). For example, the name for humans is *Homo sapiens,* and for corn, it is *Zea mays.* The first word is the genus, and the second word tells the species in that genus. (Note that the genus is capitalized, but the species is not; both are italicized.) Scientific names are universally used by biologists so as to avoid confusion. Common names tend to overlap and often are in the language of a particular country.

Taxonomists use the following categories to classify organisms as diverse, for example, as human beings and corn:

Categories	Human	Corn
kingdom	Animalia	Plantae
phylum*	Chordata	Anthophyta
class	Mammalia	Monocotyledons
order	Primates	Commelinales
family	Hominidae	Poaceae
genus	*Homo*	*Zea*
species**	*sapiens*	*mays*

* Division in kingdoms Plantae and Fungi

** To specify a particular organism, use the full name, e.g., *Homo sapiens.*

Figure 1.6

Classification of organisms. In this text, organisms are classified into the 5 kingdoms illustrated here.

Kingdom	Representative Organisms	
Monera (monerans)	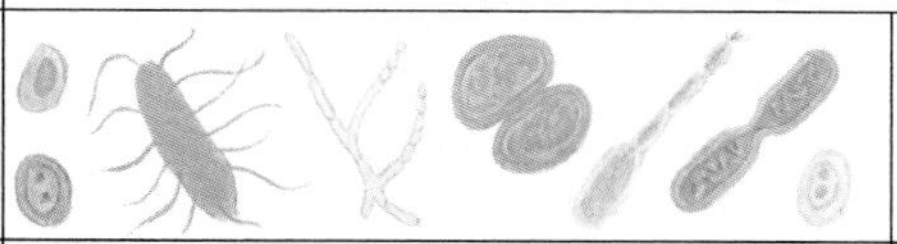	bacteria, including cyanobacteria
Protista (protists)	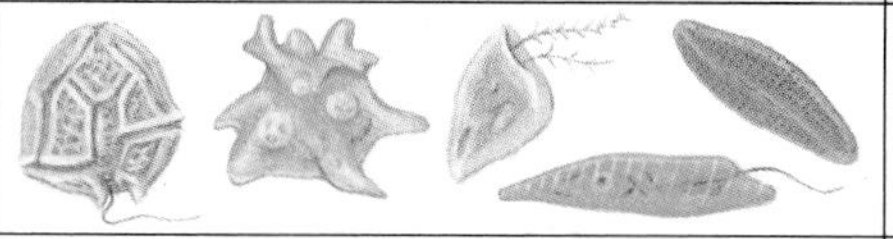	protozoans, all algae, and slime molds
Fungi (fungi)		molds and mushrooms
Plantae (plants)		mosses, ferns, various trees, and flowering plants
Animalia (animals)	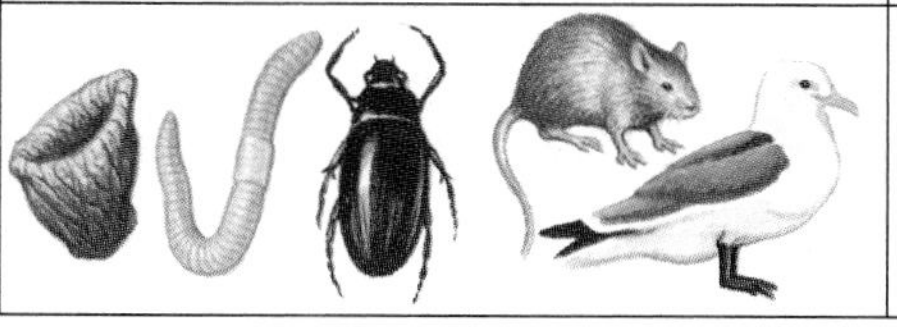	sponges, worms, insects, fishes, amphibians, reptiles, birds, and mammals

As we move from genus to kingdom, more and more different types of species are included in each successive category. Only human beings are in the genus *Homo,* but many different types of animals are in the animal kingdom. Species within the same genus share very specific characteristics, but those that are in the same kingdom have only general characteristics in common. All species in the genus *Zea* look pretty much the same—that is, like corn plants—while species in the plant kingdom are quite different, as is evident when we compare grasses to trees.

Taxonomy makes sense out of the bewildering variety of life on earth. Species are classified according to their presumed evolutionary relationship: those placed in the same genus are the most closely related, and those placed in separate kingdoms are the most distantly related. Taxonomists sometimes disagree about the number of kingdoms there are, although many today recognize the 5 kingdoms that are listed in figure 1.6.

> Scientists give known organisms a scientific name and classify them into taxonomic categories.

Ecosystems

The organization of life goes beyond separate and individual organisms. All living things on earth are part of the *biosphere,* a living network that spans the surface of the earth wherever organisms exist. The biosphere reaches up into the atmosphere and down into the soil and seas. In one area, such as a forest or a pond, all members of one species belong to a *population.* Here, the various populations, interacting with one another and with the physical environment, make up an **ecosystem** (fig. 1.7*a*). As in the case of the living organism, an ecosystem's organization usually enables it to continue to exist. The ecosystem is in a dynamic balance because of the interactions of its populations.

> Organisms are members of a population. There are many different populations within an ecosystem, a unit of the biosphere. Populations interact in such a way that the system is kept in dynamic balance.

Humans Threaten the Biosphere

The human population tends to modify existing ecosystems for its own purposes. For example, humans clear forests or grasslands in order to grow crops; later, they build houses on what was once farmland; and, finally, they convert small towns into cities. With each step, fewer and fewer original organisms remain, until at last the

Figure 1.7

Natural area versus one developed by humans. **a.** Plant and animal species thrive in an ecosystem. Their workings benefit the lives of humans, also. For example, ecosystems absorb pollutants if not overwhelmed. **b.** Humans usually live in developed areas with a limited variety of species, and these areas often produce the pollution that is harmful to all forms of life.

a.  b.

ecosystem is completely altered (fig. 1.7*b*). In the end, only humans and their domesticated plants and animals largely exist where once there were many diverse populations.

More and more ecosystems are threatened as the human population increases in size. Presently, there is great concern among scientists and laypersons about the destruction of the world's rain forests due to logging and the large numbers of persons who are starting to live and to farm there. This comes when we are beginning to realize how dependent we are on intact ecosystems and the services they perform for us. For example, the tropical rain forests act like a giant sponge, which absorbs carbon dioxide, a pollutant that pours into the atmosphere from the burning of fossil fuels, like oil and coal. An increased amount of carbon dioxide in the atmosphere is expected to have many adverse effects, such as an increase in the average daily temperature.

An ever-increasing human population size is a threat to the continued existence of *Homo sapiens* when it means that the dynamic balance of the biosphere is upset. The recognition that the workings of the biosphere need to be preserved is one of the most important developments of our new ecological awareness.

Biodiversity: Going, Going, Gone

The number of different species alive today is the highest it has been since life evolved. The biodiversity of the earth may total as many as 30 million species, of which only 1.4 million have so far been studied and classified. When ecosystems are destroyed, the species living there die off; therefore, today we are in the midst of a biodiversity-reduction crisis. For example, the existence of the species featured in figure 1.1 is threatened because tropical rain forests are being reduced in size. It is estimated that worldwide, one species becomes extinct each hour, and it is forecast that because of human activities, 50% of all species may become extinct within a few decades.

Humans are totally dependent on other species for food, clothing, medicines, and various raw materials. Therefore, it is very shortsighted of us to allow other species to become extinct. Nowhere is this more obvious than in the serious decline in the health and the biodiversity of the oceans. Off the coast of New England, 14 of the most valuable finfishes are becoming commercially extinct, meaning that too few remain to justify the cost of catching them. The reasons for this decline are overfishing, seaside development, loss of coastal wetlands, and pollution of areas, where fishes breed. The reading on page 11 discusses the plight of sharks, which may also cease to exist in the near future.

Ecosystems and the species living in them should be preserved because only then can the human species continue to exist. While the new genetic engineering can *improve* existing species, it cannot generate new ones. And it takes from 2,000 to 10,000 generations for new species to evolve and to replace the ones that have died out.

The human population tends to modify existing ecosystems and to reduce biodiversity. Because we are dependent upon the normal function and the present biodiversity of the biosphere, these should be preserved.

CAN SHARKS SURVIVE?

After 400 million years of evolution, the shark is the top predator in the ocean. At the apex of the food chain, cruising its domain in perpetual, primordial motion, it is the undisputed king of the undersea jungle.

But it has taken man—an even more ruthlessly efficient killer—just a single decade to threaten the shark's survival. Hunted for their meat and their fins, many species are in a dizzying ecological plunge. "In America, and around the world, sharks are being fished to oblivion," says University of Miami shark expert Samuel Gruber. "Without drastic conservation measures, some species will be lost."

Figure 1.A
Lemon shark. An entire population of lemon sharks in the Florida Keys was captured and used as crab bait.

Making soup. Sharks became gourmet fare in the late 1970s, after the National Marine Fisheries Service encouraged the harvesting of what had been widely regarded as an underutilized "trash" fish—good only for crab bait or sport angling. But the heaviest pressure to harvest sharks now comes from the rising demand in Asia for shark fins to make soup. Depending on the species, fins can be worth \$5 to \$30 a pound to U.S. fishermen. In parts of Asia, choice fins sell for as much as \$150 a pound. The boom has boosted the number of commercial boats targeting sharks or taking them as a bonus bycatch. It has also led to the gruesome and wasteful practice of "finning": Fishermen cut off the 2 valuable fins and toss the helpless animal back to starve.

The prospects for saving endangered shark species are complicated by the fact that the animals are slow to mature and reproduce. But for marine biologist Gruber, the idea of seas lacking sharks is deeply disturbing. Without the primary predator, fish populations would rapidly expand, possibly setting off a chain reaction that could upset the delicately balanced marine ecosystem all the way down to plankton. "We don't know enough about the true mechanics of the biosphere to predict with certainty what would happen," he says. "But we do know a diverse, balanced ecosystem is good. And for that, we need sharks."

Medical value. Some 350 species—from 8-inch cigar sharks to 35-foot whale sharks—inhabit the oceans. Beyond their food value, they show promising medical potential for humans. Their cartilage is used as artificial skin for burn victims, their corneas have been used for human replacement, and shark liver oil is a principal ingredient in many hemorrhoid ointments. Scientists are interested in studying their high resistance to cancers and their regenerative powers that allow wounds to heal rapidly. "It's amazing, the damage they suffer and how quickly they recover," Gruber says. "I've seen them with badly lacerated corneas and massive wounds from mating that heal very quickly. I've seen them with stingray spines stuck through their mouths or piercing the brain or heart cavity. They do fine."

A federal shark-recovery plan announced in January calls for an end to finning and for strict commercial- and sport-fishing limits designed to protect 39 coastal and high-seas species whose survival is threatened. Gruber says the plan is a good beginning, although for him, it comes about 4 years too late. In 1988, he was forced to abandon 30 years of behavioral studies of lemon sharks (fig. 1.A) in the Florida Keys after the population was wiped out for crab bait. "When I was a student, I thought the oceans were so wide and so vast that it was impossible to degrade the system," he says. "I thought it was a great sink for chemicals and pollution, and there was no way to put a dent in the marine populations. Boy, was I wrong."

The Process of Science

Science helps human beings understand the natural world. It is concerned solely with information gained by observing and testing that world. Science aims to be objective rather than subjective even though it is very difficult to make objective observations and to come to objective conclusions—we are often influenced by our own particular prejudices. Still, we should strive for objective observations and conclusions. Finally, scientific conclusions are subject to change whenever new findings so dictate. Quite often in science, new studies, which might utilize new techniques and equipment, tell us that previous conclusions need to be modified or changed entirely.

The ultimate goal of science is to understand the natural world in terms of **theories,** concepts based on the conclusions of experiments and observations (fig. 1.8). In a movie, a detective might claim to have a theory about the crime, or you might say that you have a theory about the win-lose record of your favorite baseball team, but in science, the word *theory* is reserved for a conceptual scheme supported by a large number of observations and not yet found lacking. Some of the unifying theories of biology are

Name of Theory	Explanation
Cell	All organisms are composed of cells.
Biogenesis	Life comes only from life.
Evolution	All living things have a common ancestor and are adapted to a particular way of life.
Gene	Organisms contain coded information that dictates their form, function, and behavior.

You can see that, in general, the theories pertain to the characteristics of life listed earlier in this chapter. Further, they apply to various aspects of living things. For example, the theory of evolution enables scientists to understand the history of life, the variety of living things, and the anatomy, physiology, and development of organisms—even their behavior. Because the theory of evolution has been supported by so many observations and experiments for over a hundred years, some biologists refer to the *principle* of evolution. They believe this is the appropriate terminology for theories that are generally accepted as valid by an overwhelming number of scientists.

Scientists ask questions and carry on investigations that pertain to the natural world. The conclusions of these investigations are tentative and subject to change. Eventually, it may be possible to arrive at a theory that is generally accepted by all.

Figure 1.8

The scientific method. Accumulated scientific data are used to formulate the hypothesis. Observations and experiments test the hypothesis. The new data allow researchers to come to a general conclusion about the phenomenon being studied. Several such conclusions (labeled 1, 2, 3) enable scientists to develop a comprehensive theory. For example, studies in comparative embryology, comparative anatomy, and paleontology all support the theory of evolution.

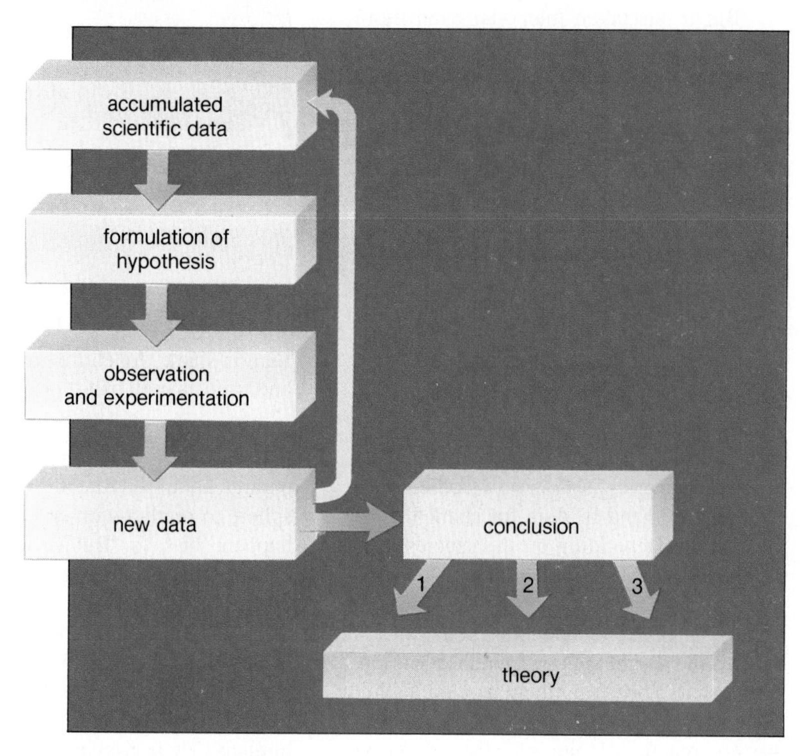

The Scientific Method: Has Steps

Scientists, including biologists, employ an approach to gathering information known as the **scientific method.** Although this approach is as varied as scientists themselves, there are still certain processes that can be identified as typical of the scientific method. Figure 1.8 outlines the essential steps in the scientific method. First, accumulated data are used to formulate a hypothesis, which becomes the basis for new observations and/or experimentation. The collected data help a scientist come to a conclusion that either supports or does not support the hypothesis. When many other observations and experiments result in conclusions that are consistent with a particular hypothesis, the hypothesis is called a theory. A theory that is accepted by many scientists sometimes is called a *principle* or a *law*.

In order to examine the scientific method in more detail, we will relate how scientists discovered the cause of Lyme disease.

Formulating the Hypothesis

A **hypothesis** is a tentative explanation of observed phenomena. To arrive at a hypothesis from accumulated **data** (factual information), scientists use various methods of reasoning, especially inductive reasoning. *Inductive reasoning* allows you to generalize after observing specific facts. For example, in 1975, a number of adults and children in the city of Lyme, Connecticut, were diagnosed as having rheumatoid arthritis. Since children rarely get rheumatoid arthritis, health officials called on Allen C. Steere, an authority on rheumatology, to investigate the matter. He found that (1) most of the victims lived in heavily wooded areas, (2) the disease was not contagious, (3) symptoms first occurred in summer, and (4) several victims remem-

Figure 1.9
This laboratory technician is testing blood samples for the presence of an infectious agent.

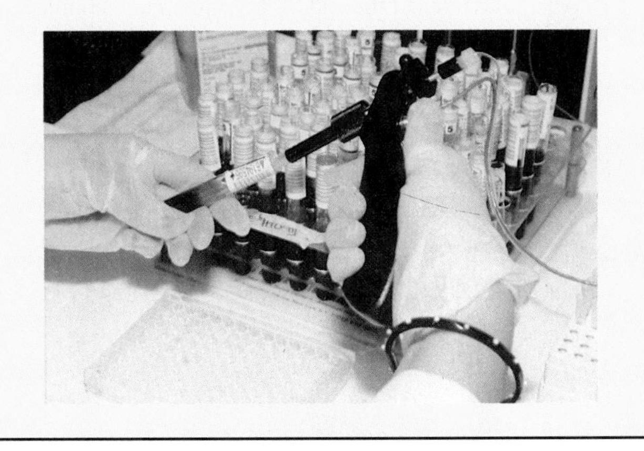

bered a strange bull's-eye rash occurring several weeks before the onset of arthritis-like symptoms. On the basis of his findings, Steere *hypothesized* that he was dealing with a disease caused by an agent (possibly a virus) transmitted by an arthropod (the phylum to which insects, spiders, and ticks belong). He named the illness Lyme disease, after the town where it was first observed.

Testing the Hypothesis

Once the hypothesis has been stated, deductive reasoning comes into play. *Deductive reasoning* begins with an "if … then" statement: If Lyme disease is caused by a virus, then it should be possible to find evidence of the virus in the blood of Lyme disease victims. Deductive reasoning allows scientists to decide which new *data* to collect. For example, Steere decided to test the blood of Lyme disease victims for the presence of every known arthropod-transmitted virus (fig. 1.9). Not a single test result was positive. Finally, in 1977, one victim who knew he had been bitten by a tick saved the arthropod and gave it to Steere. It was identified as *I. dammini,* commonly called the deer tick because adult ticks mate while feeding on a deer. Steere decided to see if the natural distribution of the deer tick corresponded to the outbreak of Lyme disease. This proved to be the case, and Steere's original hypothesis was supported. Later, Willy Burgdorfer, an authority on tick-borne diseases, isolated a spirochete (spiral) bacterium from deer ticks. The blood of Lyme disease victims tested positive for this bacterium. The new spirochete was named *Borrelia burgdorfei,* after Burgdorfer.

Because hypotheses are always subject to modification, they actually can never be proven true; however, they can be proven false; that is, falsifiable. When the data do not support the hypothesis, the hypothesis has to be rejected. It has been falsified; therefore, some think of science as what is left after alternative hypotheses have been rejected.

Even though the scientific method is quite variable, it is possible to point out certain steps that characterize it: making observations, formulating a hypothesis, testing, and coming to a conclusion. Do 1.1 Critical Thinking, found at the end of the chapter.

Reporting the Data

It is customary to report findings in a scientific journal so that the design and the results of the experiment are available to all. For example, data about tick-borne diseases are often reported in the journal *Clinical Microbiology Review.* It is necessary to give other researchers details on how experiments were conducted because results must be repeatable; that is, other scientists using the same procedures must get the same results. Otherwise, the hypothesis is no longer supported.

Often, too, authors of a report suggest what other types of experiments might clarify or broaden our understanding of the matter under study. People reading the report sometimes think of other experiments to do, also. The bull's-eye rash was later found to be due to the Lyme disease spirochete, for example.

Observations and the results of experiments are published in a journal, where they can be examined. These results are expected to be repeatable; that is, they will be obtained by anyone following the exact same procedure.

Uses Controlled Experiments

When scientists are studying a phenomenon, they often perform controlled experiments in a laboratory. A controlled experiment contains a **control sample,** which goes through all the steps of the experiment except the one being tested.

There are 2 major variables in a controlled experiment: the experimental variable and the dependent variable. The *experimental variable* is whatever is being tested, and the *dependent variable* is the result or change that is observed. The control sample is not subjected to the experimental variable.

Designing the Experiment

Suppose, for example, physiologists want to determine if sweetener S is a safe food additive (fig. 1.10). On the basis of available information, they hypothesize that sweetener S has no effect on health at a low concentration, but it does cause bladder cancer at a high concentration. They then might decide to feed sweetener S to groups of mice at ever-greater percentages of the total dietary intake of food.

Group 1: diet contains no sweetener S (the control)
Group 2: 5% of diet is sweetener S
Group 3: 10% of diet is sweetener S
↓
Group 11: 50% of diet is sweetener S

Figure 1.10

Design of a controlled experiment. From *left* to *right*: genetically identical mice are randomly divided into the control group and the test groups. All groups are exposed to the same environmental conditions, such as housing, temperature, and water supply. The control group is not tested (subjected to the experimental variable). In this case, the presence of sweetener S in the food is the variable. At the end of the experiment, all mice are examined for bladder cancer (dependent variable).

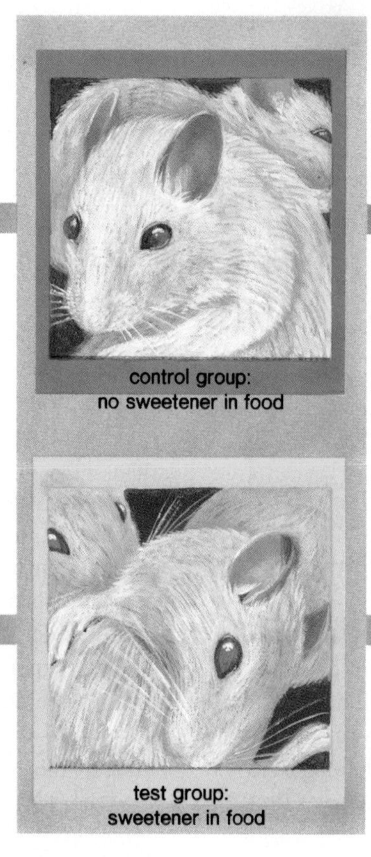

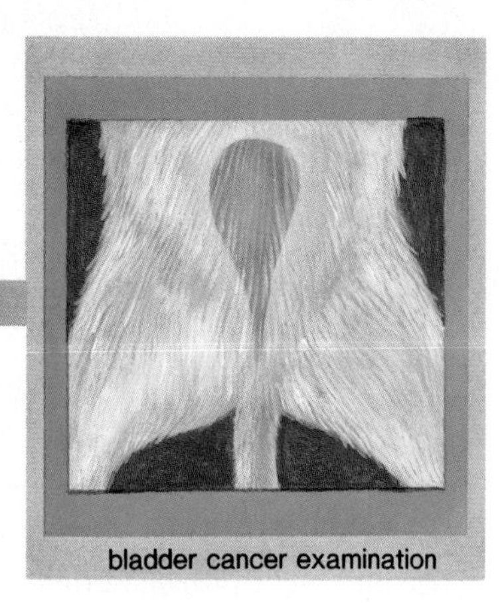

The researchers first place a certain number of randomly chosen inbred (genetically identical) mice into the various groups—say 10 mice per group. If any of the mice are different from the others, it is hoped random selection has distributed them evenly among the groups. The researchers also make sure that all conditions, such as availability of water, cage setup, and temperature of the surroundings, are the same for all groups. The food for each group is exactly the same except for the amount of sweetener S.

Analyzing the Results

Usually, data obtained from experiments such as this are presented in the form of a table or a graph (fig. 1.11). Researchers might run a statistical test to determine if the difference in the number of cases of bladder cancer between the various groups is significant. After all, if a significant number of mice in the control group develop cancer, the results are invalid. Scientists prefer *mathematical data* because they are highly objective and not subject to individual interpretation.

On the basis of the results, the experimenters try to develop a recommendation concerning the safety of sweetener S in the food of humans. They might determine, for example, that as the intake of sweetener S over 10% of food is increased, an ever-greater incidence of bladder cancer is expected.

Many scientists work in laboratories, where they carry out controlled experiments. Do 1.2 Critical Thinking, found at the end of the chapter.

Figure 1.11

Presenting the data. Scientists often report mathematical data in the form of a table or a graph. Mathematical data are more decisive and objective than visual observations. The data in this instance suggest there is a correlation between the amount of sweetener S in food and the incidence of bladder cancer. Similar experiments will be repeated many times to test these results, and the results will be statistically analyzed to determine if they are significant or due to chance alone.

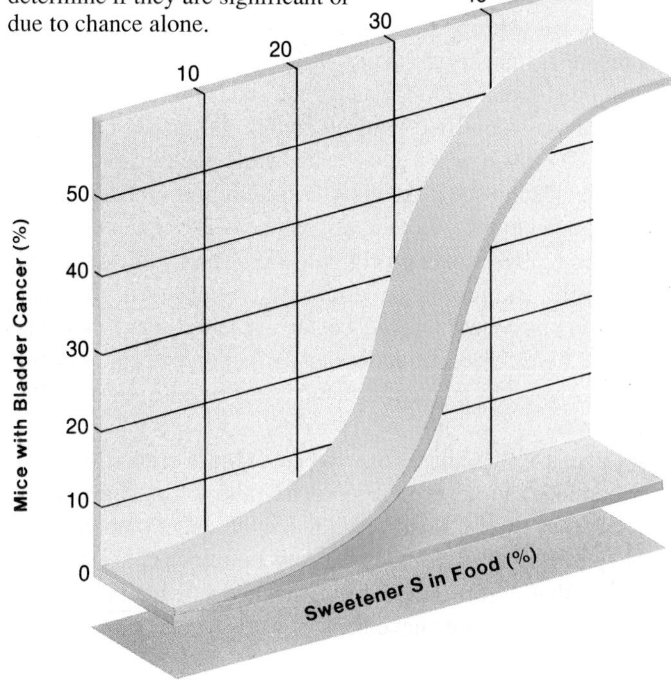

Science and Social Responsibility

There are many ways in which science has improved our lives. The most obvious examples are in the field of medicine. The discovery of antibiotics, such as penicillin, and the polio, measles, and mumps vaccines, for example, has increased our life span by decades. Cell biology research is helping us to understand the mechanisms that cause cancer. Genetic research has produced new strains of agricultural plants that have eased the burden of feeding our burgeoning world population.

Science has also had effects we find disturbing. It sometimes fosters technologies that can be ecologically disastrous if not controlled properly. Too often we blame science for these developments and think that scientists are duty bound to pursue only those avenues of research that are consistent with our present system of values and/or can never result in environmental degradation. We should understand, however, that the scientific method tests hypotheses about the natural world, and ethical and moral decisions are not testable. Therefore, all men and women have a responsibility to decide how to use scientific knowledge so that it benefits the human species and all living things.

The information presented in this text has been gathered using the scientific method. The text, while covering all aspects of biology, focuses on human biology. It is hoped that your study of biology will enable you to make wise decisions regarding your own individual well-being and also the well-being of our species.

SUMMARY

All living things share the same characteristics of life. Living things are organized; they take materials and energy from the environment; they are homeostatic; they respond to stimuli; they reproduce; they grow and develop; and they are adapted to their way of life.

There are these levels of biological organization:

atoms, molecules, cells, tissues,
organs, organ systems, organisms

Each level is more complex than the one preceding it. Cells, the lowest *living* level of organization, make up all organisms.

Because organisms are adapted to particular ways of life, their diversity is immense. Evolution accounts for both the diversity and the unity of life we see about us—all organisms share the same characteristics of life. Living things are classified according to their evolutionary relationships into these ever more specific categories:

kingdom, phylum, class, order,
family, genus, species

Organisms in different kingdoms are only distantly related; organisms in the same genus are very closely related.

When studying the world of living things, biologists and other scientists use the scientific method, which consists of these steps:

data, hypothesis, observation and
experimentation, new data, conclusions, theory

Inductive reasoning, which is based on previous and current observations and data, is used to formulate a hypothesis. Then deductive reasoning is used to decide which new experiments and observations are appropriate to test the hypothesis. The data may support a hypothesis or they may prove it false. Hypotheses cannot be proven true. All conclusions are subject to revision whenever new findings dictate. Still, there are some concepts, such as the theory of evolution, that have been supported for so long and by so many observations and experiments that they are generally accepted as true.

Science does not answer ethical questions; we must do this for ourselves. Knowledge provided by science, such as the contents of this text, can assist us in making decisions that will be beneficial to human beings and to other living things.

STUDY QUESTIONS

In order to practice **writing across the curriculum,** students should write out the answers to any or all of the study questions. The study questions are sequenced in the same order as the text.

1. Name the 7 characteristics of life, and discuss each one. (p. 5)
2. What are the levels of biological organization? What does it mean to say that living things are organized? (p. 5)
3. Food provides which 2 necessities for living things? (p. 5)
4. Give an example of homeostasis. Tell how the digestive system contributes to homeostasis in humans. (p. 7)
5. Why is the phrase "life-long developmental change" appropriate when speaking of humans? (p. 7)
6. How is an osprey adapted to its way of life? How does evolution explain both the diversity and the unity of life? (p. 7)
7. Explain the scientific name of an organism. (p. 9)
8. Name the categories of classification, from genus to kingdom. Which category contains more types of organisms having general characteristics in common? (p. 9)
9. What are ecosystems, and how are they modified by humans? (p. 9)
10. What is the ultimate goal of science? Give an example that supports your answer. (p. 12)
11. List the series of steps involved in the scientific method. (p. 12)
12. Give an example of a controlled experiment. Name the experimental variable and the dependent variable. (p. 13)

OBJECTIVE QUESTIONS

1. All living things are composed of _______, the smallest units of life.
2. All living things need a source of _______ and _______ to maintain themselves.
3. When a plant bends toward the light, it is _______ to an external stimulus.
4. When living things _______, the offspring resemble the parents.
5. Living things are suited, that is, _______, to their environment.
6. Living things are in a state of _______, that is, the internal environment stays relatively constant.
7. Scientists use _______ reasoning to formulate hypotheses.
8. Very often, the next step after formulation of the hypothesis is _______, a type of testing that usually includes a control sample.
9. Scientists try to be objective; therefore, they prefer _______ data.
10. In science, the word _______ is often used to stand for concepts based on many experiments and observations.
11. Correctly order (largest to smallest) these classification categories: genus, class, order, kingdom, phylum, species, family.
12. Correctly order (smallest to largest) these levels of biological organization: cells, organ systems, atoms, tissues, organisms, organs, molecules.

CRITICAL THINKING

In order to practice **writing across the curriculum,** students should write out the answers to any or all of the critical thinking questions. Suggested answers to the critical thinking questions are in appendix E.

1.1

Scientific hypotheses must be falsifiable.

1. Why is the hypothesis "Every human being has a guardian angel" not falsifiable?
2. Why is the hypothesis "Biotin is required for good health" falsifiable?
3. In what way are religious beliefs different from scientific beliefs?

1.2

1. A variable is an element that changes. Why is sweetener S called the experimental variable in the experiment described on page 13?
2. With reference to figure 1.10, explain this statement: A control group goes through all the steps of an experiment except the one being tested.
3. Why is bladder cancer the dependent variable in the described experiment?
4. Does the experiment have elements that are constant and invariable? What are they?
5. What is the value of a control group in an experiment?

SELECTED KEY TERMS

adaptation the fitness of an organism for its environment, including the process by which it becomes fit and is able to survive and to reproduce.

behavior all responses made by an organism to changes in the environment.

data experimentally derived facts.

development all the changes that take place during the life of an organism.

ecosystem a setting in which populations interact with each other and with the physical environment.

energy capacity to do work and bring about change; occurs in a variety of forms.

evolution changes that occur in the members of a species with the passage of time, often resulting in increased adaptation of organisms to the environment.

homeostasis the maintenance of the internal environment, such as temperature, blood pressure, and other body conditions, within narrow limits.

hypothesis a statement that is capable of explaining present data and is used to predict the outcome of future experimentation.

metabolism all of the chemical changes that occur within cells.

reproduce to make a copy similar to oneself; for example, bacteria dividing to produce more bacteria, or egg and sperm joining to produce offspring in more advanced organisms.

scientific method process by which scientists test their conclusions; consists of hypothesis generation and observation and experimentation, and results in theories.

theory a concept supported by a large number of conclusions drawn by using the scientific method.

2

THE CHEMISTRY OF LIFE

Chapter Concepts

1.
Atoms, units of matter, have a definite structure. 18, 38

2.
Atoms react with one another to form inorganic and organic molecules. 19, 38

3.
Some important inorganic molecules in living organisms are water, acids and bases, and salts. 23, 38

4.
The large organic molecules in cells are macromolecules. Macromolecules arise when their specific monomers (unit molecules) join together. 28, 38

5.
Some important organic molecules in cells are proteins, carbohydrates, lipids, and nucleic acids. 28, 38

Drosera capensis, a carnivorous plant

Chapter Outline

IT is not always easy to understand that living things, like nonliving things, are composed of chemicals. After all, it is not possible to see the chemicals that make up an organism's body. However, a few minutes' reflection regarding the dietary needs of the body usually convinces us that humans are indeed made of chemicals. For example, calcium is needed to maintain the bones, iron is necessary to prevent anemia, and adequate amino acid intake is required to build muscles.

Because living things, including humans, are composed of chemicals, it is absolutely essential for a biology student to have a basic understanding of chemistry.

Atoms

An **atom** is the smallest unit of matter nondivisible by chemical means. While it is possible to split an atom by physical means, an atom is the smallest unit to enter into chemical reactions. For our purposes, it is permissible to think of an atom as having a central **nucleus,** where subatomic particles called **protons** and **neutrons** are located, and *shells,* where **electrons** orbit about the nucleus (fig. 2.1). Electrons have varying amounts of energy. Those with the greatest amount of energy are located in the shells farthest from the nucleus. Other important features of protons, neutrons, and electrons are their charge and weight, which are indicated in table 2.1.

The periodic table of the elements in appendix A shows all the atoms that are presently known. An **element** is any substance that contains just one kind of atom. Figure 2.2 gives a simplified table highlighting the elements most common to living things. Notice that in the table, each atom has a symbol; for example, C = carbon and N = nitrogen. Each kind of atom also has an **atomic number;** for example, carbon is number 6 and nitrogen is number 7. *The atomic number equals the number of protons*. Each atom also has an **atomic weight,**[1] or mass. In the simplified table given here, carbon has an atomic weight, or mass, of 12 and nitrogen has an atomic weight of 14. *The atomic weight of an atom equals the number of protons plus the number of neutrons.*

Now, it is possible to diagram a specific electrically neutral atom (fig. 2.3). In an *electrically neutral atom,* the number of protons (+) is equal to the number of electrons (–). The first shell of an atom can contain up to 2 electrons; thereafter, each shell of those atoms in the simplified table (fig. 2.2) can contain up to 8 electrons.[2]

1. Atomic weights are relative weights. The most common isotope of carbon has been assigned an atomic weight of 12, and the other atoms are either lighter or heavier than carbon.

2. We will consider only atoms 1–20, in which all shells can have 8 electrons, except the first shell, which has 2 electrons.

Figure 2.1

Representation of an atom. The nucleus contains protons and neutrons, and the shells contain electrons.

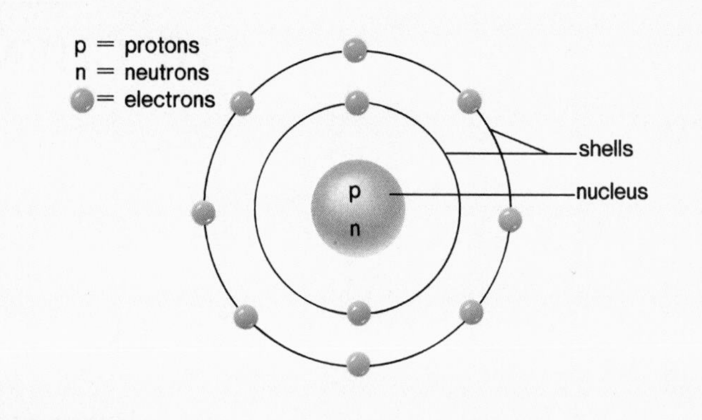

Table 2.1
Subatomic Particles

Name	Charge	Weight
Electron	One negative unit	Almost no weight
Proton	One positive unit	One atomic unit
Neutron	No charge	One atomic unit

Notice that the elements in the periodic table are horizontally arranged in order of increasing atomic number. They are also vertically arranged according to the number of electrons in the outermost shell. The Roman numerals at the top of each column indicate how many electrons there are in the outer shell of the elements listed in that column. An exception to this format is helium (He), which has only 2 electrons in the outer shell. The number of electrons in the outer shell determines the chemical properties of an atom, such as how readily it enters into chemical reactions.

Isotopes: Atomic Weights Vary

The atomic weights given in the simplified periodic table (fig. 2.2) have been rounded off; for example, the actual atomic weight of carbon (C) is 12.011. The atomic weight is the average weight of each kind of atom. Atoms of the same element can vary in weight, and when they do, they are called **isotopes.** Isotopes of carbon can be written in the following manner, where the subscript stands for the atomic number and the superscript stands for the atomic weight:

$$^{12}_{6}C \quad ^{13}_{6}C \quad ^{14}_{6}C^{*}$$

*radioactive

The number of protons (6) in these isotopes does not vary, but the number of neutrons—and therefore their weight—varies. Carbon-12 has 6 neutrons, carbon-13 has 7 neutrons, and carbon-14 has 8 neutrons.

Figure 2.2

Periodic table of the elements (simplified). See appendix A for the complete table. Each element has an atomic number, an atomic symbol, and an atomic weight. The elements most elevated and in a dark-green color are the most common, and those less elevated and in a light-green color are also common in living things. The black line separates metals (*left*) from nonmetals (*right*).

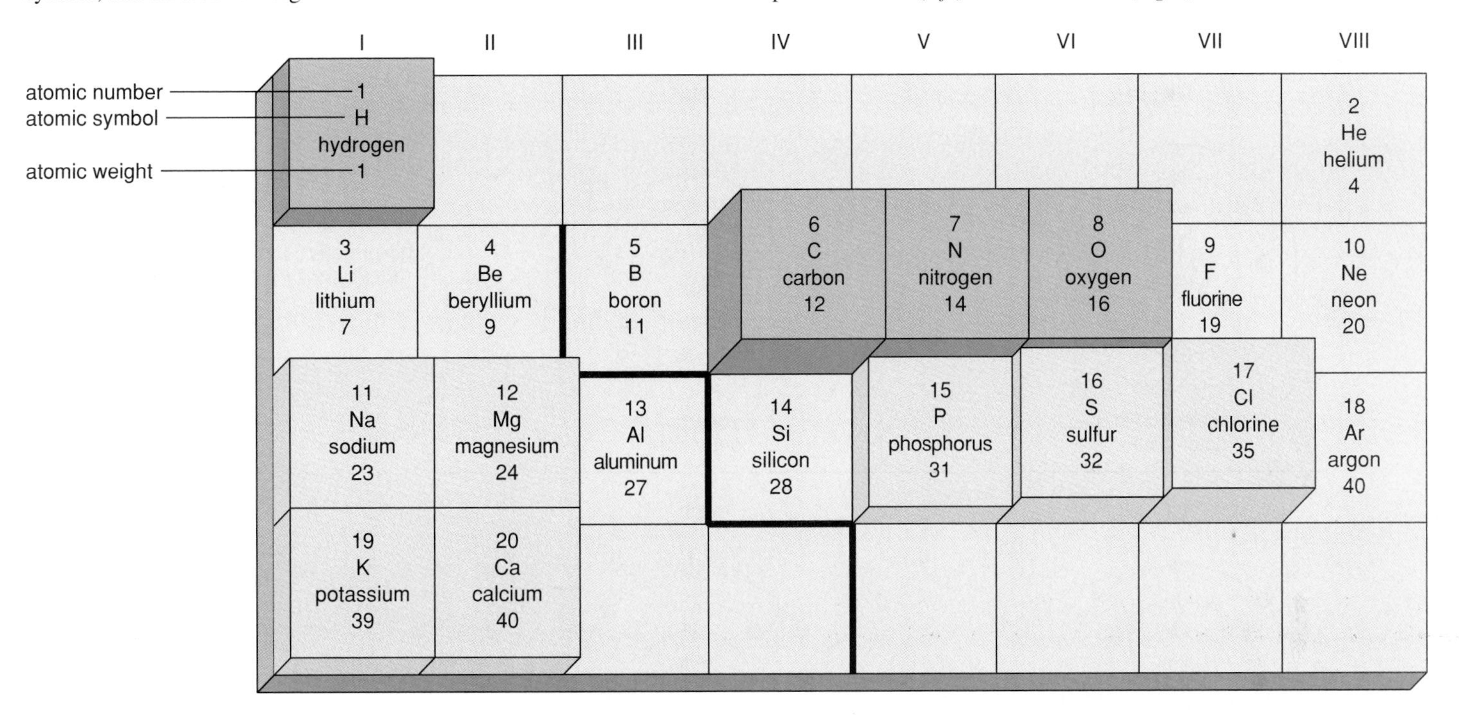

Figure 2.3

Carbon atom. The diagram shows that when the atom is electrically neutral, the number of protons (the atomic number) equals the number of electrons. Carbon can also be written in the manner shown below the diagram, where the subscript is the atomic number and the superscript is the atomic weight.

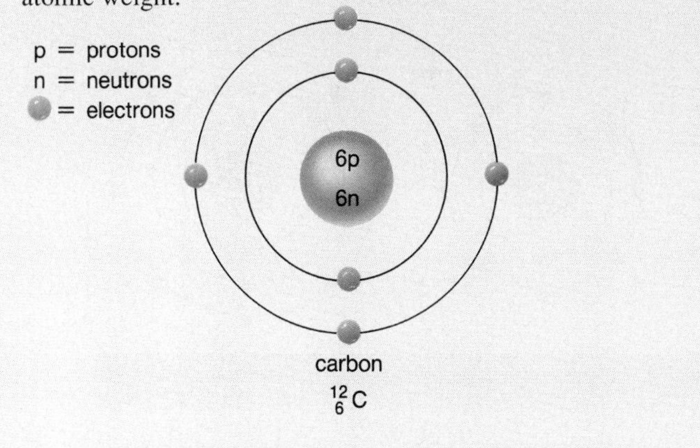

Certain isotopes, called **radioactive isotopes,** are unstable and as they decay, they emit radiation, which can be detected by using a special counter or scanner. Among those isotopes of carbon listed, only carbon-14 is radioactive, as the asterisk indicates. Radioactive isotopes are used widely in biological research and medical diagnostic procedures. For example, because the thyroid gland uses iodine, it is possible to administer a dose of radioactive iodine and then observe later that the thyroid has taken it up.

All matter is composed of atoms. The weight of an atom is dependent on the number of protons and neutrons in the nucleus. The reactivity of an atom is dependent on the number of electrons in its outermost shell.

Molecules and Compounds

Atoms often bond with each other to form a chemical unit called a **molecule.** Some molecules contain more than one kind of atom, and in those instances, a molecule is a part of a compound. A **compound** has many copies of the same type of molecule. The formula for a molecule tells the proportion of the different atoms in the molecule. For example, each water molecule contains one oxygen atom and 2 hydrogen atoms; the formula for water is, therefore, H_2O.

Octet Rule: Why Atoms React

Atoms with 8 electrons in their outermost shell do not react with one another. The *octet rule* explains the activity of other atoms. The octet rule says that atoms react with one another in order to achieve 8 electrons in their outer shell. Helium (He) and hydrogen (H) are exceptions to the octet rule because they have only one shell, which is complete with 2 electrons. Helium has 2 electrons in this shell; therefore, it does not react with other atoms. Hydrogen has only one electron and often reacts with other atoms in order to achieve another electron in the outer shell.

Figure 2.4

Ionic reactions. **a.** When the neutral atom sodium (Na) becomes an ion, it loses an electron. It then has 8 electrons in the outer shell. The sodium ion (Na^+) has a positive charge because it has one more proton than it has electrons. **b.** When the neutral atom chlorine (Cl) becomes an ion, it receives an electron. It then has 8 electrons in the outer shell. The chloride ion (Cl^-) has a negative charge because it has one less proton than it has electrons. **c.** When sodium reacts with chlorine, sodium gives an electron to chlorine, and sodium chloride (Na^+Cl^-) results. The 2 ions are held together by their opposite charge in an ionic bond.

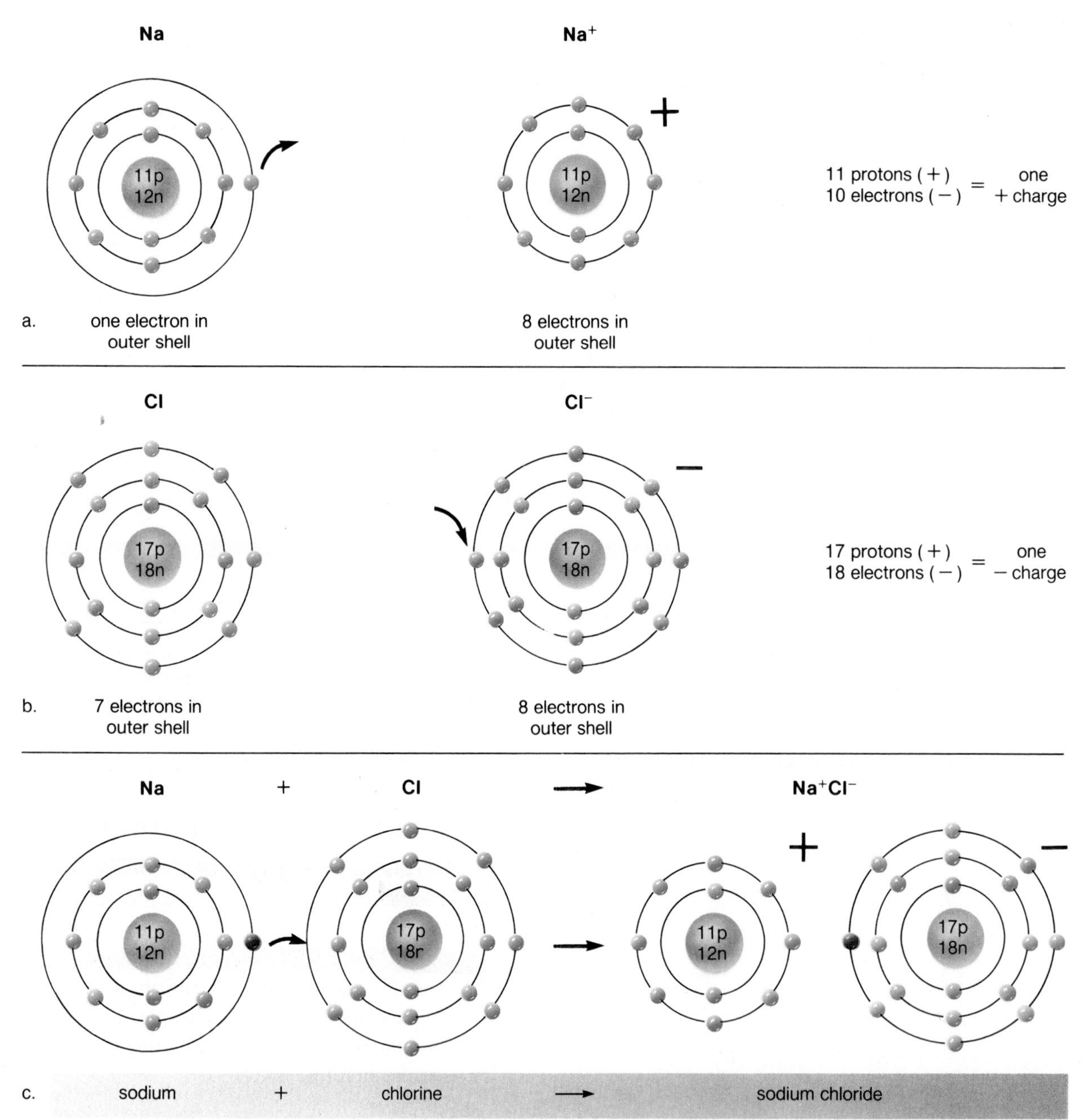

Ionic Reactions: Gain/Lose Electrons

In one type of reaction, atoms give up or take on electrons in order to achieve a completed outer shell. Such atoms, which thereafter carry a charge, are called **ions,** and the reaction is called an **ionic reaction.** In ionic reactions, atoms lose or gain electrons to produce a molecule that contains ions in a fixed ratio to one another. For example, figure 2.4 depicts a reaction between a sodium (Na) and chlorine (Cl), in which chlorine takes an electron from sodium. The resulting ions in sodium chloride (Na^+Cl^-) have 8 electrons each in the outer shell. Notice that when sodium gives up an electron, the second shell, with 8 electrons, becomes the outer shell. Now the sodium ion (Na^+)

Figure 2.5

Ionic reactions. When a metal reacts with a nonmetal, an ionic compound results—the metal has a positive charge, and the nonmetal has a negative charge. **a.** Magnesium (Mg) gives up 2 electrons to oxygen (O); in the compound magnesium oxide ($Mg^{++}O^{--}$), each magnesium ion (Mg^{++}) has 2 positive charges (why?), while each oxygen ion (O^{--}) has 2 negative charges (why?). **b.** Calcium (Ca) gives up one electron to each of 2 chlorines (Cl); in the compound calcium chloride ($Ca^{++}Cl_2^{-}$), each calcium ion (Ca^{++}) has 2 positive charges and each chloride ion (Cl^{-}) has one negative charge.

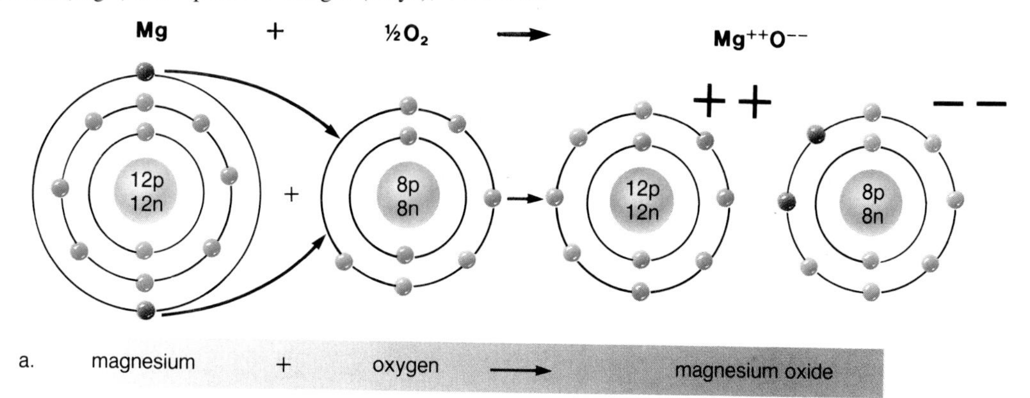

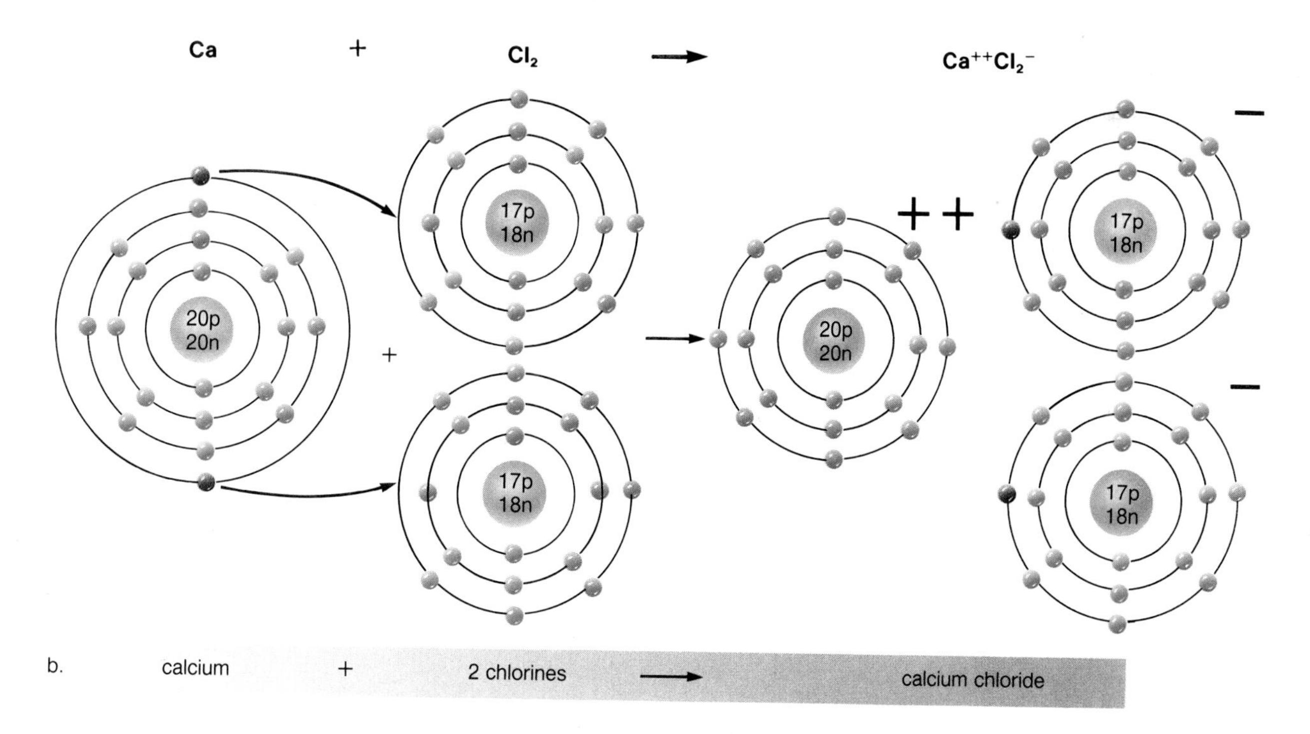

carries a positive charge because it has one more proton than electrons. Chlorine, on the other hand, receives an electron, for a total of 8 electrons in the outer shell. The chloride ion (Cl^{-}) has a negative charge because it has one more electron than protons.

The attraction between oppositely charged sodium ions and chloride ions forms an **ionic bond.** Ionic bonds are typically found in inorganic compounds. We are quite familiar with the inorganic compound sodium chloride because it is table salt, which we use to enliven the taste of foods.

Ionic-bond formation occurs when a **metal** reacts with a nonmetal. The elements that appear to the left of the black line in the simplified periodic table shown in figure 2.2 are metals, and nonmetals are those that appear to the right of this black line. Metals have atoms that lose electrons and become positively charged. In contrast, nonmetals have atoms that gain electrons and become negatively charged. This principle is illustrated in figure 2.5.

With the exception of hydrogen (H), atoms react with each other in order to achieve 8 electrons in the outermost shell. In an ionic reaction, positively and negatively charged ions form when electrons are transferred from metals to nonmetals. The attraction between ions forms an ionic bond.

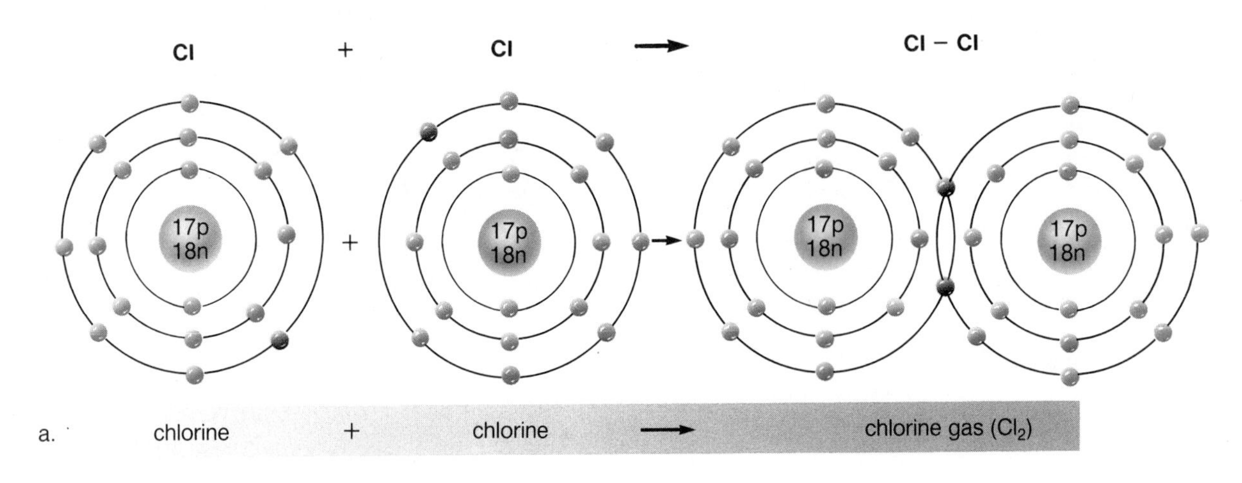

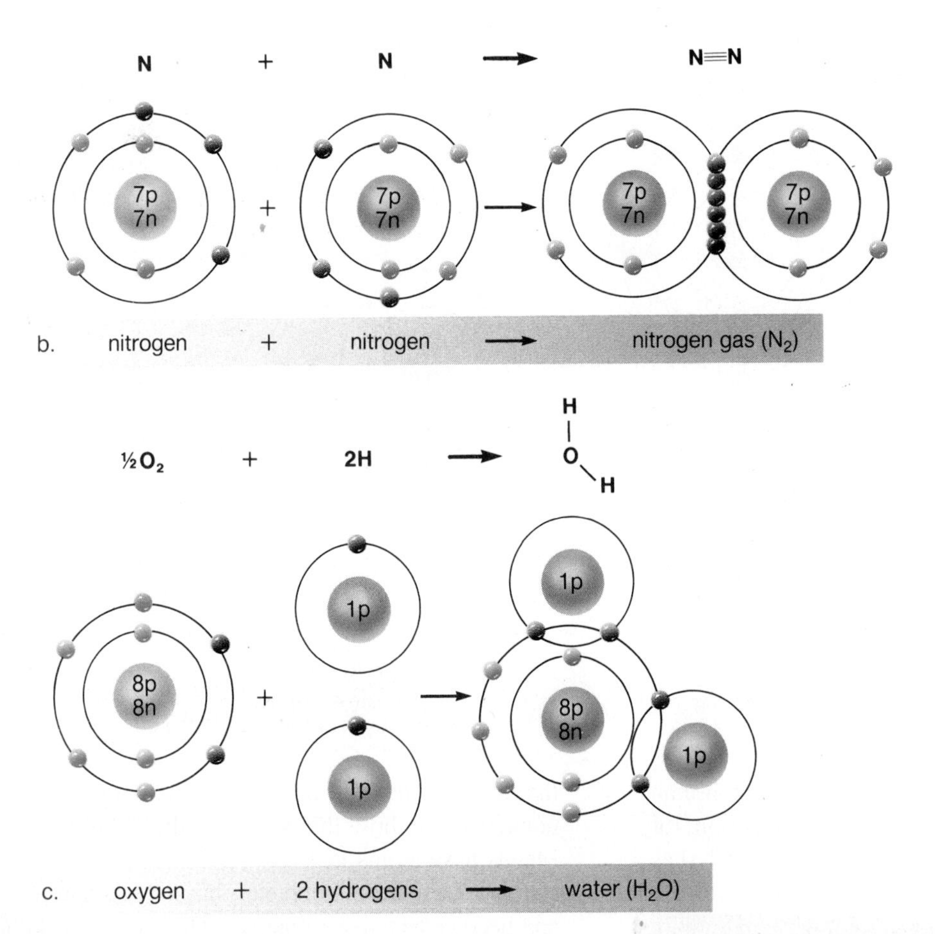

Figure 2.6
Covalent reactions. After a covalent reaction, atoms share electrons, and each atom has 8 electrons in the outer shell. To show this, it is necessary to count the shared electrons as belonging to both bonded atoms. **a.** Formation of chlorine gas (Cl_2). **b.** Formation of nitrogen gas (N_2). **c.** Formation of water (H_2O).

Covalent Reactions: Share Electrons

When nonmetals join with nonmetals, a **covalent reaction** occurs between atoms (fig. 2.6). In a **covalent bond,** the atoms share electrons instead of losing or gaining them. Covalent bonds, typically found in organic molecules, can be represented in a number of ways. The overlapping outer shells in figure 2.6 indicate that the atoms are sharing electrons. Each atom contributes one electron to each pair that is shared. These electrons spend part of their time in the outer shell of each atom; therefore, they are counted as belonging to both bonded atoms. When this is done, each atom has 8 electrons in the outermost shell.

Instead of drawing complex diagrams, electron-dot structures are sometimes used to depict covalent bonding between atoms. For example, in reference to figure 2.6*a*, each chlorine atom can be represented by its symbol, and the electrons in the outermost shell can be designated by dots. The shared electrons are placed between the 2 sharing atoms, as shown here:

$$:\ddot{\underset{..}{Cl}}\cdot + \cdot\ddot{\underset{..}{Cl}}: \rightarrow :\ddot{\underset{..}{Cl}}:\ddot{\underset{..}{Cl}}:$$

As electron-dot structures are cumbersome, other representations are often used. Structural formulas use straight lines to show the covalent bonds between the atoms. Each line represents a pair of shared electrons. Molecular formulas indicate only the number of each type of atom making up a molecule.

Structural formula: Cl — Cl

Molecular formula: Cl_2

Even if the molecule is written as Cl_2, it is easy to tell that the 2 chlorines are sharing electrons because (1) they are both nonmetals and (2) no charge is indicated. Additional examples of electron-dot, structural, and molecular formulas are shown in figure 2.7.

Double Bonds: Double and Triple

Besides a single bond, in which atoms share only a pair of electrons, a double or a triple bond can form in order for 2 atoms to complete the outer shell. In a triple bond, atoms share 3 pairs of electrons between them. For example, in figure 2.6*b*, each nitrogen atom (N) requires 3 electrons to achieve a total of 8 electrons in the outermost shell. Notice that 6 electrons are placed in the outer overlapping shells in the diagram and that 3 straight lines are in the structural formula for nitrogen gas (N_2).

In a covalent reaction, nonmetals share electrons in order to have completed outer shells. Each shared pair of electrons is a covalent bond; double and even triple bonds are possible.

Oxidation Loses, Reduction Gains

When oxygen (O) combines with a metal, oxygen receives electrons and forms ions that are negatively charged; the metal loses electrons and forms ions that are positively charged. For example, consider the reaction that is illustrated in figure 2.5*a*.

$$Mg + {}^1/_2\,O_2 \rightarrow Mg^{++}O^{--}$$

In such cases, it is obviously appropriate to say that the metal has been oxidized and that because of oxidation, the metal has lost electrons. Then we need only admit that the oxygen has been reduced because it has gained electrons, or negative charges.

Today, the terms **oxidation** and **reduction** are applied to many ionic reactions, whether or not oxygen is involved. Very simply, *oxidation refers to the loss of electrons, and reduction refers to the gain of electrons*. In the ionic reaction $Na + Cl \rightarrow Na^+Cl^-$, sodium has been oxidized (loss of electron) and chlorine has been reduced (gain of electron).

The terms *oxidation* and *reduction* also apply to certain covalent reactions. In this case, however, oxidation is the loss of hydrogen atoms (H) and reduction is the gain of hydrogen atoms. A hydrogen atom contains one proton and one electron; therefore, when a molecule loses a hydrogen atom, it has lost an electron, and when a molecule gains a hydrogen atom, it has gained an electron. We will have occasion to refer to this form of oxidation-reduction reaction again in chapter 6 because it is important to energy-conversion reactions.

When oxidation occurs, an atom is oxidized (loses electrons). When reduction occurs, an atom is reduced (gains electrons). These 2 processes occur concurrently in oxidation-reduction reactions.

Figure 2.7

Electron-dot, structural, and molecular formulas. In the electron-dot formula, only the electrons in the outer shell are designated. In the structural formula, the lines represent a pair of electrons being shared by 2 atoms. The molecular formula indicates only the number of each type of atom found within a molecule.

Electron-Dot Formula	Structural Formula	Molecular Formula
$\ddot{O}::C::\ddot{O}$ carbon dioxide	O═C═O carbon dioxide	CO_2 carbon dioxide
H : N : H (with H below, lone pair above) ammonia	H–N–H with H below (N bonded to 3 H) ammonia	NH_3 ammonia
H : O : H (lone pair below) water	H–O–H (H above O) water	H_2O water

Some Important Inorganic Molecules

Table 2.2 lists the characteristics that define **inorganic** molecules.

Water: Polar Molecule

Water is not an organic molecule because it does not contain carbon (C), but as figure 2.6*c* shows, the atoms in water (H_2O) are covalently bonded, as they are in organic molecules.

Table 2.2 Inorganic Molecules versus Organic Molecules	
Inorganic Molecules	**Organic Molecules**
Often associated with nonliving materials	Often associated with living organisms
Usually have ionic bonding between atoms	Always have covalent bonding between atoms
Usually contain metals and nonmetals	Always contain carbon and hydrogen
Always contain a small number of atoms	May be quite large, with many atoms

Both inorganic and organic molecules are involved in the proper functioning of the body.

Sometimes, covalently bonded atoms share electrons evenly, but in water, the electrons spend more time circulating the larger oxygen (O) than the smaller hydrogen (H). Therefore, there is a slight positive charge on the hydrogen atoms and a slight negative charge on the oxygen atom. Because the water molecule has charged atoms, it is called a **polar molecule.** Hydrogen bonding occurs between polar water molecules (fig. 2.8). A **hydrogen bond** occurs whenever a covalently bonded hydrogen is attracted to a negatively charged atom some distance away. The hydrogen bond is represented by a dashed line in figure 2.8 because it is relatively weak and can be broken.

Water's Characteristics

Because of hydrogen bonding, water has many characteristics beneficial to life. Hydrogen bonding causes water molecules to be cohesive—to cling together. Without hydrogen bonding between molecules, water would boil at −80°C and would freeze at −100°C, making life impossible. Instead, water boils at 100°C and freezes at 0°C; therefore, it is a liquid at body temperature. It absorbs a great deal of heat before it evaporates, and it gives off this heat as it cools down and freezes. This property allows great bodies of water, such as the oceans, to maintain a relatively constant temperature. It also helps to keep an animal's body temperature within normal limits and even accounts for the cooling effect of sweating.

The cohesiveness of water allows it to fill tubular vessels; as a result, water is an excellent medium for distributing substances and heat throughout the body. Its cohesive property is obvious whenever we observe the surface tension of bodies of water.

Also because of hydrogen bonding, liquid water is denser than ice. Therefore, ice floats on liquid water, and bodies of water always freeze from the top down, making skate sailing possible (fig. 2.9). Furthermore, the layer of ice protects the organisms below, helping them to survive the winter.

Figure 2.8

Hydrogen bonding between water molecules. Water molecules are polar: each hydrogen atom (H) carries a partial positive charge and each oxygen atom (O) carries a partial negative charge because the larger oxygen holds the electrons more tightly than the hydrogen. The polarity of the water molecules brings about hydrogen bonding between the molecules in the manner shown. The dashed lines represent hydrogen bonds.

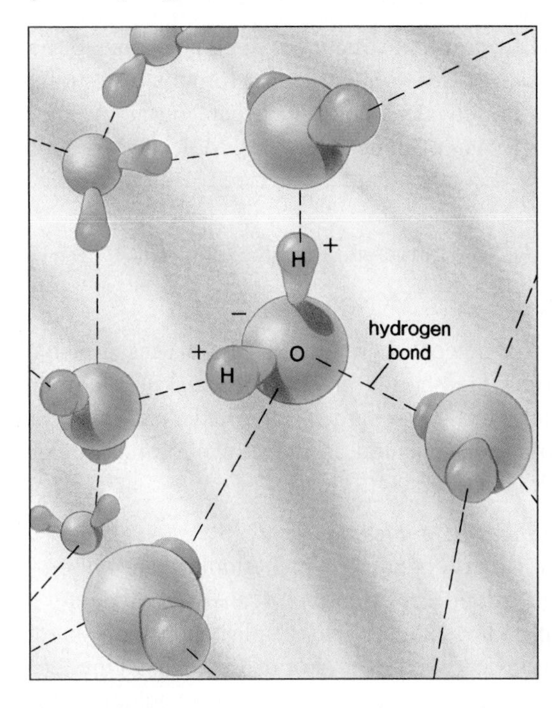

Figure 2.9

Ice. Hydrogen bonding causes liquid water to be denser than ice; therefore, bodies of water freeze from the top down. This protects the organisms that live in the water beneath the ice and also permits humans to enjoy skate sailing.

Figure 2.10
Dissociation of water molecules. In water, there are always a few molecules that have dissociated. Dissociation produces an equal number of hydrogen ions (H^+) and hydroxide ions (OH^-). (These figures are for illustration and are not mathematically accurate.)

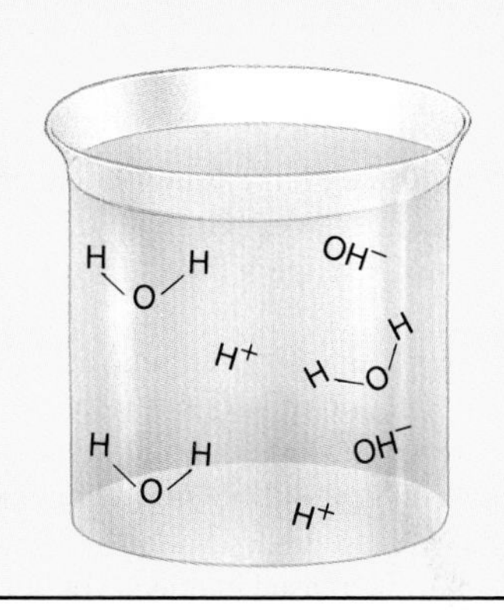

Figure 2.11
Hydrochloric acid (HCl). Hydrochloric acid releases hydrogen ions (H^+) as it dissociates. Notice that the addition of hydrochloric acid to this beaker of water has resulted in a solution with more hydrogen ions than hydroxide ions (OH^-).

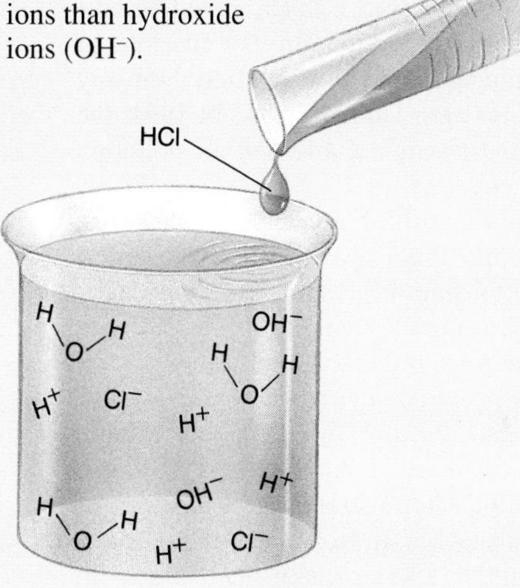

Figure 2.12
Sodium hydroxide (NaOH), a base. Sodium hydroxide releases hydroxide ions (OH^-) as it dissociates. Notice that the addition of sodium hydroxide to this beaker of water has resulted in a solution with more hydroxide ions than hydrogen ions (H^+).

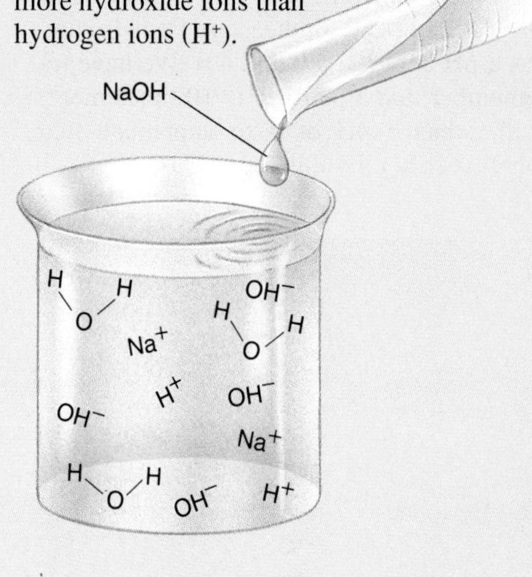

Water, being a polar molecule, acts as a solvent; it dissolves many chemical substances, particularly other polar molecules. This property of water greatly facilitates chemical reactions in cells.

Life is dependent on the various characteristics of water. Do 2.1 Critical Thinking, found at the end of the chapter.

Dissociation: Breaking Up. Polarity also causes water to tend to **dissociate,** or ionize, in this manner:

$$H—O—H \rightarrow H^+ + OH^-$$

The hydrogen ion (H^+) has lost an electron; the hydroxide ion (OH^-) has gained the electron. Very few molecules actually dissociate; therefore, few hydrogen ions and hydroxide ions result (fig. 2.10).

Acid (H^+ Up); Base (H^+ Down)

Acids are molecules that dissociate in water, releasing hydrogen ions (H^+) (or protons[3]). For example, an important inorganic acid is hydrochloric acid (HCl), which dissociates in this manner:

$$HCl \rightarrow H^+ + Cl^-$$

Dissociation is almost complete; therefore, this is called a strong acid. If hydrochloric acid is added to a beaker of water (fig. 2.11), the number of hydrogen ions (H^+) increases.

Bases are molecules that either take up hydrogen ions (H^+) or release hydroxide ions (OH^-). For example, an important inorganic base is sodium hydroxide (NaOH), which dissociates in this manner:

$$NaOH \rightarrow Na^+ + OH^-$$

Dissociation is almost complete; therefore sodium hydroxide is called a strong base. If sodium hydroxide is added to a beaker of water (fig. 2.12), the number of hydroxide ions increases.

pH Scale

The **pH[4] scale** is used to indicate the acidity and basicity (alkalinity) of a solution. A pH of exactly 7 is neutral pH. Pure water has an equal number of hydrogen ions (H^+) and hydroxide ions (OH^-), and therefore, one of each is formed when water dissociates. The fraction of water molecules that dissociate is 10^{-7} (or 0.0000001), which is the source of the pH value for neutral solutions. The pH scale was devised to simplify discussion of the hydrogen ion concentration $[H^+]$ and, consequently, of the hydroxide ion concentration $[OH^-]$; it eliminates the use of cumbersome numbers. For example,

a. $1 \times 10^{-6}\ [H^+] = \text{pH } 6$
b. $1 \times 10^{-7}\ [H^+] = \text{pH } 7$
c. $1 \times 10^{-8}\ [H^+] = \text{pH } 8$

Each lower pH unit has 10 times the amount of hydrogen ions (H^+) as the next higher unit.

3. A hydrogen atom contains one electron and one proton. A hydrogen ion has only one proton so is often called a proton.

4. pH is defined as the negative logarithm of the hydrogen ion concentration $[H^+]$.

THE HARM DONE BY ACID RAIN

Normally, rainwater has a pH of about 5.6 because the carbon dioxide in the air combines with water to give a weak solution of carbonic acid. Rain falling in northeastern United States and southeastern Canada now has a pH between 5.0 and 4.0. We have to remember that a pH of 4 is 10 times more acidic than a pH of 5 to appreciate the increase in acidity this represents.

There is very strong evidence that this observed increase in rainwater acidity is a result of the burning of fossil fuels, like coal and oil, as well as gasoline derived from oil. When fossil fuels are burned, sulfur dioxide and nitrogen oxides are produced and they combine with water vapor in the atmosphere to form acids. These acids return to earth as rain or snow, a process properly called wet deposition but more often called acid rain. Dry particles of sulfate and nitrate salts descend from the atmosphere during dry deposition.

Unfortunately, regulations that require the use of tall smokestacks to reduce local air pollution only cause pollutants to be carried far from their place of origin. Acid deposition in southeastern Canada is due to the burning of fossil fuels in factories and powerplants in the Midwest. Tensions between Canada and the United States have been eased by the 1990 Clean Air Act; it called for a 10-million-ton reduction in sulfur dioxide emissions and a 2-million-ton reduction in nitrogen oxide emissions in the United States by the year 2000. Thereafter, sulfur dioxide emissions will be capped and no increase in total emissions will be allowed. Canada agreed to make comparable reductions in its emissions of air pollutants, some of which find their way to northeastern United States. In 1991, the two countries signed a formal air pollution agreement.

Acid deposition adversely affects lakes, particularly in areas where the soil is thin and lacks limestone (calcium carbonate, $CaCO_3$), a buffer to acid deposition. It leaches aluminum from the soil, carries aluminum into the lakes, and converts mercury deposits in lake bottom sediments to soluble methyl mercury. Lakes not only become more acidic, they also show accumulation of toxic substances. In Norway and Sweden, at least 16,000 lakes contain no fish and an additional 52,000 lakes are threatened. In Canada, some 14,000 lakes are almost fishless and an additional 150,000 are in peril because of excess acidity. In the United States, about 9,000 lakes (mostly in the Northeast and Upper Midwest) are threatened, one-third of them seriously.

In forests, acid deposition weakens trees because it leaches away nutrients and releases aluminum. By 1988, most spruce, fir, and other conifers atop North Carolina's Mt. Mitchell were dead from being bathed in ozone and acid fog for years. The soil was so acidic new seedlings could not survive. Nineteen countries in Europe have reported woodland damage ranging from 5% to 15% of the forested area in Yugoslavia and Sweden to

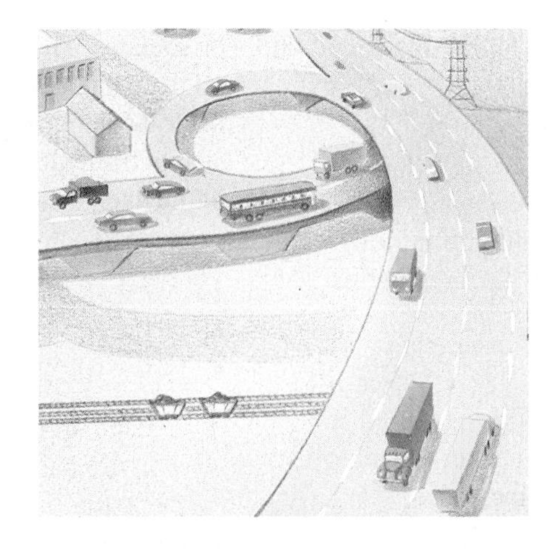

50% or more in the Netherlands, Switzerland, and the former West Germany. More than one-fifth of Europe's forests are now damaged.

These aren't the only effects of acid deposition. Reduction of agricultural yields, damage to marble and limestone monuments and buildings, and even illnesses in humans have been reported. Acid deposition has been implicated in the increased incidence of lung cancer and possibly colon cancer in residents of the East Coast. Tom McMillan, Canadian Minister of the Environment, says that acid rain is "destroying our lakes, killing our fish, undermining our tourism, retarding our forests, harming our agriculture, devastating our heritage and threatening our health."

There are of course, things that can be done. We could

a. whenever possible use alternative energy sources, such as solar, wind, hydropower, and geothermal energy.

b. use low-sulfur coal or remove the sulfur impurities from coal before it is burned.

c. require factories and power plants to use scrubbers, which remove sulfur emissions.

d. require people to use mass transit rather than driving their own automobiles.

e. reduce our energy needs through other means of energy conservation.

These measures and possibly others could be taken immediately. We only need to determine that they are necessary.

Sources: G. Tyler Miller, *Living in the Environment,* Wadsworth Publishing Company, Belmont, CA, 1985; and Lester R. Brown, et al., *State of the World,* W.W. Norton and Company, Inc., New York, NY, 1988.

Figure 2.13

The pH scale. The proportionate amount of hydrogen ions (H^+) to hydroxide ions (OH^-) is indicated by the diagonal line. Any pH above 7 is basic, while any pH below 7 is acidic.

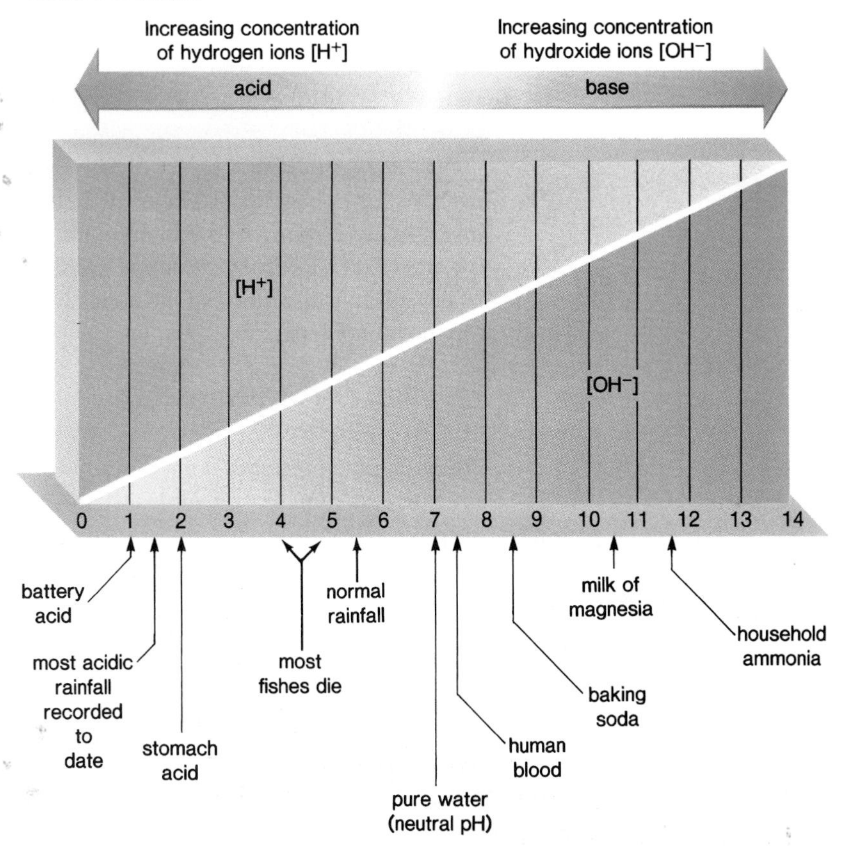

In order to understand the relationship between hydrogen ion concentration and pH, consider the following question. Of the 2 values *a* and *c* listed on page 25, which indicates a higher hydrogen ion concentration than pH 7 (neutral pH) and therefore refers to an acidic solution? A number with a smaller negative exponent indicates a greater quantity of hydrogen ions (H^+) than one with a larger negative exponent. Therefore, pH 6 is an acidic solution.

Bases add hydroxide ions (OH^-) to solutions and increase the hydroxide ion concentration [OH^-] of water. Basic solutions, then, have fewer hydrogen ions (H^+) compared to hydroxide ions. Which of the values—*a* or *c*—in the previous list refers to a basic solution? pH 8 refers to a basic solution because it indicates a lower hydrogen ion concentration [H^+] (greater hydroxide ion concentration) than pH 7.

The pH scale (fig. 2.13) ranges from 0 to 14. It uses whole numbers, instead of negative exponents of the number 10, to indicate the hydrogen ion concentration [H^+]. As we move down the pH scale, each unit has 10 times the acidity of the previous unit, and as we move up the scale, each unit has 10 times the basicity of the previous unit. A pH of 7 has an equal concentration of hydrogen ions (H^+) and hydroxide ions (OH^-). Above pH 7 there are more hydroxide ions than hydrogen ions, and below pH 7 there are more hydrogen ions than hydroxide ions.

In living things, pH needs to be maintained within a narrow range or there are health consequences. The reading on page 26 discusses the environmental consequences of rain and snow becoming more acidic.

Buffers Keep pH Steady

The pH of our blood when we are healthy is always about 7.4. **Buffers,** chemicals or combinations of chemicals that take up excess hydrogen ions (H^+) or hydroxide ions (OH^-), help to keep the pH constant. For example, carbonic acid (H_2CO_3) is a weak acid that minimally dissociates and then re-forms in the following manner:

$$\underset{\text{carbonic acid}}{H_2CO_3} \underset{\text{re-forms}}{\overset{\text{dissociates}}{\rightleftharpoons}} H^+ + \underset{\text{bicarbonate ion}}{HCO_3^-}$$

Blood always contains some carbonic acid and some bicarbonate ions. When hydrogen ions (H^+) are added to blood, the following reaction occurs:

$$H^+ + HCO_3^- \rightarrow H_2CO_3$$

When hydroxide ions (OH^-) are added to blood, this reaction occurs:

$$OH^- + H_2CO_3 \rightarrow HCO_3^- + H_2O$$

These reactions prevent any change in blood pH.

Salts: Acid + Base

When an acid reacts with a base, a salt and a water molecule result. In the case of a strong acid and a strong base, the reaction is complete:

$$HCl + NaOH \rightarrow \underset{\text{salt}}{Na^+Cl^-} + \underset{\text{water}}{HOH}$$

Further, if an equal amount of both take part in this reaction, neutralization occurs; the resulting solution is neither acidic nor basic. In the neutralization process, the hydrogen ions (H^+) from the acid and the hydroxide ions (OH^-) from the base combine to form water. The salt consists of the positive ion of the base and the negative ion of the acid.

> Acids have a pH that is less than 7, and bases have a pH that is greater than 7. Buffers, which can absorb both hydrogen ions and hydroxide ions, help to keep the pH of internal body fluids near pH 7 (neutral).

Figure 2.14

Synthesis and hydrolysis. When synthesis of a macromolecule (polymer) occurs, monomers (small molecules) join as water is released. When hydrolysis of a macromolecule occurs, water is added as bonds are broken and monomers are released.

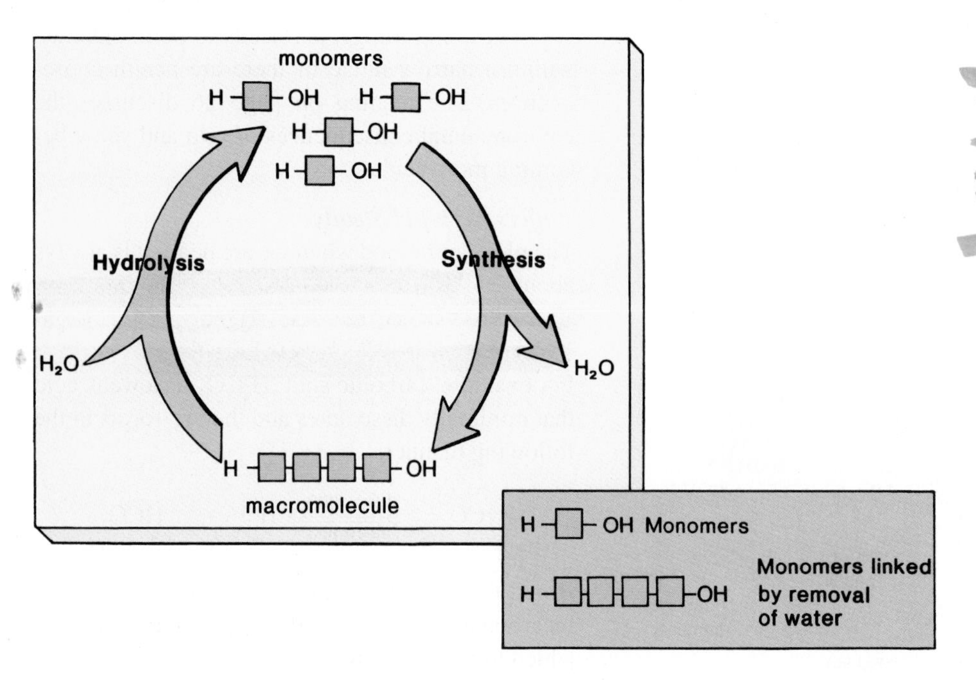

When monomers join to form a macromolecule, a bond forms between adjacent monomers. This bond forms when a hydrogen (–H) from one monomer is caused to link with a hydroxyl group (–OH) from another monomer. As a water molecule forms, dehydration **synthesis** occurs. Conversely, a macromolecule is broken down by the addition of water molecules. During this process, called **hydrolysis,** one monomer takes on a hydrogen and the adjacent monomer takes on a hydroxyl group (fig. 2.14). This leads to a disruption of the bonds linking the monomers.

Some Important Organic Molecules

Table 2.2 lists the characteristics that define **organic** molecules. Organic molecules always contain carbon (C) and hydrogen (H). The chemistry of carbon accounts for the formation of the very large variety of organic molecules we associate with living organisms. Carbon is a nonmetal, and a carbon atom has 4 electrons in the outermost shell. In order to achieve 8 electrons in the outer shell, a carbon atom shares electrons with as many as 4 other atoms, forming covalent bonds. Many times, carbon atoms share with each other to form a ring or a chain.

Cells contain very large molecules called *macromolecules.* A macromolecule forms when smaller molecules join together. Each of these smaller molecules is called a *monomer.* When monomers form a chain, the macromolecule is a *polymer.* The macromolecules of cells and their monomers are shown here:

Macromolecules	Monomers
protein	amino acid
carbohydrate	monosaccharide
lipid	glycerol and fatty acid
nucleic acid	nucleotide

Proteins: For Structure and Metabolism

Protein macromolecules sometimes have a structural function. For example, in humans, the protein keratin makes up hair and nails, while collagen is found in all types of connective tissue, including ligaments, cartilage, bones, and tendons. The muscles contain proteins, which account for their ability to contract.

Some proteins are **enzymes,** necessary contributors to the chemical workings of the cell and therefore of the body. Enzymes speed up chemical reactions; they work so quickly that a reaction that normally takes several hours or days without an enzyme takes only a fraction of a second with an enzyme. Specific enzymes in the body assist synthetic reactions, which build up macromolecules; others carry out hydrolytic reactions, which break down macromolecules (fig. 2.14).

Have Twenty Amino Acids

Proteins are polymers, or chains, of amino acids. A protein is characterized by the sequence of amino acids it contains. Most amino acids have this structural formula:

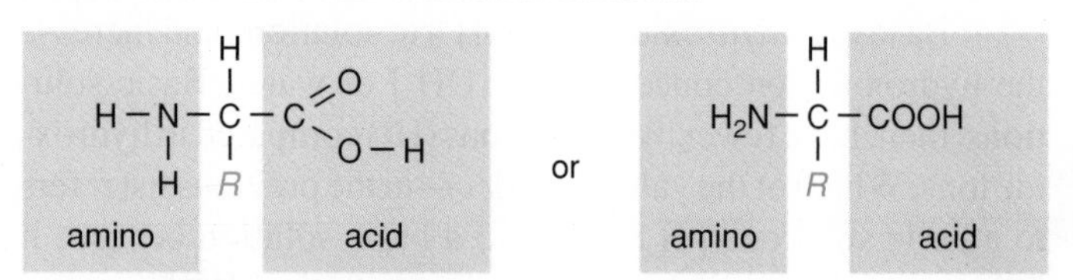

R = remainder of the molecule

The term **amino acid** is appropriate because this type of molecule has 2 functional groups: an *amino* group ($-NH_2$) and an acid (*carboxyl*) group (–COOH). Amino acids differ from one another by their *R* group, the *R*emainder of the molecule (see preceding diagram). In amino acids, the *R* group varies from a single hydrogen (H) to complicated rings (fig. 2.15). There are 20 different amino acids, and therefore, about 20 different types of *R* groups are commonly found in proteins.

Figure 2.15

Representative amino acids. Amino acids differ from one another by their *R* group; the simplest *R* group is a single hydrogen atom (H). Those that contain carbon vary as shown. Notice that each amino acid is shown twice and that the second drawing is a simplified structural formula.

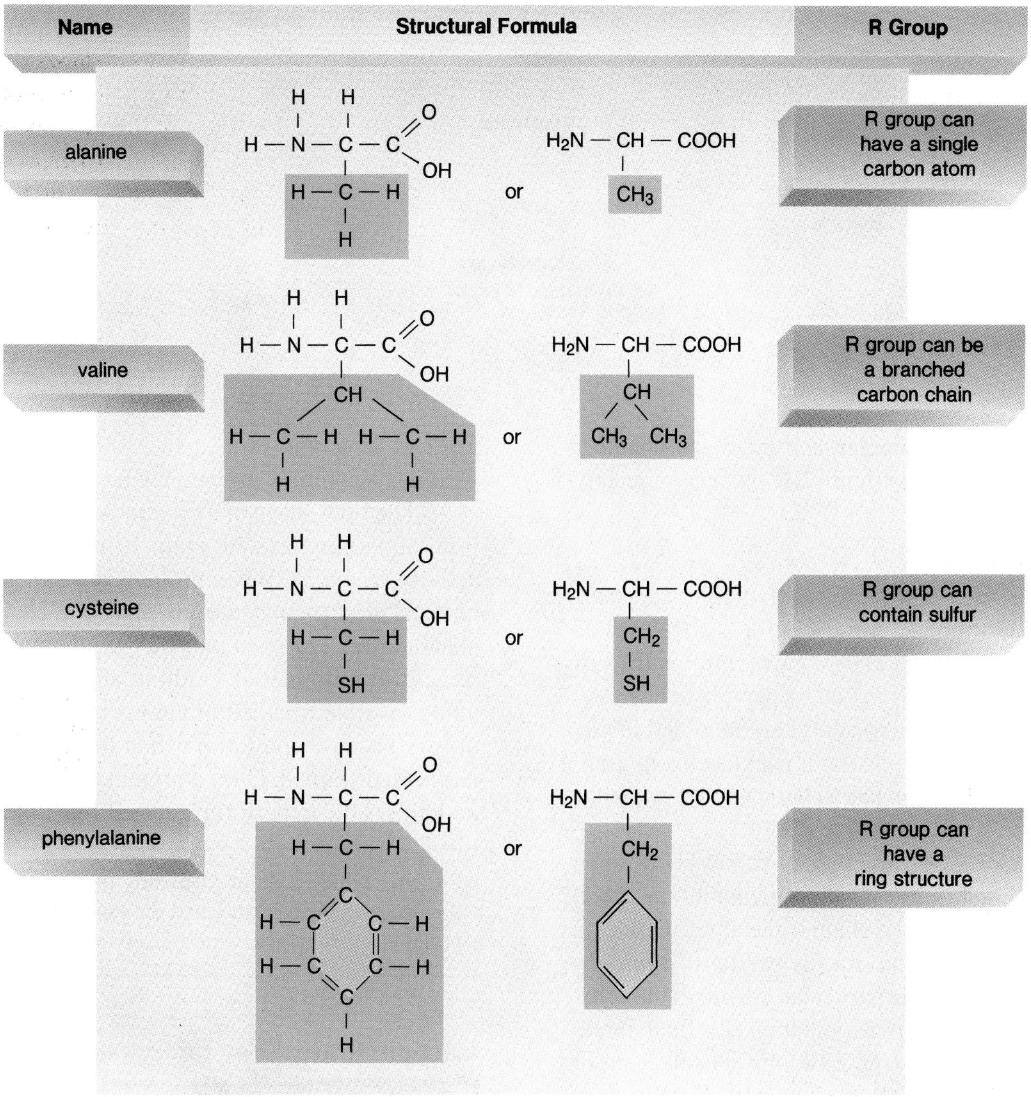

Name	Structural Formula		R Group
alanine	$H-N(H)-C(H)-C(=O)OH$, R: $H-C(H)-H$	or $H_2N-CH-COOH$, R: CH_3	R group can have a single carbon atom
valine	$H-N(H)-C(H)-C(=O)OH$, R: CH (H–C–H, H–C–H)	or $H_2N-CH-COOH$, R: CH (CH_3 CH_3)	R group can be a branched carbon chain
cysteine	$H-N(H)-C(H)-C(=O)OH$, R: $H-C-H$, SH	or $H_2N-CH-COOH$, R: CH_2, SH	R group can contain sulfur
phenylalanine	$H-N(H)-C(H)-C(=O)OH$, R: $H-C-H$, ring (C, H–C, C–H, H–C, C–H, C, H)	or $H_2N-CH-COOH$, R: CH_2, ring	R group can have a ring structure

Although many other amino acids are known, these 20 amino acids are joined in all proteins in all species of living organisms, from bacteria to humans.

Have Peptide Bonds

The bond that joins 2 amino acids is called a **peptide bond.** As you can see in figure 2.16, when dehydration synthesis occurs, the acid group of one amino acid reacts with the amino group of another amino acid, and water is given off. A **dipeptide** results when 2 amino acids join; a **polypeptide** is a string of amino acids joined by peptide bonds. A polypeptide can contain hundreds, even thousands, of amino acids and a protein consists of one or more polypeptides.

The atoms associated with a peptide bond—oxygen (O), carbon (C), nitrogen (N), and hydrogen (H)—share electrons in such a way that the oxygen carries a partial negative charge and the hydrogen carries a partial positive charge:

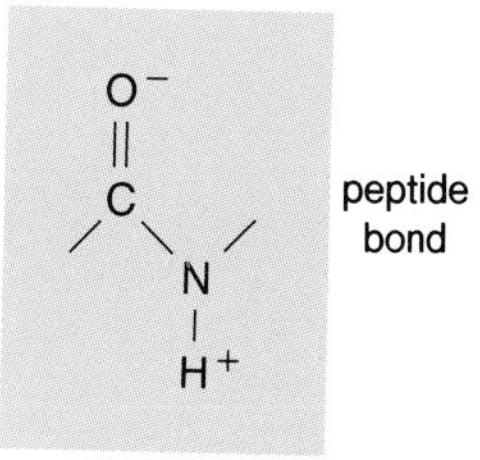

O^-
||
C
N
H^+
peptide bond

Figure 2.16

Formation of a peptide. On the left-hand side of the equation, there are 2 different amino acids, as signified by the difference in the *R*-group notations. As the peptide bond forms, water is given off—the water molecule on the right-hand side of the equation is derived from components removed from the amino acids on the left-hand side. During hydrolysis, water is added and the peptide bond is broken.

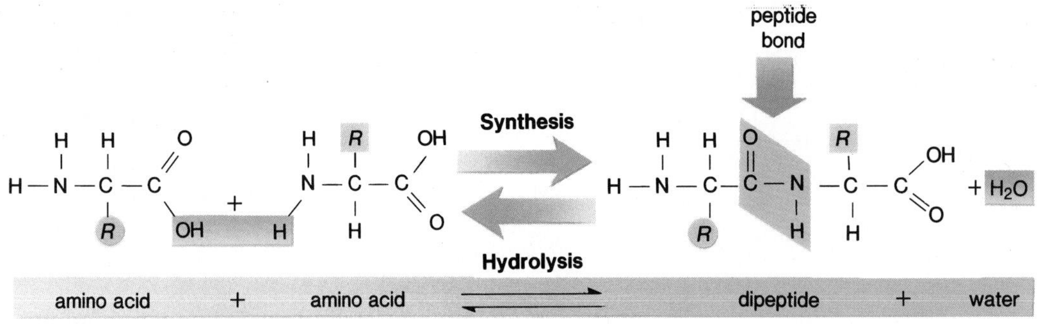

Therefore, the peptide bond is polar, and hydrogen bonding, represented by the dotted lines in figure 2.17, occurs frequently in polypeptides and proteins.

Have Levels of Organization

Proteins commonly have 3 levels of organization in their structure (fig. 2.17), although some have a fourth level as well. The first level, called the *primary structure,* is the linear sequence of the amino acids joined by peptide bonds. Any number of the 20 different amino acids can be joined in any sequence, just as if you were making a necklace from associated beads. Any given protein has a characteristic sequence of amino acids.

The *secondary structure* of a protein comes about when the polypeptide chain takes a particular orientation in space. One common arrangement of the chain is the alpha helix, or a right-handed coil, with 3.6 amino acids per turn. Hydrogen bonding between amino acids, in particular, stabilizes the helix.

The *tertiary structure* of a protein is its final three-dimensional shape. In muscles (fig. 2.18), the helical chains of myosin form a rod shape that ends in globular heads. Enzymes are globular proteins in which the helix bends and twists in different ways. The tertiary shape of a protein is maintained by various types of bonding between the *R* groups. Covalent, ionic, and hydrogen bonding are all seen.

Some proteins have more than one type of polypeptide chain, each with its own primary, secondary, and tertiary structures. These separate chains are arranged to give a fourth level of structure, termed the *quaternary structure*. Hemoglobin is a complex protein having a quaternary structure.

The final shape of a protein is very important to its function, as is emphasized again in the discussion of enzyme activity (chap. 5). When proteins are exposed to extremes in heat and pH, they undergo an irreversible change in shape called denaturation. For example, we are all aware that the addition of acid to milk causes curdling and that heating causes egg white, a protein called albumin, to coagulate. Denaturation occurs because the normal bonding between the *R* groups has been disturbed. Once a protein loses its normal shape, it is no longer able to perform its usual function.

> Proteins, which contain covalently linked amino acids, are important in the structure and the function of cells. Some proteins are enzymes, which speed up chemical reactions.

Carbohydrates: For Energy and Structure

Carbohydrate molecules are characterized by the presence of the atomic grouping CH_2O, in which the ratio of hydrogen atoms (H) to oxygen atoms (O) is approximately 2:1. Because water has this same ratio of hydrogen to oxygen, the term *carbohydrate,* which means *hydrates of carbon,* was originally thought to be appropriate. If the number of carbon atoms in a molecule is low (from 3 to 7), then the carbohydrate is a simple sugar, or monosaccharide. Thereafter, larger carbohydrates are

Figure 2.17

Levels of organization in the structure of a protein. Primary structure of a protein is the order of the amino acids; secondary structure is often an alpha (α) helix, in which hydrogen bonding occurs along the length of a polypeptide, as indicated by the dotted lines; in globular proteins, the tertiary structure is the twisting and turning of the helix, which takes place because of bonding between the *R* groups; and the quaternary structure occurs when a protein contains 2 or more linked polypeptides.

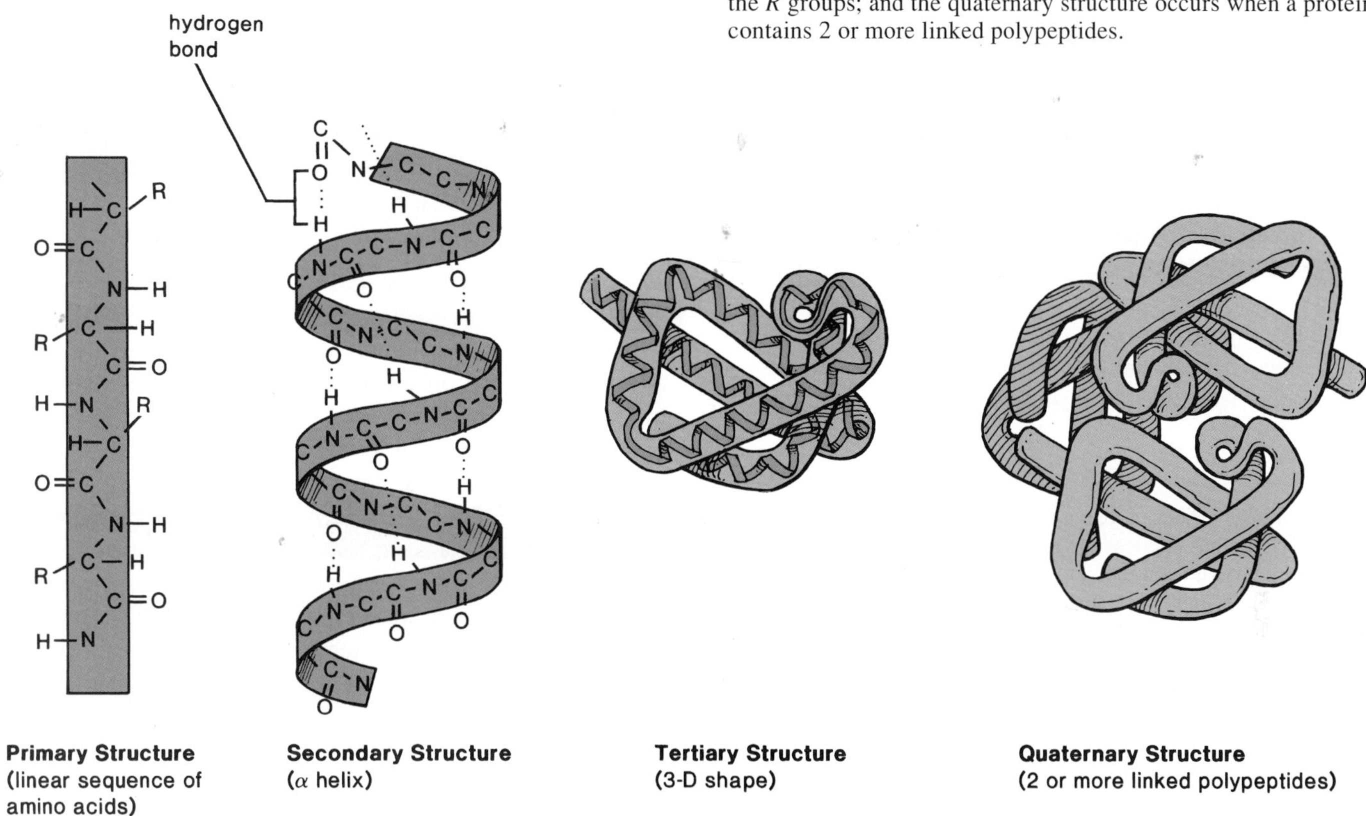

Figure 2.18

The well-developed muscles of an athlete. Muscle cells contain many protein molecules. A major concern today is steroid use to promote the buildup of muscles. This practice can be detrimental to health, as discussed in the reading on pages 360–361.

created by joining monosaccharides in the same manner described in figure 2.14 for the synthesis of organic macromolecules.

Carbohydrates are first and foremost a source of short-term energy for all organisms, including humans. Sometimes, they also join with other molecules to play a structural role.

Monosaccharides: One Sugar

As their name implies, **monosaccharides** are simple sugars of one molecule each (fig. 2.19). These molecules are often designated by the number of carbon atoms they contain; for example, **pentose** sugars, such as ribose, have 5 carbon atoms, and **hexose** sugars, such as glucose, have 6 carbon atoms. Glucose is the primary energy source of the body, and most carbohydrate polymers can be broken down into monosaccharides that either are or can be converted to glucose. Other common monosaccharides are fructose, found in fruits, and galactose, a constituent of milk. These 3 monosaccharides all have a ring structure with the molecular formula $C_6H_{12}O_6$, but they differ in the shape of the ring and/or in the arrangement of the hydrogen (–H) and the hydroxyl groups (–OH) attached to the ring.

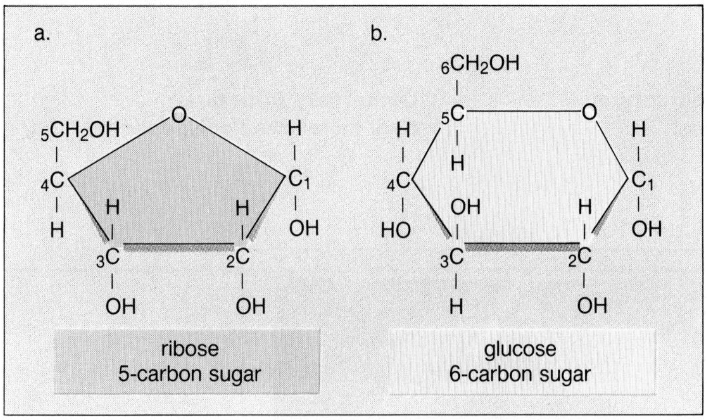

Disaccharides: Two Sugars

The term **disaccharide** tells us that the molecule contains 2 monosaccharides. When 2 glucose molecules join, **maltose** (fig. 2.20) results. When *glucose* and *fructose* join, the disaccharide **sucrose** forms. Sucrose derived from sugarcane and sugar beets is commonly known as table sugar.

Polysaccharides: Many Sugars

A **polysaccharide** is a polymer of monosaccharides. Three polysaccharides are common in organisms: starch, glycogen, and cellulose. They are all chains of glucose, just as a necklace might have only one type of bead.

Even though all 3 polysaccharides contain only glucose, they are distinguishable from one another. As figure 2.21 shows, **starch** has few side branches, or chains of glucose that branch off from the main chain. Starch is the storage form of glucose in plants. Just as we store orange juice as a concentrate, plants store starch as a concentrate of glucose. This analogy is appropriate because water is removed when glucose molecules join to form starch (fig. 2.14). The following equation also represents the synthesis of starch:

$$\text{glucose} \underset{\text{hydrolysis}}{\overset{\text{synthesis}}{\rightleftharpoons}} \text{starch} + H_2O$$

Another polysaccharide, **glycogen,** is characterized by the presence of many side chains of glucose (fig. 2.22). Glycogen is the storage form of glucose in animals. After an animal eats, the liver stores glucose as glycogen; in between eating, the liver releases glucose so that the concentration of glucose in blood is always about 0.1%.

Figure 2.19

Common monosaccharides. **a.** Ribose, a pentose—a 5-carbon sugar. Deoxyribose has one less oxygen atom (O) attached to the second carbon atom (C) compared to ribose. **b.** Glucose, a hexose—a 6-carbon sugar. The small numbers specify the carbon atoms.

a.

ribose
5-carbon sugar

b.

glucose
6-carbon sugar

Figure 2.20

Synthesis and hydrolysis of maltose, a disaccharide containing 2 glucose units. During synthesis, a bond forms between the 2 glucose molecules and the components of water are removed. During hydrolysis, the components of water are added and the bond is broken.

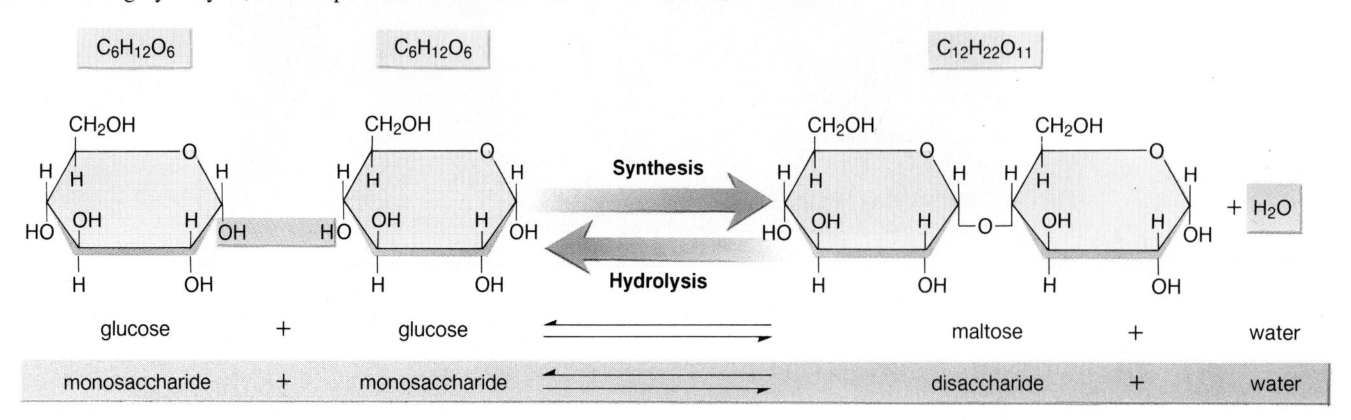

Figure 2.21

Starch structure and function. Starch is a relatively straight chain of glucose molecules. (It may also branch as shown.) The electron micrograph shows starch granules in plant cells. Starch is the storage form of glucose in plants.

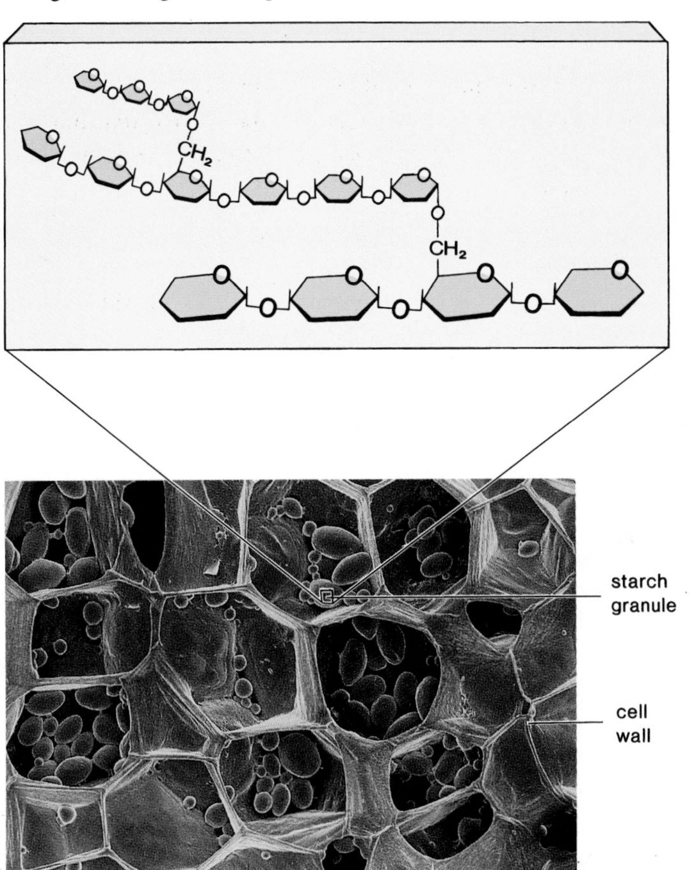

Figure 2.22

Glycogen structure and function. Glycogen is a highly branched polymer of glucose molecules. The branching allows breakdown to proceed at several points simultaneously. The electron micrograph shows glycogen granules in liver cells. Glycogen is the storage form of glucose in animals.

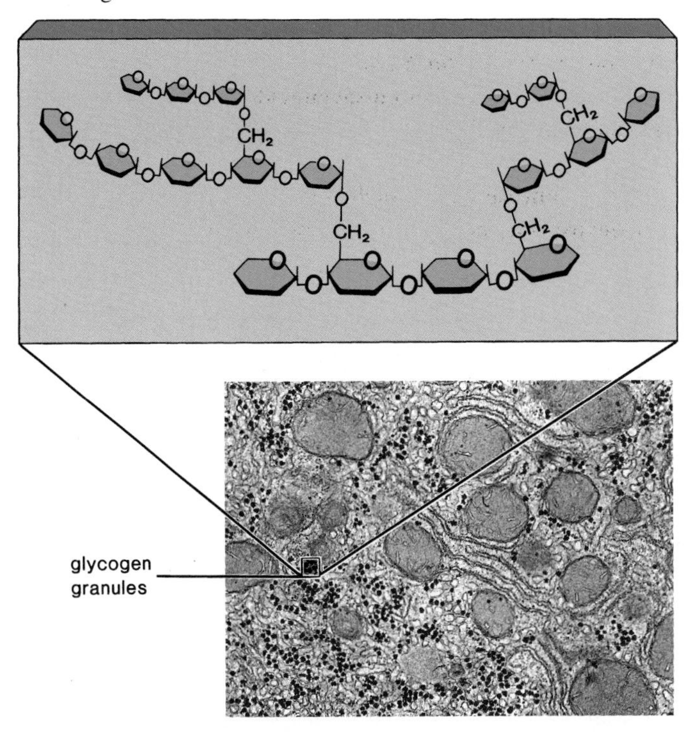

The polysaccharide **cellulose** is found in plant cell walls and accounts in part for the strong nature of these walls. In cellulose (fig. 2.23), the glucose units are joined by a slightly different type of linkage than that in starch or glycogen. (Observe the alternating position of the oxygen atoms linking with the glucose units.) While this might seem to be a technicality, actually it is important because we are unable to digest foods containing this type of linkage; therefore, cellulose passes through our digestive tract as fiber, or roughage. Recently, it has been suggested that fiber in the diet is necessary to good health and may even help to prevent colon cancer.

> Cells frequently use the monosaccharide glucose as an energy source. The polysaccharides starch and glycogen are storage compounds in plant and animal cells, respectively, and the polysaccharide cellulose is found in plant cell walls.

Figure 2.23

Cellulose structure and function. Cellulose contains a slightly different type of linkage between glucose molecules than that in starch or glycogen. Plant cell walls contain cellulose, and the rigidity of the cell walls permits nonwoody plants to stand upright as long as they receive an adequate supply of water.

Figure 2.24

Synthesis and hydrolysis of a neutral fat. During synthesis, 3 fatty acids plus one glycerol react to produce a fat molecule and 3 water molecules. During hydrolysis, one fat molecule plus 3 water molecules react to produce 3 fatty acids and one glycerol.

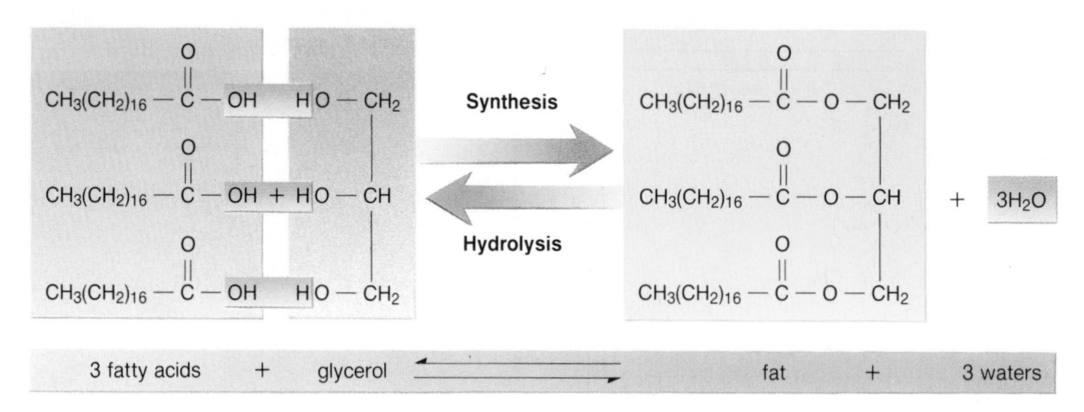

Figure 2.25

Saturated fatty acid versus unsaturated fatty acid. Fatty acids are either saturated (have no double bonds between carbon atoms) or unsaturated (have double bonds between carbon atoms). **a.** In a saturated fatty acid, the carbon atoms (C) carry all the hydrogen atoms (H) possible. **b.** In this unsaturated fatty acid, double bonds are at the third from last carbon and at other carbons.

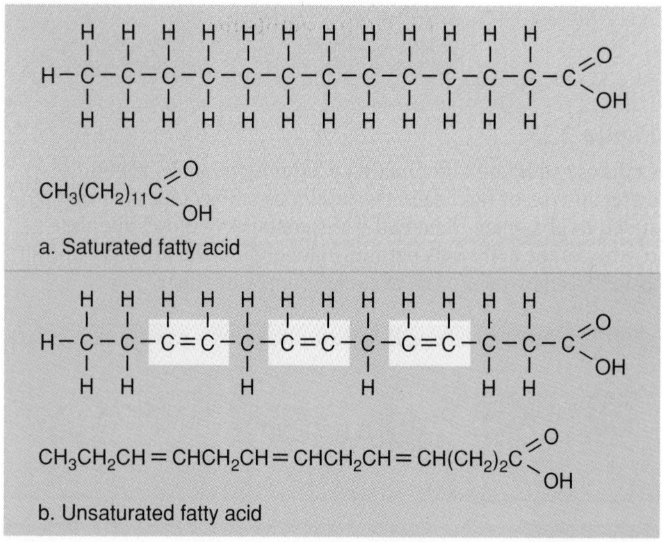

$H-C-C-C-C-C-C-C-C-C-C-C-C-COOH$ (each carbon bearing two H)

$CH_3(CH_2)_{11}COOH$

a. Saturated fatty acid

$H-C-C-C=C-C-C=C-C-C=C-C-C-COOH$

$CH_3CH_2CH=CHCH_2CH=CHCH_2CH=CH(CH_2)_2COOH$

b. Unsaturated fatty acid

Lipids: For Stored Energy and Membranes

Lipid molecules are usually water insoluble. The familiar lipids are the fats (e.g., lard and butter) and oils (e.g., corn oil and soybean oil). At room temperature, fats are solids and oils are liquids. Fat has several functions in the body: it is used for long-term energy storage, it insulates against heat loss, and it forms a protective cushion around major organs.

Fats and Oils: Glycerol + Three Fatty Acids

A fat or an oil is formed when one glycerol molecule reacts with 3 fatty acid molecules. Notice in figure 2.24 that **glycerol** has 3 hydroxyl groups –OH and that it therefore can react with 3 fatty acids to form one fat molecule and 3 water molecules. Again, the larger fat molecule is formed by dehydration synthesis in the forward direction. The backward direction shows how fat can be hydrolyzed to its components. A fat is sometimes called a *triglyceride* because of its three-part structure, and the term *neutral fat* is used because the molecule has no groups that can ionize, or become charged. Therefore, a fat is nonpolar.

Fatty Acids: Saturated/Unsaturated

A **fatty acid** has a hydrocarbon chain (a string of carbon atoms with hydrogen atoms attached) and ends with the acid group –COOH (fig. 2.25). Most of the fatty acids in cells contain 16 or 18 carbon atoms per molecule, although smaller ones with fewer carbons are also known.

Fatty acids are either saturated or unsaturated. Saturated fatty acids have no double bonds between carbon atoms. The carbon chain is saturated, so to speak, with all the hydrogens it can hold. Butter is called a saturated fat because it contains saturated hydrocarbon chains.

Unsaturated fatty acids have double bonds between carbon atoms wherever the number of hydrogens is less than 2 per carbon atom. Vegetable oils are called unsaturated fats because they contain unsaturated hydrocarbon chains. Unsaturated fatty acids account for the liquid nature of vegetable oils. Vegetable oils are hydrogenated to make margarine. Polyunsaturated margarine still contains a large number of unsaturated carbon bonds.

Figure 2.26
Emulsification. Fat molecules, being nonpolar, do not disperse in water. An emulsifier, like a soap, contains molecules that have a polar end and a nonpolar end. When an emulsifier is added to a beaker containing a layer of nondispersed fat molecules, the nonpolar ends are attracted to the nonpolar fat, and the polar ends are attracted to the water. This causes droplets of fat molecules to disperse.

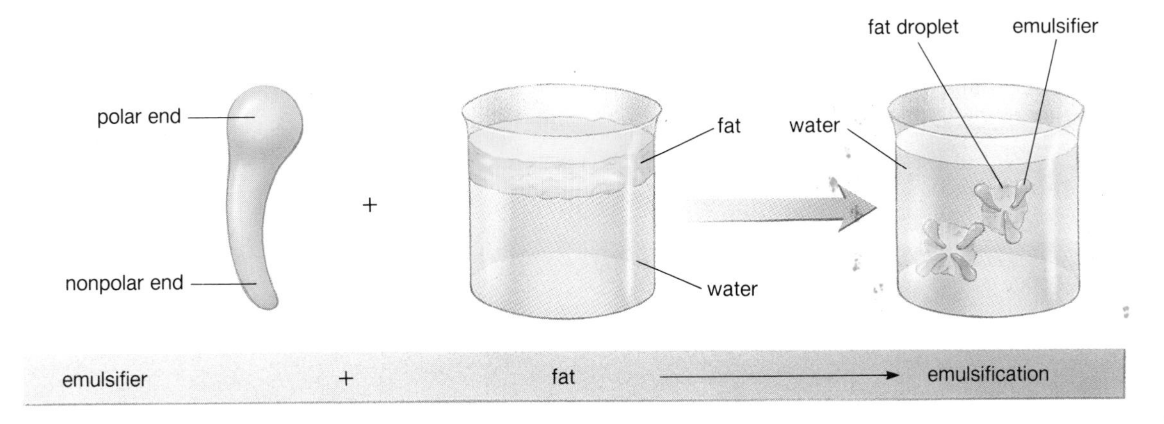

Figure 2.27
Phospholipid structure and shape. Phospholipids are constructed like fats, except that they contain a phosphate group. **a.** Lecithin, shown here, has a side chain that contains both a phosphate group and a nitrogen-containing group. **b.** The polar portion of the phospholipid molecule (head) is soluble in water, whereas the 2 hydrocarbon chains (tails) are not. This causes the molecule to arrange itself as shown.

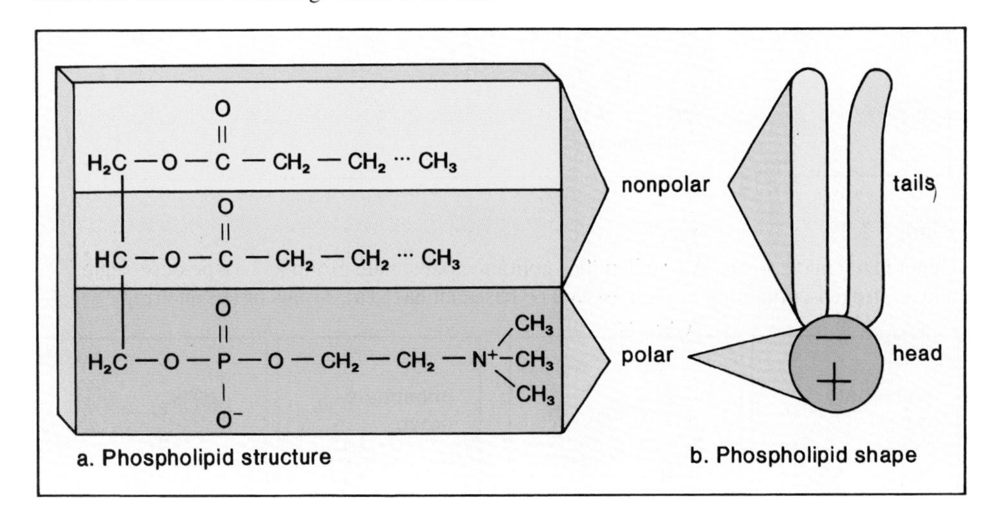

Soaps: Emulsifiers. Strictly speaking, soaps are not lipids, but they are considered here as a matter of convenience. A soap is a salt formed from a fatty acid and an inorganic base. For example,

$$\underset{\text{sodium hydroxide}}{NaOH} + \underset{\text{fatty acid}}{RCOOH} \longrightarrow \underset{\text{soap}}{RCOO^-Na^+}$$

Fats do not mix with water because they are nonpolar; a soap, however, has a nonpolar end (the hydrocarbon chain represented by *R*) and a polar end (the charged end). Therefore, a soap does mix with water. When soaps are added to oils, the oils, too, mix with water. Figure 2.26 shows how a soap positions itself about an oil droplet so that its nonpolar ends project into the fat droplet, while its polar ends project outward. Now the droplet disperses in water, and it is said that **emulsification** has occurred. Emulsification occurs when dirty clothes are washed with soaps or detergents. Also, prior to the digestion of fatty foods, fats are emulsified by bile. A person who has had the gallbladder removed may have trouble digesting fatty foods because this organ stores bile for emulsifying fats prior to the digestive process.

Phospholipids: Polar Heads and Nonpolar Tails

Phospholipids, as their name implies, contain a phosphate group:

$$^-O-\overset{\overset{\displaystyle O}{\|}}{\underset{\underset{\displaystyle O^-}{|}}{P}}-O^-$$

Essentially, phospholipids are constructed like fats, except that in place of the third fatty acid, there is a phosphate group or a grouping that contains both phosphate and nitrogen (fig. 2.27). These molecules are not electrically neutral as are fats because the phosphate group can ionize. It forms the so-called "head" of the molecule while the rest of the molecule becomes the nonpolar "tails." As we shall see in the next chapter, the cell membrane is a phospholipid bilayer in which the "heads" face outward and the tails face each other because they are water repelling.

Steroids: Four Rings

Steroids are lipids having a structure that differs entirely from that of fats. Steroid molecules have a backbone of 4 fused carbon rings, but each one differs primarily by the arrangement of the atoms in the rings and the type of functional groups attached to them (fig. 2.28). *Cholesterol* is the precursor of several other steroids, such as aldosterone, a hormone that helps to regulate the sodium level of blood, and the sex hormones, such as estrogen and testosterone, which help to maintain male and female characteristics.

Evidence has been accumulating for years that a diet high in saturated fats and cholesterol can lead to circulatory disorders. For reasons that are discussed in the reading on page 214, this type of diet leads to deposits of fatty material inside the lining of blood vessels and hence to reduced blood flow.

Lipids include nonpolar fats, long-term energy-storage molecules formed from glycerol and 3 fatty acids, and the related phospholipids, which have a charged group. Steroids have an entirely different structure, similar to that of their precursor, cholesterol. Do 2.2 Critical Thinking, found at the end of the chapter.

Figure 2.28

Steroid diversity. Like cholesterol in (**a**), all steroid molecules have 4 adjacent carbon rings, but there are structural differences in specific steroids, as noted in (**b**) and (**c**). **b.** Aldosterone is involved in the regulation of the blood level of sodium. **c.** Testosterone is the male sex hormone.

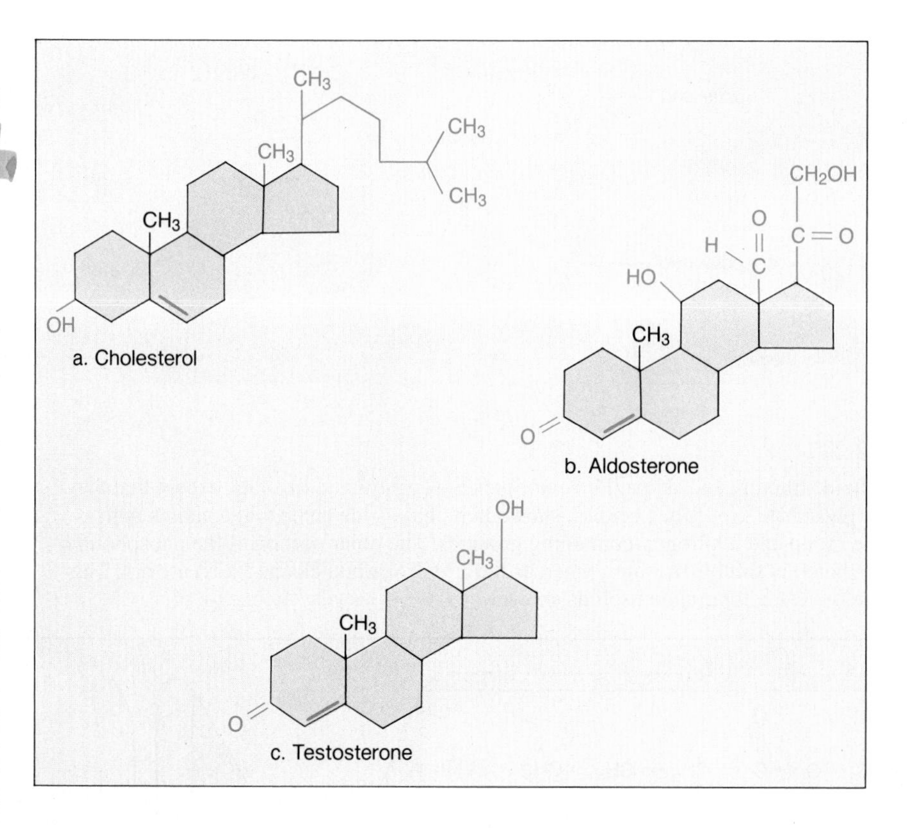

Nucleic Acids: For Reproduction and Control of Protein Synthesis

Nucleic acids are important for the growth and reproduction of cells and organisms. Human genes are composed of a nucleic acid called **DNA (deoxyribonucleic acid)**. The nucleic acid **RNA (ribonucleic acid)** works in conjunction with DNA to bring about protein synthesis.

Both DNA and RNA are polymers of nucleotides. Just like the other synthetic reactions we have studied in this section, nucleic acids are formed by dehydration synthesis:

$$\text{nucleotides} \underset{\text{hydrolysis}}{\overset{\text{synthesis}}{\rightleftharpoons}} \text{nucleic acid} + H_2O$$

Figure 2.29

Generalized nucleotides. All nucleotides contain a phosphate group (Ⓟ), a pentose sugar, and a nitrogen-containing organic base. The base can have (**a**) 2 rings or (**b**) one ring.

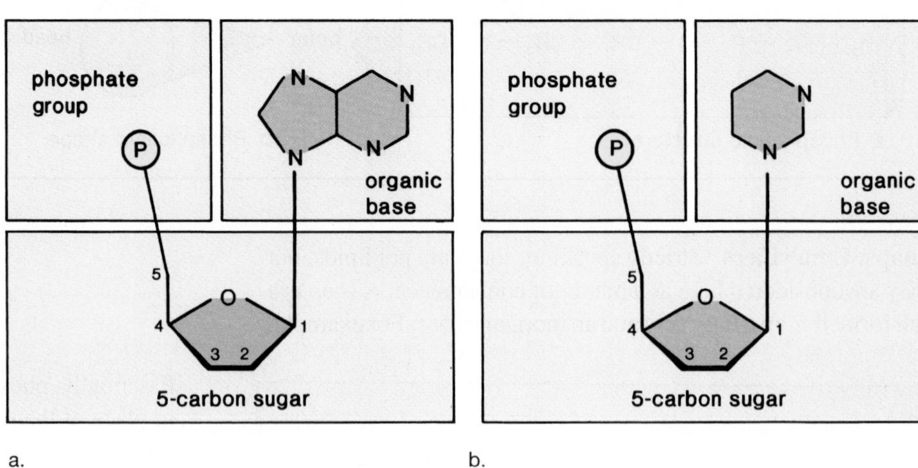

Figure 2.30

Generalized nucleic acid strand. Each nucleic acid strand is a chain of nucleotide monomers. Each strand has a backbone made of sugar molecules and phosphate groups. The bases project to the side of the backbone.

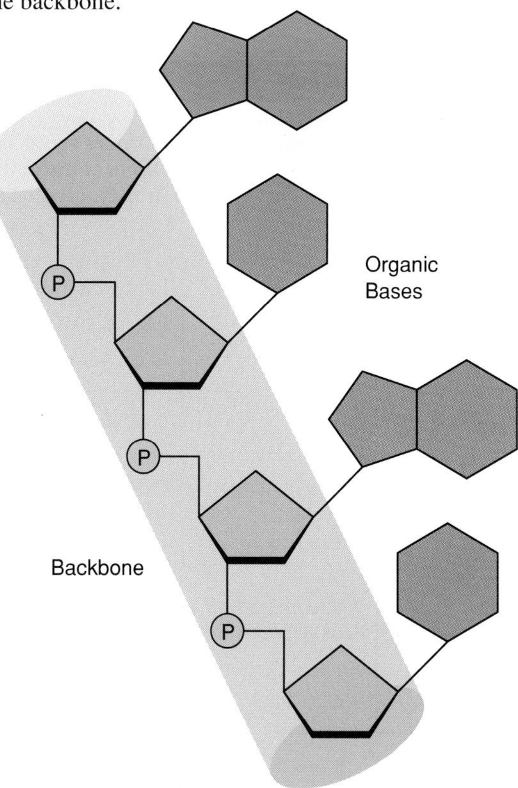

Nucleotides: Sugar + Base + Phosphate

A **nucleotide** contains 3 types of molecules: a phosphate, a pentose sugar, and a nitrogen-containing organic base. For DNA, the sugar is deoxyribose, and for RNA the sugar is ribose. For both DNA and RNA, the base can have either a single ring or a double ring. Figure 2.29 shows generalized nucleotides because the specific type of base is not designated; the phosphate group is simply represented as Ⓟ. When nucleotides join, they form a macromolecule called a strand, in which the backbone is made up of this sequence: phosphate—sugar—phosphate—sugar. The bases project to one side of the backbone (fig. 2.30). RNA is single stranded, and DNA is double stranded. The strands in DNA are held to each other by hydrogen bonding between the bases (see fig. 24.3). This is only a brief description of the structure and the function of DNA and RNA; they are considered again in greater detail in chapter 24.

Figure 2.31

Structure of ATP. ATP is a nucleotide with 3 phosphate groups (Ⓟ); 2 of the phosphate bonds are high-energy bonds, indicated here by wavy lines.

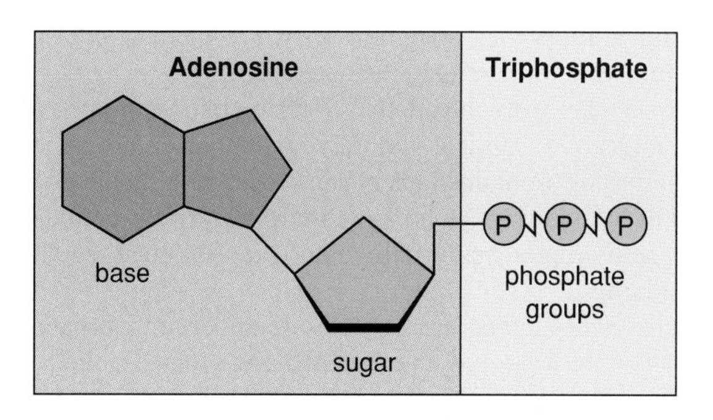

ATP: Carrier of Energy. ATP (**adenosine triphosphate**) is a nucleotide that functions as an energy carrier in cells. The structure of ATP is similar to that shown in figure 2.29*a*. The base adenine is joined to the sugar ribose (together called adenosine), and there are 3 phosphate groups instead of one. It is customary to draw the molecule as shown in figure 2.31, with the 3 phosphate groups shown on the right. ATP is known as the energy molecule because the triphosphate unit contains 2 high-energy bonds, represented in figure 2.31 by wavy lines.

> Nucleic acids function in the growth and reproduction of cells and organisms. DNA is composed of nucleotides having the sugar deoxyribose, while RNA nucleotides contain the sugar ribose. Genes are composed of DNA, and with the help of RNA, DNA controls protein synthesis.

SUMMARY

All matter is made up of atoms. Each atom has a weight, which is dependent on the number of protons and neutrons in the nucleus, and chemical properties, such as reactivity, which are dependent on the number of electrons in the outer shell.

Atoms react with one another to form either inorganic or organic molecules. Typical of nonliving things, an inorganic molecule usually contains ions bonded together by ionic bonds. Typical of living things, an organic molecule always contains carbon and hydrogen and usually other atoms bonded together by covalent bonds.

Water, acids and bases, and salts are important inorganic molecules. Pure water has a neutral pH; acids increase the hydrogen ion concentration $[H^+]$ but decrease the pH, and bases decrease the hydrogen ion concentration but increase the pH of water:

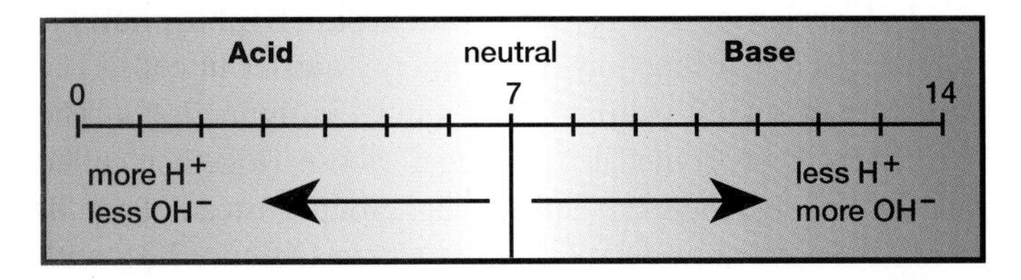

Cells contain macromolecules that arise when monomers join by dehydration synthesis, and macromolecules are broken down by a reverse process called hydrolysis. Proteins, carbohydrates, lipids, and nucleic acids are macromolecules with specific functions in cells (table 2.3).

Table 2.3
Organic Compounds Associated with Living Things

Macromolecule	Monomer	Function
Proteins	Amino acids	Enzymes speed up chemical reactions; structural components (e.g., muscle proteins)
Carbohydrates		
Starch	Glucose	Energy storage in plants
Glycogen	Glucose	Energy storage in animals
Cellulose	Glucose	Plant cell walls
Lipids		
Fats and Oils	Glycerol, 3 fatty acids	Long-term energy storage
Phospholipids	Glycerol, 2 fatty acids, phosphate group	Cell membrane structure
Nucleic Acids		
DNA	Nucleotides with deoxyribose sugar	Genetic material
RNA	Nucleotides with ribose sugar	Protein synthesis

STUDY QUESTIONS

In order to practice **writing across the curriculum,** students should write out the answers to any or all of the study questions. The study questions are sequenced in the same order as the text.

1. Name the subatomic particles of an atom; describe their charge, weight, and location in the atom. (p. 18)
2. Diagram the atomic structure of calcium. (p. 19)
3. State the octet rule, and explain how it relates to chemical reactions. (p. 19)
4. Give an example of an ionic reaction, and explain it. Mention in your explanation these terms: compound, ion, formula, and ionic bond. (pp. 20–21)
5. Give an example of a covalent reaction, and explain it. (p. 23)
6. Explain an oxidation-reduction reaction in terms of loss or gain of electrons. (p. 23)
7. Name 4 general differences between inorganic and organic compounds. (p. 24)
8. On the pH scale, which numbers indicate a basic solution? an acidic solution? Why? (p. 27)
9. What are buffers, and why are they important to life? (p. 27)
10. Explain dehydration synthesis of organic molecules and hydrolytic breakdown of organic molecules. (p. 28)
11. What are some functions of proteins? What is the monomer in proteins? What are a peptide bond, a dipeptide, and a polypeptide? (pp. 28–29)
12. Discuss the primary, secondary, and tertiary structures of proteins. (p. 30)
13. Name some monosaccharides, disaccharides, and polysaccharides, and state appropriate functions for each. What is the most common monomer for these? (pp. 32–33)
14. Name some important lipids, and state their function. What is a saturated fatty acid? an unsaturated fatty acid? How is a fat formed? (pp. 34–36)
15. What are the 2 types of nucleic acids? What is the monomer of both? (pp. 36–37)

OBJECTIVE QUESTIONS

1. The atomic number is equal to the number of _______ in an atom.
2. An ion is negatively charged when it has more _______ than _______.
3. In a covalent bond, the atoms _______ electrons.
4. _______take up either hydrogen ions (H^+) or hydroxide ions (OH^-) and therefore act to stabilize the pH.
5. When an acid is added to water, the number of hydrogen ions _______ and the pH _______.
6. _______ are organic catalysts that speed up chemical reactions.
7. A _______ bond joins 2 amino acids in a protein.
8. The sequence, or order, of amino acids in a protein is termed its _______ structure.
9. The simple sugar _______ is the primary energy source of the body.
10. The polymer _______ is found in plant cell walls.
11. Fatty acids having no double bonds between carbon atoms are said to be _______.
12. A fat results from the union of a _______ molecule and 3 fatty acid molecules.
13. Nucleic acids are composed of _______.
14. DNA is _______ stranded, while RNA is single stranded.
15. Label this diagram of synthesis and hydrolysis.

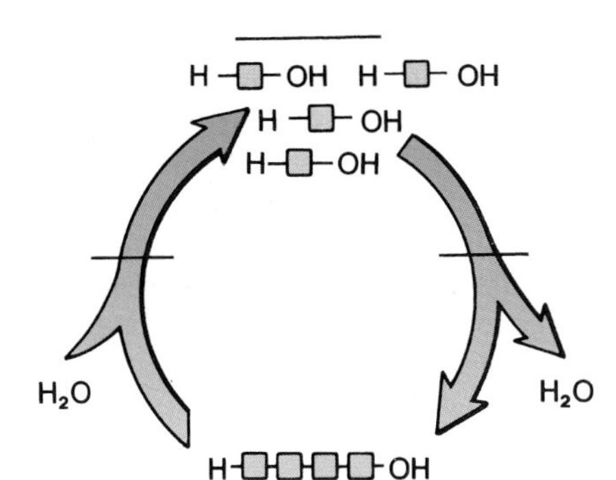

CRITICAL THINKING

In order to practice **writing across the curriculum,** students should write out the answers to any or all of the critical thinking questions. Suggested answers to the critical thinking questions are in appendix E.

2.1

1. Why isn't water termed an organic molecule?
2. Water fulfills which of the criteria for an inorganic compound? (See table 2.2.)
3. Water fulfills which of the criteria for an organic molecule? (See table 2.2.)
4. How do water's characteristics (pp. 24–25) help the body to function?

2.2

1. Actin and myosin are the proteins found in muscles. How do you predict they differ?
2. Starch and glycogen are both composed of glucose. How do they differ?
3. Oleic acid and linoleic acid are both unsaturated fatty acids. How do you predict they differ structurally?

SELECTED KEY TERMS

acid a solution in which pH is less than 7; a substance that contributes or liberates hydrogen ions (protons) in a solution.
amino acid a monomer of a protein; takes its name from the fact that it contains an amino group ($-NH_2$) and an acid group ($-COOH$).
atom smallest unit of matter that cannot be divided by chemical means.
ATP (adenosine triphosphate) a compound having a nitrogen-containing organic base, ribose, and 3 phosphate groups, which are joined by 2 high-energy bonds; a carrier of energy in cells.
base a solution in which pH is greater than 7; a substance that contributes or liberates hydroxide ions (OH^-) in a solution; alkaline; opposite of acidic. Also, a term commonly applied to one of the components of a nucleotide.
buffer a substance or compound that prevents large changes in the pH of a solution.
carbohydrate one of a class of organic compounds characterized by the presence of CH_2O groups; includes monosaccharides, disaccharides, and polysaccharides.
covalent bond a chemical bond between atoms that results from the sharing of a pair of electrons.
DNA (deoxyribonucleic acid) a nucleic acid found in cells; the genetic material that directs protein synthesis in cells.
electron a subatomic particle that has almost no weight and carries a negative charge; orbits in a shell about the nucleus of an atom.
enzyme an organic catalyst that speeds up a specific reaction or a specific type of reaction in cells.
hydrogen bond a weak attraction between a hydrogen atom carrying a partial positive charge and an atom of another molecule carrying a partial negative charge.
ion an atom or group of atoms carrying a positive or negative charge.
ionic bond a bond created by an attraction between oppositely charged ions.
isotope one of 2 or more atoms with the same atomic number that differs in the number of neutrons and therefore in weight.
lipid one of a class of organic compounds that are insoluble in water; notably fats, oils, and steroids.
neutron a subatomic particle that has a weight of one atomic mass unit, carries no charge, and is found in the nucleus of an atom.
nucleotide a monomer of a nucleic acid that forms when a nitrogen-containing organic base, a pentose sugar, and a phosphate join.
pentose a 5-carbon sugar; deoxyribose is the pentose sugar found in DNA; ribose is a pentose sugar found in RNA.
peptide bond the covalent bond that joins 2 amino acids.
pH scale a measure of the hydrogen ion concentration $[H^+]$; any pH below 7 is acidic and any pH above 7 is basic.
protein one of a class of organic compounds that are composed of either one or several polypeptides.
proton a subatomic particle found in the nucleus of an atom that has a weight of one atomic mass unit and carries a positive charge; a hydrogen ion.
RNA (ribonucleic acid) a nucleic acid found in cells that assists DNA in controlling protein synthesis.
synthesis to build up, such as the combining of 2 small molecules to form a larger molecule.

FURTHER READINGS FOR PART ONE

Baker, J. J. W., and G. E. Allen. 1981. *Matter, energy, and life*. 4th ed. Reading, Mass.: Addison-Wesley.
Drewes, F. 1992. *How to study science*. Dubuque, Iowa: Wm. C. Brown Publishers.
Fenn, J. 1982. *Engines, energy, and entropy*. New York: W. H. Freeman.
Karplus, M., and J. A. McCammon. April 1986. The dynamics of proteins. *Scientific American*.
Mathews, C. K., and K. E. van Holde. 1990. *Biochemistry*. Redwood City, Calif.: Benjamin/Cummings Publishing Co.
May, R. M. October 1992. How many species inhabit the earth? *Scientific American*.
Olson, A. J., and D. S. Goodsell. November 1992. Visualizing biological molecules. *Scientific American*.
Scientific American. October 1985. The molecules of life.
Smith, J. M. 1986. *The problems of biology*. London: Oxford University Press.
Stryer, L. 1988. *Biochemistry*. 3d ed. San Francisco: W.H. Freeman.

Part Two

THE CELL, THE SMALLEST UNIT OF LIFE

Hyaline cartilage in mammalian ear, × 36

THE CELL, THE SMALLEST UNIT OF LIFE

The cell is the smallest living thing. It is the lowest level of organization to have all the characteristics of life. A knowledge of cell structure, cell physiology, and cell biochemistry forms the foundation of an understanding of multicellular forms. In this part, we will study each of these aspects of cellular biology in detail and thereby cover the fundamental concepts of biology.

The cell is bounded by a membrane and contains organelles, many of which are also membranous. The membrane regulates the entrance and exit of molecules and determines how cellular organelles carry out their functions. Certain organelles are involved in energy metabolism, and 2 chapters in this part discuss how energy is acquired and converted to a form that cells can use.

Cell reproduction depends upon a type of cell division called mitosis, whereas both animal and plant reproduction depend upon a type of cell division called meiosis. Both types of cell division, mitosis and meiosis, are introduced.

3

CELL STRUCTURE AND FUNCTION

Chapter Concepts

1.
All organisms are composed of cells, and new cells arise only from preexisting cells. 44, 57

2.
The amount of detail they allow us to see varies from one microscope to another; therefore, their useful magnifications also vary. 44, 57

3.
The nucleus controls the metabolic functions and the structural characteristics of the cell. 46, 57

4.
A system of membranous canals and vacuoles work together to produce, store, modify, transport, and digest macromolecules. 50, 57

5.
Mitochondria and chloroplasts convert one form of energy into another and make energy-rich molecules available to cells and therefore organisms. 52, 58

6.
The cell has a cytoskeleton composed of tubules and filaments. The cytoskeleton gives the cell shape and allows it and its organelles to move. 53, 58

7.
Centrioles are related to cilia and flagella, which enable the cell to move. 53, 58

8.
There are 2 major types of cells: the eukaryotic cell, which has a nucleus; and the prokaryotic cell, which lacks a well-defined nucleus. 56, 58

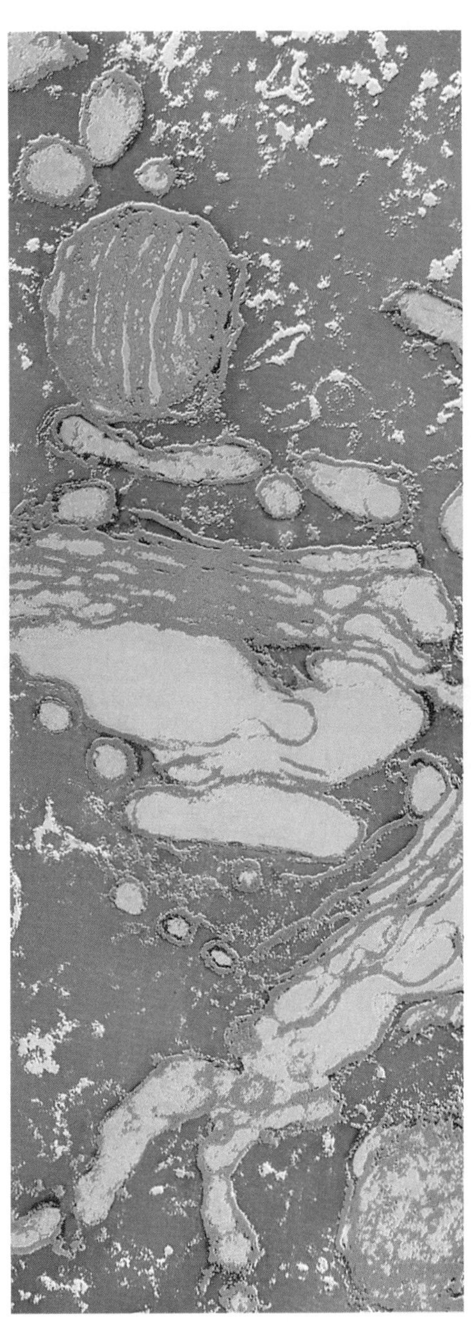

Saccules of a Golgi apparatus, ×31,800

Chapter Outline

Figure 3.1

Comparison of photomicrographs of red blood cells within a blood vessel. **a.** Light micrograph. **b.** Transmission electron micrograph (TEM). **c.** Scanning electron micrograph (SEM) with color added. **d.** Technician at a transmission electron microscope.

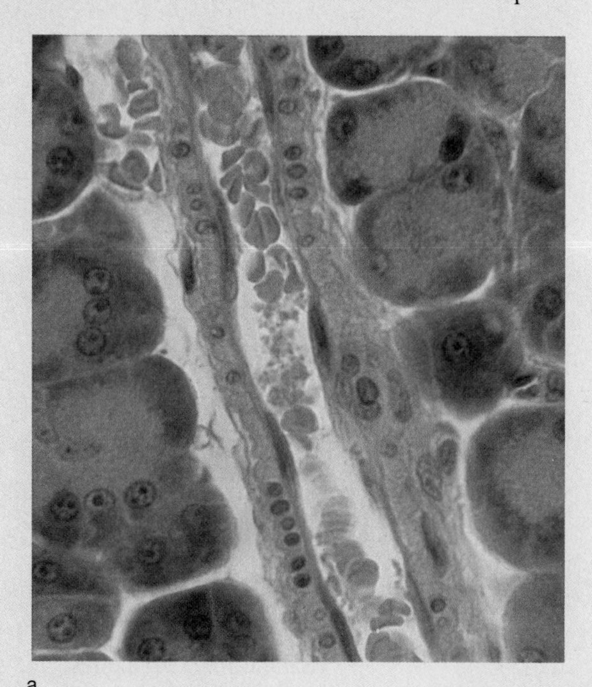

a.

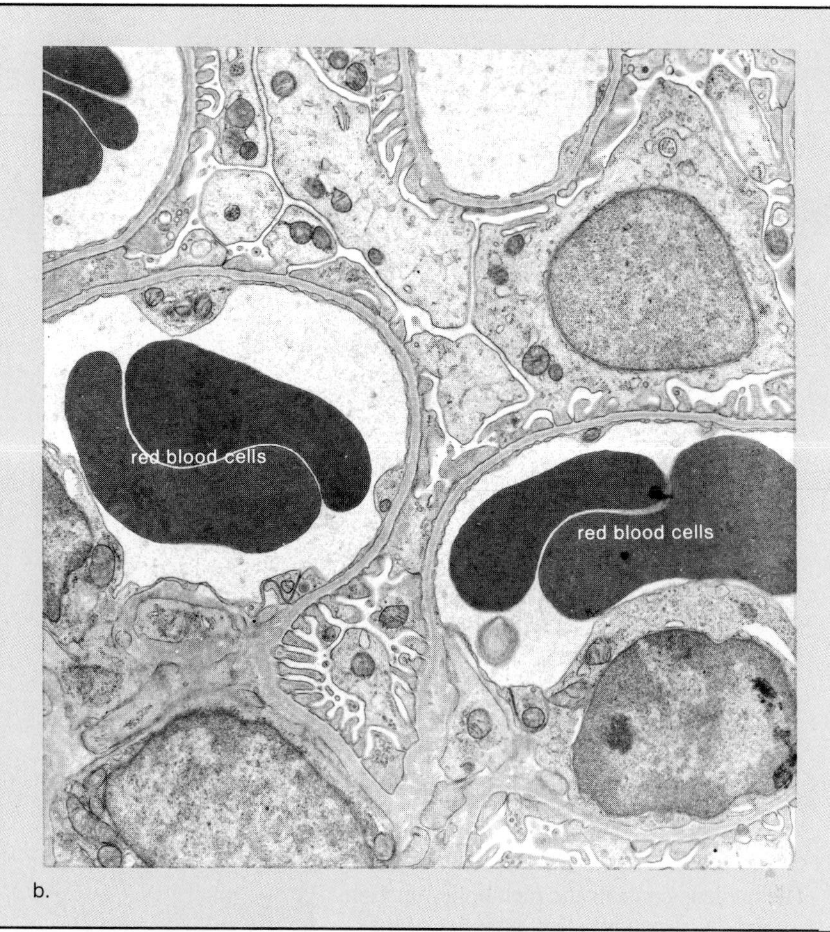

b.

The Cell Theory

All living things are made up of *cells,* the fundamental units of life. Cells come in many different shapes and sizes, but no matter what the shape or the size, each carries on the functions associated with life—interacting with the environment, obtaining chemicals and energy, growing, and reproducing.

The cell marks the boundary between the nonliving and the living. The molecules that serve as food for a cell and the organic molecules that make up a cell are not alive, and yet the cell is alive. The answer to what life is will have to be found within the cell, because the smallest living organisms are *unicellular,* while larger organisms are *multicellular*—composed of many cells. The cell theory states that all living things are composed of cells and new cells arise only from preexisting cells.

Microscopy: Reveals Cell Structure

With some exceptions, such as bird and frog eggs, cells are not readily visible to the eye; therefore, a microscope is needed to view them. Three types of microscopes are used most commonly: light microscopes, transmission electron microscopes, and scanning electron microscopes. As suggested by their names, light microscopes use light to view an object, whereas the electron microscopes utilize electrons. A transmission electron microscope gives us an image of the interior of the object, and a scanning electron microscope provides a three-dimensional view of the surface of an object.

Transmission electron microscopes have a much greater resolving power than light microscopes. *Resolving power,* the capacity to distinguish between 2 adjacent points, is dependent on the wavelength of the illumination. An electron has a much shorter wavelength than visible light. Adjacent points have to be separated by 200 nm (nanometer = 1×10^{-6} mm)[1] or more for the light microscope to resolve them as 2 separate points. The transmission electron microscope can make them out as separate if the points are only 0.5 nm apart. Therefore, a transmission electron microscope allows us to see greater cellular detail than a light microscope. This means that the useful magnification of a light microscope is much less (about ×1,000) than the useful magnification of an electron microscope (about ×30,000).

In scanning electron microscopy, a narrow beam of electrons is scanned over the surface of the specimen, which has been coated with a thin layer of metal. The metal gives off secondary

1. For units of measurement, see appendix C.

c.

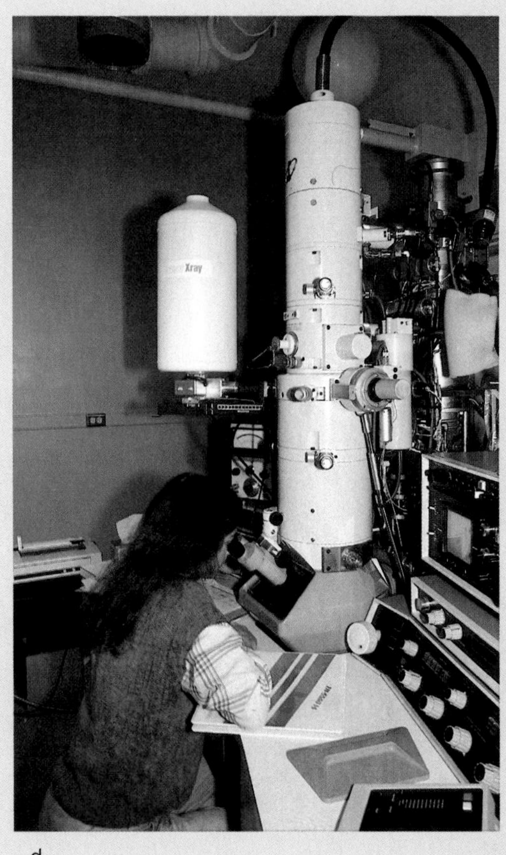

d.

electrons, which are collected to produce a television-type picture of the specimen's surface on a screen.

New types of scanning microscopes, called scanning-probe microscopes, have been recently invented. In one type, laser light is focused on a tiny probe, which is pressed against an organic macromolecule. Reflected light tells the position of the probe as it moves up and down the macromolecule, and from these data, a computer can construct a three-dimensional picture of a single macromolecule. Unlike traditional electron microscopes, scanning-probe microscopes are sensitive to the electron movements and the atomic forces that exist between the probe and the material.

A picture obtained using a light microscope sometimes is called a photomicrograph, and a picture resulting from the use of an electron microscope is called a transmission electron micrograph (TEM) or a scanning electron micrograph (SEM), depending on the type of microscope used (fig. 3.1).

All organisms are composed of cells. The internal structure of cells has been revealed by electron microscopy. Do 3.1 Critical Thinking, found at the end of the chapter.

Eukaryotic Cell Boundaries

Both animal and plant cells have a nucleus; therefore, they are called **eukaryotic cells** (*eu* means *true*; *karyon* means *nucleus*). Organisms in 4 out of the 5 kingdoms—protists, fungi, plants, and animals (see fig. 1.6)—have a nucleus.

Cell Membrane: Gatekeeper of Cell

Both animal and plant cells are surrounded by a *cell membrane*, a phospholipid bilayer in which protein molecules are embedded (fig. 3.2*a*). The cell membrane separates the contents of the cell, called the **cytoplasm,** from the surrounding environment. The membrane also compartmentalizes the cell; it divides the cell into specific regions. As we shall see, the cell contains small membranous structures called **organelles.**

Cell Wall: Supporter of Plant Cells

Plant cells (but not animal cells) have a **cell wall** in addition to a cell membrane. Many plant cells have both a primary and secondary cell wall. Figure 3.2*b* shows the cell walls of two adjacent plant cells in order to illustrate that they are separated by the middle lamella, a region that contains a sticky substance, usually pectin.

Figure 3.2

Structure of the cell membrane and the cell wall.
a. Both plant and animal cells have a cell membrane. Proteins float in a phospholipid bilayer, which has the consistency of light oil.
b. Plant cells have a cell wall, located outside the cell membrane. All plant cells have a primary cell wall, and some also have a secondary cell wall. The middle lamella, a region between cell walls, contains a sticky substance, usually pectin.

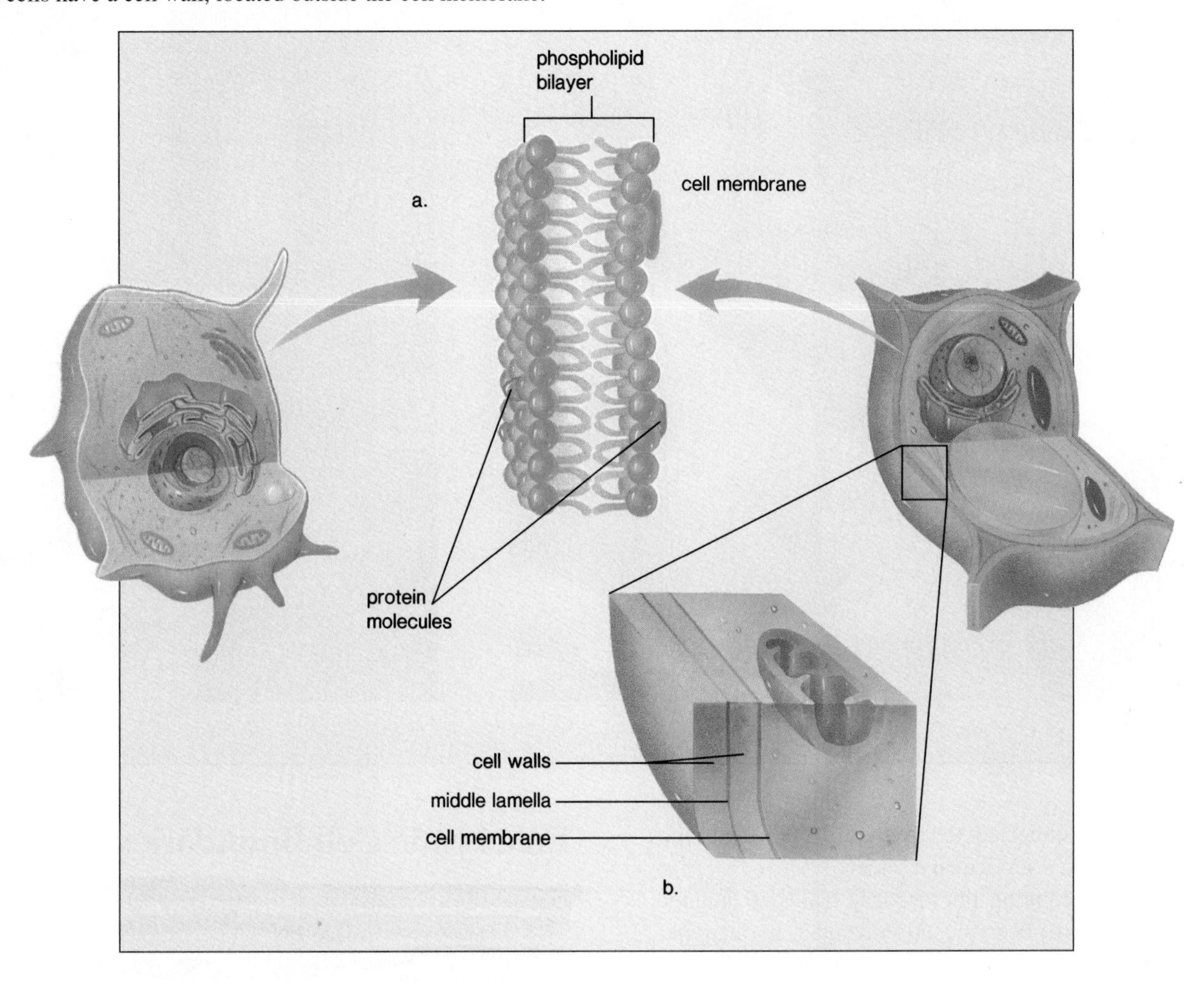

A main constituent of a primary cell wall is cellulose molecules. Cellulose molecules form threadlike microfibrils, which in turn are found in fibrils. Several layers of fibrils lying at right angles to one another are located within the cell wall. This arrangement adds strength. The secondary cell wall forms inside the primary cell wall. Secondary cell walls contain lignin, a substance that makes them even stronger than primary cell walls.

Eukaryotic Cell Organelles

Table 3.1 lists the organelles found in eukaryotic cells.

Nucleus: Controller of Cells

The **nucleus** (figs. 3.3 and 3.4) is a large organelle having a diameter of about 5 µm. A double membrane called the **nuclear envelope** keeps the contents of the nucleus separate from the cell's cytoplasm. Pores in the nuclear envelope allow large molecules to pass into and out of the nucleoplasm, the fluid interior of the nucleus. Each pore is lined by a complex of 8 somewhat cylindrical proteins that regulate the passage of materials into and out of the nucleus.

A threadlike material called **chromatin** is located within the nucleus. It is nondistinct in the nondividing cell but condenses into rodlike structures called **chromosomes** at the time of cell division. Chromatin and therefore chromosomes contain DNA, the hereditary material. DNA directs protein synthesis, and it is this function that makes the nucleus the control center of the cell. A cell's proteins help to determine its structure and function.

Nucleoli: Maker of rRNA

One or more nucleoli are present in the nucleus. Nucleoli (sing., **nucleolus**) are actually specialized parts of chromatin in which a type of RNA called ribosomal RNA (rRNA) is produced from DNA located there. Ribosomal RNA is necessary to the formation of small ribosomes, which function in the cytoplasm.

Table 3.1
Eukaryotic Cell Organelles

Category	Name	Composition	Function
Nucleus	Nucleus	Nuclear envelope surrounding the nucleoplasm, chromosomes, and nucleoli	Cellular reproduction and control of protein synthesis
	Nucleolus	Concentrated area of chromatin, RNA, and proteins	Ribosome formation
Granule-like organelle	Ribosome	Protein and RNA in 2 subunits	Protein synthesis
Membranous canals and vacuoles	Endoplasmic reticulum (ER)	Membranous flattened channels and tubular canals	Synthesis of macromolecules and transport by vesicle formation
	Rough	Studded with ribosomes	Protein synthesis
	Smooth	No ribosomes	Various; lipid synthesis in some cells
	Golgi apparatus	Stack of membranous saccules	Processing, packaging, and secretion of proteins
	Vacuole and vesicle	Membranous sacs	Storage of substances
	Lysosome	Membranous vesicle containing digestive enzymes	Intracellular digestion
Energy-converting organelles	Mitochondrion	Inner membrane (cristae) within outer membrane	Aerobic cellular respiration
	Chloroplast*	Grana within inner and outer membranes	Photosynthesis
Cytoskeleton	Microtubules	Tubulin molecules	Cell shape and movement
	Actin filaments	Actin molecules	Cell shape and contractile processes
Centrioles and other organelles	Centriole**	9 + 0 pattern of microtubules	Microtubule organization; forms basal bodies
	Cilia and flagella	9 + 2 pattern of microtubules	Movement of the cell

*Plant cells
**Animal cells

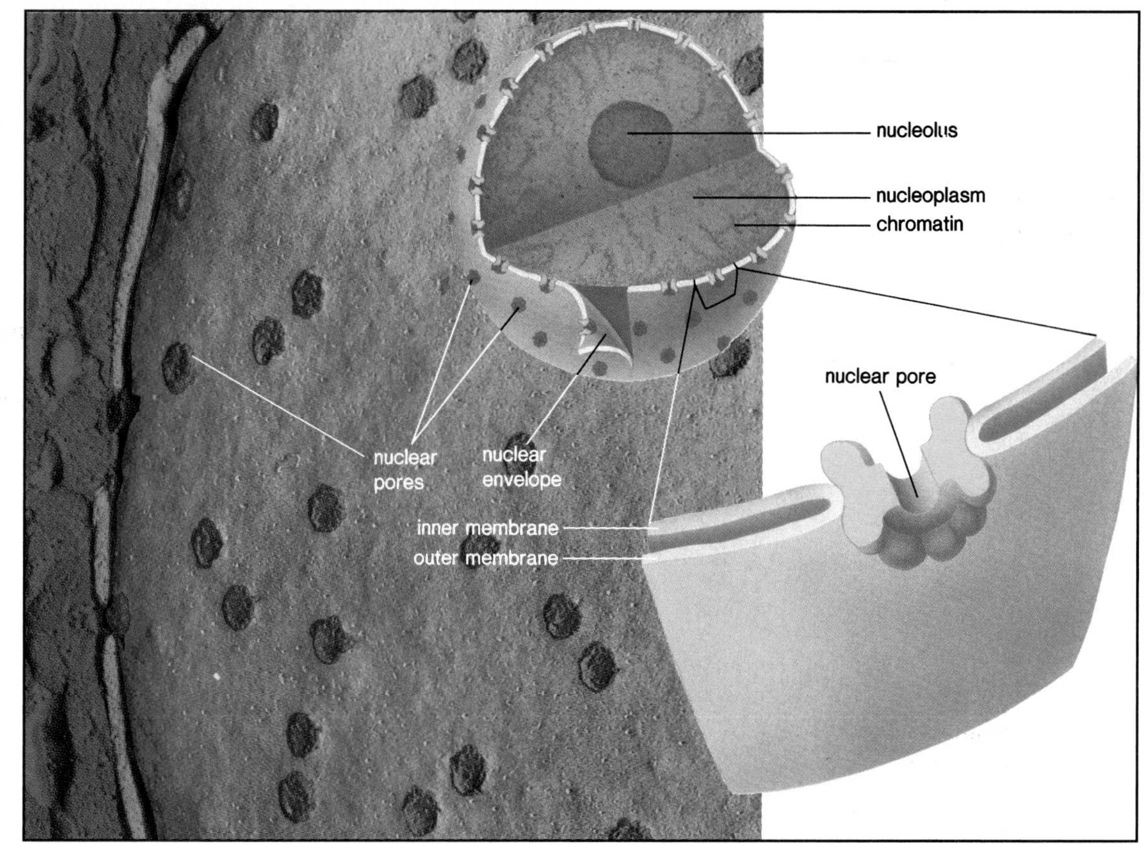

Figure 3.3
Anatomy of the nucleus. The nucleoplasm contains chromatin. Chromatin has a special region called the nucleolus, which is where rRNA is produced. The nuclear envelope contains pores, as is shown in this micrograph of a freeze-fractured nuclear envelope. Each pore is lined by a complex of 8 proteins.

Figure 3.4

Animal and plant cells. These generalized representations are based on electron micrographs, such as those shown. **a.** Animal cell. **b.** Plant cell.

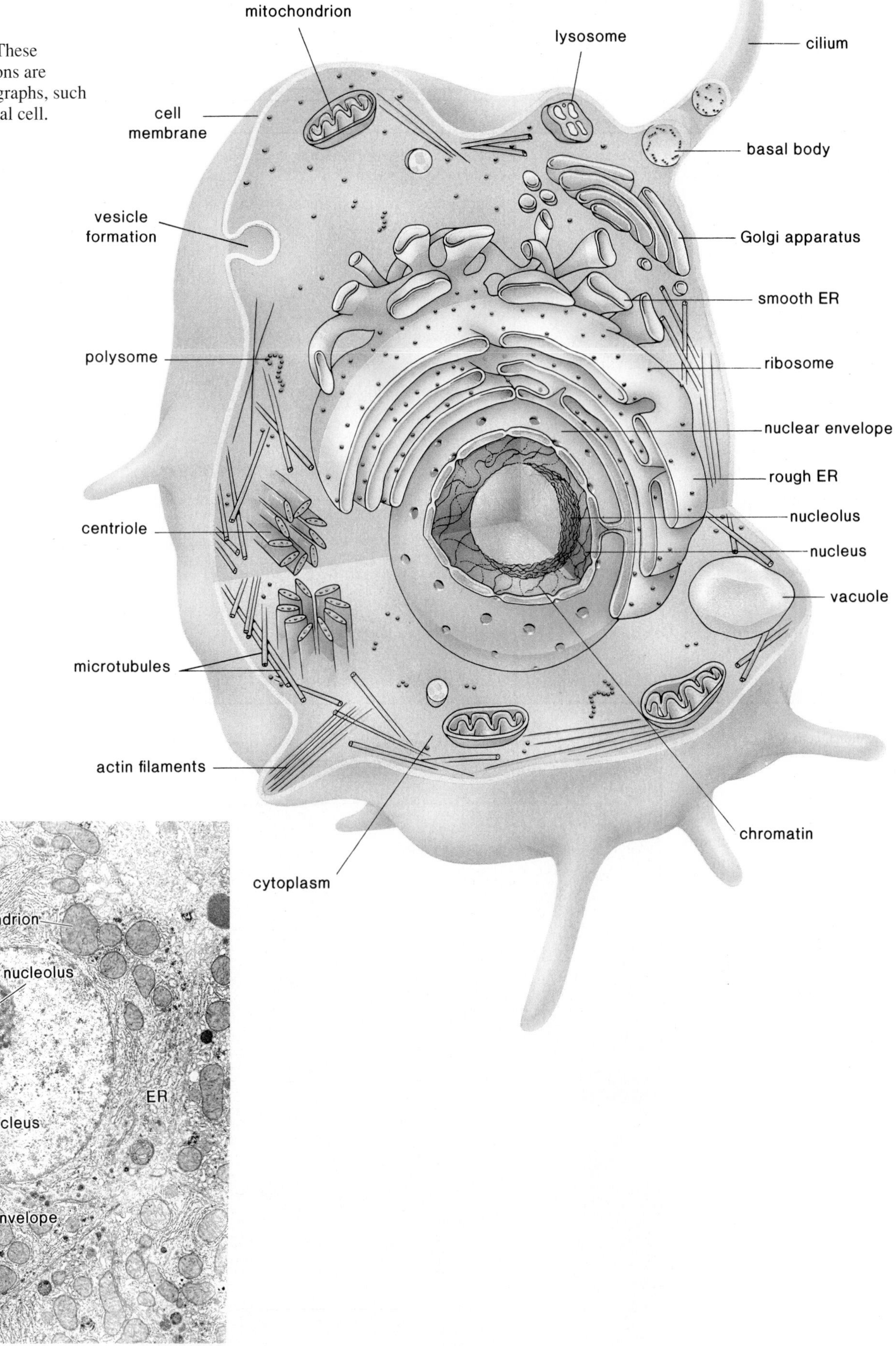

a.

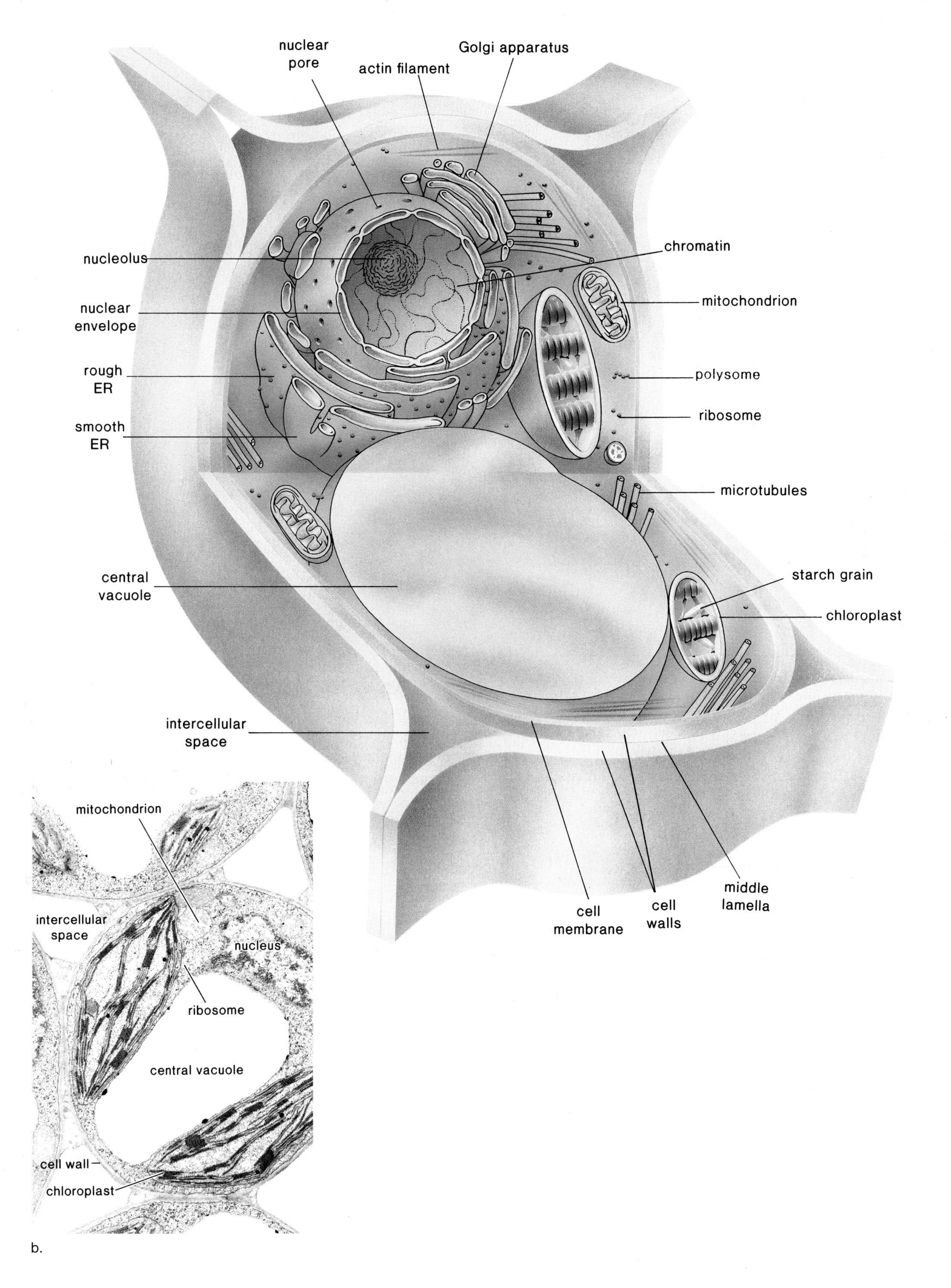
nuclear pore
actin filament
Golgi apparatus
nucleolus
chromatin
mitochondrion
nuclear envelope
rough ER
polysome
ribosome
smooth ER
microtubules
central vacuole
starch grain
chloroplast
intercellular space
cell membrane
cell walls
middle lamella
mitochondrion
intercellular space
nucleus
ribosome
central vacuole
cell wall
chloroplast
b.

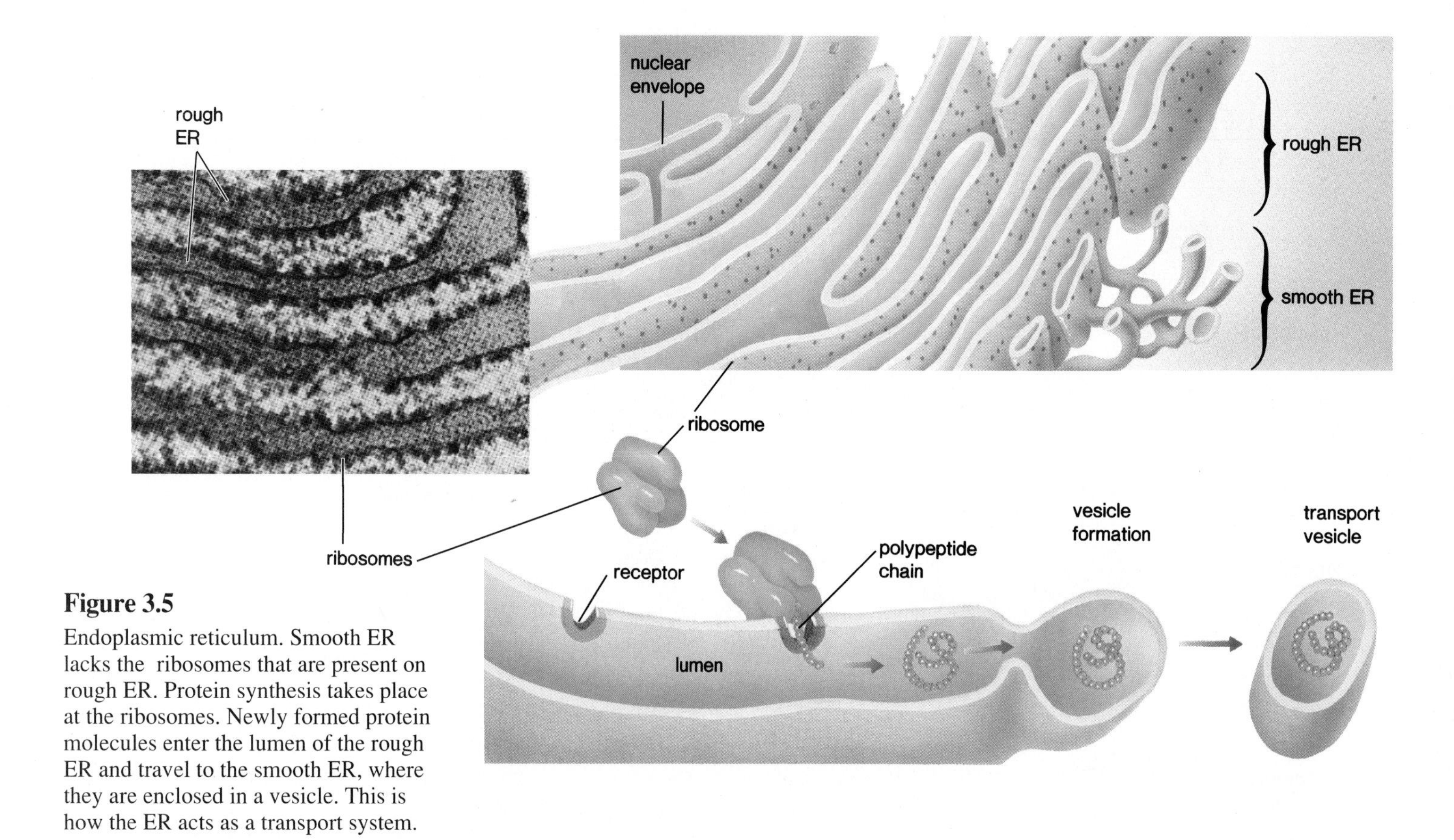

Figure 3.5

Endoplasmic reticulum. Smooth ER lacks the ribosomes that are present on rough ER. Protein synthesis takes place at the ribosomes. Newly formed protein molecules enter the lumen of the rough ER and travel to the smooth ER, where they are enclosed in a vesicle. This is how the ER acts as a transport system.

The nucleus contains chromatin, which condenses into chromosomes just prior to cell division. Chromosomes contain DNA, which directs protein synthesis in the cytoplasm. The nucleus contains one or more nucleoli, where rRNA is made from DNA located there.

Granule-like Organelles

Ribosomes look like small, dense granules in low-power electron micrographs, but they are actually composed of 2 subunits. Each of these subunits has its own particular mix of rRNA and proteins. We have already mentioned that rRNA joins with proteins within the nucleolus, but the 2 subunits are not assembled into one ribosome until they reach the cytoplasm.

Ribosomes are the site of protein synthesis in the cytoplasm. They can be attached to the endoplasmic reticulum (discussed next) or lie free within the cytoplasm. When several ribosomes are making the same protein, they are arranged in a functional group called a *polysome.*

Membranous Canals and Vacuoles

The endoplasmic reticulum (ER), the Golgi apparatus, vacuoles, and lysosomes are structurally and functionally related membranous structures. They work together to produce, transport, store, or secrete cellular products.

Endoplasmic Reticulum: Transporter of Molecules

The **endoplasmic reticulum (ER)** forms a membranous system of tubular canals, which is continuous with the nuclear envelope and branches throughout the cytoplasm (fig. 3.5). If ribosomes are attached to ER, it is called **rough ER;** if ribosomes are not present, it is called **smooth ER.** Both types of ER are involved in synthesis and modification of macromolecules.

Smooth ER produces different molecules in different cells. It is abundant in the testes and adrenal cortex, both of which produce steroid hormones. In the liver, smooth ER is involved in the detoxification of drugs, including alcohol. Special vacuoles (membrane-enclosed sacs) called *peroxisomes* are often attached to smooth ER, and these contain enzymes capable of detoxifying drugs.

Rough ER specializes in protein synthesis (fig. 3.5). The ribosomes attached to rough ER make proteins for export from the cell. The proteins enter the lumen (interior space) of rough ER, where they may be modified. After arriving at the lumen of smooth ER, a vesicle (small vacuole) pinches off and then carries the protein to the Golgi apparatus, where it is further processed. This is how the ER serves as a transport system.

Golgi Apparatus: Processor and Packager

The **Golgi apparatus** (fig. 3.6) is named for the person who first discovered its presence in cells. It is composed of a stack of about a half-dozen or more saccules (flattened vacuoles), which

Figure 3.6

Structure and function of the Golgi apparatus. The Golgi apparatus, a stack of several saccules, receives protein-containing vesicles from smooth ER. Protein molecules are modified as they move through the Golgi apparatus and are then repackaged into secretory vesicles. These vesicles take the protein to the cell membrane, where they are discharged. The Golgi apparatus also produces lysosomes. Macromolecules can enter a cell by vesicle formation, and when this vesicle fuses with the lysosome, the macromolecules are digested.

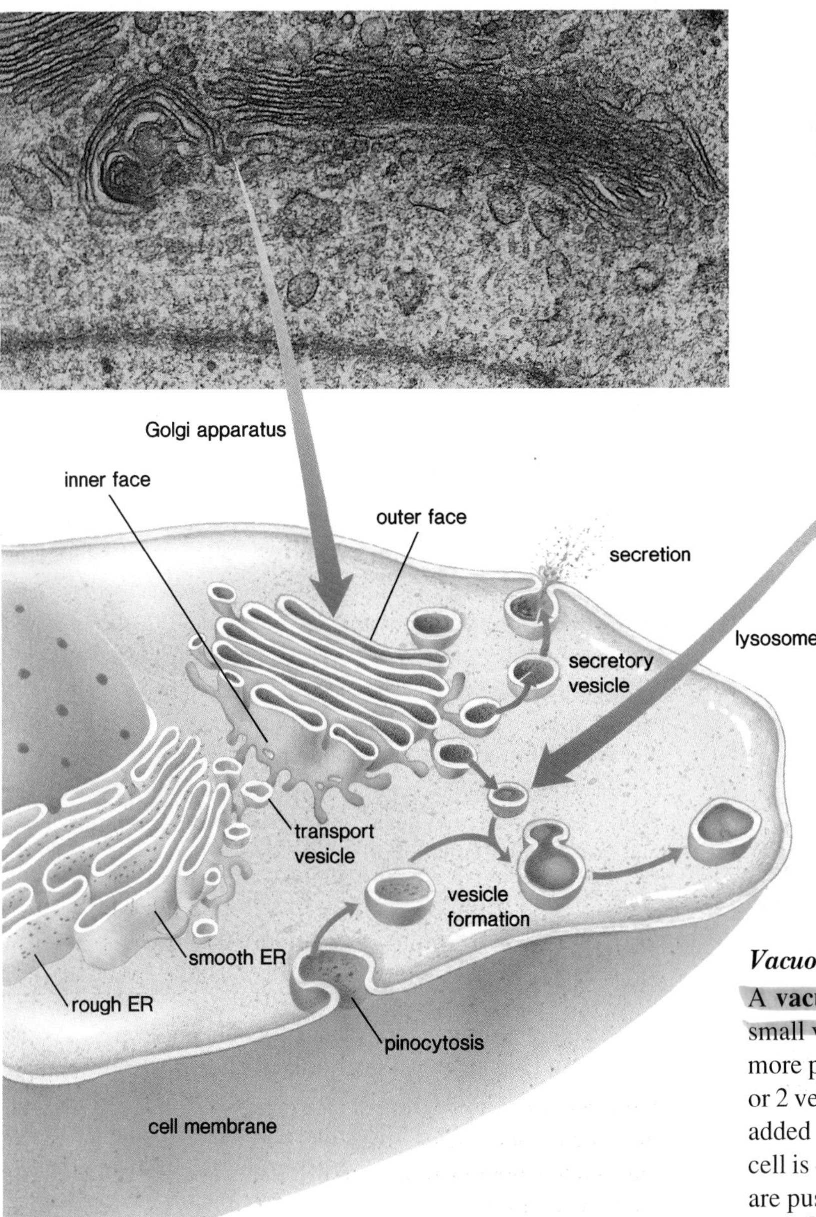

look like hollow pancakes. One side of the stack, called the inner face, is directed toward the nucleus and the ER. The other side of the stack, called the outer face, is directed toward the cell membrane. Vesicles occur at the edges of the saccules.

The Golgi apparatus functions in the packaging, storage, and distribution of molecules produced by the ER. After the molecules are received at its inner face, they move from one saccule to the next via newly formed vesicles. As they go from saccule to saccule of the Golgi apparatus, the molecules undergo many and various modifications. Their basic structure can change or be modified by the addition of a carbohydrate or a phosphate group, for example. Finally, molecules are often packaged in secretory vesicles; these move to the cell membrane and discharge their contents (fig. 3.6). For example, this is the way hormones are secreted into blood by the glands that produce them.

Vacuoles: Storer of Molecules

A **vacuole** is a large membrane-enclosed sac; a vesicle is a small vacuole. Animal cells have vacuoles, but they are much more prominent in plant cells. Typically, plant cells have one or 2 very large vacuoles filled with a watery fluid, which gives added support to the cell. Most of the central area of the plant cell is occupied by a vacuole, and the other contents of the cell are pushed to the sides (fig. 3.4*b*).

Most often, vacuoles are storage areas. Plant vacuoles contain not only water, sugars, and salts but also pigments and toxic substances. The pigments are responsible for many of the red, blue, or purple colors of flowers and some leaves. The toxic substances help protect a plant from predaceous animals. (As long as the substance is contained within a vacuole, it is not harmful to the plant.)

Lysosomes: Digester of Molecules

Lysosomes, vesicles formed by the Golgi apparatus, contain *hydrolytic enzymes.* Hydrolytic enzymes digest macromolecules in the manner described in figure 2.14, page 28. Macromolecules are sometimes brought into a cell in vesicles formed

Figure 3.7
Mitochondrion structure. A mitochondrion has a double membrane, and the cristae are infoldings of the inner membrane. The cristae project into the matrix, a space filled with fluid.

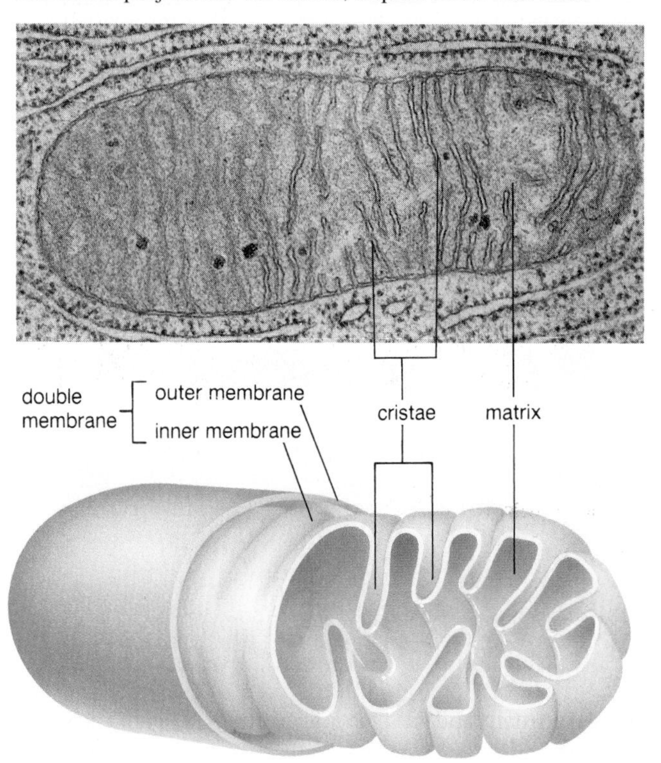

at the cell membrane (fig. 3.6). A lysosome can fuse with such a vesicle and digest its contents into simpler molecules, which then enter the cytoplasm. Some white blood cells defend the body by engulfing bacteria, a process that involves vesicle formation at the cell membrane. When lysosomes fuse with these vesicles, the bacteria are digested. Even parts of a cell are often digested by the cell's own lysosomes (called autodigestion because *auto* means *self*). Normal cell rejuvenation most likely takes place in this way, but autodigestion is also important during development. For example, when a tadpole becomes a frog, the enzymes within lysosomes digest the cells of the tail, and the fingers of a human embryo are at first webbed, but they are freed from one another by lysosomal action.

Occasionally, a child is born with a metabolic disorder involving a missing or inactive lysosomal enzyme. In these cases, the lysosomes fill to capacity with macromolecules that cannot be broken down. The cells become so full of these lysosomes that the child dies. The best known of these lysosomal storage disorders is Tay-Sachs disease, discussed on page 434.

The ER is a membranous system of tubular canals, which can be smooth or rough. Proteins synthesized at rough ER are processed and packaged in vesicles by the Golgi apparatus. Some vesicles discharge their contents at the cell membrane, and some fuse with lysosomes, which digest any material enclosed therein. Do 3.2 Critical Thinking, found at the end of the chapter.

Energy-Converting Organelles

The energy-related organelles, chloroplasts and mitochondria, convert one form of energy into another. While chloroplasts are unique to plant cells, mitochondria are found in both plant and animal cells.

Mitochondria: Producers of ATP

Most mitochondria (sing., **mitochondrion**) are between 0.5 μm and 1.0 μm in diameter and 7 μm in length, although the size and the shape can vary. Mitochondria are bounded by a double membrane. The inner membrane is folded to form little shelves called *cristae,* which project into the *matrix,* an inner space filled with a gel-like fluid (fig. 3.7).

Mitochondria produce ATP (adenosine triphosphate) molecules. Every cell uses a certain amount of ATP energy to synthesize molecules, but many cells use ATP to carry out their specialized function. For example, muscle cells use ATP for muscle contraction, which produces movement, and nerve cells use it for the conduction of nerve impulses, which make us aware of our environment.

Mitochondria are often called the powerhouses of the cell: just as a powerhouse burns fuel to produce electricity, the mitochondria convert the chemical energy of glucose products into the chemical energy of ATP molecules. In the process, mitochondria use up oxygen (O_2) and give off carbon dioxide (CO_2) and water (H_2O). The oxygen you breathe in enters cells and then mitochondria; the carbon dioxide you breathe out is released by mitochondria. Because gas exchange is involved, it is said that mitochondria carry on **aerobic cellular respiration.** A shorthand way to indicate the chemical transformation associated with cellular respiration is

$$\text{carbohydrate} + \text{oxygen} \longrightarrow \text{carbon dioxide} + \text{water} + \text{energy}$$

The energy in this equation stands for ATP molecules. As we will see in chapter 5, the matrix of a mitochondrion contains enzymes for breaking down glucose products. ATP production then occurs at the cristae. The molecules that aid in the conversion of energy are located in an assembly-line fashion on these membranous shelves.

Chloroplasts: Producers of Food

Chloroplasts, found only in plant cells and photosynthetic protists, range from about 4 μm to 6 μm in diameter and from 1 μm to 5 μm in length. They belong to a group of plant organelles known as plastids. Other plastids are the leucoplasts, which store starch, and the chromoplasts, which contain red and orange pigments.

A chloroplast is bounded by a double membrane. Inside the structure, there is even more membrane organized into flattened sacs called *thylakoids*. The thylakoids are piled up like stacks of coins, and each stack is called a **granum** (pl., grana).

Figure 3.8
Chloroplast structure. A chloroplast is bounded by a double membrane. The grana are stacks of flattened membranous sacs called thylakoids. The surrounding fluid-filled space is the stroma.

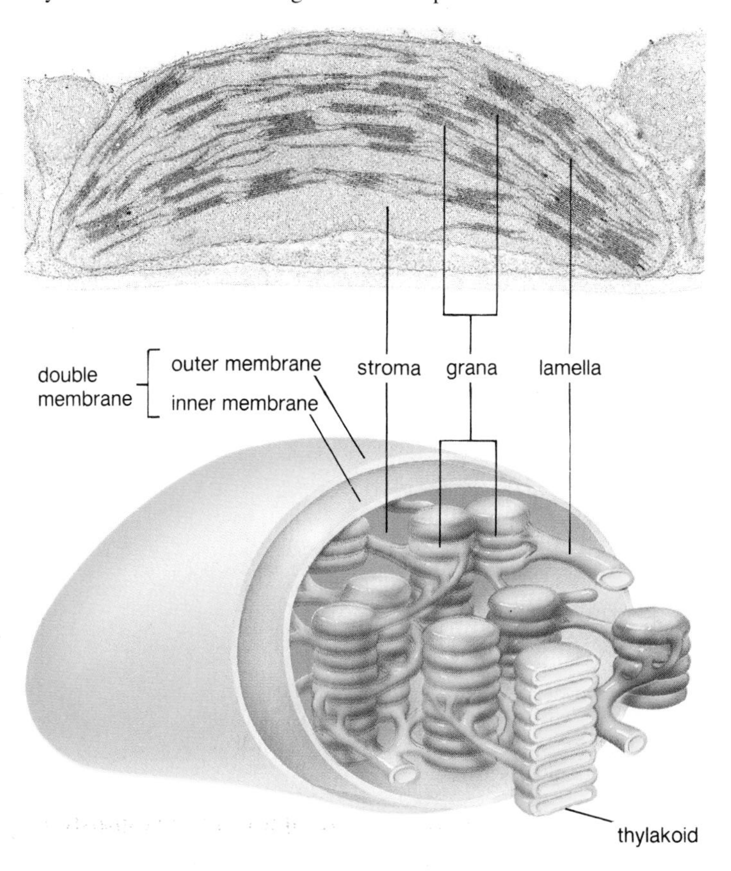

There are membranous connections between the grana called lamellae (fig. 3.8). The fluid-filled space about the grana is called the *stroma*.

Chloroplasts carry on **photosynthesis,** during which light energy (*photo*) is used to produce food molecules, like glucose (*synthesis*). Chloroplasts take in carbon dioxide (CO_2), water (H_2O), and solar (radiant) energy in order to produce glucose. They give off oxygen (O_2), which exits the leaves as a gas. Again, we can use the shorthand method to describe what has been said:

energy + carbon dioxide + water ⟶ carbohydrate + oxygen

The equation for photosynthesis is the opposite of aerobic cellular respiration, as you can see by comparing the shorthand statements for each.

Like mitochondria, chloroplasts are highly organized. The green pigment **chlorophyll,** found within the grana, makes chloroplasts and leaves green. Chlorophyll absorbs solor energy, and chloroplasts convert this energy into ATP molecules, the type of energy used by enzymes within the stroma to make the carbohydrate glucose.

Mitochondria are the powerhouses of the cell because they convert carbohydrate energy to ATP energy. Chloroplasts contain chlorophyll and carry out photosynthesis, which converts solar energy to carbohydrate energy.

Cytoskeleton: For Shape and Motion

Within the cytoplasm, several types of filamentous protein structures form a **cytoskeleton** (fig. 3.9). The cytoskeleton helps to maintain the cell's shape, anchors the organelles, and allows the cell and its organelles to move. Cells typically have a particular shape; in animals, muscle cells are tubular and nerve cells have long, skinny extensions. Sometimes cells change shape or even move about by means of cilia or flagella.

Microtubules are cylinders about 25 nm in diameter that contain 13 rows of the globular protein tubulin. Remarkably, microtubules can assemble and disassemble. When assembly occurs, 2 tubulin molecules come together as dimers, and then these join to form a microtubule (fig. 3.9*b*). In many cells, the regulation of microtubule assembly is under the control of a microtubule organizing center (MTOC), which lies near the nucleus. Microtubules radiate from the MTOC, helping to maintain the shape of the cell and acting as tracts along which organelles can move. It is well known that during cell division, microtubules form spindle fibers, which assist the movement of chromosomes (p. 110).

Actin filaments, also called microfilaments, are long, extremely thin protein fibers (about 7 nm in diameter) that usually occur in bundles (fig. 3.9*c*). In muscle cells, actin filaments interact with myosin filaments to bring about contraction. Actin filaments are seen in various types of cells, especially those in which movement occurs. For example, microvilli, which project from certain cells and which can shorten and extend, contain actin filaments. Along with myosin, they form a constriction ring, which becomes ever smaller as a cell divides into 2 cells. Actin filaments, like microtubules, can assemble and disassemble.

The cytoskeleton contains microtubules (composed of the protein tubulin) and actin protein filaments. The cytoskeleton maintains the shape of the cell and also directs the movement of the cell's organelles.

Centrioles and Other Organelles

Centrioles and related organelles are composed of microtubules arranged in specific patterns.

Figure 3.9

The cytoskeleton. **a.** The cytoskeleton both anchors the organelles and allows them to move. **b.** A microtubule is a cylinder composed of tubulin protein molecules. Microtubules radiate from the microtubule organizing center and assist intracellular movement of organelles. **c.** An actin filament contains molecules of the protein actin. Bundles of actin filaments lie close to the cell membrane. Actin filaments also form networks within the cytoplasm.

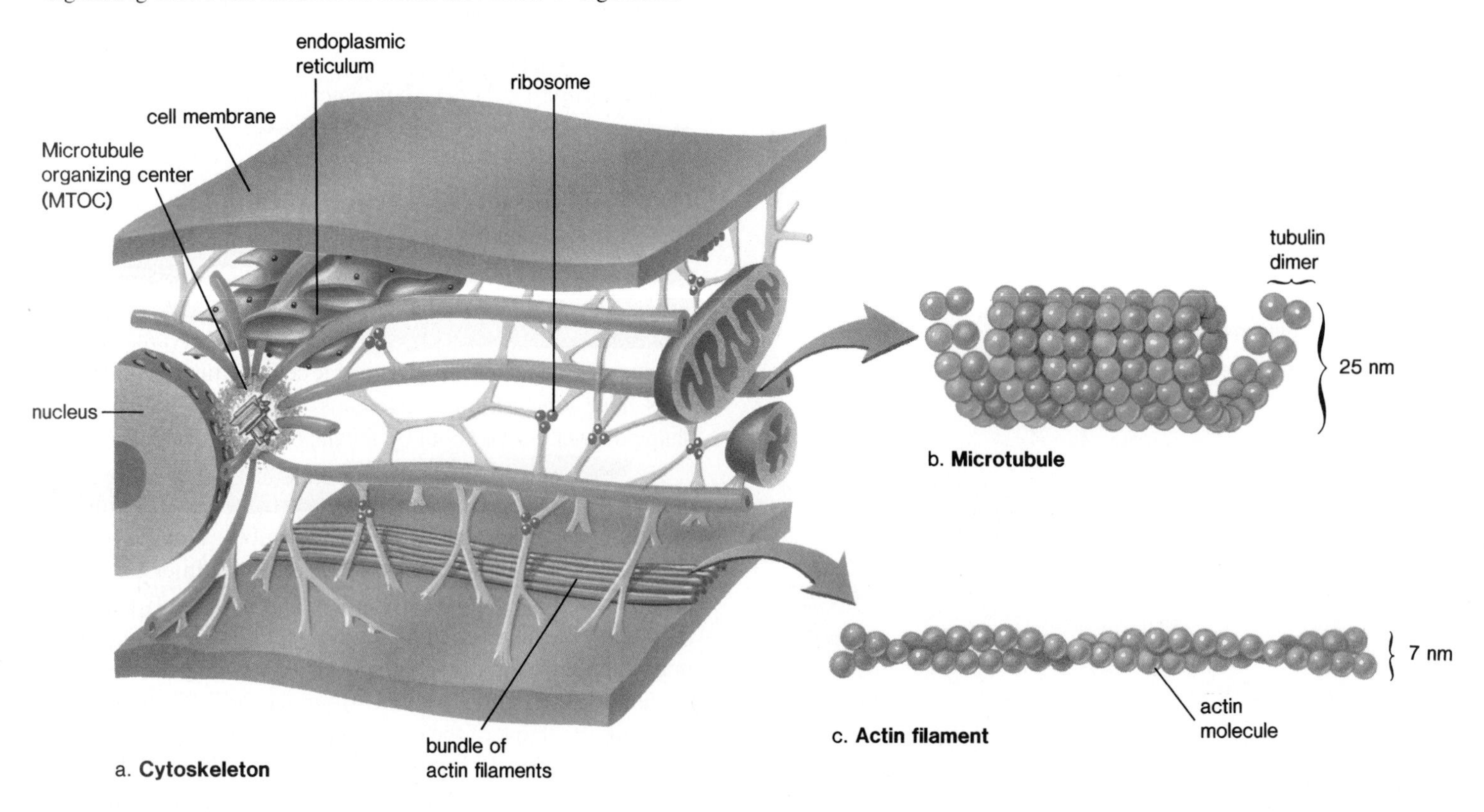

Centrioles: Organizer of Microtubules

Centrioles are short cylinders having a 9 + 0 pattern of microtubules. There is an outer ring of 9 microtubule triplets, but no microtubules are located in the center of a centriole (fig. 3.10). Animal cells, but not plant cells, have 2 centrioles, which lie at right angles to one another within the MTOC. Before an animal cell divides, the centrioles replicate; the members of each of the pairs are also at right angles to one another. During cell division, the centriole pairs separate so that each new cell receives one pair.

In cells capable of movement, centrioles give rise to **basal bodies,** which are microtubule organizing centers for the formation of cilia and flagella. In other words, a basal body does for a cilium (or a flagellum) what the MTOC does for the cell.

Cilia and Flagella: Movers of the Cell

Cilia (sing., **cilium**) and flagella (sing., **flagellum**) are hairlike extensions of cells that can move either in an undulating fashion, like a whip, or stiffly, like an oar. Cells that have these organelles are capable of movement. For example, single-celled paramecia move by means of cilia; male sperm cells move by means of flagella. The cells that line our upper respiratory tract are ciliated. These cilia sweep debris trapped within mucus back up into the throat, an action that helps to keep the lungs clean.

Cilia are much shorter than flagella, but even so they both are constructed similarly (fig. 3.11). They are both membrane-bounded cylinders enclosing a matrix having a 9 + 2 pattern of microtubules. Nine microtubule doublets are arranged in a ring around 2 central microtubules (fig. 3.11). Each doublet has arms projecting toward a neighboring doublet and radial spokes extending toward the central pair of microtubules. Recent evidence indicates that cilia and flagella move when the microtubule doublets slide along one another. The clawlike arms and the spokes seem to be involved in causing this sliding action, which requires ATP energy.

> Centrioles are small cylinders that contain microtubules. They are a part of a microtubule organizing center (MTOC) lying near the nucleus of animal cells. They also give rise to basal bodies, which organize the arrangement of microtubules within cilia and flagella. Do 3.3 Critical Thinking, found at the end of the chapter.

Figure 3.10

Centrioles. Centrioles are composed of 9 microtubule triplets. They lie at right angles to one another within the microtubule organizing center, which is believed to assemble microtubules, especially at the time of cell division.

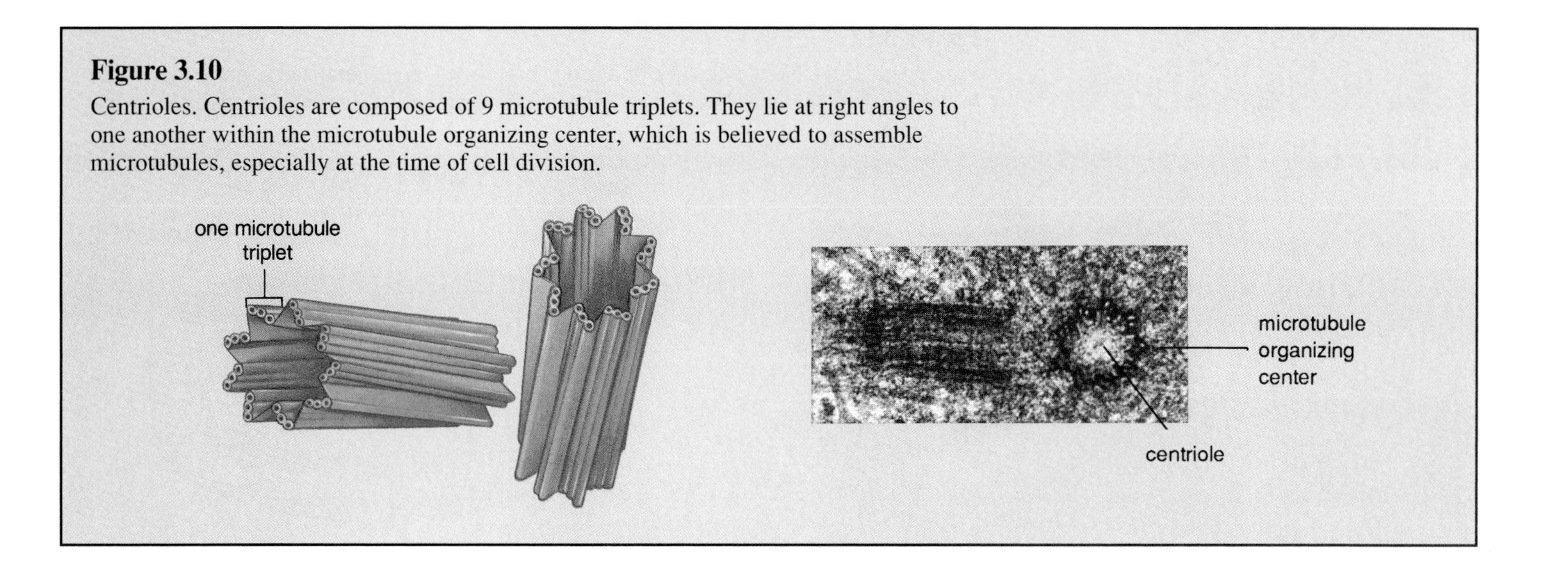

Figure 3.11

Flagellum structure. Each flagellum has a basal body, which organizes the microtubules in the shaft. A basal body has a 9 + 0 pattern of microtubules, and the shaft has a 9 + 2 pattern of microtubules. Notice the side arms of the outer microtubule doublets and the spokes, which connect them to the 2 central microtubules. Bending occurs because a doublet is held in place by the spokes as the side arms move along the length of an adjacent microtubule doublet.

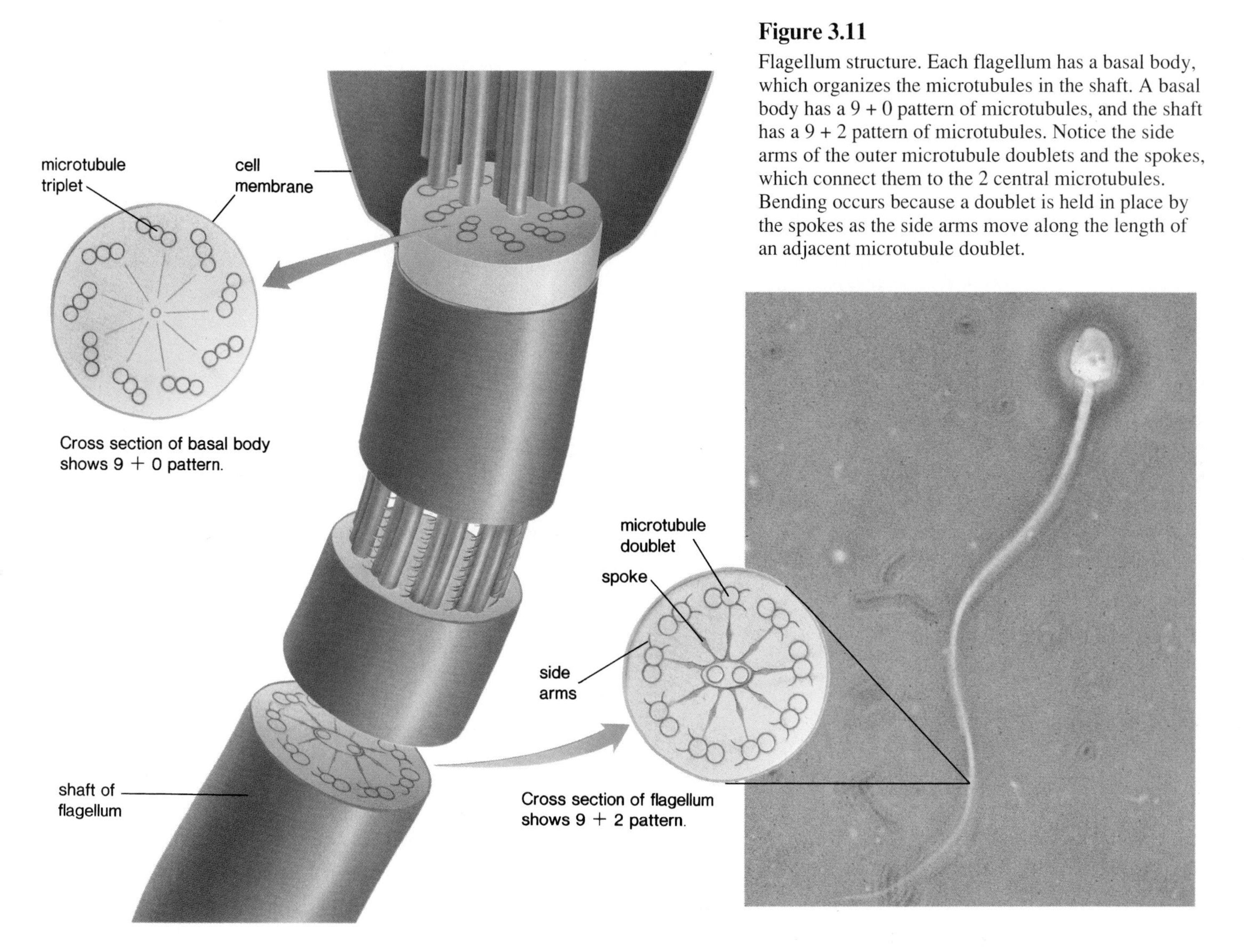

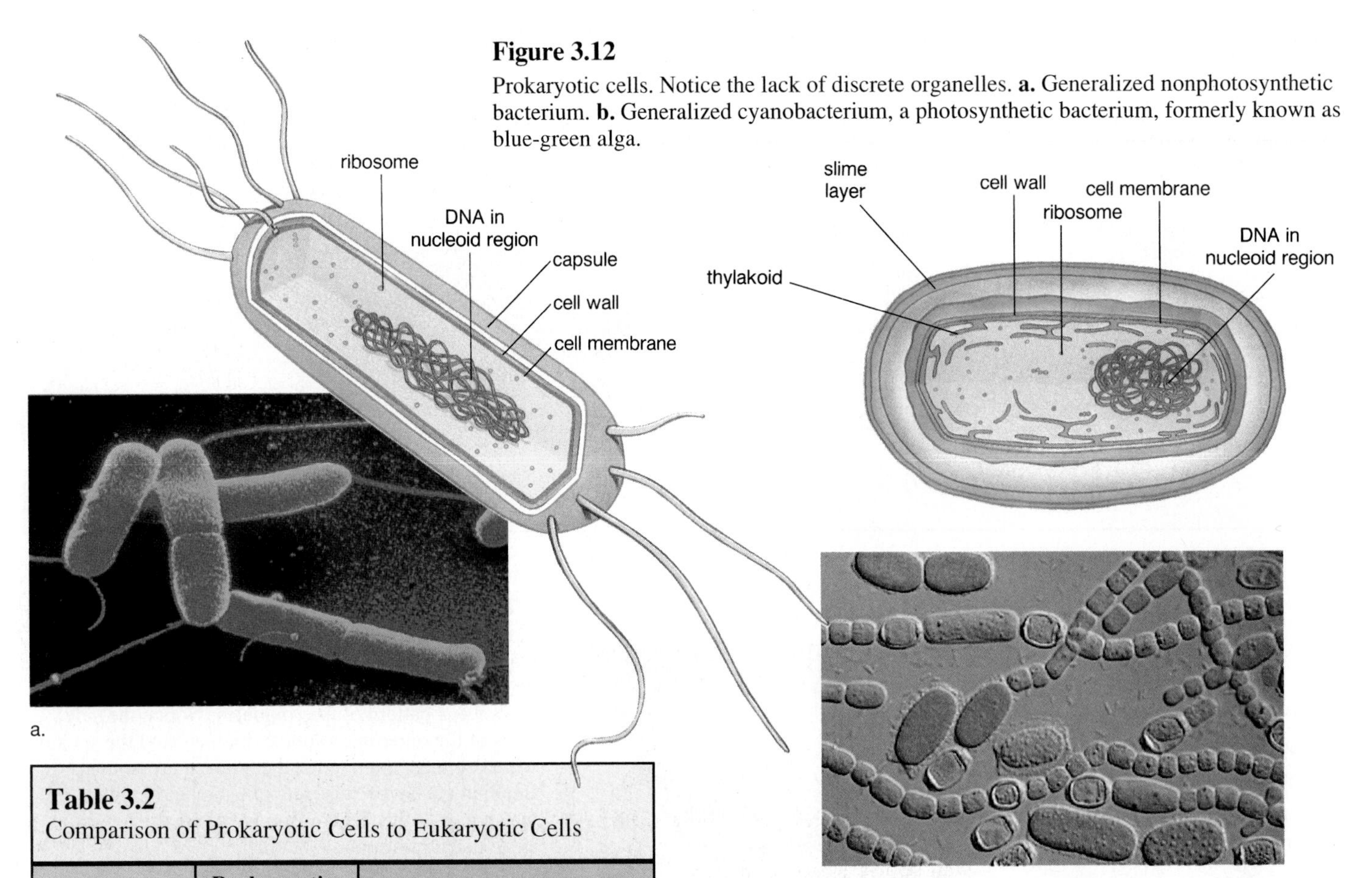

Figure 3.12
Prokaryotic cells. Notice the lack of discrete organelles. **a.** Generalized nonphotosynthetic bacterium. **b.** Generalized cyanobacterium, a photosynthetic bacterium, formerly known as blue-green alga.

Table 3.2
Comparison of Prokaryotic Cells to Eukaryotic Cells

	Prokaryotic (1–10 μm in diameter)	Eukaryotic Cells (10–100 μm in diameter)	
		Animal	*Plant*
Cell membrane	Yes	Yes	Yes
Cell wall	Yes (not cellulose)	No	Yes (cellulose)
Nuclear envelope	No	Yes	Yes
Nucleolus	No	Yes	Yes
Mitochondria	No	Yes	Yes
Chloroplasts	No	No	Yes
Endoplasmic reticulum (ER)	No	Yes	Yes
Ribosomes	Yes (smaller)	Yes	Yes
Vacuoles	Some	Yes (small)	Yes (usually large, single vacuole)
Golgi apparatus	No	Yes	Yes
Lysosomes	No	Always	Often
Cytoskeleton	No	Yes	Yes
Centrioles	No	Yes	No
Cilia or flagella	No	Often	Some male gametes

Cellular Comparisons

Table 3.2 compares prokaryotic cells (*pro* means *before*; *karyon* means *nucleus*) to the 2 types of eukaryotic cells (plant and animal cells) we have been studying in this chapter.

Prokaryotic versus Eukaryotic

In contrast to eukaryotic cells, **prokaryotic cells** do not have a well-defined nucleus. Prokaryotes, represented only by bacteria, including cyanobacteria (formerly called blue-green algae), have a chromosome composed of a single circular DNA macromolecule. It is located in an area called the *nucleoid region,* which is not enclosed by a nuclear envelope (fig. 3.12).

Prokaryotes are unicellular organisms that lack most of the organelles we have been discussing. This does not mean, however, that these cells do not carry on the functions performed by eukaryotic organelles. They have DNA in the nucleoid region, and respiratory enzymes attached to the cell membrane. In cyanobacteria, which are photosynthetic, chlorophyll is located in individual thylakoids, rather than in chloroplasts.

In addition to a cell membrane, prokaryotes have a cell wall, and if capable of movement, they possess flagella. However, the structures of these cell parts differ from those found in

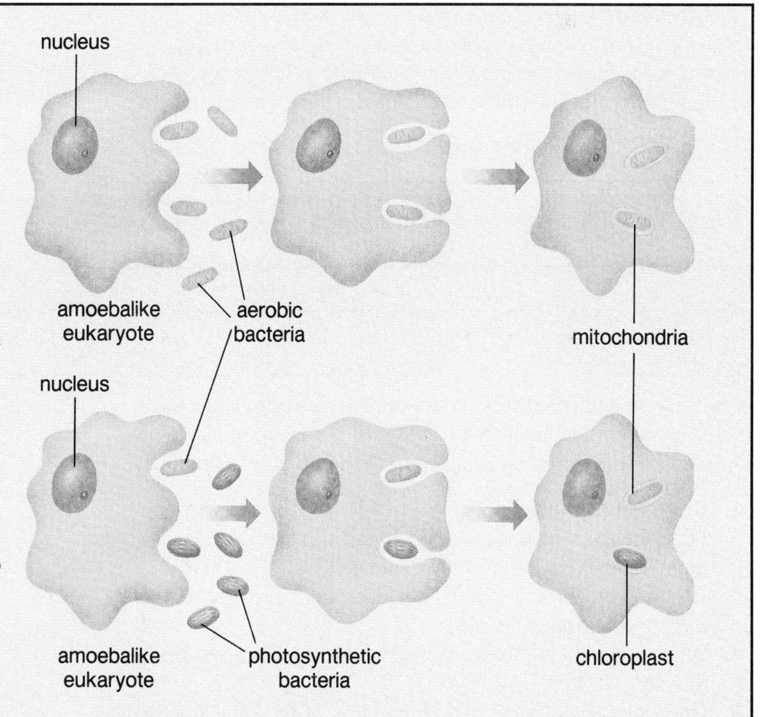

Figure 3.13

The endosymbiotic theory explains the origin of mitochondria and chloroplasts in eukaryotic cells. Mitochondria may be derived from aerobic bacteria that were engulfed by a larger amoebalike eukaryote. The evidence of the endosymbiotic theory is as follows.

- Mitochondria and chloroplasts are similar to bacteria in size and structure.
- Both organelles are membrane bounded—the outer membrane may be derived from the engulfing vesicle, and the inner one may be derived from the cell membrane of the original prokaryote.
- Mitochondria and chloroplasts contain a limited amount of genetic material and are capable of self-reproduction. Their DNA is a circular strand, which resembles the bacterial chromosome.
- Although most of the proteins within mitochondria and chloroplasts are now produced by the eukaryotic host, they do possess their own ribosomes and they do produce some proteins. Their ribosomes resemble those of bacteria.

eukaryotic cells. Outside the cell wall, there may be a polysaccharide capsule or a slime layer.

Prokaryotes are small in size (*Escherichia coli,* a bacterium, is 2 µm × 1 µm) but extremely abundant. Some even suggest that their combined number exceeds that of all other types of organisms on earth. Despite their lack of structural complexity, they are more metabolically varied than any other type of organism. They can decompose almost any organic material and therefore contribute greatly to the cycling of materials in ecosystems.

The *endosymbiotic theory,* which is widely accepted today, says that at one time in the history of life, prokaryotes entered eukaryotic cells and eventually evolved into the mitochondria and chloroplasts found in eukaryotic cells today. The data that support this theory are given in figure 3.13.

Plant versus Animal

Table 3.2, in addition to the comparison of prokaryotic cells to eukaryotic cells, compares plant cells to animal cells. Note that both plant and animal cells have a cell membrane. Plant cells also have a cell wall. Both types of cells have mitochondria, but plant cells also have chloroplasts. Typically, animal cells, not plant cells, have centrioles and basal bodies, organelles involved in the formation of cilia and flagella.

SUMMARY

All organisms are composed of cells; some organisms are unicellular and some are multicellular. The transmission electron microscope allows us to see the cellular details represented in figure 3.4.

The nucleus is a large organelle of primary importance because it controls the metabolic functions and structural characteristics of the cell. Chromatin, which condenses to become chromosomes during cell division, lies within the nucleus.

Proteins are made at rough ER before being packaged at the Golgi apparatus into vesicles for secretion. During secretion, a vesicle discharges its contents at the cell membrane. Golgi-derived lysosomes fuse with incoming vesicles to digest any enclosed material. Lysosomes also carry out autodigestion of old parts of cells.

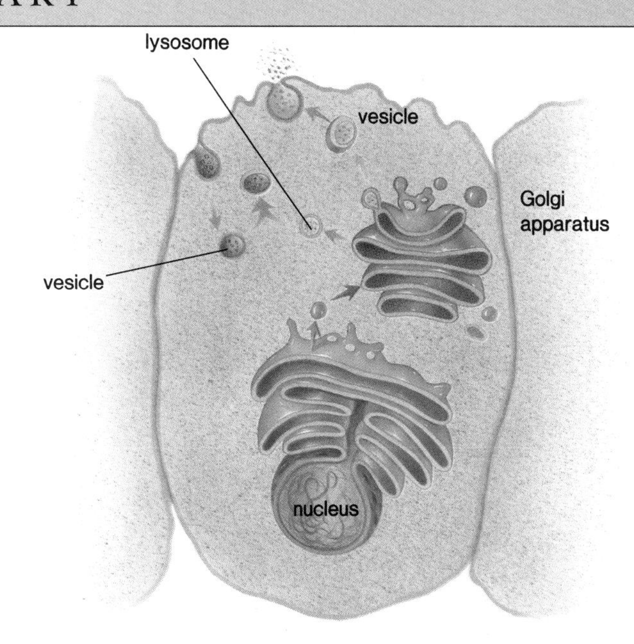

Mitochondria and chloroplasts are the energy-related organelles. During the process of aerobic cellular respiration, mitochondria convert carbohydrate energy to ATP energy, and during photosynthesis, chloroplasts convert solar energy to carbohydrate energy.

carbohydrate + oxygen ⟶ carbon dioxide + water + energy

Actin filaments and microtubules are found within the cytoskeleton, which maintains the cell's shape and permits movement of cell organelles. Centrioles are found within the MTOC, which lies just outside the nucleus of animal cells. Centrioles also produce basal bodies, which give rise to cilia and flagella.

Prokaryotic cells lack the organelles typically found in eukaryotic cells (plant and animal cells). Nevertheless, they carry on all the same functions as eukaryotic cells.

STUDY QUESTIONS

In order to practice **writing across the curriculum,** students should write out the answers to any or all of the study questions. The study questions are sequenced in the same order as the text.

1. Briefly define the cell theory. (p. 44)
2. Describe the structural and biochemical makeup of the membrane. Also describe the plant cell wall. (p. 45)
3. Describe the nucleus and its contents, including the term *DNA* in your description. (p. 46)
4. What is the nucleolus, and what function does it perform for the cell? (p. 46)
5. Describe the structure and the function of the ER. Include the terms *rough ER, smooth ER,* and *ribosomes* in your description. (p. 50)
6. Describe the structure and the function of the Golgi apparatus. Mention vacuoles and lysosomes in your description. (pp. 50–51)
7. Describe the structure and the function of mitochondria and chloroplasts. (pp. 52–53)
8. Describe the structure and the function of microtubules, actin filaments, centrioles, cilia, and flagella. (pp. 53–54)
9. What are the 2 main types of cells, and how do they differ structurally? (p. 56)
10. What are the structural differences between animal cells and plant cells? (p. 57)

OBJECTIVE QUESTIONS

1. Electron microscopes have a greater _______ power than light microscopes.
2. In a membrane, _______ molecules are embedded in a _______ bilayer.
3. Chromosomes are located within the _______, an organelle that controls the rest of the cell.
4. Rough ER has _______, but smooth ER does not.
5. Lysosomes contain _______ enzymes.
6. Vesicles derived from the ER make their way to the _______, an organelle that functions in packaging, storage, and distribution.
7. Both plant and animal cells have _______, where glucose products provide energy for ATP formation.
8. Photosynthesis takes place within _______.
9. Actin filaments and microtubules are a part of the _______, the framework of the cell, which provides its shape and regulates movement of its organelles.
10. Basal bodies, which organize the microtubules within cilia and flagella, are derived from _______.
11. Label only the parts of the cell on the right that are involved in protein synthesis and modification. Explain your choices.
12. Study the example given in (*a*) below. Then, for each other organelle listed (*b–d*), state another that is structurally and functionally related. Tell why you paired these organelles.
    a. The nucleus can be paired with *nucleoli* because nucleoli are found in the nucleus. Nucleoli occur where chromatin is producing rRNA.
    b. mitochondria
    c. centrioles
    d. endoplasmic reticulum

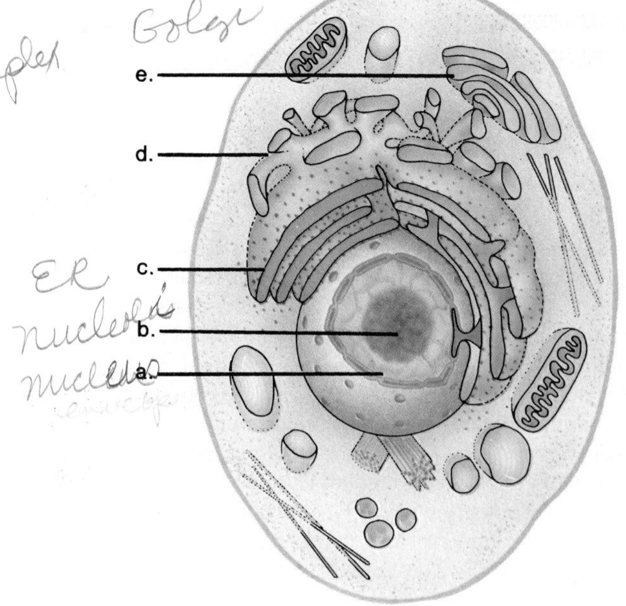

CRITICAL THINKING

In order to practice **writing across the curriculum,** students should write out the answers to any or all of the critical thinking questions. Suggested answers to the critical thinking questions are in appendix E.

3.1

1. See appendix C and note that 1 cm = 10 mm = 10,000 μm = 10,000,000 nm. A nm is equal to how many cm?
2. Some prokaryotic cells are as large as 10 μm in diameter. How many of these cells could you fit on a 1-cm line?
3. The human egg (a eukaryotic cell) is 100 μm in diameter. How many human eggs could you fit on a 1-cm line?
4. From the information provided in questions 1 and 2, would you expect a prokaryotic cell to fit inside a eukaryotic cell? Why?

3.2

1. It is possible to bathe a cell in radioactively tagged amino acids and to later detect photographically the location of radiation and therefore the amino acids in the cell. Why would you suggest using radioactive sulfur (the amino acids cysteine and methionine contain sulfur) rather than radioactive carbon?
2. An investigator uses the procedure outlined in question 1 to support the belief that proteins move from the cytoplasm into the nucleus. If so, where will radiation first appear, and where will it subsequently appear in the cell?
3. An investigator uses this same procedure to support the belief that proteins move from rough ER to secretory vesicles. Where will the radiation first appear, and where will it subsequently appear in the cell?

3.3

1. A microtubular spindle apparatus appears in plant and animal cells at the time of cell division. Why does this suggest that centrioles are not necessary to the formation of the spindle apparatus?
2. What evidence is there to suggest centrioles are necessary to microtubule organization in animal cells?
3. In animal cells, each newly formed cell receives a pair of centrioles. Why might centrioles be necessary to animal cells but not to most plant cells?

SELECTED KEY TERMS

actin filament an extremely thin fiber found within the cytoplasm that is composed of the protein actin; involved in the maintenance of cell shape and the movement of cell contents.

cell wall a protective barrier outside the cell membrane of a bacterial, fungal, algal, or plant cell.

centriole a short, cylindrical organelle in animal cells that contains microtubules in a 9 + 0 pattern; associated with the formation of basal bodies.

chloroplast (klo´ ro-plast) a membranous organelle that contains chlorophyll and is the site of photosynthesis.

chromosome rodlike structure in the nucleus seen during cell division; contains the hereditary units, or genes.

cytoplasm the contents of the cell; located between the nucleus and the cell membrane.

cytoskeleton filamentous protein structures found throughout the cytoplasm that help maintain the shape of the cell, anchor the organelles, and allow the cell and its organelles to move.

endoplasmic reticulum (ER) (en-do-plaz-mik ře-tik´ u-lum) a membranous system of tubules, vesicles, and sacs in cells sometimes having attached ribosomes. Rough ER has ribosomes; smooth ER does not.

eukaryotic cell a cell that possesses a nucleus and the other membranous organelles characteristic of complex cells.

flagellum slender, long extension used for locomotion by some protozoans, bacteria, and sperm.

Golgi apparatus an organelle consisting of concentrically folded saccules that functions in the packaging, storage, and distribution of cellular products.

lysosome (li´ so-sōm) an organelle in which digestion takes place due to the action of hydrolytic enzymes.

microtubule an organelle composed of 13 rows of globular proteins; found in multiple units within other organelles, such as the centriole, cilia, flagella, as well as spindle fibers.

mitochondrion (mi″ to-kon´ dre-on) a membranous organelle in which cellular respiration produces the energy molecule, ATP.

nucleolus (nu-kle´ o-lus) an organelle found inside the nucleus; a special region of chromatin containing DNA that produces rRNA for ribosome formation.

organelle specialized structure within cells (e.g., nucleus, mitochondria, and endoplasmic reticulum).

prokaryotic cell a cell lacking a nucleus and the membranous organelles found in complex cells; bacteria and cyanobacteria.

ribosome minute particle found attached to the ER or loose in the cytoplasm that is the site of protein synthesis.

4

THE CELL MEMBRANE

Chapter Concepts

1.
The cell membrane regulates the passage of molecules into and out of the cell. 61, 70

2.
Some types of molecules, particularly small molecules, pass freely across the cell membrane, while others do not. 63, 70

3.
Water passes freely across the cell membrane, and this can affect cell size and shape. 65, 70

4.
Carrier proteins assist the transport of some ions and molecules across the cell membrane. 68, 70

5.
Vesicle formation takes substances into the cell, and vesicle fusion with the cell membrane discharges substances from the cell. 69, 70

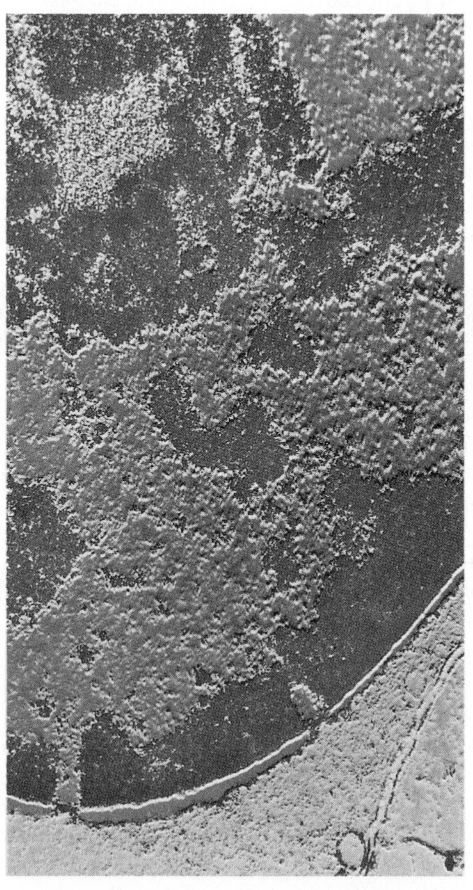

Cell nucleus and cytoplasm, ×10,000

Chapter Outline

Figure 4.1

Fluid-mosaic model of cell membrane structure. The membrane is composed of a phospholipid bilayer in which proteins are embedded. The hydrophilic heads of the phospholipids are at the surfaces of the membrane, and the hydrophobic tails make up the interior of the membrane. Note the membrane's asymmetry; for example, carbohydrate chains project externally and cytoskeleton filaments attach to proteins on the cytoplasmic side of the cell membrane.

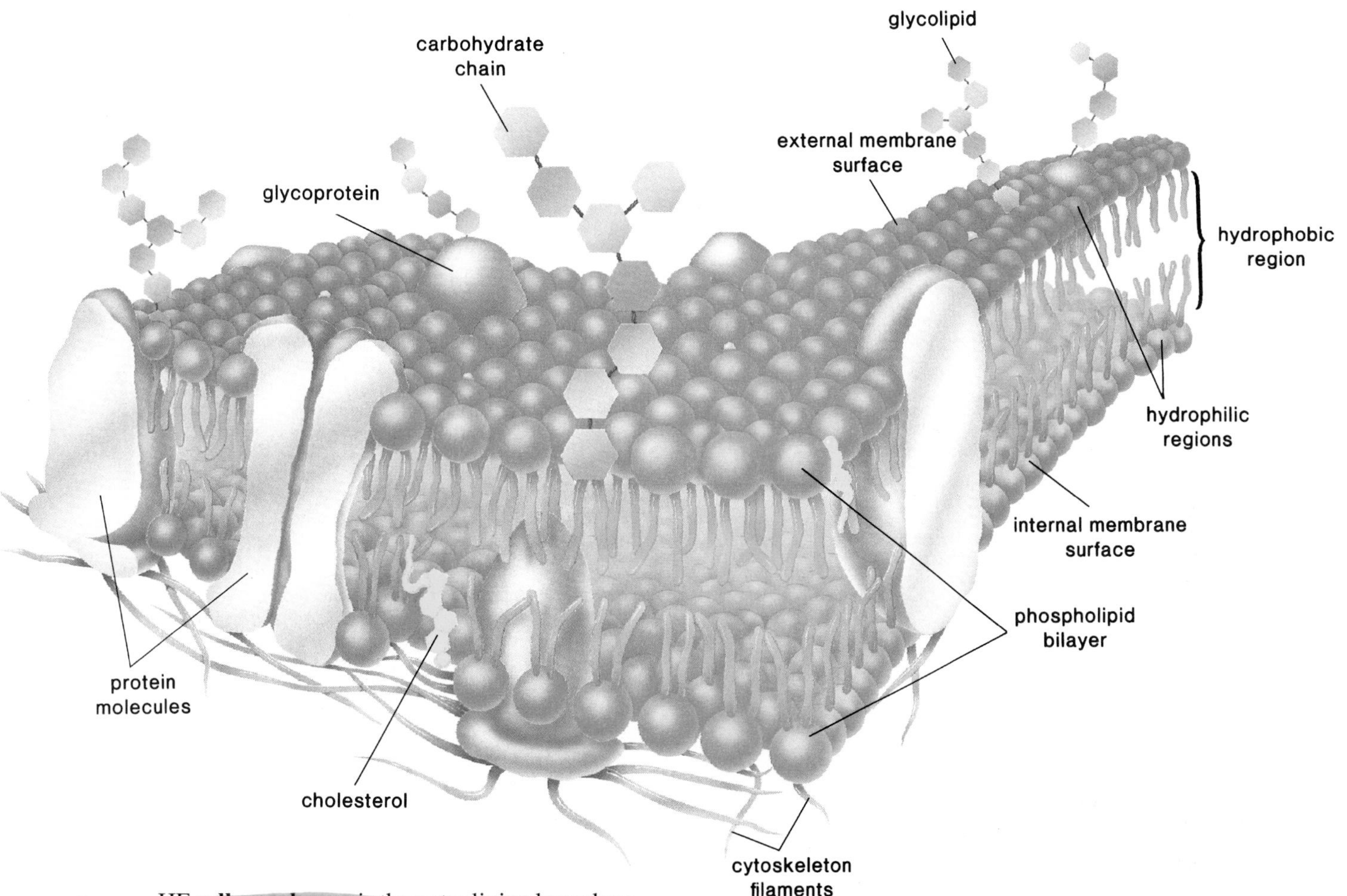

THE **cell membrane** is the outer living boundary of the cell. It helps give a cell mechanical strength and shape in addition to various other functions. In particular, the cell membrane regulates the passage of molecules into and out of the cell and, in this way, is largely responsible for maintaining cellular homeostasis. The structure of the cell membrane is intimately related to its ability to perform this and other functions.

Cell Membrane Structure and Function

The cell membrane is a phospholipid bilayer in which protein molecules are either partially or wholly embedded (fig. 4.1). The phospholipid bilayer has a *fluid* consistency, comparable to that of a light oil. The proteins are scattered throughout the membrane; therefore, they form a *mosaic* pattern. This description of the cell membrane is called the **fluid-mosaic model** of membrane structure.

Phospholipid Bilayer: Structural Element

Each phospholipid in the membrane has a polar (charged) head and 2 nonpolar tails (see fig. 2.27). When surrounded by water, phospholipid molecules form a bilayer naturally. The heads, being polar, are attracted to the water, which is also polar; therefore, the heads face outward. The polar heads are said to be hydrophilic (*water loving*). The nonpolar tails are not attracted to the water. They face inside, away from the water, and are said to be hydrophobic (*water hating*).

Some of the lipids in the external portion of the cell membrane are *glycolipids*. Glycolipids are constructed similarly to phospholipids, except the polar head consists of a chain of carbohydrate molecules. Animal cell membranes, in particular, also contain a substantial number of *cholesterol* molecules. These molecules lend stability to the lipid bilayer and prevent a drastic decrease in fluidity at low temperatures.

Figure 4.2

Membrane protein diversity. These are some of the functions performed by proteins found in the cell membrane.

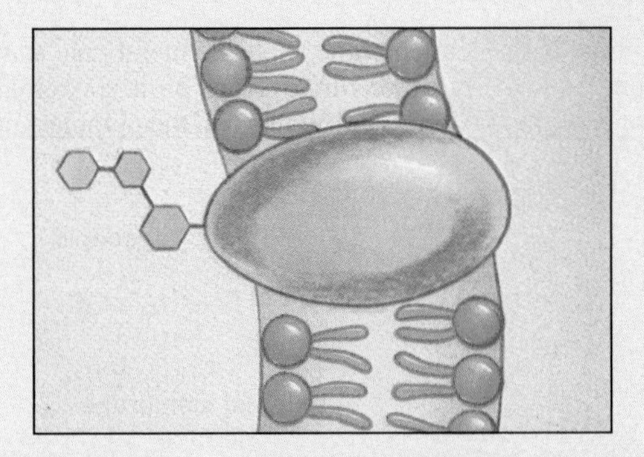

Cell-recognition protein

A glycoprotein that identifies the cell. For example, the MHC (major histocompatibility complex) glycoproteins are different for each person; thus, organ transplants are risky and relatively infrequent. Cells with foreign MHC glycoproteins are attacked by blood cells responsible for immunity.

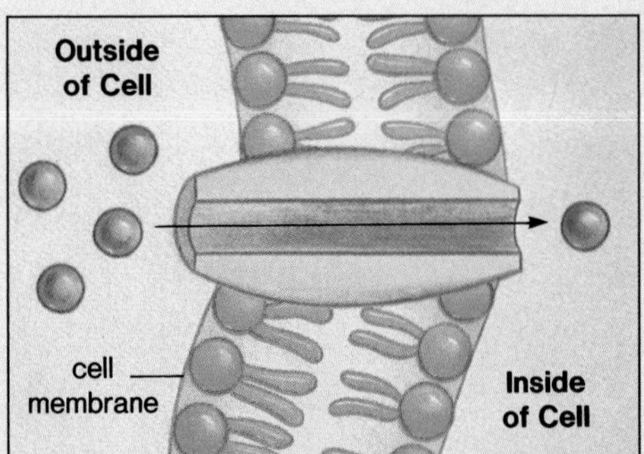

Channel protein

A protein that allows a particular molecule or ion to freely cross the cell membrane as it enters or exits the cell. Recently, it has been shown that cystic fibrosis, an inherited disorder, is caused by faulty chloride ion (Cl^-) channels. When these channels are not functioning normally, a thick mucus collects in airways and in pancreatic and liver ducts.

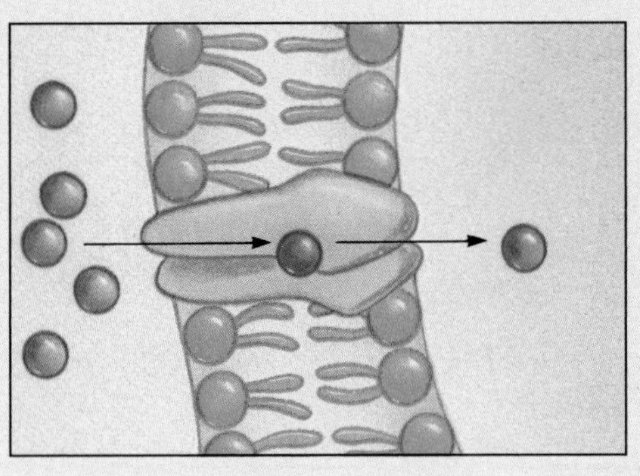

Carrier protein

A protein that selectively interacts with a specific molecule or ion so that it can cross the cell membrane to enter or exit the cell. The carrier protein that transports sodium ions (Na^+) and potassium ions (K^+) across the cell membrane requires ATP energy. The inability of some persons to use up energy for sodium-potassium transport has been suggested as the cause of their obesity.

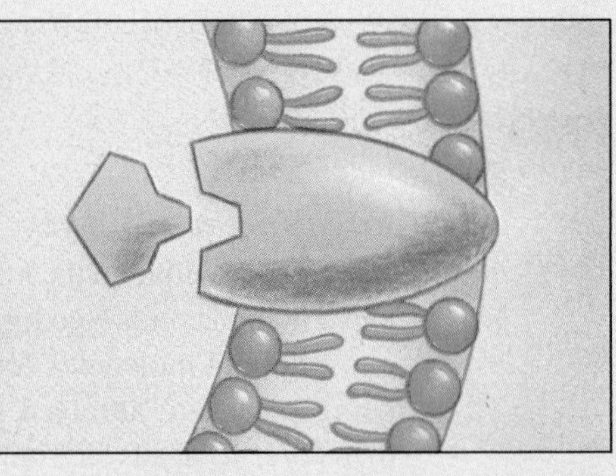

Receptor protein

A protein that is shaped in such a way that a specific molecule can bind to it. Recently, it has been shown that Pygmies are short not because they do not produce enough growth hormone, but because their cell membrane growth hormone receptors are faulty and cannot interact with the hormone.

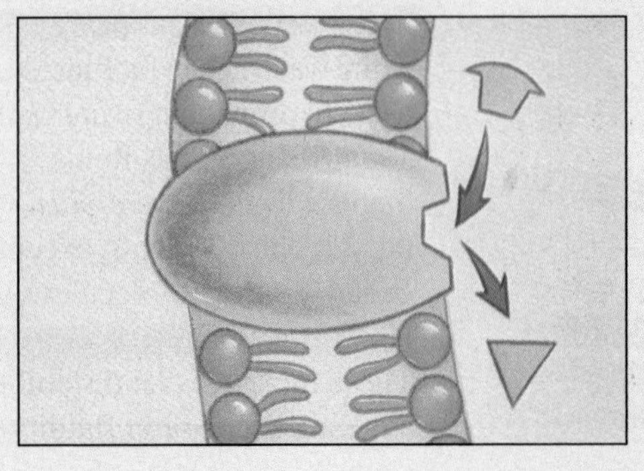

Enzymatic protein

A protein that catalyzes a specific reaction. For example, adenylate cyclase is a cell membrane protein that is involved in ATP metabolism. Polluted water may contain cholera bacteria, which release a toxin that interferes with the proper function of adenylate cyclase. Sodium ions and water leave intestinal cells in such volume that the individual dies from severe diarrhea.

Cell Membrane Proteins: Functional Elements

The lipid bilayer determines the basic structure of the cell membrane, but the various functions of the membrane are carried out by cell membrane proteins. Some protein molecules that span the membrane are glycoproteins having attached carbohydrate chains that face outward. Other proteins are located on the internal surface of the membrane, where they are held in place by cytoskeleton filaments.

The proteins of the membrane carry out various functions (fig. 4.2). Cell-recognition proteins are glycoproteins unique to the cell. They make up a cellular fingerprint by which cells can recognize one another. As we will discuss in more detail, certain proteins are involved in the passage of molecules through the membrane. Some proteins have a *channel* through which an ion or molecule can simply move across the membrane; others are *carriers,* which combine with a substance and help it to move across the membrane. Still other proteins are *receptors*; each type of receptor has a specific shape, which allows a particular molecule to bind to it. The binding of a molecule, such as a hormone, can influence the metabolism of the cell. Viruses often must attach to receptors before they enter a cell. Some proteins have an *enzymatic function* and carry out metabolic reactions.

Table 4.1
Passage of Molecules into and out of Cells

	Name	Direction	Requirements	Examples
Passive Transport Means	Simple Diffusion	Toward lower concentration	Concentration gradient	Lipid-soluble molecules, water, and gases
	Facilitated Diffusion	Toward lower concentration	Carrier and concentration gradient	Sugars and amino acids
Active Transport Means	Active Transport	Toward greater concentration	Carrier plus energy	Sugars, amino acids, and ions
	Endocytosis			
	Phagocytosis	Toward inside	Vesicle formation	Cells and subcellular material
	Pinocytosis	Toward inside	Vesicle formation	Macromolecules
	Exocytosis	Toward outside	Vesicle fuses with cell membrane	Macromolecules

All cells are surrounded by a cell membrane. The cell membrane is composed of a phospholipid bilayer in which proteins having various functions are embedded. Do 4.1 Critical Thinking, found at the end of the chapter.

How Molecules Cross the Membrane

In order for molecules to enter and exit the cell, they must cross the cell membrane. The structure of the cell membrane affects which types of molecules can freely pass through it. Small noncharged molecules, particularly if they are lipid soluble, have no difficulty crossing the membrane. Macromolecules cannot freely cross a cell membrane, and charged ions and molecules have difficulty. The membrane is usually positively charged outside and negatively charged inside. (Negatively charged ions tend to move through channels from inside the cell to outside the cell, and positively charged ions tend to move [through channels] in the opposite direction.) Because passage is restricted, the cell membrane is said to be **differentially permeable** (or selectively permeable).

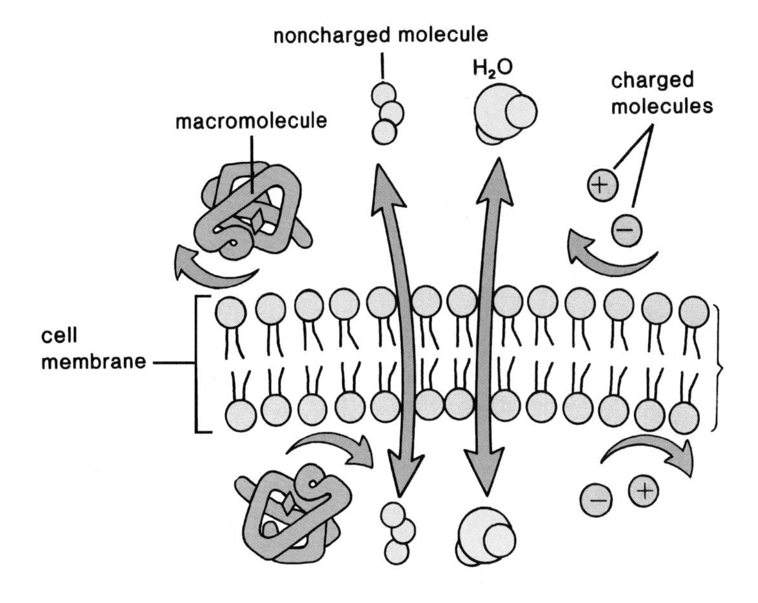

There are both passive and active ways of assisting the passage of molecules across the cell membrane. The active ways use energy (ATP molecules), while the passive ways do not. The *passive ways* involve diffusion, both simple and facilitated, and the *active ways* involve active transport, endocytosis, and exocytosis (table 4.1).

CELL IMPLANT CAPSULES RELY ON DIFFUSION

Citing results of animal experiments and a small number of tests in humans, scientists say that transplants of living cells hold great promise for attacking some of the most stubborn human disorders. . . .

Unlike conventional drug therapies, implanted cells can be targeted precisely to where they are needed in the body. And, unlike traditional drugs, living cells can provide a constant supply of the substances that are missing in diseased or damaged cells.

"This is an extremely exciting new area because the cells are the units of the body where everything happens," said Donald Wood, a cell biologist and science/technology director for the Muscular Dystrophy Association. "Many, many diseases occur when a cellular function has collapsed, and the replacement of cells to put that function back into the body is not only exciting but extremely logical". . . .

Among the most promising research reports [at the First International Congress of the Cell Transplant Society, 1992] were these:

In Sweden, brain implants of fetal tissue have improved the movement disorders of patients with Parkinson disease [see p. 295]. More work must be done, however, before the technique can be offered as therapy, said Dr. Anders Bjorklund, a leading neurosurgeon.

At McLean Hospital in Belmont [Massachusetts], researchers said rodent cells have been implanted into the brains of baboons that have a condition similar to Huntington disease [see p. 435]. The implants apparently reversed some of the damaged nerve networks that, in Huntington patients, lead to progressive loss of control over the body. Human tests may not be far off.

Cell transplants have raised hopes of treating the degeneration of muscle fibers that weakens young boys with Duchenne muscular dystrophy [see p. 452]. So far, about 30 youngsters with the ultimately fatal genetic disease have received injections of myoblasts—normal but immature muscle cells—to determine whether the cells can supply a structural protein that's missing in Duchenne patients....

Figure 4.A

Highly magnified cross section of a cell implant tube (the blue halolike structure) filled with adrenal gland cells. Dopamine and nutrients are small enough to diffuse through the pores of the tube, but immune cells are too large to enter.

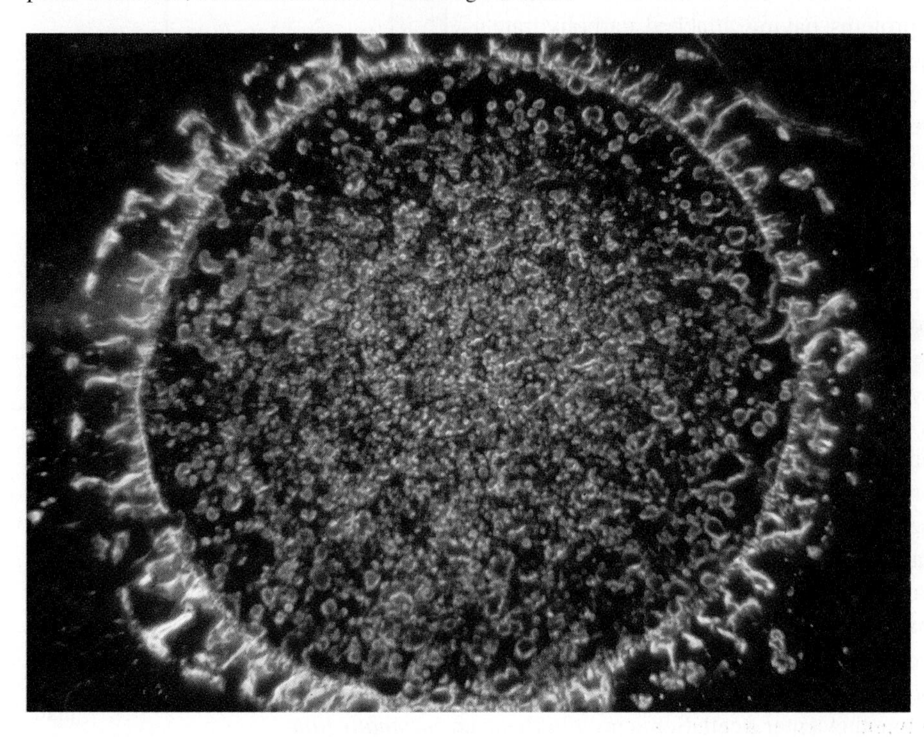

Cell transplants could be used to treat chronic pain, if research by University of Illinois scientists continues to be successful. When they implanted adrenal gland cells into the lower spine of terminal cancer patients, they told the congress, the cells manufactured natural painkillers that brought relief for as long as 11 months.

At CytoTherapeutics in Providence [Rhode Island], researchers are "hiding" implanted cells by enclosing them inside plastic capsules [fig. 4.A].

The capsules, formed of plastic membranes with tiny pores, are about the diameter of a thin pencil lead. Inside, millions of cells—cells that produce dopamine for Parkinson disease, or insulin for diabetes, for example—are safe from attack by the patient's immune cells, which are too large to enter the pores. And if they don't actually come in contact with the foreign cells, the immune defenders can't recognize them as enemies and destroy them.

The capsule membrane's pores are big enough to permit the substances made by the implanted cells to diffuse out into the body. If the implant is for Parkinson disease, for example, adrenal gland cells inside the plastic tube would continuously secrete dopamine, which is lacking in the brain of those patients, for months or even years.

Not only does the plastic capsule protect the implanted cells, it also enables doctors to remove the cells at any time.

"These implants are very safe," said Rudnick, the CytoTherapeutics CEO. "If the capsule should break, the cells would be destroyed by the patient's immune system."

Figure 4.3

Process of diffusion. Diffusion is spontaneous, and no energy is required to bring it about. **a**. When dye crystals are placed in water, they are concentrated in one area. **b.** The dye dissolves in the water, and there is a net movement of dye molecules from higher to lower concentration. There is a net movement of water molecules in the opposite direction. **c**. Eventually, the water and the dye molecules are equally distributed throughout the container.

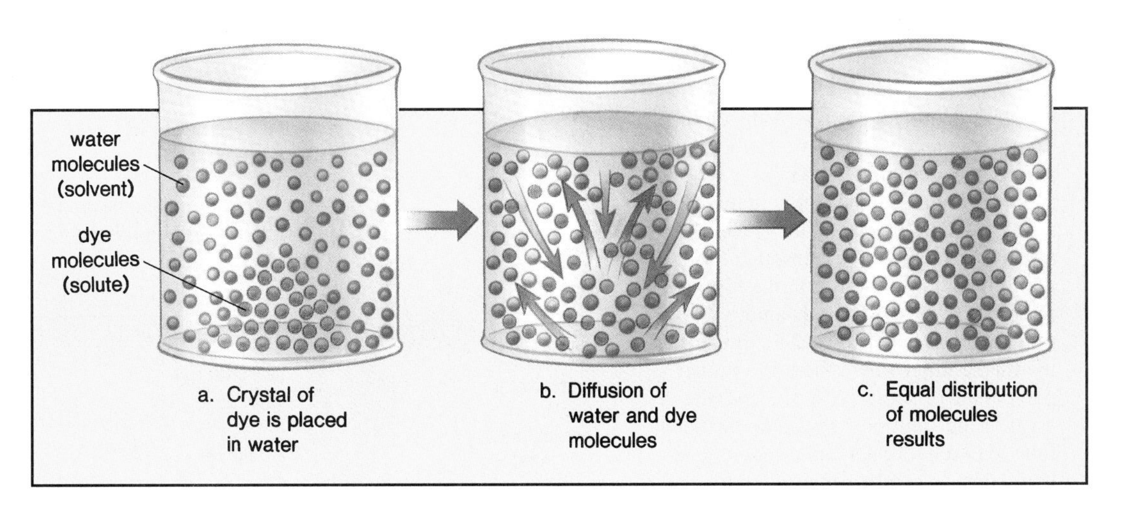

Diffusion and Osmosis

Diffusion is the net movement of a substance from a region of higher concentration to a region of lower concentration. Diffusion is a physical process that can be observed with any type of molecule. During diffusion, molecules move from higher to lower concentration—that is, down their *concentration gradient*—until they are distributed equally. For example, when a few crystals of dye are placed in water (fig. 4.3), the dye and water molecules move in various directions, but their net movement is toward the region of lower concentration. Eventually, the dye is dissolved in the water, resulting in a colored solution. A solution contains both a solute and a solvent. In this case, the **solute** is the dye molecules and the **solvent** is the water molecules. Once the solute and solvent are evenly distributed, they continue to move about, but there is no net movement of either one in any direction.

Diffusion through the Membrane

The chemical and physical properties of the cell membrane allow just a few types of molecules to enter and exit by diffusion. Lipid-soluble molecules, such as alcohols, can diffuse through the membrane because lipids are the membrane's main structural components.

Gases can also diffuse through the lipid bilayer; this is the mechanism by which oxygen enters cells and carbon dioxide exits cells. As an example, consider the movement of oxygen from the air sacs (alveoli) of the lungs to blood in the lung capillaries (fig. 4.4). After inhalation (breathing in), the concentration of oxygen in the alveoli is higher than the concentration of oxygen in blood; therefore, oxygen diffuses into blood. As discussed in the reading for this chapter, cell implants can be used to treat human disorders because needed molecules diffuse out of these cells and into the deficient cells via their cell membranes.

Water passes into and out of cells with relative ease. It probably moves through *channels* (fig. 4.2), with a pore size large enough to allow the passage of water and prevent the passage of other molecules. The fact that water can penetrate a cell membrane has important biological consequences, as described in the discussion that follows.

Figure 4.4

Gas exchange in lungs. Oxygen (O_2) diffuses into the capillaries of the lungs because there is a higher concentration of oxygen in the air sacs (alveoli) than in the capillaries.

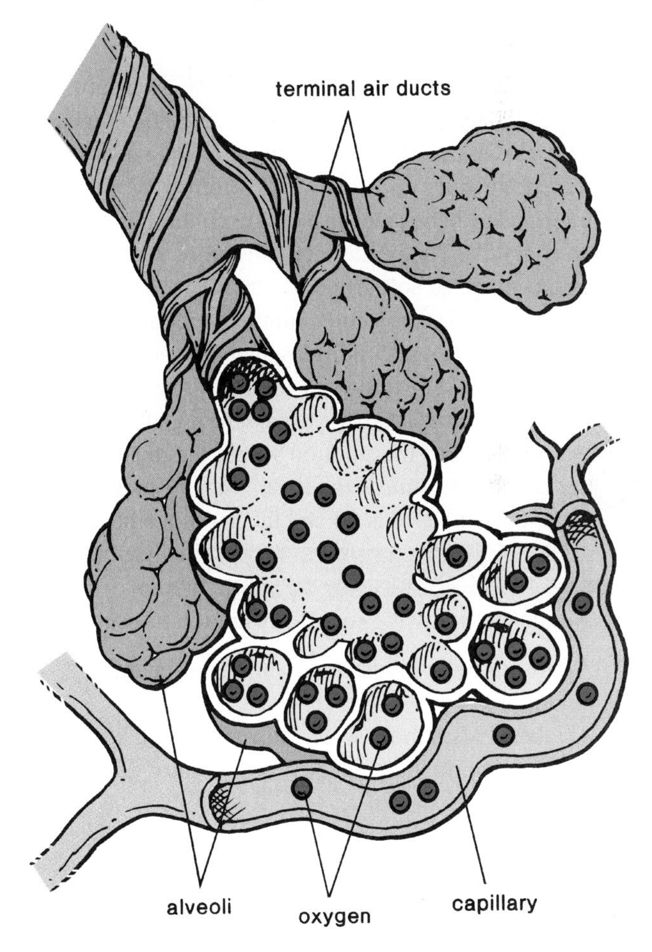

Molecules diffuse down their concentration gradients. A few types of small molecules can simply diffuse through the cell membrane.

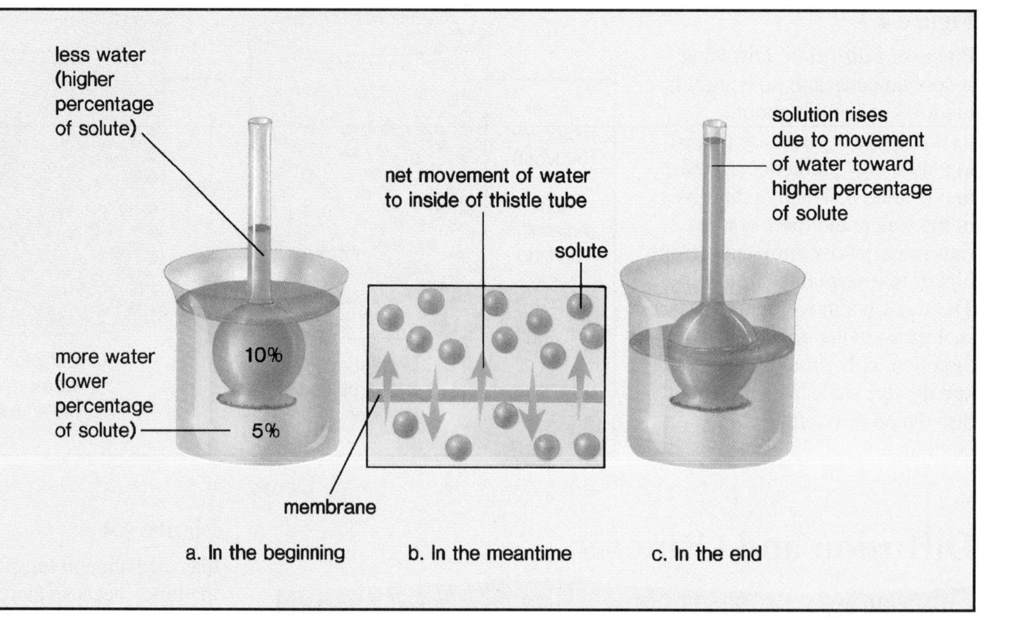

Figure 4.5
Osmosis demonstration. **a.** A thistle tube, covered at the broad end by a differentially permeable membrane, contains a 10% sugar solution. The beaker contains a 5% sugar solution. **b.** The solute (green circles) is unable to pass through the membrane, but the water passes through in both directions. There is a net movement of water toward the inside of the thistle tube, where there is a higher percentage of solute. **c.** In the end, the level of the solution rises in the thistle tube until a hydrostatic pressure equivalent to osmotic pressure builds up.

Osmosis: Water Crosses the Membrane

The diffusion of water across a differentially permeable membrane has been given a special name; it is called **osmosis.** To illustrate osmosis, a thistle tube containing a 10% sugar solution[1] is covered at one end by a differentially permeable membrane and is then placed in a beaker containing a 5% sugar solution (fig. 4.5). The beaker contains more water molecules (lower percentage of solute) per volume, and the thistle tube contains fewer water molecules (higher percentage of solute) per volume. Under these conditions, there is a net movement of water from the beaker to the inside of the thistle tube across the membrane. The solute is unable to pass through the membrane; therefore, the level of the solution within the thistle tube rises (fig. 4.5*c*). As water enters the thistle tube, a pressure called hydrostatic pressure builds up and the net movement of water ceases. The hydrostatic pressure is equivalent to the **osmotic pressure** of the solution inside the thistle tube.

Notice the following in this illustration of osmosis:

1. A differentially permeable membrane separates 2 solutions
2. The beaker has more water (lower percentage of solute), and the thistle tube has less water (higher percentage of solute)
3. The membrane does not permit passage of the solute
4. The membrane permits passage of water, and there is a net movement of water from the beaker to the inside of the thistle tube
5. An osmotic pressure is present: the amount of liquid increases on the side of the membrane with the greater percentage of solute

1. Percent solutions are grams of solute per 100 ml of solvent. Therefore, a 10% solution is 10 g sugar in 100 ml of water.

These considerations will be important as we discuss osmosis in relation to cells placed in different solutions. The cell membrane allows such solutes as sugars and salts to pass through, but the difference in permeability between water and these solutes is so great that cells in sugar and salt solutions have to cope with the osmotic movement of water.

> Osmosis is the diffusion of water across a differentially permeable membrane. Osmotic pressure is evident when there is an increased amount of water on the side of the membrane that has the higher solute concentration.

The Importance of Osmosis

Osmosis occurs constantly in living organisms. For example, due to osmosis, water is absorbed from the human large intestine, is retained by the kidneys, and is taken up by blood. Since living things contain a very high percentage of water, osmosis is an extremely important physical process that can affect health.

Tonicity: Solution Strength

Tonicity refers to the strength of a solution in relationship to osmosis. Cells can be placed in solutions that have the same percentage of solute, a higher percentage of solute, or a lower percentage of solute than the cell. These solutions are called isotonic, hypertonic, and hypotonic, respectively. Figure 4.6 depicts and describes the effects of these solutions on cells.

Isotonic Solutions: Cells Stable

In the laboratory, cells are normally placed in solutions that cause them to neither gain nor lose water. Such solutions are said to be **isotonic solutions;** that is, the solute concentration is the same on both sides of the membrane, and therefore there is no net gain or loss of water (fig. 4.6*a*). The term *iso* means *the*

Figure 4.6

Osmosis in animal and plant cells. The arrows indicate the net movement of water. In an isotonic solution, a cell neither gains nor loses water; in a hypotonic solution, a cell gains water; and in a hypertonic solution, a cell loses water.

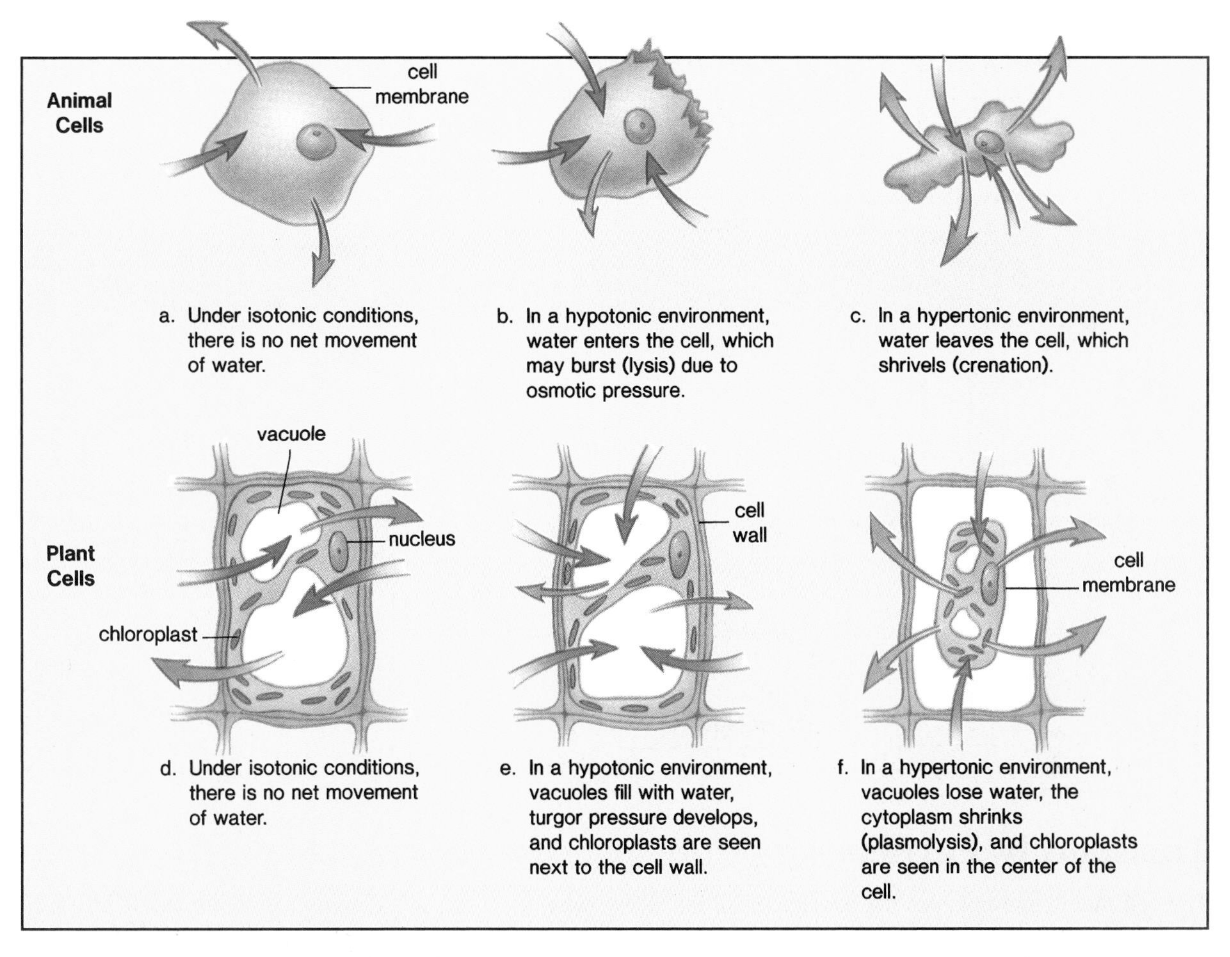

same as and the term *tonicity* refers to the strength of the solution. It is possible to determine, for example, that a 0.9% solution of the salt sodium chloride (Na^+Cl^-) is isotonic to red blood cells because the cells neither swell nor shrink when placed in this solution (fig. 4.6*a*).

Hypotonic Solutions: Cells Swell

Solutions that cause cells to swell, or even to burst, due to an intake of water are said to be **hypotonic solutions** (fig. 4.6*b*). The prefix *hypo* means *less than* and refers to a solution with a lower percentage of solute (more water) than the cell. If a cell is placed in a hypotonic solution, water enters the cell; the net movement of water is from the outside to the inside of the cell.

Any concentration of salt solution lower than 0.9% is hypotonic to red blood cells. Red blood cells placed in such a solution expand and sometimes burst due to the buildup of pressure. The term *lysis* is used to refer to disrupted cells; hemolysis, then, is disrupted red blood cells.

The swelling of a cell in hypotonic solution creates **turgor pressure.** When a plant cell is placed in a hypotonic solution, we observe expansion of the cytoplasm because the large central vacuole gains water and the cell membrane pushes against the rigid cell wall (fig. 4.6*e*). The plant cell does not burst because the cell wall does not give way. Turgor pressure in plant cells is extremely important to the maintenance of the plant's erect position.

Hypertonic Solutions: Cells Shrivel

Solutions that cause cells to shrink or to shrivel due to a loss of water are said to be **hypertonic solutions.** The prefix *hyper* means *more than* and refers to a solution with a higher percentage of solute (less water) than the cell. If a cell is placed in a hypertonic solution, water leaves the cell; the net movement of water is from the inside to the outside of the cell.

A 10% solution of sodium chloride (Na^+Cl^-) is hypertonic to red blood cells. In fact, any solution with a concentration higher than 0.9% sodium chloride is hypertonic to red blood cells. If red blood cells are placed in this solution, they shrink (fig. 4.6*c*). The term *crenation* refers to red blood cells in this condition.

A hypertonic solution brings about **plasmolysis,** shrinking of the cytoplasm due to osmosis. When a plant cell is placed in a hypertonic solution, we observe the cytoplasm shrinking because the large central vacuole loses water and the cell membrane pulls away from the cell wall (fig. 4.6*f*).

When a cell is placed in an isotonic solution, it neither gains nor loses water. When a cell is placed in a hypotonic solution (lower solute concentration than an isotonic solution) the cell gains water. When a cell is placed in a hypertonic solution (higher solute concentration than an isotonic solution), the cell loses water and the cytoplasm shrinks.

Figure 4.7

Facilitated diffusion. A carrier protein speeds the rate at which a solute crosses a membrane in the direction of lower concentration. Facilitated diffusion occurs only when there is a concentration gradient across the membrane; therefore, energy is not required.

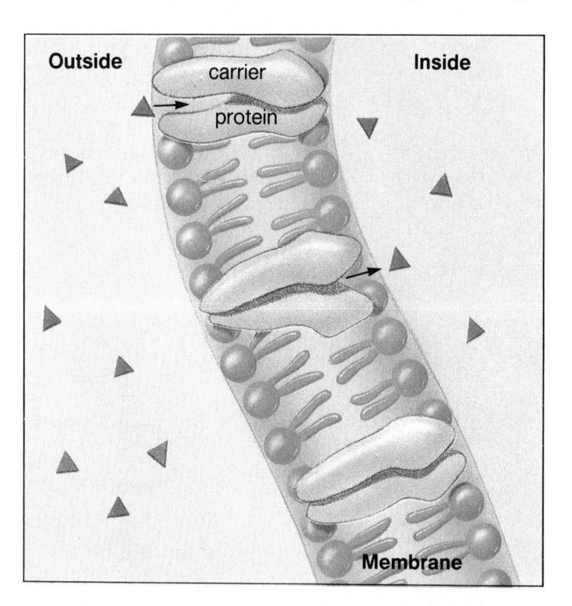

Figure 4.8

Active transport through a cell membrane. Active transport is apparent when a molecule crosses the cell membrane toward the area of higher concentration. An expenditure of ATP energy is required, presumably for a carrier protein to transport a molecule across the cell membrane against its concentration gradient.

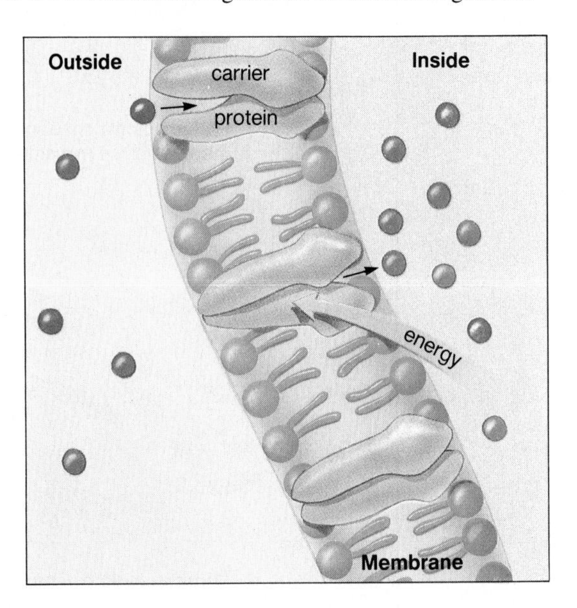

Transport by Carriers

The cell membrane impedes the passage of all but a few substances. Yet, biologically useful molecules do enter and exit the cell at a rapid rate because there are carrier proteins in the membrane. **Carrier proteins** are specific; each can combine with only a certain type of molecule, which is then transported through the membrane. It is not completely understood how carrier proteins function, but after a carrier combines with a molecule, the carrier is believed to undergo a change in shape that moves the molecule across the membrane. Carrier proteins are required for facilitated diffusion and active transport (table 4.1).

> Some of the proteins in the cell membrane are carriers. They transport biologically useful molecules into and out of the cell.

Facilitated Diffusion: With the Gradient

Facilitated diffusion explains the passage of such molecules as glucose and amino acids across the cell membrane even though they are not lipid soluble. The passage of glucose and amino acids is facilitated by their reversible combination with carrier proteins, which in some manner transport them through the cell membrane. These carrier proteins are specific. For example, various sugar molecules of identical size might be present inside or outside the cell, but glucose can cross the membrane hundreds of times faster than the other sugars. As stated earlier, this is the reason that the membrane can be called differentially permeable.

A model for facilitated diffusion (fig. 4.7) shows that after a carrier has assisted the movement of a molecule to the other side of the membrane, it is free to assist the passage of other similar molecules. Neither simple diffusion, explained previously, nor facilitated diffusion requires an expenditure of energy because the molecules are moving down their concentration gradient in the same direction they tend to move anyway.

Active Transport: Against the Gradient

During **active transport,** molecules or ions move through the cell membrane, accumulating either inside or outside the cell. For example, iodine collects in the cells of the thyroid gland; glucose is completely absorbed from the gut by the cells lining the digestive tract; and sodium is sometimes almost completely withdrawn from urine by cells lining the kidney tubules. In these instances, molecules have moved to the region of higher concentration, exactly opposite to the process of diffusion.

Both carrier proteins and an expenditure of energy (fig. 4.8) are needed to transport molecules against their concentration gradient. In this case, energy (ATP molecules) is required for the carrier to combine with the substance to be transported. Therefore, it is not surprising that cells involved primarily in active transport, such as kidney cells, have a large number of mitochondria near the membrane at which active transport is occurring (see fig. 15.10).

Figure 4.9

The sodium-potassium pump. A carrier protein actively moves sodium ions (Na^+) to the outside of the cell and potassium ions (K^+) to the inside of the cell. Note that ATP energy is required.

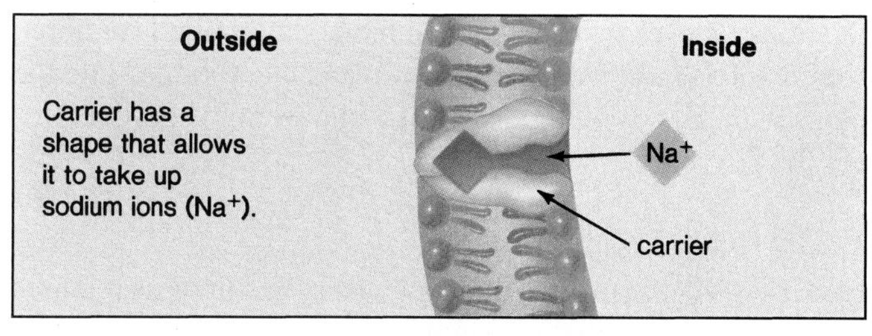

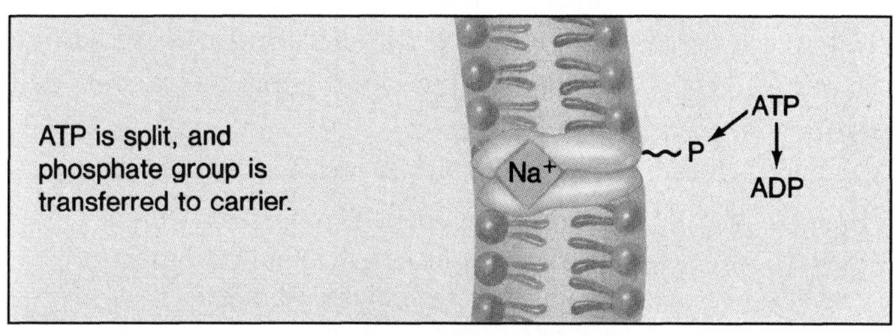

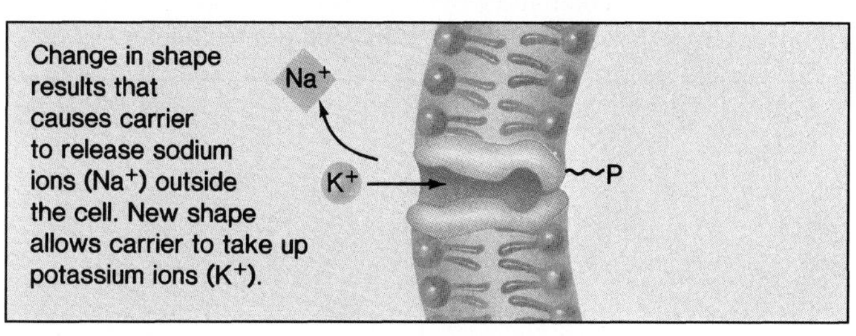

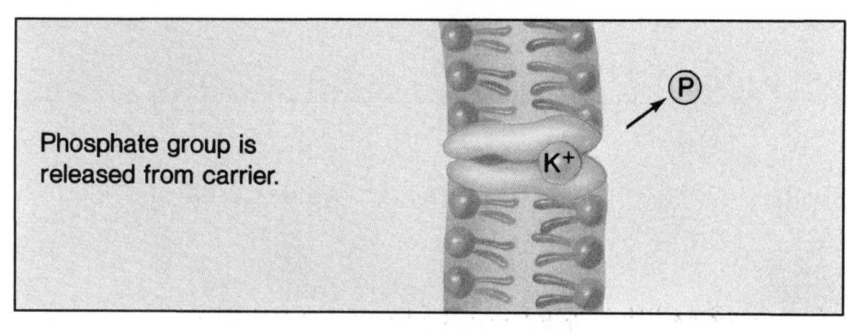

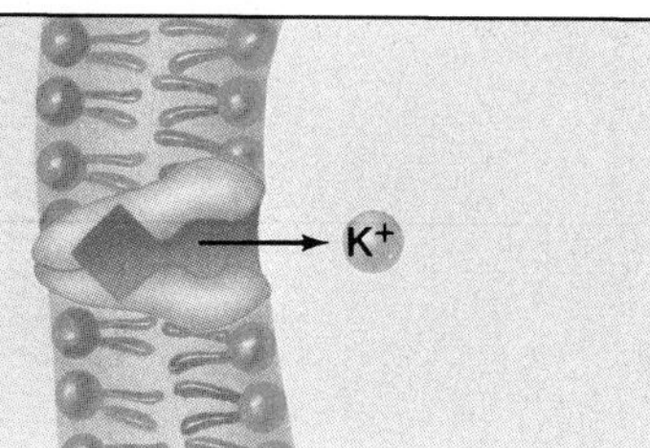

Proteins involved in active transport often are called *pumps* because just as a water pump uses energy to move water against the force of gravity, proteins use energy to move a substance against its concentration gradient. One type of pump that is active in all cells but is especially associated with nerve and muscle cells moves sodium ions (Na^+) to the outside of the cell and potassium ions (K^+) to the inside of the cell. These 2 events are presumed to be linked, and the carrier protein is called a sodium-potassium pump. A change in carrier shape after the attachment and again after the detachment of a phosphate group allows it to alternately combine with sodium ions and potassium ions (fig. 4.9). The phosphate group is donated by ATP when it is broken down enzymatically by the carrier.

The passage of salt (Na^+Cl^-) across a cell membrane is of primary importance in cells. The chloride ion (Cl^-) usually crosses the cell membrane because it is attracted by positively charged sodium ions (Na^+). First sodium ions are pumped across a membrane and then chloride ions simply diffuse through channels that allow their passage. As noted in figure 4.2, the chloride ion channels malfunction in persons with cystic fibrosis, and this leads to the symptoms of this inherited (genetic) disorder.

During facilitated diffusion (no energy required), small molecules follow their concentration gradient. During active transport (energy required), small molecules go against their concentration gradient.

Endocytosis and Exocytosis

Some molecules are too large to be transported by protein carriers. In such instances, they are transported in or out of the cell by vesicle formation.

Endocytosis: Entering by Sac

At times, macromolecules or larger substances are incorporated into cells by **endocytosis**—the process by which a vesicle is formed at the cell membrane to bring these substances into the cell (fig. 4.10). Endocytosis, even when not moving substances against a concentration gradient, requires energy.

When the material taken in by the process of endocytosis is quite large, the process is called **phagocytosis** (*cell eating*). Phagocytosis is common in such ameboid-type cells as macrophages, large cells found in humans. These cells phagocytize bacteria and worn-out red blood cells, for example.

Figure 4.10

Endocytosis and exocytosis. During endocytosis (*right* to *left*), the cell membrane forms a vesicle around the substance to be taken into the cell. During exocytosis (*left* to *right*), a vesicle fuses with the membrane, allowing an enclosed substance to leave the cell. Exocytosis occurs during secretion of substances from cells.

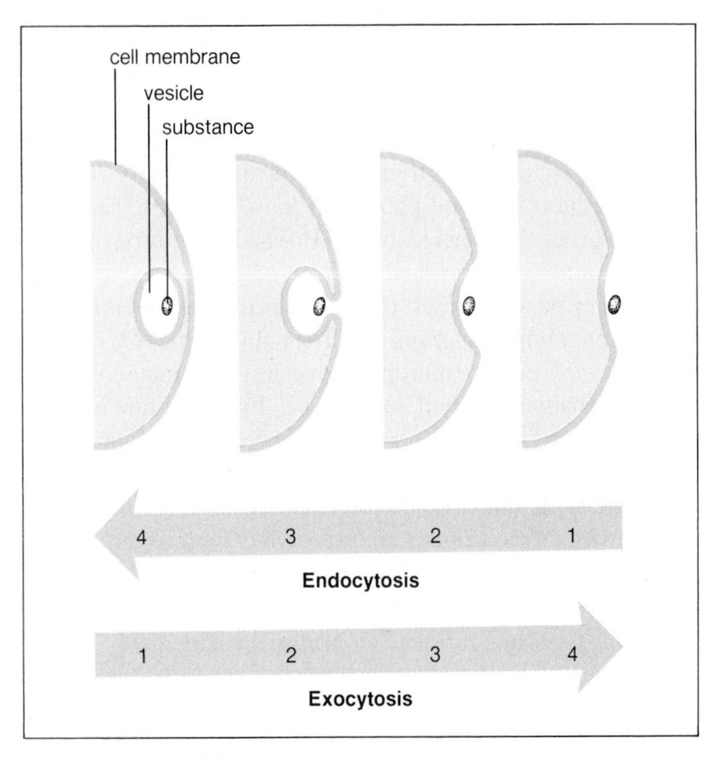

Pinocytosis (*cell drinking*) occurs when vesicles form around large-sized molecules, such as proteins. Whereas phagocytosis can be seen with the light microscope, the electron microscope must be used to observe pinocytosis.

Once formed, vacuoles or vesicles (small vacuoles) contain a substance enclosed by membrane. Digestion is required for this substance to be broken down and incorporated into the cytoplasm. Fusion with a lysosome allows this digestive process to take place.

Exocytosis: Leaving by Sac

Exocytosis is the reverse of endocytosis; this process requires a vesicle to fuse with the membrane, thereupon discharging its contents. Exocytosis is required for **secretion**. As we saw in chapter 3, vesicles formed by the Golgi apparatus secrete cell products at the cell membrane.

> Endocytosis and exocytosis are opposite processes. During endocytosis, a vesicle forms at the cell membrane; during exocytosis, a vesicle joins with the cell membrane. Macromolecules enter and exit a cell by these processes. Do 4.2 Critical Thinking, found at the end of the chapter.

SUMMARY

All cells are surrounded by a cell membrane which is a phospholipid bilayer. Proteins are either embedded in the membrane or attached to its inner surface. The proteins have various functions, but in general, the cell membrane regulates the passage of molecules across it. Substances cross cell membranes by diffusion, by carrier transport, and by vesicle formation:

Facilitated diffusion, which moves molecules down a concentration gradient, and active transport, an energy-requiring process that moves molecules against a concentration gradient, require a carrier protein. Vesicle formation during endocytosis permits molecules (pinocytosis) and debris (phagocytosis) to be taken into cells. Exocytosis is the process of cell-product secretion.

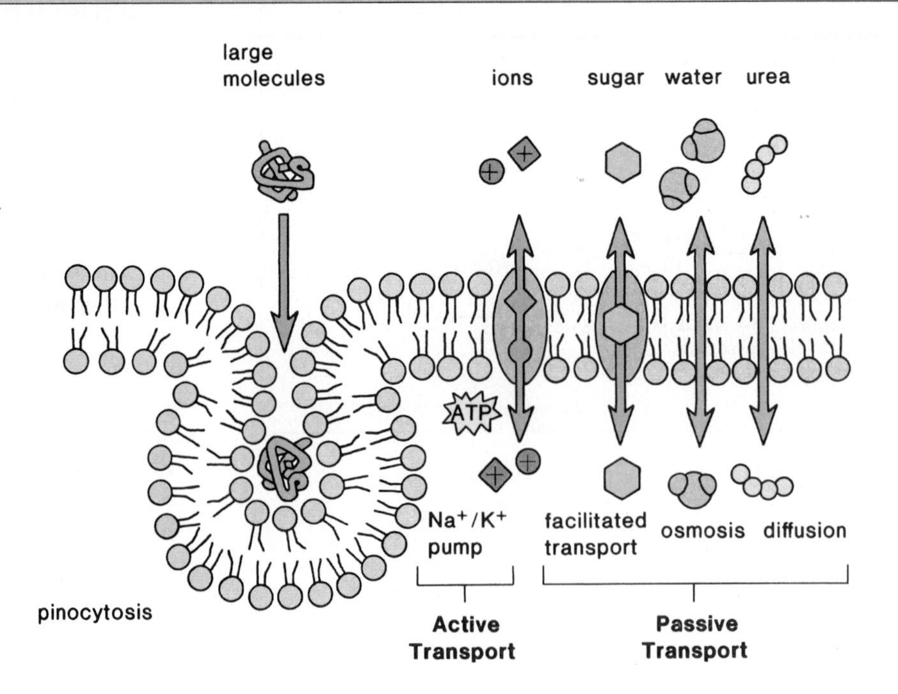

Tonicity affects the shape of cells (table 4.2). When a cell is placed in an isotonic solution, there is no net gain or loss of water. In a hypertonic solution, cells shrink, and in a hypotonic solution, cells swell. Because the cell wall does not give way, it is possible to observe plasmolysis and turgor pressure in plant cells placed in hypertonic and hypotonic solutions, respectively.

Table 4.2
Effect of Tonicity on the Cell

Tonicity of Solution	**Concentration**		**Net Movement of Water**	**Effect on Cell**
	Solute	***Water***		
Isotonic	Same as cell	Same as cell	None	None
Hypotonic	Less than cell	More than cell water	Cell gains	Swells, turgor pressure
Hypertonic	More than cell	Less than cell	Cell loses water	Shrinks, plasmolysis

STUDY QUESTIONS

In order to practice **writing across the curriculum,** students should write out the answers to any or all of the study questions. The study questions are sequenced in the same order as the text.

1. Describe the structure of the cell membrane, including the phospholipid bilayer and the various types of proteins. (pp. 61–63)
2. Why is a cell membrane called differentially permeable? (p. 63)
3. What are the mechanisms by which substances enter and exit cells? Which are active ways, and which are passive ways? (p. 63)
4. Define diffusion, and give an example. (pp. 63–65)
5. Define osmosis. Define isotonic, hypertonic, and hypotonic solutions, and give examples of how these concentrations affect red blood cells. (pp. 66–67)
6. Draw a simplified diagram of a red blood cell before and after being placed in these solutions. What terms are used to refer to the condition of the red blood cell in a hypertonic solution and in a hypotonic solution? (p. 67)
7. Draw a simplified diagram of a plant cell before and after being placed in these solutions. Describe the cell contents under these conditions. (p. 67)
8. How does facilitated diffusion differ from simple diffusion across the cell membrane? (p. 68)
9. How does active transport differ from facilitated diffusion? Give an example. (pp. 68–69)
10. Diagram endocytosis and exocytosis. Give an example for each of these. (pp. 69–70)

OBJECTIVE QUESTIONS

1. Both plant and animal cells have a cell membrane, but plant cells also have a cell _______.
2. Molecules diffuse down a _______.
3. In a hypertonic solution, cells _lose_ water and the cell contents _shrinks_
4. When plant cells are placed in a hypotonic solution, _______ is obvious because cell contents _______ against the cell wall.
5. During _______ diffusion, carriers move glucose and amino acids down their concentration gradients.
6. During active transport, a carrier moves molecules _______ their concentration gradients.
7. Sodium ions and potassium ions move across the cell membrane in opposite directions due to the action of the _______.
8. During phagocytosis, _______ formation takes large substances into the cell.
9. Label this diagram of the cell membrane.

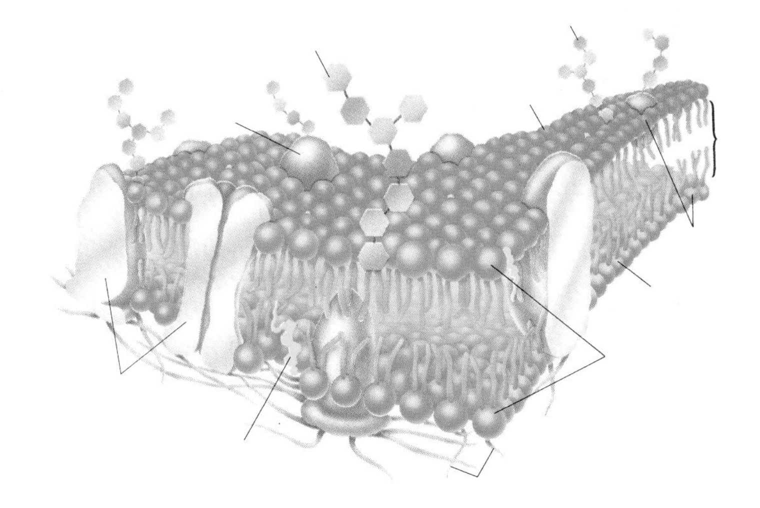

10. Label *a, b,* and *c* as hypotonic, isotonic, or hypertonic. Explain.

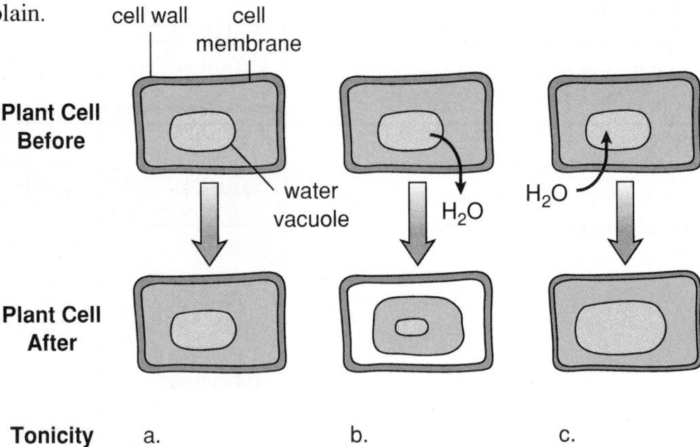

CRITICAL THINKING

In order to practice **writing across the curriculum,** students should write out the answers to any or all of the critical thinking questions. Suggested answers to the critical thinking questions are in appendix E.

4.1

1. Where in the cell would cell membrane proteins be made? Where in the cell might carbohydrate chains be added?
2. How do the proteins travel to the cell membrane?
3. In order to enter a cell, the HIV virus attaches to a cell membrane protein called the CD4 receptor, which is present on certain of our blood cells. Imagine an individual who, through a genetic defect, lacks the CD4 receptor. Do you suppose this individual could be infected by HIV?
4. Some proteins are enzymes, and some are structural proteins. The synthesis of both of these is controlled by the genes. Give an example of structural proteins found in all cells.

4.2

1. Contrast the manner in which alcohol and water enter a cell.
2. Contrast the manner in which sodium ions (Na^+) and chloride ions (Cl^-) exit a cell (fig. 4.2).
3. Contrast the manner in which amino acids and proteins enter a cell.
4. How might the proteins from question 3 be digested?

SELECTED KEY TERMS

active transport transfer of a substance into or out of a cell from a region of lower concentration to a region of higher concentration by a process that requires a carrier and an expenditure of energy.

carrier protein a protein molecule that combines with a substance and transports it through the cell membrane.

cell membrane a membrane that surrounds the cytoplasm of cells and regulates the passage of molecules into and out of the cell.

diffusion the movement of molecules from a region of higher concentration to a region of lower concentration.

endocytosis (en″ do-si-to′ sis) a process in which a vesicle is formed at the cell membrane to bring a substance into the cell.

exocytosis (eks″ o-si-to′ sis) a process in which an intracellular vesicle fuses with the cell membrane so that the vesicle's contents are released outside the cell.

facilitated diffusion passive transfer of a substance into or out of a cell along a concentration gradient by a process that requires a carrier.

hypertonic solution one that has a higher concentration of solute and a lower concentration of water than the cell.

hypotonic solution one that has a lower concentration of solute and a higher concentration of water than the cell.

isotonic solution one that contains the same concentration of solute and water as the cell.

osmosis (oz-mo′sis) the movement of water from an area of higher concentration of water to an area of lower concentration of water across a differentially permeable membrane.

osmotic pressure pressure generated by and due to the osmotic flow of water; created by the solute in a solution.

phagocytosis (fag″ o-si-to′ sis) the taking in of bacteria and/or debris by engulfing; cell eating.

pinocytosis (pin″ o-si-to′ sis) the taking in of fluid along with dissolved solutes by engulfing; cell drinking.

plasmolysis (plas-mol′ ĭ sis) contraction of the cell contents due to the loss of water.

solute a substance dissolved in a solvent to form a solution.

solvent a fluid, such as water, that dissolves solutes.

turgor pressure internal pressure that adds to the strength of the cell and builds up when water moves by osmosis into a cell.

5

METABOLISM: ENERGY AND ENZYMES

Chapter Concepts

1.
Energy cannot be created, and energy conversions result in a loss of usable energy. 74, 80

2.
The energy laws tell us that organisms need a constant input of energy. 74, 80

3.
ATP is the form of energy used by cells for any type of cellular work. 75, 80

4.
Cells have metabolic pathways in which every reaction has a specific enzyme. 75, 80

5.
Coenzymes help enzymes carry out their reactions. 79, 80

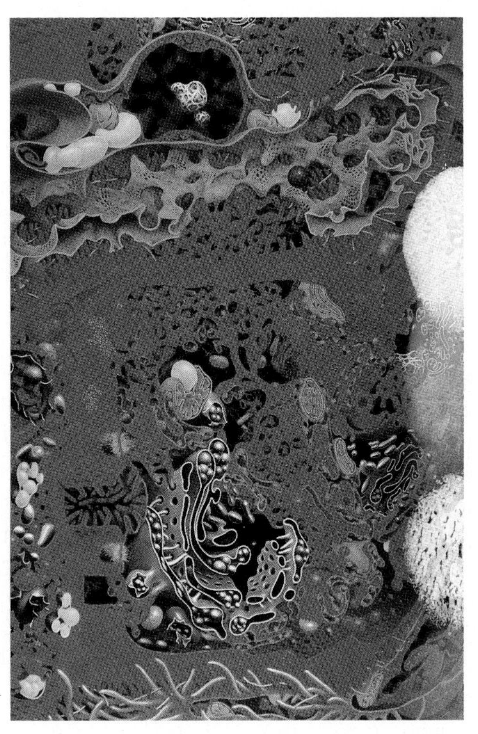

Human liver cell

Chapter Outline

Figure 5.1

Relationship of chloroplasts to mitochondria. Chloroplasts produce energy-rich carbohydrates. These carbohydrates are broken down in mitochondria, and the energy released is used for the buildup of ATP. There is a loss of usable energy during cellular respiration and photosynthesis, and eventually all ATP energy becomes heat.

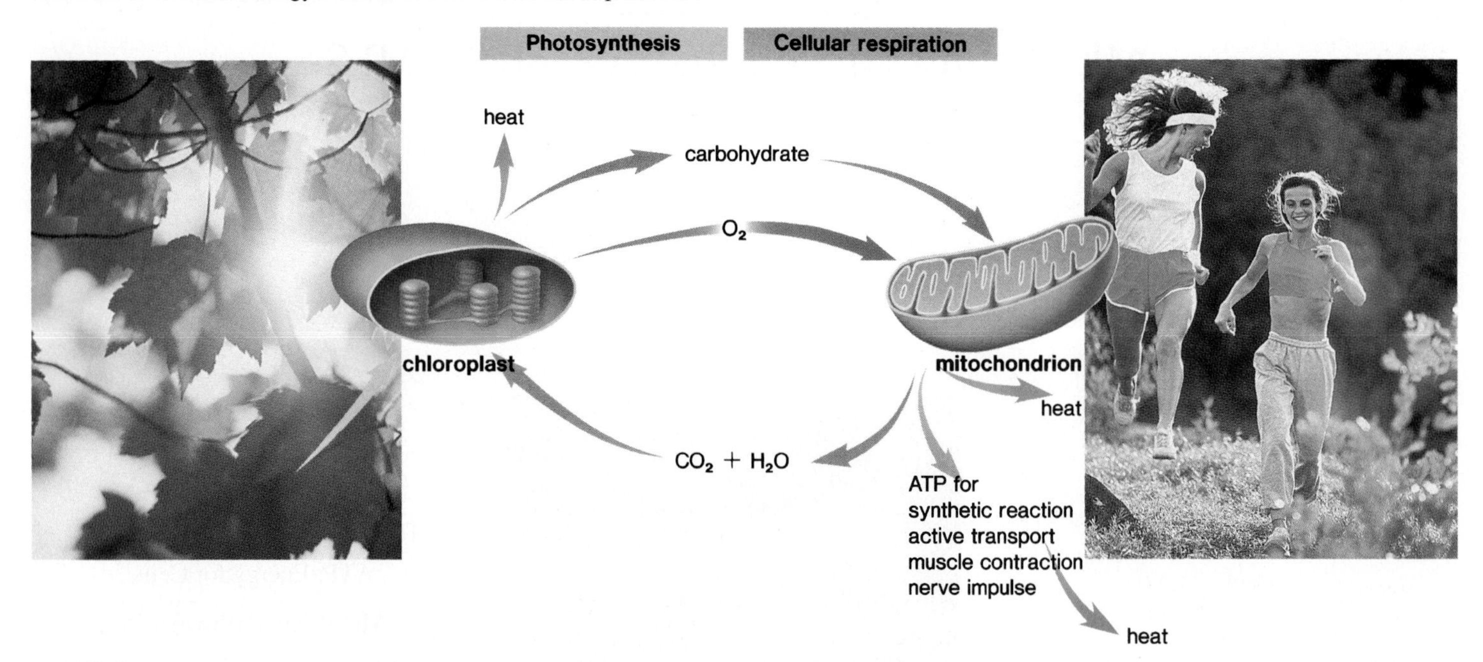

CHLOROPLASTS and mitochondria are the cellular organelles that convert one form of energy into another. When chloroplasts carry on photosynthesis, solar energy is converted into the energy of a carbohydrate, and when mitochondria carry on aerobic cellular respiration, the energy stored in carbohydrates is converted to energy temporarily held by ATP. All organisms make use of ATP. In general, photosynthesis and cellular respiration permit a flow of energy from the sun through all living things (fig. 5.1).

Energy

Energy is defined as the capacity to bring about change or to do work. A living thing must constantly perform work in order to maintain its organization, to grow, and to reproduce.

Figure 5.1 shows a way in which chemicals cycle through living things—the carbon dioxide given off by mitochondria is used by chloroplasts when carbohydrates are re-formed—but energy cannot cycle in this manner. Instead, energy conversions ultimately result in heat. The reason energy cannot cycle is apparent from a consideration of certain laws of thermodynamics.

Two Laws of Thermodynamics

Thermodynamics is the study of energy relationships and exchanges. The first law of thermodynamics says that *energy can neither be created nor destroyed.* However, it can be converted from one form to another form.

In order to demonstrate this law, we usually consider the amount of energy in a piece of the universe, termed a system. For a biologist, a system might be a cell, an organ, an organism, or even a community of organisms. All matter outside that selected system is considered to be the surroundings.

In keeping with the first law of thermodynamics, we know that energy cannot be created within a system; rather, energy must come from the surroundings. For example, the electric company does not create energy; it simply converts the energy of falling water, nuclear energy, or fossil fuel energy into electrical energy. Similarly, mitochondria do not create ATP energy. Rather, they convert the energy of carbohydrates into ATP energy. We also know from the first law of thermodynamics that a system cannot destroy energy, although it can lose energy to its surroundings. A cell loses energy to its surroundings; therefore, it must constantly take in energy (i.e., carbohydrates) in order to continue living.

The second law of thermodynamics says that *one usable form of energy cannot be completely converted into another usable form.* For example, muscles convert the chemical-bond energy within ATP to the mechanical energy of contraction, though not completely; some chemical-bond energy of ATP becomes heat. Because heat is the form of energy that living systems easily lose to the surroundings, it is termed nonusable energy. Because of energy conversions in living things, all the energy in an ATP molecule eventually becomes heat that is lost to the environment (fig. 5.1). Now, we can see that living systems continually take usable energy (i.e., carbohydrates) from the environment and return to it heat, a nonusable form of energy.

Figure 5.2

The ATP cycle. ATP is continually made and remade in cells. The average male needs about 8 kg (17.6 lb) of ATP an hour, yet the body has on hand only about 50 grams (1.8 oz) at any time. The answer to this paradox is that the entire supply of ATP is recycled about once each minute.

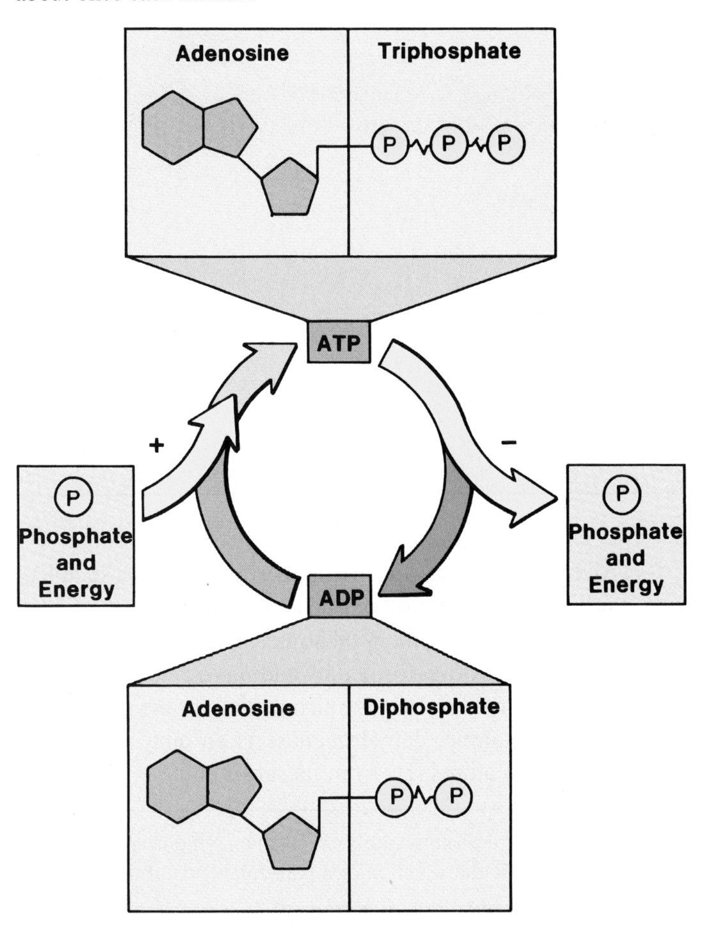

Living things need a constant supply of energy from their surroundings because (1) energy cannot be created and (2) when one form of energy is converted to another form, some is lost as heat.

ATP: Energy for Cells

ATP (adenosine triphosphate) is the immediate source of energy in cells. Figures 2.31 and 5.2 show that ATP is a nucleotide composed of the nitrogenous base adenine and the sugar ribose (together called adenosine) and 3 phosphate groups. The wavy lines in the formula for ATP indicate high-energy phosphate bonds; when these bonds are broken, an unusually large amount of energy is released.[1] Because of this property, ATP is the energy currency of cells. When cells "need" something, they "spend" ATP. ATP is the common energy currency for synthesis of molecules, active transport of molecules from one site to another, conduction of nerve impulses, and muscle contraction. When energy is required for these processes, the end phosphate group is removed from ATP and the molecule breaks down to ADP (adenosine diphosphate) and Ⓟ (fig. 5.2).

ATP Cycle: Add/Drop Ⓟ

Cells constantly remake ATP molecules from ADP + Ⓟ (fig. 5.2). The rebuilding process, however, requires an input of energy, and this energy comes from the metabolism of glucose products in mitochondria. This rejoining of ADP and Ⓟ provides a constant supply of ATP in cells.

ATP is the energy currency in cells because it has high-energy phosphate bonds. ATP breaks down to ADP + Ⓟ + energy and is rebuilt from these same components. Do 5.1 Critical Thinking, found at the end of the chapter.

Metabolic Pathways

Metabolism is the sum of all chemical reactions occurring inside a living cell. Reactions do not occur haphazardly in cells; they are usually a part of a metabolic pathway. *Metabolic pathways* begin with a particular reactant and terminate with an end product. While it is possible to write an overall equation for a pathway as if the beginning *reactant* went to the end *product* in one step, there are actually many minute steps in between. In the pathway, one reaction leads to the next reaction, which leads to the next reaction, and so forth in an organized, highly structured manner. This arrangement makes it possible for one pathway to lead to several others, especially since various pathways have several molecules in common. Also, metabolic energy is captured and utilized more easily if it is released in small increments rather than all at once.

Metabolic pathways can be represented by the following diagram:

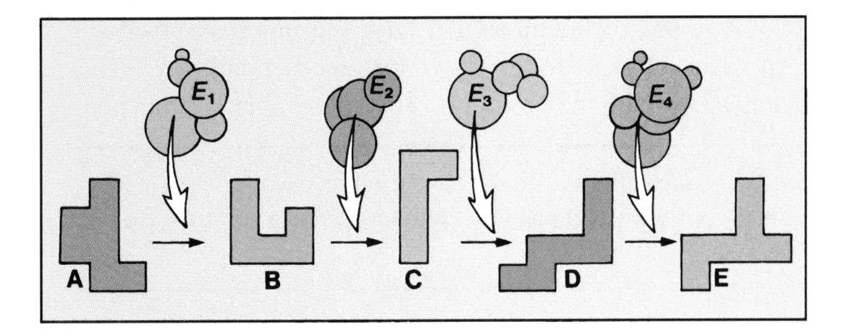

In this diagram, the angular shapes lettered *A–E* are reactants and products in the various reactions. (A reactant is a substance that participates in a reaction. A product is a substance that is formed by a chemical reaction.) The circular shapes numbered E_1–E_4 are enzymes. The reactants in an enzymatic reaction are called the **substrates** for that enzyme. In the first reaction, *A* is the substrate for E_1 and *B* is the product. Now *B* becomes the

1. A high-energy phosphate bond liberates about 7 Kcal/gram-molecular weight compared to about 2 Kcal for a low-energy phosphate bond. Kcal (kilocalorie) is a common way to measure heat.

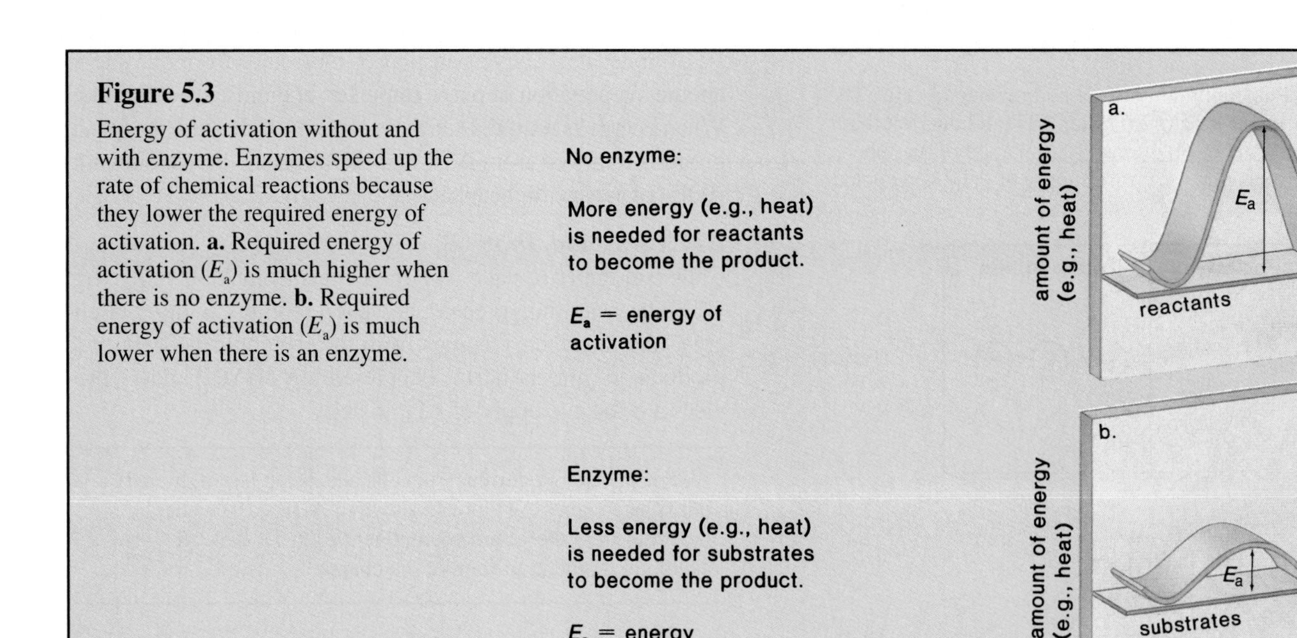

Figure 5.3
Energy of activation without and with enzyme. Enzymes speed up the rate of chemical reactions because they lower the required energy of activation. **a.** Required energy of activation (E_a) is much higher when there is no enzyme. **b.** Required energy of activation (E_a) is much lower when there is an enzyme.

Table 5.1
Enzymes Named for Their Substrate

Substrate	Enzyme
Lipid	Lipase
Urea	Urease
Maltose	Maltase
Ribonucleic acid	Ribonuclease
Lactose	Lactase

substrate for E_2, and C is the product. This process continues until the final product E forms.

Any one of the molecules (A–E) in this linear pathway could be a substrate for an enzyme in another pathway. A diagram showing all the possibilities would be highly branched.

A metabolic pathway is a series of reactions that proceed in an orderly, step-by-step manner. Each reactant is a substrate for a particular enzyme.

Enzymes: Speed Reactions

Every reaction in a cell requires a specific enzyme. In most instances, an **enzyme** is a protein molecule[2] that functions as an organic catalyst to speed up a chemical reaction. In a crowded ballroom, a mutual friend can cause particular people to interact. In the cell, an enzyme brings together particular molecules and causes them to react with one another.

An enzyme is very specific in its action and can speed up only one particular reaction or one type of reaction; therefore, enzymes are named for their substrates. Table 5.1 shows that enzyme names are often formed by adding *-ase* to the name of its substrate. Some enzymes are also named for the action they perform; for example, dehydrogenase is an enzyme that removes hydrogen atoms (H) from its substrate.

Most enzymes are protein catalysts. Each enzyme is specific; it speeds up a particular reaction or a particular type of reaction.

Enzymes are absolutely necessary to the continued existence of a cell because they allow reactions to occur at moderate temperatures. Molecules usually have to be activated in some way for a reaction to occur. For example, wood does not burn unless it is heated to a high temperature. Likewise, in the laboratory, activation very often is achieved by heating the reaction flask. This increases the number of effective collisions between molecules, allowing them to react with one another.

Enzymes lower the energy of activation—the amount of heat needed for a reaction to occur. Figure 5.3 gives an example: the amount of energy needed when there is no enzyme is much higher than the amount of energy needed when there is an enzyme present. It is known, for example, that the hydrolysis of casein (the protein found in milk) requires 20,600 Kcal/gram-molecular weight to occur when there is no enzyme and only 12,000 Kcal with the enzyme. Enzymes lower the energy of activation by binding with their substrates in such a way that a reaction can occur more readily.

2. The recent discovery of catalytic RNA molecules suggests that not all enzymes are proteins.

Figure 5.4
Enzymatic action. An enzyme has an active site, which is where the substrates and enzyme fit together in such a way that the substrates are oriented to react. Following the reaction, the products are released and the enzyme assumes its original shape. **a.** Enzymes carry out synthetic reactions. **b.** Enzymes carry out degradative reactions.

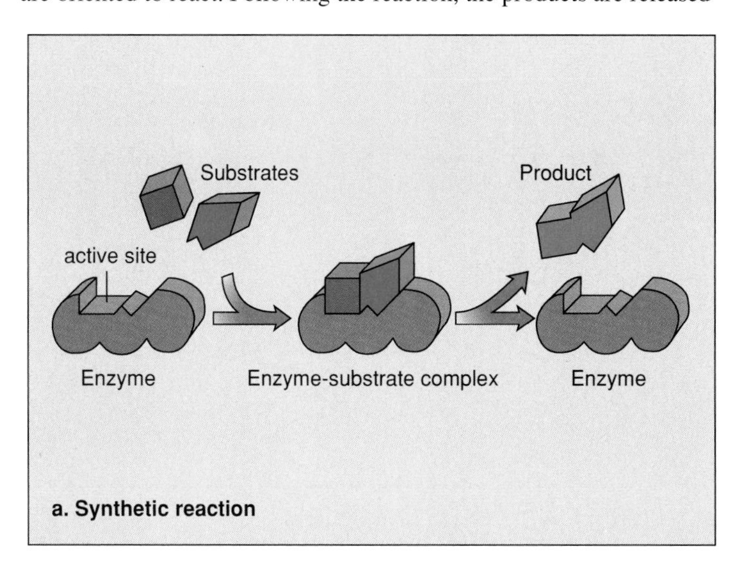

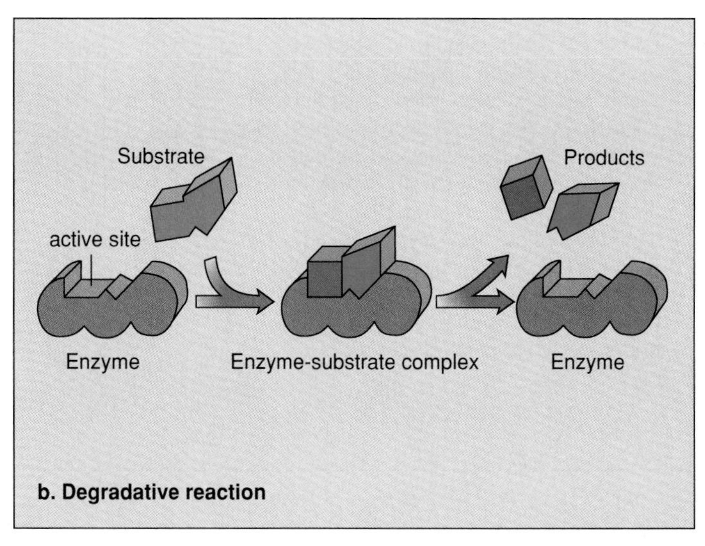

Enzyme-Substrate Complex: Induced Fit

The reaction that is pictorially shown in figure 5.4 is often used to show that enzymes form a complex with their substrates:

$$E + S \rightarrow ES \rightarrow E + P$$

In this reaction, E = enzyme, S = substrate, ES = enzyme-substrate complex, and P = product.

Notice in figure 5.4 that the enzyme has an **active site,** a place where the substrates fit onto the enzyme in such a way that they are oriented to react. The enzyme, not heat, facilitates the progress of the reaction. The enzyme does not undergo a permanent change, so it can be used over and over again. Therefore, only a small amount of enzyme actually is found in a cell.

The shape of an enzyme allows an enzyme-substrate complex to form, which explains the specificity of an enzyme. The substrates are seemingly specific to the enzyme because their shapes fit together as a *key fits a lock.* However, it is now thought that the active site may very well undergo a slight change in shape in order to more perfectly accommodate the substrates. This is called the **induced-fit model** because as binding occurs, the active site is induced (undergoes a slight alteration) to achieve the best fit. After the reaction is complete, the product is released, and the active site returns to its original state.

The Product: Determined by the Reaction

An enzymatic reaction can bring about synthesis—2 smaller molecules may be joined to form a larger molecule. For example, 2 amino acids can be joined to form a dipeptide. An enzymatic reaction can also bring about degradation—a larger molecule can be broken down to smaller molecules. For example, a disaccharide can be hydrolyzed to monosaccharides.

In a given reaction, only certain products are produced by any particular reactants, and the presence of an enzyme does not change the outcome of a reaction. However, the presence of an active enzyme does determine whether or not a reaction takes place. For example, if substance A can react to form either substance B or substance C, then the enzyme that is active—E_1 or E_2—determines which product (substance D or substance F) is produced:

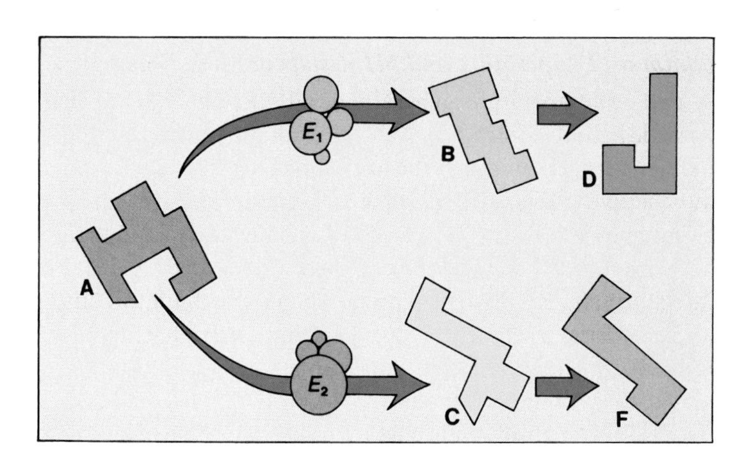

Each enzyme has an active site, where specific substrates are oriented to facilitate a particular reaction. Once the reaction is complete, the products are released from the enzyme, which can then speed up another reaction.

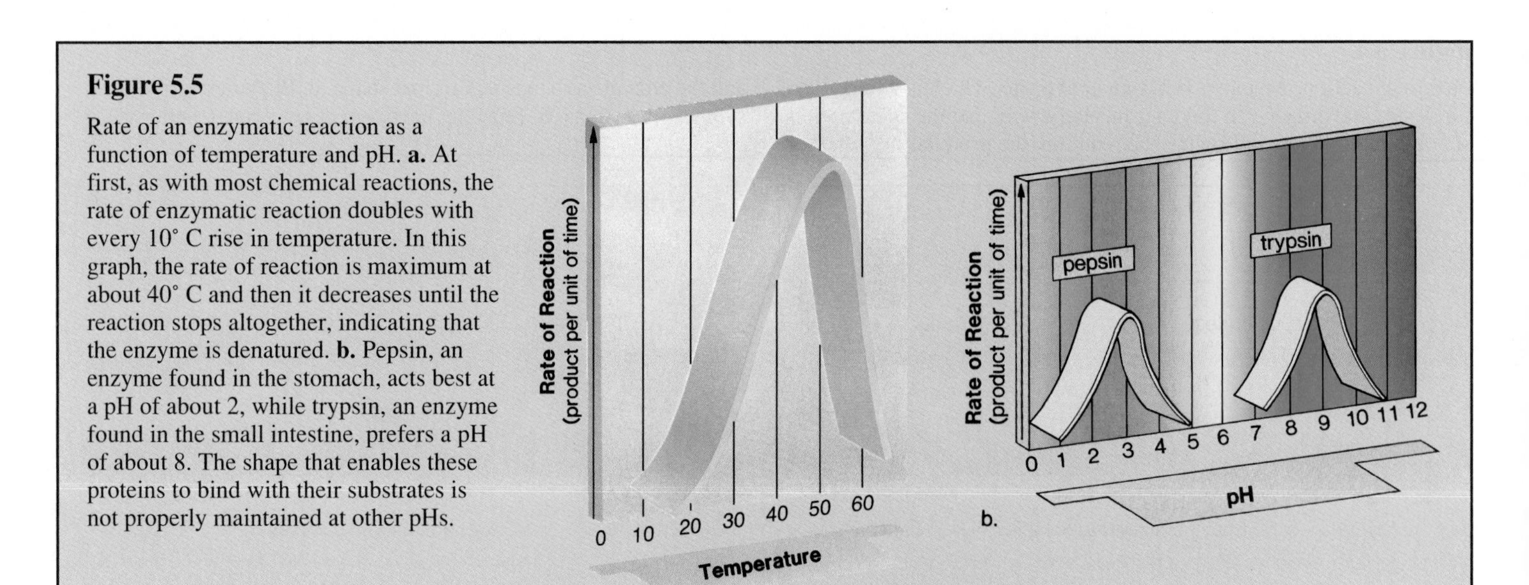

Figure 5.5

Rate of an enzymatic reaction as a function of temperature and pH. **a.** At first, as with most chemical reactions, the rate of enzymatic reaction doubles with every 10° C rise in temperature. In this graph, the rate of reaction is maximum at about 40° C and then it decreases until the reaction stops altogether, indicating that the enzyme is denatured. **b.** Pepsin, an enzyme found in the stomach, acts best at a pH of about 2, while trypsin, an enzyme found in the small intestine, prefers a pH of about 8. The shape that enables these proteins to bind with their substrates is not properly maintained at other pHs.

What Affects Enzyme Productivity

Enzymes normally allow reactions to proceed quite rapidly. For example, the breakdown of hydrogen peroxide into water and oxygen can occur 600,000 times a second when the enzyme catalase is present. How quickly an enzyme works, however, is affected by certain conditions.

Adding Substrate Results in More Product

Generally, enzyme activity increases as substrate concentration increases because there are more collisions between substrate molecules and the enzyme. As more substrate molecules fill active sites, more product results per unit time. But when the enzyme's active sites are filled almost continuously with substrate, the enzyme's rate of activity cannot increase anymore. Maximum rate has been reached.

Optimum Temperature and pH Results in More Product

A rise in temperature generally results in an increase in enzyme activity (fig. 5.5*a*). As the temperature increases, the movement of enzyme and substrate molecules increases, and more active sites are filled. If the temperature rises beyond a certain point, however, the reaction time levels off and then declines rapidly because the enzyme is denatured. A denatured protein has lost its normal shape and therefore its ability to form an enzyme-substrate complex. High temperatures disrupt the hydrogen bonding between amino acids; that is, the secondary structure of the protein is affected (see fig. 2.17). Denaturation is observed when the white of an egg (albumin) is cooked, for example.

A change in pH can also affect enzyme activity (fig. 5.5*b*). Each enzyme has an optimal pH. This pH helps to maintain the enzyme's tertiary structure because it helps to maintain the normal interactions between *R* groups of the amino acids within the enzyme. A change in pH can alter the ionization of *R* groups and disrupt the normal interactions of the *R* groups, and a change in shape, or denaturation, eventually occurs. Without its normal shape, the enzyme is unable to combine efficiently with its substrate.

Inhibition Results in Less Product

In *competitive inhibition,* another molecule is so close in shape to an enzyme's substrate that this molecule can compete with the true substrate for the active site of the enzyme. Such a molecule is designated as *I*, for inhibitor, in these reactions:

$$\text{Irreversible: } I + E \rightarrow EI \rightarrow \text{no further reaction}$$

$$\text{Reversible: } I + E \rightarrow EI \rightarrow E + I$$

Any molecule that binds with an enzyme other than its substrate is an inhibitor of the reaction because only the binding of the substrate results in a product. The first reaction given represents irreversible inhibition, and the second represents reversible inhibition. Irreversible inhibition of an enzyme is not common and usually is caused by a poison. For example, penicillin causes the death of bacteria due to the irreversible inhibition of an enzyme needed to form the bacterial cell wall. In humans, hydrogen cyanide is an inhibitor of a very important enzyme (cytochrome oxidase) present in all cells, which accounts for its lethal effect.

Reversible inhibition is common in cells, and an enzyme can be inhibited competitively by its own product if the product closely resembles the enzyme's substrate. One interesting example of reversible inhibition, though, concerns 2 foreign substances in cells. The enzyme alcohol dehydrogenase combines with either alcohol or ethylene glycol (automobile antifreeze). Because the breakdown product of the latter damages the kidneys, the medical remedy for accidental ingestion of antifreeze is administration of alcohol to the point of intoxication. Most of the ethylene glycol then is harmlessly excreted.

Figure 5.6

Noncompetitive inhibition. The end product of this metabolic pathway can bind to enzyme E_1. When it does, the pathway is shut down because the binding of the end product changes the enzyme's shape so that the substrate cannot bind to the active site. In this way, noncompetitive inhibition helps to regulate the output of metabolic pathways.

One type of reversible inhibition is called *noncompetitive inhibition* because a molecule binds to an enzyme at a site other than the active site. The binding of the molecule causes the enzyme to assume a shape that prevents it from binding with its substrate. Noncompetitive reversible inhibition is the normal way by which metabolic pathways are regulated in cells. Consider, for example, that it would be possible to slow down a metabolic pathway if the end product inhibited the first enzyme (*E1*) noncompetitively (fig. 5.6). When there is an adequate quantity of end product, the pathway is shut down, and when more end product is needed, the pathway is active.

An enzyme's shape is appropriate for its substrate. Any environmental factor that affects the shape of a protein also affects the ability of an enzyme to speed up its reaction. Do 5.2 Critical Thinking, found at the end of the chapter.

Coenzymes: Helpers of Enzymes

Many enzymes require a nonprotein cofactor to assist them in carrying out their function. Some cofactors are ions; for example, magnesium ions (Mg^{++}), potassium ions (K^+), and calcium ions (Ca^{++}) are very often involved in enzymatic reactions.

Some other cofactors, called **coenzymes,** are organic molecules that bind to enzymes and serve as carriers for chemical groups or electrons. In this case, the protein portion of the enzyme accounts for its specificity, that is, the ability of the enzyme to form an enzyme-substrate complex and to speed up only one particular reaction. The coenzyme portion of the enzyme participates in the reaction.

A coenzyme is generally a large molecule that the body is incapable of synthesizing without the ingestion of a vitamin. **Vitamins** are organic dietary requirements needed in small amounts only. Niacin (or nicotinate), thiamin (or vitamin B_1), riboflavin, folate, and biotin are just a few examples of well-known vitamins that are parts of coenzymes.

Figure 5.7

The NAD cycle. NAD is reduced and becomes $NADH_2$ when it accepts hydrogen atoms (H). $NADH_2$ is oxidized to NAD again when the hydrogen atoms are passed to another acceptor.

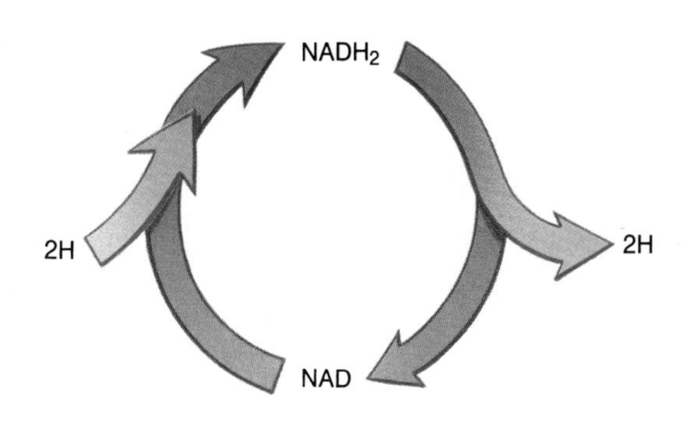

NAD Helps Oxidations

NAD[3] is a coenzyme that contains the vitamin niacin and works in conjunction with enzymes called dehydrogenases. An enzyme of this type removes 2 hydrogen atoms (H) from its substrate, and these are accepted by NAD.

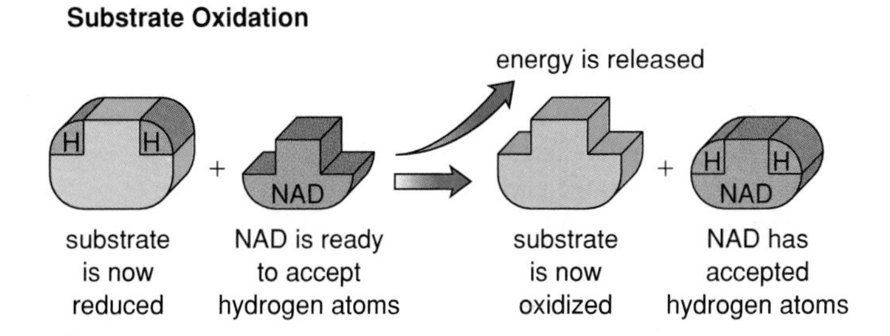

This is an oxidation reaction, and energy is released because a substrate bond has been broken. **Oxidation** is usually defined as the removal of electrons. Note that when hydrogen atoms are removed from a substrate, electrons are also removed since a hydrogen atom consists of a hydrogen ion (H^+) and an electron (e^-).

Just as there is an ATP cycle, there is also an NAD cycle in cells (fig. 5.7). Only a small amount of NAD is needed in cells because the same molecule is used over and over again. After NAD accepts hydrogen atoms and is reduced to $NADH_2$, $NADH_2$ carries the hydrogen atoms to another acceptor, becoming oxidized to NAD again.

NAD is involved in cellular respiration, a metabolic pathway that breaks down glucose to carbon dioxide (CO_2) and water (H_2O). As mentioned, the energy from glucose breakdown is used to rebuild ATP from ADP + Ⓟ. This topic is discussed further in chapter 6.

3. Nicotinamide adenine dinucleotide

NADP Helps Reductions

NADP,[4] also a coenzyme, has a structure similar to NAD, but it contains a phosphate group that is not found in NAD. This may signal that it has a slightly different function. NADP carries 2 hydrogen atoms (H) as does NAD, but NADP is used to bring about substrate reduction:

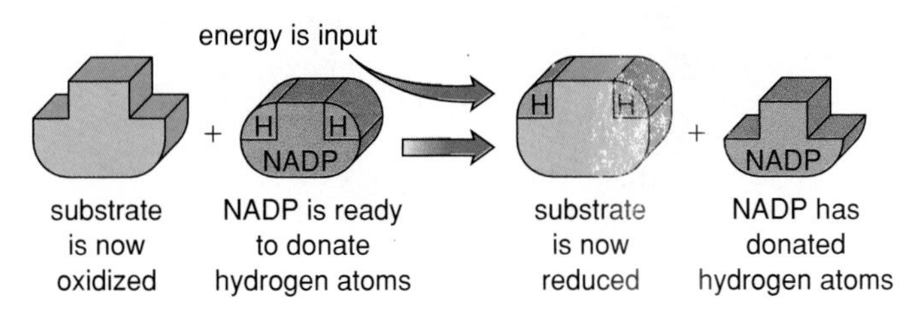

4. Nicotinamide adenine dinucleotide phosphate

Reduction is usually defined as the addition of electrons. Note that when hydrogen atoms are accepted by a substrate, electrons are also accepted because a hydrogen atom consists of a hydrogen ion (H^+) and an electron (e^-). Reduction is an energy-requiring process because a substrate bond is formed.

There is also an $NADPH_2$–NADP cycle in cells. During photosynthesis within chloroplasts, carbon dioxide (CO_2) is reduced to a carbohydrate. $NADPH_2$ supplies the necessary hydrogen atoms and ATP supplies the necessary energy to bring about the reduction. Both of these are produced within chloroplasts after solar energy is absorbed.

> Enzymes have helpers called coenzymes, which participate in the reaction. When a substrate is oxidized (as in cellular respiration), energy is released and the coenzyme NAD becomes $NADH_2$. When a substrate is reduced (as in photosynthesis), energy is required and the coenzyme $NADPH_2$ becomes NADP.

SUMMARY

All living things need a constant input of energy because they cannot create energy and because all energy conversions result in a loss of usable energy in the form of heat. After absorbing solar energy, chloroplasts produce the carbohydrates that are eventually broken down to provide the energy for ATP buildup in mitochondria. ATP is the energy currency of cells; energy-requiring reactions use ATP molecules. Eventually, usable energy in living things is converted to heat.

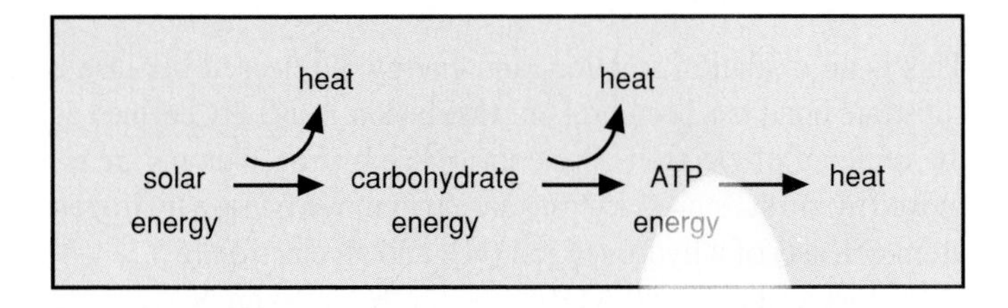

A metabolic pathway is a series of reactions that proceed in an orderly, step-by-step manner. Each reaction requires a specific enzyme. Reaction rates increase when enzymes form a complex with their substrates. Any environmental factor that affects the shape of a protein also affects the ability of an enzyme to do its job. Sometimes enzymes require coenzymes. NAD is a coenzyme that, by accepting electrons, oxidizes substrates with a concomitant release of energy. During cellular respiration, this energy is used to build up ATP. $NADPH_2$ is a coenzyme that gives up electrons and so reduces substrates. Reduction requires ATP energy. This occurs during photosynthesis.

Cellular Respiration	Photosynthesis
$C_6H_{12}O_6$	$CO_2 + H_2O$
ADP + (P) → ATP; NAD → $NADH_2$	ATP → ADP + (P); $NADPH_2$ → NADP
$CO_2 + H_2O$	$C_6H_{12}O_6$

STUDY QUESTIONS

In order to practice **writing across the curriculum,** students should write out the answers to any or all of the study questions. The study questions are sequenced in the same order as the text.

1. How do chloroplasts and mitochondria permit a flow of energy between organisms? (p. 74)
2. Use the 2 laws of thermodynamics to explain why a living system needs a constant input of energy from its surroundings. (p. 74)
3. Why is ATP called the energy currency of cells? What is the ATP cycle? (p. 75)
4. Diagram a metabolic pathway. Label the substrates and the enzymes. (p. 75)
5. Why are enzymes specific, and why can't each one speed up many different reactions? (p. 76)
6. Why is less energy needed for a reaction to occur when an enzyme is present? (p. 76)
7. Name and explain the manner in which at least 3 factors can influence the speed of an enzymatic reaction. (p. 78)
8. What are coenzymes, and how do they help enzymes? (p. 79)
9. NAD participates in what kind of reaction? NAD participates in which cellular process? (p. 79)
10. NADP participates in what kind of reaction? NADP participates in which cellular process? (p. 80)

OBJECTIVE QUESTIONS

1. Life requires a constant supply of energy. Ultimately, the energy comes from the _______.
2. Cells make use of the chemical energy found in the high-energy bonds of _______ molecules.
3. One form of energy cannot be transformed completely into another type; some usable energy is always lost as _______.
4. ADP, Ⓟ, and _______ are needed to re-form ATP.
5. The shape of an enzyme allows only its _______ to fit at the active site.
6. After substrates come together on the surface of an enzyme, less _______ is needed to make the reaction occur.
7. For an enzyme to work most efficiently, the solution should have the correct _______ and the temperature should be moderate.
8. Enzymes often have _______, which participate in the reaction.
9. NAD _______ a substrate and becomes $NADH_2$.
10. $NADPH_2$ _______ a substrate and becomes NADP.
11. Use these terms to label this diagram: substrate, enzyme, active site, product, and enzyme-substrate complex. Explain the importance of an enzyme's shape to its activity.

CRITICAL THINKING

In order to practice **writing across the curriculum,** students should write out the answers to any or all of the critical thinking questions. Suggested answers to the critical thinking questions are in appendix E.

5.1

1. At each link of a food chain, such as when cows eat grass and humans eat cows, there is a loss of energy. Explain.
2. Considering this or any other food chain, why is it correct to say that humans are dependent on the energy of the sun?
3. Explain the meaning of the word *drives* in the expression "glucose breakdown drives ATP buildup."

5.2

1. Pepsin is an enzyme that breaks down protein. A student has a test tube that contains pepsin, egg white, and water. What conditions would you recommend to ensure digestion of the egg white?
2. If all the conditions are perfect, how could you increase the yield (i.e., amount of product—amino acids—per unit of time)?
3. The instructor adds an inhibitor to the test tube. How could the student tell if reversible or irreversible inhibition is taking place?

SELECTED KEY TERMS

active site the region on the surface of an enzyme where the substrate binds and where the reaction occurs.
ATP (adenosine triphosphate) a compound containing adenine, ribose, and 3 phosphates, 2 of which are high-energy phosphates; the "common currency" of energy for most cellular processes.
coenzyme a nonprotein molecule that aids the action of the enzyme to which it is loosely bound.
energy the capacity to do work.
enzyme a protein catalyst that speeds up a specific reaction or a specific type of reaction.
metabolism all of the chemical changes that occur within a cell.
NAD a coenzyme of oxidation; accepts hydrogen atoms (H) from a substrate and carries them to another acceptor.
NADP a coenzyme of reduction; $NADPH_2$ donates hydrogen atoms (H) to a substrate.
oxidation the loss of electrons (usually inorganic); the removal of hydrogen atoms (H) (usually organic).
reduction the gain of electrons (inorganic); the addition of hydrogen atoms (H) (organic).
substrate a reactant in a reaction controlled by an enzyme.
vitamin essential requirement in the diet needed in small amounts. They are often part of coenzymes.

6

PHOTOSYNTHESIS AND CELLULAR RESPIRATION

Chapter Concepts

1.
Photosynthetic organisms (plants, algae, and a few bacteria) produce the organic molecules that are the primary source of nutrients and chemical energy for all organisms. 84, 99

2.
Photosynthesis in plants and algae takes place in chloroplasts. 84, 99

3.
Photosynthesis has 2 sets of reactions. The light-dependent reactions provide the energy (ATP) and the hydrogen atoms (H) that are used by light-independent reactions to reduce carbon dioxide (CO_2) to a carbohydrate (R–CH_2O). 85, 99

4.
Cellular respiration, which occurs in all organisms, converts the energy of carbohydrates to the energy of ATP. 90, 99

5.
Mitochondria complete the process of aerobic cellular respiration: glucose ($C_6H_{12}O_6$) becomes carbon dioxide (CO_2) and water (H_2O). 90, 99

6.
Fermentation, an anaerobic process that occurs in the cytoplasm, results in the incomplete breakdown of glucose. 95, 100

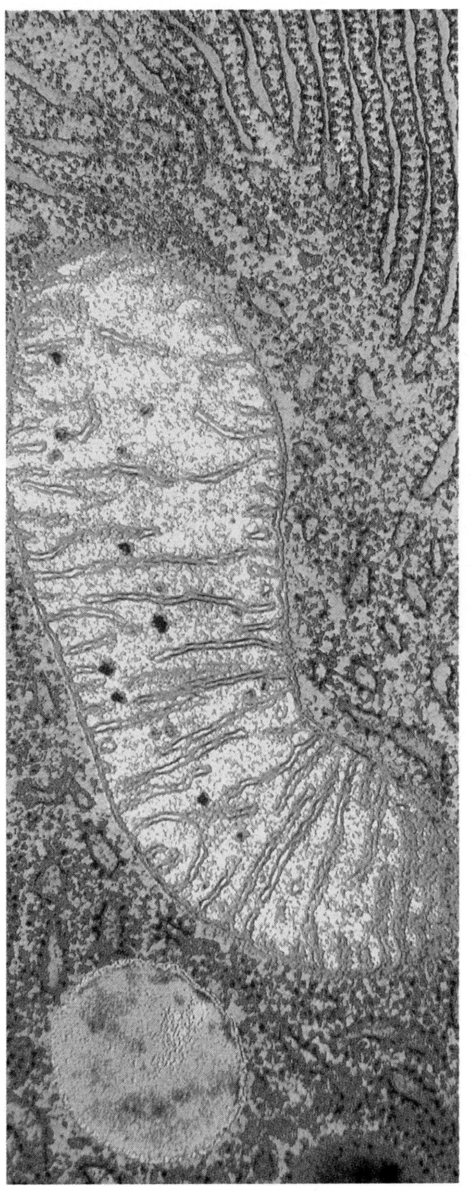

Mitochondrion and endoplasmic reticulum

Chapter Outline

Figure 6.1

Chloroplast structure and function. Thylakoids (flattened sacs) are stacked like poker chips into grana (one stack is a granum). Each thylakoid consists of a thylakoid membrane surrounding a thylakoid space. Thylakoid membranes contain chlorophyll and other pigments that absorb solar energy. The light-dependent reactions occur here. The fluid-filled stroma contains enzymes that, with the help of $NADPH_2$, reduce carbon dioxide (CO_2) to carbohydrate (R–CH_2O) during the light-independent reactions.

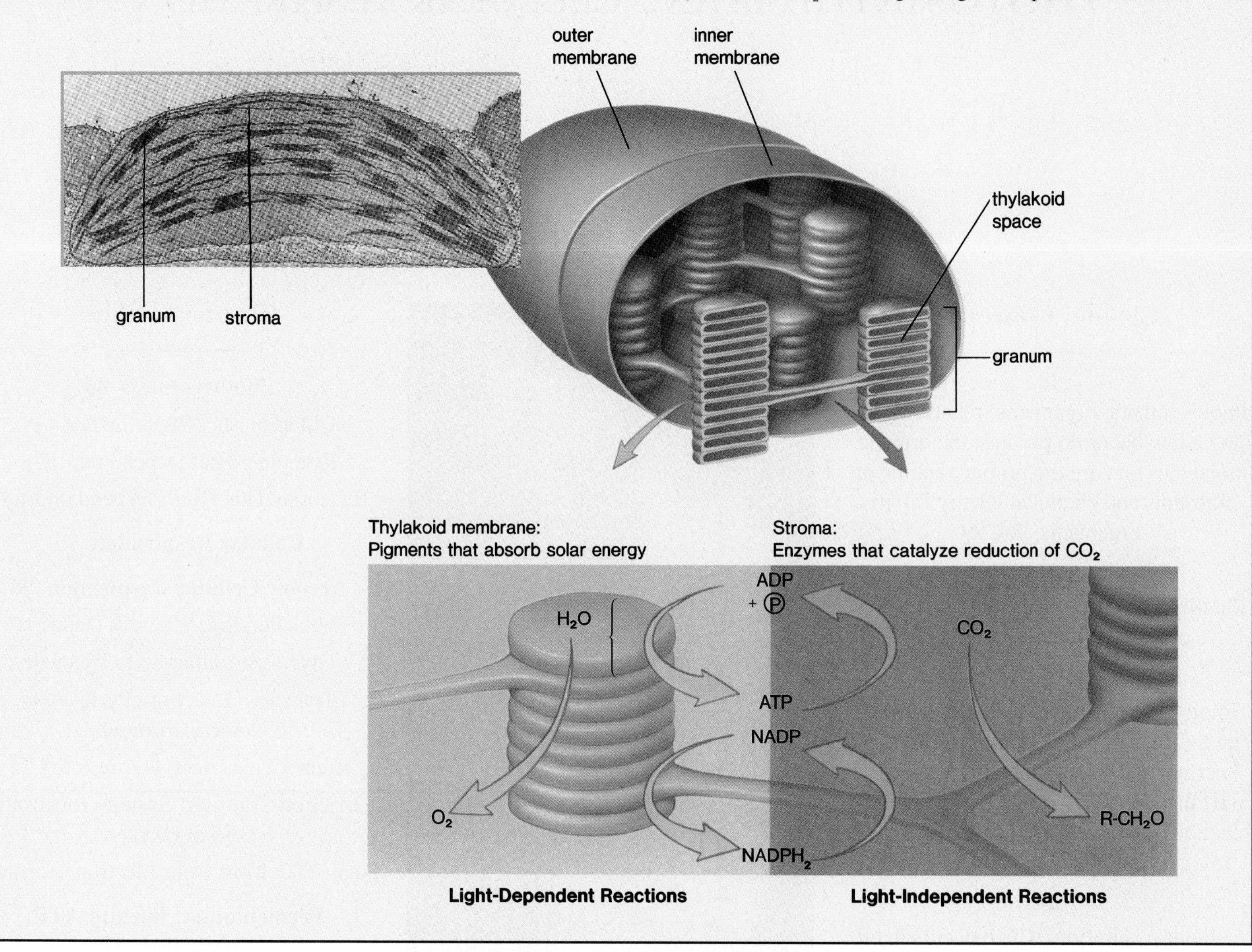

WE have noted that chloroplasts and mitochondria carry out energy transformations that result in a supply of ATP molecules for the cell. Now it is time to look a little more closely at the actual metabolic pathways of photosynthesis and cellular respiration. We will see that several steps are needed to transform solar energy into carbohydrate-bond energy; similarly, several steps are needed to transform carbohydrate-bond energy into ATP-bond energy.

Photosynthesis

Only plants, algae, and a few kinds of bacteria carry on photosynthesis. As its name implies, *photosynthesis* refers to the ability of these organisms to make their own food in the presence of sunlight. Plants and some algae carry on photosynthesis within chloroplasts.

Chloroplasts: Where It Happens

The interior of a chloroplast contains stacks of membranous compartments called grana (sing., **granum**) (fig. 6.1). The individual flattened sacs within each granum are called **thylakoids.**

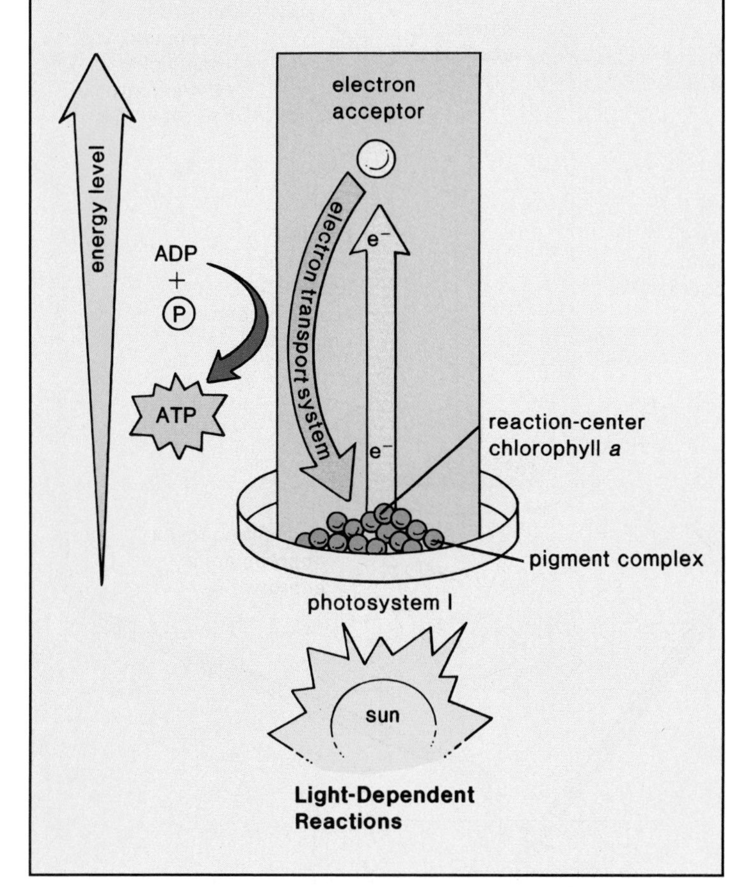

Figure 6.2

Cyclic electron pathway involves only Photosystem I. Excited electrons (e^-) leave the PS I reaction-center chlorophyll and are taken up by an electron acceptor, which passes them down an electron transport system before they return to PS I. Only ATP production results from this pathway.

The green pigment **chlorophyll** is found within the membrane of the thylakoids, and it is here that solar energy is captured. Surrounding the grana is a fluid-filled space called the **stroma.** The stroma contains enzymes that participate in photosynthesis.

Photosynthesis involves 2 sets of reactions known as the light-dependent reactions and the light-independent reactions. The **light-dependent reactions** begin after the thylakoids absorb solar energy. Then, water (H_2O) is split and oxygen (O_2) is released. $NADPH_2$ molecules and ATP molecules (see chap. 5) are also made. The **light-independent reactions** involve the enzymes of the stroma. The $NADPH_2$ and ATP produced by the light-dependent reactions are used to reduce carbon dioxide (CO_2) to form a carbohydrate (R–CH_2O) (fig. 6.1).

The existence of the 2 sets of reactions is consistent with the observation that even when light is being maximally absorbed by a photosynthetic system, a rise in temperature still increases the rate of photosynthesis. This occurs because only the light-dependent reactions are sensitive to the amount of light; the light-independent reactions utilize temperature-sensitive enzymes.

Photosynthesis takes place in chloroplasts. The light-dependent reactions occur in thylakoids, which capture solar energy and produce $NADPH_2$ and ATP. The light-independent reactions occur in the stroma, where the $NADPH_2$ and ATP are used to reduce carbon dioxide (CO_2) to a carbohydrate (R–CH_2O).

Reactions That Depend on Light

The thylakoid membrane is highly organized into 2 groups of light-gathering units called Photosystem I (PS I) and Photosystem II (PS II). Each **photosystem** contains a pigment complex composed of chlorophyll *a* and chlorophyll *b* molecules (green pigments) and accessory pigments, such as **carotenoid** molecules (primarily red and orange but also yellow pigments). The carotenoids give tomatoes and carrots their color (accounting for the name *carrot*) and also give leaves their color after they lose chlorophyll in the fall. The closely packed pigment molecules in the photosystems serve as an "antenna" for gathering solar energy. Solar energy is passed from one pigment to the other until it is concentrated into one particular chlorophyll *a* molecule, the *reaction-center chlorophyll.* Electrons in the reaction-center chlorophyll molecule become so excited that they escape and move to a nearby *electron acceptor molecule.*

Cyclic Electron Pathway: ATP Only

The *cyclic electron pathway* (fig. 6.2) begins after the PS I pigment complex absorbs solar energy. In this pathway, excited electrons (e^-) leave the PS I reaction-center chlorophyll but eventually return to it. Before they return, however, the electrons enter an **electron transport system,** a series of carriers that pass electrons from one to the other. Some of the carriers are cytochrome molecules; for this reason, the electron transport system is sometimes called a cytochrome system. As the electrons pass from one carrier to the next, energy that will be used to produce ATP molecules is released and stored.

Some photosynthetic bacteria utilize the cyclic electron pathway only; therefore, this pathway probably evolved early in the history of life. It is believed that in plants, the cyclic flow of electrons is utilized only when carbon dioxide (CO_2) is in such limited supply that carbohydrate is not being produced. At this time, there would be no need for additional $NADPH_2$, which is produced by the noncyclic electron pathway.

The cyclic electron pathway, from PS I back to PS I, has only this one effect: production of ATP.

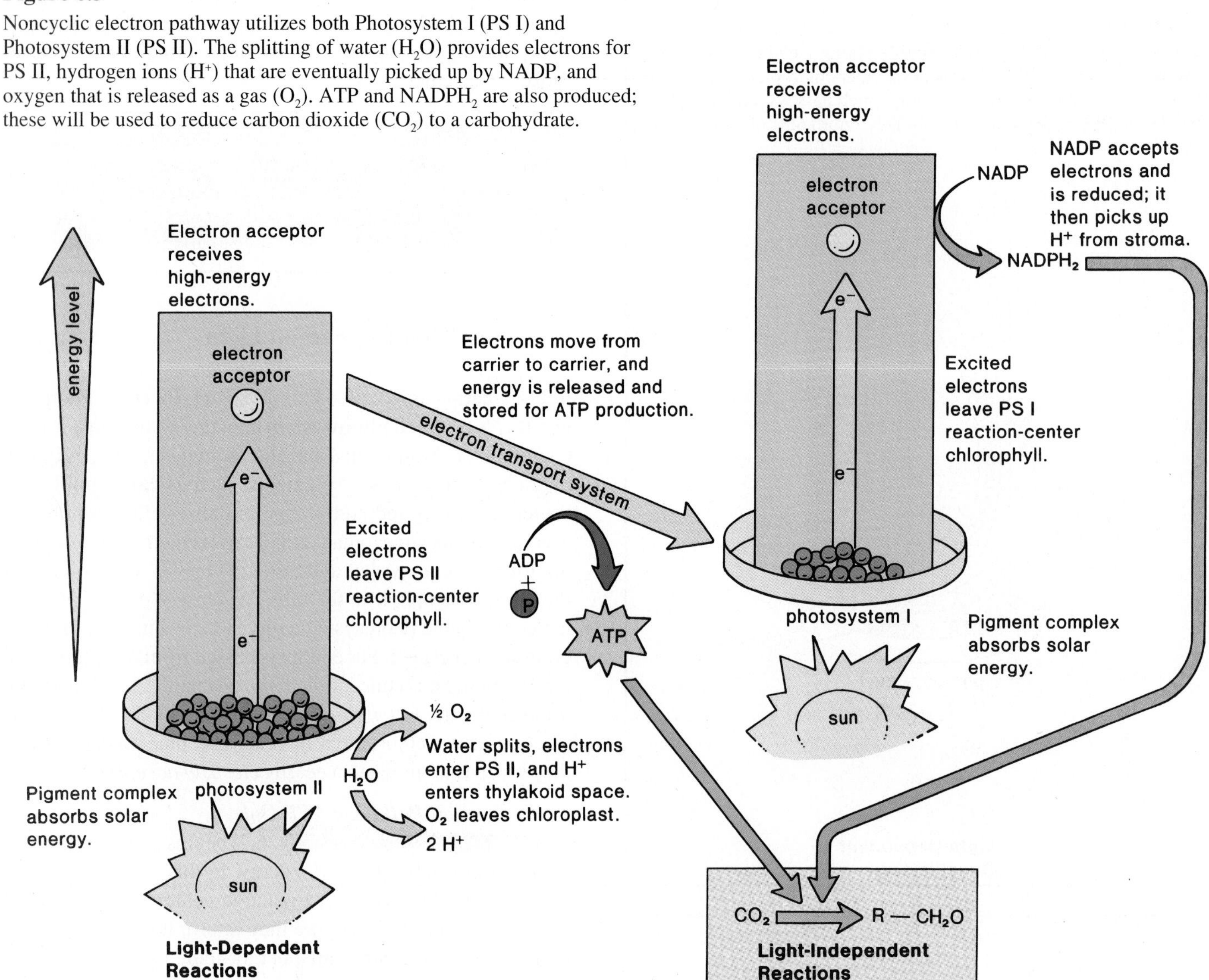

Figure 6.3

Noncyclic electron pathway utilizes both Photosystem I (PS I) and Photosystem II (PS II). The splitting of water (H_2O) provides electrons for PS II, hydrogen ions (H^+) that are eventually picked up by NADP, and oxygen that is released as a gas (O_2). ATP and $NADPH_2$ are also produced; these will be used to reduce carbon dioxide (CO_2) to a carbohydrate.

Noncyclic Electron Pathway: ATP and $NADH_2$

During the *noncyclic electron pathway*, electrons move from water (H_2O) through PS II to PS I and then on to NADP (fig. 6.3). This pathway begins when the PS II pigment complex absorbs solar energy and excited electrons (e^-) leave the reaction-center chlorophyll. PS II receives replacement electrons from water, which splits, releasing oxygen:

$$H_2O \longrightarrow 2\,H^+ + 2\,e^- + 1/2\,O_2$$

This oxygen evolves from the chloroplast and the plant as oxygen gas. The hydrogen ions (H^+) temporarily stay in the thylakoid space.

The high-energy electrons that leave PS II are captured by an electron acceptor, which sends them to an electron transport system. As the electrons pass from one carrier to the next, energy that will be used to produce ATP molecules is released and stored. Low-energy electrons leaving the electron transport system enter PS I.

The PS I pigment complex absorbs solar energy, and excited electrons leave the reaction-center chlorophyll and are

Figure 6.4

Electron transport system and chemiosmotic ATP synthesis. **a.** The electron transport system is located within the thylakoid membrane, which surrounds the thylakoid space. **b.** As electrons are passed from carrier to carrier of the electron transport system, energy is released and is used to pump hydrogen ions (H^+) from the stroma into the thylakoid space. The flow of H^+ down the resulting electrochemical gradient through a channel within a protein (ATP synthase complex) provides the energy for ATP formation.

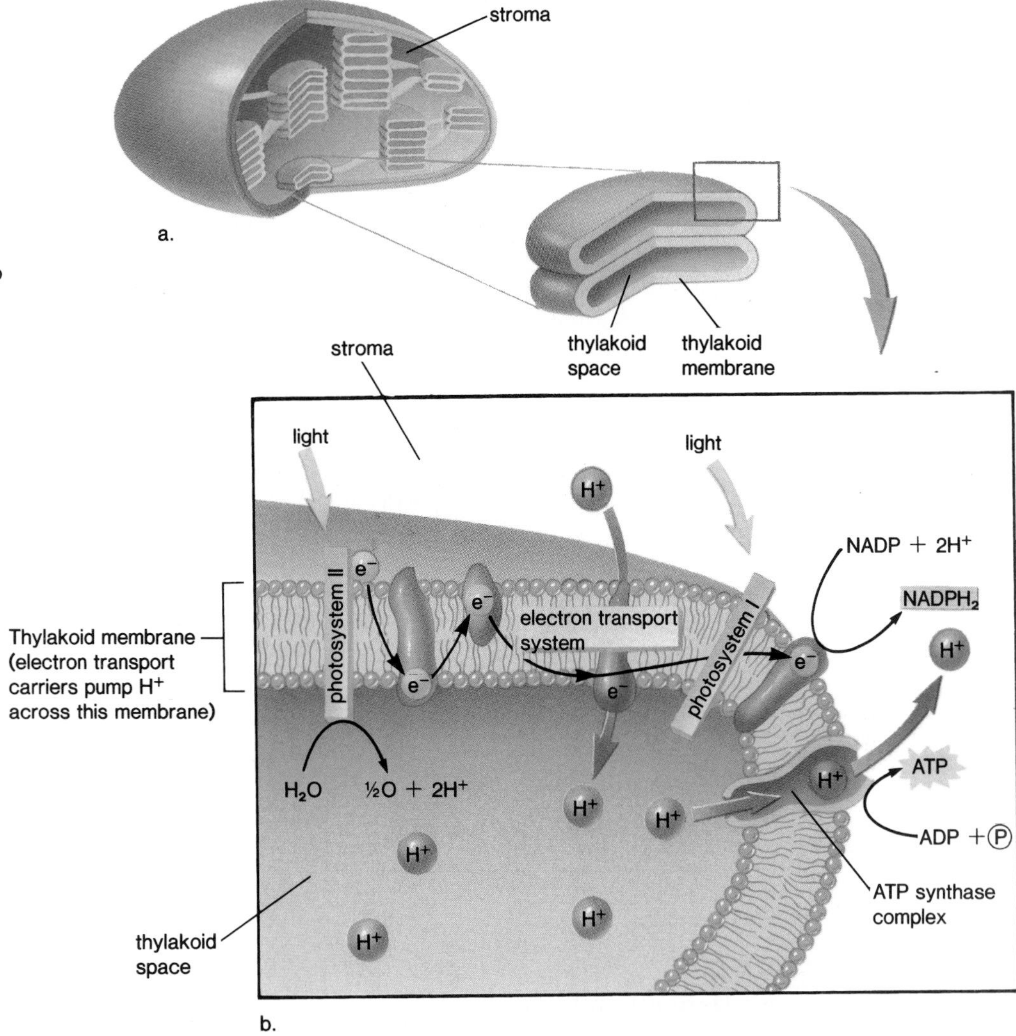

captured by an electron acceptor. This time, the electron acceptor passes the electrons on to NADP. NADP now takes on $2H^+$ and becomes $NADPH_2$.

The $NADPH_2$ and ATP produced by the noncyclic flow of electrons in the thylakoid membrane are used by enzymes in the stroma during the light-independent reactions.

The noncyclic flow of electrons from water to NADP has these effects:

Water is split, yielding H^+, e^-, and O_2;
ATP is produced from ADP + Ⓟ;
NADP becomes $NADPH_2$.

Chemiosmosis: The Making of ATP

As mentioned before, the *thylakoid space* acts as a reservoir for hydrogen ions (H^+). First, each time water is split, 2 H^+ remain in the thylakoid space. Second, as the electrons move from carrier to carrier in the electron transport system, they give up energy, which is used to pump H^+ from the stroma into the thylakoid space (fig. 6.4). Therefore, there is a large number of hydrogen ions in the thylakoid space compared to the number in the stroma. The hydrogen ions diffuse down their concentration gradient through special channels within proteins, each one called an *ATP synthase complex*. This flow of H^+ from high to low concentration provides the energy that allows the ATP synthase complex to enzymatically produce ATP from ADP + Ⓟ.

This method of producing ATP is called **chemiosmotic ATP synthesis** because ATP production is tied to an electrochemical gradient. And it is said that hydrogen ions flow down their electrochemical gradient, through the ATP synthase complex. (Hydrogen ions in this context are often referred to as *protons*.)

Figure 6.5
Light-independent reactions. Each turn of the Calvin cycle uses one CO_2, 3 ATP, and 2 $NADPH_2$ to produce 2 PGAL. Because 5 PGAL are needed to re-form 3 RuBP, it takes 3 turns of the cycle to have a net gain of one PGAL, which can be used to form glucose ($C_6H_{12}O_6$).

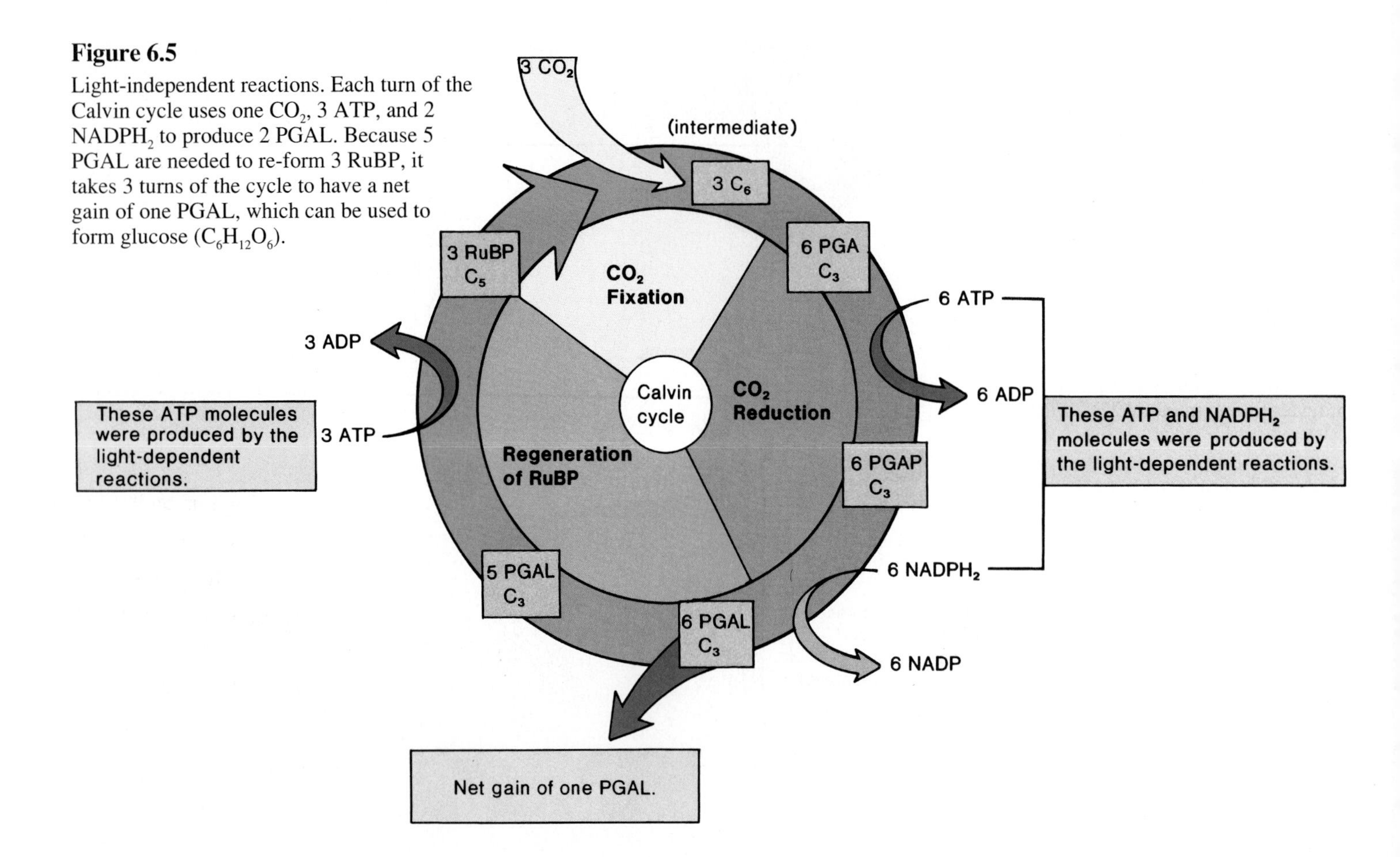

Reactions That Don't Depend on Light

The **light-independent reactions** use $NADPH_2$ and ATP from the light-dependent reactions to reduce carbon dioxide to form a carbohydrate:

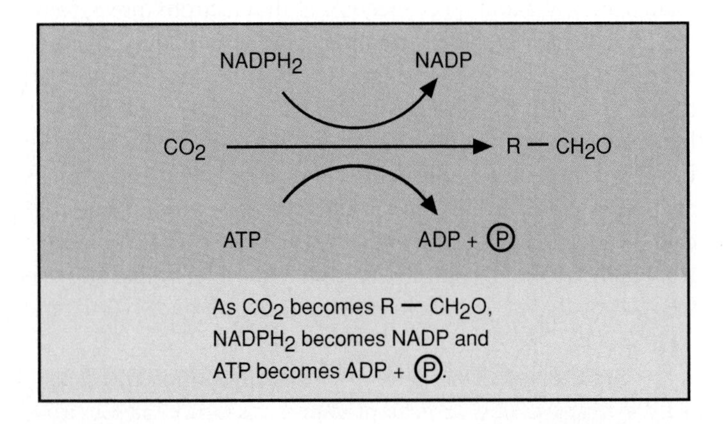

The reduction of carbon dioxide occurs in the stroma of the chloroplast by means of a series of reactions known as the Calvin cycle. Although this cycle does not require light, it is most likely to occur during the day, when a plant is producing high levels of ATP and $NADPH_2$.

Calvin Cycle: CO_2 Becomes Carbohydrate

The **Calvin cycle** is named for one of the individuals who was instrumental in discovering the reactions that make up the cycle. The cycle can be divided into 3 stages: carbon dioxide fixation, carbon dioxide reduction, and **RuBP (ribulose bisphosphate)** regeneration (fig. 6.5).

Fixing Carbon Dioxide. Carbon dioxide fixation, which is the incorporation of carbon dioxide (CO_2) into organic compounds, occurs during the first reaction of the Calvin cycle. At that time RuBP, a 5-carbon molecule, combines with carbon dioxide. The enzyme that catalyzes this reaction is RuBP carboxylase, often called rubisco for short. Rubisco makes up more than 15% of all the protein in a chloroplast, and some say it is the most abundant protein on earth. Rubisco works slowly compared to other enzymes, which accounts for its abundance in plant cells.

The 6-carbon molecule resulting from carbon dioxide fixation immediately breaks down to form 2 *PGA* (*phosphoglycerate*), 3-carbon molecules. Because the first detectable molecule in the Calvin cycle is a 3-carbon (C_3) molecule, it is also known as the C_3 cycle.

Reducing Carbon Dioxide. Each of the 2 PGA molecules undergoes reduction to **PGAL (phosphoglyceraldehyde)** in 2 steps:

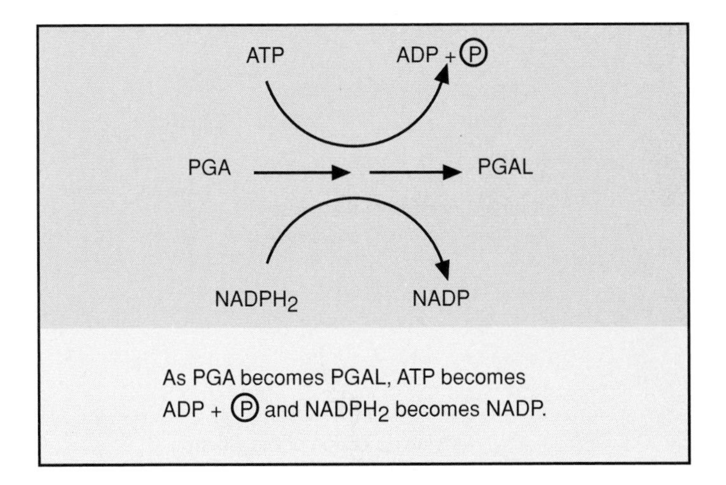

As PGA becomes PGAL, ATP becomes ADP + Ⓟ and $NADPH_2$ becomes NADP.

This is the actual reaction that uses $NADPH_2$ and ATP from the light-dependent reactions, and it signifies the reduction of carbon dioxide (CO_2) to a carbohydrate (R–CH_2O). The reduction of carbon dioxide is a synthetic process because it requires the formation of new bonds. Hydrogen atoms and energy are needed for reduction synthesis, and these are supplied by $NADPH_2$ and ATP, respectively.

Regenerating RuBP. For every 3 turns of the Calvin cycle, 5 molecules of PGAL are used to re-form 3 molecules of RuBP so that the cycle can continue:

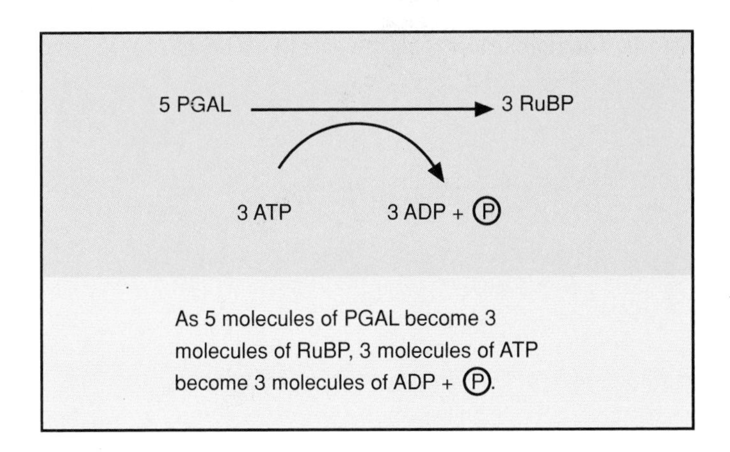

As 5 molecules of PGAL become 3 molecules of RuBP, 3 molecules of ATP become 3 molecules of ADP + Ⓟ.

The net gain of 3 turns of the Calvin cycle is one PGAL molecule.

This reaction also utilizes some of the ATP produced by the light-dependent reactions.

PGAL: The End Product

PGAL, the end product of the Calvin cycle, is converted to all sorts of organic molecules. In comparison to animal cells, algae and plants have enormous biochemical capabilities. They use PGAL for these purposes:

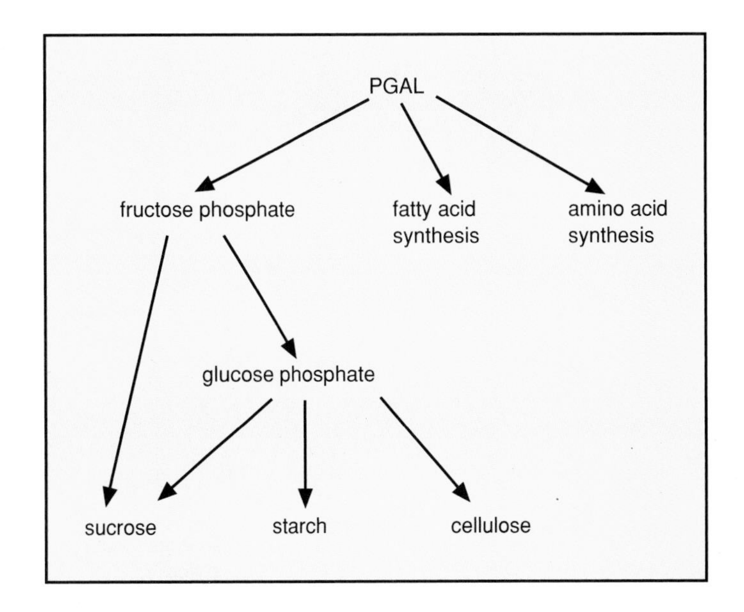

As shown here, *glucose phosphate* is among the organic molecules that result from PGAL metabolism. This is of interest to us because glucose is the molecule that plants and animals most often use to produce the ATP molecules they require for their night and day energy needs.

During the light-independent reactions, ATP and $NADPH_2$ are utilized to reduce carbon dioxide (CO_2) to PGAL within the stroma. It takes 2 PGAL to form glucose phosphate. To obtain these 2 PGAL molecules requires 6 turns of the Calvin cycle. Do 6.1 Critical Thinking, found at the end of the chapter.

Figure 6.6

Mitochondrion structure and function. A mitochondrion is bounded by a double membrane, with an intermembrane space. The inner membrane invaginates to form the shelflike cristae. The matrix is the fluid-filled interior of the mitochondrion. Note that glycolysis occurs outside the mitochondrion, while the Krebs cycle occurs in the matrix and the electron transport system is located on the cristae, both within the mitochondrion.

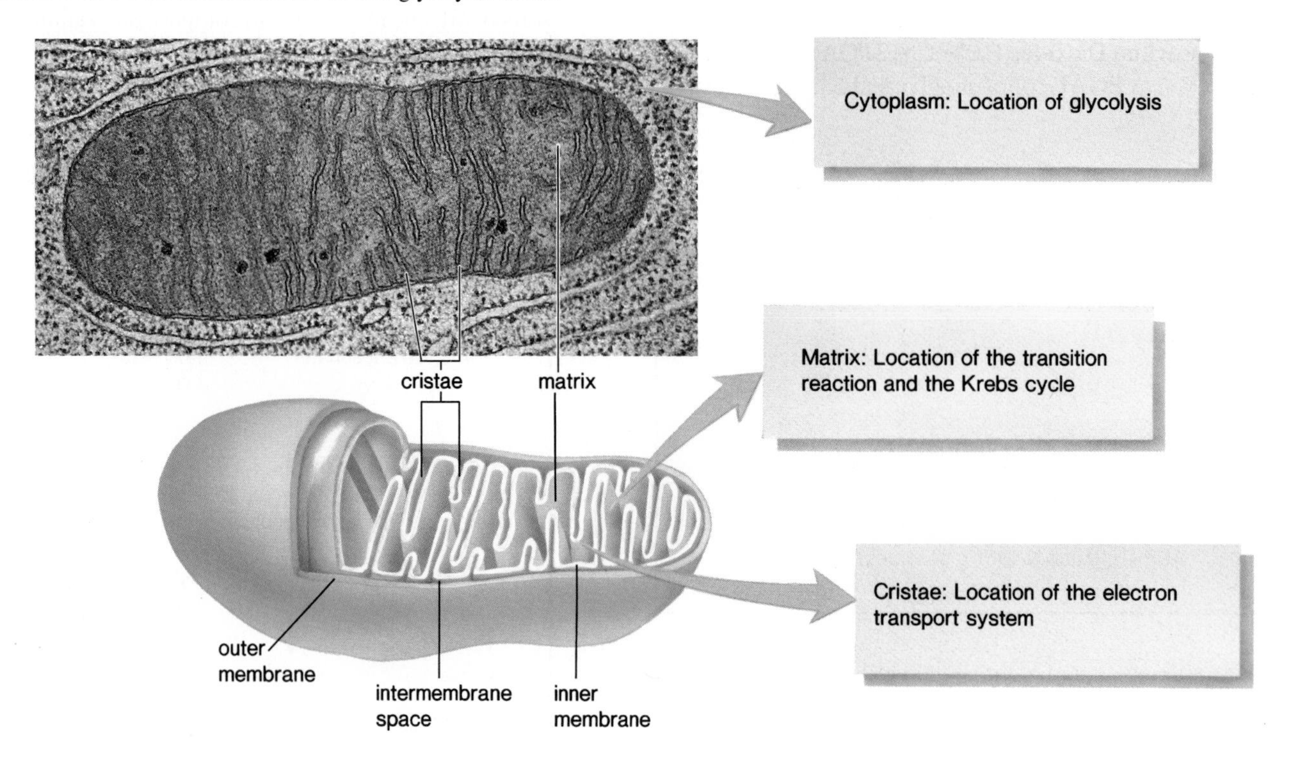

Cellular Respiration

Cellular respiration includes 2 metabolic processes: aerobic cellular respiration and fermentation. **Aerobic cellular respiration** is the series of enzymatic reactions by which glucose is completely metabolized to carbon dioxide and water with the resultant buildup of 38 ATP molecules. The complete breakdown of glucose requires oxygen, as implied by the term **aerobic.**

Fermentation is a series of enzymatic reactions by which glucose is incompletely metabolized to lactate (animals) or carbon dioxide and alcohol (yeast). Fermentation is **anaerobic** and does not require oxygen; however, it results in a net gain of only 2 ATP.

Aerobic Cellular Respiration

Cells would lose a tremendous amount of energy if they used glucose directly—it contains far too much chemical energy for individual reactions and therefore much energy would become nonusable heat. Also, the cell is far too delicate to utilize all the energy released in one burst. Most organisms, including all plants and animals, carry on aerobic cellular respiration, a process that requires many steps. Gradual glucose breakdown releases energy slowly and allows for the simultaneous buildup of many ATP.

Aerobic cellular respiration requires 3 individual pathways: **glycolysis,** the **Krebs cycle,** and an **electron transport system.** The **transition reaction** acts like a bridge connecting glycolysis with the Krebs cycle.

Mitochondria: Where It Happens

A mitochondrion has a double membrane, with an *intermembrane space* (between the outer and inner membrane) (fig. 6.6). Cristae are folds of the inner membrane that jut out into the *matrix,* the innermost compartment, which is filled with a gel-like fluid.

The Krebs cycle and the electron transport system occur in mitochondria. The Krebs cycle enzymes are located in the matrix and the electron transport system carriers are located on the cristae. The electron transport system is directly involved in ATP production; therefore, mitochondria are often called the powerhouses of the cell.

Figure 6.7

Relationship of aerobic cellular respiration to the body proper. Glucose and oxygen are delivered to the cells by the bloodstream. Carbon dioxide and water are removed by the bloodstream. The water can remain in the cell or leave as needed. ATP remains in the cytoplasm as a source of energy for the cell to do work.

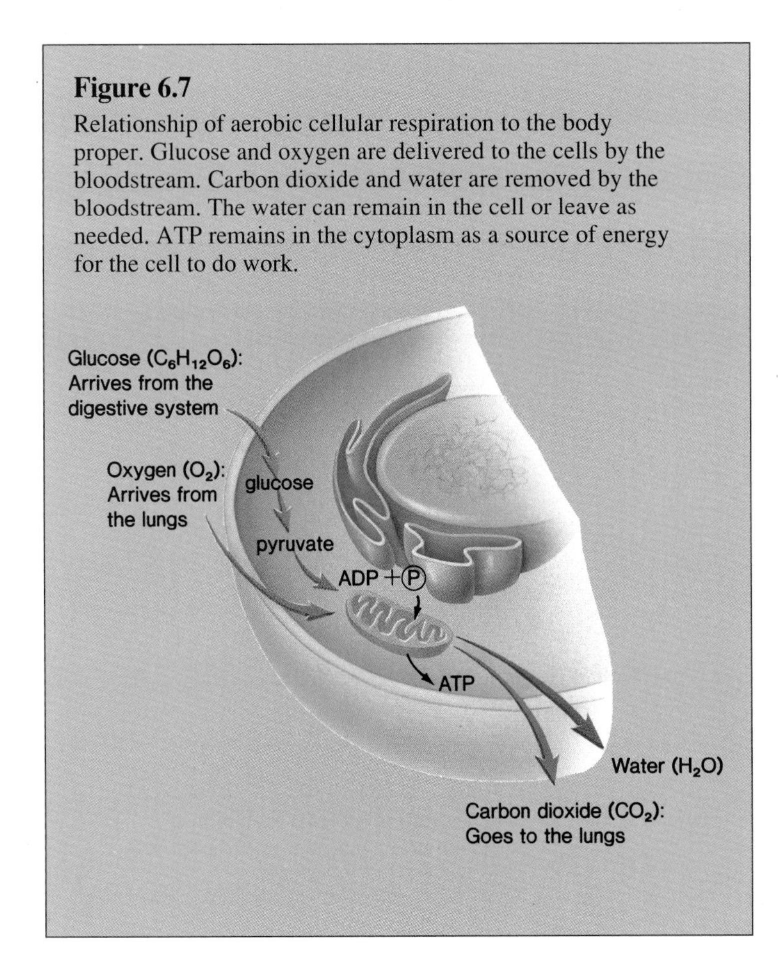

It is interesting to think about how our bodies provide the reactants for complete glucose breakdown and how they dispose of the products (fig. 6.7). The air we breathe contains *oxygen* (O_2), and the food we eat contains *glucose*. These enter the bloodstream, which carries them about the body, and they diffuse into each and every cell. Glycolysis occurs in the cytoplasm, and the end product, pyruvate, enters the mitochondria. In mitochondria, pyruvate is broken down to carbon dioxide (CO_2) and water (H_2O) as ATP is produced. All 3 of these diffuse out of the mitochondria into the cytoplasm. The *ATP* is utilized inside the cytoplasm for energy-requiring processes. Carbon dioxide diffuses out of the cell and enters the bloodstream. The bloodstream takes the *carbon dioxide* to the lungs, where it is exhaled. The *water* molecules, called metabolic water, become important if they are the organism's only supply of water. In these cases, metabolic water can help to prevent dehydration of the organism.

Glycolysis: Glucose to Pyruvate

Glycolysis is a series of enzymatic reactions. Glucose is broken down to **pyruvate,** and this releases enough energy to produce small amounts of ATP and $NADH_2$ (fig. 6.8). Since glycolysis is universally found in organisms, it most likely evolved before the Krebs cycle and the electron transport system. This may be why glycolysis occurs in the cytoplasm and does not require oxygen. Bacteria evolved before other organisms, and there are some bacteria today that are anaerobic; they die in the presence of oxygen.

As glycolysis begins, the addition of 2 phosphate groups readies *glucose*, a 6-carbon molecule, to react. This requires 2 separate reactions and uses 2 ATP. Also, at one point, hydrogen atoms (H) are removed from the substrates of the pathway and are picked up by NAD. Altogether, 2 $NADH_2$ are produced as substrate oxidation occurs. This releases enough energy to allow the formation of 4 ATP. Subtracting the 2 ATP that were used to get started, glycolysis results in a net gain of 2 ATP.

When glycolysis is part of aerobic cellular respiration, pyruvate enters the mitochondria, where oxygen (O_2) is utilized. However, glycolysis does not need to be a part of *aerobic* cellular respiration. As we will see on page 95, glycolysis is also part of fermentation, an *anaerobic* process—it does not require oxygen.

Glycolysis, which takes place within the cytoplasm, breaks down glucose to smaller fragments and produces small amounts of ATP and $NADH_2$. The following are the result of glycolysis:

Glucose becomes 2 pyruvate;
net gain of 2 ATP results;
2 $NADH_2$ are formed.

Transition Reaction: Pyruvate to Acetyl Groups

The **transition reaction,** which occurs within the matrix of mitochondria (fig. 6.6), produces an activated product and additional $NADH_2$. Specifically, pyruvate is converted to a 2-carbon *acetyl group,* and carbon dioxide (CO_2) is given off in the process:

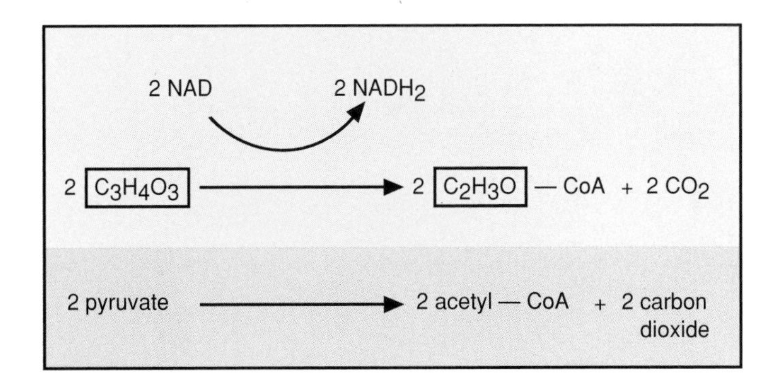

This is an oxidation reaction in which hydrogen atoms (H) are removed from pyruvate by a dehydrogenase that uses NAD as a coenzyme. The activated acetyl group is picked up by another coenzyme called **coenzyme A,** or CoA, and *acetyl-CoA* forms.

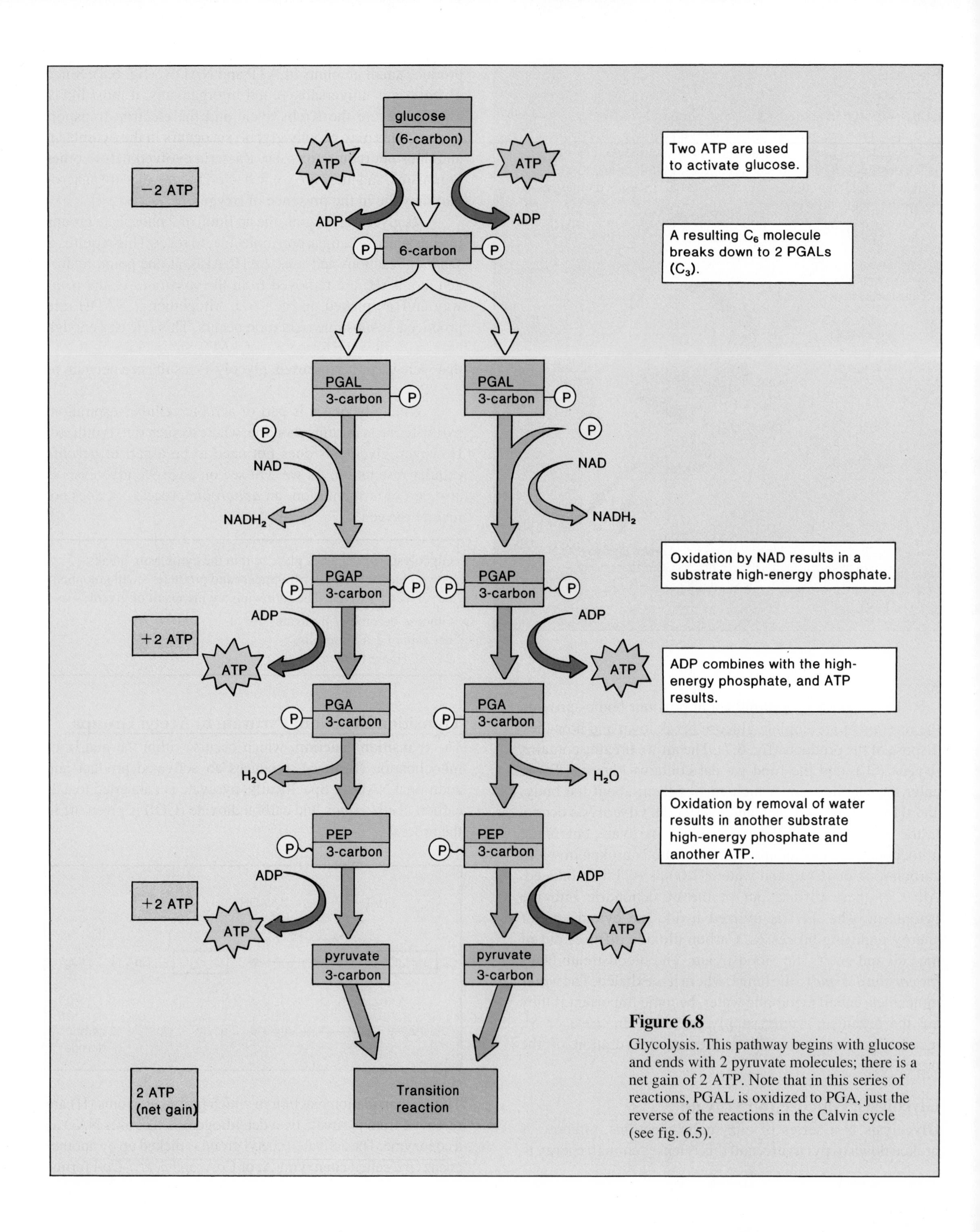

Figure 6.8

Glycolysis. This pathway begins with glucose and ends with 2 pyruvate molecules; there is a net gain of 2 ATP. Note that in this series of reactions, PGAL is oxidized to PGA, just the reverse of the reactions in the Calvin cycle (see fig. 6.5).

Figure 6.9

The Krebs cycle. The cycle begins and ends with citrate. The cycle turns twice per glucose molecule.

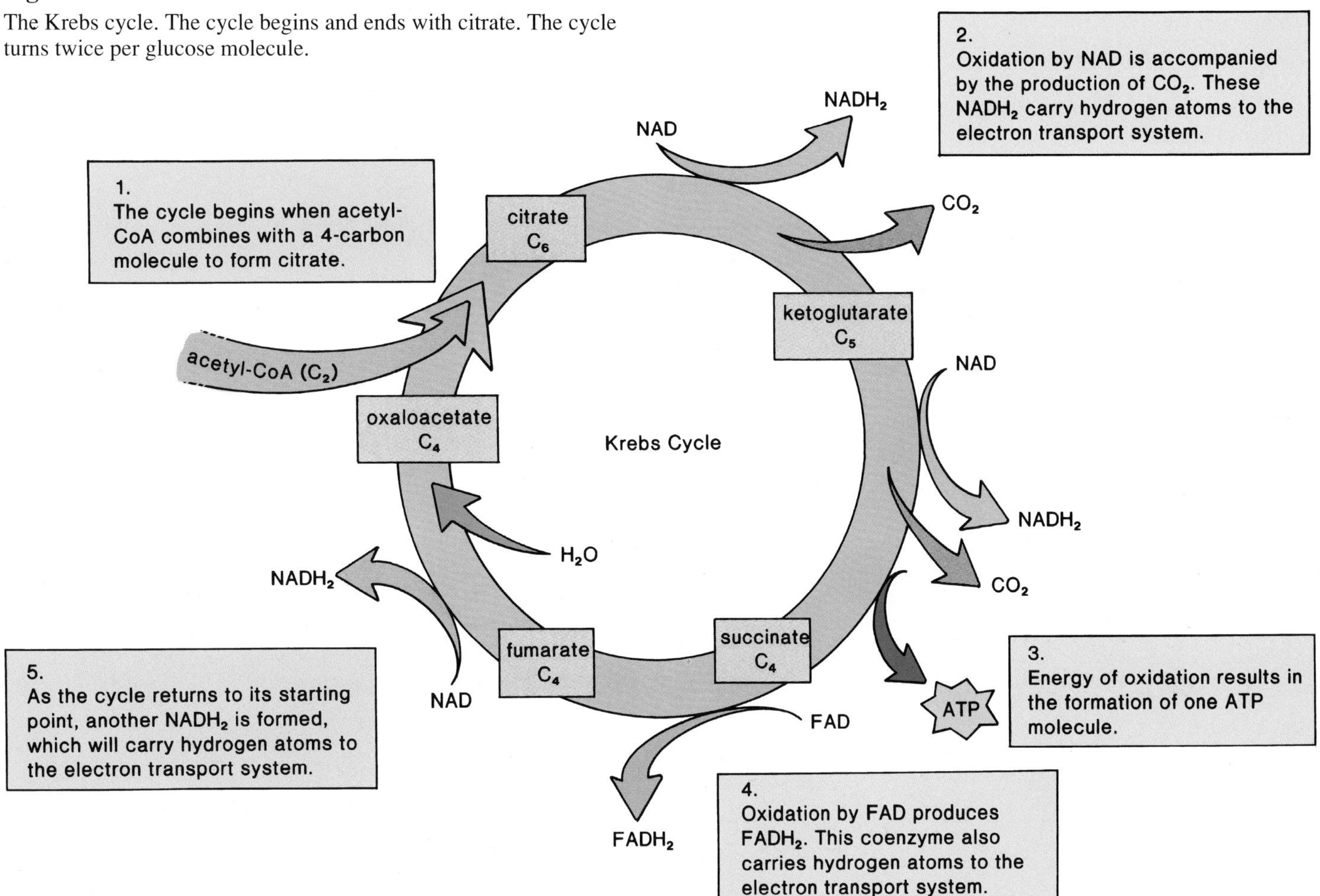

The carbon dioxide is a metabolic waste and is excreted by cells. The reaction occurs twice per glucose molecule.

The transition reaction, which occurs in the matrix of mitochondria, follows glycolysis. The following are the result of the transition reaction (per glucose molecule):

Two pyruvate become 2 acetyl-CoA;
2 CO_2 are given off;
2 $NADH_2$ are formed.

Krebs Cycle: Acetyl Groups to CO_2

During the **Krebs cycle,** as each acetyl group from a transition reaction is completely oxidized, activated products and carbon dioxide (CO_2) result. The Krebs cycle, named for the person who discovered it, is called a cycle because it involves a series of enzymatic reactions that begin and end with *citrate* (citric acid).

The hydrogen atoms (H) removed during the Krebs cycle are accepted by NAD and **FAD,**[1] a coenzyme that is used infrequently compared to NAD (fig. 6.9). Oxidation releases the energy used to form one ATP per turn. Because there are 2 transition reactions per glucose molecule, the Krebs cycle must turn twice per glucose molecule.

The Krebs cycle, which occurs in the matrix of mitochondria, takes up each acetyl group (acetyl-CoA) from a transition reaction and oxidizes it to CO_2. The Krebs cycle turns twice per glucose molecule with these results:

Two acetyl groups are oxidized to 4 CO_2;
6 $NADH_2$ are formed;
2 $FADH_2$ are formed;
2 ATP are produced.

---

1. FAD (flavin adenine dinucleotide)

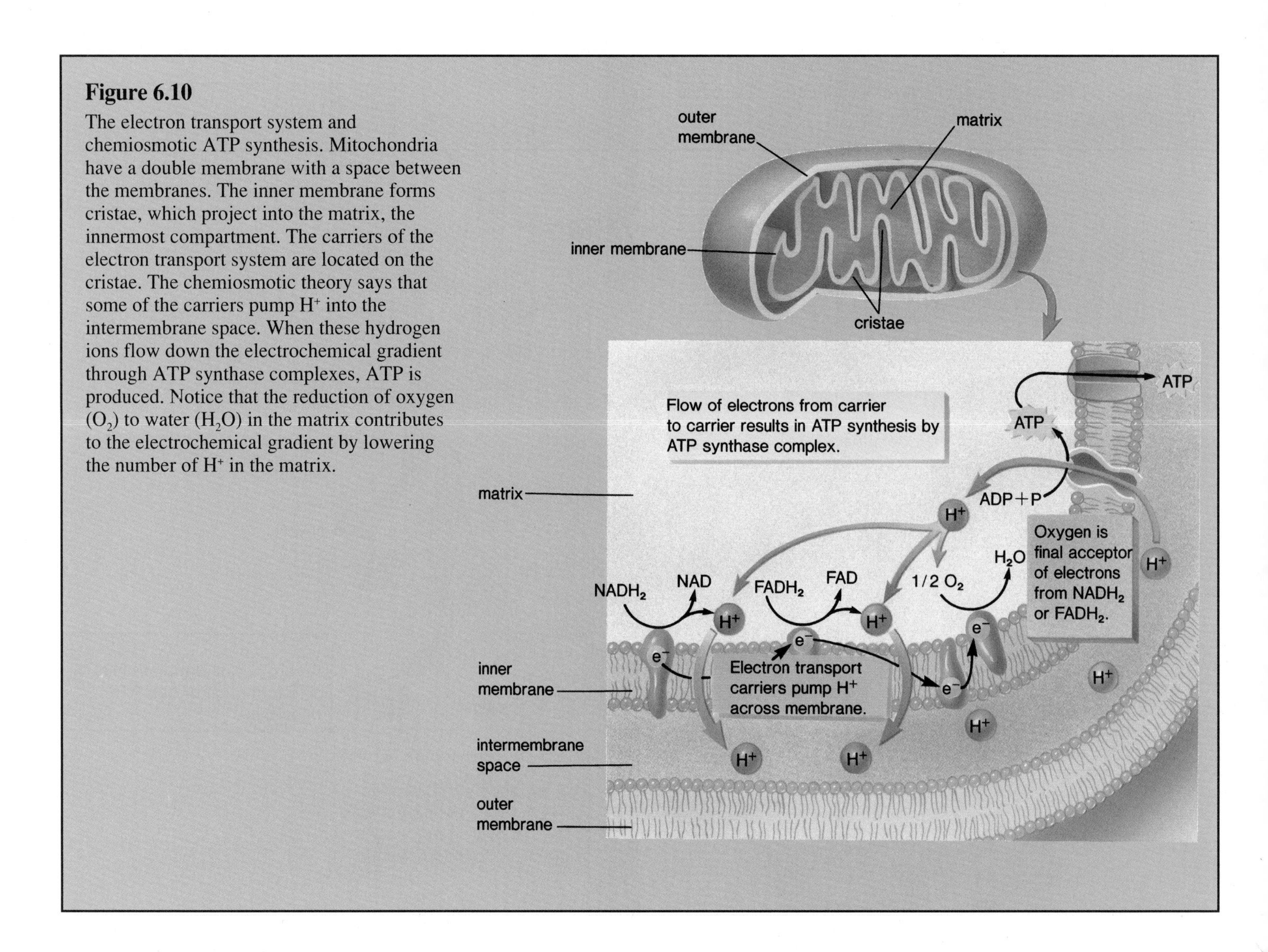

Figure 6.10
The electron transport system and chemiosmotic ATP synthesis. Mitochondria have a double membrane with a space between the membranes. The inner membrane forms cristae, which project into the matrix, the innermost compartment. The carriers of the electron transport system are located on the cristae. The chemiosmotic theory says that some of the carriers pump H^+ into the intermembrane space. When these hydrogen ions flow down the electrochemical gradient through ATP synthase complexes, ATP is produced. Notice that the reduction of oxygen (O_2) to water (H_2O) in the matrix contributes to the electrochemical gradient by lowering the number of H^+ in the matrix.

Electron Transport System: Electrons Pass to Oxygen

The **electron transport system,** located on the cristae of the mitochondria, establishes a large electrochemical gradient that leads to ATP production (fig. 6.10). As in chloroplasts, the electron transport system is a series of carriers that pass electrons from one carrier to another. Some of the carriers are cytochrome molecules, and the system sometimes is called a cytochrome system. The electrons that enter the electron transport system are at first a part of the hydrogen atoms ($e^- + H^+$) attached to NAD or FAD. These are the same hydrogen atoms that were removed from the metabolites of glycolysis, the transition reaction, and the Krebs cycle.

$NADH_2$ and $FADH_2$ carry hydrogen atoms ($e^- + H^+$) to the electron transport system, but certain carriers accept only the electrons. The hydrogen ions (H^+) are deposited in the intermembrane space, which lies between the outer and inner membranes surrounding the mitochondrion. The accumulation of hydrogen ions in this space, which acts as a *hydrogen ion reservoir,* results in a large electrochemical gradient. As the hydrogen ions move down their concentration gradient through channels within proteins embedded in the membrane of cristae, energy is provided for ATP production. Each of the channel proteins is an *ATP synthase complex,* which can enzymatically join ADP with Ⓟ to form ATP. As in photosynthesis, this method of producing ATP is called **chemiosmotic ATP synthesis** because ATP production is tied to an electrochemical gradient.

Oxygen (O_2) is the final acceptor of electrons at the end of the electron transport system. After accepting electrons, oxygen combines with hydrogen ions from the intermembrane space, and water forms:

$$1/2\ O_2 + 2\,e^- + 2\,H^+ \longrightarrow H_2O$$

Table 6.1
ATP Produced per Glucose Molecule

	Direct	By Way of Electron Transport System	
Glycolysis	2 ATP	2 $NADH_2$ =	6 ATP*
Transition reaction		2 $NADH_2$ =	6 ATP
Krebs cycle	2 ATP	6 $NADH_2$ =	18 ATP
		2 $FADH_2$ =	4 ATP
Subtotals	4 ATP		34 ATP
Grand Total		38 ATP*	

*The numbers in this column and the total number of ATP are usually less because the electron transport system does not always produce the maximum possible number of ATP per $NADH_2$.

Total ATP: Most from Electron Transport

In order to calculate the total number of ATP produced per glucose molecule, it is customary to first consider the number of ATP that are made directly during glycolysis and the Krebs cycle (first column, table 6.1). Then, we must realize that for every $NADH_2$ that enters the electron transport system, 3 ATP result. For every $FADH_2$ that enters the electron transport system, 2 ATP result. (Examine figure 6.10 and note that $FADH_2$ enters the electron transport system after $NADH_2$. Therefore, fewer hydrogen ions are transported into the intermembrane space following delivery by $FADH_2$ than $NADH_2$.) Using this method of calculation, we arrive at a total of 38 ATP per glucose molecule (table 6.1).

Thirty-eight is the maximum number of ATP produced per glucose molecule. In many cells, $NADH_2$ from glycolysis cannot cross the mitochondrial membrane, and its hydrogen atoms are shuttled across by a carrier that sometimes delivers them to an FAD instead of an NAD. In that case, for every cytoplasmic $NADH_2$ formed, only 2 ATP are produced (instead of 3), and the grand total of ATP per glucose molecule is then 36 instead of 38.

> The electron transport system is located in the cristae of mitochondria. The following are the result of the electron transport system (per glucose molecule):
>
> O_2 accepts electrons and becomes H_2O;
> 32–34 ATP are produced.
> Do 6.2 Critical Thinking, found at the end of the chapter.

Figure 6.11 summarizes our discussion of aerobic cellular respiration.

Figure 6.11
The pathways of aerobic cellular respiration. The process requires these 3 pathways plus the transition reaction.

Glycolysis: Glucose is broken down to 2 pyruvate (PYR) for a net gain of 2 ATP. The hydrogen atoms (H) removed are taken to the electron transport system via $NADH_2$.

Transition reaction: Carbon dioxide (CO_2) and hydrogen atoms (H) are removed from pyruvate. Acetyl-CoA results. The hydrogen atoms are taken to the electron transport system via $NADH_2$.

Krebs cycle: Acetyl-CoA enters the cycle; carbon dioxide (CO_2) and hydrogen atoms (H) exit the cycle. There is a gain of 2 ATP per glucose.

Electron transport system: The electron transport system receives hydrogen atoms ($e^- + H^+$). The electrons pass from carrier to carrier (cytochromes are carriers), releasing energy, and 34 ATP result. Oxygen (O_2) acts as the final acceptor of electrons, and then water (H_2O) results.

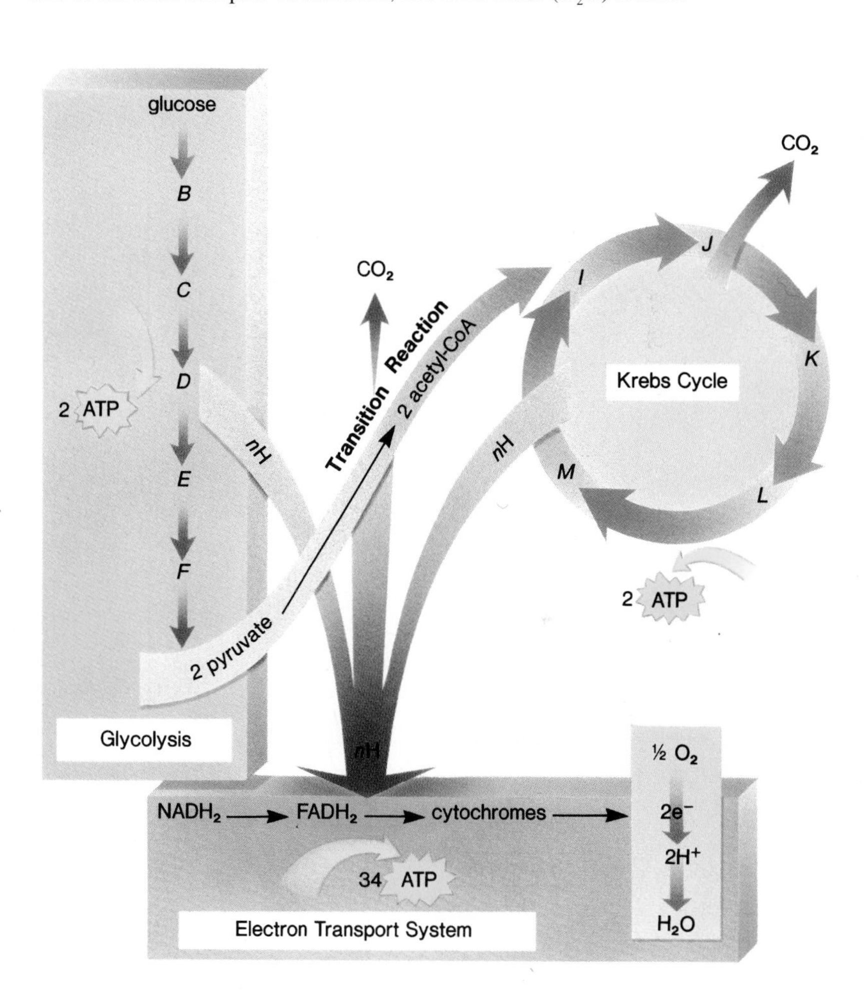

Fermentation: Backup ATP Production

Fermentation is an anaerobic process that does not require oxygen. It consists of glycolysis plus one other reaction—the reduction of pyruvate to either lactate (animal cells) or alcohol and carbon dioxide (yeast cells) (fig. 6.12). Other end products are possible, depending on the type of enzyme in the particular organism.

Figure 6.12

Fermentation, an anaerobic process. Fermentation in animals consists of glycolysis plus an additional reaction in which the end product of glycolysis, pyruvate, accepts hydrogen atoms (H) and is reduced. This "frees" NAD so that it can return to pick up more hydrogen atoms.

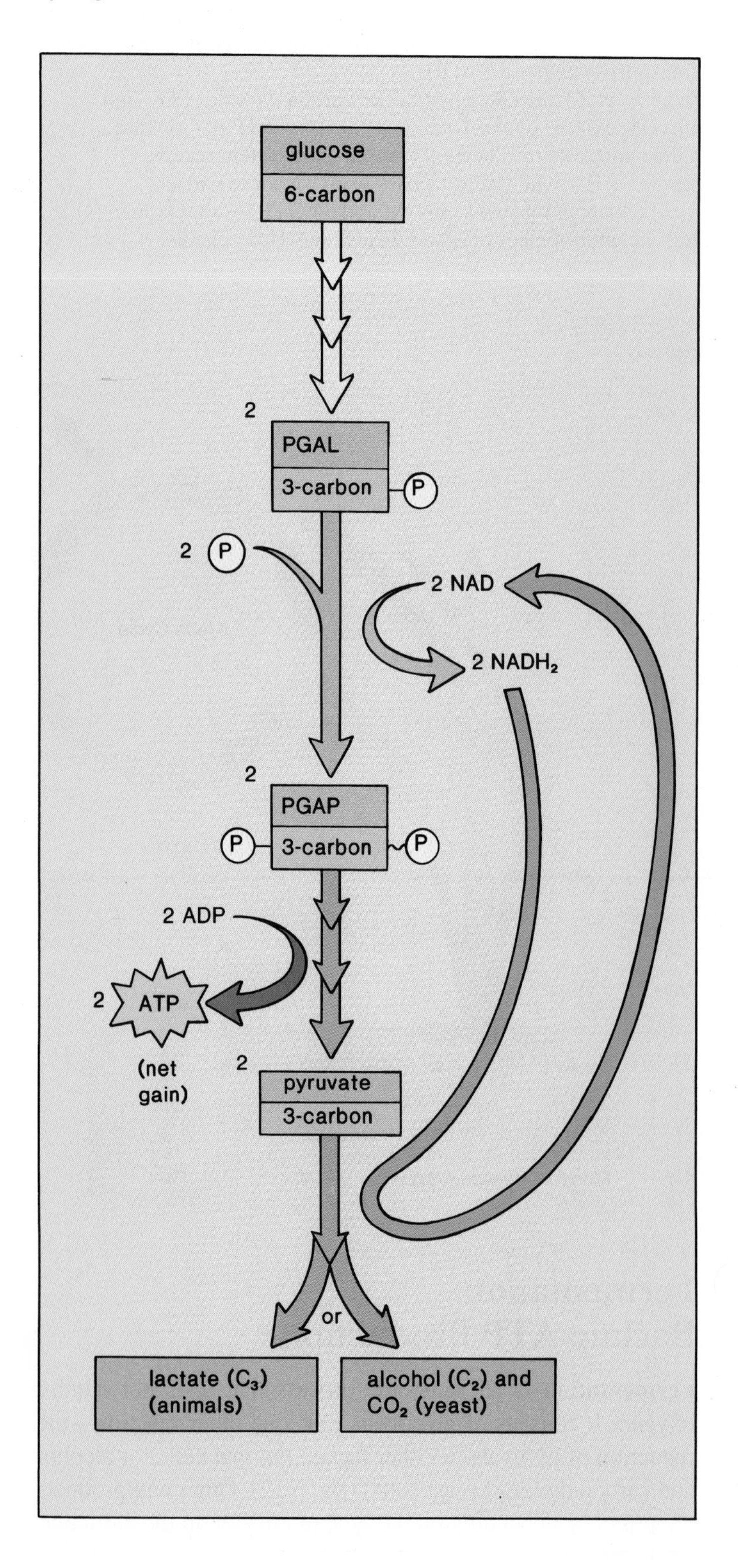

If oxygen (O_2) is not available to cells, the electron transport system soon becomes inoperative. Electrons back up in the system when oxygen, the final acceptor, is not present. In this case, cells still produce ATP because they resort to fermentation. However, fermentation results in only 2 ATP per glucose molecule. The end products of the process (lactate or alcohol) are toxic to cells, and high concentrations can result in the death of the organism.

Glycolysis still occurs during fermentation because $NADH_2$ passes its hydrogen atoms to pyruvate instead of to the electron transport system. Now NAD is "free" to return and pick up more hydrogen atoms.

Advantages and Disadvantages

Despite its low yield, fermentation is important because it can provide a rapid burst of ATP. In our own bodies, muscle cells more than other cells are apt to carry on fermentation. When our muscles are working vigorously over a short period of time, as when we run, fermentation is a way to produce ATP even though oxygen (O_2) is temporarily in limited supply. At first, blood carries away all the lactate formed in the muscles. Eventually, however, lactate begins to build up in the muscles, changing the pH and causing the muscles to fatigue so that they no longer contract. When we stop running, our bodies are in oxygen debt, as signified by the fact that we keep on breathing very heavily for a time. Recovery is complete when the lactate is transported to the liver, where it is reconverted to pyruvate. Some of the pyruvate is respired completely, and the rest is converted back to glucose ($C_6H_{12}O_6$).

Because of fermentation, yeast cells are capable of growing and dividing anaerobically for a time. If the initial glucose level is high, however, the cells are eventually killed by the very alcohol they produce. Presumably, human beings were delighted to discover this form of fermentation, as the ethyl alcohol produced has been consumed in great quantity for thousands of years.

Fermentation is anaerobic; it requires glycolysis and then the reduction of pyruvate by $NADH_2$, which frees NAD molecules. Per glucose molecule, the results of fermentation are as follows:

Two lactate or 2 alcohol and 2 CO_2 are produced;
2 ATP are produced.

Metabolic Pool and Biosynthesis

Some reactions, called catabolic reactions, break down molecules, and others, called anabolic (or synthetic) reactions, build them up. It is correct to say that *catabolic reactions drive anabolic reactions* because the ATP that results from catabolism is used during anabolism when molecules and macromolecules are synthesized. Anabolism produces all the substances that a cell produces whether they are structural components, such as membranous organelles, or secretory products, such as hormones or digestive enzymes.

Catabolic Reactions: Breaking Down

We already know that glucose is broken down during aerobic cellular respiration. Other molecules can also undergo catabolism, however. When a fat is used as an energy source, it breaks down to glycerol and 3 fatty acids. As figure 6.13 indicates, glycerol is converted to PGAL, a metabolite in glycolysis. The fatty acids are converted to acetyl-CoA, which enters the Krebs cycle. An 18-carbon fatty acid results in 9 acetyl-CoA. Calculation shows that respiration of these can produce a total of 216 ATP. For this reason, fats are an efficient form of stored energy—there are 3 long fatty acid chains per fat molecule.

The carbon skeleton of amino acids can also be broken down. The skeleton is produced in the liver when an amino acid undergoes **deamination,** or the removal of the amino group. The amino group becomes ammonia (NH_3), which enters the urea cycle and becomes part of urea, the primary excretory product of humans. Just where the carbon skeleton begins degradation is dependent on the length of the *R* group, since this determines the number of carbons left after deamination.

Anabolic Reactions: Building Up

We have already mentioned that the ATP produced during catabolism drives anabolism. But there is another way catabolism is related to anabolism. The substrates making up the pathways in figure 6.13 can be used as starting materials for synthetic reactions. In other words, compounds that enter the pathways are oxidized to substrates that can be used for biosynthesis. This is the cell's **metabolic pool,** in which one type of molecule can be converted to another. In this way, carbohydrate intake can result in the formation of fat. PGAL molecules can be converted to glycerol molecules, and acetyl groups can be joined to form fatty acids. Fat synthesis follows. This explains why you gain weight from eating too much candy, ice cream, and cake.

Some metabolites of the Krebs cycle can be converted to amino acids through transamination, the transfer of an amino group from one amino acid to another. Plants are able to synthesize all of the amino acids they need. Animals, however, lack some of the enzymes necessary for synthesis of all amino acids. Adult humans, for example, can synthesize 11 of the common amino acids, but they cannot synthesize the other 9. The amino acids that cannot be synthesized must be supplied by the diet; they are called the essential amino acids. The nonessential amino acids can be synthesized. It is quite possible for animals to suffer from protein deficiency if their diet does not contain adequate quantities of all the essential amino acids.

All the reactions involved in aerobic cellular respiration are a part of a metabolic pool; the metabolites from the pool can be used for catabolism or for anabolism.

Figure 6.13

Interrelationship of metabolic pathways.

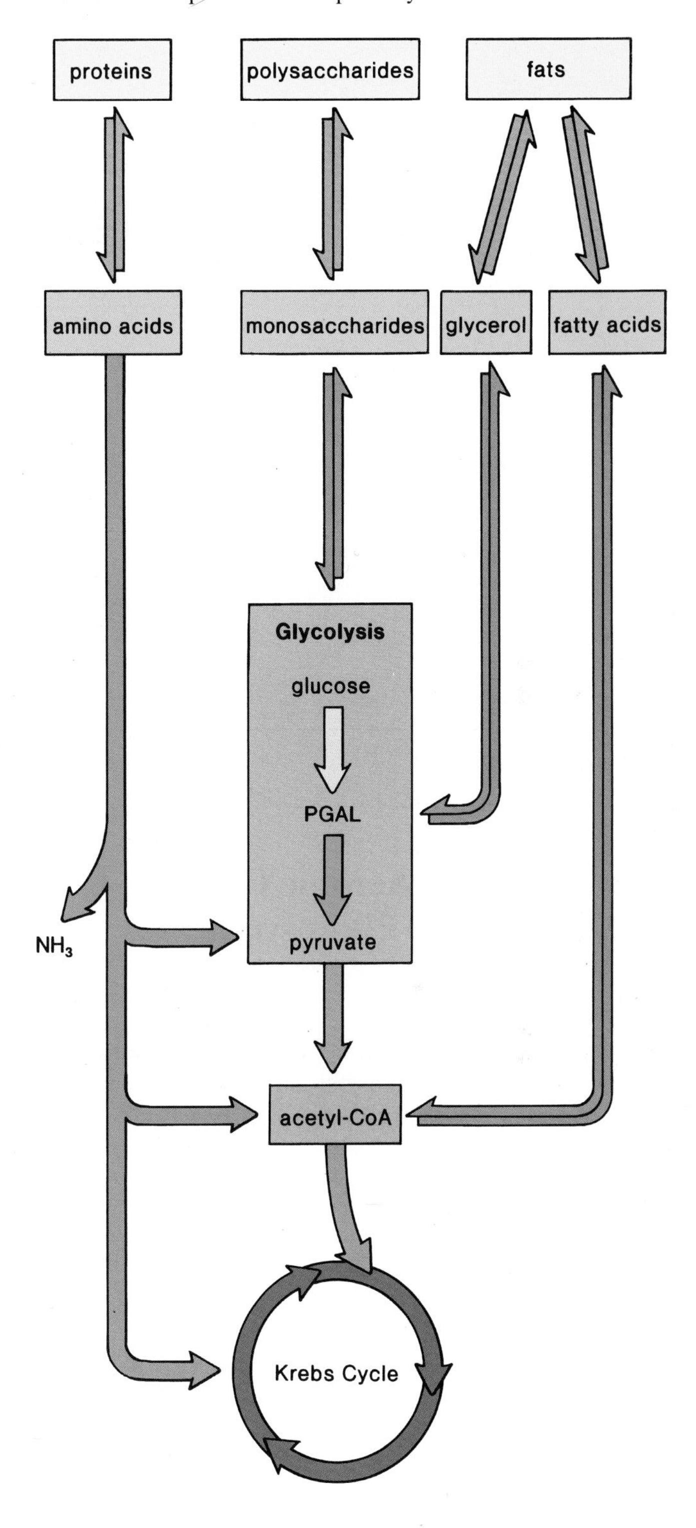

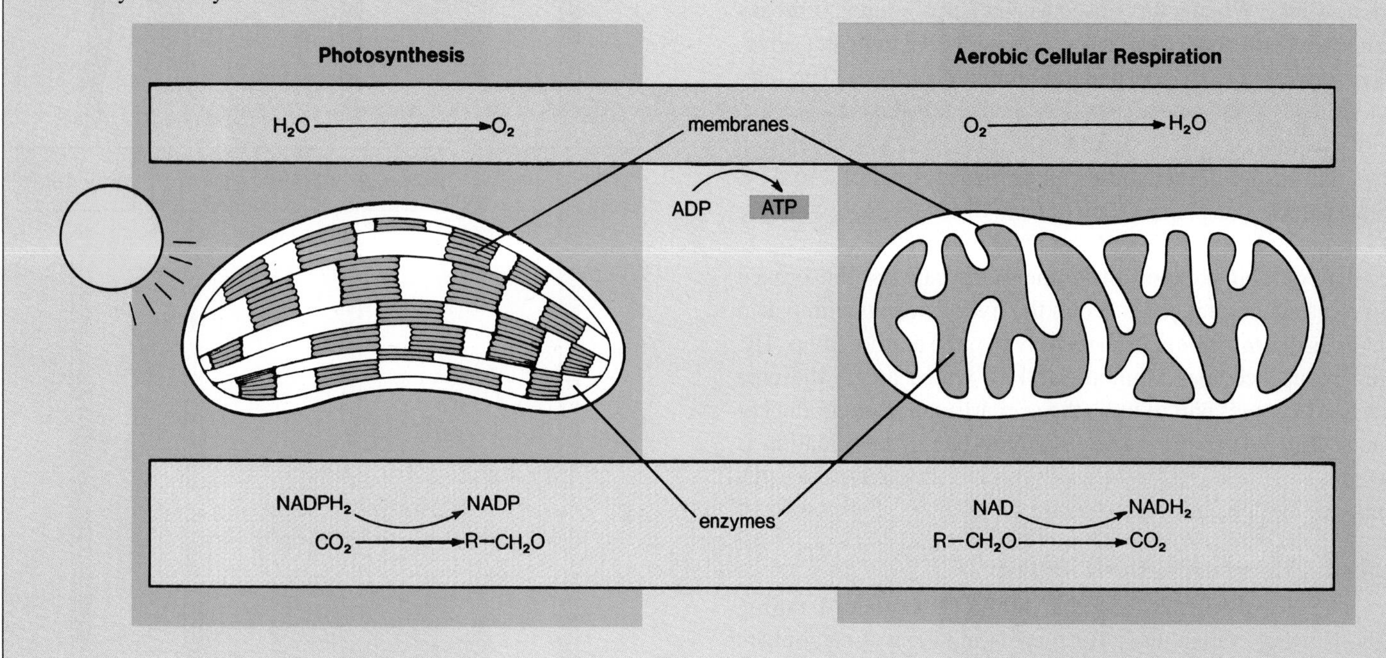

Figure 6.14
Diagram illustrating similarities and differences of photosynthesis, which takes place in chloroplasts, and aerobic cellular respiration, which takes place in mitochondria. Both have an electron transport system, located within membranes, where ATP is produced. Both have enzyme-catalyzed reactions within the fluid interior. The coenzymes NADP (chloroplasts) and NAD (mitochondria) operate in the membrane and within the fluid interior. Photosynthesis releases oxygen (O_2) and reduces carbon dioxide (CO_2) to a carbohydrate; aerobic cellular respiration reduces oxygen and releases carbon dioxide.

Comparison of Aerobic Cellular Respiration and Photosynthesis

Differences

Both plant and animal cells carry on aerobic cellular respiration, but only plant cells photosynthesize. The cellular organelle for aerobic cellular respiration is the mitochondrion, while the cellular organelle for photosynthesis is the chloroplast.

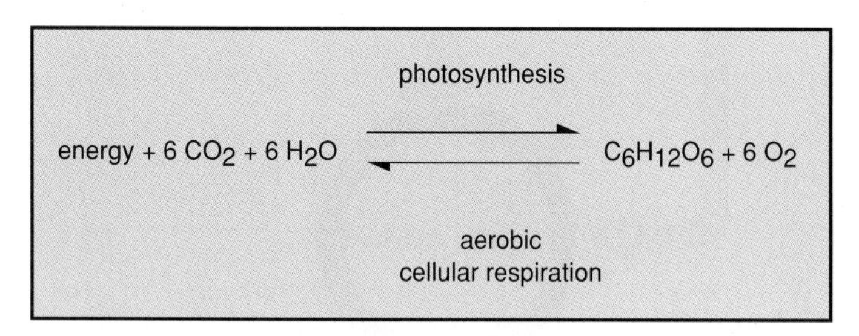

The overall equation for photosynthesis is the opposite of that for aerobic cellular respiration: The reaction in the forward direction represents photosynthesis, and the energy is solar energy. The reaction in the opposite direction represents aerobic cellular respiration, and the energy is ATP.

Obviously, photosynthesis is the building up of glucose, while aerobic cellular respiration is the breaking down of glucose. See table 6.2 for a summarized list of differences between these processes.

Table 6.2
Aerobic Cellular Respiration versus Photosynthesis

Aerobic Cellular Respiration	Photosynthesis
Mitochondrion	Chloroplast
Oxidation	Reduction
Releases energy	Requires energy
Requires O_2	Releases O_2
Releases CO_2	Requires CO_2

Similarities

Both photosynthesis and aerobic cellular respiration are metabolic pathways within cells and therefore consist of a series of reactions that the overall reaction does not indicate. Both pathways make use of an electron transport system located in membranes (fig. 6.14) to generate a supply of ATP, and both make use of a hydrogen carrier—aerobic cellular respiration uses NAD, and photosynthesis uses NADP.

Both pathways utilize this reaction, but in opposite directions. For photosynthesis, read from left to right in the following diagram, and for aerobic cellular respiration, read from right to left.

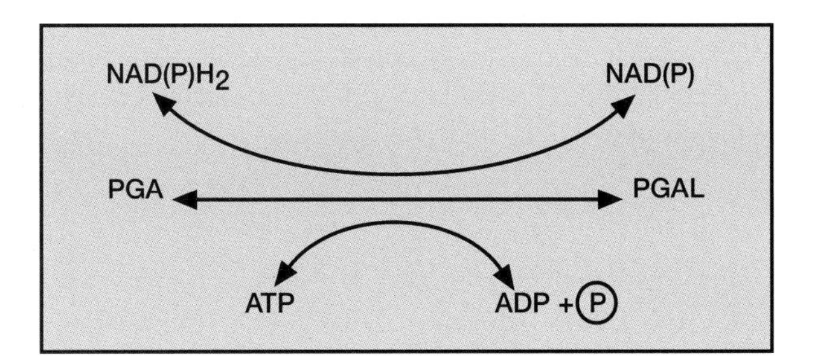

Both photosynthesis and aerobic cellular respiration occur in plant cells. While both of these occur during the daylight hours, only aerobic cellular respiration occurs at night. During daylight hours, the rate of photosynthesis exceeds the rate of aerobic cellular respiration, resulting in a net increase and storage of glucose. The stored glucose is used to support cellular metabolism, which continues during the night.

Aerobic cellular respiration and photosynthesis can be compared; both similarities and differences exist. Do 6.3 Critical Thinking, found at the end of the chapter.

SUMMARY

Solar energy is converted to carbohydrate energy during photosynthesis. Plants and some algae have chloroplasts, where the light-dependent reactions occur in the thylakoid membranes and the light-independent reactions occur in the stroma. The light-dependent reactions produce the ATP and $NADPH_2$ needed to reduce carbon dioxide (CO_2) during the light-independent reactions. Carbon dioxide enters the Calvin cycle directly from the atmosphere, and PGAL molecules exit from it. Plants convert PGAL to many different organic molecules. Among these, glucose phosphate is of special interest because glucose is used by all organisms to produce ATP molecules.

Aerobic cellular respiration (the breakdown of glucose to carbon dioxide and water) requires 3 sets of reactions: glycolysis, the Krebs cycle, and an electron transport system. Glycolysis takes place in the cytoplasm, but the Krebs cycle occurs in the matrix and the electron transport system occurs on the cristae of the mitochondrion. The $NADH_2$ molecules produced during

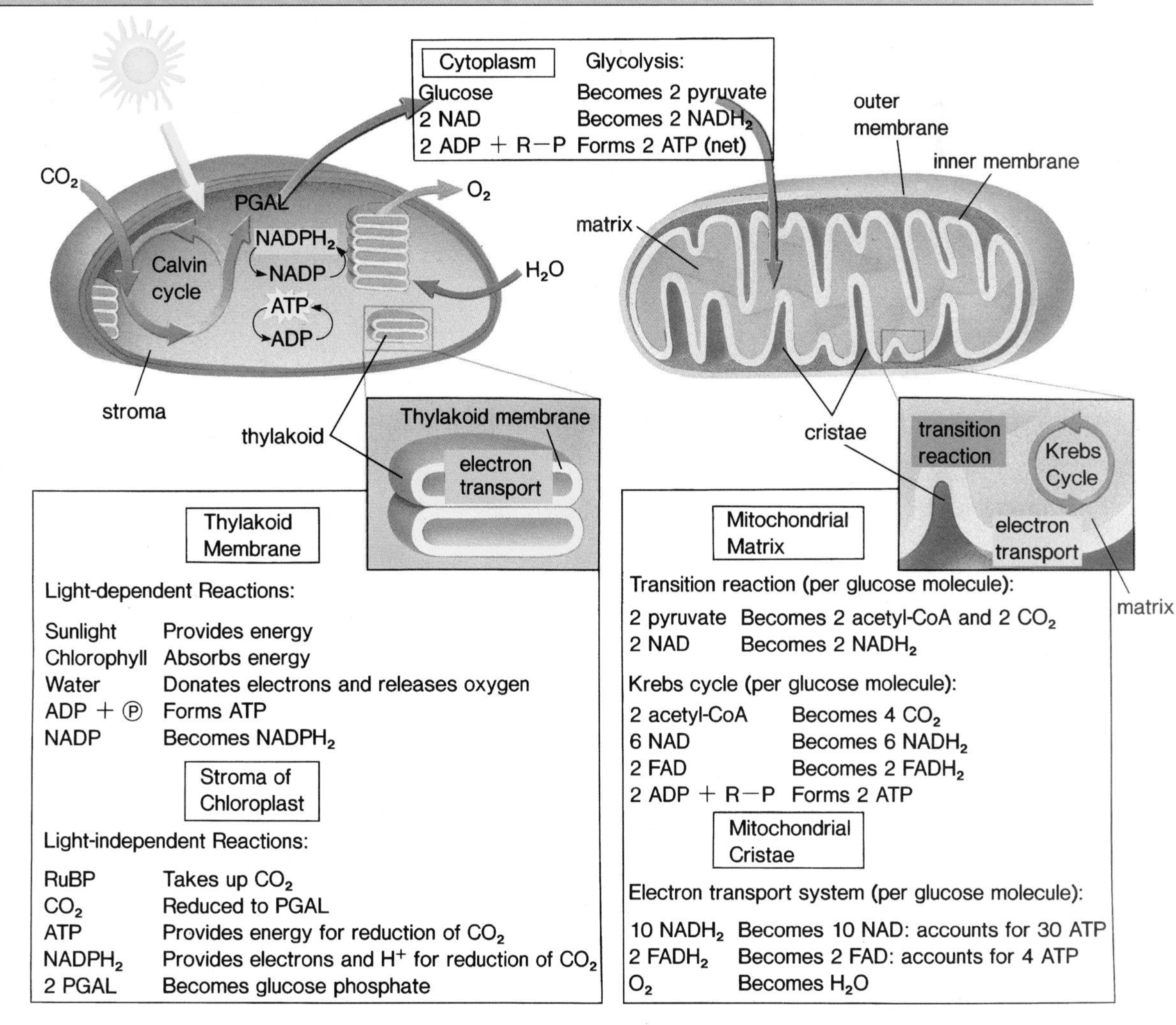

glycolysis and the Krebs cycle take hydrogen atoms (H) to the electron transport system. Energy is released as electrons are passed from carrier to carrier; this energy is used to pump hydrogen ions (H^+) into the intermembrane space. When the hydrogen ions flow down their electrochemical gradient through an ATP synthase complex, energy is provided for ATP production. The final acceptor for electrons is oxygen (O_2), which then takes on hydrogen ions, forming water. The electron transport system produces a maximum of 34 ATP (for technical reasons, most often 32 ATP). Glycolysis and the Krebs cycle (2 turns) directly produce another 4 ATP.

If oxygen is not available in cells, the electron transport system is inoperative, and fermentation (an anaerobic process) occurs. Pyruvate from glycolysis is reduced by $NADH_2$ to either lactate (animal cells) or alcohol and carbon dioxide (yeast). This frees NAD so that once again glycolysis can occur. Fermentation gives a net gain of 2 ATP. Buildup of lactate or alcohol eventually causes the death of animal and yeast cells.

STUDY QUESTIONS

In order to practice **writing across the curriculum,** students should write out the answers to any or all of the study questions. The study questions are sequenced in the same order as the text.

1. Why are almost all living things dependent upon the process of photosynthesis and the energy of the sun? (p. 84)
2. Describe the structure of a chloroplast, and indicate where the light-dependent and the light-independent reactions occur. In what way are these 2 processes connected? (pp. 84–85)
3. Trace the path of electrons during both the cyclic and the noncyclic electron pathway for the light-dependent reactions. (pp. 85–86)
4. List and discuss the 3 stages of the Calvin cycle. Which stage represents the reduction of carbon dioxide (CO_2) to a carbohydrate? (pp. 88–89)
5. Why is it correct to say that a plant cell is biochemically more competent than an animal cell? (p. 89)
6. Relate the pathways of aerobic cellular respiration (except glycolysis) to the structure of a mitochondrion. (pp. 90–91)
7. How do our body cells obtain glucose and oxygen? What happens to carbon dioxide given off by these cells? (p. 91)
8. Name and describe the events within the 3 pathways and the transition reaction that make up aerobic cellular respiration. Be sure to include chemiosmotic ATP synthesis to calculate the number of ATP produced. (pp. 94–95)
9. Describe the events of fermentation. Be sure to include where and when it occurs and to calculate the number of ATP produced. (pp. 95–96)
10. Outline the manner in which the body can make use of proteins and fats, in addition to carbohydrates, as energy sources. (p. 97)
11. Contrast cellular respiration to photosynthesis in at least 5 ways. How are the 2 cellular processes similar? (pp. 98–99)

OBJECTIVE QUESTIONS

1. Life requires a continual supply of energy. Ultimately, this energy comes from _______.
2. In the noncyclic electron pathway, both _______ and _______ are produced.
3. The source of electrons for the noncyclic electron pathway is ultimately _______.
4. In the Calvin cycle, the molecule RuBP _______ carbon dioxide.
5. The useful end product of the Calvin cycle is _______.
6. During aerobic cellular respiration, glucose is broken down completely to _______.
7. Glycolysis ends in 2 _______ molecules and produces _______ $NADH_2$ and a net gain of _______ ATP.
8. Carbon dioxide is given off by the _______ reaction and the _______ cycle.
9. Most of the ATP produced during aerobic cellular respiration comes from _______.
10. The thylakoid space in the chloroplast compares to the _______ space of the mitochondrion because they both serve as hydrogen ion (H^+) reservoirs.

11. Label these diagrams of a chloroplast and a mitochondrion. Then relate the following metabolic pathways to the correct diagram labels. The pathways may be used for both organelles or for no organelle as needed.

electron transport system
Krebs cycle
transition reaction
light-independent reactions
light-dependent reactions
reduction of carbon dioxide
chemiosmotic ATP synthesis
Calvin cycle

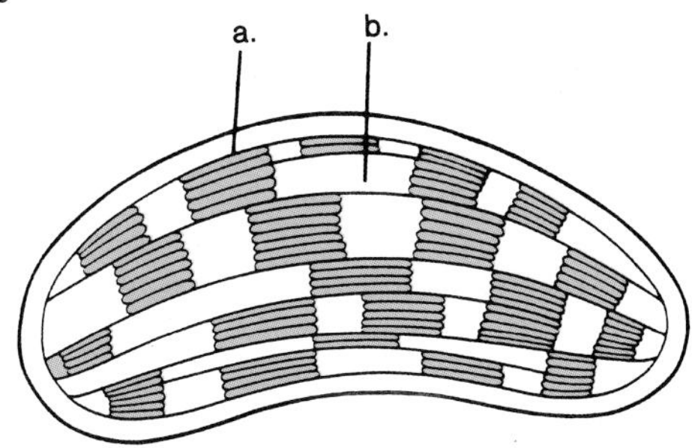

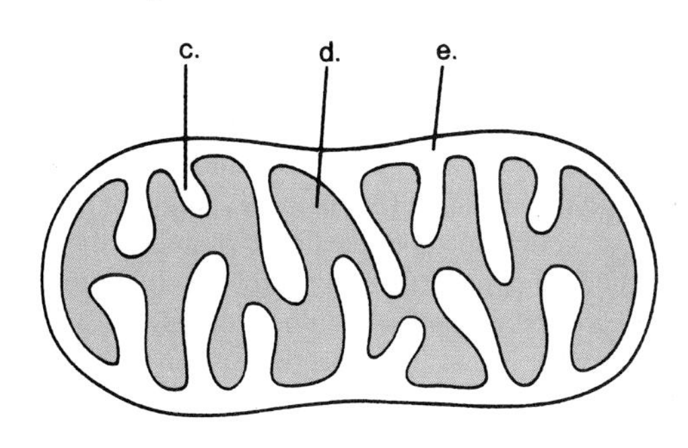

CRITICAL THINKING

In order to practice **writing across the curriculum,** students should write out the answers to any or all of the critical thinking questions. Suggested answers to the critical thinking questions are in appendix E.

6.1

1. A plant exchanges materials with its environment (surroundings). What substances does it take from the environment, and what substances does it give to the environment?
2. What part of a chloroplast gives off oxygen (O_2)? What is the source of this oxygen? What part of a chloroplast takes up carbon dioxide (CO_2)? What happens to the carbon dioxide?
3. How do plants and animals use the oxygen given off by a plant? What is the source of carbon dioxide used by a plant?

6.2

1. The body breathes in oxygen (O_2). What exact function does oxygen perform in the body?
2. Just before a competition, athletes eat carbohydrates. Why does the intake of carbohydrates ensure them of a supply of energy for the athletic event?
3. The body breathes out carbon dioxide (CO_2). Exactly how does the body produce carbon dioxide?

6.3

The structure of chloroplasts and of mitochondria can be compared.

1. What part of a mitochondrion compares to the stroma of a chloroplast?
2. Are the biochemical processes in each of these parts the same or the opposite? Explain.
3. What part of a mitochondrion compares to the thylakoid membrane of a chloroplast? to the thylakoid space of a chloroplast?
4. Are the biochemical processes in each of these parts the same or the opposite? Explain.

SELECTED KEY TERMS

Calvin cycle the primary (C_3) pathway of the light-independent reactions of photosynthesis; converts CO_2 to carbohydrate.

carotenoid (ka-rot´en-oid˝) an orange or yellow pigment that serves as an accessory to chlorophyll in photosynthesis.

chlorophyll the green pigment that converts solar energy to chemical energy during photosynthesis.

electron transport system a chain of electron carriers in the thylakoid membranes of chloroplasts and the cristae of mitochondria that utilize released energy to produce ATP.

glycolysis (gli-kol´i-sis) a metabolic pathway found in the cytoplasm that participates in aerobic cellular respiration and fermentation: it converts glucose to 2 molecules of pyruvate.

granum (pl., grana) a stack of thylakoids within chloroplasts.

Krebs cycle a cyclical metabolic pathway found in the matrix of mitochondria that participates in aerobic cellular respiration; breaks down acetyl groups to carbon dioxide. Also called the citric acid cycle because the reactions begin and end with citric acid (citrate).

light-dependent reactions the first stage of photosynthesis, in which solar energy is stored temporarily as ATP and often $NADPH_2$.

light-independent reactions the second stage of photosynthesis, in which energy produced by light-dependent reactions converts CO_2 to carbohydrates.

photosystem a cluster of light-absorbing pigment molecules within thylakoid membranes.

pyruvate the end product of glycolysis; pyruvic acid.

ribulose bisphosphate (RuBP) (ri-bu-lōs´ bis-fos´-fāt) the 5-carbon molecule that unites with CO_2 in the Calvin cycle.

stroma the fluid component of chloroplasts surrounding the grana.

thylakoid one of the flattened sacs within chloroplasts, the walls (thylakoid membranes) of which are sites of the light-dependent reactions of photosynthesis.

transition reaction a reaction within aerobic cellular respiration during which hydrogen atoms and carbon dioxide are removed from pyruvate; results in acetyl groups and connects glycolysis to the Krebs cycle.

7

CELL DIVISION

Chapter Concepts

1.
Each species has a characteristic number of chromosomes. 104, 119

2.
Animals typically have a life cycle that includes 2 types of cell division, one (mitosis) that maintains the chromosome number and another (meiosis) that halves the number. 105, 119

3.
Mitosis is a part of the cell cycle. First cells get ready to divide, and then they divide. 106, 119

4.
During mitosis, a complete set of chromosomes is distributed equally to 2 daughter cells. 107, 119

5.
During meiosis, half the total number of chromosomes is distributed to 4 daughter cells. 112, 119

6.
The occurrence, events, and daughter cells of mitosis and meiosis can be compared. 116, 119

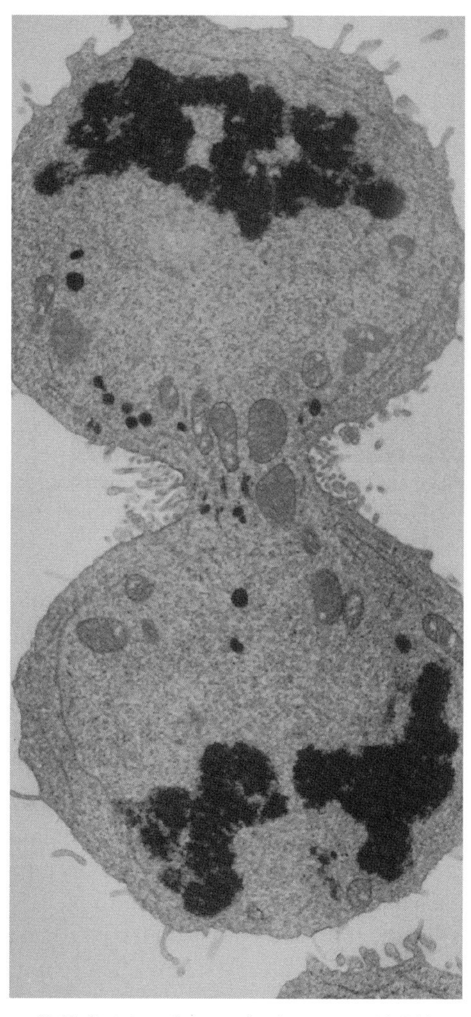

Cell division, human leukocyte, ×12,000

Chapter Outline

Figure 7.1

Human karyotype preparation. As illustrated here, the stain used can result in chromosomes with a banded appearance. The bands help researchers identify and analyze the chromosomes. The enlargement of a chromosome on the far *left* shows that the chromosomes in a karyotype consist of 2 chromatids held together at a centromere. This is the appearance of chromosomes just before they divide.

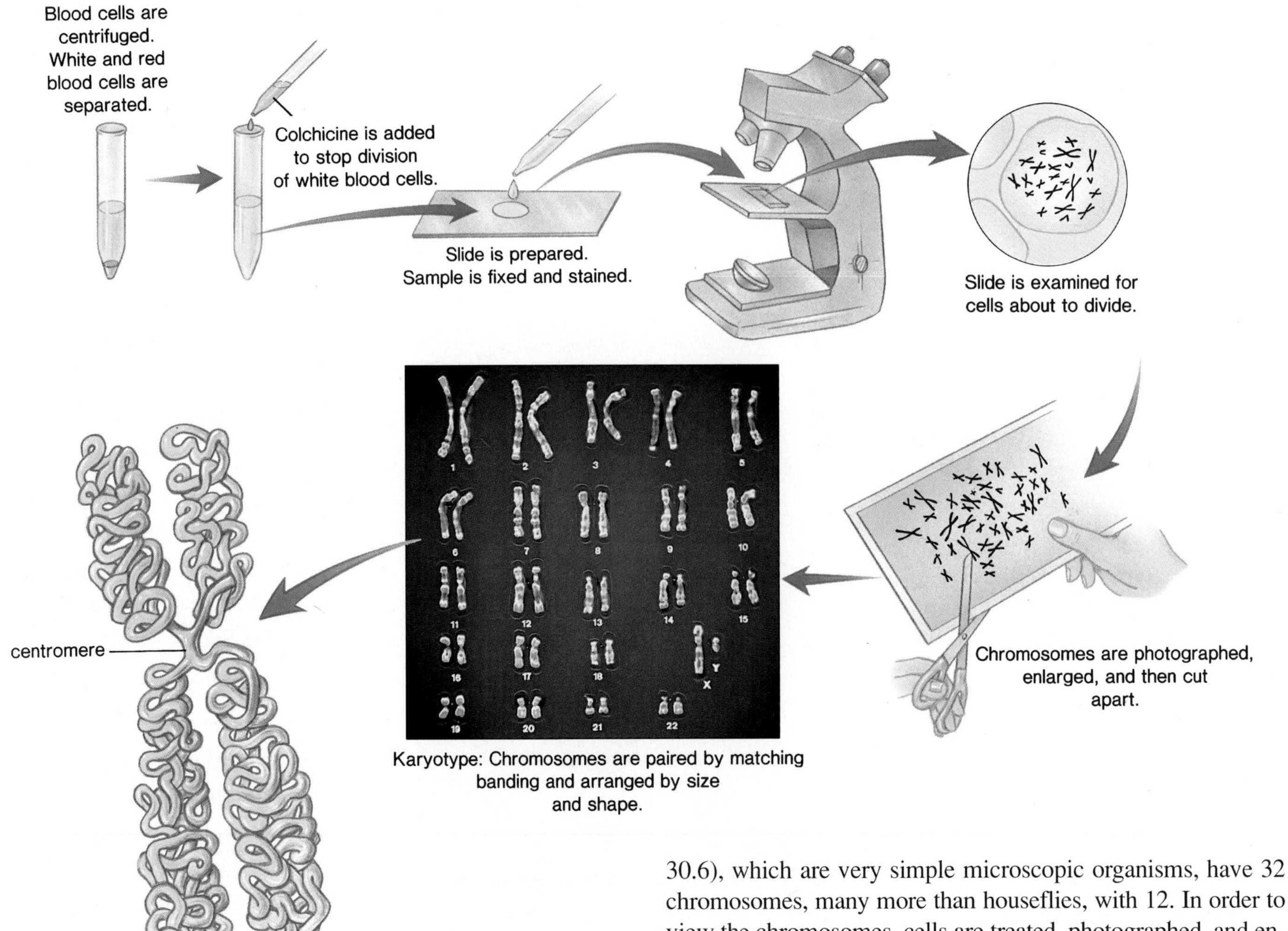

CELL division is necessary for growth and repair of multicellular organisms and for reproduction of all organisms. Cell division requires division of both the nucleus and the cytoplasm. We learned in chapter 3 that the nucleus contains *chromatin,* long threads made up of the DNA double helix and associated proteins. At the time of cell division, chromatin coils, loops, and condenses to give a highly compacted structure. Individual *chromosomes* can be seen just as cell division is about to occur. The reading for this chapter describes the transition from chromatin to chromosomes in greater detail.

An examination of multicellular organisms shows that each type of organism has a characteristic number of chromosomes—corn plants have 20 chromosomes, houseflies have 12, and humans have 46. The particular number has nothing to do with the complexity of the organism. For example, hydras (see fig. 30.6), which are very simple microscopic organisms, have 32 chromosomes, many more than houseflies, with 12. In order to view the chromosomes, cells are treated, photographed, and enlarged. From photographs of cells during division, chromosomes can be cut out and arranged by pairs. (Pairs of chromosomes have the same size and general appearance.) The resulting display of chromosome pairs is called a **karyotype** (fig. 7.1).

Although both males and females have 23 pairs of chromosomes, one pair is of unequal length in the male. The larger chromosome of this pair is called the X chromosome, and the smaller chromosome is called the Y chromosome. Females have 2 X chromosomes in their karyotype. The X and Y chromosomes are called the **sex chromosomes** because they carry the genes that determine sex. All other chromosomes except X and Y are called **autosomes.**

Notice, as the enlargement in figure 7.1 shows, prior to division each chromosome is composed of 2 identical parts called sister **chromatids.** Sister chromatids are genetically identical, and they contain the same *genes,* the units of heredity that control the cell. The chromatids are held together at a region called the **centromere.**

Each organism has a characteristic number of chromosomes; humans have 46. A human karyotype shows 22 pairs of autosomes and 1 pair of sex chromosomes. Males have an X and Y chromosome; females have 2 X chromosomes. Each chromosome in a karyotype is composed of 2 sister chromatids held together at the centromere.

Figure 7.2
Life cycle of humans. Meiosis in males is a part of sperm production, and meiosis in females is a part of egg production. When a haploid sperm fertilizes a haploid egg, the zygote is diploid. The zygote undergoes mitosis as it develops into a newborn child. Mitosis continues after birth until the individual reaches maturity; then the life cycle begins again.

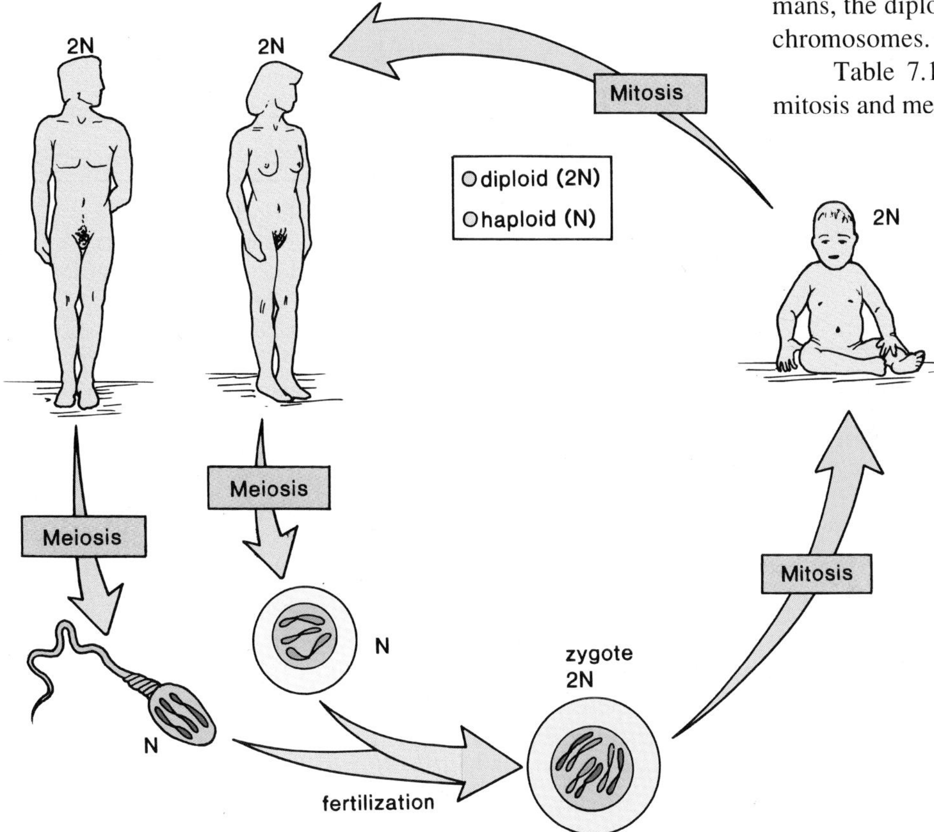

The Life Cycle of Animals

Advanced multicellular animals, including humans, typically have a life cycle (fig. 7.2) that requires 2 types of cell division: meiosis and mitosis.

Meiosis occurs in the sex organs. In males, it produces the cells that become sperm; in females, it produces the cells that become eggs. The sperm and the egg are the sex cells, or **gametes.** Gametes contain half the number of chromosomes in a karyotype—one chromosome from each of the pairs of chromosomes. This is called the **N**, or **haploid**, number of chromosomes; the haploid number of chromosomes in humans is 23.

A new individual comes into existence when a sperm fertilizes an egg. The resulting **zygote** has the **2N**, or **diploid**, number of chromosomes. Each parent contributes one chromosome to each of the pairs of chromosomes present. As the individual develops, mitosis occurs and each **somatic** (body) **cell** has the diploid number of chromosomes. In humans, the diploid number is 46 because there are 23 pairs of chromosomes.

Table 7.1 summarizes the major differences between mitosis and meiosis in multicellular animals.

The life cycle of humans requires 2 types of cell divisions: mitosis and meiosis. Do 7.1 Critical Thinking, found at the end of the chapter.

Mitosis

Because of mitosis, each body cell is genetically identical; they all have the same number and kinds of chromosomes. Mitosis is important to the growth and the repair of multicellular organisms. For example, mitosis occurs as a baby develops in the womb, as the body generates new skin and blood cells, and as a wound heals and the damage is repaired.

Mitosis also occurs when unicellular organisms reproduce asexually. Division results in 2 organisms where before there was only one organism.

Table 7.1
Mitosis versus Meiosis in Animals

Location	Cell Division	Description	Result
Somatic (body) cells	Mitosis	2 N (diploid) → 2 N (diploid)	Growth and repair
Sex organs	Meiosis	2 N (diploid) → N (haploid)	Gamete production

Figure 7.3

The cell cycle. Immature cells go through a cycle that consists of interphase and mitosis. During interphase, DNA replication occurs, preceded and followed by growth. Eventually some daughter cells "break out" of the cell cycle and become specialized cells, performing a specific function.

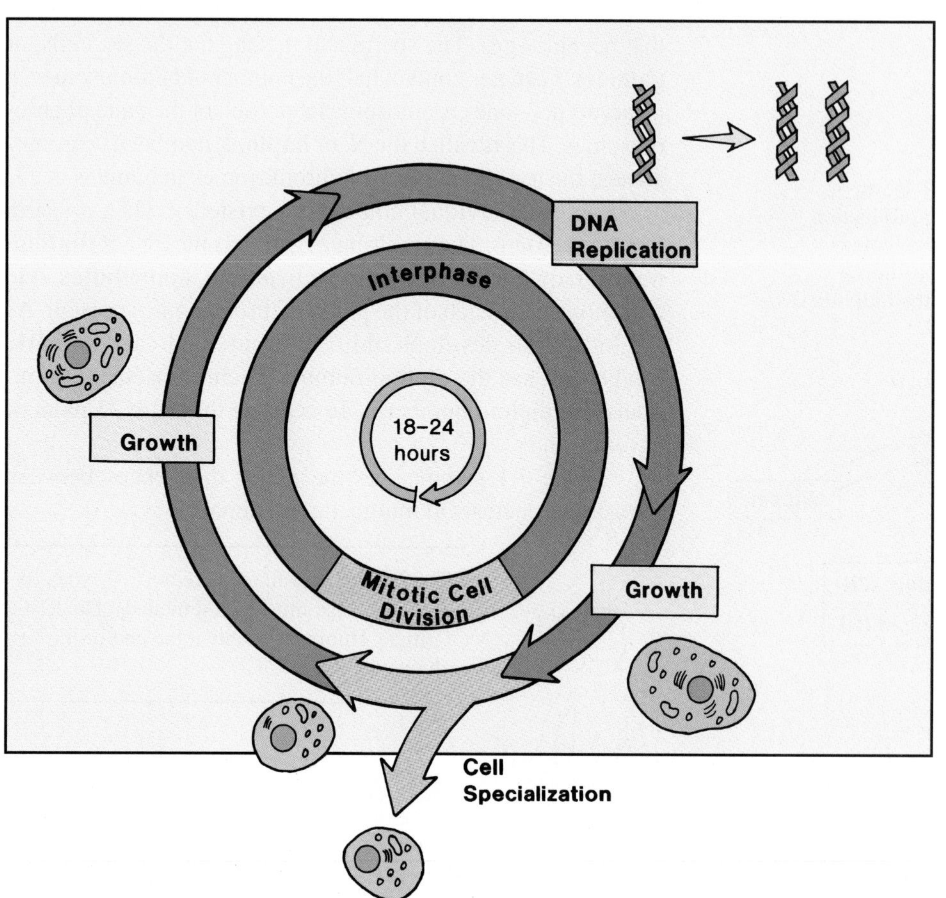

During **interphase,** an animal cell resembles that shown in figure 3.4*a* (p. 48). The nuclear envelope and the nucleoli are visible. The chromosomes, however, are not visible because the chromosome material is not compacted; only indistinct chromatin is seen.

It used to be said that interphase was a resting stage, but we now know that this is not the case. The organelles are metabolically active and are carrying on their normal functions. If the cell is going to divide, *DNA replication* occurs. During replication, DNA is copied and then each chromosome has sister chromatids. *Sister chromatids* are genetically identical—they contain the same genes. Also, organelles, including the *centrioles,* duplicate. A nondividing cell has one pair of centrioles, but in a cell that is going to divide, this pair duplicates, and there are 2 pairs of centrioles outside the nucleus.

> The cell cycle includes mitosis and interphase. During interphase, DNA replication results in each chromosome having sister chromatids. The organelles, including centrioles, also duplicate during interphase. Do 7.2 Critical Thinking, found at the end of the chapter.

Cell Cycle: DNA Replication, Then Cell Division

The **cell cycle** consists of mitosis,[1] which includes division of the cytoplasm, and interphase (fig. 7.3). The length of time required for the entire cell cycle varies according to the organism and even the type of cell within the organism, but 18–24 hours is typical for animal cells. Mitosis lasts less than an hour to slightly more than 2 hours; for the rest of the cycle, the cell is in interphase.

There is usually a limit to the number of times an animal cell enters the cell cycle and divides before degenerative changes lead to the cell's death. Normal cells divide only about 50 times. Usually, they break out of the cell cycle and become specialized even before this number of divisions is reached. This can be contrasted to cancer cells, which can continue to divide indefinitely. Cancer cells have abnormal chromosomes and/or cell structure irregularities associated with their ability to keep repeating the cell cycle.

Today, researchers are actively investigating the control of the cell cycle. They have found that in the early stages, the

1. The term *mitosis* technically refers only to nuclear division, but for convenience, it is used here to refer to division of the entire cell.

cell releases molecules that control later stages, and later stages release molecules that feed back to control early stages. A critical regulatory event occurs during the first stage—if the nutrient and control signals are positive, the cell always completes the rest of the cycle.

Overview of Mitosis: Chromosome Number Stays the Same

Mitosis is cell division that produces 2 *daughter cells, each with the same number and kinds of chromosomes as the parent cell,* the cell that divides. Therefore, the parent cell and the daughter cells are genetically identical.

Figure 7.4 gives an overview of mitosis; for simplicity, only 4 chromosomes are depicted. Because DNA replication has occurred during interphase, each chromosome has 2 sister chromatids held together at a centromere. During mitosis, the centromeres divide, the sister chromatids separate, and a complete set of chromosomes goes to each newly forming cell. (Following separation, each chromatid is called a chromosome.) Each daughter cell receives the same number and kinds of chromosomes as the parent cell and therefore is genetically identical to the parent cell.

Figure 7.4

Mitosis overview. Following DNA replication during interphase, each chromosome in the parent cell consists of 2 sister chromatids. During mitotic division, the sister chromatids separate so that daughter cells have the same number and kinds of chromosomes as the parent cell. (The blue chromosomes were inherited from one parent, and the red chromosomes were inherited from the other parent.)

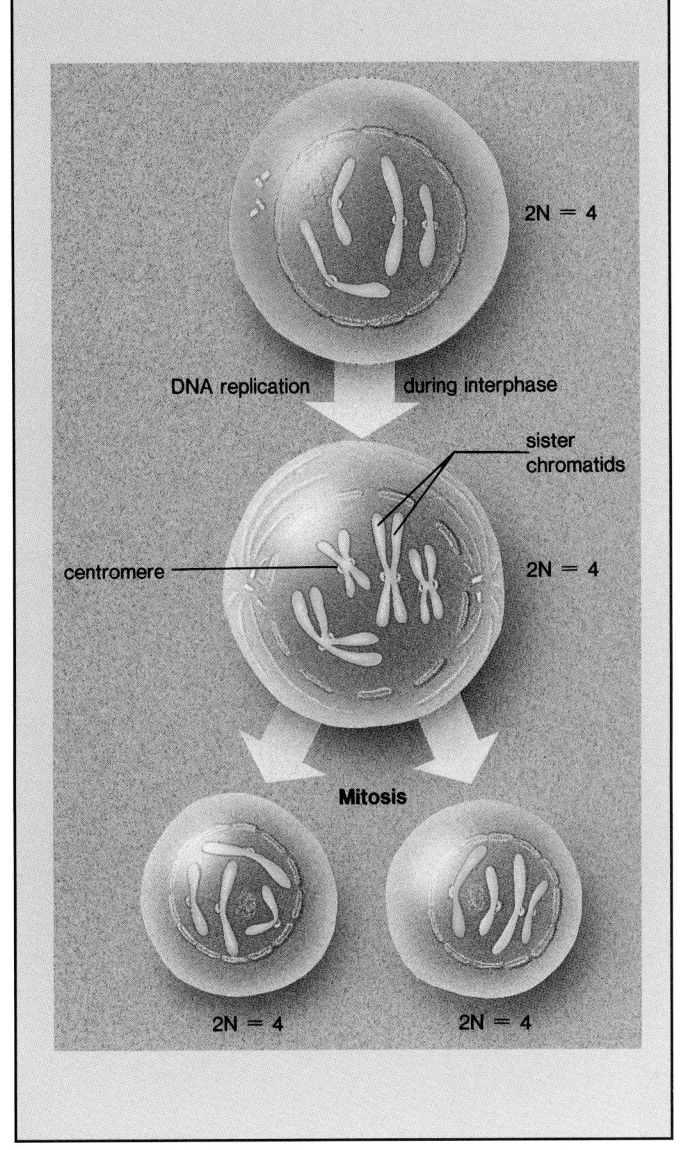

The Stages: One Set

As an aid in describing the events of mitosis, the process is divided into 4 phases: prophase, metaphase, anaphase, and telophase (fig. 7.5). Although it is helpful to depict the stages of mitosis as if they could be separate, they are actually continuous and flow from one stage to another with no noticeable interruption.

Prophase

It is apparent during prophase that cell division is about to occur. The 2 pairs of centrioles outside the nucleus begin moving away from each other toward opposite ends of the nucleus. *Spindle fibers* appear between the separating centriole pairs, the nuclear envelope begins to fragment, and the nucleolus begins to disappear.

The chromosomes are now visible. Each is composed of sister chromatids held together at a centromere. As the chromosomes continue to shorten and to thicken, they become attached by way of their centromeres to spindle fibers. At **prophase,** chromosomes are randomly placed and have not yet aligned at the equator of the spindle.

Figure 7.5

Mitotic cell division. Mitosis is a continuous process but is conveniently divided into 4 stages, excluding interphase and daughter cells. Notice that these drawings and micrographs are for mitosis in animal cells. Because centrioles duplicate in animal cells about to undergo mitosis, there are 2 pairs of centrioles in the late-interphase cell at the start of mitosis. (The blue chromosomes were inherited from one parent, and the red chromosomes were inherited from the other parent.)

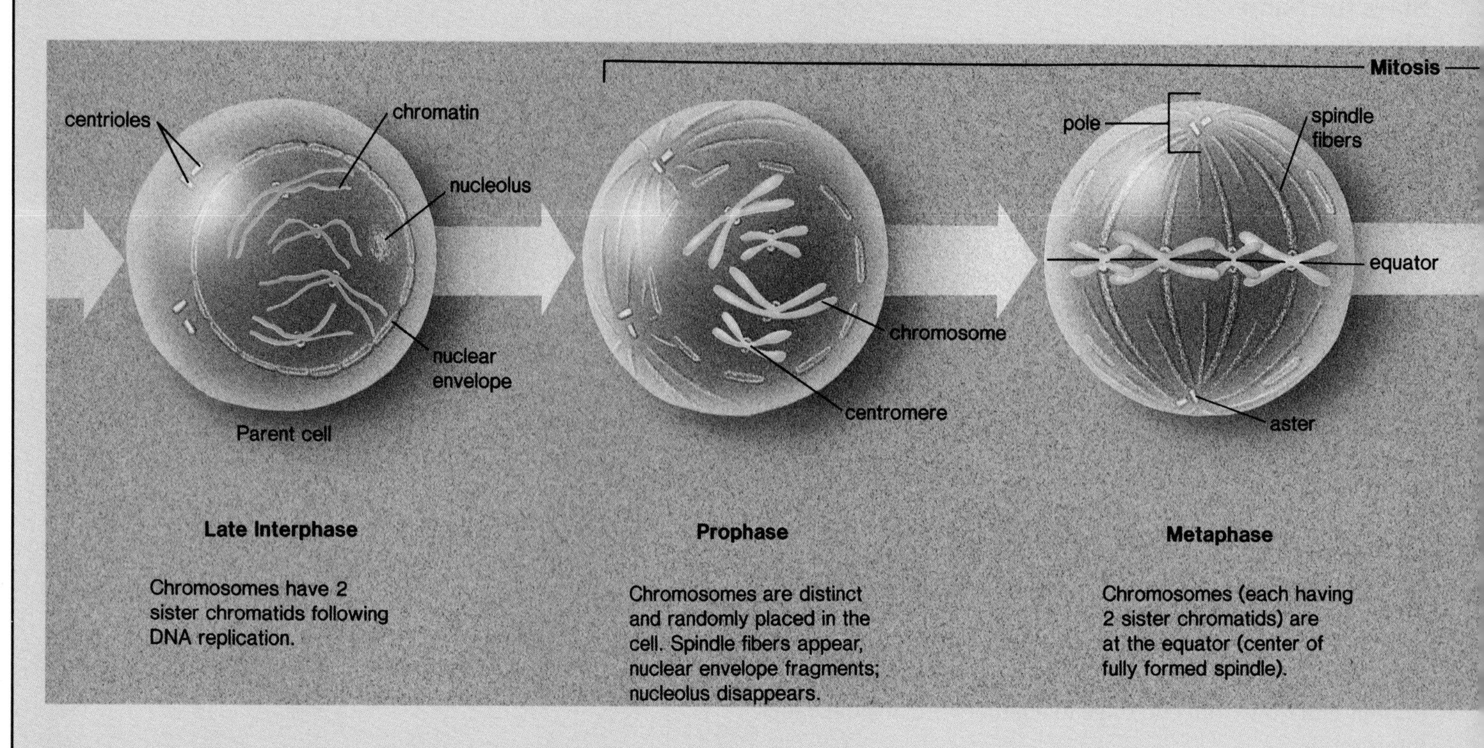

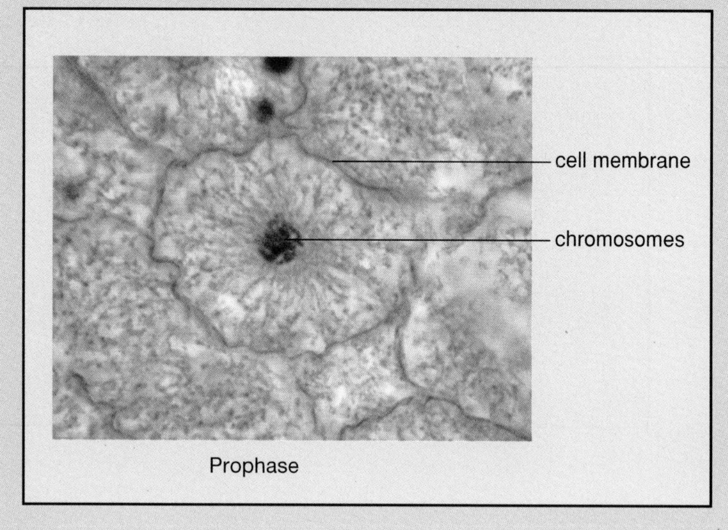

Prophase

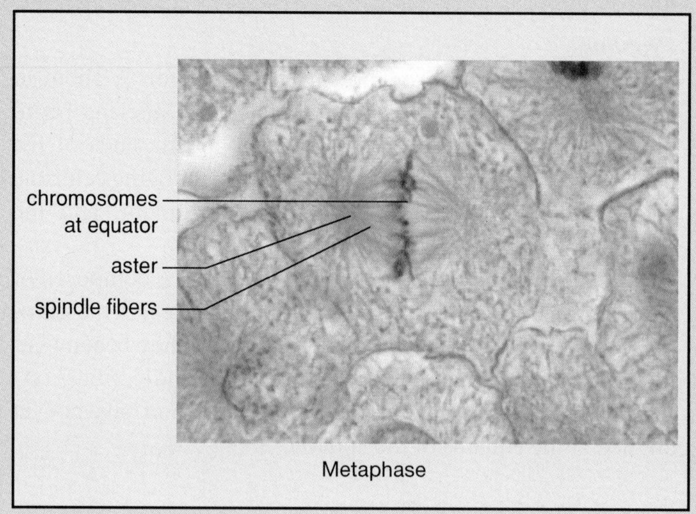

Metaphase

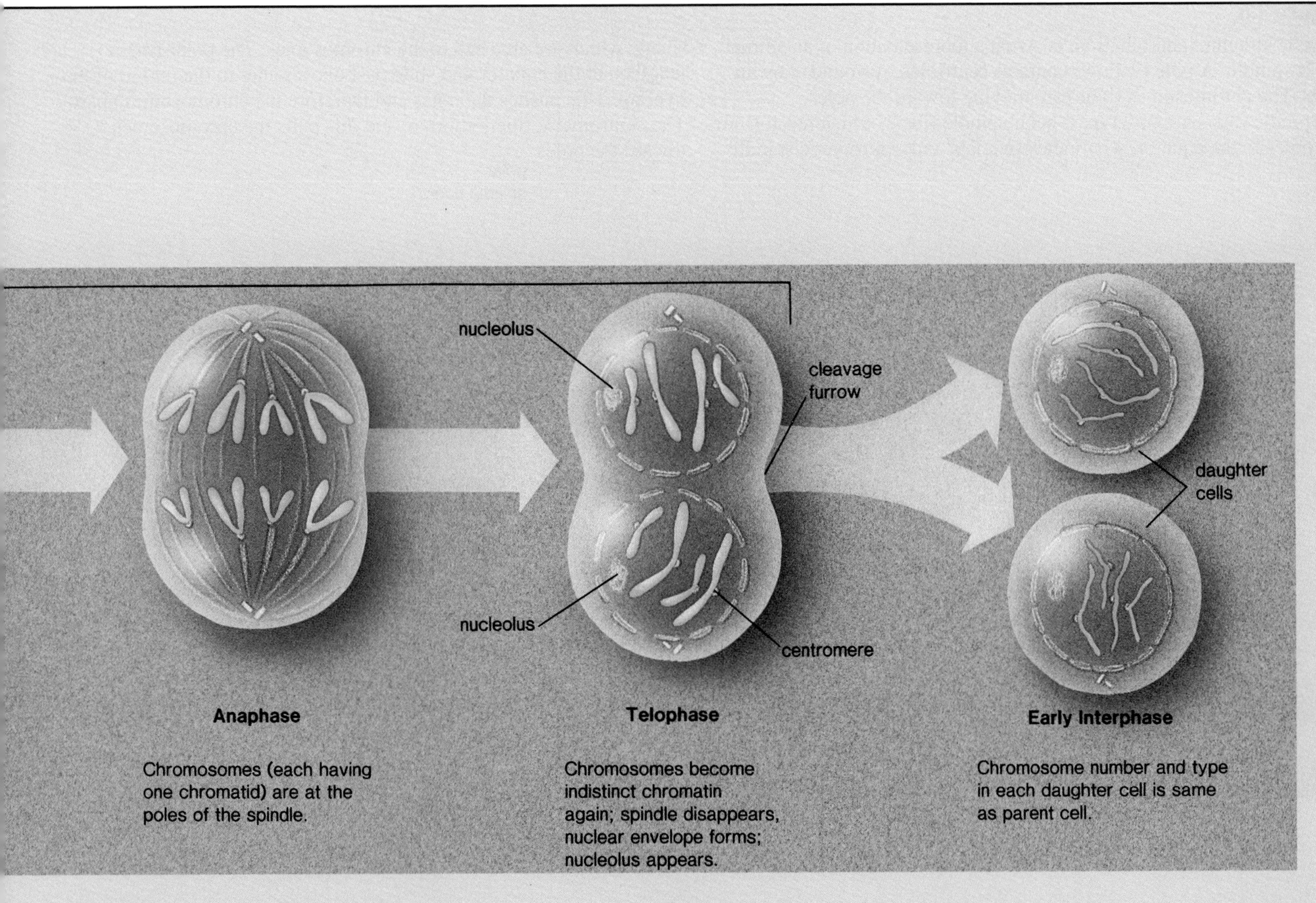

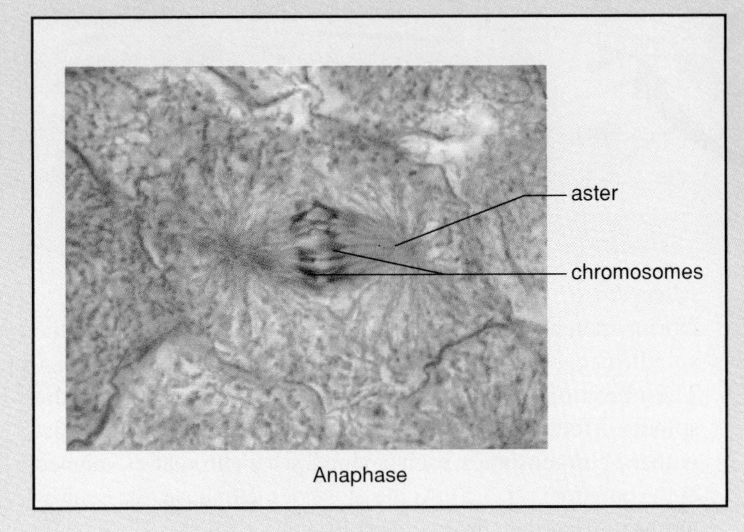

Anaphase

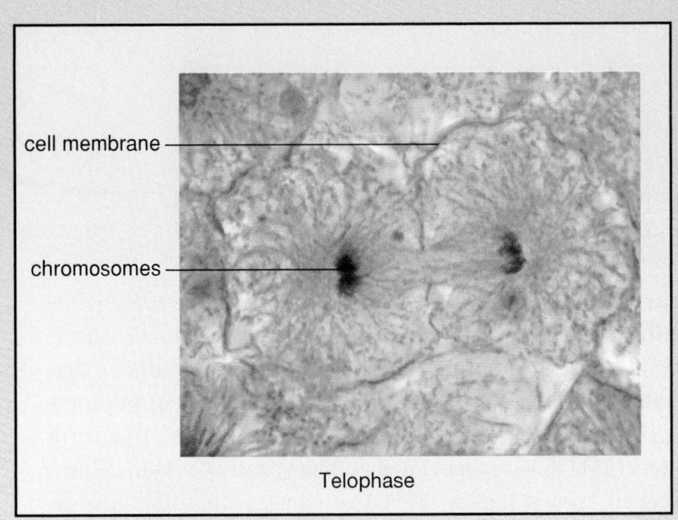

Telophase

Figure 7.6

Spindle structure and function. **a.** Artist's representation of an animal cell's spindle. A pole (yellow) contains centrioles, surrounded by an aster. The chromosomes (blue) are moving toward the poles. **b.** Spindle fibers are of 2 types: polar spindle fibers, which reach from the poles to the equator, where they overlap, and centromeric spindle fibers, which are attached to the chromosomes. The polar fibers lengthen at the equator and slide past one another in the region of the overlap. This pushes the poles and therefore the chromosomes apart. The centromeric fibers shorten, and this pulls the chromosomes toward the poles.

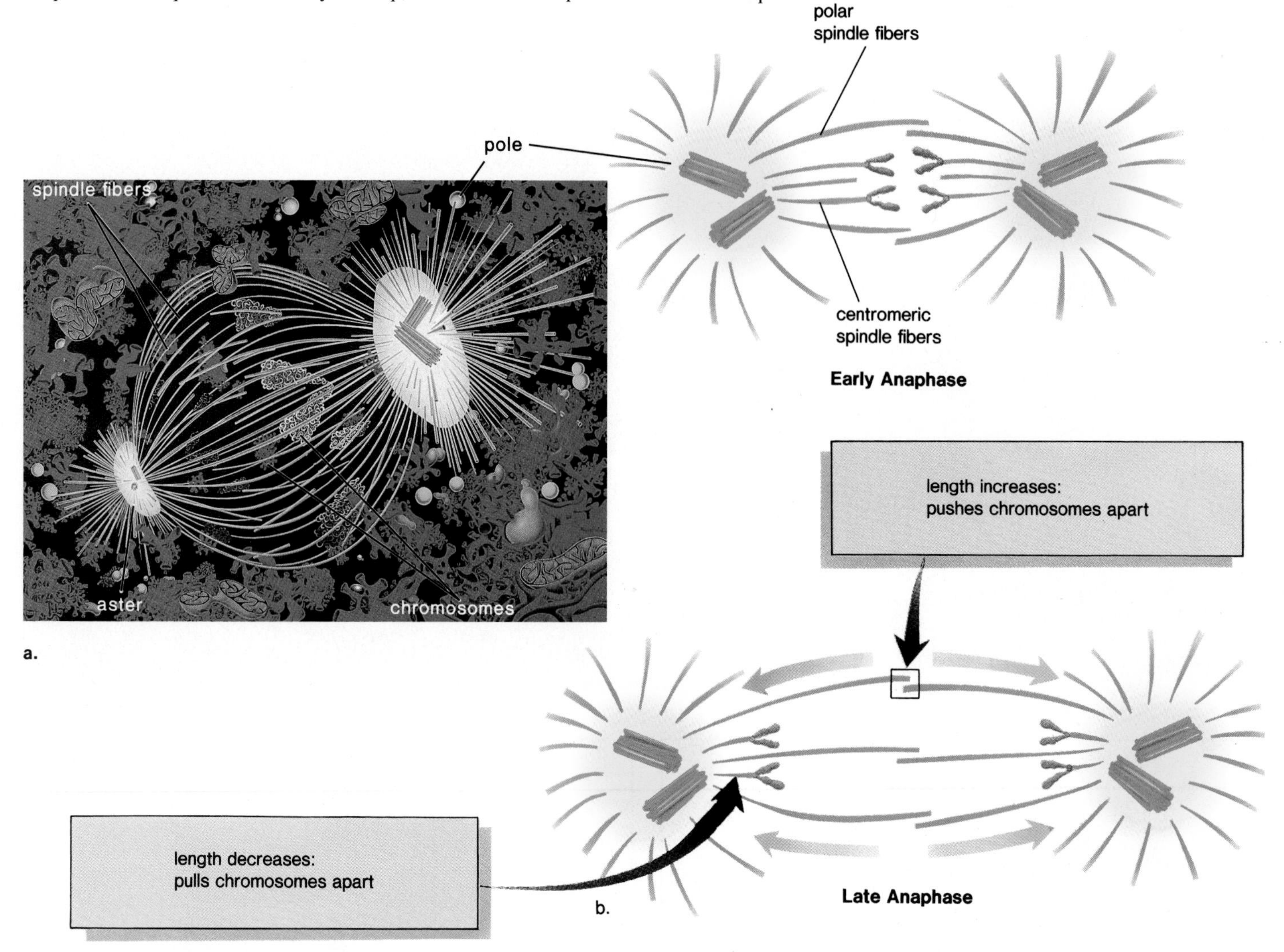

Structure of the Spindle. At the end of prophase, a cell has a fully formed spindle (fig. 7.6). A **spindle** has poles, asters, and fibers. The **asters** are arrays of short microtubules that radiate from the poles, and the fibers are bundles of microtubules that stretch between the poles. Microtubule organizing centers (MTOC) are associated with the centrioles at the poles. It is well known that an MTOC organizes microtubules, including, presumably, those of the spindle. It is possible that the centrioles assist in this function, but it could also be that their location at the poles of a spindle simply ensures that each daughter cell receives a pair of centrioles.

Metaphase

During metaphase, the nuclear envelope is fragmented and the spindle occupies the region formerly occupied by the nucleus. The chromosomes are now at the equator (center) of the spindle. **Metaphase** is characterized by a fully formed spindle, with the chromosomes, each having 2 sister chromatids, aligned at the equator (fig. 7.5). At the close of metaphase, the centromeres uniting the chromatids split.

Anaphase

At the start of anaphase, the sister chromatids separate (fig. 7.5). *Once separated, the chromatids are called chromosomes.* Separation of the sister chromatids ensures that each cell receives a copy of each type of chromosome and thereby has a full complement of genes. During anaphase, the daughter chromosomes move up (*ana* means *up*) to the poles of the spindle. **Anaphase** is characterized by the diploid number of chromosomes at each pole.

Function of the Spindle. The spindle brings about chromosome movement. Two types of spindle fibers are involved in the movement of chromosomes during anaphase. The polar spindle fibers, one type, extend from the poles to the equator of the spindle; there they overlap one another (fig. 7.6). As mitosis proceeds, polar spindle fibers increase in length and then slide past one another. As the spindle gets longer, the poles and the chromosomes get pushed apart.

The chromosomes themselves are not attached to the polar fibers. They are attached to a second type of spindle fiber, the centromeric spindle fibers, which extend from their centromeres to the poles. The centromeric spindle fibers get shorter and shorter as the chromosomes move toward the poles, and eventually the centromeric spindle fibers disappear.

Spindle fibers, as stated, are composed of microtubules. Microtubules can assemble and disassemble by the addition or subtraction of tubulin (protein) subunits. This is what enables spindle fibers to lengthen and shorten and what ultimately causes the movement of the chromosomes.

Telophase

Telophase begins when the chromosomes, each having one chromatid, arrive at the poles. During telophase, the chromosomes become indistinct chromatin again. The spindle disappears as nucleoli appear in each cell, and nuclear envelopes form. **Telophase** is characterized by the presence of 2 daughter nuclei.

Following nuclear division, cytoplasmic division, called **cytokinesis,** usually occurs. In animal cells, a slight indentation called a **cleavage furrow** passes around the circumference of the cell. Actin filaments form a contractile ring, and as the ring gets smaller and smaller, the cleavage furrow pinches the cell in two. Now each cell is enclosed by its own cell membrane (fig. 7.7).

Animal Mitosis versus Plant Mitosis

There are 2 main differences between plant cell mitosis and animal cell mitosis. First of all, in plant cells, centrioles and asters are not seen during mitosis. However, spindle fibers do appear (fig. 7.8). It is interesting to note that animal cells deprived of centrioles will also form a spindle. It may be, therefore, that centrioles do *not* contribute to the function of an MTOC.

The second difference pertains to cytokinesis. The rigid cell wall that surrounds plant cells does not permit division of the cytoplasm by means of a cleavage furrow. Instead, vesicles largely derived from the Golgi apparatus travel down the polar spindle fibers to the region of the equator. These vesicles fuse to form a double-layered membrane called a **cell plate,** which spreads to the sides and marks the boundary of the 2 daughter cells. Cell wall material accumulates between the 2 membranes of the cell plate. When this material splits, each plant cell has its own cell membrane and cell wall.

Mitosis ensures that each somatic cell is 2N. When the sister chromatids separate during anaphase, each newly forming cell receives the same number and kinds of chromosomes as the original cell. Division of the cytoplasm involves a cleavage furrow in animal cells and a cell plate in plant cells.

Figure 7.7
Cytokinesis in an animal cell. A single cell becomes 2 cells by a furrowing process. A contractile ring composed of actin filaments gradually gets smaller, and the cleavage furrow pinches the cell into 2 cells.

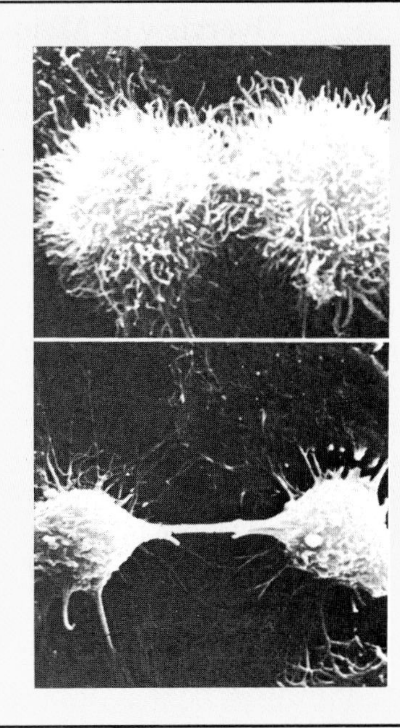

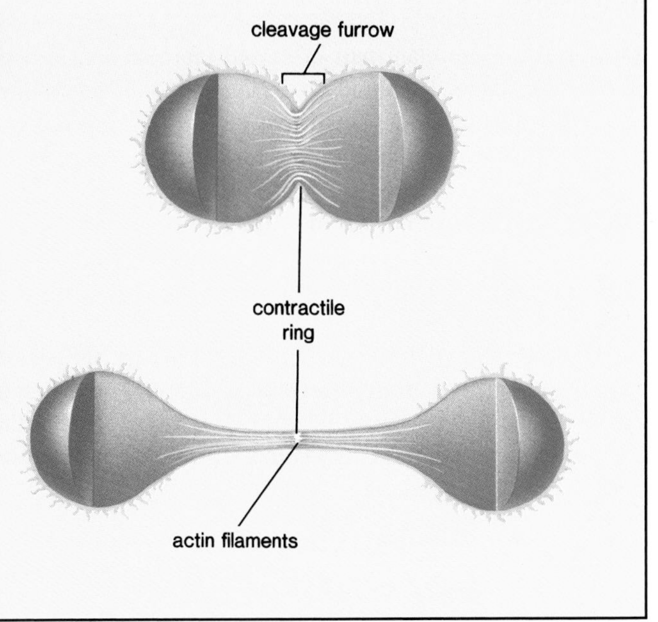

Figure 7.8

Plant cell mitosis. Note the absence of centrioles and asters and the presence of the cell wall. In telophase, a cell plate develops in between the 2 daughter cells. The cell plate marks the boundary of the new daughter cells, where new cell membrane and new cell wall will form for each cell.

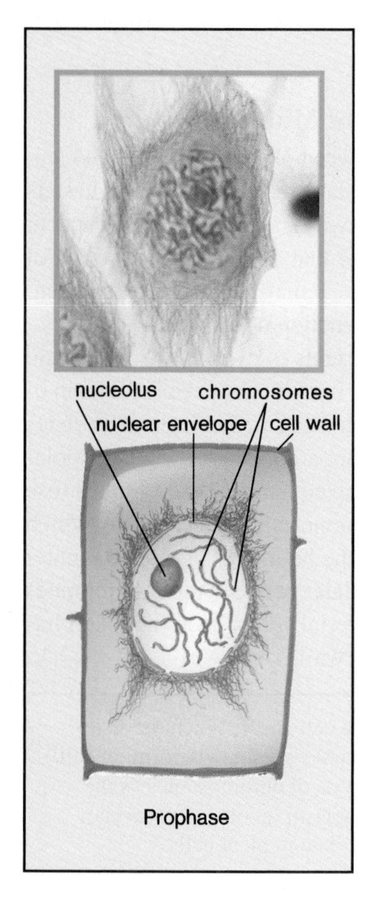

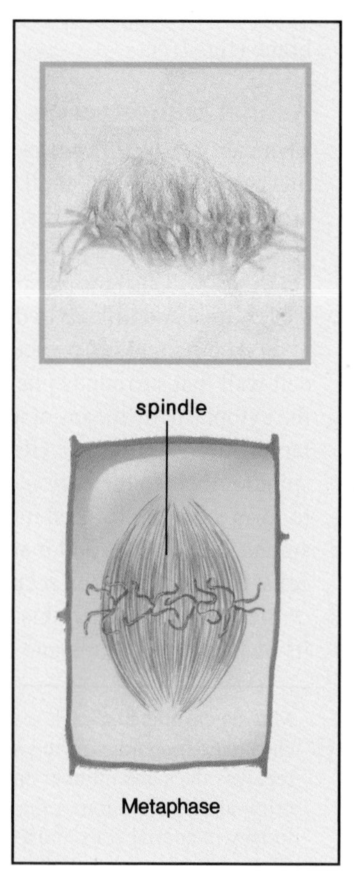

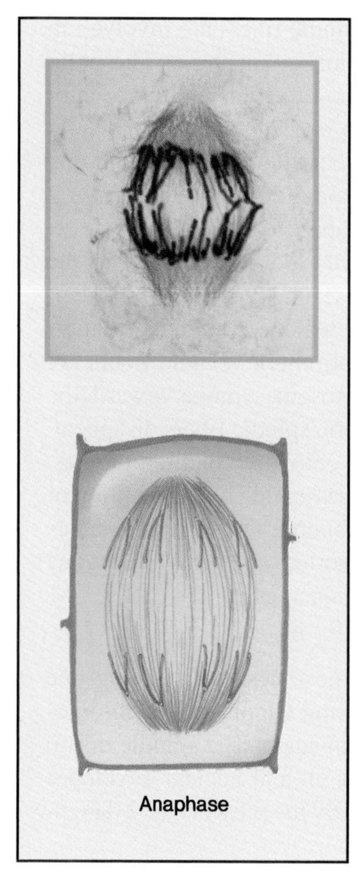

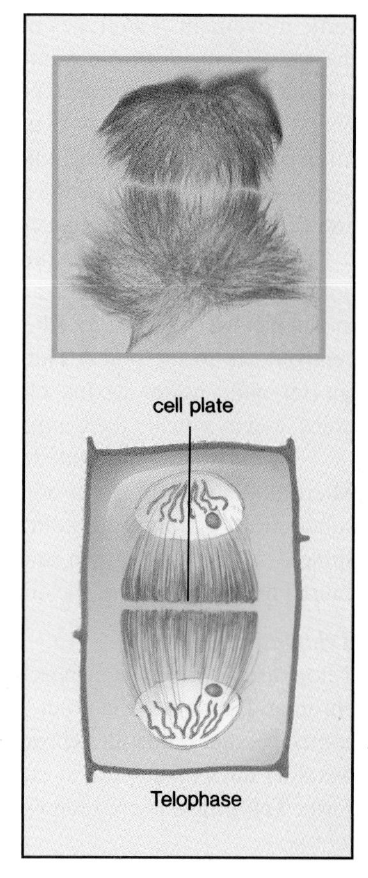

Meiosis

Meiosis is an important part of sexual reproduction in both animals and plants. In animals, it occurs during the production of the egg and the sperm. In plants, meiosis occurs during spore formation. Spores are specialized structures that precede the formation of the sex cells.

Meiosis keeps the chromosome number constant generation after generation; it also ensures that each generation has a different genetic makeup than the previous one. For example, in animals, the egg carries one-half the chromosomes (and genes) from the female parent and the sperm carries one-half the chromosomes (and genes) from the male parent. When the sperm fertilizes the egg, the zygote has the diploid number of chromosomes and a different combination of genes than either parent.

Overview of Meiosis: Chromosome Number Cut in Half

Meiosis, which requires 2 cell divisions, results in *4 daughter cells, each having one of each kind of chromosome and therefore half the number of chromosomes as the parent cell.*[2] The parent cell has the diploid number of chromosomes, while the daughter cells have the haploid number of chromosomes. Therefore, meiosis is often called reduction division.

Recall that when a cell is 2N, or diploid, the chromosomes occur in pairs. For example, the 46 chromosomes of humans occur in 23 pairs of chromosomes. These pairs are called **homologous chromosomes.** During meiosis, homologous chromosomes separate. Because of this, each daughter cell receives one member from each homologous pair and therefore half the total number of chromosomes.

2. The term *meiosis* technically refers only to nuclear division, but for convenience, it is used here to refer to the division of the entire cell.

Figure 7.9

Overview of meiosis. Following DNA replication, each chromosome is a dyad. During meiosis I, the homologous chromosomes pair, forming tetrads (4 chromatids), and then separate. During meiosis II, the chromatids separate, and the 4 daughter cells are haploid. (The blue chromosomes were inherited from one parent, and the red chromosomes were inherited from the other parent.)

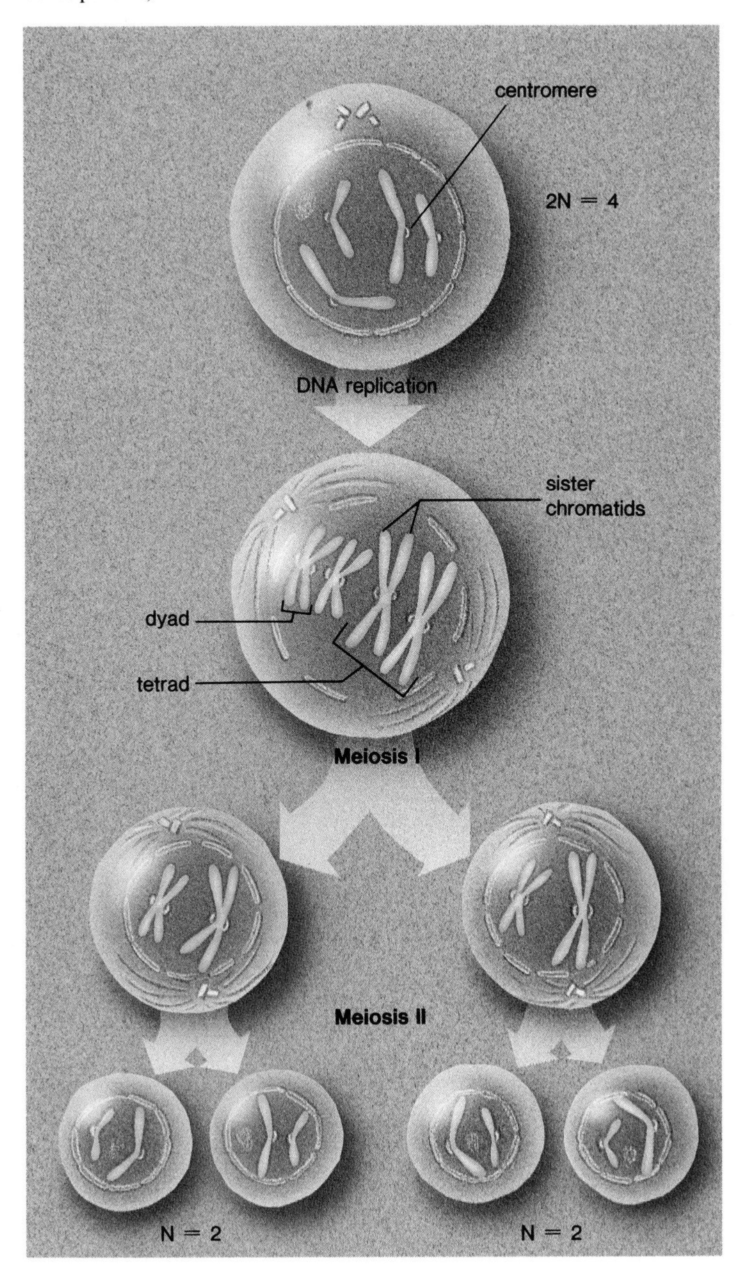

Figure 7.9 presents an overview of meiosis, indicating the 2 cell divisions, **meiosis I** and **meiosis II.** Prior to meiosis I, DNA replication occurs, and each chromosome has 2 sister chromatids. Therefore, each chromosome can be called a **dyad.** During meiosis I, the homologous chromosomes come together and line up side by side due to a means of attraction still unknown. This so-called **synapsis** results in **tetrads,** associations

Figure 7.10

Crossing-over. When homologous chromosomes are in synapsis, the nonsister chromatids exchange genetic material. Following crossing-over, each chromatid has a different combination of genes.

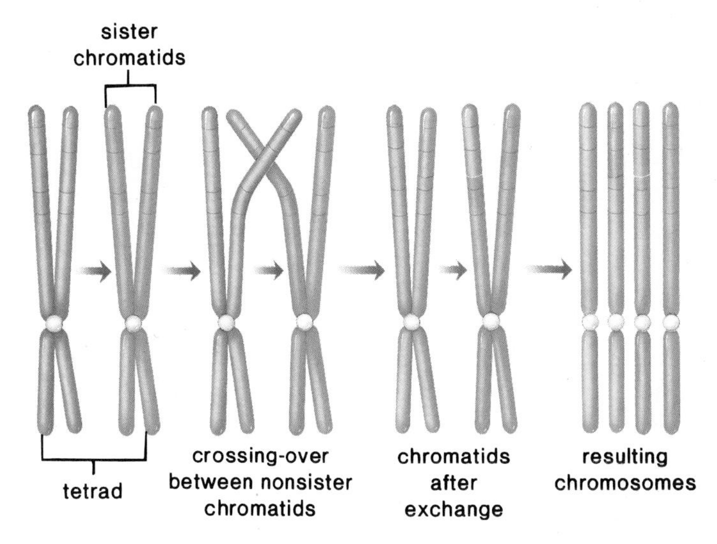

of 4 chromatids that stay in close proximity during the first 2 phases of meiosis I. During synapsis, nonsister chromatids exchange genetic material, as illustrated in figure 7.10. Now each chromatid of the tetrad has a different combination of genes. The exchange of genetic material between chromatids is called **crossing-over** and is an additional means by which offspring receive a different genetic makeup than either of its parents.

Following synapsis during meiosis I, the homologous chromosomes separate. This separation means that one chromosome from each homologous pair reaches each haploid daughter cell. There are no restrictions on the separation process; either chromosome of a homologous pair can occur in a daughter cell with either chromosome of any other pair. Therefore, all possible combinations of chromosomes and genes occur within the gametes, which result after meiosis is complete.

No replication of DNA is needed between meiosis I and meiosis II because the chromosomes are dyads; they already have 2 sister chromatids. During meiosis II, the sister chromatids separate in each of the cells from meiosis I, and therefore meiosis produces 4 daughter cells. Each daughter cell has the haploid number of chromosomes. You can count the number of centromeres to verify that the parent cell has the diploid number of chromosomes and each daughter cell has the haploid number.

The Stages: Two Sets

The same 4 stages seen in mitosis—prophase, metaphase, anaphase, and telophase—occur during both meiosis I and meiosis II. The complete designation indicates the stage and the division; for example, prophase I is the first stage of meiosis I.

The First Division

The stages of meiosis I as they appear in an animal cell are diagrammed in figure 7.11. During *prophase I,* the spindle appears while the nuclear envelope fragments and the nucleolus disappears. The homologous chromosomes, each having 2 sister chromatids, undergo synapsis, forming tetrads. Crossing-over occurs now, but for simplicity, this event has been omitted from figure 7.11. In *metaphase I,* tetrads line up at the equator of the spindle. During *anaphase I,* homologous chromosomes separate and dyads move to opposite poles of the spindle. During *telophase I,* nuclear envelopes form, and nucleoli appear as the spindle disappears. In certain species, the cell membrane furrows to give 2 cells, and in others, the second division begins without benefit of complete furrowing. Regardless, each daughter cell contains only one chromosome from each homologous pair. The chromosomes are dyads and each has 2 sister chromatids. No replication of DNA occurs during a period of time called *interkinesis.*

The Second Division

The stages of meiosis II for an animal cell are diagrammed in figure 7.12. At the beginning of *prophase II,* a spindle appears while the nuclear envelope fragments and the nucleolus disappears. Dyads (one dyad from each pair of homologous chromosomes) are present, and each attaches to the spindle independently. During *metaphase II,* the dyads are lined up at the equator. At the close of metaphase II, the centromeres split. During *anaphase II,* the sister chromatids of each dyad separate and move toward the poles. Each pole receives the same number of chromosomes. In *telophase II,* the spindle disappears as nuclear envelopes form. The cell membrane furrows to give 2 complete cells, each of which has the haploid, or N, number of chromosomes. Each chromosome has one chromatid. Since each cell from meiosis I undergoes meiosis II, there are 4 daughter cells altogether.

Figure 7.11 shows an animal cell undergoing meiosis. Centrioles are present, the cells have no cell walls, and a cleavage furrow forms to divide the cells. In animals, meiosis is involved in either spermatogenesis or oogenesis.

> Meiosis involves 2 cell divisions. During meiosis I, tetrads form and crossing-over occurs. Homologous chromosomes, each consisting of 2 sister chromatids, separate and each daughter cell receives one chromosome from each pair. During meiosis II, separation of chromatids in daughter cells from meiosis I results in 4 daughter cells, each with the haploid number of chromosomes. Do 7.3 Critical Thinking, found at the end of the chapter.

Figure 7.11

Meiosis I. During meiosis I, homologous chromosomes undergo synapsis and then separate so that each daughter cell has only one chromosome from each original homologous pair. For simplicity's sake, the results of crossing-over have not been depicted. Notice that each daughter cell is haploid and each chromosome has 2 chromatids. (The blue chromosomes were inherited from one parent, and the red chromosomes were inherited from the other parent.)

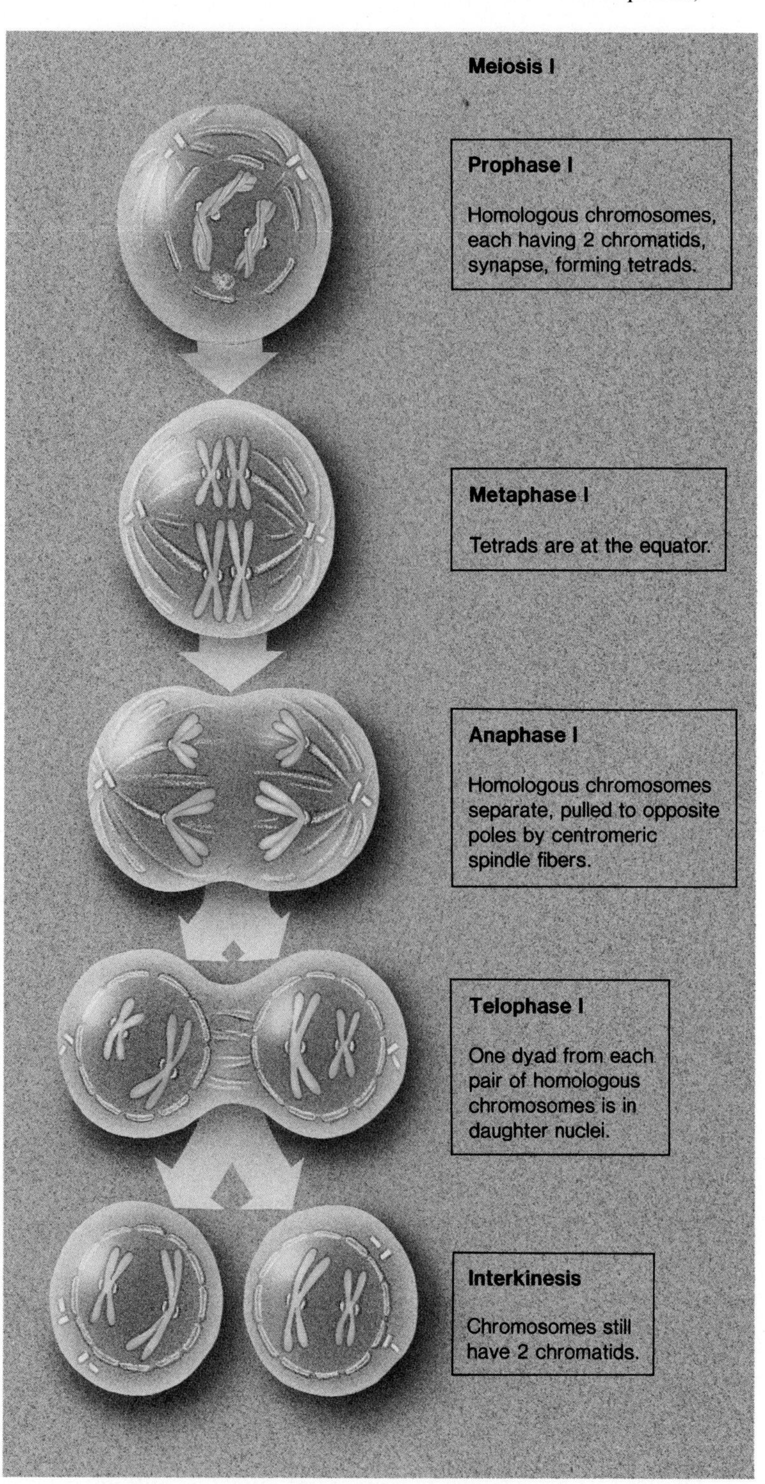

Figure 7.12
Meiosis II. During meiosis II, chromatids separate. Each daughter cell is haploid, and each chromosome has one chromatid.

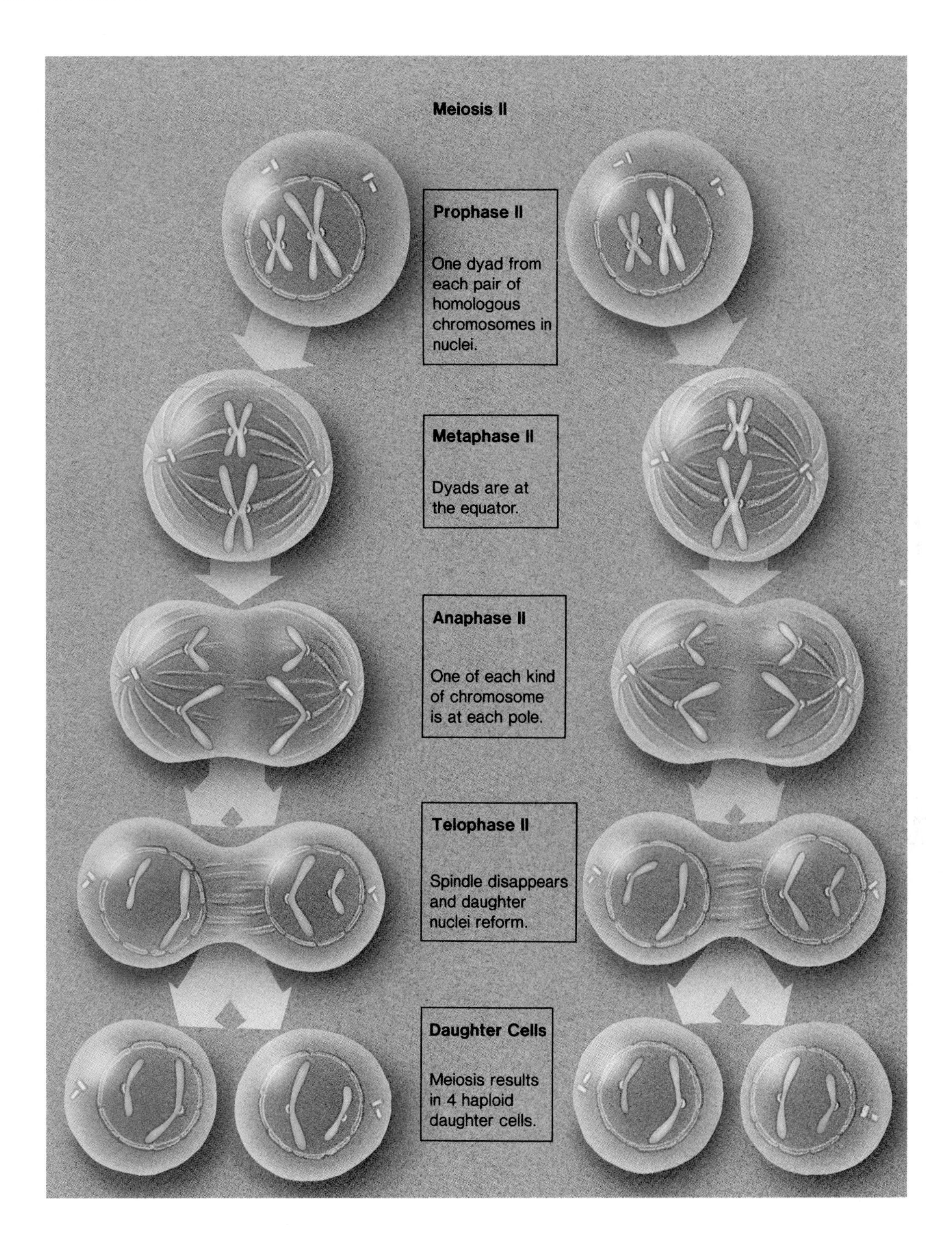

Animal Meiosis: Sperm and Egg Production

Spermatogenesis and oogenesis occur in the sex organs, the testes in males and the ovaries in females. During **spermatogenesis,** sperm are produced, and during **oogenesis,** eggs are produced. The gametes appear differently in the 2 sexes (fig. 7.13), and meiosis is different, too. The process of meiosis in males always results in 4 cells that become sperm. Meiosis in females produces only one cell that goes on to become an egg. Meiosis I results in one large cell called a secondary oocyte and one polar body. After meiosis II, there are one egg and 3 polar bodies. The **polar bodies** are a way to discard unnecessary chromosomes while retaining much of the cytoplasm in the egg. The cytoplasm serves as a source of nutrients for the developing embryo.

Spermatogenesis, once started, continues to completion and mature sperm result. In contrast, oogenesis does not necessarily go to completion. Only if a sperm fertilizes the secondary oocyte does it undergo meiosis II and become an egg. Regardless of this complication, however, both the sperm and the egg contribute the haploid number of chromosomes to the fertilized egg. In humans, each contributes 23 chromosomes. In addition, the egg contributes most of the cytoplasm. Figure 7.13 shows how the sperm and the egg are adapted to their function. The sperm is a tiny flagellated cell, while the egg is stationary and quite large.

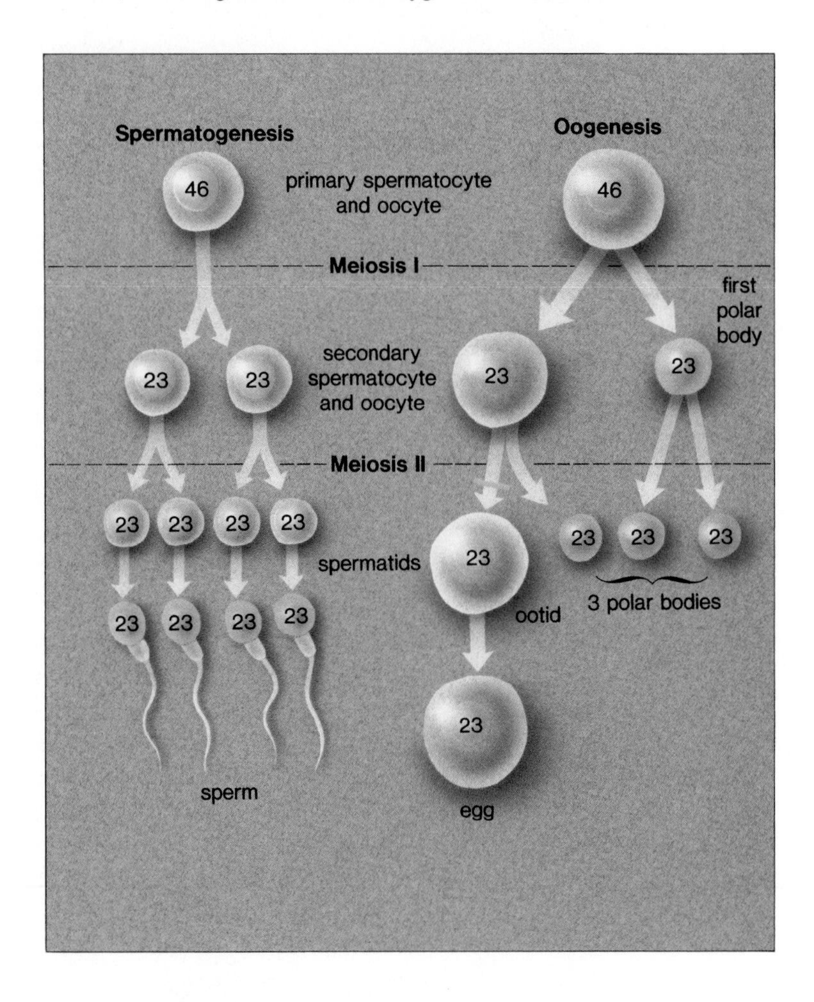

Figure 7.13

Spermatogenesis and oogenesis. Spermatogenesis produces 4 viable sperm, whereas oogenesis produces one egg and 3 polar bodies. Notice that oogenesis does not go to completion unless the secondary oocyte is fertilized. In humans, both sperm and egg have 23 chromosomes each; therefore, following fertilization, the zygote has 46 chromosomes.

Comparison of Mitosis to Meiosis

Figure 7.14 compares mitosis to meiosis. The differences between these cellular divisions can be categorized according to occurrence and events.

Occurrence of Each

In animals, meiosis occurs only in cells that eventually give rise to the gametes. Mitosis occurs in somatic cells—all the other types of cells in the body.

Events of Each

Mitosis can be compared to meiosis I in this manner:

Mitosis	**Meiosis I**
Prophase No pairing of chromosomes	*Prophase I* Pairing of homologous chromosomes
Metaphase Dyads at equator	*Metaphase I* Tetrads at equator
Anaphase Sister chromatids separate	*Anaphase I* Homologous chromosomes separate
Telophase Chromosomes have one chromatid.	*Telophase I* Chromosomes have 2 sister chromatids.

Mitosis can be compared with meiosis II in this manner:

Mitosis	**Meiosis II**
Metaphase Dyads at equator	*Metaphase II* Dyads at equator
Anaphase Sister chromatids separate	*Anaphase II* Sister chromatids separate
Telophase 2 diploid daughter cells	*Telophase II* 4 haploid daughter cells

Figure 7.14
Mitosis compared to meiosis. Crossing-over during meiosis is shown.

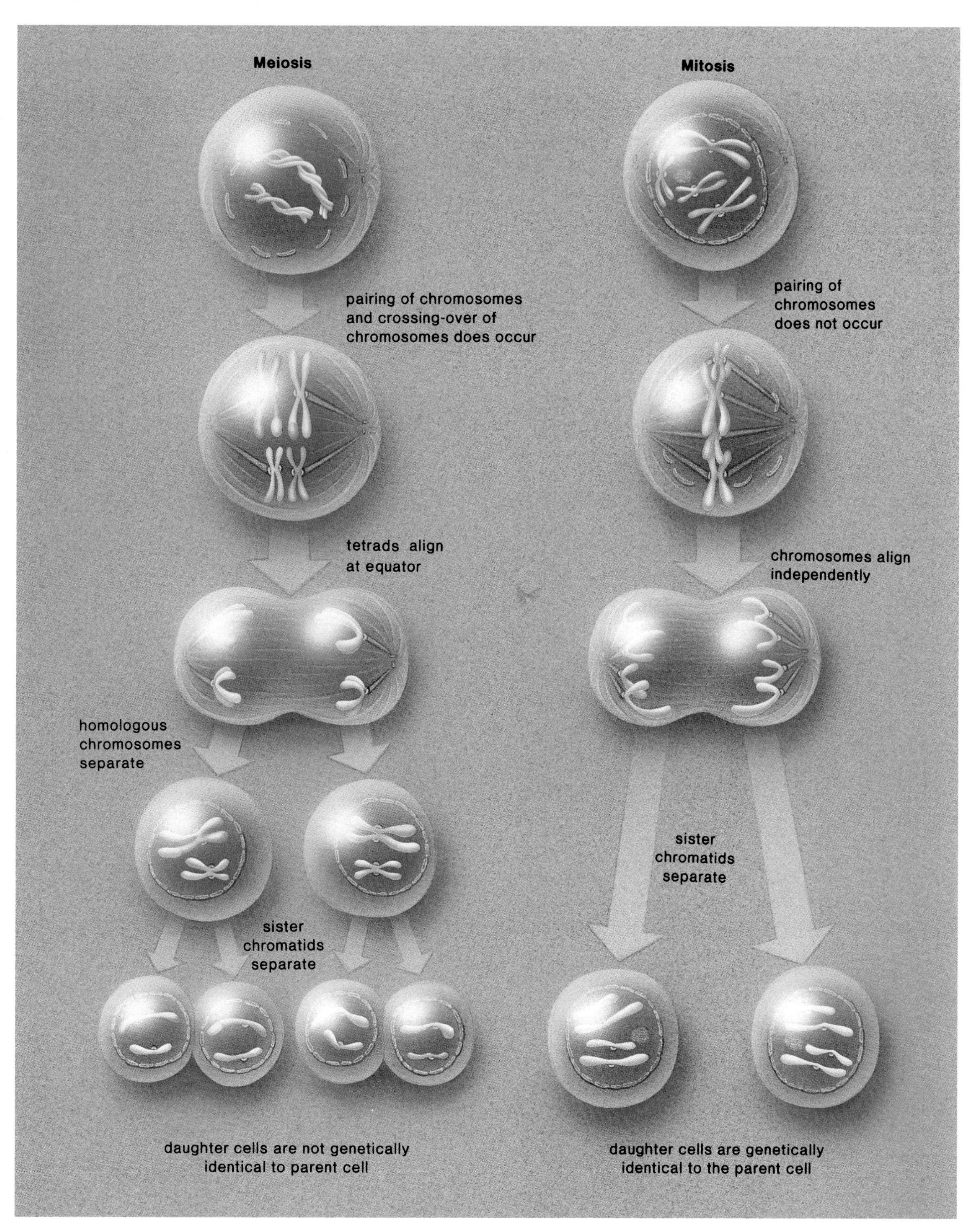

WHAT'S IN A CHROMOSOME?

What's in a chromosome? When early investigators decided that the genes are on the chromosomes, they had no idea of chromosome composition. By the mid-1900s, it was known that chromosomes are made up of both DNA and protein. Only in recent years, however, have investigators been able to produce models suggesting how chromosomes are organized.

A eukaryotic chromosome is more than 50% protein. Many of these proteins are concerned with DNA and RNA synthesis, but a large proportion, termed histones, seem to play primarily a structural role. A human cell contains 46 chromosomes, and the length of the DNA in each chromosome is about 5 cm. Therefore, a human cell contains at least 2 m of DNA. Yet, all of this DNA is packed into a nucleus that is about 5 μm in diameter. The histones seem to be responsible for packaging the DNA so that it can fit into such a small space. The packing unit, termed a nucleosome, gives chromatin a beaded appearance in certain electron micrographs.

Figure 7.A shows that the DNA double helix is wound around a core of histone molecules in a nucleosome. Notice how DNA stretches between the nucleosomes at the location of H_1 histone molecules. Whenever the H_1 histone molecules make contact, the chromatin shortens. At the time of cell division, the entire structure coils, folds, and condenses to give the highly compact form of the chromosome seen during the stage of cell division called metaphase. No doubt, compact chromosomes are easier to move about than extended chromatin.

Figure 7.A
Chromosome versus chromatin structure.

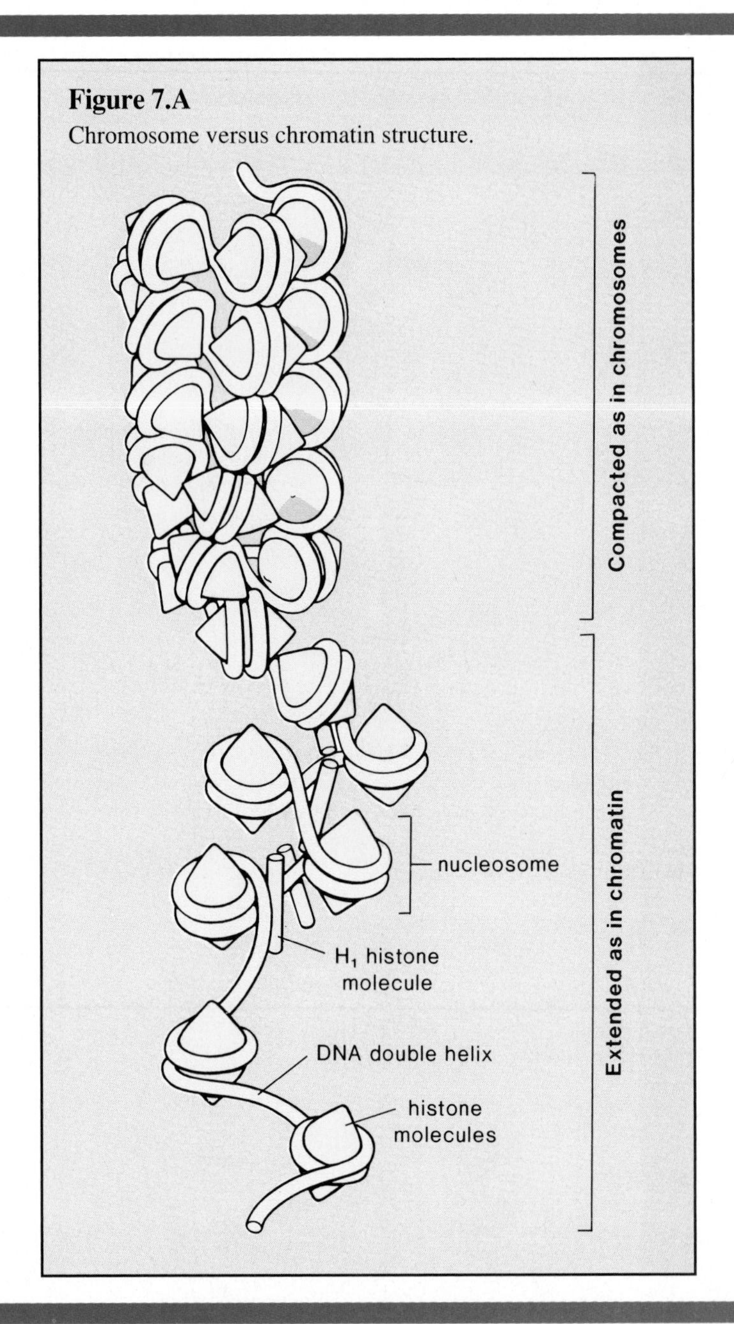

Results of Each

The genetic consequences of mitosis and meiosis are quite different.

1. Mitosis results in 2 daughter cells; meiosis results in 4 daughter cells.
2. The 2 daughter cells from mitosis are diploid; the 4 daughter cells from meiosis are haploid.
3. The mitotic daughter cells are genetically identical to each other and to the parent cell. The daughter cells from meiosis are not genetically identical to the parent cell. (Crossing-over and independent assortment ensure that no daughter cell is identical to another.)

SUMMARY

Humans have 23 pairs of homologous chromosomes, or 46 chromosomes altogether. The typical life cycle of animals includes these events:

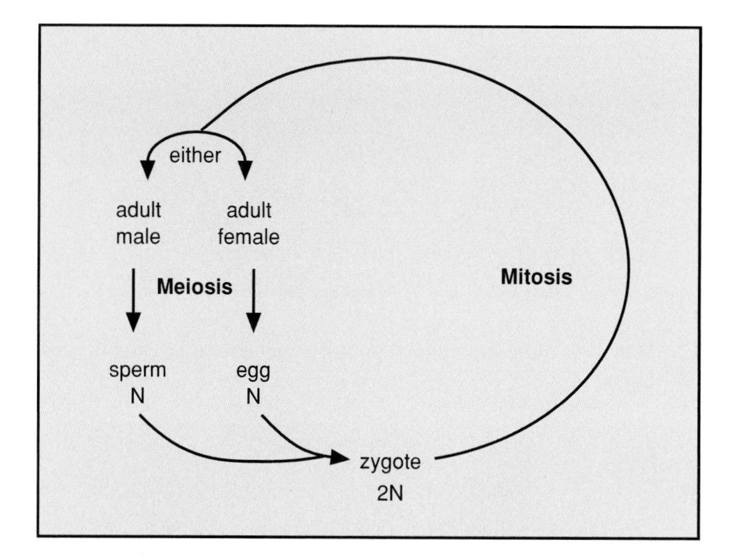

Mitosis ensures that all somatic cells have the diploid number and same kinds of chromosomes. Meiosis occurs during the production of gametes and ensures that the gametes have half the number of chromosomes—all possible combinations of chromosomes can occur in the gametes.

Mitosis is part of the cell cycle, which also includes the events of interphase. During interphase, chromosomes and organelles duplicate before mitosis begins.

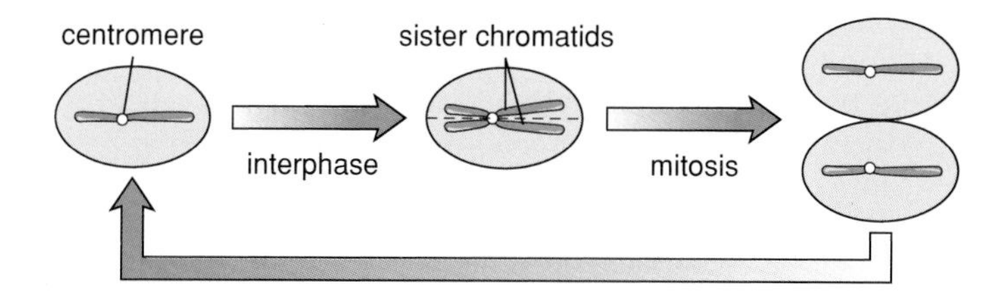

During mitosis, sister chromatids separate; then they are called daughter chromosomes. The cell cycle begins again.

For study purposes, mitosis can be divided into 4 stages: prophase (chromosomes move toward equator of spindle), metaphase (chromosomes are at equator), anaphase (chromosomes move to the poles), and telophase (chromosomes become chromatin). The cytoplasm is partitioned by a cleavage furrow in animals and by the formation of a cell plate in plants.

Meiosis involves 2 cell divisions. During meiosis I, the homologous chromosomes (following crossing-over between nonsister chromatids) separate, and during meiosis II the sister chromatids separate.

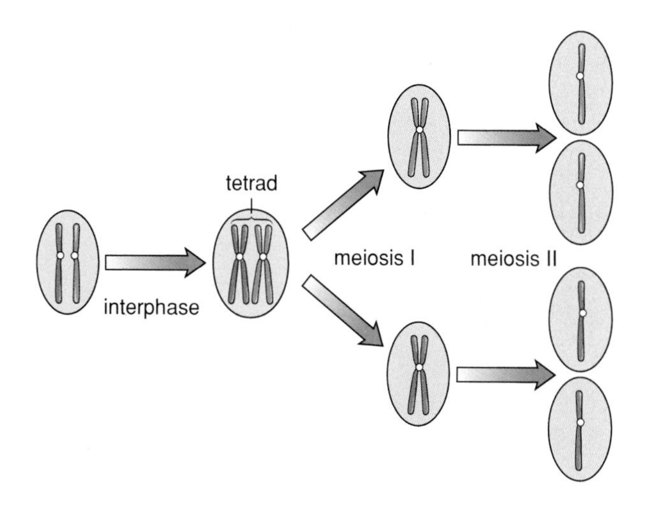

Meiosis results in 4 daughter cells, each having the haploid number of chromosomes. Meiosis is part of gamete formation in animals and spore formation in plants.

Mitosis can be compared to meiosis, as depicted in figure 7.14.

STUDY QUESTIONS

In order to practice **writing across the curriculum,** students should write out the answers to any or all of the study questions. The study questions are sequenced in the same order as the text.

1. Describe the normal karyotype of a human being. What is the difference between a male karyotype and a female karyotype? (p. 104)
2. Relate the terms *diploid (2N)* and *haploid (N)* to mitosis in somatic (body) cells and to meiosis in sex organs. (pp. 105–107)
3. Explain the makeup of a chromosome just prior to mitosis and meiosis. (p. 104)
4. Give several instances of when mitosis occurs in humans. (p. 106)
5. Describe the stages of animal mitosis, including in your description the terms *centriole, nucleolus, spindle,* and *cleavage furrow.* (pp. 107–111)
6. Name 2 differences between plant cell mitosis and animal cell mitosis. (p. 111)
7. What is the importance of meiosis in the life cycle of any organism? (p. 112)
8. Describe the stages of meiosis I, including in your description the term *tetrad.* (p. 113)
9. Describe the stages of meiosis II, including in your description the term *dyad.* (p. 113)
10. Explain why oogenesis produces one mature egg, but spermatogenesis results in 4 sperm. (p. 116)
11. Compare mitosis to both meiosis I and meiosis II. (p. 116)

OBJECTIVE QUESTIONS

1. During interphase, the chromosomes are not visible because they are extended into fine threads called _______.
2. If an organism has 12 chromosomes, it has _______ homologous pairs.
3. Just prior to division, every chromosome is composed of 2 identical parts called _______.
4. If the parent cell has 24 chromosomes, following mitosis the daughter cells have _______ chromosomes.
5. As the organelles called _______ separate and move to the poles, the spindle fibers appear.
6. _______ is the stage of mitosis during which the chromatids separate and become chromosomes.
7. Cytokinesis in an animal cell requires a _______, while in a plant cell it involves the formation of a _______.
8. Whereas mitosis results in 2 daughter cells, meiosis produces _______ daughter cells.
9. During anaphase I of meiosis, the _______ separate. This means that eventually the gametes will have the haploid number of chromosomes.
10. Meiosis ensures that the zygote will have a _______ combination of genes than either parent.
11. Label this diagram of a cell in prophase of mitosis.

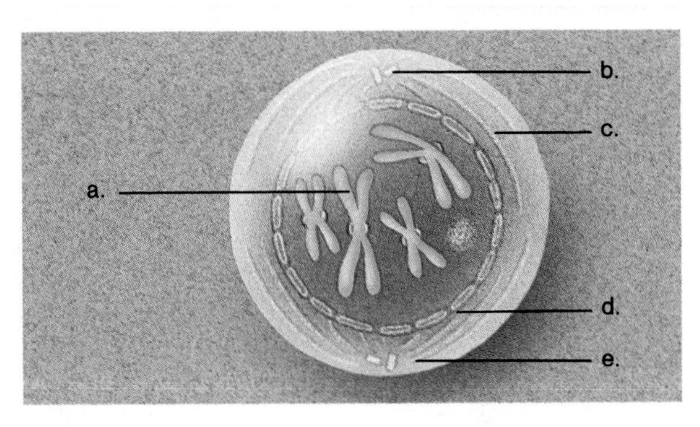

12. Which of these drawings represents metaphase I? How do you know?

CRITICAL THINKING

In order to practice **writing across the curriculum,** students should write out the answers to any or all of the critical thinking questions. Suggested answers to the critical thinking questions are in appendix E.

7.1

1. Which form of the genetic material—chromatin or chromosomes—would you expect to be actively directing protein synthesis? Why?
2. The genes are on the chromosomes. Ordinarily, a person inherits 2 copies of a gene—one from the mother and one from the father. Does it seem reasonable that the 2 copies of the gene might be different forms of the gene?
3. The genes direct protein (enzyme) synthesis. Why do you suppose a person with 3 chromosomes of the same kind might suffer from various disorders?
4. Not all genes are active in all cells. For example, why might a defective gene not adversely affect the maturation of a sperm or an egg?

7.2

1. As cells grow, there is more cytoplasm per amount of cell membrane. Cell division restores the original cell membrane-to-cytoplasm ratio. Why is this useful to the cell?
2. During the last growth portion of the cell cycle, cellular organelles duplicate. Why is this useful?
3. Specialized cells "break out" of the cell cycle and do not divide anymore. Give an example of a specialized cell you would not expect to divide again.

7.3

1. The drug colchicine halts cell division (fig. 7.1). What part of a dividing cell do you suppose it disrupts?
2. Asexual reproduction ordinarily does not produce genetic variation. Why not?
3. Sexual reproduction does produce genetic variation. Why?
4. Would you expect sexual reproduction to aid the evolutionary process? Why?

SELECTED KEY TERMS

aster short microtubule that extends outward from a spindle pole in animal cells during cell division.

autosome chromosome other than a sex chromosome.

cell cycle a repeating sequence of events in eukaryotic cells consisting of interphase, when growth and DNA synthesis occurs, and mitosis, when cell division occurs.

cell plate a double-layered membrane that precedes the formation of the cell wall as a part of cytokinesis in plant cells.

centromere (sen´tro-mēr) a region of attachment of a chromosome to spindle fibers that is generally seen as a constricted area.

chromatid (kro´ma-tid) one of the 2 identical parts of a chromosome following replication of DNA.

cleavage furrow an indentation that begins the process of cleavage, by which animal cells undergo cytokinesis.

crossing-over the exchange of corresponding segments of genetic material between nonsister chromatids of homologous chromosomes during synapsis of meiosis I.

cytokinesis (si″to-ki-ne´sis) division of the cytoplasm of a cell following telophase of mitosis and meiosis I and II.

diploid (dĭp´loid) the **2N** number of chromosomes; twice the number of chromosomes found in gametes.

gamete (gam´et) one of 2 types of reproductive cells that join in fertilization to form a zygote; most often an egg or a sperm.

haploid the **N** number of chromosomes; half the diploid number; the number characteristic of gametes that contain only one set of chromosomes.

homologous chromosome (ho-mol´o-gus kro´mo-sōm) similarly constructed; homologous chromosomes have the same shape and contain genes for the same traits.

karyotype (kar´e-o-tīp) the arrangement of all the chromosomes within a cell by pairs in a fixed order.

meiosis (mi-o´sis) a type of cell division occurring during the production of gametes in animals by means of which the 4 daughter cells have the haploid number of chromosomes.

mitosis (mi-to´sis) type of cell division in which daughter cells receive the exact chromosome and genetic makeup of the parent cell; occurs during growth and repair.

oogenesis (o″o-jen´ĕ-sis) production of an egg in females by the process of meiosis and maturation.

sex chromosome chromosome responsible for the development of characteristics associated with gender; an X or Y chromosome.

somatic cell in animals, any cell other than those that undergo meiosis and become a sperm or egg; body cell.

spermatogenesis (sper″mah-to-jen´ĕ-sis) production of sperm in males by the process of meiosis and maturation.

spindle the structure that brings about the movement of chromosomes during cell division; microtubular fibers stretch between poles, which are surrounded by microtubular asters.

synapsis (sĭ-nap´sis) the attracting and pairing of homologous chromosomes during prophase I of meiosis.

tetrad a set of 4 chromatids resulting from the pairing of homologous chromosomes during prophase I of meiosis.

zygote (zi´gōt) diploid cell formed by the union of 2 gametes; the product of fertilization.

FURTHER READINGS FOR PART TWO

Albersheim, P., and G. A. Darvill. September 1985. Oligosaccharins. *Scientific American.*

Alberts, B., et al. 1989. *Molecular biology of the cell.* 2d ed. New York: Garland Publishing.

Allen, R. D. February 1987. The microtubule as an intracellular engine. *Scientific American.*

Atkins, P. W. October 1987. Molecules. New York: *Scientific American.*

Baker, J. J. W., and G. E. Allen. 1987. *Matter, energy, and life.* 4th ed. Reading, Mass.: Addison-Wesley.

Becker, W. M., and D. W. Deamer. 1991. *The world of the cell.* 2d ed. Redwood City, Calif.: Benjamin/Cummings Publishing Co.

Bubel, A. 1989. *Microstructure and function of cells.* New York: John Wiley and Sons.

Darnell, J., H. Lodish, and D. Baltimore. 1986. *Molecular cell biology.* New York: Scientific American Books.

Dautry-Varsat, A., and H. F. Lodish. May 1984. How receptors bring proteins and particles into cells. *Scientific American.*

deDuve, C. May 1983. Microbodies in the living cell. *Scientific American.*

—. 1986. *A guided tour of the living cell.* Vols. 1 and 2. New York: Scientific American.

Glover, D. M., C. Gonzalez, and J. W. Raff. June 1993. The centrosome. *Scientific American.*

Govindjee, and W. J. Coleman. February 1990. How plants make oxygen. *Scientific American.*

Grivell, L. A. March 1983. Mitochondrial DNA. *Scientific American.*

Hall, D., and K. Rao. 1987. *Photosynthesis.* 4th ed. Baltimore: Arnold.

Harold, F. 1986. *The vital force: A study of bioenergetics.* San Francisco: W.H. Freeman.

Karplus, M., and J. A. McCannon. April 1986. The dynamics of proteins. *Scientific American.*

Kleinsmith, L. J., and V. M. Kish. 1988. *Principles of cell biology.* New York: Harper and Row.

Lienhard, G. E., et al. January 1992. How cells absorb glucose. *Scientific American.*

McIntosh, J. R., and K. L. McDonald. October 1989. The mitotic spindle. *Scientific American.*

Mathews, C. K., and K. E. van Holde. 1990. *Biochemistry.* Redwood City, Calif.: Benjamin/Cummings Publishing Co.

Ostro, M. J. January 1987. Liposomes. *Scientific American.*

Prescott, D. M. 1988. *Cells.* Boston: Jones and Bartlett.

Richards, F. M. January 1991. The protein folding problems. *Scientific American.*

Rothman, J. September 1985. The compartmental organization of the Golgi apparatus. *Scientific American.*

Scientific American. October 1985. The molecules of life.

Sharon, N., and H. Lis. January 1993. Carbohydrates in cell recognition. *Scientific American.*

Sheeler, P., and D. E. Bianchi. 1987. *Cell and molecular biology.* 3d ed. New York: John Wiley and Sons.

Stryer, L. 1988. *Biochemistry.* 3d ed. San Francisco: W.H. Freeman.

Unwin, N., and R. Henderson. February 1984. The structure of proteins in biological membranes. *Scientific American.*

Welch, W. J. May 1993. How cells respond to stress. *Scientific American.*

Wickramasinghe, H. K. October 1989. Scanning-probe microscopes. *Scientific American.*

Youvan, D. C., and B. I. Marrs. June 1987. Molecular mechanisms of photosynthesis. *Scientific American.*

Leaves of *Coleus blumei*

PLANT BIOLOGY

Flowering plants, representative of plants in general, are composed of a root system and a shoot system; the latter includes stems, leaves, and flowers. The organization of these organs suits their individual functions and a plant's ability to carry on photosynthesis. There are special transport pathways that take water and minerals from the roots to the leaves.

Other pathways take organic substances, produced in the leaves, throughout the plant. Plants grow and reproduce their entire life, and these processes are regulated to be in tune with the seasons.

8

PLANT ORGANIZATION AND GROWTH

Chapter Concepts

1.
The vegetative organs of a plant are the roots, the stems, and the leaves. 126, 135

2.
Flowering plants are divided into 2 groups, the monocots and the dicots. 126, 135

3.
Each organ in a plant contains specific tissues and specialized cell types. 127, 135

4.
Water and minerals enter a plant at the roots, and the organization of a root suits this function. 129, 135

5.
Stems continue to grow in length and girth; therefore, they experience both primary and secondary growth. 130, 135

6.
Leaves carry on photosynthesis, and the organization of a leaf suits this function. 134, 135

Western bleeding heart
(*Dicentra formosa*)

Chapter Outline

Figure 8.1

Root and shoot systems of a plant. A plant has 2 main divisions: the root system below ground and the shoot system containing the stems and the leaves above ground. Vascular tissue transports materials from the roots to the leaves and vice versa. The production of new cells at the terminal bud (shoot tip) and the root tip results in a plant's lengthwise growth.

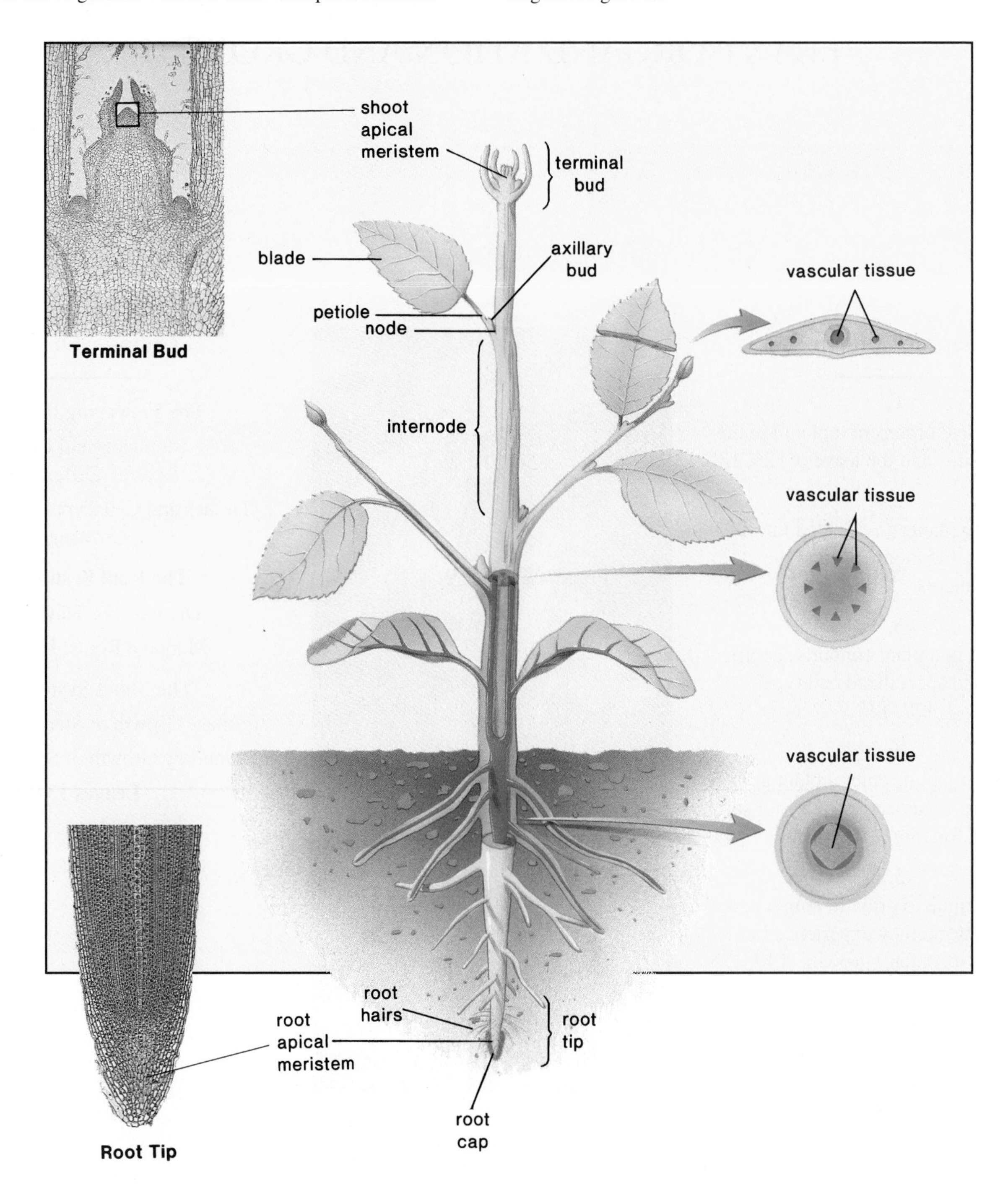

The Flowering Plant

The body of a flowering plant has 2 divisions: the root system and the shoot system (fig. 8.1). The root anchors the plant in the soil, and a stem holds the leaves aloft to catch the rays of the sun. The root absorbs water (H_2O) and minerals; then these nutrients are transported in the stem to the leaves, which receive carbon dioxide (CO_2) from the air and carry on photosynthesis. A carbohydrate (CH_2O) produced by photosynthesis is transported in the stem to the rest of the plant and then finally to the root, where it is stored as starch.

Monocots and Dicots: Several Differences

Flowering plants are divided into the **monocots** and the **dicots,** depending on the number of cotyledons, or "seed leaves," the embryonic plant has. A **cotyledon** provides nutrient molecules for growing embryos before the true leaves begin photosynthesizing. Adult monocots (e.g., corn and other grasses) and dicots (e.g., beans and peas) have several structural differences, which we will study in this chapter.

Tissues and Cell Types: Plants Keep Growing

Plants grow throughout their entire life because **meristem,** which contains embryonic cells, is located in the shoot and root apexes. Meristem tissue divides and produces the cells that go on to differentiate into various types of cells. Apical meristem gives rise to 3 tissue systems: dermal, ground, and vascular (fig. 8.2). The *dermal tissue* system forms the outer, protective covering of a plant. For example, **epidermis,** composed of epidermal cells, covers the entire body of nonwoody and young woody plants. Epidermis protects inner body parts and prevents the plant from drying out. In addition, the epidermis has specialized structures and functions, which are examined when each vegetative (nonreproductive) organ is considered. The *ground tissue* system fills the interior of a plant. Representative cells in this system are parenchyma and sclerenchyma cells. **Parenchyma** cells are relatively unspecialized and correspond best to the generalized cell of a plant (see fig. 3.4*b*). **Sclerenchyma** cells are often hollow and nonliving with extremely strong walls. They support other plant tissues and organs.

The *vascular (transport) tissue* system conducts water and nutrients in a plant. **Xylem** transports water (H_2O) and minerals from the roots to the leaves. Xylem contains 2 types of conducting cells, *tracheids* and *vessel elements.* Both of these types of conducting cells are hollow and nonliving at maturity. Water flows from tracheid to tracheid through pits, depressions where the secondary wall does not form. Water flows even more freely from one vessel element to the next because they have no end walls. **Phloem** transports organic nutrients, usually from the leaves to the roots. Phloem contains *sieve-tube cells,* each of which is associated with at least one *companion cell.* Sieve-tube cells, which have perforated end walls called sieve plates, contain cytoplasm but no nuclei. Strands of cytoplasm called plasmodesmata extend from one cell to the next through the sieve plates. Companion cells are much smaller than sieve-tube cells, but each has all the cellular components, including a nucleus. It is believed that a companion cell helps a sieve-tube cell perform its function of transporting organic nutrients. Chapter 9 discusses phloem transport in more detail.

Figure 8.2

Some plant cell types. Meristem produces new cells, and these cells differentiate into the specialized cells shown. The dermal tissue system consists of the epidermal cells of epidermis, the outermost tissue in all organs of the plant. The ground tissue system contains parenchyma and sclerenchyma cells found in cortex and pith, for example. The vascular tissue system contains the vessel elements and the tracheids found in xylem (see fig. 9.1) and the sieve-tube cells found in phloem (see fig. 9.4).

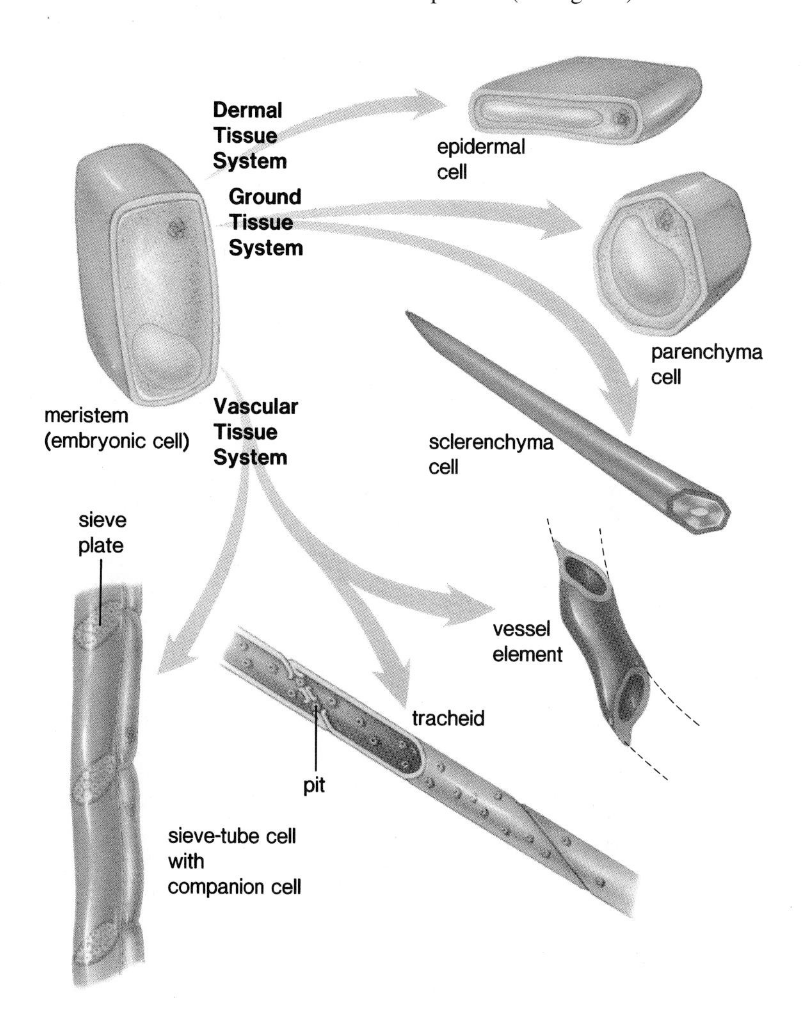

The body of a plant is composed of tissues containing specialized cells, which are differentiated to perform various functions. Meristem remains undifferentiated and is capable of continually dividing and producing new cells.

Figure 8.3

Dicot root tip. **a.** A root tip has 3 zones plus the root cap. **b.** The vascular cylinder of a dicot root contains vascular tissue. Xylem is typically star-shaped, and phloem lies between the points of the star. **c.** Because of the Casparian strip (waxy substance), water and solutes must pass through the endodermal cells. Therefore, endodermal cells regulate the passage of materials into the vascular cylinder.

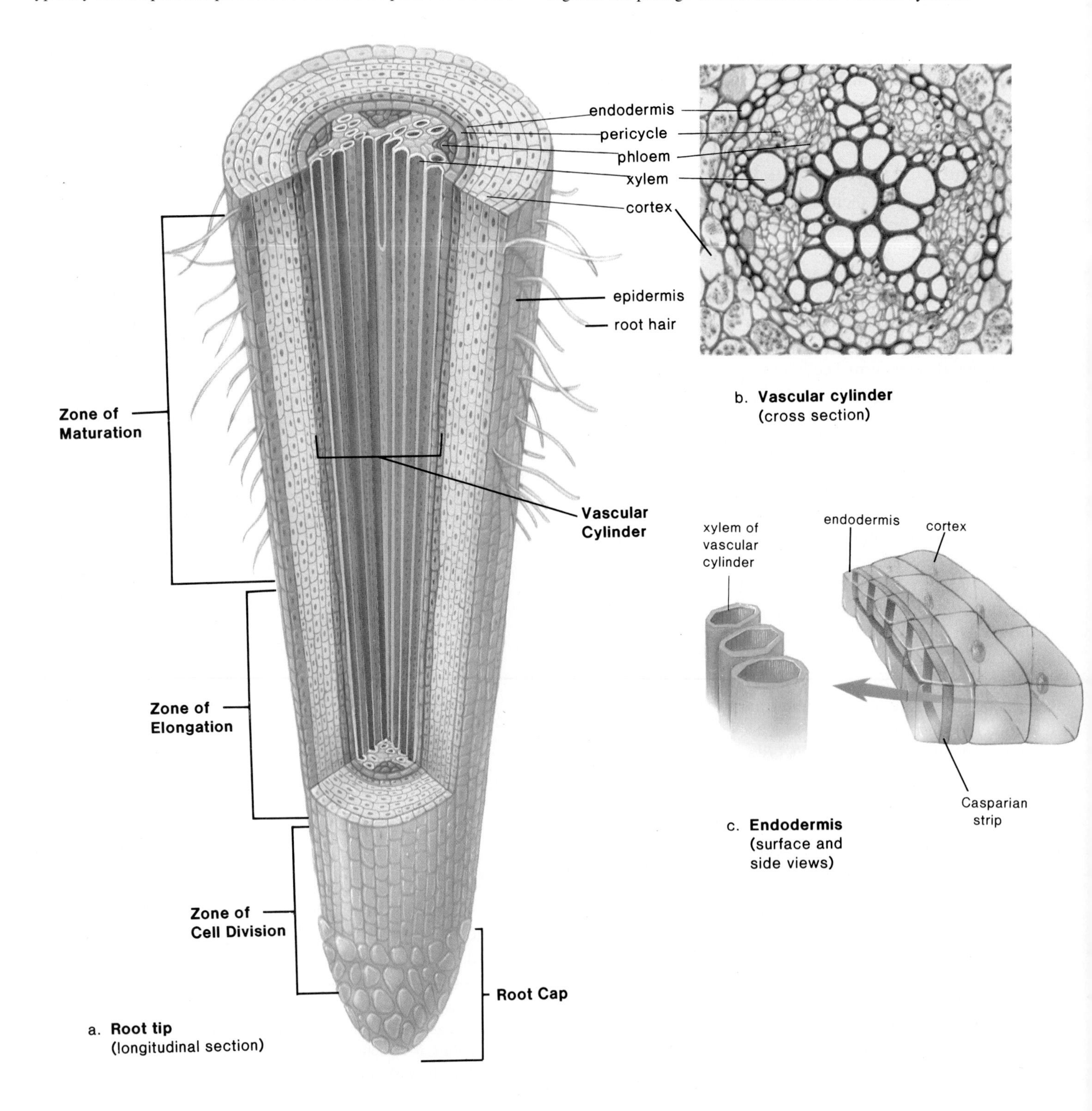

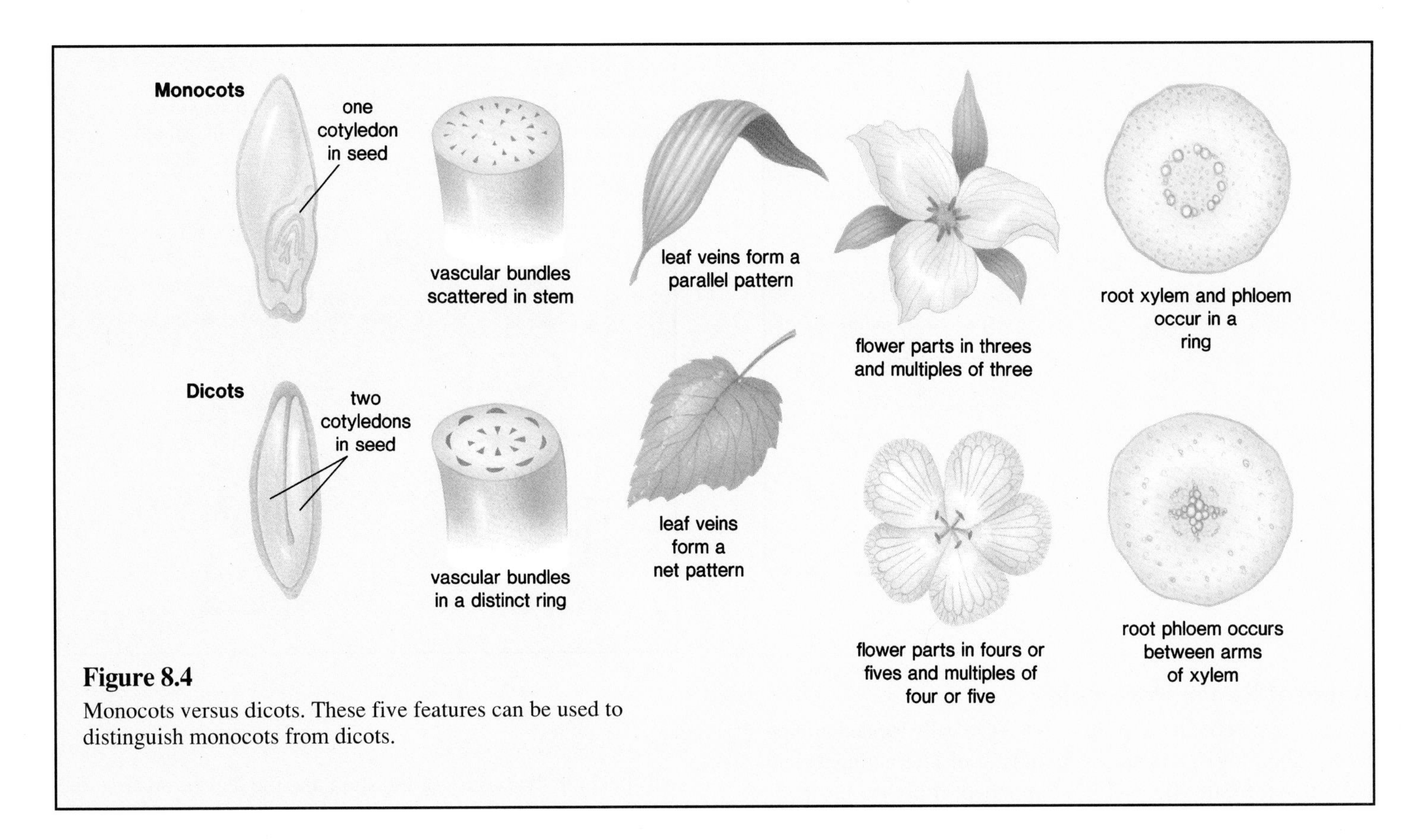

Figure 8.4
Monocots versus dicots. These five features can be used to distinguish monocots from dicots.

The Root System

The root system anchors a plant in the soil. It also absorbs water and minerals from the soil and stores the products of photosynthesis received from the leaves. Figure 8.3*a* is a longitudinal section of a dicot root. Monocots and dicots are contrasted in figure 8.4.

Dicot Roots: Have Zones

As primary growth (growth in length) occurs, the cells in the root enter zones that correspond to the various stages of differentiation. The *zone of cell division* contains apical meristem. Here cells are continuously added to the root cap below and to the zone of elongation above. The **root cap** is a protective cover for the root tip. The cells in the root cap are replaced constantly, because they are ground off as the root pushes through rough soil particles. In the *zone of elongation,* the cells get longer as they specialize. In the *zone of maturation,* the cells are mature and fully differentiated. This zone is recognizable even in a whole root because root hairs are borne by many of the epidermal cells. Root hairs add tremendously to the total absorptive surface area of roots, as described in the reading on page 134.

A cross section of a root at the zone of maturation is shown in figure 8.3*a* and *b*. These specialized tissues are present:

Epidermis The epidermis, which forms the outer layer of the root, consists of only a single layer of cells. The majority of epidermal cells are thin walled and rectangular, but in the zone of maturation, many epidermal cells have root hairs. These project as far as 5–8 mm into the soil.

Cortex Moving inward from the epidermis, large, thin-walled parenchyma cells make up the cortex. These irregularly shaped cells are loosely packed, making it possible for water and minerals to move through the cortex without entering the cells. The cells contain starch granules, and the cortex functions in food storage.

Endodermis The endodermis is a single layer of rectangular cells that forms a boundary between the cortex and the vascular cylinder. The endodermal cells fit snugly together and are bordered on 4 sides (fig. 8.3*c*) by a ring of waxy material known as the **Casparian strip.** Because of the Casparian strip, water and minerals must pass through endodermal cells in order to reach the vascular cylinder. Therefore, the endodermis regulates the entrance of minerals into the vascular cylinder.

Vascular cylinder The **pericycle,** the first layer of cells within the vascular cylinder, retains the capacity to divide and start the development of branch or secondary roots. Most of the vascular cylinder contains vascular tissue, xylem and phloem. Xylem appears star-shaped in dicots (fig. 8.3*b*) because several arms of tissue radiate from a common center. The phloem is found in separate regions between the arms of the xylem.

Figure 8.5

Cross section of the root of a corn plant (a monocot), showing the xylem and phloem tissue surrounding the pith.

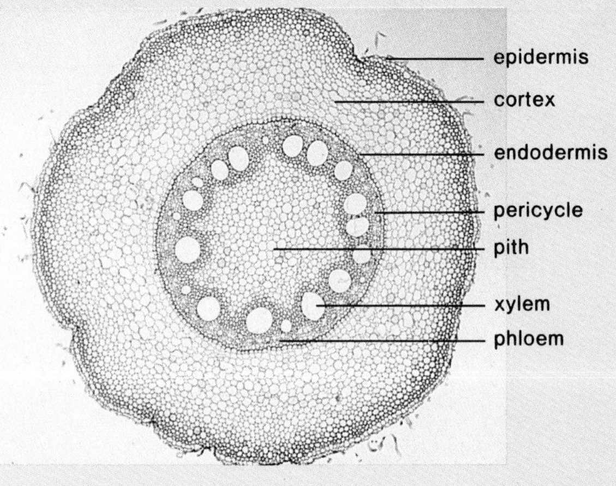

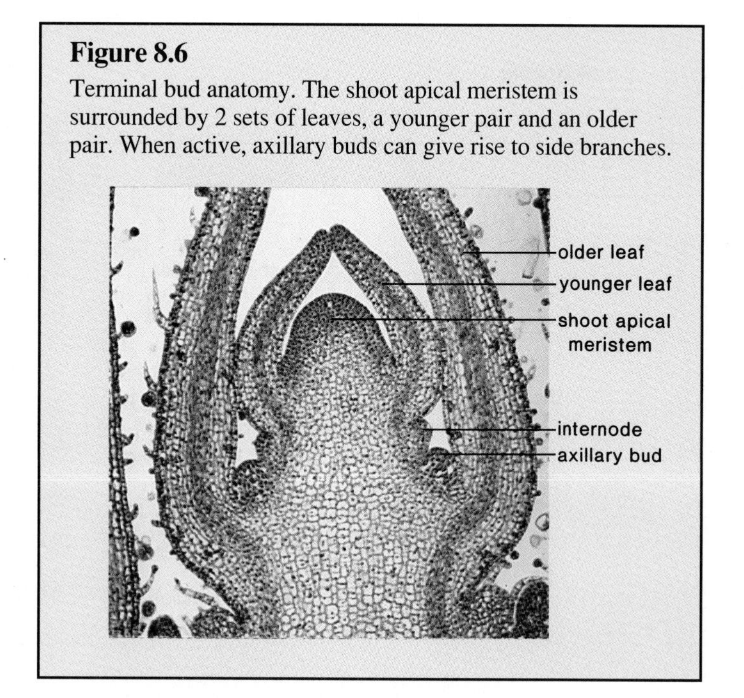

Figure 8.6

Terminal bud anatomy. The shoot apical meristem is surrounded by 2 sets of leaves, a younger pair and an older pair. When active, axillary buds can give rise to side branches.

Monocot Roots: Have Pith

A monocot root often has **pith,** which is centrally located ground tissue. The pith is surrounded by a ring of alternating xylem and phloem tissue (fig. 8.5). The monocot root also has pericycle, endodermis, cortex, and epidermis with root hairs.

> The root system of a plant absorbs water and minerals, which cross the epidermis and cortex before entering the endodermis, the tissue that regulates the entrance of molecules into the vascular cylinder.

The Shoot System

The shoot system of a flowering plant includes the stem and the leaves. Here, we will consider stem anatomy, and in the next section, we will examine leaf anatomy. A stem supports leaves, flowers, and fruits; it conducts substances to and from the roots and the leaves; and it helps to store water and the products of photosynthesis.

Even highly modified stems can be recognized by the presence of nodes, internodes, buds, and leaves (fig. 8.1). A *node* is the point at which leaves or buds are attached to a stem, and an *internode* is a segment between nodes.

Primary Growth of Stems: From Buds

The *terminal bud* of a growing stem contains shoot apical meristem (fig. 8.6). Whereas root apical meristem is protected by a root cap, shoot apical meristem is protected by newly formed leaves within a bud. Shoot apical meristem produces cells that will become the tissues of the stem and the leaves. At first, the nodes are very close together; then, internode growth separates them. This pattern complicates the growth of stems, and it is not possible to divide the entire stem into zones of cell division, elongation, and maturation as with the root.

Axillary buds, which are usually dormant but may develop into branch shoots, occur at axils, an angle between the stem and the leaves. Inactive buds are covered by protective bud scales. In the temperate zone, the terminal bud stops growing in the winter and is then protected by bud scales. In the spring, when growth resumes, these scales fall off and leave a scar. You can tell the age of a stem by counting these bud-scale scars.

Herbaceous Stems: Only Primary Growth

Mature nonwoody (**herbaceous**) stems exhibit only primary growth. The outermost tissue of herbaceous stems is the epidermis, which is covered by a waxy cuticle to prevent water loss. These stems have distinctive **vascular bundles,** where xylem and phloem are found enclosed by a sheath of support cells. In each bundle, xylem is typically found toward the inside and phloem is found toward the outside of the stem.

In the dicot herbaceous stem (fig. 8.7), the bundles are grouped in a distinct ring that separates the *cortex* from the central *pith.* The cortex is sometimes green and carries on photosynthesis, and the pith may function as a storage site for the products of photosynthesis. In the monocot herbaceous stem (fig. 8.8), the vascular bundles are scattered throughout the stem, and there is no well-defined cortex nor well-defined pith.

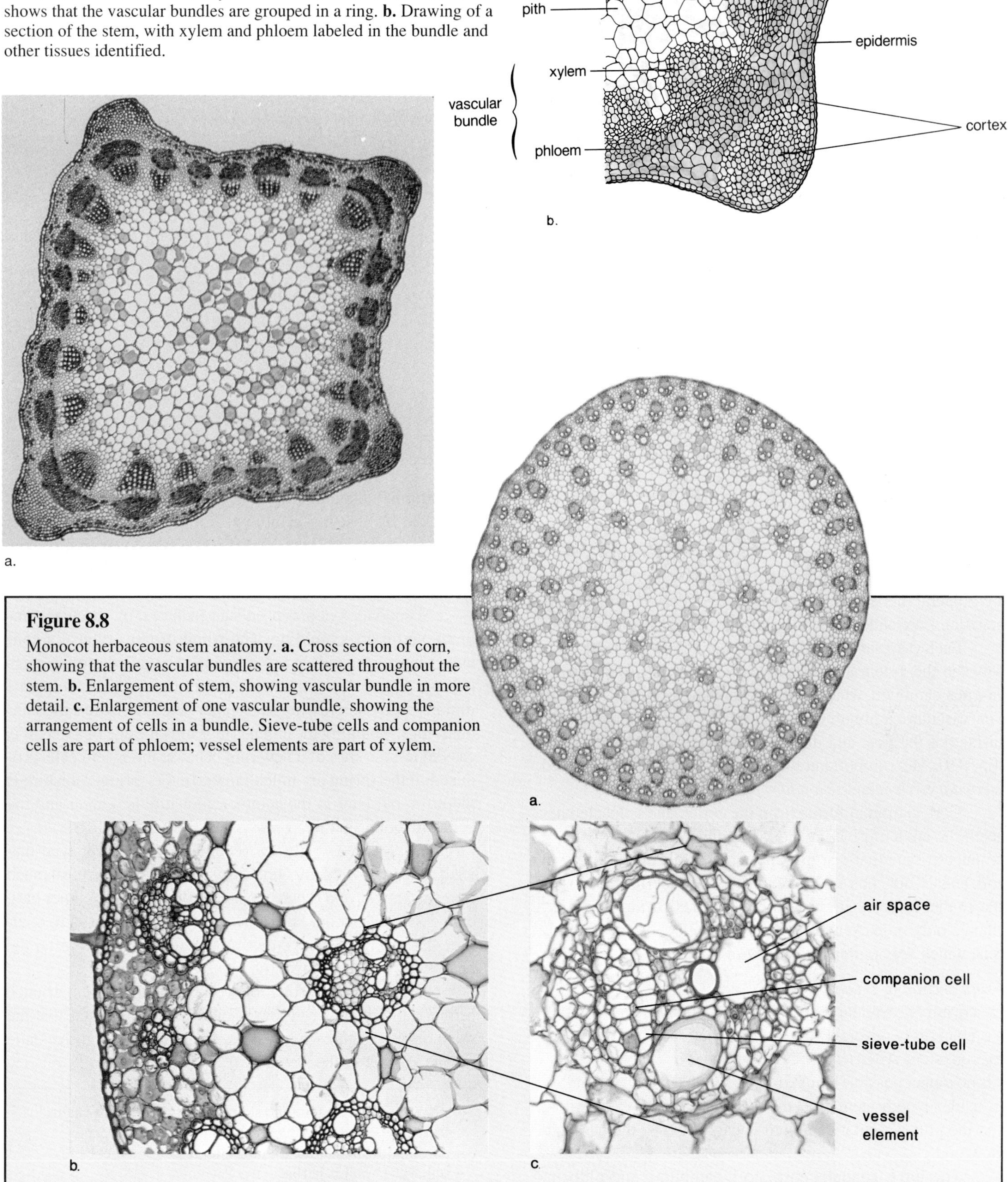

Figure 8.7

Dicot herbaceous stem anatomy. **a.** Cross section of an alfalfa stem shows that the vascular bundles are grouped in a ring. **b.** Drawing of a section of the stem, with xylem and phloem labeled in the bundle and other tissues identified.

Figure 8.8

Monocot herbaceous stem anatomy. **a.** Cross section of corn, showing that the vascular bundles are scattered throughout the stem. **b.** Enlargement of stem, showing vascular bundle in more detail. **c.** Enlargement of one vascular bundle, showing the arrangement of cells in a bundle. Sieve-tube cells and companion cells are part of phloem; vessel elements are part of xylem.

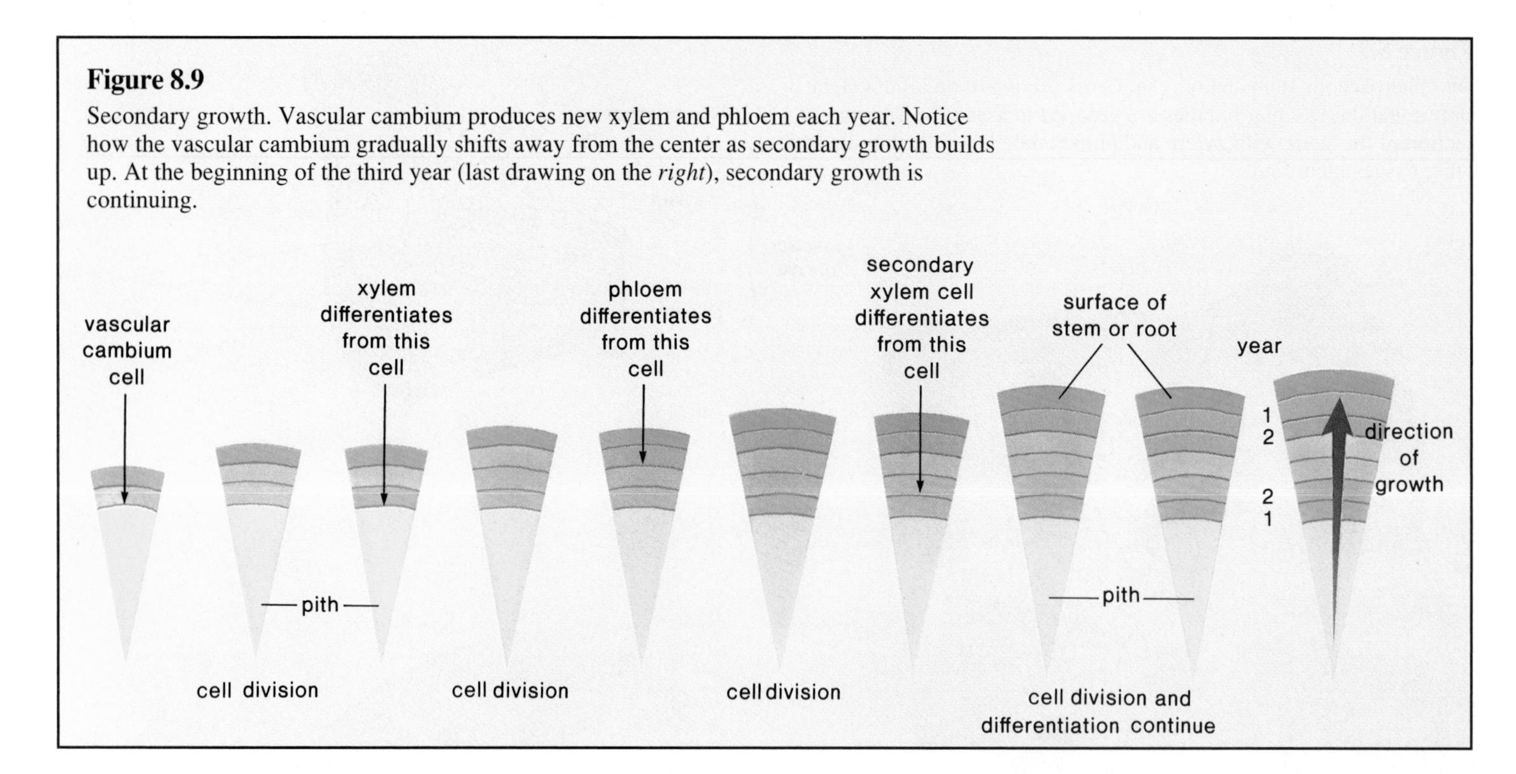

Figure 8.9
Secondary growth. Vascular cambium produces new xylem and phloem each year. Notice how the vascular cambium gradually shifts away from the center as secondary growth builds up. At the beginning of the third year (last drawing on the *right*), secondary growth is continuing.

Secondary Growth of Stems: In Trees

Apical meristem produces the primary tissues and is responsible for primary growth, which increases the length of a stem. Particularly in woody dicots, such as many types of trees, there are 2 lateral meristems that produce secondary growth. These meristems are called **vascular cambium** and **cork cambium.**

Each vascular bundle in a stem contains meristematic cells between the xylem and the phloem. These cells join to form vascular cambium, which stretches completely around the stem. The vascular cambium cells divide in a plane parallel to the surface of the tree, and this causes a tree to increase in girth (fig. 8.9). More important, the cells become the tissues called *secondary xylem* and *secondary phloem.*

Cork cambium forms from the cortex and is located beneath the epidermis. When cork cambium begins to divide, it produces tissue that disrupts and replaces the epidermis with cork cells. Cork cells are impregnated with suberin, a fatty substance that makes them waterproof. Dead cork allows gas exchange only in pockets of loosely arranged cells called **lenticels,** which are not impregnated with suberin.

Cross Section of Woody Stem

As a result of secondary growth, a dicot woody stem (fig. 8.10) has an entirely different type of organization than a dicot herbaceous stem. After secondary growth has continued for a time, it is no longer possible to make out individual vascular bundles. Instead, a woody stem has 3 distinct areas: the bark, the wood, and the pith. Vascular cambium occurs between the bark and the wood.

The **bark** contains cork, cork cambium, and phloem. Because phloem is in the bark, even partial removal of bark can seriously damage a tree. Secondary phloem is produced each year by vascular cambium. In older stems, the previous year's secondary phloem contributes to the formation of new cork cambium. Therefore, secondary phloem does not build up year after year.

Secondary xylem is produced each year by vascular cambium, and it does build up. The **wood** of trees typically contains rings of secondary xylem, which are called growth rings because there is one for each season of growth. In temperate regions with one growing season per year, these are called annual rings. Counting the annual rings tells the age of a tree. It is easy to tell the start of a new ring because the xylem cells produced in the spring are much larger. In the spring, moisture is plentiful, but later in the summer, moisture is scarcer and the cells are much smaller.

In large trees, only the secondary xylem produced that year, called sapwood, functions in water transport. Deposits such as resins, gums, and other substances plug the older inner part, called the heartwood. Heartwood may help support a tree, although some trees stand erect and live for many years after the heartwood has rotted away.

Like stems, the roots of woody plants also experience primary and secondary growth. Figure 8.11 illustrates that the secondary growth of roots arises and progresses in the same manner as that of the stem.

> Within the shoot system, the stem transports water and nutrients between the leaves and the roots. Stem anatomy differs according to whether the plant is a monocot or a dicot and whether the plant is herbaceous or woody.

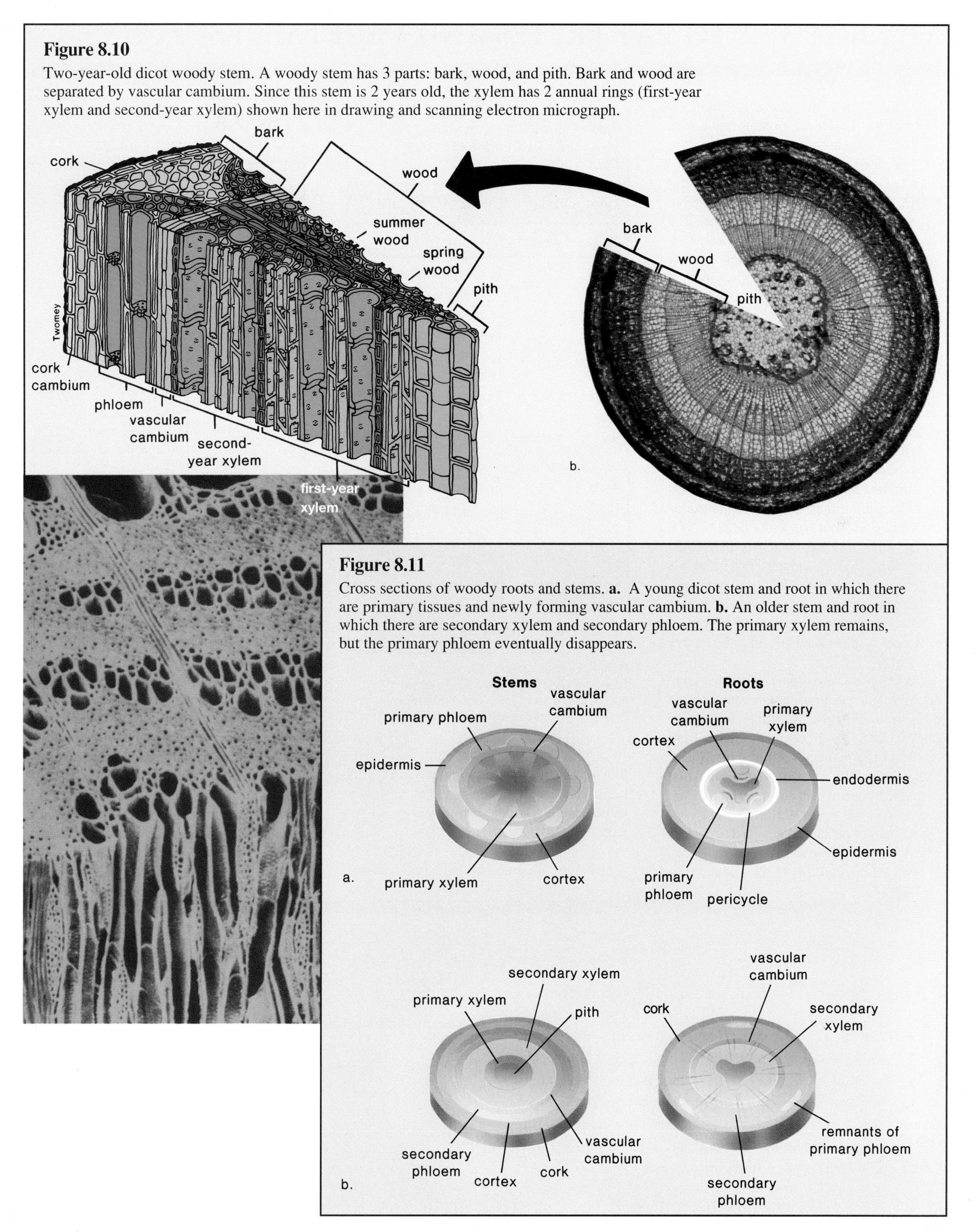

Figure 8.10

Two-year-old dicot woody stem. A woody stem has 3 parts: bark, wood, and pith. Bark and wood are separated by vascular cambium. Since this stem is 2 years old, the xylem has 2 annual rings (first-year xylem and second-year xylem) shown here in drawing and scanning electron micrograph.

Figure 8.11

Cross sections of woody roots and stems. **a.** A young dicot stem and root in which there are primary tissues and newly forming vascular cambium. **b.** An older stem and root in which there are secondary xylem and secondary phloem. The primary xylem remains, but the primary phloem eventually disappears.

WHY BOTHER TO EAT FRUITS AND VEGETABLES?

Plants need only inorganic nutrients in order to produce all the organic molecules that make up their body. Aside from carbon (C), hydrogen (H), and oxygen (O_2) (obtained from carbon dioxide [CO_2] and water [H_2O]), the mineral elements listed in table 8.A are required nutrients for plants. The mineral elements are classified as *macronutrients* when they are used by plants in great amounts and as *micronutrients* when they are needed by plants in very small amounts. Both types of minerals are found in the soil but in low concentrations; therefore, not only must a plant be able to take them up, it must also be able to concentrate them. Fortunately, the root system of a plant is designed for just this purpose. As the root system grows, it branches and branches again so that the roots are exposed to a tremendous amount of soil. It has been estimated that a rye plant has roots totaling about 900 km (more than 650 mi) in length. Further, because of the extensive number of root hairs, the total surface area is about 635 m^2, or more than 7,000 square feet! Water, and possibly minerals too, enters root hairs by diffusion, but eventually active transport is used to concentrate the minerals within the organs of a plant. A plant uses a great deal of ATP for active transport.

When plants are organically grown, the soil is enriched with organic matter instead of inorganic fertilizer. Contrary to popular belief, it makes absolutely no difference from which source a plant gets its nutrients. Sometimes organically grown also means that mechanical means instead of herbicides are used to control weeds and biological controls instead of pesticides are used to control pests. For example, mail-order ladybugs can be released to control aphid populations.

Table 8.A
Inorganic Nutrients Necessary for Plant Life

Compound	Mineral Supplied
Macronutrients	
KNO_3	K, N (potassium, nitrogen)
$CaNO_3$	Ca, N (calcium, nitrogen)
$NH_4H_2PO_4$	N, P (nitrogen, potassium)
$MgSO_4$	Mg, S (magnesium, sulfur)
Micronutrients	
KCl	Cl (chlorine)
H_3BO_3	B (boron)
$MnSO_4$	Mn (manganese)
$ZnSO_4$	Zn (zinc)
$CuSO_4$	Cu (copper)
H_2MoO_4	Mo (molybdenum)
Fe-EDTA	Fe (iron)

Since herbicides and pesticides are harmful to human health if consumed in quantity, some benefit is expected if they are not used to grow foods.

Human beings are lucky that plants can concentrate minerals, for we often are dependent on them for our basic supply of such minerals as sodium to maintain blood pressure, calcium to build bones and teeth, and iron to help carry oxygen to our cells. Once plants have taken the minerals up, they are incorporated into proteins, fats, and vitamins. When we eat plants, we are supplied with minerals and all types of organic molecules, some of which become building blocks for our own cells and some of which we use as an energy source.

Table 8.B lists examples of plant foods that humans consume. Each type of food is associated with a particular organ of a flowering plant: root, stem, leaf, or flower.

Table 8.B
Food from Flowering Plants

Plant Organ	Foods
Roots	Sweet potato, radish, turnip, parsnip
Stems	White potato, sugar cane, asparagus
Leaves	Cabbage, kale, spinach, lettuce, tea leaves
Petioles*	Celery, rhubarb
Seeds†	Pea, navy bean, lima bean, nuts, coffee bean
Fruits†	Wheat, rice, corn, oat, rye, string bean, apple, orange, peach, tomato, squash

*Part of a leaf

†Derived from flower parts

Leaves

Leaves are the organs of photosynthesis in flowering plants. A leaf usually consists of a flattened **blade** and a **petiole,** which connects the blade to the stem. The blade may be single or it may be composed of several leaflets. Externally, it is possible to see the pattern of the **leaf veins,** which are the final extensions of vascular tissue. Leaf veins have a net pattern in dicot leaves and a parallel pattern in monocot leaves (fig. 8.4). Because leaf veins contain vascular tissue, they transport both water and nutrients to and from the leaves.

The cross section of a typical dicot leaf belonging to a temperate-zone plant is shown in figure 8.12. There is a layer of *epidermis* at the top and the bottom of the leaf. The epidermis may bear protective hairs and glands that produce irritating substances; these may prevent the leaf from being eaten by insects. The epidermis is covered by a waxy *cuticle,* which keeps the leaf from drying out. Unfortunately, it also prevents gas exchange because the cuticle is not gas permeable. However, the epidermis, particularly the lower epidermis, contains openings called *stomata* (sing., **stoma**), which allow gases to move

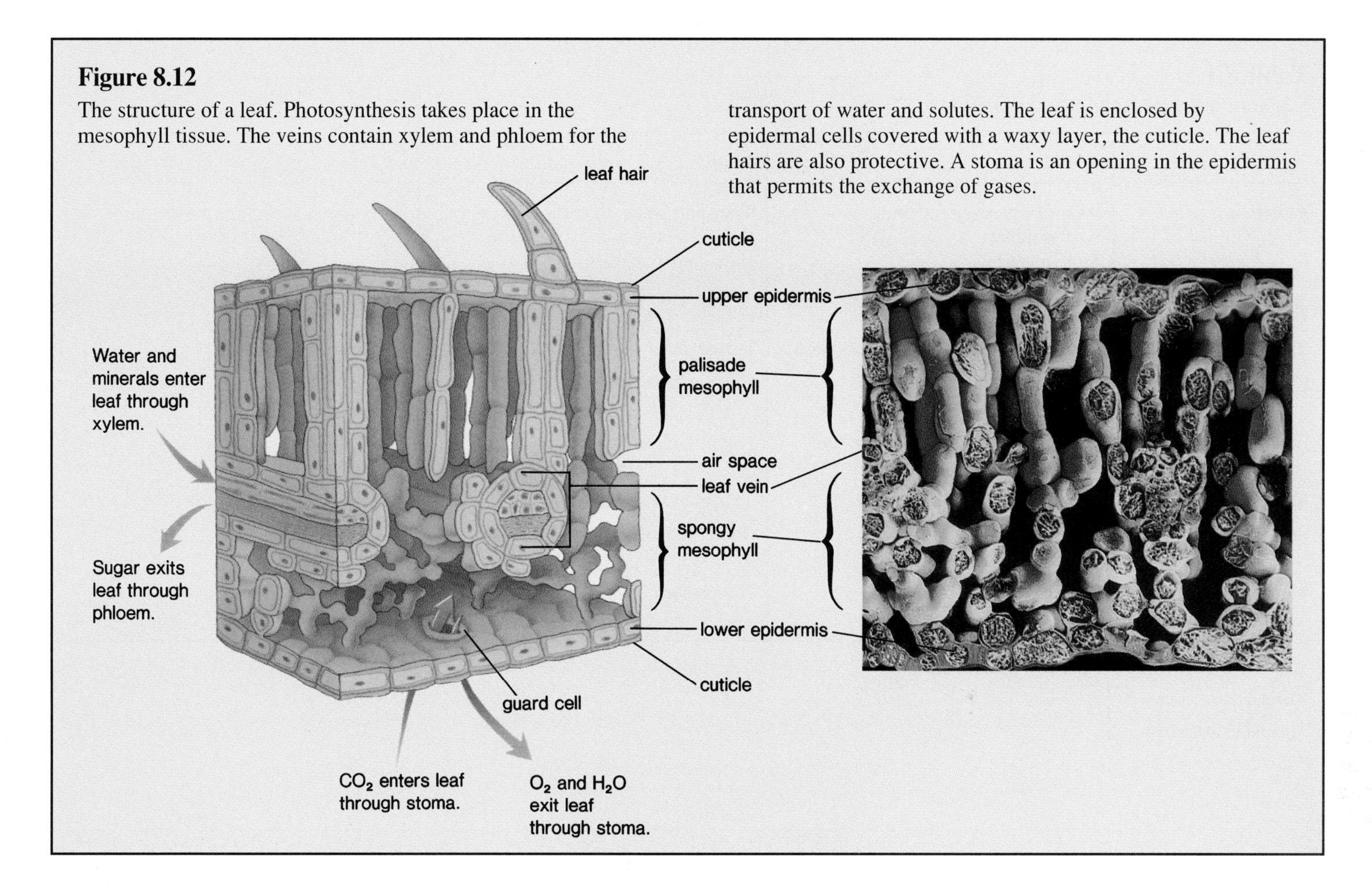

Figure 8.12
The structure of a leaf. Photosynthesis takes place in the mesophyll tissue. The veins contain xylem and phloem for the transport of water and solutes. The leaf is enclosed by epidermal cells covered with a waxy layer, the cuticle. The leaf hairs are also protective. A stoma is an opening in the epidermis that permits the exchange of gases.

into and out of the leaf. Each stoma has 2 **guard cells,** which regulate its opening and closing.

The body of a leaf is composed of **mesophyll,** which has 2 layers: the *palisade mesophyll,* an upper layer that contains elongated cells, and *spongy mesophyll,* a lower layer that contains irregular cells bounded by air spaces. The parenchyma cells of the palisade and spongy layers have many chloroplasts and carry on most of the photosynthesis for the plant. The loosely packed arrangement of the cells in the spongy layer increases the surface area for gas exchange.

Within the shoot system, the leaves carry on photosynthesis. The leaf is covered by epidermis and contains mesophyll tissue and leaf veins. Mesophyll contains a palisade layer and a spongy layer of cells. Do 8.1 Critical Thinking, found at the end of the chapter.

SUMMARY

The body of a flowering plant is divided into the root system and the shoot system. The shoot system contains the stems and the leaves. There are 2 categories of flowering plants (monocots and dicots), which are distinguished especially by whether there are one or 2 cotyledons.

The apical meristems produce cells that undergo specialization, forming 3 tissue systems: dermal tissue (i.e., epidermal cells); ground tissue (i.e., parenchyma and sclerenchyma cells); and vascular tissue. Vascular tissue consists of xylem and phloem. Xylem transports water and minerals, and phloem transports organic nutrients.

The root anchors a plant, absorbs water and minerals, and stores the products of photosynthesis. A root tip has 3 zones of primary growth: the zone of cell division (apical meristem), the zone of elongation, and the zone of maturation. A cross section of an herbaceous dicot root has the specific tissues listed in table 8.1.

Stems support leaves, conduct materials to and from roots and leaves, and help store plant products. Primary growth of a stem, due to the activity of the apical meristem, results in the specific tissues listed in table 8.1. In herbaceous dicot stems, the vascular bundles are arranged in a ring; in monocots, the vascular bundles are scattered. Secondary growth of a temperate woody stem is due to vascular cambium, which produces new secondary xylem and secondary phloem every year. The cross section of a woody stem shows bark, wood, and pith. Wood contains annual rings of xylem.

A leaf carries on photosynthesis and has the specific tissues listed in table 8.1. Gas exchange occurs at the stomata; the leaf veins contain xylem and phloem. Xylem brings water to and phloem transports photosynthetic products away from leaves.

Table 8.1
Vegetative Organs and Major Tissues

	Roots	Stems	Leaves
Function	Absorb water and minerals Anchor plant Store materials	Transport water and nutrients Support leaves Help store materials	Carry on photosynthesis
Tissue			
Epidermis*	Root hairs absorb water and minerals	Protect inner tissues	Stomata carry on gas exchange
Cortex†	Store water and products of photosynthesis	Carry on photosynthesis, if green	Not present
Endodermis†	Regulate passage of minerals into vascular cylinder	Not usually present	Not present
Vascular‡	Transport water and nutrients	Transport water and nutrients	Transport water and nutrients
Pith†	Store water and products of photosynthesis	Store products of photosynthesis	Not present
Mesophyll†	Not present	Not present	Carry on gas exchange and photosynthesis

Note: Plant tissues belong to one of 3 tissue systems:
*Dermal tissue system
†Ground tissue system
‡Vascular tissue system

STUDY QUESTIONS

In order to practice **writing across the curriculum,** students should write out the answers to any or all of the study questions. The study questions are sequenced in the same order as the text.

1. Contrast the root system with the shoot system. (p. 126)
2. Name and discuss the zones of a root tip. (p. 128)
3. Contrast monocots with dicots in 4 ways. (p. 129)
4. Name several cell types found in flowering plants. (p. 129)
5. Describe the anatomy of a dicot root tip, both longitudinal and cross section. How does the anatomy of a monocot root differ from this? (pp.128–130)
6. Describe the anatomy of a dicot herbaceous stem and of a dicot woody stem in cross section. How does the anatomy of a monocot stem differ from that of a dicot? (pp. 130–132)
7. Contrast the primary growth of stems and roots with the secondary growth. (pp. 132–133)
8. Describe the anatomy of a leaf in cross section. (pp. 134–135)

OBJECTIVE QUESTIONS

1. _______ is tissue that is unspecialized and capable of continual cell division.
2. In the roots, many epidermal cells have _______, and in the leaves, the epidermis contains openings called _______.
3. The conducting cells in xylem are called _______ and _______.
4. In a dicot root, mature cells are found in the zone of _______.
5. In a dicot root, the _______ regulates the passage of minerals into the vascular cylinder.
6. In a monocot stem, the vascular bundles are said to be _______.
7. In a woody stem, the secondary xylem builds up and forms the _______, which can be counted to tell the age of a tree.
8. Mesophyll contains 2 layers, the _______ layer and the spongy layer.
9. Label the following on this simplified drawing of a root: endodermis, phloem, xylem, cortex, and epidermis.

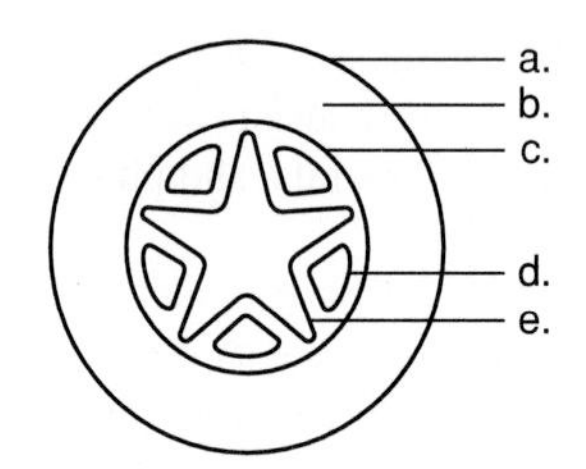

10. Label the following on this simplified drawing of a woody stem: phloem, pith, vascular cambium, cork, and xylem (wood).

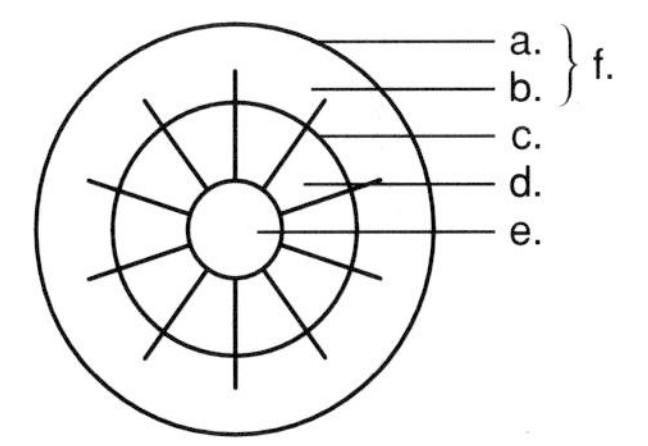

11. Label the following on this simplified drawing of a leaf: leaf vein, spongy mesophyll, palisade mesophyll, lower epidermis, and upper epidermis.

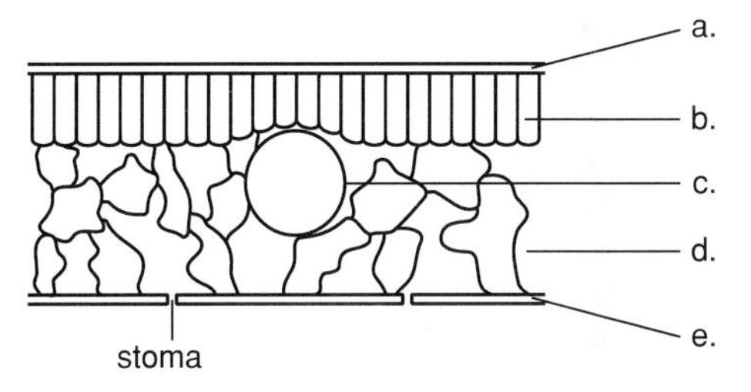

CRITICAL THINKING

In order to practice **writing across the curriculum,** students should write out the answers to any or all of the critical thinking questions. Suggested answers to the critical thinking questions are in appendix E.

8.1

1. What is the function of leaf epidermis, and how does the structure suit the function?
2. What is the function of the spongy layer of mesophyll, and how does the structure suit the function?
3. What is the function of leaf veins, and how does the structure suit the function?

SELECTED KEY TERMS

Casparian strip a waxy ring around endodermal cells that prevents passage of water and minerals other than through the cells.

cortex a tissue composed mainly of parenchyma cells, found between the vascular tissue and the epidermis in stems and roots.

cotyledon (kot˝ĭ-lédon) the seed leaf of the embryonic plant.

dicot dicotyledon; a type of flowering plant distinguished particularly by the presence of 2 cotyledons in the seed, such as beans and geraniums.

endodermis plant tissue consisting of a single layer of cells that surrounds and regulates the entrance of minerals, particularly into the vascular cylinder of roots.

epidermis the outer layer of cells of plants and other organisms.

herbaceous nonwoody.

meristem plant tissue that always remains undifferentiated and capable of dividing to produce new cells.

mesophyll the middle portion of a leaf made up of parenchyma cells, which carries on photosynthesis and gas exchange.

monocot (mon´o-kot) monocotyledon; a type of flowering plant in which the seed has only one cotyledon, such as corn and lily.

phloem (flóem) the vascular tissue in plants that transports organic nutrients.

pith central tissue composed of parenchyma cells that occurs in dicot stems.

stoma (pl., stomata) opening in the leaves of plants through which gas exchange takes place.

vascular bundle structure that includes xylem and phloem; typically found in herbaceous plant stems.

vascular cambium a meristem that produces secondary phloem and secondary xylem, which add to the girth of plants.

vascular cylinder a central region of roots; contains vascular and other tissues.

xylem (zɪ´-lem) the vascular tissue in plants that transports water and minerals.

9

PLANT PHYSIOLOGY AND REPRODUCTION

Chapter Concepts

1.
Water transport is dependent upon the chemical and physical properties of water. 139, 150

2.
Organic transport is dependent on the physical properties of water. 142, 150

3.
Plants respond to outside stimuli by changing their growth patterns. 143, 150

4.
Some plant responses are controlled by the length of daylight (photoperiod). 143, 150

5.
Plants reproduce both asexually and sexually. 146, 150

6.
Seeds within fruits are the products of sexual reproduction in flowering plants. 148, 150

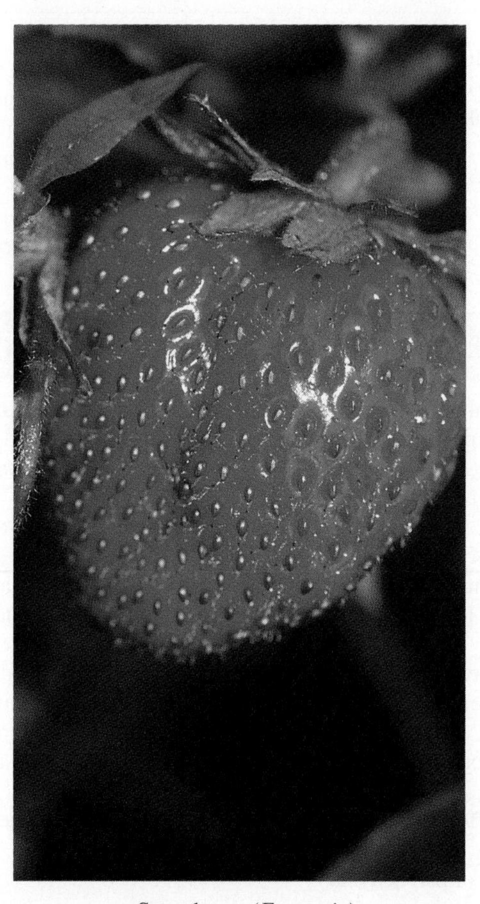

Strawberry (*Fragaria*)

Chapter Outline

LIKE all organisms, plants grow and reproduce. In order to grow, plants must acquire the inorganic nutrients they need to carry on photosynthesis. Water is transported from the roots to the leaves in xylem, a vascular tissue that is continuous throughout the body of the plant (fig. 9.1). Carbon dioxide enters a leaf through the stomata, openings in the lower epidermis. Following photosynthesis, sugars are transported throughout the plant by another vascular tissue, phloem. As discussed in chapter 6, plants have the metabolic capability of changing sugars into all the other types of organic molecules they require in order to grow.

Plant growth is seasonal—in the temperate zone, plants begin to grow in the spring and become dormant in the fall. As we shall see, plants have a means to detect the photoperiods—the relative periods of light and dark—that mark the seasons. The reproductive organs of flowering plants are located in flowers; some plants even produce flowers only during certain seasons of the year. Flowers attract motile pollinators (e.g., insects) that help bring the sperm and egg together so reproduction can occur. When all the members of a species flower at the same time, reproductive success is more likely.

The topics covered in this chapter tell us how plant structures function to bring about growth and reproduction.

Water Transport

Water enters a root primarily through the root hairs, makes its way across the cortex, and finally enters xylem within the vascular cylinder (see fig. 8.3). As also discussed in chapter 8, xylem contains 2 types of conducting cells: tracheids and vessel elements. The **tracheids** have pitted end walls, but the **vessel elements** have no end walls. The vessel elements, therefore, form a completely hollow pipeline from the roots to the leaves (fig. 9.1).

Figure 9.1

Xylem structure. Light micrograph shows generalized organization of xylem, and the drawings show the internal structure of tracheids and vessel elements. Tracheids and vessel elements are stacked one on top of the other; therefore, they conduct water from the roots to the leaves.

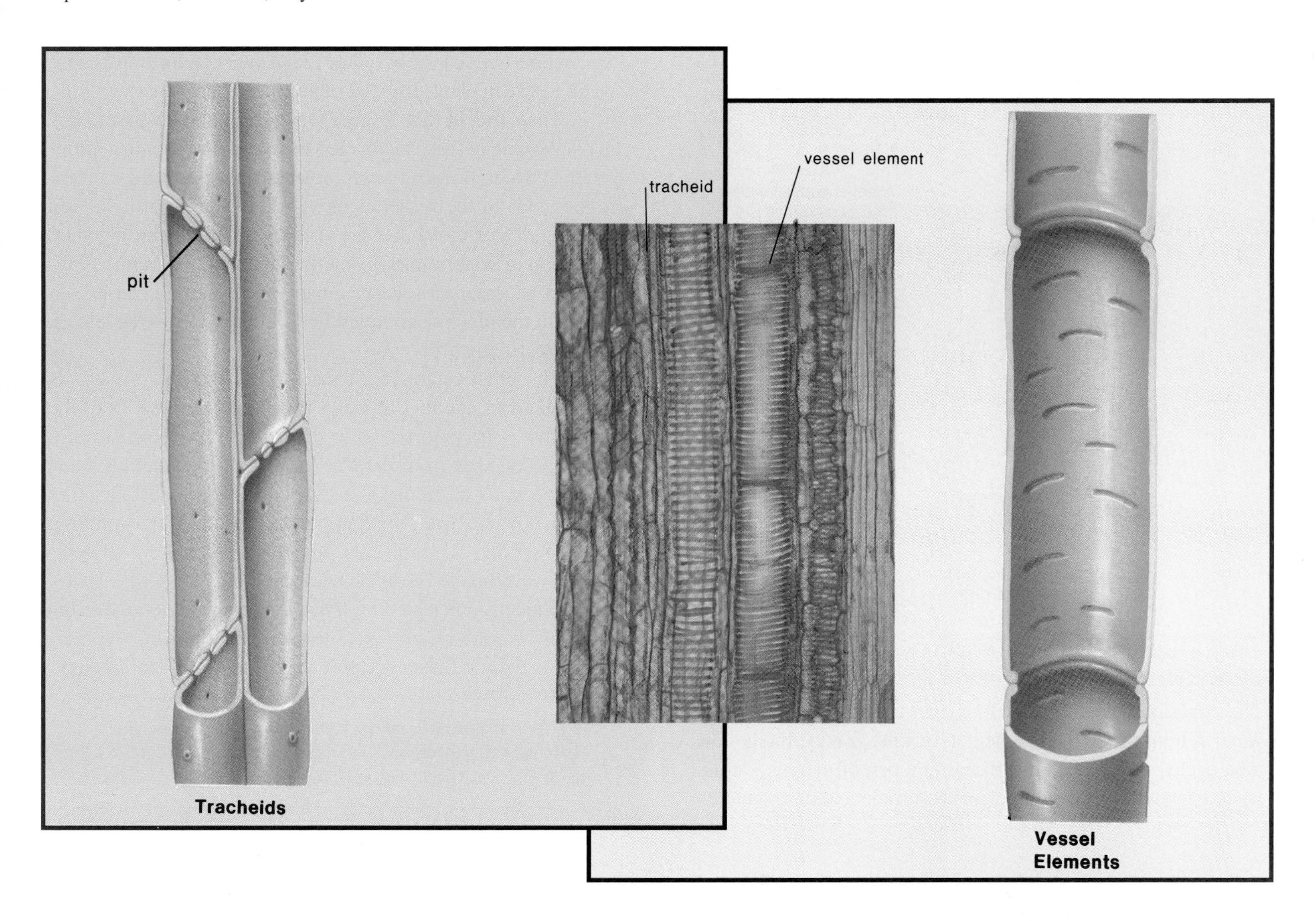

Figure 9.2

Cohesion-tension theory of xylem transport. Water (H_2O) enters a plant at the root hairs and transpires (evaporates) at the leaves. Vessel elements form a continuous pipeline from the roots to the leaves, and this pipeline is completely full of water. Transpiration exerts a pull on this water column, causing it to move upward.

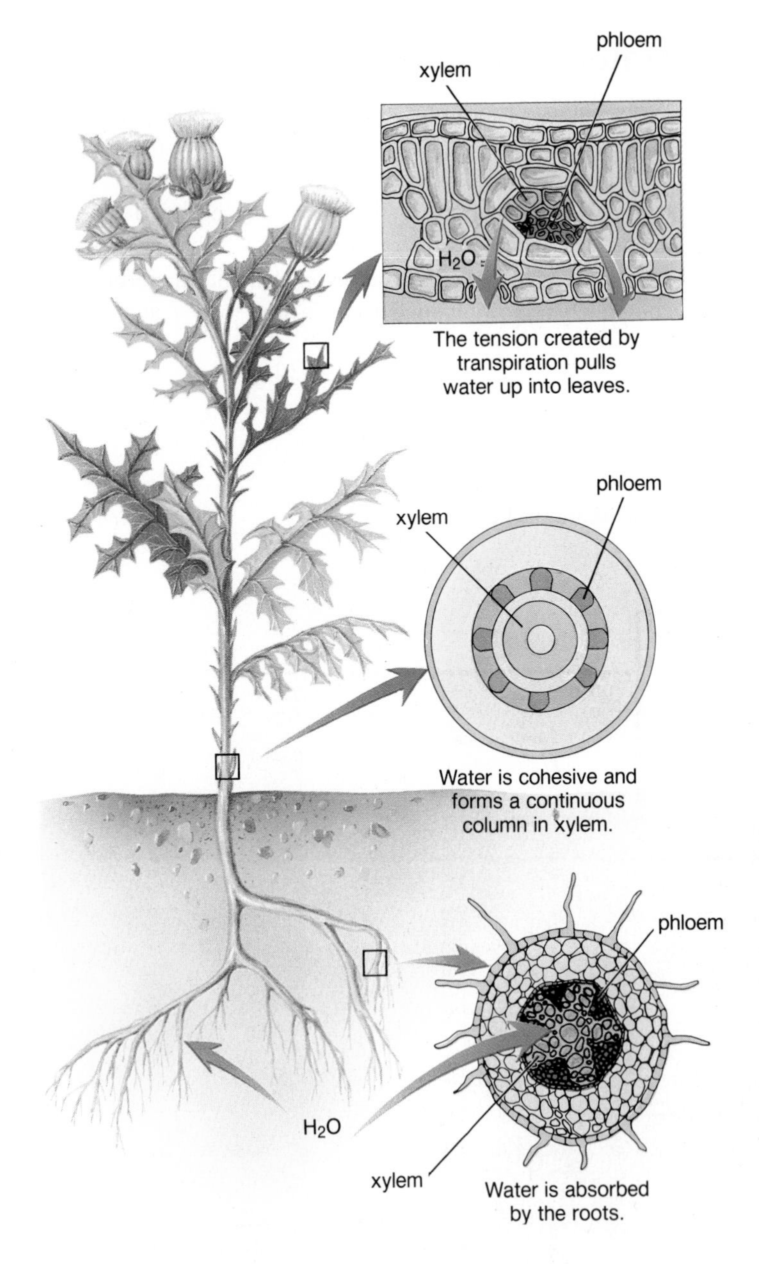

The Cohesion-Tension Theory: How Xylem Moves Water

The cohesion-tension theory of xylem transport explains how water is transported to great heights against gravity. If a hollow tube enclosed at one end is placed in a pan of mercury, atmospheric pressure (the weight of air) is sufficient to raise mercury to 76 cm. This height expressed in terms of water is 10.4 m. Since some trees can be as tall as 120 m, other factors must cause water to rise in plants (fig. 9.2) above the 10.4 m pressure limit.

Table 9.1
Transpiration Rates

Per Day Midsummer		Liters
Ragweed		6–7
3.5m apple tree		10–20
4m cactus		0.02
Coconut palm		70–80
Date palm		400–500
Per Growing Season		**Liters**
Tomato	100 days	113
Sunflower	90 days	450
Apple tree	188 days	6,800
Coconut	365 days	15,900
Date palm	365 days	132,100

Table from *Botany: A Human Concern,* Second Edition, by David L. Rayle and Hale L. Wedberg, copyright © 1980 by Saunders College Publishing, reprinted by permission of the publisher.

There are, in fact, 2 other factors that allow water to rise in plants. One of these factors has to do with the chemical properties of water. Because water molecules are polar, they adhere to the walls of the vessel elements, and because of hydrogen bonding, water molecules are cohesive—they cling together. *Cohesion* of water molecules within the xylem pipeline is absolutely necessary for water transport in a plant. It causes water to fill the pipeline completely, from the roots to the leaves, and to resist any separation.

The other factor that causes water to rise in plants is **transpiration,** the loss of water by evaporation. Much of the water that is transported from the roots to the leaves evaporates and escapes from the leaf by way of the stomata (table 9.1). The water molecules that evaporate are replaced by other water molecules from the leaf veins. Therefore, transpiration exerts a pulling effect by creating *tension,* which draws a column of water up the vessel elements from the roots to the leaves.

The tension created by transpiration is only effective because of the cohesive property of water. Therefore, this explanation for xylem transport is called the **cohesion-tension theory.**

Water is transported from the roots to the leaves in xylem. Transport is dependent on the ability of a water column to remain intact as transpiration occurs at the leaves.

Figure 9.3
Opening and closing of stomata due to water pressure changes in the guard cells.

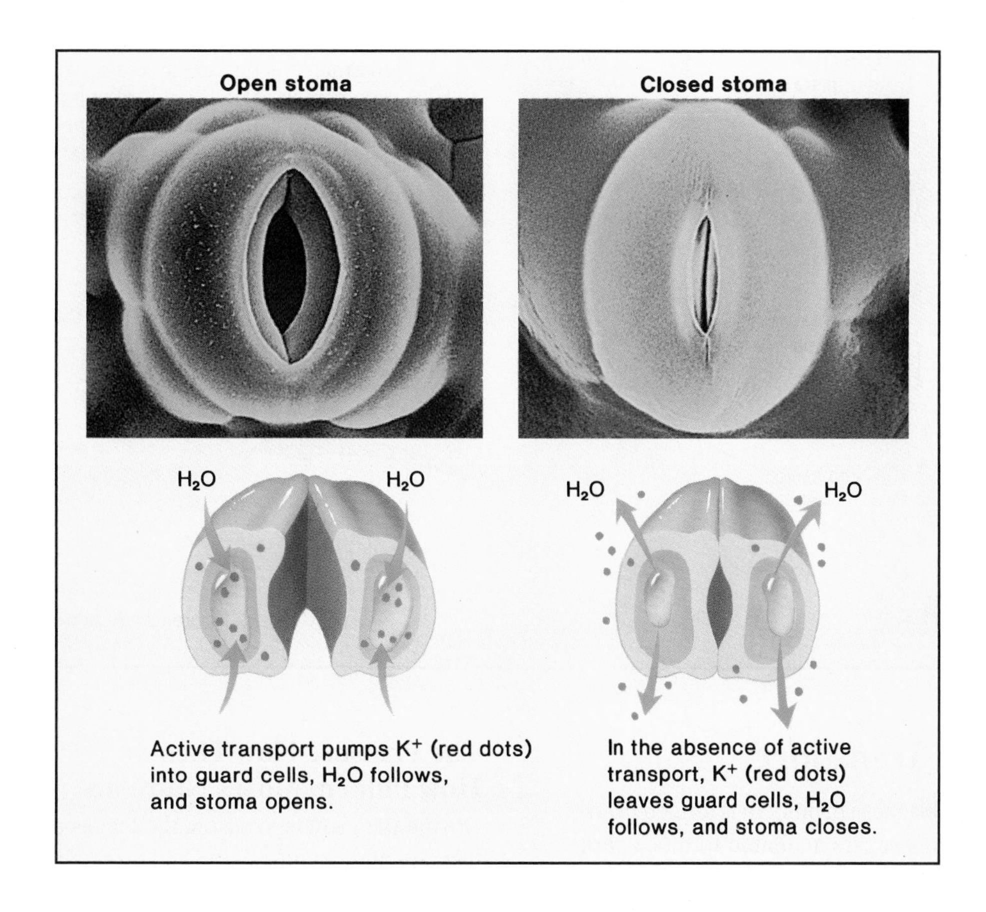

Xylem not only transports water, it also transports minerals. As discussed in the previous chapter, minerals are needed in order for a plant to produce the many organic molecules needed in cells (table 8.A). For example, the absorption and concentration of nitrates allow a plant to produce its own amino acids.

When Stomata Open and Close

A plant photosynthesizes when the stomata are open because carbon dioxide (CO_2) is entering the leaves. But we have to remember that water is also exiting the stomata because of transpiration. When a plant is water stressed, the stomata close to conserve water and then photosynthesis ceases. In order to keep the stomata open, photosynthesis requires an especially abundant supply of water for transpiration to occur.

Each stoma has 2 guard cells. Notice in figure 9.3 that the guard cells are attached to each other at their ends and that the inner walls are thicker than the outer walls. When guard cells take up water, they buckle out from their region of attachment, and the stoma opens. When a plant is photosynthesizing (note also that the guard cells contain chloroplasts), ATP is available. At that time, an ATP-driven pump actively transports potassium ions (K^+) into the guard cells. Now the guard cells take up water by osmosis, and the stoma opens. Without active transport, potassium ions move into surrounding cells, the guard cells lose water, and the stoma closes.

Stomata are open during the day and are closed during the night. Furthermore, they tend to open and close at the same time each day. An event that occurs every 24 hours like this is called a circadian rhythm. Biological clocks, which are discussed on page 631, control circadian rhythms.

A stoma is usually open during the day when a plant is photosynthesizing. At that time, potassium ions (K^+) are pumped into the 2 guard cells, which then take up water, causing the stoma to open. Do 9.1 Critical Thinking, found at the end of the chapter.

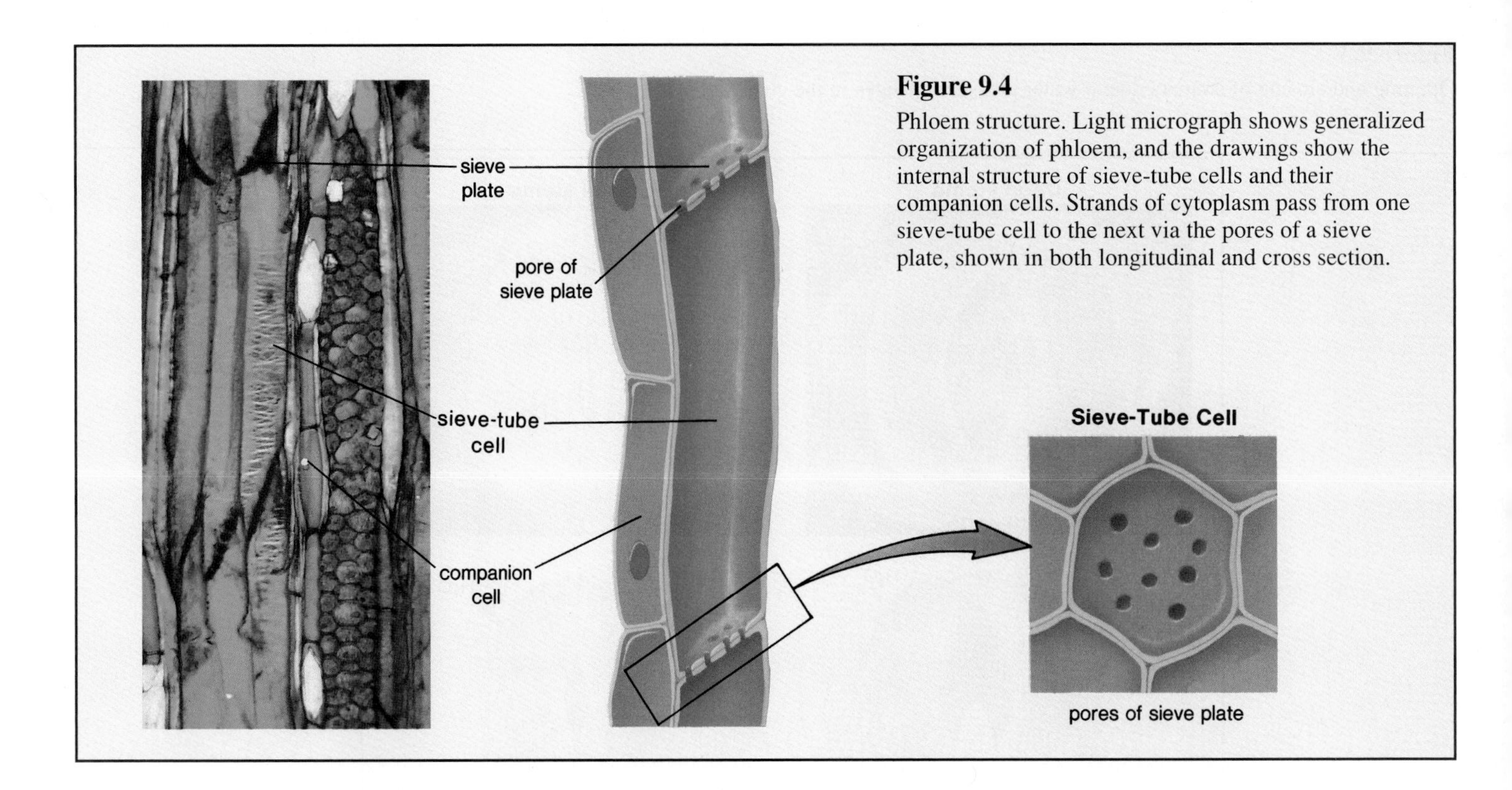

Figure 9.4

Phloem structure. Light micrograph shows generalized organization of phloem, and the drawings show the internal structure of sieve-tube cells and their companion cells. Strands of cytoplasm pass from one sieve-tube cell to the next via the pores of a sieve plate, shown in both longitudinal and cross section.

Organic Nutrient Transport

The movement of organic substances in phloem is termed *translocation.* Translocation makes sugars available to those parts of a plant that are actively metabolizing and growing. The conducting cells in phloem are sieve-tube cells, each of which typically has a companion cell (fig. 9.4). **Sieve-tube cells** contain cytoplasm but have no nucleus. Their end walls have pores and resemble a sieve; therefore, the end walls are said to be sieve plates. The sieve-tube cells are aligned vertically, and strands of cytoplasm called *plasmodesmata* extend from one cell to the other through the sieve plates. Therefore, there is a continuous pathway for organic nutrient transport throughout the plant.

The smaller **companion cell,** which does have a nucleus, is a more generalized cell than the sieve-tube cell. It is speculated that the companion-cell nucleus controls and maintains the lives of both cells and helps a sieve-tube cell perform its translocating function.

Chemical analysis of phloem sap shows that it is composed chiefly of sugar and that the concentration of organic nutrients is 10–13% by volume. Samples for chemical analysis most often are obtained by using aphids, small insects that are phloem feeders. The aphid drives its stylet, a short mouthpart functioning like a hypodermic needle, between the epidermal cells and withdraws phloem sap from a sieve-tube cell. The body of the aphid can be cut away carefully, leaving the stylet, which exudes phloem sap for collection and analysis.

The Pressure-Flow Theory: How Phloem Moves Nutrients

During the growing season, the leaves are a source of sugar—they are photosynthesizing and producing sugar (fig. 9.5). This sugar is actively transported into sieve-tube cells, and water follows passively by osmosis. Active transport is possible because sieve-tube cells have a living cell membrane and the necessary energy is provided by the companion cells. The buildup of water within the sieve-tube cells creates *pressure,* which starts a *flow* of phloem sap. The roots (and other plant organs) are a sink for sugar, and it is actively transported out of the sieve-tube cells. When water follows passively by osmosis, phloem sap flows from the leaves (source) to the roots (sink). This explanation for translocation in phloem is called the **pressure-flow theory.**

The pressure-flow theory can account for the observed reversal of flow in phloem, for example, in the spring before the leaves are out. At that time, when a plant is not photosynthesizing, the roots serve as a source of sugar, and the other parts of the plant serve as a sink. Water again enters sieve-tube cells at the source and flows toward the sink, carrying sugar with it.

Phloem transports organic nutrients in a plant. Typically, sugar and then water enter sieve-tube cells in the leaves. This creates pressure, which causes water to flow to the roots, carrying sugar with it.

Figure 9.5

Pressure-flow theory of phloem transport. Sugar and water enter sieve-tube cells at a source. This creates pressure, which causes phloem contents to flow. Sieve-tube cells form a continuous pipeline from a source to a sink, where sugar and water exit sieve-tube cells.

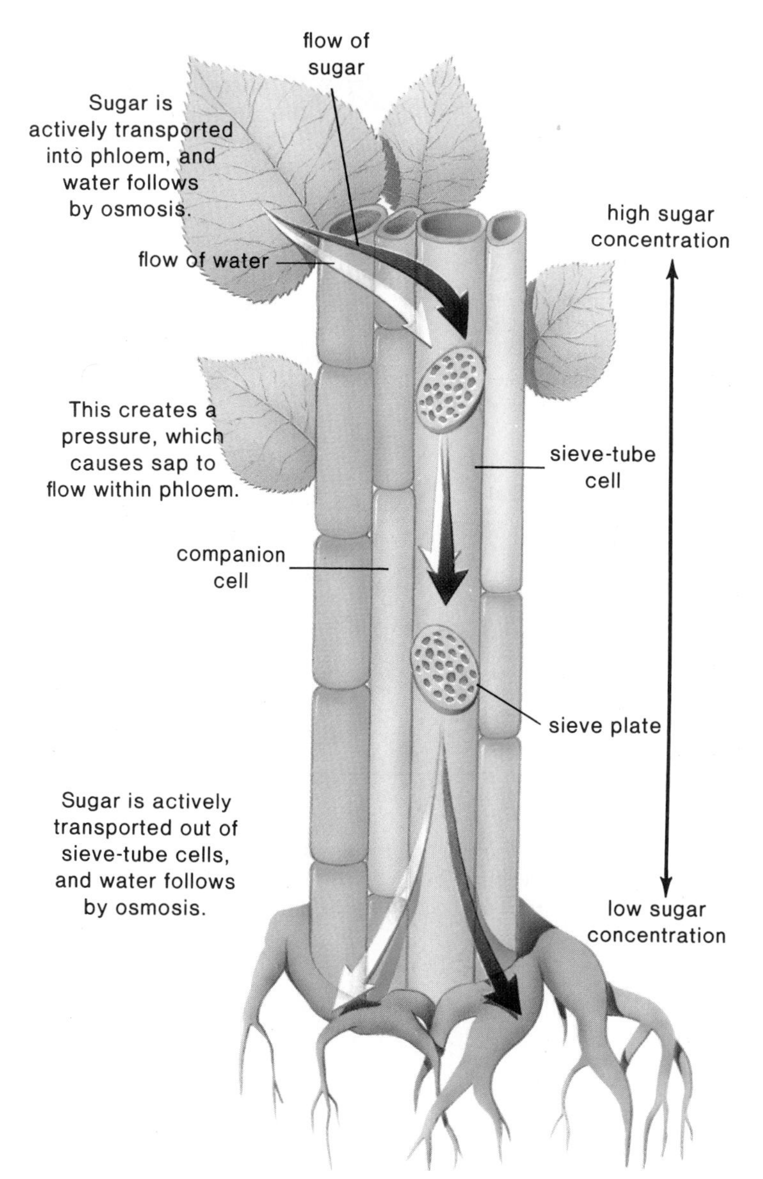

Plant Responses to Environmental Stimuli

Plants respond to environmental stimuli (e.g., light, day length, gravity, and temperature) usually by changing their growth pattern. Among the principal internal factors that regulate such responses are plant hormones. A *plant hormone* is an organic molecule synthesized by the plant that has physiological and/or developmental effects at very low concentrations. Some plant hormones are made in meristem and transported to other tissues; others move directly from the tissue of origin to another tissue; and still others are used where they are made. Table 9.2 lists the major types of plant hormones and their function. Some of these hormones are promoters of growth and some are inhibitors of growth. These hormones often interact to control physiological responses.

Table 9.2
Plant Hormones

Type	Primary Example	Notable Function
Growth Promoters		
Auxins	Indoleacetic acid (IAA)	Cell elongation
Gibberellins	Gibberellic acid (GA)	Stem elongation
Cytokinins	Zeatin	Cell division
Growth Inhibitors		
Abscisic acid	Abscisic acid (ABA)	Dormancy
Ethylene	Ethylene	Leaf and fruit drop

Each naturally occurring hormone has a specific chemical structure. Other synthetic chemicals, some of which differ only slightly from the natural hormones, also affect the growth of plants. These and the naturally occurring hormones are sometimes grouped and called plant growth regulators. The reading on page 145 discusses the various uses of plant growth regulators.

Photoperiodism: Night/Day Responses

A response based on the proportion of light to darkness in a 24-hour cycle is called **photoperiodism.** Photoperiodic responses in plants are particularly obvious in the temperate zone. In the spring, plants respond to increasing day length by initiating growth; in the fall, they respond to decreasing day length by ceasing growth processes. Day length controls flowering in some plants; for example, violets and tulips flower in the spring, asters and goldenrods flower in the fall. Flowers contain the structures that produce the gametes; when all the members of a species flower at the same time, reproduction is more likely to be successful.

Long-day plants initiate flowering when the days get longer than a certain minimum value, or critical length. Short-day plants initiate flowering when the days get shorter than a critical length. Day-length-neutral plants are insensitive to the length of the day. The cocklebur is a short-day plant; if a long night is interrupted by a flash of light, it will not flower (fig. 9.6). Clover, on the other hand, is a long-day plant; if a long night is interrupted by a flash of light, it will still flower. Interrupting the day with darkness has no effect. This shows that the length of continuous darkness, not the day length, actually controls flowering.

In order to flower, short-day plants require a period of darkness longer than a critical length and long-day plants require a period of darkness shorter than a critical length.

Figure 9.6

Day-length effect on 2 types of plants.Some plants flower only during a particular season. The cocklebur flowers when days are short, and clover flowers when days are long. The length of the night is the determining factor, proven by interrupting a longer-than-critical-length night with a flash of light.

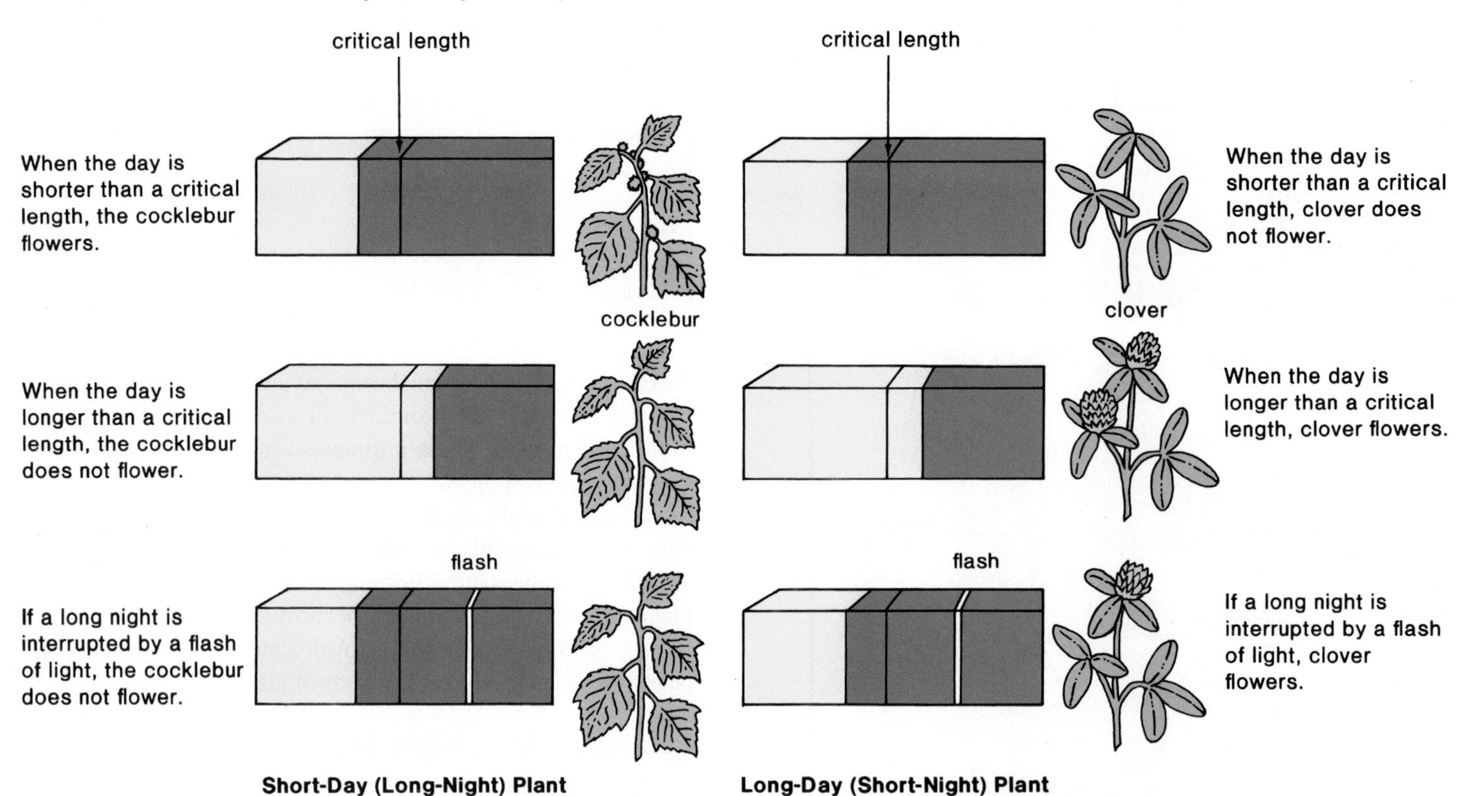

Phytochrome: A Necessary Pigment

Phytochrome is a plant pigment having a chemical structure that changes when it is exposed to light and when it is exposed to darkness. It could be that flowering is initiated in some plants according to the time of day phytochrome changes its structure. The pigment could be part of a biological clock system that controls flowering.

Phytochrome is also believed to be involved in other plant responses. Some seeds do not germinate if planted too close to the surface; others do not germinate if planted too deep in the ground. These and other experiments indicate that the detection of sunlight by phytochrome affects seed germination. Seedlings grown in the dark *etiolate*; that is, the stem increases in length and the leaves remain small (fig. 9.7). Etiolation allows a seedling to reach sunlight before its stored food runs out. Once the seedling is exposed to sunlight, however, it begins to grow normally—the leaves expand and the stem branches. Evidently, the exposure of phytochrome to sunlight brings about these changes.

The use of the plant pigment phytochrome to detect the photoperiod and the presence of sunlight is only one mechanism by which plants regulate their life cycle and growth. The reading on the next page tells how hormones are also involved in the regulation of plant responses to maximize chances of survival.

Figure 9.7

Phytochrome. The plant pigment phytochrome is sensitive to sunlight and seems to be involved in controlling aspects of physiology, such as the growth response in a shady versus a sunny environment.

PLANTS HAVE HORMONES, TOO

Plant growth involves production of cells by means of cell division, enlargement of these cells, and finally, differentiation as the cells take on specific functions. Three types of hormones are known to promote plant growth: the cytokinins, which stimulate cell division; the auxins, which bring about cell division and enlargement; and the gibberellins, which promote enlargement of cells and to a lesser extent cell division. Two hormones, ABA and ethylene, are known to inhibit plant growth in general. Plant growth regulators include natural hormones and related synthetic hormones. Today, plant growth regulators are used to bring about an increase in crop yields just as fertilizers, irrigation, and pesticides have done in the past.

Bending toward the light is an adaptation of plants that promotes photosynthesis. Experiments with oat seedlings have shown that bending occurs because auxin is transported to the shady side of the shoot. This can be proven by removing the tip of a shoot and placing an auxin-containing agar block on one side of the stump. The cells on this side elongate, causing bending to occur.

Since the time these experiments were first performed, many commercial uses for auxins have been discovered. Auxins can cause the base of a shoot to form new roots so that new plants can be started from cuttings. When sprayed on trees, auxins can prevent fruit from dropping too soon. Auxins also inhibit the growth of axillary buds; potatoes sprayed with auxin will not sprout and therefore can be stored longer. In high concentrations, auxins are used as herbicides, which prevent the growth of weeds—broad-leaved plants. The synthetic auxins known as 2,4D and 2,4,5T were used as defoliants during the Vietnam War. The latter, called Agent Orange, is sometimes contaminated with dioxin, a by-product that causes cancer in humans.

Gibberellins (fig. 9.A) cause the entire plant to grow larger. Before World War II, the Japanese studied a disease they called "foolish seedling disease" because the young plants grew rapidly, became spindly, and fell over. They found that this disease was caused by gibberellins secreted by a fungus that had infected the plants. Since then, it has been discovered that the application of gibberellins can cause seeds to germinate and plants such as cabbages to bolt (meaning rapid stem elongation) and flower. Gibberellins are used commercially to increase the size of plants. Treatment of sugarcane fields with as little as 56 g per acre increases the cane yield by more than 5 t.

Figure 9.A

The effect of gibberellic acid (GA) on Thompson seedless grapes (*Vitis vinifera*). Control grapes (*left*). GA sprayed at bloom and at fruit set (*right*). Almost all grapes sold in stores are now treated with gibberellic acid.

Cytokinins were discovered when mature carrot and tobacco plant cells began to divide when grown in coconut milk. Testing revealed the presence of cytokinins in the milk. Later, scientists were able to grow entire plants from single cells in test tubes when various plant hormones were present in correct proportions.This was the first use of the technique called tissue culture.

Nurseries now culture all sorts of plants with assembly-line efficiency. Plant breeders are extremely interested in utilizing a modification of the tissue culture technique, in which most often leaf cells are treated to produce *protoplasts,* cells that have been chemically stripped of their outer wall. (A single protoplast will give rise to a new plant identical to the original plant.) It is faster and easier to test protoplasts instead of entire plants for desired characteristics, such as resistance to bacteria and fungi, high temperatures, and drought. Also, selected protoplasts have been genetically engineered, as described in chapter 25, and the resulting plants have shown the characteristics dictated by the inserted genes. Perhaps protoplast technology will someday allow botanists to alter the genetic makeup of a variety of plants.

The hormone ethylene, which causes fruit to ripen, is classified as an inhibitor. Fruits commonly are kept in cold storage to prevent the release of ethylene, which is a gas. Many synthetic inhibitors simply oppose the action of the natural stimulatory hormones (auxins, gibberellins, and cytokinins). The application of synthetic inhibitors can cause leaf and fruit drop. Removal of the leaves of cotton plants by chemical means aids harvesting; thinning the fruit of young fruit trees produces larger fruit from the trees as they mature; and retarding the growth of some plants increases their hardiness. For example, an inhibitor has been used to reduce stem length in wheat plants so that they do not fall over in heavy winds and rain. Other synthetic inhibitors mimic the action of ethylene and cause fruit and other crops to ripen. Fields and orchards now are sprayed with synthetic growth regulators, just as they are sprayed with pesticides.

Reproduction in Flowering Plants

Flowering plants can reproduce asexually (without gametes) or sexually (with gametes).

Asexual Reproduction: Potato Eyes and Such

In a form of asexual reproduction known as *vegetative propagation,* a portion of one plant gives rise to a completely new plant. Both plants, the original and the new, have identical genes. For example, strawberry plants grow from the nodes of runners, or horizontal stems; violets grow from the nodes of rhizomes (horizontal stems); and white potato plants grow from the "eyes" of a potato. Sometimes asexual reproduction has great commercial importance. Once a desired plant has been produced through hybridization (when different species are crossed) or bioengineering, cuttings can be treated with hormones to encourage them to grow roots. The plants are then ready for sale.

Sexual Reproduction: Sex among Flowers

Plants reproduce sexually. This may come as a surprise to those who never thought of plants as being male and female. Sexual reproduction is defined properly as reproduction requiring gametes, often an egg and a sperm. In flowering plants, the sex organs are located in the flower.

Sexual Anatomy of Flowers

Figure 9.8 shows the parts of a typical flower. The **sepals,** most often green, form a whorl about the **petals,** the color of which accounts for the attractiveness of many flowers. In the center of the flower is a small vaselike structure, the **pistil,** which usually has 3 parts: the **stigma,** an enlarged sticky knob; the **style,** a slender stalk; and the **ovary,** an enlarged base. The ovary contains a number of **ovules,** which play a significant role in reproduction. Grouped about the pistil are the **stamens,** each of which has 2 parts: the **anther,** a saclike container, and the **filament,** a slender stalk.

The Life Cycle: Generations Alternate

Plants have a life cycle called **alternation of generations** because 2 generations are involved: the sporophyte and the gametophyte. The *sporophyte* is a diploid (2N) generation that produces haploid (N) **spores** by meiosis. A flower produces 2 types of spores: **microspores** and **megaspores.** Microspores are produced in the anthers of stamens, and megaspores are produced within ovules (fig. 9.8). A microspore becomes a pollen grain, which upon maturity is a sperm-containing *male gametophyte.* A megaspore becomes an egg-containing embryo sac, which is the *female gametophyte.* Following fertilization, the zygote develops into an embryo located within a seed. When the seed germinates, the new sporophyte plant begins to grow.

Figure 9.8

Alternation of generations in a flowering plant. Flowering plants are heterosporous; they produce microspores and megaspores. A microspore becomes a pollen grain, or male gametophyte, which upon maturity has sperm nuclei. A megaspore becomes the embryo sac, or female gametophyte, which produces an egg. When a sperm nucleus joins with an egg nucleus, the zygote develops into an embryo, retained within a seed.

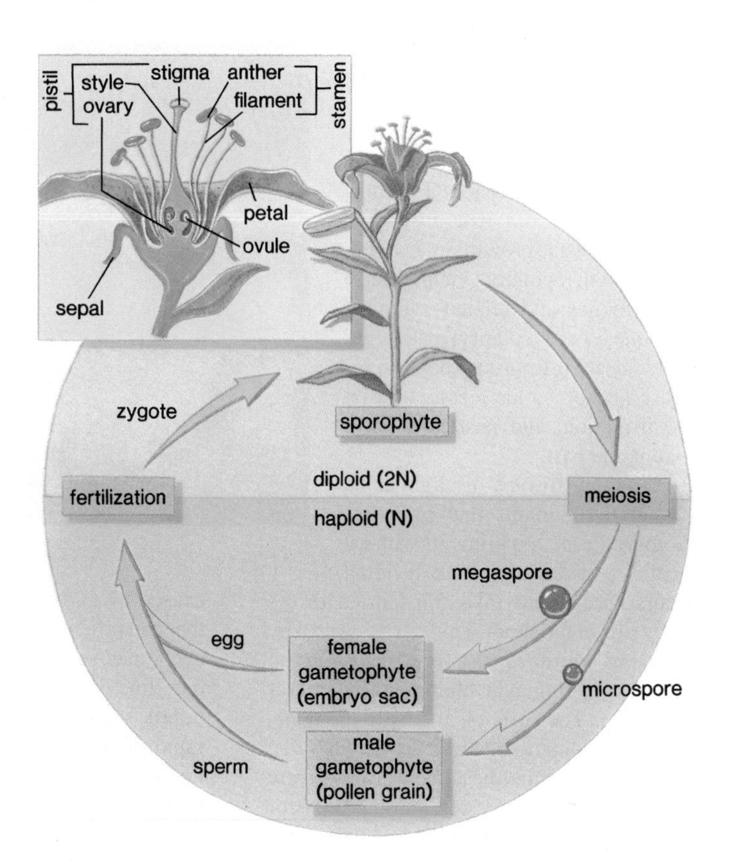

Figure 9.9 shows these same steps in greater detail. Within an ovule, a megaspore (*mega* means *large*) parent cell undergoes meiosis to produce 4 haploid megaspores. Three of these megaspores disintegrate, leaving one functional megaspore, which divides mitotically. The result is the female gametophyte, or **embryo sac,** which typically consists of 8 haploid nuclei embedded in a mass of cytoplasm. The cytoplasm differentiates into cells, one of which is an *egg* and another of which is the endosperm cell with 2 nuclei (called the *polar nuclei*).

The anther has **pollen sacs,** which contain numerous microspore (*micro* means *small*) parent cells. Each parent cell undergoes meiosis to produce 4 haploid cells called micro-

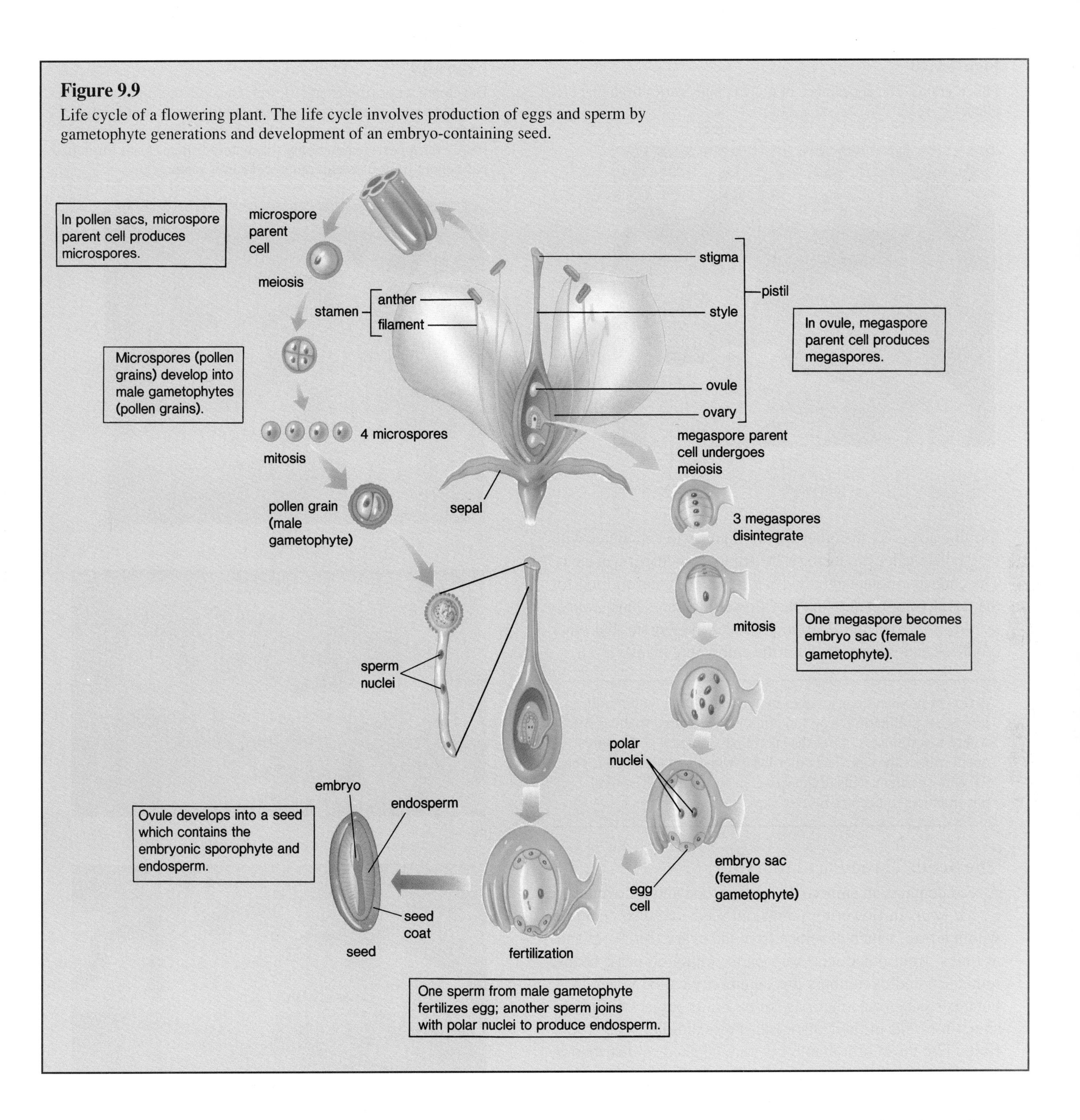

Figure 9.9
Life cycle of a flowering plant. The life cycle involves production of eggs and sperm by gametophyte generations and development of an embryo-containing seed.

spores. The microspores usually separate, and each one becomes a **pollen grain** or male gametophyte (fig. 9.10). At this point, the young male gametophyte contains 2 nuclei, the *generative nucleus* and the *tube nucleus*.

Pollination occurs when pollen is windblown or carried by insects, birds, or bats to the stigma of the same type of plant. Only then does a pollen grain germinate and produce a long pollen tube. This pollen tube grows within the style until it reaches an ovule in the ovary. Before fertilization occurs, the generative nucleus divides, producing 2 sperm, which have no flagella. On reaching the ovule, the pollen tube discharges the sperm. One of the 2 sperm migrates to and fertilizes the egg,

Figure 9.10

Pollen grains. The appearance of pollen grains varies from plant to plant.

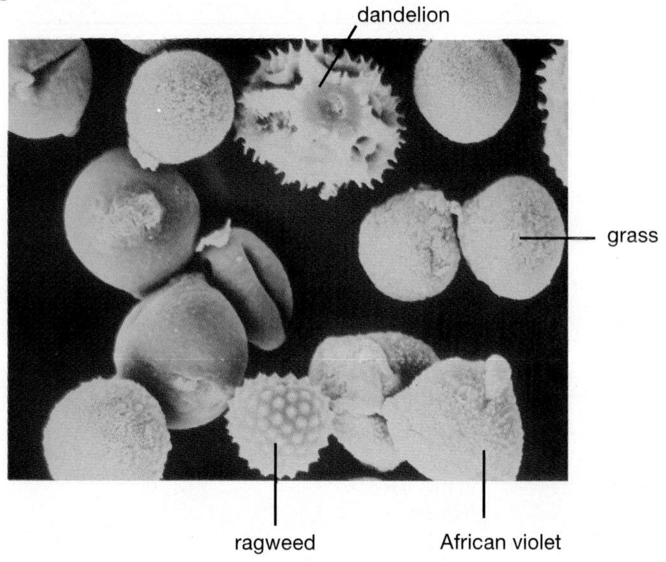

forming a zygote; the other sperm migrates to and unites with the 2 polar nuclei, producing a 3N (triploid) endosperm nucleus. The endosperm nucleus divides to form endosperm, food for the developing plant. Note that flowering plants have *double fertilization.* One fertilization produces the zygote; the other produces endosperm, food for the embryonic plant.

A flower produces eggs within ovules and sperm within pollen grains. At pollination, a pollen grain is transported to the pistil, where it germinates. After the sperm pass down a pollen tube, one fertilizes the egg. The other fuses with the polar nuclei. The 3N endosperm nucleus divides to give endosperm. The ovule now matures to become a seed.

The Seed: Three-Part Embryo

A seed contains an embryo and stored food and is covered by a *seed coat.* In flowering plants, all seeds are also enclosed within a fruit, which develops from the ovary (fig. 9.11) and, at times, from other accessory parts. Although peas, beans, tomatoes, and cucumbers are commonly called vegetables, botanists categorize them as fruits. Fruits protect seeds and sometimes aid in their dispersal. For example, winged dry fruits, like those of a maple tree, are adapted to distribution by the wind, while fleshy fruits, like those of a cherry, are eaten by birds and the seeds are deposited some distance away. Wide dispersal exposes plants to more resources than limited dispersal does.

Monocots have seeds with one *cotyledon* (seed leaf), and dicots have seeds with 2 cotyledons (see fig. 8.4). Cotyledons typically provide nutrient molecules for the growing embryo. In monocot embryos, the cotyledon rarely stores food; rather, it absorbs food molecules from the endosperm and passes them

Figure 9.11

Development of fruit from a flower. Once seed formation begins, the ovary and the accessory parts begin to enlarge, until only remnants of the other flower parts remain. Even these remnants finally disappear, leaving only the mature fruit. **a.** Fruit formation has begun. **b.** Fruit is enlarging. **c.** Fruit is mature.

a.

b.

c.

Figure 9.12

Monocot seed and germination. **a.** Longitudinal section of mature seed. Notice the large amount of endosperm, in addition to the cotyledon, in the monocot seed. **b.** When the seed germinates, the plumule becomes the leaves and the radicle becomes the roots.

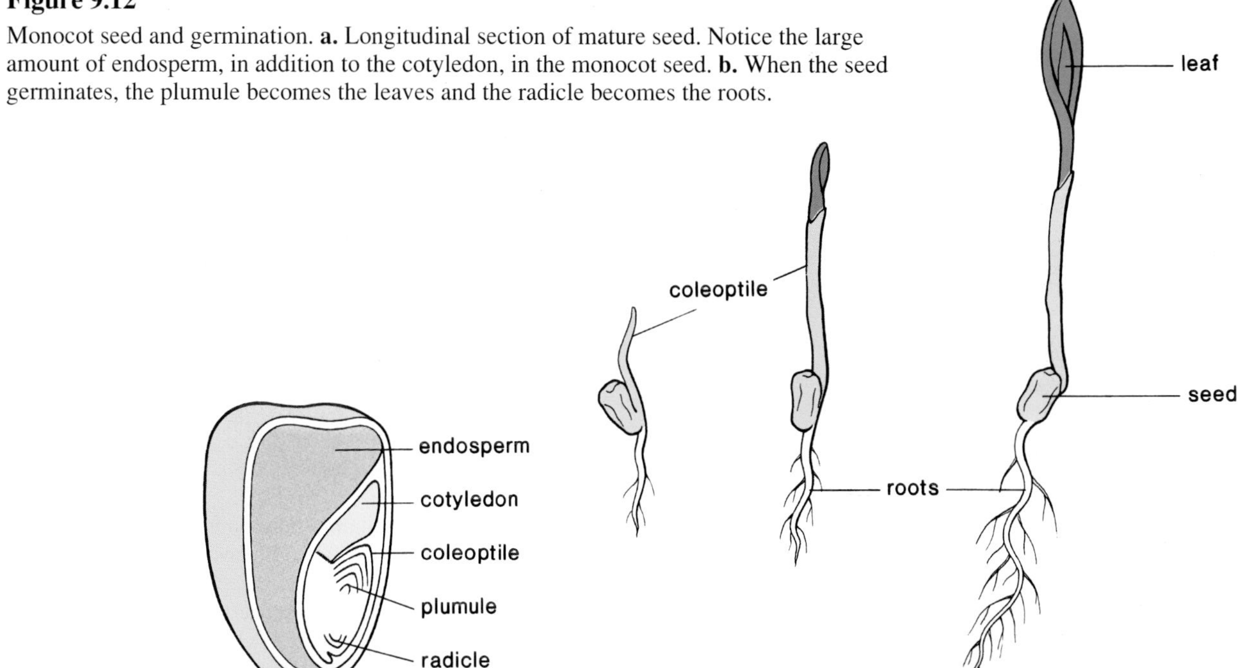

Figure 9.13

Dicot seed and germination. **a.** Longitudinal section of dicot seed shows 2 large cotyledons, one on either side of the embryo. **b.** Germination of a dicot seed makes it easier to detect that the epicotyl gives rise to the leaves, the hypocotyl becomes a portion of the stem, and the radicle becomes the roots.

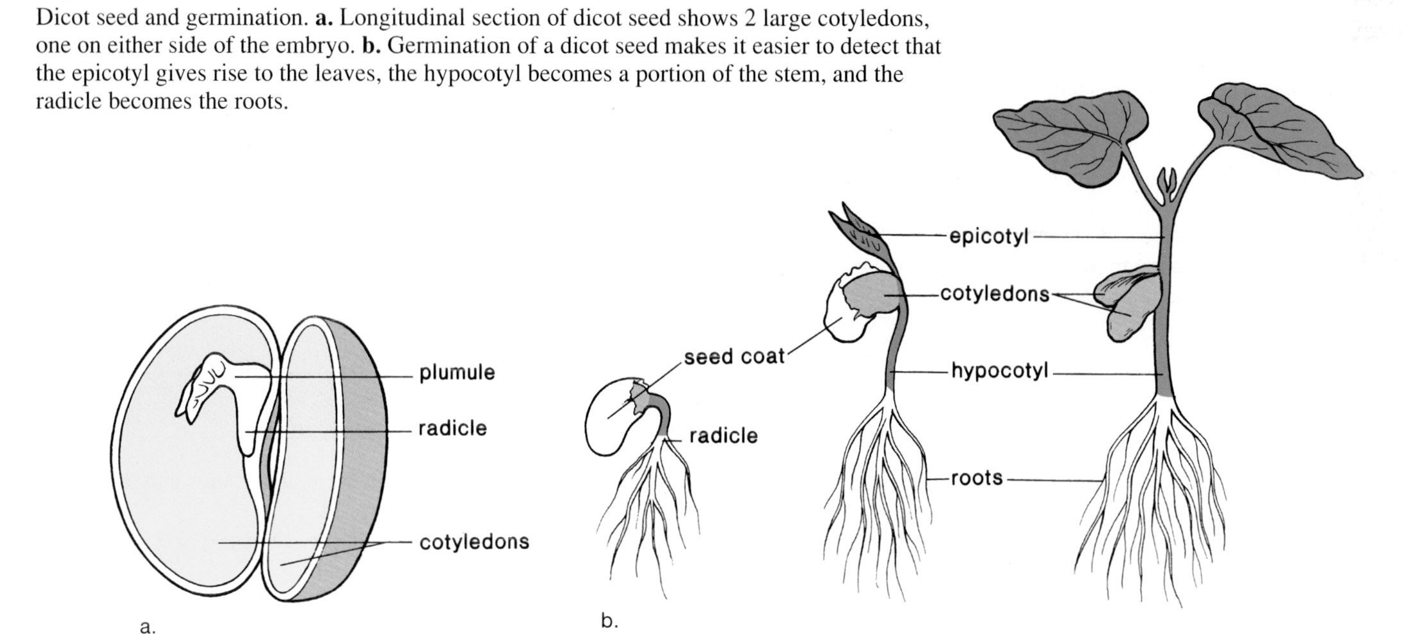

to the embryo. In many dicot embryos, the cotyledons replace the endosperm, which typically has already transferred its nutrients to the cotyledons.

Aside from the *cotyledon*(s), a seed contains the **epicotyl,** that portion of the embryo above the attachment of the cotyledon(s), and the **hypocotyl,** which lies below the attachment of the cotyledons and becomes a portion of the stem. The epicotyl contains apical meristem and sometimes bears young leaves, in which case it is called a *plumule*. A **radicle,** the embryonic root at the lower end of the hypocotyl, also contains apical meristem (figs. 9.12 and 9.13).

> The mature seed typically contains an embryo consisting of the cotyledon(s), the epicotyl, and the hypocotyl. Upon germination, a radicle (root) appears, and the epicotyl gives rise to the leaves.

SUMMARY

Water transport in plants occurs within xylem. The cohesion-tension model of xylem transport states that transpiration creates tension, which pulls water upward in xylem. This transport works only because water molecules are cohesive. Most of the water taken in by a plant is lost through stomata by transpiration.

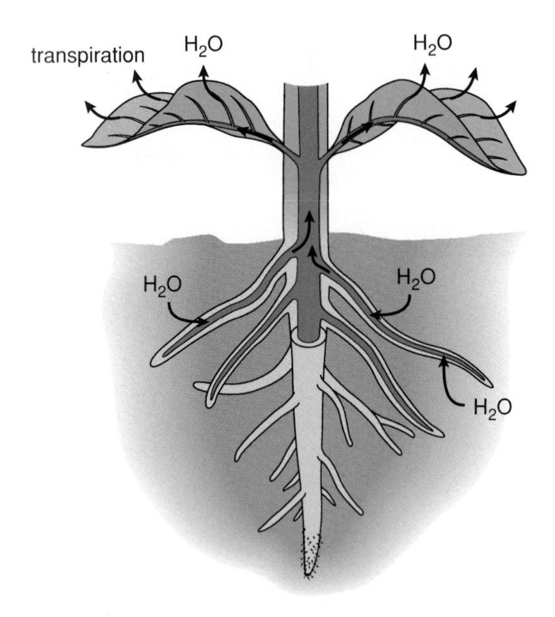

Stomata open when guard cells take up water (H_2O), stretching their thin side walls. Water follows potassium ions (K^+) into the guard cells.

Transport of organic nutrients in plants occurs within phloem. The pressure-flow theory of phloem transport states that sugar is actively transported into phloem at a source and water follows by osmosis. The resulting increase in pressure creates a flow, which moves water and sucrose to a sink.

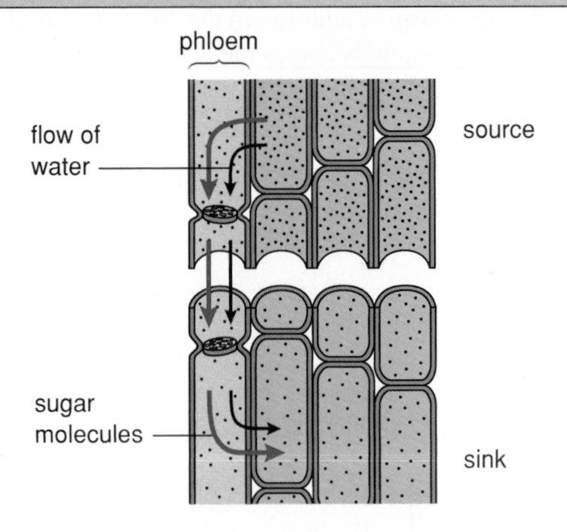

Both stimulatory and inhibitory hormones help to control certain plant growth patterns. Some hormones stimulate growth (auxins, gibberellins, and cytokinins), while others inhibit growth (ethylene and abscisic acid). Plant hormones most likely control photoperiodism. Short-day plants flower when the days are shorter (nights are longer) than a critical length, and long-day plants flower when the days are longer (nights are shorter) than a critical length. Some plants are day-length neutral. Phytochrome, a plant pigment that responds to daylight, is believed to be a part of a biological clock system that in some unknown way brings about flowering.

Flowering plants have an alternation of generations life cycle, which includes separate male and female gametophytes. The pollen grain, the male gametophyte, is produced within the stamens of a flower. The female gametophyte is produced within the ovule of a flower. Following pollination and fertilization, the ovule matures to become the seed and the ovary becomes the fruit. The enclosed seeds contain the embryo (hypocotyl, epicotyl, plumule, radicle) and stored food (endosperm and/or cotyledons). When a seed germinates, the root appears below and the shoot appears above.

STUDY QUESTIONS

In order to practice **writing across the curriculum,** students should write out the answers to any or all of the study questions. The study questions are sequenced in the same order as the text.

1. Explain the cohesion-tension theory of water transport. (pp. 139–140)
2. What events precede the opening and closing of stomata by guard cells? (pp. 140–141)
3. Explain the pressure-flow theory of phloem transport. (pp. 142–143)
4. Name 5 plant hormones, and state their functions. (p. 143)
5. Define photoperiodism, and discuss its relationship to flowering in certain plants. (p. 143)
6. What is phytochrome, and what are some possible functions of phytochrome in plants? (p. 144)
7. How do plants reproduce asexually? sexually? (p. 146)
8. Describe how a female gametophyte forms in flowering plants. (p. 147)
9. Describe how a male gametophyte forms in flowering plants. (p. 147)
10. Contrast the monocot seed and seedling with the dicot seed and seedling. (pp. 148–149)

OBJECTIVE QUESTIONS

1. The transport of water is dependent upon _______, which occurs whenever the stomata are open.
2. Stomata open when _______, followed by _______, enters guard cells.
3. The _______ theory explains the transport of sugar in sieve-tube cells.
4. Short-day plants _______ (will, will not) flower when a longer-than-critical-length night is interrupted by a flash of light.
5. _______ is the pigment that is believed to signal a biological clock in plants that exhibit photoperiodism.
6. Plants have a life cycle called _______.
7. The female gametophyte develops within the _______ of a flower, and the male gametophyte develops within the _______.
8. Monocots have seeds with one _______, while dicots have seeds with 2.
9. Label this diagram of the alternation of generations life cycle and the reproductive parts of a flower.

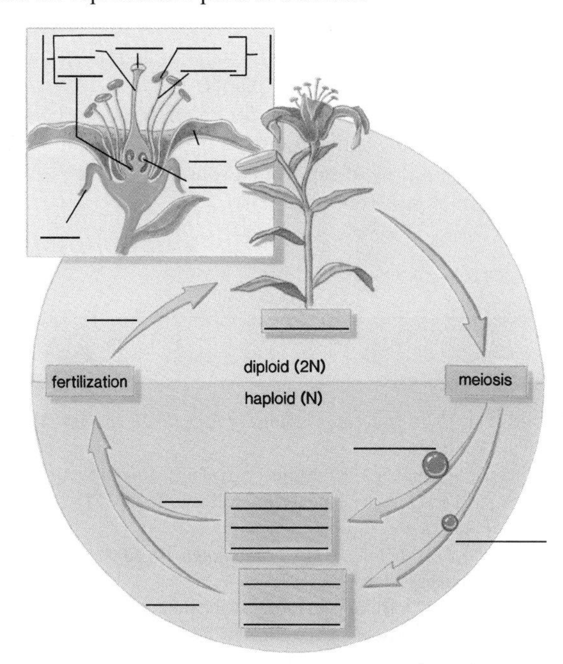

CRITICAL THINKING

In order to practice **writing across the curriculum,** students should write out the answers to any or all of the critical thinking questions. Suggested answers to the critical thinking questions are in appendix E.

9.1

A twig with leaves is placed in the top of an open tube. The tube contains water above mercury:

1. Atmospheric pressure alone is sufficient to raise mercury (Hg) only 760 mm (760 mm Hg = 10.4 m water). What is atmospheric pressure, and of what significance is this finding for a tree that is 120 m high?
2. Why does the mercury rise higher than 760 mm Hg when a twig with leaves is placed in the top of the tube?
3. What does the experiment suggest about the ability of transpiration to raise water to the top of tall trees?

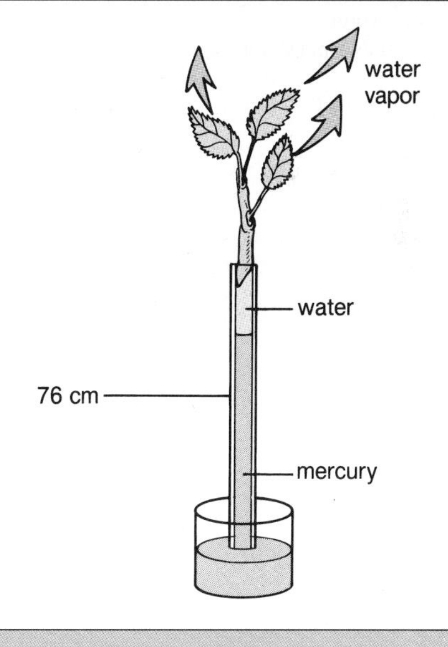

SELECTED KEY TERMS

alternation of generations a life cycle typical of plants in which a diploid sporophyte alternates with a haploid gametophyte.

cohesion-tension theory explanation for upward transportation of water in xylem based upon transpiration-created tension and the cohesive properties of water molecules.

companion cell a small nucleated cell that lies adjacent to and helps with the activities of a sieve-tube cell.

embryo sac the female gametophyte of flowering plants that contains an egg cell.

epicotyl (ep″ĭ-kot′il) the plant embryo portion above the cotyledons; contributes to stem development.

hypocotyl (hi″po-kot′il) the plant embryo portion below the cotyledons; contributes to stem development.

photoperiodism a response to light and darkness; particularly in reference to flowering in plants.

phytochrome (fi´to-krōm) a plant pigment that enables a photoperiodic response in plants.

pollen grain the male gametophyte of flowering plants that, upon maturity, contains sperm.

pollination the delivery of pollen by wind or animals to the stigma of a pistil in flowering plants.

pressure-flow theory explanation for phloem transport; osmotic pressure following active transport of sugar into phloem brings about a flow of sap from a source to a sink.

radicle the embryonic root of a plant.

sieve-tube cell a phloem cell that functions in transport of organic nutrients. During development, sieve-tube cells align vertically and form a continuous pathway for transport.

spore a haploid reproductive cell produced by the diploid sporophyte of a plant; asexually gives rise to the haploid gametophyte.

tracheid (tra´ke-id) a component of xylem made of long, tapered nonliving cells.

transpiration the evaporation of water from a leaf; pulls water from the roots through a stem to leaves.

vessel element a conducting cell in xylem. During development, vessel elements lose their contents and end walls so that they form a continuous vertical pipeline for transport of water.

FURTHER READINGS FOR PART THREE

Alberts, B., et al. 1989. *Molecular biology of the cell*. 2d ed. New York: Garland Publishing.

Barrett, S. C. H. September 1987. Mimicry in plants. *Scientific American*.

Bazzar, F. A., and E. D. Fajer. January 1992. Plant life in a CO_2-rich world. *Scientific American*.

Bold, H. C. 1980. *Morphology of plants and fungi*. 4th ed. New York: Harper & Row, Publishers, Inc.

Brill, W. J. March 1977. Biological nitrogen fixation. *Scientific American*.

Cronquist, A. 1982. *Basic botany*. 2d ed. New York: Harper & Row, Publishers, Inc.

Epel, D. November 1977. The program of fertilization. *Scientific American*.

Hesslop-Harrison, Y. February 1978. Carnivorous plants. *Scientific American*.

Jansen, W., and F. B. Salisbury. 1971. *Botany: An ecological approach*. Belmont, Calif.: Wadsworth.

Niklas, K. J. July 1987. Aerodynamics of wind pollination. *Scientific American*.

Raven, H., et al. 1986. *Biology of plants*. 4th ed. New York: Worth Publishers, Inc.

Rayle, D., and H. L. Wedberg. 1980. *Botany: A human concern*. Boston: Houghton Mifflin.

Rost, R., et al. 1984. *Botany: A brief introduction to plant biology*. 2d ed. New York: John Wiley and Sons.

Salisbury, F. B., and C. W. Ross. 1985. *Plant physiology*. 3d ed. Belmont, Calif.: Wadsworth.

Shepard, J. F. May 1982. The regeneration of potato plants from leaf-cell protoplasts. *Scientific American*.

Stern, K. 1991. *Introductory plant biology*. 5th ed. Dubuque, Iowa: Wm. C. Brown Publishers.

Zimmerman, M. H. March 1963. How sap moves in trees. *Scientific American*.

Part Four

HUMAN ANATOMY AND PHYSIOLOGY

Pancreatic cell adjoins a blood capillary

HUMAN ANATOMY AND PHYSIOLOGY

Human anatomy and physiology is the study of the structure and the function of the human body. A limited number of tissues make up organs, which form systems to carry out the functions assumed by the cell in less complex animals.

All body systems help to maintain a relatively stable internal environment. The digestive system provides nutrients, and the excretory system rids the body of metabolic wastes. The respiratory system supplies oxygen (O_2), but it also eliminates carbon dioxide (CO_2). The circulatory system carries nutrients and oxygen to and wastes from the cells so that tissue fluid composition remains constant. The immune system helps to protect the body from disease. The nervous and hormonal systems control body functions. The nervous system directs body movements, allowing the organism to manipulate the external environment, an important life-sustaining function.

10

HUMAN ORGANIZATION

Chapter Concepts

1.
Animal tissues can be categorized into 4 major types: epithelial, connective, muscular, and nervous tissues. 156, 168

2.
Organs usually contain several types of tissues. For example, although skin is composed primarily of epithelial tissue and connective tissue, it also contains muscle and nerve fibers. 162, 168

3.
Organs are grouped into organ systems, each of which has specialized functions. 164, 168

4.
Humans have a marked ability to maintain a relatively stable internal environment. All organ systems contribute to homeostasis. 165, 168

Ciliated epithelial cells lining the trachea, ×3,840

Chapter Outline

N the chapters to follow, human anatomy and physiology is studied as representative of vertebrate anatomy and physiology. Our study will be more meaningful if we first review human organization. Figure 1.2 shows that the human body, like the bodies of other organisms, has levels of biological organization. Cells of the same type are joined to form a tissue. Different tissues are found in an organ, and various types of organs are arranged into an organ system. Finally, the organ systems make up the organism.

As you study this chapter, note that the structure and the function of an organ system are dependent upon the structure and the function of the organ, the tissue, and the cell type contained therein. For example, the structure and the function of the skeletal muscle system are the same as that of the skeletal muscles, the muscular tissue, and the muscle cells.

Types of Tissues

The tissue of the human body can be categorized into 4 major types: *epithelial tissue,* which covers body surfaces and lines body cavities; *connective tissue,* which binds and supports body parts; *muscular tissue,* which moves body parts; and *nervous tissue,* which responds to stimuli and conducts impulses from one body part to another.

Epithelial Tissue: Covers Body and Lines Cavities

Epithelial tissue, also called epithelium, consists of tightly packed cells that form a continuous sheet over the entire body surface and most of the body's inner cavities. Externally, like the epidermis in plants, it protects an animal from injury and drying out. Internally, epithelial tissue may be specialized for other functions in addition to protection. For example, it secretes mucus along the digestive tract; it sweeps up impurities from the lungs by means of cilia; and it efficiently absorbs molecules from kidney tubules and the intestine because of minute cellular extensions called microvilli.

There are 3 types of epithelial tissue (fig. 10.1): **squamous epithelium,** which *is composed of flattened cells;* **cuboidal epithelium,** which contains *cube-shaped cells;* and **columnar epithelium,** in which *the cells resemble pillars or columns.* An epithelium can be simple or *stratified.* Simple means the tissue has a single layer of cells, and stratified means that the tissue has layers of cells piled one on top of the other. The outer layer of skin is stratified squamous epithelium. One type of columnar epithelium is *pseudostratified*—it appears to be layered, but actually true layers do not exist because each cell touches a basement membrane.

A so-called *basement membrane* often joins an epithelium to underlying connective tissue. We now know that the basement membrane is glycoprotein reinforced by fibers supplied by the connective tissue, the type of tissue discussed next.

An epithelium sometimes secretes a product, in which case it is described as glandular. A **gland** can be a single epithelial cell, as in the case of mucus-secreting goblet cells, found within the columnar epithelium lining the digestive tract, or a gland can contain many cells. Glands that secrete their product into ducts are called *exocrine glands,* and those that secrete their product directly into the bloodstream are called *endocrine glands.* The pancreas is both an exocrine gland, because it secretes digestive juices into the small intestine via ducts, and an endocrine gland, because it secretes insulin into the bloodstream.

Junctions: Help Communication

Three types of junctions join the tightly packed epithelial cells (fig. 10.2). A *tight junction* forms an impermeable barrier because adjacent cell membrane proteins actually join, producing a zipperlike fastening. Tight junctions prevent substances from passing between the cells; for example, the epithelial lining of the digestive tract prevents digestive juices and microbes (infectious agents such as bacteria and viruses) from entering blood.

A *desmosome* is an adhesion junction between 2 cells. The adjacent cell membranes do not touch but are held together by intercellular filaments firmly attached to buttonlike thickenings. Filaments of the cytoskeleton attach to the inner sides of the desmosome. The presence of desmosomes makes an epithelial tissue, such as the outer layer of skin, sturdy yet flexible.

A *gap junction* is a communication junction. It forms when 2 identical cell membrane channels join. This lends strength, but it also allows ions, sugars, and other small molecules to pass between the 2 cells. Their presence in heart and smooth muscle ensures synchronized contraction.

> Epithelial tissue is named according to the shape of the cell. These tightly packed protective cells can occur in more than one layer, and the cells lining a cavity can be ciliated and/or secretory.

Connective Tissue: Connects Organs

Connective tissue (table 10.1) binds organs together, provides support and protection, insulates, stores fat, and produces blood cells. As a rule, connective tissue cells are widely separated by a **matrix,** noncellular material found between cells. The matrix may have fibers of 2 types. White fibers contain *collagen,* a substance that gives them flexibility and strength. Yellow fibers contain *elastin,* a substance that is not as strong as collagen but is more elastic.

Loose Connective Tissue: Two Types of Fibers

Loose connective tissue binds structures together (fig. 10.3*a*). The cells of this tissue, which are mainly **fibroblasts,** are located some distance from one another and are separated by a jellylike matrix, which contains many white collagen fibers and yellow elastic fibers. The collagen fibers occur in bundles and are strong and flexible. The elastic fibers form networks that when stretched return to their original length. As discussed later,

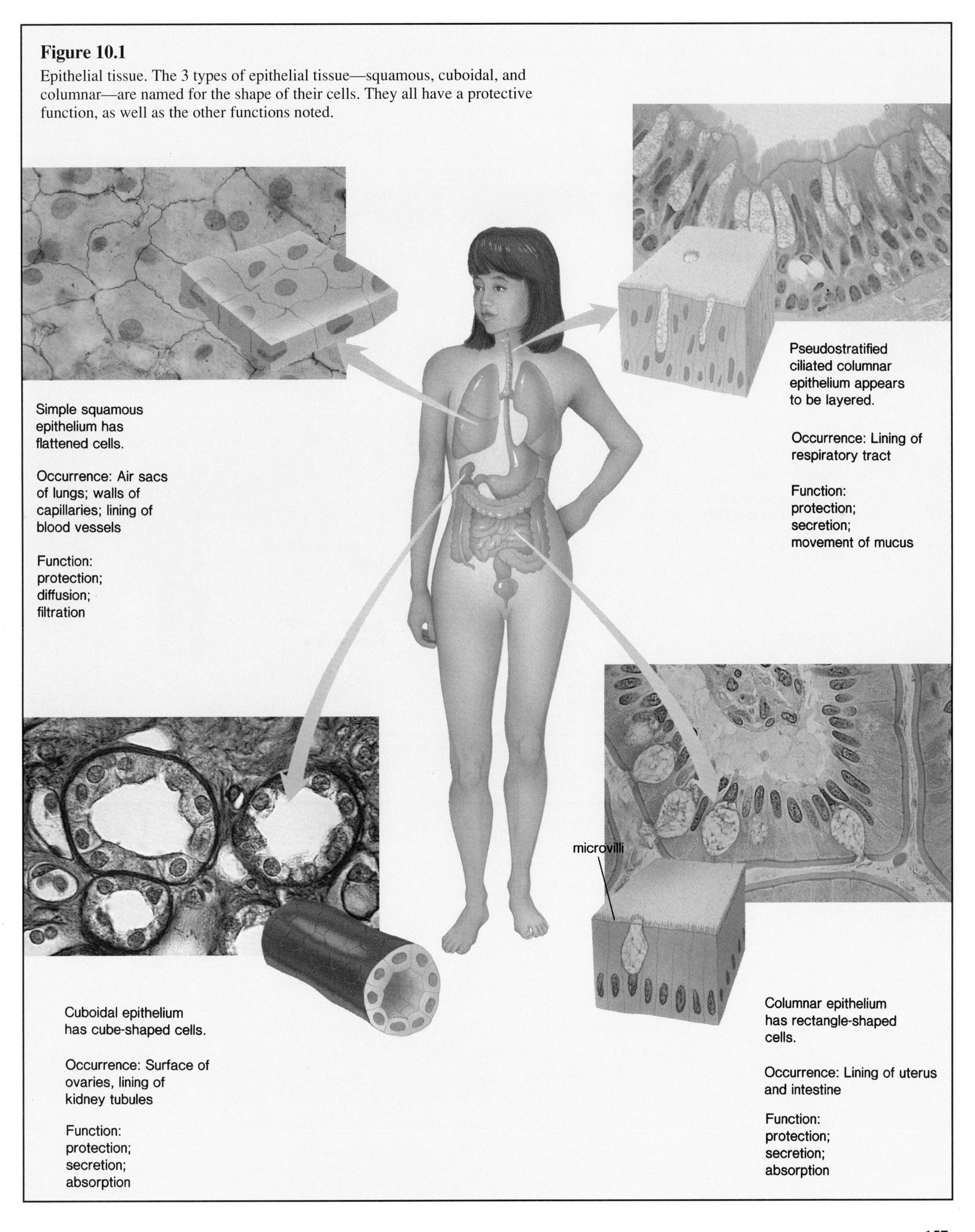

Figure 10.1

Epithelial tissue. The 3 types of epithelial tissue—squamous, cuboidal, and columnar—are named for the shape of their cells. They all have a protective function, as well as the other functions noted.

Figure 10.2

Junctions between epithelial cells. Tight junctions form an impermeable barrier because adjacent cell membrane proteins join; desmosomes are adhesion junctions, where intercellular filaments run between 2 cells; and gap junctions are communication junctions because 2 identical cell membrane channels join.

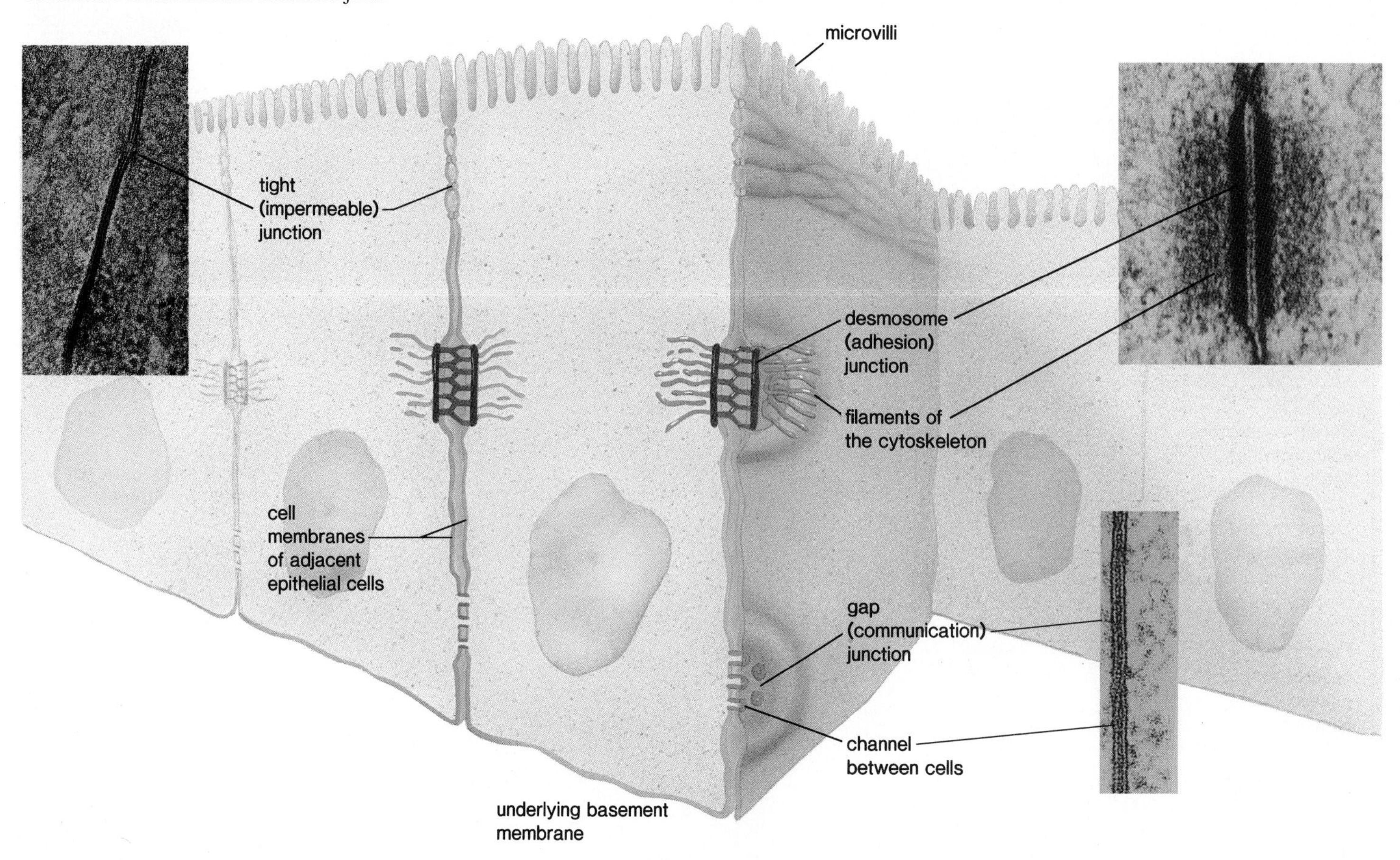

Table 10.1
Connective Tissue

Type	Function	Location
Loose connective tissue	Supports and binds organs	Beneath skin; beneath most epithelial layers
Adipose tissue	Insulates; stores fat	Beneath skin; around organs; heart
Fibrous connective tissue	Binds organs	Tendons; ligaments
Cartilage		
Hyaline cartilage	Supports; protects	Ends of bones; nose; rings in walls of respiratory passages
Elastic cartilage	Supports; protects	External ear; part of the larynx
Fibrocartilage	Supports; protects	Between bony parts of backbone and knee
Compact bone	Supports; protects; produces blood cells	Bones of skeleton
Blood	Transports gases, nutrients, and wastes about body; fights infection; forms blood clots	Blood vessels

Figure 10.3

Connective tissue examples. **a.** In loose connective tissue, cells called fibroblasts are separated by a jellylike matrix, which contains both collagen and elastic fibers. **b.** Adipose tissue cells have nuclei (arrow) pushed to one side because the cells are filled with fat. **c.** In hyaline cartilage, the flexible matrix is glassy in appearance. **d.** In compact bone, the hard matrix contains calcium salts. The concentric rings of cells in lacunae are part of a Haversian system, an elongated cylinder with central canals that contain blood vessels and nerve fibers.

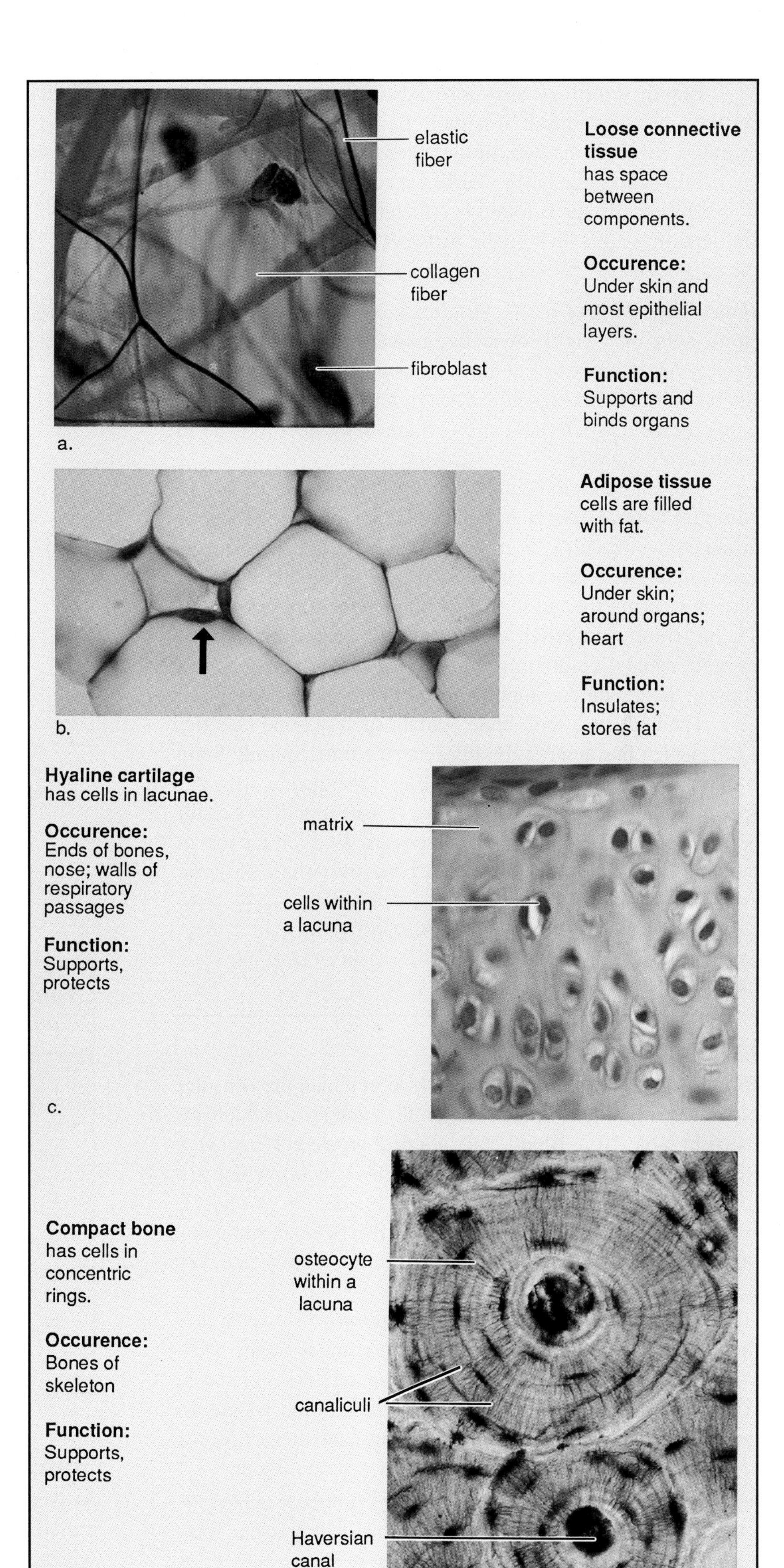

loose connective tissue commonly lies beneath epithelial layers. In certain instances, epithelium and its underlying connective tissue form body membranes (p. 165). In addition, adipose tissue (fig. 10.3*b*) is a type of loose connective tissue in which the fibroblasts enlarge with and store fat and in which the intercellular matrix is reduced.

Fibrous Connective Tissue: Tendons and Ligaments

Fibrous connective tissue contains many white collagen fibers packed closely together. This type of tissue has more specific functions than loose connective tissue. For example, fibrous connective tissue is found in **tendons,** which connect muscles to bones, and in **ligaments,** which connect bones to other bones at joints. Tendons and ligaments take a long time to heal following an injury because their blood supply is relatively poor.

> Loose connective tissue and fibrous connective tissue, which bind and support body parts, differ according to the type and the abundance of fibers in the matrix.

Cartilage: Less Rigid Matrix

In **cartilage,** the cells lie in small chambers called lacunae (sing., **lacuna**), separated by a matrix that is solid yet flexible. Unfortunately, because this tissue lacks a direct blood supply, it heals very slowly. There are 3 types of cartilage, distinguished by the type of fiber in the matrix.

Hyaline cartilage (fig. 10.3*c*), the most common type of cartilage, contains only very fine collagen fibers. The matrix has a milk glass appearance. Hyaline cartilage is found in the nose and at the ends of the long bones and the ribs, and it forms rings in the walls of respiratory passages. The fetal skeleton also is made of this type of cartilage. Later, the cartilaginous fetal skeleton is replaced by bone.

Elastic cartilage has more elastic fibers than hyaline cartilage. For this reason, it is more flexible and is found, for example, in the framework of the outer ear.

Fibrocartilage has a matrix containing strong collagen fibers. Fibrocartilage is found in structures that withstand tension and pressure, such as the pads between the vertebrae in the backbone and the wedges found in the knee joint.

Bone: Rigid Matrix

Bone is the most rigid connective tissue. It consists of an extremely hard matrix of calcium salts deposited around protein fibers. The calcium salts give bone rigidity, and the protein fibers provide elasticity and strength, much as steel rods do in reinforced concrete.

The shaft of a long bone is compact bone (fig. 10.3*d*). In **compact bone,** osteocytes (bone cells) are located in lacunae arranged in concentric circles around tiny tubes called Haversian canals. Nerve fibers and blood vessels are in these canals. The latter bring the nutrients that allow bone to renew itself. The nutrients can reach all of the cells because canaliculi (minute canals) containing thin processes of the osteocytes connect them with one another and with the Haversian canals.

The ends of a long bone contain spongy bone (see fig. 17.6), which has an entirely different structure. **Spongy bone** contains numerous bony bars and plates, separated by irregular spaces. Although lighter than compact bone, spongy bone still is designed for strength. Just as braces are used for support in buildings, the solid portions of spongy bone follow lines of stress.

Cartilage and bone are support tissues. Cartilage is more flexible than bone because the matrix is rich in protein, not calcium salts like that of bone.

Blood: Liquid Matrix

Blood (fig. 10.4) is a connective tissue in which the cells are separated by a liquid called *plasma,* the contents of which are listed in table 10.2. Blood cells are of 2 types: **erythrocytes** (red), which carry oxygen, and **leukocytes** (white), which aid in fighting infection. Also present in plasma are *platelets,* which are important to blood clotting. Platelets are not complete cells; rather, they are fragments of giant cells found in bone marrow.

Blood is unlike other types of connective tissue in that the intercellular matrix (i.e., plasma) is not made by the cells. Plasma (table 10.2) is a mixture of different types of molecules, which enter the blood at various locations. Some people do not classify blood as connective tissue; instead, they suggest a separate tissue category for blood called vascular tissue.

Blood is a connective tissue in which the matrix is plasma.

Figure 10.4

Blood, a liquid tissue. Blood is classified as connective tissue because the cells are separated by a matrix—plasma. Plasma, the liquid portion of blood, contains red blood cells, white blood cells, and platelets.

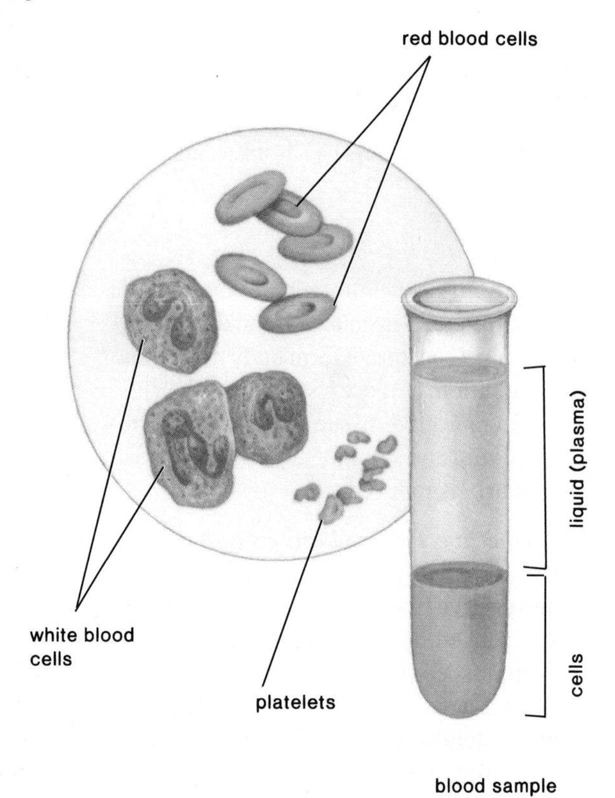

Table 10.2
Blood Plasma

Water (92% of Total)	
Solutes (8% of Total)	
Inorganic ions (salts)	Na^+, Ca^{++}, K^+, Mg^{++}; Cl^-, HCO_3^-, HPO_4^-, SO_4^-
Gases	O_2, CO_2
Plasma proteins	Albumin, globulins, fibrinogen
Organic nutrients	Glucose, fats, phospholipids, amino acids, etc.
Nitrogenous waste products	Urea, ammonia, uric acid
Regulatory substances	Hormones, enzymes

Muscular Tissue: Contracts

Muscular (contractile) tissue is composed of cells that are called *muscle fibers.* Muscle fibers contain actin filaments and myosin filaments, whose interaction accounts for the movements we associate with animals. There are 3 types of vertebrate muscles: *skeletal, smooth,* and *cardiac.*

Skeletal muscle (fig. 10.5*a*) is attached to the bones of the skeleton; it moves body parts. It is under voluntary control and contracts faster than all the other muscle types. Skeletal muscle fibers are cylindrical and quite long—sometimes they run the length of the muscle. They arise during development when several cells fuse, giving one cell with multiple nuclei. The nuclei are located at the periphery of the cell, just inside the cell membrane.

Skeletal muscle cells are **striated.** Light and dark bands run perpendicular to the length of the cell. These bands are due to the placement of actin filaments and myosin filaments in the cell.

Smooth muscle is so named because the cells lack striations. The spindle-shaped cells form layers in which the thick middle portion of one cell is opposite the thin ends of adjacent cells. Consequently, the nuclei form an irregular pattern in the tissue (fig. 10.5*b*). Smooth muscle is not under voluntary control and therefore is said to be involuntary. Smooth muscle, found in walls of viscera (intestine, stomach, and other internal organs) and blood vessels, contracts more slowly than skeletal muscle but can remain contracted for a longer time. When the smooth muscle of the intestine contracts, it moves the food along, and when the smooth muscle of the blood vessels contracts, it constricts the blood vessels, helping to raise blood pressure.

Cardiac muscle (fig. 10.5*c*), which is found only in the wall of the heart, is responsible for the heartbeat, which pumps blood. Cardiac muscle seems to combine features of both smooth muscle and skeletal muscle. It has striations like skeletal muscle, but the contraction of the heart is involuntary for the most part. Cardiac muscle cells also differ from skeletal muscle cells in that they have a single, centrally placed nucleus. The cells are branched and seemingly fused one with the other, and the heart appears to be composed of one large interconnecting mass of muscle cells. Actually, cardiac muscle cells are separate and individual, but they are bound end to end at *intercalated disks,* areas where folded cell membranes between 2 cells contain desmosomes and gap junctions.

All muscular tissue contains actin filaments and myosin filaments; these form a striated pattern in skeletal and cardiac muscle, but not in smooth muscle.

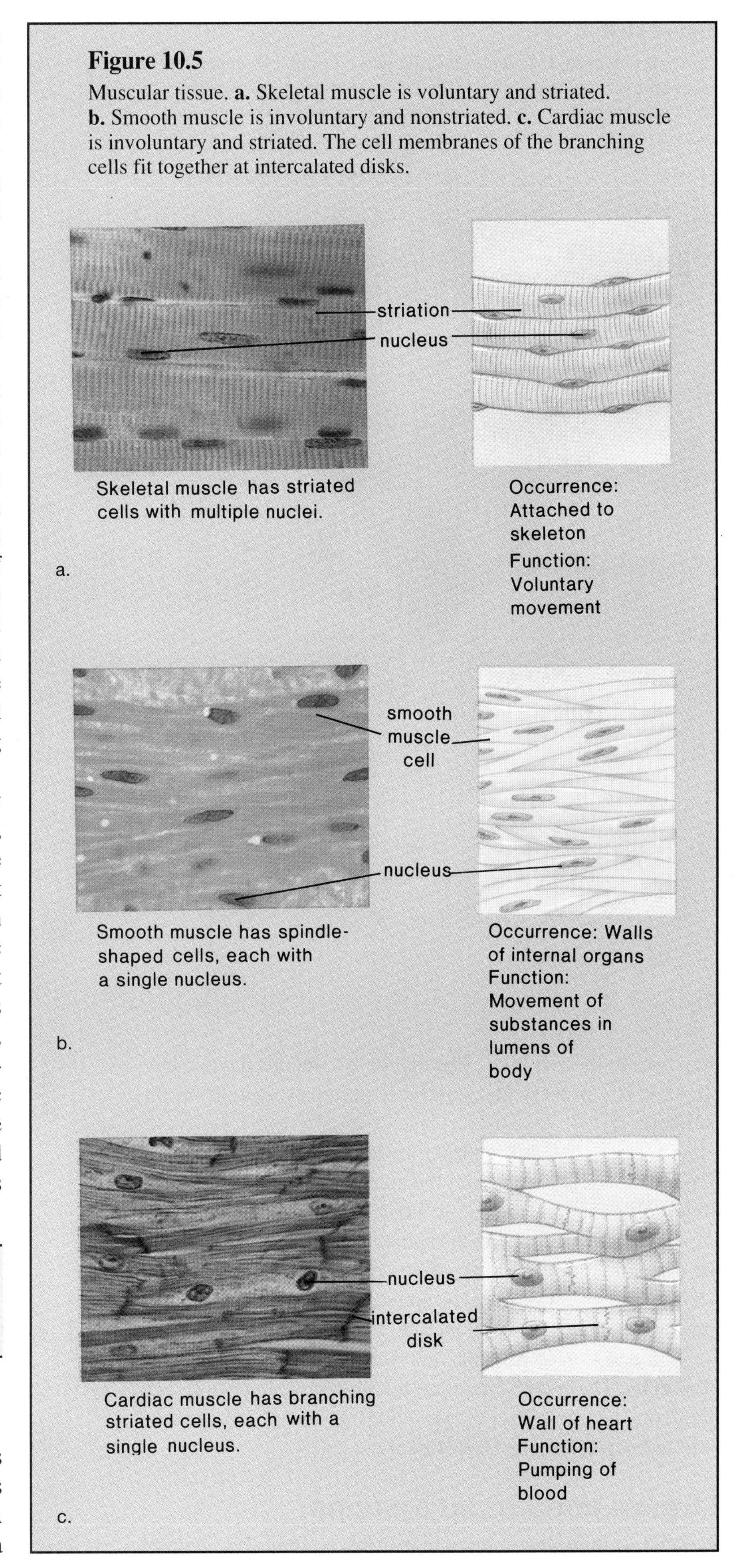

Figure 10.5
Muscular tissue. **a.** Skeletal muscle is voluntary and striated. **b.** Smooth muscle is involuntary and nonstriated. **c.** Cardiac muscle is involuntary and striated. The cell membranes of the branching cells fit together at intercalated disks.

Nervous Tissue: Conducts Impulses

The brain and the spinal cord contain conducting cells termed neurons. A **neuron** is a specialized cell that has 3 parts: dendrites, cell body, and an axon (fig. 10.6). A dendrite is a process that conducts impulses (sends a

Figure 10.6
Photo of a neuron. Conduction of the nerve impulse is dependent on neurons, each of which has the 3 parts indicated. A dendrite takes nerve impulses to the cell body, and an axon takes them away from the cell body. The nucleus is in the cell body.

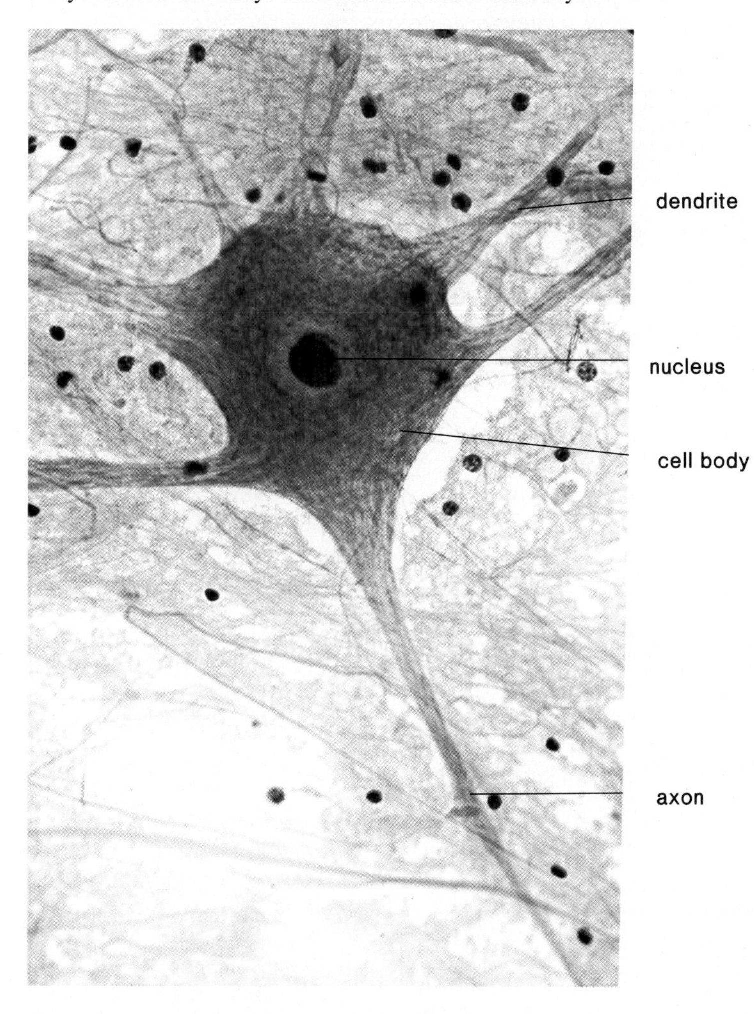

message) to the cell body. The cell body contains the nucleus. An axon is a process that conducts impulses away from the cell body.

When axons and dendrites are long, they are called *nerve fibers.* Outside the brain and the spinal cord, nerve fibers are bound by connective tissue to form **nerves.** Nerves conduct impulses from receptors to the spinal cord and the brain, where the phenomenon called sensation occurs. They also conduct nerve impulses away from the spinal cord and the brain to the muscles, causing them to contract.

In addition to neurons, nervous tissue contains **neuroglial cells.** These cells maintain the tissue by supporting and protecting neurons. They also provide nutrients to neurons and help to keep the tissue free of debris.

Organs and Organ Systems

We tend to think that a particular organ contains one type of tissue. For example, we associate muscular tissue with muscles and nervous tissue with the brain. However, these organs also contain other types of tissue; for example, they contain loose connective tissue and blood. An **organ** is a structure that is composed of 2 or more types of tissues that work together to perform particular functions. An **organ system** contains many different organs that cooperate to carry out a process such as digestion of food.

We are going to consider skin as an example of an organ. Some people even like to call skin the *integumentary system,* especially since it cannot be placed in one of the other organ systems. However, skin does not really have distinct organs.

Skin as an Organ

Skin (fig. 10.7) covers the body, protecting underlying parts from physical trauma, microbes, and water loss. Skin helps to regulate body temperature (p. 164), and because it contains sense organs, skin also helps us to be aware of our surroundings and to communicate with others.

Skin has an outer epidermal layer (the epidermis) and an inner dermal layer (the dermis). Beneath the dermis, a subcutaneous layer binds skin to the underlying organs.

Skin Layers: Structure and Function

The **epidermis** is the outer, thinner layer of the skin. It is made up of stratified squamous epithelium, which is continually produced by a bottom layer of cells termed basal cells. Newly formed cells push to the surface, then gradually flatten and harden. Eventually, they die and are sloughed off. Hardening is caused by cellular production of a waterproof protein called *keratin.* When you have *dandruff,* the rate of keratinization is 2 or 3 times the normal rate in certain areas of the scalp. Over much of the body, keratinization is minimal, but the palm of the hand and the sole of the foot have a particularly thick outer layer of dead keratinized cells arranged in spiral and concentric patterns. These patterns form fingerprints and footprints.

Specialized cells in the epidermis called *melanocytes* produce melanin, the pigment responsible for skin color in dark-skinned persons. When you sunbathe, the melanocytes become more active, producing melanin in an attempt to protect the skin from the damaging effects of the ultraviolet (UV) radiation in sunlight.

> The epidermis, the outer layer of skin, is made up of stratified squamous epithelium. New cells, continually produced in the innermost layer of the epidermis, push outward, become keratinized, die, and are sloughed off.

The **dermis,** a layer of fibrous connective tissue, is thicker than the epidermis. It contains elastic fibers and collagen fibers. The collagen fibers form bundles, which interlace with each other and run, for the most part, parallel to the skin surface. As a person ages and is exposed to the sun, the number of these fibers decreases, and those remaining have characteristics that make the skin less supple and prone to wrinkling.

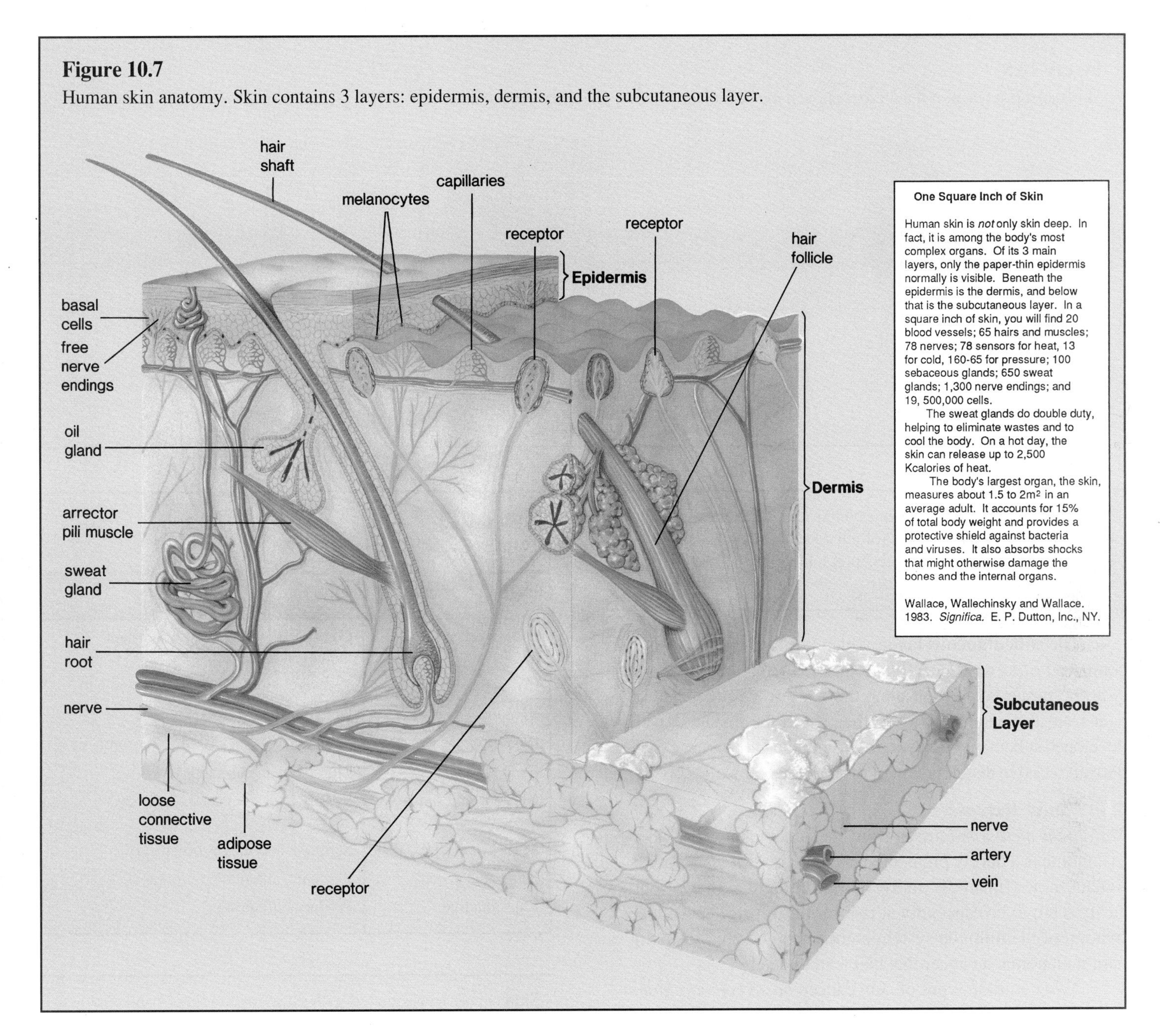

Figure 10.7
Human skin anatomy. Skin contains 3 layers: epidermis, dermis, and the subcutaneous layer.

There are several types of structures in the dermis. A hair shaft, except for the root, contains dead, hardened epidermal cells; the root is alive and resides at the base of a *hair follicle.* Each follicle has one or more *oil* (sebaceous) *glands,* which secrete sebum, an oily substance that lubricates the hair and the skin. Particularly on the nose and the cheeks, the sebaceous glands may fail to discharge, and the secretions collect and form "whiteheads" or "blackheads." The color of blackheads is due to oxidized sebum. If pus-inducing bacteria also are present, a boil or a pimple may result.

The *arrector pili muscle,* a smooth muscle, is attached to the hair follicle in such a way that when contracted, the muscle causes the hair to "stand on end." When you have had a scare or are cold, goose bumps develop due to the contraction of these muscles.

Sweat (sudoriferous) *glands* are quite numerous and are present in all regions of skin. A sweat gland begins as a coiled tubule within the dermis, but then it straightens out near its opening. Some sweat glands open into hair follicles, but most open onto the surface of the skin.

Small receptors are present in the dermis. *Receptors* are specialized nerve endings, present in all organs and parts of the body, that respond to either external or internal stimuli as appropriate. In skin there are different receptors for touch, pressure, pain, and temperature. The fingertips contain the most touch receptors, and these add to our ability to use our fingers for delicate tasks. The dermis also contains nerve fibers and blood vessels. When blood rushes into these vessels, a person blushes, and when blood is reduced in them, a person turns "blue."

Figure 10.8
Skin cancer. In each of the 3 types shown, the skin clearly has an abnormal appearance.

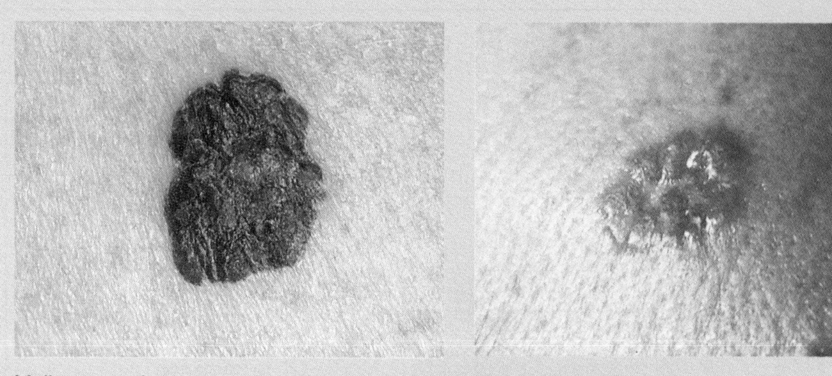

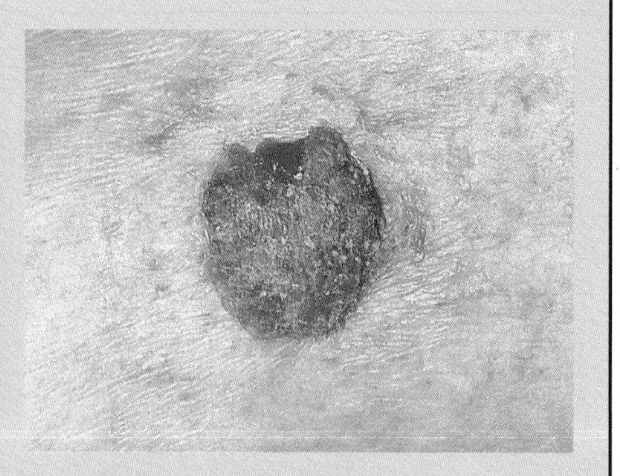

Malignant melanoma — Basal cell carcinoma — Squamous cell caracinoma

> The dermis, composed of fibrous connective tissue, lies beneath the epidermis. It contains hair follicles, sebaceous glands, and sweat glands. It also contains receptors, blood vessels, and nerve fibers.

The **subcutaneous layer,** which lies below the dermis, is composed of loose connective tissue, including adipose tissue. Adipose tissue helps to insulate the body from either gaining heat from the outside or losing heat from the inside. A well-developed subcutaneous layer gives a rounded appearance to the body. Excessive development of this layer accompanies obesity.

Skin Cancer Increases

Malignant melanoma arises from melanocytes in the skin. A melanoma is a darkly pigmented spot that resembles a nonmalignant mole (fig. 10.8). Individuals with light skin who burn easily seem to be especially at risk for this type of cancer. Activation of the immune system is the latest therapy for malignant melanoma, a cancer that can lead to death.

Two common types of skin cancer, *basal cell carcinoma* and *squamous cell carcinoma,* are likely to occur in all persons exposed to sunlight. Ultraviolet radiation causes epidermal cells to become cancerous. Precancerous dark patches of skin become rough and scaly with a reddish base. In basal cell carcinoma, epidermal cells invade the dermis and form ulcers. Both types of cancer can usually be surgically removed.

In recent years, there has been a great increase in the number of persons with skin cancer, and physicians believe this is due to sunbathing or even to the use of tanning machines, as discussed in the reading on page 165. These professionals strongly recommend that everyone stay out of the sun and refrain from using tanning machines. If persons must be in the sun, sunscreens should be used. A lotion with a sun protection factor (SPF) of 15 means that 15 minutes in the sun with protection is equivalent to one minute without protection. Even higher SPF strengths are available.

Table 10.3
Human Organ Systems

Name	Function
Digestive	Converts food particles to nutrient molecules
Circulatory	Transports molecules to and from cells
Immune	Defends against disease
Respiratory	Exchanges gases with the environment
Excretory	Eliminates metabolic wastes
Nervous and sensory	Regulates systems and response to environment
Musculoskeletal	Supports and moves organism
Hormonal	Regulates internal environment
Reproductive	Produces offspring

> Skin cancer is associated with ultraviolet radiation and appears in 3 forms. Malignant melanoma is the most dangerous skin cancer. Squamous cell carcinoma and basal cell carcinoma can usually be removed surgically.

Studying Organ Systems

In this text, we will study the organ systems listed in table 10.3. Each of these systems has a specific location within the body. The central nervous system is located dorsally (toward the back); the brain is protected by the skull; and the spinal cord, which gives off spinal nerves, is protected by the vertebrae (fig. 10.9). The repeating units of vertebrae and spinal nerves show that humans are segmented animals, meaning that body parts reoccur at regular intervals.

Within the musculoskeletal system, the skeleton provides the surface area for attachment of well-developed and power-

THAT OH, SO NICE, TAN

Are tanning machines safe? Most dermatologists feel they are not. Tanning booths (fig. 10.A) almost certainly are capable of causing skin to age, to degenerate, or to develop cancer.

On the whole, people who use tanning machines probably get away with it, but they are taking a chance. Tanning machines expose their patrons to intense ultraviolet light, which is capable of producing the acute and chronic side effects of exposure to sunlight. Tanning machines use mainly UVA light, which is somewhat less potent in producing biological changes than the slightly shorter UVB waves, but sufficient doses of UVA can be deleterious. Also, extra UVA may enhance the carcinogenic potential of exposure to natural sunlight. People taking medications that induce photosensitivity may have severe reactions to the light from tanning machines. Certain diseases, notably lupus erythematosus, which are exacerbated by sunlight, also are made worse by tanning machines.

The light in these booths is intense so as to achieve in a few minutes the skin reaction that would otherwise take a longer period of baking in natural sunlight. People who tan in booths may do so without any clothing; thus they expose areas of skin that lack protective pigmentation built up from long-term exposure to sunlight. Ultraviolet light has the potential to injure the retina; people who patronize tanning parlors always should wear protective goggles.

The American Academy of Dermatology has formed a task force on photobiology. Its chairman, Dr. Leonard C. Harber, also is chairman of the Department of Dermatology at Columbia University College of Physicians and Surgeons. He says, "We want health warnings to appear in tanning parlors as they do on cigarette packs."

Excerpted from the March 1988 issue of the *Harvard Health Letter* © 1988, President and Fellows of Harvard College. Reprinted by permission.

Figure 10.A

Tanning booth. Tanning the skin most likely contributes to skin aging, skin degeneration, and skin cancer.

ful striated muscles. The musculoskeletal system makes up most of the body weight and is specialized for locomotion.

The other internal organs are found within a body cavity called the **coelom.** In humans, the coelom is divided by the muscular diaphragm, which assists breathing. The heart, a pump for the closed circulatory system, and the lungs are located in the upper (thoracic, or chest) cavity. The major portion of the digestive system, the entire excretory system, and much of the reproductive system are located in the lower (abdominal) cavity. The major organs of the excretory system are the paired kidneys, and the accessory organs of the digestive system are the liver and the pancreas. Each sex has characteristic sex organs.

Body Membranes: Two-part Lining

At the organ level, the term *membrane* generally refers to a thin lining or covering composed of an epithelium overlying a layer of loose connective tissue. For example, mucous membrane lines the organs of the respiratory and digestive systems. This type of membrane, as its name implies, secretes mucus. Serous membrane lines enclosed cavities and covers the organs that lie within these cavities, such as the heart, the lungs, and the kidneys. This type of membrane secretes a watery lubricating fluid.

The body is divided into cavities, within which the organs are found. These cavities usually are lined with membrane. Do 10.1 Critical Thinking, found at the end of the chapter.

Homeostasis

Homeostasis means that the internal environment remains relatively constant, regardless of the conditions in the external environment. In humans, for example;

1. Glucose concentration of blood remains at about 0.1%.
2. The pH of blood is always near 7.4.
3. Blood pressure in the brachial artery averages near 120/80.
4. Blood temperature averages around 37° C (98.6° F).

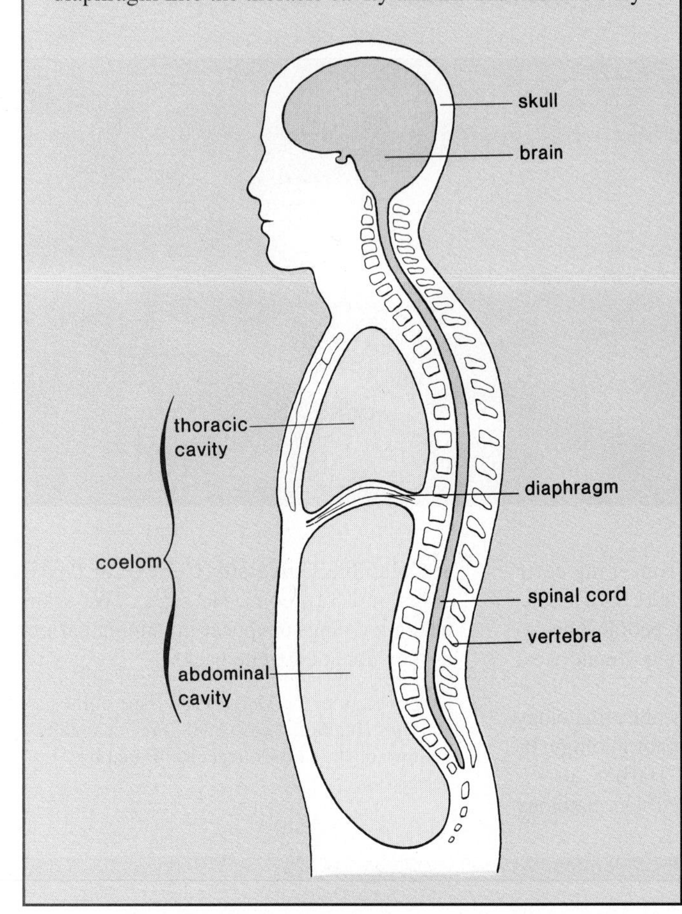

Figure 10.9

Organization of the human body. Humans have a dorsal nervous system and a well-developed coelom, which contains the internal organs. The coelom is divided by the diaphragm into the thoracic cavity and the abdominal cavity.

The ability of the body to keep the internal environment within a certain range allows humans to live in a variety of habitats, such as the arctic regions, the deserts, or the tropics.

This internal environment includes tissue fluid, which bathes all the tissues of the body. Tissue fluid is refreshed when such molecules as oxygen and nutrients exit blood and wastes enter blood (fig. 10.10). Tissue fluid remains constant only as long as blood composition remains constant. Although we are accustomed to using the word *environment* to mean the external environment of the body, it is important to realize that it is the internal environment of tissues that is ultimately responsible for our health and well-being.

The internal environment of the body consists of blood and tissue fluid, which bathes the cells.

Most systems of the body contribute to maintenance of a constant internal environment. The digestive system takes in and digests food, providing nutrient molecules that enter blood

Figure 10.10

Tissue fluid composition. Cells are surrounded by tissue fluid, which is continually refreshed because oxygen and nutrient molecules constantly exit and waste molecules continually enter the bloodstream as shown.

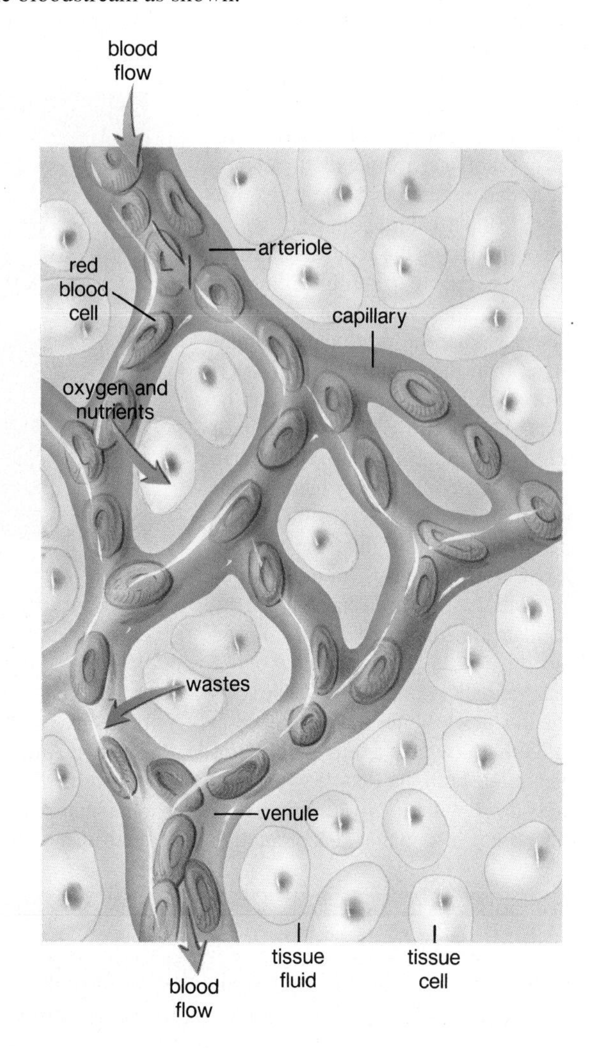

and replace the nutrients that are constantly being used by the body cells. The respiratory system adds oxygen to and removes carbon dioxide from blood. The amount of oxygen taken in and carbon dioxide given off can be increased to meet body needs. The chief regulators of blood composition, however, are the liver and the kidneys. They monitor the chemical composition of plasma (table 10.2) and alter it as required. Immediately after glucose enters blood, it can be removed by the liver for storage as glycogen. Later, the glycogen can be broken down to replace the glucose used by the body cells; in this way, the glucose composition of blood remains constant. The hormone insulin, secreted by the pancreas, regulates glycogen storage. The liver also removes toxic chemicals, such as ingested alcohol, drugs, and cellular wastes from blood. These are converted to molecules that can be excreted by the kidneys. The kidneys are also under hormonal control as they excrete wastes and salts, substances that can affect blood pH.

All the systems of the body contribute to homeostasis, that is, maintaining the relative constancy of the internal environment.

Although homeostasis is, to a degree, controlled by hormones, it is ultimately controlled by the nervous system. The brain contains centers that regulate factors such as temperature and blood pressure. Maintenance of body conditions requires receptors, which detect unacceptable levels and signal a regulator center (fig. 10.11). If a correction is required, the center then directs an effector (muscles or glands) to bring about a response. The results of the response feed back into the system to influence the stimulus, usually depressing it. This is called **negative feedback** control because the response dampens or even cancels the stimulus that brought about the response. This type of homeostatic regulation results in fluctuation above and below a mean, as illustrated by temperature control (fig. 10.12). Notice that negative feedback control is a self-regulatory mechanism.

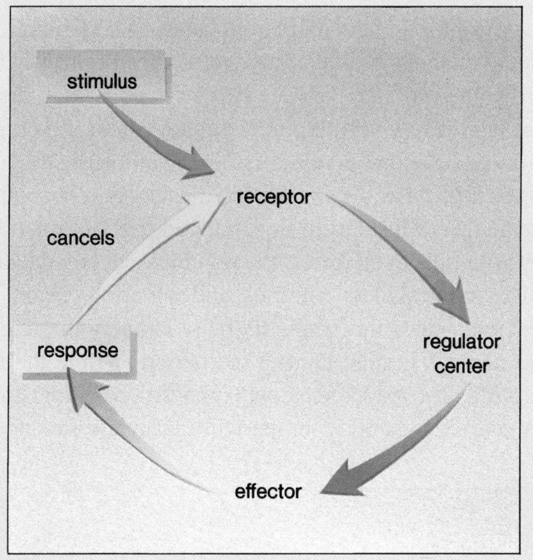

Figure 10.11

Negative feedback control. A receptor is stimulated by a change such as low temperature and signals a regulator center (in the brain), which directs effectors (muscles or glands) to react. The response, constriction of blood vessels in the skin and perhaps shivering, raises the temperature and cancels the original stimulus.

Figure 10.12

Temperature control. When body temperature rises, the regulator center directs blood vessel dilation and sweat gland activation. As a result, body temperature lowers. Then the regulator center directs the blood vessels to constrict, hairs to stand on end, and even shivering to occur if needed. The body temperature then rises again. Because the regulator center is activated only by extremes, body temperature fluctuates above and below normal.

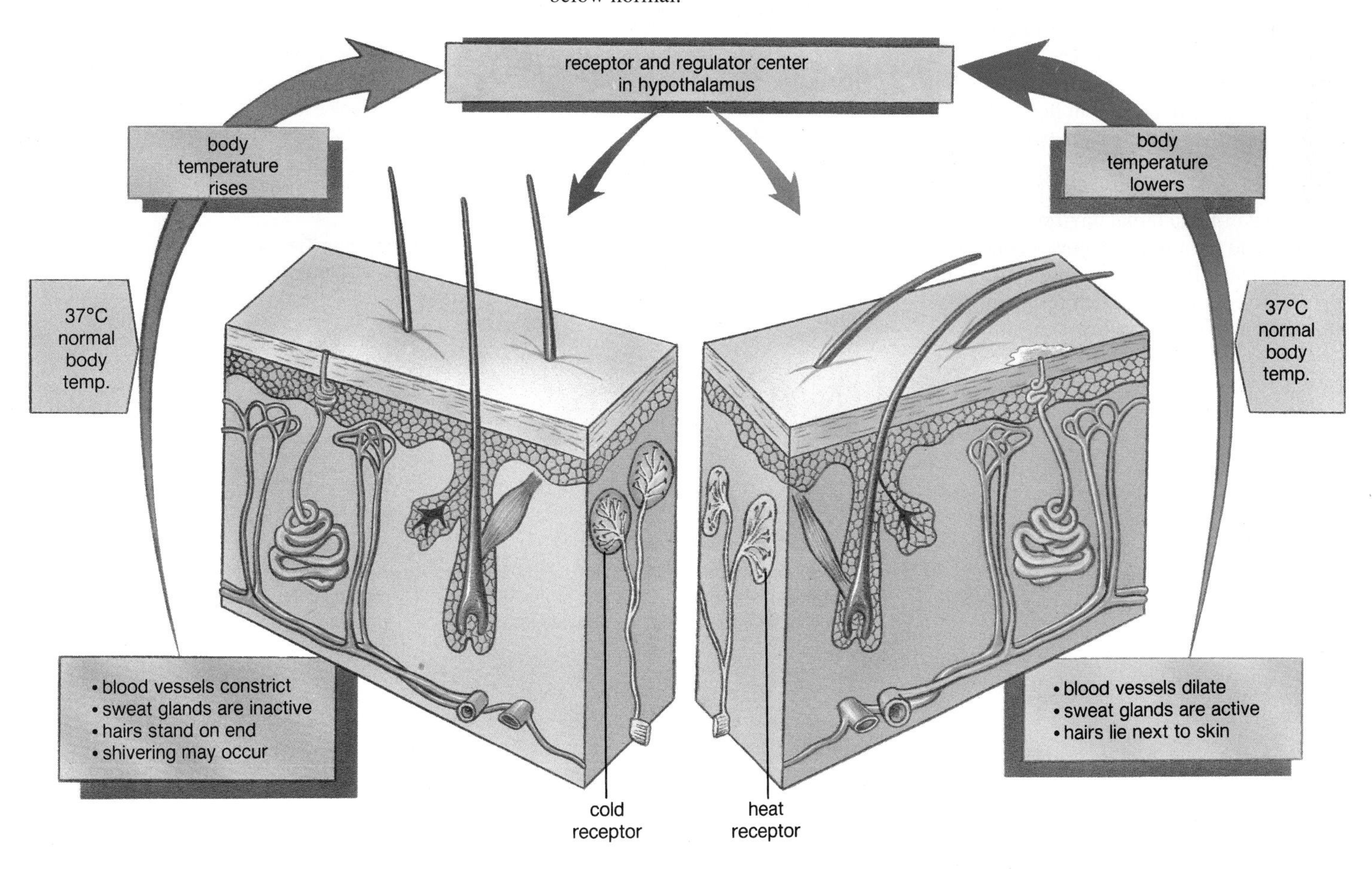

Controlling Body Temperature

The receptor and the regulator center for body temperature are located in the hypothalamus, a part of the brain. The receptor is sensitive to the temperature of blood, and when the temperature falls below normal, the regulator center directs (via nerve impulses) the blood vessels of the skin to constrict. This conserves heat. Also, the arrector pili muscles in skin pull hairs erect, and a layer of insulating air is trapped next to skin. If body temperature falls even lower, the regulator center sends nerve impulses to the skeletal muscles, and shivering occurs. Shivering generates heat, and gradually body temperature rises to 37° C and perhaps higher. During the period of time the body temperature is normal, the receptor and the regulator center are not active, but once body temperature is higher than normal, they are reactivated. Now the regulator center directs the blood vessels of the skin to dilate. This allows more blood to flow near the surface of the body, where heat is lost to the environment. The regulator center also activates the sweat glands because the evaporation of sweat helps to lower body temperature. Gradually, body temperature decreases to 37° C and perhaps lower. Once body temperature is below normal, the cycle begins again.

> Homeostasis of internal conditions is a self-regulatory mechanism that results in slight fluctuations above and below a mean. For example, body temperature rises above and drops below a normal temperature of 37° C. Do 10.2 Critical Thinking, found at the end of the chapter.

SUMMARY

Human tissues are categorized into 4 major types. Epithelial tissue covers the body and lines its cavities. Connective tissue often binds body parts. Contraction of muscular tissue moves the body and its parts. Nerve impulses conducted by neurons within nervous tissue help to bring about coordination of body parts.

Different types of tissues are joined to form organs, each one having a specific function. Organs are grouped into organ systems. In vertebrates, the brain and the spinal cord are dorsally located, and the internal organs are located in the coelom, divided into the thoracic and abdominal cavities.

All organ systems contribute to the constancy of the internal environment. The nervous and hormonal systems regulate the other systems. Both of these are controlled by negative feedback, which results in fluctuation above and below a mean.

Epithelial Tissue
(example: cuboidal epithelium)

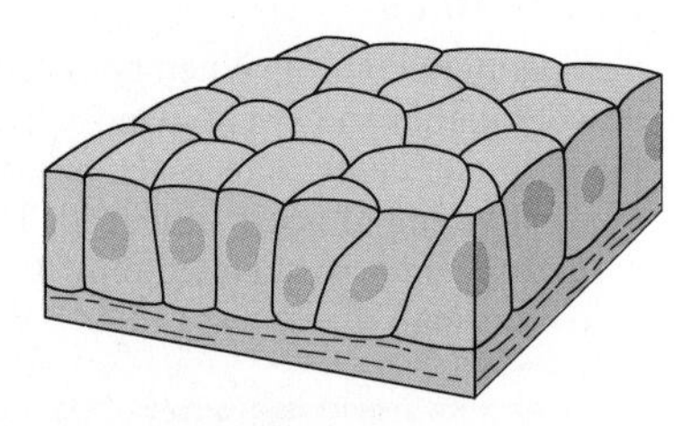

- named for shape of cell
- located on body surface or lining a cavity
- various functions, including protection

Connective Tissue
(example: loose connective tissue)

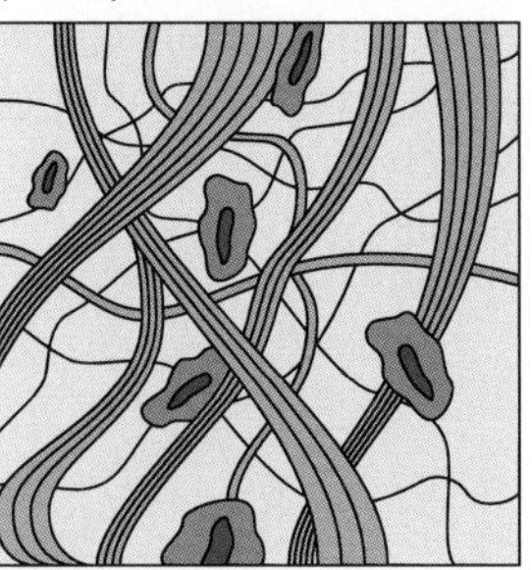

- cells separated by a matrix that contains fibers
- located between and around body parts
- functions to support, protect, and hold things together

Muscular Tissue
(example: cardiac muscle)

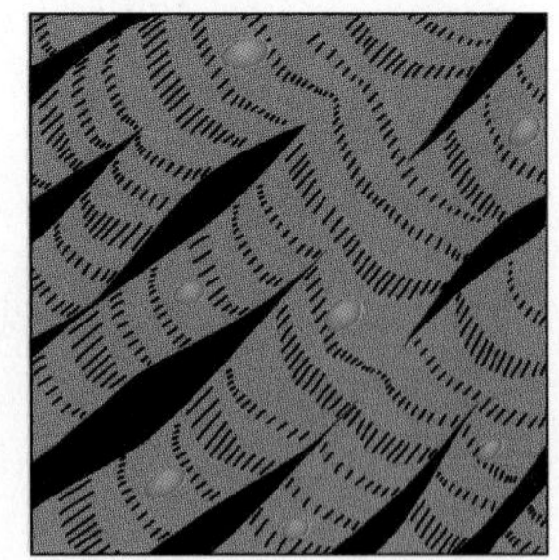

- cells contain contractile filaments
- located in skeletal muscles and organ walls
- functions to move the body and body parts

Nervous Tissue
(example: neuron)

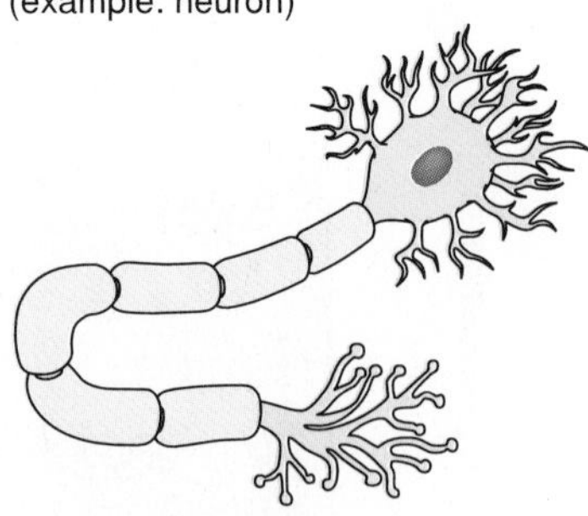

- cells have nerve fibers
- located in brain, spinal cord, and ganglia
- functions to transmit nerve impulses

STUDY QUESTIONS

In order to practice **writing across the curriculum,** students should write out the answers to any or all of the study questions. The study questions are sequenced in the same order as the text.

1. Name the 4 major types of tissues. (p.156)
2. What are the functions of epithelial tissue? Name the different kinds, and give a location for each. (p. 156)
3. What are the functions of connective tissue? Name the different kinds, and give a location for each. (pp. 156–159)
4. What are the functions of muscular tissue? Name the different kinds, and give a location for each. (p. 161)
5. Nervous tissue contains what type of cell? Which organs in the body are made up of nervous tissue? (p. 161)
6. Describe the structure of skin, and state at least 2 functions of this organ. (p. 162)
7. In general terms, describe the location of the human organ systems. (pp. 164–165)
8. Distinguish between cell membrane and body membrane. (p. 165)
9. What is homeostasis, and how is it achieved in the human body? (pp. 165–166)
10. Specifically describe how body temperature is maintained at about 37° C. (p. 168)

OBJECTIVE QUESTIONS

1. Most organs contain several different types of _______.
2. Kidney tubules are lined by cube-shaped cells called _______ epithelium.
3. Pseudostratified ciliated columnar epithelium contains cells that appear to be _______, have projections called _______, and are _______ in shape.
4. Both cartilage and blood are classified as _______ tissue.
5. Cardiac muscle is _______ but involuntary.
6. Nerve cells are called _______.
7. Skin has 3 layers: epidermis, _______, and the subcutaneous layer.
8. Outer skin cells are filled with _______, a waterproof protein that strengthens them.
9. Mucous membrane contains _______ tissue overlying _______ tissue.
10. Homeostasis is maintenance of the relative _______ of the internal environment, that is, blood and _______ fluid.
11. Give the name, the location, and the function for each of these tissues.
    a. Type of epithelial tissue
    b. Type of muscular tissue
    c. Type of connective tissue

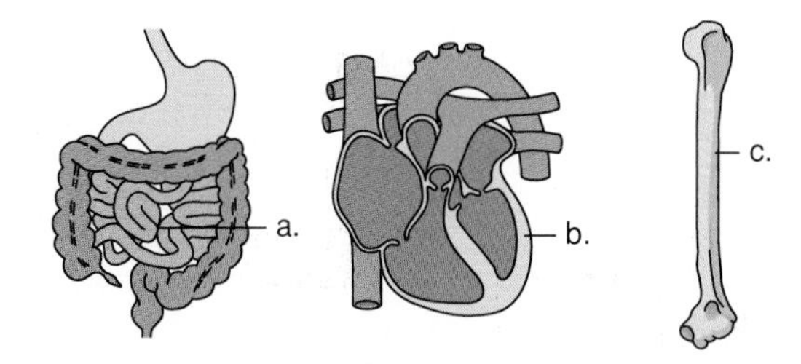

CRITICAL THINKING

In order to practice **writing across the curriculum,** students should write out the answers to any or all of the critical thinking questions. Suggested answers to the critical thinking questions are in appendix E.

10.1

1. How is the structure of an epithelial cell suited to its function?
2. Tight junctions between epithelial cells are equivalent to which feature of endodermal cells in plants (p. 142)? How are they equivalent?
3. How is the structure of a skeletal muscle cell suited to muscle contraction? If, upon contraction, muscular tissue always shortens from right to left, what would happen to an object attached at the right?
4. How is the structure of a nerve cell suited to its function?

10.2

1. Normal body temperature is said to be 37° C. Is body temperature always exactly 37° C, or does it fluctuate?
2. Why does the text use the phrase "*relative constancy* of the internal environment" when referring to homeostasis?
3. What type of stimuli activate the receptor and the regulator center shown in figure 10.11?
4. Does this account for fluctuation of body temperature above and below a certain temperature?

SELECTED KEY TERMS

bone connective tissue having a hard matrix of calcium salts deposited around protein fibers.

cartilage a connective tissue in which the cells lie within lacunae separated by a flexible matrix.

coelom (se´lom) a body cavity of higher animals that contains internal organs, such as those of the digestive system.

compact bone hard bone consisting of Haversian systems cemented together.

connective tissue a type of tissue characterized by cells separated by a matrix that often contains fibers.

dermis (der´mis) the layer of thick skin that lies beneath the epidermis.

epidermis (ep˝ ĭ-der´mis) the outer layer of skin, composed of stratified squamous epithelium.

epithelial tissue (ep˝ ĭ-the´le-al tish´u) a type of tissue that covers the external surface of the body and lines its cavities.

gland a cell or group of epithelial cells that are specialized to secrete a substance.

hyaline cartilage (hi´ah-līn kar´tĭ-lij) cartilage composed of very fine collagen fibers and a matrix having a milk glass appearance.

lacuna (lah-ku´nah) a small pit or hollow cavity, as in bone or cartilage, where a cell or cells are located.

ligament dense fibrous connective tissue that joins bone to bone at a joint.

muscular (contractile) tissue a type of tissue that contains cells capable of contracting; skeletal muscles are attached to the skeleton, smooth muscle is found within walls of internal organs, and cardiac muscle makes up the heart.

negative feedback a self-regulatory mechanism that is activated by an imbalance and results in a fluctuation above and below a mean.

neuron (nu´ron) nerve cell that characteristically has 3 parts: dendrites, cell body, axon.

spongy bone porous bone found at the ends of long bones.

striated having bands; cardiac and skeletal muscle are striated with bands of light and dark.

subcutaneous layer (sub˝ku-ta´ne-us la´er) a tissue layer found in vertebrate skin that lies just beneath the dermis and tends to contain adipose tissue.

tendon dense, fibrous connective tissue that joins muscle to bone.

11

DIGESTION

Chapter Concepts

1.
The human digestive system is a tube with specialized parts between the 2 openings, the mouth and the anus. 172, 194

2.
The liver and the pancreas are the accessory organs of digestion because their secretions assist the digestive process. 179, 194

3.
The digestive enzymes are specific and have an optimum temperature and pH. 182, 194

4.
The products of digestion are small molecules, such as amino acids and glucose, that can cross cell membranes. 182, 194

5.
Proper nutrition supplies the body with energy and nutrients, including vitamins, minerals, and the essential amino acids. 184, 194

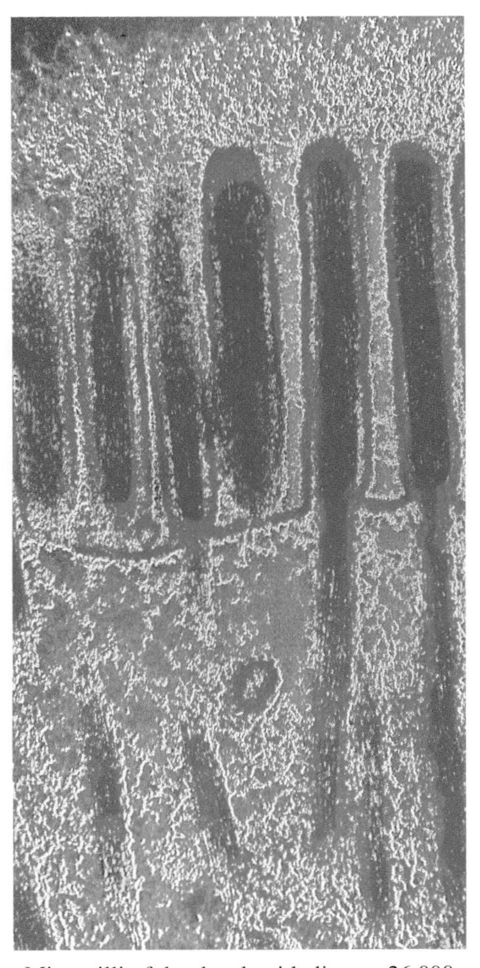

Microvilli of duodenal epithelium, ×26,000

Chapter Outline

DIGESTION takes place within a tube called the digestive tract, which begins with the mouth and ends with the anus (table 11.1 and fig. 11.1). Digestion of food in humans is an extracellular (outside cells) process. Digestive enzymes are secreted into the tract by glands located either in the tract or nearby. Food is never found within these *accessory glands,* only within the tract itself.

While strictly speaking the term *digestion* means the breakdown of food by enzymatic action, in this text the term is expanded to include both physical and chemical processes that reduce food to small soluble molecules. Only small molecules can cross cell membranes and be absorbed by the tract lining. Too often we are inclined to think that since we eat meat (protein), potatoes (carbohydrate), and butter (fat), these are the substances that nourish our bodies. Instead, it is, for example, the amino acids from the protein and the sugars from the carbohydrate that actually enter blood and are transported throughout the body to nourish the cells. Any component of food, such as cellulose, that cannot be digested to small molecules leaves the tract as waste.

Digestion of food requires a cooperative effort between different parts of the body. We will see that the production of hormones and the performance of the nervous system achieve the cooperation of body parts.

The Digestive System

The functions of the digestive system are to ingest the food, to digest it to small molecules that can cross cell membranes, to absorb these nutrient molecules, and to eliminate nondigestible wastes.

Mouth: Food Receiver

The mouth, sometimes called the oral cavity, receives the food in humans. Most people enjoy eating mainly because they find food tasty. Receptors called taste buds are found primarily on the tongue. These are activated by the presence of food in the mouth. Taste buds initiate nerve impulses, which travel by way of cranial nerves to the brain. Still, what we call taste is largely due to stimulation of olfactory (smell) receptors in the nose. If we have a cold and our nose is blocked, food has little taste. The sense of taste is discussed further in chapter 18.

The teeth chew the food into pieces convenient for swallowing. During the first 2 years of life, the 20 deciduous, or baby, teeth appear. These are eventually replaced by the adult teeth. Normally, adults have 32 teeth (fig. 11.2). One-half of each jaw has teeth of 4 different types: 2 chisel-shaped *incisors* for biting; one pointed *canine* for tearing; 2 fairly flat *premolars* for grinding; and 3 *molars,* more flattened than the other teeth, for crushing. The last molar, called a wisdom tooth, may fail to erupt, or if it does, it is sometimes crooked and useless. Oftentimes, dentists recommend the extraction of the wisdom teeth.

Each tooth (fig. 11.3) has 2 main divisions, a crown and a root. The crown is composed of a layer of enamel, an extremely

Figure 11.1

Trace the path of food from the mouth to the anus. The large intestine consists of three parts: the transverse colon, the ascending colon, and the descending colon, plus the rectum and anal canal. Note the placement of the accessory organs of digestion, the liver and the pancreas.

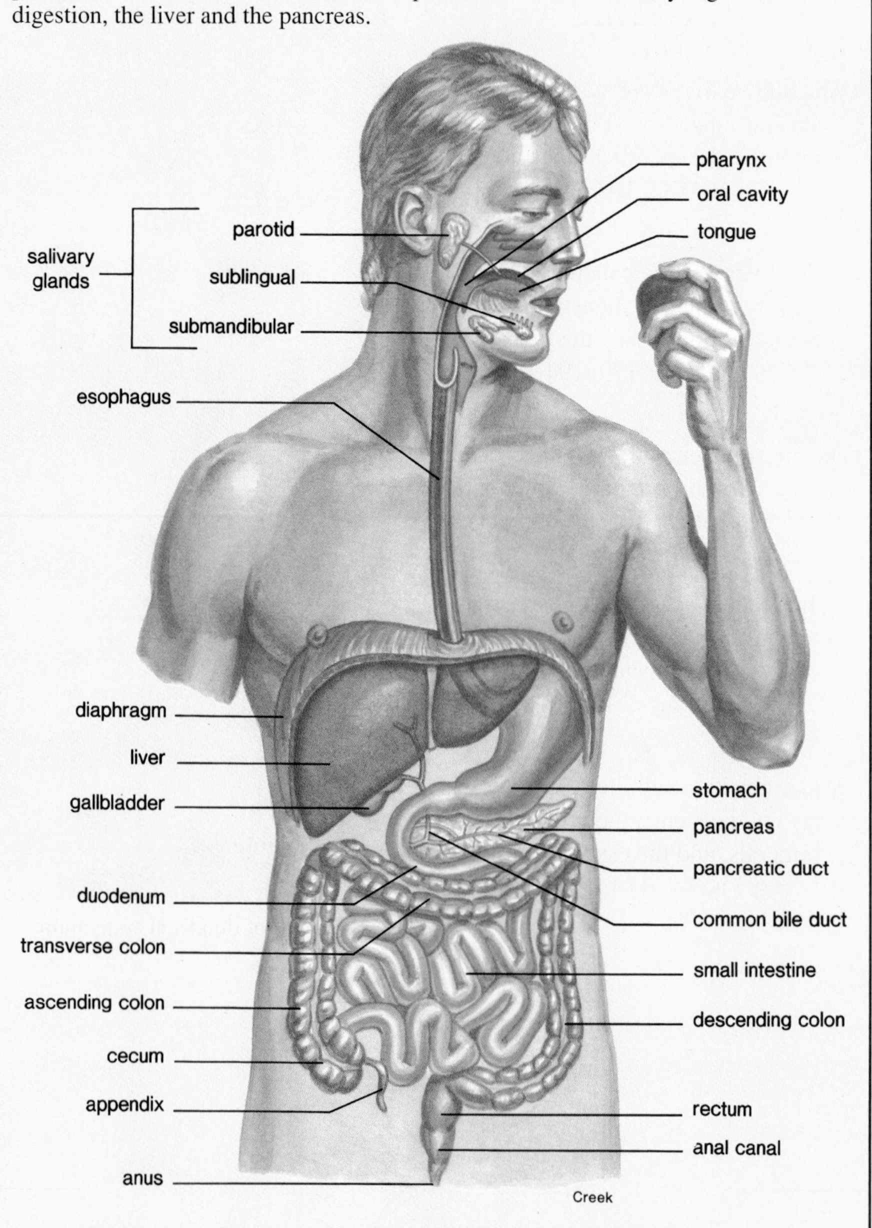

A N I M A L • B I O L O G Y

DIGESTIVE TRACTS

Comparatively speaking, some animal digestive tracts are incomplete and some are complete. The planarian has an incomplete digestive tract because the only opening is used as both an entrance and an exit; an earthworm has a complete digestive tract, with both a mouth and an anus. The branched digestive tract in a planarian delivers nutrients directly to the body cells. You can see that a complete digestive tract, such as in earthworms, leads to specialization of parts—mouth, pharynx, esophagus, crop, and gizzard precede the earthworm's intestine. Both the planarian and the earthworm draw food in through a muscular pharynx, but of the 2, the planarian is the predator. The earthworm feeds on decayed organic matter in dirt, but a planarian grasps and clings to its live prey, even partially predigesting it before sucking it in whole or in pieces.

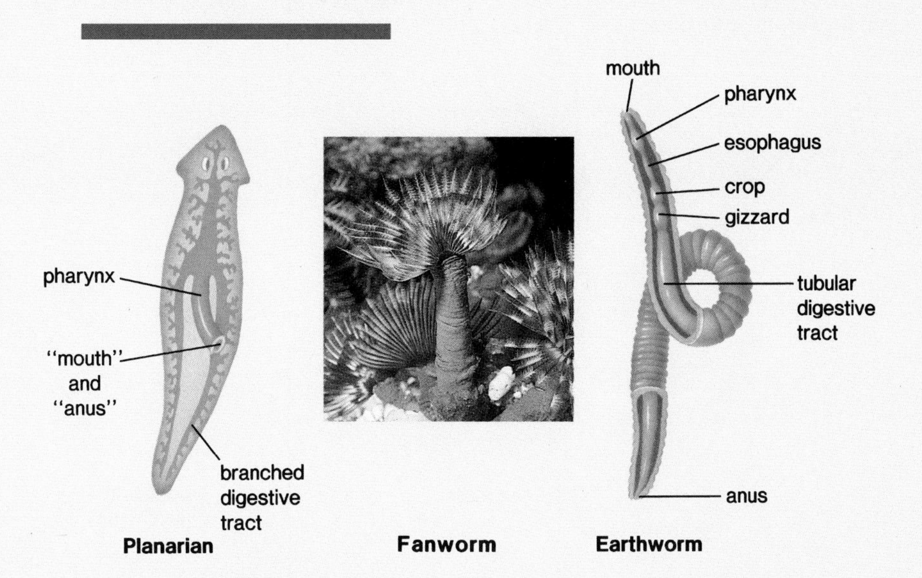

Among the annelids, earthworms are discontinuous feeders, as witnessed by the presence of a crop, the storage organ. The marine fanworms, however, are continuous filter feeders. These animals lack a head entirely; featherlike structures arranged in a circle sift food particles out of water currents and pass them to the mouth.

Table 11.1
Path of Food

Organ	Special Feature	Function
Mouth	Teeth, tongue	Chewing of food; digestion of starch
Esophagus		Movement of food by peristalsis
Stomach	Gastric glands	Storage of food; acidity kills some bacteria; digestion of protein
Small intestine	Villi	Digestion of all foods; absorption of nutrients
Large intestine		Absorption of water; storage of nondigestible remains
Anus		Defecation

hard outer covering of calcium compounds; a thick layer of dentin, a bonelike material; and inner pulp, which contains the nerves and the blood vessels. Dentin and pulp are also found in the root. Tooth decay, or *caries,* commonly called a cavity, occurs when the bacteria within the mouth metabolize sugar and give off acids that erode the tooth. Two measures can prevent tooth decay: eating a limited amount of sweets, and daily brushing and flossing of teeth. It also has been found that fluoride treatments, particularly in children, can make the enamel stronger and more resistant to decay.

Gum disease is more apt to occur with aging. Inflammation of the gums (gingivitis) can spread to the periodontal

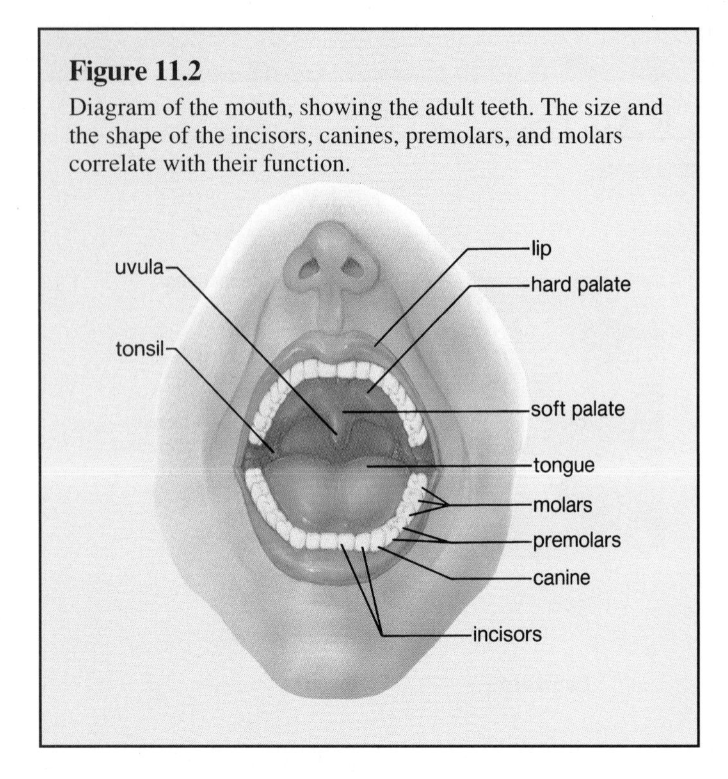

Figure 11.2
Diagram of the mouth, showing the adult teeth. The size and the shape of the incisors, canines, premolars, and molars correlate with their function.

membrane (fig. 11.3), which lines the tooth socket. The individual then has **periodontitis,** characterized by a loss of bone and loosening of the teeth so that extensive dental work may be required. Stimulation of the gums in a manner advised by dentists is helpful in controlling this condition.

In humans, the roof of the mouth separates the nasal cavities from the oral cavity. The roof has 2 parts: an anterior **hard palate** and a posterior **soft palate** (fig. 11.2). The hard palate contains several bones, but the soft palate is composed only of muscle. The soft palate ends in the *uvula,* a suspended process often mistaken by the layperson for the tonsils. In fact, the tonsils are at the sides of the oral cavity (fig. 11.2), at the base of the tongue, and in the nose (called adenoids). The tonsils play a minor role in protecting the body from microbes (microscopic infectious agents, such as bacteria and viruses), as is discussed in chapter 13.

The three pairs of **salivary glands** are exocrine glands that send their juices (saliva) by way of ducts to the mouth (fig. 11.1). The *parotid glands* lie at the sides of the face immediately below and in front of the ears. These glands swell when a person has the mumps, a viral infection most often seen in children. Each parotid gland has a duct, which opens on the inner surface of the cheek at the location of the second upper molar. The *sublingual glands* lie beneath the tongue, and the *submandibular glands* lie beneath the lower jaw. The ducts from these glands open into the mouth under the tongue. You can locate all these openings if you use your tongue to feel for small flaps on the inside of your cheek and under your tongue. An enzyme within saliva begins the process of digesting food, specifically starch.

Figure 11.3
Longitudinal section of a canine tooth. Nerves and blood vessels are found within the pulp of a tooth.

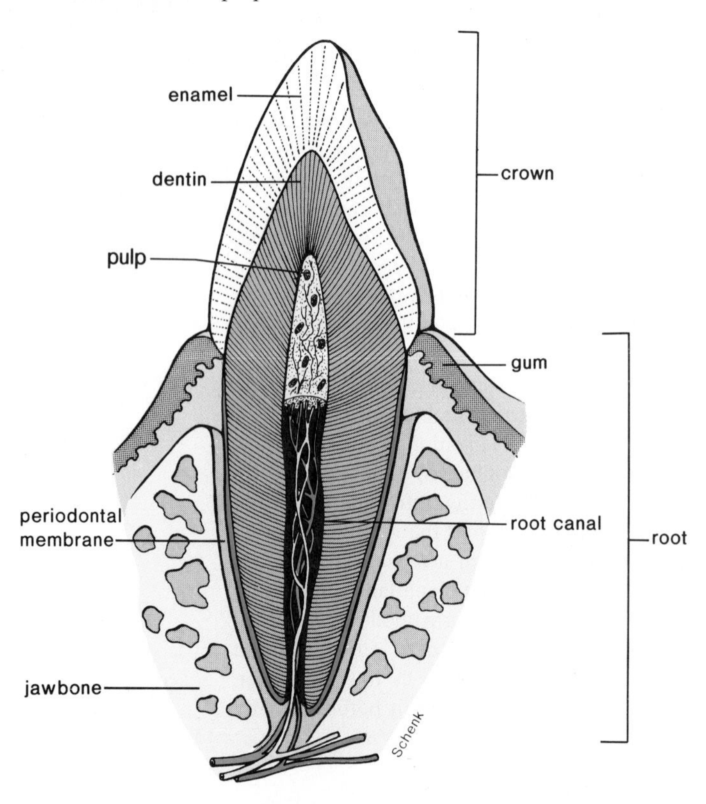

The tongue, which is composed of striated muscle with an outer layer of mucous membrane, mixes the chewed food with saliva. It then forms this mixture into a mass called a *bolus* in preparation for swallowing.

The salivary glands send saliva into the mouth, where the teeth chew the food and the tongue forms it into a bolus for swallowing.

Pharynx: A Crossroad

Swallowing (fig. 11.4) occurs in the **pharynx,** a region between the mouth and the esophagus, which is a long muscular tube leading to the stomach. Swallowing is a *reflex action,* which means the action is usually performed automatically, without conscious thought. During swallowing, food normally enters the esophagus because the air passages are blocked. Unfortunately, we have all had the unpleasant experience of having food "go the wrong way." The wrong way may be either into the nose or into the trachea (windpipe). If it is the latter, coughing forces the food up out of the trachea and into the pharynx again. Usually the opening to the *nasopharynx,* which leads to the nasal cavities, is covered when the soft palate moves back. The opening to the larynx (voice box) at the top of the trachea, called the **glottis,** is covered when the trachea moves up under a flap of tissue called the **epiglottis.** This is easy to observe in the up-and-down movement of the *Adam's apple,* a part of the larynx, when a person eats. Notice that breathing does not occur during swallowing because air passages are closed off.

> The air passage and the food passage cross in the pharynx. When you swallow, the air passage is usually blocked off, and food must enter the esophagus.

Esophagus: Food Conductor

After swallowing occurs, the *esophagus* conducts the bolus through the thoracic cavity and into the abdominal cavity. The wall of the esophagus (fig. 11.5) in the abdominal cavity is representative of the digestive tract in general. A *mucosa* lines the **lumen** (space within the tube); this is followed by a *submucosa* of loose connective tissue, which contains nerve and blood vessels; a muscularis, a smooth muscle layer having both longitudinal and circular muscles; and finally, a *serosa.*

A rhythmic contraction of the esophageal wall, **peristalsis** (fig. 11.9), pushes the food along. Occasionally, peristalsis begins even though there is no food in the esophagus. This produces the sensation of a lump in the throat.

Figure 11.4
Swallowing. When food is swallowed, the soft palate covers the nasopharynx and the epiglottis covers the glottis, forcing the bolus to pass down the esophagus. Therefore, you do not breathe when swallowing.

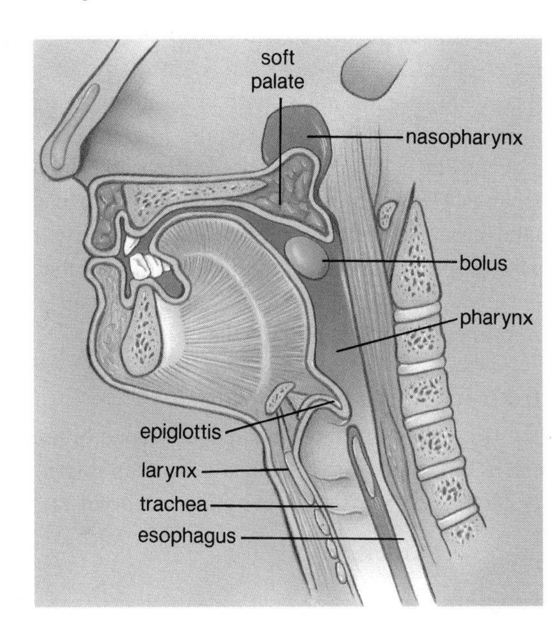

Figure 11.5
Wall of the esophagus in abdominal cavity. Like the rest of the digestive tract, several different types of tissues are found in the wall of the esophagus. Note the placement of circular muscle inside longitudinal muscle. This arrangement ensures that the action of the circular muscle does not interfere with that of the longitudinal muscle. **a.** Diagrammatic drawing. **b.** Scanning electron micrograph (Lu = central lumen; Mu = mucosa; Su = submucosa; Me = muscularis; and Ad = adventitia, or serosa).
Source: U.S. Department of Agriculture.

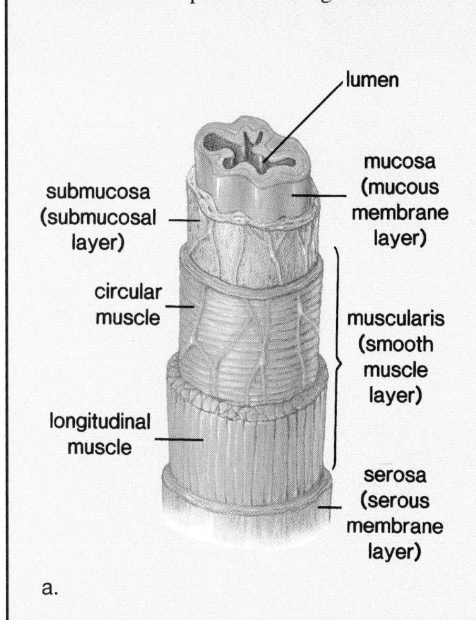

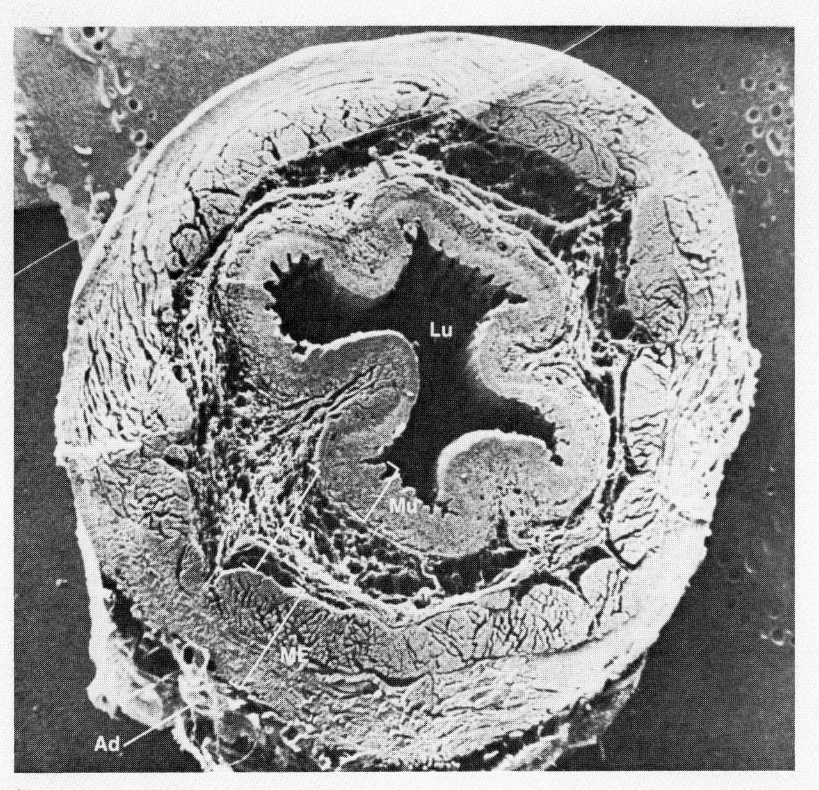

The esophagus extends from the back of the pharynx to just below the diaphragm, where it meets the stomach at an angle. The entrance of the esophagus to the stomach is marked by a constrictor called the lower esophageal sphincter, although the muscle in this sphincter is not as developed as in a true sphincter. **Sphincters** are muscles that encircle tubes and act as valves; tubes close when sphincters contract, and they open when sphincters relax. When food is swallowed, the sphincter relaxes, allowing the bolus to pass into the stomach. Normally, the lower esophageal sphincter prevents the acidic contents of the stomach from entering the esophagus. Heartburn, which feels like a burning pain rising up into the throat, occurs when some of the stomach contents escape into the esophagus. When vomiting occurs, a reverse peristaltic wave causes the sphincter to relax, and the contents of the stomach are propelled upward through the esophagus.

Stomach: Food Storer and Grinder

The *stomach* (fig. 11.6) is a thick-walled, J-shaped organ that lies on the left side of the body beneath the diaphragm. The stomach is continuous with the esophagus above and the duodenum of the small intestine below. The stomach stores food. The wall of the stomach has 3 layers of muscle and also contains deep folds that disappear as the stomach fills. The muscular wall of the stomach churns, mixing the food with gastric secretions. When food leaves the stomach, it is a pasty material called *chyme*.

The columnar epithelial lining of the stomach, which contains many goblet cells, has millions of gastric pits, which lead into **gastric glands** (the term *gastric* always refers to the stomach). The gastric glands produce gastric juice, which contains hydrochloric acid (HCl) and a protein-digesting enzyme called pepsin. The high acidity of the stomach (about pH 2) is beneficial because it kills most bacteria present in food. Although hydrochloric acid does not digest food, it does break down the connective tissue of meat and activates the digestive enzyme in gastric juice called pepsin. Normally, the wall of the stomach is protected by a thick layer of mucus, but if by chance hydrochloric acid does penetrate this mucus, autodigestion of the wall can begin, and an ulcer results (fig. 11.6). An **ulcer** is an open sore in the wall caused by the gradual disintegration of tissue. It is

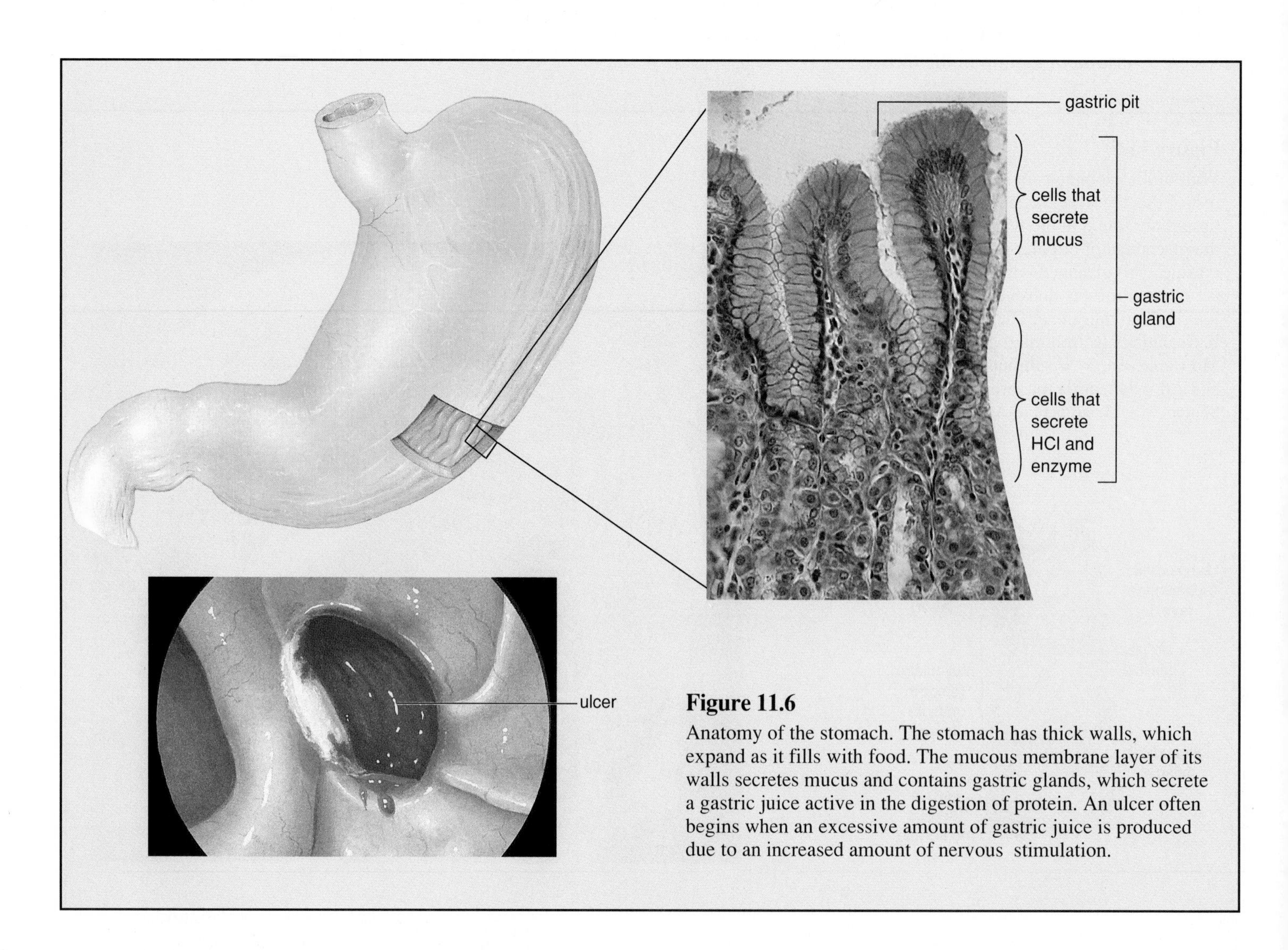

Figure 11.6
Anatomy of the stomach. The stomach has thick walls, which expand as it fills with food. The mucous membrane layer of its walls secretes mucus and contains gastric glands, which secrete a gastric juice active in the digestion of protein. An ulcer often begins when an excessive amount of gastric juice is produced due to an increased amount of nervous stimulation.

believed that the most frequent cause of an ulcer is oversecretion of gastric juice due to too much nervous stimulation; persons under stress tend to have a greater incidence of ulcers. However, there is now evidence that a bacterial (*Helicobacter pylori*) infection may impair the ability of cells to produce protective mucus.

Normally, the stomach empties in about 2–6 hours. Chyme leaves the stomach and enters the small intestine by way of the *pyloric sphincter.* The pyloric sphincter repeatedly opens and closes, allowing chyme to enter the small intestine in small squirts only.

> The stomach can expand to accommodate large amounts of food. While food is in the stomach, it churns, mixing food with acidic gastric juice.

Small Intestine: Food Processor

The **small intestine** gets its name from its small diameter (compared to that of the large intestine), but perhaps it should be called the long intestine because it averages about 6.0 m (20 ft) in length compared to the large intestine, which is about 1.5 m (5 ft) long. The small intestine receives bile (p. 180) from the gallbladder and secretions from the pancreas, chemically and mechanically breaks down chyme, absorbs nutrient molecules, and transports undigested material to the large intestine.

The first 25 cm (10 in) of the small intestine is the **duodenum.** Duodenal ulcers sometimes occur because the gastric juice within chyme digests the intestinal wall in this region. Ducts from the gallbladder and the pancreas join and then enter into the duodenum (fig. 11.1). The juices of the small intestine are normally basic because pancreatic juice, which enters the duodenum, contains sodium bicarbonate ($NaHCO_3$). This normally neutralizes the acidity of chyme from the stomach.

The wall of the small intestine contains fingerlike projections called **villi** (fig. 11.7). Because they are so numerous, the villi give the intestinal wall a soft, velvety appearance. Each villus has an outer layer of columnar epithelium and contains blood vessels and a small lymphatic vessel called a **lacteal.** The lymphatic system is an adjunct to the circulatory system; its vessels carry a fluid called lymph to the circulatory veins.

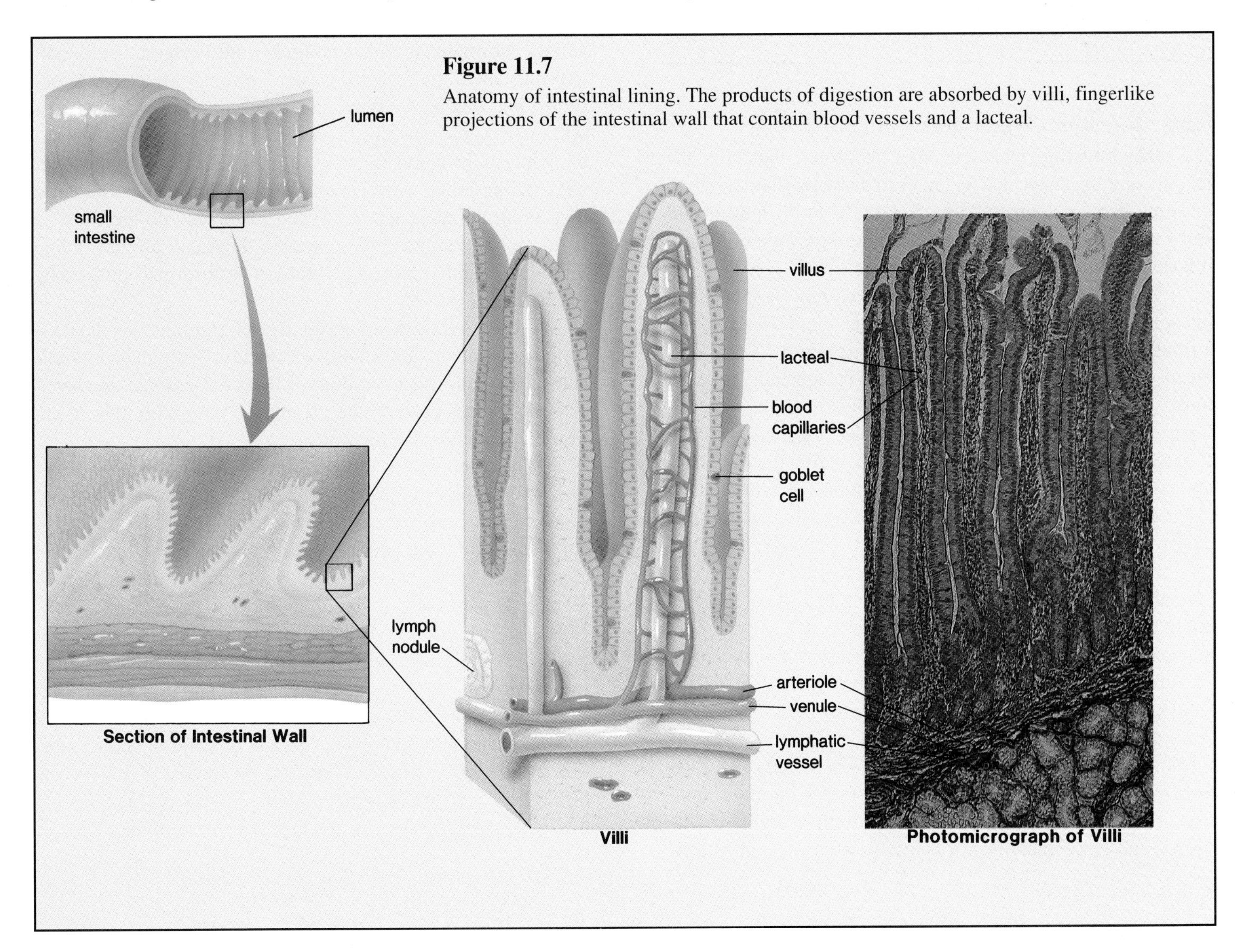

Figure 11.7
Anatomy of intestinal lining. The products of digestion are absorbed by villi, fingerlike projections of the intestinal wall that contain blood vessels and a lacteal.

The epithelial cells of the villi contain microvilli, microscopic projections that extend into the lumen of the small intestine. In electron micrographs, the microvilli give the cells a fuzzy border, collectively called a brush border. The microvilli bear the intestinal digestive enzymes, which are therefore referred to as brush-border enzymes. These enzymes finish the digestion of chyme to small molecules that can cross cell membranes and be absorbed. The microvilli greatly increase the surface area of the small intestine for absorption of nutrients.

Absorption of nutrient molecules occurs across the wall of each villus and continues until all small molecules have been absorbed. Therefore, absorption is an active process involving active transport of molecules across cell membranes and requiring an expenditure of cellular energy. Sugars and amino acids cross the columnar epithelial cells to enter blood. The components of fats rejoin in epithelial cells and are packaged as lipoprotein droplets, which enter the lacteals.

The small intestine is specialized to absorb the products of digestion. It is quite long (6.0 m) and has fingerlike projections called villi, where nutrient molecules are absorbed into the circulatory and lymphatic systems.

Large Intestine: Water and Salt Processor

The **large intestine,** which includes the cecum, the colon, the rectum, and the anal canal, is larger in diameter than the small intestine (6.5 cm compared to 2.5 cm). The large intestine absorbs water and salts. It also stores nondigestible material until it is defecated (expelled) at the anus.

The *cecum,* which lies below the entrance of the small intestine, is the blind end of the large intestine. The cecum has a small projection called the vermiform **appendix** (*vermiform* means *wormlike*) (fig. 11.8). In humans, the appendix, like the tonsils, may play a role in immunity. This organ, however, is subject to inflammation, a condition called appendicitis. If inflamed, it is wise to remove the appendix before the fluid content rises to the point that the appendix bursts, a situation that can lead to generalized infection of the serosa of the abdominal cavity.

Peristalsis (fig. 11.9), which began in the esophagus, occurs along the entire digestive tract, including the colon. The **colon** has 3 parts: the *ascending colon*, which goes up the right side of the body to the level of the liver; the *transverse colon*, which crosses the abdominal cavity just below the liver and the stomach; and the *descending colon*, which passes down the left side of the body to the rectum, the last 20 cm of the large intestine. The rectum opens at the **anus** where *defecation*, expulsion of *feces*, occurs (fig. 11.10). Feces contains nondigestible remains, bile pigments (which account for its color), and large quantities of bacteria (which account for its

Figure 11.8
The anatomical relationship between the small intestine and the ascending colon. The cecum is the blind end of the ascending colon. The appendix is attached to the cecum.

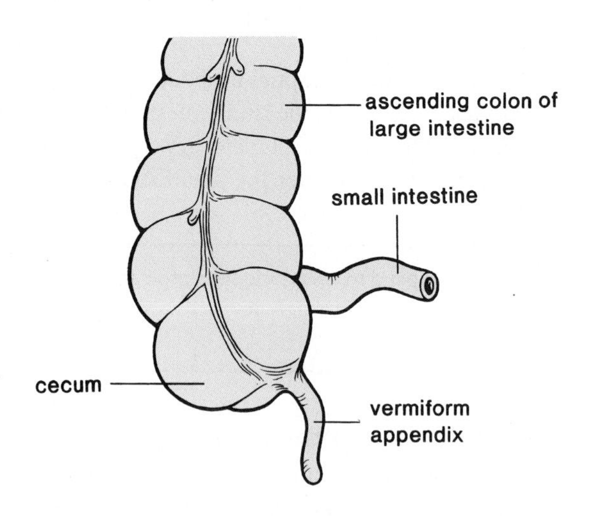

smell). Normally, these are noninfectious bacteria that live off any substances not digested earlier. For many years, it was believed that facultative bacteria (bacteria that can live with or without oxygen), such as *Escherichia coli*, were the major inhabitants of the colon, but new culture methods show that over 99% of the colon bacteria are obligate anaerobes (bacteria that die in the presence of oxygen). Not only do the bacteria break down nondigestible material, they also produce some vitamins and other molecules that can be absorbed and used by our bodies.

Water is considered unsafe for swimming when the coliform bacterial count reaches a certain level. A high count indicates that a significant amount of feces has entered the water. The more feces present, the greater the possibility that infectious bacteria are also present.

The colon is subject to the development of polyps, small growths arising from the epithelial lining. Polyps, whether benign or cancerous, can be removed individually. If colon cancer is detected while still confined to a polyp, the outcome is expected to be a complete cure. Some investigators believe dietary fat increases the likelihood of colon cancer because it causes an increase in bile secretion. It could be that intestinal bacteria convert bile salts (p. 180) into substances that promote the development of cancer. On the other hand, fiber in the diet seems to inhibit the development of colon cancer. Dietary fiber absorbs water and adds bulk, thereby diluting the concentration of bile salts and facilitating the movement of substances through the intestine. Regular elimination reduces the time that the colon wall is exposed to any cancer-promoting agents in feces.

Figure 11.9
Peristalsis in the digestive tract. Rhythmic waves of muscle contraction move material along the digestive tract. The 3 drawings show how a peristaltic wave moves through a single section of the tract over time.

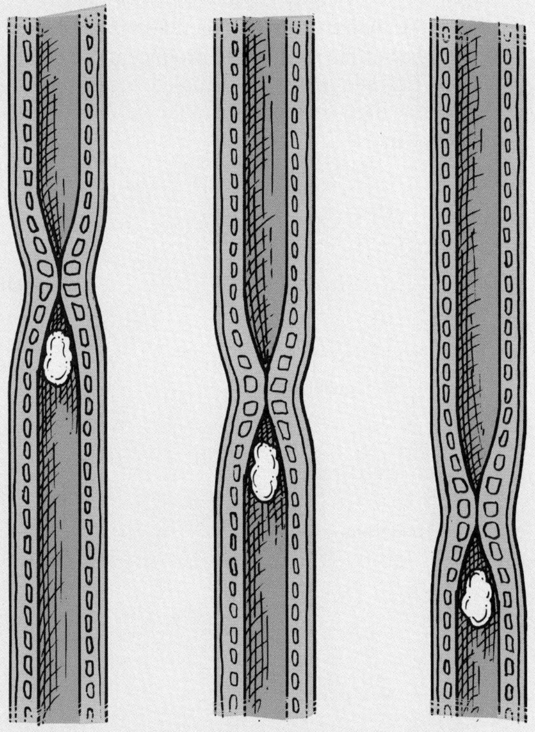

Figure 11.10
Defecation reflex. The accumulation of feces in the rectum causes it to stretch, which initiates a reflex action resulting in rectal contraction and expulsion of the fecal material.

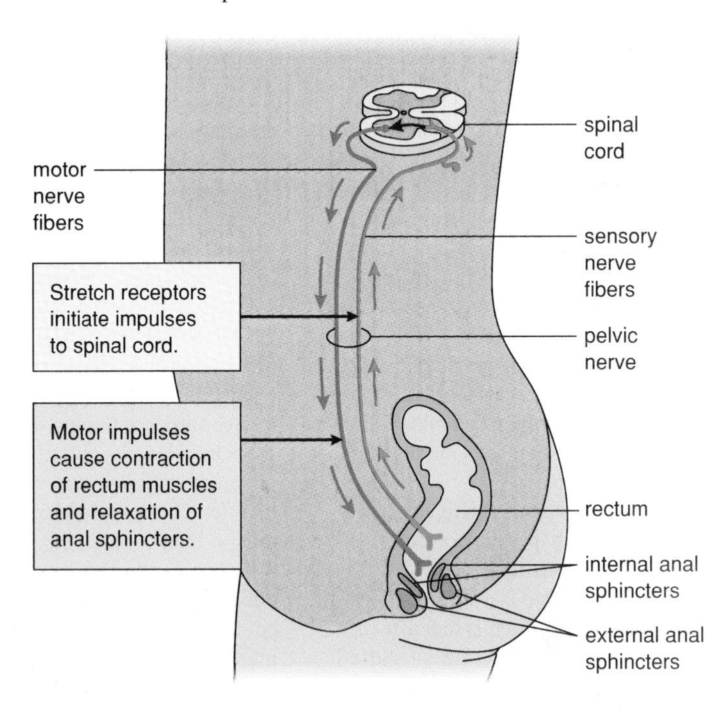

Diarrhea and Constipation

Two common everyday complaints associated with the large intestine are *diarrhea* and *constipation.* The major causes of diarrhea are infection of the lower tract and nervous stimulation. In the case of infection, such as food poisoning caused by eating contaminated food, the intestinal wall becomes irritated and peristalsis (fig. 11.9) increases. As a protective measure, water is not absorbed, and the diarrhea that results rids the body of the microbes. In nervous diarrhea, the nervous system stimulates the intestinal wall and diarrhea results. Prolonged diarrhea can lead to dehydration because of water loss and to disturbances in the heart's contraction due to an imbalance of salts in blood.

When a person is constipated, the feces are dry and hard. One reason for this condition is that socialized persons have learned to inhibit defecation to the point that the desire to defecate is ignored. Two components of the diet can help to prevent constipation: water and fiber (roughage). Water intake prevents drying out of the feces, and fiber provides the bulk needed for elimination. The frequent use of laxatives is discouraged. If, however, it is necessary to take a laxative, a bulk laxative is the most natural because, like fiber, it produces a soft mass of cellulose in the colon. Lubricants, like mineral oil, make the colon slippery, and saline laxatives, like milk of magnesia, act osmotically—they prevent water from being absorbed and may even cause water to enter the colon, depending on the dosage. Some laxatives are irritants; they increase peristalsis to the degree that the contents of the colon are expelled.

Chronic constipation is associated with the development of hemorrhoids, a condition that is discussed on page 217.

The large intestine does not produce digestive enzymes; it does absorb water and salts. In diarrhea, too little water has been absorbed by the large intestine; in constipation, too much water has been absorbed.

Two Accessory Organs

The pancreas and the liver are the accessory organs of digestion. Figure 11.1 shows how ducts conduct pancreatic juice from the pancreas and bile from the liver to the duodenum.

The Pancreas: For Digestive Enzymes

The **pancreas** lies deep in the abdominal cavity, resting on the posterior abdominal wall. It is an elongated and somewhat

HORMONAL CONTROL OF DIGESTIVE JUICES

The study of the control of digestive gland secretion began in the late 1800s. At that time, Ivan Pavlov showed that dogs would begin to salivate at the ringing of a bell because they had learned to associate the sound of the bell with being fed. Pavlov's experiments demonstrated that even the thought of food can cause the nervous system to order the secretion of digestive juices. If food is present in the mouth, the stomach, and the small intestine, digestive secretion occurs because of simple reflex action. The presence of food sets off nerve impulses that travel to the brain. Thereafter, the brain stimulates the digestive glands to secrete.

In this century, investigators have discovered that specific control of digestive secretions is achieved by hormones. A *hormone* is a substance produced by one set of cells that affects a different set of cells, the so-called target cells. Hormones are transported by the bloodstream. For example, when a person has eaten a meal particularly rich in protein, the gastric glands of the stomach wall produce the hormone gastrin. Gastrin enters the bloodstream, and soon stomach churning and the secretory activity of gastric glands increase.

Cells of the duodenal wall produce hormones, 2 of which are of particular interest, secretin and CCK (cholecystokinin). Acid, especially hydrochloric acid (HCl) present in chyme, stimulates the release of secretin, while partially digested protein and fat stimulate the release of CCK. Soon after these hormones enter the bloodstream, the pancreas increases its output of pancreatic juice and the liver increases its output of bile. The gallbladder contracts to release bile.

Another hormone produced by the duodenal wall, GIP (gastric inhibitory peptide), works opposite to gastrin—it inhibits gastric gland secretion and stomach motility. This is not surprising, because the hormones of the body often have opposite effects.

Figure 11.A

Hormonal control of digestive gland secretions. Gastrin, produced by the lower part of the stomach, enters the bloodstream and thereafter stimulates the upper part of the stomach to produce more digestive juices. Secretin and CCK stimulate the pancreas to secrete its digestive juices and the gallbladder to release bile.

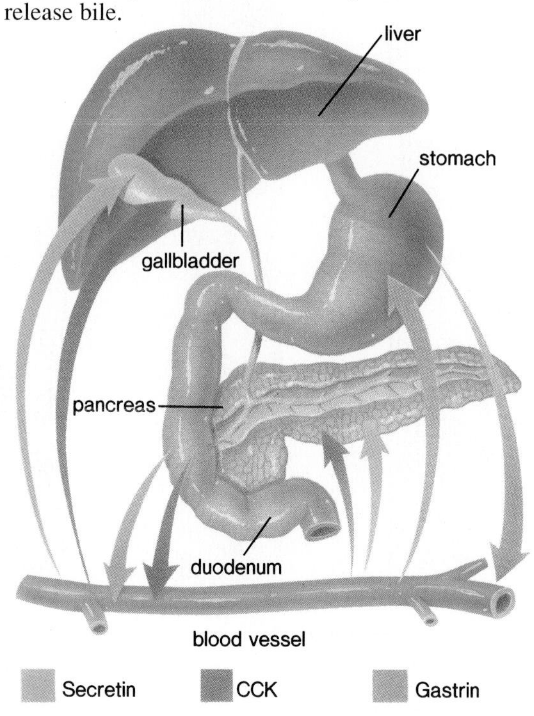

flattened organ that has both an endocrine and an exocrine function. We are now interested in its exocrine function—most of its cells produce pancreatic juice, which contains enzymes for digestion of carbohydrate, protein, and fat. In other words, the pancreas secretes enzymes for the digestion of all types of food. The enzymes travel by way of ducts to the duodenum of the small intestine (fig. 11.1). Regulation of pancreatic secretion is discussed in the reading on this page.

The Liver: Makes Bile

The **liver,** which is the largest gland in the body, lies mainly in the upper right section of the abdominal cavity, under the diaphragm.

Bile Production. Bile is a yellowish green fluid because it contains the bile pigments bilirubin and biliverdin, which are derived from the breakdown of hemoglobin, the pigment found in red blood cells. Bile also contains bile salts (derived from cholesterol), which emulsify fat in the duodenum of the small intestine. When fat is emulsified, it breaks up into droplets that can be acted upon by a digestive enzyme from the pancreas (fig. 2.26, p. 35).

The liver produces up to 1,500 ml of bile each day. This bile is sent by way of bile ducts to the gallbladder, where it is stored. The **gallbladder** is a pear-shaped, muscular sac attached to the undersurface of the liver. Here, water is absorbed, and bile then becomes a thick, mucuslike material. Bile leaves the gallbladder and proceeds to the duodenum via ducts (fig. 11.1).

The Liver as Gatekeeper. In some ways, the liver acts as the gatekeeper to blood. Once nutrient molecules have been absorbed by the small intestine, they enter the **hepatic portal vein,** pass through the blood vessels of the liver, and then enter the hepatic vein (fig. 11.11).

As blood passes through the liver, it removes poisonous substances and works to keep the contents of blood constant. For example, the liver removes excess glucose present in the hepatic portal vein and stores it as glycogen:

$$\text{glucose} \longrightarrow \text{glycogen} + H_2O$$

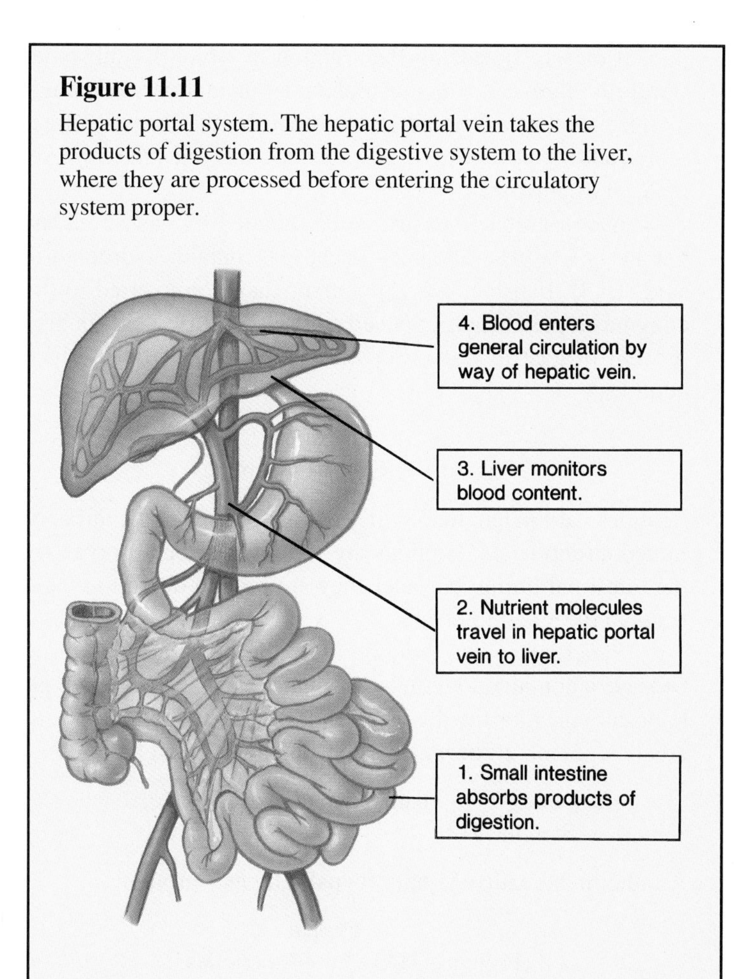

Figure 11.11
Hepatic portal system. The hepatic portal vein takes the products of digestion from the digestive system to the liver, where they are processed before entering the circulatory system proper.

Between eating periods, when the glucose level of blood falls below 0.1%, glycogen is broken down to glucose, which enters the hepatic vein. In this way, the glucose content of blood remains near 0.1%. It is interesting to note that glycogen is sometimes called animal starch because both starch and glycogen are made up of glucose molecules (p. 33).

If, by chance, the supply of glycogen or glucose runs short, the liver converts amino acids to glucose molecules:

$$\text{amino acids} \longrightarrow \text{glucose} + \text{amino groups}$$

Recall that amino acids contain nitrogen in the form of amino groups, whereas glucose contains only carbon, oxygen, and hydrogen. Therefore, before amino acids can be converted to glucose molecules, *deamination,* or the removal of amino groups from the amino acids, must take place. By an involved metabolic pathway, the liver converts these amino groups to urea:

$$H_2N\text{—}\overset{\overset{\displaystyle O}{||}}{C}\text{—}NH_2$$

Urea is the usual nitrogenous waste product of humans; after its formation in the liver, it is excreted by the kidneys.

The liver also makes blood proteins from amino acids. These proteins are not used as food for cells; rather, they serve important functions within blood itself.

Altogether, we have mentioned the following functions in the liver:

1. Converts hemoglobin from red blood cells to break down products (bilirubin and biliverdin) excreted along with bile salts in bile.
2. Produces bile, which is stored in the gallbladder before entering the small intestine, where bile salts emulsify fats.
3. Detoxifies blood by removing and metabolizing poisonous substances.
4. Stores glucose as glycogen after eating and breaks down glycogen to glucose to maintain the glucose concentration of blood between eating periods.
5. Produces urea from the breakdown of amino acids.
6. Makes blood proteins from amino acids.

Two accessory organs of digestion, the pancreas and the liver, send secretions to the duodenum via ducts. The pancreas produces pancreatic juice, which contains enzymes for the digestion of carbohydrate, protein, and fat. The liver produces bile, which is stored in the gallbladder and is used to emulsify fats.

Serious Liver Disorders. Jaundice, hepatitis, and cirrhosis are 3 serious diseases that affect the entire liver and hinder its ability to repair itself. Therefore, they are life-threatening diseases. When a person is jaundiced, there is a yellowish tint to the skin and the whites of the eyes. Bilirubin has been deposited in the skin due to an abnormally large amount in blood. In *hemolytic jaundice,* red blood cells are broken down in abnormally large amounts; in *obstructive jaundice,* the bile duct is blocked or the liver cells are damaged. Obstructive jaundice often occurs when crystals of cholesterol come out of solution and form gallstones. The stones may be so numerous that passage of bile along a bile duct is blocked, and the gallbladder must be removed (fig. 11.12).

Jaundice can also result from *viral hepatitis,* a collective term that includes several types of hepatitis. Hepatitis A is most often caused by eating shellfish from polluted waters. Other types are commonly spread by blood transfusions, kidney dialysis, or injection with inadequately sterilized needles. Hepatitis can also be acquired by sexual contact.

Cirrhosis is a chronic disease of the liver in which the organ first becomes fatty. Liver tissue is then replaced by inactive fibrous scar tissue. In alcoholics, who often develop cirrhosis of the liver, the condition most likely is caused by the excessive amounts of alcohol (a toxin) the liver is forced to break down.

Figure 11.12

Gallstones. After removal, this gallbladder was cut open to show its contents—numerous gallstones. The dime was added later to indicate the size of the stones.

The liver is a very critical organ. Any malfunction is a matter of considerable concern. Do 11.1 Critical Thinking, found at the end of the chapter.

Digestive Enzymes

The digestive enzymes are **hydrolytic enzymes,** which catalyze breakdown by the introduction of water at specific bonds (see fig. 2.14). Digestive enzymes are like other enzymes we have studied (fig. 5.4, p. 77). They are proteins having a particular shape that fits their substrate. They also have an optimum pH, which maintains their shape, thereby enabling them to speed up their specific reaction.

The various digestive enzymes are present in the digestive juices mentioned previously. We now consider each of the enzymes listed in table 11.2 as we discuss the digestion of carbohydrates, proteins, and lipids, the major components of food.

In the mouth, saliva from the salivary glands has a neutral pH and contains **salivary amylase,** an enzyme that acts on starch:

$$\text{Starch} + H_2O \xrightarrow{\text{salivary amylase}} \text{maltose}$$

In this equation, salivary amylase is written above the arrow to indicate that it is neither a reactant nor a product in the reaction. It merely speeds up the reaction in which its substrate, starch, is digested to many molecules of maltose. Although maltose molecules cannot be absorbed by the digestive tract lining, additional digestive action in the small intestine converts maltose to glucose.

In the stomach, gastric juice secreted by gastric glands has a very low pH—about 2—because it contains hydrochloric acid (HCl). Pepsinogen, a precursor that is converted to the enzyme **pepsin** when exposed to hydrochloric acid, is also present in gastric juice. Pepsin acts on protein to produce peptides:

$$\text{Protein} + H_2O \xrightarrow{\text{pepsin}} \text{peptides}$$

Peptides vary in length, but they always consist of a number of linked amino acids. Peptides are too large to be absorbed by the intestinal lining, but later they are broken down to amino acids in the small intestine.

Pancreatic juice, which enters the duodenum, is basic because it contains sodium bicarbonate ($NaHCO_3$). It also contains enzymes for the digestion of all types of food. One pancreatic enzyme, **pancreatic amylase,** digests starch:

$$\text{Starch} + H_2O \xrightarrow{\text{pancreatic amylase}} \text{maltose}$$

Another pancreatic enzyme, **trypsin,** digests protein:

$$\text{Protein} + H_2O \xrightarrow{\text{trypsin}} \text{peptides}$$

Trypsin is secreted as trypsinogen, which is converted to trypsin in the duodenum.

Lipase, a third pancreatic enzyme, digests fat droplets after they have been emulsified by bile salts:

$$\text{Fat} \xrightarrow{\text{bile salts}} \text{fat droplets}$$

$$\text{Fat droplets} + H_2O \xrightarrow{\text{lipase}} \text{glycerol} + \text{fatty acids}$$

The end products of lipase digestion, glycerol and fatty acid molecules, are small enough to cross the cells of the intestinal villi, where absorption takes place. As mentioned previously, glycerol and fatty acids enter the cells of the villi, and within these cells, they are rejoined and packaged as lipoprotein droplets, which enter the lacteals (fig. 11.7).

Peptidases and **maltase,** 2 enzymes present in the mucosa of the intestinal villi, complete the digestion of protein and starch to small molecules that cross into the cells of the villi. Peptides, which result from the first step in protein digestion, are digested to amino acids by peptidases:

$$\text{Peptides} + H_2O \xrightarrow{\text{peptidases}} \text{amino acids}$$

Table 11.2
Comparison of Digestive Enzymes

Enzyme	Source	Optimum pH	Type of Food Digested	Product
Salivary amylase	Salivary glands	Neutral	Starch	Maltose
Pepsin	Stomach	Acidic	Protein	Peptides
Pancreatic amylase	Pancreas	Basic	Starch	Maltose
Trypsin	Pancreas	Basic	Protein	Peptides
Lipase	Pancreas	Basic	Fat	Glycerol, fatty acids
Peptidases	Small intestine	Basic	Peptides	Amino acids
Maltase	Small intestine	Basic	Maltose	Glucose

Table 11.3
Major Digestive Enzymes

Reaction	Enzyme	Produced by	Site of Action
Starch + $H_2O \longrightarrow$ maltose	{ Salivary amylase	Salivary glands	Mouth
	Pancreatic amylase	Pancreas	Small intestine
Maltose + $H_2O \longrightarrow$ glucose*	Maltase	Small intestine	Small intestine
Protein + $H_2O \longrightarrow$ peptides	{ Pepsin	Gastric glands	Stomach
	Trypsin	Pancreas	Small intestine
Peptides + $H_2O \longrightarrow$ amino acids*	Peptidases	Small intestine	Small intestine
Fat droplets + $H_2O \longrightarrow$ glycerol + fatty acids*	Lipase	Pancreas	Small intestine

*Absorbed by villi.

Note: Food is made up largely of carbohydrate (starch), protein, and fat. These very large macromolecules are broken down by digestive enzymes to small molecules that can be absorbed by intestinal villi. This table indicates the steps needed for carbohydrate digestion (starch and maltose), protein digestion (protein and peptides), and fat digestion (fat) and shows that they are all hydrolytic reactions.

Maltose, which results from the first step in starch digestion, is digested to glucose by maltase:

$$\text{Maltose} + H_2O \xrightarrow{\text{maltase}} \text{glucose}$$

Other disaccharides, each of which has its own enzyme, are digested in the small intestine. The absence of any one of these enzymes can cause illness. For example, many people, including as many as 75% of African Americans, cannot digest lactose, the sugar found in milk, because they do not produce lactase, the enzyme that converts lactose to its components, glucose and galactose. Drinking untreated milk often gives these individuals the symptoms of *lactose intolerance* (diarrhea, gas, cramps), caused by a large quantity of undigested lactose in the intestine. In most areas, it is possible to purchase milk made lactose-free by the addition of synthetic lactase.

Table 11.3 lists some of the major digestive enzymes produced by the digestive tract, salivary glands, or pancreas. Each type of food is broken down by specific enzymes.

Digestive enzymes present in digestive juices break down food to the nutrient molecules—glucose, amino acids, fatty acids, and glycerol. The first 2 are absorbed into the blood capillaries of the villi and the last 2 re-form and enter the lacteals as lipoprotein droplets. Do 11.2 Critical Thinking, found at the end of the chapter.

Best Conditions for Digestion

Laboratory experiments can define the necessary conditions for digestion (fig. 11.13). For example, the following 4 test tubes can be prepared and observed for the digestion of egg white, or the protein albumin.

Figure 11.13

An experiment to demonstrate that enzymes digest food when the environmental conditions are correct. Tube number 1 lacks the enzyme pepsin, and no digestion occurs; tube number 2 has too high a pH, and little or no digestion occurs; tube number 3 has the proper pH because of the presence of hydrochloric acid, but still no digestion occurs because the enzyme is missing; tube number 4 contains the enzyme, and the environmental conditions are correct for digestion.

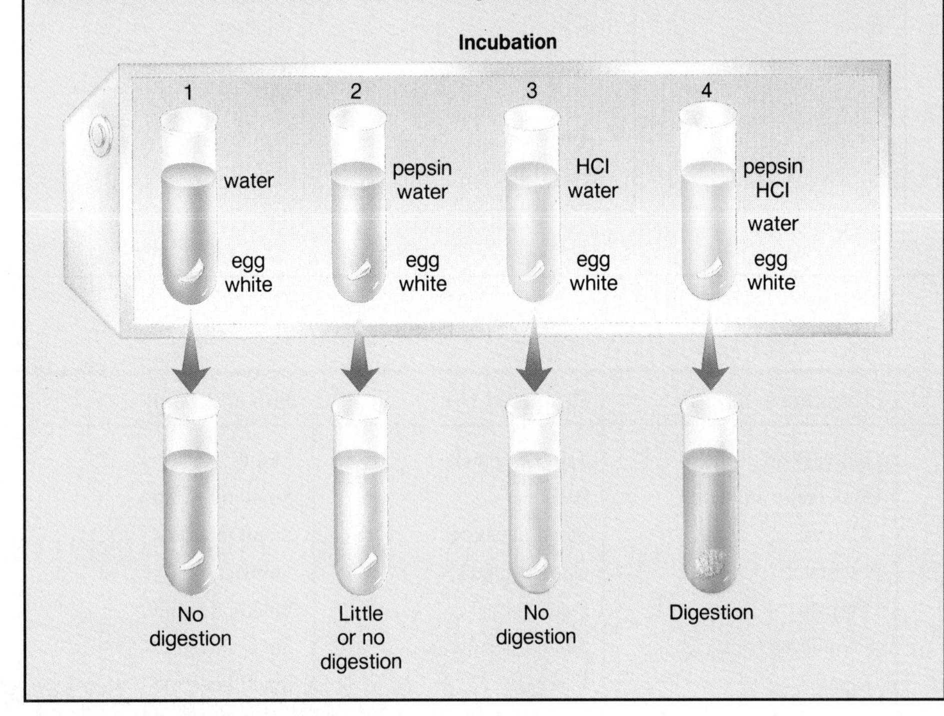

1. H_2O + a small sliver of egg white (protein)
2. Pepsin + H_2O + a small sliver of egg white
3. HCl + H_2O + a small sliver of egg white
4. Pepsin + HCl + H_2O + a small sliver of egg white

All tubes are then placed in an incubator at body temperature for at least one hour. At the end of this time, we can predict that tube number 4 will show the best digestive action because the environmental conditions are appropriate. Tube number 3 does not contain the enzyme (pepsin) and tube number 2 has too high a pH (hydrochloric acid is lacking), so these 2 tubes are expected to show little or no digestion. Tube number 1 is a control tube, and no digestion is expected to occur in this tube. This experiment shows that for digestion to occur, the enzyme, as well as the substrate and hydrochloric acid, must be present.

Nutrition

The body requires many different types of organic molecules and a smaller number of various types of inorganic ions and compounds from the diet each day. Nutrition involves an interaction between food and the living organism, and a **nutrient** is a substance in food that is used by the body for the maintenance of health. In order for the diet to contain all the essential nutrients, it must be balanced. A *balanced diet* includes a variety of foods proportioned as shown in figure 11.14.

In order to get a daily supply of essential nutrients, it is necessary to have a balanced diet.

Food consists largely of proteins, carbohydrates, and lipids (fats and cholesterol). Therefore, we begin by considering these substances.

Proteins: Supply Building Blocks

Foods rich in protein include red meat, fish, poultry, dairy products, legumes, nuts, and cereals. Following digestion of protein, amino acids enter the bloodstream and are transported to the tissues. Ordinarily, amino acids are not used as an energy source. Most are incorporated into structural proteins found in muscles, skin, hair, and nails. Others are used to synthesize such proteins as hemoglobin, plasma proteins, enzymes, and hormones.

Protein formation requires 20 different types of amino acids. Of these, 9 are required in the diet because the body is unable to produce them. These are termed the **essential amino acids.** The body produces the other 11 amino acids by simply transforming one type into another type. Some protein sources, such as meat, are *complete;* they provide all 20 types of amino acids. Vegetables and grains supply us with amino acids, but each vegetable or grain alone is an *incomplete* protein source because at least one of the essential amino acids is absent. However, it is possible to combine nonmeat foods in order to acquire all the essential amino acids. For example, the combinations of cereal with milk or beans with rice provide all the essential amino acids.

A complete source of protein is absolutely necessary to ensure a sufficient supply of the essential amino acids. Do 11.3 Critical Thinking, found at the end of the chapter.

Because amino acids are not stored in the body, a daily supply is needed. However, it does not take very much protein to meet the daily requirement. In the United States, the *required dietary (daily) allowances (RDAs)* are determined by the National Research Council, a part of the National Academy of Sciences. The RDA for a woman (120 lb) is 44 g of protein per day. For a man (154 lb), the RDA is 56 g of protein per day. A single serving of roast beef (3 oz) provides 25 g of protein, and a cup of milk provides 8 g.

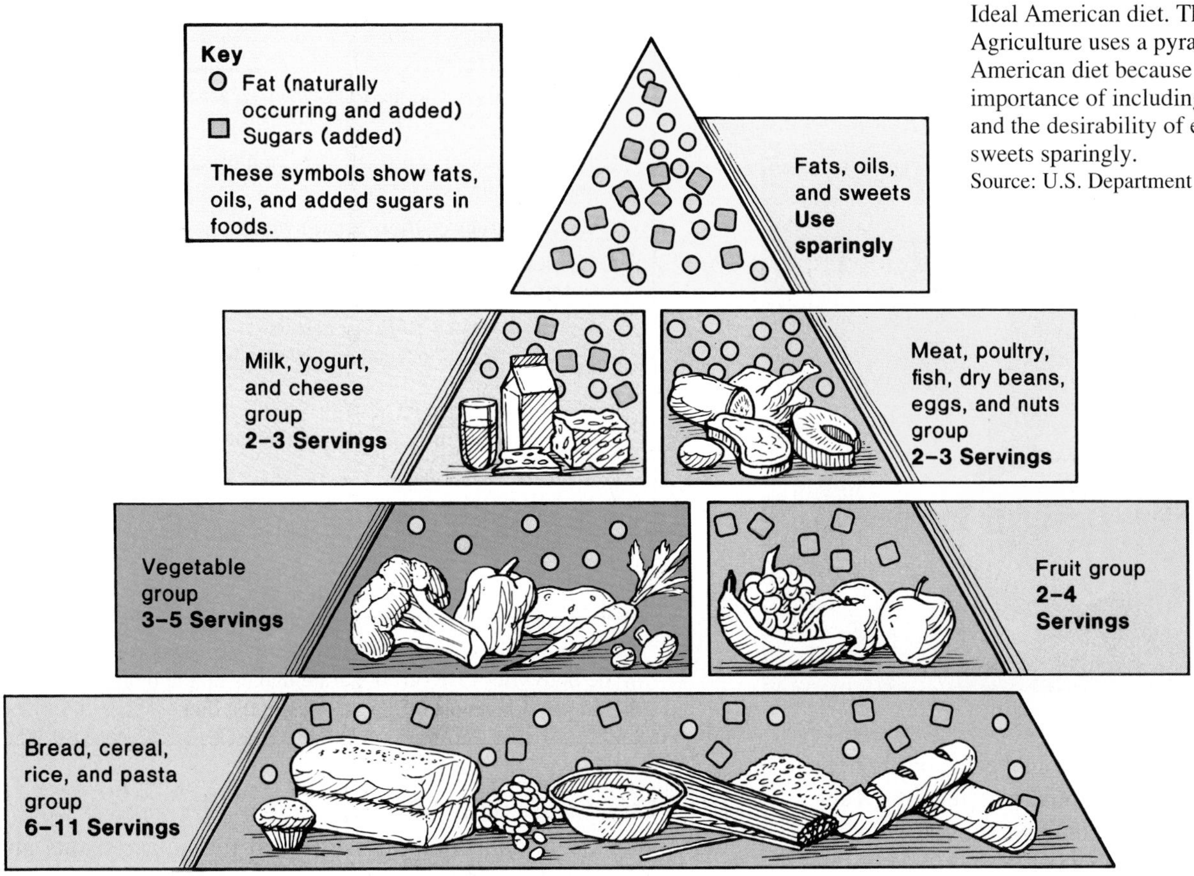

Key
O Fat (naturally occurring and added)
□ Sugars (added)
These symbols show fats, oils, and added sugars in foods.

Fats, oils, and sweets **Use sparingly**

Milk, yogurt, and cheese group **2–3 Servings**

Meat, poultry, fish, dry beans, eggs, and nuts group **2–3 Servings**

Vegetable group **3–5 Servings**

Fruit group **2–4 Servings**

Bread, cereal, rice, and pasta group **6–11 Servings**

Food Guide Pyramid: A Guide to Daily Food Choices

Figure 11.14

Ideal American diet. The U.S. Department of Agriculture uses a pyramid to show the ideal American diet because it emphasizes the importance of including grains in the diet and the desirability of eating fats, oils, and sweets sparingly.
Source: U.S. Department of Agriculture.

While it is very important to meet the RDA for protein, consuming more can actually be detrimental. Calcium loss, detected in the urine, has been noted when dietary protein intake is over twice the RDA. Some meats, hamburger, for example, are high in protein but also fat. Everything considered, it is probably a good idea to depend on protein from plant origins (e.g., whole-grain cereals, dark breads, legumes) to a greater extent than is the custom in this country.

Carbohydrates: Eat More Complex

The quickest, most readily available source of energy for the body is carbohydrates, which can be complex (i.e., polysaccharides), as in breads and cereals, or simple (i.e., monosaccharides or disaccharides), as in candy, ice cream, and soft drinks. As mentioned previously, starches are digested to glucose, which is stored by the liver in the form of glycogen. Between eating periods, the blood glucose level is maintained at about 0.1% by the breakdown of glycogen or by the conversion of amino acids to glucose. While body cells can utilize fat as an energy source, brain cells require glucose. If necessary, amino acids are taken from the muscles, even from heart muscle. To avoid this situation, it is suggested that the daily diet contain at least 100 g of carbohydrate. As a point of reference, a slice of bread contains approximately 14 g of carbohydrate.

> Carbohydrates are needed in the diet to maintain the blood glucose level.

Actually, the dietary guidelines produced jointly by the U.S. Department of Agriculture and the Department of Health and Human Services recommend that we increase the proportion of carbohydrates per total energy content of the diet:

	Typical Diet (%)	Recommended Diet (%)
Proteins	12	12
Carbohydrates	46	58
Fats	42	30

Further, it is assumed that these carbohydrates are complex, not simple (fig. 11.15). Simple carbohydrates (e.g., sugars) are labeled "empty calories" by some dieticians because they

Figure 11.15
Complex carbohydrates. To meet our energy needs, dieticians recommend complex carbohydrates, like those shown here, rather than simple carbohydrates, like candy and ice cream. Simple carbohydrates are more likely to cause weight gain.

contribute to energy needs and weight gain and are not part of foods that supply other nutritional requirements. Table 11.4 gives suggestions on how to cut down the consumption of dietary sugar (simple carbohydrates).

In contrast to simple sugars, complex carbohydrates are likely to be accompanied by a wide range of other nutrients and by fiber, which is nondigestible plant material. Insoluble fiber, such as that found in wheat bran, has a laxative effect and therefore may reduce the risk of colon cancer. Soluble fiber, such as that found in oat bran, may possibly reduce cholesterol in blood because it combines with cholesterol in the digestive tract and prevents it from being absorbed.

A word of caution, however: while the diet should have an adequate amount of fiber, a high-fiber diet can be detrimental. Some evidence suggests that the absorption of iron, zinc, and calcium is impaired by a diet too high in fiber.

> Complex carbohydrates, along with fiber, are considered beneficial to health.

Carbohydrates usually provide most of the dietary calories,[1] even though they have fewer calories per gram than fats:

Food Component	Kcal[2]/g
Carbohydrate	4.1
Protein	4.1
Fat	9.3

1. A calorie is the amount of heat required to raise one gram of water one degree.

2. 1,000 calories = one Kcalorie (Kcal)

Table 11.4
Reducing Dietary Sugar

To reduce dietary sugar, the following suggestions are recommended.

1. Eat fewer sweets, such as candy, soft drinks, ice cream, and pastry.
2. Eat fresh fruits or fruits canned without heavy syrup.
3. Use less sugar—white, brown, or raw—and less honey and syrup.
4. Avoid sweetened breakfast cereals.
5. Eat less jelly.
6. Drink pure fruit juices, not imitations.
7. When cooking, use spices like cinnamon instead of sugar to flavor foods.
8. Do not put sugar in tea or coffee.

> All types of foods can be used as energy sources (see fig. 6.13) in the body. Because carbohydrate makes up the bulk of the diet, it provides most of the calories in the diet.

Lipids: Risky Excess

Our discussion of lipids is divided into 2 parts: fats and cholesterol.

Fats: Beware High-Fat Diets

Fats are present not only in butter, margarine, and oils, but also in many foods high in protein. After being absorbed, the products of fat digestion are packaged as lipoproteins, enter the lymph, and are transported by blood to the tissues. The liver can alter ingested fats to suit the body's needs, except it is unable to produce the fatty acid linoleic acid. Since this is required for phospholipid production, linoleic acid is considered an essential fatty acid. Although fats have the highest caloric content, they should not be avoided entirely because they do contain this essential fatty acid.

While we need to be sure to ingest some fat in order to satisfy our need for linoleic acid, recent dietary guidelines (p. 185) suggest that we should reduce the amount of fat per total energy content of the diet from 40% to 30%. Dietary fat has been implicated in cancer of the colon, pancreas, ovary, prostate, and breast (fig. 11.16). Many animal studies have shown that a high-fat diet stimulates the development of breast cancer, while a low-fat diet does not. It has also been found that women who have a low-fat diet are less likely to develop breast cancer. Specifically, a reduction in the amount of linoleic acid appears to prevent breast cancer. Linoleic acid is found in corn oil, safflower oil, sunflower oil, and other common plant oils but is not abundant in olive oil or in fatty fishes and marine animals.

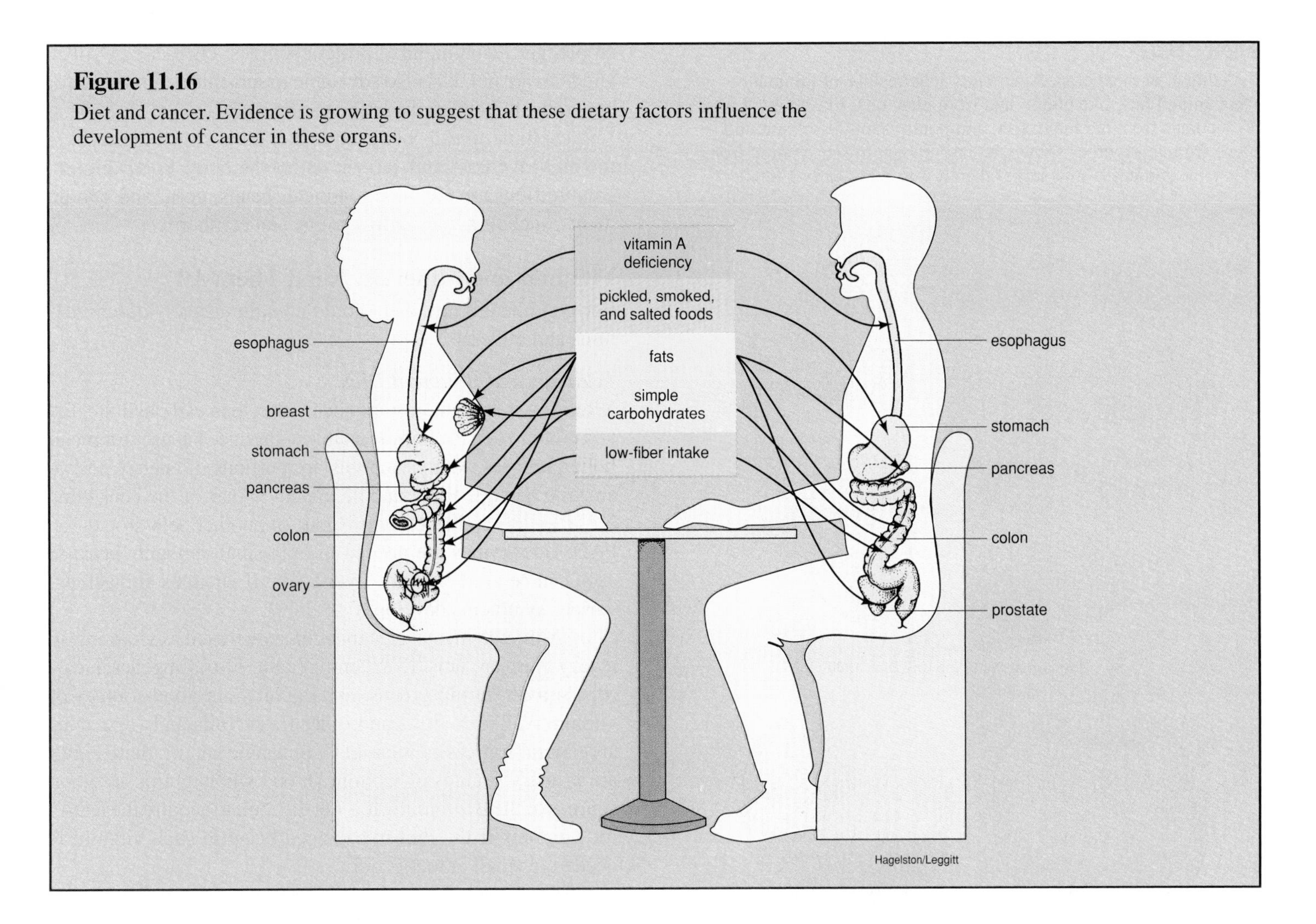

Figure 11.16
Diet and cancer. Evidence is growing to suggest that these dietary factors influence the development of cancer in these organs.

> There is very strong evidence that a high-fat diet increases chances of cancer development.

Fat is the component of food that has the highest energy content (9.3 Kcal/g compared to 4.1 Kcal/g for carbohydrate). Raw potatoes, which contain roughage, have about 0.9 Kcal per gram, but when they are cooked in fat, the number of Kcalories jumps to 6 Kcal per gram. Another problem for those trying to limit their caloric intake is that fat is not always highly visible: butter melts on toast or potatoes. Table 11.5 gives suggestions for cutting down on the amount of fat in the diet.

As a nation, we have increased our consumption of fat from plant sources and have decreased our consumption from animal sources, such as red meat and butter (fig. 11.17). Most likely, this is due to recent publicized studies linking diets high in saturated fats and cholesterol to hypertension and heart attack.

Cholesterol and Heart Disease

The risk of cardiovascular disease includes the many factors discussed on page 214. One of these factors, according to the

Table 11.5
Reducing Dietary Fat

To reduce dietary fat, the following suggestions are recommended.

1. Choose lean red meat, poultry, fish, or dry beans and peas as a protein source.
2. Trim fat off meat and remove skin from poultry before cooking.
3. Cook meat or poultry on a rack so that fat will drain off.
4. Broil, boil, or bake, rather than fry.
5. Limit your intake of butter, cream, hydrogenated oils, shortenings, and coconut and palm oils.*
6. Use herbs and spices to season vegetables instead of butter, margarine, or sauces. Use lemon juice instead of salad dressing.
7. Drink skim milk instead of whole milk, and use skim milk in cooking and baking.
8. Eat nonfat and low-fat foods.

*Although coconut and palm oils are from plant sources, they are saturated fats.

Figure 11.17

Fat content in the diet. **a.** Americans acquire 43% of fat intake from animal fats, like butter, and from plant oils, like cooking oils; 36% comes from red meat, fish, and poultry; and lesser amounts come from the sources shown. **b.** The amount of fat acquired from vegetable sources is now larger than it was in the early 1900s.

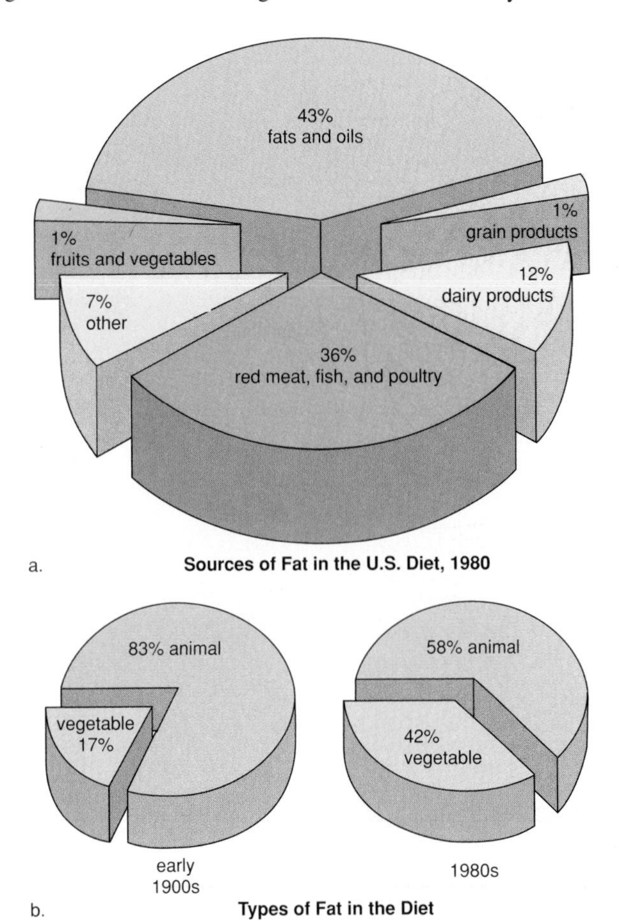

National Heart, Lung, and Blood Institute, is a blood cholesterol level of 240 mg/100 ml or higher. If the cholesterol level is this high, additional testing can determine how much of each of 2 important subtypes of cholesterol is in blood. Cholesterol is carried from the liver to the cells by plasma proteins called *low-density lipoprotein (LDL)* and is carried away from the cells to the liver by *high-density lipoprotein (HDL)*. Therefore, LDL is the type of lipoprotein that apparently contributes to formation of plaque (p. 214), which can clog the arteries, while HDL protects against the development of clogged arteries.

A diet low in saturated fat and cholesterol decreases the blood cholesterol level (LDL level) in some individuals. The suggestions in table 11.5 promote this type of diet, as do certain protein sources. White fish, poultry, and shellfish, for example, are recommended protein sources; cheese, egg yolks, and liver are not. It is also advisable to substitute egg whites for egg yolks in both cooking and eating.

Soluble fiber is believed to combine with cholesterol in the digestive tract and carry it out of the body. Foods high in soluble fiber are oat bran, oatmeal, beans, corn, and certain fruits, such as apples, citrus fruits, and cranberries.

Vitamins and Minerals: Need Them All

For good health, the diet should contain all the various vitamins and minerals.

Vitamins and Balanced Diets

Vitamins are organic compounds (other than carbohydrate, fat, and protein) that the body is unable to produce but uses for metabolic purposes. Many vitamins are portions of coenzymes, or enzyme helpers. For example, niacin is part of the coenzyme NAD (p. 79) and riboflavin is part of another dehydrogenase, FAD. Coenzymes are needed in only small amounts because each can be used over and over again. If vitamins are lacking, various symptoms develop (fig. 11.18).

Although many substances are advertised as vitamins, in reality there are only 13 vitamins (table 11.6). In general, carrots, squash, turnip greens, and collards are good sources of vitamin A. Citrus fruits and other fresh fruits and vegetables are natural sources of vitamin C. Sunshine and irradiated milk are primary sources of vitamin D, and whole grains are good sources of the B vitamins. It is not difficult to acquire the RDAs for vitamins if the diet is balanced because each vitamin is needed in small amounts only.

The National Academy of Sciences suggests that we eat more fruits and vegetables in order to acquire a good supply of vitamins C and A because these 2 vitamins may help to prevent cancer. Nevertheless, the intake of excess vitamins by way of pills is discouraged because this practice can possibly lead to illness. For example, excess vitamin C can cause kidney stones or it can be converted to oxalic acid, a molecule that is toxic to the body. Vitamin A taken in excess over long periods can cause hair loss, bone and joint pains, and loss of appetite. Excess vitamin D can cause an overload of calcium in the blood; in children, this leads to loss of appetite and retarded growth. Megavitamin therapy should always be supervised by a physician.

Minerals: Macro and Trace

In addition to vitamins, various **minerals** are required by the body (table 11.7). Minerals are divided into the macrominerals, which are recommended in amounts more than 100 mg per day, and the microminerals (trace elements), which are recommended in amounts less than 20 mg per day. The macrominerals sodium, magnesium, phosphorus, chlorine, potassium, and calcium are structural components of tissues. For example,

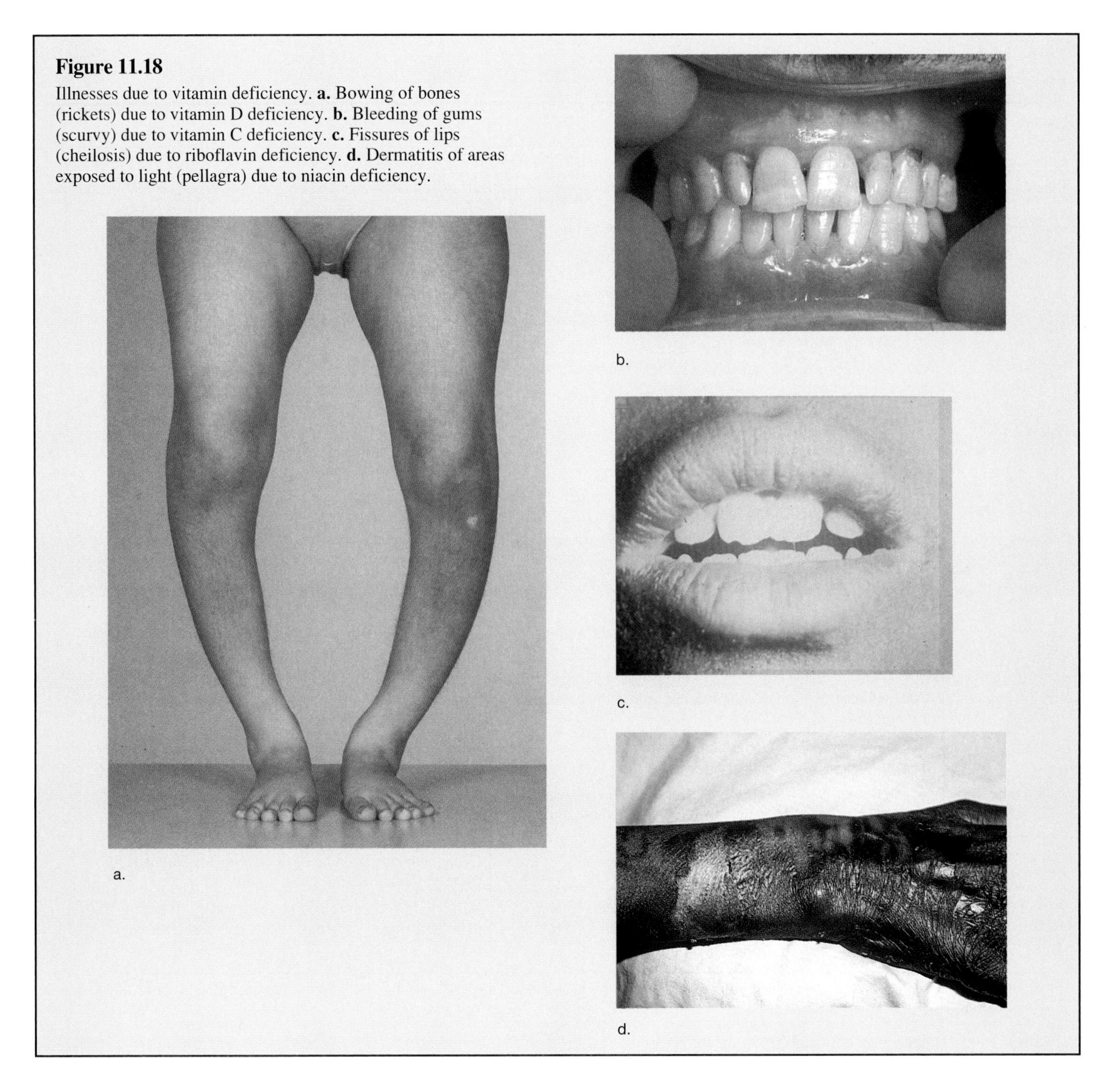

Figure 11.18
Illnesses due to vitamin deficiency. **a.** Bowing of bones (rickets) due to vitamin D deficiency. **b.** Bleeding of gums (scurvy) due to vitamin C deficiency. **c.** Fissures of lips (cheilosis) due to riboflavin deficiency. **d.** Dermatitis of areas exposed to light (pellagra) due to niacin deficiency.

calcium is needed for the construction of bones and teeth and for nerve conduction and muscle contraction.

The microminerals seem to have very specific functions. For example, iron is needed for the production of hemoglobin, and iodine is used in the production of thyroxin, a hormone produced by the thyroid gland. As research continues, more and more elements are added to the list of microminerals considered to be essential. During the past 3 decades, for example, very small amounts of molybdenum, selenium, chromium, nickel, vanadium, silicon, and even arsenic have been found to be essential to good health.

Occasionally, individuals do not receive enough iron, calcium, magnesium, or zinc in their diet. Adult females need more iron in the diet than males (RDA of 18 mg compared to 10 mg) because they lose hemoglobin each month during menstruation. Stress can bring on a magnesium deficiency, and due

Table 11.6
Vitamins: Their Role in the Body and Their Food Sources

Vitamins	Role in Body	Good Food Sources
Fat Soluble		
Vitamin A	Assists in the formation and maintenance of healthy skin, hair, and mucous membranes; aids in the ability to see in dim light (night vision); essential for proper bone growth, tooth development, and reproduction	Deep yellow/orange and dark green vegetables and fruits (carrots, broccoli, spinach, cantaloupe, sweet potatoes), cheese, milk, and fortified margarines
Vitamin D	Aids in the formation and maintenance of bones and teeth; assists in the absorption and use of calcium and phosphorus	Milks fortified with vitamin D, tuna, salmon, or cod liver oil; also made in the skin when exposed to sunlight
Vitamin E	Protects vitamin A and essential fatty acids from oxidation; prevents cell membrane damage	Vegetable oils and margarine, nuts, wheat germ and whole-grain breads and cereals, and green leafy vegetables
Vitamin K	Aids in synthesis of substances needed for clotting of blood; helps maintain normal bone metabolism	Green leafy vegetables, cabbage, and cauliflower; also made by bacteria in intestines of humans, except for newborns
Water Soluble		
Vitamin C	Important in forming collagen, a protein that gives structure to bones, cartilage, muscle, and vascular tissue; helps maintain capillaries, bones, and teeth; aids in absorption of iron; helps protect other vitamins from oxidation	Citrus fruits, berries, melons, dark green vegetables, tomatoes, green peppers, cabbage, and potatoes
Thiamin	Helps in release of energy from carbohydrates; promotes normal function of nervous system	Whole-grain products, dried beans and peas, sunflower seeds, and nuts
Riboflavin	Helps body transform carbohydrate, protein, and fat into energy	Nuts, yogurt, milk, whole-grain products, cheese, poultry, and leafy green vegetables
Niacin	Helps body transform carbohydrate, protein, and fat into energy	Nuts, poultry,whole-grain products, dried fruit, leafy greens, and beans; can be formed in the body from tryptophan, an essential amino acid found in protein
Vitamin B_6	Aids in the use of fats and amino acids; aids in the formation of protein	Sunflower seeds, beans, poultry, nuts, leafy green vegetables, bananas, and dried fruit
Folic acid	Aids in the formation of hemoglobin in red blood cells; aids in the formation of genetic material	Dark green leafy vegetables, nuts, beans, whole-grain products, and fruit juices
Pantothenic acid	Aids in the formation of hormones and certain nerve-regulating substances; helps in the metabolism of carbohydrate, protein, and fat	Nuts, beans, seeds, dark green leafy vegetables, poultry, dried fruit, and milk
Biotin	Aids in the formation of fatty acids; helps in the release of energy from carbohydrate	Occurs widely in foods, especially eggs; also made by bacteria in the intestines of humans
Vitamin B_{12}	Aids in the formation of red blood cells and genetic material; helps the function of the nervous system	Milk, yogurt, cheese, fish, poultry, and eggs; not found in plant foods unless fortified (such as in some breakfast cereals)

to its high-fiber content, a vegetarian diet may make zinc less available to the body. However, a varied and complete diet usually supplies the RDAs for minerals.

Calcium and Bone Disease. There has been much public interest in calcium supplements (fig. 11.19), to counteract osteoporosis, a degenerative bone disease that afflicts an estimated one-fourth of older men and one-half of older women in the United States. These individuals have porous bones that break easily because they lack sufficient calcium. Studies have shown, however, that calcium supplements cannot prevent osteoporosis even when the dosage is 3,000 mg a day. In women who have ceased to menstruate, bone-eating cells called osteoclasts are known to be more active than bone-forming cells called osteoblasts. Until now, the most effective defense against osteoporosis in older women has been estrogen hormone therapy and exercise, which encourage the work of osteoblasts. Recently, however, studies have shown that the drug etidronate disodium, which inhibits osteoclast activity, is effective in osteoporotic women when administered at the proper dosage.

Women can guard against osteoporosis when they are older by forming strong, dense bones when they are younger. Eighteen-year-old women on the average get only 679 mg of calcium a day when the RDA is at least 800 mg. They should consume more calcium-rich foods, such as milk and dairy

Table 11.7
Minerals: Their Role in the Body and Their Food Sources

Minerals	Role in Body	Good Food Sources
Macrominerals		
Calcium	Used for building bones and teeth and maintaining bone strength; also involved in muscle contraction, blood clotting, and maintenance of cell membranes	All dairy products, dark green leafy vegetables, beans, nuts, sunflower seeds, dried fruit, molasses, and canned fish
Phosphorus	Used to build bones and teeth, release energy from carbohydrate, proteins, and fats, and form genetic material, cell membranes, and many enzymes	Beans, sunflower seeds, milk, cheese, nuts, poultry, fish, and lean meats
Magnesium	Used to build bones, produce proteins, release energy from muscle carbohydrate stores (glycogen), and regulate body temperature	Sunflower and pumpkin seeds, nuts, whole-grain products, beans, dark green vegetables, dried fruit, and lean meats
Sodium	Regulates body-fluid volume and blood acidity; aids in transmission of nerve impulses	Most of the sodium in the American diet is added to food as salt (sodium chloride) in cooking, at the table, or in commercial processing; animal products contain some natural sodium
Chloride	A component of gastric juice; aids in acid-base balance	Table salt, seafood, milk, eggs, and meats
Potassium	Assists in muscle contraction, the maintenance of fluid and electrolyte balance in the cells, and the transmission of nerve impulses; also aids in the release of energy from carbohydrate, proteins, and fats	Widely distributed in foods, especially fruits and vegetables, beans, nuts, seeds, and lean meats
Microminerals		
Iron	Involved in the formation of hemoglobin in the red blood cells of blood and myoglobin in muscles; also part of several enzymes and proteins	Molasses, seeds, whole-grain products, fortified breakfast cereals, nuts, dried fruits, beans, poultry, fish, and lean meats
Zinc	Involved in the formation of protein (growth of all tissues), wound healing, and prevention of anemia; a component of many enzymes	Whole-grain products, seeds, nuts, poultry, fish, beans, and lean meats
Iodine	Integral component of thyroid hormones	Table salt (fortified), dairy products, shellfish, and fish
Fluoride	Maintenance of bone and tooth structure	Fluoridated drinking water is the best source; also found in tea, fish, wheat germ, kale, cottage cheese, soybeans, almonds, onions, and milk
Copper	Vital to enzyme systems and in manufacturing red blood cells; needed for utilization of iron	Nuts, oysters, seeds, crab, wheat germ, dried fruit, whole-grain products, and legumes
Selenium	Functions in association with vitamin E and may assist in protecting tissues and cell membranes from oxidative damage; may also aid in preventing cancer	Nuts, whole-grain products, lean pork, cottage cheese, milk, molasses, and squash
Chromium	Required for maintaining normal glucose metabolism; may assist insulin function	Nuts, prunes, vegetable oils, green peas, corn, whole-grain products, orange juice, dark green vegetables, and legumes
Manganese	Needed for normal bone structure, reproduction, and the normal function of nervous system; a component of many enzyme systems	Whole-grain products, nuts, seeds, pineapple, berries, legumes, dark green vegetables, and tea
Molybdenum	Component of enzymes; may help prevent dental caries	Tomatoes, wheat germ, lean pork, legumes, whole-grain products, strawberries, winter squash, milk, dark green vegetables, and carrots

Figure 11.19
Calcium in the diet. Many over-the-counter calcium supplements are now available to boost the amount in the diet.

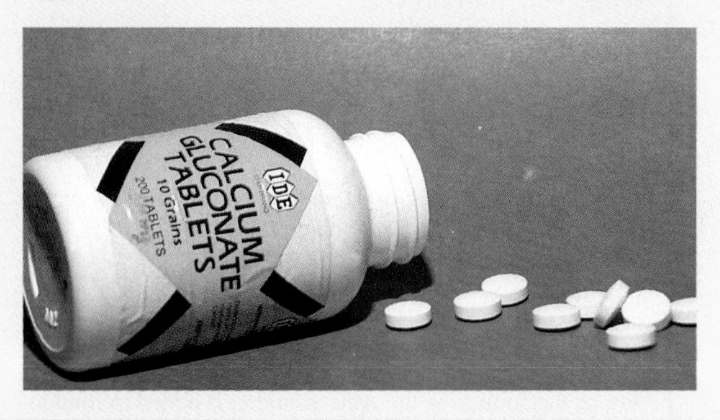

Table 11.8
Reducing Dietary Sodium

To reduce dietary sodium, the following suggestions are recommended.

1. Use spices instead of salt to flavor food.
2. Add little or no salt to foods at the table, and add only small amounts of salt when you cook.
3. Eat unsalted crackers, pretzels, potato chips, nuts, and popcorn.
4. Avoid hot dogs, ham, bacon, luncheon meat, smoked salmon, sardines, and anchovies.
5. Avoid processed cheese and canned or dehydrated soups.
6. Avoid prepared catsup and horseradish.
7. Avoid brine-soaked foods, such as pickles and olives.

products. Taking calcium supplements may not be as effective; a cup of milk supplies 270 mg of calcium, while a 500-mg tablet of calcium carbonate provides only 200 mg. The excess supplemental calcium is not taken up by the body, as it is not in a form that is *bioavailable*. However, an excess of bioavailable calcium can lead to kidney stones.

Dietary calcium and exercise, plus estrogen hormone therapy in older women if needed, are the best safeguards against osteoporosis.

Too Much Sodium. The recommended amount of sodium intake per day is 400–3,300 mg, although the average American takes in 4,000–4,700 mg every day. In recent years, this imbalance has caused concern because high sodium intake has been linked to hypertension (high blood pressure) in some people. About one-third of the sodium we consume occurs naturally in foods; another one-third is added during commercial processing; and we add the last one-third either during home cooking or at the table in the form of table salt.

Clearly, it is possible for us to cut down on the amount of sodium in the diet. Table 11.8 gives recommendations for doing so.

Excess sodium in the diet can lead to hypertension; therefore, excess sodium intake should be avoided.

Dieting

Figure 11.20 indicates that weight loss occurs if the caloric intake is reduced while the same level of activity is maintained. Nutritionists say that the best way to lose weight is to modify behavior and to lose the weight slowly. Behavior modification requires us to examine our eating behaviors (fig. 11.21), to identify the situations that cause us to snack unnecessarily, and to work at changing these. For example, if you are used to having cake and ice cream for dessert, substitute fruit. If you pass an ice cream shop every day and tend to stop in, change your route and do not go by this shop.

Physical activity is a very important addition to the daily routine. Aside from helping to firm up the body after weight loss, exercise also puts you in the mood to keep the weight off and it burns calories.

Excessive dieting and rapid weight loss are dangerous to health and do not work because the weight loss is simply regained once the diet ceases. The body is not adapted for rapid weight loss. First, if calories are severely and suddenly restricted, the body burns protein rather than fat. This protein is removed from the muscles, even the heart muscle. Second, the body apparently has a "set point" for its usual amount of fat. If the amount of stored fat falls below this point, the fat cells are believed to signal the brain by the release of a chemical substance. In response, the metabolic rate is lowered so that fewer calories are needed to stay at the same weight. This set-point hypothesis also explains why people tend to quickly regain any weight that has been lost through dieting. There is some evidence that exercise and avoidance of fatty foods lowers the set point for body fat.

The overall conclusion, then, is that in order to keep unwanted pounds off, a long-term program is needed. This program should include regular exercise and a balanced diet that avoids fatty foods.

Fad diets can be dangerous to health. Long-term weight control is achieved when a balanced diet containing a reduced number of calories and an exercise program are adopted.

Three Eating Disorders

Authorities recognize 3 primary eating disorders: obesity, bulimia, and anorexia nervosa. Although these exist in a con-

Figure 11.20

Diagram illustrating the relationship between caloric intake and weight gain or loss. In each instance, energy needs are divided between basal metabolism (the work the body does to maintain such normal functions as heartbeat, breathing, and blood circulation) and physical activity. **a.** Energy content of food is greater than energy needs of the body—weight gain occurs. **b.** Energy content of food is less than energy needs of the body—weight loss occurs. **c.** Energy content of food equals energy needs of the body—no weight change occurs.

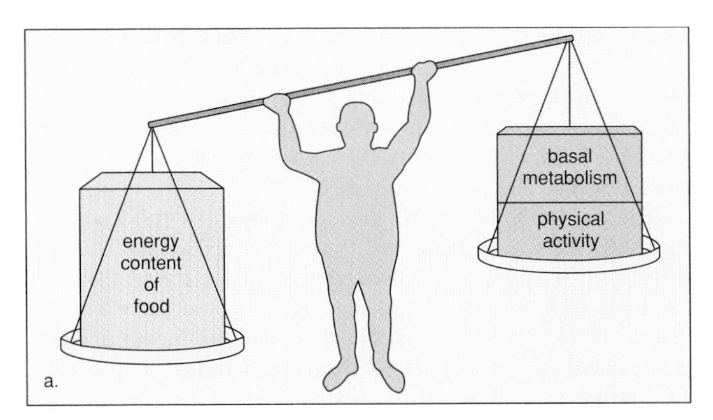

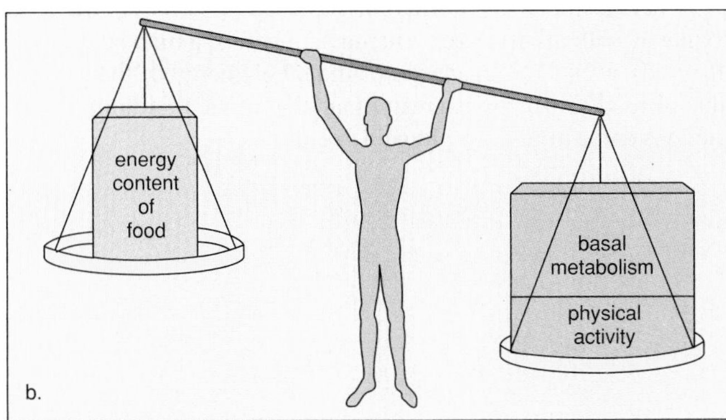

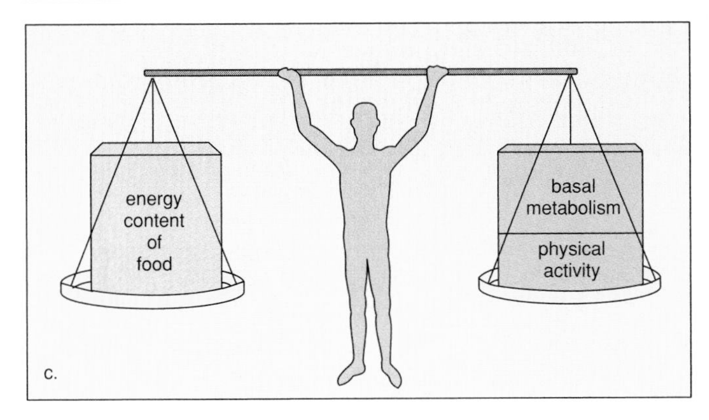

tinuum as far as body weight is concerned, there is much overlap among them.

Obesity is defined as a body weight of more than 20% above the ideal weight and is most likely caused by a combination of factors, including hormonal, metabolic, and social factors. The social factors include the eating habits of other family members. Obese individuals need to consult a physician if they want to bring their body weight down to normal and to keep it there permanently.

Figure 11.21

Behavior and food intake. Social situations sometimes cause us to indulge in fattening foods.

A physician may be able to determine the many complex causes of obesity.

Bulimia can coexist with either obesity or anorexia nervosa. People, usually young women, who are afflicted with bulimia have the habit of eating to excess and then purging themselves by some artificial means, such as induced vomiting or laxatives. These individuals usually are depressed, but whether the depression causes or is caused by the bulimia cannot be determined. While individual psychological help does not seem to be effective, there is some indication that group therapy does help the bulimia patient. The possibility of a hormonal disorder has not been ruled out, however.

Anorexia nervosa is diagnosed when an individual is extremely thin but still claims to "feel fat" and continues to diet. It is possible these individuals have an incorrect body image, which makes them think they are fat. It is also possible they have various psychological problems, including a desire to suppress their sexuality. Menstruation ceases in very thin women.

Both bulimia and anorexia nervosa are serious disorders that require the attention of competent medical personnel.

SUMMARY

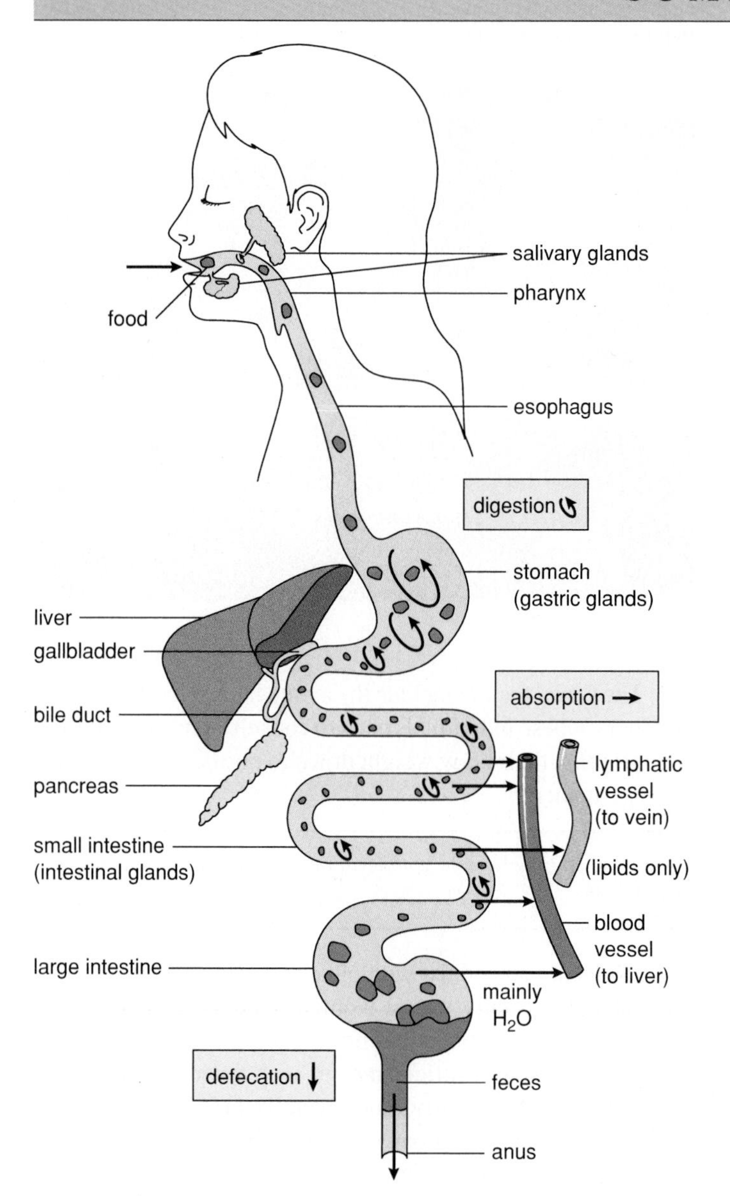

In the mouth, food is chewed and starch is acted upon by salivary amylase. After swallowing, peristaltic action moves the food along the entire digestive tract. In the stomach, pepsin, in the presence of hydrochloric acid (HCl), acts on protein. In contrast, the small intestine has a basic pH environment. Here, fat is emulsified by bile salts to fat droplets before being acted upon by lipase. Protein is digested by trypsin, and starch is digested by pancreatic amylase. Intestinal enzymes present on the surface of the intestinal mucosa finish the digestion of protein and carbohydrate. Only nondigestible material passes from the small intestine to the large intestine. The large intestine absorbs water from this material. It also contains a large population of bacteria that can use the material as food. In the process, the bacteria produce vitamins that can be absorbed and used by the body.

The walls of the small intestine have fingerlike projections called villi, within which are blood capillaries and a lymphatic lacteal. Amino acids and glucose enter blood, and glycerol and fatty acids re-form to give fat packaged as a lipoprotein before entering the lacteal. Blood from the small intestine moves into the hepatic portal vein, which goes to the liver, an organ that monitors and contributes to blood composition.

A balanced diet is required for good health. Food should provide us with all necessary vitamins, minerals, amino acids, and fatty acids, as well as an adequate amount of energy. If the caloric value of food consumed is greater than that needed for body functions and activity, weight gain occurs.

STUDY QUESTIONS

In order to practice **writing across the curriculum,** students should write out the answers to any or all of the study questions. The study questions are sequenced in the same order as the text.

1. List the parts of the digestive tract, anatomically describe them, and state the contribution of each to the digestive process. (p. 173)
2. Discuss the absorption of the products of digestion into the circulatory system. (p. 177)
3. What functions do intestinal bacteria perform? (p. 178)
4. List the accessory glands, and describe the part they play in the digestion of food. (pp. 179–180)
5. What are gastrin, secretin, and CCK? Where are they produced? What are their functions? (p. 180)
6. List 6 functions of the liver. How does the liver maintain a constant blood glucose level? (p. 181)
7. What is jaundice? cirrhosis of the liver? (p. 181)
8. Discuss the digestion of starch, protein, and fat, listing all the steps that occur to bring about digestion of each of these. (p. 182)
9. Give reasons why carbohydrates, fats, proteins, vitamins, and minerals are all necessary for good nutrition. (pp. 184–188)
10. What factors determine how many calories should be ingested? (p. 193)

OBJECTIVE QUESTIONS

1. In the mouth, salivary _______ digests starch to _______.
2. When swallowing, the _______ covers the opening to the larynx.
3. The _______ takes food to the stomach, where _______ is primarily digested.
4. The gallbladder stores _______, a substance that _______ fat.
5. The pancreas sends digestive juices to the _______, the first part of the small intestine.
6. Pancreatic juice contains _______ for digesting protein, _______ for digesting starch, and _______ for digesting fat.
7. Whereas pepsin works best at a _______ pH, the enzymes found in pancreatic juice works best at a _______ pH.
8. The products of digestion are absorbed into the cells of the _______, fingerlike projections of the intestinal wall.
9. After eating, the liver stores glucose as _______.
10. The diet should include a complete protein source, one that includes all the _______.
11. Label the diagram of the digestive tract, and give a function for each organ.
12. Predict and explain the results of this experiment for each test tube.
    a. Test tube 1: water, bile salts, oil
    b. Test tube 2: water, bile salts, pepsin, oil
    c. Test tube 3: water, bile salts, pancreatic lipase, oil

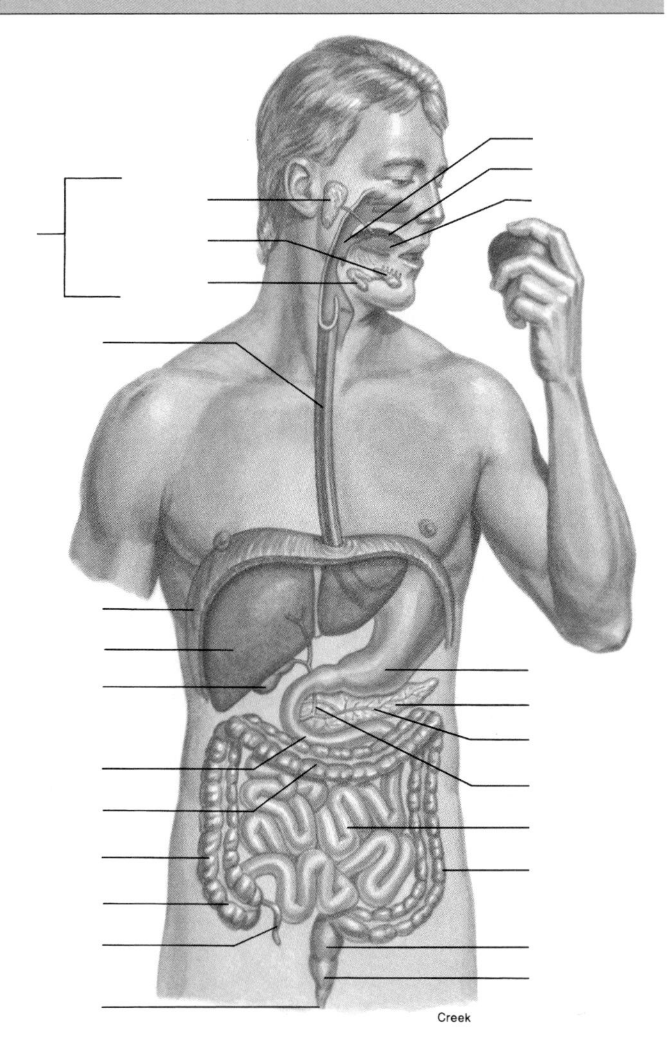

CRITICAL THINKING

In order to practice **writing across the curriculum,** students should write out the answers to any or all of the critical thinking questions. Suggested answers to the critical thinking questions are in appendix E.

11.1

1. The digestive tract is lined by mucous membrane (p. 156). What type of tissue found in the mucous membrane of the stomach and the small intestine produces digestive enzymes?
2. All systems of the body contribute to homeostasis. How does the digestive system contribute?
3. The liver "fine tunes" the contribution of the digestive system to homeostasis. How?
4. The nutrients from digestion enter the bloodstream and are transported to the cells. What does a cell do with the glucose? with the amino acids?

11.2

1. Does the experiment described in figure 11.13 show that the enzyme pepsin prefers an acidic rather than a basic pH? Describe an experiment that would show this.
2. Does the experiment described in figure 11.13 show that pepsin digests only protein and not starch, for example? Describe an experiment that would show this.
3. Does the experiment described in figure 11.13 show that better digestion occurs at a warm temperature rather than a cold temperature? Describe an experiment that would show this.
4. Describe the contents of a test tube that would give the best digestion of fat. Use test tube 4 in figure 11.13 as a guide, but change the contents to suit the digestion of fat.

11.3

1. Essential amino acids cannot be made by the body. From your knowledge of metabolic pathways, what is lacking in the cell?
2. Genes (DNA) control the production of enzymes. What happened to a gene if it no longer codes for a functioning enzyme?
3. Plants are able to make all types of amino acids. What would happen to a plant that was unable to make a particular amino acid? Why?
4. Why can animals survive even though they cannot make all types of amino acids?

SELECTED KEY TERMS

amylase (am´ĭ-lās) starch-digesting enzyme secreted by the salivary glands (salivary amylase) and the pancreas (pancreatic amylase).

bile a secretion of the liver that is temporarily stored in the gallbladder before being released into the small intestine, where it emulsifies fat.

colon the large intestine.

epiglottis (ep″ĭ-glot´is) a structure that covers the glottis during the process of swallowing.

gallbladder a saclike organ associated with the liver that stores and concentrates bile.

gastric gland gland within the stomach wall that secretes gastric juice.

glottis slitlike opening to the larynx between the vocal cords.

hard palate bony anterior portion of the roof of the mouth.

hydrolytic enzyme an enzyme that catalyzes a reaction in which the substrate is broken down by the addition of water.

lipase (li´pās) a fat-digesting enzyme secreted by the pancreas.

lumen (lu´men) the cavity inside any tubular structure, such as the lumen of the digestive tract.

pepsin a protein-digesting enzyme secreted by gastric glands.

peristalsis (per″ĭ-stal´sis) a rhythmic contraction that serves to move the contents along in tubular organs, such as the digestive tract.

pharynx (far´ingks) a common passageway (throat) for both food intake and air movement.

salivary gland a gland associated with the mouth that secretes saliva.

soft palate entirely muscular posterior portion of the roof of the mouth.

sphincter (sfingk´ter) a muscle that surrounds a tube and closes or opens the tube by contracting and relaxing.

trypsin (trip´sin) a protein-digesting enzyme secreted by the pancreas.

villus (vil´us) a fingerlike projection from the wall of the small intestine that functions in absorption.

vitamin essential requirement in the diet, needed in small amounts, that is often a part of a coenzyme.

12

CIRCULATION

Chapter Concepts

1.
The human heart is a double pump; the right side pumps blood to the lungs, and the left side pumps blood to the rest of the body. 201, 217

2.
A series of vessels delivers blood to the capillaries, where exchange of molecules takes place, and then another series of vessels delivers blood back to the heart. 206, 217

3.
Blood is composed of cells and a fluid containing proteins and various other molecules and ions. 209, 217

4.
Exchange of molecules between blood and tissue fluid across capillary walls supplies cells with nutrients and removes wastes. 212, 217

5.
Blood clotting is a series of reactions that produces a clot—fibrin threads in which red blood cells are trapped. 213, 217

6.
Although the circulatory system is very efficient, it is still subject to degenerative disorders. 215, 217

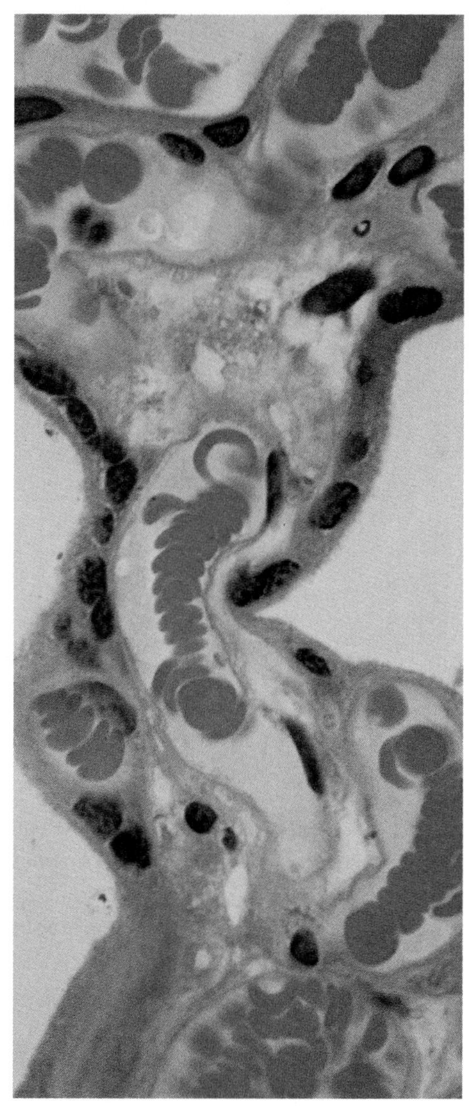

Capillaries with red blood cells, ×1,000

Chapter Outline

Figure 12.1

Blood vessels. **a.** Blood leaving the heart moves from an artery to arterioles to capillaries to venules and then returns to the heart by way of a vein. **b.** Arteries have well-developed walls with a thick middle layer of elastic tissue and smooth muscle. **c.** Capillary walls are one cell thick. **d.** Veins have flabby walls, particularly because the middle layer is not as thick as in arteries. Veins have valves, which point toward the heart.

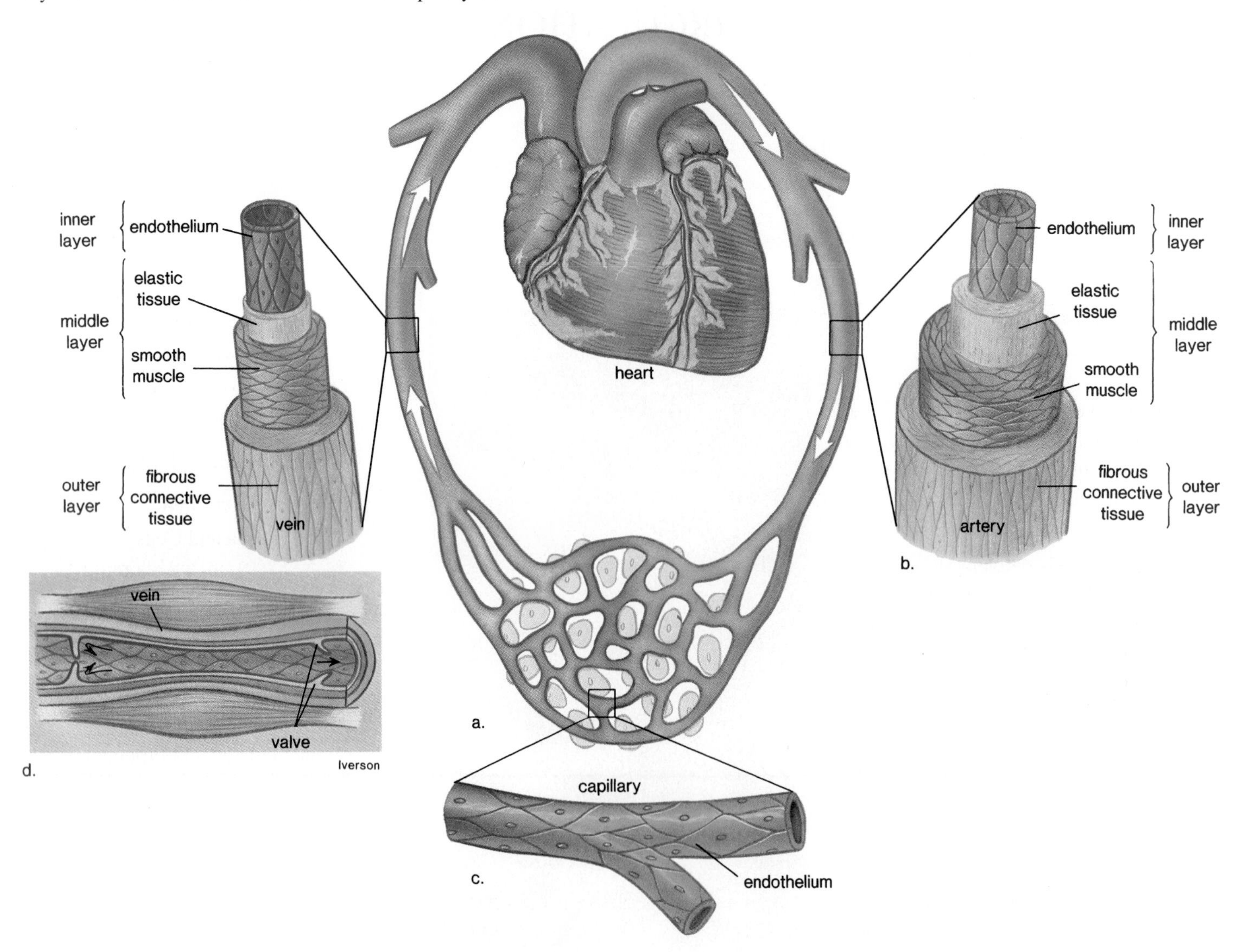

BLOOD moving through tubular vessels brings our cells their daily supply of nutrients, such as amino acids and glucose, and takes away their wastes, such as carbon dioxide. The heart (fig. 12.1) keeps blood moving along its predetermined circular path. Circulation of blood is so important that if the heart stops beating for only a few minutes, death results.

The Circulatory System

The circulatory system has 3 types of blood vessels: the **arteries** (and arterioles), which carry blood away from the heart; the **capillaries,** which exchange material with the tissues; and the **veins** (and venules), which return blood to the heart.

Arteries have thick walls (fig. 12.1*b*). The walls are composed of an inner membranous layer called endothelium, which contains squamous epithelial cells; a thick middle layer of elastic tissue and smooth muscle; and an outer fibrous connective tissue layer. Even though an artery can expand to accommodate the sudden increase in blood volume after each heartbeat, the walls are so thick that they are supplied with blood vessels. **Arterioles** are small arteries, just visible to the naked eye, that can constrict or dilate. The greater the number of vessels dilated, the lower the blood pressure.

Arterioles branch into the smaller capillaries. Each capillary is an extremely narrow, microscopic tube with one-cell-thick walls composed only of endothelium. *Capillary beds* (networks of many capillaries) are present in all regions of the body;

Figure 12.2

External heart anatomy. **a.** The venae cavae bring deoxygenated blood to the right side of the heart from the body, and the pulmonary arteries take this blood to the lungs. The pulmonary veins bring oxygenated blood from the lungs to the left side of the heart, and the aorta takes this blood to the body. **b.** The coronary arteries and cardiac veins pervade cardiac muscle. The coronary arteries are the first blood vessels to branch off the aorta. They bring oxygen and nutrients to cardiac cells.

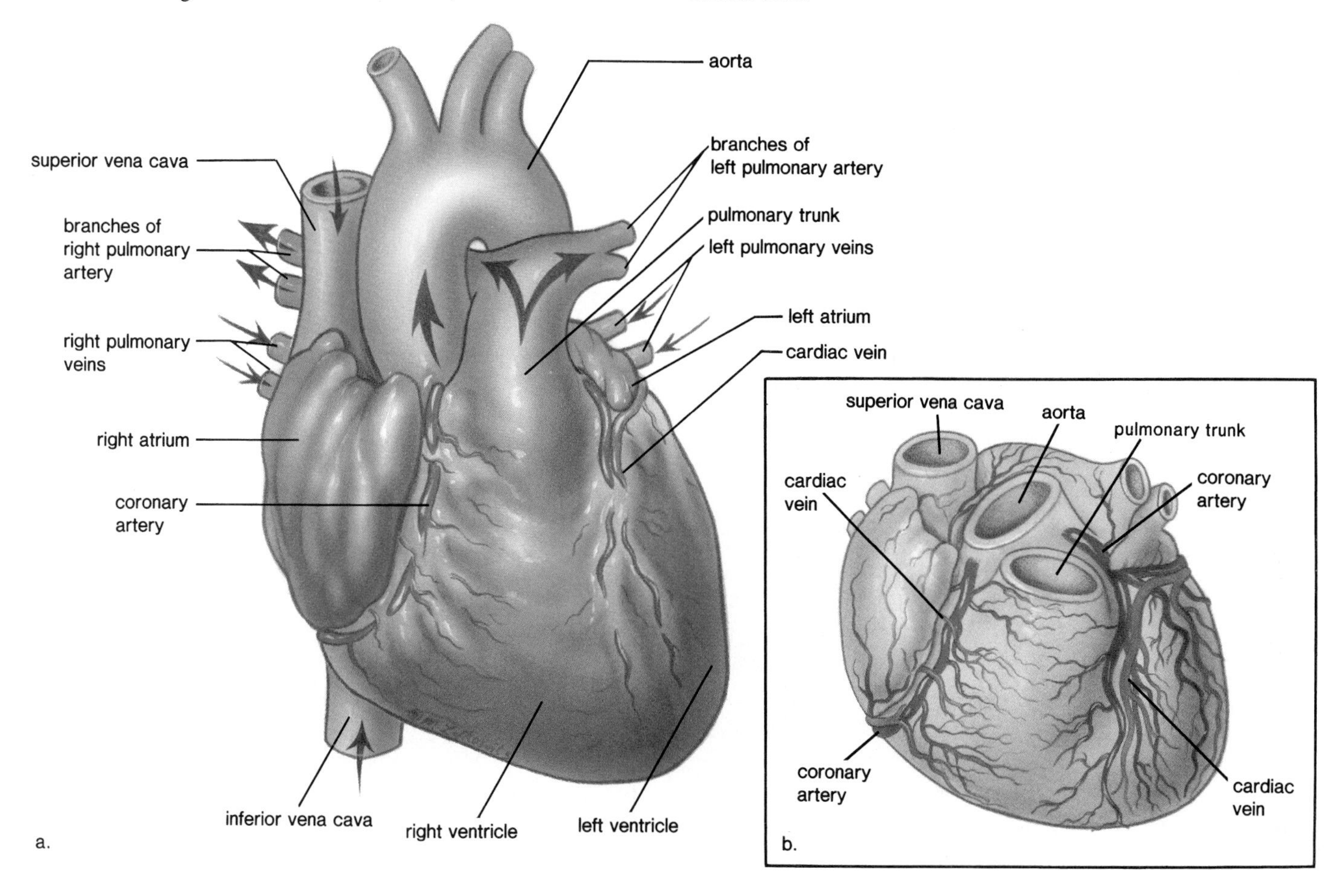

consequently, a cut to any body tissue draws blood. The capillaries are the most important part of a closed circulatory system because the exchange of nutrient and waste molecules takes place across their thin walls. Since the capillaries serve the cells, the heart and the other vessels of the circulatory system can be thought of as the means by which blood is conducted to and from the capillaries. Only certain capillaries are open at any given time. Shunting of blood is possible because each capillary bed has a thoroughfare channel, which allows blood to go directly from arteriole to venule. After eating, blood is shunted through the muscles of the body and diverted to the digestive system. This is why swimming after a heavy meal may cause cramping.

Veins and venules take blood from the capillary beds to the heart. First, the **venules** (small veins) drain blood from the capillaries and then join to form a vein. The walls of venules (and veins) have the same 3 layers as the walls of arteries, but the middle layer is poorly developed and therefore the walls are thinner. Veins often have **valves,** which allow blood to flow only toward the heart when open and prevent the backward flow of blood when closed.

Arteries and arterioles carry blood away from the heart; veins and venules carry blood to the heart; and capillaries join arterioles to venules.

The Heart: Pumps Blood

The **heart** is a cone-shaped (fig. 12.2), muscular organ about the size of a fist. It is located between the lungs directly behind the sternum (breastbone) and is tilted so that the apex is directed to the left. The major portion of the heart, called the **myocardium,** consists largely of cardiac muscle. The muscle fibers within the myocardium are branched and tightly joined to one another (see p. 161). The heart lies within the pericardium, a sac which contains a small quantity of lubricating liquid.

Figure 12.3

Internal view of the heart. **a.** The right side of the heart contains deoxygenated blood. The venae cavae empty into the right atrium, and the pulmonary trunk leaves the right ventricle. The left side of the heart contains oxygenated blood. The pulmonary veins enter the left atrium, and the aorta leaves from the left ventricle. **b.** A diagrammatic representation of the heart, which allows you to trace the path of blood. On the right side of the heart: venae cavae, right atrium, right ventricle, pulmonary arteries to lungs. On the left side of the heart: pulmonary veins, left atrium, left ventricle, aorta to body. Restate this and put in the name of the valves where appropriate.

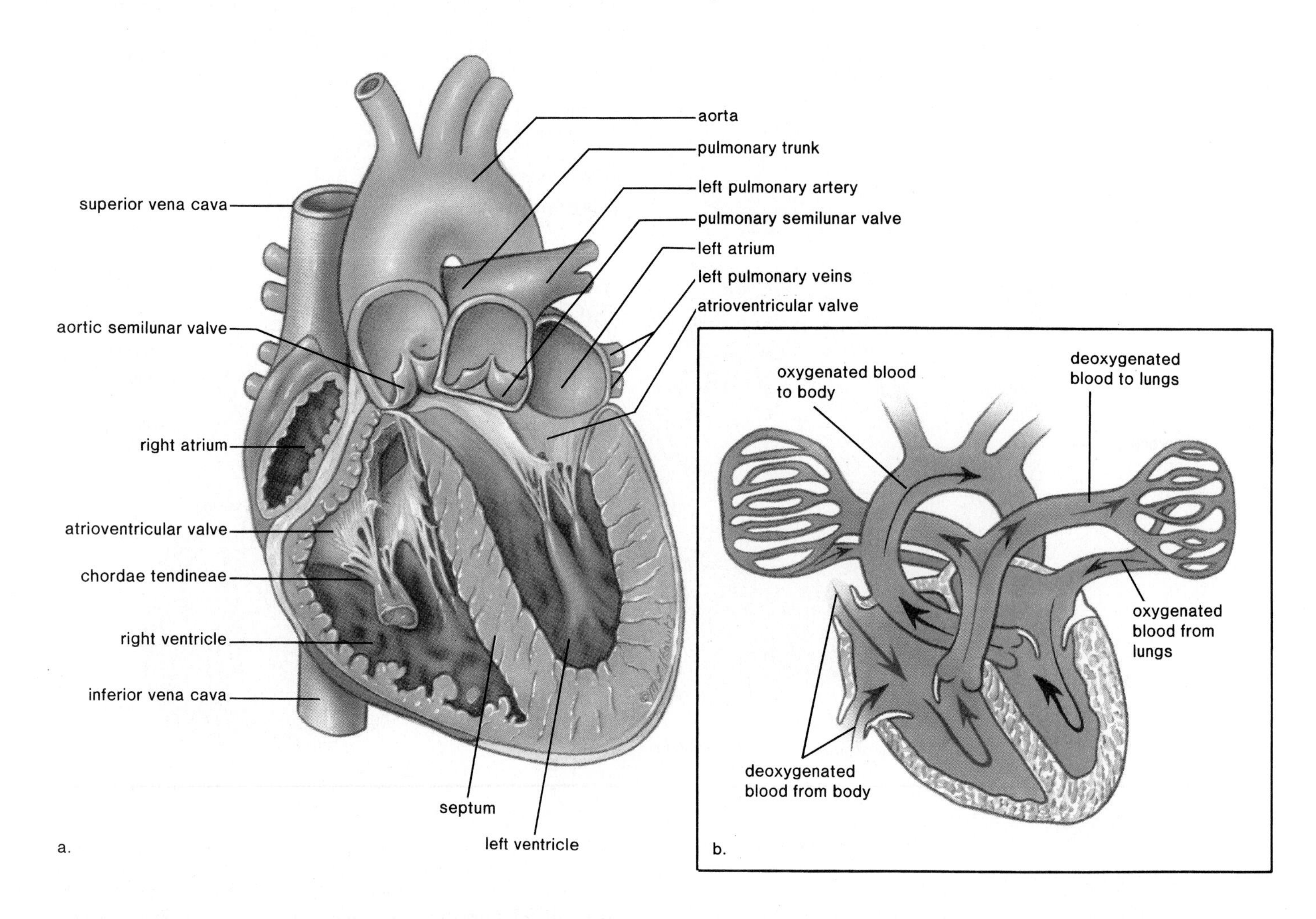

Internally (fig. 12.3), a wall called the **septum** separates the heart into a right side and a left side. The heart has 4 chambers: 2 upper, thin-walled atria (sing., **atrium**), sometimes called auricles, and 2 lower, thick-walled **ventricles.** The atria are much smaller and weaker than the muscular ventricles. The heart also has valves, which direct the flow of blood and prevent its backward movement. The valves that lie between the atria and the ventricles are called the *atrioventricular valves.* These valves are supported by strong, fibrous strings called chordae tendineae. The chordae, which are attached to muscular projections of the ventricular walls, support the valves and prevent them from inverting. The atrioventricular valve on the right side is called the tricuspid valve because it has 3 cusps, or flaps. The valve on the left side is called the bicuspid (mitral) because it has 2 flaps. There are also *semilunar valves,* which resemble half moons, between the ventricles and their attached vessels. The pulmonary semilunar valve lies between the right ventricle and the pulmonary trunk. The aortic semilunar valve lies between the left ventricle and the aorta.

Humans have a four-chambered heart (2 atria and 2 ventricles). A septum separates the right side from the left side. Do 12.1 Critical Thinking, found at the end of the chapter.

Figure 12.4

Cardiac cycle. During systole, the 4 chambers of the heart contract: the atria contract for 0.15 sec and the ventricles contract for 0.30 sec. During diastole (0.70 sec) all chambers rest. Note when the semilunar and atrioventricular valves are open or closed.

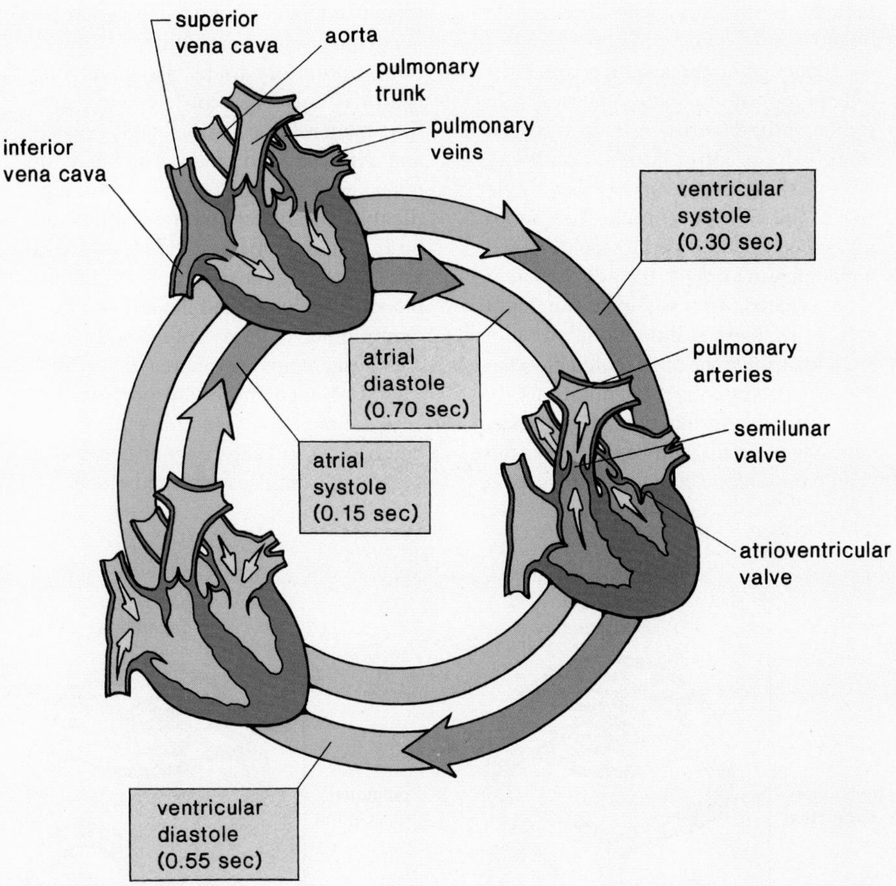

Path through the Heart

We can trace the path of blood through the heart (fig. 12.3) in the following manner:

The superior (anterior) **vena cava** and the inferior (posterior) vena cava, both carrying deoxygenated blood (low in oxygen and high in carbon dioxide), enter the right atrium.

The right atrium sends blood through an atrioventricular valve (the tricuspid valve) to the right ventricle.

The right ventricle sends blood through the pulmonary semilunar valve into the pulmonary trunk and the pulmonary arteries to the lungs.

The pulmonary veins, carrying oxygenated blood (high in oxygen and low in carbon dioxide) from the lungs, enter the left atrium.

The left atrium sends blood through an atrioventricular valve (the bicuspid, or mitral, valve) to the left ventricle.

The left ventricle sends blood through the aortic semilunar valve into the aorta to the body proper.

From this description, you can see that deoxygenated blood never mixes with oxygenated blood and that blood must go through the lungs in order to pass from the right side to the left side of the heart. In fact, the heart is a *double pump* because the right side of the heart sends blood through the lungs, and the left side sends blood throughout the body. Since the left ventricle has the harder job of pumping blood to the rest of the body, its walls are thicker than those of the right ventricle, which pumps blood to the lungs.

> The right side of the heart pumps blood to the lungs, and the left side of the heart pumps blood throughout the body.

Cardiac Cycle: 0.85 Seconds

Each heartbeat is called a cardiac cycle (fig. 12.4). First, the 2 atria contract at the same time; then the 2 ventricles contract at the same time. Then all chambers relax. The word **systole** refers to contraction of heart muscle, and the word **diastole** refers to relaxation of heart muscle. The heart contracts, or beats, about 70 times a minute, and each heartbeat lasts about 0.85 sec.

Heart Sounds

When the heart beats, the familiar lub-dub sound occurs as the valves of the heart close. The lub is caused by vibrations occurring when the atrioventricular valves close, and the dub is heard when the semilunar valves close. A heart murmur, or a slight slush sound after the lub, is often due to ineffective valves, which allow blood to pass back into the atria after the atrioventricular valves have closed.

Pulse: Swell and Recoil

The surge of blood entering the arteries causes their elastic walls to swell, but then they almost immediately recoil. This alternating expansion and recoil of an arterial wall can be felt as a **pulse** in any artery that runs close to the body's surface. It is customary to feel the pulse by placing several fingers on a radial artery, which lies near the outer border of the palm side of the wrist. A carotid artery, on either side of the trachea in the neck, is another good location to feel the pulse. Normally, the pulse rate indicates the rate of the heartbeat because the arterial walls pulse whenever the left ventricle contracts.

ANIMAL CIRCULATORY SYSTEMS

Every cell requires a supply of oxygen and nutrient molecules and must rid itself of waste molecules. The sac body plan of cnidarians and flatworms makes a circulatory system unnecessary. Each cell can exchange materials either with the fluid within the gastrovascular cavity or with the external environment.

Arthropods and many mollusks have an open circulatory system. In this system, a heart pumps blood into vessels that empty into body cavities or into sinuses located within organs themselves. Vertebrates and several invertebrates, including annelids, squids, and octopuses, have a closed circulatory system. Blood, which usually consists of cells and plasma, is pumped by the heart into a system of blood vessels; it never runs free.

Among vertebrates, there are 3 different types of circulatory pathways. In fishes, blood follows a one-circuit (single-loop circulatory) pathway through the body. The heart has a single atrium and a single ventricle. The pumping action of the ventricle sends blood under pressure only to the gills, where it is oxygenated. After passing through gill capillaries, there is little blood pressure left to distribute the oxygenated blood to the tissues (systemic circulation).

Amphibians, like other vertebrates, have a two-circuit pathway. Because the heart pumps blood both to the lungs (pulmonary circulation) and the tissues, there is adequate blood pressure and flow to both circulatory loops. However, the heart has only one ventricle, and therefore, there is some mixing of oxygenated and deoxygenated blood. The hearts of other vertebrates are partially (most reptiles) or completely (some reptiles, all birds, and mammals) divided into right and left halves. The right ventricle pumps blood to the lungs, and the left ventricle pumps blood to the rest of the body. This arrangement not only increases the likelihood of adequate blood pressure for both the pulmonary and systemic circulations, it also keeps oxygenated blood separated from deoxygenated blood.

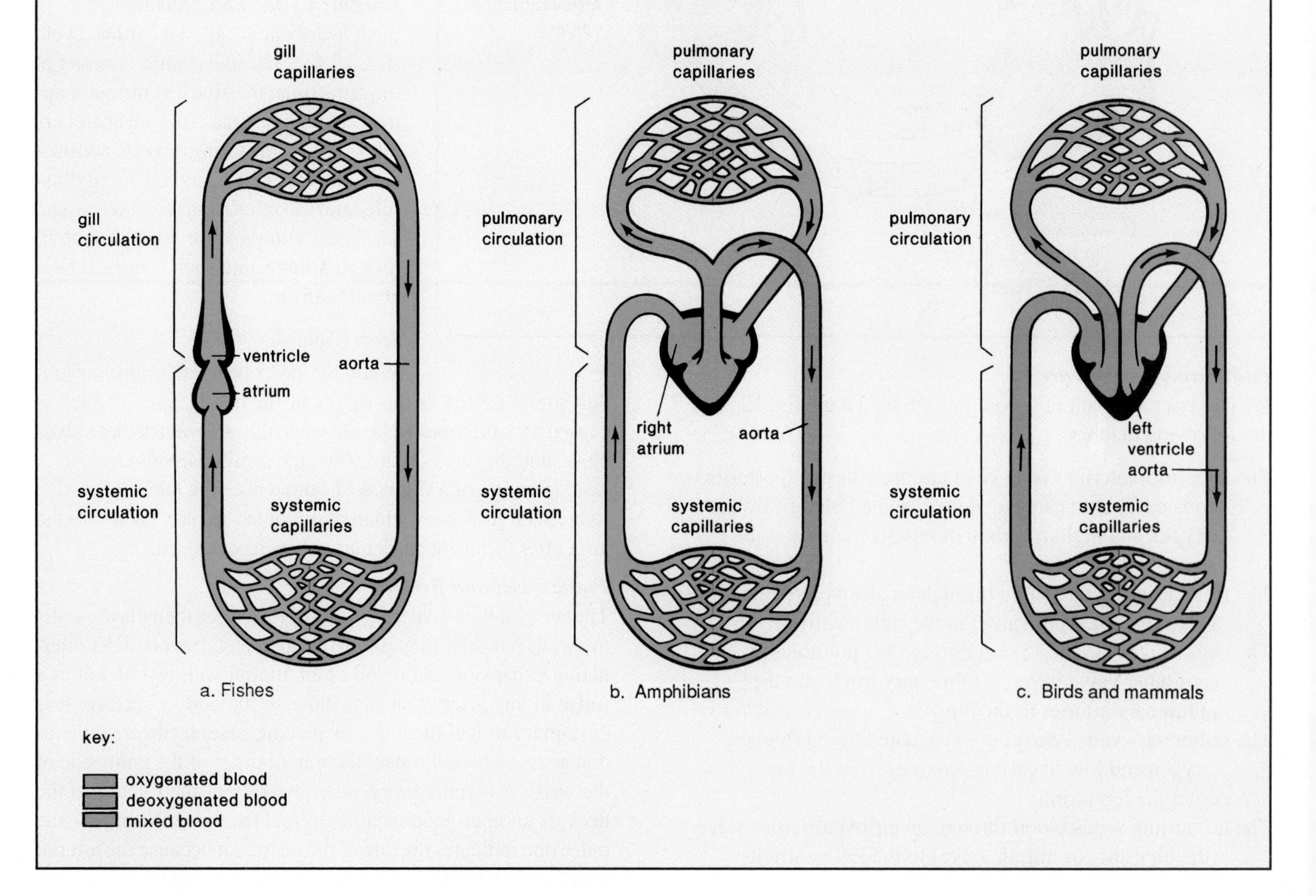

Figure 12.5

Control of the cardiac cycle. **a.** The SA node sends out a stimulus that causes the atria to contract. When this stimulus reaches the AV node, the AV node signals the ventricles to contract by way of the Purkinje fibers. **b.** A normal ECG indicates that the heart is functioning properly. The *P* wave occurs as the atria contract; the *QRS* wave occurs as the ventricles contract; and the *T* wave occurs when the ventricles are recovering from contraction. **c.** Abnormal ECGs: sinus tachycardia is an abnormally fast heartbeat due to a fast pacemaker; ventricular fibrillation is an irregular heartbeat due to irregular stimulation of the ventricles; and mitral stenosis occurs because the bicuspid (mitral) valve is obstructed.

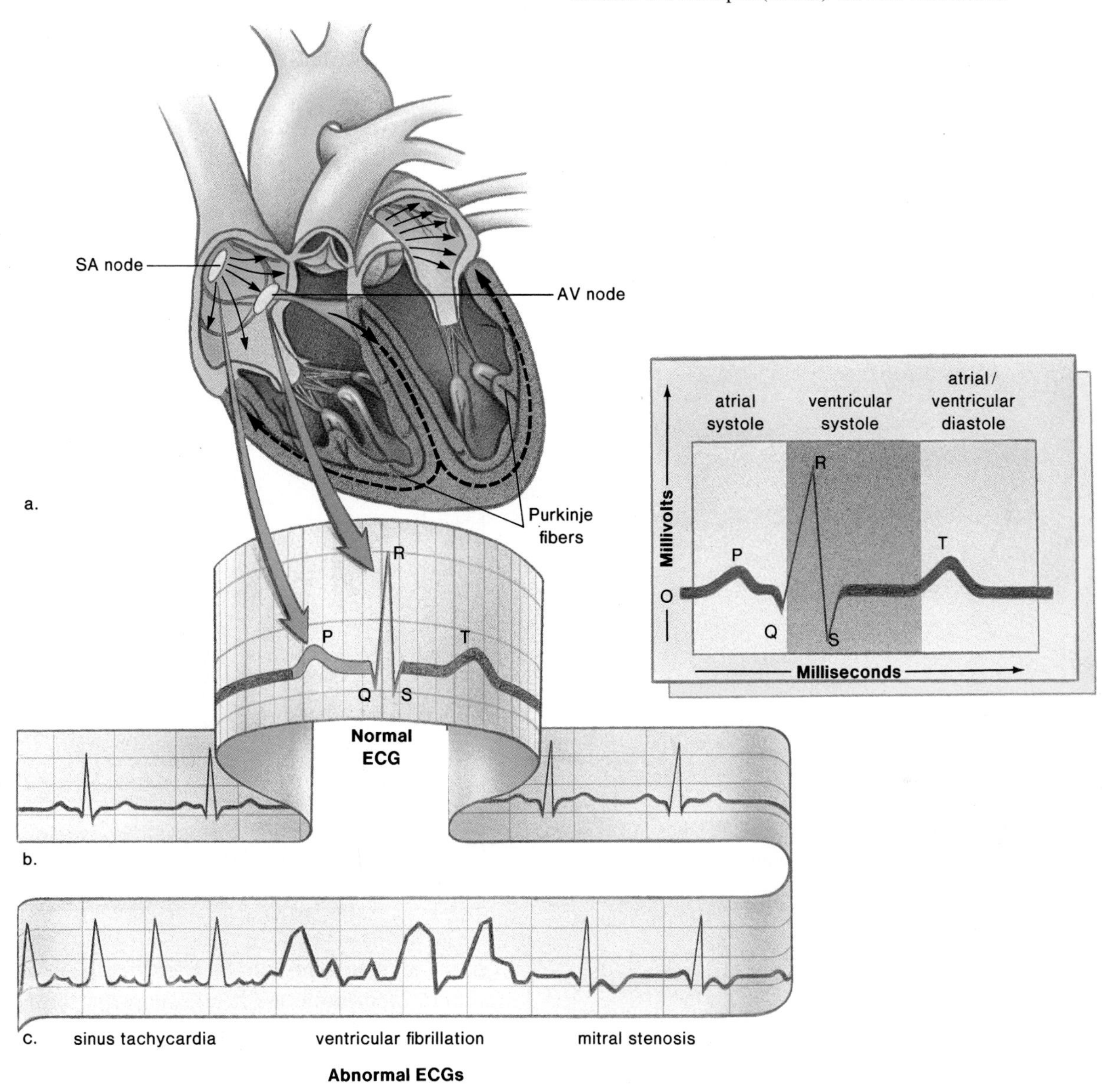

Conduction System: Unique Tissue

Nodal tissue, a unique type of tissue located in 2 regions of the heart, has both muscular and nervous characteristics. The **SA (sinoatrial) node** is found in the upper dorsal wall of the right atrium; the other, the **AV (atrioventricular) node,** is found in the base of the right atrium very near the septum (fig. 12.5*a*). The SA node initiates the heartbeat and automatically sends out an excitation impulse every 0.85 sec; this causes the atria to contract. When the impulse reaches the AV node, the AV node signals the ventricles to contract by way of two large fibers terminating in the more numerous and smaller Purkinje fibers. The SA node is called the **pacemaker** because it usually keeps the heartbeat regular. If the SA node fails to work properly, the heart still beats, but irregularly. To correct this condition, it is possible to implant an artificial pacemaker, which automatically gives an electric shock to the heart every 0.85 sec. The heart then beats regularly again.

With the contraction of any muscle, including the myocardium, ionic changes occur; these can be detected with

Figure 12.6

Determination of blood pressure using a sphygmomanometer. The technician inflates the cuff with air, and then, as the pressure is gradually reduced, she or he listens with a stethoscope for the sounds that indicate blood is moving past the cuff in an artery. This is systolic blood pressure. The pressure in the cuff is further reduced until no sound is heard, indicating that blood is flowing freely through the artery. This is diastolic pressure.

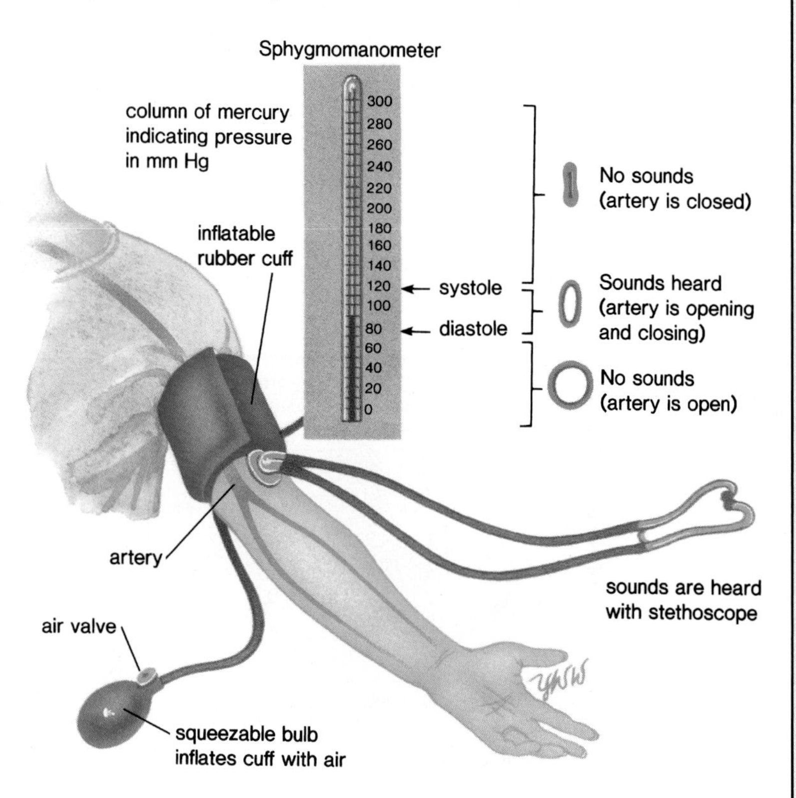

electrical recording devices. The pattern that results, called an **electrocardiogram (ECG** or **EKG)** (fig. 12.5*b*), has an atrial phase and a ventricular phase. The first wave in the electrocardiogram, called the *P* wave, represents the excitation and contraction of the atria. The second wave, or the *QRS* wave, occurs during ventricular excitation and contraction. The third, or *T*, wave is caused by the recovery of the ventricles. An examination of the electrocardiogram indicates whether the heartbeat has a normal or an irregular pattern.

The conduction system of the heart includes the SA node, the AV node, and the Purkinje fibers. With an ECG, it is possible to determine if the conduction system, and therefore the beat of the heart, is regular.

As we will discuss in chapter 16, the nervous system can modify the heartbeat rate, which is set by the SA node.

Blood Pressure: High/Low

Blood pressure is the pressure of blood against the wall of a blood vessel. A sphygmomanometer is used to measure blood pressure, as described in figure 12.6. The highest arterial pressure, called the *systolic pressure,* is reached during ejection of blood from the heart. The lowest arterial pressure is called the *diastolic pressure.* Diastolic pressure occurs while the heart ventricles are relaxing. Normal resting blood pressure for a young adult is said to be 120 mm of mercury (Hg) over 80 mm, or simply 120/80. The higher number is the systolic pressure, and the lower number is the diastolic pressure. Actually, 120/80 is the expected blood pressure in the brachial artery of the arm; blood pressure decreases with distance from the left ventricle. Blood pressure is, therefore, higher in the arteries than in the arterioles. Further, there is a sharp drop in blood pressure when the arterioles reach the capillaries. The decrease can be correlated with the increase in the total cross-sectional area of the vessels as blood moves through arteries, arterioles, and then into capillaries. There are more arterioles than arteries and many more capillaries than arterioles (fig. 12.7).

Figure 12.7

Diagram illustrating how velocity and blood pressure are related to the total cross-sectional area of blood vessels. Capillaries have the greatest cross-sectional area and the least pressure and velocity. Skeletal muscle contraction, not blood pressure, accounts for the velocity of blood in the veins.

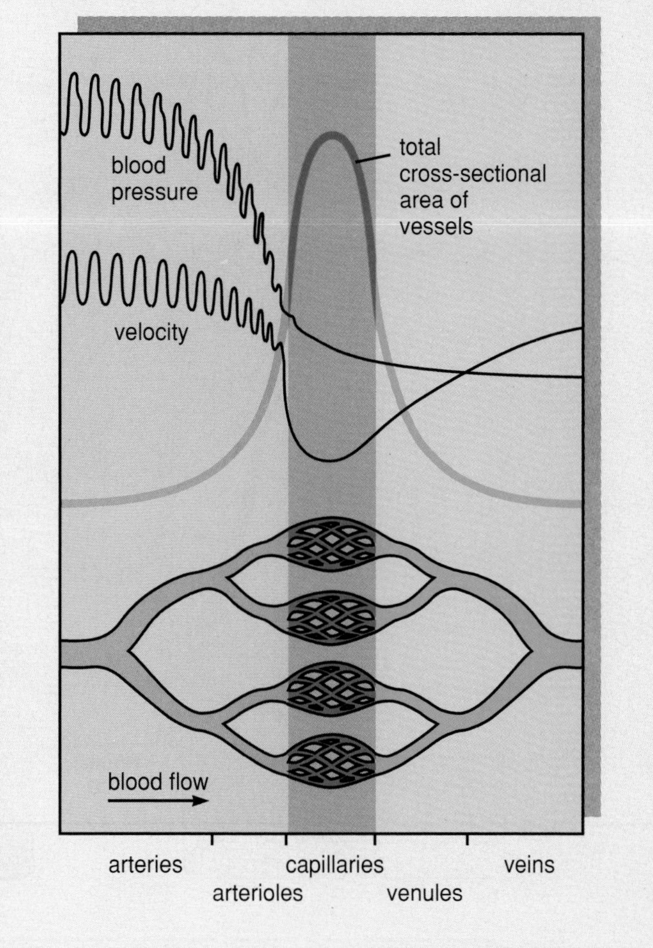

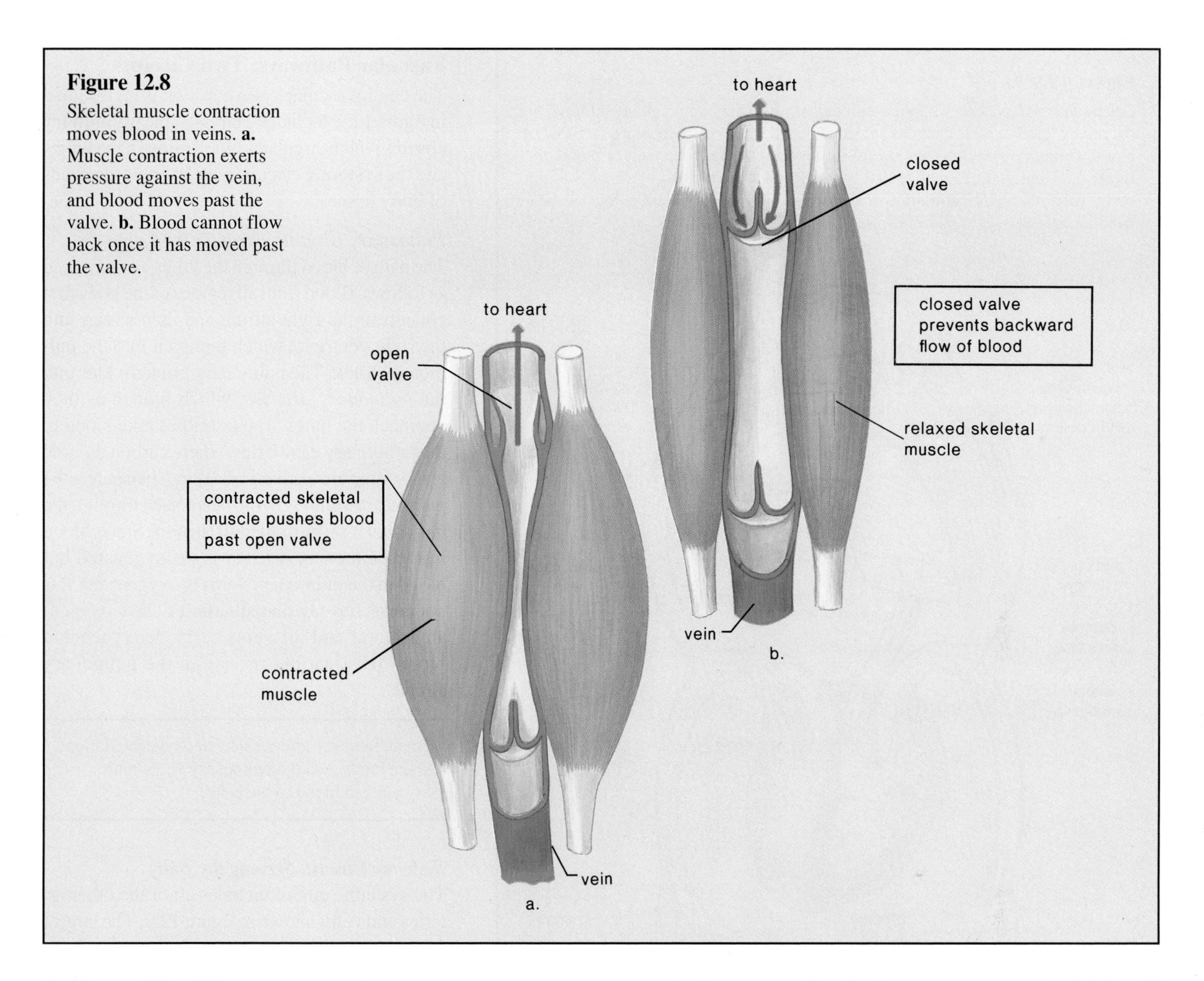

Figure 12.8
Skeletal muscle contraction moves blood in veins. **a.** Muscle contraction exerts pressure against the vein, and blood moves past the valve. **b.** Blood cannot flow back once it has moved past the valve.

Swiftness of Blood Flow

The velocity of blood flow varies in different parts of the circulatory system (fig. 12.7). Blood pressure accounts for the velocity of the blood flow in the arterial system, and therefore, as blood pressure decreases due to the increased cross-sectional area of the arterial system, so does velocity. Blood moves more slowly through the capillaries than it does through the aorta. This is important because the slow progress allows time for the exchange of molecules between blood and tissues.

Blood pressure cannot account for the movement of blood through the venules and the veins since they lie on the other side of the capillaries. Instead, movement of blood through the venous system is due to skeletal muscle contraction. When the skeletal muscles contract, they put pressure against the weak walls of the veins. This causes blood to move past a *valve* (fig. 12.8). Once past the valve, blood cannot return. The importance of muscle contraction in moving blood in the venous system can be demonstrated by forcing a person to stand rigidly still for a number of hours. Frequently, fainting occurs because blood collects in the limbs, robbing the brain of oxygen. In this case, fainting is beneficial because the resulting horizontal position aids in moving blood to the head.

Blood flow gradually increases in the venous system (fig. 12.7) due to a progressive reduction in the cross-sectional area as small venules join to form veins. The 2 venae cavae together have a cross-sectional area only about double that of the aorta. The blood pressure is lowered in the thoracic cavity whenever the chest expands during inspiration. This also aids the flow of venous blood into the thoracic cavity because blood flows in the direction of reduced pressure.

Blood pressure accounts for the flow of blood in the arteries and the arterioles; skeletal muscle contraction accounts for the flow of blood in the venules and the veins.

Figure 12.9

Cardiovascular system. The blue-colored vessels carry deoxygenated blood, and the red-colored vessels carry oxygenated blood; the arrows indicate the flow of blood. Compare this diagram, useful for learning to trace the path of blood, to figure 12.10 in order to realize that both arteries and veins go to all parts of the body. Also, there are capillaries in all parts of the body. No cell is located far from a capillary.

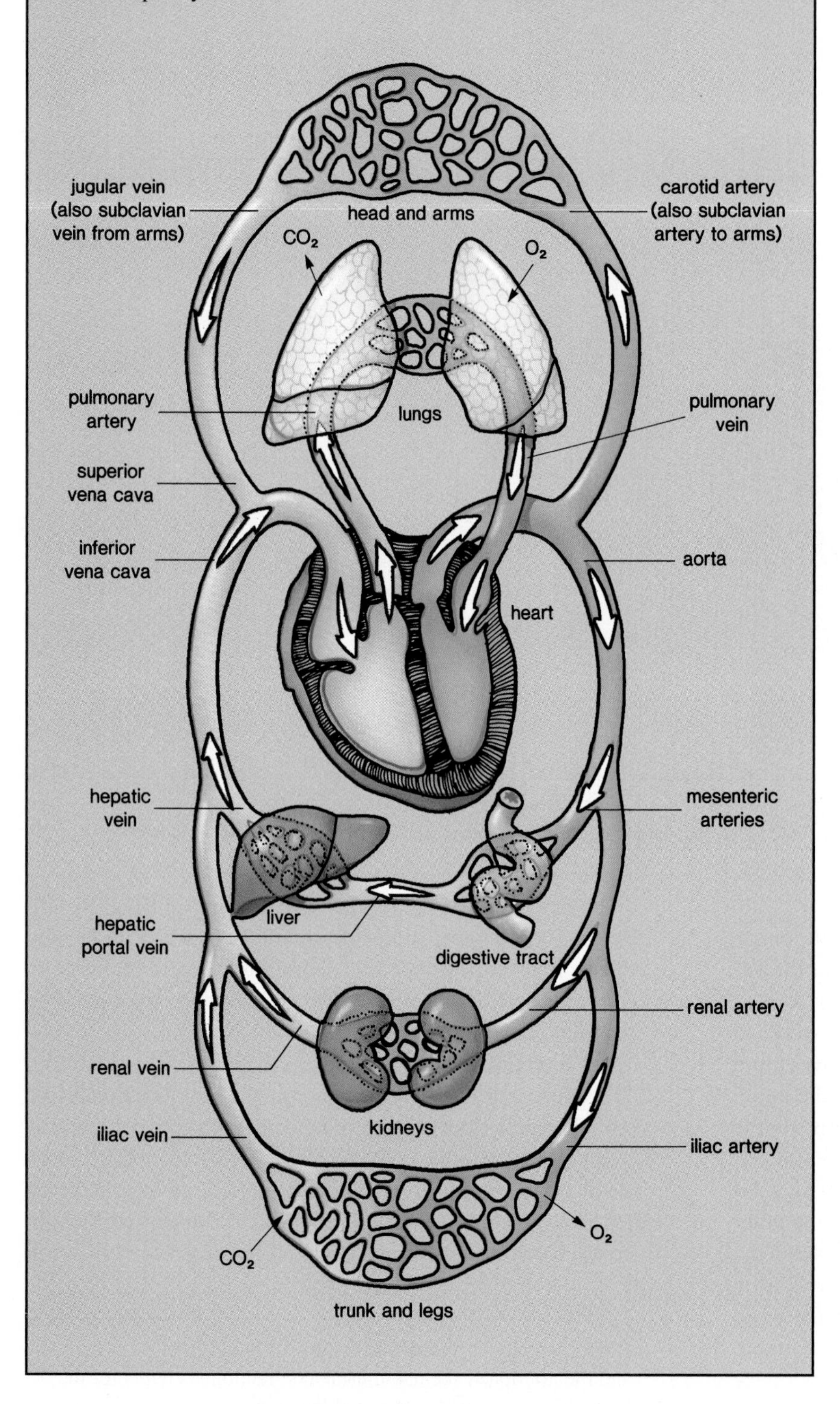

Vascular Pathways: Two Circuits

The cardiovascular system, which is represented in figure 12.9, includes 2 circuits: the **pulmonary circuit,** which circulates blood through the lungs, and the **systemic circuit,** which serves the needs of body tissues.

Pulmonary Circuit: Through the Lungs

The path of blood through the lungs can be traced as follows. Blood from all regions of the body first collects in the right atrium and then passes into the right ventricle, which pumps it into the pulmonary trunk. The pulmonary trunk divides into the *pulmonary arteries,* which branch as they approach the lungs. The arterioles take blood to the pulmonary capillaries, where carbon dioxide and oxygen are exchanged. Blood then enters the pulmonary venules, which lead back through the *pulmonary veins* to the left atrium. Since blood in the pulmonary arteries is deoxygenated but blood in the pulmonary veins is oxygenated, it is not correct to say that all arteries carry oxygenated blood and all veins carry deoxygenated blood. It is just the reverse in the pulmonary circuit.

The pulmonary arteries take deoxygenated blood to the lungs, and the pulmonary veins return oxygenated blood to the heart.

Systemic Circuit: Serving the Body

The systemic circuit includes all of the other arteries and veins shown in figure 12.9. The largest artery in the systemic circuit is the **aorta,** and the largest veins are the *superior* and *inferior venae cavae.* The superior vena cava collects blood from the head, the chest, and the arms, and the inferior vena cava collects blood from the lower body regions. Both enter the right atrium. The aorta and the venae cavae serve as the major pathways for blood in the systemic circuit.

The path of systemic blood to any organ in the body begins in the left ventricle, which pumps blood into the aorta. Branches from the aorta go to the major body regions and organs. For example, the path of blood to the kidneys can be traced as follows:

Left ventricle—aorta—renal artery—renal arterioles, capillaries, venules—renal vein—inferior vena cava—right atrium

To trace the path of blood to any organ in the body, you need only mention the aorta, the proper branch of the aorta, the organ, and the vein returning blood to the vena cava. In most instances, the artery and the vein that serve the same organ are given the same name (fig. 12.10). In the systemic circuit, unlike the pulmonary system, arteries contain oxygenated blood and have a bright red color, but veins contain deoxygenated blood and appear a dark purplish color.

The **coronary arteries** (fig. 12.2*b*), which are a part of the systemic circuit, are extremely important because they serve the heart muscle itself. (The heart is not nourished by blood in its chambers.) The coronary arteries are the first branches off the aorta. They are just above the aortic semilunar valve and they lie on the exterior surface of the heart, where they divide into diverse arterioles. The coronary capillary beds join to form venules. The venules converge to form the cardiac veins, which empty into the right atrium. The coronary arteries have a very small diameter and may become blocked, as discussed on page 214.

The body has a portal system called the *hepatic portal system,* which is associated with the liver. A portal system begins and ends in capillaries; in this instance, the first set of capillaries occurs at the villi of the small intestine and the second occurs in the liver. Blood passes from the capillaries of the villi into venules that join to form the *hepatic portal vein,* a vessel that connects the villi of the intestine with the liver. The *hepatic vein* leaves the liver and enters the inferior vena cava.

While figure 12.9 is helpful in tracing the path of blood, remember that all parts of the body receive both arteries and veins, as illustrated in figure 12.10.

The systemic circuit takes blood from the left ventricle of the heart to the right atrium of the heart. It serves the body proper.

Figure 12.10

Human circulatory system. A more realistic representation of major blood vessels in the body shows that arteries and veins go to all parts of the body. The superior and inferior venae cavae take their names from their relationship to which organ?

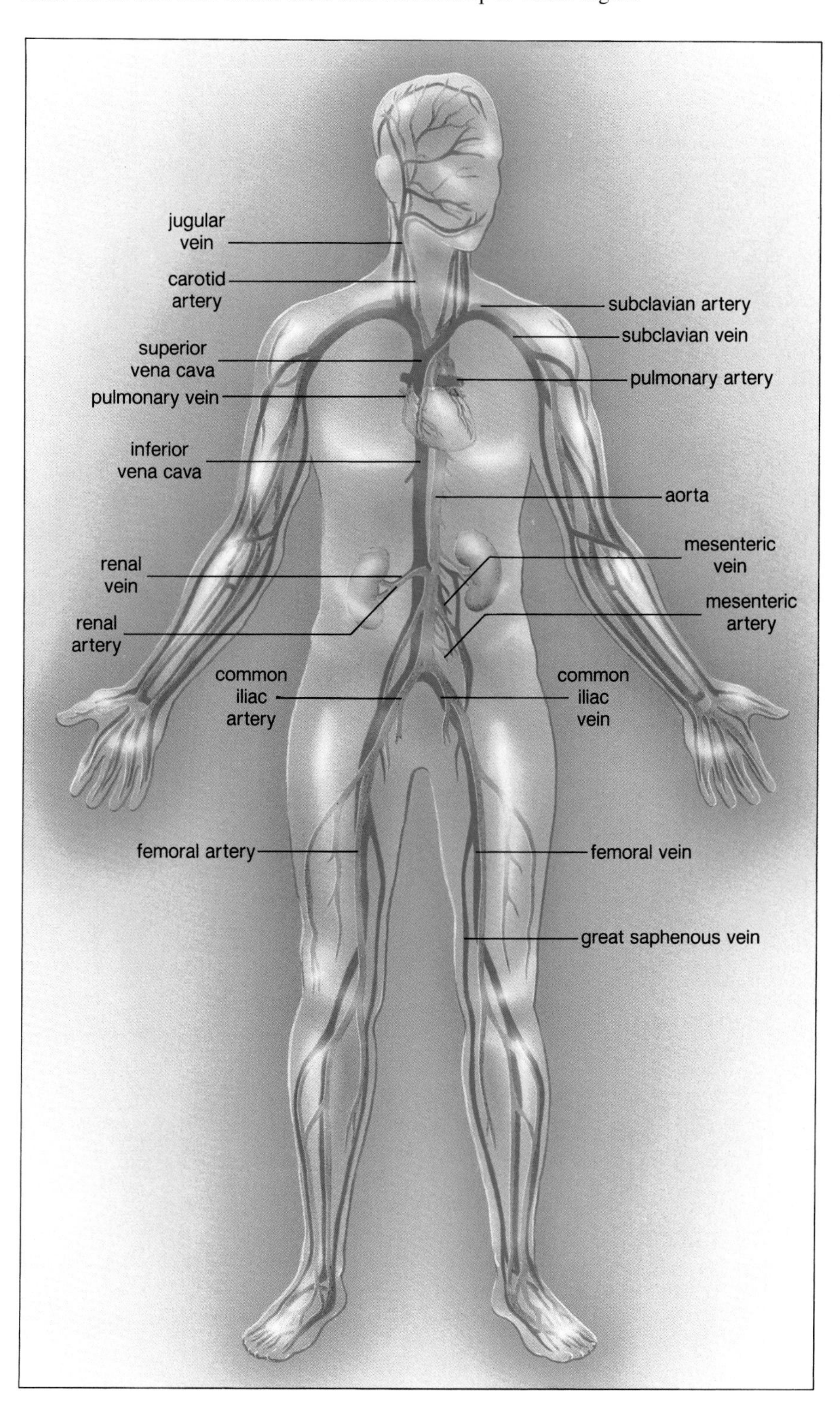

Figure 12.11

Composition of blood. When blood is transferred to a test tube and is prevented from clotting, it forms 2 layers. The transparent, yellow, top layer is plasma, the liquid portion of blood. The formed elements are in the bottom layer. The tables describe these components in detail.

FORMED ELEMENTS	Function and Description	Source
Red blood cells (erythrocytes) 4 million–6 million per mm³ blood	Transport O_2 and help transport CO_2 7–8 μm in diameter Bright-red to dark-purple biconcave disks without nuclei	Bone marrow
White blood cells (leukocytes)	Fight infection	Bone marrow
Granular leukocytes: Basophil 20–50 per mm³ blood	10–12 μm in diameter Spherical cells with lobed nuclei; large, irregularly shaped, deep-blue granules* in cytoplasm	
Eosinophil 100–400 per mm³ blood	10–14 μm in diameter Spherical cells with bilobed nuclei; coarse, deep-red, uniformly-sized granules* in cytoplasm	
Neutrophil 3,000–7,000 per mm³ blood	10–14 μm in diameter Spherical cells with multilobed nuclei; fine, pink granules* in cytoplasm	
Agranular leukocytes: Lymphocyte 1,500–3,000 per mm³ blood	5–17 μm in diameter (average 9–10 μm) Spherical cells with large, round nuclei	
Monocyte 100–700 per mm³ blood	14–24 μm in diameter Large spherical cells with kidney-shaped, round, or lobed nuclei	
Platelets (thrombocytes) 250,000–500,000 per mm³ blood	Initiate clotting 2–4 μm in diameter Disk-shaped cell fragments with no nuclei; purple granules in cytoplasm	Bone marrow

Plasma 55%

Formed Elements 45%

PLASMA	Function	Source
Water (90–92% of plasma)	Maintains blood volume; transports molecules	Absorbed from intestine
Plasma proteins (7–8% of plasma)	Maintain blood osmotic pressure and pH	Liver
Albumin	Maintain blood volume and pressure	
Fibrinogen	Clotting	
Immunoglobulins	Transport; fight infection	
Salts (less than 1% of plasma)	Maintain blood osmotic pressure and pH; aid metabolism	Absorbed from intestinal villi
Gases		
Oxygen	Cellular respiration	Lungs
Carbon dioxide	End product of metabolism	Tissues
Nutrients Fats Glucose Amino acids	Food for cells	Absorbed from intestinal villi
Urea	Nitrogenous waste	Liver
Hormones, vitamins, etc.	Aid metabolism	Varied

*with Wright's stain

Blood

Blood has numerous functions that help maintain homeostasis. Blood transports molecules to and from the capillaries, where exchanges with tissue fluids take place. It helps guard the body against invasion by **microbes** (microscopic infectious agents, such as bacteria and viruses), and it clots, preventing a potentially life-threatening loss of blood.

Plasma and Cells

If blood is transferred from a person's vein to a test tube and is prevented from clotting, it separates into 2 layers (fig. 12.11).

Figure 12.12

Physiology of red blood cells. **a.** Red blood cells move single file through the capillaries. **b.** Each red blood cell is a biconcave disk containing many molecules of hemoglobin, the respiratory pigment. **c.** Hemoglobin contains 4 polypeptide chains, 2 of which are alpha (α) chains and 2 of which are beta (β) chains. The plane in the center of each chain represents an iron-containing heme group. Oxygen combines loosely with iron when hemoglobin is oxygenated.

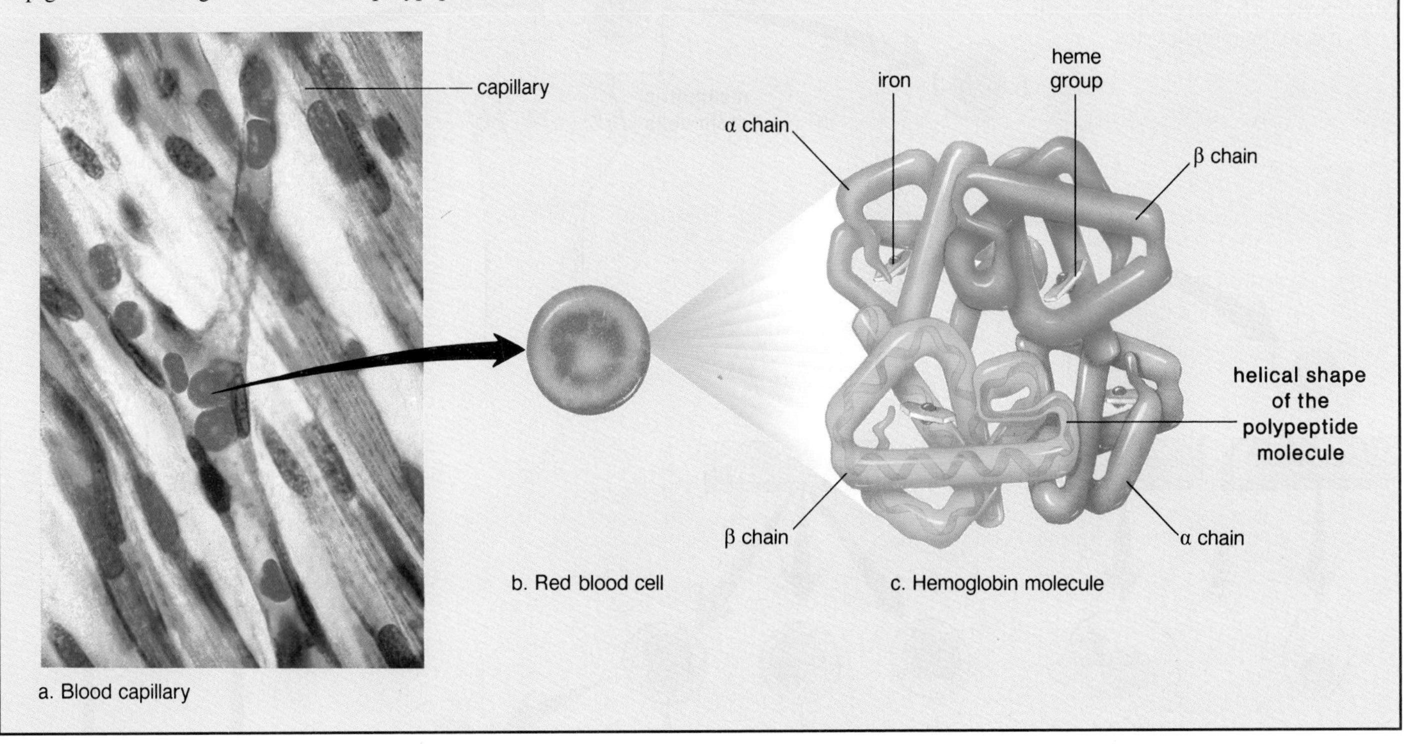

The lower layer consists of red blood cells (erythrocytes), white blood cells (leukocytes), and blood platelets (thrombocytes). Collectively, these are called the **formed elements.** Formed elements make up about 45% of the total volume of whole blood. The upper layer, called **plasma,** is the liquid portion of blood. Plasma, which accounts for about 55% of the total volume of whole blood, contains a variety of inorganic and organic substances dissolved or suspended in water.

Plasma proteins, which make up 7–8% of plasma, assist in transporting large organic molecules in blood. For example, **albumin** transports bilirubin, a breakdown product of hemoglobin. The lipoproteins that transport cholesterol are a type of protein called globulins. Plasma proteins also maintain blood volume because they are too large to pass through a capillary wall. Therefore, capillaries are always areas of lesser water concentration compared to tissue fluid, and water automatically diffuses into capillaries. Certain plasma proteins have specific functions. Fibrinogen is necessary to blood clotting, and immunoglobulins are antibodies, which help fight infection.

Plasma proteins transport molecules in blood and help maintain blood volume. Some have more specific functions.

Red Blood Cells: Carry Oxygen

Red blood cells (erythrocytes) are continuously manufactured in the bone marrow of the skull, the ribs, the vertebrae, and the ends of the long bones. Normally, there are 4 million–6 million red blood cells per mm^3 of whole blood.

Each red blood cell contains about 200 million hemoglobin molecules tightly joined to one another. Since hemoglobin is a red pigment, the cells appear red, and their color also makes blood red. Each hemoglobin molecule (fig. 12.12) contains 4 polypeptide chains, which make up the protein globin. Each chain is associated with heme, a complex iron-containing group. The iron portion of hemoglobin acquires oxygen in the lungs and gives it up in the tissues. Plasma carries only about 0.3 ml of oxygen per 100 ml, but whole blood carries 20 ml of oxygen per 100 ml. This shows that hemoglobin increases the oxygen-carrying capacity of blood more than 60 times. Hemoglobin also assists in the transport of carbon dioxide, as we will discuss in chapter 14.

The number of red blood cells produced increases whenever arterial blood carries a reduced amount of oxygen, as happens when an individual first takes up residence at a high altitude. Under these circumstances, the kidneys (and probably other organs as well) produce a growth factor called erythropoietin, which stimulates erythrocyte stem cells in the red

Figure 12.13

Blood-cell formation in red bone marrow. Multipotent stem cells give rise to specialized stem cells. The myeloid stem cell gives rise to still other cells, which become red blood cells, platelets, and all the white blood cells except lymphocytes. The lymphoid stem cell gives rise to the lymphocytes.

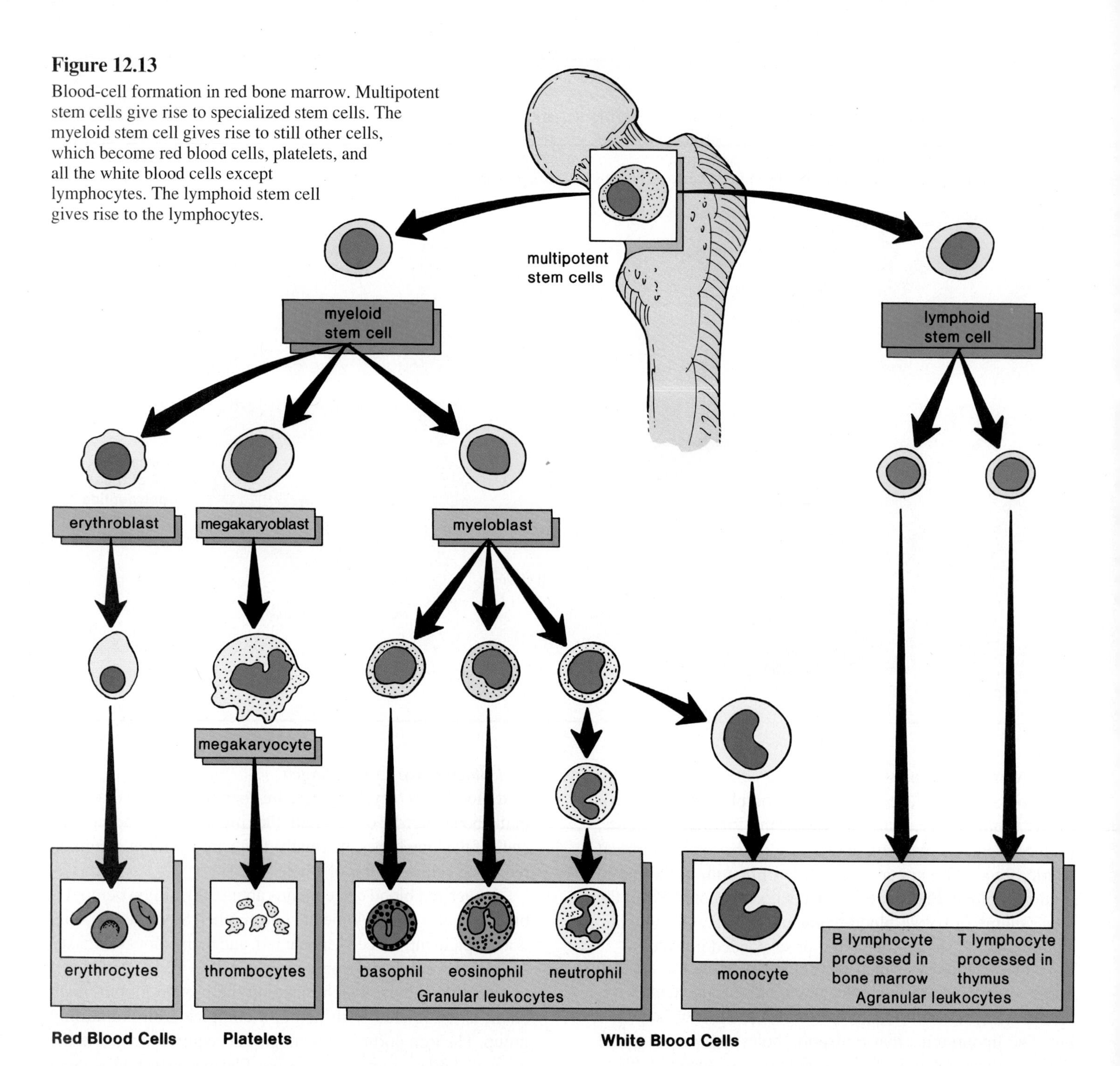

bone marrow (fig. 12.13). A *stem cell* is ever capable of dividing and producing new cells that differentiate. The stem cell called an erythroblast produces red blood cells. Before they are released from the bone marrow into blood, red blood cells lose their nucleus and acquire the respiratory pigment **hemoglobin.**

Possibly because they lack a nucleus, red blood cells only live about 120 days. They are destroyed chiefly in the *liver* and the *spleen,* where they are engulfed by large phagocytic cells. When red blood cells are broken down, the hemoglobin is released. The iron is recovered and returned to the bone marrow for reuse. The heme portion of hemoglobin undergoes chemical degradation and is excreted by the liver in the bile as bile pigments. These bile pigments contribute to the color of feces.

When there is an insufficient number of red blood cells or the red blood cells do not have enough hemoglobin, the individual suffers from **anemia** and has a tired, run-down feeling. In iron-deficiency anemia, the hemoglobin level is low, probably due to a diet that does not contain enough iron. Certain foods that are rich in iron, such as raisins and liver, can be added to the diet to help prevent iron-deficiency anemia.

In another type of anemia called pernicious anemia, the digestive tract is unable to absorb enough vitamin B_{12}. This vitamin is essential to the proper formation of red blood cells;

Figure 12.14

Macrophage (red) engulfing bacteria (green). Monocyte-derived macrophages are the body's scavengers. They engulf microbes and debris in the body's fluids and tissues.

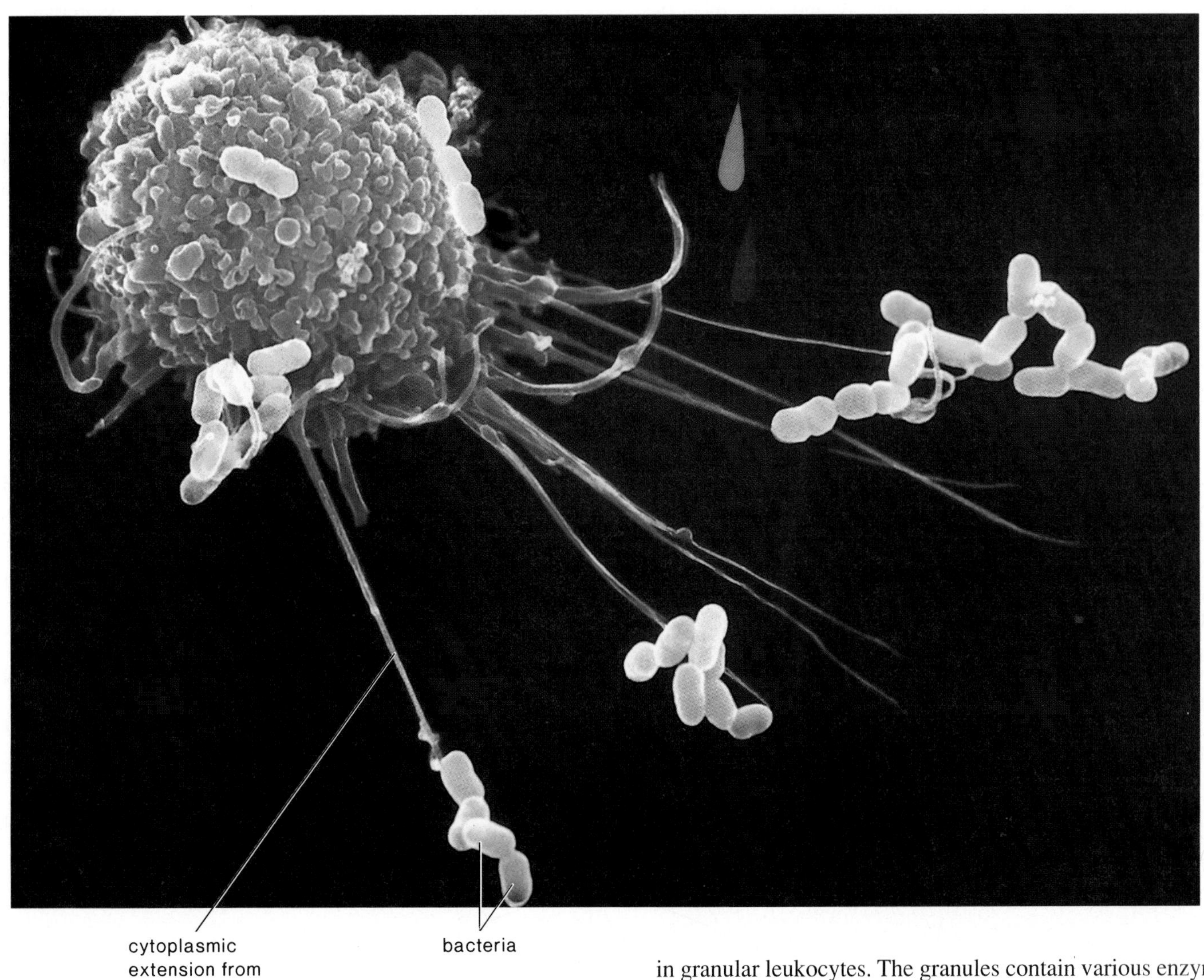

without it, immature red blood cells tend to accumulate in the bone marrow in large quantitites. A special diet and administration of vitamin B_{12} by injection is an effective treatment for pernicious anemia.

White Blood Cells: Fight Infection

White blood cells (leukocytes) differ from red blood cells in that they are usually larger, have a nucleus, lack hemoglobin, and without staining, appear white in color. White blood cells are not as numerous as red blood cells, with only 5,000–11,000 cells per mm^3. White blood cells fight infection, and therefore they will be discussed at greater length in the next chapter, which concerns immunity.

On the basis of structure, it is possible to divide white blood cells into *granular leukocytes* and *agranular leukocytes*. Both types of cells have granules, but they are more prominent in granular leukocytes. The granules contain various enzymes and antibiotic-like proteins, which help white blood cells do their job of defending the body against microbes. Neutrophils are granular leukocytes with a multilobed nucleus joined by nuclear threads; therefore, they are also called *polymorphonuclear.* They are the most abundant of the white blood cells and are able to phagocytize and digest bacteria. The agranular leukocytes (monocytes and lymphocytes) typically have a spherical or kidney-shaped nucleus. After *monocytes,* the largest of the white blood cells, take up residence in the tissues, they differentiate into the even larger macrophages (fig. 12.14). *Macrophages* phagocytize microbes and stimulate other white blood cells to defend the body. The *lymphocytes* are of 2 types, B lymphocytes and T lymphocytes, and each type has a specific role to play in immunity, as discussed in chapter 13.

If the total number of white blood cells increases beyond normal, leukemia or an infection may be present. *Leukemia* is a form of cancer characterized by uncontrolled production of abnormal white blood cells. Sometimes an infection results in the increase of only one type of white blood cell, a condition that is detected with a differential white blood cell count. A blood sample is examined microscopically, and the number of each type of white blood cell is counted up to a total of 100 cells. A person with *infectious mononucleosis,* caused by the Epstein-Barr virus, has an excessive number of lymphocytes of the B type. A person with AIDS, caused by an HIV infection, has an abnormally low number of lymphocytes of the T type.

White blood cells are produced in the bone marrow (fig. 12.13), and various growth factors have been discovered. The best known is GM-CSF (granulocyte-macrophage colony-stimulating factor). Many white blood cells live only a few days and are believed to die combating invading microbes. Others live months or even years.

White blood cells fight infection. They attack microbes that have invaded the body.

Exchanges with Tissue Fluid

At the arterial end of a capillary, blood pressure (40 mm Hg) is higher than the osmotic pressure of blood (25 mm Hg) (fig. 12.15). Osmotic pressure is created by the presence of salts and in particular by the plasma proteins. Because blood pressure is higher than osmotic pressure, water exits a capillary at the arterial end. This is a *filtration* process in that such large substances as red blood cells and plasma proteins remain in the capillaries, but small substances, such as water molecules, leave. *Tissue fluid,* created by this process, consists of all the components of plasma except the proteins.

Along the length of the capillary, molecules follow their concentration gradient as diffusion occurs. Blood always has a greater concentration of oxygen and nutrients than tissue fluid. Therefore, they diffuse out of a capillary. Cells use glucose ($C_6H_{12}O_6$) and oxygen (O_2) in the process of aerobic cellular respiration, and they use amino acids for protein synthesis. Following cellular respiration, the cells give off carbon dioxide (CO_2) and water (H_2O). Tissue fluid always has the greater concentration of waste materials; therefore, they diffuse into blood at the capillary.

Because blood pressure is much reduced (10 mm Hg) at the venous end of the capillary, osmotic pressure (25 mm Hg) tends to pull water back into the capillary. Retrieving water by means of osmotic pressure is not completely effective; there is

Figure 12.15

Diagram of a capillary, illustrating the exchanges that take place and the forces that aid the process. At the arterial end of a capillary, the blood pressure is higher than the osmotic pressure, and therefore water (H_2O), oxygen (O_2), amino acids, and glucose ($C_6H_{12}O_6$) tend to leave the bloodstream.

At the venous end of a capillary, the osmotic pressure is higher than the blood pressure, and therefore water, carbon dioxide, and other waste molecules tend to enter the bloodstream. Notice that the red blood cells and the plasma proteins are too large to exit a capillary.

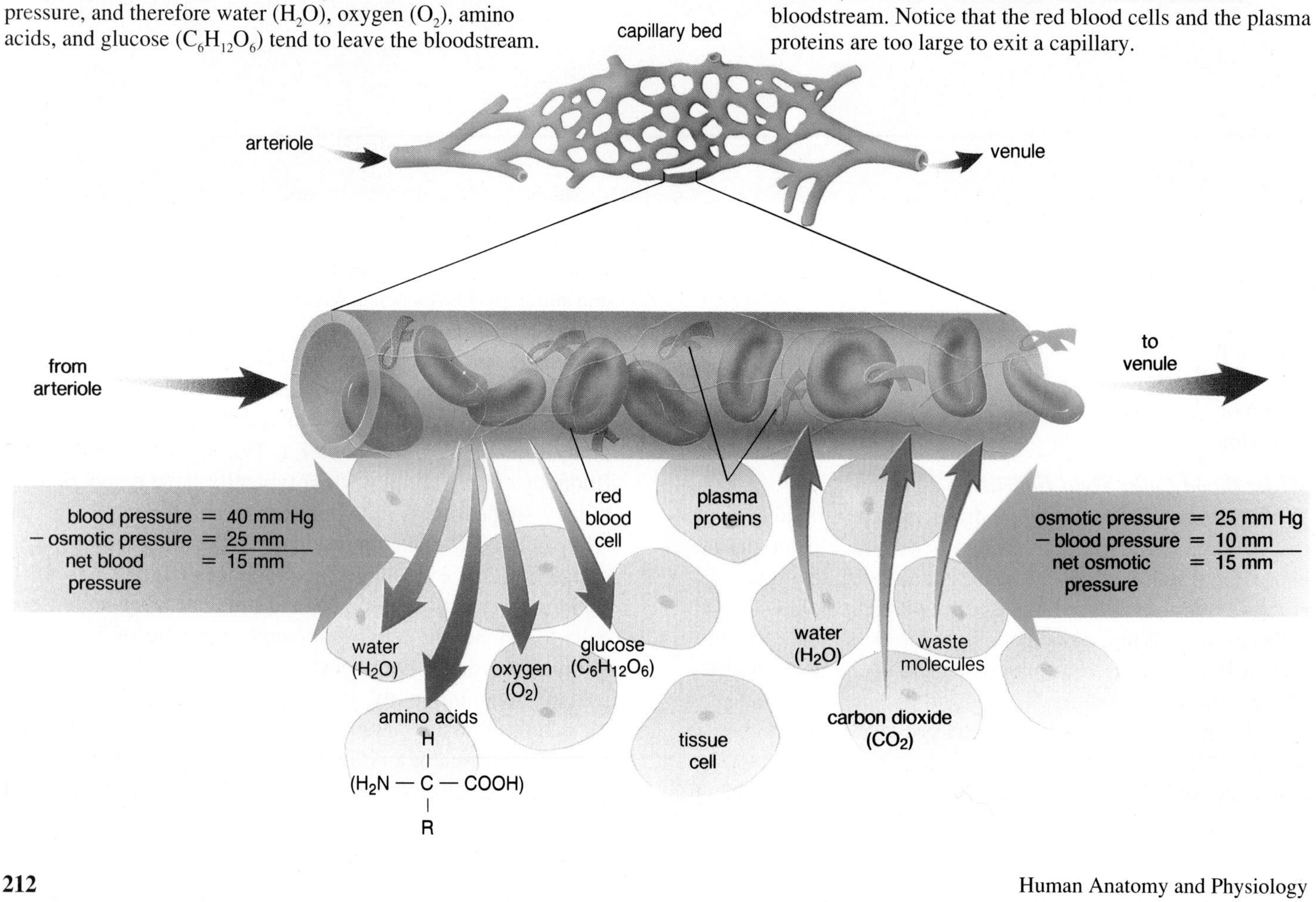

always some fluid that is not picked up at the venous end. This excess tissue fluid enters the lymphatic capillaries (fig. 12.16). Lymph is tissue fluid contained within lymphatic vessels. Lymph is returned to the systemic venous blood when the major lymphatic vessels enter the subclavian veins in the shoulder region (see fig. 13.1).

Water, oxygen, and nutrient molecules (e.g., glucose and amino acids) exit a capillary near the arterial end; water and waste molecules (e.g., carbon dioxide) enter a capillary near the venous end.

Figure 12.16

Lymphatic vessels. Arrows indicate that lymph is formed when lymphatic capillaries take up excess tissue fluid. Lymphatic capillaries lie near blood capillaries.

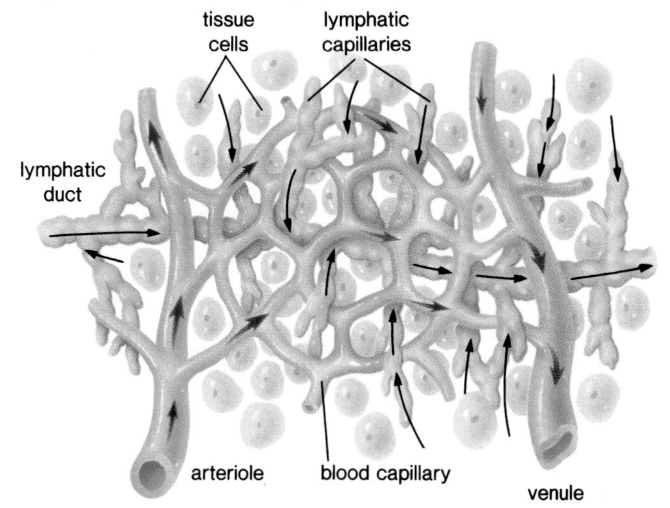

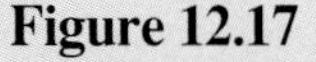

How Blood Clots

When a blood vessel is injured, **clotting,** or coagulation, of blood takes place. At least 12 clotting factors in blood participate in the formation of a blood clot. We will discuss only the roles played by platelets, prothrombin, and fibrinogen. **Platelets** result from fragmentation of certain large cells, called megakaryocytes, in the bone marrow (fig. 12.13). Platelets are produced at a rate of 200 billion a day, and the bloodstream carries more than a trillion. **Fibrinogen** and **prothrombin** are proteins manufactured and deposited in blood by the liver. Vitamin K is necessary to the production of prothrombin, and if by chance this vitamin is missing from the diet, hemorrhagic disorders develop.

If blood is allowed to clot in a test tube, a yellowish fluid develops above the clotted material. This fluid is called **serum,** and it contains all the components of plasma except fibrinogen. (Because we have now used a number of different terms to refer to various body fluids related to blood, table 12.1 reviews these terms for you.)

When a blood vessel in the body is damaged, platelets clump at the site of the puncture and partially seal the leak. They and the injured tissues release a clotting factor called prothrombin activator, which converts prothrombin to thrombin. This reaction requires calcium ions (Ca^{++}). **Thrombin,** in turn, acts as an enzyme that severs 2 short amino acid chains from each fibrinogen molecule. These activated fragments then join end to end, forming long threads of **fibrin.** Fibrin threads wind around the platelet plug in the damaged area of the blood vessel and provide the framework for the clot. Red blood cells are also trapped within the fibrin threads; these cells make a clot appear red (fig. 12.17).

Figure 12.17

Blood clotting. **a.** Prothrombin and fibrinogen are components of blood. Prothrombin activator is an enzyme that speeds up the conversion of prothrombin to thrombin. Thrombin is an enzyme that speeds up the conversion of fibrinogen to fibrin. **b.** Scanning electron micrograph showing a red blood cell (erythrocyte) caught in the fibrin threads of a clot.

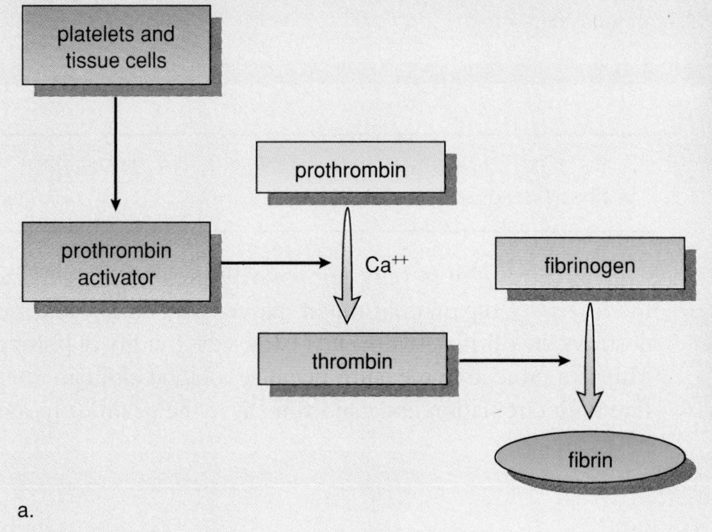

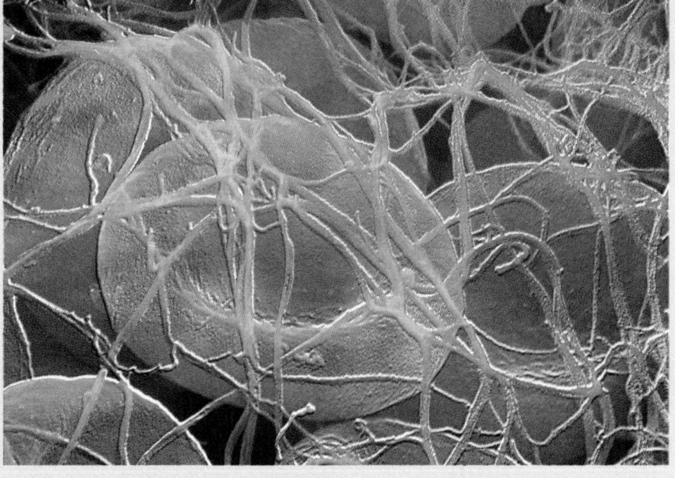

THE CHOLESTEROL CONNECTION

Because sudden cardiac death happens once every 72 sec in the United States, it is good to identify the factors that predispose an individual to cardiovascular disease. The risk factors for cardiovascular disease include the following:

- Male gender
- Family history of heart attack under age 55
- Smoking more than 10 cigarettes a day
- Severe obesity (30% or more overweight)
- Hypertension
- Unfavorable blood HDL and LDL cholesterol levels
- Impaired circulation to the brain or legs
- Diabetes mellitus

Hypertension is well recognized as a major factor in cardiovascular disease, and 2 controllable behaviors contribute to hypertension. Smoking cigarettes, including filtered cigarettes, causes hypertension, as does obesity. It is best to never take up the habit of smoking cigarettes, but most of the detrimental effects are reversed when the habit is broken. Since it is very difficult for obese individuals to lose weight, it is recommended that weight control be a lifelong endeavor.

Investigators have identified several behaviors that may help reduce the possibility of heart attack and stroke. Exercise seems to be critical. Sedentary individuals have a risk of cardiovascular disease about double that of those who are very active. One physician, for example, recommends that his patients walk for one hour, 3 times a week. Stress reduction also is desirable. The same investigator recommends everyday meditation and yoga-like stretching and breathing exercises to reduce stress.

Another behavior that is much in the news of late is the adoption of a diet low in saturated fats and cholesterol (see p. 187). Many believe such a diet protects against the development of cardiovascular disease. Cholesterol is ferried in the blood by 2 types of plasma proteins called LDL (low-density lipoprotein) and HDL (high-density lipoprotein). LDL (called "bad" lipoprotein) takes cholesterol to the tissues from the liver, and HDL (called "good" lipoprotein) transports cholesterol out of the tissues to the liver. When the LDL level in blood is abnormally high or the HDL level is abnormally low, cholesterol accumulates in the cells. When cholesterol-laden cells line the arteries, plaque develops, and this interferes with circulation.

Cholesterol guidelines have been established by the National Heart, Lung, and Blood Institute. According to the institute, everyone should know his or her blood cholesterol level. Individuals with a borderline-high blood cholesterol level (200–239 mg/100 ml) should be further tested if they already have heart disease or if they have 2 known risk factors for cardiovascular disease (see list). Individuals with a high blood cholesterol level (240 mg/100 ml) always should be further tested. Persons with an LDL cholesterol level of over 130 mg/100 ml should be treated if they have other risk factors, and those with an LDL cholesterol of 160 mg/100 ml should be treated even if this is the only risk factor.

Persons with a total-to-HDL cholesterol ratio higher than 4.5 also are considered to be at risk. Individuals with a normal total cholesterol level, but with an unfavorable total-to-HDL cholesterol ratio, have had heart attacks. For example, if a person's total blood cholesterol level is 200 mg/100 ml, but the HDL level is only 25 mg/100 ml, then the total-to-HDL cholesterol ratio is 8.0, and circulatory difficulties most likely will develop.

First and foremost, treatment for unfavorable cholesterol levels consists of adopting a diet low in saturated fat and cholesterol. If the prescribed diet does not lower blood cholesterol sufficiently, then drugs can be prescribed. Some of the drugs act in the intestine to remove cholesterol, and others act in the body to prevent its production. Lovastatin dramatically lowers cholesterol and is easy to administer, but muscle and kidney damage have been reported as side effects.

A diet low in saturated fat and cholesterol may lower the total blood cholesterol level and the LDL level of some individuals, but this diet is not expected to raise the HDL level. Aside from certain drugs that apparently raise HDL level, exercise is sometimes effective.

There is nothing that can be done about some of the cardiovascular risk factors, such as male gender and family history. However, other risk factors likely can be controlled if the individual believes it is worth the effort. It is clear that the 4 great admonitions for a healthy life—heart-healthful diet, regular exercise, proper weight maintenance, and no smoking—all contribute to acceptable blood pressure and blood cholesterol levels.

Table 12.1
Body Fluids

Name	Composition
Blood	Formed elements and plasma
Plasma	Liquid portion of blood
Serum	Plasma minus fibrinogen
Tissue fluid	Plasma minus proteins
Lymph	Tissue fluid within lymphatic vessels

A blood clot consists of platelets and red blood cells entangled in fibrin threads.

A fibrin clot is only temporarily present. As soon as blood vessel repair is initiated, an enzyme called plasmin destroys the fibrin network and restores the fluidity of plasma. This is a protective measure because a blood clot can interfere with circulation and can even cause the death of tissues in the area.

Figure 12.A

Plaque. **a.** Plaque (yellow) in the coronary artery of a heart patient. **b.** Cross section of plaque shows its composition and indicates how it bulges out into the lumen of an artery, obstructing blood flow.

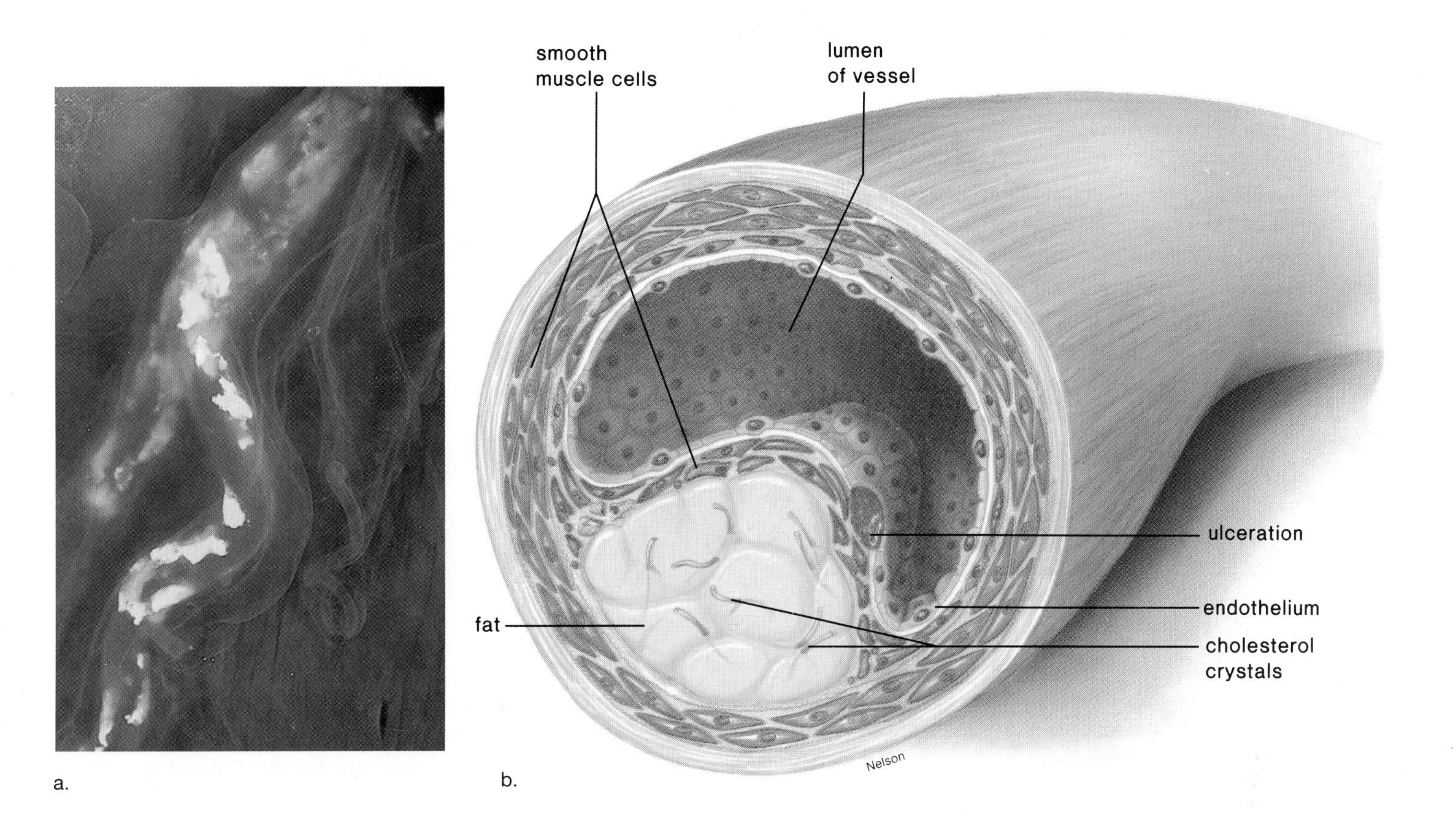

Circulatory Disorders

During the past 30 years, the number of deaths due to cardiovascular disease has declined more than 30%. Even so, more than 50% of all deaths in the United States still are attributable to cardiovascular disease. The number of deaths due to hypertension, stroke, and heart attack is greater than the number due to cancer and accidents combined.

Cardiovascular disease is the number-one killer in the United States.

Hypertension: Silent Killer

It is estimated that about 20% of all Americans suffer from *hypertension,* high blood pressure as indicated by a blood pressure reading. Women of any age are considered to have hypertension if their blood pressure reading is 160/95 or above. For a man under age 45, a reading above 130/90 is hypertensive, and beyond age 45, a reading above 140/95 is considered hypertensive. While both systolic and diastolic pressures are considered important, it is the diastolic pressure that is emphasized when medical treatment is being considered.

Hypertension is sometimes called a silent killer because it may not be detected until a stroke or heart attack occurs. Therefore, it is important to have regular blood pressure checks and to adopt a life-style that protects against the development of hypertension.

Atherosclerosis: Fatty Arteries

Hypertension is also seen in individuals who have *atherosclerosis* (formerly called arteriosclerosis), an accumulation of soft masses of fatty materials, particularly cholesterol, beneath the inner linings of arteries. Such deposits are called *plaque,* and as it develops, plaque tends to protrude into the vessel, interfering with the flow of blood. Atherosclerosis begins in early adulthood and develops progressively through middle age, but symptoms may not appear until an individual is 50 or older. To prevent its onset and development, the American Heart Association and other organizations recommend a diet low in saturated fat and cholesterol, as discussed in the reading on page 214.

Plaque can cause a clot to form on the irregular arterial wall. As long as the clot remains stationary, it is called a *thrombus,* but when and if it dislodges and moves along with blood, it is called an *embolus.* If *thromboembolism* is not treated, complications can arise, as mentioned in the following section.

> Development of atherosclerosis, which is associated with a high blood cholesterol level, can lead to thromboembolism.

Stroke and Heart Attack: Lack of Oxygen

Both strokes and heart attacks are associated with hypertension and atherosclerosis. A *stroke* occurs when a portion of the brain dies due to a lack of oxygen. A stroke, characterized by paralysis or death, often results when a small arteriole bursts or is blocked by an embolus. A person sometimes is forewarned of a stroke by a feeling of numbness in the hands or the face, difficulty in speaking, or temporary blindness in one eye. A *heart attack* occurs when a portion of the heart muscle dies because of a lack of oxygen. Due to atherosclerosis, the coronary artery may be partially blocked. The individual may then suffer from *angina pectoris,* characterized by a radiating pain in the left arm. Nitroglycerin or related drugs dilate blood vessels and help relieve the pain. When a coronary artery is completely blocked, perhaps because of thromboembolism, a heart attack occurs.

> Stroke and heart attack are associated with both hypertension and atherosclerosis. Do 12.2 Critical Thinking, found at the end of the chapter.

Treating Blocked Coronary Arteries

Medical and surgical treatments now are available for blocked coronary arteries.

Figure 12.18

Coronary bypass operation. During this operation, the surgeon grafts segments of another vessel, usually a vein, between the aorta and the coronary vessels, bypassing areas of blockage. Patients who are ill enough to require surgery often receive 2 or 3 bypasses in a single operation.

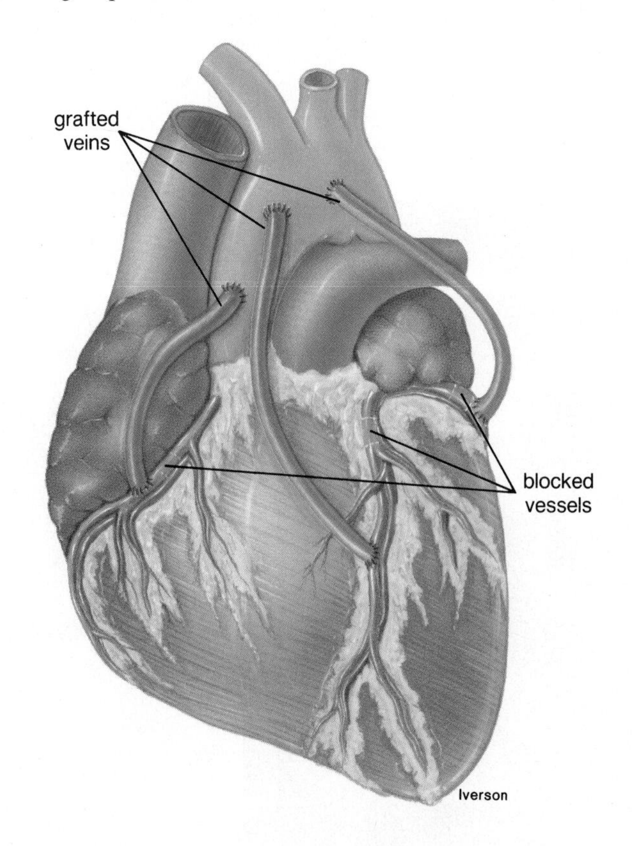

Dissolving Blood Clots. Medical treatment for thromboembolism includes 2 drugs that can be given intravenously to dissolve a clot: streptokinase, normally produced by bacteria, and tPA, which is bioengineered (see p. 490). Both drugs convert plasminogen, a molecule found in blood, into plasmin, an enzyme that dissolves blood clots. In fact, tPA, which stands for *tissue plasminogen activator,* is the body's own way of converting plasminogen to plasmin. Streptokinase and tPA are used particularly when it is known that a clot is present.

If a person has symptoms of angina or a thrombolytic stroke, then an anticoagulant drug, such as aspirin, may be given. Aspirin reduces the stickiness of platelets and therefore lowers the probability that a clot will form. There is evidence that aspirin protects against first heart attacks, but there is no clear support for taking aspirin every day to prevent strokes in symptom-free people. Physicians warn that long-term use of aspirin might have harmful effects, including bleeding in the brain.

Clearing Clogged Arteries. Surgical procedures are available to clear clogged arteries. In *angioplasty,* the cardiologist threads a plastic tube into an artery of an arm or a leg and guides it through a major blood vessel toward the heart. When the tube reaches the region of plaque in a coronary artery (fig. 12.A), a balloon attached to the end of the tube is inflated, forcing the vessel open. The problem with this procedure is the vessel may not remain open, and worse, it may cause clots to form. Various alternatives, including a laser technique, are available.

Each year, thousands of persons have coronary bypass surgery. During this operation, surgeons take a segment of another blood vessel from the patient's body and stitch one end to the aorta and the other end to a coronary artery past the point of obstruction (fig. 12.18).

Once the heart is exposed, some physicians use lasers to open up clogged coronary vessels. Presently, this technique is used in conjunction with coronary bypass operations, but eventually it may be possible to use lasers independently, without opening the thoracic cavity.

Donated and Artificial Hearts. Persons with weakened hearts may eventually suffer from *congestive heart failure,* meaning the heart is no longer able to pump blood adequately. These individuals, depending on their age, are candidates for a donor heart transplant. The difficulties with a donor heart transplant are, first, one of availability and, second, the tendency of the body to reject foreign organs. Sometimes, it is possible to repair the heart instead of replacing it. For example, a back muscle can be wrapped around a heart too weak to pump adequately. An artificial pacemaker causes the muscle to contract regularly and helps pump blood.

On December 2, 1982, Barney Clark became the first person to receive a Jarvik-7, an artificial heart. The Jarvik-7, no longer approved for experimental use in humans, is driven by bursts of air received from a large external machine. Today, investigators are experimenting with a totally internal electric heart. Radio signals transmit power from a portable battery pack through the skin to an orange-sized mechanical heart, consisting of 2 motor-driven plastic sacs.

Veins: Dilated and Inflamed

Varicose veins are abnormal and irregular dilations in superficial (near the surface) veins, particularly those in the lower legs. Varicose veins in the rectum, however, are commonly called piles, or more properly, *hemorrhoids.* Varicose veins develop when the valves of the veins become weak and ineffective due to the backward pressure of blood. The problem can be aggravated when venous blood flow is obstructed by crossing the legs or by sitting in a chair so that its edge presses against the backs of the knees.

Phlebitis, or inflammation of a vein, is a more serious condition, particularly when a deep vein is involved. Blood in the inflamed vessel may clot, in which case thromboembolism occurs. An embolus that originates in a systemic vein eventually may come to rest in a pulmonary arteriole, blocking circulation through the lungs. This condition, termed *pulmonary embolism,* can result in death.

SUMMARY

The movement of blood is dependent on the beat of the heart, an organ with 4 chambers. The right side of the heart pumps blood to the lungs. In the pulmonary circuit, the pulmonary artery takes blood from the right ventricle to the lungs, and the pulmonary veins return it to the left atrium. In the systemic circuit, the aorta takes blood from the left ventricle to the body cells, and the venae cavae return blood to the right atrium. The end result of transport is exchange of materials between tissue fluid and blood at the capillaries.

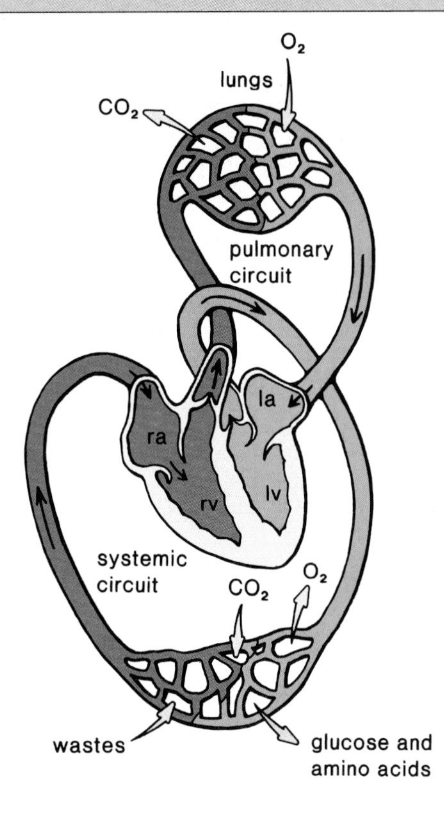

Blood pressure accounts for the flow of blood in the arteries but not the veins. Skeletal muscle contraction pushes blood past a venous valve, which then shuts, preventing backward flow. The velocity of blood is lowest in the capillaries, allowing time for exchange to take place.

Blood is composed of plasma and formed elements. The plasma proteins have various functions (fig. 12.11). The red blood cells contain hemoglobin, which transports oxygen and assists in the transport of carbon dioxide. White blood cells and immunoglobulins fight infection. Platelets and fibrinogen are involved in blood clotting. A blood clot consists of red blood cells trapped in fibrin threads.

Hypertension and atherosclerosis are 2 circulatory disorders that lead to heart attack and to stroke. Medical and surgical procedures are available to control cardiovascular disease, but the best policy is prevention by following a heart-healthful diet, getting regular exercise, maintaining a proper weight, and not smoking cigarettes.

STUDY QUESTIONS

In order to practice **writing across the curriculum,** students should write out the answers to any or all of the study questions. The study questions are sequenced in the same order as the text.

1. What types of blood vessels are there? Discuss their structure and function. (pp. 198–199)
2. Trace the path of blood in the heart, mentioning the vessels attached to and the valves within the heart. (p. 201)
3. Describe the cardiac cycle (using the terms *systole* and *diastole*), and explain the heart sounds. (p. 201)
4. Describe the cardiac conduction system and an ECG. Tell how an ECG is related to the cardiac cycle. (pp. 203–204)
5. In what type of vessel is blood pressure highest? lowest? Velocity is lowest in which type of vessel, and why is it lowest? Why is this beneficial? What factors assist venous return of blood? (pp. 204–205)
6. Trace the path of blood in the pulmonary circuit. Trace the path of blood to and from the kidneys in the systemic circuit. (p. 206)
7. State the major components of blood, and give a function for each. (p. 208)
8. What forces operate to facilitate exchange of molecules across the capillary wall? (pp. 212–213)
9. Name the steps that take place when blood clots. Which substances are present in the blood at all times, and which appear during the clotting process? (pp. 213–214)
10. What is atherosclerosis? (p. 216) Name 2 illnesses associated with hypertension and thromboembolism. (p. 216)
11. Discuss the medical and surgical treatment of cardiovascular disease. (pp. 216–217)

OBJECTIVE QUESTIONS

1. Arteries are blood vessels that take blood _______ the heart.
2. When the left ventricle contracts, blood enters the _______.
3. The pulmonary veins carry blood _______ in oxygen.
4. The _______ node is known as the pacemaker.
5. Blood moves in the arteries due to _______ and in the veins due to _______ _______.
6. Red blood cells _______ and white blood cells _______ _______.
7. When a blood clot occurs, fibrinogen has been converted to _______ threads.
8. The most common granular leukocyte is the _______, a phagocytic white blood cell.
9. Reducing the amount of _______ and _______ in the diet reduces the chances of plaque buildup in the arteries.
10. Label this diagram of the heart.

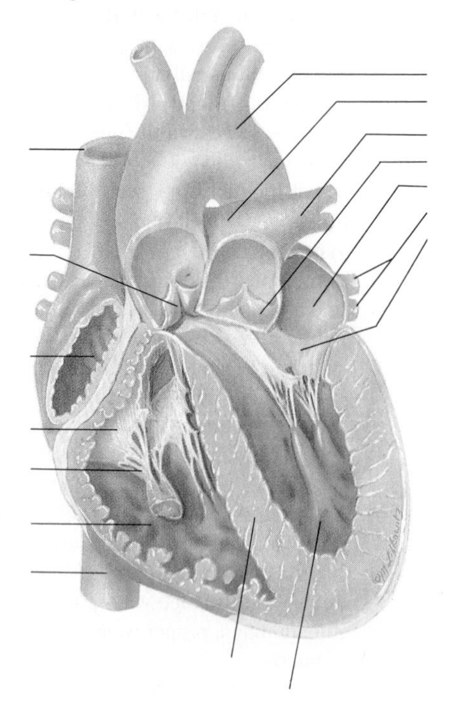

11. Add these labels to this diagram of a capillary: arterial end, plasma proteins, venous end, oxygen, nutrients, carbon dioxide, water.

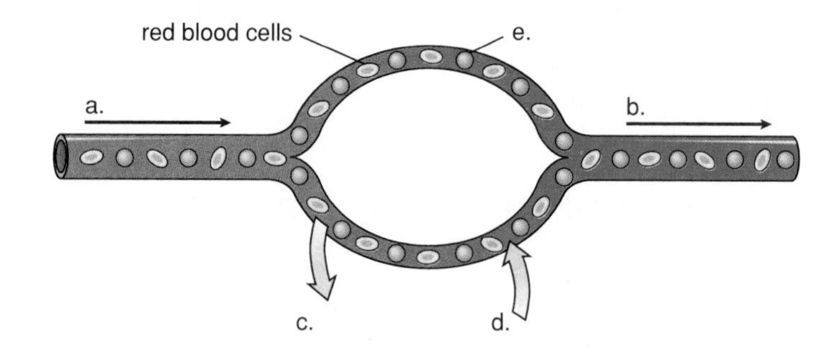

CRITICAL THINKING

In order to practice **writing across the curriculum,** students should write out the answers to any or all of the critical thinking questions. Suggested answers to the critical thinking questions are in appendix E.

12.1

1. Diagram the heart, placing the atria above and the ventricles below. Imagine that the vessels attached to the atria enter from above the heart and the vessels attached to the ventricles exit from below. Where would you place (a) the superior vena cava, (b) the aorta, (c) the pulmonary vein, and (d) the pulmonary artery in your diagram of the heart?
2. During fetal development, a blood vessel connects the pulmonary trunk to the aorta. What effect does this have on circulation to the lung? Under these circumstances, would the aorta carry fully oxygenated blood?

12.2

1. Assume that an embolus is most apt to lodge in a capillary. If an embolus has formed in the iliac vein, in which organ would you expect it to lodge?
2. If an embolus has formed in the carotid artery, in which organ would you expect it to lodge?
3. If an embolus has formed in the hepatic portal vein, in which organ would you expect it to lodge?
4. Explain the term *coronary thrombosis* by telling why you would expect a thrombus and not an embolus formed in the veins to cause a heart attack.

SELECTED KEY TERMS

aorta (a-or´tah) major systemic artery that receives blood from the left ventricle.
arteriole (ar-te´re-ōl) vessel that takes blood from an artery to capillaries.
artery vessel that takes blood away from the heart to arterioles; characteristically possessing thick elastic and muscular walls.
atrium (a´tre-um) chamber; particularly an upper chamber of the heart lying above the ventricles; either the left atrium or the right atrium.
AV (atrioventricular) node a small region of neuromuscular tissue that transmits impulses received from the SA node to the ventricular walls.
capillary (kap´ĭ-lar″ē) microscopic vessel connecting arterioles to venules through the thin walls of which molecules either exit or enter blood.
clotting process of blood coagulation, usually when injury occurs.
coronary artery artery that supplies blood to the wall of the heart.
diastole (di-as´to-le) relaxation of a heart chamber.
fibrinogen plasma protein that is converted into fibrin threads during blood clotting.
formed element a constituent of blood that is either cellular (red blood cells and white blood cells) or at least cellular in origin (platelets).
hemoglobin a red iron-containing pigment in blood that combines with and transports oxygen.
microbe microscopic infectious agent, such as a bacterium or a virus.
plasma the liquid portion of blood, consisting of all components except the formed elements.
platelet component of blood that is necessary to blood clotting.
prothrombin plasma protein that is converted to thrombin during the process of blood clotting.
pulmonary circuit that part of the circulatory system that takes deoxygenated blood to and oxygenated blood away from the gas-exchanging surfaces in the lungs.
SA (sinoatrial) node small region of neuromuscular tissue that initiates the heartbeat; also called the pacemaker.
serum light yellow liquid left after clotting of blood.
systemic circuit that part of the circulatory system that serves body parts other than the gas-exchanging surfaces in the lungs.
systole (sis´to-le) contraction of a heart chamber.
thrombin an enzyme that converts fibrinogen to fibrin threads during blood clotting.
valve membranous extension of a vessel or the heart wall that opens and closes, ensuring one-way flow.
vein vessel that takes blood to the heart from venules; characteristically having nonelastic walls.
vena cava (ve´nah ka´vah) a large systemic vein that returns blood to the right atrium of the heart; either the superior or inferior vena cava.
ventricle cavity in an organ, such as a lower chamber of the heart; either the left ventricle or the right ventricle.
venule vessel that takes blood from capillaries to a vein.

13

THE LYMPHATIC SYSTEM AND IMMUNITY

Chapter Concepts

1.
The lymphatic vessels form a one-way system that transports lymph from the tissues and fat from the lacteals to certain cardiovascular veins. 221, 235

2.
Lymphocytes are especially abundant in the lymphoid organs.
222, 235

3.
The body has various general (nonspecific) ways to protect itself from disease. 223, 235

4.
Immunity is specific and requires 2 types of lymphocytes, B lymphocytes and T lymphocytes. Both of these are produced in the bone marrow. 225, 235

5.
Immunotherapy involves the use of vaccines to achieve long-lasting immunity and the use of antibodies to provide temporary immunity. 229, 235

6.
While immunity preserves our existence, it is also responsible for certain undesirable effects, such as allergies, tissue rejection, and autoimmune diseases. 234-236

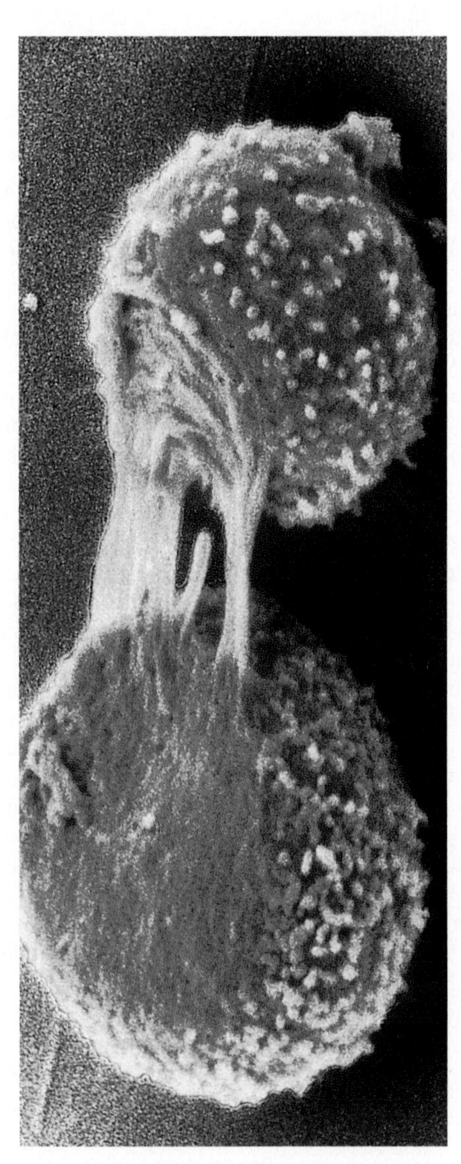

Hybridoma cells, ×10,000

Chapter Outline

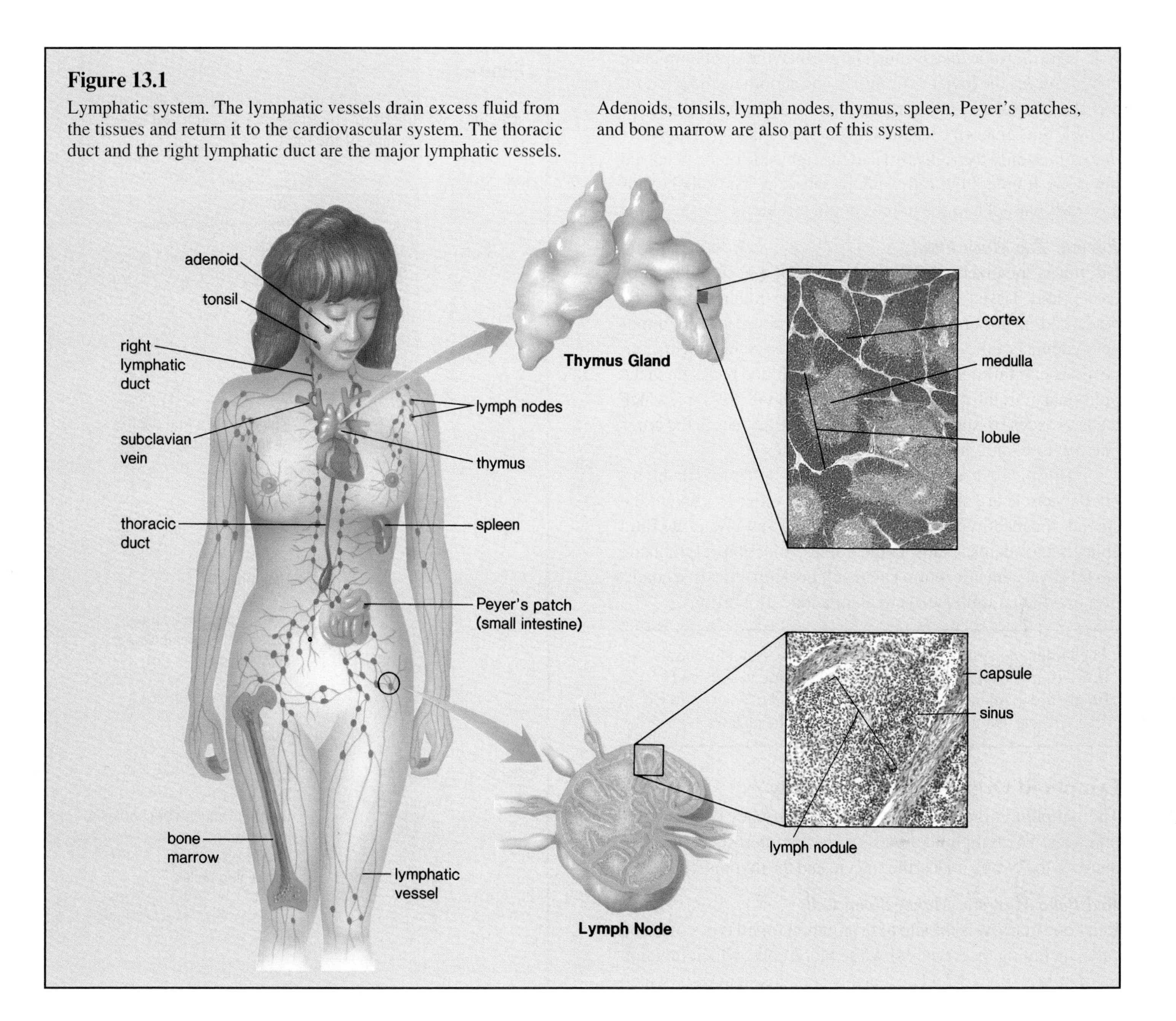

Figure 13.1
Lymphatic system. The lymphatic vessels drain excess fluid from the tissues and return it to the cardiovascular system. The thoracic duct and the right lymphatic duct are the major lymphatic vessels. Adenoids, tonsils, lymph nodes, thymus, spleen, Peyer's patches, and bone marrow are also part of this system.

The Lymphatic System

The **lymphatic system** consists of lymphatic vessels and lymphoid organs (fig. 13.1). This system, which is closely associated with the cardiovascular system, has 3 main functions: (1) lymphatic vessels take up excess tissue fluid and return it to the bloodstream; (2) lymphatic capillaries absorb fat molecules at the lacteals and transport them to the bloodstream (p. 177); and (3) the lymphatic system helps to defend the body against disease.

Lymphatic Vessels: One-Way Transport

Lymphatic vessels are quite extensive; every region of the body is richly supplied with lymphatic capillaries. The construction of the larger lymphatic vessels is similar to that of cardiovascular veins, including the presence of valves. Also, the movement of lymph within lymphatic vessels is dependent upon skeletal muscle contraction. When the muscles contract, lymph is squeezed past a valve, which closes, preventing lymph from flowing backwards.

The lymphatic system is a one-way transport system that begins with lymphatic capillaries. These capillaries take up fluid that has diffused from and has not been reabsorbed by the blood capillaries (see fig. 12.16). Once tissue fluid enters the lymphatic vessels, it is called **lymph.** The lymphatic capillaries join to form lymphatic vessels, which merge in the thoracic cavity before entering one of 2 ducts: the thoracic duct or the right lymphatic duct.

The *thoracic duct* is much larger than the right lymphatic duct. It serves the lower extremities, the abdomen, the left arm, and the left side of the head and the neck. In the thorax, the thoracic duct enters the left subclavian vein. The *right lymphatic duct* serves only the right arm and the right side of the head and the neck. It enters the right subclavian vein. The subclavians are cardiovascular veins in the shoulder regions.

Edema: Too Much Fluid

Edema is localized swelling caused by the accumulation of tissue fluid. Tissue fluid accumulates if too much of it is being made and/or not enough of it is being drained away. Pulmonary edema is a life-threatening condition associated with congestive heart failure. Due to a weak heart, blood backs up in the pulmonary circuit, causing an increase in blood pressure, which leads to excess tissue fluid. The walls of the air sacs in the lungs may rupture, and the patient may suffocate.

During a breast cancer operation, lymph nodes and lymphatic vessels are sometimes removed in order to prevent the spread of cancer. Without these lymphatic vessels, tissue fluid collects and edema results. In the tropics, infection of lymphatic vessels by a parasitic worm can result in elephantiasis, a condition in which a limb swells to elephantine proportions.

> The lymphatic system is a one-way transport system. Lymph flows from a capillary to ever-larger lymphatic vessels and finally to a lymphatic duct, which enters a subclavian vein. Do 13.1 Critical Thinking, found at the end of the chapter.

Lymphoid Organs: Assist Immunity

The lymphoid organs are so-called because they contain lymphocytes. The lymphoid organs of special interest are the bone marrow, the lymph nodes, the spleen, and the thymus (fig. 13.1).

Red Bone Marrow: Makes Blood Cells

Red bone marrow is the site of origination for all types of blood cells, including the 5 types of white blood cells, which function in immunity (fig. 13.2). The marrow contains stem cells, which are ever capable of dividing and producing cells that go on to differentiate into the various types of blood cells. All the white blood cells originate from the same multipotent stem cell but eventually have their own particular stem cell (see fig. 12.13). In the child, most bones have red bone marrow, but in the adult, it is present only in the bones of the skull, the sternum (breastbone), the ribs, the clavicle, the spinal column, and the ends of the long bones (p. 209). The red bone marrow consists of a network of connective tissue fibers called reticular fibers, which are produced by cells called reticular cells. These and the stem cells and their progeny are packed about thin-walled sinuses (open spaces) filled with venous blood. Differentiated blood cells enter the bloodstream at these sinuses.

Lymph Nodes: Cleanse Lymph

At certain points along lymphatic vessels, small (about 1–25 mm), ovoid or round structures called lymph nodes occur. A lymph node has a fibrous connective tissue capsule. Connective tissue also divides a node into nodules. Each nodule contains a sinus, filled with many lymphocytes and macrophages. As lymph passes through the sinuses, the macrophages purify it of microbes and any other debris.

While nodules usually occur within lymph nodes, they can also occur singly or in groups. The *tonsils,* located in back of the mouth on either side of the tongue, and the adenoids, located on the posterior wall above the border of the soft pal-

Figure 13.2

There are 5 types of white cells, which differ according to structure and function. The frequency of each cell type is given as a percentage of the total.

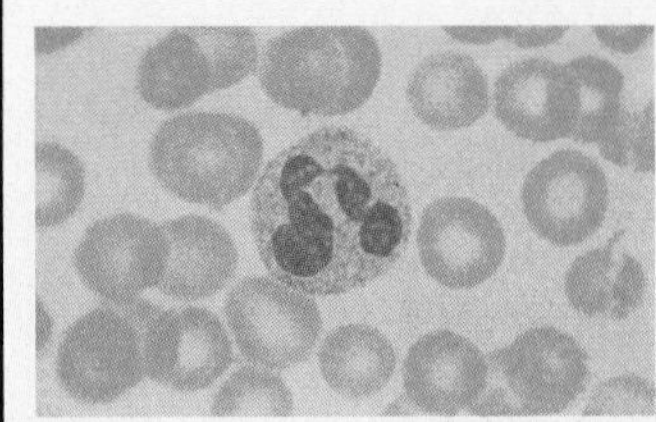

Neutrophil
40 – 70%
Phagocytizes primarily bacteria

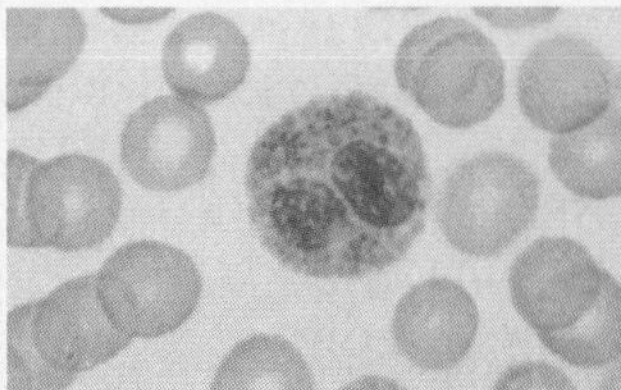

Eosinophil
1 – 4%
Phagocytizes and destroys antigen-antibody complexes

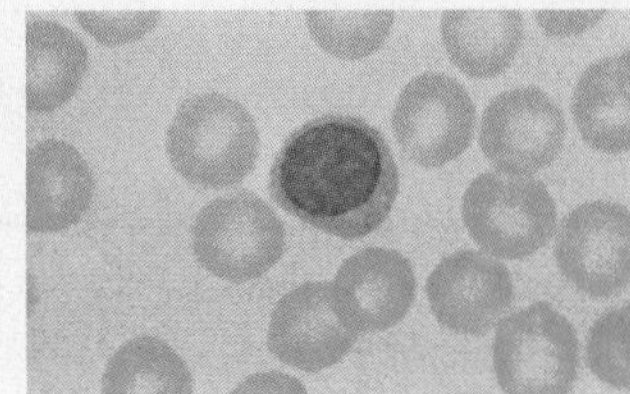

Basophil
0 – 1%
Congregates in tissues; releases histamine when stimulated

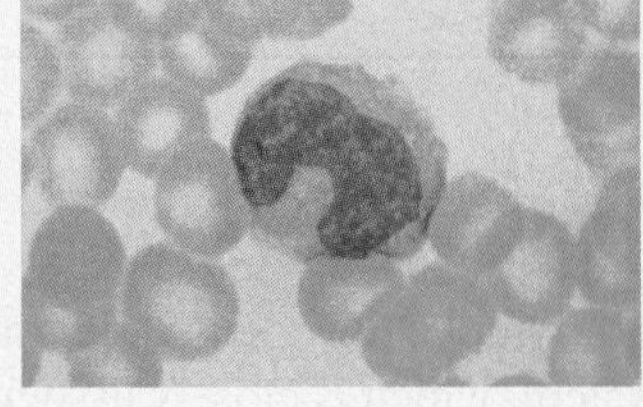

Lymphocyte
20 – 45%
B type produces antibodies in blood and lymph;
T type kills virus-containing cells

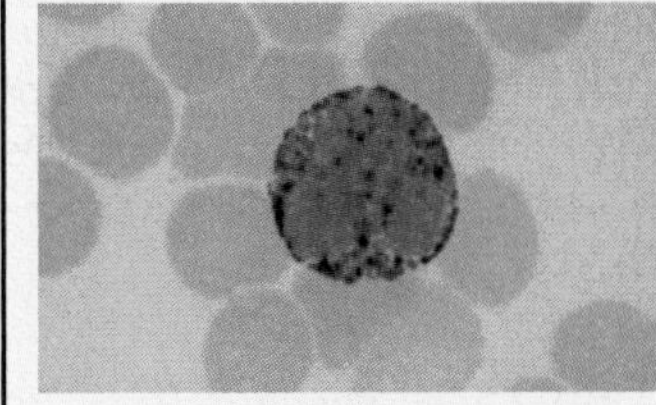

Monocyte
4 – 8%
Becomes macrophage—phagocytizes bacteria and viruses

ate, are composed of partly encapsulated lymph nodules. Also, nodules called *Peyer's patches* are found within the intestinal wall. The lymph nodes occur in groups in certain regions of the body. For example, the inguinal nodes are in the groin and the axillary nodes are in the armpits.

Spleen: Cleanses Blood

The spleen is located in the upper left abdominal cavity just beneath the diaphragm. The construction of the spleen is similar to that of a lymph node. Outer connective tissue divides the organ into lobules, which contain sinuses. In the spleen, however, the sinuses are filled with blood instead of lymph. Especially since the blood vessels of the spleen can expand, this organ serves as a blood reservoir and makes blood available in times of low pressure or when the body needs extra oxygen in the blood.

A spleen nodule contains red pulp and white pulp. Red pulp contains red blood cells, lymphocytes, and macrophages. White pulp contains only lymphocytes and macrophages. Both types of pulp help to purify blood that passes through the spleen. If the spleen ruptures due to injury, it can be removed, and although its functions are duplicated by other organs, the individual is expected to be slightly more susceptible to infections and may have to take antibiotic therapy indefinitely.

Thymus: T Lymphocytes Mature

The thymus (fig. 13.1) is located along the trachea atop the heart and behind the sternum in the upper thoracic cavity. This gland varies in size, but it is larger in children than in adults and may eventually almost disappear. Some believe this contributes to the increased incidence of cancer as we age.

The thymus is divided into lobules by connective tissue. The T lymphocytes mature in these lobules. Those in the interior (medulla) are more mature than those in the exterior (cortex) of a lobule. Mature T lymphocytes have survived an arduous test: If any show the ability to react with "self" cells, they die. If they have the ability to attack a foreign cell, they leave the thymus. The thymus secretes thymosin, a molecule that is believed to be an inducing factor; that is, it causes pre-T lymphocytes to become T lymphocytes. Thymosin may also have other functions in immunity.

> The lymphoid organs have specific functions that assist immunity. White blood cells are made in the bone marrow; lymph is cleansed in lymph nodes; blood is cleansed in the spleen; and T lymphocytes mature in the thymus.

Immunity

Immunity is the ability of the body to defend itself against microbes, foreign cells, and even body cells that have gone astray, such as cancer cells. Immunity includes nonspecific and specific defenses.

Nonspecific Defenses: Three Types

The 3 nonspecific defenses—barriers to entry, the inflammatory reaction, and protective proteins—are useful against all types of infectious agents.

Barriers to Entry: From Skin to Bacteria

Skin and the mucous membrane lining the respiratory and digestive tracts serve as mechanical barriers to entry by microbes. Oil gland secretions contain chemicals that weaken or kill bacteria in skin. The respiratory tract is lined by cells that sweep mucus and trapped particles up into the throat, where they can be swallowed. The stomach has an acidic pH, which inhibits the growth of many types of bacteria. The various bacteria that normally reside in the intestine and other organs, such as the vagina, prevent microbes from taking up residence.

Inflammatory Reaction: Call to Arms

Whenever skin is broken due to a minor injury, a series of events occurs that is known as the **inflammatory reaction** because the site becomes inflamed—it reddens and swells. Figure 13.3 illustrates the participants in the inflammatory reaction. The mast cells, one type of participant, are derived from basophils (fig. 13.2), which take up residence in the tissues.

When an injury occurs, a capillary and several tissue cells are apt to rupture and to release **histamine** and **bradykinin.** Histamine causes the capillary to dilate and become more permeable. Bradykinin potentiates this effect and initiates nerve impulses, resulting in the sensation of pain.The enlarged capillary causes the skin to redden, and its increased permeability allows proteins and fluids to escape, so that swelling results.

Any break in skin allows microbes to enter the body. Neutrophils and monocytes are amoeboid; they can change shape and squeeze through capillary walls to enter tissue fluid. Neutrophils phagocytize bacteria. When phagocytosis occurs, an endocytic vesicle is formed (see fig. 4.10). The engulfed bacterium is destroyed by hydrolytic enzymes when this vesicle combines with a lysosome.

Monocytes differentiate into **macrophages,** large phagocytic cells that are able to devour dozens of microbes and still survive. Some organs, like the liver, kidney, spleen, and brain, have resident macrophages, which routinely act as scavengers, devouring old blood cells, bits of dead tissue, and other debris. Macrophages also are capable of bringing about an explosive increase in the number of leukocytes by liberating a growth factor that passes by way of blood to the red bone marrow, where it stimulates the production and the release of white blood cells, usually neutrophils.

As the infection is being overcome, some neutrophils die. These, along with dead tissue, cells, and bacteria and living white blood cells, form **pus,** a thick, yellowish fluid. Pus indicates that the body is trying to overcome the infection.

> The inflammatory reaction is a "call to arms"—it marshals phagocytic white blood cells to the site of invasion by bacteria.

Figure 13.3

Inflammatory reaction. When a blood vessel is injured, histamine is released by mast cells, dilates blood vessels, and bradykinin stimulates the pain nerve endings. Neutrophils and monocytes congregate at the injured site and squeeze through the capillary wall. The neutrophils begin to phagocytize bacteria. The monocytes become macrophages, large cells that are especially good at phagocytosis and that stimulate other white blood cells to action.

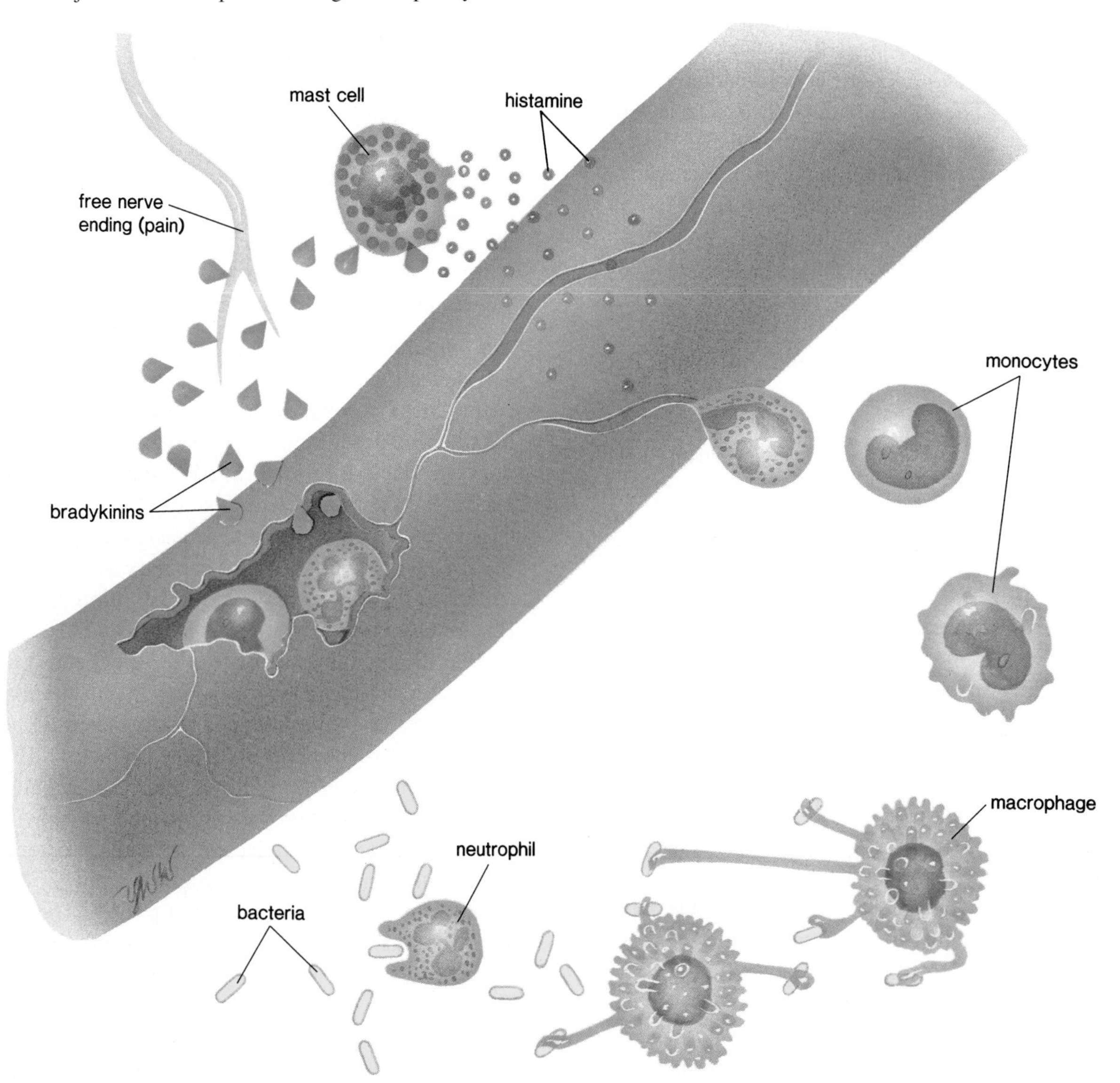

Protective Proteins: Plasma Proteins

The **complement system,** often simply called complement, consists of a number of plasma proteins designated by the letter *C* and a number or letter. Once a complement protein is activated, it activates another protein, and the result is a set series of reactions. A limited amount of activated protein is needed because a cascade occurs; each protein in the series is capable of activating many proteins next in line.

Complement is activated when microbes enter the body. One series of reactions is complete when complement proteins form pores in bacterial cell walls and membranes. These pores allow fluids and salts to enter the bacterial cell to the point that it bursts (fig. 13.4).

Complement also releases chemicals that attract phagocytes to the site and induce inflammation. It "complements" certain immune responses, and this accounts for its name. For example, some complement proteins bind to the surface of microbes already coated with antibodies; this ensures that the microbes will be phagocytized by a neutrophil or a macrophage.

When viruses infect a tissue cell, the affected cell produces and secretes interferon. **Interferon** binds to receptors on the surface of noninfected cells, and this action causes the cells to prepare for possible attack by producing substances that interfere with viral replication. Interferon is specific to the species; therefore, only human interferon can be used in humans. Whereas before it was a problem to collect enough

Figure 13.4

Action of the complement system against a bacterium. When complement proteins in the plasma are activated by an immune reaction, they form pores in bacterial cell walls and membranes, allowing fluids and salts to enter until the cell eventually bursts.

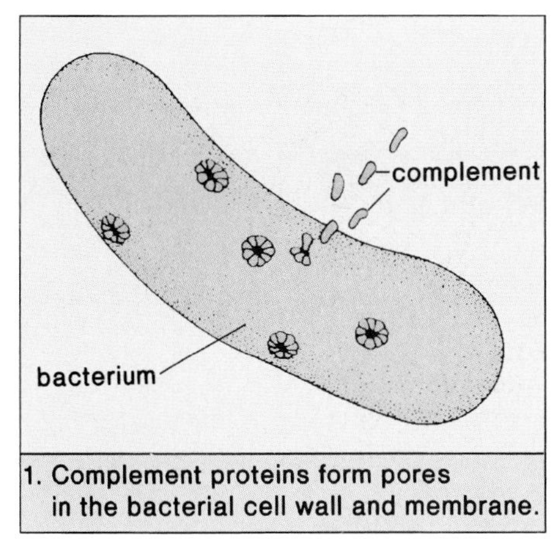

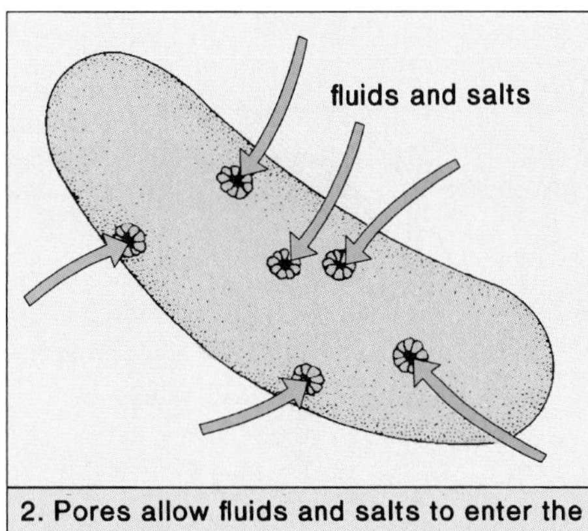

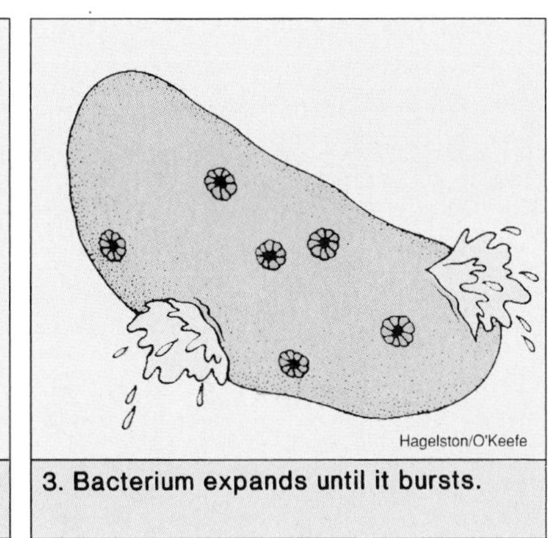

interferon for clinical and research purposes, interferon is now made by recombinant DNA technology.

> Immunity includes these nonspecific defenses: barriers to entry, the inflammatory reaction, and protective proteins.

Specific Defense: Lock and Key

Sometimes, we are threatened by an invasion of microbes that cannot be successfully counteracted by a nonspecific defense mechanism. In such cases, it is necessary to rely on a specific defense against a particular antigen. An **antigen** is a protein (or polysaccharide) molecule that the body recognizes as nonself. Microbes are antigenic, but antigens can also be part of a foreign cell or a cancer cell. Because we do not ordinarily become immune to our own cells, it is said that the immune system is able to tell self from nonself.

Immunity usually lasts for some time. For example, once we recover from the measles, we usually cannot be infected by the measles virus a second time. Immunity is primarily the result of the action of the B lymphocytes and the T lymphocytes. In humans, B (for bone marrow) lymphocytes mature in the bone marrow, and T (for thymus) lymphocytes mature in the thymus gland. **B lymphocytes,** also called B cells, give rise to plasma cells, which produce **antibodies,** proteins that are capable of combining with and neutralizing antigens. These antibodies are secreted into blood, lymph, and mucus. In contrast, **T lymphocytes,** also called T cells, do not produce antibodies. Instead, certain T cells directly attack cells that bear antigens. Other T cells regulate the immune response.

Lymphocytes are capable of recognizing an antigen because they have receptor molecules on their surface. The shape of the receptors on any particular lymphocyte is complementary to a specific antigen. It is often said that the receptor and the antigen fit together like *a lock and a key.* It is estimated that during our lifetime, we encounter a million different antigens, so we need the same number of different lymphocytes for protection against those antigens. It is remarkable that so much diversification occurs during the maturation process that in the end there is a different lymphocyte type for each possible antigen. Despite this great diversity, none of the lymphocytes ordinarily attacks the body's own cells. It is believed that if by chance a lymphocyte arises that is equipped to respond to the body's own proteins, it is normally suppressed and develops no further.

> There are 2 types of lymphocytes. B cells produce and secrete antibodies, which combine with antigens. Certain T cells directly attack antigen-bearing cells, and others regulate the immune response.

B Cells: Make Memory and Plasma Cells

The receptor on a B cell is called a membrane-bound antibody because antibodies are the secreted form of the B-cell receptor. When a B cell encounters a bacterial cell or a toxin bearing an appropriate antigen, it divides (if stimulated by a helper T cell) many times and changes into many **plasma cells,** which secrete antibodies against this antigen (fig. 13.5). All of the plasma cells derived from one parent lymphocyte are called a clone, and a clone produces many copies of the same antibody. The *clonal selection theory* states that the antigen selects which B cell will produce a clone of plasma cells. (Notice that a B cell does not clone until the appropriate antigen is present.)

Once antibody production is sufficient, the antigen disappears from the system, and the development of plasma cells ceases. However, some members of a clone do not participate in antibody production; instead, they remain in the bloodstream

Figure 13.5

Clonal selection theory as it applies to B cells. An antigen activates the appropriate B cell, which undergoes clonal expansion if stimulated by a helper T cell. During the process, many plasma cells, which produce specific antibodies against this antigen, are produced. Memory cells, which retain the ability to secrete these antibodies, are also produced.

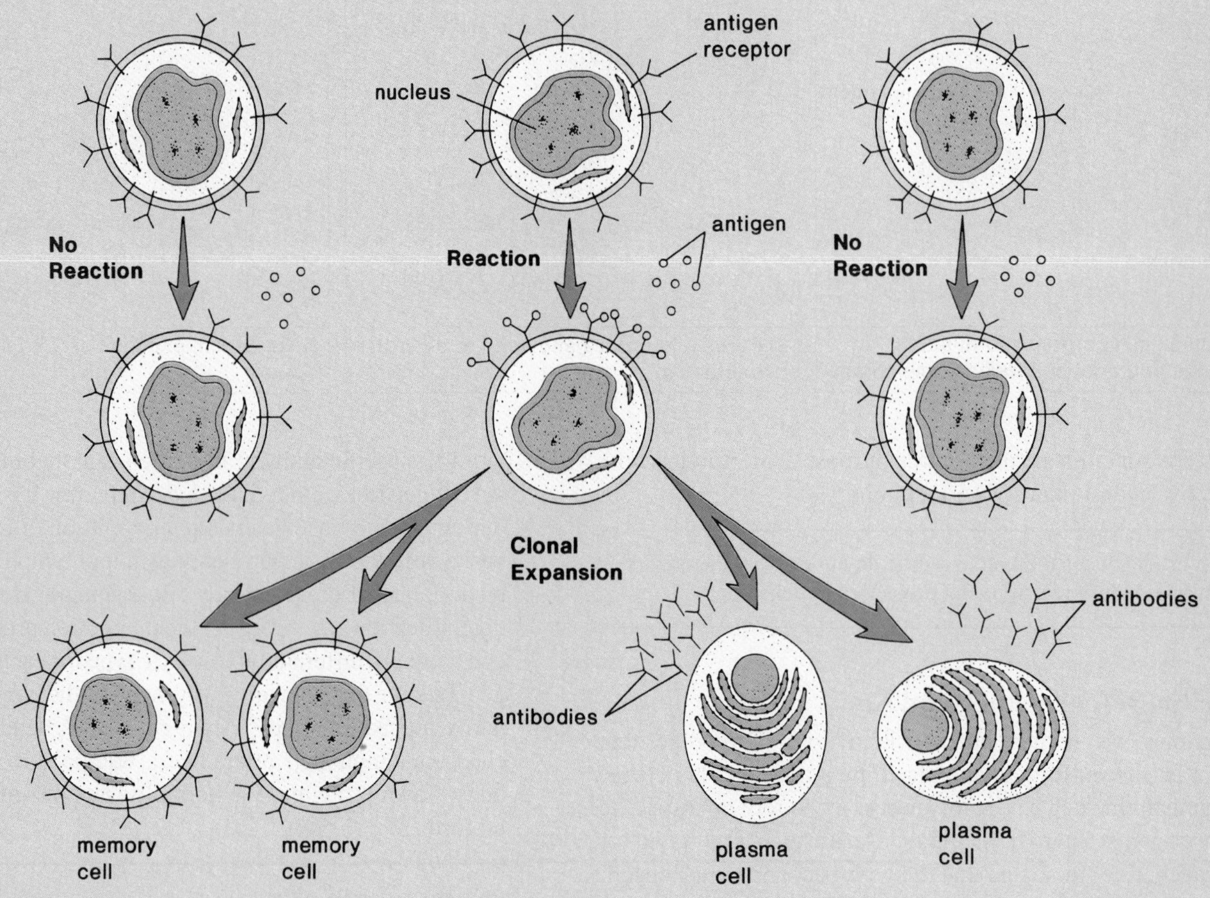

as **memory B cells.** Memory B cells are capable of producing a specific antibody for some time. Also, they will divide and grow into plasma cells if the same antigen invades the system again. Therefore, memory B cells make the individual actively immune.

Defense by B cells is called **antibody-mediated immunity** because the various types of B cells produce antibodies. It is also called humoral immunity because these antibodies are present in the bloodstream. A humor is any fluid occurring normally in the body.

B cells

- Antibody-mediated immunity
- Produced and mature in bone marrow
- Direct recognition of antigen
- Clonal expansion produces antibody-secreting plasma cells as well as memory cells

Antibodies and Antigens Often Form Complexes. The most common type of antibody (IgG) in the blood is a Y-shaped protein molecule having 2 arms. Each arm has a heavy (long) chain and a light (short) chain of amino acids. These chains have *constant regions,* where the sequence of amino acids is set, and *variable regions,* where the sequence of amino acids varies (fig. 13.6). The constant regions are not identical among all the antibodies (table 13.1). Instead, they are the same for different classes of antibodies. The variable regions form an antigen-combining site because their shape is specific to a particular antigen. The antigen combines with the antigen-combining site in a lock-and-key manner.

The antigen-antibody reaction can take several forms, but quite often the antigen-antibody reaction produces complexes of antigens combined with antibodies (fig. 13.6). When viruses and bacterial toxins have combined with specific antibodies, they cannot attach to target cells. Antigen-antibody complex,

Figure 13.6

Antigen-antibody reaction. An IgG antibody has 2 heavy (long) amino acid chains and 2 light (short) amino acid chains arranged to give 2 variable regions, where a particular antigen is capable of combining with the antibody. Quite often the antigen-antibody reaction produces complexes of antigens combined with antibodies.

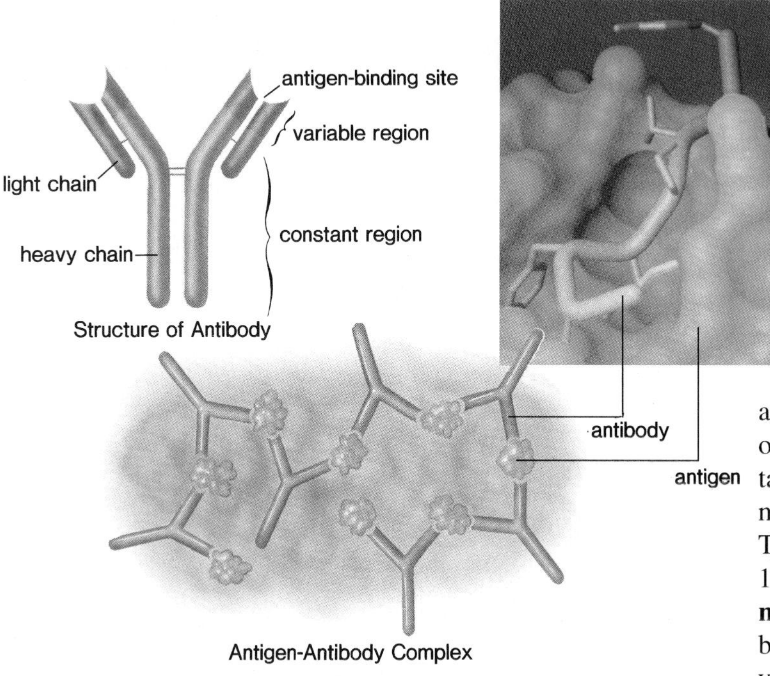

Table 13.1
Antibodies

Class	Presence	Function
IgG	Main antibody type in circulation	Attacks microbes* and bacterial toxins; enhances phagocytosis
IgA	Main antibody type in secretions such as saliva and milk	Attacks microbes and bacterial toxins
IgE	Antibody type found as membrane-bound receptor on basophils in blood and mast cells in tissues	Responsible for allergic reactions
IgM	Antibody type found in circulation, largest antibody	Activates complement; clumps cells
IgD	Antibody type found as a membrane-bound receptor	Functions unknown

* Viruses and bacteria

sometimes called the immune complex, marks the antigen for destruction by other forces. For example, the complex may be engulfed by neutrophils or macrophages or it may activate complement. Complement makes microbes more susceptible to phagocytosis, as discussed.

An antibody combines with its antigen-combining site in a lock-and-key manner. The reaction can produce antigen-antibody complexes, which contain several molecules of antibody and antigen. These complexes are then destroyed by phagocytic cells or by complement.

T Cells: Cytotoxic, Helper, Memory, Suppressor

There are 4 different types of T cells: cytotoxic T cells, helper T cells, memory T cells, and suppressor T cells. All 4 types look alike but can be distinguished by their functions.

Cytotoxic T cells sometimes are called killer T cells. They attack and destroy antigen-bearing cells, such as virus-infected or cancer cells. Cytotoxic T cells have storage vacuoles containing perforin molecules. Perforin molecules perforate a cell membrane, forming a pore that allows water and salts to enter. The cell under attack then swells and eventually bursts (fig. 13.7). It often is said that T cells are responsible for **cell-mediated immunity,** characterized by destruction of antigen-bearing cells. Of all the T cells, only cytotoxic T cells are involved in this type of immunity.

Helper T cells regulate immunity by enhancing the response of other immune cells. When exposed to an antigen, they enlarge and secrete **lymphokines,** messenger proteins that stimulate helper T cells to clone and other immune cells to perform their functions. For example, lymphokines stimulate macrophages to phagocytize and B cells to manufacture antibodies. Because the HIV virus, which causes AIDS, attacks helper T cells, it inactivates the immune response. AIDS is discussed on pages 386–388.

When an activated helper T cell divides, the clone contains **suppressor T cells** and **memory T cells.** Once there is a sufficient number of suppressor T cells, the immune response ceases. Following suppression, however, a population of memory T cells persists, perhaps for life. These cells are able to secrete lymphokines and to stimulate macrophages and B cells whenever the same antigen reenters the body.

Activating Cytotoxic (Killer) and Helper T Cells. T cells have receptors just as B cells do. Unlike B cells, however, cytotoxic T cells and helper T cells are unable to recognize an antigen that simply is present in lymph or blood. Instead, the antigen must be presented to them by an *antigen-presenting cell* (*APC*). When an APC, usually a macrophage, engulfs a microbe, it is enclosed within an endocytic vesicle. Here it is broken down to peptide fragments, which are

Figure 13.7

Cell-mediated immunity. The scanning electron micrographs show cytotoxic T cells attacking and destroying a cancer cell. During the killing process, the vacuoles in a cytotoxic T cell fuse with the cell membrane and release perforin molecules. These molecules combine to form pores in the target cell membrane. Thereafter, fluid and salts enter so that the target cell eventually bursts.

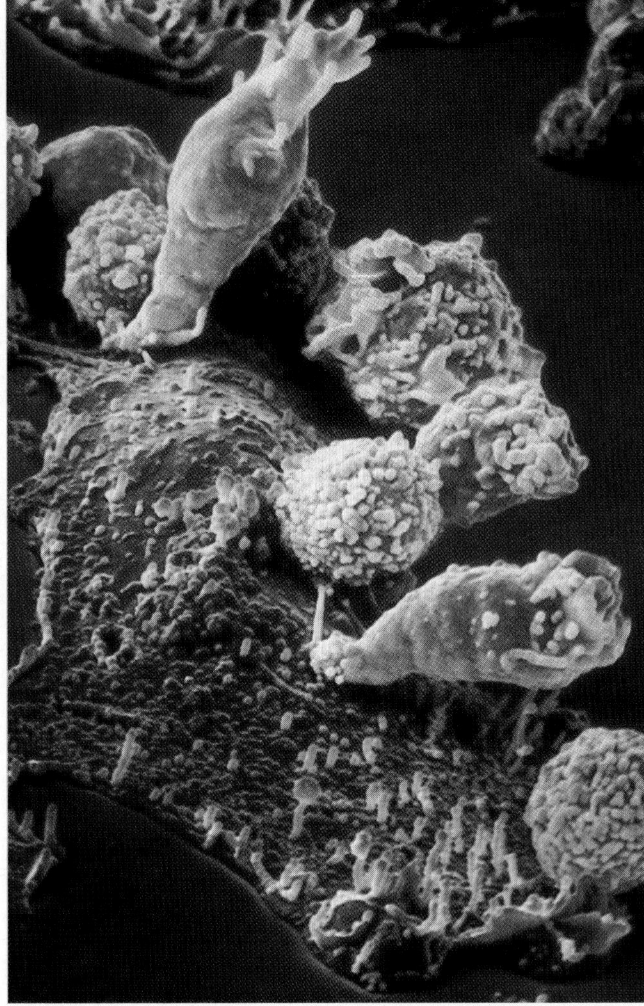

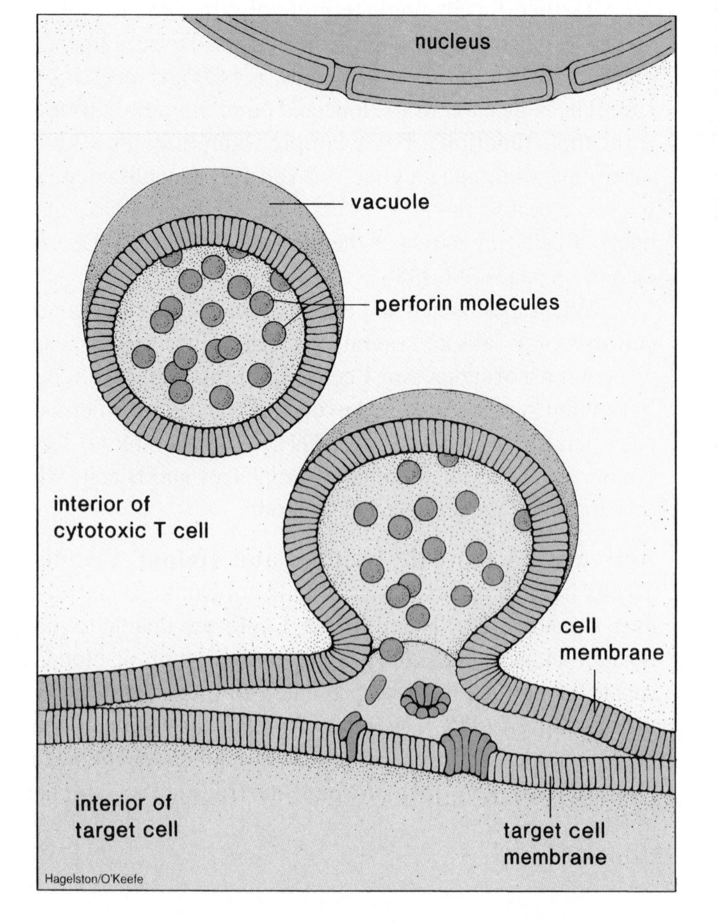

antigenic. The antigenic peptide fragment is linked to an **MHC (major histocompatibility complex) protein,** and together they are displayed at the cell membrane and presented to a T cell.

The importance of MHC proteins in cell membranes was first recognized when it was discovered that they contribute to the specificity of tissues and make it difficult to transplant tissue from one person to another. In other words, the donor and the recipient must be histo-(tissue)compatible (the same or nearly so) for a transplant to be successful without the administration of immunosuppressive drugs.

Figure 13.8 shows a macrophage presenting an antigen to a helper T cell. Once a helper T cell recognizes an antigen, it undergoes clonal expansion, producing suppressor T cells and memory T cells, which can also recognize this same antigen. Once a cytotoxic T cell recognizes an antigen, it attacks and destroys any cell that is infected with the same antigen.

Figure 13.8

T cell activation. T cell receptors are combining with macrophage receptors—a combination of an MHC protein and an antigenic peptide. Thereafter, cytotoxic T cells are activated to destroy cells displaying this antigen.

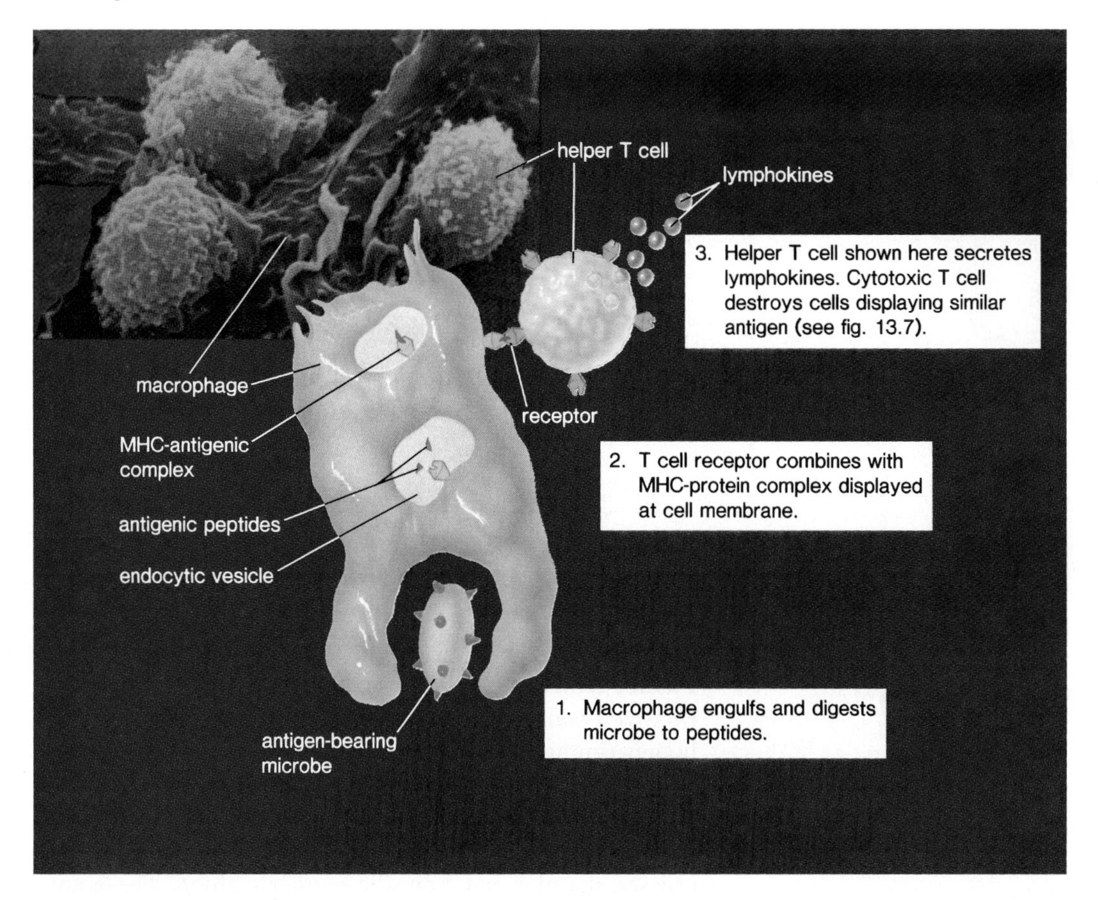

T Cells

- Cell-mediated immunity
- Produced in bone marrow; mature in thymus
- Antigen must be presented by macrophage
- Cytotoxic T cells search and destroy antigen-bearing cells
- Helper T cells secrete lymphokines and stimulate other immune cells

Do 13.2 Critical Thinking, found at the end of the chapter.

Immunotherapy

The immune system can be manipulated to help people avoid or recover from diseases. Some of these techniques have been utilized for a long time, and some are relatively new.

Induced Immunity: Active and Passive

Induced immunity is immunity brought about artificially by medical intervention. There are 2 types of induced immunity: active and passive.

Active Immunity: Long-lived

Active immunity sometimes develops naturally after a person is infected with a microbe. Today, however, active immunity is often induced when a person is well so that future infection is avoided. To prevent infections, people can be artificially immunized against them. One recommended immunization schedule for children is given in figure 13.9. The importance of following one of these recommended schedules is being demonstrated at this time. There have been outbreaks of childhood communicable diseases among college-aged people because they were not immunized properly when they were younger.

Immunization involves the use of **vaccines,** which traditionally are microbes that have been treated so that they are no longer virulent (able to cause disease). New methods of producing vaccines are being developed. For example, it is possible to use the recombinant DNA technique to mass-produce a protein that can be used as a vaccine. This method is being used to prepare a vaccine against hepatitis B.

After a vaccine is given, it is possible to determine the amount of antibody present in a sample of serum—this is called the *antibody titer.* After the first exposure to a vaccine, a primary response occurs. For a period of several days, no antibodies are present; then, there is a slow rise in the titer, followed by a gradual decline (fig. 13.10). After a second exposure, a secondary response may occur. If so, the titer rises rapidly to a level much greater than before. The second exposure in that case often is called the "*booster*" because it boosts the antibody titer to a high level. The antibody titer now may be high enough to prevent disease symptoms even if the individual is exposed to the disease. If so, the individual is now immune to that particular disease. A good secondary response is related to the number of plasma and memory cells in serum. Upon the second exposure, these cells are already present, and antibodies can be produced rapidly. Therefore, active immunity is long-lived.

Active (long-lasting) immunity can be induced by the use of vaccines when a person is well and in no immediate danger of contracting an infectious disease.

Figure 13.9

Suggested immunization schedule for infants and young children. Children who are not immunized are subject to childhood diseases that can cause serious health consequences.

Source: U.S. Department of Health and Human Services, Parent's Guide to Childhood Immunization, revised May 1991, p. 26.

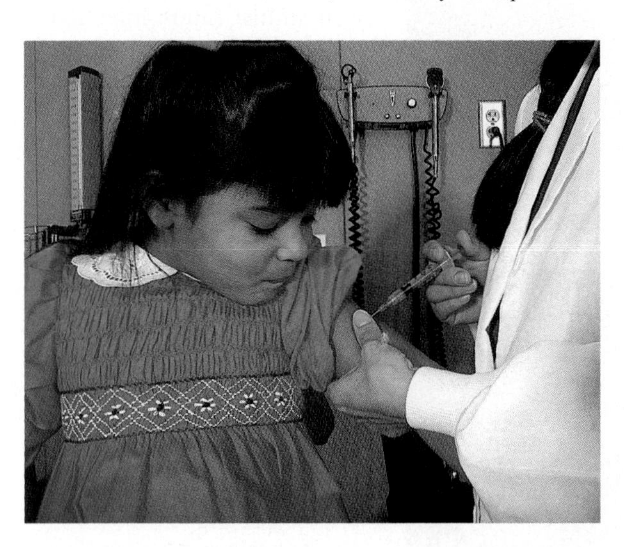

Vaccine	Months of Age				Years of Age
DTP (diphtheria, tetanus, whooping cough)	2	4	6	15*	4-6
OPV (oral polio vaccine)	2	4		15*	4-6
MMR (measles, mumps, rubella)				15**	4-6 #
Hib (Haemophilus influenza, type b)	2	4	6†	15††	
Td (tetanus and less diphtheria than DTP)					14-16

*Many experts recommend this dose of vaccine at 18 months.

**In some areas, this dose of MMR vaccine may be given at 12 months.

†This dose may not be required, depending on which Hib vaccine is used.

††This dose may be given at 12 months, depending on which Hib vaccine is used.♣

#Some experts recommend that this dose of MMR vaccine be given at entry to middle or junior high school.

♣And every 10 years thereafter.

Figure 13.10

Development of active immunity due to immunization. The primary response, after the first exposure to a vaccine, is minimal, but the secondary response, which may occur after the second exposure, shows a dramatic rise in the amount of antibody present in serum.

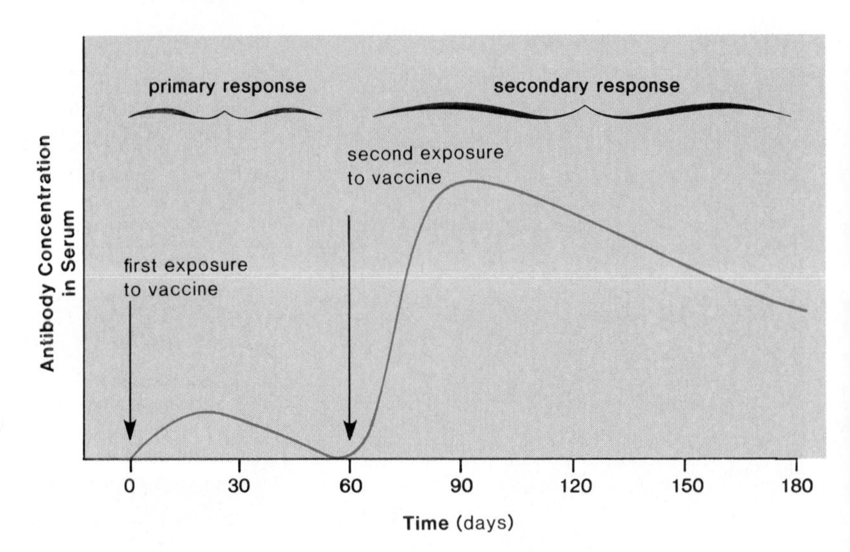

Passive Immunity: Short-lived

Passive immunity occurs when an individual is given antibodies (immunoglobulins) to combat a disease. Since these antibodies are not produced by the individual's B cells, passive immunity is short-lived. For example, newborn infants are passively immune to disease because antibodies have crossed the placenta from the mother's blood. These antibodies soon disappear, however, so that within a few months, infants become more susceptible to infections. Breast feeding (fig. 13.11) prolongs the passive immunity an infant receives from the mother because antibodies are present in the mother's milk.

Even though passive immunity does not last, it sometimes is used to prevent illness in a patient who has been unexpectedly exposed to an infectious disease. Usually, the person receives a gamma globulin injection (serum that contains antibodies), perhaps taken from donors who have recovered from the illness. In the past, horses were immunized and serum was taken from them to provide the needed antibodies against such diseases as diphtheria, botulism, and tetanus. Occasionally, a patient who received these antibodies became ill because the serum contained proteins that the individual's immune system recognized as foreign. This was called serum sickness.

> Passive immunity is needed when an individual is in immediate danger of succumbing to an infectious disease. Passive immunity is short-lived because the antibodies are administered to and not made by the individual.

Figure 13.11

Example of passive immunity. Breast feeding is believed to prolong the passive immunity an infant receives from the mother because antibodies are present in the mother's milk.

Lymphokines: Boost White Blood Cells

Lymphokines, the messenger proteins used by lymphocytes, are being investigated as possible adjunct therapy for cancer and AIDS because they stimulate white blood cell formation and/or function. Both interferon and various other types of lymphokines called *interleukins* have been used as immunotherapeutical drugs, particularly to potentiate the ability of the individual's own T cells (and possibly B cells) to fight cancer.

Interferon, discussed previously on page 224, is a substance produced by leukocytes, fibroblasts, and probably most cells in response to a viral infection. When it is produced by T cells, interferon is called a lymphokine. Interferon still is being investigated as a possible cancer drug, but so far it has proven to be effective only in certain patients, and the exact reasons as yet cannot be discerned.

When and if cancer cells carry an altered protein on their cell surface, they should be attacked and destroyed by cytotoxic T cells. Whenever cancer does develop, it is possible that the cytotoxic T cells have not been activated. In that case, lymphokines might awaken the immune system and lead to the destruction of the cancer. In one technique being investigated, researchers first withdraw T cells from the patient and activate the cells by culturing them in the presence of an interleukin. The cells then are reinjected into the patient, who is given doses of interleukin to maintain the killer activity of the T cells.

Those who are actively engaged in interleukin research believe that interleukins soon will be used as adjuncts for vaccines, for the treatment of chronic infectious diseases, and perhaps for the treatment of cancer. Interleukin antagonists also may prove helpful in preventing skin and organ rejection, autoimmune diseases, and allergies.

The interleukins and other lymphokines show some promise of potentiating the individual's own immune system.

Monoclonal Antibodies: Same Specificity

As previously discussed, every plasma cell derived from the same B cell secretes antibodies against a specific antigen. These are **monoclonal antibodies** because all of them are the same type (*mono*) and because they are produced by plasma cells derived from the same B cell (*clone*).

One method of producing monoclonal antibodies in vitro (in laboratory glassware) is depicted in figure 13.12. B lymphocytes are removed from the body (today, usually a mouse) and are exposed to a particular antigen. The activated B lymphocytes are fused with myeloma cells (malignant plasma cells that live and divide indefinitely). The fused cells are called hybridomas—*hybrid* because they result from the fusion of 2 different cells, and *oma* because one of the cells is a cancer cell.

At present, monoclonal antibodies are being used for quick and certain diagnosis of various conditions. For example, a particular hormone is present in the urine of a pregnant woman. A monoclonal antibody can be used to detect the hormone and so indicate that the woman is pregnant. Monoclonal antibodies also are used to identify infections. They are so accurate they can even sort out the different types of T cells in a blood sample. And because they can distinguish between cancer cells and normal tissue cells, they are used to carry radioactive isotopes or toxic drugs only to tumors so these can be selectively destroyed.

Monoclonal antibodies are considered to be a biotechnology product because the production process makes use of a living system to mass-produce the product.

Monoclonal antibodies are produced in pure batches—they all react to just one type of molecule (antigen); therefore, they can distinguish one cell, or even one molecule, from another.

Figure 13.12
One possible method of producing human monoclonal antibodies.

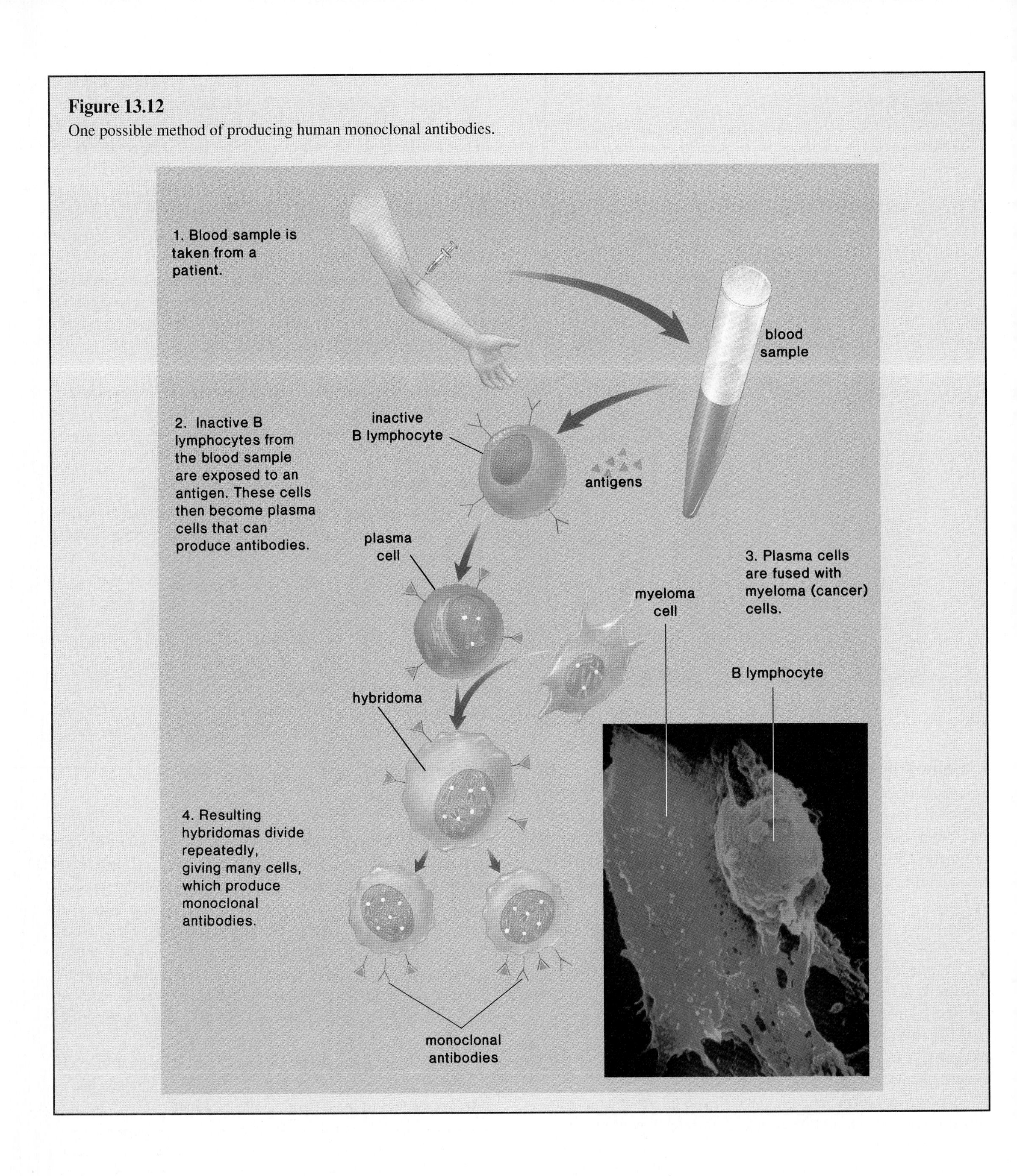

Table 13.2
The ABO System

Blood Type	Antigen or Red Blood Cells	Antibody in Plasma	% U.S. African-American	% U.S. Caucasian	% U.S. Asian	% North American Indians	% Americans of Chinese Descent
A	A	Anti-B	27	41	28	8	25
B	B	Anti-A	20	9	27	1	35
AB	A,B	None	4	3	5	0	10
O	None	Anti-A and anti-B	49	47	40	92	30

Figure 13.13
Blood typing. The standard test to determine ABO and Rh blood type consists of putting a drop of anti-A antibodies, anti-B antibodies, and anti-Rh antibodies on a slide. To each of these, a drop of the person's blood is added. **a.** If agglutination occurs, as seen in the photo on the right, the person has this antigen on red blood cells. **b.** Several possible results.

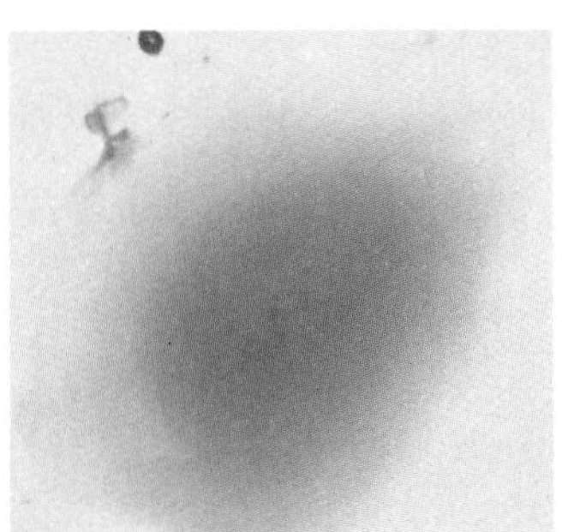

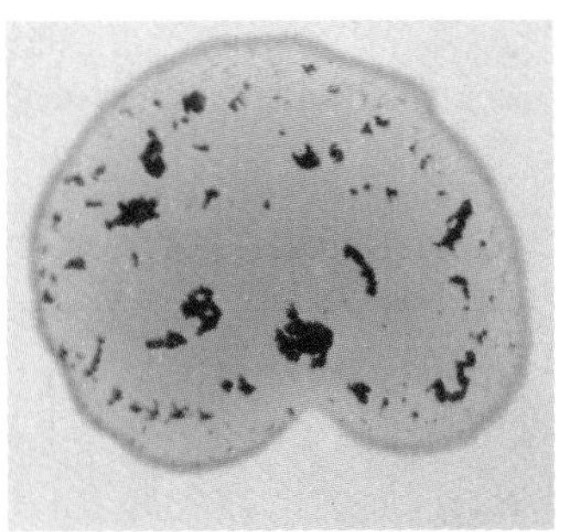

a.

anti-A	anti-B	anti-Rh	type blood
			O+
			A−
			B+
			AB−

b.

Immunological Side Effects and Illnesses

The immune system protects us from disease because it can tell self from nonself. Sometimes, however, the immune system is underprotective, as when an individual develops cancer, or is overprotective, as when an individual cannot receive certain types of blood.

Finding Compatible Blood

When blood transfusions were first attempted, illness and even death sometimes resulted. Eventually, it was discovered that only certain types of blood are compatible because red blood cell membranes carry proteins that are antigens to blood recipients. The ABO system of typing blood is based on this principle.

ABO System: Two Antigens

Blood typing in the ABO system is based on 2 antigens known as antigen A and antigen B. There are 4 blood types: O, A, B, and AB. Type O has neither the A antigen nor the B antigen on red blood cells; the other types of blood have antigen A, B, or both A and B, respectively (table 13.2).

Within plasma, there are antibodies to the antigens *not* present on the person's red blood cells. This is reasonable, because if the same antigen and antibody are present in blood, **agglutination,** or clumping of red blood cells, occurs. Agglutination causes blood to stop circulating and red blood cells to burst.

Figure 13.13 shows a way to use the antibodies derived from plasma to determine the blood type. If agglutination occurs after a sample of blood is exposed to a particular antibody, the person has that type of blood.

Figure 13.14
Diagram describing the development of hemolytic disease of the newborn (HDN).

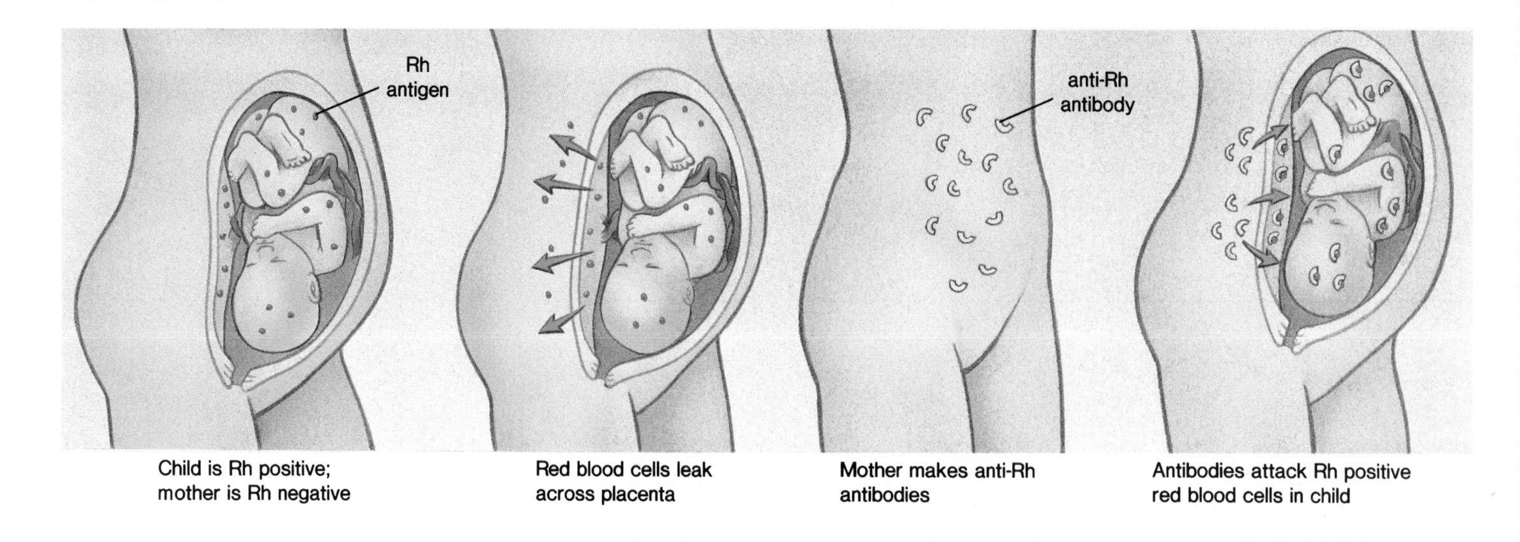

Rh System: Pregnancy Risks
Another important antigen in matching blood types is the Rh factor. Persons with the Rh factor on their red blood cells are Rh positive (Rh^+); those without it are Rh negative (Rh^-). Rh-negative individuals normally do not have antibodies to the Rh factor, but they may make them when exposed to the Rh factor.

If a mother is Rh negative and a father is Rh positive, a child may be Rh positive (fig. 13.14). The Rh-positive red blood cells of the child may begin leaking across the placenta into the mother's circulatory system, as placental tissues normally break down before and at birth. This causes the mother to produce anti-Rh antibodies. In this or a subsequent pregnancy with another Rh-positive child, anti-Rh antibodies may cross the placenta and destroy the child's red blood cells. This condition is called hemolytic disease of the newborn (HDN).

The Rh problem has been solved by giving Rh-negative women an Rh-immunoglobulin injection (often an Rho-Gam injection) either midway through the first pregnancy or no later than 72 hours after giving birth to any Rh-positive child. This injection contains anti-Rh antibodies, which attack any of the child's red blood cells in the mother's blood before these cells can stimulate her immune system to produce her own antibodies. This injection is not beneficial if the woman has already begun to produce antibodies; therefore, the timing of the injection is most important.

A common way of typing blood is to use the ABO system combined with the Rh system. The possibility of hemolytic disease of the newborn exists when the mother is Rh negative and the father is Rh positive. Do 13.3 Critical Thinking, found at the end of the chapter.

Allergies: Overactive Immune System

Allergies are caused by an overactive immune system, which forms antibodies to substances that usually are not recognized as foreign substances. Unfortunately, allergies usually are accompanied by coldlike symptoms, or even at times, by severe systemic reactions, such as anaphylactic shock, a sudden drop in blood pressure, and respiratory difficulties that can lead to death.

Of the 5 varieties (table 13.1) of antibodies—IgG, IgA, IgE, IgM, and IgD—IgE antibodies cause allergies. IgE antibodies are found on the membrane of basophils in the blood and on the membrane of *mast cells,* which are found in the tissues. As mentioned, mast cells are basophils that have left the bloodstream and taken up residence in the tissues. When an *allergen,* an antigen that provokes an allergic reaction, attaches to the IgE antibodies on mast cells, these cells release histamine and other substances that cause mucus secretion and airway constriction, resulting in the characteristic symptoms of allergy. On occasion, basophils and other white blood cells release these substances into the bloodstream. The increased capillary permeability that results from this can lead to fluid loss and shock.

Allergy shots sometimes prevent the onset of allergic symptoms. Injections of the allergen cause the body to build up high quantities of IgG antibodies, and these combine with allergens received from the environment before they have a chance to reach the IgE antibodies located in the membrane of mast cells.

> Histamine and other substances released by mast cells cause allergic symptoms.

Tissue Rejection: Foreign MHC Proteins

Certain organs, such as skin, the heart, and the kidneys, could be transplanted easily from one person to another if the body did not attempt to *reject* them. Rejection occurs because cytotoxic T cells bring about disintegration of foreign tissue in the body.

Organ rejection can be controlled in 2 ways: careful selection of the organ to be transplanted and the administration of immunosuppressive drugs. It is best if the transplanted organ has the same type of MHC proteins as those of the recipient, because cytotoxic T cells can recognize foreign MHC proteins. The immunosuppressive drug cyclosporine has been in use for some years. A new experimental drug, FK-506, eventually may replace cyclosporine as the drug of choice for transplant patients. In more than 100 patients taking FK-506, the rate of organ rejection was one-sixth that of patients taking cyclosporine. The mechanism of action of immunosuppressive drugs is not known.

> When an organ is rejected, the immune system is attacking cells that bear different MHC proteins from those of the individual.

Autoimmune Diseases: The Body Attacks Itself

Certain human illnesses are autoimmune diseases, in which the body's own antibodies and T cells attack tissues. Exactly what causes autoimmune diseases is not known, but they seem to appear after the individual has recovered from an infection. Some bacteria have been observed to produce toxic products that can cause T cells to bind prematurely to macrophages. Perhaps it is at this time that T cells learn to recognize the body's own tissues. This might be the cause of at least some autoimmune diseases. In myasthenia gravis, neuromuscular junctions do not work properly and muscular weakness results. In MS (multiple sclerosis), the myelin sheath of nerve fibers is attacked, and this causes various neuromuscular disorders. A person with SLE (systemic lupus erythematosus) has various symptoms including fever and malaise. At some point, kidney failure usually develops. In rheumatoid arthritis, the joints are affected. It is suspected heart damage following rheumatic fever and the loss of pancreatic cells following type I diabetes are also autoimmune illnesses. There are no cures for autoimmune diseases.

> Autoimmune diseases seem to be preceded by an infection that results in cytotoxic T cells attacking the body's own organs.

SUMMARY

The lymphatic system consists of lymphatic vessels and lymphoid organs. The lymphatic vessels collect excess tissue fluid and absorb fat molecules at lacteals and carry these to the cardiovascular system. Lymphocytes are produced in the bone marrow and accumulate in other lymphoid organs (thymus, lymph nodes, and spleen).

The body is prepared to defend itself in both nonspecific and specific ways. Barriers to entry, the inflammatory reaction, and protective proteins react to any threat. The immune response is specific to a particular antigen and requires 2 types of lymphocytes, both of which are produced in the bone marrow. B cells mature in the bone marrow, and T cells mature in the thymus.

B cells directly recognize an antigen and give rise to antibody-secreting plasma cells if stimulated to do so by helper T cells. In order for the T cell to recognize an antigen, the antigen must be presented by an APC, usually a macrophage. There are 4 types of T cells. Cytotoxic T cells kill infected cells on contact; helper T cells stimulate other immune cells and produce lymphokines; suppressor T cells suppress the immune response. There are also memory T and memory B cells, which remain in the body and provide long-lasting immunity.

Immunity can be fostered by immunotherapy. Vaccines are available to promote long-lived active immunity, and antibodies sometimes are available to provide an individual with short-lived passive immunity. Lymphokines, notably interferon and

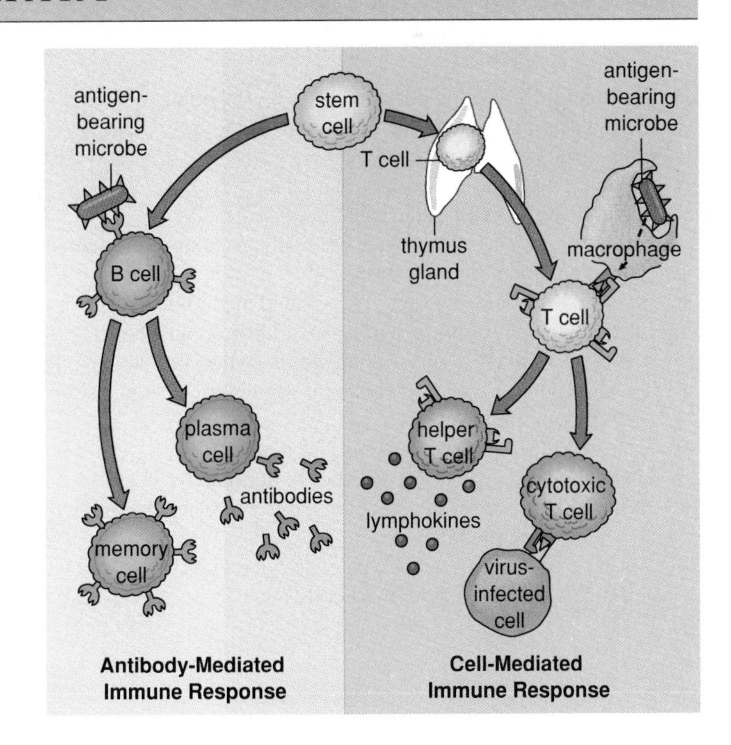

interleukins, are used to promote the body's ability to recover from cancer and to treat AIDS.

Immunity has certain undesirable side effects. Blood transfusions require compatible blood types. Of consideration are the antigens (A and B) on the red blood cells and the antibodies (anti-A and anti-B) in the plasma. The Rh antigen is also important, particularly because an Rh-negative mother may produce anti-Rh antibodies, which will attack the red blood cells of an Rh-positive fetus. Allergies result when an overactive immune system forms antibodies to substances not normally recognized as foreign. Cytotoxic T cells attack transplanted organs, although immunosuppressive drugs are available. Autoimmune illnesses occur when antibodies and T cells attack the body's own tissues.

STUDY QUESTIONS

In order to practice **writing across the curriculum,** students should write out the answers to any or all of the study questions. The study questions are sequenced in the same order as the text.

1. What is the lymphatic system, and what are its 3 functions? (p. 221)
2. Describe the structure and the function of the bone marrow, lymph nodes, the spleen, and the thymus. (pp.222–223)
3. What are the body's nonspecific defense mechanisms? (p. 223)
4. Describe the inflammatory reaction, and give a role for each type of cell and molecule that participates in the reaction. (pp. 223–224)
5. What is the clonal selection theory? (p. 225) B cells are responsible for which type of immunity? (p. 226)
6. Describe the structure of an antibody, including the terms *variable regions* and *constant regions.* (p. 226)
7. Name the 4 types of T cells, and state their functions. (p. 227)
8. Explain the process by which a T cell is able to recognize an antigen. (pp. 227–228)
9. How is active immunity achieved? How is passive immunity achieved? (pp. 229–230)
10. What are lymphokines, and how are they used in immunotherapy? (p. 231)
11. How are monoclonal antibodies produced, and what are their applications? (p. 231)
12. What are the 4 ABO blood types in humans? For each, state the antigen(s) on the red blood cells and the antibody(ies) in the plasma. (p. 233)
13. What are the Rh types of the parents if Rh-related problems develop during childbearing? Explain why this is so. (p. 234)
14. Discuss allergies, tissue rejection, and autoimmune diseases as they relate to the immune system. (pp. 234–235)

OBJECTIVE QUESTIONS

1. Lymphatic vessels collect excess _______ and return it to the _______ veins.
2. The function of lymph nodes is to _______ lymph.
3. The most common granular leukocyte is the _______, a phagocytic white blood cell.
4. T lymphocytes have passed through the _______.
5. A stimulated B cell produces antibody-secreting _______ cells and _______ cells, which are ready to produce the same type of antibody at a later time.
6. B cells are responsible for _______-mediated immunity.
7. In order for a T cell to recognize an antigen, it must be presented by a(n) _______ along with an MHC protein.
8. T cells produce _______, which are stimulatory chemicals for all types of immune cells.
9. Type AB blood has the antigens _______ and _______ on red blood cells and _______ antibodies in plasma.
10. Hemolytic disease of the newborn can occur when the mother is _______ and the father is _______.
11. Allergic reactions are associated with the release of _______ and other substances from mast cells.
12. Immunization with _______ brings about active immunity.
13. Hybridomas produce _______ antibodies.
14. Give the function of the 4 types of T cells shown.

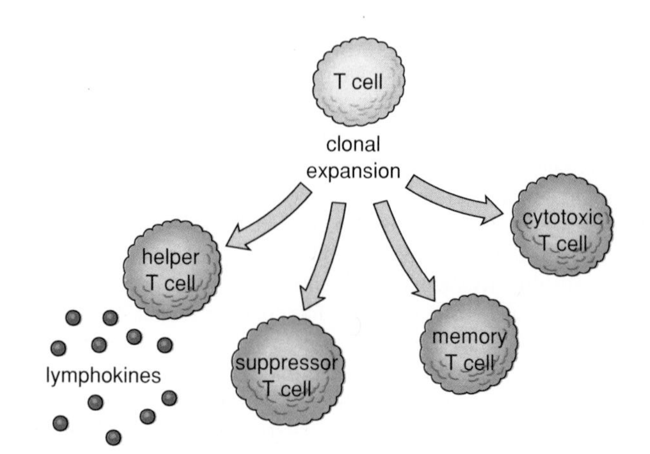

CRITICAL THINKING

In order to practice **writing across the curriculum,** students should write out the answers to any or all of the critical thinking questions. Suggested answers to the critical thinking questions are in appendix E.

13.1

1. Why would you expect edema when blood pressure rises, but not when it decreases?
2. Why would you expect edema when blood osmotic pressure decreases, but not when it rises?
3. Suppose blood proteins leak into the tissues so that an equal amount occurs on both sides of the capillary wall. Under these circumstances, what will happen to the osmotic pressure of blood? What will happen to the blood pressure? What will happen to tissue-fluid formation?

13.2

1. A mouse is irradiated so that its bone marrow and thymus are destroyed. It then is resupplied only with bone marrow. The mouse is unable to form antibodies. Why?
2. A mixture of B cells is exposed to a specific radiolabeled antigen in vitro (within laboratory glassware). Would you expect all B cells to bind with the antigen and to be radiolabeled?
3. When B cells and T cells are incubated in vitro with a radiolabeled antigen, binding to certain B cells occurs but not to T cells. Why?
4. Human beings communicate by sight, sound, and touch. How do immune cells communicate with one another?

13.3

1. Type O blood used to be called the universal donor for blood transfusions. Why?
2. Type AB blood used to be called the universal recipient. Why?
3. Newborns already have a particular blood type. What does this tell you about your blood type?
4. Other systems aside from the ABO system exist for typing blood. Why is this to be expected?

SELECTED KEY TERMS

agglutination (ag-gloo″tĭ-na′shun) clumping of cells, particularly in reference to red blood cells involved in an antigen-antibody reaction.

antibody a protein produced in response to the presence of an antigen; an antibody combines with the antigen to produce a nonharmful complex.

antigen (ant′i-jen) a foreign substance, usually a protein, that stimulates the immune system to react, such as to produce antibodies.

B lymphocyte a lymphocyte that matures in the bone marrow and when stimulated by the presence of a specific antigen, gives rise to antibody-producing plasma cells.

complement system a series of proteins in plasma that counteracts a microbe invasion in a variety of ways; complements the antigen-antibody reaction.

cytotoxic T cell T lymphocyte that attacks and kills antigen-bearing cells; killer T cell.

helper T cell T lymphocyte that releases lymphokines and stimulates certain other immune cells to perform their respective functions.

inflammatory reaction a tissue response to injury that is characterized by dilation of blood vessels and accumulation of fluid in the affected region.

interferon (in″ter-fē′on) a protein formed by a cell infected with a virus that can increase the resistance of other cells to the virus.

lymph fluid having the same composition as tissue fluid and carried in lymphatic vessels.

lymphatic system a one-way vascular system that takes up excess fluid in the tissues and transports it to cardiovascular veins in the shoulders.

lymphokine (lim′fo-kīn) molecule secreted by T lymphocytes that has the ability to affect the activity of all types of immune cells.

memory B cell one of a persistent population of B cells ready to produce antibodies specific to a particular antigen; accounts for the development of active immunity.

memory T cell a T cell that is ready to recognize an antigen that previously invaded the body.

MHC (major histocompatibility complex) protein a membrane protein that serves to identify the cells of a particular individual.

monoclonal antibody an antibody of the same type produced by a hybridoma—a lymphocyte that has fused with a cancer cell.

plasma cell a cell derived from a B cell lymphocyte that is specialized to mass-produce antibodies.

pus thick, yellowish fluid composed of dead phagocytes, dead tissue, and bacteria.

suppressor T cell T lymphocyte that suppresses certain other T and B lymphocytes from continuing to divide and perform their respective functions.

T lymphocyte a lymphocyte that matures in the thymus and occurs in 4 varieties, one of which kills antigen-bearing cells outright.

vaccine antigens prepared in such a way that they can promote active immunity without causing disease.

14

RESPIRATION

Chapter Concepts

1.
As air passes along the respiratory tract, it is filtered, warmed, and saturated with water before gas exchange takes place across a very extensive moist surface. 241, 256

2.
During inspiration, the pressure in the lungs decreases and then air comes rushing in. During expiration, increased pressure in the thoracic cavity causes air to leave the lungs. 245, 256

3.
During external respiration, the respiratory pigment hemoglobin combines with oxygen in the lungs, and during internal respiration, hemoglobin gives up oxygen. 248, 256

4.
Hemoglobin also aids in the transport of carbon dioxide from the tissues to the lungs. 252, 256

5.
The respiratory tract is especially subject to disease because it is exposed to microbes. Polluted air contributes to 2 major lung disorders—emphysema and cancer. 252, 256

False color SEM of ciliated epithelial cells lining the bronchi, ×1,000

Chapter Outline

BREATHING is more continuously necessary than eating. While it is possible to stop eating altogether for several days, it is not possible to remain alive for longer than several minutes without breathing. Breathing supplies the body with the oxygen (O_2) needed for aerobic cellular respiration, as indicated in the following equation.

$$38\ ADP + 38\ Ⓟ \rightarrow 38\ ATP$$

$$C_6H_{12}O_6 + 6\ O_2 \longrightarrow 6\ H_2O + 6\ CO_2$$

As glucose is broken down to carbon dioxide and water, ATP molecules are formed.

This equation tells us that the body requires oxygen to convert the energy within glucose ($C_6H_{12}O_6$) to phosphate-bond energy.[1] Therefore, the more energy expended, the greater the need for oxygen (fig. 14.1). The average young adult male utilizes about 250 ml of oxygen per minute in a basal, or restful, state. Exercise and digestion of food raise the oxygen need. The average amount of oxygen needed with mild exercise is 500 ml of oxygen per minute.

The equation for aerobic cellular respiration also shows that carbon dioxide (CO_2) is an end product of the process. Carbon dioxide is eliminated from the body by the breathing process.

Breathing is necessary to supply the body with oxygen so that ATP can be formed by aerobic cellular respiration.

Altogether, the term *respiration* can be used to refer to the complete process of supplying oxygen to body cells for aerobic cellular respiration and the reverse process of ridding the body of carbon dioxide given off by cells. Respiration can be said to include the following components.

1. **Breathing:** entrance and exit of air into and from the lungs
2. **External respiration:** exchange of the gases oxygen (O_2) and carbon dioxide (CO_2) between air and blood
3. **Internal respiration:** exchange of the gases oxygen and carbon dioxide between blood and tissue fluid
4. **Cellular respiration:** production of ATP in cells

This chapter discusses the first 3 listed components of the respiratory process. Aerobic cellular respiration was discussed in detail in chapter 6.

1. The body requires oxygen (O_2) for the respiration of fats and amino acids as well as for glucose. Thirty-six ATP molecules are sometimes produced instead of 38 (p. 95).

Figure 14.1

The body's oxygen need. Exercising increases the body's need for oxygen (O_2) because aerobic cellular respiration is sped up in order to provide ATP for muscle contraction. The heart pumps faster to deliver oxygen to the tissues in a more timely manner.

Table 14.1
Composition of Inspired and Expired Air

Component of Air	Inspired Air (%/Vol)	Expired Air (%/Vol)
Nitrogen (N_2)	79.00	79.60
Oxygen (O_2)	20.96	16.02*
Carbon dioxide (CO_2)	0.04	4.38

*We exhale some oxygen, making mouth-to-mouth resuscitation possible.

Breathing

The normal breathing rate is about 14–20 times per minute. Breathing consists of taking air in, called **inspiration** (inhalation), and forcing air out, called **expiration** (exhalation). Expired air contains less oxygen (O_2) and more carbon dioxide (CO_2) than inspired air, indicating that the body takes in oxygen and gives off carbon dioxide (table 14.1). Nitrogen gas is in air but plays no role in respiration.

ANIMAL RESPIRATORY ORGANS

Animals do not have a storage area for gases; therefore, they must continually acquire oxygen and rid the body of carbon dioxide. Some animals, like hydras and planarians, are small and shaped in a way that allows their body cells to carry out gas-exchange. Among larger animals, the earthworm's shape provides an extensive outer surface for respiration. Glands keep the surface moist, and the worm is behaviorally adapted to remain in damp soil during the day.

In most complex animals, vascularization—close association with an extensive capillary system—enhances the effectiveness of the respiratory organ. Aquatic animals take oxygen from the water. Quite often they have gills, which are finely divided and vascularized outgrowths of either an outer or inner body surface. In the crayfish and similar crustaceans, the gills are located on the thorax, just beneath the exoskeleton. In many fishes, the gills are outward extensions of the pharynx. When the mouth opens, water is drawn in. When the mouth closes, water flows through the gill slits located between the gill arches.

Terrestrial animals take oxygen from the air. Among invertebrates, insects have a tubular respiratory system. Air enters the system at valvelike openings, and then the tubes called trachea branch and rebranch, until finally tracheoles are in direct contact with the body cells. Terrestrial vertebrates, in particular, have evolved lungs, which are outgrowths from the lower pharyngeal region. The lungs of amphibians are simple sacs, and most amphibians also make use of the skin as a respiratory surface. The lungs are more finely divided in reptiles and are especially divided in birds and mammals. It has even been estimated that human lungs have a total surface area that is at least 40 times the surface area of skin.

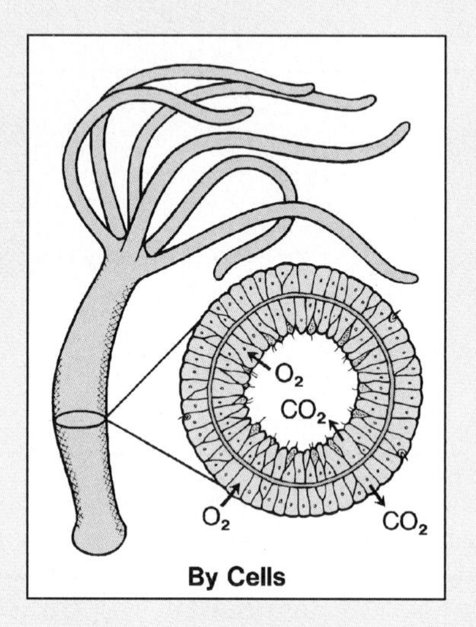

By Cells

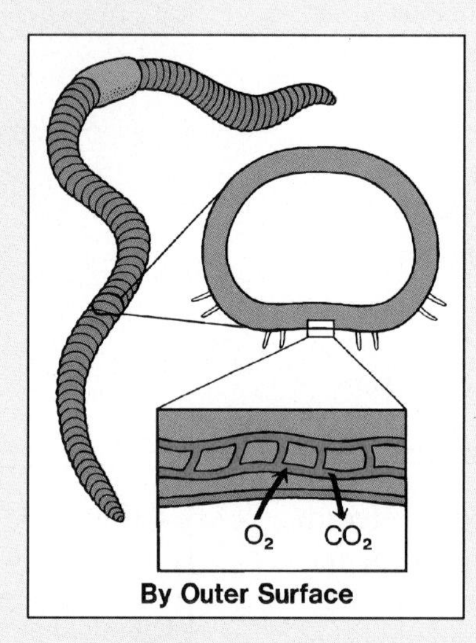

By Outer Surface

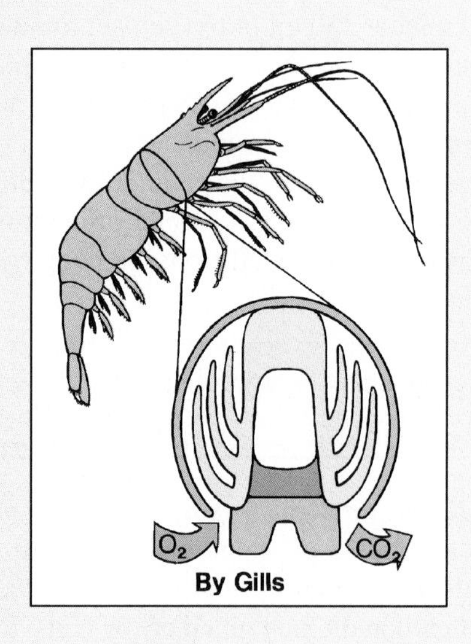

By Gills

By Gills with Arches

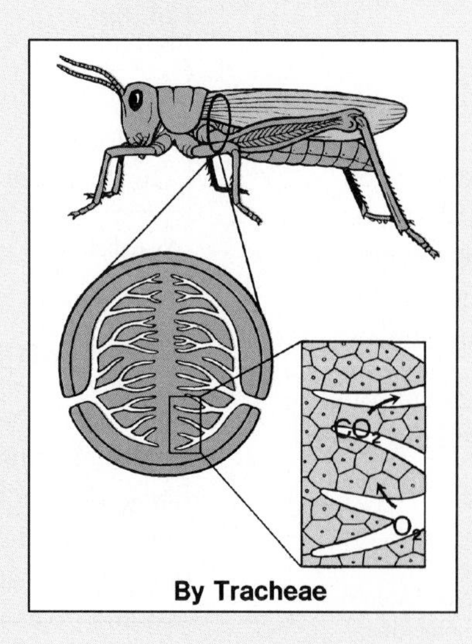

By Tracheae

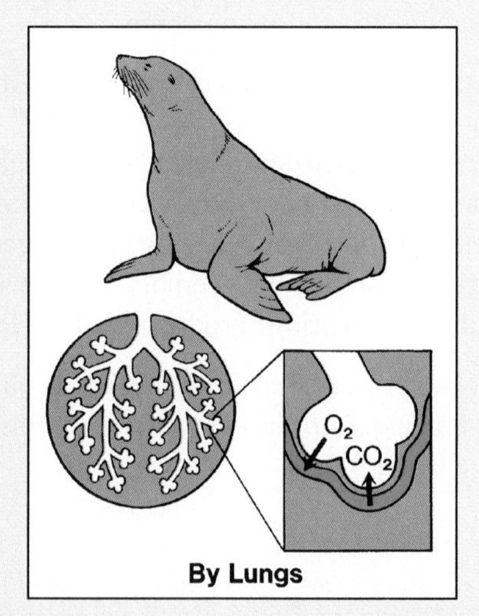

By Lungs

Air Passageway: Filters, Warms, Moistens

During inspiration and expiration, air is conducted toward or away from the lungs by a series of cavities, tubes, and openings, listed in order in table 14.2 and illustrated in figure 14.2.

As air moves in along the air passages, it is filtered, warmed, and moistened. Filtering is accomplished by coarse hairs and cilia in the region of the nostrils and by cilia alone in the rest of the nose and the trachea. In the nose, the hairs and the cilia act as a screening device. In the trachea, cilia beat upward, carrying mucus, dust, and occasional bits of food that "went down the wrong way" into the pharynx, where the accumulation can be swallowed or expectorated. The air is warmed by heat given off by the blood vessels lying close to the surface of the lining of the air passages, and it is moistened by the wet surface of these passages.

Conversely, as air moves out during expiration, it cools and loses its moisture. As the air cools, it deposits its moisture on the lining of the trachea and the nose, and the nose may even drip as a result of this condensation. The air still retains so much moisture, however, that upon expiration on a cold day, it condenses and forms a small cloud.

Table 14.2
Path of Air

Structure	Description	Function
Nasal cavities	Hollow spaces in nose	Filter, warm, and moisten air
Pharynx	Chamber behind oral cavity and between nasal cavity and larynx	Connection to surrounding regions
Glottis	Opening into larynx	Passage of air into larynx
Larynx	Cartilaginous organ that contains vocal cords (voice box)	Sound production
Trachea	Flexible tube that connects larynx with bronchi (windpipe)	Passage of air to bronchi
Bronchi	Major divisions of trachea that enter lungs	Passage of air to each lung
Bronchioles	Branched tubes that lead from the bronchi to the alveoli	Passage of air to each alveolus
Lungs	Soft, cone-shaped organs that occupy a large portion of the thoracic cavity	Gas exchange

Air is warmed, filtered, and moistened as it moves from the nose toward the lungs.

Each portion of the air passage has its own structure and function, as described in the sections that follow.

The Nose: Two Cavities

The nose contains 2 *nasal cavities,* narrow canals with convoluted lateral walls. The cavities are separated from one another by the septum, a wall composed of bone and cartilage. Special ciliated cells in the narrow upper recesses of the nasal cavities (see fig. 18.4) act as odor receptors. Nerves lead from these cells to the brain, where the impulses generated by the odor receptors are interpreted as smell.

The tear (lacrimal) glands drain into the nasal cavities by way of tear ducts. For this reason, crying produces a runny nose. The nasal cavities also communicate with the cranial sinuses, air-filled, mucous membrane-lined spaces in the skull. If these membranes are inflamed due to a cold or an allergic reaction, mucus can accumulate in the sinuses, causing a sinus headache.

The nasal cavities empty into the nasopharynx, the upper portion of the pharynx. The *eustachian tubes* lead from the nasopharynx to the middle ears (see fig. 18.12).

The nasal cavities, which receive air, open into the nasopharynx.

The Pharynx: A Crossroad

The **pharynx** is essentially the throat, and air taken in by either the nose or the mouth enters the pharynx. In the pharynx, the air (trachea) and food (esophagus) passages temporarily join. The trachea, which lies in front of the esophagus, is normally open, allowing the passage of air, but the esophagus is normally closed and opens only when swallowing occurs. The larynx lies at the top of the trachea.

Air from either the nose or the mouth enters the pharynx, as does food. The passage of air continues in the larynx and then the trachea proper.

The Larynx: Voice Box

The **larynx** can be imagined as a triangular box whose apex, the Adam's apple, is located at the front of the neck. At the top of the larynx is a variable-sized opening called the **glottis.** When food is being swallowed, the glottis is covered by a flap of tissue called the **epiglottis** so that no food passes into the larynx. If, by chance, food or some other substance does enter the larynx, reflex coughing usually expels the substance. If this reflex is not sufficient, it may be necessary to resort to the Heimlich maneuver (fig. 14.3).

Figure 14.2

Diagram of the human respiratory tract, with internal structure of one lung revealed in an enlargement of a section. Gas exchange occurs in the alveoli, which are surrounded by a capillary network. Notice that the pulmonary arteriole carries deoxygenated blood (colored blue) and the pulmonary venule carries oxygenated blood (colored red). (See page 206.)

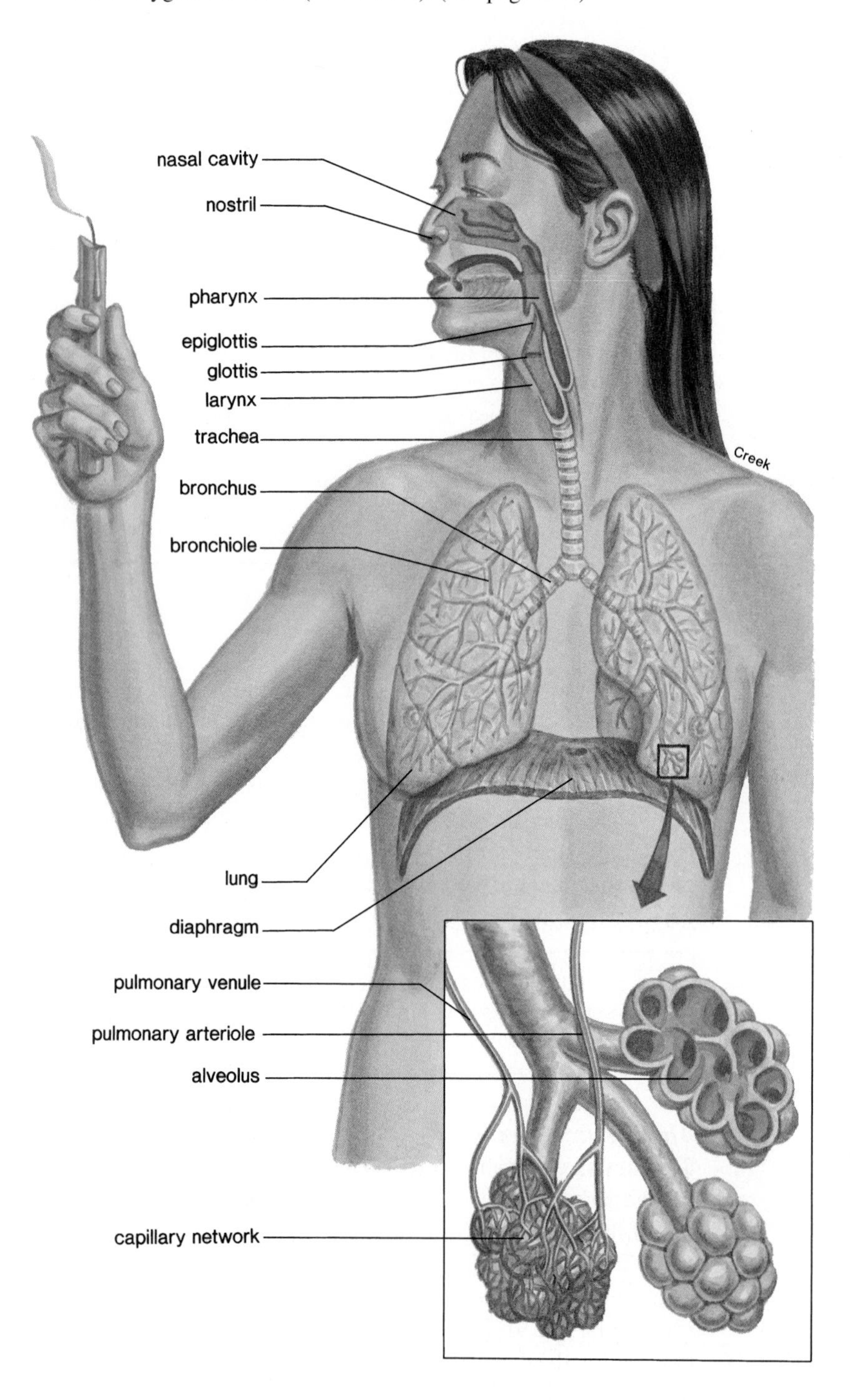

Figure 14.3

The Heimlich maneuver. More than 8 Americans choke to death each day on food lodged in the trachea. A simple process termed the abdominal thrust or Heimlich maneuver can save the life of a person who is choking. If the person is standing or sitting: (1) Stand behind the person or the person's chair, and wrap your arms around the choking person's waist; (2) grasp your fist with your other hand, and place the fist against the abdomen, slightly above the navel and below the rib cage; (3) press your fist into the abdomen with a quick upward thrust; (4) repeat several times if necessary. If the choking person is lying down: (1) Position this person so that the face is directed upward; (2) face the choking person, and kneel astride the hips; (3) with one of your hands on top of the other, place the heel of your bottom hand on the abdomen, slightly above the navel and below the rib cage; (4) press into the abdomen with a quick upward thrust; (5) repeat several times if necessary. If you are alone and choking, use anything that applies force just below your diaphragm. Press into a table or a sink, or use your own fist.

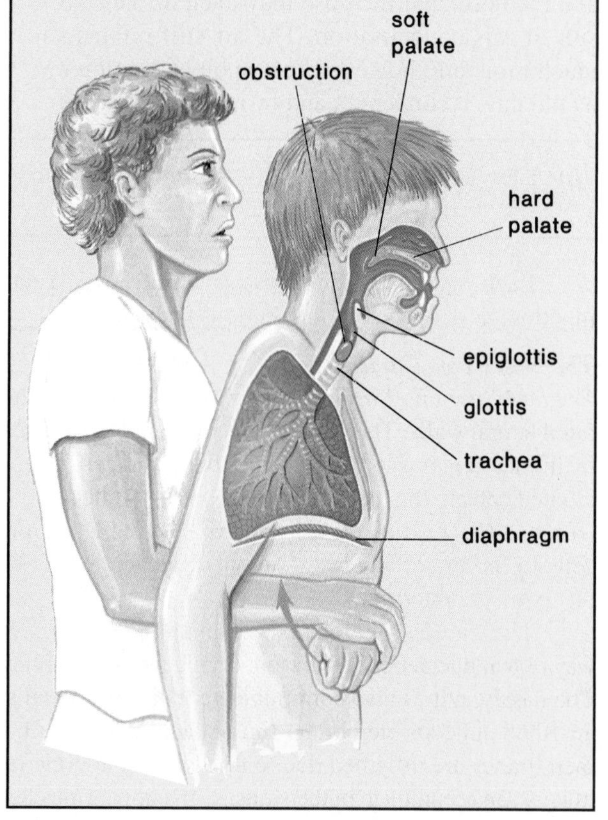

Figure 14.4

Placement of the vocal cords. The vocal cords are on either side of the glottis, an opening into the larynx. The epiglottis covers the glottis during swallowing. **a.** Longitudinal section of the larynx, showing the placement of the vocal cords. When air is expelled from the larynx, the cords vibrate. **b.** The glottis is narrow when we produce a high-pitched sound (*top*) and widens as the pitch deepens (*bottom*).

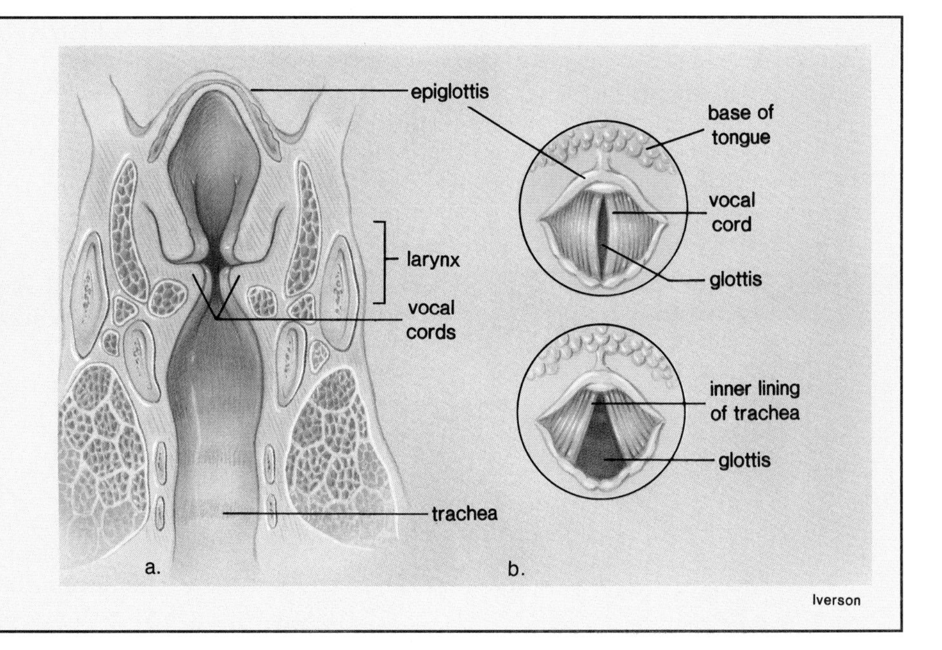

The **vocal cords,** mucous membrane folds supported by elastic ligaments, occur at the edges of the glottis (fig. 14.4). These cords vibrate when air is expelled past them through the glottis, and their vibration produces sound. The high or low pitch of the voice depends upon the length, the thickness, and the degree of elasticity of the vocal cords and the tension at which they are held. The loudness, or intensity, of the voice depends upon the amplitude of the vibrations, or the degree to which vocal cords vibrate.

At the time of puberty, the growth of the larynx and the vocal cords is much more rapid and accentuated in the male than in the female, causing the male to have a more prominent Adam's apple and a deeper voice. The voice "breaks" in the young male due to his inability to control the longer vocal cords.

> The larynx is the voice box because it contains the vocal cords, which occur at the sides of the glottis, an opening covered by the epiglottis during swallowing.

The Trachea: Windpipe

The **trachea** is a tube held open by C-shaped cartilaginous rings. Ciliated mucous membrane lines the trachea, and normally these cilia keep it free of debris. Smoking is known to destroy the cilia, and consequently the soot in cigarette smoke collects in the lungs. Smoking is discussed more fully at the end of this chapter.

If the trachea is blocked because of illness or accidental swallowing of a foreign object, it is possible to insert a tube by way of an incision made in the trachea. This tube acts as an artificial air intake and exhaust duct. The operation is called a *tracheotomy.*

Bronchi: Air Tubes

The trachea divides into 2 *bronchi* (sing., **bronchus**), which lead respectively into the right and left lungs (fig. 14.5). The bronchi branch into a great number of smaller passages called **bronchioles.** The bronchi resemble the trachea in structure, but as the bronchiolar tubes divide and subdivide, their walls become thinner and the small rings of cartilage are no longer present. During an asthma attack, the bronchioles constrict even to the point of closing, and movement of air through the narrowed tubes may result in the wheezing characteristic of asthma. Each bronchiole terminates in an elongated space enclosed by a multitude of air sacs, or pockets, called *alveoli* (sing., **alveolus**) (fig. 14.2). The alveoli make up the lungs.

Lungs

The lungs are cone-shaped organs; the right lung has 3 lobes and the left lung has 2 lobes. A lobe is further divided into lobules, each of which has a bronchiole serving many alveoli. The lungs lie on either side of the heart in the thoracic (chest) cavity. The base of each lung is broad and concave so that it fits the convex surface of the diaphragm. The other surfaces of the lungs follow the contours of the ribs and the organs in the thoracic cavity.

Alveoli: 300 Million Air Sacs. Each alveolar sac is made up of simple squamous epithelium surrounded by blood capillaries. Gas exchange occurs between air in the alveoli and blood in the capillaries (fig. 14.2).

A film of lipoprotein lining the alveoli of mammalian lungs lowers the surface tension and prevents them from closing. The lungs collapse in some newborn babies, especially premature infants, who lack this film. This condition, called infant respiratory distress syndrome, is now treatable.

Figure 14.5

Airways to the lungs. The trachea divides into the bronchi, which give rise to the bronchioles. The bronchioles have many branches and terminate in the alveoli. Each bronchiole and its branches are similarly colored.

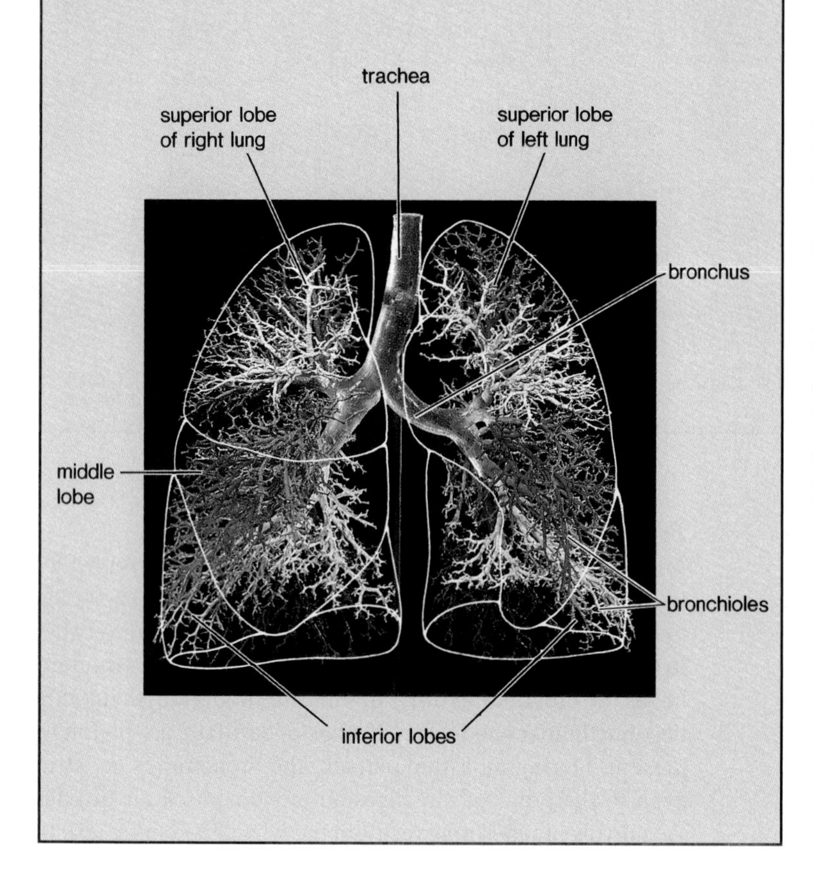

There are approximately 300 million alveoli, having a total cross-sectional area of 50–70 m^2. This is at least 40 times the surface area of the skin. Because of their many air spaces, the lungs are very light; normally, a piece of lung tissue dropped in a glass of water floats.

Air moves from the trachea and the 2 bronchi, which are held open by cartilaginous rings, into the lungs. The lungs are composed of air sacs called alveoli.

The Mechanism of Breathing

In order to understand **ventilation,** the manner in which air is drawn into and expelled from the lungs, it is necessary to remember first that when we are breathing, there is a continuous column of air from the pharynx to the alveoli of the lungs; that is, the air passages are open.

Secondly, the lungs lie within the sealed-off thoracic cavity. The **rib cage** forms the top and sides of the thoracic cavity. It contains the ribs, hinged to the vertebral column at the back and to the sternum (breastbone) at the front, and the intercostal muscles, which lie between the ribs. The **diaphragm,** a dome-shaped horizontal muscle, forms the floor of the thoracic cavity.

The lungs are enclosed by the **pleural membranes.** An infection of the pleural membranes is called pleurisy. The outer pleural membrane adheres to the rib cage and the diaphragm, and the inner membrane is fused to the lungs. The 2 pleural layers lie very close to one another, separated only by a thin film of fluid. Normally, the intrapleural pressure is lower than atmospheric pressure by 4 mm Hg.

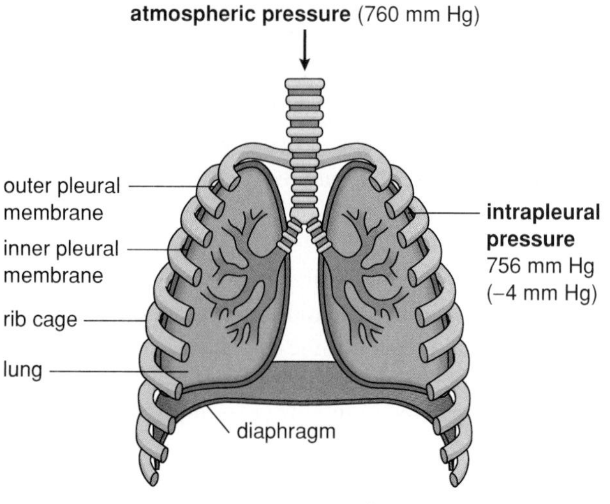

The importance of the reduced intrapleural pressure is demonstrated when, by design or accident, air enters the intrapleural space. The lungs collapse, and inspiration is impossible.

The lungs are completely enclosed and by way of the pleural membranes adhere to the thoracic cavity walls. The intrapleural pressure is lower than atmospheric pressure.

Inspiration Is Active

Carbon dioxide (CO_2) and hydrogen ions (H^+) are the primary stimuli that cause us to breathe. When the concentration of carbon dioxide and subsequently the concentration of hydrogen ions reach a certain level in blood, the *respiratory center* in the medulla oblongata, the stem portion of the brain, is stimulated. This center is not affected by low oxygen (O_2) levels. There are also chemoreceptors in the *carotid bodies,* located in the carotid arteries, and in the *aortic bodies,* located in the aorta, that respond primarily to hydrogen ion concentration [H^+] but also to the level of carbon dioxide and oxygen in blood. These bodies communicate with the respiratory center. When the level of carbon dioxide and hydrogen rise, the rate and depth of breathing increase.

> In humans, carbon dioxide and hydrogen ions are the primary stimuli causing us to breathe.

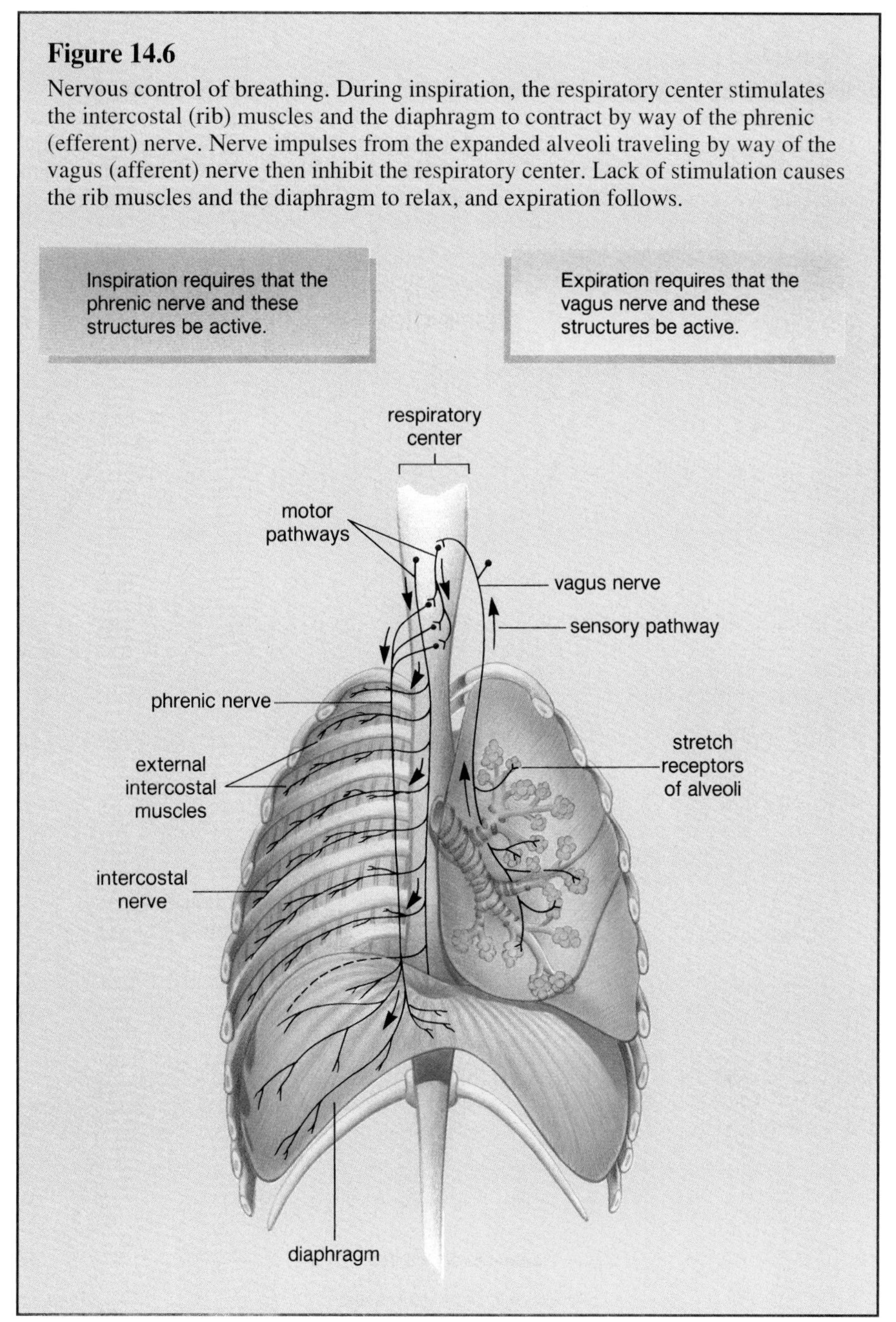

Figure 14.6

Nervous control of breathing. During inspiration, the respiratory center stimulates the intercostal (rib) muscles and the diaphragm to contract by way of the phrenic (efferent) nerve. Nerve impulses from the expanded alveoli traveling by way of the vagus (afferent) nerve then inhibit the respiratory center. Lack of stimulation causes the rib muscles and the diaphragm to relax, and expiration follows.

When the respiratory center is stimulated, a nerve impulse goes out by way of nerves to the diaphragm and the rib cage (fig. 14.6). In its relaxed state, the *diaphragm* is dome shaped, but upon stimulation, it contracts and lowers. When the external intercostal muscles contract, the *rib cage* moves upward and outward. Both of these contractions serve to increase the size of the thoracic cavity. As the thoracic cavity increases in size, the lungs expand. When the lungs expand, air pressure within the enlarged alveoli lowers and is immediately rebalanced by air rushing in through the nose or the mouth.

Inspiration (fig. 14.7) is the active phase of breathing. During this time, the diaphragm and the intercostal muscles contract, the intrapleural pressure decreases even further, the lungs expand, and air comes rushing in. Note that air comes in because the lungs already have opened up; air does not force the lungs open. This is why it is sometimes said that *humans breathe by negative pressure.* The creation of a partial vacuum sucks air into the lungs.

> Stimulated by nervous impulses, the rib cage lifts up and out and the diaphragm lowers to expand the thoracic cavity and the lungs; this allows inspiration to occur.

Expiration Is Usually Passive

When the lungs are expanded, the alveoli stretch. This stimulates stretch receptors in the alveolar walls, and they initiate nerve impulses that travel from the inflated lungs to the respiratory center. When the impulses arrive at the medulla oblongata, the center is inhibited, and it stops sending signals to the diaphragm and the rib cage. The *diaphragm* relaxes and resumes its dome shape (fig. 14.7). The abdominal organs press up against the diaphragm. The *rib cage* moves down and inward. The elastic lungs recoil, and air is pushed out.

Figure 14.7

Inspiration versus expiration. During inspiration, the rib cage lifts up and out, the diaphragm lowers, the lungs expand, and air is drawn in. This sequence of events is only possible because the pressure within the intrapleural space, containing a thin film of fluid, is less than atmospheric pressure. During expiration, the rib cage lowers, the diaphragm rises, the lungs recoil, and air is forced out.

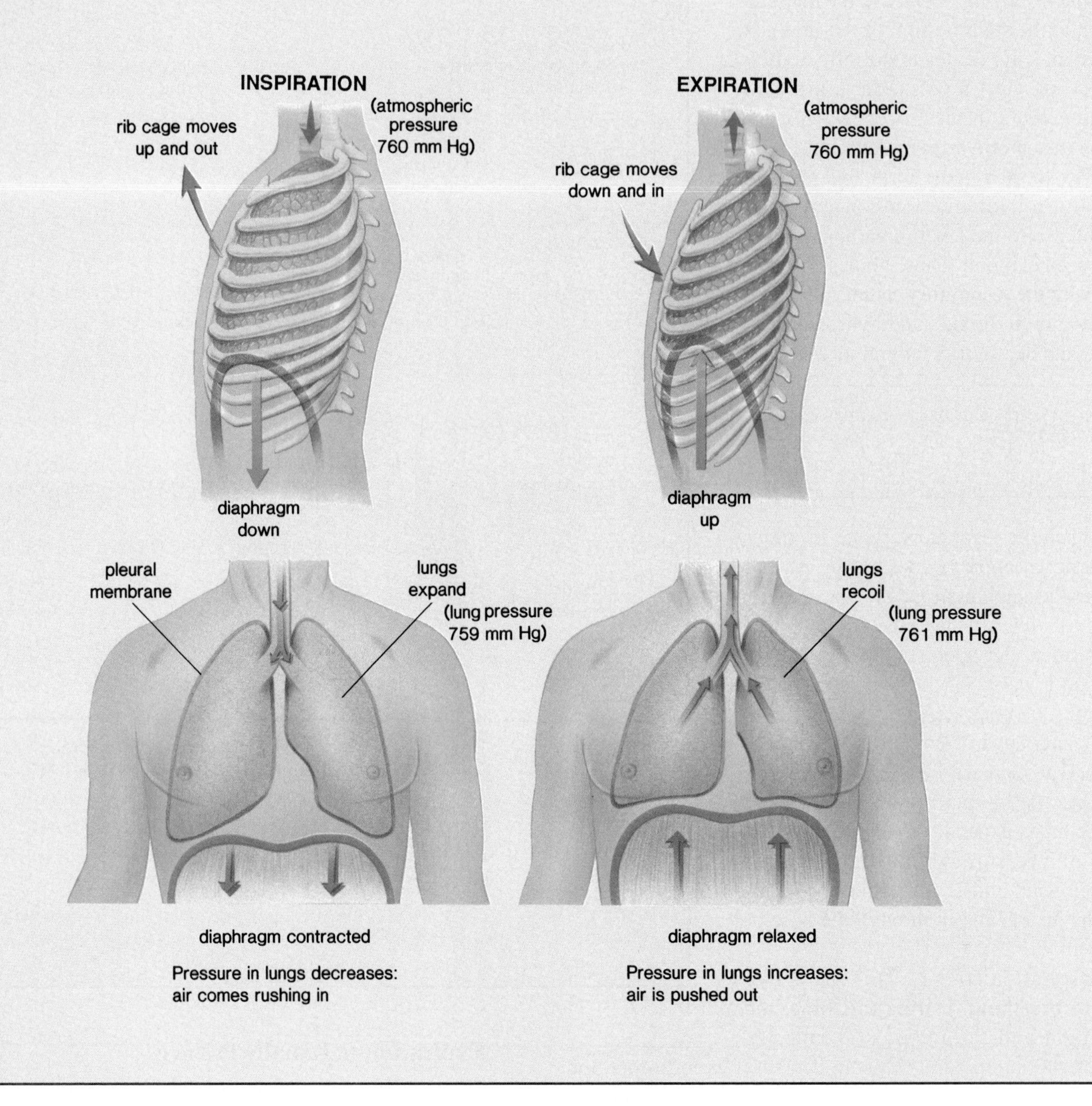

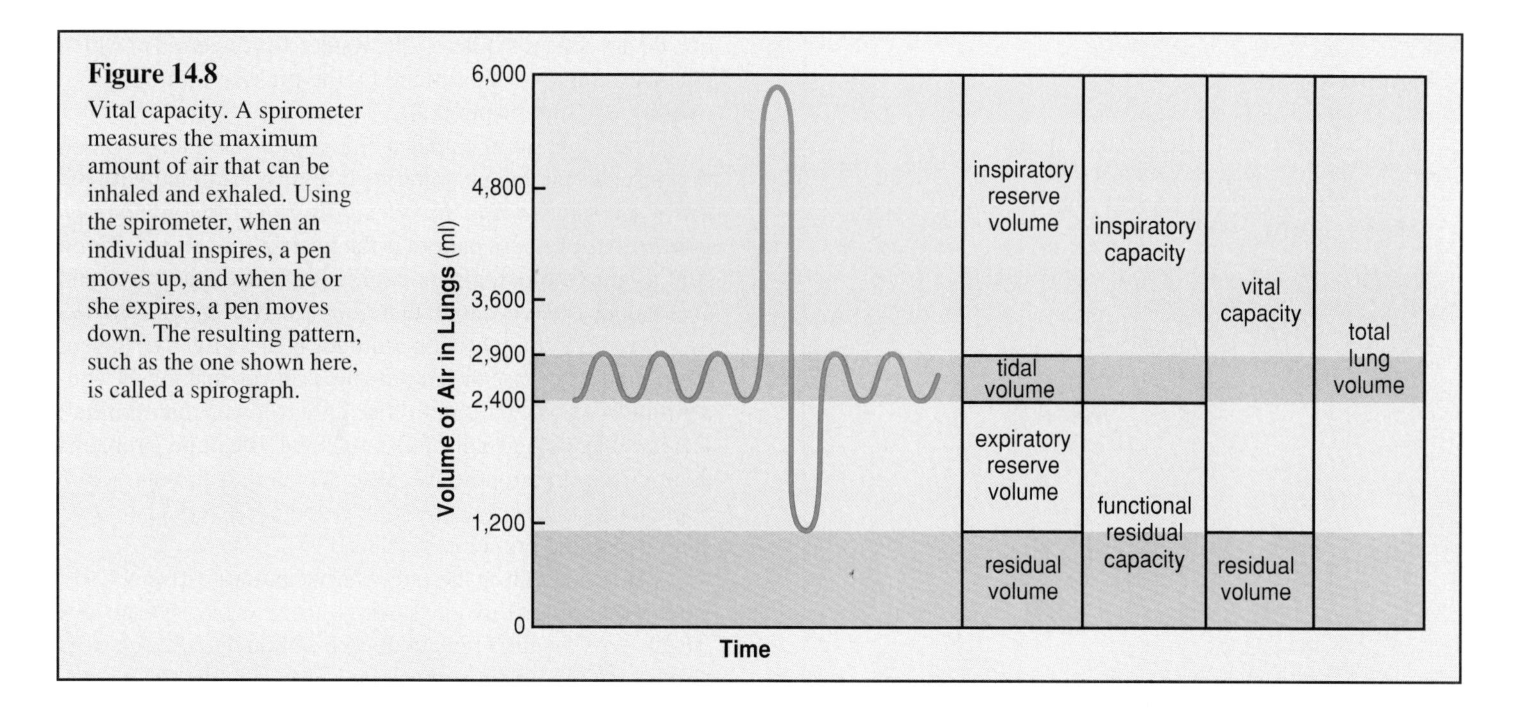

Figure 14.8
Vital capacity. A spirometer measures the maximum amount of air that can be inhaled and exhaled. Using the spirometer, when an individual inspires, a pen moves up, and when he or she expires, a pen moves down. The resulting pattern, such as the one shown here, is called a spirograph.

It is clear that while inspiration is an active phase of breathing, expiration is normally passive because the breathing muscles automatically relax following contraction. It is possible, in deeper and more rapid breathing, for both phases to be active because the contraction of intercostal muscles can force the rib cage to move downward and inward. Also, when the abdominal wall muscles are contracted, increased pressure helps to expel air.

When nervous stimulation ceases, the rib cage lowers and the diaphragm rises, allowing the lungs to recoil and expiration to occur.

How Much Do Lungs Hold?

When we breathe, the amount of air moved in and out with each breath is called the **tidal volume.** Normally, the tidal volume is about 500 ml, but we can increase the amount inhaled and exhaled by deep breathing. The maximum volume of air that can be moved in and out during a single breath is called the **vital capacity** (fig. 14.8). First, we can increase inspiration by as much as 3,100 ml of air. This is called the *inspiratory reserve volume.* Similarly, we can increase expiration by contracting the thoracic muscles. This is called the *expiratory reserve volume,* and it measures approximately 1,400 ml of air. Vital capacity is the sum of tidal, inspiratory reserve, and expiratory reserve volumes.

Note in figure 14.8 that even after very deep breathing, some air (about 1,000 ml) remains in the lungs; this is called the **residual volume.** This air is not too useful for gas exchange purposes. In some lung diseases, such as emphysema (p. 255), the residual volume builds up because the individual has difficulty emptying the lungs. This means that the lungs tend to be filled with useless air, and as you can see from examining figure 14.8, the vital capacity is markedly reduced.

Dead Space in Airways

Some of the inspired air never reaches the lungs; instead it fills the conducting airways (fig. 14.9). These passages are not used for gas exchange and therefore are said to contain *dead space.* To ventilate the lungs, then, it is better to breathe slowly and deeply because this ensures that a greater percentage of the tidal volume reaches the lungs.

Breathing through a very long tube increases the amount of dead space beyond maximum inspiratory capacity. Thereafter, death will occur because the air inhaled never reaches the alveoli.

The volume of fresh air reaching the lungs can vary. Do 14.1 Critical Thinking, found at the end of the chapter.

Figure 14.9

Distribution of air in the lungs. The air colored blue does not reach the alveoli immediately; therefore, this is called dead space. The air colored purple represents the amount of residual air that has not left the lungs. Only the air colored pink brings additional oxygen (O_2) for respiration.

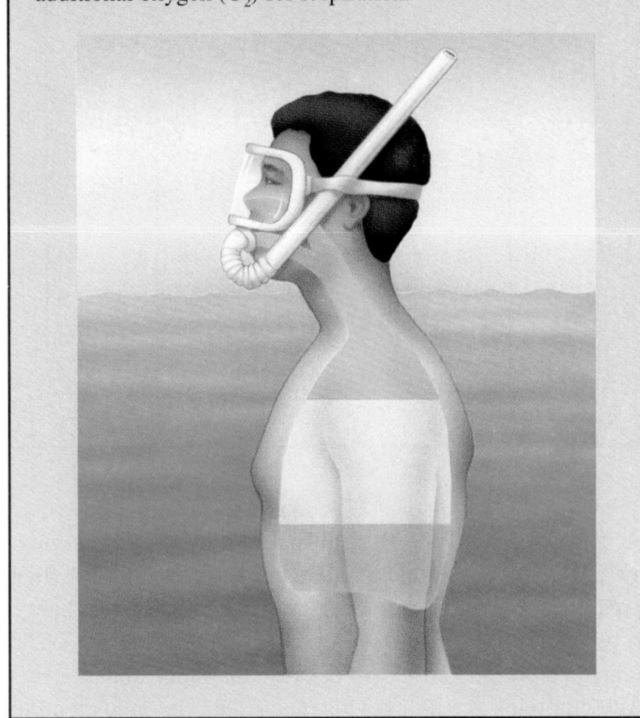

External and Internal Respiration

Figure 14.10 shows both external respiration and internal respiration. The term *external respiration* refers to the exchange of gases between air in the alveoli and blood in the pulmonary capillaries. The term *internal respiration* refers to the exchange of gases between blood in systemic capillaries and tissue fluid.

External Respiration: Air Sacs and Blood

The walls of both an alveolus and a blood capillary consist of a thin, single layer of cells. Since neither wall offers resistance to the passage of gases, *diffusion* alone governs the exchange of oxygen (O_2) and carbon dioxide (CO_2) between alveolar air and blood. Active cellular absorption and secretion do not appear to play a role. Rather, the direction in which the gases move is determined by the pressure gradients between blood and inspired air.

Atmospheric air contains little carbon dioxide, but blood flowing into the lung capillaries is almost saturated with the gas. Therefore, *carbon dioxide diffuses out of blood into the alveoli*. The pressure pattern is the reverse for oxygen. Blood coming into the pulmonary capillaries is deoxygenated, and alveolar air is oxygenated; therefore, *oxygen diffuses into the capillary*. Breathing at high altitudes is less effective than at low altitudes because the air pressure is lower, making the concentration of oxygen (and other gases) lower than normal; therefore, less oxygen diffuses into blood. Breathing problems do not occur in airplanes because the cabin is pressurized to maintain an appropriate pressure. Emergency oxygen is available in case the pressure is reduced.

As blood enters the pulmonary capillaries (fig. 14.10), most of the carbon dioxide is being carried as bicarbonate ions (HCO_3^-). As the little remaining free carbon dioxide begins to diffuse out, the following reaction is driven to the right:

$$H^+ + HCO_3^- \longrightarrow H_2CO_3 \longrightarrow H_2O + CO_2\uparrow$$

bicarbonate ion

"Up" arrow indicates carbon dioxide is leaving the body.

The enzyme carbonic anhydrase (p. 253), present in red blood cells, speeds up the reaction. As the reaction proceeds, the respiratory pigment **hemoglobin** gives up the hydrogen ions (H^+) it has been carrying; HHb becomes Hb. Hb is called deoxyhemoglobin.

Now, hemoglobin more readily takes up oxygen and becomes oxyhemoglobin.

$$Hb + \downarrow O_2 \longrightarrow HbO_2$$

deoxyhemoglobin oxyhemoglobin

"Down" arrow indicates that oxygen is entering the body.

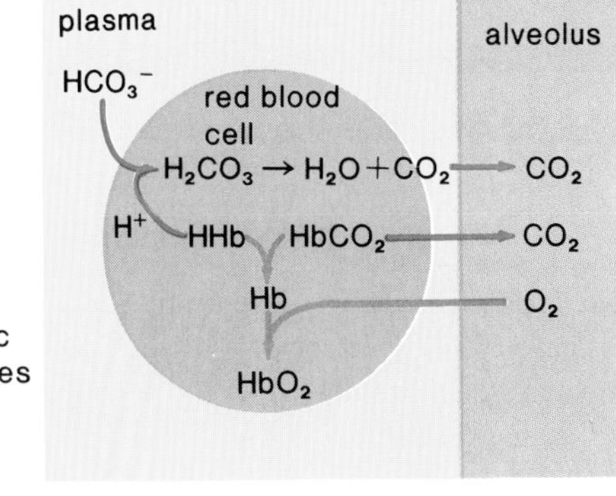

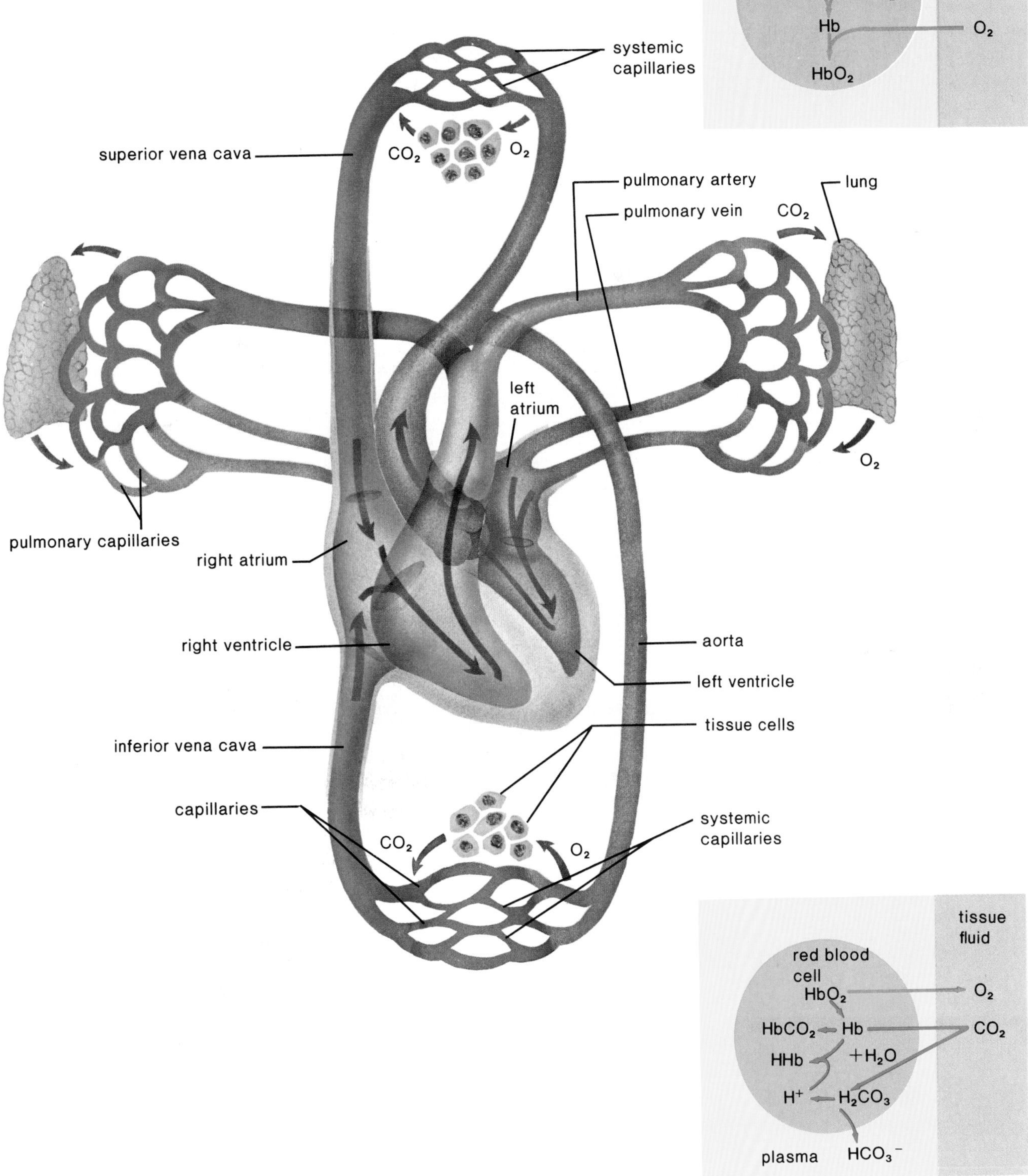

Figure 14.10

Diagram illustrating external and internal respiration. During external respiration in the lungs, carbon dioxide (CO_2) leaves blood and oxygen (O_2) enters blood. During internal respiration in the tissues, oxygen leaves blood and carbon dioxide enters blood. Steps necessary for gas exchange are shown for the lungs (*top*) and for the tissues (*bottom*).

THE RISKS OF SMOKING AND THE BENEFITS OF QUITTING

Based on available statistics, the American Cancer Society informs us of the risks of smoking and the benefits of quitting.

Risks of Smoking	Benefits of Quitting
Shortened life expectancy. Twenty-five-year-old, 2-pack-a-day smokers have a life expectancy of 8.3 years shorter than nonsmoking contemporaries. Other smoking levels: proportional risk.	***Reduced risk of premature death.*** After 10–15 years, exsmokers' risk approaches that of those who have never smoked.
Lung cancer. Smoking cigarettes is "major cause in both men and women."	***Gradual decrease in risk.*** After 10–15 years, risk approaches that of those who never smoked.
Larynx cancer. In all smokers (including pipe and cigar), it is 2.9–17.7 times that of nonsmokers.	***Gradual reduction of risk.*** Reaches normal after 10 years.
Mouth cancer. Cigarette smokers have 3–10 times as many oral cancers as nonsmokers. Pipes, cigars, chewing tobacco are also major risk factors. Alcohol seems to be synergistic carcinogen with smoking.	***Risk is reduced in first few years.*** Risk drops to level of nonsmoker in 10–15 years if both smoking/drinking are eliminated.
Cancer of esophagus. Cigarettes, pipes, and cigars increase risk of dying of esophageal cancer about 2–9 times. Synergistic relationship exists between smoking and alcohol.	***Risk reduction is expected.*** Risk is dose-related to smoking/drinking.
Cancer of bladder. Cigarette smokers have 7–10 times risk of bladder cancer as nonsmokers. Also synergistic with certain exposed occupations (dyestuffs, etc).	***Gradual decrease in risk.*** After 7 years, risk is same as nonsmokers.
Cancer of pancreas. Cigarette smokers have 2–5 times risk of dying of pancreatic cancer as nonsmokers.	***Risk reduction is expected.*** Evidence suggests risk is dose-related to smoking/drinking.
Coronary heart disease. Cigarette smoking is major factor; responsible for 120,000 excess U.S. deaths from coronary heart disease (CHD) each year.	***Sharply decreased risk after one year of not smoking.*** After 10 years, exsmokers' risk is same as that of those who never smoked.
Chronic bronchitis and pulmonary emphysema. Cigarette smokers have 4–25 times risk of death from these diseases as nonsmokers. Damage seen in lungs of even young smokers.	***Cough and sputum disappear during first few weeks after quitting.*** Lung function may improve, and rate of deterioration may slow down.
Stillbirth and low birth weight. Smoking mothers have more stillbirths and babies of low birth weight, who are more vulnerable to disease and death.	***Risk of stillbirth and low birth weight caused by smoking is eliminated*** in women who stop smoking before fourth month of pregnancy.
Childhood development. Children of smoking mothers are smaller, underdeveloped physically and socially 7 years after birth.	***Proper childhood development is expected.*** Since children of nonsmoking mothers are bigger and more advanced socially, inference is that not smoking during pregnancy might avoid underdeveloped children.

Figure 14.A
Normal lung versus cancerous lung. **a.** Normal lung with heart in place. Notice the healthy red color. **b.** Lungs of a heavy smoker. Notice how black the lungs are except where cancerous tumors have formed.

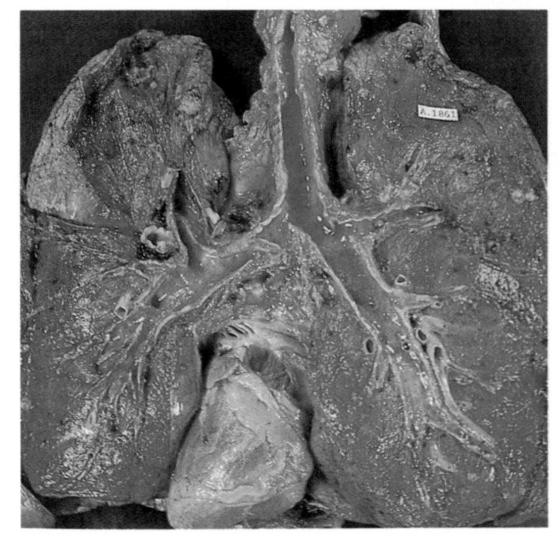

a.

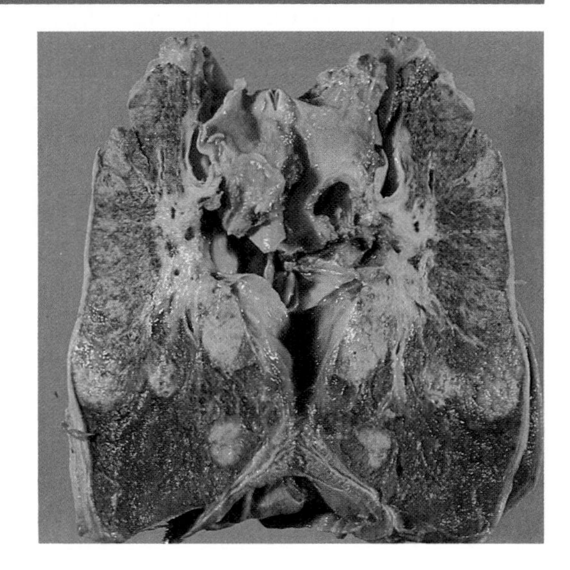

b.

Risks of Smoking	Benefits of Quitting
Peptic ulcer. Cigarette smokers get more peptic ulcers and die more often of them; cure is more difficult in smokers.	***More rapid and better healing is expected.*** Exsmokers get ulcers, but these are likely to heal more rapidly and completely than those of smokers.
Immune system. In smokers, allergies are present and there is impairment of immune system.	***Effects due to smoking are avoided.*** Impairment of the immune system was due to smoking in the first place. Since these are direct, immediate effects of smoking, they are obviously avoidable by not smoking.
Medicines. Alters pharmacologic effects of many medicines, diagnostic tests, and greatly increases risk of thrombosis with oral contraceptives.	***A return to normality.*** Majority of blood components elevated by smoking return to normal after cessation. Nonsmokers on the Pill have much lower risks of thrombosis.

Figure 14.11

Effect of conditions on hemoglobin saturation. Hemoglobin becomes more saturated in the lungs because the partial pressure of oxygen (PO_2) increases. Oxygen take-up is also enhanced because (**a**) temperature and (**b**) acidity decrease in the lungs. Hemoglobin is about 60–70% saturated in the tissues, and about 98–100% saturated in the lungs.

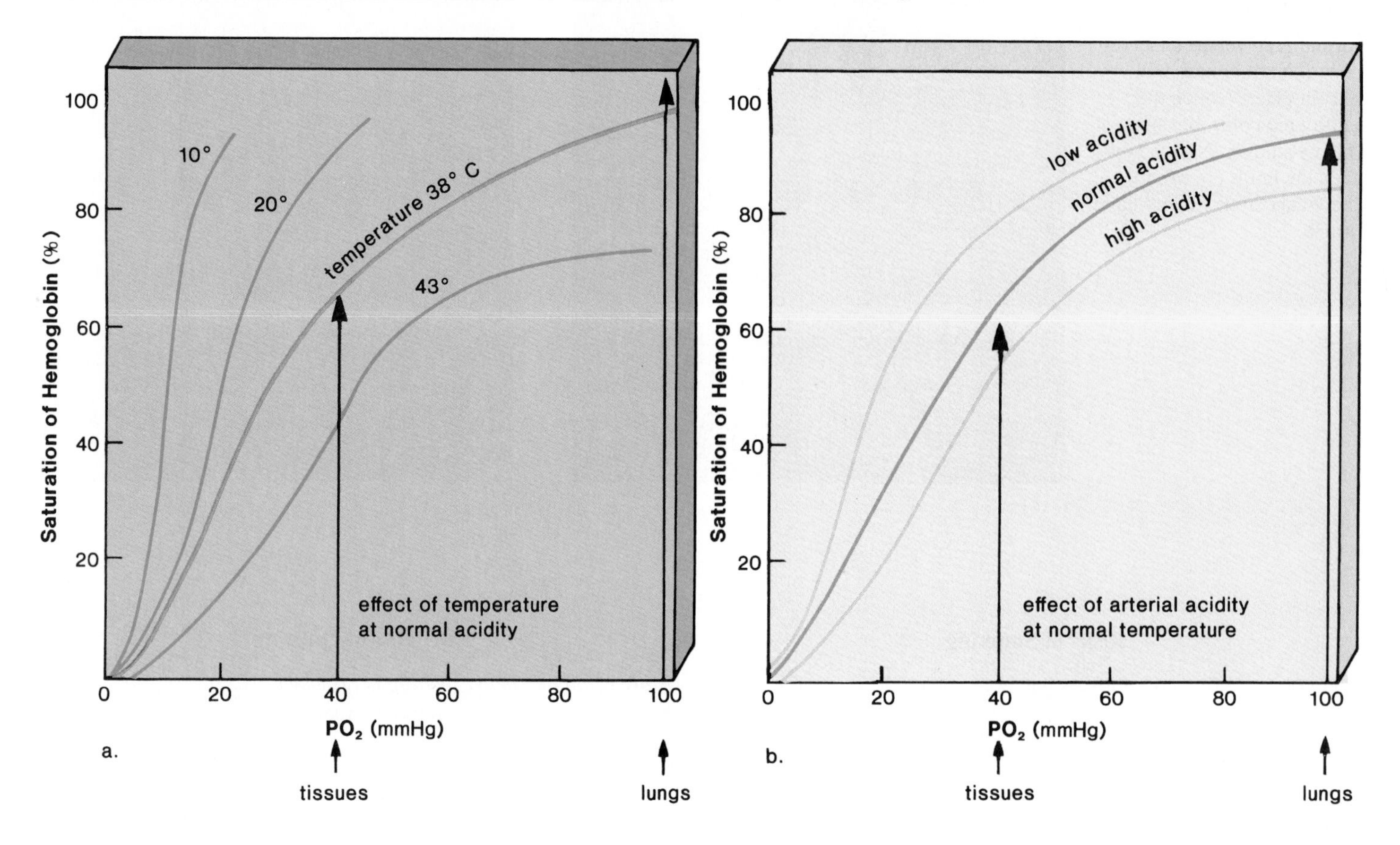

It is remarkable that at the partial pressure[2] of oxygen in the lungs (PO_2 = about 100 mm Hg), hemoglobin is about 98% saturated (fig. 14.11). Hemoglobin takes up oxygen in increasing amounts as the PO_2 increases and likewise gives it up as the PO_2 decreases. The curve begins to level off at about 90 mm Hg. This means that hemoglobin easily retains oxygen in the lungs but tends to release it in the tissues. This effect is potentiated by the fact that hemoglobin takes up oxygen more readily in the cool temperature (fig. 14.11*a*) and neutral pH (fig. 14.11*b*) of the lungs. On the other hand, it gives up oxygen more readily at the warmer temperature and more acidic pH of the tissues.[3]

External respiration, the exchange of oxygen (O_2) and carbon dioxide (CO_2) between air within alveoli and blood in pulmonary capillaries, is dependent on the process of diffusion.

Internal Respiration: Blood and Tissue Fluid

Blood that enters the systemic capillaries is bright red in color because red blood cells contain oxyhemoglobin. Oxyhemoglobin gives up oxygen, which diffuses out of blood into the tissues (fig. 14.10):

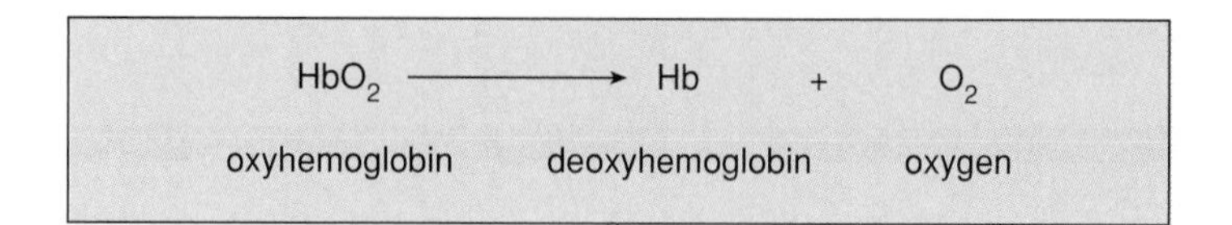

2. Air exerts pressure, and the amount of pressure each gas exerts in air is called its partial pressure, symbolized by a capital P.

3.

	pH	Temperature
Lungs	7.40	37°C (98.6° F)
Body	7.38	38°C (100.4° F)

Oxygen diffuses out of blood into the tissues because the oxygen concentration of tissue fluid is low—the cells continuously use up oxygen in cellular respiration. *Carbon dioxide diffuses into blood from the tissues* because the carbon dioxide concentration of tissue fluid is high. Carbon dioxide, produced continuously by cells, collects in tissue fluid.

After carbon dioxide diffuses into blood, it enters the red blood cells, where a small amount is taken up by hemoglobin, forming **carbaminohemoglobin.** Most of the carbon dioxide combines with water, forming carbonic acid (H_2CO_3), which dissociates to hydrogen ions (H^+) and bicarbonate ions (HCO_3^-). The enzyme **carbonic anhydrase,** present in red blood cells, speeds up the reaction:

$$\underset{\text{carbon dioxide}}{CO_2} + \underset{\text{water}}{H_2O} \rightleftharpoons \underset{\text{carbonic acid}}{H_2CO_3} \rightleftharpoons \underset{\text{hydrogen ion}}{H^+} + \underset{\text{bicarbonate ion}}{HCO_3^-}$$

The globin portion of hemoglobin combines with excess hydrogen ions produced by the reaction, and Hb becomes HHb called **reduced hemoglobin.** In this way, the pH of blood remains fairly constant. Bicarbonate ions diffuse out of red blood cells and are carried in the plasma. Blood that leaves the capillaries is deep purple in color because red blood cells contain reduced hemoglobin.

> Internal respiration, the exchange of oxygen and carbon dioxide between blood in the systemic capillaries and tissue fluid, is dependent on the process of diffusion. Do 14.2 Critical Thinking, found at the end of the chapter.

Respiration and Health

We have seen that the entire respiratory tract has a warm, wet mucous membrane lining, which is constantly exposed to environmental air. The quality of this air, determined by the pollutants and the microbes it contains, can affect our health.

Respiratory Tract Infections

Microbes frequently spread from one individual to another by way of the respiratory tract. Droplets from one single sneeze can be loaded with billions of bacteria or viruses. The mucous membranes are protected by mucus and by the constant beating of the cilia, but if the number of infective agents is large and/or our resistance is reduced, respiratory infections such as colds and influenza (flu) can result. Other more serious infections and disorders are discussed here.

Bronchitis: Acute and Chronic

Viral infections can spread from the nasal cavities to the sinuses (sinusitis), to the middle ears (otitis media), to the larynx (laryngitis), and to the bronchi (bronchitis). Acute bronchitis (fig. 14.12) is usually caused by a secondary bacterial infection of the bronchi, resulting in a heavy mucus discharge with much coughing. Acute bronchitis usually responds to antibiotic therapy. Chronic bronchitis, on the other hand, is not necessarily due to infection. It is often caused by constant irritation of the lining of the bronchi, which as a result undergo degenerative changes, including the loss of cilia and their normal cleansing action. There is frequent coughing, and the individual is more susceptible to respiratory infections. Chronic bronchitis is most often seen in cigarette smokers.

Strep Throat: Risks Rheumatic Fever

Strep throat is a very severe throat infection caused by the bacterium *Streptococcus pyogenes.* Swallowing may be difficult, and there is fever. Unlike a viral infection, strep throat should be treated with antibiotics. If not treated, it can lead to complications such as rheumatic fever, which can permanently damage the heart valves.

Lung Disorders

Pneumonia and tuberculosis are 2 serious infections of the lungs ordinarily controlled by antibiotics. Two other illnesses discussed, emphysema and lung cancer, are not due to infections; in most instances, they are due to cigarette smoking.

Pneumonia: Lobules Fill and Breathing Ceases

Most forms of pneumonia are caused by a microbe that has infected the lungs. AIDS patients are subject to a particularly rare form of pneumonia caused by the protozoan *Pneumocystis carinii.* Sometimes, pneumonia is localized in specific lobules of the lungs. These lobules become nonfunctional as they fill with mucus and pus. Obviously, the more lobules involved, the more serious the infection.

Figure 14.12

Common bronchial and pulmonary infectious diseases and disorders. Exposure to microbes and/or polluted air, including cigarette and cigar smoke, causes the diseases and disorders pictured here.

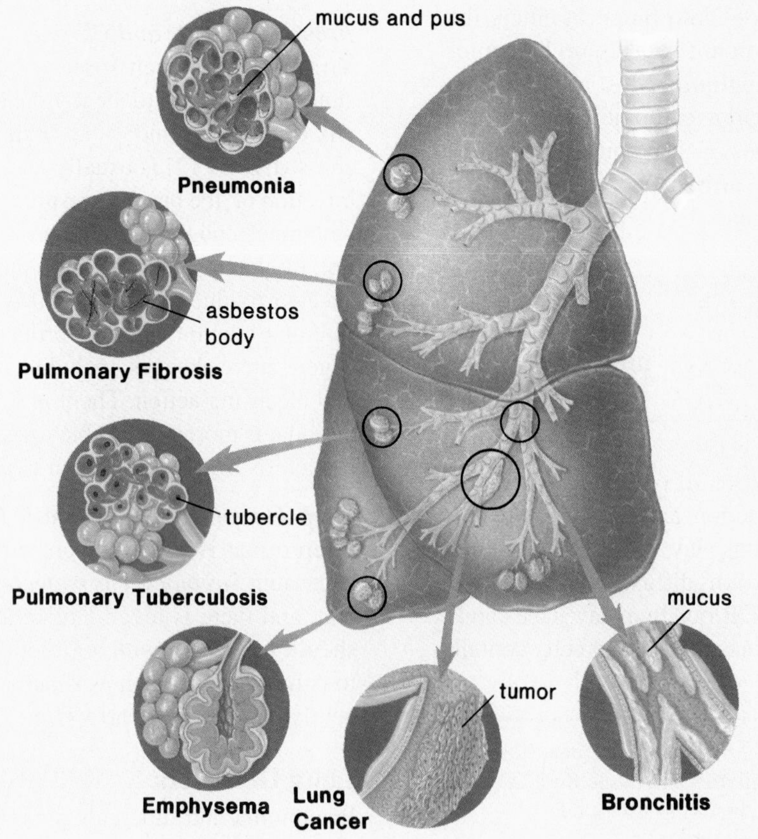

mucus and pus

Pneumonia

asbestos body

Pulmonary Fibrosis

tubercle

Pulmonary Tuberculosis

mucus

tumor

Emphysema

Lung Cancer

Bronchitis

Pulmonary Tuberculosis: Past and Recent Threat

Pulmonary tuberculosis is caused by the tubercle bacillus, a type of bacterium. When a person has tuberculosis, the alveoli burst and are replaced by inelastic connective tissue. It is possible to tell if a person has ever been exposed to tuberculosis with a skin test in which a highly diluted extract of the bacilli is injected into the skin of the patient. A person who has never been in contact with the bacillus shows no reaction, but one who has developed immunity to the organism shows an area of inflammation that peaks in about 48 hours. If these bacilli invade the lung tissue, the cells build a protective capsule about the foreigners to isolate them from the rest of the body. This tiny capsule is called a *tubercle*. If the resistance of the body is high, the imprisoned organisms die, but if the resistance is low, the organisms eventually can be liberated. If a chest X ray detects tubercles, the individual is put on appropriate drug therapy to ensure the localization of the disease and the eventual destruction of any live bacterial organisms.

Tuberculosis killed about 100,000 people in the United States each year before the middle of this century, when antibiotic therapy brought it largely under control. In recent years, however, the incidence of tuberculosis is on the rise, particularly among AIDS patients, the homeless, and the rural poor. Worse, the new strains are resistant to the usual antibiotic therapy. Therefore, some physicians would like to again make use of sanatoriums to quarantine patients during treatment.

Emphysema: Bronchioles Collapse and Alveoli Burst

Emphysema refers to the destruction of lung tissue, with accompanying ballooning or inflation of the lungs due to trapped air. The trouble stems from the destruction and collapse of the bronchioles. When this occurs, the alveoli are cut off from renewed oxygen supply, and the air within them is trapped. The trapped air very often causes the alveolar walls to rupture (fig. 14.12), with a fibrous thickening of associated blood vessel walls. The victim is breathless and may have a cough. Since the surface area for gas exchange is reduced, not enough oxygen reaches the heart and the brain. Even so, the heart works furiously to force more blood through the lungs, which can lead to a heart condition. Lack of oxygen to the brain can make the person feel depressed, sluggish, and irritable.

Pulmonary Fibrosis: Inhaling Particles

Inhaling particles such as silica (sand), coal dust, and asbestos (fig. 14.12) can lead to pulmonary fibrosis, a condition in which fibrous connective tissue builds up in the lungs. Breathing capacity can be seriously impaired, and the development of cancer is common. Since asbestos has been used so widely as a fireproofing and insulating agent, unwarranted exposure has occurred. It is projected that 2 million deaths could be caused by asbestos exposure—mostly in the workplace—between 1990 and 2020.

Lung Cancer: Women Catch Up

Lung cancer used to be more prevalent in men than in women, but recently it has surpassed breast cancer as a cause of death in women. This can be linked to an increase in the number of women who smoke today. Autopsies on smokers have revealed the progressive steps by which the most common form of lung cancer develops. The first event appears to be thickening and callusing of the cells lining the bronchi. (Callusing occurs whenever cells are exposed to irritants.) Then there is a loss of cilia so that it is impossible to prevent dust and dirt from settling in the lungs. Following this, cells with atypical nuclei appear in the callused lining. A tumor consisting of disordered cells with atypical nuclei is considered to be cancer in situ (at one location). A final step occurs when some of these cells break loose and penetrate other tissues, a process called metastasis. Now the cancer has spread. The tumor may grow until the bronchus is blocked, cutting off the supply of air to that lung. The entire lung then collapses, the secretions trapped in the lung spaces become infected, and pneumonia or a lung abscess (localized area of pus) results. The only treatment that offers a possibility of cure is to remove the lung completely before secondary growths have had time to form. This operation is called a *pneumonectomy*.

> The incidence of lung cancer is much higher in individuals who smoke than in those who do not.

Current research indicates that *involuntary smoking*, simply breathing in air filled with cigarette smoke, can also cause lung cancer and other illnesses associated with smoking. The reading on pages 250–251 lists the various illnesses associated with smoking. If a person stops both voluntary and involuntary smoking and if the body tissues are not already cancerous, they usually return to normal over time.

SUMMARY

Air enters and exits the lungs by way of the respiratory tract (table 14.2). Inspiration begins when the respiratory center in the medulla oblongata sends excitatory nerve impulses to the diaphragm and the rib cage. As they contract, the diaphragm lowers and the rib cage moves upward and outward; the lungs expand, creating a partial vacuum, which causes air to rush in. Nerves within the expanded lungs then send inhibitory impulses to the respiratory center. As the diaphragm relaxes, it resumes its dome shape, and as the rib cage retracts, air is pushed out of the lungs during expiration.

External respiration occurs when carbon dioxide (CO_2) leaves blood and oxygen (O_2) enters blood at the alveoli. Oxygen is transported to the tissues in combination with hemoglobin. Internal respiration occurs when oxygen leaves blood and carbon dioxide enters blood at the tissues. Carbon dioxide is carried to the lungs within the plasma as the bicarbonate ion (HCO_3^-). Hemoglobin combines with hydrogen ions and becomes reduced hemoglobin (HHb).

There are a number of illnesses associated with the respiratory tract. In addition to colds and flu, the lungs may be infected by the more serious pneumonia and tuberculosis. Two illnesses that have been attributed to breathing polluted air are emphysema and lung cancer.

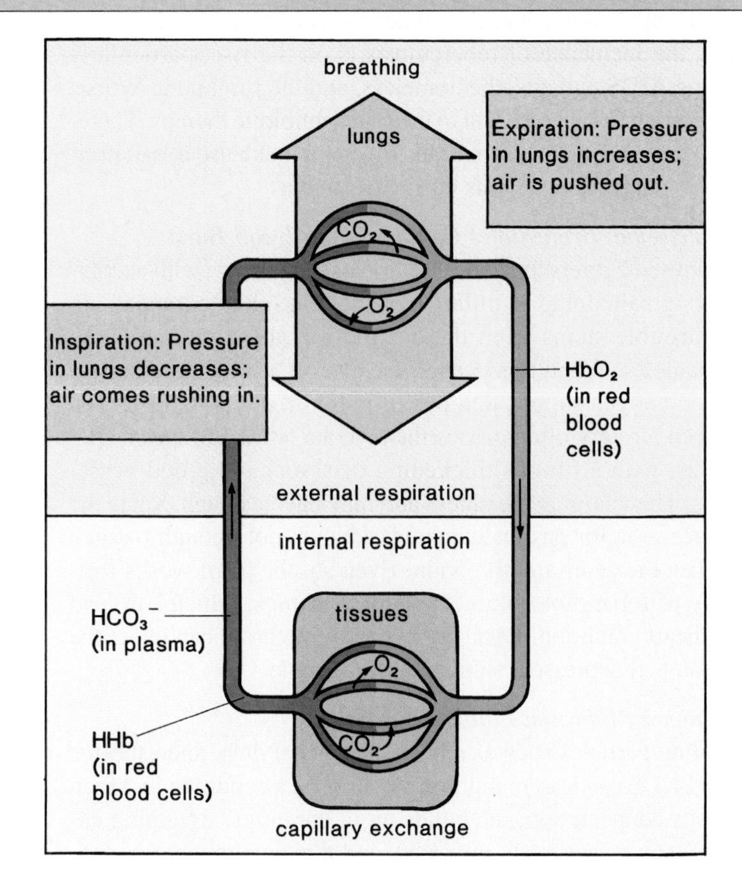

STUDY QUESTIONS

In order to practice **writing across the curriculum,** students should write out the answers to any or all of the study questions. The study questions are sequenced in the same order as the text.

1. Name and explain the 4 parts of respiration. (p. 239)
2. List the parts of the respiratory tract. What are the special functions of the nasal cavity, the larynx, and the alveoli? (p. 241)
3. What are the steps in inspiration and expiration? How is breathing controlled? (p. 245)
4. Why can't we breathe through a very long tube? (p. 247)
5. What physical process is believed to explain gas exchange? (p. 248)
6. What 2 equations are needed to explain external respiration? (pp. 248, 252)
7. How is hemoglobin remarkably suited to its job? (p. 248)
8. What 2 equations are needed to explain internal respiration? (pp. 252–253)
9. Name and discuss some respiratory tract infections. (p. 253)
10. What are emphysema and pulmonary fibrosis, and how do they affect a person's health? (p. 255)
11. By what steps is cancer believed to develop in the person who smokes? (p. 255)

OBJECTIVE QUESTIONS

1. In tracing the path of air, the _______ immediately follows the pharynx.
2. The lungs contain air sacs called _______.
3. The breathing rate is primarily regulated by the amount of _______ and _______ in blood.
4. Air enters the lungs after they have _______.
5. Carbon dioxide (CO_2) is carried in blood as the _______ ion.
6. The hydrogen ions (H^+) given off when carbonic acid (H_2CO_3) dissociates are carried by _______.
7. Gas exchange is dependent on the physical process of _______.
8. Reduced hemoglobin becomes oxyhemoglobin in the _______.
9. The most likely cause of emphysema and chronic bronchitis is _______.
10. Most cases of lung cancer actually begin in the _______.
11. Label this diagram of the human respiratory tract.

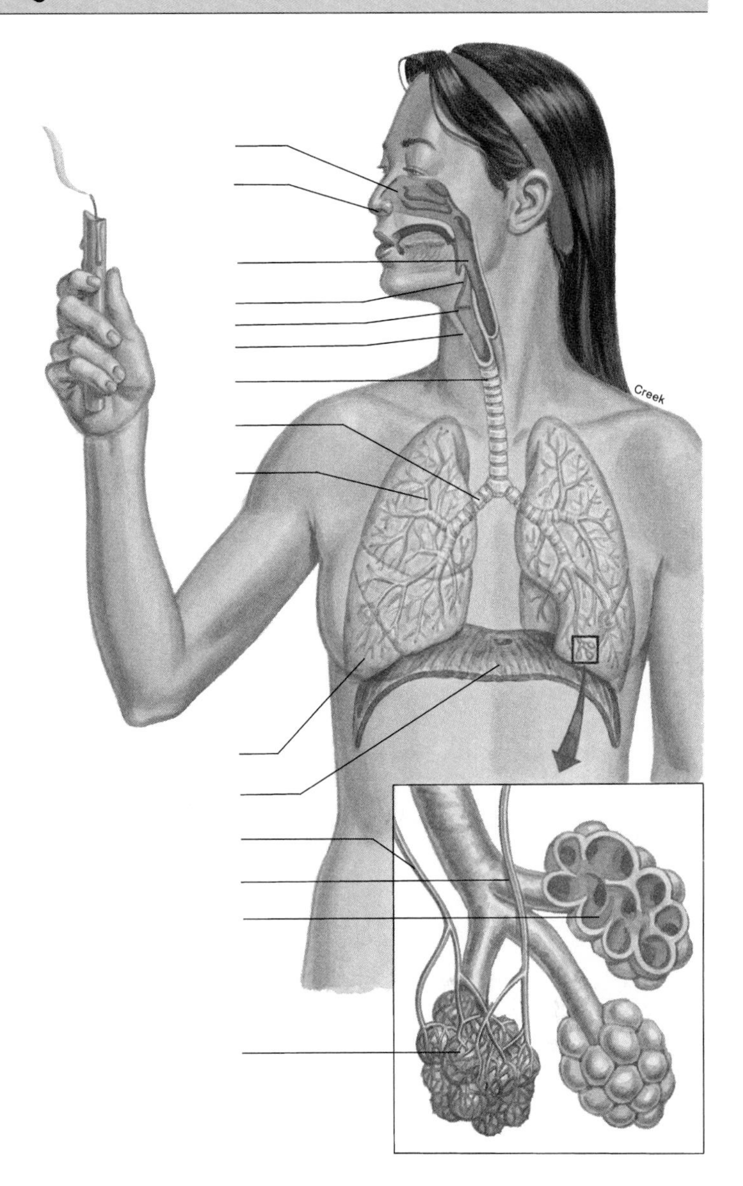

CRITICAL THINKING

In order to practice **writing across the curriculum,** students should write out the answers to any or all of the critical thinking questions. Suggested answers to the critical thinking questions are in appendix E.

14.1

1. Humans use negative pressure to fill the lungs, and frogs use positive pressure. What is the difference between negative pressure and positive pressure?
2. Birds have a flow-through system—the air flows from the lungs into air sacs and then out by way of a separate set of tubes. In what way is this more efficient than human breathing?
3. Frogs have thin, moist skin, and reptiles have thick skin. Which animal do you predict practices skin breathing? Which animal has better developed lungs?
4. Humans have a diaphragm, reptiles do not. In what way does a diaphragm assist breathing?

14.2

1. Why is it better to give a person a mixture of oxygen (O_2) and carbon dioxide (CO_2) rather than pure oxygen to stimulate breathing?
2. Why is it impossible for a person to commit suicide by holding his or her breath? (Hint—What gas builds up?)

3. Why is it workable for the carotid and aortic bodies to be sensitive to the hydrogen ion concentration (H^+) of blood rather than to oxygen?

4. Why would you predict that the respiratory center is sensitive to the *presence* of carbon dioxide rather than to the *absence* of oxygen in blood?

SELECTED KEY TERMS

alveolus (al-ve´ o-lus) (pl., alveoli) air sac of a lung.
bronchiole (brong´ke ol) the smaller air passages in the lungs that eventually terminate in alveoli.
bronchus (brong´ kus) (pl., bronchi) one of 2 major divisions of the trachea leading to the lungs.
diaphragm (di´ ah-fram) a sheet of muscle that separates the thoracic cavity from the abdominal cavity in higher animals.
expiration process of expelling air from the lungs; exhalation.
external respiration exchange between blood and alveoli of carbon dioxide and oxygen.
hemoglobin a red iron-containing pigment in blood that combines with and transports oxygen.
inspiration the act of breathing in.
internal respiration exchange between blood and tissue fluid of oxygen and carbon dioxide.
larynx cartilaginous organ located between the pharynx and the trachea that contains the vocal cords; voice box.
rib cage the top and sides of the thoracic cavity; contains ribs and intercostal muscles.
trachea (tra´ ke ah) in vertebrates, a tube supported by C-shaped cartilaginous rings that lies between the larynx and the bronchi; also called the windpipe.
ventilation breathing; the process of moving air into and out of the lungs.
vocal cord fold of tissue within the larynx; creates vocal sounds when it vibrates.

15

EXCRETION

Chapter Concepts

1.
Excretion rids the body of unwanted substances, particularly the end products of metabolism. 260, 275

2.
Several organs assist in the process of excretion, but the kidneys, which are a part of the urinary system, are the primary organs of excretion. 262, 275

3.
The kidneys contain nephrons, which produce urine in several steps. The formation of urine rids the body of nitrogenous wastes and regulates the salt/water content and the pH of blood. 267, 275

4.
The kidneys, important organs of homeostasis, are under the control of various hormones. 272, 275

5.
The malfunction of the kidneys causes illness and even death. 274, 275

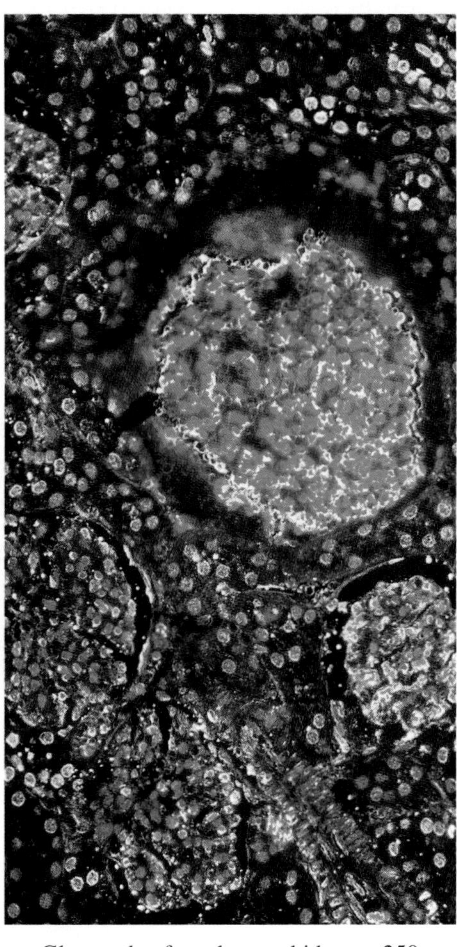

Glomerulus from human kidney, ×250

Chapter Outline

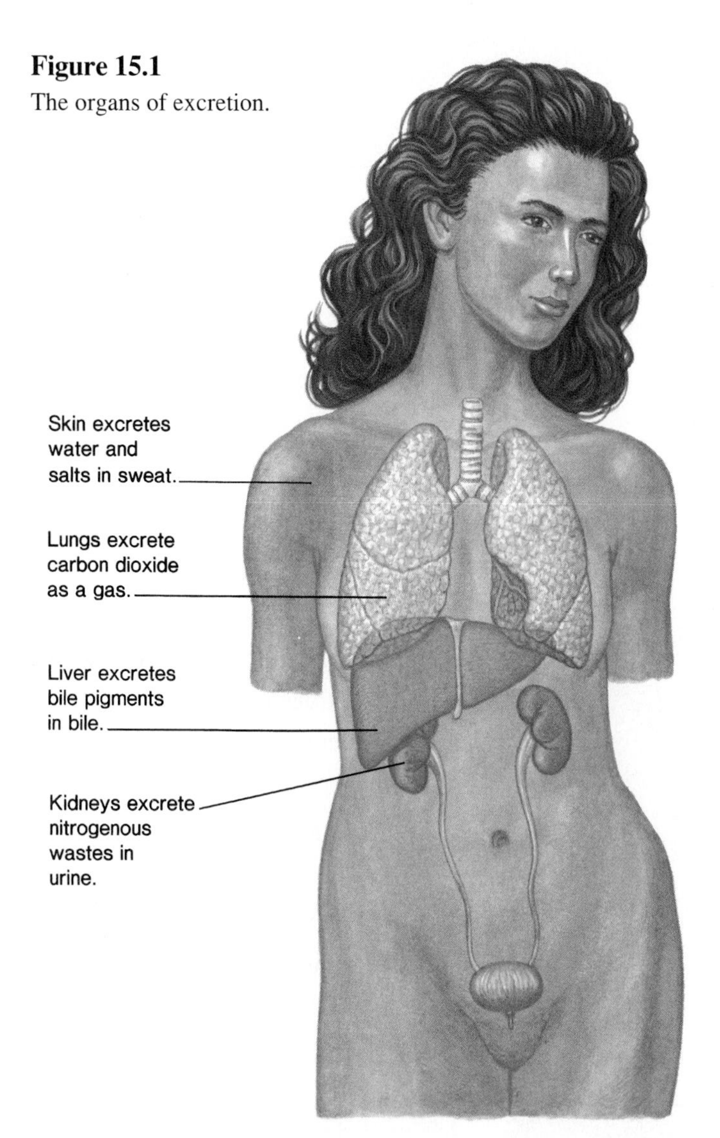

Figure 15.1
The organs of excretion.

Table 15.1
Some Metabolic End Products

Name	Process	Primary Excretory Organs
Nitrogenous Wastes		
Ammonia	Amino acid metabolism	Kidneys
Urea	Amino acid metabolism	Kidneys, skin
Uric acid	Nucleotide metabolism	Kidneys
Creatinine	Creatine phosphate metabolism	Kidneys
Other Wastes		
Water	Various	Kidneys
Salts	Various	Kidneys
Carbon dioxide	Cellular respiration	Lungs
Bile pigments	Hemoglobin metabolism	Liver

THE composition of blood serving the tissues remains relatively constant due to the action of several organs. In previous chapters, we discussed how the digestive tract and the lungs add nutrients and oxygen to blood. In this chapter, we discuss how the organs of excretion (fig. 15.1) remove substances from blood and thereby contribute to homeostasis. Most other animals also have organs of excretion.

At this point, it is helpful to remember that the term *defecation,* not *excretion,* is used to refer to the elimination of feces from the body. Substances that are excreted are waste products of metabolism. Undigested food and bacteria, which make up feces, have never entered the body proper.

Excretory Substances and Organs

Excretion rids the body of metabolic wastes. Amino acids, nucleotides, and creatine phosphate all contain nitrogen, and their metabolism results in nitrogenous wastes (table 15.1). In addition to nitrogenous wastes, water, salts, carbon dioxide, and bile pigments are end products that are excreted.

Figure 15.2
Marine fishes. Most aquatic animals, including most bony fishes, excrete ammonia as their nitrogenous waste. Marine fishes, such as these big-eye jacks, excrete excess salt by way of their gills.

Nitrogenous Wastes

Ammonia (NH_3) arises from the deamination, or removal, of amino groups from amino acids. Ammonia is extremely toxic, and only animals that live in water and continually flush out their body with water excrete ammonia (fig. 15.2). In our body, deamination occurs in the liver, where ammonia is converted to urea.

ANIMAL ORGANS OF EXCRETION

Most animals have tubular organs that function in the excretion of nitrogenous wastes and osmotic regulation. Planarians have 2 strands of blind-end excretory tubules, which branch throughout the body before opening at excretory pores. Tissue fluid enters the excretory system at the flame cells, where cilia propel it into, through, and out the system. The beating cilia look like a flickering flame under the microscope, which accounts for the name of these cells.

Nearly every segment of an earthworm's body has a pair of tubules called nephridia, which begin with a ciliated opening and end at an excretory pore. As tissue fluid moves through the tubules, certain substances are reabsorbed and carried away by a capillary network surrounding the tubules. The urine contains metabolic wastes, salts, and water.

Insects have blind-end tubules called Malpighian tubules, which are attached to the gut. Uric acid, along with tissue fluid, enters these long, thin tubules before moving into the gut. Here, water and other useful substances are reabsorbed, but uric acid eventually passes out of the gut. Insects in dry environments reabsorb most of the water and excrete a dry, semisolid mass.

Osmotic regulation must be suited to the environment of the animal. For example, fishes that live in fresh water need to rid the body of water while retaining salt. They produce large quantities of hypotonic urine and actively transport salt into blood at the gills. On the other hand, fishes that live in salt water need to acquire water while ridding the body of salt. They drink water constantly and excrete the excess salt by way of the gills. Their urine is isotonic to blood.

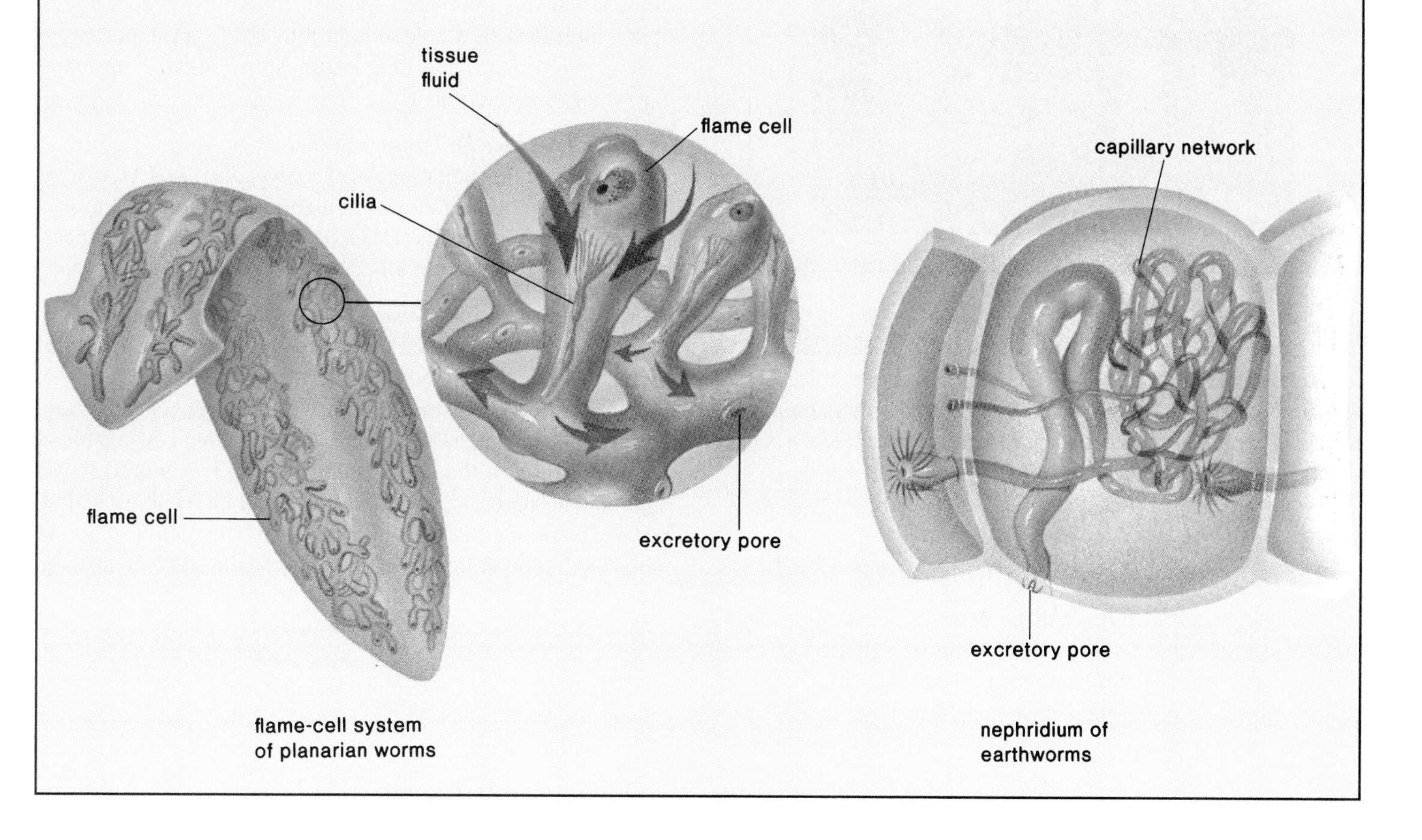

flame-cell system of planarian worms

nephridium of earthworms

Figure 15.3
Seabirds on cliff. Birds excrete uric acid, a solid material, as their nitrogenous waste. It is mixed with feces in a common repository for the urinary, digestive, and reproductive systems. Seabirds congregate in such numbers that their droppings build up to give a nitrogen-rich substance called guano. At one time, guano was harvested for natural fertilizer.

Urea is produced in the liver by a complicated series of reactions called the urea cycle. In this cycle, carrier molecules take up carbon dioxide (CO_2) and 2 molecules of ammonia to release a combined product, urea, which is a much less toxic substance:

$$H_2N-\overset{\overset{\displaystyle O}{\|}}{C}-NH_2$$

Many terrestrial animals that need to conserve water excrete **uric acid** as their nitrogenous end product (fig. 15.3). In humans, uric acid only occurs when nucleotides are broken down; if uric acid is present in excess, it precipitates out of the plasma. Crystals of uric acid sometimes collect in the joints, producing a painful ailment called *gout.*

Creatinine is an end product of muscle metabolism. It results when creatine phosphate, a molecule that serves as a reservoir of high-energy phosphate, breaks down.

Water and Other Wastes

Other excretory substances are water (H_2O), salts (ions), carbon dioxide (CO_2), and bile pigments.

Water: Affects Blood Pressure

Water (H_2O) is an end product of metabolism, notably aerobic cellular respiration. Water, taken into the body when we consume food or liquids, is often excreted. The volume of blood, determined by its fluid content, affects blood pressure. Treatment of hypertension sometimes includes the administration of a diuretic drug, which increases the excretion of sodium and water by the kidneys.

Salts: Important Balance

Various salts (ions) that participate in metabolism are excreted, largely by the kidneys. The blood level of salts is important to the pH, the osmotic pressure, and the electrolyte balance of blood. Various ions have specific functions. For example, potassium ions (K^+) and sodium ions (Na^+) are important to nerve conduction, and magnesium ions (Mg^{++}) help many enzymes function properly.

Carbon Dioxide: To Lungs and Out

The lungs are the major organs of *carbon dioxide* (CO_2) excretion, although the kidneys are also important. The kidneys excrete bicarbonate ions (HCO_3^-), the form in which most carbon dioxide is carried in blood.

Bile Pigments: From Heme

Bile pigments are derived from the heme portion of hemoglobin and are incorporated into bile within the liver (fig. 15.4). (The iron portion of the heme molecule is recycled; it is returned to the bone marrow to be used in making new red blood cells.) Although the liver produces bile, it is stored in the gallbladder before passing into the small intestine by way of ducts. If for any reason a bile duct is blocked, bile spills out into blood and collects in the skin, causing it to be colored yellow (p. 180).

Excretion rids the body of metabolic wastes, such as urea, a nitrogenous waste, carbon dioxide, salts, and water.

Four Excretory Organs

The kidneys are the primary excretory organs, but other organs also function in excretion (fig. 15.1), including those described in the discussion that follows.

Figure 15.4

The liver. This is an organ of excretion because it adds end products of hemoglobin breakdown to bile. These are the bile pigments. The liver makes bile, which is stored in the gallbladder before being sent to the small intestine by way of ducts.

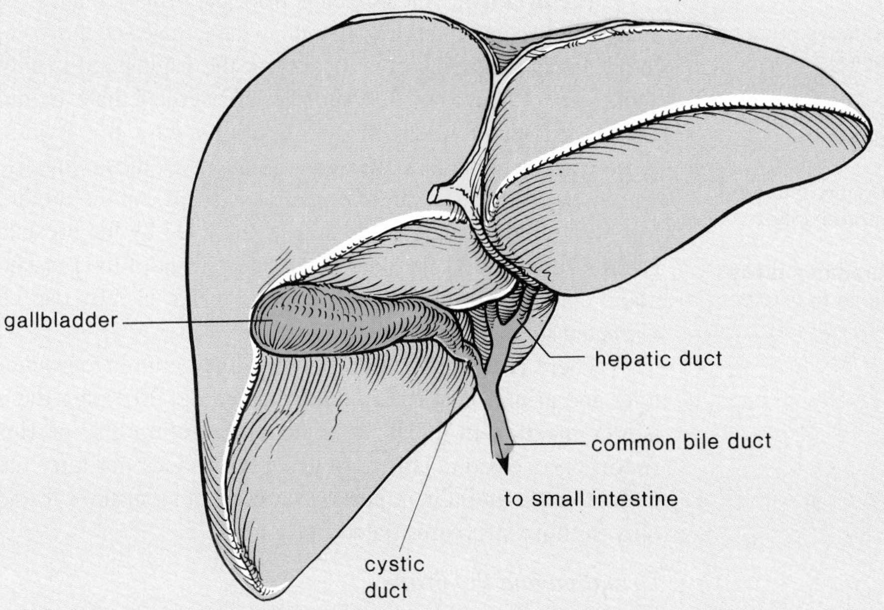

Table 15.2
Composition of Urine: 95% Water, 5% Solids

Organic Wastes	per 1,500 ml of Urine
Urea	30 g
Creatinine	1–2 g
Uric acid	1 g
Salts	25 g
Positive Ions	**Negative Ions**
Sodium	Chlorides
Potassium	Sulfates
Magnesium	Phosphates
Ammonium	Bicarbonate
Calcium	

Skin Sweats

The sweat glands in the skin (see fig. 10.7) excrete perspiration, which is a solution of water, salt, and some urea. In the dermis, a sweat gland is a coiled tubule, but then it straightens as it passes through and exits the epidermis. Although perspiration is a form of excretion, we perspire not so much to rid the body of wastes as to cool it. The body cools because heat is lost as perspiration evaporates. Sweating keeps the body temperature within normal range during muscular exercise or when the outside temperature rises. In times of kidney failure, urea is excreted by the sweat glands and forms a so-called urea frost on the skin.

Liver Excretes Bile Pigments

The liver excretes bile pigments, which are incorporated into bile, a substance stored in the gallbladder before it passes into the small intestine by way of ducts (fig. 15.4). The yellow pigment found in urine, called urochrome, also is derived from the breakdown of heme, but this pigment is deposited in blood and is subsequently excreted by the kidneys.

Lungs Remove CO_2 and H_2O

The process of expiration (breathing out) not only removes carbon dioxide (CO_2) from the body, it also results in the loss of water (H_2O). The air we exhale contains moisture, as demonstrated by breathing onto a cool mirror.

Kidneys Produce Urine

The kidneys produce urine, which ordinarily contains the organic wastes and inorganic salts listed in table 15.2. The kidneys are part of the urinary system.

> There are various organs that excrete metabolic wastes, but ordinarily the kidneys primarily rid the body of urea.

Figure 15.5
The urinary system. Urine is found only within the kidneys, the ureters, the urinary bladder, and the urethra.

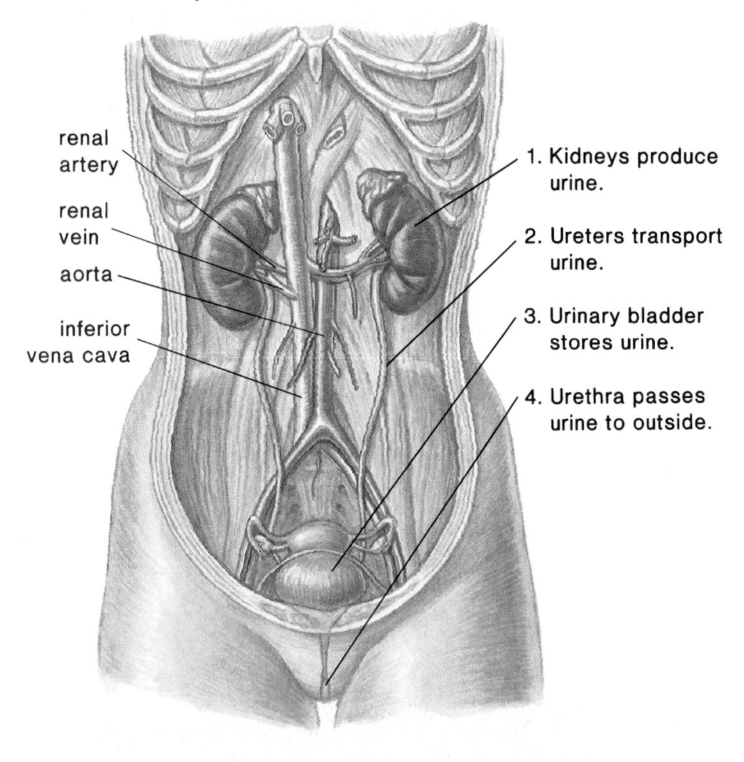

The Urinary System

The urinary system includes the structures illustrated in figure 15.5. The organs are listed in order, according to the path of urine.

The Path of Urine

Urine is made by the **kidneys,** bean-shaped, reddish brown organs, each about the size of a fist. One kidney is found on either side of the vertebral column, just below the diaphragm. The kidneys lie in depressions against the deep muscles of the back, beneath the membranous lining of the abdominal cavity, where they also receive some protection from the lower rib cage. Each is covered by a tough fibrous capsule of connective tissue overlaid by adipose tissue.

The **ureters** are muscular tubes that convey the urine from the kidneys toward the bladder by peristalsis. Urine enters the bladder by peristaltic contractions, in jets that occur at the rate of 5 per minute.

The **urinary bladder,** which can hold up to 600 ml of urine, is a hollow, muscular organ that gradually expands as urine enters. In the male, the urinary bladder lies in front of the rectum, the seminal vesicles, and the vas deferens. In the female, the urinary bladder is in front of the uterus and the upper vagina.

The **urethra,** which extends from the urinary bladder to an external opening, differs in length in the female and the male. In the female, the urethra is in front of the vagina and is only about 4 cm (1.6 in) long. The short length of the female urethra invites bacterial invasion and explains why the female is more prone to bladder infections. In the male, the urethra averages 20 cm (8 in) when the penis is flaccid. As the urethra leaves the male urinary bladder, it is encircled by the prostate gland (see fig. 20.1). In older men, enlargement of the prostate gland can prevent urination, a condition that usually can be corrected surgically.

There is no connection between the genital (reproductive) and urinary systems in females (see fig. 20.5), but there is a connection in males. In males, the urethra also carries sperm on occasion. This double function does not alter the path of urine, and it is important to realize that urine is found only in those structures noted in figure 15.5.

Urination and the Brain

When the urinary bladder fills with urine, stretch receptors send nerve impulses to the spinal cord; nerve impulses leaving the cord then cause the urinary bladder to contract and the sphincters to relax so that urination is possible. In older children and adults, it is possible for the brain to control this reflex, delaying urination until a suitable time.

Only the urinary system, consisting of the kidneys, the urinary bladder, the ureters, and the urethra, ever holds urine.

Kidneys: Three Regions

On the concave side of each kidney there is a depression where the renal blood vessels and the ureters enter (fig. 15.6*a*). When a kidney is sliced lengthwise, it is possible to make out 3 regions. The **renal cortex** is an outer granulated layer that dips down in between a radially striated, or lined, layer called the **renal medulla.** The renal medulla contains cone-shaped tissue masses called *renal pyramids* (fig. 15.6*b*). The **renal pelvis** is an inner space, or cavity, that is continuous with the ureter.

Over One Million Nephrons

Microscopically, the kidney is composed of over one million **nephrons,** sometimes called renal or kidney tubules (figs. 15.6*c* and 15.7). Each nephron is made up of several parts. The blind end of the nephron is pushed in on itself to form a

Figure 15.6

Microscopic anatomy of the kidney. **a.** A longitudinal section of the kidney, showing the blood supply. Note that the renal artery divides to give smaller arteries. Smaller veins join to form the renal vein. **b.** Same section without the blood supply. Now it is easier to make out the renal cortex, the renal medulla, and the renal pelvis, which connects with the ureter. The renal pyramids make up the renal medulla. **c.** An enlargement of a lobe, showing the placement of nephrons in a renal pyramid.

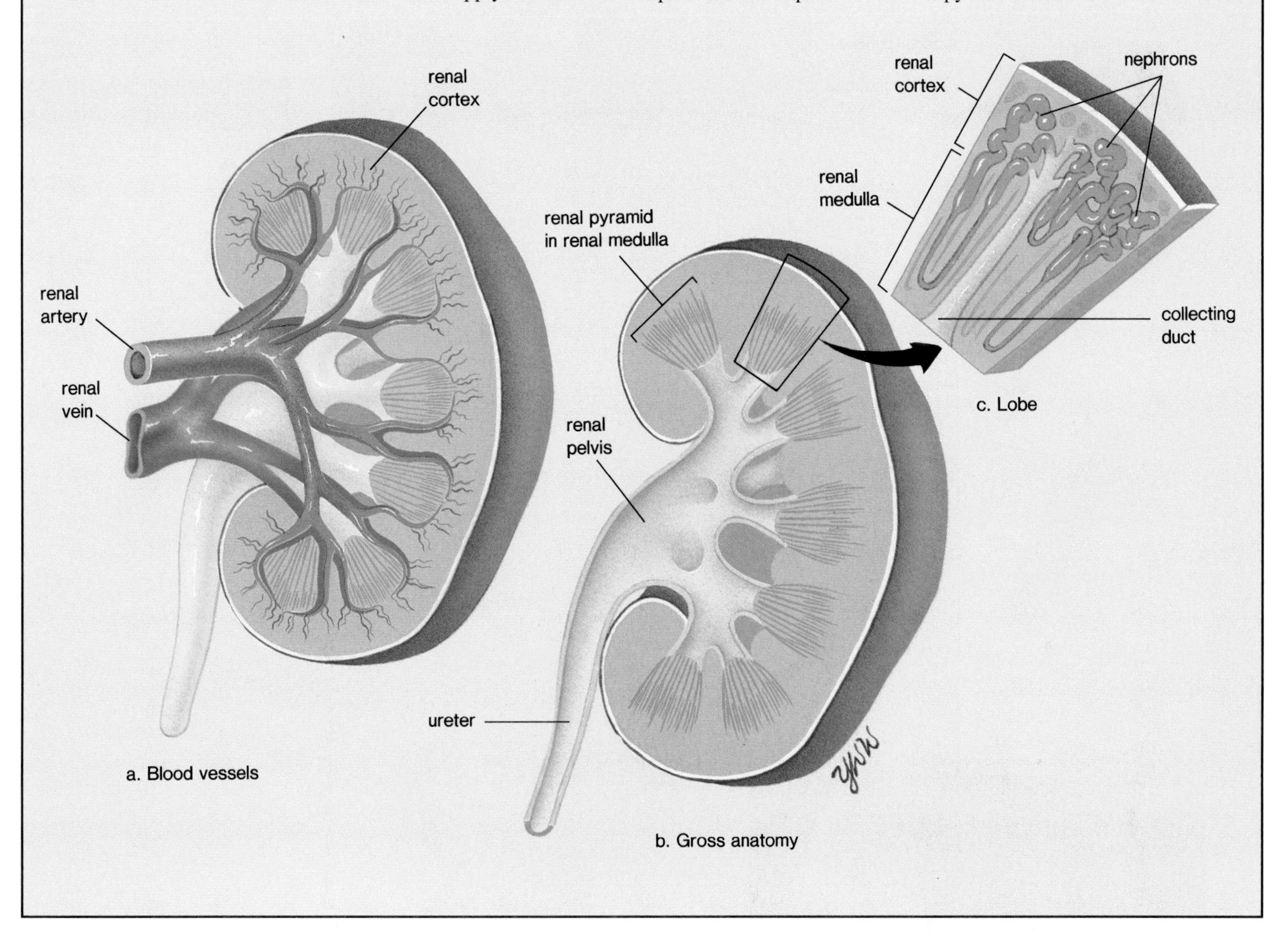

cuplike structure called **Bowman's capsule** (also called the glomerular capsule). The outer layer of Bowman's capsule is composed of squamous epithelial cells; the inner layer is composed of specialized cells that allow easy passage of molecules. Next, there is a **proximal** (meaning near the Bowman's capsule) **convoluted tubule,** in which the cells are composed of cuboidal epithelium, with many mitochondria and an inner brush border (tightly packed microvilli). Then simple squamous epithelium appears as the tube narrows and makes a U-turn to form the portion of the tubule called the **loop of Henle** (also called the loop of the nephron). This leads to the **distal** (far from Bowman's capsule) **convoluted tubule,** where the cells are composed of cuboidal epithelium, again with mitochondria, but without a brush border. The distal convoluted tubules of several nephrons join to form one **collecting duct.** A kidney contains many collecting ducts, which enter the renal pelvis.

As shown in figure 15.6*c*, Bowman's capsule and the convoluted tubules always lie within the renal cortex. The loop of Henle dips down into the renal medulla; a few nephrons have a very long loop of Henle, which penetrates deep into the renal medulla. The collecting ducts are also located in the renal medulla, and they give the renal pyramids their lined appearance.

Figure 15.7

Nephron macroscopic and microscopic anatomy. Note the portions of a nephron that are in the renal cortex and the portions that are in the renal medulla. The insets show the microscopic anatomy. The large cells adhering to the glomerulus are actually from the inner lining of Bowman's capsule. They are spaced wide apart; these spaces are called filtration slits.

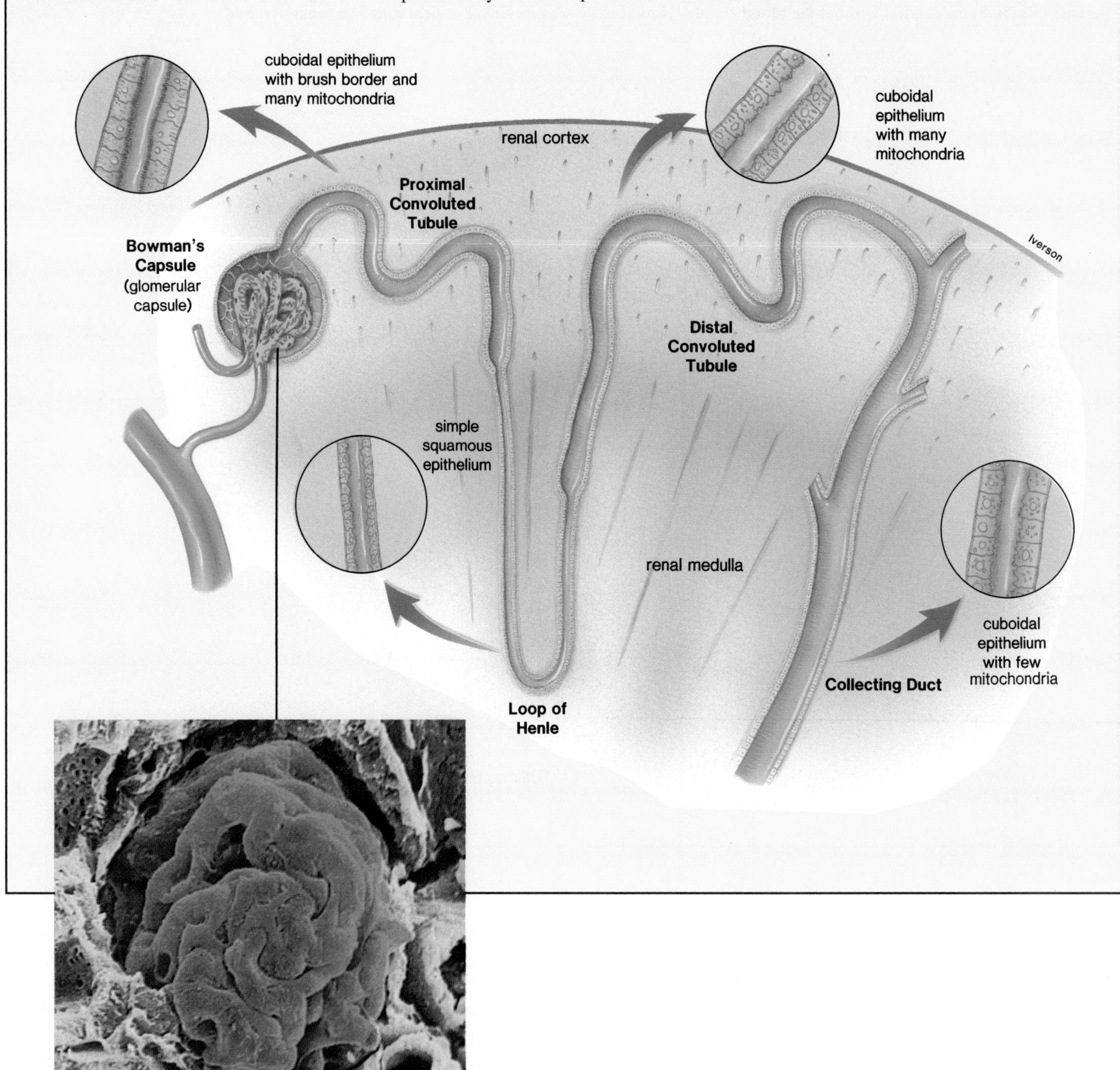

Figure 15.8

Circulation about a nephron. Trace the path of blood by following the arrows. Use the table provided to assist you in this task.

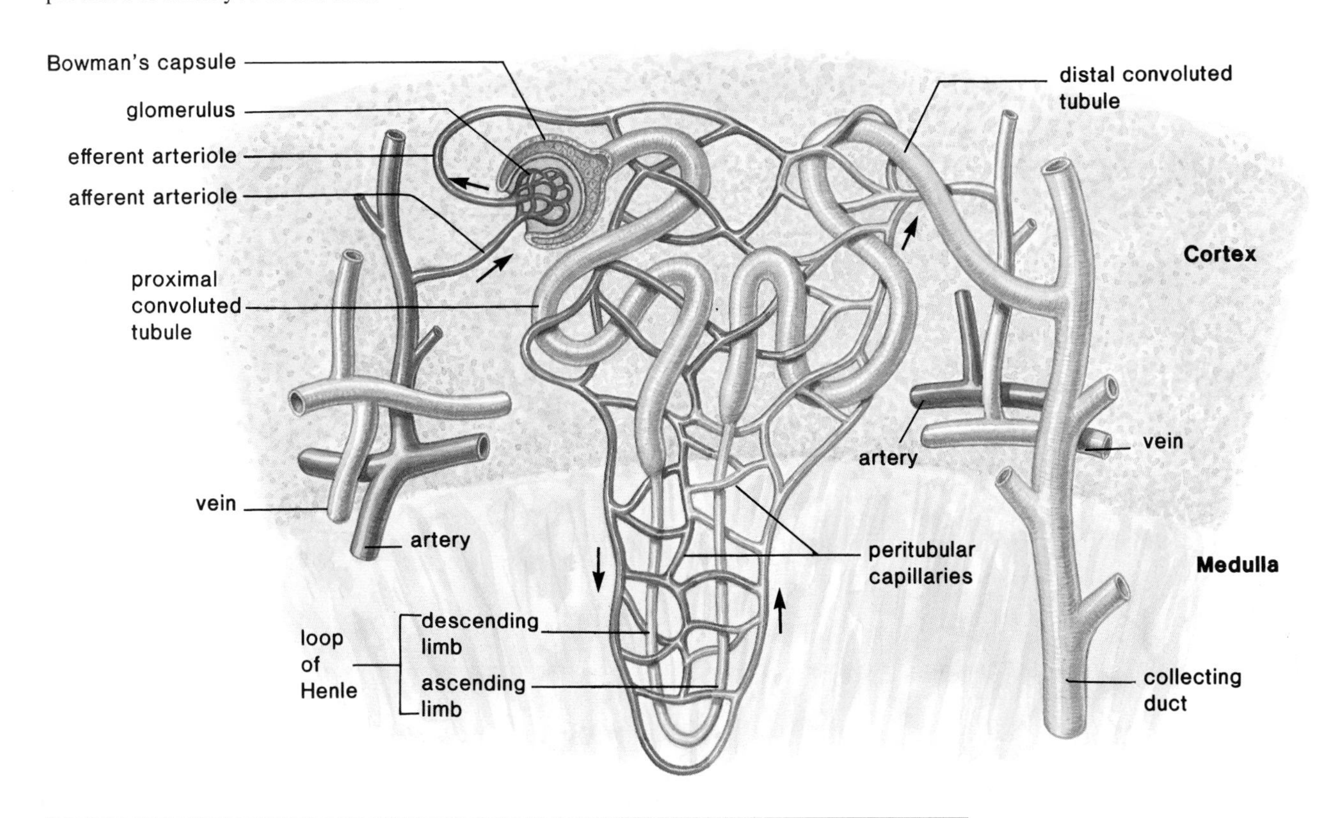

Circulation about a Nephron	
Name of Structure	**Significance**
Afferent arteriole	Brings arteriolar blood toward Bowman's capsule
Glomerulus	Capillary tuft enveloped by Bowman's capsule
Efferent arteriole	Takes arteriolar blood away from Bowman's capsule
Peritubular capillary network	Capillary bed that envelops the rest of the tubule
Venule	Takes venous blood away from tubule

Urine Formation

Each nephron has its own blood supply, including 2 capillary regions. The **glomerulus** is a capillary tuft inside Bowman's capsule, and the **peritubular capillary** surrounds the rest of the nephron (fig. 15.8). Urine formation requires the movement of molecules between these capillaries and the nephron. Three steps are involved: *pressure filtration, selective reabsorption,* and *tubular excretion.*

The pattern of blood flow about the nephron is critical to urine formation.

Pressure Filtration: Divides Blood

Figure 15.9 gives a simple overview of urine formation. Whole blood, of course, enters the glomerulus (fig. 15.9*a*). Under the influence of glomerular blood pressure, which is usually about 60 mm Hg, small molecules move from the glomerulus to the inside of Bowman's capsule, across the thin walls of each. This is a **pressure filtration** process because large molecules and formed elements are unable to pass through these thin walls. In effect, then, blood that enters the glomerulus is divided into 2 portions: the filterable components and the nonfilterable components.

Filterable Blood Components	Nonfilterable Blood Components
Water	Formed elements (blood cells and platelets)
Nitrogenous wastes	Proteins
Nutrients	
Salts (ions)	

The filterable components form the **glomerular filtrate,** which contains small dissolved molecules in approximately the same concentration as plasma. The filtrate

Figure 15.9

Steps in urine formation (simplified). **a.** The steps are noted where they occur within the nephron. **b.** Steps, including the molecule and the processes involved, are listed.

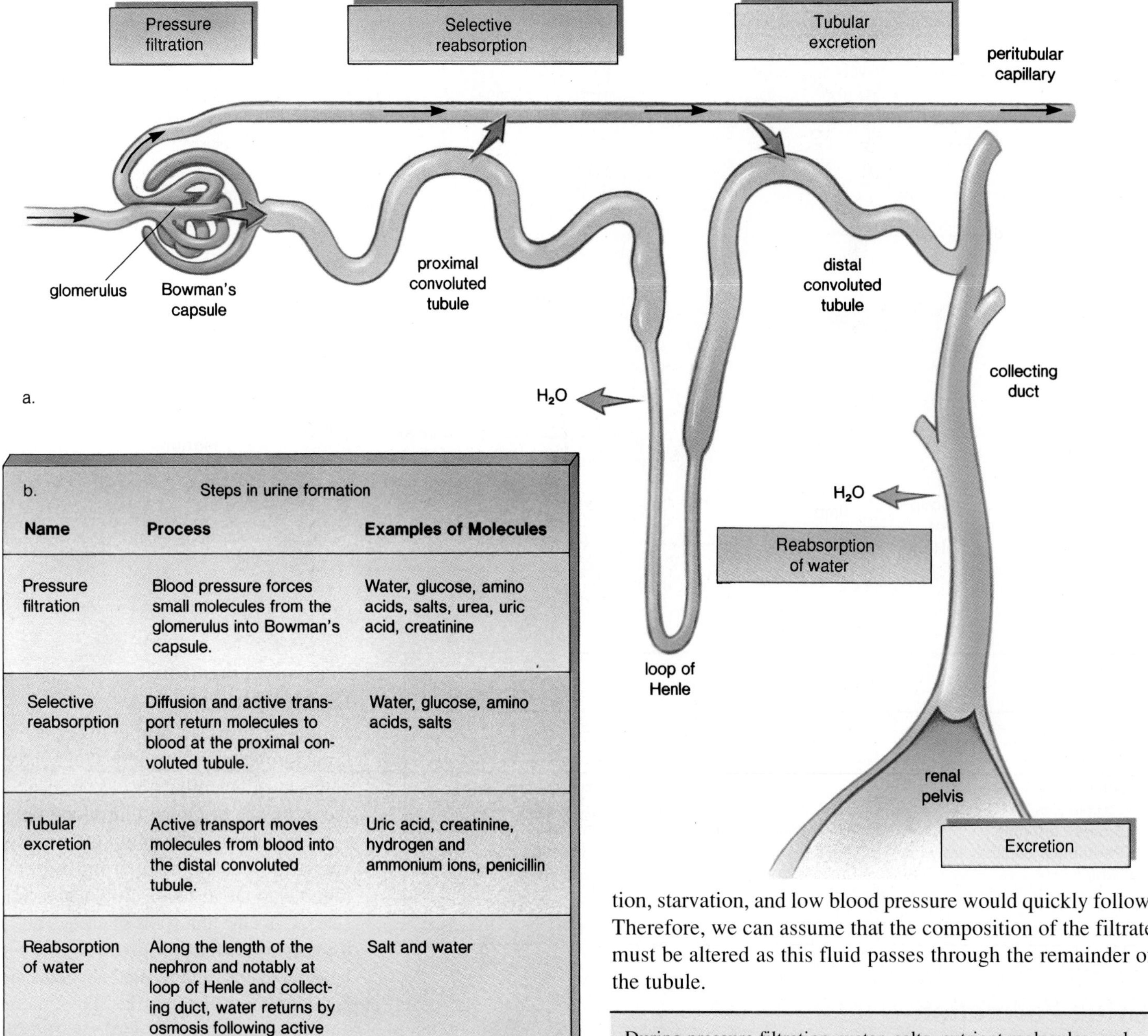

b. Steps in urine formation

Name	Process	Examples of Molecules
Pressure filtration	Blood pressure forces small molecules from the glomerulus into Bowman's capsule.	Water, glucose, amino acids, salts, urea, uric acid, creatinine
Selective reabsorption	Diffusion and active transport return molecules to blood at the proximal convoluted tubule.	Water, glucose, amino acids, salts
Tubular excretion	Active transport moves molecules from blood into the distal convoluted tubule.	Uric acid, creatinine, hydrogen and ammonium ions, penicillin
Reabsorption of water	Along the length of the nephron and notably at loop of Henle and collecting duct, water returns by osmosis following active reabsorption of salt.	Salt and water
Excretion	Urine formation rids body of metabolic wastes.	Water, salts, urea, uric acid, ammonium chloride, creatinine

stays inside Bowman's capsule, and the nonfilterable components stay within the glomerulus.

A consideration of the preceding filterable substances leads us to conclude that if the composition of urine were the same as that of the glomerular filtrate, the body would continually lose nutrients, water, and salts. Death from dehydration, starvation, and low blood pressure would quickly follow. Therefore, we can assume that the composition of the filtrate must be altered as this fluid passes through the remainder of the tubule.

During pressure filtration, water, salts, nutrient molecules, and waste molecules move from the glomerulus to the inside of Bowman's capsule. The filtered substances are called the glomerular filtrate. Do 15.1 Critical Thinking, found at the end of the chapter.

Selective Reabsorption: Passive and Active

Both passive and active reabsorption of molecules from the nephron to blood in a peritubular capillary occur as the filtrate moves along the *proximal convoluted tubule.*

Because of *passive reabsorption,* even some urea is reabsorbed (table 15.3). However, we are particularly interested in the passive reabsorption of water (H_2O). Two factors aid this process: the nonfilterable proteins remain in blood, and salt is returned to blood. Following active reabsorption of sodium ions (Na^+), chloride ions (Cl^-) follow passively, as does water. Therefore, water moves to the area of greater solute concentration. This process occurs along the length of the nephron, until eventually nearly all water and sodium ions have been reabsorbed into the blood (table 15.3).

Table 15.3
Reabsorption from Nephron

Substance	Amount Filtered (per Day)	Amount Excreted (per Day)	Reabsorption (%)
Water (L)	180	1.8	99.0
Sodium (g)	630	3.2	99.5
Glucose (g)	180	0.0	100.0
Urea (g)	54	30.0	44.0

The cells lining the proximal convoluted tubule are anatomically adapted for *active reabsorption* (fig. 15.10). These cells have numerous microvilli, each about 1 µm in length, which increase the surface area for reabsorption. In addition, the cells contain numerous mitochondria, which produce the energy necessary for active transport. Reabsorption by active transport is **selective reabsorption** because only molecules recognized by carrier molecules are actively reabsorbed. After passing through the tubule cells, the molecules enter blood in the peritubular capillary.

Glucose is an example of a molecule that ordinarily is completely reabsorbed because there is a plentiful supply of carrier molecules for it. However, every substance has a maximum rate of transport, and after all its carriers are in use, any excess in the filtrate will appear in the urine. For example, as reabsorbed levels of glucose approach 400 mg/100 ml plasma, the rest will appear in the urine. In diabetes mellitus, excess glucose occurs in the glomerulus and then in the filtrate because the liver fails to store glucose as glycogen.

Figure 15.10
The cells that line the lumen (inside) of the proximal convoluted tubule, where selective reabsorption takes place. **a.** This photomicrograph shows that the cells have a brushlike border composed of microvilli (mv), which greatly increases the surface area exposed to the lumen. The peritubular capillary surrounds the cells (nu = nucleus). **b.** Each cell has many mitochondria, which supply the energy needed for active transport, the process that moves molecules (green) from the lumen to the capillary.

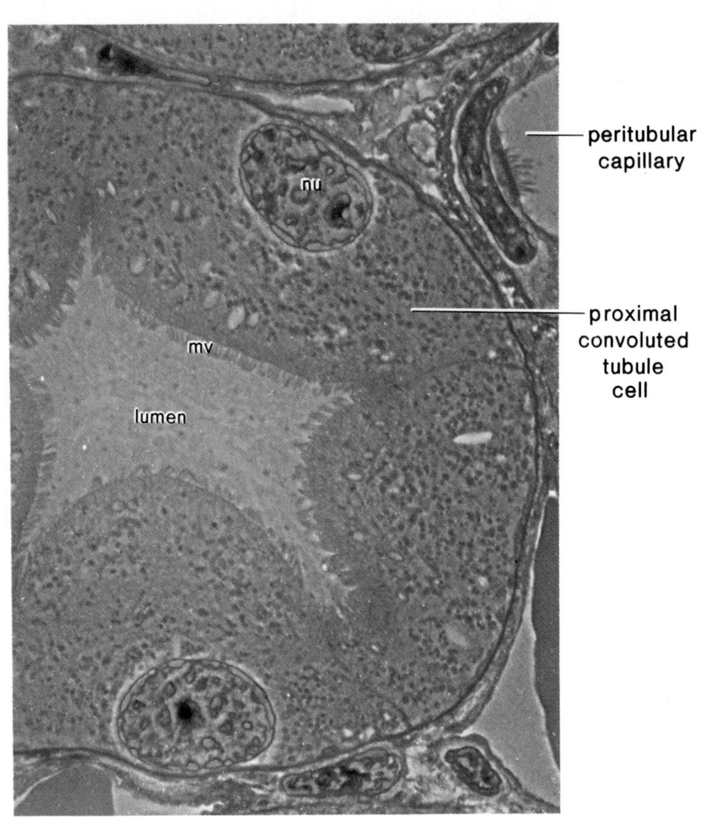

a.

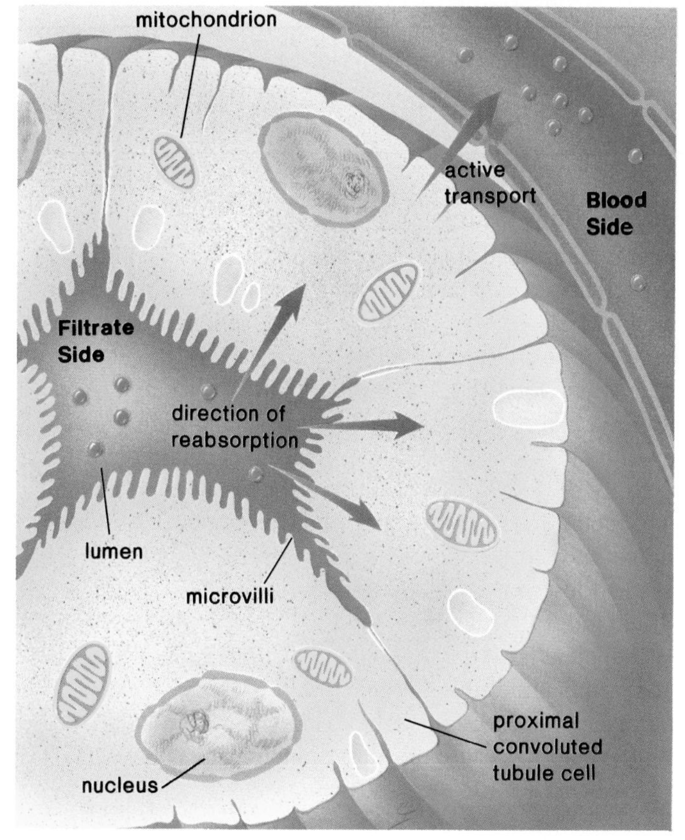

b.

We have seen that the filtrate that enters the proximal convoluted tubule is divided into 2 portions: the components that are reabsorbed from the tubule into blood and the components that are nonreabsorbed.

Reabsorbed Filtrate Components	Nonreabsorbed Filtrate Components
Most water	Some water
Nutrients	Much nitrogenous waste
Required salts (ions)	Excess salts (ions)

The substances that are not reabsorbed become the tubular fluid, which enters the loop of Henle.

During selective reabsorption, nutrient and salt molecules are actively reabsorbed from the proximal convoluted tubule into the peritubular capillary, and water follows passively.

Tubular Excretion: Second Addition of Wastes

Tubular excretion[1] is a second way substances can be added to tubular fluids. Hydrogen and ammonium ions, creatinine, and drugs like penicillin move *from* blood in the peritubular capillaries *into* the distal convoluted tubule. The cells that line this portion of the tubule also have numerous mitochondria because tubular excretion is an active process that requires ATP. In the end, urine contains substances that underwent pressure filtration and substances that underwent tubular excretion. However, pressure filtration is the more important of the 2 processes.

During tubular excretion, certain molecules are actively secreted from the peritubular capillary into the fluid of the distal convoluted tubule. These molecules are found in urine.

Reabsorbing Water

Water is reabsorbed along the whole length of the nephron, but the excretion of a hypertonic urine (one that is more concentrated than blood) is dependent upon the action of the loop of Henle and the collecting duct.

A long *loop of Henle,* which typically penetrates deep into the renal medulla, is made up of a *descending* (going down) limb and an *ascending* (going up) limb. Salt (Na^+Cl^-) passively diffuses out of the lower portion of the ascending limb, but the upper, thick portion of the limb actively transports salt out into the tissue of the outer renal medulla (fig. 15.11). Less and less salt is available for transport from the tubule as fluid moves up the thick portion of the ascending limb. In the end, there is an osmotic gradient within the tissues of the renal medulla: the concentration of salt is greater in the direction of the inner medulla. (Note that water cannot leave the ascending limb because the limb is impermeable to water.)

1. The term *tubular excretion* is used instead of *tubular secretion* because it better represents the end result of the process.

Figure 15.11

Reabsorption of water at the loop of Henle and the collecting duct. Salt (Na^+Cl^-) diffuses and is actively transported out of the ascending limb of the loop of Henle into the renal medulla; also, urea is believed to leak from the collecting duct and to enter the tissues of the renal medulla. This creates a hypertonic environment, which draws water out of the descending limb and the collecting duct. This water is returned to the circulatory system.

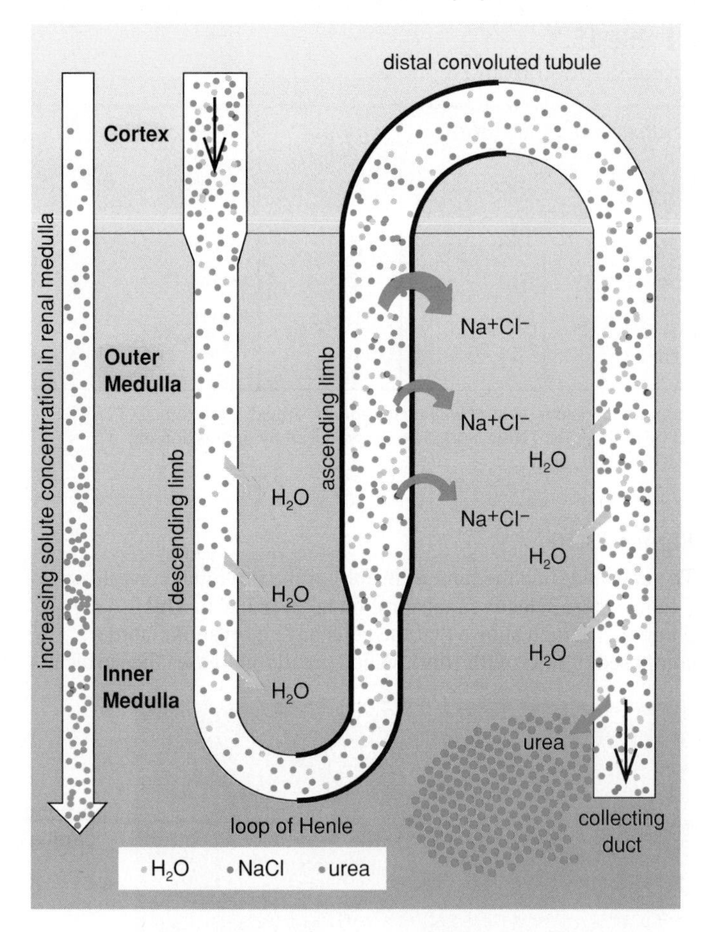

If you examine figure 15.11 carefully, you can see that the *innermost* portion of the inner renal medulla has the highest concentration of solutes. This cannot be due to salt because active transport of salt does not start until the thick portion of the ascending limb. Urea is believed to leak from the lower portion of the collecting duct, and it is this molecule that contributes to the high solute concentration of the inner renal medulla.

Because of the solute concentration gradient of the renal medulla, water leaves the descending limb of the loop of Henle along its length. This is a *countercurrent mechanism*—the increasing concentration of solute encounters the decreasing number of water molecules in the descending limb, ensuring that water continues to leave the descending limb from the top to the bottom.

Fluid entering a *collecting duct* comes from the distal convoluted tubule. This fluid is isotonic to the cells of the cortex. This means that to this point, the net effect of reabsorption

SPARE PARTS

Transplantation of the kidney, heart, liver, pancreas, lung, and other organs is now possible due to 2 major breakthroughs. First, solutions have been developed that preserve donor organs for several hours. This made it possible for one young boy to undergo surgery for 16 hours, during which time he received 5 different organs. Second, rejection of transplanted organs is now prevented by immunosuppressive drugs; therefore, organs can be donated by unrelated individuals, living or dead. Living individuals can donate one kidney, a portion of their liver, and certainly bone marrow, which quickly regenerates.

After death, it is still possible to give the "gift of life" to someone else—over 25 organs and tissues from the same person can be used for transplants at that time. A liver transplant, for example, can save the life of a child born with biliary atresia, a congenital defect in which the bile ducts do not form. Dr. Thomas Starzl, a pioneer in this field, reports there is a 90% chance of complete rehabilitation among children who survive a liver transplant (fig. 15.A). (He has also tried animal-to-human liver transplants, but so far these have not been successful.) So many heart recipients are now alive and healthy they have formed basketball and softball teams, demonstrating the normalcy of their lives after surgery.

Figure 15.A

Children receiving a liver transplant to cure a birth defect have a good chance of survival and complete recovery.

One problem persists. Although there are as many Americans waiting for organs as ever, only a small percentage of us signify that we are willing to donate organs at the time of our death. To become an organ and tissue donor, a donor card must be signed and carried at all times (fig. 15.B). In many states, the back of the driver's license acts as a donor card. Age is no drawback, but the donor should have been in good health prior to death.

Organ and tissue donation will not interfere with funeral arrangements, and most religions do not object to the donation. Family members should know ahead of time about the desire to become a donor because they will be asked to sign permission papers at the time of death. No money is received for the gift organs, which are removed by a team of surgeons from the nearest organ procurement center.

The United Network for Organ Sharing (UNOS), based in Richmond, Va., which was established after the 1984 National Organ Transplant Act, has a computerized system for matching needy patients with available organs. The patients are ranked according to various medical criteria, and UNOS notifies the appropriate hospital of the availability of an organ. Donor and recipient identities are confidential.

Today, there are over 27,000 Americans waiting for a gift of life. Will they wait in vain?

Source: Data from Arkansas Regional Organ Recovery Agency (AURORA), Little Rock, AR.

ORGAN DONOR CARD

____
Print or type name of donor

In the hope that I may help others, I hereby make this anatomical gift, if medically acceptable, to take effect upon my death. The words and marks below indicate my desires.

I give: (a) ______ any needed organs or parts
(b) ______ only the following organs or parts

____
Specify the organ(s) or part(s)
for the purposes of transplantation, therapy, medical research or education;
(c) ______ my body for anatomical study if needed.
Limitations or special wishes, if any: ______________

Signed by the donor and the following witnesses in the presence of each other:

Signature of Donor | Date of Birth of Donor

Date Signed | City & State

Witness | Witness

This is a legal document under the Uniform Anatomical Gift Act or similar laws. For further information consult your physician or

UNOS P.O. Box 13770 Richmond, Virginia 23225-8770

Figure 15.B

An organ donor card gives the name of the donor and witnesses, one of whom should be a close family member.

of water and salt is the production of a fluid that has the same tonicity as blood. Now, however, the collecting duct passes through the renal medulla, which is increasingly hypertonic, as previously explained (fig. 15.11). Therefore, water diffuses out of the collecting duct into the renal medulla, and the urine within the collecting duct becomes hypertonic to blood plasma.

Urine, the composition of which is listed in table 15.2, now passes out of the collecting duct into the renal pelvis of the kidney. Urine contains all the molecules that were not reabsorbed, as well as those that underwent tubular excretion at the distal convoluted tubule.

Water diffuses from both the descending limb of the loop of Henle and the collecting duct due to an increasingly hypertonic renal medulla. Urine (table 15.2) formation is complete.

Regulatory Functions of the Kidneys

The kidneys regulate the pH, the salt balance, and the volume of blood.

Maintaining Blood pH and Salt Balance

The kidneys help to maintain the pH level of blood within a narrow range, and the whole nephron takes part in this process. The excretion of hydrogen ions (H^+) and ammonia (NH_3), together with the reabsorption of sodium ions (Na^+) and bicarbonate ions (HCO_3^-), is adjusted to keep the pH within normal bounds. If blood is acidic, hydrogen ions are excreted in combination with ammonia, while sodium ions and bicarbonate ions are reabsorbed. This restores the pH because $NaHCO_3$ is a base. If blood is basic, fewer hydrogen ions are excreted and fewer sodium ions and bicarbonate ions are reabsorbed.

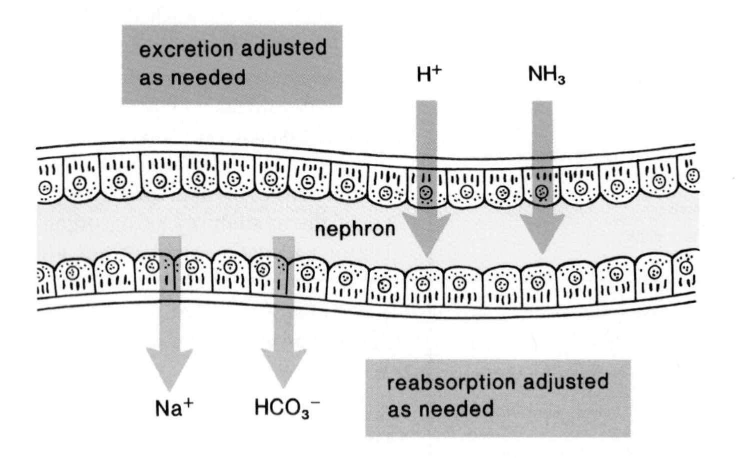

These examples also show that the kidneys regulate the salt balance in blood by controlling the excretion and the reabsorption of various ions. Sodium (Na^+) is an important ion in plasma that must be regulated, but the kidneys also excrete or reabsorb other ions, such as bicarbonate ions, potassium ions, and magnesium ions, as needed.

Maintaining Blood Volume

Maintenance of blood volume and salt balance is under the control of hormones. **Antidiuretic hormone (ADH),** secreted by the posterior pituitary, primarily maintains blood volume. ADH increases the permeability of the collecting duct so that more water can be reabsorbed into the blood. In order to understand the function of this hormone, consider its name. *Diuresis* means *increased amount of urine,* and *antidiuresis* means *decreased amount of urine*. When ADH is present, more water is reabsorbed into the blood and a decreased amount of urine results.

The secretion of ADH is dependent on whether blood volume needs to be increased or decreased. When water is reabsorbed at the collecting duct, blood volume increases, and when water is not reabsorbed, blood volume decreases. In practical terms, if an individual does not drink much water on a certain day, the *posterior lobe of the pituitary* releases ADH, more water is reabsorbed, blood volume is maintained at a normal level, and consequently, there is less urine. On the other hand, if an individual drinks a large amount of water and does not perspire much, the posterior lobe of the pituitary does not release ADH, more water is excreted, blood volume is maintained at a normal level, and a greater amount of urine is formed.

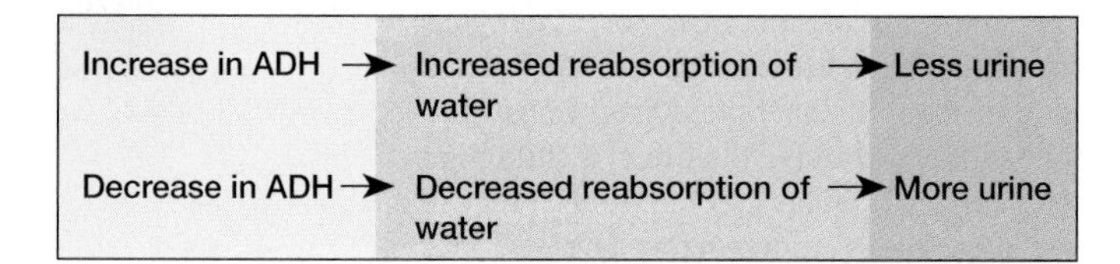

Drinking alcohol causes diuresis because it inhibits the secretion of ADH. The dehydration that follows is believed to contribute to the symptoms of a "hangover." Drugs called diuretics often are prescribed for high blood pressure. The drugs cause salts and water to be excreted; therefore, they reduce blood volume and blood pressure. Concomitantly, any *edema* (p. 274) that is present is also reduced.

Aldosterone, secreted by the adrenal cortex, is a hormone that primarily maintains sodium ion (Na^+) and potassium ion (K^+) balance. It causes reabsorption of sodium ions into the blood at the distal convoluted tubule and the excretion of potassium ions. The increase of sodium ions in blood causes water to be reabsorbed, leading to an increase in blood volume and blood pressure.

Blood pressure is constantly monitored by the afferent arteriole cells within the juxtaglomerular apparatus. The juxtaglomerular apparatus (fig. 15.12) occurs at a region of contact between the afferent arteriole and the distal convoluted tubule. The afferent arteriole cells in the region secrete *renin* when blood pressure is insufficient to promote efficient filtration in the glomerulus. Renin is an enzyme that changes angiotensinogen, a large plasma protein made by the liver, into angiotensin I (fig. 15.12*c*). Then an enzyme called angiotensin-converting enzyme, present in the lining of the capillaries of the lungs, changes angiotensin I to angiotensin II. Angiotensin II, a powerful vasoconstrictor, stimulates the adrenal cortex to release aldosterone. Now, blood pressure rises.

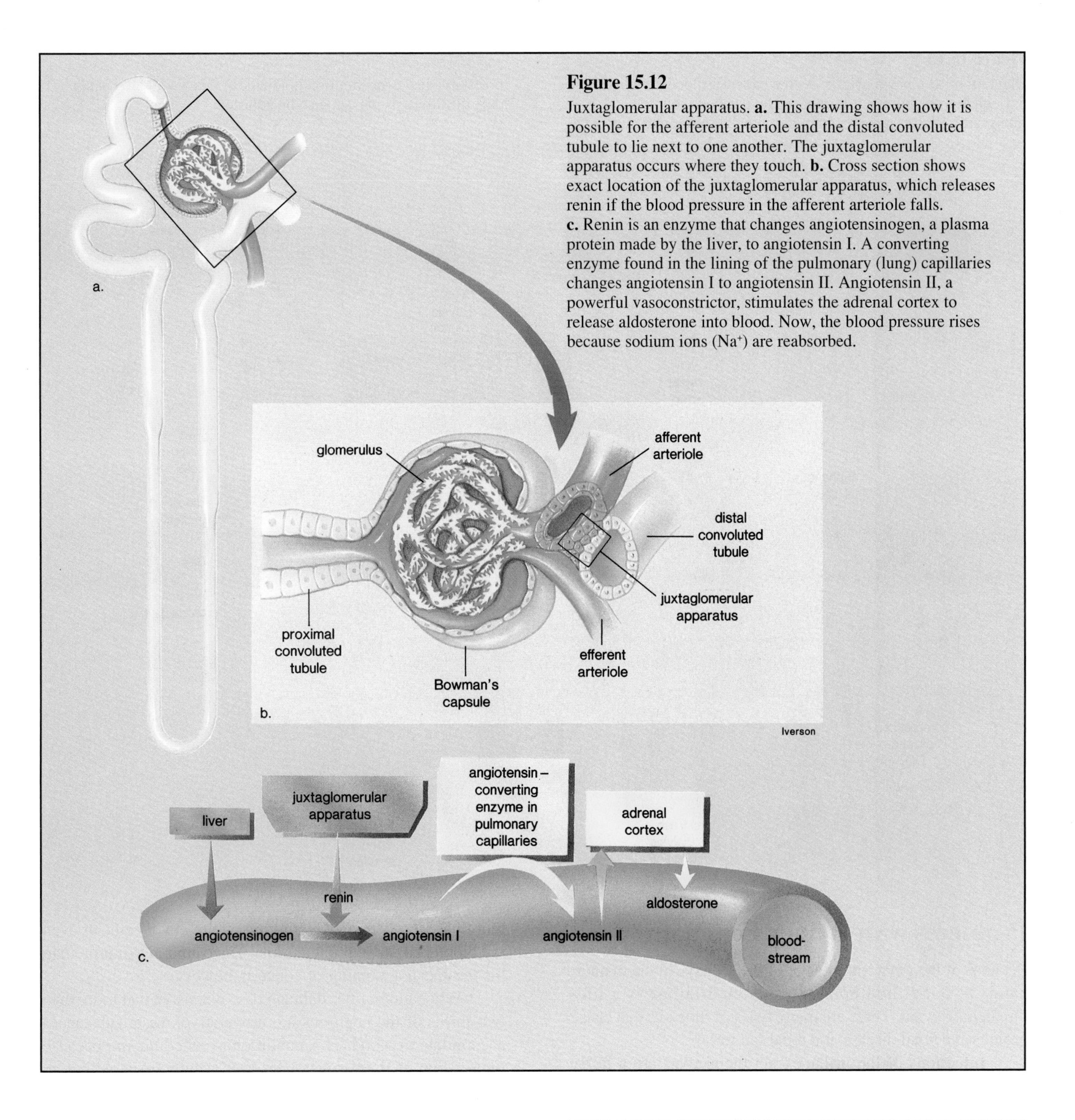

Figure 15.12

Juxtaglomerular apparatus. **a.** This drawing shows how it is possible for the afferent arteriole and the distal convoluted tubule to lie next to one another. The juxtaglomerular apparatus occurs where they touch. **b.** Cross section shows exact location of the juxtaglomerular apparatus, which releases renin if the blood pressure in the afferent arteriole falls. **c.** Renin is an enzyme that changes angiotensinogen, a plasma protein made by the liver, to angiotensin I. A converting enzyme found in the lining of the pulmonary (lung) capillaries changes angiotensin I to angiotensin II. Angiotensin II, a powerful vasoconstrictor, stimulates the adrenal cortex to release aldosterone into blood. Now, the blood pressure rises because sodium ions (Na^+) are reabsorbed.

The renin-angiotensin-aldosterone system always seems to be active in some patients with hypertension. In response to this possibility, a drug for hypertension that inhibits angiotensin-converting enzyme is available. The drug is called ACE (for angiotensin-converting enzyme) inhibitor.

The kidneys contribute to homeostasis by excreting urea. They also maintain both the pH and the salt balance of blood and regulate the volume of blood, three very important functions. Do 15.2 Critical Thinking, found at the end of the chapter.

Figure 15.13

Diagram of a kidney machine. As the patient's blood circulates through dialysis tubing, it is exposed to a dialysis solution (dialysate). Wastes move from blood into the solution because of a preestablished concentration gradient. In this way, blood is not only cleansed, its pH can also be adjusted.

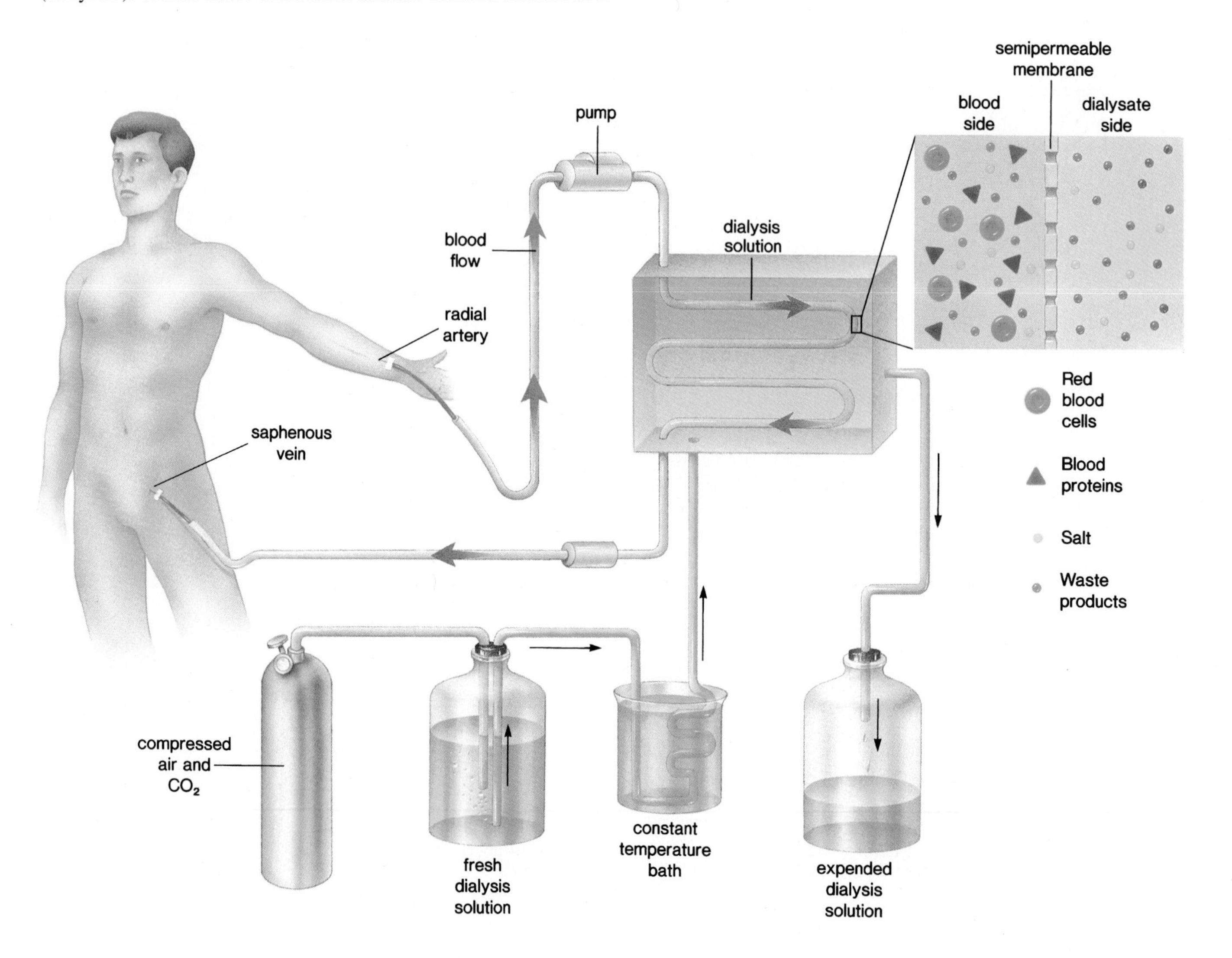

Problems with Kidney Function

Because of the great importance of the kidneys to the maintenance of body fluid homeostasis, renal failure is a life-threatening event. There are many types of illnesses that cause progressive renal disease and renal failure.

Infections of the urinary tract themselves are a fairly common occurrence, particularly in females since the urethra is considerably shorter than the male urethra. If the infection is localized in the urethra, it is called *urethritis*. If it invades the urinary bladder, it is called *cystitis*. Finally, if the kidneys are affected, the infection is called *pyelonephritis*.

Glomerular damage sometimes leads to blockage of the glomeruli so that no fluid moves into the tubules, or damage can cause the glomeruli to become more permeable than usual. This is detected when a **urinalysis** is done. If the glomeruli are too permeable, albumin, white blood cells, or even red blood cells appear in the urine. Trace amounts of protein in the urine is not a matter of concern, however.

When glomerular damage is so extensive that more than two-thirds of the nephrons are inoperative, waste substances accumulate in blood. This condition is called *uremia* because urea is one of these substances that accumulates. Although nitrogenous wastes can cause serious damage, the retention of water and salts (ions) is of even greater concern. The latter causes edema, fluid accumulation in the body tissues. Imbalance in the ionic composition of body fluids can even lead to loss of consciousness and to heart failure.

When Kidneys Fail

Treatment options for failed kidneys include kidney transplant or dialysis.

Replacing the Kidney

Patients with renal failure sometimes undergo a kidney transplant operation, during which a functioning kidney from a donor is received (see reading, page 271). As with all organ transplants, there is the possibility of organ rejection. Receiving a kidney from a close relative has the highest chance of success. The current one-year survival rate is 97% if the kidney is received from a relative and 90% if it is received from a nonrelative.

Dialysis: Treating Blood

If a satisfactory donor cannot be found for a kidney transplant, which is frequently the case, the patient can undergo dialysis, utilizing either a kidney machine or continuous ambulatory peritoneal (abdominal) dialysis, or CAPD. *Dialysis* is defined as the diffusion of dissolved molecules through a semipermeable membrane. These molecules, of course, move across a membrane from the area of greater concentration to one of lesser concentration.

During hemodialysis (fig. 15.13), the patient's blood is passed through a semipermeable membranous tube, which is in contact with a balanced salt (dialysis) solution. Substances more concentrated in blood diffuse into the dialysis solution, which is also called the dialysate, and substances more concentrated in the dialysate diffuse into blood. Accordingly, the artificial kidney can be utilized either to extract substances from blood, including waste products or toxic chemicals and drugs, or to add substances to blood—for example, bicarbonate ions (HCO_3^-) if blood is acidic. In the course of a six-hour hemodialysis, from 50 g to 250 g of urea can be removed from a patient, which greatly exceeds the amount excreted by normal kidneys. Therefore, a patient must undergo treatment only about twice a week.

In the case of CAPD, a fresh amount of dialysate is introduced directly into the abdominal cavity from a bag attached to a permanently implanted plastic tube. Waste and water molecules pass into the dialysate from the surrounding organs before the fluid is collected either 4 hours or 8 hours later. The individual can go about his or her normal activities during CAPD, unlike during hemodialysis.

> Kidney transplants and hemodialysis are corrective procedures that are available for persons who have suffered renal failure. Do 15.3 Critical Thinking, found at the end of the chapter.

SUMMARY

The end products of metabolism for the most part are nitrogenous wastes, such as urea, uric acid, and creatinine, all of which are primarily excreted by the kidneys, organs that have many nephrons. Urine formation requires 3 steps: during pressure filtration, small components of plasma pass into Bowman's capsule from the glomerulus due to blood pressure; during selective reabsorption, nutrients and sodium are actively reabsorbed from the proximal convoluted tubule into blood; and during tubular excretion, a few types of substances are actively transported into the distal convoluted tubule from blood.

Water is reabsorbed along the length of the nephron, but it is the loop of Henle and the collecting duct that enable humans to secrete a hypertonic urine. ADH, a hormone produced by the posterior pituitary, directly controls the reabsorption of water, and aldosterone, from the adrenal cortex, controls it indirectly by affecting sodium ion (Na^+) reabsorption. The whole nephron participates in maintaining the pH of blood by regulating the pH of urine. In practice, hydrogen ions (H^+) are excreted from and sodium bicarbonate ions (HCO_3^-) are reabsorbed by the nephron to maintain the pH.

Various problems can lead to kidney failure. In such cases, the person can either receive a kidney from a donor or undergo dialysis, by means of either the kidney machine or CAPD.

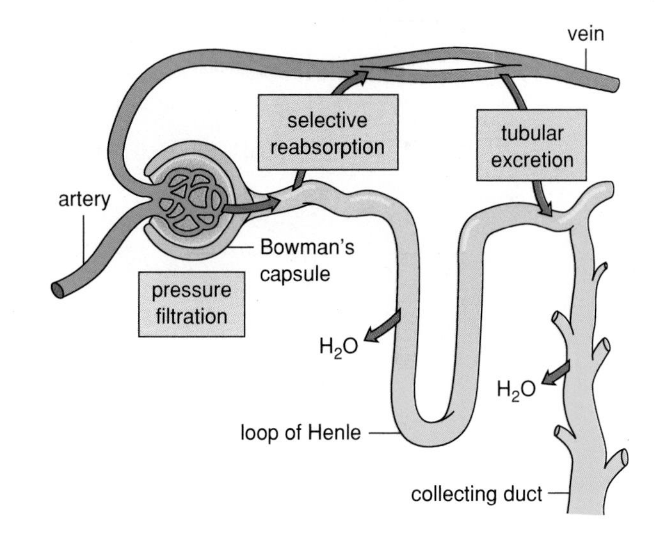

STUDY QUESTIONS

In order to practice **writing across the curriculum,** students should write out the answers to any or all of the study questions. The study questions are sequenced in the same order as the text.

1. Name 4 nitrogenous end products, and explain how each is formed in the body. (p. 260)
2. Name several excretory organs, as well as the substances they excrete. (p. 260)
3. What is the composition of urine? (p. 263)
4. Give the path of urine. (p. 264)
5. Name the parts of a nephron. (pp. 264–265)
6. Trace the path of blood about the nephron. (p. 267)
7. Describe how urine is made by telling what happens at each part of the tubule. (p. 268)
8. Explain these terms: *pressure filtration, selective reabsorption,* and *tubular excretion.* (p. 268)
9. How does the nephron regulate the blood volume and the pH of blood? What hormones are involved? (p. 272)
10. Explain how the kidney machine and CAPD work. (p. 275)

OBJECTIVE QUESTIONS

1. The primary nitrogenous end product of humans is _______.
2. The liver excretes _______, which are derived from the breakdown of _______.
3. Urine leaves the urinary bladder in the _______.
4. The capillary tuft inside Bowman's capsule is called the _______.
5. _______ is a substance that is found in the filtrate, is reabsorbed, and is still in urine.
6. _______ is a substance that is found in the filtrate, is minimally reabsorbed, and is concentrated in the urine.
7. Tubular excretion takes place at the _______, a portion of the nephron.
8. Reabsorption of water from the collecting duct is regulated by the hormone _______.
9. In addition to excreting nitrogenous wastes, the kidneys adjust the _______ and _______ of blood.
10. Persons who have nonfunctioning kidneys often have their blood cleansed by _______ machines.
11. Label this diagram of a nephron.

CRITICAL THINKING

In order to practice **writing across the curriculum,** students should write out the answers to any or all of the critical thinking questions. Suggested answers to the critical thinking questions are in appendix E.

15.1

1. Urine, not urea, is made by the kidneys. What is the difference between urine and urea?
2. Blood pressure in the glomerulus favors filtration of molecules. What force in the glomerulus opposes filtration (see fig. 12.15)?
3. The efferent arteriole is narrower than the afferent arteriole. What effect does this have on blood pressure in the glomerulus?
4. If there is a loss of proteins from blood into Bowman's capsule, would the filtration rate increase or decrease? (Assume constant blood pressure.) Explain.

15.2

1. If 99% of water is reabsorbed (table 15.3), how can urine be 95% water (table 15.2)?
2. Carrier molecules work as fast as they can to return glucose to blood. Explain why excess glucose is not returned.
3. When CO_2 is excreted by the lungs, does blood become more acidic or more basic? If bicarbonate ions are excreted by the kidneys, does blood become more acidic or more basic?
4. The maintenance of normal blood pH is a very important function of the kidneys. What molecules in the cells are affected by pH changes?

15.3

1. Which of the steps given in figure 15.9 are also part of hemodialysis? Which are absent?
2. Why don't patients lose plasma proteins during hemodialysis?
3. Why is more urea excreted during hemodialysis than during urine formation?
4. What would you add to the dialysate to prevent the loss of glucose from blood?

SELECTED KEY TERMS

antidiuretic hormone (ADH) (an″ tĭ-di″ u-ret′ik hōr′ mōn) sometimes called vasopressin, a hormone secreted by the posterior pituitary that controls the degree to which water is reabsorbed by the kidneys.

Bowman's capsule a double-walled cup that surrounds the glomerulus at the beginning of the nephron.

collecting duct a tube that receives urine from the distal convoluted tubules of several nephrons.

distal convoluted tubule highly coiled region of a nephron that is distant from Bowman's capsule.

excretion removal of metabolic wastes from the body.

glomerular filtrate (glo-mer′ u-lar fil′ trāt) the filtered portion of blood contained within Bowman's capsule.

glomerulus (glo-mer′ u-lus) a cluster; for example, the cluster of capillaries surrounded by Bowman's capsule in a nephron.

kidney an organ in the urinary system that produces and excretes urine.

nephron (nef′ ron) the anatomical and functional unit of the vertebrate kidney; kidney tubule.

peritubular capillary capillary that surrounds a nephron and functions in reabsorption during urine formation.

pressure filtration the movement of small molecules from the glomerulus into Bowman's capsule due to the action of blood pressure.

proximal convoluted tubule highly coiled region of a nephron near Bowman's capsule.

renal pelvis a hollow chamber in the kidney that lies inside the renal medulla and receives freshly prepared urine from the collecting ducts.

selective reabsorption the movement of nutrient molecules, as opposed to waste molecules, from the contents of the nephron into blood at the proximal convoluted tubule.

tubular excretion the movement of certain molecules from blood into the distal convoluted tubule so that they are added to urine.

urea (u-re′ ah) primary nitrogenous waste of mammals derived from amino acid breakdown.

ureter(u-re′ ter) one of 2 tubes that take urine from the kidneys to the urinary bladder.

urethra (u-re′ thrah) tube that takes urine from the bladder to the outside.

uric acid waste product of nucleotide metabolism.

urinary bladder an organ where urine is stored before being discharged by way of the urethra.

16

THE NERVOUS SYSTEM

Chapter Concepts

1.
The nervous system is made up of cells called neurons, which are specialized to carry nerve impulses. 279, 300

2.
Transmission of impulses between neurons is accomplished by means of chemicals called neurotransmitter substances. 284, 300

3.
The peripheral nervous system contains nerves, which conduct nerve impulses between body parts and the central nervous system. 286, 300

4.
The central nervous system, made up of the spinal cord and the brain, is highly organized. In the brain, consciousness is a function only of the cerebrum, which is more highly developed in humans than in other animals. 292, 300

5.
Drugs that affect the psychological state of the individual, such as alcohol, marijuana, cocaine, and heroin, are abused, usually to the detriment of the body. 296, 300

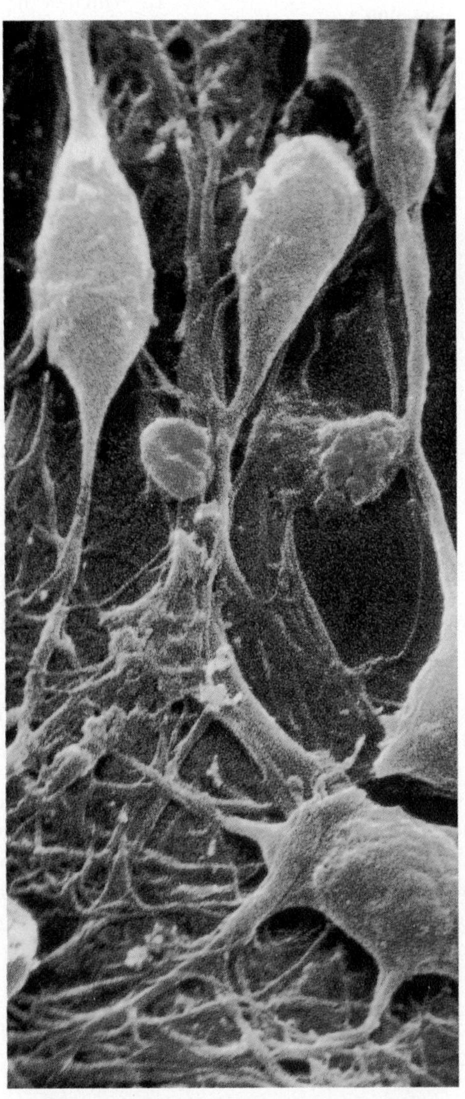

Neurons from the cerebral cortex, ×480

Chapter Outline

Figure 16.1

Organization of the nervous system. The nervous system consists of the central nervous system (CNS) and the peripheral nervous system (PNS). Nerve impulses flow from one neuron to another much as a baton is passed from one runner to another in a relay race.

THE nervous system tells us that we exist and, along with the muscles and sense organs, accounts for our distinctly animal characteristic of quick reaction to environmental stimuli. The nerve cell is called a **neuron,** and it is neurons that carry nerve impulses (messages) from sense organs to a central receiving station (brain and spinal cord) before other impulses go to the muscles or the glands, which then react. As figure 16.1 indicates, the nervous system has 2 main divisions, the central nervous system and the peripheral nervous system. The **central nervous system (CNS),** which includes the brain and the spinal cord, lies in the midline of the body, while the **peripheral nervous system (PNS)** includes the nerves, which project from the CNS.

Neurons

Neurons are the principal type of nerve cell in the body.

Neuron Structure: Three Parts

All neurons (fig. 16.2) have 3 parts: dendrite(s), cell body, and axon. A **dendrite** conducts nerve impulses toward the **cell body,** the part of a neuron that contains the nucleus. An **axon** conducts nerve impulses away from the cell body.

There are 3 types of neurons: sensory neurons, motor neurons, and interneurons (fig. 16.2). A **sensory neuron** takes a message from a **receptor** in a sense organ to the CNS and typically has a long dendrite and a short axon. A **motor neuron** takes a message away from the CNS to an **effector,** a muscle or a gland, and has short dendrites and a long axon. Because motor neurons cause muscle fibers and glands to react, they are said to **innervate** these structures. Sometimes a sensory neuron is referred to as the *afferent neuron,* and a motor neuron is called the *efferent neuron.* These words, which are derived from Latin, mean *running to* and *running away from,* respectively. Obviously, they refer to the relationship of these neurons to the CNS.

An **interneuron** (also called association neuron or connector neuron), which is always found completely within the CNS, conveys messages between parts of the system. An interneuron has short dendrites and either a long or a short axon. Table 16.1 summarizes the 3 types of neurons, which are also illustrated in figure 16.2.

> Although all neurons have the same 3 parts, each is specialized in structure and in function. Specialization is dependent on the location of the neuron in relation to the CNS.

The dendrites and the axons of neurons are sometimes called fibers or processes. Most long fibers, whether dendrites or axons, are covered by tightly packed spirals of *Schwann cells* (neurolemmocytes) (fig. 16.3). Schwann cells are one of several types of neuroglial cells in the nervous system. Neuroglial cells service the neurons—they have supportive and nutritive functions. Schwann cells encircle a fiber, and as they wrap themselves around the axon many times, they lay down several layers of cellular membrane containing myelin, a lipid substance that is an excellent insulator. Myelin gives nerve fibers their white, glistening appearance. Because of the manner in which Schwann cells wrap themselves about nerve fibers, 2 sheaths are formed (fig. 16.3). The outermost sheath is called the **neurilemma,** or the cellular sheath, and the inner one is called the **myelin sheath.** (Scarlike patches

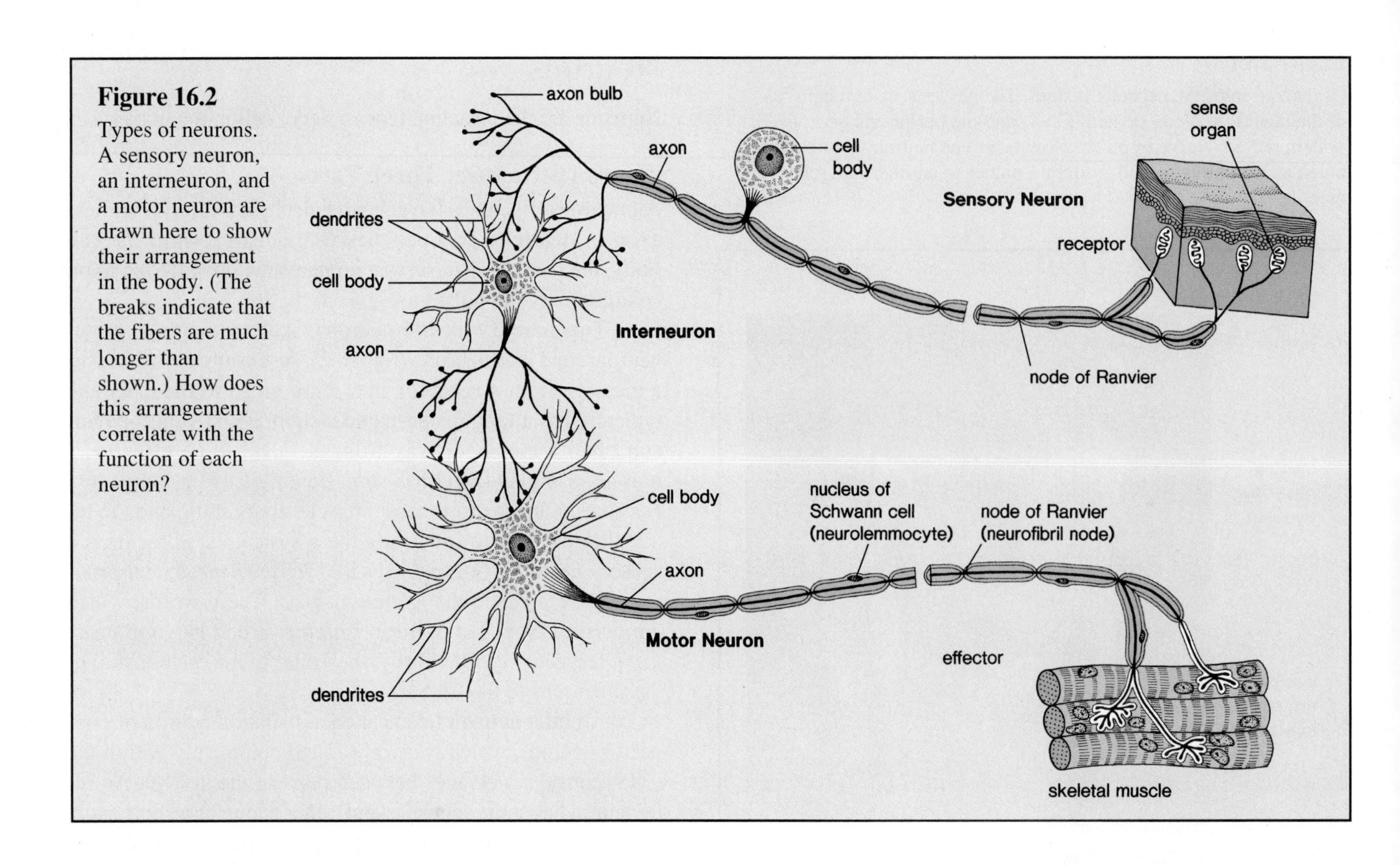

Figure 16.2
Types of neurons. A sensory neuron, an interneuron, and a motor neuron are drawn here to show their arrangement in the body. (The breaks indicate that the fibers are much longer than shown.) How does this arrangement correlate with the function of each neuron?

Table 16.1
Neurons

Neuron	Structure	Function
Sensory neuron (afferent)	Long dendrite, short axon	Carries nerve impulses (messages) from a receptor to the CNS*
Motor neuron (efferent)	Short dendrite, long axon	Carries nerve impulses (messages) from the CNS to an effector
Interneuron	Short dendrites, long or short axon	Carries nerve impulses (messages) within the CNS

*CNS = central nervous system

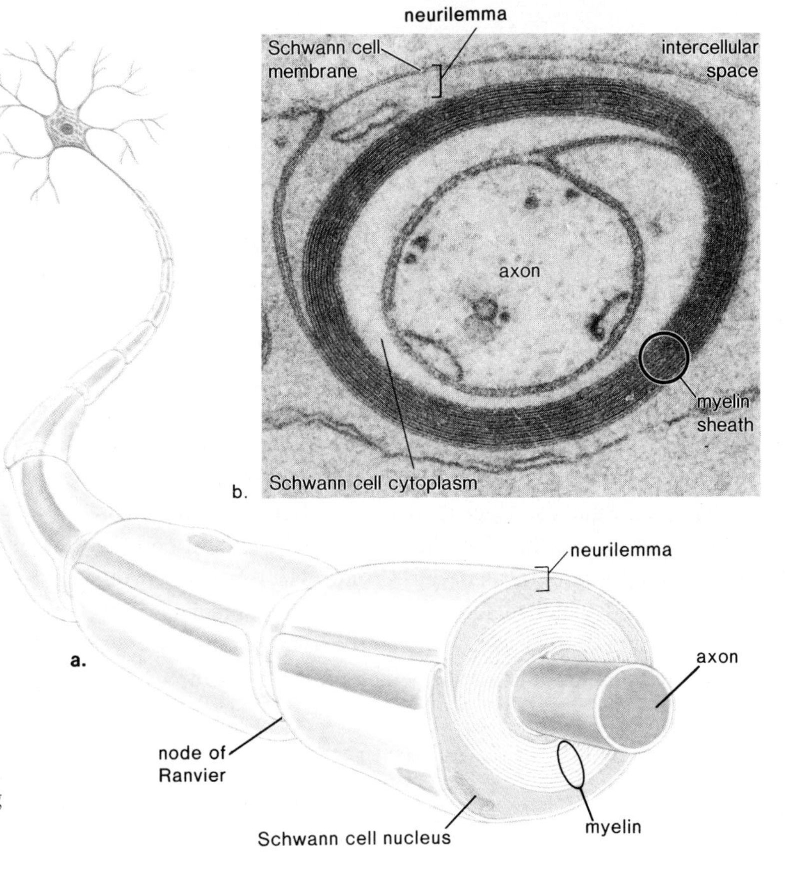

Figure 16.3
Neurilemma and myelin sheath. **a.** Axon of a motor neuron ending in a cross section of neurilemma and myelin sheath, which enclose the long fibers of all neurons. The myelin sheath is composed of many layers of Schwann cell membrane and has a white, glistening appearance in the body. **b.** Electron micrograph of a cross section of an axon, surrounded by neurilemma and myelin sheath.

A N I M A L • B I O L O G Y

ANIMAL NERVOUS SYSTEMS

A comparative study of animal nervous systems indicates the steps that may have led to the complex system of vertebrates. In hydras, the simple nervous system looks like a net of threads extending throughout the radially symmetrical body. The net is composed of neurons, which contact one another within the mesoglea. Bilaterally symmetrical planarians have the rudiments of a central nervous system (CNS) and a peripheral nervous system (PNS). A "brain" receives sensory information from the eyespots and other sensory cells and sends messages along parallel nerve cords. The transverse nerves between the nerve cords keep the movement of the 2 sides coordinated.

The annelids and the arthropods have what is usually considered the typical invertebrate nervous system. The CNS consists of a brain and a single ventral solid nerve cord; the PNS consists of ganglia and lateral nerves in each segment. The brain most likely receives sensory information and controls the activity of the ganglia so that the entire animal is coordinated.

In contrast to these invertebrates, vertebrates have a much larger brain and a dorsal hollow nerve cord. Vertebrates exhibit a vast increase in the number of neurons within the CNS and the PNS. An insect's nervous system may contain a total of about one million neurons, while there may be many thousand to several billion times that number in a vertebrate nervous system.

Among vertebrates, there is a progressive increase in the size of the forebrain from fishes to humans. In fishes and amphibians, the forebrain has largely an olfactory function, but in reptiles, birds, and mammals, the forebrain receives information from other parts of the brain and coordinates sensory data and motor functions. The forebrain is the part of the brain responsible for consciousness, intelligence, and reason.

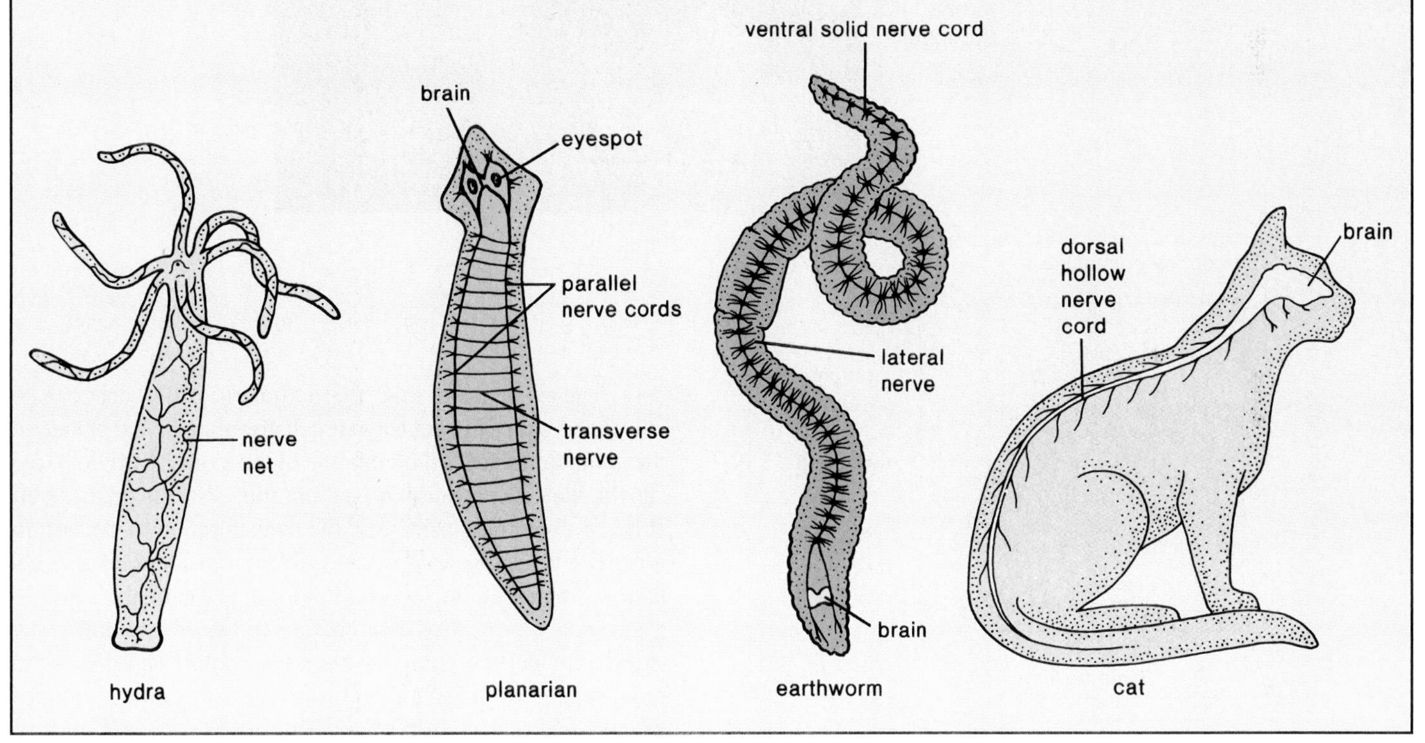

begin to replace myelin in multiple sclerosis, and motor difficulties result.) In the PNS, neurilemma plays an important role in nerve regeneration. If a nerve fiber is accidently severed, the part on the far side of the cut, away from the cell body, except for the neurilemma, degenerates. The neurilemma serves as a passageway for new growth. The myelin sheath affects nerve conduction and is discussed in the following section.

Nerve Impulse: Same for All Neurons

A **nerve impulse** is the way a neuron transmits information. The nature of a nerve impulse has been studied using giant axons from the squid and an instrument called a voltmeter. Voltage is a measure of the electrical potential difference between 2 points, which in this case are the inside and the outside of the axon. The change in voltage is displayed on an *oscilloscope,* an instrument with a screen that shows a trace, or pattern, indicating a change in voltage with time.

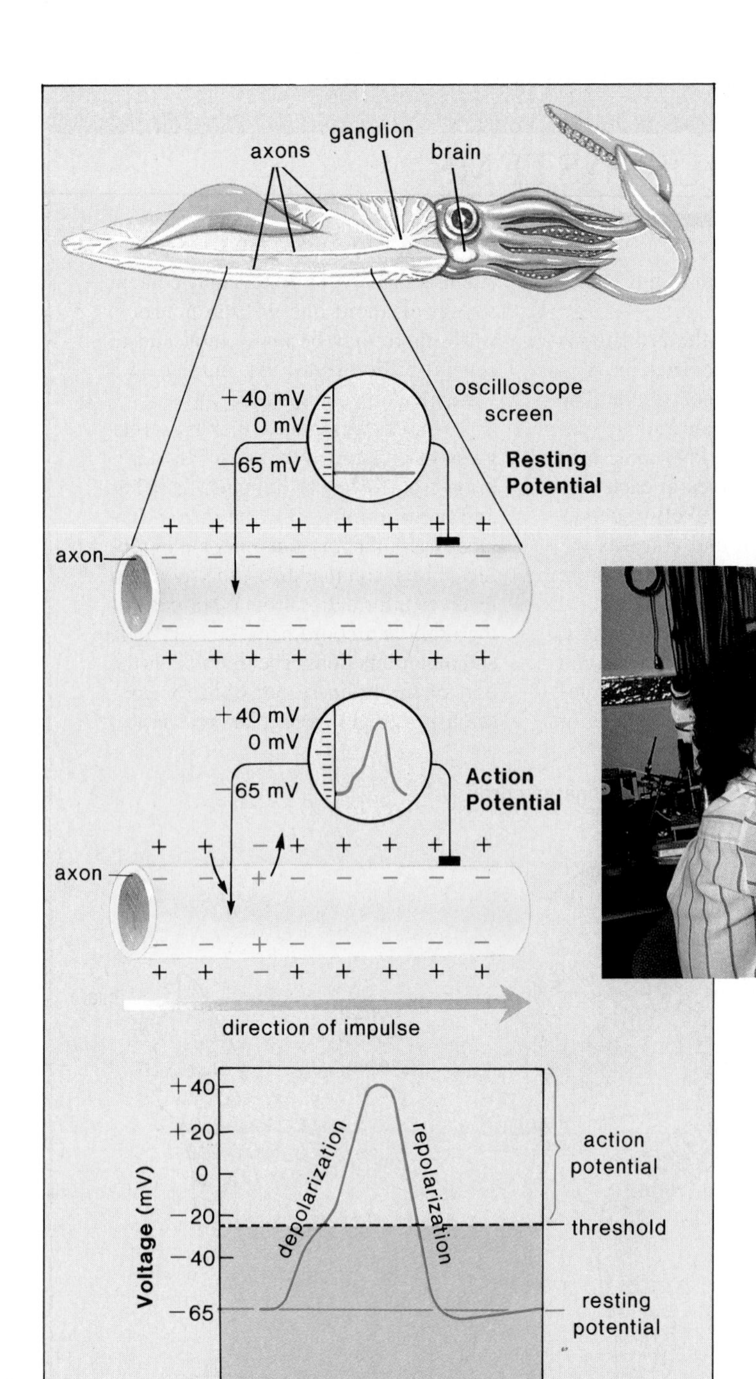

Figure 16.4

Study of nerve impulse requires the use of an oscilloscope and a nerve fiber, such as a giant squid axon. These squid axons are so large (about 1 mm in diameter) that a microelectrode can be inserted into them. When the axon is not conducting a nerve impulse, the electrode registers and the oscilloscope shows a resting potential of –65 mV. When the axon is conducting a nerve impulse, the threshold for an action potential is achieved, and there is a rapid change in potential from –65 mV to +40 mV (called depolarization), followed by a return to –65 mV (called repolarization).

Resting Potential: Inside Is Negative

In the experimental setup shown in figure 16.4, an oscilloscope is wired to 2 electrodes, one inside and one outside a giant axon of the squid. The axon is essentially a membranous tube filled with axoplasm (cytoplasm of the axon). When the axon is not conducting an impulse, the oscilloscope records a *membrane potential* (potential difference across a membrane) equal to about −65 mV. This reading indicates that the inside of the neuron is negative compared to the outside. This is called the **resting potential** because the axon is not conducting an impulse.

The existence of this polarity (charge difference) can be correlated with a difference in ion distribution on either side of the axomembrane (cell membrane of the axon). As figure 16.5 shows, the concentration of sodium ions (Na^+) is greater outside the axon than inside and the concentration of potassium ions (K^+) is greater inside the axon than outside. The unequal distribution of these ions is due to the action of the sodium-potassium pump. This is an active transport system in the axomembrane that pumps sodium ions out of and potassium ions into the axon (p. 283). The work of the pump maintains the unequal distribution of sodium ions and potassium ions across the axomembrane.

The pump is always working because the membrane is somewhat permeable to these ions and they tend to diffuse toward their lesser concentration. Since the membrane is more permeable to potassium than to sodium, there are always more positive ions outside the axomembrane than inside; this accounts for the polarity recorded by the oscilloscope. There are also large, negatively charged proteins in the axoplasm, which are termed immobile in figure 16.5 because they are too large to cross the axomembrane.

Figure 16.5

Action potential and resting potential. The action potential is the result of an exchange of sodium ions (Na^+) and potassium ions (K^+), and it is shown (*right*) as a change in polarity by an oscilloscope. So few ions are exchanged for each action potential that it is possible for a nerve fiber to repeatedly conduct nerve impulses. Whenever the fiber rests, the sodium-potassium pump restores the original distribution of ions.

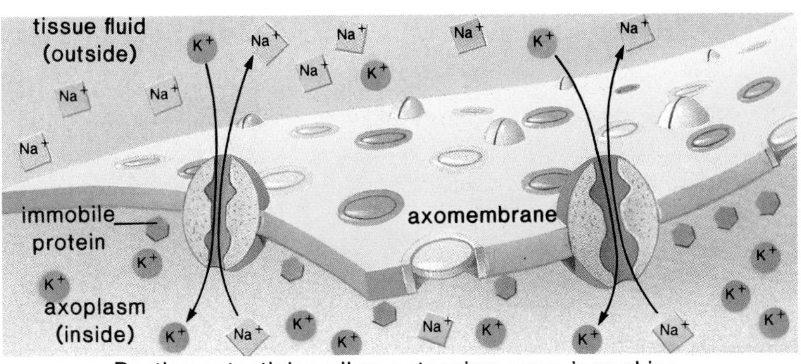

Resting potential: sodium-potassium pump is working.

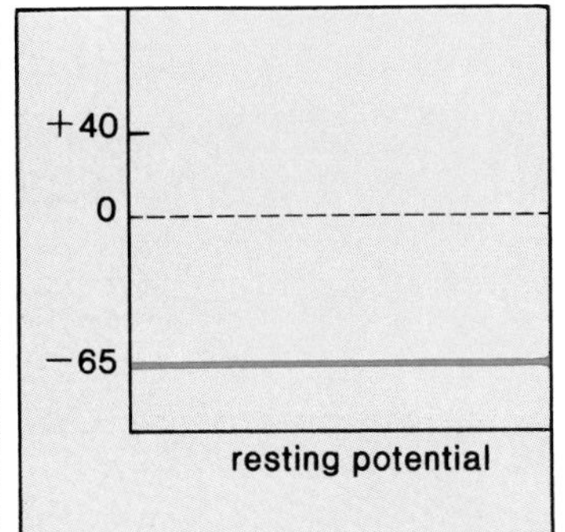

When a neuron is not conducting a nerve impulse, the sodium (Na^+) and potassium (K^+) gates are closed. The sodium-potassium pump maintains the uneven distribution of these ions across the axomembrane. The oscilloscope registers a resting potential of −65 mV inside compared to outside.

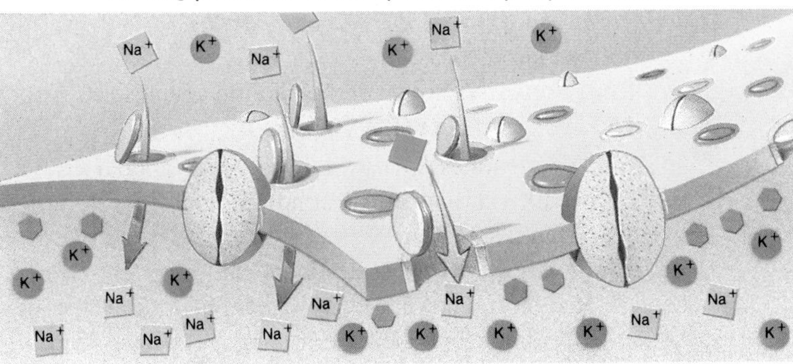

Action potential: sodium gates are open.

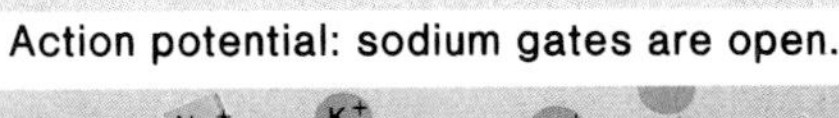

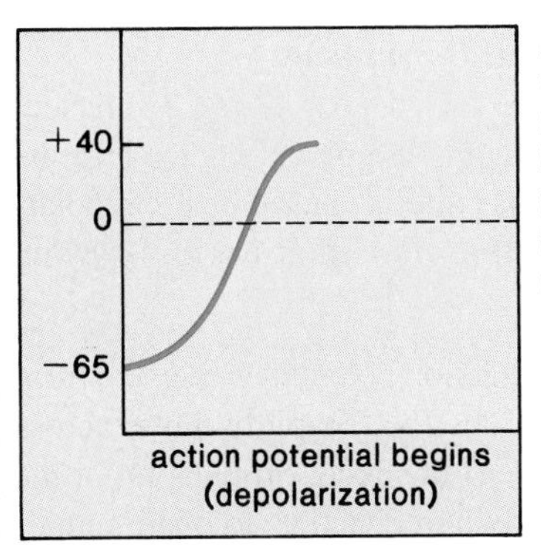

An action potential begins when the sodium gates open and sodium ions move to the inside. The oscilloscope registers a depolarization as the axoplasm reaches +40 mV compared to tissue fluid.

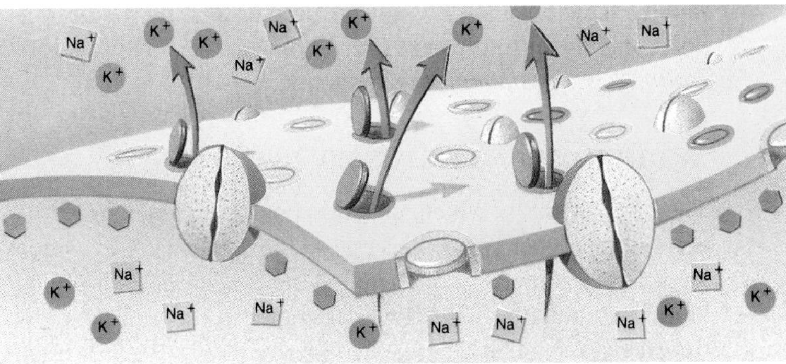

Action potential: potassium gates are open.

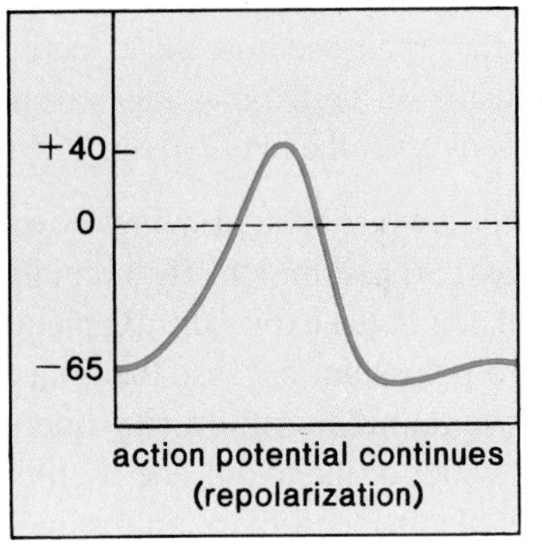

Action potential continues as the sodium gates close and the potassium gates open, allowing potassium ions to move to the outside. The oscilloscope registers repolarization as the axoplasm again becomes −65 mV compared to tissue fluid.

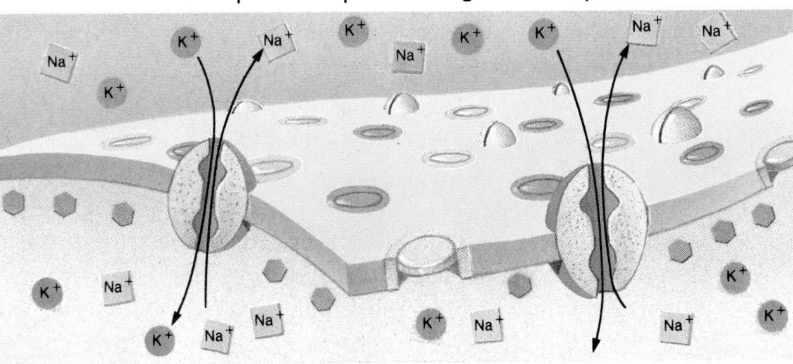

Retractory period: sodium-potassium pump is working.

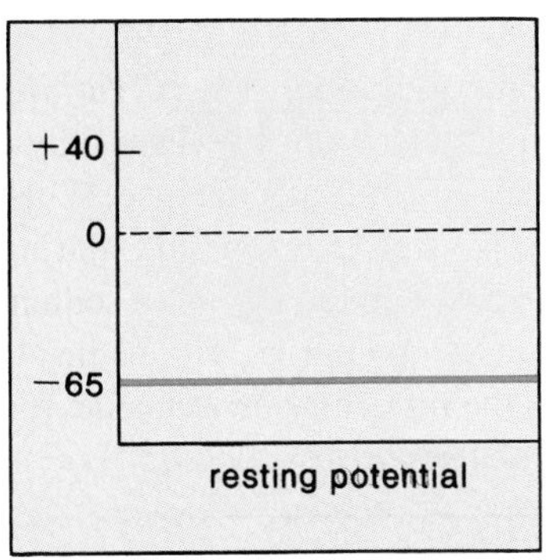

The oscilloscope registers −65 mV again, but the sodium-potassium pump is working to restore the original sodium and potassium ion distribution. The sodium and potassium gates are now closed but will open again in response to another stimulus.

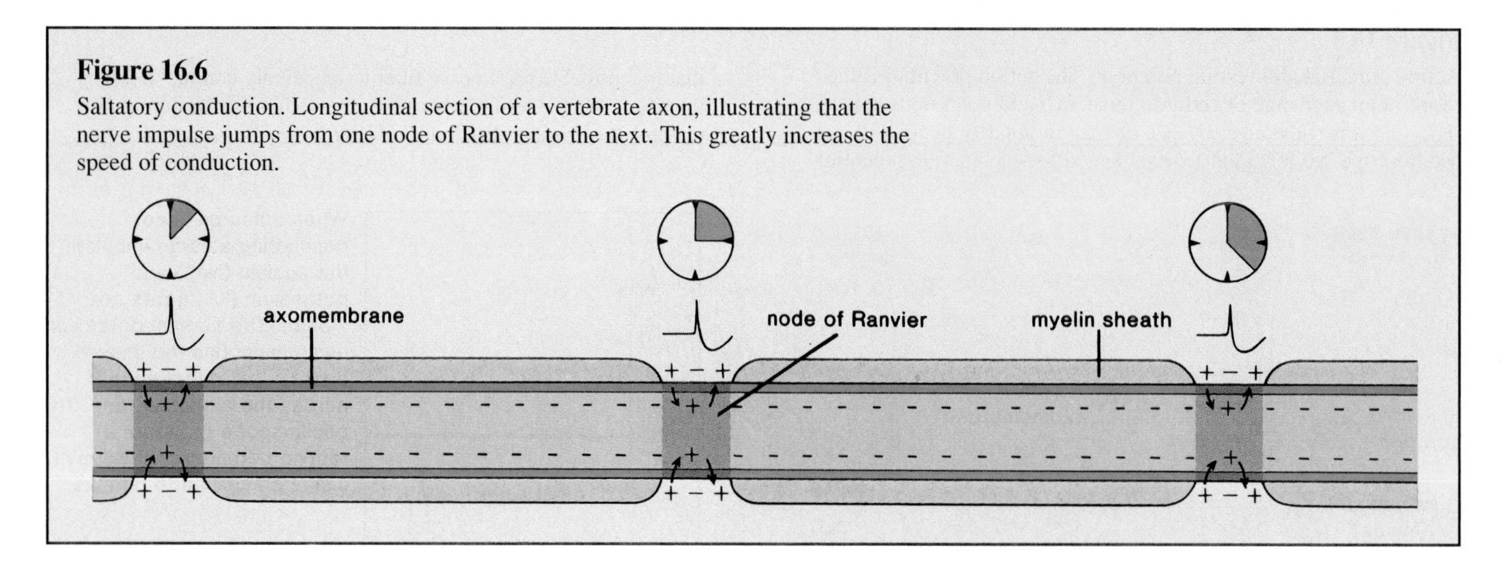

Figure 16.6
Saltatory conduction. Longitudinal section of a vertebrate axon, illustrating that the nerve impulse jumps from one node of Ranvier to the next. This greatly increases the speed of conduction.

Action Potential: Upswing and Downswing
If the axon is stimulated to conduct a nerve impulse by an electric shock, a sudden change in pH, or a pinch, a trace appears on the oscilloscope screen. This pattern, caused by rapid polarity changes and called the **action potential,** has an upswing and a downswing.

Sodium Gates Open. As the action potential swings up from −65 mV to +40 mV, sodium ions (Na^+) rapidly move across the axomembrane to the inside of the axon. Stimulation of the axon causes the gates of the sodium channels to open temporarily, allowing sodium to flow into the axon. This sudden permeability of the axomembrane causes the oscilloscope to record a *depolarization*: the charge inside of the axon changes from negative to positive as sodium ions enter the interior (fig. 16.5).

Potassium Gates Open. As the action potential swings down from +40 mV to at least −65 mV, potassium ions (K^+) rapidly move from the inside to the outside of the axon. The axomembrane is suddenly permeable to potassium because the potassium gates of the potassium channels temporarily open, allowing potassium ions to flow out of the axon. The oscilloscope records a *repolarization*: the inside of the axon resumes a negative charge (fig. 16.5).

Refractory Period: Sodium-Potassium Pump Keeps Working
A fiber can conduct a volley of nerve impulses because only a small number of ions are exchanged with each impulse. When the fiber rests, however, there is a refractory period, during which the sodium-potassium pump continues to return sodium ions (Na^+) to the outside and potassium ions (K^+) to the inside of the axon (fig. 16.5). During the refractory period, a neuron is unable to conduct a nerve impulse.

All neurons, whether sensory or motor, transmit the same type of nerve impulse—an electrochemical change that is propagated along the nerve fiber(s).

Conduction Speed: From 20 m/sec to 200 m/sec
The oscilloscope records changes at only one location in a nerve fiber. Actually, however, the action potential travels along the length of a fiber (fig. 16.6).

Unlike vertebrates, invertebrate nerve fibers do not have a myelin sheath. The speed of conduction in invertebrate nerve fibers can reach 20 m per second, but the speed of conduction in myelinated vertebrate fibers can reach 200 m per second. Notice in figure 16.3 that the myelin sheath has gaps between Schwann cells called the **nodes of Ranvier** (neurofibril node). The speed of conduction in myelinated fibers is much faster because the action potential jumps from one node of Ranvier to the next (fig. 16.6). This is called saltatory (*saltatory* means *jumping*) conduction.

Transmission across a Synapse

The mechanism by which an action potential passes from one neuron to another is not the same as the mechanism by which an action potential is conducted along a neuron. Every axon branches into many fine terminal branches, each of which is tipped by a small swelling, the axon bulb (fig. 16.7*a*). Each bulb lies very close to the dendrite (or the cell body) of another neuron. This region of close proximity is called a **synapse.** At a synapse, the axomembrane is called the **presynaptic membrane** and the membrane of the next neuron is called the **postsynaptic membrane.** The small gap between is the **synaptic cleft.**

Transmission of nerve impulses across a synaptic cleft is carried out by **neurotransmitter substances,** which are stored in synaptic vesicles (fig. 16.7*b*) before their release. When nerve impulses traveling along an axon reach a synaptic ending, the axomembrane becomes permeable to calcium ions (Ca^{++}). These ions then interact with microfilaments, causing the microfilaments to pull the synaptic vesicles to the inner surface of the presynaptic membrane. When the vesicles

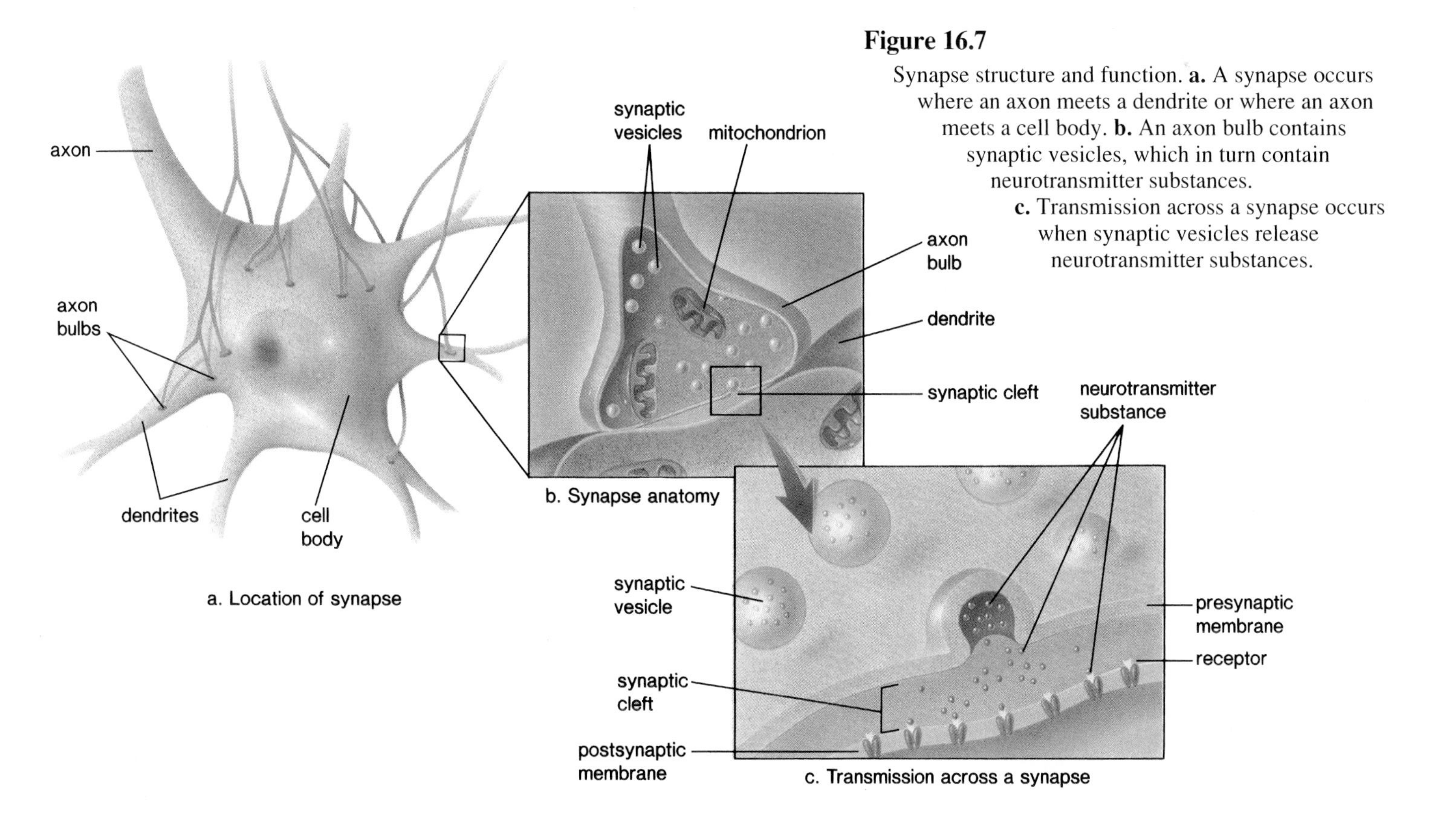

Figure 16.7

Synapse structure and function. **a.** A synapse occurs where an axon meets a dendrite or where an axon meets a cell body. **b.** An axon bulb contains synaptic vesicles, which in turn contain neurotransmitter substances. **c.** Transmission across a synapse occurs when synaptic vesicles release neurotransmitter substances.

merge with this membrane, a neurotransmitter substance is discharged into the synaptic cleft. The neurotransmitter molecules diffuse across the cleft to the postsynaptic membrane, where they bind with a receptor in a lock-and-key manner (fig. 16.7*c*).

Depending on the type of neurotransmitter and/or the type of receptor, the response to a neurotransmitter can be excitation or inhibition. If excitation occurs, the membrane potential of the postsynaptic membrane decreases, the sodium ion (Na^+) channels open at that locale, and the likelihood of the neuron firing (transmitting a nerve impulse) increases. If inhibition occurs, the membrane potential of the postsynaptic membrane increases as the inside becomes more negative, and the likelihood of a nerve impulse decreases.

Transmission across a synapse is dependent on the release of neurotransmitters, which diffuse across the synaptic cleft, a small space separating neuron from neuron. Do 16.1 Critical Thinking, found at the end of the chapter.

Neurotransmitters: Quick Acting

Acetylcholine (ACh) and **norepinephrine (NE)** are well-known neurotransmitters active in both the PNS and the CNS. They are excitatory or inhibitory according to the type of receptor at the postsynaptic membrane. Examples of other neurotransmitters known to be active only in the CNS are given on page 295.

Once a neurotransmitter substance has been released into a synaptic cleft, it has only a short time to act. In some synapses, the cleft contains enzymes that rapidly inactivate the neurotransmitter. For example, the enzyme **acetylcholinesterase (AChE),** or simply cholinesterase, breaks down acetylcholine. In other synapses, the synaptic ending rapidly absorbs the neurotransmitter substance, possibly for repackaging in synaptic vesicles or for chemical breakdown. The enzyme monoamine oxidase breaks down norepinephrine after it is absorbed. The short existence of neurotransmitters in the synapse prevents continuous stimulation (or inhibition) of postsynaptic membranes.

Transmission of nerve impulses across a synapse is dependent on a neurotransmitter substance, which changes the permeability of the postsynaptic membrane.

Summation and Integration: On the Receiving End

A dendrite or a cell body is on the receiving end of many synapses. Whether or not a neuron fires depends on **summation,** the net effect of all the excitatory and inhibitory neurotransmitters received. If enough sodium ion (Na^+) channels open, excitation is sufficient to raise the membrane potential above threshold level (fig. 16.4), and the neuron fires. Otherwise, it does not fire.

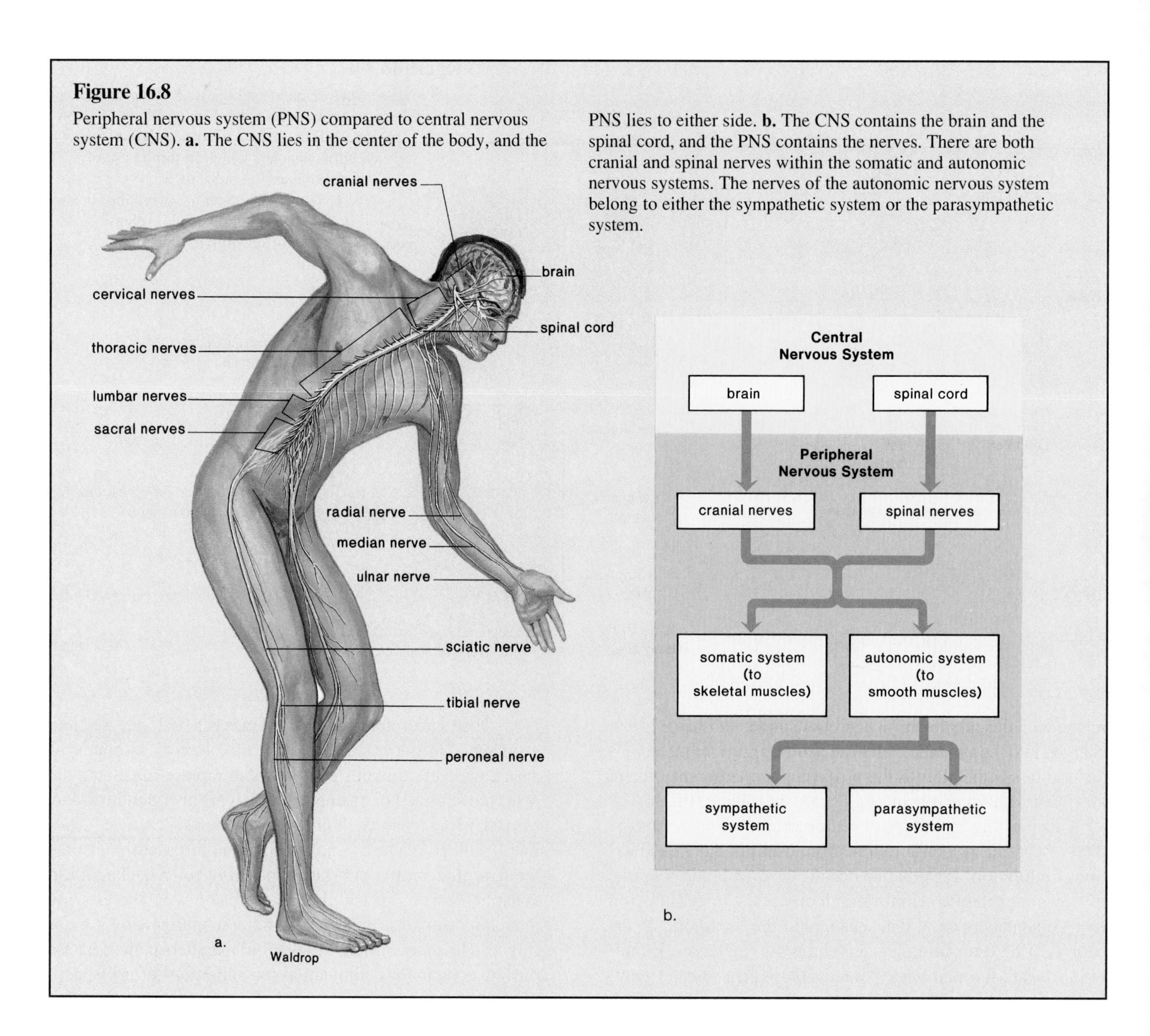

Figure 16.8
Peripheral nervous system (PNS) compared to central nervous system (CNS). **a.** The CNS lies in the center of the body, and the PNS lies to either side. **b.** The CNS contains the brain and the spinal cord, and the PNS contains the nerves. There are both cranial and spinal nerves within the somatic and autonomic nervous systems. The nerves of the autonomic nervous system belong to either the sympathetic system or the parasympathetic system.

The CNS *integrates* (sums up) the information it receives from all over the body. Summation in a neuron is integration at the cellular level. Integration in the brain allows us to make decisions about the body in general.

The Peripheral Nervous System

The peripheral nervous system (PNS) lies outside the central nervous system (CNS) (fig. 16.8*a*). The PNS is made up of nerves, which are part of either the somatic system or the autonomic system (fig. 16.8*b*). The somatic system contains nerves that control skeletal muscles, skin, and joints. The autonomic system contains nerves that control the smooth muscles of the internal organs and the glands.

Nerves: Only Long Fibers

Nerves are structures that contain many long fibers—long dendrites and/or long axons. The cell bodies of these neurons are found in the brain, the spinal cord, or the ganglia. Ganglia (sing., **ganglion**) are collections of cell bodies within the PNS. An individual nerve fiber obeys an all-or-none law, meaning that it fires maximally or it does not fire. A nerve does not obey the all-or-none law—a nerve can have degrees of performance because it contains many fibers, any number of which can be carrying nerve impulses (fig. 16.9).

There are 3 types of nerves (table 16.2). **Sensory nerves** contain only the long dendrites of sensory neurons. **Motor nerves** contain only the long axons of motor neurons. **Mixed**

Figure 16.9

Nerve anatomy and physiology. A nerve contains many nerve fibers, any number of which can be carrying a nerve impulse. Therefore, a nerve has degrees of response and does not obey an all-or-none law.

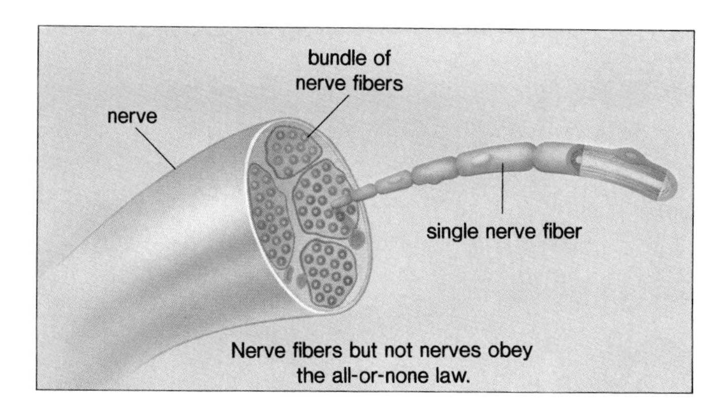

Table 16.2
Nerves

Type	Structure	Function
Sensory nerve	Long axons of sensory neurons only	Carries message from a receptor to the CNS
Motor nerve	Long axons of motor neurons only	Carries message from the CNS to an effector
Mixed nerve	Both long dendrites of sensory neurons and long axons of motor neurons	Carries message in dendrite to and away from the CNS in axons

nerves, however, contain both the long dendrites of sensory neurons and the long axons of motor neurons. Each nerve fiber within a nerve is surrounded by myelin (fig. 16.3), and therefore nerves have a white, glistening appearance.

Cranial Nerves: Attached to Brain

Humans have 12 pairs of **cranial nerves** attached to the brain. Some of these are sensory, some are motor, and others are mixed. Notice that although the brain is part of the CNS, the cranial nerves are part of the PNS. All cranial nerves, except the vagus, travel to or from the head, neck, and face. The vagus nerve has many branches serving the internal organs.

Spinal Nerves: Two Roots from the Spinal Cord

Humans have 31 pairs of **spinal nerves.** Each spinal nerve emerges from the spinal cord (fig. 16.10) by 2 short branches, or roots, which lie within the vertebral column. The *dorsal root* contains the dendrites of sensory neurons, which conduct impulses to the cord. The *ventral root* contains the axons of motor neurons, which conduct impulses away from the cord. These 2 roots join just before a spinal nerve leaves the vertebral column. Therefore, all spinal nerves are mixed nerves, which contain many sensory dendrites and motor axons. Each spinal nerve serves the particular region of the body in which it is located.

In the PNS, cranial nerves take impulses to and/or from the brain, and spinal nerves take impulses to and from the spinal cord.

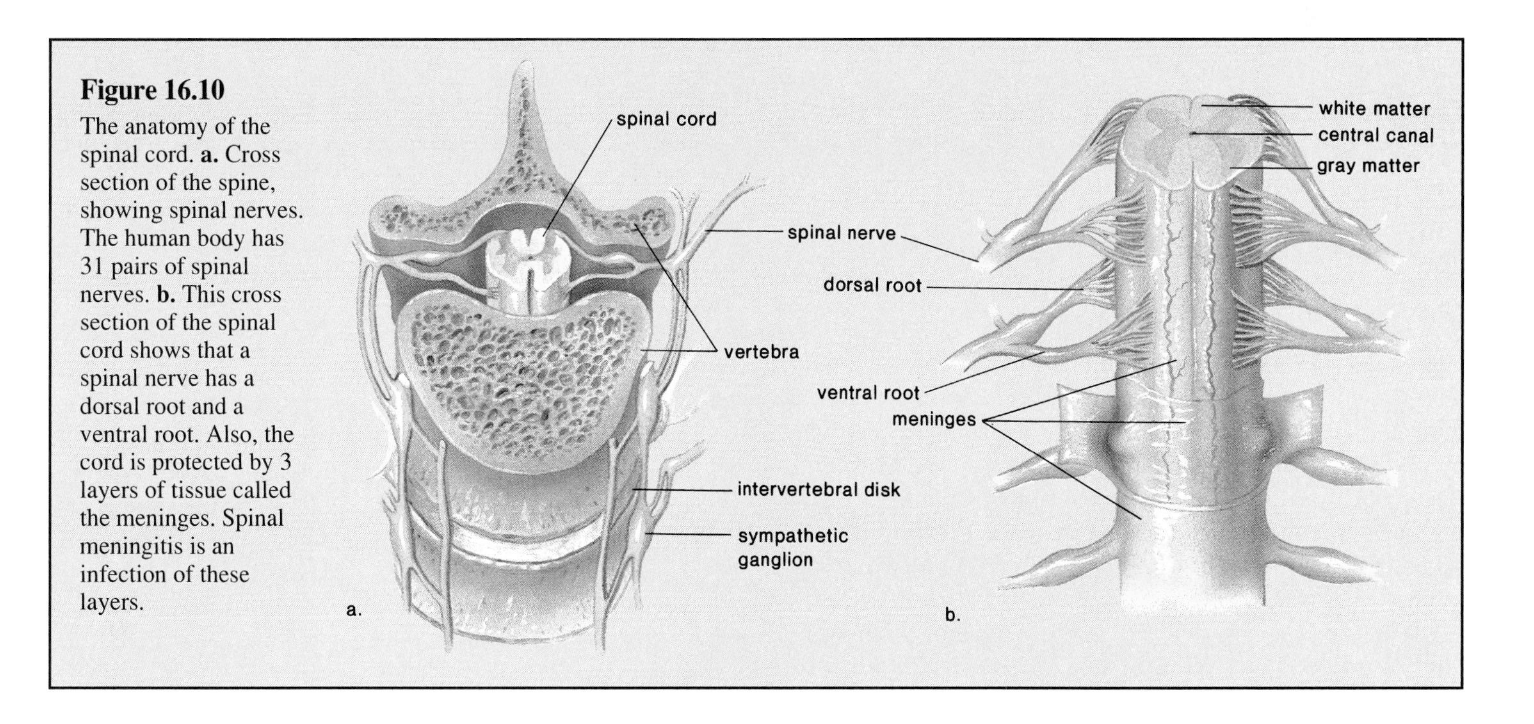

Figure 16.10

The anatomy of the spinal cord. **a.** Cross section of the spine, showing spinal nerves. The human body has 31 pairs of spinal nerves. **b.** This cross section of the spinal cord shows that a spinal nerve has a dorsal root and a ventral root. Also, the cord is protected by 3 layers of tissue called the meninges. Spinal meningitis is an infection of these layers.

Figure 16.11

Diagram of a reflex arc showing the detailed composition of a spinal nerve. When a receptor in skin is stimulated, nerve impulses (see arrows) move along a sensory neuron to the spinal cord. (Note that the cell body of a sensory neuron is in a ganglion outside the cord.) The nerve impulses are picked up by an interneuron, which lies completely within the cord, and pass to the dendrites and the cell body of a motor neuron that lies ventrally within the cord. The nerve impulses then move along the axon of the motor neuron to an effector, such as a muscle, which contracts. The brain receives information concerning sensory stimuli by way of other interneurons, with long fibers in tracts that run up and down the cord within the white matter.

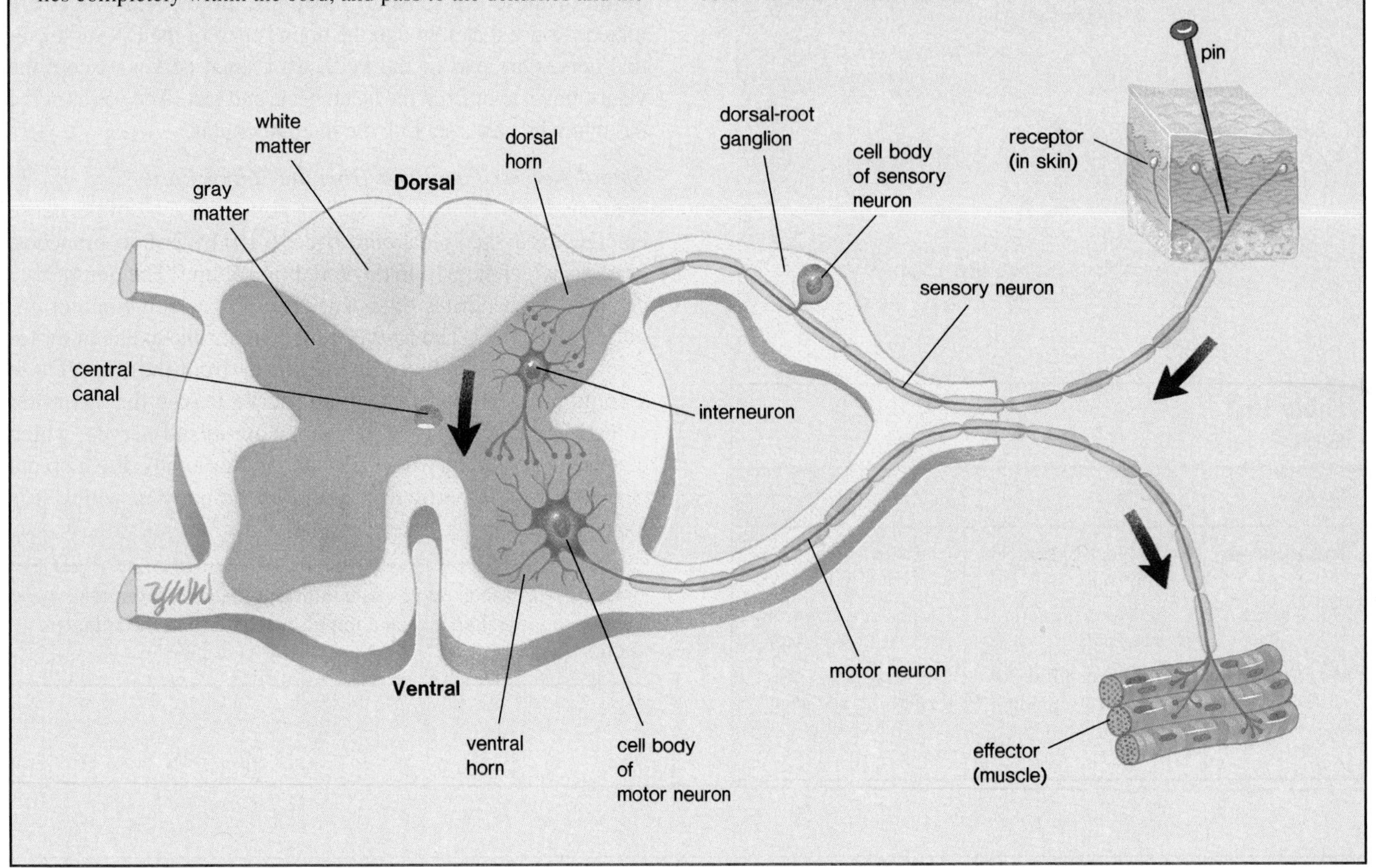

Somatic Nervous System: For Muscles

The **somatic nervous system** includes all nerves that serve the musculoskeletal system and the receptors, including those in skin. Receptors receive environmental stimuli and then initiate nerve impulses. Muscles are effectors, which bring about a reaction to the stimulus. Receptors are studied in chapter 18, and muscle effectors are studied in chapter 17.

The Reflex Arc: Main Functional Unit

Reflexes are automatic, involuntary responses to changes occurring inside or outside the body. In the somatic nervous system, outside stimuli often initiate a reflex action. Some reflexes, such as blinking the eye, involve the brain, while others, such as withdrawing the hand from a hot object, do not necessarily involve the brain. Figure 16.11 illustrates the path of the second type of reflex action. Whenever a person touches a very hot object, a *receptor* in skin generates nerve impulses, which move along the dendrite of a *sensory neuron* toward the cell body and the CNS. The cell body of a sensory neuron is located in the **dorsal-root ganglion,** just outside the cord. From the cell body, the impulses travel along the axon of the sensory neuron and enter the cord by the dorsal root of a spinal nerve. The impulses then pass to many interneurons, one of which connects with a motor neuron. The short dendrites and the cell body of the *motor neuron* lead to the axon, which leaves the cord by way of the ventral root of a spinal nerve. The nerve impulses travel along the axon to *muscle fibers,* which then contract so that the hand is withdrawn from the hot object. (See table 16.3 for a listing of these events.)

Various other reactions usually accompany a reflex response: the person may look in the direction of the object, jump back, and utter appropriate exclamations. This whole series of responses is explained by the fact that the sensory neuron stimulates several interneurons, which take impulses to all parts of the CNS, including the cerebrum, which in turn, makes the person conscious of the stimulus and his or her reaction to it.

Table 16.3
Path of a Simple Reflex

1. Receptor (formulates message)*	Generates nerve impulses
2. Sensory neuron (takes message to CNS)	Impulses move along dendrite (spinal nerve)† and proceed to cell body (dorsal-root ganglion) and then go from cell body to axon (spinal cord)
3. Interneuron (passes message to motor neuron)	Impulses picked up by dendrites and pass through cell body to axon (spinal cord)
4. Motor neuron (takes message away from CNS)	Impulses travel through short dendrites and cell body (spinal cord) to axon (spinal nerve)
5. Effector (receives message)	Receives nerve impulses and reacts: glands secrete and muscles contract

*Phrases within parentheses state overall function.

†Words within parentheses indicate location of structure.

The reflex arc is the main functional unit of the nervous system. It allows us to react to internal and external stimuli. Do 16.2 Critical Thinking, found at the end of the chapter.

Autonomic Nervous System: For Internal Organs

The autonomic nervous system (fig. 16.12), a part of the PNS, is made up of motor neurons that control the internal organs automatically and usually without need for conscious intervention. The sensory neurons that come from the internal organs allow us to feel internal pain. The cell bodies for these sensory neurons are in dorsal root ganglia along with the cell bodies of somatic sensory neurons.

There are 2 divisions of the autonomic nervous system: the sympathetic system and the parasympathetic system. Both of these (1) function automatically and usually subconsciously in an involuntary manner; (2) innervate all internal organs; and (3) utilize 2 motor neurons and one ganglion for each impulse. The first of these 2 neurons has a cell body within the CNS and a **preganglionic axon.** The second neuron has a cell body within the ganglion and a **postganglionic axon.**

The autonomic nervous system subconsciously controls the function of internal organs.

Sympathetic System: Fight or Flight

The preganglionic fibers of the **sympathetic nervous system** arise from the *thoraco-lumbar* (middle) portion of the spinal cord and almost immediately terminate in ganglia that lie near the cord. Therefore, in this system, the preganglionic fiber is short, but the postganglionic fiber that makes contact with an organ is long.

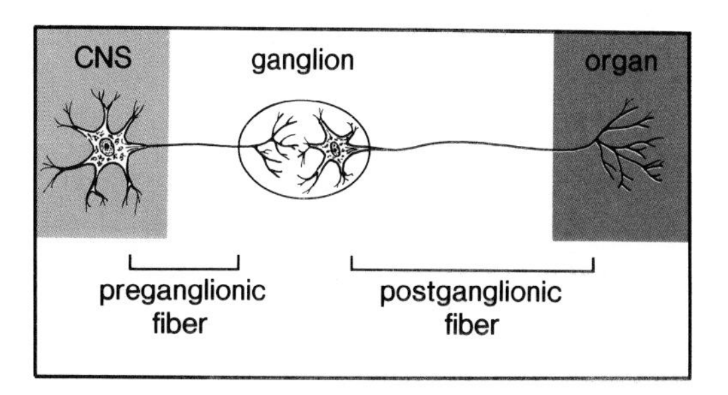

The sympathetic system is especially important during emergency situations and is associated with "fight or flight." For example, it inhibits the digestive tract, but it dilates the pupil, accelerates the heartbeat, and increases the breathing rate. The neurotransmitter released by postganglionic axons is primarily norepinephrine, a chemical close in structure to epinephrine (adrenalin), a medicine used as a heart stimulant.

The sympathetic system brings about those responses we associate with "fight or flight."

Parasympathetic System: Housekeeper

A few cranial nerves, including the vagus nerve, together with fibers that arise from the sacral (bottom) portion of the spinal cord, form the **parasympathetic nervous system** (fig. 16.12). Therefore, this system is often referred to as the *craniosacral portion* of the autonomic nervous system. In the parasympathetic nervous system, the preganglionic fiber is long and the postganglionic fiber is short because the ganglia lie near or within the organ.

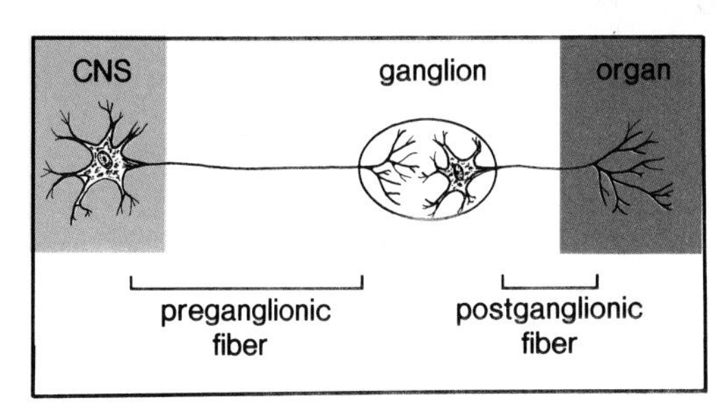

The parasympathetic system, sometimes called the "housekeeper system," promotes all the internal responses we associate with a relaxed state; for example, it causes the pupil of the eye to contract, promotes digestion of food, and retards the heartbeat. The neurotransmitter utilized by the parasympathetic system is primarily acetylcholine.

The parasympathetic system brings about the responses we associate with a relaxed state.

Figure 16.12

Structure and function of the autonomic nervous system. The sympathetic fibers arise from the thoracic and lumbar portions of the spinal cord; the parasympathetic fibers arise from the brain and the sacral portion of the cord. Each system innervates the same organs although they have contrary effects. For example, the sympathetic system speeds up the beat of the heart, while the parasympathetic system slows it down.

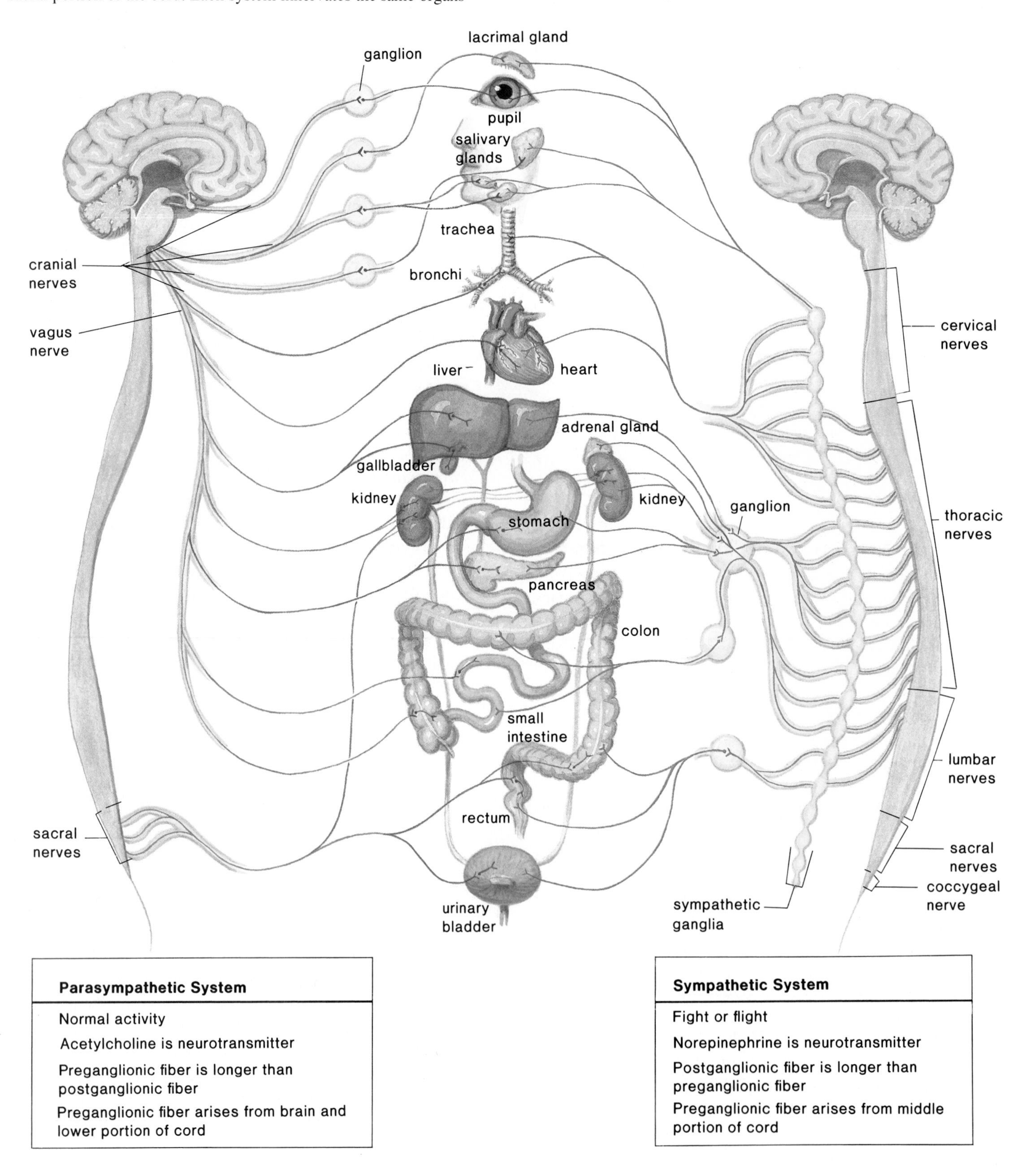

Figure 16.13
The human brain. Note how large the cerebrum is compared to the rest of the brain.

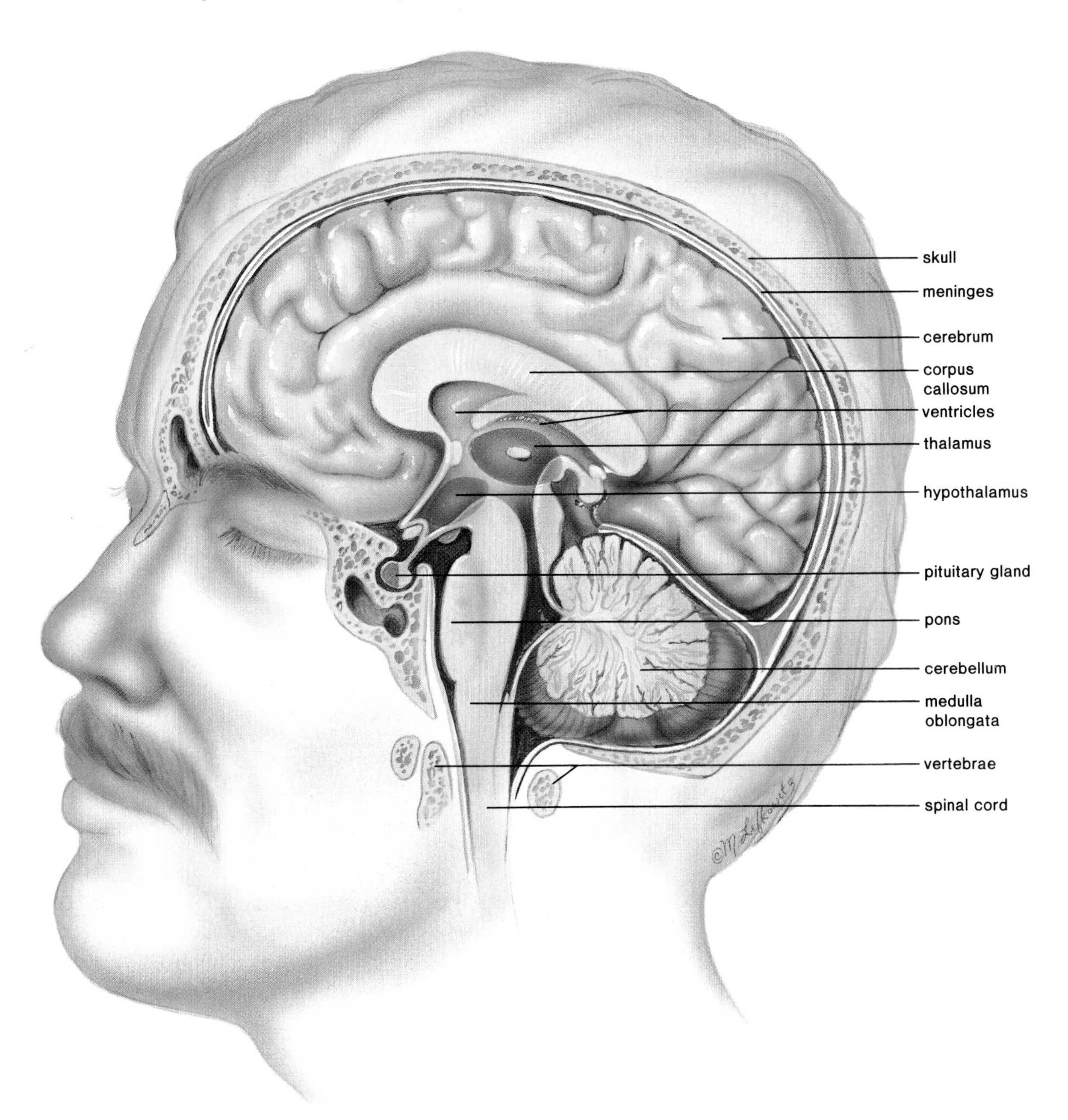

The Central Nervous System

The central nervous system (CNS) consists of the spinal cord and the brain. As figures 16.10 and 16.13 illustrate, the CNS is protected by bone: the brain is enclosed by the skull and the spinal cord is surrounded by vertebrae. Also, both the brain and the spinal cord are wrapped in 3 protective membranes known as **meninges** (sing., meninx); meningitis is an infection of these coverings (fig. 16.10*b*). The spaces between the meninges are filled with **cerebrospinal fluid,** which cushions and protects the CNS. Cerebrospinal fluid is contained within the *ventricles* of the brain, which are interconnecting spaces that produce and serve as a reservoir for cerebrospinal fluid, and the **central canal** of the spinal cord. A small amount of this fluid is sometimes withdrawn for laboratory testing when a spinal tap (i.e., lumbar puncture) is performed.

Spinal Cord: Two Main Functions

The spinal cord lies along the middorsal line of the body. It has 2 main functions: (1) it is the center for many reflex actions, and (2) it provides a means of communication between the brain and the spinal nerves, which leave the spinal cord.

The path of a spinal reflex passes through the gray matter of the cord (fig. 16.11). Unmyelinated cell bodies and short fibers give this area its gray color. In cross section, the gray matter looks like a butterfly or the letter H. The axons of sensory neurons are found in the dorsal regions (horns) of the gray matter, and the dendrites and the cell bodies of motor neurons are found in the ventral regions (horns) of the gray matter. Short interneurons connect sensory neurons to motor neurons on the same and the opposite sides of the spinal cord.

The white matter of the spinal cord is found in between the regions of the gray matter (fig. 16.10*b*). Myelinated long fibers of interneurons that run together in bundles called *tracts* give white matter its color. These tracts connect the spinal cord to the brain. Dorsally, there are primarily ascending tracts taking information to the brain, and ventrally, there are primarily descending tracts carrying information from the brain. Because the tracts at one point cross over, the left side of the brain controls the right side of the body and the right side of the brain controls the left side of the body.

> The CNS lies in the midline of the body and consists of the brain and the spinal cord. Sensory information is received and motor control is initiated in the CNS.

Brain: Subconscious and Conscious

The largest and most prominent portion of the human brain (fig. 16.13) is the cerebrum. Consciousness resides only in the cerebrum; the rest of the brain functions below the level of consciousness. In addition to the portions mentioned next, the subconscious brain contains many tracts that relay messages to and from the spinal cord.

Subconscious Brain: Controls, Channels

The **medulla oblongata** lies closest to the spinal cord and contains centers for heartbeat, breathing, and blood pressure. It also contains reflex centers for vomiting, coughing, sneezing, hiccuping, and swallowing.

The **hypothalamus** forms the lower walls and floor of the third ventricle. This part of the brain is concerned with homeostasis, or the constancy of the internal environment, and contains centers for hunger, sleep, thirst, body temperature, water balance, and blood pressure. The hypothalamus controls the pituitary gland and thereby serves as a link between the nervous and endocrine systems.

> The medulla oblongata and the hypothalamus are both concerned with control of the internal organs.

The *midbrain* and the *pons* contain tracts that connect the cerebrum with other parts of the brain. In addition, the pons functions with the medulla oblongata in regulating respiration, and the midbrain has reflex centers concerned with head movements in response to visual and auditory stimuli.

Figure 16.14

The reticular activating system (RAS). This system sorts out information received from receptors and thereafter only certain data reach the cerebrum.

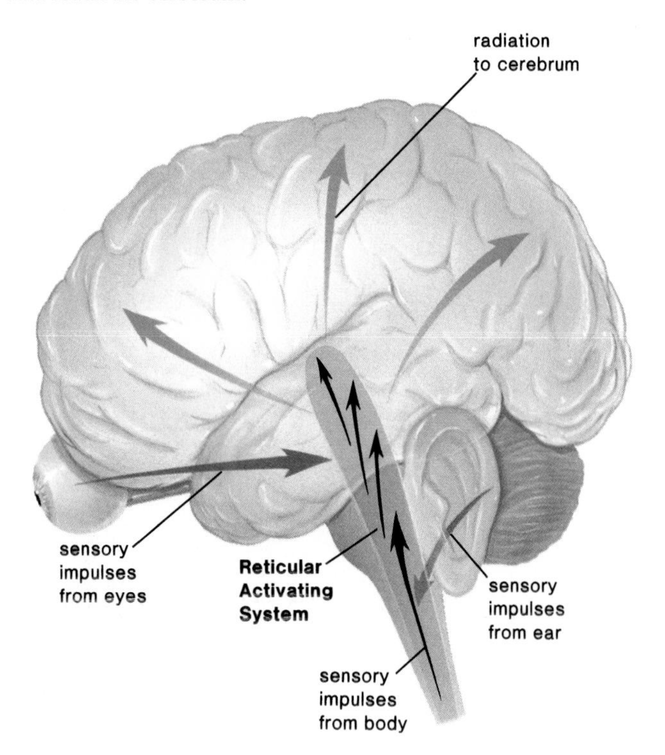

The **thalamus** is an egg-shaped structure in the third ventricle. It is a central relay station for sensory impulses traveling upward from other parts of the cord and the brain to the cerebrum. It receives all sensory impulses (except those associated with the sense of smell) and channels them to appropriate regions of the cerebrum. In other words, the thalamus is the last portion of the brain for sensory input before the cerebrum.

The thalamus has connections to various parts of the brain by way of nerve fibers that radiate from the upper part of the *reticular activating system* (*RAS*) (fig. 16.14). The RAS, which extends from the medulla oblongata to the thalamus, sorts out stimuli received from sense organs, including the eyes and the ears, and passes on only those that require immediate attention. The thalamus sometimes is called the "gatekeeper to the cerebrum" because it alerts the cerebrum to only certain sensory input. We are not aware of most of the sensory impulses received by the CNS.

> The thalamus receives sensory impulses from other parts of the CNS and channels only certain of these to the cerebrum.

The **cerebellum,** a bilobed, butterfly-shaped structure, is the second largest portion of the brain. It is located dorsal to the pons and the medulla oblongata. The cerebellum functions in muscle coordination (fig. 16.15), integrating impulses received from higher centers to ensure that all the skeletal muscles work together to produce smooth and graceful motions. The cerebellum also is responsible for maintaining normal muscle tone and transmitting impulses that maintain posture. It receives information from the inner ear indicating the position of the body and sends impulses to those muscles whose contraction maintains or restores balance.

The cerebellum controls balance and complex muscular movements.

Figure 16.15

Roller blading. This sport requires muscle coordination and balance, which are controlled by the cerebellum.

Conscious Brain: Largest Part

The **cerebrum,** the foremost part of the brain, is the only area responsible for consciousness. It is the largest portion of the brain in humans. The outer layer of the cerebrum, called the *cerebral cortex,* is gray in color and contains cell bodies and unmyelinated short fibers. The cerebrum is divided into halves, known as the right and left **cerebral hemispheres.** Each hemisphere contains 4 surface lobes: **frontal, parietal, temporal,** and **occipital** (fig. 16.16). Little is known about the functions of a fifth lobe, the insula, which lies beneath the surface.

Figure 16.16

The convoluted cortex of the cerebrum is divided into 4 surface lobes: frontal, temporal, parietal, and occipital. It is possible to map the cerebral cortex since each area has a particular function.

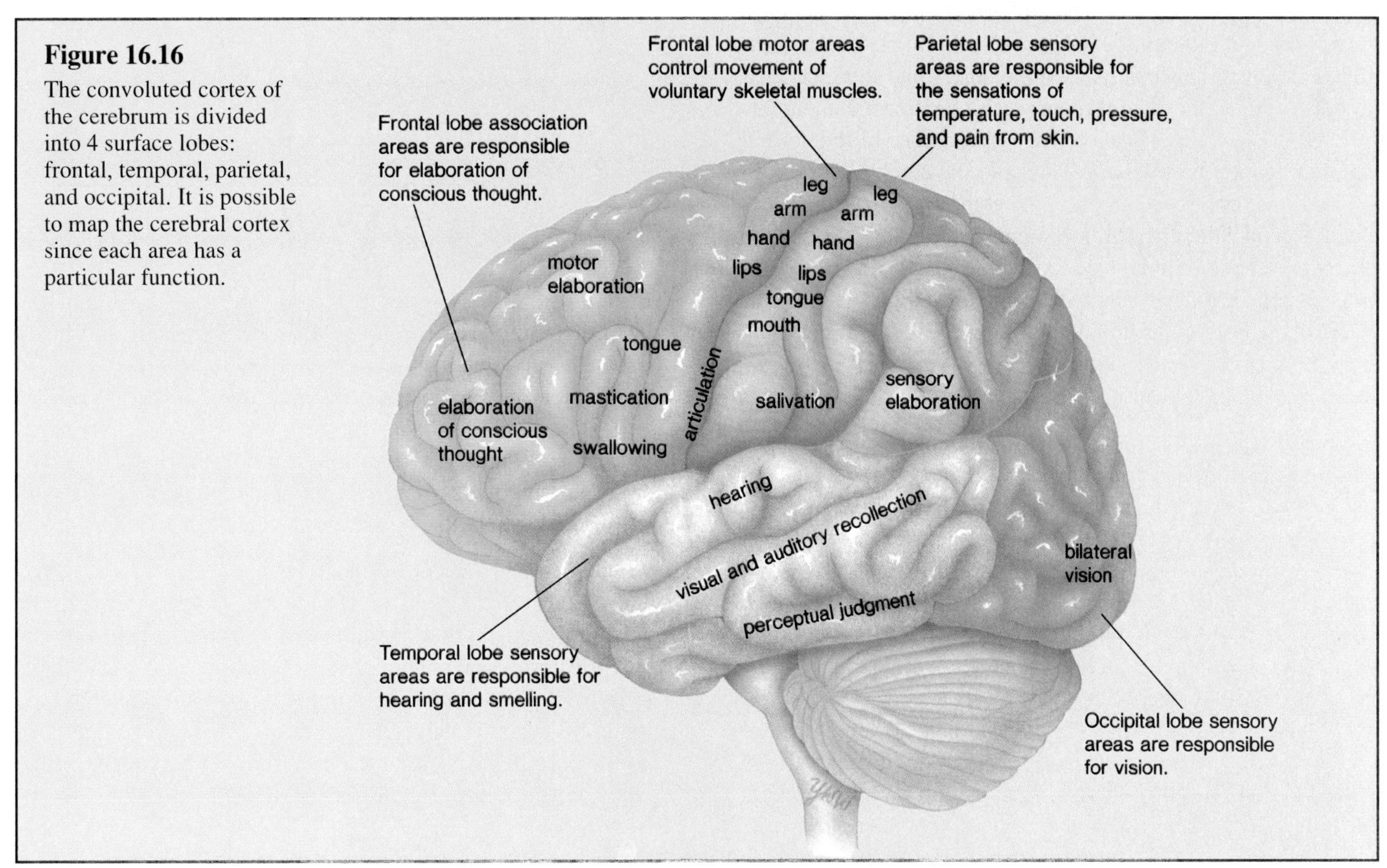

Certain areas of the cerebral cortex have been "mapped" in great detail. Physiologists have identified the *motor areas* of the frontal lobe, which initiate contraction of skeletal (voluntary) muscles; the *sensory areas* of the parietal lobe, which receive impulses from receptors; and the *association areas*, which receive information from the other lobes and integrate it into higher, more complex levels of consciousness. Association areas are concerned with intellect, artistic and creative abilities, learning, and memory.

Consciousness is the province of the cerebrum, the most developed portion of the human brain. The cerebrum is responsible for higher mental processes, including the interpretation of sensory input and the initiation of voluntary muscular movements.

There has been a great deal of testing to determine whether the right and left halves of the cerebrum serve different functions. These studies tend to suggest that the left half of the brain is the verbal (word) half and the right half of the brain is the visual (spatial relation) and artistic half. However, other results indicate that such a strict dichotomy does not always exist between the 2 halves. In any case, the 2 cerebral hemispheres normally share information because they are connected by a horizontal tract called the **corpus callosum** (fig. 16.13).

Severing the corpus callosum can control severe epileptic seizures, but then the 2 halves of the brain no longer communicate; each half has its own memories and thoughts. Today, use of the laser permits more precise treatment without this side effect. *Epilepsy* is caused by a disturbance of the normal communication between the RAS and the cerebral cortex. In a grand mal epileptic seizure, the cerebrum is extremely excited. Due to a reverberation of signals within the RAS and the cerebrum, the individual loses consciousness, even while convulsions are occurring. Finally, the neurons fatigue and the signals cease. Following an attack, the brain is so fatigued the person must sleep for a while.

EEG: Shows Brain Waves. The electrical activity of the brain can be recorded in the form of an **electroencephalogram (EEG).** Electrodes are taped to different parts of the scalp, and an instrument called the electroencephalograph records the so-called brain waves (fig. 16.17).

When the subject is awake, 2 types of waves are usual. *Alpha waves,* with a frequency of about 6–13 per second and a potential of about 45 μV, predominate when the eyes are closed. *Beta waves,* with higher frequencies but lower voltage, appear when the eyes are open.

During sleep the waves become larger, slower, and more erratic. The eyes move back and forth rapidly during periods called **REM** (rapid eye movement) **sleep.** When subjects are awakened during REM, they always report that they had been dreaming. The significance of REM sleep is still being debated, but some studies indicate that REM sleep is needed for memory development.

Figure 16.17

Electroencephalograms (EEGs), recordings of the electrical activity of the brain. The alpha waves, which appear when the subject is awake with eyes closed, are the most common. Second most common are the beta waves, recorded when the subject is awake with eyes open. Sleep has various stages, as indicated.

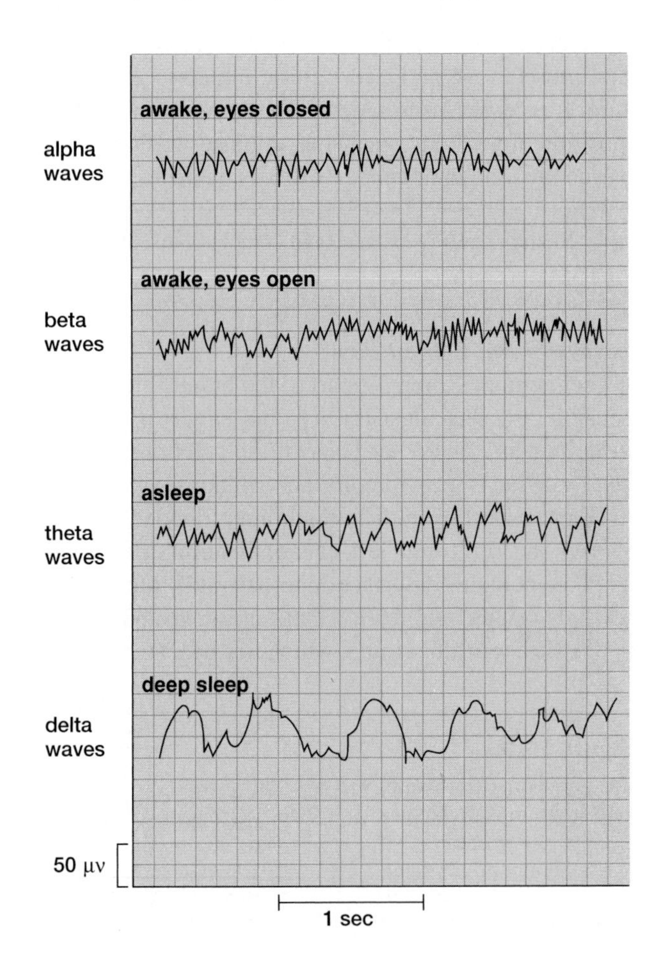

The EEG is a diagnostic tool; for example, an irregular pattern can signify epilepsy or a brain tumor. A flat EEG signifies lack of electrical activity of the brain, or brain death; therefore, it can be used to determine the precise time of death.

Limbic System: Good and Bad Feelings

The **limbic system** (fig. 16.18) involves portions of both the subconscious and the conscious brain. It lies just beneath the cerebral cortex and contains neural pathways that connect portions of the frontal lobes, the temporal lobes, the thalamus, and the hypothalamus. Several masses of gray matter lying deep within each hemisphere of the cerebrum, termed the *basal nuclei,* are also a part of the limbic system.

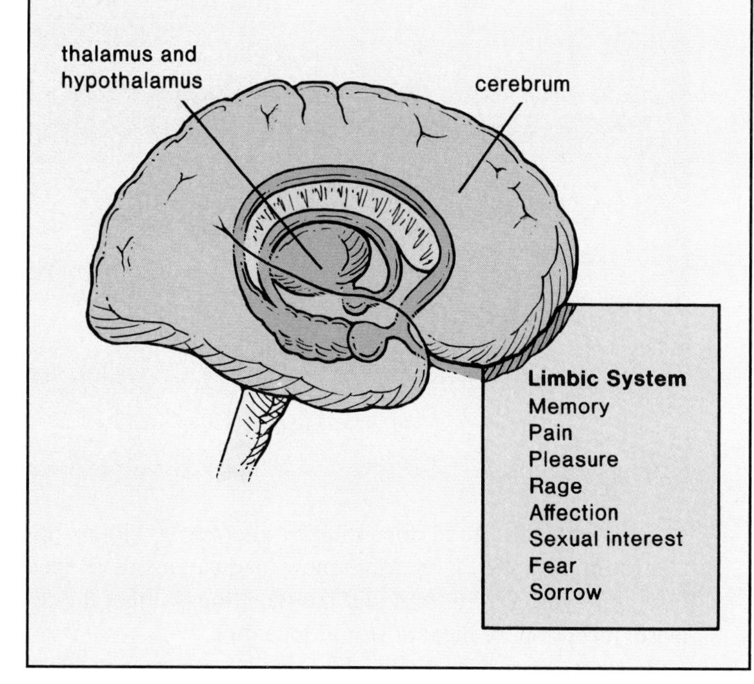

Figure 16.18

The limbic system. The limbic system, which includes portions of the cerebrum, the thalamus, and the hypothalamus, is sometimes called the emotional brain because it seems to control the emotions listed.

Stimulation of different areas of the limbic system causes the subject to experience pain, pleasure, rage, affection, sexual interest, fear, or sorrow. By causing pleasant or unpleasant feelings about experiences, the limbic system apparently guides the individual into behavior that is likely to increase the chance of survival.

Essential for Learning and Memory. The limbic system also is involved in the processes of learning and memory. Learning requires memory, but just what permits memory development is not definitely known. Experimentation with invertebrates such as slugs and snails indicates that learning is accompanied by an increase in the number of synapses in the brain, while forgetting involves a decrease in the number of synapses.

Experiments with monkeys have led to the conclusion that the limbic system is absolutely essential to both short-term and long-term memory. An example of short-term memory in humans is the ability to recall a telephone number long enough to dial it; an example of long-term memory is the ability to recall the events of the day. It is believed that at first, impulses move only within the limbic circuit, but eventually the basal nuclei transmit the neurotransmitter acetylcholine to the sensory areas where memories are stored. The involvement of the limbic system certainly explains why emotionally charged events result in our most vivid memories. The fact that the limbic system communicates with the sensory areas for touch, smell, vision, hearing, and taste accounts for the ability of any particular sensory stimulus to awaken a complex memory.

The limbic system is particularly involved in emotions and in memory and learning.

Brain Neurotransmitters and Runner's High

As discussed previously, neurotransmitters released at the axon bulbs affect the membrane potential of postsynaptic membranes. Acetylcholine (ACh), as well as the chemically related norepinephrine (NE), serotonin, and dopamine, is found in the brain. The excitatory or inhibitory effect of ACh and NE varies according to the type of receptor. *Serotonin* is generally inhibitory, and *dopamine* is generally excitatory. Both are associated with behavior states, such as mood, sleep, learning, and memory. Increasingly, it appears that feelings of pleasure are accompanied by the release of dopamine.

Certain amino acids and peptides are also neurotransmitters in the brain. The amino acids gamma-aminobutyrate (GABA) and glycine are inhibitory, while glutamate is excitatory. *Endorphins* are peptides. They are called the body's own opiates because morphine and heroin, both of which are derived from opium, utilize receptors for endorphin in the CNS. When endorphins are present, neurons do not release substance P, a neurotransmitter that brings about the sensation of pain. Exercise has been associated with the presence of endorphins, and this may account for the so-called runner's high.

Neurotransmitter Disorders. It has been discovered that several neurological illnesses, such as *Parkinson disease* and *Huntington disease,* are due to a neurotransmitter imbalance. Parkinson disease is a condition characterized by a wide-eyed, unblinking expression, an involuntary tremor of the fingers and the thumbs, muscular rigidity, and a shuffling gait. Implanting fetal grafts of dopamine-secreting neurons has alleviated the symptoms of patients with Parkinson disease, but some object to such operations on moral grounds.

Huntington disease is characterized by a progressive deterioration of the individual's nervous system, which eventually leads to constant thrashing and writhing movements and finally to insanity and death. Huntington disease is believed to be the malfunction of the inhibitory neurotransmitter GABA.

Alzheimer disease, a severe form of senility with marked memory loss in 5–10% of all people over age 65, is recognized by the presence of numerous plaques of abnormal protein in the brain. Some patients are taking experimental drugs that prevent the breakdown of acetylcholine because researchers have found that acetylcholine secretion is substantially below normal in the brain of persons with Alzheimer disease.

Some neurological illnesses are associated with the deficiency of a particular neurotransmitter in the brain.

Figure 16.19
Drug action at synapses.

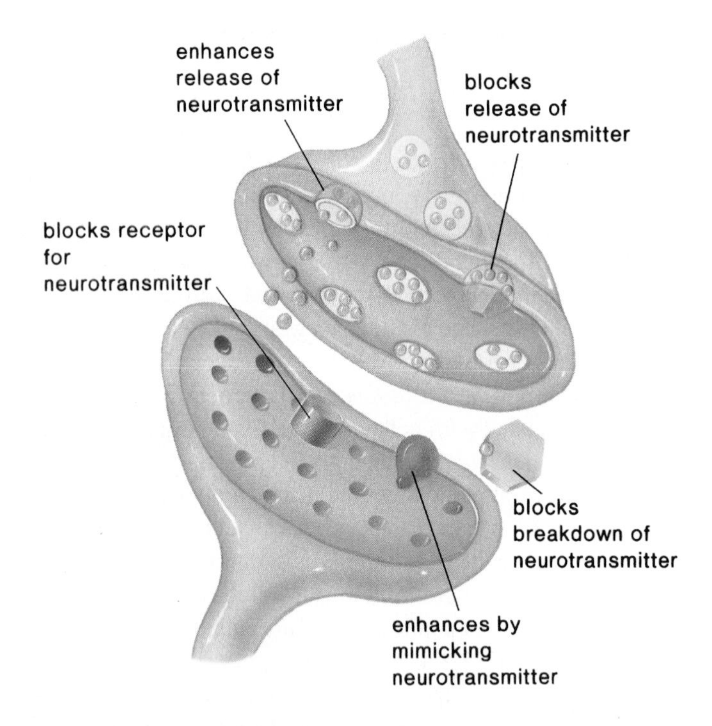

Drug Action	Psychological Effect of Drug
Blocks excitatory neurotransmitter	Depression
Enhances excitatory neurotransmitter	Stimulation
Blocks inhibitory neurotransmitter	Stimulation
Enhances inhibitory neurotransmitter	Depression

Figure 16.20
Drug abuse involvement in accidents. Those who abuse drugs, including alcohol, are more likely to be involved in automobile accidents. Unfortunately, others in addition to the abuser often suffer the consequences.

Drug Abuse

A wide variety of drugs can be used to alter the mood and/or emotional state (see appendix B). Drugs that affect the nervous system have 2 general effects: (1) they impact the RAS (p. 292) and the limbic system, and (2) they either promote or decrease the action of a particular neurotransmitter (fig. 16.19). Stimulants can either enhance the action of an excitatory neurotransmitter or block the action of an inhibitory neurotransmitter. Depressants can either enhance the action of an inhibitory neurotransmitter or block the action of an excitatory neurotransmitter. Increasingly, researchers believe that dopamine secretion primarily affects mood. Cocaine is known to potentiate the effects of dopamine by interfering with its uptake from synaptic clefts. Many new medications developed to counter drug addiction and mental illness affect the release, reception, or breakdown of dopamine.

Drug abuse occurs when a person takes a drug at a dose level and under circumstances that increase the potential for a harmful effect. Drug abusers are apt to display either a psychological and/or a *physical dependence* on the drug. Dependence has developed when the person spends much time thinking about the drug or arranging to get it and often takes more of the drug than was intended. With physical dependence, formerly called an addiction to the drug, the person is *tolerant* to the drug—that is, must increase the amount of the drug to get the same effect—and has *withdrawal symptoms* when he or she stops taking the drug.

> Drugs that affect the nervous system can cause physical dependence and withdrawal symptoms.

Alcohol: Most Abused Drug

The type of alcohol in beer, wine, and liquor is ethanol. While it is possible to drink alcohol in moderation, the drug is often abused. Alcohol use becomes "abuse," or an illness, when alcohol ingestion impairs an individual's social relationships, health, job performance, or ability to avoid legal difficulties (fig. 16.20). Table 16.4 lists some of the questions that are used to identify the alcohol-dependent person.

Tape Here

Tape Here

NO POSTAGE
NECESSARY
IF MAILED
IN THE
UNITED STATES

BUSINESS REPLY MAIL

FIRST- CLASS MAIL PERMIT NO. 47 EDWARDSVILLE, KS

POSTAGE WILL BE PAID BY ADDRESSEE

MEDI-SIM, INC.
P.O. BOX 13267
EDWARDSVILLE KS 66113-9989

Table 16.4
Some Questions to Identify the Alcohol-Dependent Person

1. Do you occasionally drink heavily after a disappointment, a quarrel, or when the boss gives you a bad time?
2. When you are having trouble or feel under pressure, do you always drink more heavily than usual?
3. Have you noticed that you are able to handle more liquor than you did when you were first drinking?
4. Did you ever wake up the "morning after" and discover that you could not remember part of the evening before, even though your friends tell you that you did not "pass out"?
5. When drinking with other people, do you try to have a few extra drinks when others will not know it?
6. Are there certain occasions when you feel uncomfortable if alcohol is not available?
7. Have you recently noticed that when you begin drinking you are in more of a hurry to get the first drink than you used to be?
8. Do you sometimes feel a little guilty about your drinking?
9. Are you secretly irritated when your family or friends discuss your drinking?
10. Have you recently noticed an increase in the frequency of your memory "blackouts"?
11. When you are sober, do you often regret things you have done or said while drinking?
12. Have you often failed to keep the promises you have made to yourself about controlling or cutting down on your drinking?
13. Do more people seem to be treating you unfairly without good reason?
14. Do you eat very little or irregularly when you are drinking?
15. Do you get terribly frightened after you have been drinking heavily?

Source: National Council on Alcoholism.

How Alcohol Works

Alcohol effects on the brain are biphasic. After consuming several drinks, blood alcohol concentration rises rapidly and the drinker reports feeling "high" and happy (euphoric). Ninety minutes later—and lasting some 300–400 minutes after consumption—the drinker feels depressed and unhappy (dysphoric) (fig. 16.21). On the other hand, if the drinker continues to drink in order to maintain a high blood alcohol level, he or she will experience ever-increasing loss of control. Coma and death are even possible if a substantial amount of alcohol (1 1/4 pt of whiskey) is consumed within an hour. In the short run, research seems to indicate that alcohol potentiates GABA, an inhibitory neurotransmitter. Exactly how this leads to motor incoordination and poor judgment is not known. In the long run, alcohol causes the death of neurons, permanent brain damage, and cirrhosis of the liver.

Cirrhosis of the Liver: Fat-Filled and Scarred

The stomach and the liver contain the enzyme alcohol dehydrogenase, which begins the breakdown of alcohol to acetic acid. A new study reports that women have less of this enzyme in their stomach, and this may explain why they show a greater sensitivity to alcohol, including a greater chance of liver damage. Acetic acid can be used in the liver to produce energy, but the calories provided are termed "empty" because they contribute to energy needs and weight gain without supplying any other nutritional requirements. Worse still, the molecules (glucose and fatty acids) that the liver ordinarily uses as an energy source are converted to fats (see fig. 6.13). Eventually, the liver cells become engorged with fat droplets. After a few years of being overtaxed, the liver cells begin to die, causing an inflammatory condition known as alcoholic hepatitis. Finally, scar tissue appears in the liver, and it is no longer able to perform its vital functions. This condition is called cirrhosis of the liver, a frequent cause of death among drinkers. Brain impairment and generalized deterioration of other vital organs also are seen in heavy drinkers.

It should be stressed that the early signs of deterioration can be reversed if the habit of drinking to excess is given up.

> Alcohol is the most abused drug in the United States. Its abuse often results in well-recognized illnesses and early death.

Figure 16.21
Typical blood alcohol curve for a normal drinker after intake of 1 ml of alcohol per kilogram of body weight. As blood alcohol concentration increases, the user often feels euphoric (happy), but as blood alcohol concentration declines, the user feels dysphoric (unhappy). In most states, a person is considered legally drunk when the blood alcohol content is 0.1%. This usually requires imbibing 3 mixed drinks within 90 minutes.

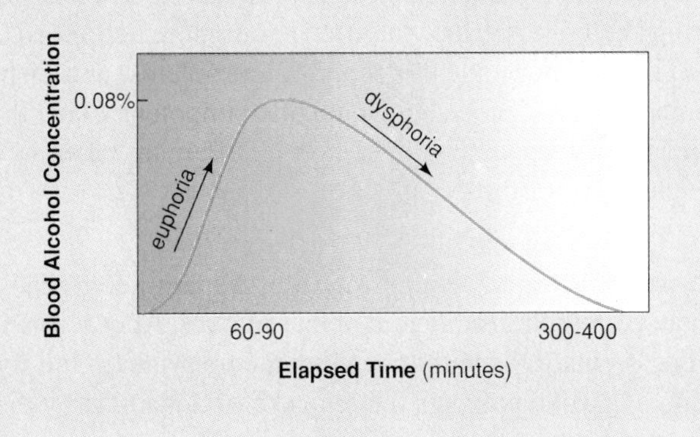

Figure 16.22

Cannabis sativa. This plant, which is used to make marijuana, often is smoked in the same manner as tobacco.

Marijuana: May Act on Serotonin

The dried flowering tops, leaves, and stems of the Indian hemp plant *Cannabis sativa* (fig. 16.22) contain and are covered by a resin that is rich in THC (tetrahydrocannabinol). The names *cannabis* and *marijuana* apply to either the plant or THC.

The effects of marijuana differ depending upon the strength and the amount consumed, the expertise of the user, and the setting in which it is taken. Usually, the user reports experiencing a mild euphoria along with alterations in vision and judgment, which result in distortions of space and time. The inability to concentrate and to speak coherently and motor incoordination also can be involved.

Intermittent use of low-potency marijuana generally is not associated with obvious symptoms of toxicity, but heavy use can produce chronic intoxication. Intoxication is recognized by the presence of hallucinations, anxiety, depression, rapid flow of ideas, body image distortions, paranoid reactions, and similar psychotic symptoms. The terms *cannabis psychosis* and *cannabis delirium* refer to such reactions.

Classified as Hallucinogen

Marijuana is classified as an hallucinogen. It is possible that, like LSD (lysergic acid diethylamide), it has an effect on the action of serotonin, an excitatory neurotransmitter.

The use of marijuana does not seem to produce physical dependence, but a psychological dependence on the euphoric and sedative effects can develop. Craving or difficulty in stopping use also can occur as a part of regular, heavy use.

Marijuana has been called a *gateway drug* because adolescents who have used marijuana also tend to try other drugs. For example, in a study of 100 cocaine abusers, 60% had smoked marijuana for more than 10 years.

Links to Illnesses

Usually marijuana is smoked in a cigarette form called a joint. Since this allows toxic substances, including carcinogens, to enter the lungs, chronic respiratory disease and lung cancer are considered dangers of long-term, heavy use. Some researchers claim that marijuana use leads to long-term brain impairment. Others report that males and females suffer reproductive dysfunctions. *Fetal cannabis syndrome,* which resembles fetal alcohol syndrome, has been reported.

Some psychologists are very concerned about the use of marijuana among adolescents. Marijuana can be used as a means to avoid coming to grips with the personal problems that often develop during this maturational phase.

> Although marijuana does not produce physical dependence, it does produce psychological dependence.

Cocaine: Affects Dopamine in the Brain

Cocaine is an alkaloid derived from the shrub *Erythroxylum cocoa.* Cocaine is sold in powder form and as *crack,* a more potent extract (fig. 16.23). Users often use the word *rush* to describe the feeling of euphoria that follows intake of the drug. Snorting (inhaling) produces this effect in a few minutes, injection, within 30 seconds, and smoking, in less than 10 seconds. Persons dependent upon the drug are, therefore, most likely to smoke cocaine. The rush only lasts a few seconds and then is replaced by a state of arousal, which lasts from 5 minutes to 30 minutes. Then the user begins to feel restless, irritable, and depressed. To overcome these symptoms, the user is apt to take more of the drug, repeating the cycle over and over again until there is no more drug left. A binge of this sort can go on for days, after which the individual suffers a *crash.* During the binge period, the user is hyperactive and has little desire for food or sleep, but he or she has an increased sex drive. During the crash period, the user is fatigued, depressed, irritable, has memory and concentration problems, and displays no interest in sex. Indeed, men are often impotent. Other drugs, such as marijuana, alcohol, or heroin, often are taken to ease the symptoms of the crash.

How Cocaine Works

Cocaine affects the concentration of dopamine, a generally excitatory neurotransmitter, in brain synapses. After release, dopamine ordinarily is withdrawn into the presynaptic cell for recycling. Cocaine prevents the reuptake of dopamine by the presynaptic membrane; this causes an excess of dopamine in the synaptic cleft so that the user experiences the sensation of a rush. The epinephrine-like effects of dopamine account for the state of arousal that lasts for some minutes after the rush experience.

Figure 16.23

Cocaine use. **a.** Crack, the ready-to-smoke form of cocaine, is a more potent and more deadly form than the powder. **b.** Users often smoke crack in a glass water pipe. The high produced consists of a "rush" lasting a few seconds, followed by a few minutes of euphoria. Continuous use makes the user extremely dependent on the drug.

a.

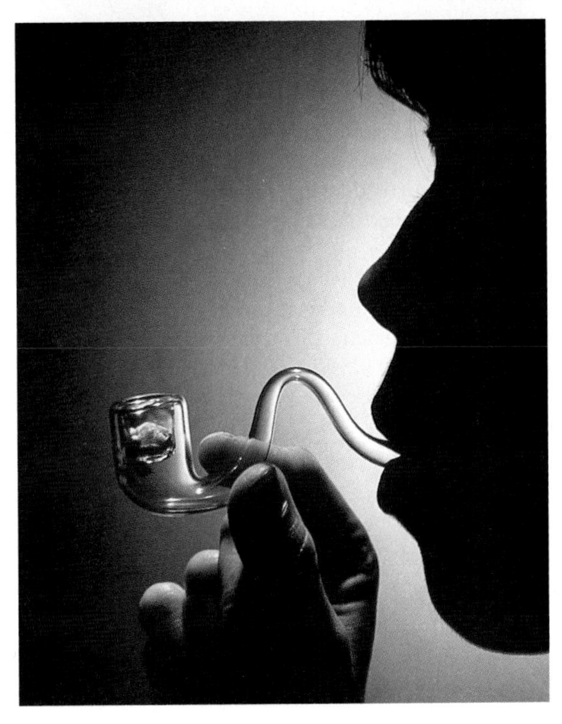

b.

With continued cocaine use, the body begins to make less dopamine to compensate for a seemingly excess supply. The user, therefore, now experiences *tolerance, withdrawal symptoms,* and an intense *craving* for cocaine. These are indications that the person is highly dependent upon the drug or, in other words, that cocaine is extremely addictive.

Links to Illnesses

Overdosing on cocaine is a real possibility. The number of deaths from cocaine and the number of emergency-room admissions for drug reactions involving cocaine have increased greatly. High doses can cause seizures and cardiac and respiratory arrest.

Individuals who snort the drug can suffer damage to the nasal tissues and even perforation of the septum, the membrane between the nostrils. Whether long-term cocaine abuse causes brain damage is not yet known, but this possibility is under investigation. It is known that babies born to addicts suffer withdrawal symptoms and may suffer neurological and developmental problems.

Heroin: Mimics Body's Endorphins

Heroin is derived from morphine, an alkaloid of *opium.* Heroin usually is injected. After intravenous injection, the onset of action is noticeable within one minute and reaches its peak in 3–6 minutes. There is a feeling of euphoria along with relief from pain. Side effects can include nausea, vomiting, dysphoria, and respiratory and circulatory depression leading to death.

How Heroin Works

Heroin binds to receptors meant for the body's own opioids, the endorphins. As mentioned previously, the opiates are believed to alleviate pain by preventing the release of a neurotransmitter termed substance P from certain sensory neurons in the region of the spinal cord. When substance P is released, pain is felt, and when substance P is not released, pain is not felt. Evidence also indicates that there are opioid receptors in neurons that travel from the spinal cord to the limbic system and that stimulation of these can cause a feeling of pleasure. This explains why opium and heroin not only kill pain but also produce a feeling of tranquility.

Individuals who inject heroin become physically dependent on the drug. With time, the body's production of endorphins decreases. *Tolerance* develops so that the user needs to take more of the drug just to prevent *withdrawal* symptoms. The euphoria originally experienced upon injection is no longer felt.

Heroin withdrawal symptoms include perspiration, dilation of pupils, tremors, restlessness, abdominal cramps, gooseflesh, defecation, vomiting, and increase in systolic pressure and respiratory rate. Those who are excessively dependent may experience convulsions, respiratory failure, and death. Infants born to women who are physically dependent also experience these withdrawal symptoms.

Cocaine and heroin produce a very strong physical dependence. An overdose of these drugs can cause death.

Methamphetamine (Ice): Acts like Cocaine

Methamphetamine is related to amphetamine, a well-known stimulant. Both methamphetamine and amphetamine have been drugs of abuse for some time, but a new form of methamphetamine known as "ice" is now being used as an alternative to cocaine. Ice is a pure, crystalline hydrochloride salt that has the appearance of sheetlike crystals. Unlike cocaine, ice can be illegally produced in this country in laboratories and does not need to be imported.

Ice, like crack, will vaporize in a pipe, so it can be smoked, avoiding the complications of intravenous injections. After rapid absorption into the bloodstream, the drug moves quickly to the brain. It has the same stimulatory effect as cocaine, and subjects report they cannot distinguish between the 2 drugs after intravenous administration. Methamphetamine effects, however, persist for hours instead of a few seconds. Therefore, it is the preferred drug of abuse by many.

Designer Drugs: Slightly Altered Structure

Designer drugs are analogues; that is, they are chemical compounds of controlled substances slightly altered in molecular structure. One such drug is MPPP (1-methyl-4-phenylprionoxy piperidine), an analogue of the narcotic fentanyl. Even small doses of the drug are very toxic; MPPP already has caused many deaths on the West Coast.

SUMMARY

The cell bodies of neurons are found in the CNS and the ganglia. Axons and dendrites are nerve fibers. The nerve impulse is a change in permeability of the axomembrane so that sodium ions (Na^+) move to the inside of a neuron and potassium ions (K^+) move to the outside. The nerve impulse is transmitted across the synapse by neurotransmitter substances.

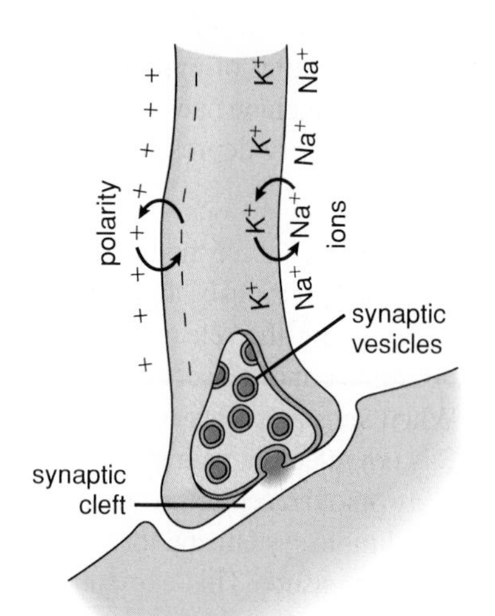

During a spinal reflex, a sensory neuron transmits nerve impulses from a receptor to an interneuron, which in turn transmits impulses to a motor neuron, which conducts them to an effector. Reflexes are automatic, and some do not require involvement of the brain.

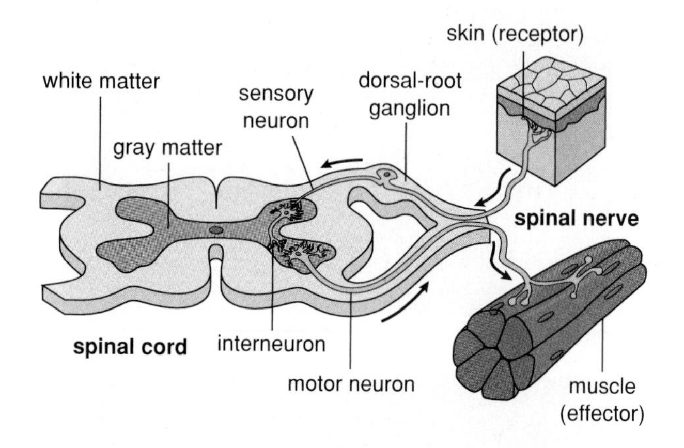

Long fibers of sensory and/or motor neurons make up cranial and spinal nerves of the somatic and autonomic divisions of the PNS. While the somatic division controls skeletal muscle, the autonomic division controls smooth muscle of the internal organs and glands.

The CNS consists of the spinal cord and the brain. Only the cerebrum is responsible for consciousness; the other portions of the brain have their own function. The cerebrum can be mapped, and each lobe also seems to have particular functions. Neurological drugs, although quite varied, have been found to affect the RAS and the limbic system by either promoting or preventing the action of neurotransmitters.

STUDY QUESTIONS

In order to practice **writing across the curriculum,** students should write out the answers to any or all of the study questions. The study questions are sequenced in the same order as the text.

1. What are the 2 main divisions of the nervous system? Explain why these names are appropriate. (p. 279)
2. What are the 3 types of neurons? How are they similar, and how are they different? (p. 279)
3. What does the term *resting potential* mean, and how is it brought about? (p. 282) Describe the 2 parts of an action potential and the change that can be associated with each part. (p. 284)
4. What is the sodium-potassium pump, and when is it active? (p.282)
5. What is a neurotransmitter substance, where is it stored, how does it function, and how is it destroyed? (pp. 284–285) Name 2 well-known neurotransmitters. (p. 285)
6. What are the 3 types of nerves, and how are they anatomically different? functionally different? Distinguish between cranial and spinal nerves. (p. 287)
7. Trace the path of a reflex action after discussing the structure and the function of the spinal cord and the spinal nerve. (pp. 287–289)
8. What is the autonomic nervous system, and what are its 2 major divisions? (p. 289) Give several similarities and differences between these divisions. (p. 289)
9. Name the major parts of the brain, and give a function for each. (pp. 292–293)
10. Describe the EEG, and discuss its importance. (p. 294)
11. Describe the physiological effects and mode of action of alcohol, marijuana, cocaine, and heroin. (pp. 296–299)

OBJECTIVE QUESTIONS

1. A(n) _______ carries nerve impulses away from the cell body.
2. During the upswing of the action potential, _______ ions are moving to the _______ of the nerve fiber.
3. The space between the axon of one neuron and the dendrite of another is called the _______.
4. Acetylcholine is broken down by the enzyme _______ after it has altered the permeability of the postsynaptic membrane.
5. Motor nerves innervate _______.
6. The vagus nerve is a(n) _______ nerve that controls the _______.
7. In a reflex arc, only the neuron called the _______ is completely within the CNS.
8. The brain and the spinal cord are covered by 3 protective layers called _______.
9. The _______ is the part of the brain that allows us to be conscious.
10. The _______ is the part of the brain responsible for coordination of body movements.
11. Label this diagram showing the 3 types of neurons.

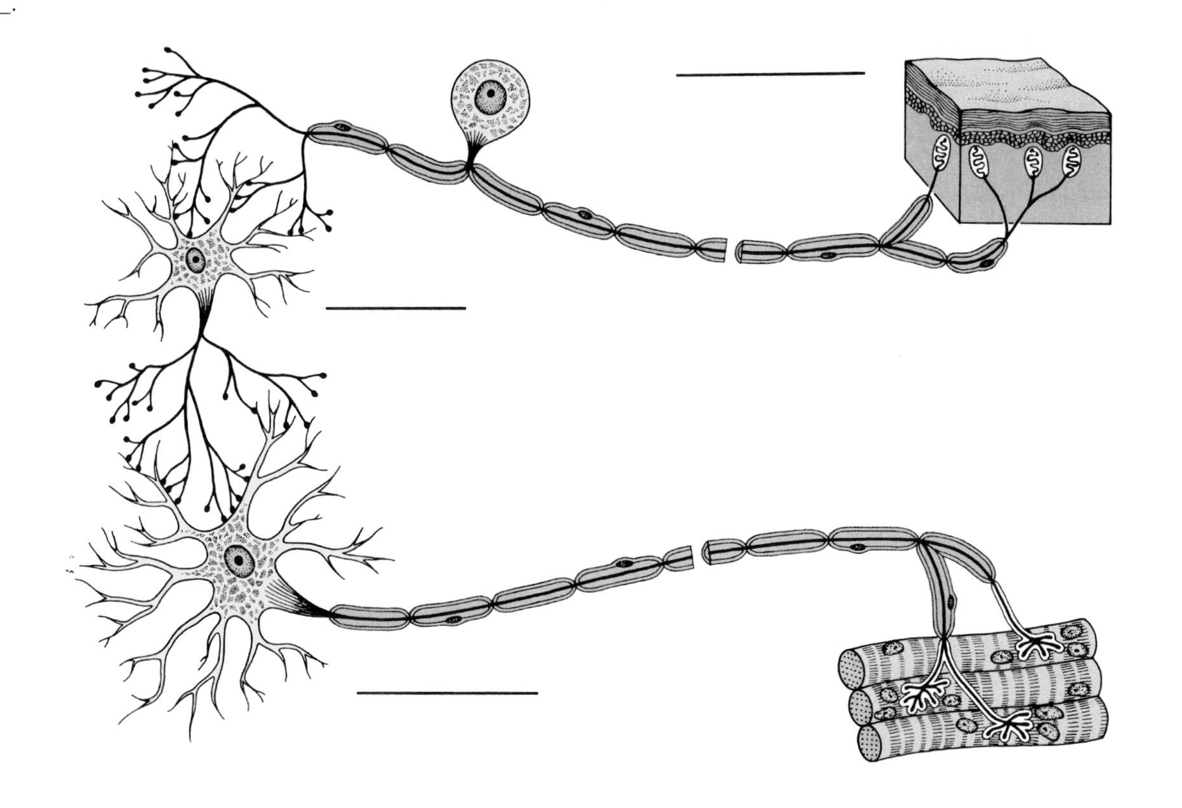

CRITICAL THINKING

In order to practice **writing across the curriculum,** students should write out the answers to any or all of the critical thinking questions. Suggested answers to the critical thinking questions are in appendix E.

16.1

1. Electricity is the flow of electrons within a wire. How is the nerve impulse different from this?
2. If a neurotransmitter substance is inhibitory, would you expect a higher or lower voltage reading compared to –65 mV on the oscilloscope?
3. In the laboratory, an axon segment can conduct a nerve impulse in either direction. Why do nerve impulses travel only from axon to dendrite or cell body across a synapse in the body?

16.2

1. If you applied acid to the left leg of a frog, why might both legs respond?
2. If you severed just the dorsal-root ganglion of the sciatic nerve serving the left leg, would either leg be able to respond?
3. If you severed just the ventral root of the left sciatic nerve, would either leg be able to respond?
4. If you destroyed just the spinal cord, would either leg be able to respond?

SELECTED KEY TERMS

axon fiber of a neuron that conducts nerve impulses away from the cell body.

cell body portion of a neuron that contains the nucleus and from which the nerve fibers extend.

central nervous system (CNS) the brain and the spinal cord in vertebrate animals.

cerebral hemisphere (ser´ ĕ-bral hem´ ĭ sfēr) one of the large paired structures that together constitute the cerebrum of the brain.

cerebrospinal fluid fluid found in the ventricles of the brain, the central canal of the spinal cord, and in association with the meninges.

dendrite fiber of a neuron, typically branched, that conducts nerve impulses toward the cell body.

effector a structure such as a muscle or a gland that allows an organism to respond to environmental stimuli.

ganglion (gang´ gle-on) a collection of neuron cell bodies within the PNS.

innervate (in´ er-vāt) to activate an organ, muscle, or gland by motor neuron stimulation.

interneuron a neuron found within the CNS that takes nerve impulses from one portion of the system to another.

limbic system a portion of the brain concerned with memory and emotions.

motor neuron a neuron that takes nerve impulses from the CNS to an effector.

myelin sheath (mi´ ĕ-lin shēth) the Schwann cell membranes that cover long neuron fibers and give them a white, glistening appearance.

nerve impulse an electrochemical change due to increased membrane permeability that is propagated along a neuron from the dendrite to the axon following excitation.

neurotransmitter substance a chemical found at the ends of axons that is responsible for transmission across a synapse.

parasympathetic nervous system that part of the autonomic nervous system that usually promotes activities associated with a normal state.

peripheral nervous system (PNS) nerves and ganglia that lie outside the CNS.

receptor a dendrite ending of a sensory neuron specialized to receive stimuli and generate a nerve impulse.

sensory neuron a neuron that takes nerve impulses to the CNS, typically has a long dendrite and a short axon; afferent neuron.

somatic nervous system that part of the PNS containing motor neurons that control skeletal muscles.

sympathetic nervous system that part of the autonomic nervous system that usually causes effects associated with emergency situations.

synapse (sin´ aps) the region between 2 nerve cells where the nerve impulse is transmitted from one to the other, usually from axon to dendrite.

17

THE MUSCULOSKELETAL SYSTEM

Chapter Concepts

1.
The skeleton provides a framework for the body and permits flexible movement, among various other functions. 304, 319

2.
The jointed human skeleton is divided into the axial skeleton and the appendicular skeleton. 304, 319

3.
Bone is a rigid but living material that is continually rejuvenated. 308, 319

4.
Macroscopically, skeletal muscles work in antagonistic pairs and exhibit certain physiological characteristics. 310, 319

5.
Microscopically, muscle fiber contraction is dependent on filaments of both actin and myosin and a ready supply of calcium ions (Ca^{++}) and ATP. 314, 319

Striated muscle fibers, ×1,300

Chapter Outline

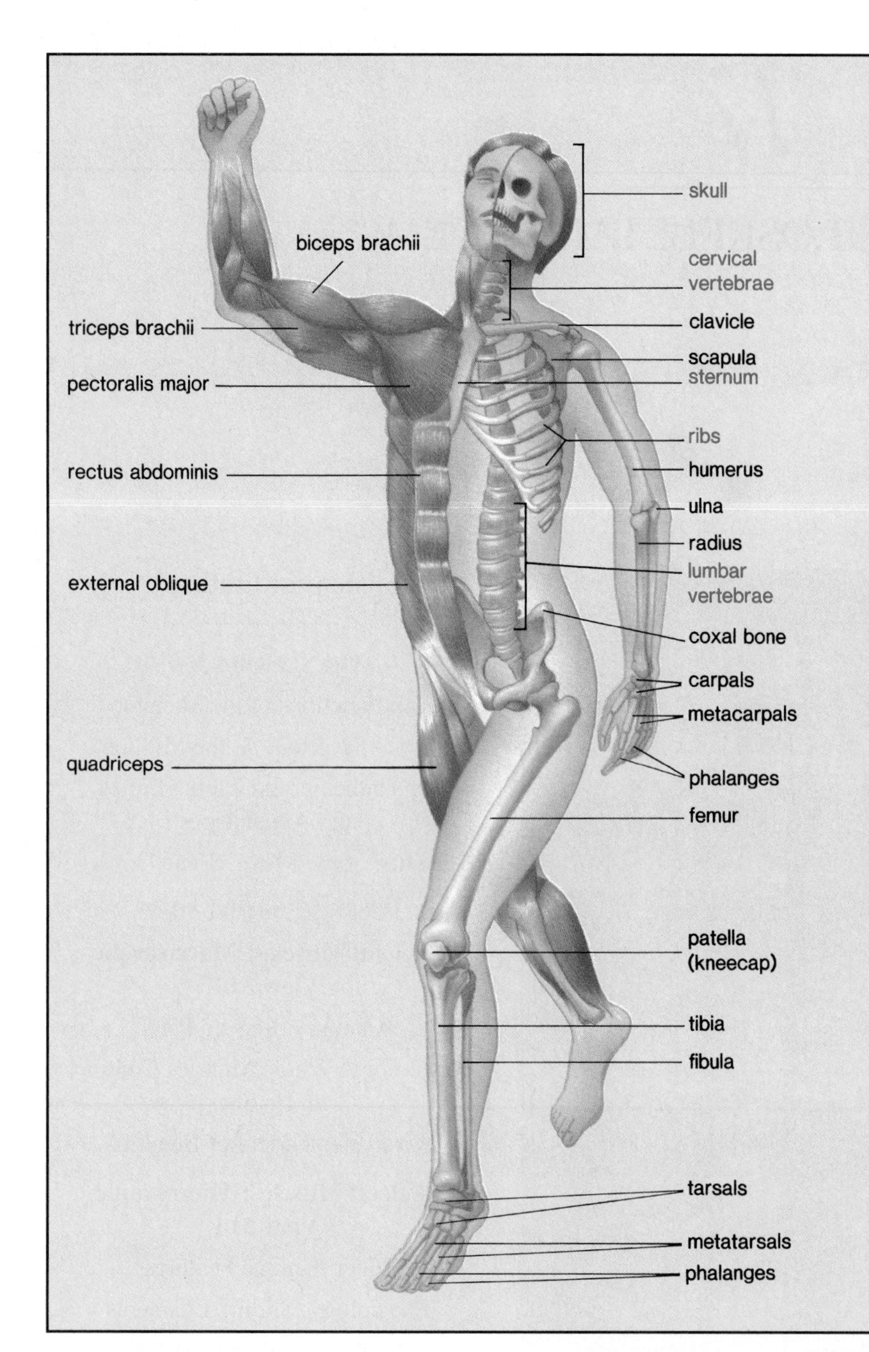

Figure 17.1
Major bones (*right*) and skeletal muscles (*left*) of the human body. The axial skeleton, composed of the skull, the vertebral column, the sternum, and the ribs (red leaders and labels), lies in the midline; the rest of the bones belong to the appendicular skeleton (black leaders and labels).

Axial skeleton

Skull

Vertebral column

Sternum

Ribs

Appendicular skeleton

Pectoral girdle: Clavicle, scapula

Arm: Humerus, ulna, radius

Hand: Carpals, metacarpals, phalanges

Pelvic girdle: Coxal bones

Leg: Femur, tibia, fibula, patella

Foot: Tarsals, metatarsals, phalanges

USCLES and bones largely account for body weight and appearance. Working together, they allow us to perform many mechanical tasks, some of which require grace and agility.

The Skeleton

The human skeleton, which has many functions, is divided into the axial skeleton and the appendicular skeleton (fig. 17.1).

Six Functions of the Skeleton

The skeleton (fig. 17.1), notably the large, heavy bones of the legs, *supports the body* against the pull of gravity. The skeleton also *protects soft body parts*. For example, the skull forms a protective encasement for the brain, as does the rib cage for the heart and the lungs. Flat bones, such as those of the skull, the ribs, and the sternum (breastbone), *produce red blood cells* in both adults and children. All bones are *storage areas* for inorganic calcium and phosphorus salts. Bones also provide *sites for muscle attachment*. The long bones, particularly those of the legs and the arms, *permit flexible body movement*.

The skeleton not only permits flexible movement, it also supports and protects the body, produces red blood cells, and serves as a storehouse for certain inorganic salts.

A N I M A L • B I O L O G Y

ANIMAL SKELETAL SYSTEMS

The skeleton is the framework of the body. It helps to protect the internal organs and assists in movement. To produce body movements, the force of muscle contractions must be specifically directed against other parts of the body.

Hydras and planarians use their fluid-filled gastrovascular cavity and annelids use their fluid-filled coelom as a hydrostatic skeleton. In these animals, muscle contraction is applied against the fluid-filled cavity. In the earthworm, the contractions of first the circular muscle and then the longitudinal muscle of the body wall enable the worm to alternately extend and shorten. In this way, the earthworm can move forward.

Other members of the animal kingdom have an exoskeleton. The rigid calcium carbonate ($CaCO_3$) exoskeleton of a clam (the bivalve shell) serves primarily for protection. It grows as the animal grows. The chitinous exoskeleton of arthropods is jointed and movable. Chitin is a strong, flexible, nitrogenous polysaccharide. Arthropods molt to rid themselves of an exoskeleton that has become too small.

Vertebrates have an endoskeleton composed of bone and cartilage that grows with the animal. It is also jointed, as is the arthropod skeleton. A rigid but flexible skeleton helped the arthropods and vertebrates successfully colonize the terrestrial environment.

Axial Skeleton: At the Midline

The **axial skeleton** lies in the midline of the body and consists of the skull (cranium and facial bones), the vertebral column, the sternum, and the ribs.

The bones of the skull contain the *sinuses,* air spaces lined by mucous membrane. Two of these, called the mastoid sinuses, drain into the middle ear. Mastoiditis, a condition that can lead to deafness, is an inflammation of these sinuses. The *cranium* is composed of 8 bones that fit tightly together in adults. In newborns, certain cranial bones are not completely formed and instead are joined by membranous regions called *fontanels,* all of which usually close by the age of 16 months. Whereas the cranium protects the brain, the numerous facial bones join to support and protect the special sense organs and to form the jawbones.

The *vertebral column* extends from the skull to the pelvis and forms a dorsal backbone, which protects the spinal cord (see fig. 16.10). Normally, the vertebral column has 4 curvatures, which provide more resiliency and strength than a straight column could. The column is composed of many parts called *vertebrae,* which are held together by bony facets, muscles, and strong ligaments. The vertebrae are named according to their location in the body (fig. 17.2).

Intervertebral disks, located between the vertebrae, allow motion between the vertebrae so that we can bend forward, backward, and from side to side. They act as a kind of padding, preventing the vertebrae from grinding against one another and absorbing shock caused by such movements as running, jumping, and even walking. Unfortunately, these disks weaken with age and may slip or even rupture. Pain results when the

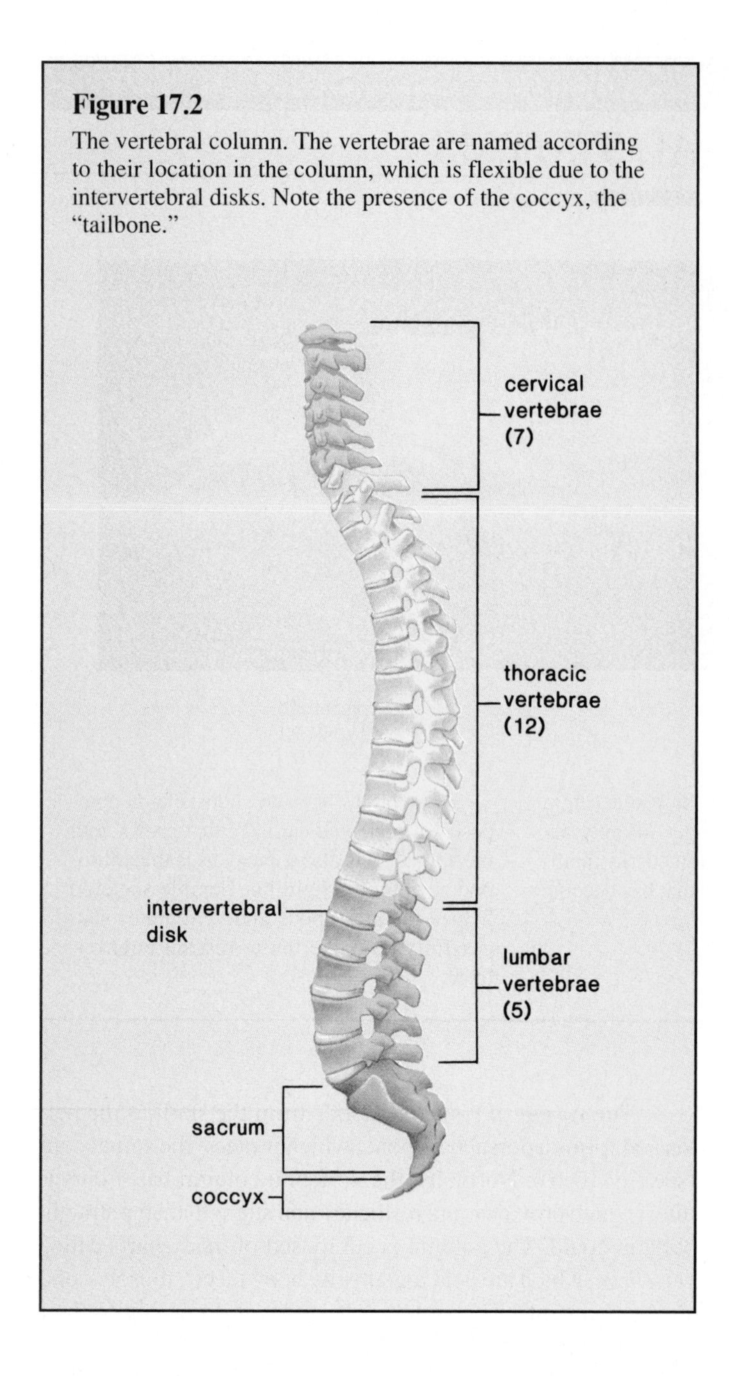

Figure 17.2

The vertebral column. The vertebrae are named according to their location in the column, which is flexible due to the intervertebral disks. Note the presence of the coccyx, the "tailbone."

damaged disk presses up against the spinal cord and/or spinal nerves. The body may heal itself or the disk can be removed surgically. If surgery is necessary, the 2 adjacent vertebrae can be fused together, but this limits the flexibility of the body.

The vertebral column, directly or indirectly, serves as an anchor for all the other bones of the skeleton (fig. 17.1). All 12 pairs of *ribs* connect directly to the thoracic vertebrae, and all but 2 pairs connect either directly or indirectly via shafts of cartilage to the *sternum.* The lower 2 pairs of ribs are called "floating ribs" because they do not attach to the sternum.

Appendicular Skeleton: Girdles and Appendages

The **appendicular skeleton** consists of the bones within the pectoral and pelvic girdles and the attached appendages. The pectoral girdles (shoulders) and appendages (arms and hands) are specialized for flexibility, but the pelvic girdle (coxal bones) and appendages (legs and feet) are specialized for strength.

The components of the *pectoral girdle* (fig. 17.3) are loosely linked by ligaments rather than firm joints. Each *clavicle* (collarbone) connects with the sternum and a *scapula* (shoulder blade). The scapula is held in place by muscles and can follow the movements of the arm freely. The single long bone in the upper arm (fig. 17.3), the *humerus,* has a smooth, round head, which fits into a socket on the scapula. The socket, however, is very shallow and much smaller than the head. Although this means that the arm can move in almost any direction, there is little stability. Therefore, this is the joint that is most apt to dislocate. The opposite end of the humerus meets the 2 bones of the lower arm, the *ulna* and the *radius,* at the elbow. (The prominent bone in the elbow is the topmost part of the ulna.) When the arm is held so that the palm is turned frontward, the radius and the ulna are about parallel to one another. When the arm is turned so that the palm is next to the body, the radius crosses in front of the ulna, a feature that contributes to the easy twisting motion of the lower arm.

The numerous bones of the hand increase its flexibility. The wrist has 8 *carpal* bones, which look like small pebbles. From these, 5 *metacarpal* bones fan out to form a framework for the palm. The metacarpal bone that leads to the thumb is positioned so that the thumb can reach out and touch the other digits. (*Digits* is a term that refers to either fingers or toes.) Beyond the metacarpals are the *phalanges,* the bones of the digits. The phalanges of the hand are long, slender, and lightweight.

The *pelvic girdle* (fig. 17.4) consists of 2 heavy, large *coxal bones* (hipbones). The coxal bones are anchored to the *sacrum,* and together these bones form a hollow cavity, the **pelvis**. The weight of the body is transmitted through the pelvis to the legs and then onto the ground. The largest bone in the body is the *femur,* or thighbone. Although the femur is a strong bone, there is a limit to the weight it can support. A giant 10 times taller than an ordinary human being would also be about 10 times wider and thicker, making him or her weigh about one thousand times as much. This amount of weight would break even giant-sized femurs.

In the lower leg, the larger of the 2 bones, the *tibia* (fig. 17.4), has a ridge we call the shin. Both of the bones of the lower leg have a prominence that contributes to the ankle—the tibia on the inside of the ankle and the *fibula* on the outside of

Figure 17.3

The bones of the pectoral girdle, the arm, and the hand. The humerus becomes the "funny bone" of the elbow, the sensation upon bumping it due to the activation of a nerve that passes across its end.

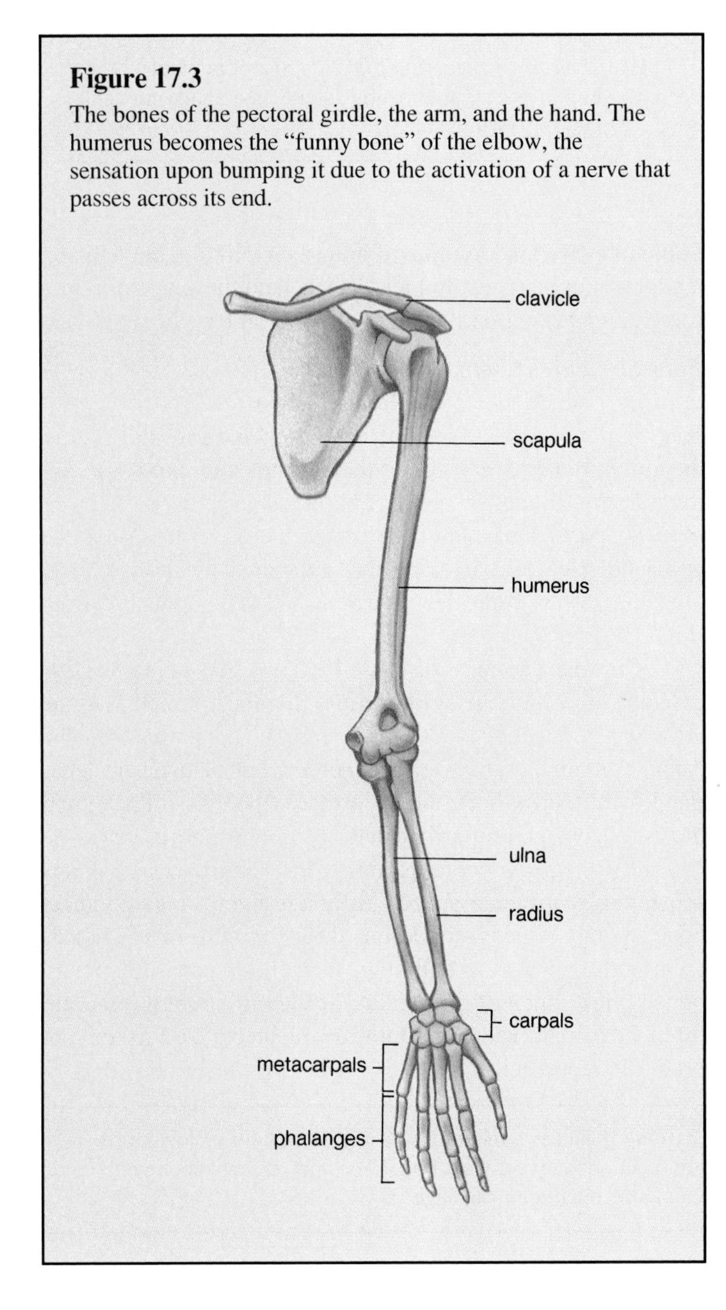

the ankle. Although there are 7 *tarsal* bones in the ankle, only one bone receives the weight of the body and passes it on to the heel and the ball of the foot. If you wear high-heeled shoes, weight is thrown even farther toward the front of the foot. The *metatarsal* bones form the arches of the foot—a longitudinal arch from the heel to the toes, and a transverse arch across the foot. These provide a stable, springy base for the body. If the tissues that bind the metatarsals together weaken, flatfeet are apt to result. The bones of the toes are called *phalanges,* just like those of the fingers, but in the foot, phalanges are stout and extremely sturdy.

The axial and appendicular skeletons contain the bones listed in figure 17.1.

Figure 17.4

The bones of the pelvic girdle, the leg, and the foot. The femur is our strongest bone—it withstands a pressure of 540 kg per 2.5 cm^3 when we walk.

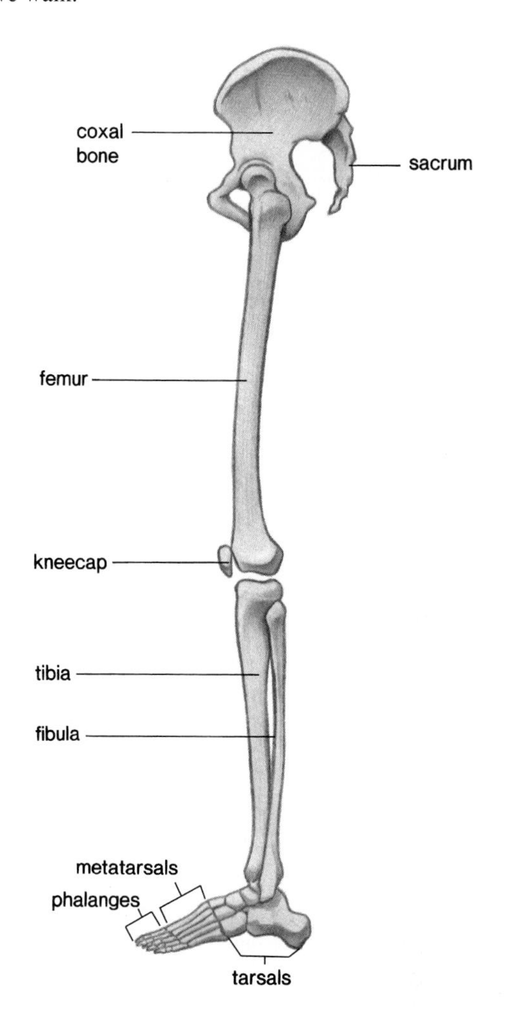

Joints: Some Move, Some Don't

Bones are linked at the joints, which are often classified according to the amount of movement they allow. Some bones, such as those that make up the cranium, are sutured together; they are *immovable.* Other joints are *slightly movable,* such as the joints between the vertebrae. The vertebrae are separated by disks, described earlier, which increase their flexibility. Similarly, the 2 coxal bones are slightly movable where they are joined ventrally by cartilage. Owing to hormonal changes, this joint becomes more flexible during late pregnancy, which allows the female pelvis to expand during childbirth.

Most joints are *freely movable,* or **synovial joints,** in which the 2 bones are separated by a cavity. The bones are held in place by **ligaments,** which form a capsule. In a "double-jointed" individual, the ligaments are unusually loose. The joint capsule is lined by synovial membrane, which produces *synovial fluid,* a lubricant for the joint.

Figure 17.5
The knee joint, an example of a synovial joint. Notice the cavity between the bones, which is encased by ligaments and lined by synovial membrane. The kneecap protects the joint and keeps the tendons of the quadriceps in proper alignment.

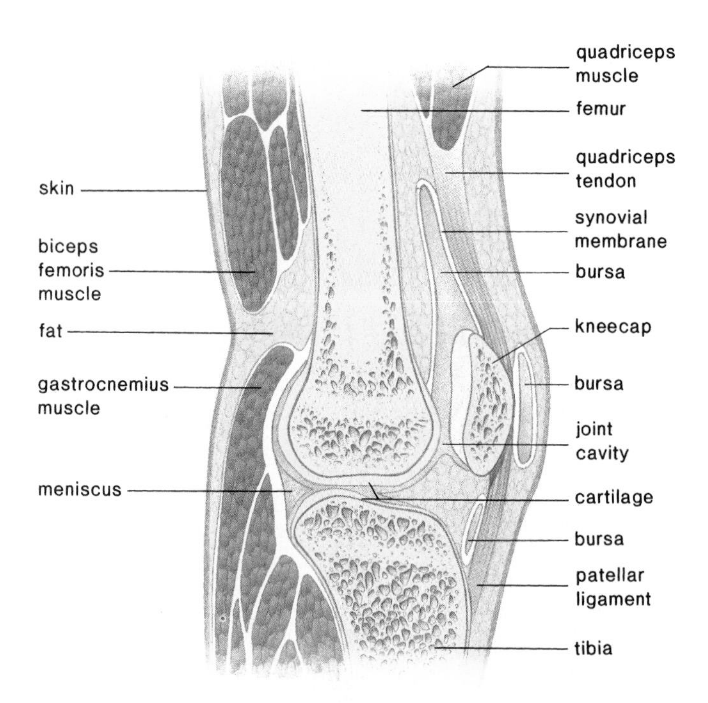

The knee is an example of a synovial joint (fig. 17.5). As in other freely movable joints, the bones of the knee are capped by cartilage, but the knee also has crescent-shaped pieces of cartilage between the bones called menisci (sing., *meniscus*). These add stability, helping to support the weight placed on the knee joint. Unfortunately, athletes often injure the menisci, an injury known as torn cartilage. The knee joint also contains 13 sacs called bursae (sing., *bursa*), which are filled with synovial fluid. These ease friction between tendons and ligaments and between tendons and bones. Inflammation of bursae is called bursitis, one form of which is tennis elbow.

There are different types of synovial joints. The knee and elbow joints are *hinge joints,* because like a hinged door, they permit movement largely in one direction only. More versatile are the ball-and-socket joints; for example, the ball of the femur fits into a socket on the coxal bone. *Ball-and-socket joints* allow movement in all planes and even a rotational movement.

Synovial joints are subject to *arthritis.* In rheumatoid arthritis, the synovial membrane thickens and becomes inflamed. Degenerative changes take place that make the joint almost immovable and painful to use. There is evidence that rheumatoid arthritis is brought on by an autoimmune reaction. In old-age arthritis, or osteoarthritis, the cartilage at the ends of the bones disintegrates so that the 2 bones become rough and irregular. This type of arthritis is apt to affect the joints that have received the greatest use over the years, such as finger joints.

> Joints are classified according to the degree of movement they afford. Some joints are immovable, some are slightly movable, and some are freely movable.

Bones Are Living Tissue

Sometimes we tend to think of bone as nonliving, but actually it contains living cells. In fact, the size and the shape of bones can change even after they stop growing in length.

Bone Structure: Compact and Spongy

A long bone, such as the femur, illustrates principles of bone anatomy. When the bone is split open, as in figure 17.6, the longitudinal section shows that it is not solid but has a cavity, called the medullary cavity. This cavity usually contains *yellow bone marrow,* which is a fat-storage tissue. It is bounded at the sides by compact bone and at the ends by spongy bone. Over the spongy bone, there is a thin shell of compact bone and finally a layer of cartilage.

Compact bone, as discussed on page 160, contains Haversian systems—bone cells in tiny chambers called lacunae arranged in concentric circles around Haversian canals. The canals contain blood vessels and nerves. The lacunae are separated by a matrix that contains protein fibers of collagen and mineral deposits, primarily calcium and phosphorus salts.

Spongy bone contains numerous bony bars and plates separated by irregular spaces. Although lighter than compact bone, spongy bone is still designed for strength. Just as braces are used for support in buildings, the solid portions of spongy bone follow lines of stress. The spaces in spongy bone are often filled with **red bone marrow,** a specialized tissue that produces blood cells.

> A long bone has a medullary cavity filled with yellow marrow and bounded by compact bone. The ends contain spongy bone and are covered by cartilage.

Bone Growth: Constant Renewal

Most of the bones of the skeleton are cartilaginous when first formed during development, and then they are ossified as bone. When ossification of a long bone begins, there is only a primary ossification center at the middle of the bone; later, secondary centers form at the ends of the bone. A cartilaginous disk remains between the primary ossification center and each secondary center. The length of a bone is dependent on how long the cartilage cells within the disk continue to divide. Eventually, though, the disks disappear, and the bone stops growing when the individual attains adult height.

In the adult, bone is continually broken down and built up again. Bone-absorbing cells, called *osteoclasts,* are derived from cells carried in the bloodstream. As they break down bone, they remove worn cells and deposit calcium in blood. Apparently, osteoclasts disappear after about 3 weeks and the

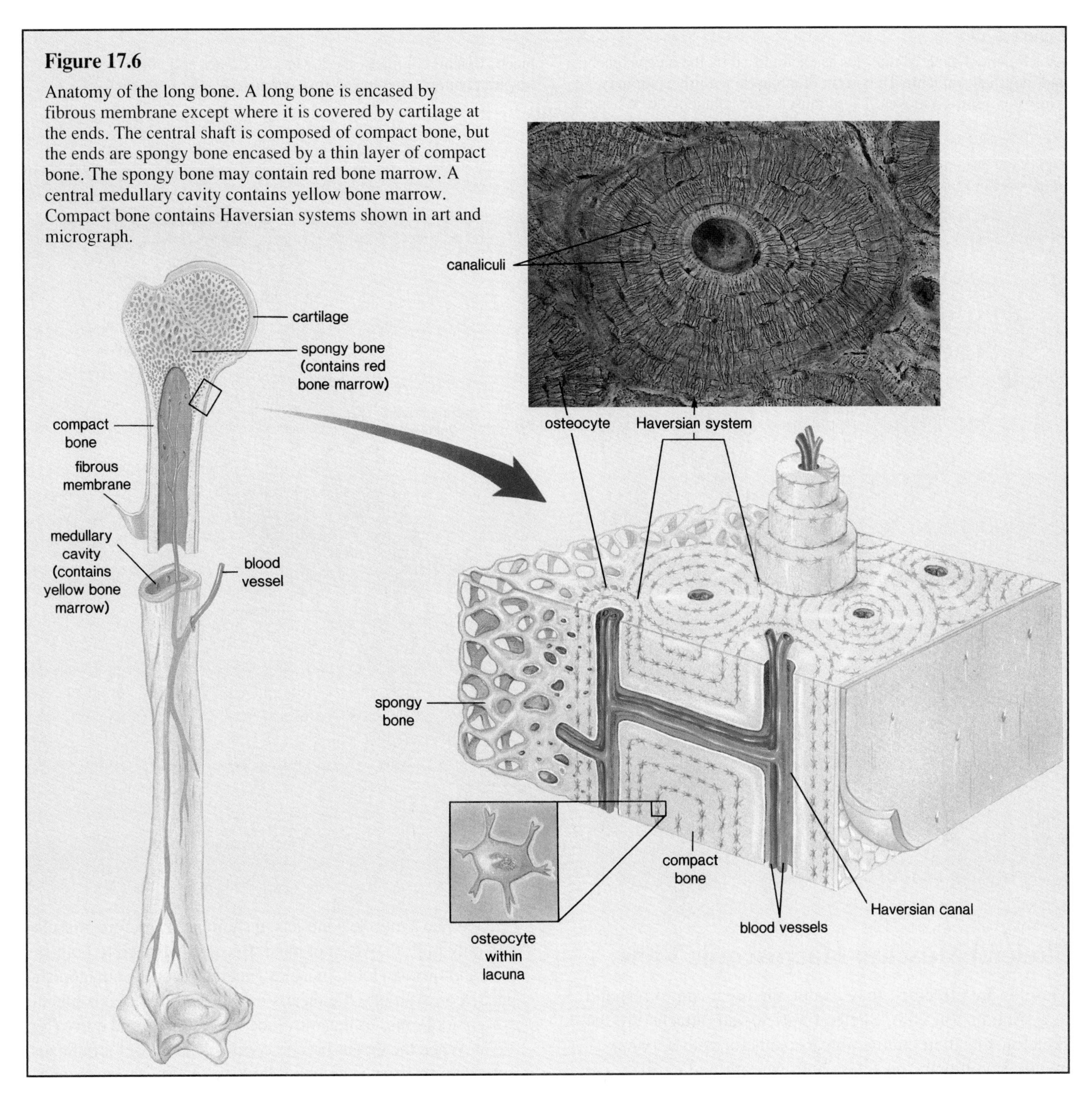

Figure 17.6

Anatomy of the long bone. A long bone is encased by fibrous membrane except where it is covered by cartilage at the ends. The central shaft is composed of compact bone, but the ends are spongy bone encased by a thin layer of compact bone. The spongy bone may contain red bone marrow. A central medullary cavity contains yellow bone marrow. Compact bone contains Haversian systems shown in art and micrograph.

destruction they caused is repaired by osteoblasts. As they form new bone, the osteoblasts take calcium from blood. Eventually, some of these cells get caught in the matrix they secrete and are converted to **osteocytes,** the cells found within Haversian systems (fig. 17.6). Osteocytes maintain healthy bone and its mineral content until the cycle is repeated.

Because of continual renewal and depending on the amount of physical activity or change in certain hormonal balances, the thickness of bones can change. In postmenopausal women, osteoclasts are more active than osteoblasts due to a lack of estrogen. To prevent osteoporosis, a condition in which weak and thin bones cause aches and pains and fracture easily, some women take estrogen. Also, the drug etidronate disodium has been shown to be effective in alleviating osteoporosis. Exercise and sufficient calcium in the diet also encourage osteoblast activity.

Bone is living tissue and is always being rejuvenated. Do 17.1 Critical Thinking, found at the end of the chapter.

Figure 17.7

Attachment of skeletal muscles as exemplified by the biceps brachii and the triceps brachii. The origin of a muscle is fairly stationary, while the insertion moves. These muscles are antagonistic. When the biceps brachii contracts, the lower arm flexes, and when the triceps brachii contracts, the lower arm extends.

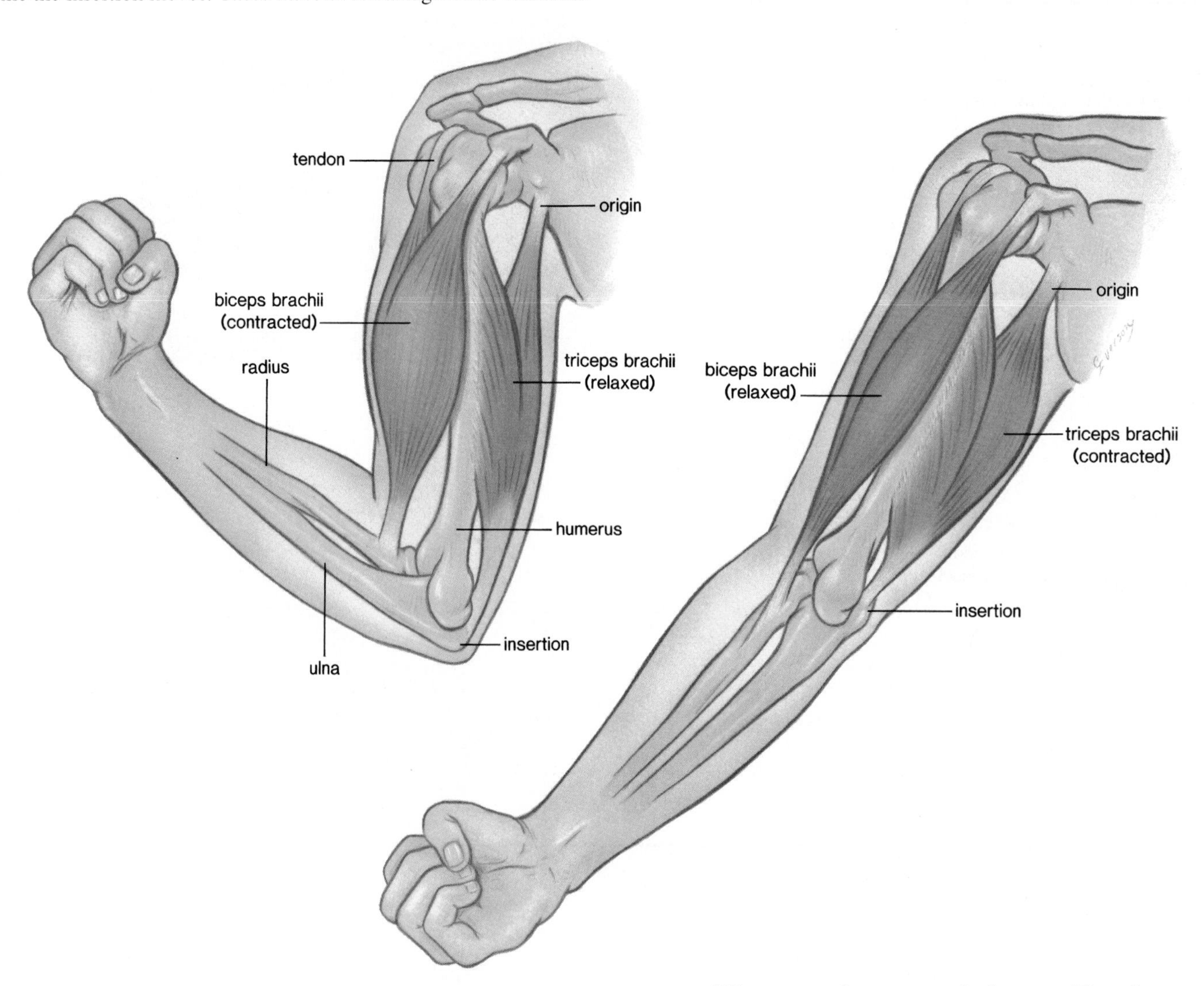

Skeletal Muscles: Macroscopic View

Muscles are effectors; they enable the organism to respond to a stimulus (p. 288). Skeletal muscles are attached to the skeleton, and their contraction accounts for voluntary movements. Involuntary muscles, both smooth and cardiac, are discussed on page 161.

Anatomy: Muscle Pairs

Muscles typically are attached to bone by **tendons,** which are made of fibrous connective tissue. Tendons most often attach muscles to the far side of a joint, so that the muscle extends across the joint (fig. 17.7). When the central portion of the muscle, called the belly, contracts, one bone remains fairly stationary and the other one moves. The **origin** of the muscle is on the stationary bone, and the **insertion** of the muscle is on the bone that moves.

When a muscle contracts, it shortens. Therefore, muscles can only pull; they cannot push. Because we need to both extend and flex at a joint, muscles generally work in antagonistic pairs. For example, the biceps brachii and the triceps brachii are a pair of muscles that move the lower arm up and down (fig. 17.7). When the biceps brachii contracts, the lower arm flexes, and when the triceps brachii contracts, the lower arm extends.

Whole skeletal muscles are attached to bones and work in antagonistic pairs. Do 17.2 Critical Thinking, found at the end of the chapter.

Physiology: Whole Muscles Contract by Degrees

It is possible to study the contraction of individual whole muscles in the laboratory. Customarily, a calf muscle from a frog is loosened at the tendon of Achilles and attached to a muscle lever. The muscle is stimulated and the mechanical force of contraction is transduced (changed) into an electrical current recorded by a *physiograph* (fig. 17.8*a*). The resulting visual pattern is called a *myogram.*

All-or-None Response: For Fibers Only

Like a nerve fiber, a *single* muscle fiber (muscle cell, p. 160) either responds to a stimulus and contracts or it does not. At first, the stimulus may be so weak that no contraction occurs, but as soon as the strength of the stimulus reaches the *threshold stimulus,* the muscle fiber contracts completely. Therefore, a muscle fiber obeys the *all-or-none law.*

Contrary to that of an individual fiber, the strength of contraction of a whole muscle can increase according to the strength of the stimulus beyond the threshold stimulus. A whole muscle contains many fibers, and the degree of contraction is dependent on the total number of fibers contracting. The *maximum stimulus* is the one beyond which the degree of contraction does not increase.

> Although muscle fibers obey the all-or-none law, whole muscles do not. The degree of contraction is dependent on the total number of fibers contracting.

Muscle Twitch

If a muscle is attached to a physiograph and is given a maximum stimulus, it contracts and then relaxes. This action—a single contraction that lasts only a fraction of a second—is called a muscle *twitch.* Figure 17.8*b* is a myogram of a twitch, which is customarily divided into the *latent period,* or the period of time between stimulation and initiation of contraction, the *contraction period,* and the *relaxation period.*

If a muscle is exposed to 2 maximum stimuli in quick succession, it responds to the first but not to the second stimulus. This is because it takes an instant following a contraction for the muscle fibers to recover in order to respond to the next stimulus. The very brief moment following stimulation, during which a muscle is unresponsive, is called the refractory period.

Summation and Tetanus: Blending Twitches

If a muscle is given a rapid series of threshold stimuli, it can respond to the next stimulus without relaxing completely. In this way, muscle tension *summates* until maximal sustained contraction, called **tetanus,** is achieved (fig. 17.8*c*). The myogram no longer shows individual twitches; rather, they are fused and blended completely into a straight line. Tetanus continues until the muscle fatigues due to depletion of energy reserves. *Fatigue* is apparent when a muscle relaxes even though stimulation continues.

Tetanic contractions occur whenever skeletal muscles are actively used. Ordinarily, however, only a portion of any particular muscle is involved—while some fibers are contracting, others are relaxing. Because of this, intact muscles rarely fatigue completely.

Figure 17.8

Physiology of skeletal muscle contraction. **a.** A physiograph is an apparatus used to record a myogram, a visual representation of the contraction of a muscle that has been dissected from an animal. **b.** Simple muscle twitch is composed of 3 periods: latent, contraction, and relaxation. **c.** Summation and tetanus. When a muscle is not allowed to relax completely between stimuli, the contraction gradually increases in intensity until the muscle is in tetanus. Eventually, the muscle fatigues.

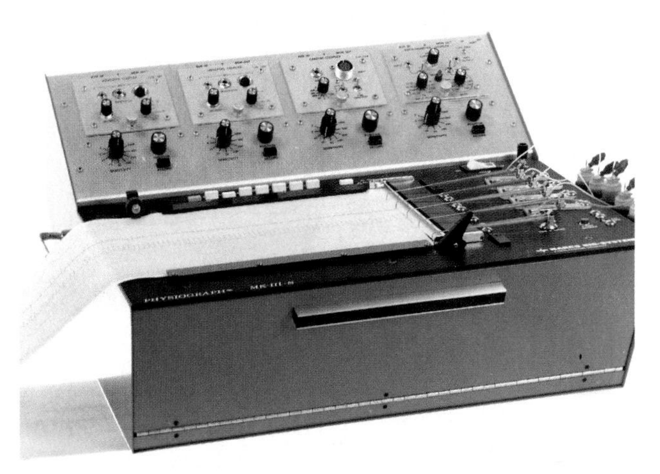

a.

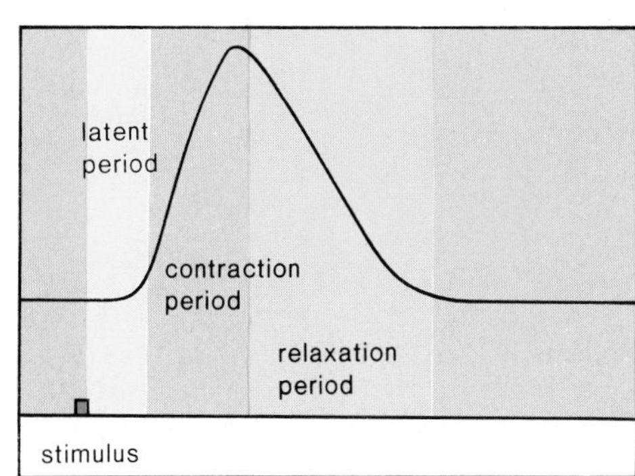

b.

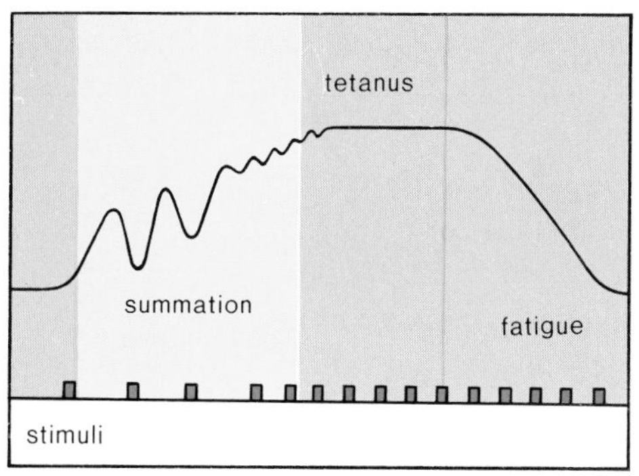

c.

> Muscle twitch, summation, and tetanus are related to the frequency with which a muscle is stimulated.

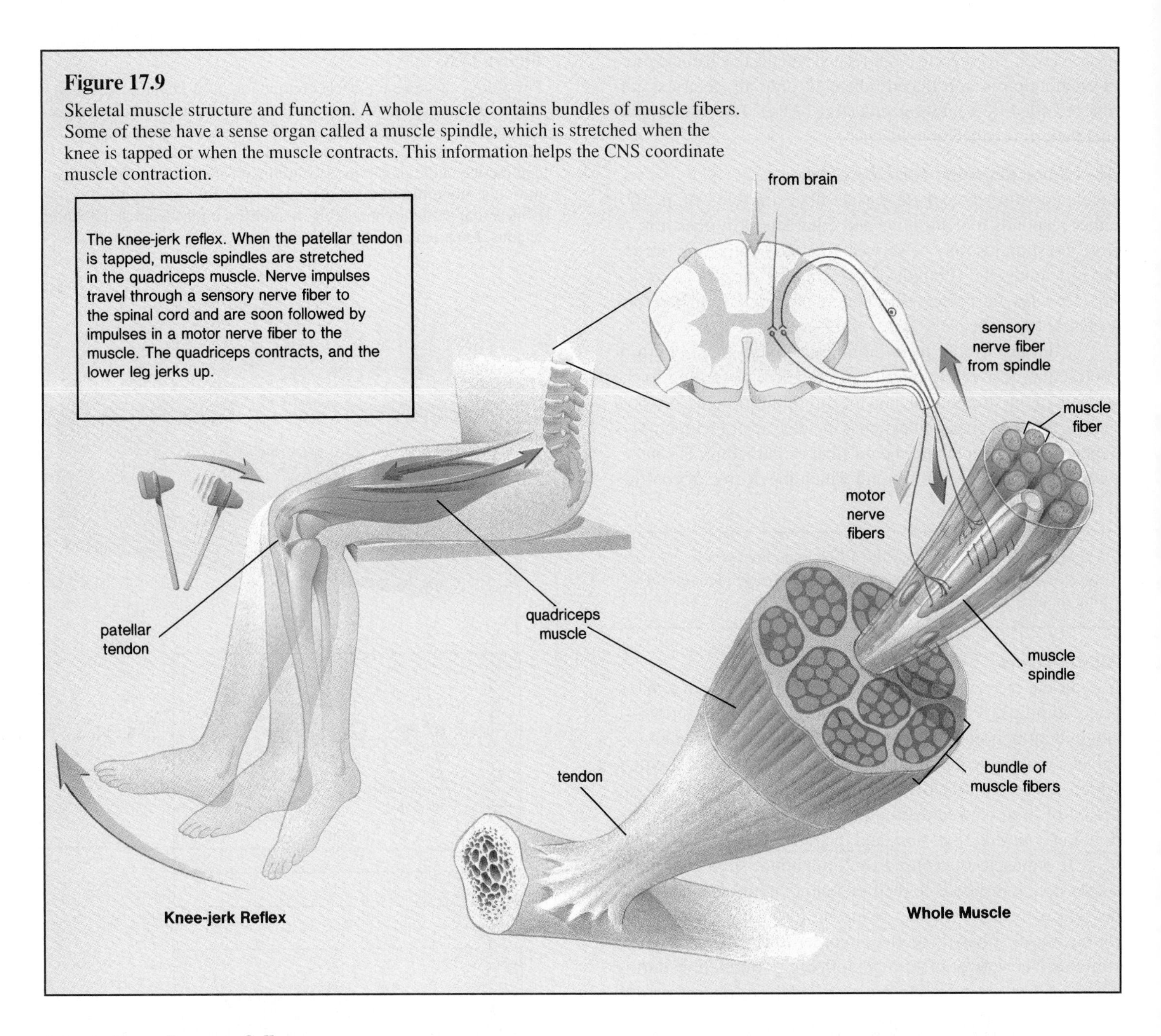

Figure 17.9
Skeletal muscle structure and function. A whole muscle contains bundles of muscle fibers. Some of these have a sense organ called a muscle spindle, which is stretched when the knee is tapped or when the muscle contracts. This information helps the CNS coordinate muscle contraction.

Muscle Tone: Prevents Collapse

Skeletal muscles exhibit **tone,** a condition in which some fibers are always contracted. Muscle tone is particularly important in maintaining posture. If all the fibers within the neck, trunk, and leg muscles suddenly relaxed, the body would collapse.

A *muscle spindle* is a sense organ that consists of several modified muscle fibers. These muscle fibers have sensory nerve fibers wrapped around a specialized region near the middle of their length (fig. 17.9). Muscle spindles send nerve impulses to the central nervous system (CNS) when they are stretched by muscle contraction. These nerve impulses help the CNS coordinate the contraction of many muscles at a time.

Exercise: Variety of Benefits

A regular exercise program, such as the one described in table 17.A in the reading on the next page, has many benefits. Increased endurance and strength of muscles are 2 possible benefits. Endurance is measured by the length of time the muscle can work before fatiguing, and strength is the force a muscle (or a group of muscles) can exert against a resistance.

A regular exercise program brings about physiological changes that build endurance, such as increased stores of ATP in the muscles and increased tolerance to lactate buildup. Muscle strength increases as muscle enlargement occurs due to exercise. When a muscle enlarges, the number of muscle fibers does not usually increase, but the protein content of the muscle does. This happens because the contractile elements in

EXERCISE, EXERCISE, EXERCISE!!

. . . If a single researcher could be considered responsible for the fitness craze, it would be Dr. Ralph Paffenbarger, a specialist in heart disease and exercise with posts at both Stanford and Harvard universities. His landmark research, involving 17,000 Harvard alumni and 6,000 San Francisco longshoremen, showed that the men who exercised vigorously and burned at least 2,000 calories a week doing so cut their risk of dying from heart disease by half. The reduction was dramatic even if the men smoked or if their parents both had died of heart disease.

The problem was the public—and many health professionals—interpreted these findings as meaning that anything less didn't do any good. If you weren't up for running 20 miles a week, or for an hour of tennis 5 times a week, or for cross-country skiing for a half hour a day, well, you'd be just as well off sitting in front of the tube with a beer. The 2,000-calorie threshold was transformed into a "magic number"—above it, you were fit; below it, you were a basket case.

Paffenbarger says he never believed that the 2,000-calorie figure was carved in stone. Neither did Dr. Arthur Leon, an epidemiologist at University of Minnesota's School of Public Health. Both were surprised, however, when Leon's research showed that a far more modest level of activity could exert a very powerful protective effect on the heart. "People who don't want to do formal, sweaty exercise can be told that less can be beneficial," says Leon. "Just moving around more is a big help."

A nationwide study involving 12,138 men at 22 medical centers found that men who were only moderately active—spending an average of 48 minutes a day on leisure-time physical activity—had one-third fewer heart attacks than their peers who moved around during leisure time an average of 16 minutes each day. And the moderately active group didn't spend all, or even most, of their exercise time huffing and puffing. Mostly, the report found, their activities were in the light-to-moderate range: lawn and garden work, bowling, ballroom dancing—activities we don't often even think of as exercise.

Indeed, some of the most dramatic gains are made by the sedentary folks whose initial efforts fall short of the magic 2,000 calories and whose activities never include much bouncing around. For example, after only 4 months of attending a twice-weekly, low-impact aerobics class at Northwestern University, men and women with rheumatoid arthritis—a chronic, severe form of joint disease—reported much less pain, swelling, fatigue, and depression than before they began exercising.

Figure 17.A

Fitness walkers. The enclosed shopping mall, intended to be a boon for buyers, now has also become a haven for fitness walkers. Climate controlled and its passages unimpeded by curbs or stoplights, the mall provides a perfect environment for those determined to put in their daily mileage, rain or shine, while eliminating many of the outdoor hazards that may deter older pedestrians. Some mall walkers, like these Galleria Mall GoGetters in Glendale, California, have formed their own clubs. A few malls now issue special walking maps, while others open on holidays, even when the stores are closed, just to accommodate the local ramblers.

Leon and Paffenbarger point out that it's relatively easy to program greater amounts of activity into your normal day. Walk down to the accounting department at work instead of calling the accountant. Take stairs instead of elevators. Stroll during your lunchtime. A brisk walk—especially if you swing your arms—can get your heart rate up without the jouncing of jogging that many people dislike.

Leon's study showed that the benefits of exercise start to level off at a certain point—at least as far as fatal heart attack is concerned. The rate of fatal heart attacks among the moderately active men was the same as the most active men in the study, who devoted an average of 134 minutes a day to leisure-time physical activity.

Does this mean the most active group was no more fit than the moderately active group? Of course not. Fitness, stresses Bud Getchell, executive director of the nonprofit National Institute for Fitness and Sport in Indianapolis, includes flexibility, muscular strength, endurance, and body composition (lean vs. fat), as well as cardiovascular health. To achieve the type of fitness that would allow you to take on most vigorous activities with ease—from lifting a heavy box at work, to shoveling snow without hurting your back, to enjoying a game of pickup basketball—most fitness experts say you should spend about 3 hours (table 17.A), spaced out during the week, doing activities that strengthen muscles and enhance flexibility as well as challenge the heart.

Your body will reward your efforts. Although most people think the main benefit of

–Continued

–Continued from previous page

Table 17.A
A Checklist for Staying Fit

Children, Ages 7–12	Teenagers, Ages 13–18	Adults, Ages 19–55	Seniors, Ages 55+
Vigorous activity 1–2 hours daily	Vigorous activity 3–5 times a week	Vigorous activity for one-half hour, 3 times a week	Moderate exercise 3 times a week
Free play	Build muscle with calisthenics	Exercise to prevent lower back pain: aerobics, stretching, yoga	Plan a daily walk
Build motor skills through team sports, dance, swimming	Plan aerobic exercise to control buildup of fat cells	Take active vacations: hike, bicycle, cross-country ski	Daily stretching exercises
Encourage more exercise outside of physical education classes	Pursue tennis, swimming, riding—sports that can be enjoyed for a lifetime	Find exercise partners: join a running club, bicycle club, outing group	Learn a new sport: golf, fishing, ballroom dancing
Initiate family outings: bowling, boating, camping, hiking	Continue team sports, dancing, hiking, swimming		Try low-impact aerobics
			Before undertaking new exercises, consult your doctor

fitness is reducing the risk of heart disease, there's mounting evidence that other tissues and organs benefit as well. Several studies at the Institute for Aerobics Research in Dallas, Harvard, and other institutions have found that more active people have lower rates of colon, brain, kidney, and reproductive cancers, as well as leukemia, than their more sedentary counterparts. And studies that looked at exercise along with other variables found this still held true when factors like age, diet, and socioeconomic background were taken into account. Researchers speculate this may be because activity promotes the delivery of more nutrients and oxygen to these organs and tissues.

Indeed, according to Dr. Everett L. Smith, director of the Biogerontology Laboratory at the University of Wisconsin–Madison, half of the functional decline between ages 30 and 70 could be prevented if we simply used our body more. Bone density, nerve functions, and kidney efficiency, as well as overall strength and flexibility, can be largely preserved into our later years simply by keeping up an active life.

muscles, the myofibrils, which contain the protein filaments actin and myosin, increase in number.

Aside from improved endurance and strength, an exercise program also helps many other organs of the body. Cardiac muscle enlarges, and the heart can work harder than before. The resting heart rate decreases. Lung and diffusion capacity increase. Body fat decreases, but bone density increases so that breakage is less likely. Blood cholesterol and fat levels decrease, as does blood pressure. The reading on this and the previous page discusses in particular studies showing that an exercise program lowers the risk of heart attack.

Skeletal Muscles: Microscopic View

A whole skeletal muscle (fig. 17.10) is composed of a number of *muscle fibers* in bundles.

Fiber: Unique Features

Each muscle fiber is a cell containing the usual cellular components, but special names have been assigned to some of these components. The cell membrane is called the **sarcolemma,** the cytoplasm is the **sarcoplasm,** and the endoplasmic reticulum is the **sarcoplasmic reticulum** (fig. 17.10). A muscle fiber also has some unique anatomical characteristics. For one thing, it has a *T* (for transverse) *system*: the sarcolemma forms *T tubules,* which penetrate, or dip down, into the cell so that they come into contact—but do not fuse—with expanded portions of the sarcoplasmic reticulum. The expanded portions of the sarcoplasmic reticulum, called calcium-storage sacs, contain calcium ions (Ca^{++}), which are essential for muscle contraction. The sarcoplasmic reticulum encases hundreds and sometimes even thousands of **myofibrils,** which are the contractile portions of the fibers.

Myofibrils and Sarcomeres: Banded

Myofibrils are cylindrical in shape and run the length of the muscle fiber. The light microscope shows that a myofibril has light and dark bands called *striations.* It is these bands that cause skeletal muscle to appear striated (see fig. 10.5*a*). The electron microscope shows that the striations of myofibrils are formed by the placement of protein filaments within contractile units called **sarcomeres.** A sarcomere extends between 2 dark lines called the *Z lines.*

Figure 17.10

Skeletal muscle fiber structure and function. A muscle fiber contains many myofibrils, divided into sarcomeres, which are contractile. A sarcomere contains actin (thin) filaments and myosin (thick) filaments. When a muscle fiber contracts, the actin filaments move toward the center so that the H zone gets smaller, to the point of disappearing.

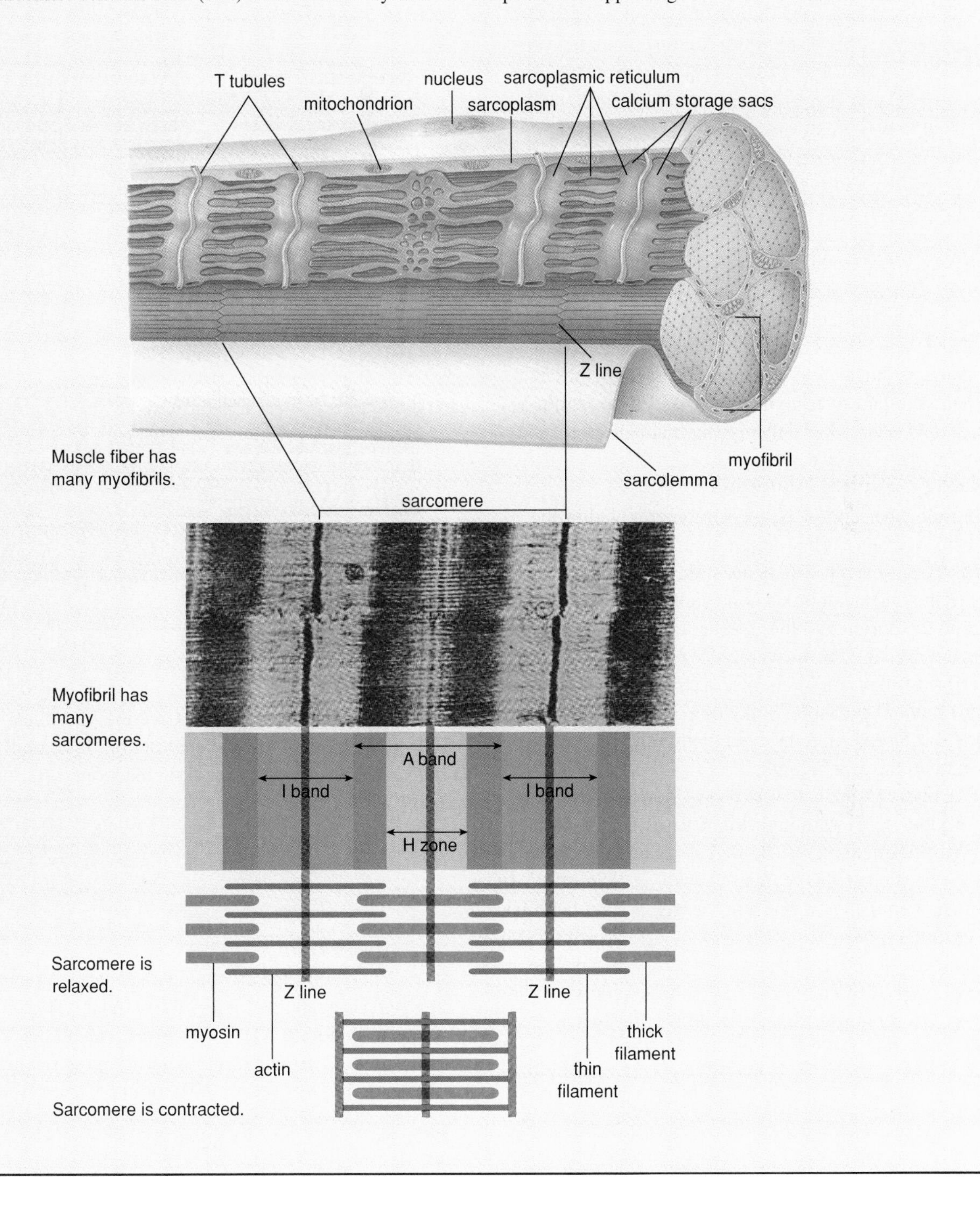

Table 17.1
Muscle Contraction

Name	Function
Actin filaments	Slide past myosin, causing contraction
Ca^{++}	Needed for myosin to bind to actin
Myosin filaments	a. Pull actin filaments by means of cross-bridges b. Are enzymatic and split ATP
ATP	Supplies energy for muscle contraction

A sarcomere contains 2 types of protein filaments. The thick filaments are made up of a protein called **myosin,** and the thin filaments are made up of a protein called **actin.** The I band is light colored because it contains only actin filaments attached to a Z line. The dark regions of the A band contain overlapping actin and myosin filaments, and its H zone has only myosin filaments.

Physiology: Sliding Filaments

As a muscle fiber contracts, the sarcomeres within the myofibrils shorten. When a sarcomere shortens (fig. 17.11*b*), the actin (thin) filaments slide past the myosin (thick) filaments and approach one another. This causes the I band to shorten and the H zone to almost or completely disappear. The movement of actin filaments in relation to myosin filaments is called the **sliding filament theory** of muscle contraction. During the sliding process, the sarcomere shortens even though the filaments themselves remain the same length.

The participants in muscle contraction have the functions listed in table 17.1. Although the actin filaments slide past the myosin filaments, it is the myosin filaments that do the work. In the presence of calcium ions (Ca^{++}), portions of a myosin filament called *cross-bridges* (fig. 17.11) bend backward and attach to an actin filament. After attaching to the actin filament, the cross-bridges bend forward and the actin filament is pulled along. Now, ATP is broken down by myosin, and detachment occurs. Note, therefore, that myosin is not only a structural protein, it is also an ATPase enzyme. The cross-bridges attach and detach some 50–100 times as the thin filaments are pulled to the center of a sarcomere.

> The sliding filament theory states that actin filaments slide past myosin filaments because myosin has cross-bridges, which pull the actin filaments inward.

ATP provides the energy for muscle contraction to continue. In order to ensure a ready supply of ATP, muscle fibers contain **creatine phosphate** (phosphocreatine), a storage form of high-energy phosphate. Creatine phosphate does not directly participate in muscle contraction. Instead, it is used to regenerate ATP by the following reaction:

Figure 17.11
Sliding filament theory. **a.** Relaxed sarcomere. **b.** Contracted sarcomere. Note that during contraction, the I band and the H zone decrease in size. This indicates that the actin filaments slide past the myosin filaments. Even so, the myosin filaments do the work by pulling the actin filaments by means of cross-bridges.

a.

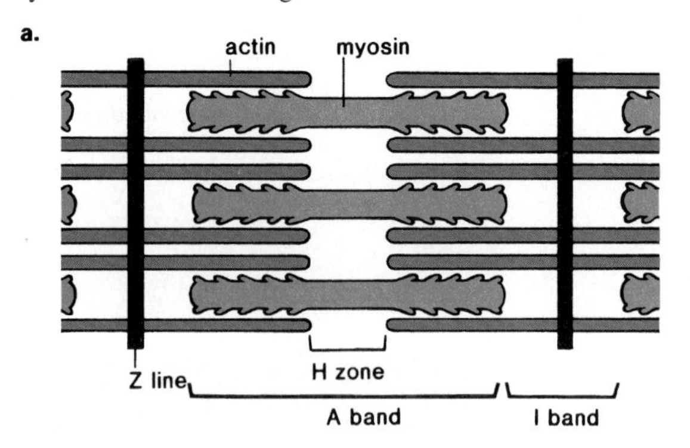

b.

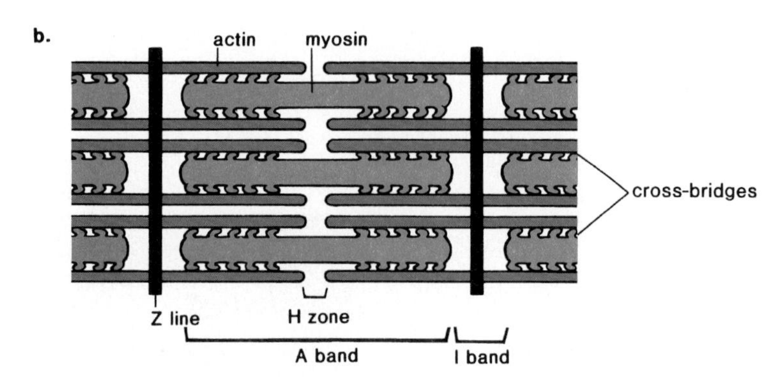

$$\text{creatine} \sim \text{P} + \text{ADP} \longrightarrow \text{ATP} + \text{creatine}$$

Oxygen Debt

When all of the creatine phosphate is depleted and no oxygen (O_2) is available for aerobic cellular respiration, a muscle fiber can generate ATP using fermentation, an anaerobic process (p. 95). Fermentation, which is apt to occur during strenuous exercise, can supply ATP for only a short time because of lactate buildup. The buildup is noticeable when it produces muscle aches and fatigue upon exercising.

We all have had the experience of having to continue deep breathing following strenuous exercise. This continued intake of oxygen is required to complete the metabolism of the lactate that has accumulated during exercise and represents an **oxygen debt** that the body must pay to rid itself of lactate. The lactate is transported to the liver, where one-fifth of it is completely broken down to carbon dioxide (CO_2) and water

(H_2O) by means of the Krebs cycle and the respiratory chain (see chap. 6). The ATP gained by this respiration then is used to reconvert four-fifths of the lactate to glucose.

Muscle contraction requires a ready supply of ATP. Creatine phosphate is used to generate ATP rapidly. If oxygen is in limited supply, fermentation produces ATP but results in oxygen debt.

Innervation: Releasing Ca^{++}

Muscles are innervated; that is, nerve impulses cause muscles to contract. A motor axon branches to several muscle fibers; collectively, this is called a motor unit. Each branch ends in several axon bulbs, where there are synaptic vesicles filled with the neuromuscular transmitter acetylcholine (ACh). The region where an axon bulb lies in close proximity to the sarcolemma of a muscle fiber is called a **neuromuscular junction.** A neuromuscular junction (fig. 17.12) has the same components as a synapse: a presynaptic membrane, a synaptic cleft, and a postsynaptic membrane. In this case, however, the postsynaptic membrane is a portion of the sarcolemma of a muscle fiber.

Nerve impulses cause synaptic vesicles to merge with the presynaptic membrane and to release acetylcholine into the synaptic cleft. When acetylcholine reaches the sarcolemma, the sarcolemma is depolarized. The result is a **muscle action potential,** which spreads over the sarcolemma and down the T system (fig. 17.10) to the calcium ions (Ca^{++}) stored in the calcium-storage sacs of the sarcoplasmic reticulum. When the muscle action potential reaches a sac, calcium ions are released. They diffuse into the sarcoplasm, where they participate in muscle contraction, as discussed previously.

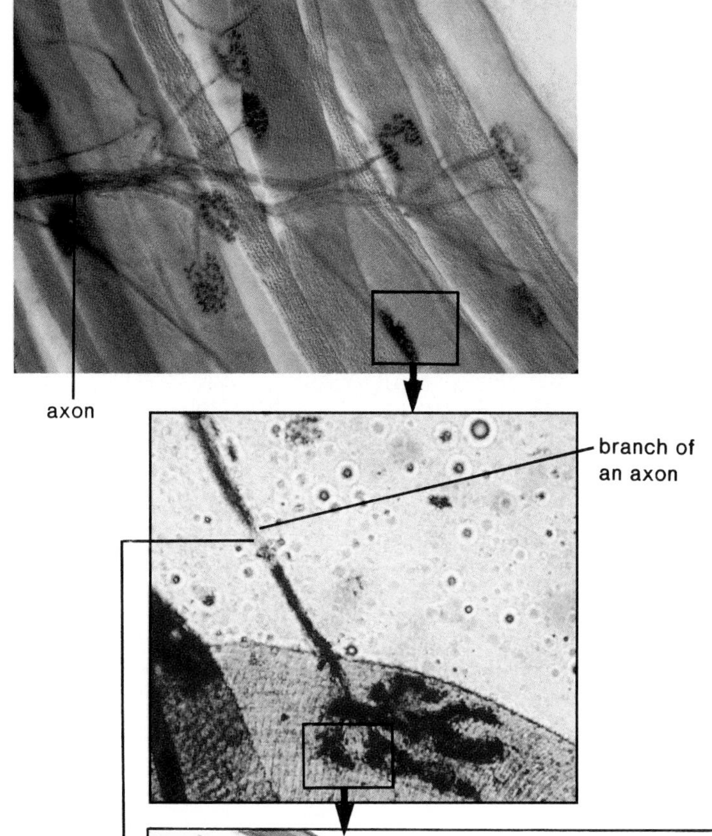

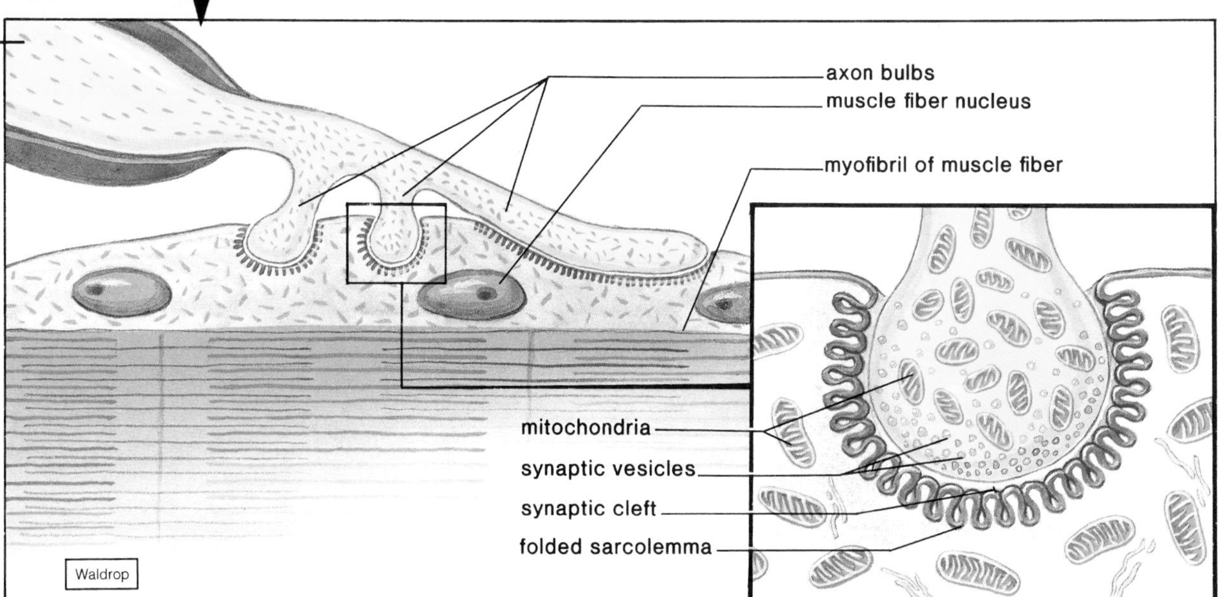

Figure 17.12

Muscle innervation. As shown, a motor neuron innervates several muscle fibers. Each axon branch ends in axon bulbs, and a neuromuscular junction occurs where an axon bulb comes in close proximity to a sarcolemma. The bulb contains synaptic vesicles filled with acetylcholine, which diffuses across the synaptic cleft when released. Muscle contraction follows.

A neuromuscular junction functions like a synapse except that a muscle action potential causes calcium ions (Ca^{++}) to be released from calcium-storage sacs, and thereafter muscle contraction occurs. Do 17.3 Critical Thinking, found at the end of the chapter.

It is possible to consider muscle contraction in greater detail. Figure 17.13 shows the placement of 2 other proteins associated with a thin filament (the double row of twisted globular actin molecules). Threads of tropomyosin wind about a thin filament, and troponin occurs at intervals along the threads. Calcium ions (Ca^{++}) that have been released from their storage sac combine with troponin. After binding occurs, the tropomyosin threads shift their position, and the cross-bridge binding sites are exposed.

The thick filament is a bundle of myosin molecules, each having a globular head. Each head is a cross-bridge, which has an ATP-binding site. After ATP attaches to ATP-binding sites, the cross-bridges bend backward and attach to the cross-bridge binding sites on the actin filaments. The cross-bridges then bend forward, pulling the actin filaments a short distance. Then the myosin heads break down ATP, and detachment of the cross-bridges occurs. The actin filaments move nearer the center of the sarcomere each time the cycle is repeated.

The movement of the actin filaments causes muscle contraction. Contraction ceases when nerve impulses no longer stimulate the muscle fiber. With the cessation of a muscle action potential, calcium ions are pumped back into their storage sac by active transport. Relaxation then occurs.

Figure 17.13

Detailed structure and function of sarcomere contraction. Following a muscle action potential, calcium ions (Ca^{++}) are released from their storage sacs. They combine with troponin, a protein that occurs periodically along tropomyosin threads. Then tropomyosin threads shift their position so that cross-bridge binding sites are revealed along the actin (thin) filaments. The myosin (thick) filaments extend globular heads, forming cross-bridges, which bind to these sites. The breakdown of ATP by myosin causes the cross-bridges to detach and to reattach farther along the actin filaments. In this way, the actin filaments are pulled along past the myosin filaments.

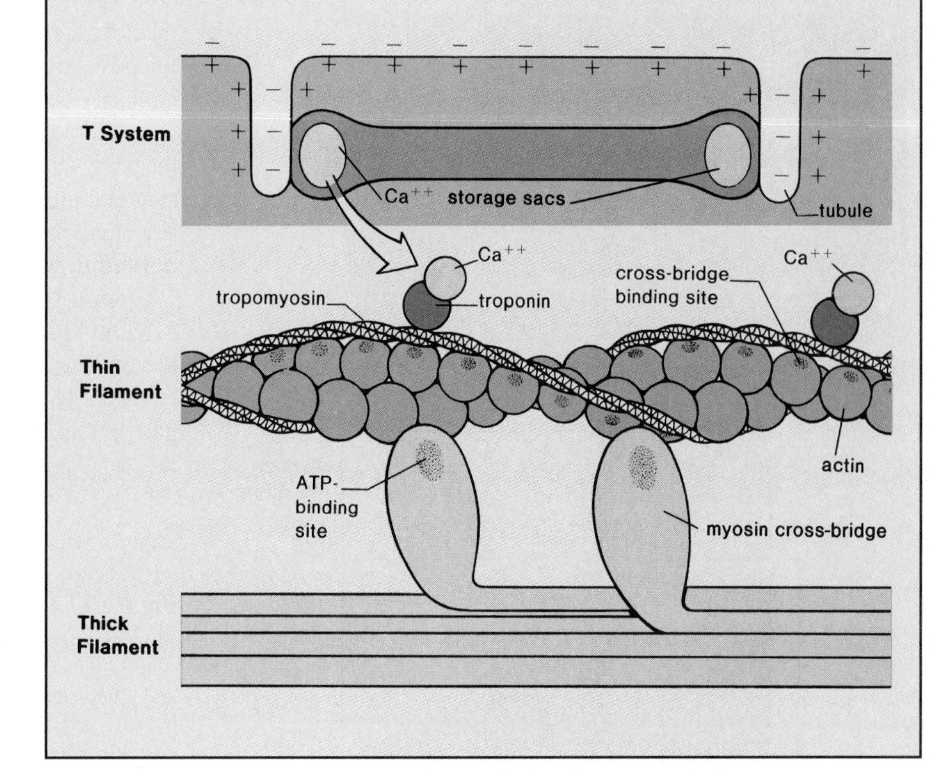

SUMMARY

The skeleton aids movement while it supports and protects the body. Bones serve as deposits for inorganic salts, and some bones are sites for blood-cell production. The skeleton is divided into 2 parts: (1) the axial skeleton, which is made up of the skull, the ribs, the sternum, and the vertebrae; and (2) the appendicular skeleton, which is composed of the girdles and their appendages. Joints are regions where bones are linked. Bone is constantly being renewed: osteoclasts break down bone and osteoblasts build new bone. Osteocytes are found in the lacunae of Haversian systems.

Whole skeletal muscles get shorter when they contract; therefore, muscles work in antagonistic pairs. For example, biceps brachii contraction flexes the forearm and triceps brachii contraction extends the forearm.

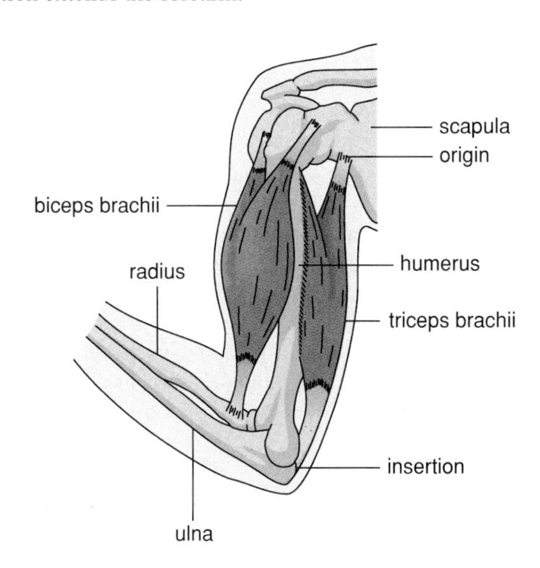

Muscle fibers obey the all-or-none law; it is possible to study a single contraction (muscle twitch) and sustained contraction (summation and tetanus) using a physiograph.

Muscle fibers are cells that contain myofibrils in addition to the usual cellular components. Longitudinally, myofibrils are divided into sarcomeres, where it is possible to note the arrangement of actin filaments and myosin filaments. When a sarcomere contracts, the actin filaments slide past the myosin filaments and the H zone all but disappears. Myosin has cross-bridges, which attach to and pull the actin filaments along. ATP breakdown by myosin is necessary for detachment to occur.

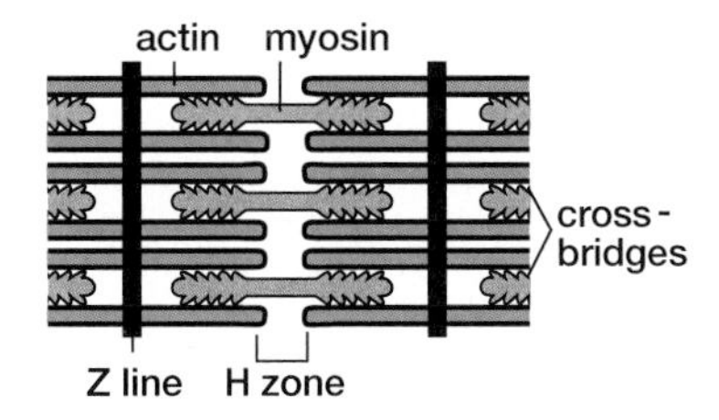

Innervation of a muscle fiber begins at a neuromuscular junction. Here, synaptic vesicles release acetylcholine into the synaptic cleft. When the sarcolemma receives acetylcholine, a muscle action potential moves down the T system to calcium-storage sacs. Contraction occurs when calcium ions (Ca^{++}) are released, and when calcium ions are actively transported back into the storage sacs, muscle relaxation occurs.

STUDY QUESTIONS

In order to practice **writing across the curriculum,** students should write out the answers to any or all of the study questions. The study questions are sequenced in the same order as the text.

1. Distinguish between the axial skeleton and the appendicular skeleton. (pp. 305–306)
2. List the bones that form the pectoral and pelvic girdles. (p. 306)
3. Describe the anatomy of a freely movable joint and of a long bone. (p. 308)
4. Describe how muscles are attached to bones. Why do muscles act in antagonistic pairs? (p. 310)
5. Describe the significance of threshold and maximum stimuli, muscle twitch, summation, and tetanic contraction. (p. 311)
6. How is the tone of a muscle maintained, and how do muscle spindles contribute to the maintenance of tone? (p. 312)
7. Discuss the microscopic anatomy of a muscle fiber and the structure of a sarcomere. What is the sliding filament theory? (pp. 314–316)
8. What is the role of creatine phosphate? (p. 316)
9. Discuss the availability and the specific role of ATP during muscle contraction. What is oxygen debt, and how is it repaid? (pp. 316–317)
10. What causes a muscle action potential? How does the muscle action potential bring about sarcomere and muscle fiber contraction? (pp. 317–318)

OBJECTIVE QUESTIONS

1. The skull, the ribs, and the sternum are all in the _______ skeleton.
2. The vertebral column protects the _______.
3. The 2 bones of the lower arm are the _______ and the _______.
4. Most joints are freely movable _______ joints, in which the 2 bones are separated by a cavity.
5. Muscles work in _______ pairs; the biceps brachii flexes and the triceps brachii extends the lower arm.
6. Maximal sustained contraction of a muscle is called _______.
7. Actin filaments and myosin filaments are found within cell inclusions called _______, which are divided into units called _______.
8. The molecule _______ serves as an immediate source of high-energy phosphate for ATP production in muscle cells.
9. The juncture between axon ending and muscle cell sarcolemma is called a _______ junction.
10. A muscle action potential causes _______ ions to be released from storage sacs; this signals the muscle fiber to contract.
11. Label this diagram of a muscle fiber using these terms: myofibril, mitochondrion, T tubules, sarcomere, sarcolemma, sarcoplasmic reticulum, sarcoplasm.

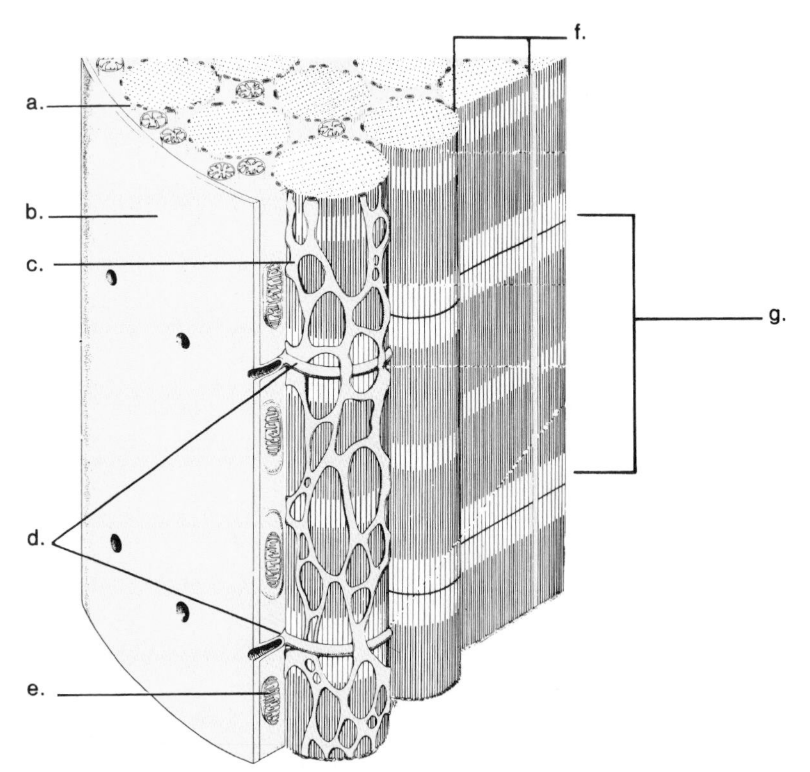

CRITICAL THINKING

In order to practice **writing across the curriculum,** students should write out the answers to any or all of the critical thinking questions. Suggested answers to the critical thinking questions are in appendix E.

17.1

1. What evidence do you have that bone is living tissue?
2. Why would you expect persons with stronger muscles to have stronger bones?
3. Bones have all sorts of grooves and protuberances. What purpose might these have?
4. The female pelvis is wider than the male pelvis. With what might this be associated?

17.2

1. Explain why the biceps brachii causes the lower arm and not the upper arm (where it is located) to flex.
2. Why can't the contraction of the biceps brachii both flex and extend the lower arm?
3. Why do you suppose the muscles of the legs are larger than the muscles of the arms?

17.3

1. A muscle fiber obeys the all-or-none law. Therefore, do all myofibrils in a muscle fiber contract at the same time?
2. When a sarcomere contracts, does the Z line move? If so, in which direction?
3. A respiratory pigment in muscle called myoglobin receives oxygen (O_2) from hemoglobin. Which of these 2 respiratory pigments has the higher affinity for oxygen?
4. When we exercise, blood brings more oxygen to the muscles. What do the muscles specifically do with all this oxygen?

SELECTED KEY TERMS

actin one of 2 major proteins of muscle; makes up thin filaments in myofibrils of muscle fibers. *See* myosin.

appendicular skeleton portion of the skeleton forming the upper extremities, the pectoral girdles, the lower extremities, and the pelvic girdle.

axial skeleton portion of the skeleton that supports and protects the organs of the head, the neck, and the trunk.

compact bone bone in which cells, separated by a matrix of collagen and mineral deposits, are located within Haversian systems.

creatine phosphate compound unique to muscles that contains a high-energy phosphate bond.

insertion the end of a muscle that is attached to a movable bone.

ligament dense fibrous connective tissue that joins bone to bone at a joint.

muscle action potential an electrochemical change due to increased sarcolemma permeability that is propagated down the T system and results in muscle contraction.

myofibril the contractile portion of a muscle fiber.

myosin one of 2 major proteins of muscle; makes up thick filaments in myofibrils and is capable of breaking down ATP. *See* actin.

neuromuscular junction the point of contact between a nerve cell and a muscle fiber.

origin end of a muscle that is attached to a relatively immovable bone.

osteocyte a mature bone cell.

oxygen debt oxygen that is needed to metabolize lactate, a compound that accumulates during vigorous exercise.

red bone marrow tissue located in the cavity of bones that forms blood cells.

sarcomere structural and functional unit of a myofibril; contains actin and myosin filaments.

synovial joint a freely movable joint.

tendon dense fibrous connective tissue that joins muscle to bone.

tetanus sustained muscle contraction without relaxation.

tone the continuous partial contraction of muscle.

18

SENSES

Chapter Concepts

1.
Receptors respond to mechanical stimuli, chemical stimuli, and light energy. Our knowledge of the outside world is largely dependent on interpretation of these stimuli. 323, 340

2.
Each receptor responds to one type of stimulus, but they all initiate nerve impulses. The sensation realized is the prerogative of the region of the cerebrum receiving the nerve impulses. 323, 340

3.
The general sense organs are present throughout the body in skin, joints, and muscles. 323, 340

4.
The special sense organs, such as the eyes and the ears, contain groupings of specialized receptors. 324, 340

5.
The microvilli of taste cells have receptors for the molecules in food. Certain receptors are responsible for bitter, sour, salty, or sweet tastes. 325, 340

6.
The receptors for sight contain visual pigments, which respond to light rays. 331, 340

7.
The receptors for hearing are hair cells, which respond to pressure waves. 337, 340

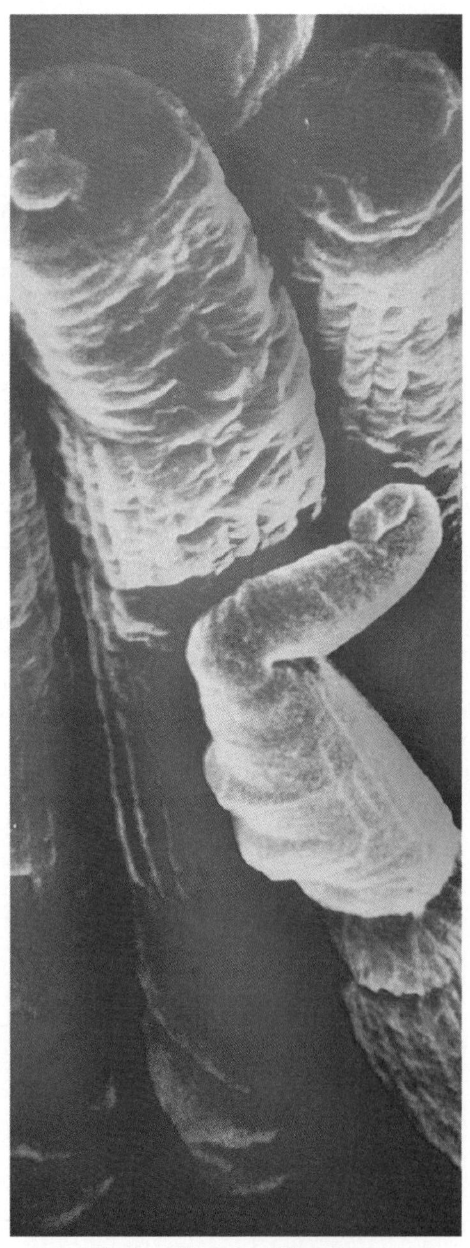

Rods and cones of the retina

Chapter Outline

Figure 18.1
Sense perception. Do you see a vase in this picture or a man and a woman looking at each other? Our sense organs are dependent on the stimuli received, and they can be fooled. This vase was specially designed to commemorate the 25th year of the queen of England's reign. The man and the woman are Prince Philip and Queen Elizabeth.

SENSE organs have receptors, which receive external and internal stimuli (fig. 18.1). Receptors are the first components of the reflex arc described in chapter 16. When a receptor is stimulated, it generates nerve impulses that are transmitted to the spinal cord and/or the brain. We are conscious of a sensation, however, only if the impulses reach the cerebrum. The sensory portion of the cerebrum can be mapped according to the parts of the body and the type of sensation realized at different loci (see fig. 16.16).

Sensation actually occurs in the brain and not in the sense organ itself. If the nerve fiber from the sense organ is cut, there is no sensation. Also, since a nerve impulse is always the same electrochemical charge, the particular sensation realized does not have to do with the nerve impulse. The brain is responsible for the type of sensation felt and for the localization of the sensation. For example, if we connect a pain receptor in the foot to a nerve normally receiving impulses from a heat receptor in the hand and then proceed to stick the pain receptor in the foot, the subject will report the feeling of warmth in the hand. The brain indicates the sensation and the localization. This realization is mildly disturbing because it makes us aware of how dependent we are on the anatomical wholeness of the body in order to be properly aware of our surroundings.

Table 18.1
General Sense Organs

Receptor	Main Location	Sense
Free nerve endings	Skin and other epithelial layers	Pain, possibly touch and temperature
Meissner corpuscles	Skin (notably fingertips)	Touch
Ruffini endings	Skin and subcutaneous tissue	Heat and touch
Merkel disks	Skin and subcutaneous tissue	Touch
Pacini corpuscles	Skin and subcutaneous tissue (notably joints, mammary glands, and external genitals)	Pressure
Krause end bulbs	Skin and subcutaneous tissue (notably lips, eyelids, and external genitals)	Cold and touch
Muscle spindles	Skeletal muscles	Proprioception
Golgi tendon receptors	Near junction of tendons and muscles	Proprioception

General Sense Organs

Microscopic receptors (table 18.1) are present in skin, the visceral organs, and the muscles and the joints.

Skin: Mosaic of Receptors

Skin (fig. 18.2) contains receptors for touch, pressure, pain, and temperature. It is a mosaic of these tiny receptors, as you can determine by slowly passing a metal probe over your skin. At certain points, there is a feeling of pressure, and at others, there is a feeling of hot or cold (depending on the temperature of the probe). Certain parts of the skin contain more receptors for a particular sensation; for example, the fingertips have an abundance of touch receptors.

Adaptation occurs when the receptor becomes so accustomed to stimulation that it stops generating impulses, even though the stimulus is still present. For example, the touch receptors adapt because soon after we put on an article of clothing, we are no longer aware of the feel of clothes against our skin.

Muscles and Joints: Sensing Position of Limbs

Proprioception is the sense of knowing the position of the limbs; for example, if you close your eyes and move your arm about slowly, you still have a sense of your arm's location. Muscle spindles, discussed in chapter 17, contribute to our sense of proprioception. Contraction of associated muscle fibers stretches muscle spindles and causes them to generate nerve impulses; for this reason, they are sometimes called **stretch receptors.**

Figure 18.2

Receptors in human skin. The classical view shown here is that each receptor has the main function indicated. However, investigations in this century indicate that matters are not so clear cut. For example, microscopic examination of the skin of the ear shows only free nerve endings (pain receptors), and yet the skin of the ear is sensitive to all sensations. Therefore, it appears that the receptors of the skin are somewhat but not completely specialized.

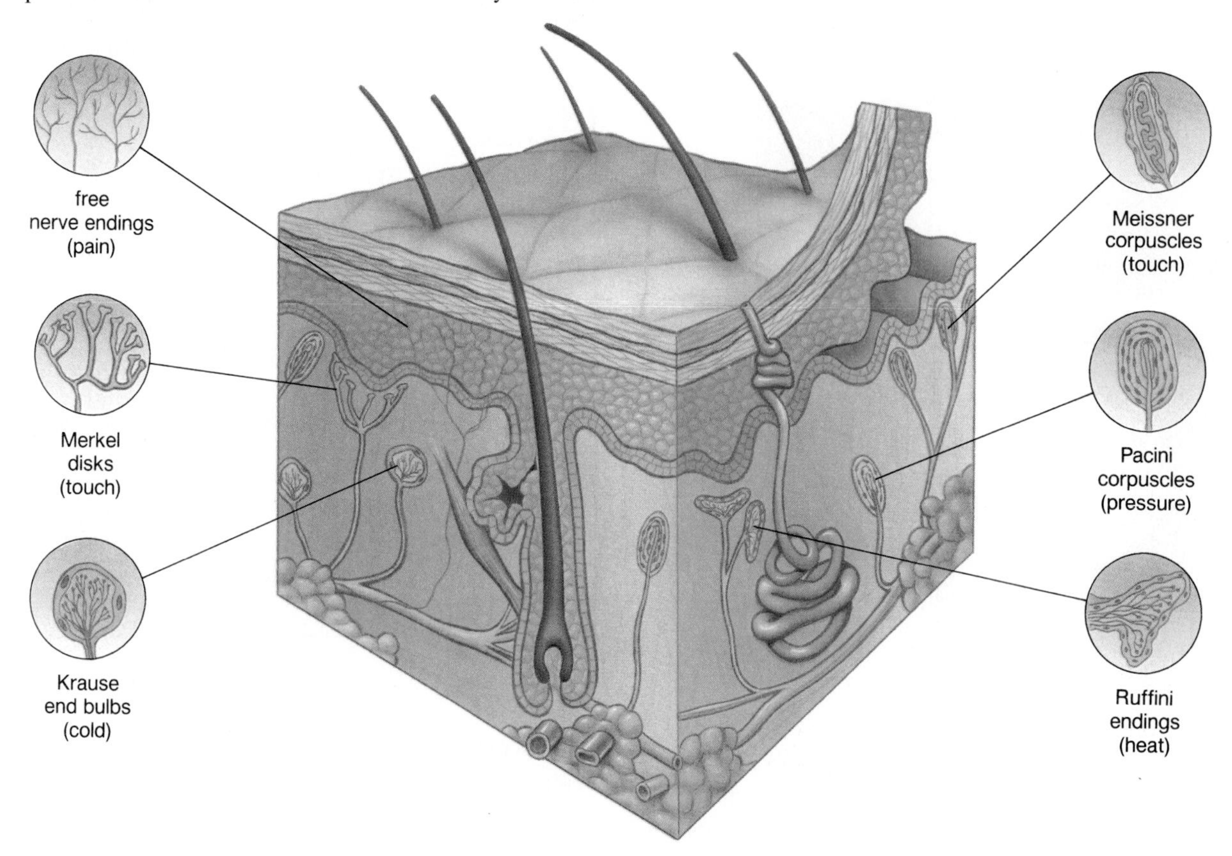

The *knee-jerk* reflex is a common example of the manner in which muscle spindles act (see fig. 17.9). When the legs are crossed at the knee and the tendon at the knee is tapped, both the tendon and the muscle spindles in the thigh are stretched. Then muscle spindles transmit impulses to the spinal cord, and thereafter the thigh muscles contract. This causes the lower leg to jerk upward in a kicking motion.

Receptors such as Golgi tendon receptors are located in the joints and associated ligaments and tendons. They respond to stretching, pressure, and pain. The brain integrates this information with that received from other types of receptors so that we know the position of body parts.

Special Sense Organs

The special sense organs (table 18.2) include the taste buds, the nose, the eye, and the ear. The nose and the taste buds contain chemoreceptors, the eye contains photoreceptors, and the ear contains mechanoreceptors.

Table 18.2
Special Sense Organs

Sense Organ	Type of Receptor	Specific Receptor	Senses
Taste buds	Chemoreceptor	Taste cells	Taste
Nose	Chemoreceptor	Olfactory cells	Smell
Eye	Photoreceptor	Rods and cones in retina	Vision
Ear	Mechanoreceptor	Hair cells in utricle, saccule, and semicircular canals	Equilibrium
	Mechanoreceptor	Hair cells in organ of Corti	Hearing

Figure 18.3

Taste buds. **a.** Elevations on the tongue are called papillae. The location of those containing taste buds responsive to sweet, sour, salt, and bitter is indicated. **b.** Enlargement of papillae. **c.** The taste buds occur along the walls of the papillae. **d.** Drawing shows the various cells that make up a taste bud. Taste cells in a bud end in microvilli that are sensitive to the chemicals exhibiting the tastes noted in (**a**). When the chemicals combine with membrane-bound receptors, nerve impulses are generated.

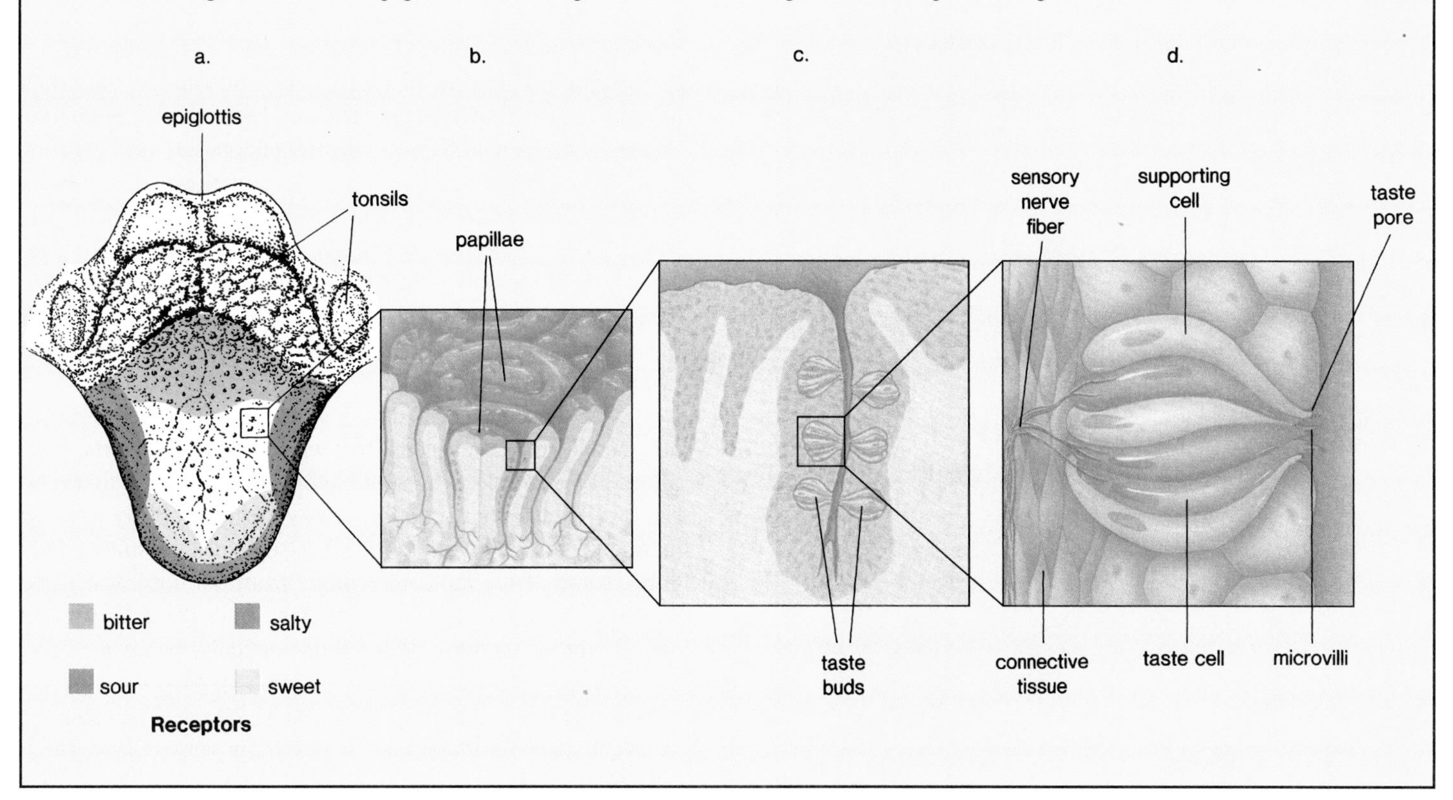

Taste Buds and the Nose

Taste and smell are called the *chemical senses* because the receptors for these senses are sensitive to chemical substances in the food we eat and the air we breathe. Therefore, they are called chemoreceptors.

Taste Buds: For Tasting

Taste buds are located primarily on the tongue (fig. 18.3). Many lie along the walls of the papillae, the small elevations on the tongue that are visible to the naked eye. Isolated ones are also present on the hard palate, the pharynx, and the epiglottis.

Taste buds are pockets of cells that extend through the tongue epithelium and open at a taste pore. Taste buds have supporting cells and a number of elongated taste cells that end in microvilli. Taste cells, which have associated sensory nerve fibers, are sensitive to chemicals. Nerve impulses most probably are generated when the chemicals bind to receptor sites found on the microvilli.

It is believed that there are 4 types of tastes (bitter, sour, salty, sweet) and that taste buds for each are concentrated on the tongue in particular regions (fig. 18.3*a*). Sweet receptors are most plentiful near the tip of the tongue. Sour receptors occur primarily along the margins of the tongue. Salt receptors are most common on the tip and the upper front portion of the tongue. Bitter receptors are located toward the back of the tongue.

Figure 18.4

Olfactory cell location and anatomy. **a.** The olfactory area in humans is located high in the nasal cavity. **b.** Enlargement of the olfactory cells shows they are modified neurons located between supporting cells. When olfactory cells are stimulated by chemicals, olfactory nerve fibers conduct nerve impulses to the olfactory bulb. An olfactory tract within the bulb takes the nerve impulses to the brain.

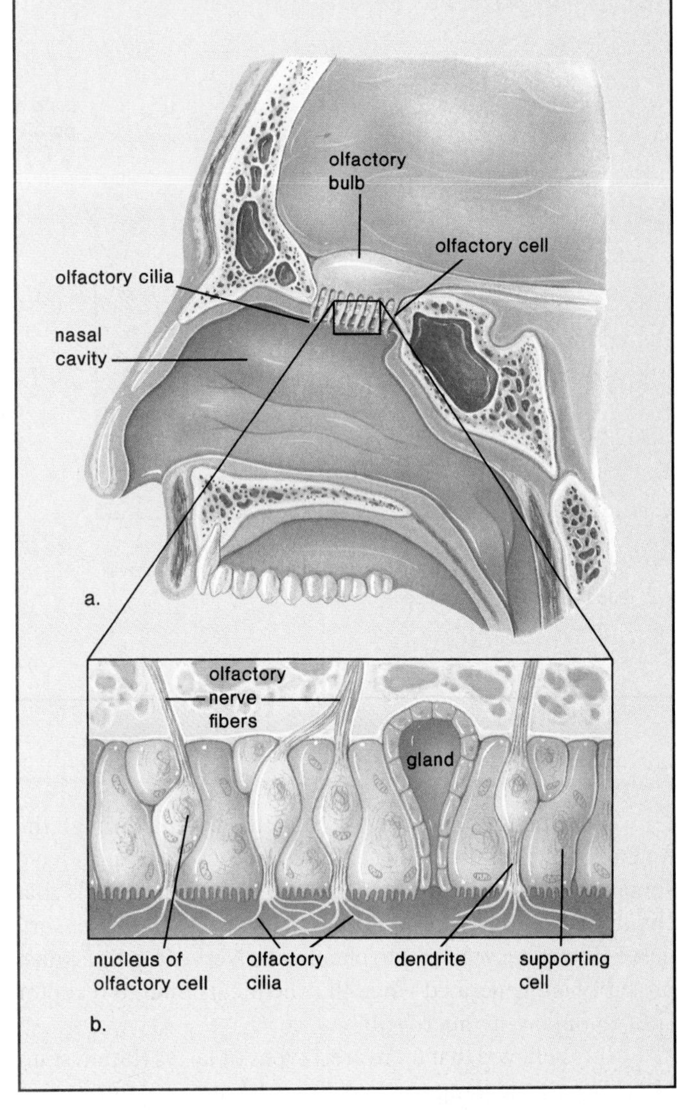

The Nose: For Smelling

Our sense of smell is dependent on olfactory cells (fig. 18.4), located high in the roof of the nasal cavity. Each cell ends in a tuft of about 5 cilia, which bear receptor sites for various chemicals. Research resulting in the stereochemical theory of smell suggests that different types of smell are related to the various shapes of molecules. When molecules combine with the receptor sites, nerve impulses are generated in the olfactory nerve fibers. Within the olfactory bulbs, which are paired masses of gray matter beneath the frontal lobes of the cerebrum, an olfactory tract takes this sensory information to an olfactory area of the cerebrum.

The olfactory receptors, like touch and temperature receptors, adapt to outside stimuli. In other words, after a while, the presence of a particular chemical no longer causes the olfactory cells to generate nerve impulses, and we are no longer aware of a particular smell.

The sense of taste and the sense of smell supplement each other, creating a combined effect when interpreted by the cerebrum. For example, when we have a cold, we think that food has lost its taste, but actually we have lost the ability to sense its smell. This may work in reverse also. When we smell something, some of the molecules move from the nose down into the mouth region and stimulate the taste buds there. Therefore, part of what we refer to as smell may actually be taste.

The receptors for taste (taste cells) and the receptors for smell (olfactory cells) work together to give us our sense of taste and our sense of smell.

The Eye

The anatomy and the physiology of the eye are complex, and we will consider each topic separately.

How the Eye Looks

The eyeball (fig. 18.5 and table 18.3), an elongated sphere about 2.5 cm in diameter, has 3 layers, or coats: the sclera, the choroid, and the retina. The outer **sclera** is a white, fibrous, protective layer except for the transparent cornea, the window of the eye. The middle, thin, dark brown layer, the **choroid,** contains many blood vessels and absorbs stray light rays. Toward the front, the choroid thickens and forms the ring-shaped ciliary body. The **ciliary body** contains the *ciliary muscle,* which controls the shape of the lens for near and far vision. Finally, the choroid becomes a thin, circular, muscular, and pigmented diaphragm, the **iris,** which regulates the size of the **pupil,** a hole in the center through which light enters the eyeball. The **lens,**

Figure 18.5

Anatomy of the human eye. Notice that the sclera becomes the cornea and the choroid becomes the ciliary body and the iris. The retina contains the receptors for vision; the fovea centralis is the region where vision is most acute. A blind spot occurs where the optic nerve leaves the retina. There are no receptors for light at this location.

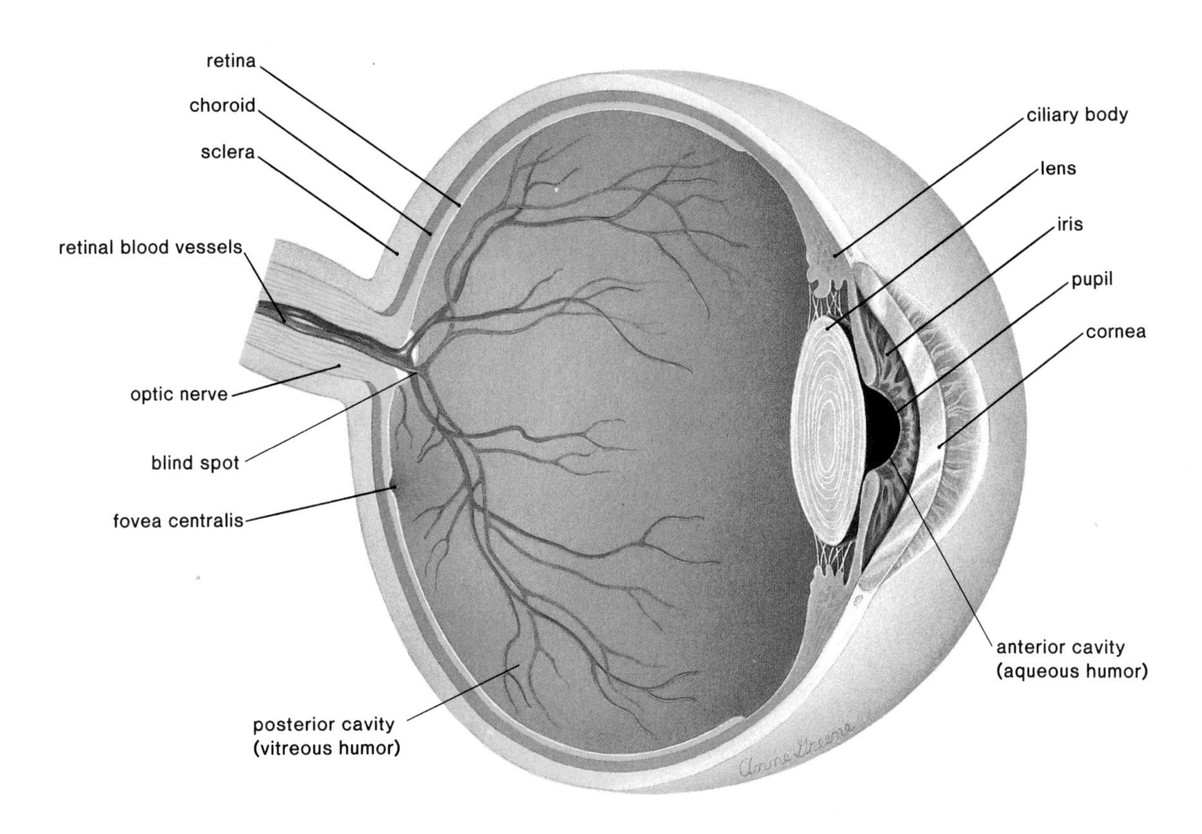

Table 18.3
Function of Parts of the Eye

Part	Function
Lens	Refracts and focuses light rays
Iris	Regulates light entrance
Pupil	Admits light
Choroid	Absorbs stray light
Sclera	Protects eyeball
Cornea	Refracts light rays
Humors	Refract light rays
Ciliary body	Holds lens in place, accommodation
Retina	Contains receptors for sight
Rods	Make black-and-white vision possible
Cones	Make color vision possible
Optic nerve	Transmits impulse to brain
Fovea centralis	Makes acute vision possible

attached to the ciliary body by ligaments, divides the cavity of the eye into 2 smaller cavities. A viscous, gelatinous material, the **vitreous humor,** fills the posterior cavity behind the lens. The anterior cavity between the cornea and the lens is filled with an alkaline, watery solution secreted by the ciliary body and called the **aqueous humor.**

A small amount of aqueous humor is continually produced each day. Normally, it leaves the anterior cavity by way of tiny ducts located where the iris meets the cornea. When a person has glaucoma, these drainage ducts are blocked, and aqueous humor builds up. If glaucoma is not treated, the resulting pressure compresses the arteries that serve the nerve fibers of the retina, where the receptors for sight are located. The nerve fibers begin to die due to lack of nutrients, and the person becomes partially blind. Over time, total blindness can result.

Figure 18.6

Anatomy of the retina. The retina is the inner layer of the eye. Rods and cones are located at the back of the retina, followed by the bipolar cells and the ganglionic cells, whose fibers become the optic nerve. (Notice that rods share bipolar cells but cones do not. Cones, therefore, distinguish more detail.) The optic nerve carries impulses to the occipital lobe of the cerebrum.

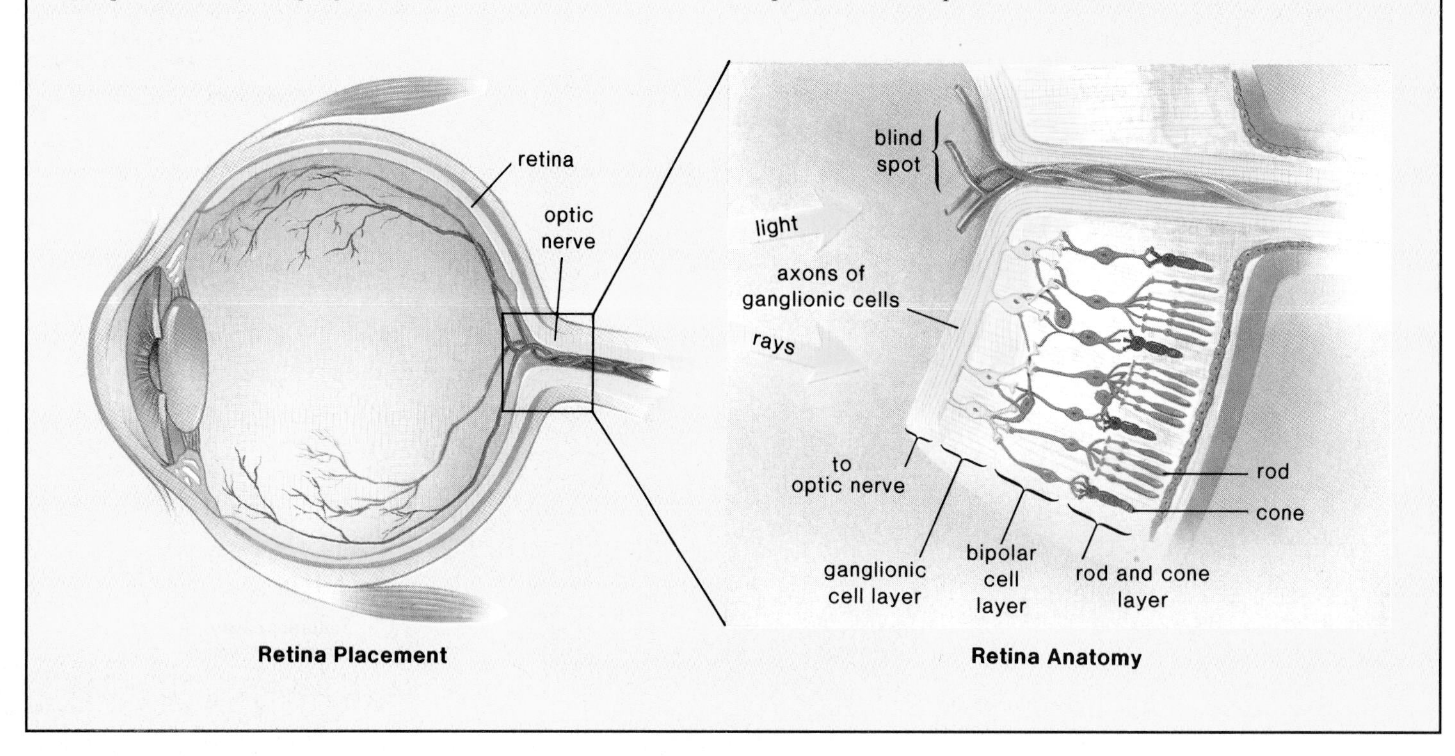

Retina: Three Layers of Cells

The inner layer of the eye, the **retina,** has 3 layers of cells (fig. 18.6). The layer closest to the choroid contains the sense receptors for vision, the **rods** and the **cones;** the middle layer contains bipolar cells; and the innermost layer contains ganglionic cells, which have axons that become the **optic nerve.** Only the rods and the cones contain light-sensitive pigments, and therefore light must penetrate to the back of the retina before nerve impulses are generated.

Nerve impulses initiated by the rods and the cones are passed to the bipolar cells, which in turn pass them to the ganglionic cells. The axons of the ganglionic cells pass in front of the retina, forming the optic nerve, which turns to pierce the layers of the eye. Notice in figure 18.6 that there are many more rods and cones than ganglionic cells. In fact, the retina has as many as 150 million rods but only one million ganglionic cells and optic nerve fibers. This means that there is considerable mixing of messages and a certain amount of integration before nerve impulses are sent to the visual cortex of the cerebrum. There are no rods or cones where the optic nerve passes through the retina; therefore, this is a **blind spot,** where vision is impossible.

The retina contains a very special region called the **fovea centralis** (fig. 18.5), an oval, yellowish area with a depression in which there are only cones. Vision is most acute in the fovea centralis.

The eye has 3 layers: the outer sclera, the middle choroid, and the inner retina. Only the retina contains receptors for vision.

ANIMAL EYES AND EARS

Some photoreceptors are very simple. The eyespots of planaria only detect light, enabling the animal to remain in the shadows and avoid the light. Image-forming eyes are found among 4 invertebrate groups: cnidarians, annelids, mollusks, and arthropods. Arthropods have compound eyes; each of the many individual visual units has its own lens and views a separate portion of an object. How well the brain combines this information is not known, but compound eyes seem especially well suited to detecting motion, as anyone who has tried to catch a fly knows.

Vertebrates and certain mollusks, like the squid, have a camera type of eye. A single lens focuses an image of the visual field on the photoreceptors, which are packed closely together. All of the photoreceptors taken together can be compared to a piece of film in a camera.

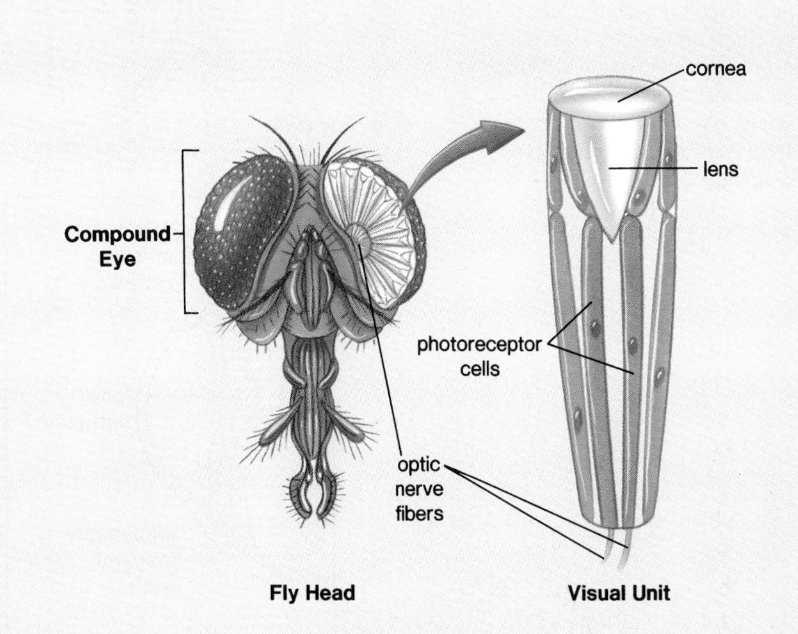

Crustaceans, spiders, and insects have a simple type of ear. In grasshoppers, the tympanum passes sound vibrations to sensory cells, which initiate nerve impulses. Among the vertebrates, the ears have varying complexity. Fishes have an inner ear, which detects primarily body position. This ear is associated with a lateral-line system containing hair cells that respond to pressure waves in the water. The cochlea, which contains the receptors for hearing, is first seen in amphibians, which have both a middle ear and an inner ear. The cochlea is somewhat elongated in birds and highly coiled in mammals. The extended length allows humans to distinguish a range in pitch from high to low.

How the Eye Sees

Focusing: Bending Light

When we look at an object, light rays are **focused** on the retina (fig. 18.7*a*). In this way, an *image* of the object appears on the retina. The image on the retina occurs when the rods and the cones in a particular region are excited. Obviously, the image is much smaller than the object. In order to produce this small image, light rays must be bent (refracted) and brought into focus. They are initially bent as they pass through the cornea, and further bending occurs as the rays pass through the lens and the humors.

Light rays are reflected from an object in all directions. For distant objects, only nearly parallel rays enter the eye, and the cornea alone is needed for focusing. For close objects, many of the rays are at sharp angles to one another, and

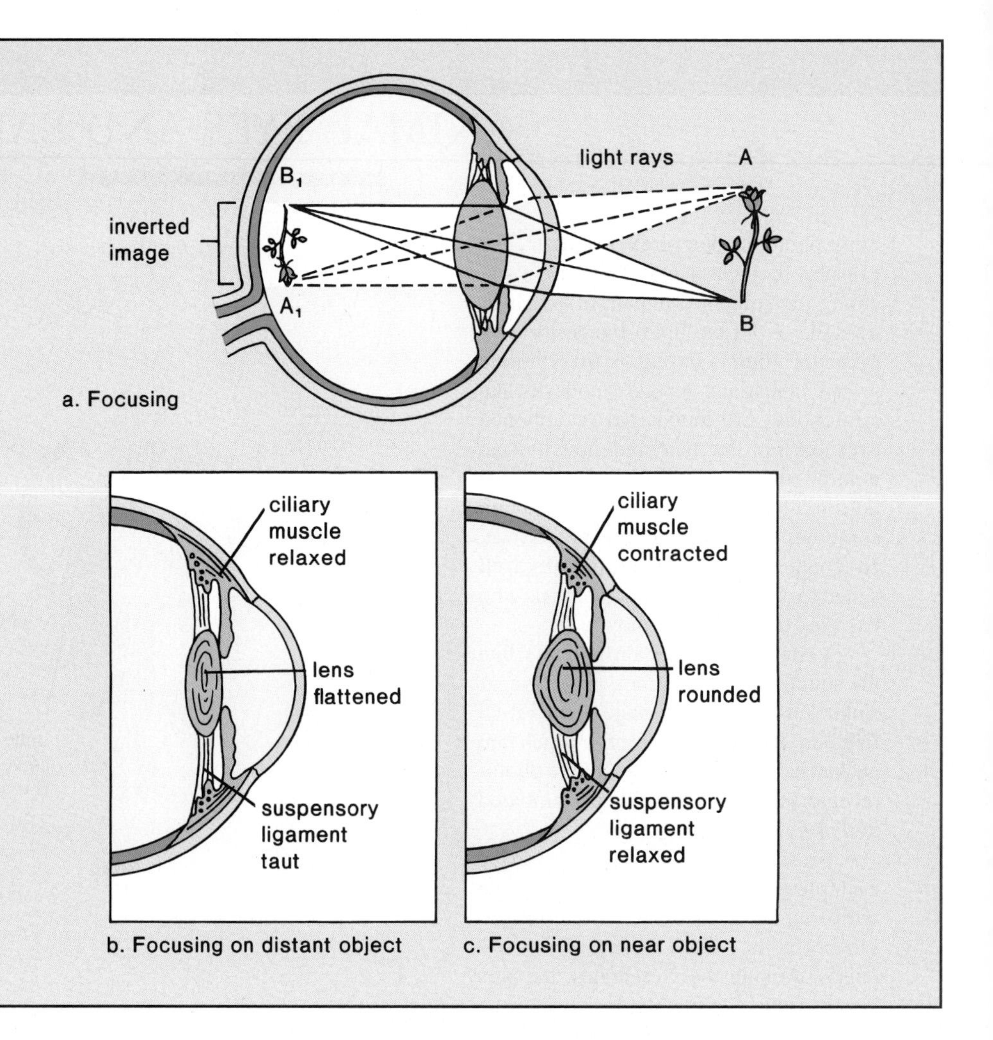

Figure 18.7
Focusing. **a.** Light rays from each point on an object are bent by the cornea and the lens in such a way that they are directed to a single point after emerging from the lens. By this process, an inverted image of the object forms on the retina. **b.** When focusing on a distant object, the lens is flat because the ciliary muscle is relaxed and the suspensory ligament is taut. **c.** When focusing on a near object, the lens accommodates: it rounds up because the ciliary muscle contracts, causing the suspensory ligament to relax.

additional focusing is required. The lens provides this additional focusing power as **accommodation** occurs: the lens remains flat when we view distant objects, but it rounds up when we view near objects.

The shape of the lens is controlled by the ciliary muscle within the ciliary body. When we view a distant object, the ciliary muscle is relaxed, causing the suspensory ligaments attached to the ciliary body to be taut; therefore, the lens remains relatively flat (fig. 18.7*b*). When we view a close object, the ciliary muscle contracts, releasing the tension on the suspensory ligaments, and the lens rounds up due to its natural elasticity (fig. 18.7*c*). Because close work requires contraction of the ciliary muscle, it very often causes eyestrain.

With aging, the lens loses some of its elasticity and is unable to accommodate. Corrective lenses are then usually necessary, as discussed on page 333. Also with aging, the lens is subject to *cataracts;* it can become opaque and therefore incapable of transmitting rays of light. Recent research suggests that cataracts develop when crystallin proteins, contained by special cells within the interior, oxidize, causing the three-dimensional shape to change. If so, researchers believe that eventually they may be able to find ways to restore the normal configuration of crystallin so that cataracts can be treated medically instead of surgically.

For the present, however, surgery is the only viable treatment for cataracts. First, a surgeon opens the eye near the rim of the cornea. Zonulysin, an enzyme, may be used to digest away the ligaments holding the lens in place. Most surgeons then use a cryoprobe, an instrument that freezes the lens for easy removal. An intraocular lens attached to the iris can then be implanted so that the patient does not need to wear thick glasses or contact lenses.

The lens, assisted by the cornea and the humors, focuses images on the retina.

The Upside-Down Image. The image formed on the retina is inverted (fig. 18.7*a*), and it is thought that perhaps this image is righted in the brain by experience. In one experiment, scientists wore glasses that inverted the field of vision. At first, they had difficulty adjusting to the placement of the objects, but they soon became accustomed to their inverted world. Experiments such as this one suggest that if we see the world upside down, the brain learns to see it right side up.

Combining the Data

We can see well with either eye alone, but the 2 eyes functioning together provide us with **stereoscopic vision.** Normally, the 2 eyes are directed by the eye muscles toward the same object, and therefore the object is focused on corresponding points of the 2 retinas. Each eye, however, sends its own data to the brain about the placement of the object because each forms an image from a slightly different angle. These data are pooled to produce depth perception by a two-step process. First, because the optic nerves cross at the optic chiasma (fig. 18.8), one-half of the brain receives information from both eyes about the same part of an object. Later, the 2 halves of the brain communicate to arrive at a three-dimensional interpretation of the whole object.

> The anatomy and the physiology of the brain allow us to see the world right side up and in 3 dimensions.

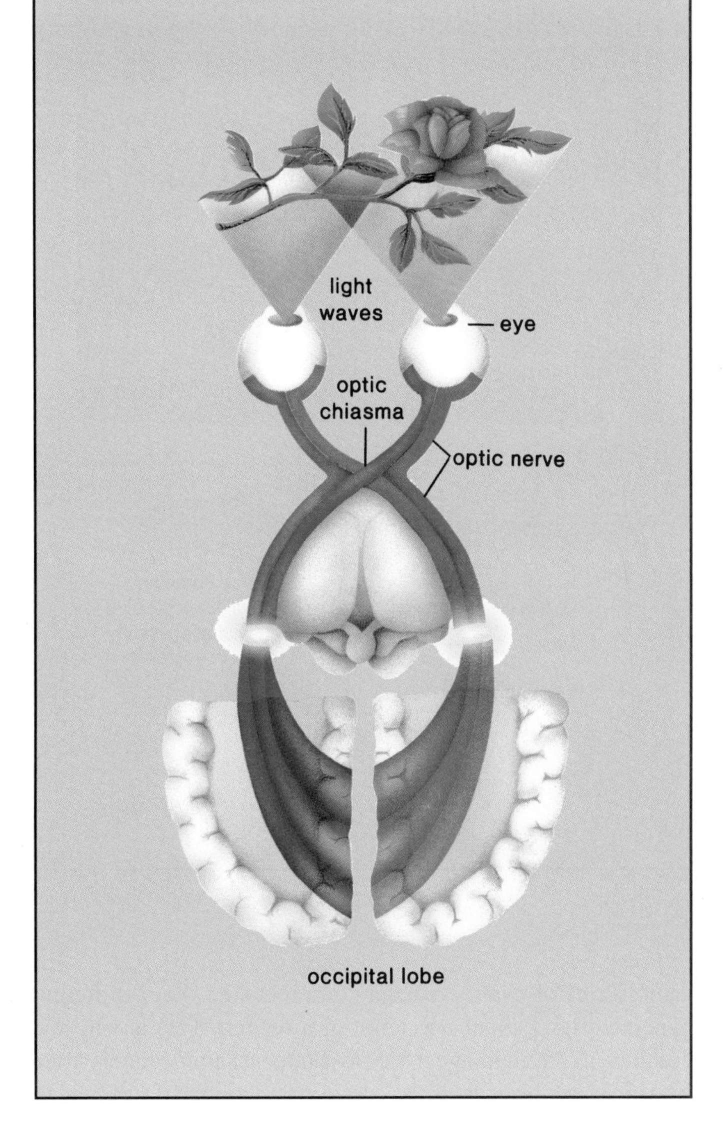

Figure 18.8
Optic chiasma. Both eyes "see" the entire object, but data from the right half of each retina goes to the right occipital lobe of the cerebrum and information from the left half of each retina goes to the left occipital lobe because of the optic chiasma. When the data are combined, the brain "sees" the entire object in depth.

Seeing: Is Chemical

In *dim light,* the iris causes the pupils to enlarge so that more rays of light can enter the eyes. As the rays of light enter, they strike the rods and the cones, but only the 150 million rods located in the periphery, or sides, of the eyes are sensitive enough to be stimulated by this faint light. The rods do not detect fine detail or color, so at night, for example, all objects appear to be blurred and a shade of gray. Rods do detect even the slightest motion, however, because of their abundance and position in the eyes.

Although it has been known for some time that vision generated by the rods is dependent on the presence of a visual pigment called **rhodopsin,** the actual events have only recently been worked out in detail. Many molecules of rhodopsin, also called visual purple, are located within the membrane of the disks found in the outer segment of the rods (fig. 18.9). Rhodopsin is a complex molecule that contains a protein (*opsin*) and a pigment molecule called *retinal,* which is a derivative of vitamin A. When retinal absorbs light energy, it changes shape and disengages from opsin (fig. 18.9). A complex series of reactions follows, which leads to the excitation of bipolar cells (fig. 18.6). In dim light, enzymes re-form rhodopsin, and the

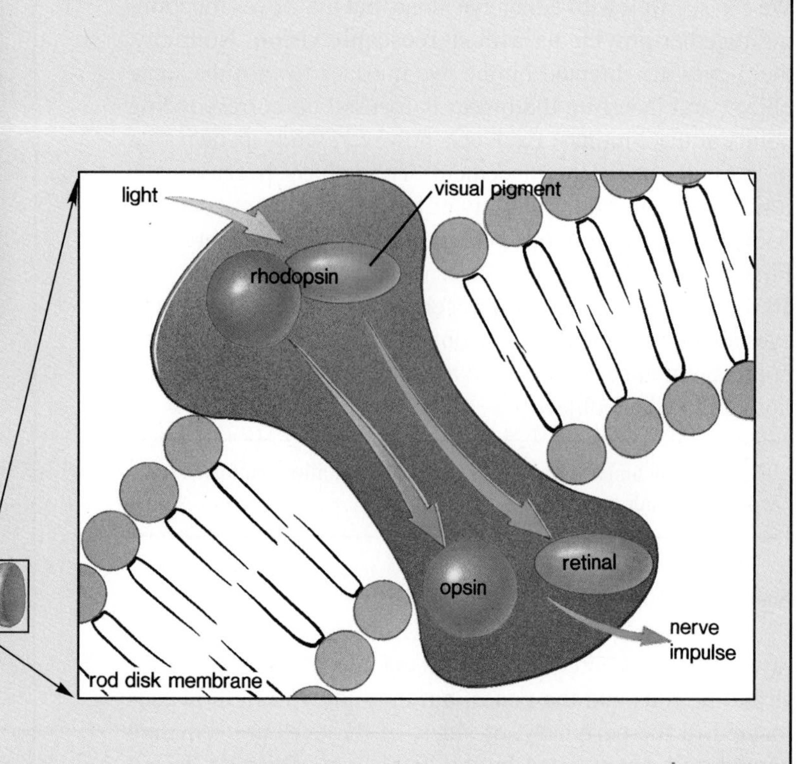

Figure 18.9

Structure and function of rods and cones. The outer segment of rods and cones is composed of stacks of membranous disks that contain visual pigments. In rods, the membrane of each disk contains rhodopsin, a complex molecule containing the protein opsin and the pigment retinal. When retinal absorbs energy, it changes shape and disengages from opsin. This leads to nerve impulses in bipolar cells (fig. 18.6).

same series of events reoccurs and reoccurs. Each stimulus generated lasts about one-tenth of a second. This is why we continue to see an image if we close our eyes immediately after looking at an object. It also allows us to see motion if still frames are presented at a rapid rate, as in "movies." In bright light, rhodopsin is not re-formed, and the cones become active. This is the reason vision is limited when we first walk into a dark place. It takes a few minutes for rhodopsin to re-form.

The cones, located primarily in the fovea centralis, detect the fine detail and the color of an object. In order to perceive depth, as well as to see color, we turn our eyes so that reflected light from the object strikes the fovea centralis. Color vision has been shown to depend on the 3 kinds of cones that contain pigments sensitive to blue, green, or red light. The nerve impulses generated from one type of cone not only stimulate certain cells in the visual cortex of the cerebrum, they also inhibit the reception of impulses from other types of cones. For example, when we see red, certain cells in the brain are prohibited from receiving impulses from green cones. Similarly, impulses sent through blue cones tend to oppose the combination of signals sent by red and green cones—which together produce yellow. This process assists integration and enables the brain to tell the location of various colors in the environment.

Complete color blindness is extremely rare. In most instances, a particular type of cone is lacking or deficient in number. The lack of red or green cones is the most common, affecting about 5% of human males. If the eye lacks red cones, the green colors are accentuated, and vice versa (fig. 18.10).

Figure 18.10

Test plates for color blindness. When looking at the top plate, the person with normal color vision sees the number 8, and when looking at the bottom plate, the person with normal color vision sees the number 12. The most common form of color blindness involves an inability to distinguish reds from greens.

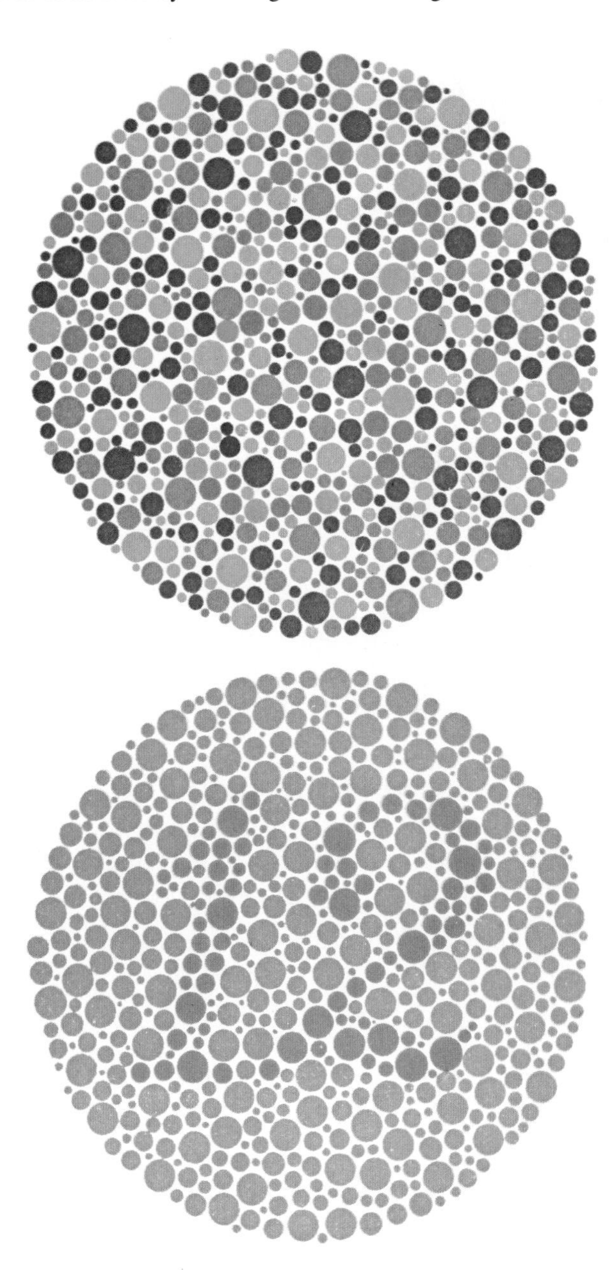

These plates have been reproduced from Ishihara's Tests for Colour Blindness published by Kanehara & Co., Ltd., Tokyo, Japan, but tests for color blindness cannot be conducted with this material. For accurate testing, the original plates should be used.

The sense receptors for sight are the rods and the cones. The rods are responsible for vision in dim light, and the cones are responsible for vision in bright light and for color vision. When either is stimulated, nerve impulses are transmitted in the optic nerve to the brain. Do 18.1 Critical Thinking, found at the end of the chapter.

Helping the Eye

The majority of people can see what is designated as a size 20 letter 20 ft away and so are said to have 20/20 vision. Persons who can see close objects but cannot see the letters from this distance are said to be nearsighted. Nearsighted people can see near better than they can see far. These individuals often have an elongated eyeball, and when they attempt to look at a far object, the image is brought to focus in front of the retina (fig. 18.11). They can see near because they can adjust the lens so that the image is focused on the retina, but to see far, these people must wear concave lenses, which diverge the light rays so that the image can be focused on the retina. There is a new treatment for nearsightedness called radial keratotomy, or radial K. From 4 to 8 cuts are made in the cornea so that they radiate out from the center like spokes in a wheel. When the cuts heal, the cornea is flattened. Although some patients are satisfied with the result, others complain of glare and varying visual acuity.

Persons who can easily see the optometrist's chart but cannot see close objects well are farsighted; these individuals can see far away better than they can see near. They often have a shortened eyeball, and when they try to see near, the image is focused behind the retina. When the object is far away, the lens can compensate for the short eyeball, but when the object is close, these persons must wear a convex lens to increase the bending of light rays so that the image can be focused on the retina.

When the cornea or lens is uneven, the image is fuzzy. The light rays cannot be evenly focused on the retina. This condition, called **astigmatism,** can be corrected by an unevenly ground lens to compensate for the uneven cornea.

As We Age

With normal aging, the lens loses some of its ability to change shape in order to focus on close objects. Because nearsighted individuals still have difficulty clearly seeing objects in the distance, they must wear bifocals, which means that the upper part of the lens is for distant vision and the remainder is for near vision.

The shape of the eyeball determines the need for corrective lenses; the inability of the lens to accommodate as we age also requires corrective lenses for close vision.

Figure 18.11

Common abnormalities of the eye, with possible corrective lenses. **a**. The cornea and the lens function in bringing light rays (lines) to focus, but sometimes they are unable to compensate for the shape of the eyeball or for an uneven cornea. **b.** In these instances corrective lenses enable the individual to see normally.

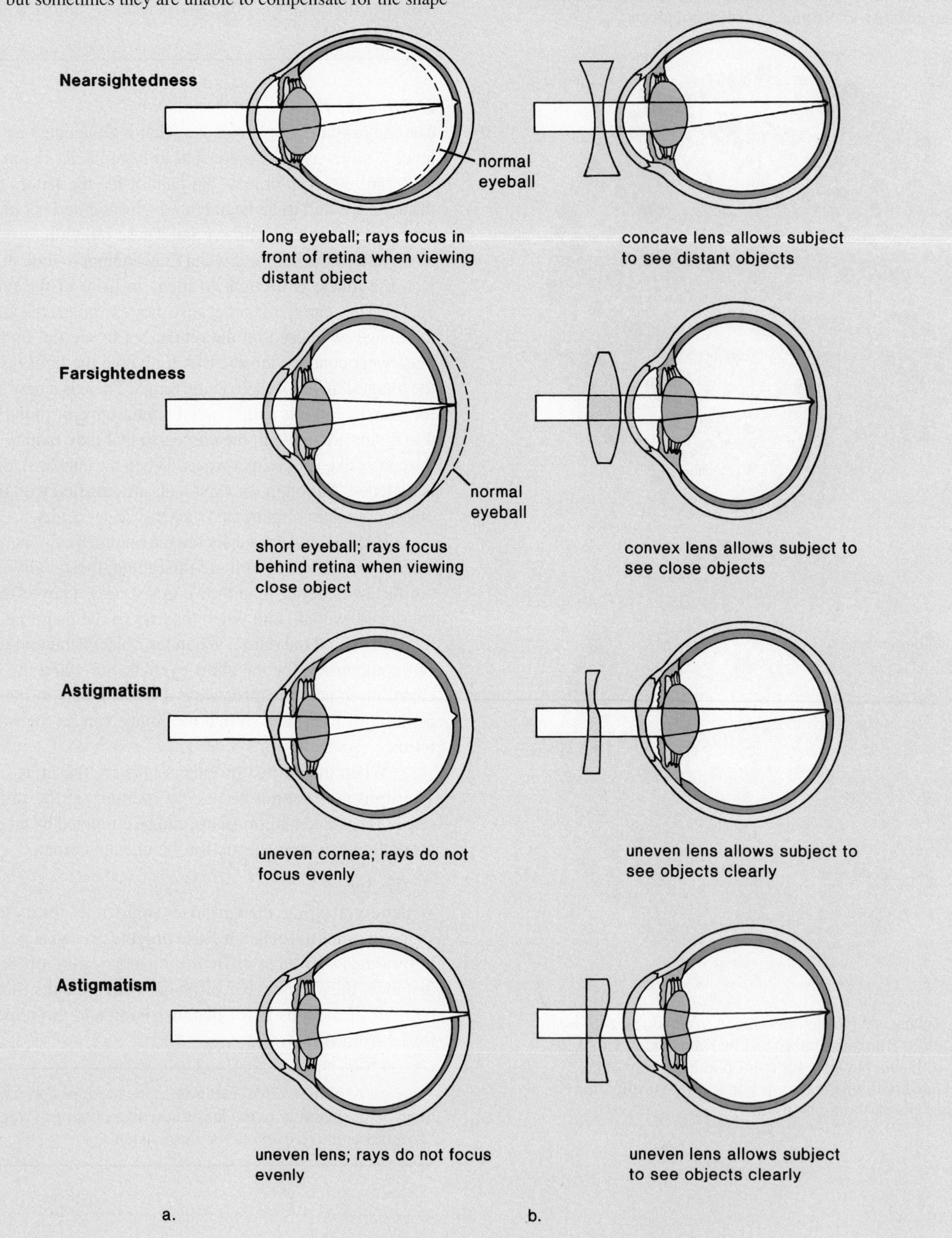

Figure 18.12

Anatomy of the human ear. In the middle ear, the hammer (malleus), the anvil (incus), and the stirrup (stapes) amplify sound waves. Otosclerosis is a condition in which the stirrup becomes attached to the inner ear and is unable to carry out its normal function. It can be replaced by a plastic piston, and thereafter the individual hears normally because sound waves are transmitted as usual to the cochlea, which contains the receptors for hearing.

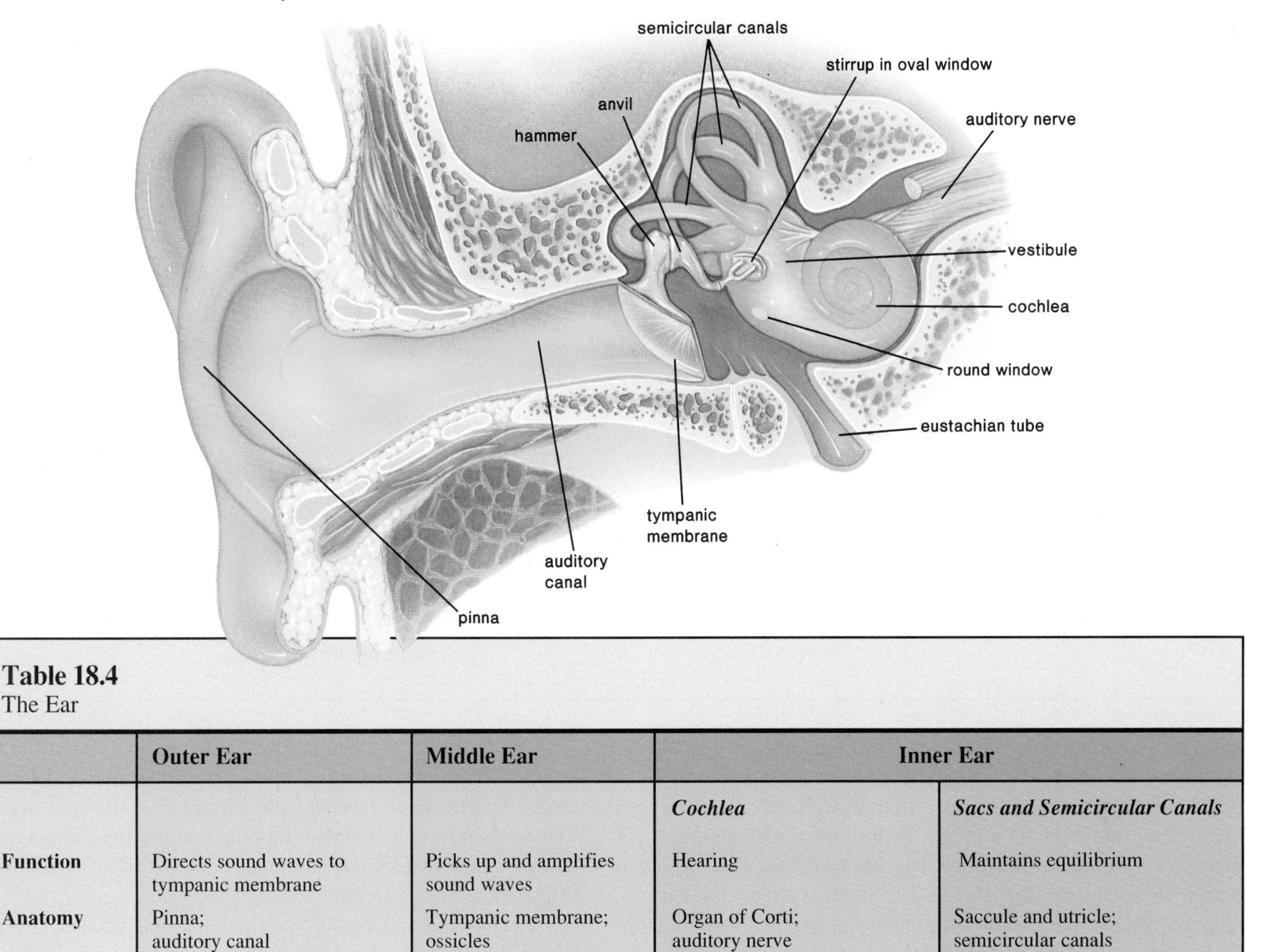

Table 18.4
The Ear

	Outer Ear	Middle Ear	Inner Ear	
			Cochlea	*Sacs and Semicircular Canals*
Function	Directs sound waves to tympanic membrane	Picks up and amplifies sound waves	Hearing	Maintains equilibrium
Anatomy	Pinna; auditory canal	Tympanic membrane; ossicles	Organ of Corti; auditory nerve	Saccule and utricle; semicircular canals
Medium	Air	Air (eustachian tube)	Fluid	Fluid

Path of vibration: Sound waves—vibration of tympanic membrane—vibration of hammer, anvil, and stirrup—vibration of oval window—fluid pressure waves in canals of inner ear lead to stimulation of hair cells—bulging of round window

The Ear

The ear accomplishes 2 sensory functions: equilibrium (balance) and hearing. The receptors for both of these are located in the inner ear and consist of hair cells with cilia that respond to mechanical stimulation. Each hair cell has from 30 to 150 cilia. When the cilia of any particular hair cell are displaced in a certain direction, the cell generates nerve impulses that are sent along a cranial nerve to the brain.

How the Ear Appears

Figure 18.12 is a drawing of the ear, and table 18.4 lists the parts of the ear. The ear has 3 divisions: outer ear, middle ear, and inner ear. The **outer ear** consists of the **pinna** (external flap) and the **auditory canal.** The opening of the auditory canal is lined with fine hairs and sweat glands. Modified sweat glands that secrete earwax, a substance that helps to guard the ear against the entrance of foreign materials such as air pollutants, are in the upper wall of the canal.

The **middle ear** begins at the **tympanic membrane** (eardrum) and ends at a bony wall containing 2 small openings

Figure 18.13

Anatomy of the inner ear. **a.** The inner ear contains the semicircular canals, the utricle and the saccule within a vestibule, and the cochlea. The cochlea has been cut to show the location of the organ of Corti. **b.** An ampulla at the base of each semicircular canal contains the receptors (hair cells) for dynamic equilibrium. **c.** The utricle and the saccule are small sacs that contain the receptors (hair cells) for static equilibrium. **d.** The receptors for hearing (hair cells) are in the organ of Corti.

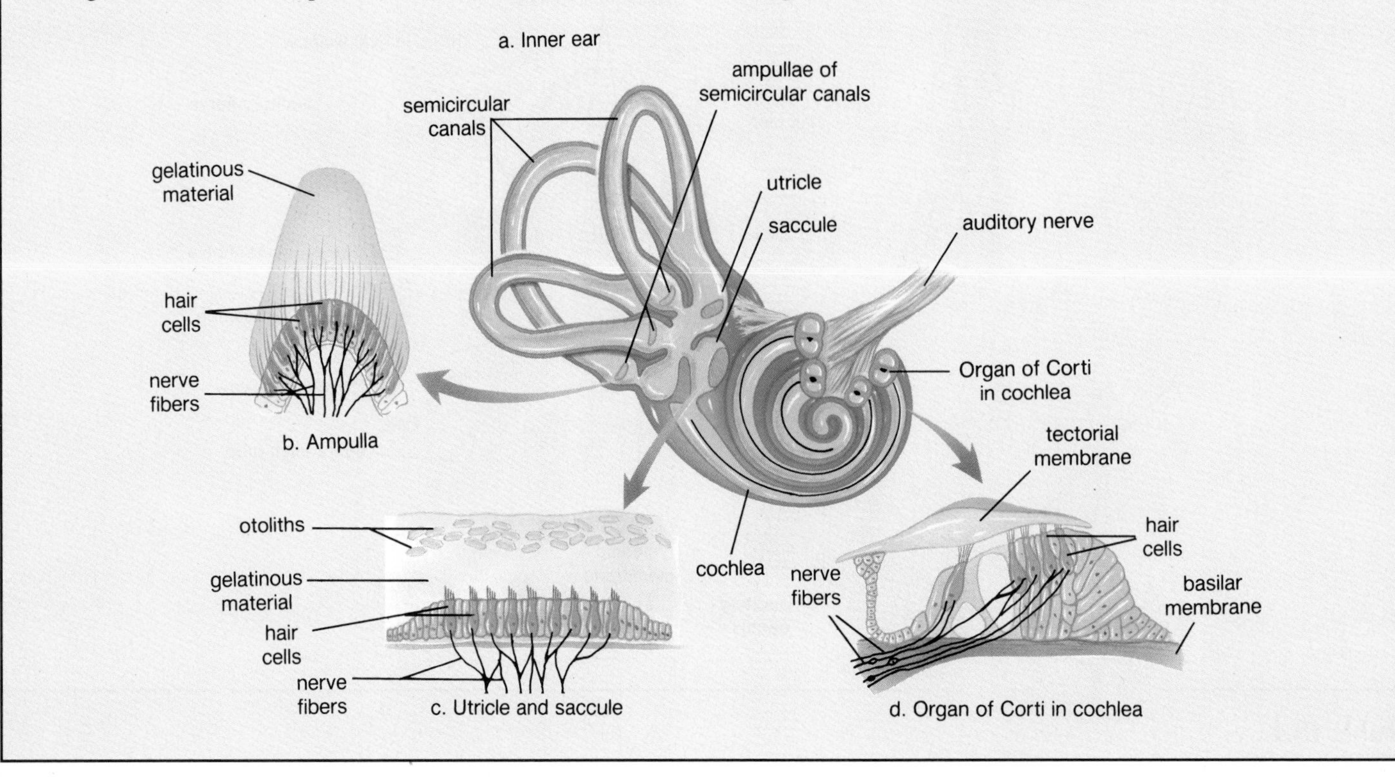

covered by membranes. These openings are called the **oval** and **round windows.** Three small bones are found between the tympanic membrane and the oval window. Collectively called the **ossicles,** individually they are the **hammer** (malleus), the **anvil** (incus), and the **stirrup** (stapes) because their shapes resemble these objects (fig. 18.12). The hammer adheres to the tympanic membrane, and the stirrup touches the oval window. The posterior wall has an opening that leads to the mastoid sinuses of the skull.

A **eustachian tube,** which extends from each middle ear to the nasopharynx, permits equalization of air pressure. Chewing gum, yawning, and swallowing in elevators and airplanes help to move air through the eustachian tubes upon ascent and descent.

Whereas the outer ear and the middle ear contain air, the inner ear is filled with fluid. The **inner ear** (fig. 18.13*a*), anatomically speaking, has 3 areas: the first and the second, the semicircular canals and the vestibule, are concerned with equilibrium; the third, the cochlea, is concerned with hearing.

The **semicircular canals** are arranged so that there is one in each dimension of space. The base of each of the 3 canals, called the **ampulla** (fig. 18.13*b*), is slightly enlarged. Little hair cells with cilia inserted into a gelatinous material are found within the ampullae.

A vestibule, or chamber, lies between the semicircular canals and the cochlea. It contains 2 small membranous sacs called the **utricle** and the **saccule** (fig. 18.13*c*). Both of these sacs contain little hair cells with cilia that protrude into a gelatinous material. Calcium carbonate ($CaCO_3$) granules, or **otoliths,** rest on this material.

The **cochlea** resembles the shell of a snail because it spirals. Three canals are located within the tubular cochlea: the vestibular canal, the **cochlear canal,** and the tympanic canal. Along the length of the basilar membrane, which forms the lower wall of the cochlear canal, are little hair cells with cilia that come into contact with another membrane, the tectorial membrane. The hair cells of the cochlear canal plus the **tectorial membrane** are called the **organ of Corti** (fig. 18.13*d*). This organ sends the nerve impulses to the auditory cortex of the cerebrum, where they are interpreted as sound.

> The ear has 3 major divisions: outer ear, middle ear, and inner ear. The outer ear contains the auditory canal; the middle ear contains the ossicles; and the inner ear contains the semicircular canals, a vestibule, and the cochlea.

Figure 18.14

Receptors for balance. **a.** The ampullae of the semicircular canals contain hair cells with cilia embedded in a gelatinous material. When the head rotates, the material is displaced, and bending of cilia initiates nerve impulses in sensory nerve fibers. This permits dynamic equilibrium. **b.** The utricle and the saccule contain hair cells with cilia embedded in a gelatinous material. When the head bends, otoliths are displaced, causing the gelatinous material to sag and the cilia to bend. The bending initiates nerve impulses in sensory nerve fibers. This permits static equilibrium.

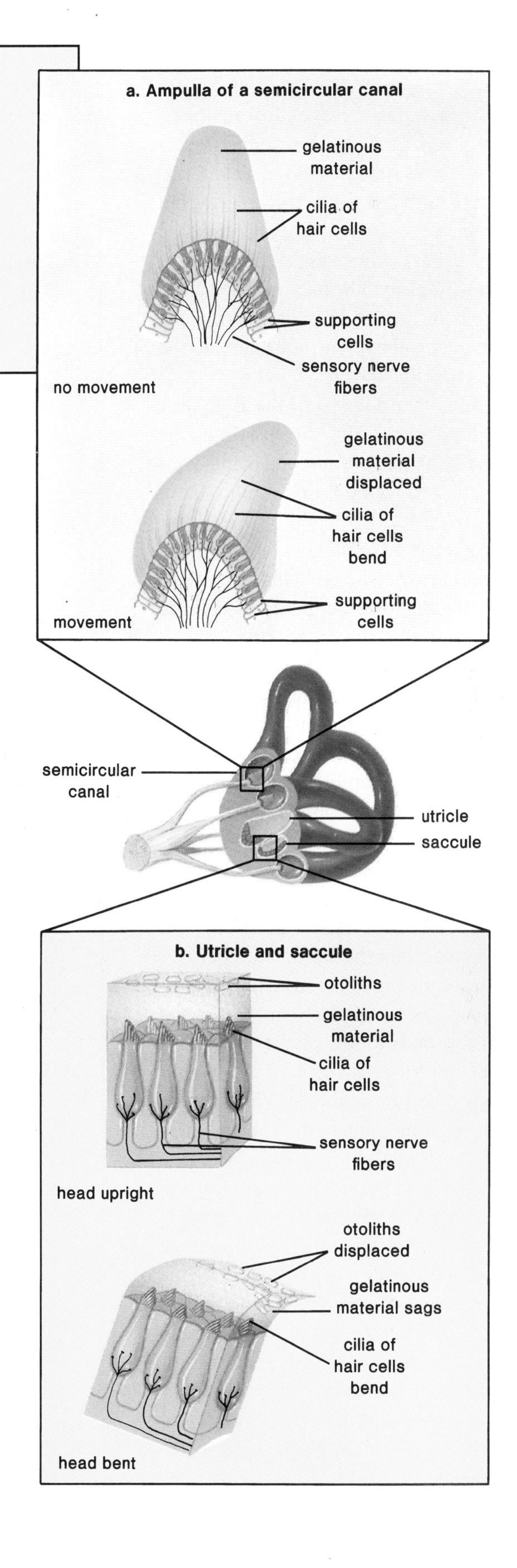

How the Ear Balances and Hears

Balance: Two Kinds

The sense of balance has been subdivided into 2 senses: *dynamic equilibrium,* requiring a knowledge of angular and/or rotational movement, and *static equilibrium,* requiring a knowledge of movement in one plane, either vertical or horizontal.

Dynamic equilibrium is required when the body is moving. At that time, the fluid within the semicircular canals flows over and displaces the gelatinous material within the ampullae (fig. 18.14*a*). This causes the cilia of the hair cells to bend, which initiates nerve impulses that travel to the brain. Continuous movement of the fluid in the semicircular canals causes one form of motion sickness.

When the body is still, the otoliths in the utricle and the saccule rest on the gelatinous material above the hair cells (fig. 18.14*b*). Static equilibrium is required when the body moves horizontally or vertically. At that time, the otoliths are displaced and the gelatinous material sags, bending the cilia of the hair cells beneath. The hair cells then generate nerve impulses that travel to the brain.

Movement of fluid within the semicircular canals contributes to the sense of dynamic equilibrium. Movement of the otoliths within the utricle and the saccule is important for static equilibrium.

Hearing: Rubbing Cilia

The process of hearing begins when sound waves enter the auditory canal (fig. 18.12). Just as ripples travel across the surface of a pond, sound travels by the successive vibrations of molecules. Ordinarily, sound waves do not carry much energy, but when a large number of waves strike the tympanic membrane (eardrum), it moves back and forth (vibrates) ever so slightly. The hammer then takes the pressure from the inner surface of the tympanic membrane and passes it by way of the anvil to the stirrup in such a way that the pressure is multiplied about 20 times as it moves from the tympanic membrane to the stirrup. The stirrup strikes the oval window, causing it to vibrate, and in this way, the pressure is passed to the fluid within the cochlea (fig. 18.15).

If the cochlea is unwound, as shown in figure 18.15, you can see that the vestibular canal connects with the tympanic canal and that pressure waves move from one canal to the other toward the round window, a membrane that bulges to absorb the pressure. As a result of the movement of the fluid within the cochlea, the basilar membrane moves up and down, and the cilia of the hair cells rub against the tectorial membrane. This bending of the cilia initiates nerve impulses that pass by way of the **auditory nerve** to the temporal lobe of the cerebrum, where the impulses are interpreted as a sound.

The organ of Corti is narrow at its base but widens as it approaches the tip of the cochlear canal. Each part of the organ is sensitive to different wave frequencies, or pitch. Near the tip, the organ of Corti responds to low pitches, such as a tuba, and near the base, it responds to higher pitches, such as a bell or a whistle. The neurons from each region along the length of the cochlea lead to slightly different areas in the brain. The pitch sensation we experience depends upon which of these areas of the brain is stimulated.

Volume is a function of the amplitude of sound waves. Loud noises cause the fluid of the cochlea to vibrate to a greater degree, and this, in turn, causes the basilar membrane to move up and down to a greater extent. The resulting increased stimulation is interpreted by the brain as volume. It is believed that tone is an interpretation of the brain based on the distribution of the hair cells stimulated.

Figure 18.15

Receptors for hearing. The organ of Corti is located within the cochlea. In the unwound cochlea, note that the organ of Corti consists of hair cells resting on the basilar membrane with the tectorial membrane above. The arrows represent the pressure waves that move from the oval window to the round window due to the motion of the stirrup. These pressure waves cause the basilar membrane to vibrate and the cilia of at least a portion of the 16,000 hair cells to bend against the tectorial membrane. The generated nerve impulses result in hearing. The micrograph shows the hair cells in the organ of Corti.

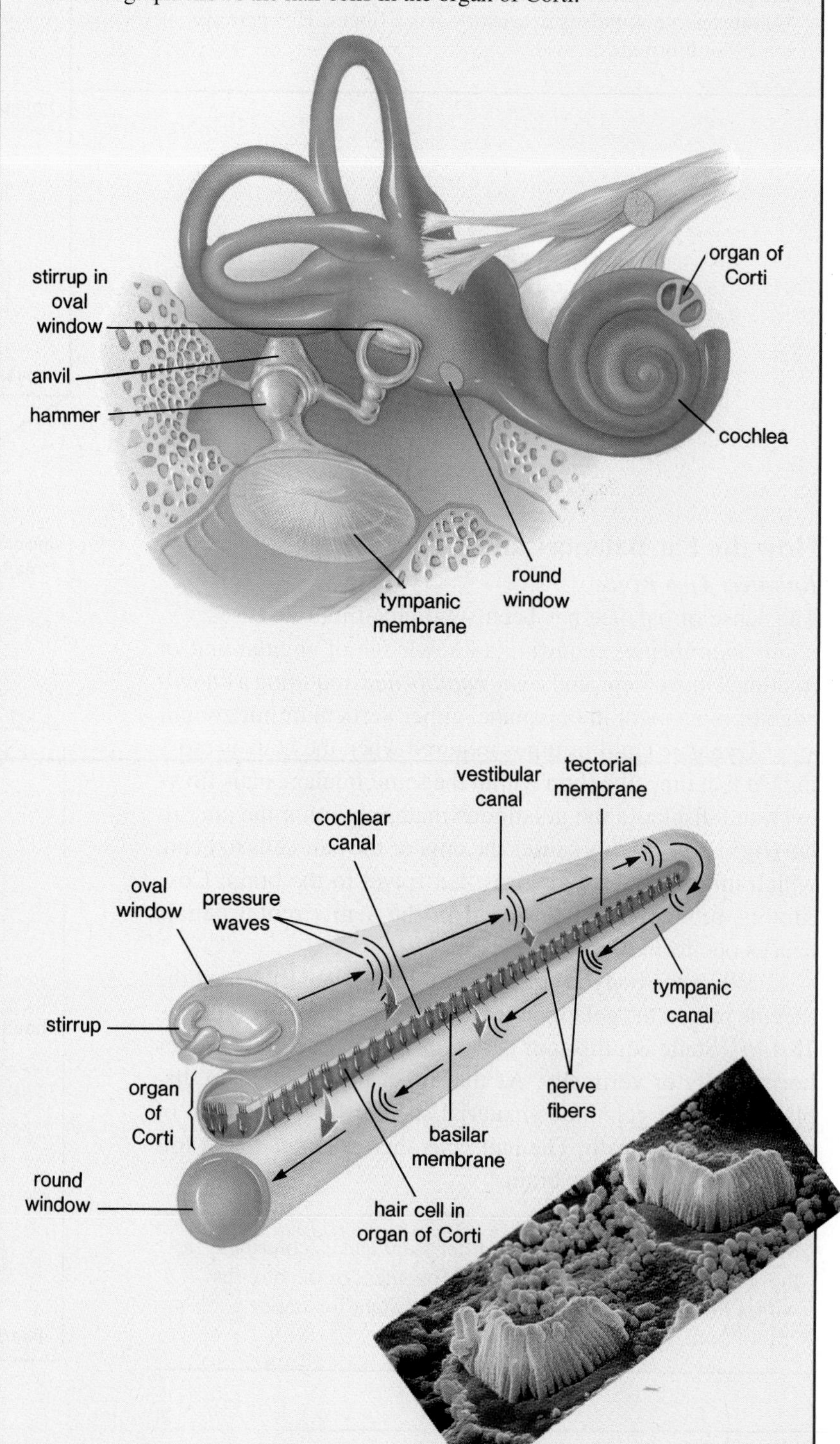

PICTURING THE EFFECTS OF NOISE

"We have an idea of what noise does to the ear," David Lipscomb (of the University of Tennessee Noise Laboratory) says. "There's a pretty clear cause-effect relationship." And these scanning electron micrographs of the cochlea's tiny structures graphically document noise trauma to the inner ear.

Hair cells transmit the mechanical energy of sound waves into those neural impulses that the brain interprets as sound. Loud noise can damage or destroy hair cells, as these scanning electron micrographs illustrate (fig. 18.A).

Hair cells come in 2 varieties: a single row of inner cells and a triple row of outer ones. "Outer cells degenerate before inner cells," notes Clifton Springs, New York, otolaryngologist Stephen Falk. The most subtle change wrought by noise is the development of vesicles, or blisterlike protrusions, along the walls of the hair cells' cilia. Continued assault by noise leads to the rupture of the vesicles and to damage. In addition, the "cuticular plate"—base tissue supporting the cilia—may soften, followed by swelling and ultimate degeneration of hair cells.

But sensory hair cells are not the only structures at risk. Adjacent inner ear cells . . . may undergo vacuolation—development of degenerative empty spaces in cells. Even nerve fibers synapsing at the hair cells' roots may die. In the final phase of noise-induced cochlear damage, the organ of Corti—of which hair cells and supporting cells are a part—is completely denuded of its natural components and is covered by a layer of scar tissue.

Figure 18.A
The hair cells of the organ of Corti are damaged when we listen to loud noises. The electron micrographs show normal hair cells in the organ of Corti before exposure, and hair cells after 24-hour exposure to a noise level typical of rock music (120 db). Note scars where cilia have worn away.

hair cells

Organ of Corti

Outer Hair Cells

before exposure to loud noise

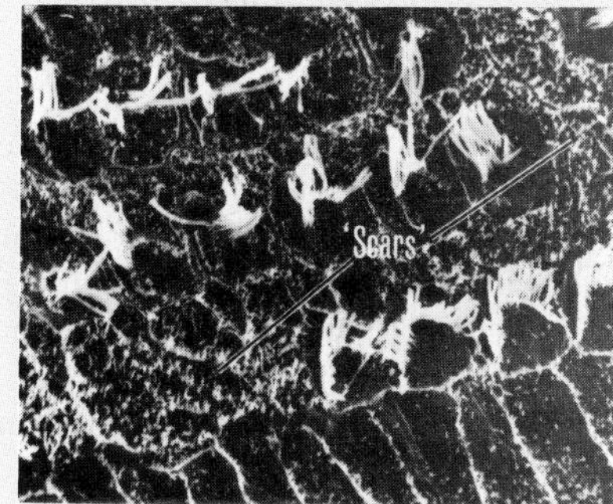

after 24-hours of exposure to loud noise

The sense receptors for sound are hair cells on the basilar membrane (the organ of Corti). When the basilar membrane vibrates, the cilia of the delicate hair cells touch the tectorial membrane, initiating nerve impulses that are transmitted in the auditory nerve to the brain. Do 18.2 Critical Thinking, found at the end of the chapter.

Deafness: Two Main Kinds

There are 2 major types of deafness: *conduction deafness* and *nerve deafness*. Conduction deafness can be due to a congenital defect, such as that which occurs when a pregnant woman contracts rubella (German measles) during the first trimester of pregnancy. (For this reason, every female should be immunized against rubella before the childbearing years.) Conduction deafness can also be due to infections that have caused the ossicles to fuse, restricting their ability to magnify sound waves. Because respiratory infections can spread to the ear by way of the eustachian tubes, every cold and ear infection should be taken seriously.

Nerve deafness most often occurs when cilia on the sense receptors within the cochlea have worn away (see reading above). Since this can happen with normal aging, old people are more likely than young persons to have trouble hearing; however, nerve deafness also occurs when people listen to loud music amplified to or above 130 db. Because the usual types of hearing aids are not helpful for nerve deafness, it is wise to avoid subjecting the ears to any type of continuous loud noise. Costly cochlear implants, which stimulate the auditory nerve directly, are available, but those who have these electronic devices report that the speech they hear is like that of a robot.

SUMMARY

All receptors are the first part of a reflex arc: they initiate nerve impulses that eventually reach the cerebrum, where sensation occurs. General sense organs include those located in skin and proprioceptors located in joints. Special sense organs include the taste buds, nose, eyes, and ears where receptors are concentrated in specific regions.

Vision is dependent on the eye, the optic nerve, and the visual cortex of the cerebrum. The rods, receptors for vision in dim light, and the cones, receptors that depend on bright light and provide color and detailed vision, are located in the retina, the inner layer of the eyeball. The cornea, the humors, and especially the lens bring the light rays to focus on the retina. To see a close object, accommodation occurs as the lens rounds up. Due to the optic chiasma, both sides of the brain must function together to give three-dimensional vision.

Hearing is a specialized sense dependent on the ear, the auditory nerve, and the auditory cortex of the cerebrum. The outer and middle portions of the ear simply convey and magnify the sound waves that strike the oval window. Its vibrations set up pressure waves within the cochlea, which contains the organ of Corti, consisting of hair cells with the tectorial membrane above. When the cilia of the hair cells strike this membrane, nerve impulses are initiated that finally result in hearing.

The ear also contains receptors for our sense of balance. Dynamic equilibrium is dependent on the stimulation of hair cells within the ampullae of the semicircular canals. Static equilibrium relies on the stimulation of hair cells by otoliths within the utricle and the saccule.

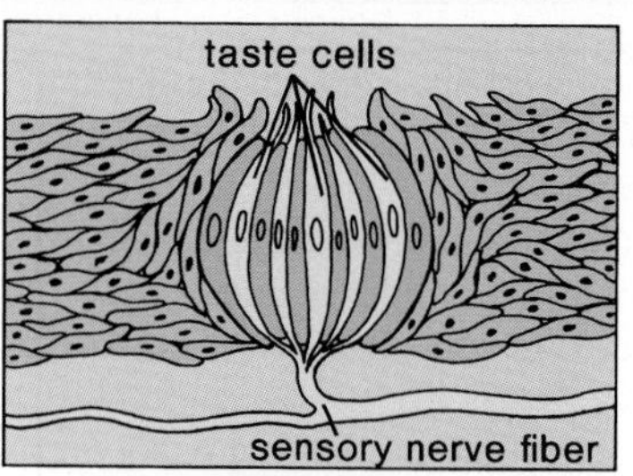

Taste Bud

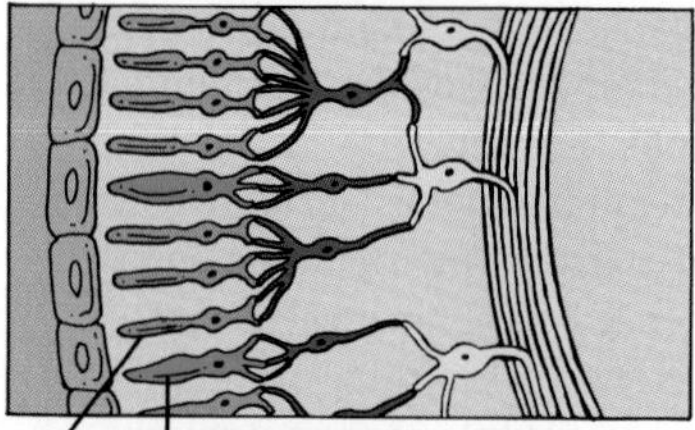

Retina

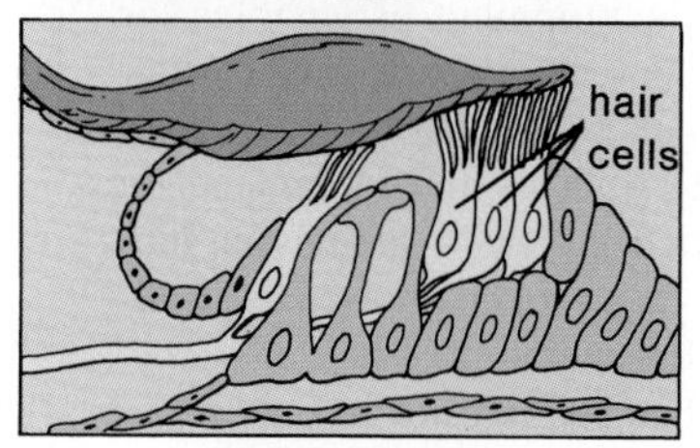

Organ of Corti

STUDY QUESTIONS

In order to practice **writing across the curriculum,** students should write out the answers to any or all of the study questions. The study questions are sequenced in the same order as the text.

1. Which sense organs are categorized as general, and which are categorized as special? (pp. 323–324)
2. Discuss the receptors of the skin and the joints. (pp. 323–324)
3. Discuss the chemoreceptors. (pp. 325–326)
4. Describe the anatomy of the eye (p. 326), and explain focusing and accommodation. (p. 330)
5. Describe vision in dim light. What chemical reaction is responsible for vision in dim light? (p. 331) Discuss color vision. (p. 332)
6. Relate the need for corrective lenses to 3 possible eye shapes. (p. 333) Discuss bifocals. (p. 333)
7. Describe the anatomy of the ear and how we hear. (pp. 335–338)
8. Describe the role of the utricle, the saccule, and the semicircular canals in balance. (p. 337)
9. Discuss the 2 causes of deafness, including why young people frequently suffer loss of hearing. (p. 339)

OBJECTIVE QUESTIONS

1. _______ is the sense of knowing the location of body parts.
2. Taste cells and olfactory cells are _______ because they are sensitive to chemicals in the air and in food.
3. The receptors for vision, the _______ and the _______, are located in the _______, the inner layer of the eye.
4. The cones give us _______ vision and work best in _______ light.
5. The lens _______ for viewing close objects.
6. People who are nearsighted cannot see objects that are _______. A(n) _______ lens restores this ability.
7. The ossicles are the _______, the _______, and the _______.
8. The semicircular canals are involved in our sense of _______.
9. The organ of Corti is located in the _______ canal of the _______.
10. Vision, hearing, taste, and smell do not occur unless nerve impulses reach the proper portion of the _______.
11. Label this diagram of the eye.

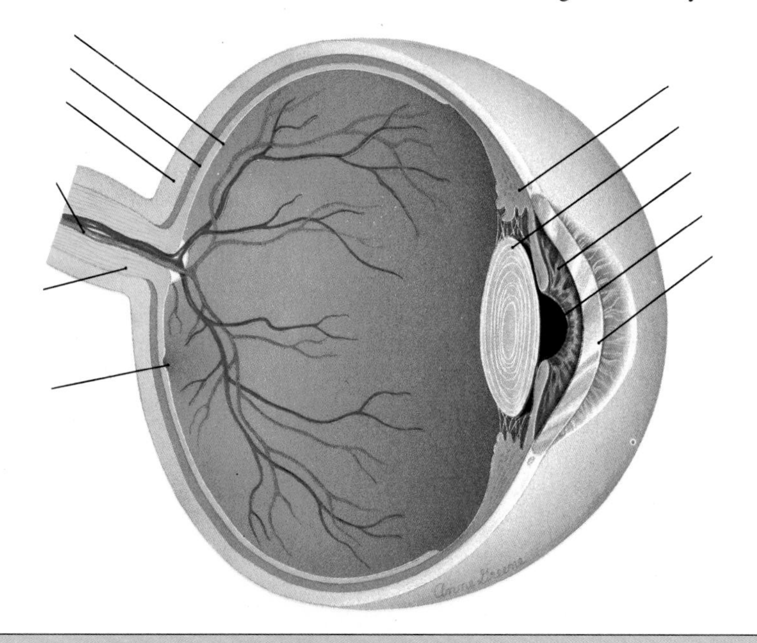

CRITICAL THINKING

In order to practice **writing across the curriculum,** students should write out the answers to any or all of the critical thinking questions. Suggested answers to the critical thinking questions are in appendix E.

18.1

1. Devise a categorization for the parts of the eye (table 18.3). Justify your system of categorization. When a person needs glasses, which of your categories is deficient?
2. Phytochrome (p. 144), chlorophyll (p. 85), and retinal (p. 331) are all pigments. What features do they have in common?
3. The pineal gland, located on the edge of the third ventricle of the brain, is sensitive to the length of both the day and the night. Hypothesize how the pineal gland detects when it is day or night.

18.2

1. Fishes do not have ears, but they do have a lateral line, which contains mechanoreceptors that are sensitive to fluid pressure waves. What fossil evidence is needed to support the hypothesis that the lateral line of fishes evolved into the mammalian ear?
2. What part of the human ear (outer, middle, inner) would you expect to have evolved from the lateral line? What parts of the human ear would you expect to have evolved from other structures or to have been "added on" considering that humans are terrestrial? Why?

SELECTED KEY TERMS

accommodation lens adjustment in order to see close objects.
choroid (ko´roid) the vascular, pigmented middle layer of the eyeball.
ciliary body (sil´ e-er˝e) structure associated with the choroid layer that contains the ciliary muscle, which controls the shape of the lens of the eye.
cochlea that portion of the inner ear that resembles a snail's shell and contains the organ of Corti, the sense organ for hearing.
cone bright-light receptor in the retina of the eye that detects color and provides visual acuity.
fovea centralis (fo´ ve-ah sen-tral´ is) region of the retina consisting of densely packed cones that is responsible for the greatest visual acuity.
lens a clear membranelike structure found in the eye behind the iris; brings objects into focus.
organ of Corti a portion of the inner ear that contains the receptors for hearing.
ossicle one of the small bones of the middle ear—hammer, anvil, stirrup.
otolith calcium carbonate granule associated with ciliated cells in the utricle and the saccule.
proprioception the sense of knowing the position of the limbs.
retina (ret´ ĭ-nah) the innermost layer of the eyeball that contains the rods and the cones.
rhodopsin (ro-dop´sin) visual purple, a pigment found in the rods of a type of receptor in the retina of the eye.
rod dim-light receptor in the retina of the eye that detects motion but not color.
saccule (sak´ ūl) a saclike cavity that makes up part of the membranous labyrinth of the inner ear; contains receptors for static equilibrium.
sclera white, fibrous outer layer of the eyeball.
semicircular canal tubular structure within the inner ear that contains the receptors responsible for the sense of dynamic equilibrium.
tympanic membrane (tim-pan´ ik mem´brān) membrane located between the outer ear and the middle ear that receives sound waves; the eardrum.
utricle (u´ tre-k´l) saclike cavity that makes up part of the membranous labyrinth of the inner ear; contains receptors for static equilibrium.

19

HORMONES

Chapter Concepts

1.
The endocrine system utilizes chemical messengers called hormones to bring about coordination of body parts. 344, 364

2.
The hypothalamus, a part of the brain, controls the function of the pituitary gland. 346, 364

3.
A negative feedback mechanism controls the secretions of the pituitary gland and other glands. 351, 364

4.
The malfunction of the endocrine glands can bring about a dramatic change in appearance and cause early death. 352, 364

5.
Environmental signals are not unique to the endocrine system and most likely are present whenever one biological element influences the behavior or the chemistry of another element. 362, 364

Polarized light micrograph of progesterone crystals, ×10

Chapter Outline

Endocrine Glands

Two major systems, the nervous system (see chap. 16) and the **endocrine system,** coordinate the various activities of body parts. Both systems utilize chemical messengers (chemical signals) to fulfill their functions, and therefore, a certain amount of overlap between the nervous and endocrine systems is to be expected. The systems evolved together, no doubt making occasional use of the same chemical messengers and communicating not only with other systems but with each other as well.

The nervous system, as discussed in chapter 16, utilizes neurotransmitter substances, which are released by one neuron and influence the excitability of other neurons. The human endocrine system utilizes **hormones**, chemical messengers that are produced in one body region but affect a different body region (fig. 19.1). These glands, called **endocrine glands,** can be contrasted with exocrine glands. Exocrine glands secrete their products into ducts for transport into body cavities; for example, the salivary glands send saliva into the mouth by way of the salivary ducts. Endocrine glands are ductless; they secrete their hormones directly into the bloodstream for distribution throughout the body. The cells of these glands abut capillaries, the thin walls of which allow easy entrance of hormones.

The nervous system reacts quickly to external and internal stimuli; you rapidly pull your hand away from a hot stove, for example. The endocrine system is slower than the nervous system because it takes time for a hormone to travel through the circulatory system to its *target organ.* You might think from the use of this terminology that the hormone is seeking out a particular organ. Quite the contrary is true, however; the organ is awaiting the arrival of the hormone. The cells that can react to a hormone have specific receptors that combine with the hormone in a lock-and-key manner. Therefore, certain cells respond to one hormone and not to another, depending on their receptors.

Endocrine glands secrete hormones into the bloodstream for transport to target organs.

Figure 19.1
Anatomical location of major endocrine glands in the body.

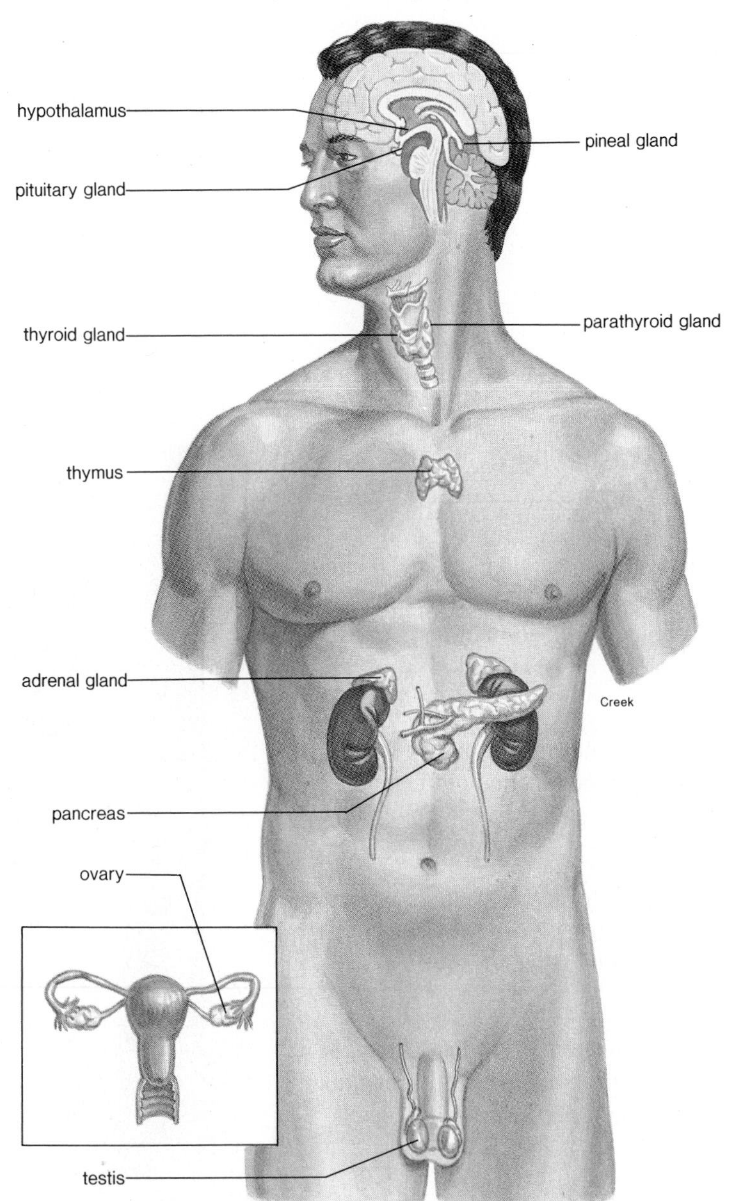

How Hormones Work

Some hormones are either *peptides or proteins.* These molecules are coded for by genes and are synthesized at the ribosomes. Eventually, they are packaged into vesicles at the Golgi apparatus and are secreted at the cell membrane (see fig. 3.6). There are some other hormones, such as norepinephrine, that are derived from the amino acid tyrosine; their production requires only a series of metabolic reactions within the cytoplasm.

For the purpose of discussing hormonal action, we can group the hormones that are peptides and proteins with those derived from amino acids. We will call all of these peptide hormones.

Still other hormones are *steroids.* Steroid hormones are produced by the adrenal glands, the ovaries, and the testes (fig. 19.1). Steroids are derived from cholesterol (see fig. 2.28) by a series of metabolic reactions. These hormones are stored in fat droplets in the cell cytoplasm until their release at the cell membrane.

A N I M A L • B I O L O G Y

INSECT GROWTH HORMONES

All animals use chemical messengers to regulate the activities of their cells. These messengers are now being identified in a wide variety of invertebrate animals. The work on insect growth hormones began in the 1930s, and therefore they have been studied for some time.

Insects, like other arthropods, have an exoskeleton that hinders growth unless it is shed periodically. Typically, the larvae of insects molt (shed their skeleton) a number of times before they pupate and metamorphose into an adult (see figure). V. B. Wigglesworth showed in the 1930s that the insect brain was necessary for maturation to take place because it produced a hormone appropriately called *brain hormone*. Brain hormone stimulates the prothoracic gland, which lies in the region just behind the head, to secrete *ecdysone*, a steroid hormone. This hormone is also called the molt-and-maturation hormone because it promotes both molting and maturation. During larval stages, however, response to ecdysone is modified by the action of *juvenile hormone*. If juvenile hormone is present, the insect molts into another larval form; if it is minimally present, the insect pupates; and if it is absent, the insect undergoes metamorphosis into an adult.

Some plants produce compounds that are similar or identical to either ecdysone or juvenile hormone. These compounds apparently protect the plants by disrupting the development of the insects that feed upon them. Work is underway to extract these compounds for use as insecticides.

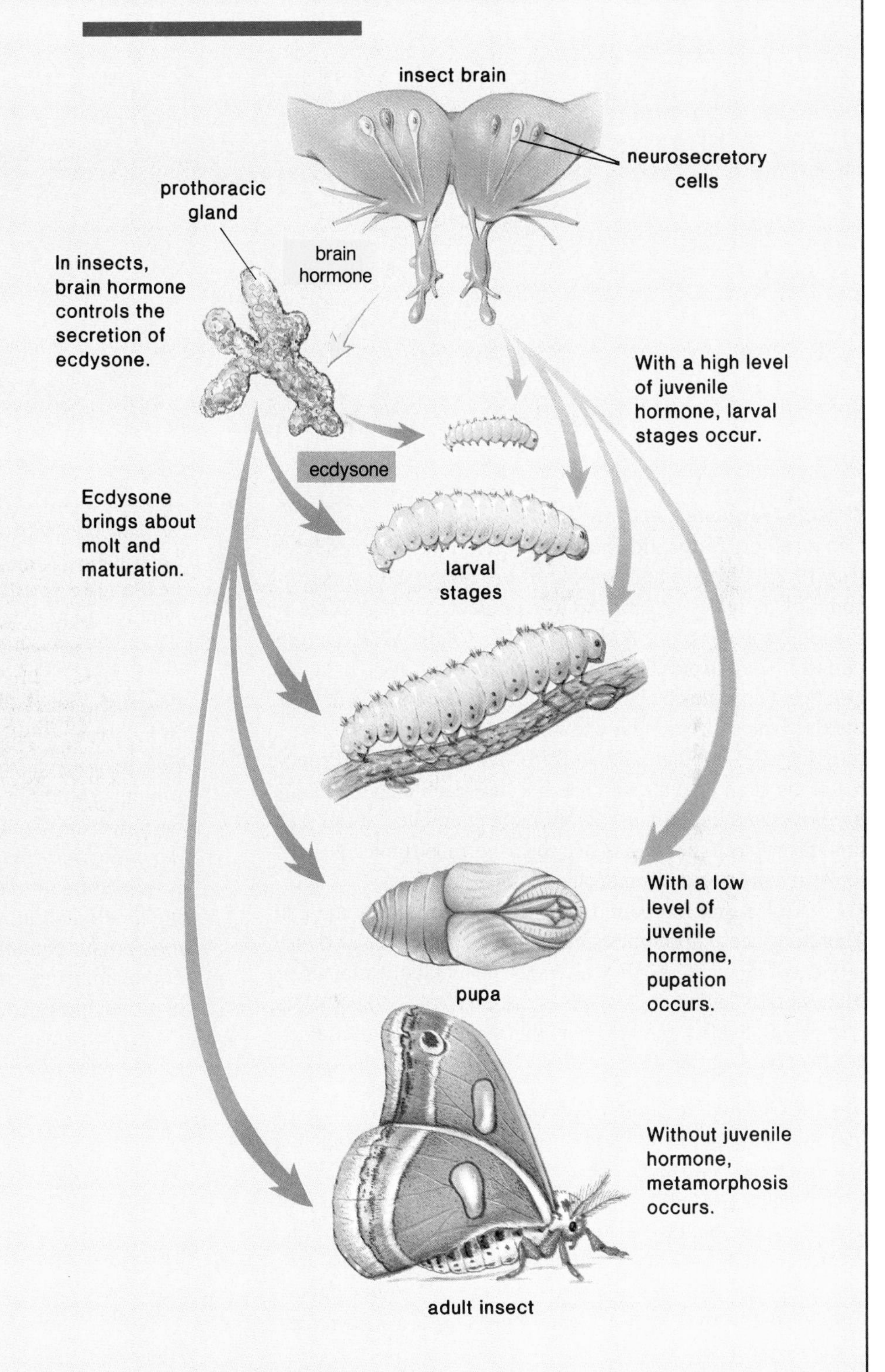

Figure 19.2

Cellular activity of hormones. **a.** Peptide hormones combine with receptors located on the cell membrane. This promotes the production of cyclic AMP, which in turn leads to activation of a particular enzyme. **b.** Steroid hormones pass through the cell membrane to combine with receptors; the complex activates certain genes, leading to protein synthesis.

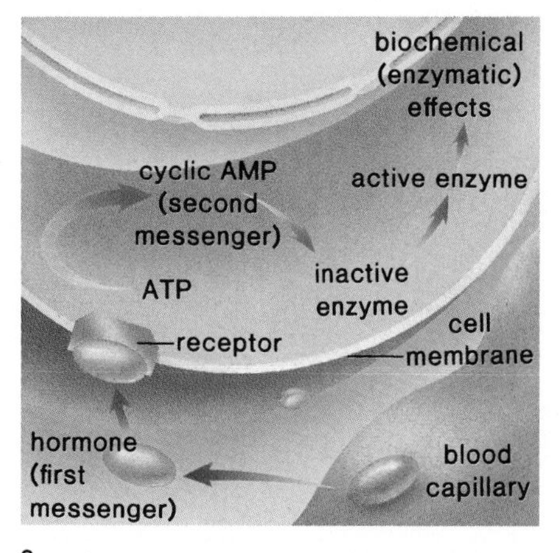

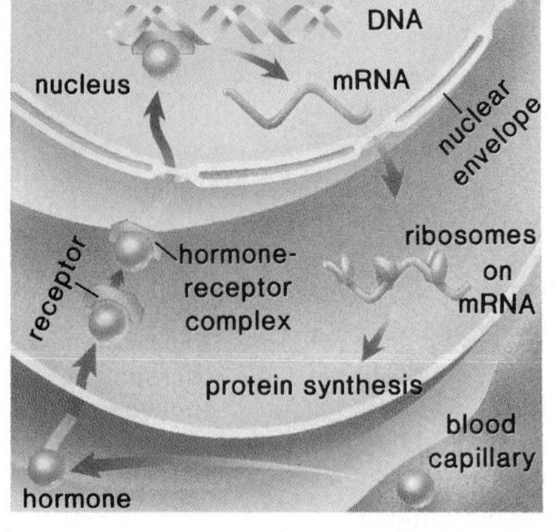

Peptide Hormones: Activate Existing Enzymes

The receptors for peptide hormones are on the cell membrane (fig. 19.2*a*). When a hormone binds to a receptor, the resulting complex activates an enzyme that produces cyclic adenosine monophosphate (cyclic AMP or cAMP). Cyclic AMP is a compound made from ATP, but it contains only one phosphate group, which is attached to adenosine at 2 locations. Cyclic AMP then activates a particular enzyme in the cell; this enzyme in turn activates another, and so forth. This series of enzymatic reactions is an enzyme cascade because each enzyme can be used over and over again; therefore, at every step, more reactions take place. The binding of a single hormone molecule eventually results in a thousandfold response.

Notice that the peptide hormone never enters the cell. Therefore, these hormones sometimes are called the *first messenger,* while cyclic AMP, which sets the metabolic machinery in motion, is called the *second messenger* (fig. 19.2*a*). Other types of molecules can also play the role of a second messenger in cells.

Steroid Hormones: Call for New Enzymes

Steroid hormones do not bind to cell membrane receptors; they can enter the cell freely because they are lipids (fig. 19.2*b*). Once inside, steroid hormones bind to receptors in the cytoplasm (or the nucleus). The hormone-receptor complex then enters the nucleus, where it binds with chromatin at a location that promotes activation of particular genes. Protein synthesis follows. In this manner, steroid hormones lead to the synthesis of new enzymes.

Steroids act more slowly than peptides because more time is needed to synthesize new proteins than to activate enzymes that are already present in the cell. Steroids have a more sustained effect on the metabolism of the cell than peptide hormones.

Hormones are chemical messengers that influence the metabolism of the receiving cell. Peptide hormones activate existing enzymes in the cell, and steroid hormones bring about the synthesis of new enzymes.

Negative Feedback: Dampens Stimulus

In a self-regulating negative feedback mechanism, an adaptive response dampens or even cancels the stimulus that brought about the response (see fig. 10.11). The function of an endocrine gland is usually controlled by negative feedback. An endocrine gland can be sensitive to either the condition it is regulating or to the blood level of the hormone it is producing. For example, when blood is concentrated, the hypothalamus produces a hormone that causes blood dilution. The hypothalamus then stops producing the hormone. On the other hand, the pituitary gland produces a hormone that stimulates the thyroid gland. When the blood level of a hormone produced by the thyroid rises, the pituitary gland no longer stimulates the thyroid gland.

The Hypothalamus and the Pituitary Gland

The hypothalamus, located beneath the thalamus in the third ventricle of the brain, regulates the internal environment. For example, it helps to control heart rate, body temperature, and water balance, as well as the activity of the **pituitary gland.** The pituitary gland (only about 1 cm in diameter) lies just below the hypothalamus (fig. 19.1) and is divided into 2 portions: the **posterior pituitary** and the **anterior pituitary.**

Figure 19.3

The hypothalamus produces 2 hormones, ADH and oxytocin, which are stored in and secreted by the posterior pituitary.

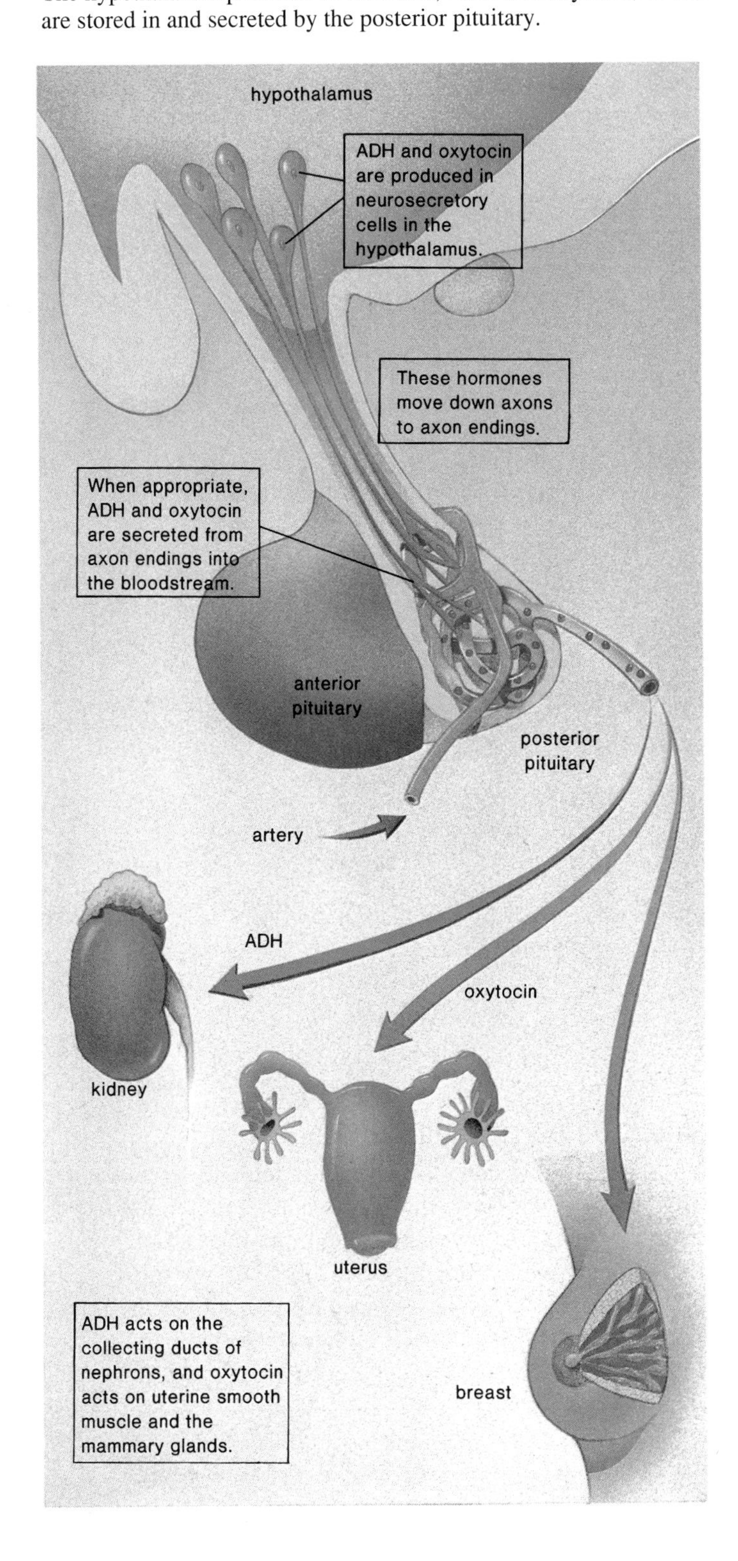

Figure 19.4

Regulation of ADH secretion. Neurons in the hypothalamus are sensitive to the osmolarity of blood. When blood is concentrated, they send signals to the neurosecretory cells in the hypothalamus. These release ADH from their axon endings. ADH increases the permeability of the collecting ducts in the kidneys so that more water is reabsorbed. Once blood is diluted, ADH is no longer secreted. This is an example of control by a negative feedback mechanism.

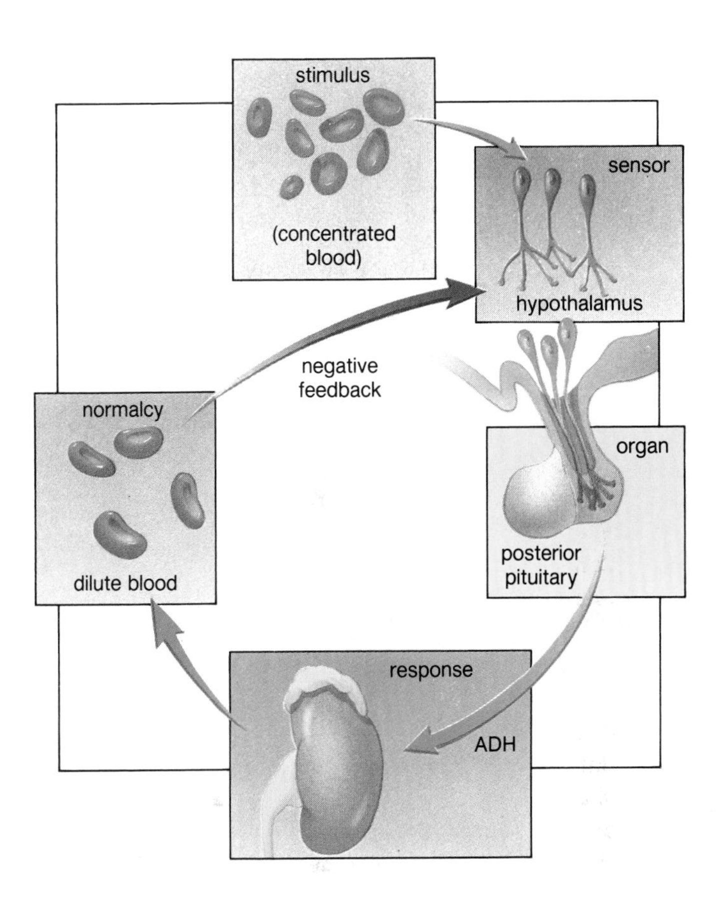

Posterior Pituitary: Stores 2 Hormones

The posterior pituitary is connected to the hypothalamus by means of a stalklike structure. Neurons in the hypothalamus, called *neurosecretory cells,* respond to neurotransmitter substances and produce the hormones that are stored in and released from the posterior pituitary (fig. 19.3). These hormones are *antidiuretic hormone* (*ADH*), sometimes called vasopressin, and *oxytocin.* ADH, as discussed in chapter 15, promotes the reabsorption of water from the collecting duct, a portion of the nephron. When osmoreceptors in the hypothalamus determine that blood is too concentrated, ADH is released into the bloodstream from the axon endings in the posterior pituitary. Once blood is diluted, the hormone is no longer released. This is an example of control by a negative feedback mechanism (fig. 19.4).

Figure 19.5
Hypothalamus and anterior pituitary. The hypothalamus controls the secretions of the anterior pituitary.

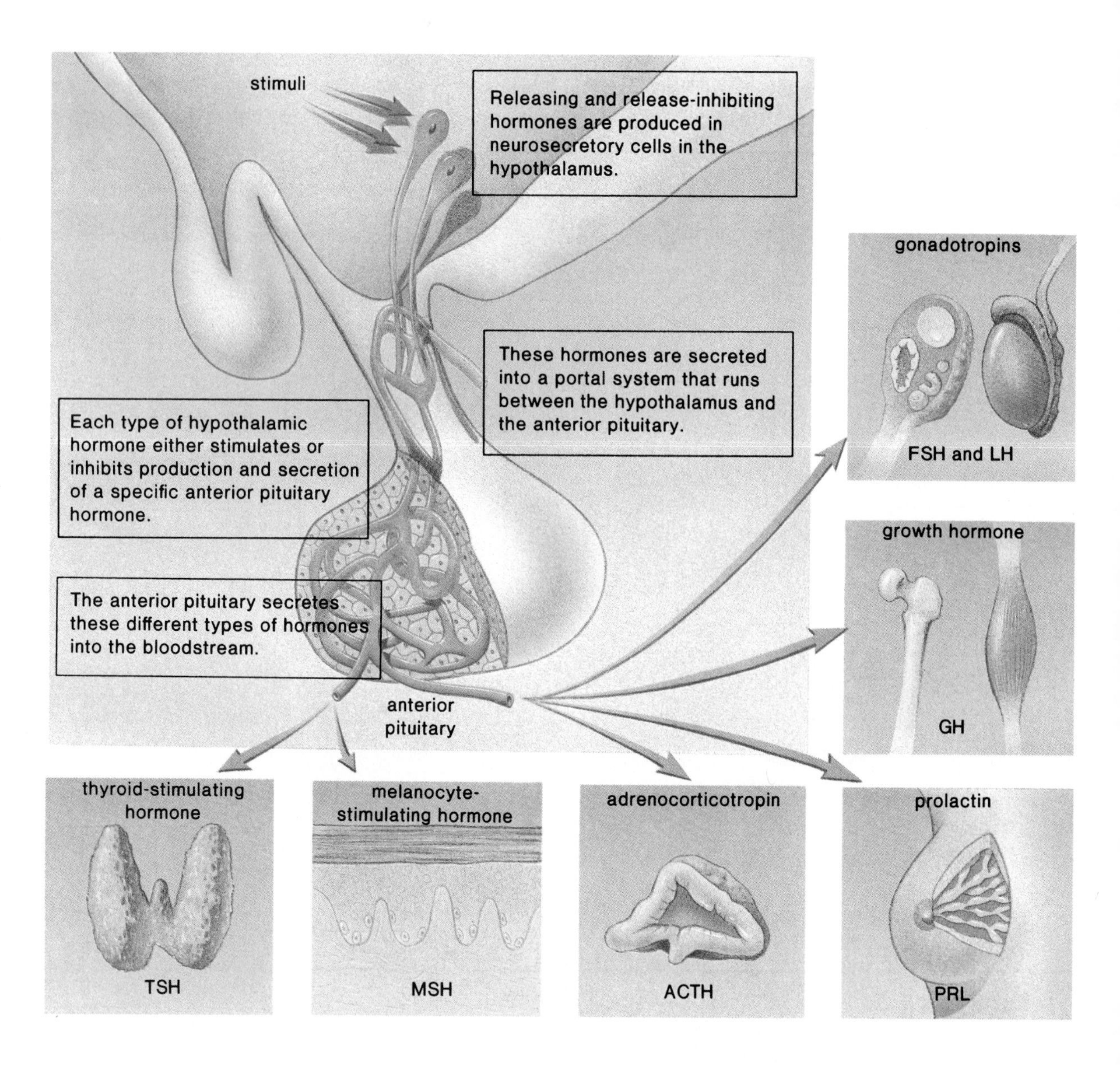

The inability to produce **antidiuretic hormone (ADH)** causes **diabetes insipidus** (watery urine), in which a person produces copious amounts of urine with a resultant loss of salts from blood. This condition can be corrected with the administration of ADH.

Oxytocin is the other hormone that is made in the hypothalamus and stored in the posterior pituitary. Oxytocin causes the uterus to contract and is used to artificially induce labor. It also stimulates the release of milk from the mother's mammary glands when her baby is nursing.

It is appropriate to note that the neurosecretory cells in the hypothalamus provide an example of a way the nervous system and the endocrine system are joined. This topic is discussed again later.

The posterior pituitary stores and releases 2 hormones, ADH and oxytocin, both of which are produced by neurosecretory cells in the hypothalamus.

Anterior Pituitary: Master Gland

The hypothalamus controls the anterior pituitary by producing hypothalamic-releasing and release-inhibiting hormones. For example, there is a thyroid-releasing hormone (TRH) and a thyroid release-inhibiting hormone (TRIH).

Releasing and release-inhibiting hormones are transported from the hypothalamus to the anterior pituitary by way of a portal system connecting the 2 organs (fig. 19.5).

The anterior pituitary is controlled by the hypothalamic-releasing hormones and hypothalamic release-inhibiting hormones. These are produced in neurosecretory cells in the hypothalamus and pass to the anterior pituitary by way of a portal system.

Hormones produced by the anterior pituitary are listed in table 19.1. Three of these hormones have a direct effect on the body. **Growth hormone (GH),** or somatotropin, dramatically

Table 19.1
The Principal Endocrine Glands and Their Hormones

Endocrine Gland	Hormone Released	Target Tissues/ Organ	Chief Function of Hormone	Disorders (Too Much/ Too Little)
Hypothalamus	Hypothalamic-releasing and release-inhibiting hormones	Anterior pituitary	Regulate anterior pituitary hormones	*See* anterior pituitary
Posterior pituitary (storage of hypothalamic hormones)	Antidiuretic hormone (ADH, vasopressin)	Kidneys	Stimulates water reabsorption by kidneys	Diverse*/diabetes insipidus
	Oxytocin	Uterus, mammary glands	Stimulates uterine muscle contraction and release of milk by mammary glands	
Anterior pituitary	Thyroid-stimulating hormone (TSH, thyrotropic)	Thyroid	Stimulates thyroid	*See* thyroid
	Adrenocorticotropic hormone (ACTH)	Adrenal cortex	Stimulates adrenal cortex	*See* adrenal cortex
	Gonadotropic hormones	Gonads		
	Follicle-stimulating hormone (FSH)		Controls egg and sperm production	
	Luteinizing hormone (LH)		Controls sex hormone production	*See* testis and ovary
	Prolactin (PRL)	Mammary glands	Stimulates milk production and secretion	
	Growth hormone (GH, somatotropin)	Soft tissues, bones	Stimulates cell division, protein synthesis, and bone growth	Giantism, acromegaly/ dwarfism
	Melanocyte-stimulating hormone (MSH)	Melanocytes in skin	Regulates skin color in lower vertebrates; unknown function in humans	
Thyroid	Thyroxin	All tissues	Increases metabolic rate; helps to regulate growth and development	Exophthalmic goiter/ simple goiter, myxedema, cretinism
	Calcitonin	Bones, kidneys, intestine	Lowers blood calcium level	Tetany/weak bones
Parathyroids	Parathyroid hormone (PTH)	Bones, kidneys, intestine	Raises blood calcium level	Weak bones/tetany
Adrenal medulla	Epinephrine and norepinephrine	Cardiac and other muscles	Stimulate fight or flight reactions; raise blood glucose level	
Adrenal cortex	Glucocorticoids (e.g., cortisol)	All tissues	Raise blood glucose level; stimulate breakdown of protein	
	Mineralocorticoids (e.g., aldosterone)	Kidneys	Stimulate kidneys to reabsorb sodium and to excrete potassium	Cushing syndrome/ Addison disease
	Sex hormones	Sex organs, skin, muscles, bones	Stimulate development of secondary sex characteristics (particularly in male)	
Pancreas	Insulin	Liver, muscles, adipose tissue	Lowers blood glucose level; promotes formation of glycogen, proteins, and fats	Shock/diabetes mellitus
	Glucagon	Liver, muscles, adipose tissue	Raises blood glucose level; promotes breakdown of glycogen, proteins, and fats	
Gonads				
Testes	Androgens (testosterone)	Sex organs, skin, muscles, bones	Stimulate spermatogenesis; develop and maintain secondary male sex characteristics	Diverse/feminization
Ovaries	Estrogen and progesterone	Sex organs, skin, muscles, bones	Stimulate growth of uterine lining; develop and maintain secondary female sex characteristics	Diverse/masculinization
Thymus	Thymosins	T lymphocytes	Stimulates maturation of T lymphocytes	
Pineal gland	Melatonin	Circadian rhythms	Involved in circadian and circannual rhythms; possibly involved in maturation of sex organs	

*The word *diverse* in this table means that the symptoms have not been described as a syndrome in the medical literature.

Figure 19.6
Giantism. Sandy Allen, one of the world's tallest women due to a higher than usual amount of GH, produced by the anterior pituitary.

affects physical appearance since it determines the height of the individual (fig. 19.6). If little or no GH is secreted by the anterior pituitary during childhood, the person will probably become a pituitary dwarf—although of perfect proportions, quite small in stature. If too much GH is secreted, the person will probably become a giant. Giants usually have poor health, primarily because GH has a secondary effect on blood sugar level, promoting an illness called diabetes (sugar) mellitus, discussed later.

GH promotes cell division, protein synthesis, and bone growth. It stimulates the transport of amino acids into cells and increases the activity of ribosomes, both of which are essential to protein synthesis. In bones, GH promotes growth of the cartilaginous disks and causes osteoblasts to form bone (p. 308). Evidence suggests that the effects on cartilage and bone may actually be due to hormones called somatomedins, which are released by the liver. GH causes the liver to release somatomedins.

If the production of GH increases in an adult after full height has been attained, only certain bones respond. These are the bones of the jaw, the eyebrow ridges, the nose, the fingers, and the toes. When these begin to grow, the person takes on a slightly grotesque look, with enlargement of facial features and huge fingers and toes, a condition called **acromegaly** (fig. 19.7).

Figure 19.7
Acromegaly. The condition is caused by the overproduction of GH in the adult. It is characterized by an enlargement of the bones in the face, the fingers, and the toes of an adult. **a.** At age 20, this individual was normal. **b.** At age 24, there is some enlargement of the nose, the jaw, and the fingers.

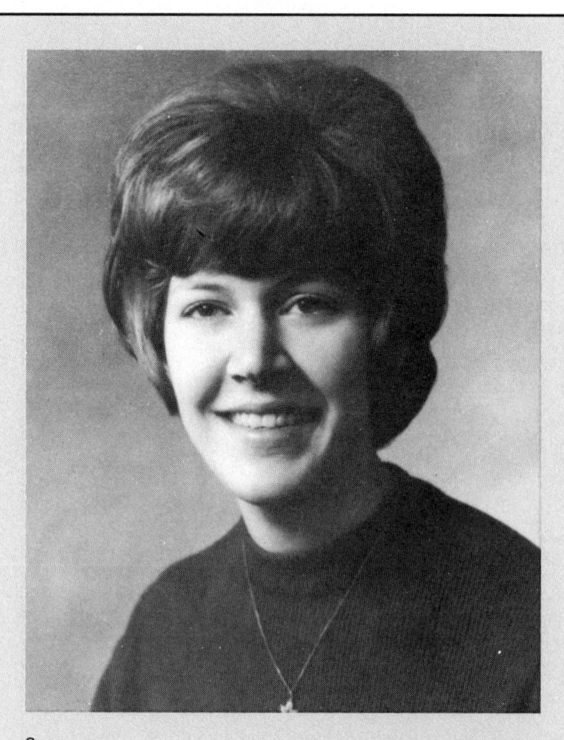

a.

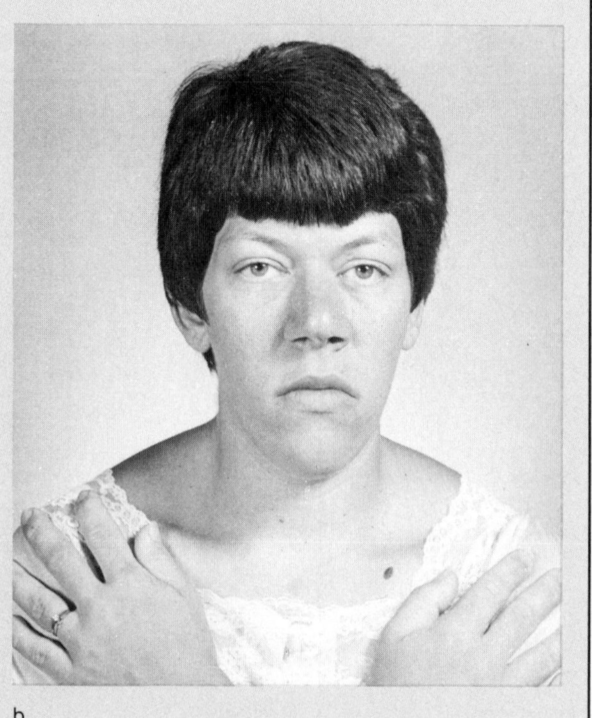

b.

Prolactin (PRL) is produced in quantity only after childbirth. It causes the mammary glands in the breasts to develop and to produce milk.

Melanocyte-stimulating hormone (MSH) regulates skin color in lower vertebrates, but its function in humans is obscure. However, it is derived from a molecule that is also the precursor for both adrenocorticotropic hormone (ACTH) and the anterior pituitary endorphins. These endorphins are structurally and functionally similar to the endorphins produced in brain cells.

GH and PRL are 2 hormones produced by the anterior pituitary. GH influences the height of children, and overproduction brings about a condition called acromegaly in adults. PRL promotes milk production after childbirth.

Other Hormones Produced by the Anterior Pituitary

The anterior pituitary sometimes is called the *master gland* because it controls the secretion of certain other endocrine glands (fig. 19.5). As indicated in table 19.1, the anterior pituitary secretes the following hormones, all of which have an effect on other glands:

1. **Thyroid-stimulating hormone (TSH),** a hormone that stimulates the thyroid
2. **Adrenocorticotropic hormone (ACTH),** a hormone that stimulates the adrenal cortex
3. **Gonadotropic hormones** (FSH and LH), which stimulate the gonads—the testes in males and the ovaries in females.

TSH causes the thyroid to produce thyroxin; ACTH causes the adrenal cortex to produce cortisol; and gonadotropic hormones cause the gonads to secrete sex hormones. Notice that it is now possible to indicate a three-tiered relationship among the hypothalamus, the anterior pituitary, and the other endocrine glands. The hypothalamus produces hormones that control the anterior pituitary, and the anterior pituitary produces hormones that control the thyroid, the adrenal cortex, and the gonads. Figure 19.8 illustrates the negative feedback mechanism that controls the activity of all these glands.

The hypothalamus, the anterior pituitary, and the other endocrine glands controlled by the anterior pituitary are all involved in a self-regulating negative feedback system.

Figure 19.8

The hypothalamus-pituitary-thyroid control relationship. TRH (thyroid-releasing hormone) stimulates the anterior pituitary, and TSH (thyroid-stimulating hormone) stimulates the thyroid to secrete thyroxin. The level of thyroxin in the body is negatively controlled in 3 ways: (**a**) The level of TSH exerts feedback control over the hypothalamus; (**b**) the level of thyroxin exerts feedback control over the anterior pituitary; and (**c**) the level of thyroxin exerts feedback control over the hypothalamus. In this way, thyroxin controls its own secretion. Cortisol and sex hormone levels are controlled similarly.

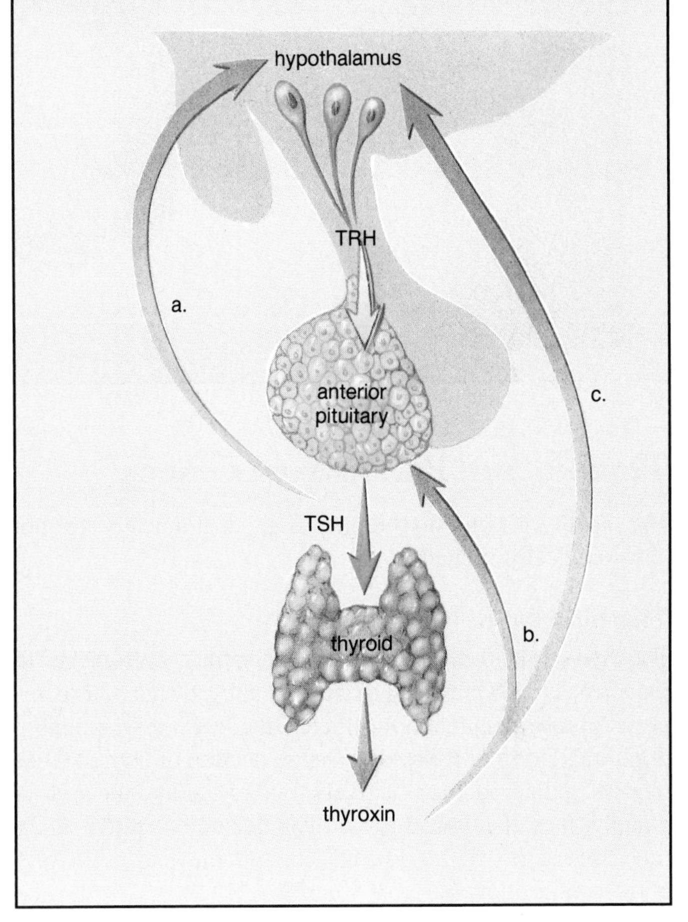

Figure 19.9

Simple goiter. An enlarged thyroid gland often is caused by a lack of iodine in the diet. Without iodine, the thyroid is unable to produce thyroxin, and continued anterior pituitary stimulation causes the gland to enlarge.

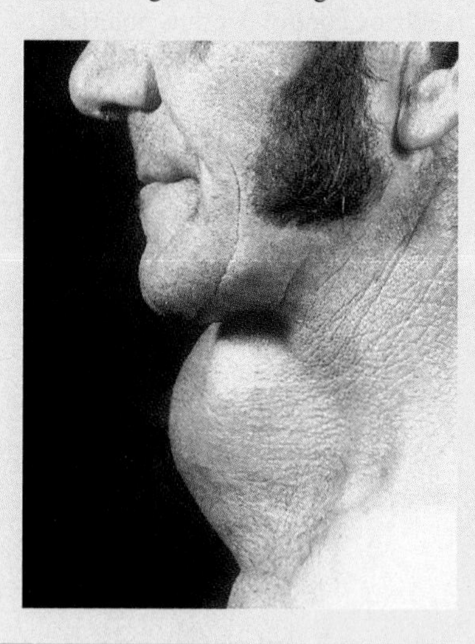

Figure 19.10

Cretinism. Cretins are individuals who have suffered from thyroxin insufficiency since birth or early childhood. Skeletal growth is usually inhibited to a greater extent than soft tissue growth; therefore, the child appears short and stocky. Sometimes, the tongue becomes so large that it obstructs swallowing and breathing.

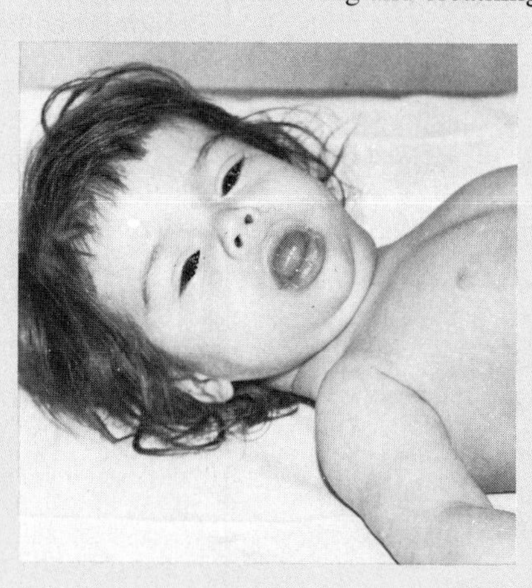

Figure 19.11

Myxedema. This condition is caused by thyroxin insufficiency in the older adult. An unusual type of edema leads to swelling of the face and bagginess under the eyes.

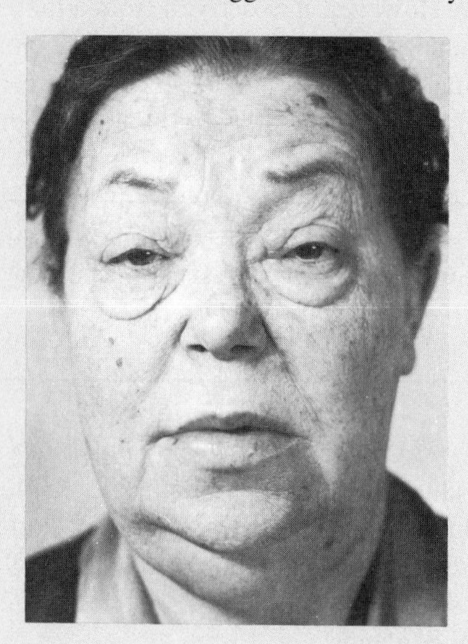

Thyroid and Parathyroid Glands

The parathyroid and thyroid glands are anatomically, but not physiologically, related.

Thyroid Gland: Needs Iodine

The thyroid gland (fig. 19.1) is located in the neck, attached to the trachea just below the larynx. Internally, the gland is composed of a large number of follicles filled with thyroglobulin, the storage form of **thyroxin.** The production of both of these requires iodine. Iodine is actively transported into the thyroid gland, where the concentration can become as much as 25 times that of blood. If iodine is lacking in the diet, the thyroid gland enlarges, producing a goiter (fig. 19.9). (In the United States, salt is iodized for this reason.)

The cause of goiter becomes clear if we refer to figure 19.8. When there is a low level of thyroxin in blood, a condition called hypothyroidism, the anterior pituitary is stimulated to produce *TSH.* TSH causes the thyroid to increase in size so that enough thyroxin usually is produced. In this case, enlargement continues because enough thyroxin is never produced. An enlarged thyroid that produces some thyroxin is called a **simple goiter.**

Thyroxin: Speeds Up Metabolism

Thyroxin does not have a target organ; instead, it stimulates most of the cells of the body to metabolize at a faster rate. The number of respiratory enzymes in the cell increases, as does oxygen (O_2) uptake.

If the thyroid fails to develop properly, a condition called **cretinism** results. Cretins (fig. 19.10) are short, stocky persons who have had extreme hypothyroidism since infancy and/or childhood. Thyroxin therapy can initiate growth, but unless treatment is begun within the first 2 months of life, mental retardation results. The occurrence of hypothyroidism in adults produces the condition known as **myxedema** (fig. 19.11), which is characterized by lethargy, weight gain, hair loss, slowed pulse rate, decreased body temperature, and thick and puffy skin. The administration of adequate doses of thyroxin restores normal function and appearance.

In the case of hyperthyroidism (too much thyroxin), the thyroid gland is enlarged and overactive, causing a goiter to form and the eyes to protrude because of edema in the tissues of the eye sockets and swelling of muscles that move the eyes. This type of goiter is called **exophthalmic goiter** (fig. 19.12). The patient usually becomes hyperactive, nervous, irritable, and suffers from insomnia. Removal or destruction of a portion of the thyroid by means of radioactive iodine sometimes is effective in curing the condition.

Calcitonin: Regulates Calcium

In addition to thyroxin, the thyroid gland produces the hormone **calcitonin.** This hormone helps to regulate the calcium level in blood and opposes the action of parathyroid hormone. The interaction of these 2 hormones is discussed in the following paragraphs.

Figure 19.12
Exophthalmic goiter. In hyperthyroidism, the eyes protrude because of edema in the tissues of the eye sockets.

The anterior pituitary produces TSH, a hormone that promotes the production of thyroxin by the thyroid, a gland subject to goiters. Thyroxin, which speeds up metabolism, can affect the body as a whole, as exemplified by cretinism and myxedema.

Parathyroid Glands: Affect Calcium, Too

The parathyroid glands are embedded in the posterior surface of the thyroid gland, as shown in figure 19.13*b*. Many years ago, these 4 small glands were sometimes removed by mistake during thyroid surgery. Under the influence of **parathyroid hormone (PTH),** the blood calcium level increases and the phosphate level decreases. The hormone stimulates the absorption of calcium from the digestive tract, the retention of calcium by the kidneys, and the demineralization of bone. In other words, PTH promotes the activity of osteoclasts, the bone-resorbing cells. Although this also raises the level of phosphate in blood, PTH acts on the kidneys to excrete phosphate in the urine. When a woman stops producing the female sex hormone estrogen following menopause, she is at risk of developing osteoporosis, characterized by thinning bones. How and whether estrogen affects PTH has not been determined.

If insufficient PTH is produced, the blood calcium level drops, resulting in **tetany.** In tetany, the body shakes from continuous muscle contraction. The effect really is brought about by increased excitability of the nerves, which

Figure 19.13
Thyroid and parathyroid glands. **a.** The thyroid gland is located in the neck in front of the trachea. **b.** The 4 parathyroid glands are embedded in the posterior surface of the thyroid gland.
c. Regulation of parathyroid hormone (PTH) secretion. A low blood calcium level causes the parathyroids to secrete PTH, which causes the kidneys and the gut to retain calcium and osteoclasts to break down bone. The end result is an increased level of calcium in blood. A high blood calcium level inhibits secretion of PTH.

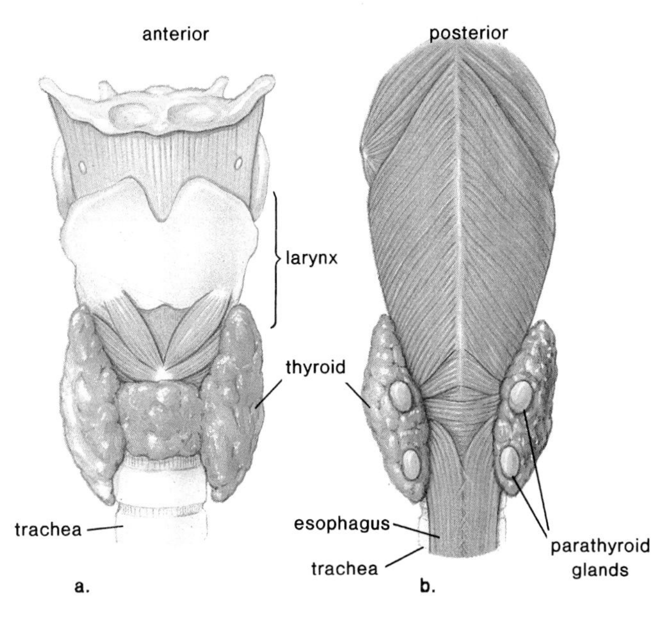

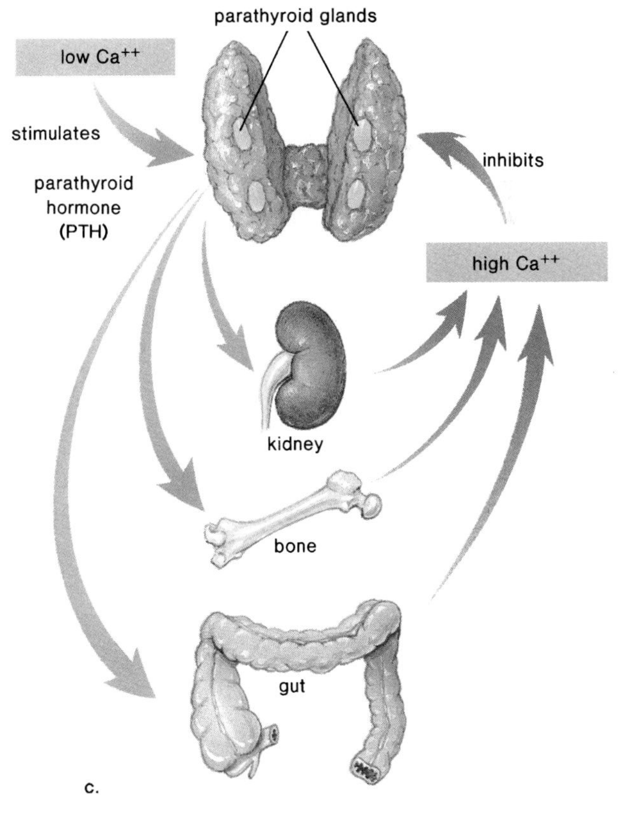

initiate nerve impulses spontaneously and without rest. Calcium plays an important role in both nervous conduction and muscle contraction.

The level of PTH secretion is controlled by a negative feedback mechanism involving calcium (fig. 19.13*c*). When the blood calcium level increases, PTH secretion is inhibited, and when the blood calcium level decreases, PTH secretion is stimulated.

As mentioned previously, the thyroid secretes calcitonin, which also influences blood calcium level. Although calcitonin has the opposite effect of PTH, particularly on the bones, its action is not believed to be as significant. Still, the 2 hormones function together to regulate the level of calcium in blood.

> PTH maintains a high blood calcium level by promoting calcium absorption in the digestive tract, calcium reabsorption by the kidneys, and the demineralization of bone. These actions are opposed by calcitonin, produced by the thyroid.

Adrenal Glands

The adrenal glands, as their name implies (*ad* means *near*; *renal* means *kidney*), lie atop the kidneys (fig. 19.1). Each consists of an inner portion called the *medulla* and an outer portion called the *cortex*. These portions, like the anterior pituitary and the posterior pituitary, have no functional connection with one another.

Adrenal Medulla: Secretes for Fight or Flight

The adrenal medulla secretes **norepinephrine** and epinephrine under conditions of stress. They bring about all the responses we associate with the "fight or flight" reaction: the blood glucose level and the metabolic rate increase, as do breathing and the heart rate. People can perform feats of strength, such as lifting up a car to free a child. The blood vessels in the intestine constrict, and those in the muscles dilate. This increased circulation to the muscles causes them to have more stamina than usual. In times of emergency, the sympathetic nervous system *initiates* these responses, but they are maintained by secretions from the adrenal medulla.

The adrenal medulla is innervated by only one set of sympathetic nerve fibers. Recall from chapter 16 that usually there are pre- and postganglionic nerve fibers for each organ stimulated. In this instance, what happened to the postganglionic neurons? It appears that the adrenal medulla may have evolved from a modification of the postganglionic neurons. Like the neurosecretory neurons in the hypothalamus, these neurons also secrete hormones into the bloodstream.

> The adrenal medulla releases norepinephrine and epinephrine into the bloodstream. These hormones help us and other animals to cope with situations that threaten survival.

Adrenal Cortex: Absolutely Necessary

Although the adrenal medulla can be removed with no ill effects, the adrenal cortex is absolutely necessary to life. The 2 major classes of hormones made by the adrenal cortex are the *glucocorticoids* and the *mineralocorticoids*. The adrenal cortex also secretes a small amount of male sex hormone and an even smaller amount of female sex hormone. All of these hormones are steroids.

Glucocorticoids: Cortisol, for Example

Of the various glucocorticoids, the hormone responsible for the greatest amount of activity is **cortisol.** Cortisol promotes the hydrolysis of muscle protein to amino acids, which then enter blood. This leads to an increased level of glucose when the liver converts these amino acids to glucose. Cortisol also favors metabolism of fatty acids rather than carbohydrates. In opposition to insulin, therefore, cortisol raises the blood glucose level. Cortisol also counteracts the inflammatory response, which leads to the pain and swelling of joints in arthritis and bursitis. The administration of cortisol aids these conditions because it reduces inflammation.

The secretion of cortisol by the adrenal cortex is under the control of the anterior pituitary hormone ACTH (adrenocorticotropic hormone). Using the same kind of system shown in figure 19.8, the hypothalamus produces a releasing hormone (CRH) that stimulates the anterior pituitary to release ACTH. ACTH, in turn, stimulates the adrenal cortex to secrete cortisol, which regulates its own synthesis by negative feedback of both CRH and ACTH synthesis.

Figure 19.14

Renin-angiotensin-aldosterone system. If the blood sodium level is low, the kidneys secrete renin. Thereafter, there are increased levels of angiotensin I and II in blood. Angiotensin II stimulates the adrenal cortex to release aldosterone. Aldosterone promotes reabsorption of sodium by the kidneys. When the blood sodium level rises, the kidneys stop secreting renin.

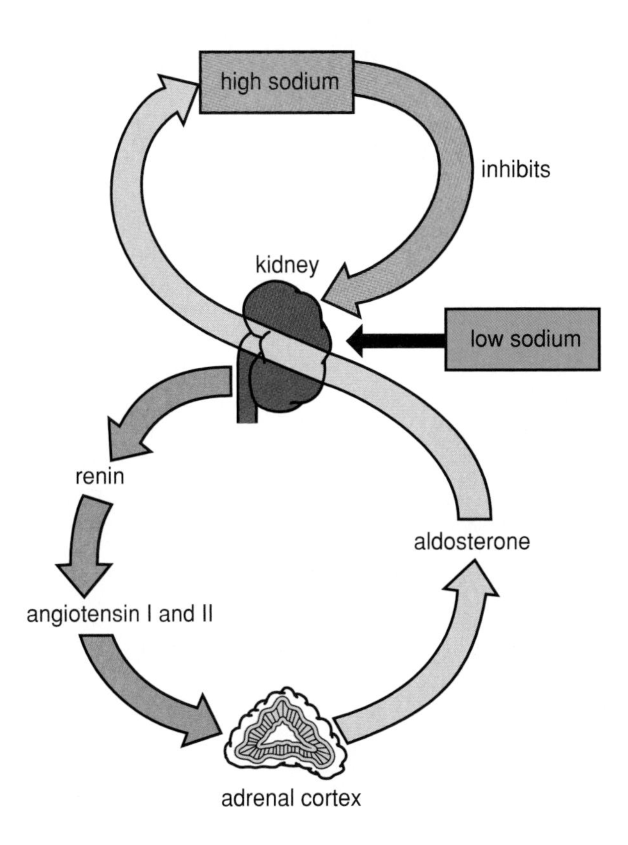

Mineralocorticoids: Aldosterone, for Example

The secretion of mineralocorticoids, the most significant of which is **aldosterone,** is not under the control of the anterior pituitary. Aldosterone regulates the level of sodium and potassium in blood. Its primary target organ is the kidney, where it promotes renal absorption of sodium and renal excretion of potassium (fig. 19.14). The level of sodium is particularly important to the maintenance of blood pressure because its concentration indirectly regulates the secretion of aldosterone. When the blood sodium level is low, the kidneys secrete renin. Renin is an enzyme that converts the plasma protein angiotensinogen to angiotensin I, which is changed to angiotensin II by a converting enzyme found in the lungs. Angiotensin II stimulates the adrenal cortex to release aldosterone (see fig. 15.12). The effect of this system, called the renin-angiotensin-aldosterone system, is to raise blood pressure in 2 ways. First, angiotensin II constricts the arteries directly, and second, aldosterone causes the kidneys to reabsorb sodium. When the blood sodium level is high, water is reabsorbed, and blood volume and blood pressure are maintained.

The heart produces a hormone that acts contrary to aldosterone. This hormone is called the atrial natriuretic hormone because (1) it is produced by the atria of the heart and (2) it causes natriuresis, the excretion of sodium. Once sodium is excreted, so is water; therefore, blood volume and blood pressure decrease.

> Cortisol, which raises the blood glucose level, and aldosterone, which raises the blood sodium level, are 2 hormones secreted by the adrenal cortex.

Sex Hormones: Male for Females and Vice Versa

The adrenal cortex produces a small amount of both male and female sex hormones. In males, the cortex is a source of female sex hormones, and in females, it is a source of male hormones. A tumor in the adrenal cortex can cause the production of an excess of sex hormones, which can lead to feminization in males and masculinization in females.

Disorders: Bronze Skin or Moon Face

When the level of adrenal cortex hormones is low, Addison disease results. When the level of adrenal cortex hormones in the body is high, Cushing syndrome results.

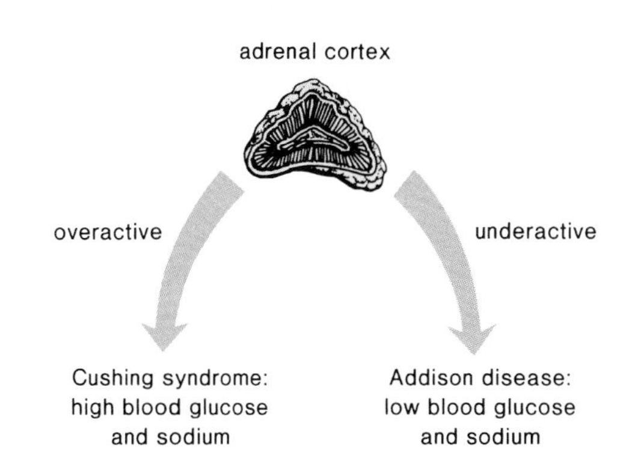

Figure 19.15

Addison disease. This condition is characterized by a peculiar bronzing of the skin, as seen in the face and the thin skin of the nipples of this patient.

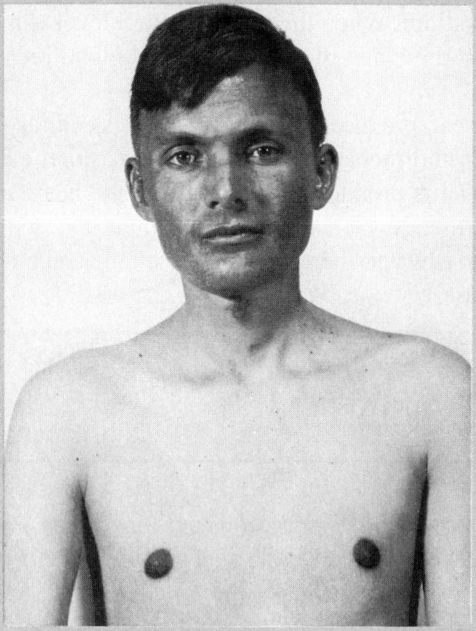

Figure 19.16

Cushing syndrome. Persons with this condition tend to have an enlarged trunk and a moonlike face. Masculinization may occur in women due to the excess male sex hormones in the body.

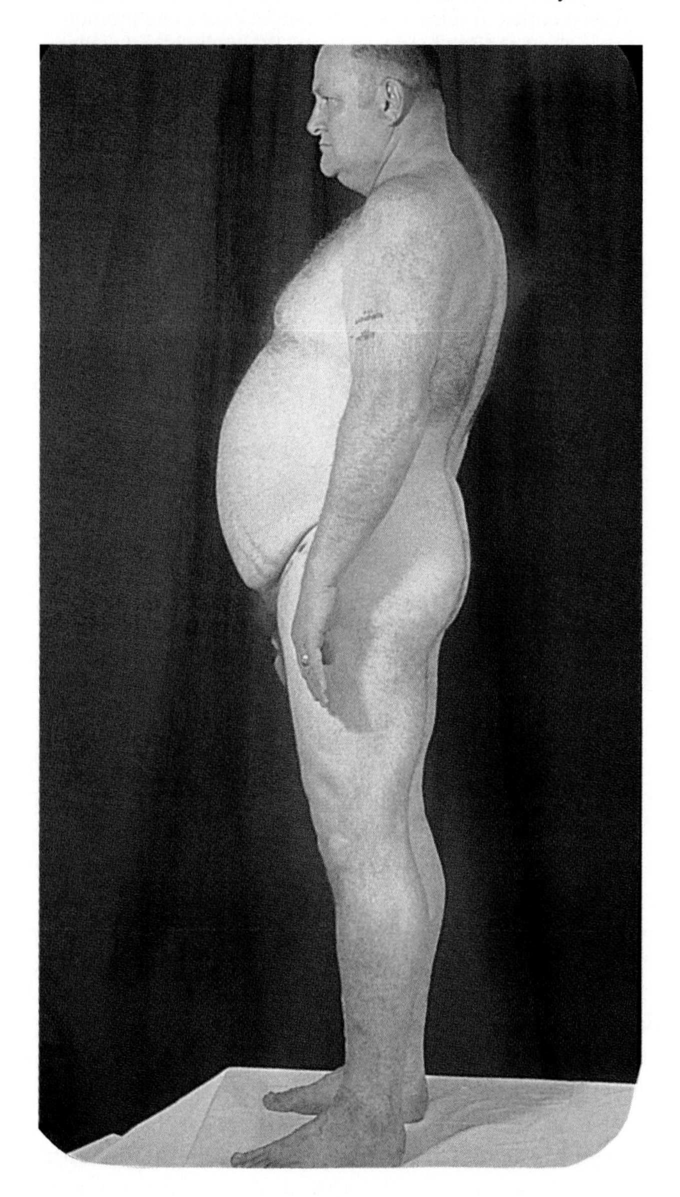

Addison Disease. Because of the lack of cortisol, the Addison disease patient is unable to maintain the blood glucose level, tissue repair is suppressed, and there is a high susceptibility to any kind of bodily stress. Even a mild infection can cause death. Due to the lack of aldosterone, the blood sodium level is low, and the person experiences low blood pressure along with low blood pH, a condition called acidosis. In addition, the patient's skin has a peculiar bronze cast (fig. 19.15).

Cushing Syndrome. In Cushing syndrome, a high level of cortisol causes a tendency toward diabetes mellitus, a decrease in muscle protein, and an increase in subcutaneous fat. Because of these effects, the person usually develops thin arms and legs and an enlarged trunk. Due to the high level of sodium, blood is basic and the patient is hypertensive and has edema of the face, which gives it a moon shape (fig. 19.16).

Addison disease is due to adrenal cortex hyposecretion, and Cushing syndrome is due to adrenal cortex hypersecretion.

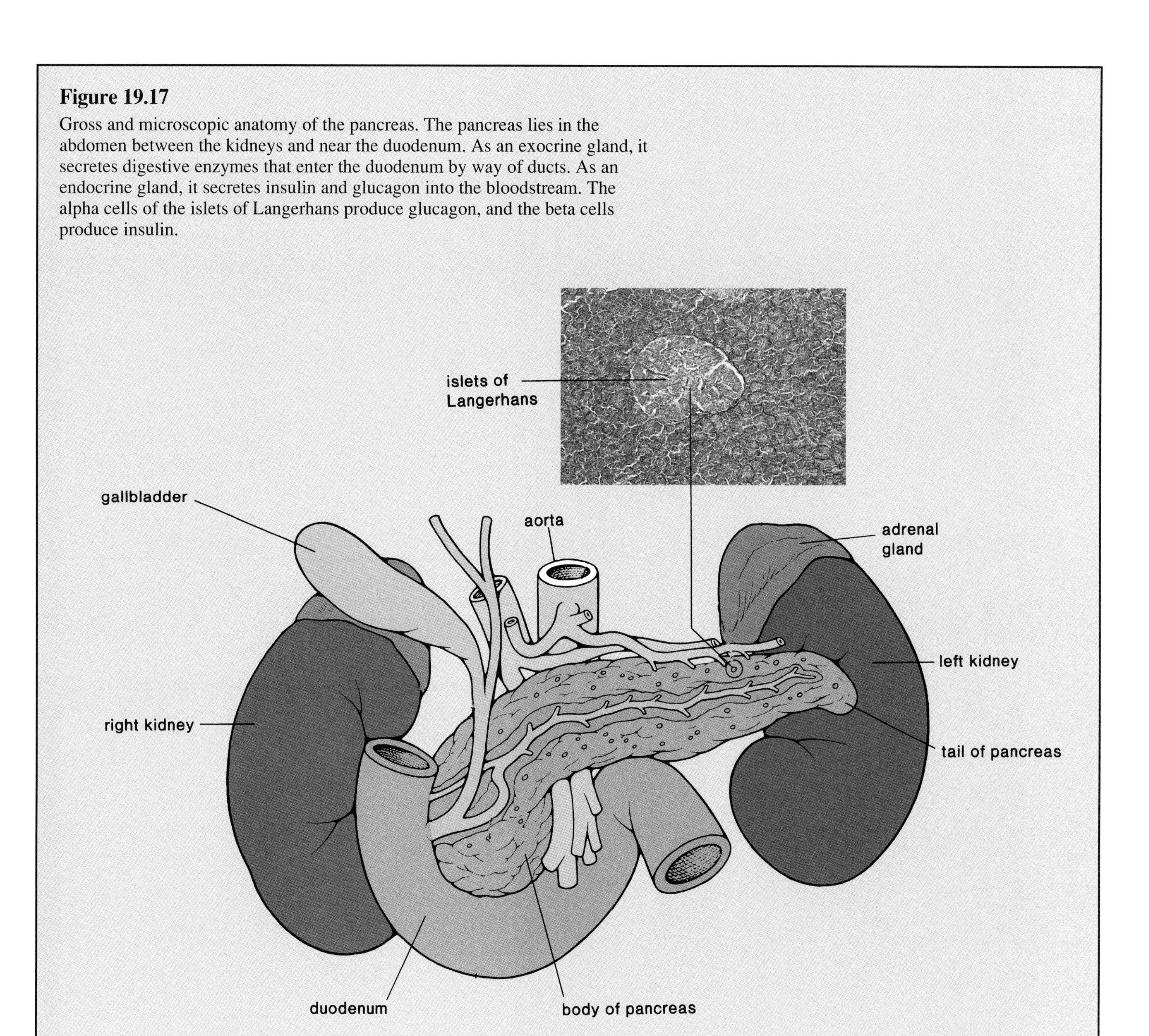

Figure 19.17

Gross and microscopic anatomy of the pancreas. The pancreas lies in the abdomen between the kidneys and near the duodenum. As an exocrine gland, it secretes digestive enzymes that enter the duodenum by way of ducts. As an endocrine gland, it secretes insulin and glucagon into the bloodstream. The alpha cells of the islets of Langerhans produce glucagon, and the beta cells produce insulin.

The Pancreas

The **pancreas** is a long organ that lies transversely in the abdomen (fig. 19.17) between the kidneys and near the duodenum of the small intestine. It is composed of 2 types of tissue—exocrine, which produces and secretes *digestive* juices that go by way of ducts to the small intestine, and endocrine, called the **islets of Langerhans** (pancreatic islets), which produces and secretes the hormones **insulin** and **glucagon** directly into blood.

All the cells of the body use glucose as an energy source; in order to preserve the health of the body, it is important that the glucose concentration remain within normal limits. Insulin is secreted when there is a high level of glucose in blood, which usually occurs just after eating. Insulin has 3 different actions: (1) it stimulates liver, fat, and muscle cells to take up and metabolize glucose; (2) it stimulates the liver and the muscles to store glucose as glycogen; and (3) it promotes the buildup of

Figure 19.18

Contrary effects of insulin and glucagon. When the blood glucose level is high, the pancreas secretes insulin. Insulin promotes the storage of glucose as glycogen and the synthesis of proteins and fats as opposed to their use as energy sources. Therefore, insulin lowers the blood glucose level. When the blood glucose level is low, the pancreas secretes glucagon. Glucagon acts in opposition to insulin in all respects; therefore, glucagon raises the blood glucose level.

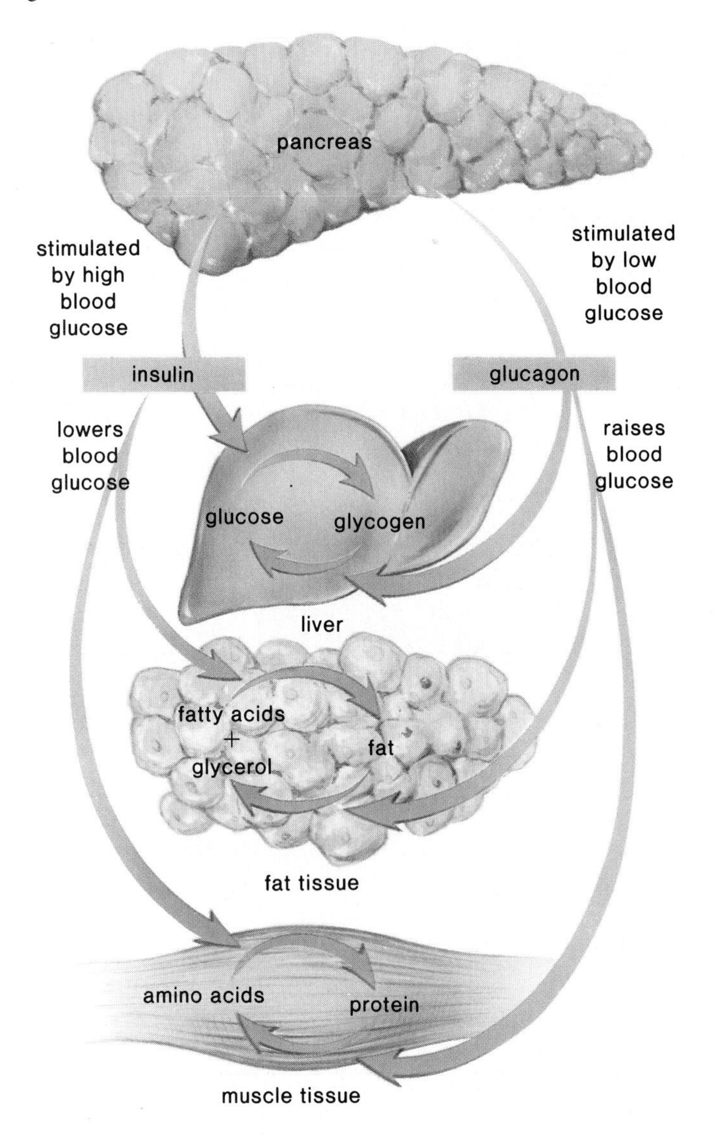

fats and proteins and inhibits their use as an energy source. Therefore, insulin is a hormone that promotes storage of nutrients so that they are on hand during leaner times. It also helps to lower the blood glucose level.

Glucagon is secreted from the pancreas in between eating, and its effects are opposite to those of insulin. Glucagon stimulates the breakdown of stored nutrients and causes the blood glucose level to rise (fig. 19.18).

Table 19.2
Symptoms of Hyperglycemia and Hypoglycemia

Hyperglycemia	Hypoglycemia
Slow, gradual onset	Sudden onset
Dry, hot skin	Perspiration, pale skin
No dizziness	Dizziness
No palpitation	Heart palpitation
No hunger	Hunger
Excessive urination	Normal urination
Excessive thirst	Normal thirst
Deep, labored breathing	Shallow breathing
Fruity breath odor	Normal breath odor
Large amounts of urinary sugar	Urinary sugar absent or slight
Ketones in urine	No ketones in urine
Drowsiness and great lethargy leading to stupor	Confusion, disorientation, strange behavior

Diabetes Mellitus: Insulin Lack or Insensitive Cells

The symptoms of **diabetes mellitus** (sugar diabetes) include the following:

Sugar in the urine
Frequent, copious urination
Abnormal thirst
Rapid weight loss
General weakness
Drowsiness and fatigue
Itching of the genitals and the skin
Visual disturbances, blurring
Skin disorders, such as boils, carbuncles, and infection

Many of these symptoms develop because sugar is not being metabolized by the cells. The liver fails to store glucose as glycogen, and all the cells fail to utilize glucose as an energy source. This means that the blood glucose level rises very high after eating, causing glucose to be excreted in the urine. More water than usual is therefore excreted, so that the diabetic is extremely thirsty.

Since carbohydrate is not being metabolized, the body turns to the breakdown of protein and fat for energy. Unfortunately, the breakdown of these molecules leads to the buildup of ketones in blood and urine. The resulting reduction in blood volume and acidosis (acid blood) can eventually lead to coma and death. The symptoms of hyperglycemia (table 19.2) develop slowly and there is time for intervention and reversal of symptoms.

There are 2 types of diabetes. In *type I (insulin-dependent) diabetes,* the pancreas is not producing insulin. The condition is believed to be brought on by exposure to an environmental agent, most likely a virus, whose presence causes cytotoxic T cells to destroy the islets of Langerhans (pancreatic islets). As a result, the patient must have daily insulin injections. These injections control the diabetic symptoms but still can cause inconveniences since either an overdose of insulin or the absence of regular eating can bring on the symptoms of hypoglycemia (table 19.2). These symptoms appear when the blood glucose level falls below normal levels. Since the brain requires a constant supply of sugar, unconsciousness can result. The cure is quite simple: an immediate source of sugar, such as a sugar cube or fruit juice, can very quickly counteract hypoglycemia.

Obviously, insulin injections are not the same as a fully functioning pancreas, which responds on demand to a high glucose level by supplying insulin. For this reason, some doctors advocate an islet transplant for type I diabetes.

Of the 12 million people who now have diabetes in the United States, at least 10 million have *type II (insulin-independent) diabetes.* This type of diabetes usually occurs in people of any age who are obese and inactive. The pancreas produces insulin, but the cells do not respond to it. At first, cells lack the receptors necessary to detect the presence of insulin, and later, the organs and tissues listed in figure 19.18 are even incapable of taking up glucose. If type II diabetes is untreated, the results can be as serious as type I diabetes. (Diabetics are prone to blindness, kidney disease, and circulatory disorders, including strokes. Pregnancy carries an increased risk of diabetic coma, and the child of a diabetic is somewhat more likely to be stillborn or to die shortly after birth.) It is important, therefore, to prevent or to at least control type II diabetes. The best defense is a nonfattening diet and regular exercise. If this fails, oral drugs that make the cells more sensitive to the effects of insulin or that stimulate the pancreas to make more insulin are available.

Diabetes mellitus is caused by the lack of insulin or the insensitivity of cells to insulin. Insulin lowers blood glucose levels by causing the cells to take up glucose and the liver to convert it to glycogen. Do 19.1 Critical Thinking, found at the end of the chapter.

Other Endocrine Glands

There are some other glands in the body that produce hormones. We will discuss some of these here.

Gonads: For Sex Characteristics

The gonads are endocrine glands that produce the hormones that determine sexual characteristics. As is discussed in detail in the following chapter, the *testes* produce the androgens—the most important of which is testosterone—which are the male sex hormones, and the *ovaries* produce estrogen and progesterone, the female sex hormones. The secretion of these hormones is under the control of the gonadotropic hormones produced by the anterior pituitary.

The sex hormones bring about the secondary sex characteristics of males and females. Among other traits, males have greater muscular strength than females. As discussed in the reading on the next page, athletes and others sometimes take so-called **anabolic steroids,** which are synthetic steroids that mimic the action of testosterone, in order to improve their strength and physique. Unfortunately, this practice is accompanied by harmful side effects.

Thymus: Most Active in Children

The **thymus** is a lobular gland that lies in the upper thoracic cavity (fig. 19.1). This organ reaches its largest size and is most active during childhood; with aging, the organ gets smaller and becomes fatty. Certain lymphocytes that originate in the bone marrow and then pass through the thymus are transformed into T cells (p. 225). The thymus produces various hormones called *thymosins,* which aid the differentiation of T cells and may stimulate immune cells in general. There is hope that these hormones can be used in conjunction with lymphokine therapy to restore or to stimulate the function of T cells in patients suffering from AIDS or cancer.

Pineal Gland: Hormone at Night

The **pineal gland** produces the hormone called melatonin, primarily at night. In fishes and amphibians, the pineal gland is located near the surface of the body and is a "third eye," which receives light rays directly. In mammals, the pineal gland is located in the third ventricle of the brain and cannot receive direct light signals. However, it does receive nerve impulses from the eyes by way of the optic tract.

The pineal gland and melatonin are involved in daily cycles called **circadian rhythms**. Normally, we grow sleepy at night, when melatonin levels are high, and awaken once daylight returns and melatonin levels are low. Shift work is

THE TRAGEDY OF STEROIDS

Inside a glitzy health club in an affluent North Shore suburb of Boston, 19-year-old Matthew Creighton stands in front of a mirror that covers an entire wall and inspects his bulging biceps.

Creighton has been pumping iron for 6 years, going through the same routine 5 mornings a week. He began entering weight-lifting tournaments last year and quickly realized that good training alone was not going to make him very competitive. On a September morning last year, Creighton met a man outside a health club and for $50 brought his first bottle of steroids. He has been a regular user since....

While much of the attention surrounding the nation's drug epidemic has focused on cocaine and heroin, muscle-enhancing anabolic drugs such as steroids are being used in steadily increasing numbers, and especially alarming is their growing popularity among youngsters.

Because steroids by and large have not been the target of federal law-enforcement authorities, there are no accurate figures on the number of steroid users. But the best estimate, according to federal officials, is that one million–3 million Americans use steroids, a figure that has increased steadily since the early 1970s.

The uproar surrounding the use of steroids by Canadian sprinter Ben Johnson in the Seoul Olympics has underscored an issue that has received little public attention in the past. For many years, steroids were used primarily by bodybuilders, weight lifters, and other athletes, such as professional football players, to help them in competition. But in recent years, steroid use has expanded to fitness buffs, high school and college athletes, and skinny youngsters wanting to gain bulk quickly.

Anabolic steroids are synthetic hormones taken either orally or by injection. They were developed in the 1930s to prevent muscle atrophy in patients with debilitating illness. In some cases, they also were given to burn victims and surgery patients to speed recovery....

Steroids, which can be dispensed legally only through a prescription, also have been used to treat certain forms of anemia and breast cancer.

But those using steroids illegally to enhance their appearance take them in large quantities, sometimes as much as 40 or 50 times the recommended dosage, according to FDA officials. Prolonged use of such a large quantity can lead to stunted growth in youngsters and high blood cholesterol. In men, it can result in baldness, acne, shrunken testes, feminized breasts, and infertility.

In women, who normally produce very low levels of testosterone and therefore gain much more from steroids, such overuse can promote facial hair, deepening of the voice, and an enlarged clitoris (fig. 19.A).

While scientific research has not been exhaustive, some studies have linked steroid abuse to cancer of the liver, prostate, and testes, as well as to kidney diseases and atherosclerosis.

"No one has really conclusively determined the long-term effects of steroid use because most doctors don't want to appear to condone a practice many consider unhealthy," said Dr. Gloria Troendle, senior medical officer at the FDA (Federal Drug Administration).

But another FDA official in Washington, who asked not to be identified, said, "It seems the evidence is becoming clear that youngsters who use anabolic steroids in large doses for 2 or 3 months face the possibility of dying in their 30s or 40s."

Growing evidence also points to another conclusion that steroids, which some in the medical community say appear to be addictive, can damage the mind. In psychotic side effects, sometimes referred to as " 'roid mania," users of large quantities of steroids have experienced mania, wild aggression, and delusions, said Dr. Harrison Pope, a psychiatrist at McLean Hospital in Belmont who has studied the effects of steroids on the mind.

One patient Pope examined had a friend videotape him while he deliberately drove a car into a tree at 35 mph. "My hunch is that we are seeing only a small part of . . . the magnitude of the psychiatric effects of steroids," Pope said in a recent interview in *Currents*, a medical trade journal.

Steroids had their first nonmedical use in World War II when Nazi doctors gave them to soldiers in an attempt to make them more aggressive. Following the war, the Soviet Union and other Eastern European nations began dispensing steroids to their athletes. In the 1950s, an American doctor working with the U.S. weight-lifting team learned that Soviet athletes were using steroids and markedly improving their performance. The doctor introduced steroids to lifters at the York (Pa.) Barbell Club, where they were an immediate hit.

In the 1960s, as their dangerous side effects started being documented, steroids fell into disfavor in the medical community. Since then, the FDA has banned most steroids. Currently, only about a dozen are approved for limited medical use.

The limited availability of steroids by prescription has helped trigger a flourishing black market. Smuggled into the United States from Mexico and to a lesser degree from Eastern Europe, the illicit trafficking has become a $100 million-a-year business, according to federal authorities.

Steroids are sold through the mail or in gyms and health clubs, federal authorities said. "You've got kids in the club making maybe $5 or $6 an hour, but if they're selling steroids, they can make as much as $900–1,000 a week," said the operator of several health clubs in the Los Angeles area who did not want to give his name....

Congress includes a provision raising the penalty for the distribution of steroids from a misdemeanor to a felony, with jail terms of 3 years for dispensing to adults and 6 years for dispensing to children. Still, most medical, sports, and law-enforcement officials agree that the nation's craving for steroids will not subside soon without more stringent law enforcement and better education about the dangers of the drugs....

Despite the health risks, steroids will continue to be popular among younger us-

Figure 19.A

Steroid's effects on the body in general and muscle cell in particular.

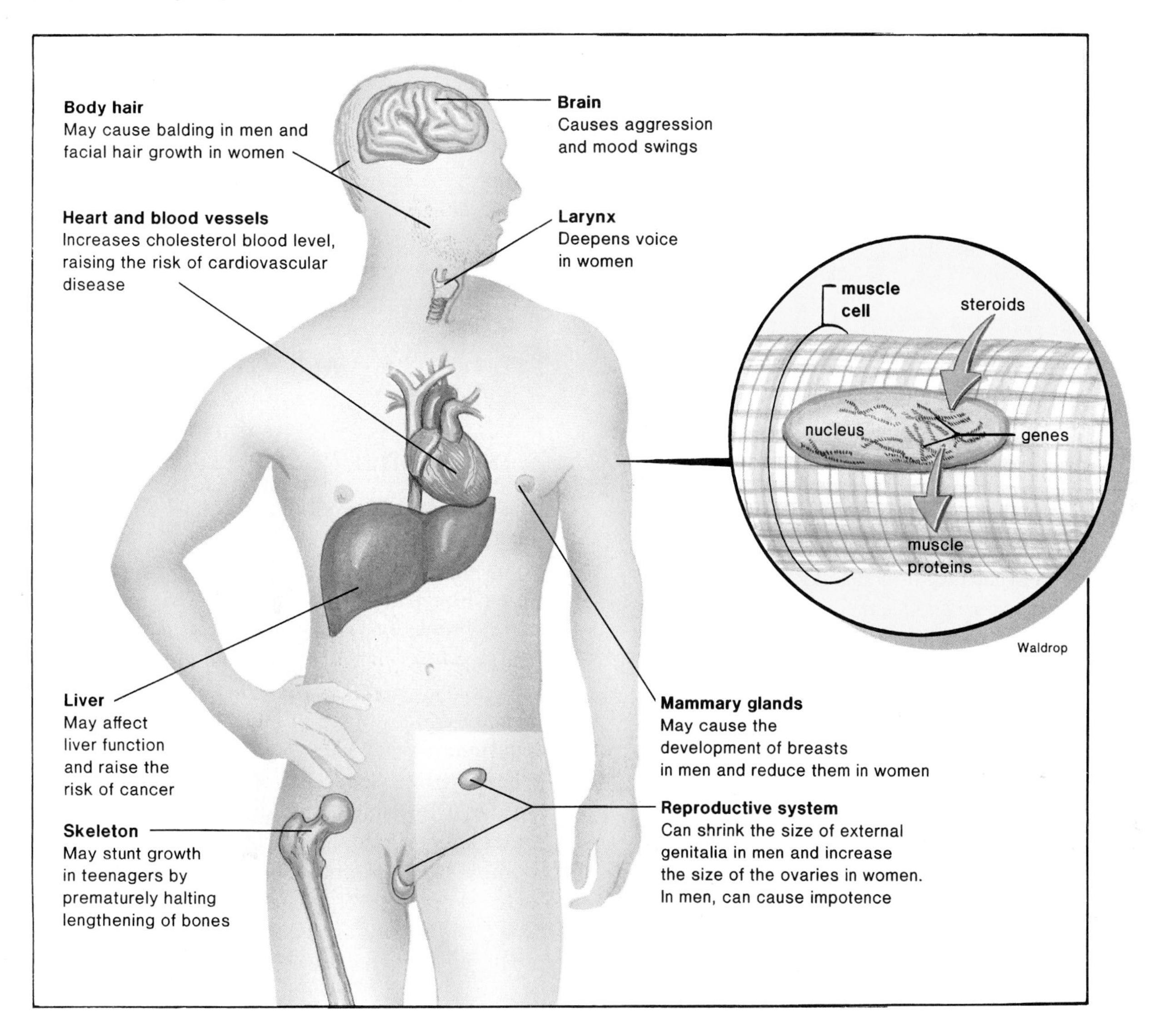

ers who take them to "feel better about themselves," said Dr. Jack Freinhar, a psychiatrist at Del Amo Hospital in Torrance, California, who treats adolescent steroid abusers.

"Physical risk doesn't matter when you're talking about saving the self," said Freinhar, who believes steroids can become an addiction he calls "reverse anorexia." "With anorexia, one is never thin enough. In our culture, there's this push to be muscular. Most of the youngsters I treat have had a deprived childhood and they have holes in the self. They need acceptance. One way is to look good, and steroids provide a fast method. But they can never get enough because each time they look in the mirror they still see themselves as too thin."

From John Powers and Diego Rabadeneira, "The Growing Threat of Steroids" in *The Boston Globe*, October 1988.

usually troublesome because it upsets this normal daily rhythm. Similarly, travel to another time zone, as when going to Europe from the United States, results in jet lag because the body is still producing melatonin according to the old schedule. Some people even have Seasonal Affective Disorder (SAD); they become depressed and have an uncontrollable desire to sleep with the onset of winter. Melatonin administration makes their symptoms worse, but exposure to a bright light improves them.

Many animals go through a yearly cycle that includes enlargement of reproductive organs during the summer, when melatonin levels are low. Mating occurs in the fall and young are born in the spring. It is of interest that children with a brain tumor that destroys the pineal gland experience early puberty. It's possible that the pineal gland is also involved in human sexual development.

Nontraditional Sources

Even organs that are not usually considered to be endocrine glands, such as the heart, the stomach, and the small intestine, have been found to secrete hormones. The heart produces *atrial natriuretic hormone,* which helps to regulate the sodium and water balance of the body. It lowers blood pressure by promoting renal excretion of sodium and water, and it also inhibits the release of renin and the hormones aldosterone and ADH. Atrial natriuretic factor is a peptide that is released not only by the atria but also by the aortic arch, ventricles, lungs, and pituitary gland in response to increases in blood pressure. The stomach and the small intestine produce the peptide hormones discussed in chapter 11.

A number of different types of organs and cells produce peptide *growth factors,* which stimulate cell division and mitosis. They are like hormones in that they act on cell types having specific receptors to receive them. Some, including lymphokines (p. 231) and blood cell growth factors (p. 212), are released into blood, others diffuse to nearby cells, and still others are self-stimulating. The latter are of interest because they play a role in the formation of cancer (p. 481). Other growth factors are described in the following listing.

Platelet-derived growth factor is released from platelets and from many other cell types. It helps in wound healing and causes an increase in the number of fibroblasts, smooth muscle cells, and certain cells of the nervous system.

Epidermal growth factor and *nerve growth factor* stimulate the cells indicated by their names as well as many others. These growth factors are also important in wound healing.

Tumor angiogenesis factor stimulates the formation of capillary networks and is released by tumor cells. One treatment for cancer is to prevent the activity of this growth factor.

Prostaglandins (PG) are another class of chemical messengers that also are produced and act locally. They are derived from fatty acids stored in cell membranes as phospholipids. When a cell is stimulated by reception of a hormone or even by trauma, a series of synthetic reactions takes place in the cell membrane, and PG is first released into the cytoplasm and then secreted from the cell. There are many different types of prostaglandins produced by many different tissues. In the uterus, certain prostaglandins cause muscles to contract; therefore, they are implicated in the pain and discomfort of menstruation in some women. (Antiprostaglandin therapy is useful in these cases.) On the other hand, certain prostaglandins are used to treat ulcers because they reduce gastric secretion, to treat hypertension because they lower blood pressure, and to prevent thrombosis because they inhibit platelet aggregation. Because the different prostaglandins can have contrary effects, however, it has been very difficult to standardize their use, and in most instances, prostaglandin therapy is still considered experimental.

Environmental Signals

In this chapter, we concentrated on the functions of the human endocrine glands and their hormonal secretions. We already know that hormones are only one type of chemical messenger or environmental signal between cells. In fact, the concept of the environmental signal has now been broadened to include at least the following 3 categories of messengers (fig. 19.19).

Environmental signals that act at a distance between individuals. Many organisms release chemical messengers called **pheromones** into the air or in externally deposited body fluids. These are intended to be messages for other members of the species. For example, ants lay down a pheromone trail to direct other ants to food, and the female silkworm moth releases bombykol, a sex attractant that is received by male moth antennae even several miles away. This chemical is so potent that it has been estimated that only 40 out of 40,000 receptors on the male antennae need to be activated in order for the male to respond. Mammals, too, release pheromones; the urine of dogs serves as a territorial marker, for example. Some studies are being conducted to determine if humans have pheromones, and other studies have suggested that humans respond to pheromones. For example, investigators have observed that women who live in close quarters tend to have coinciding menstrual cycles. They reason that this might be caused by a pheromone, but they don't know what the pheromone is or how it exerts its effect.

Environmental signals that act at a distance between body parts. This category includes the endocrine secretions, which traditionally have been called hormones. It also includes the

Figure 19.19

The 3 categories of environmental signals. Pheromones are chemical messengers that act at a distance between individuals. Endocrine hormones and neurosecretions typically are carried in the bloodstream and act at a distance within the body of a single organism. Some chemical messengers have local effects only; they pass between cells that are adjacent to one another. This, of course, includes neurotransmitter substances.

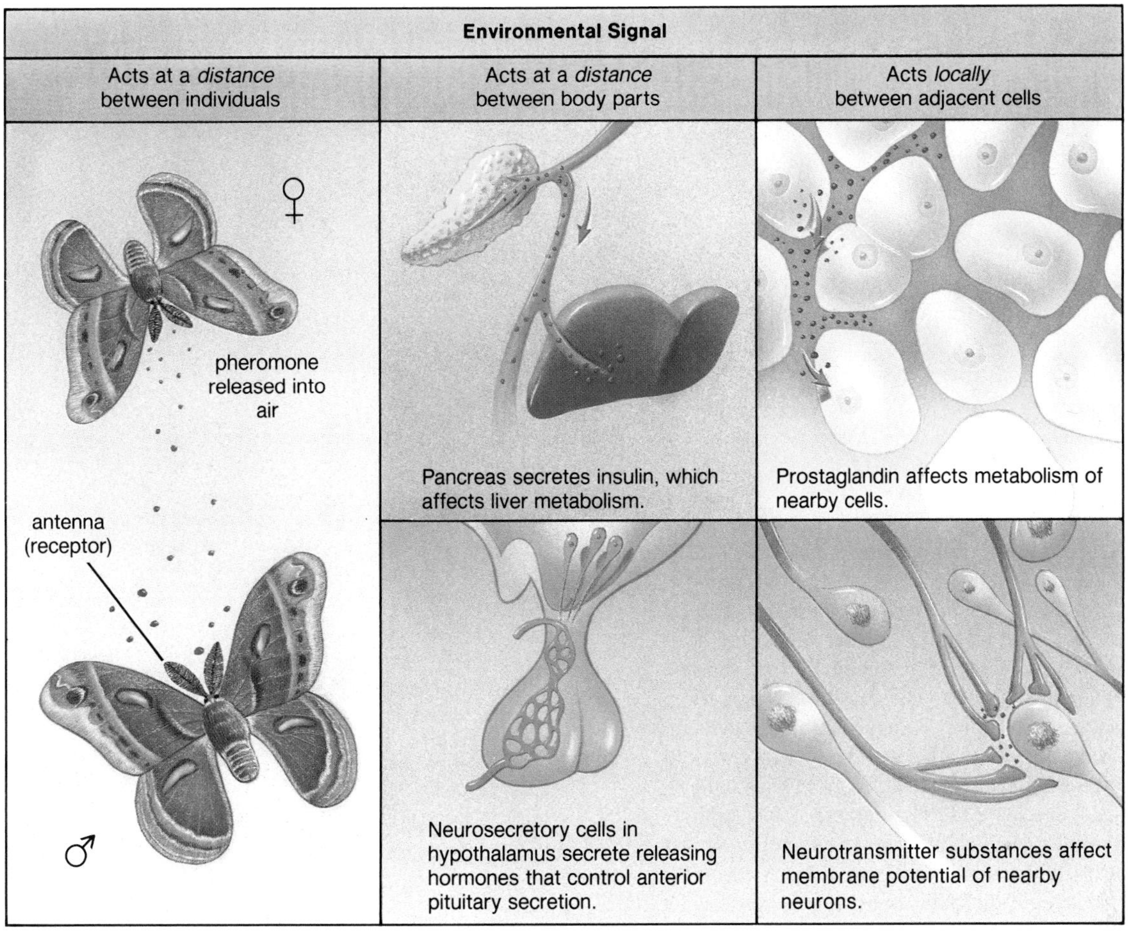

secretions of the neurosecretory cells in the hypothalamus—the production and action of ADH and oxytocin illustrate the close relationship between the nervous system and the endocrine system. Neurosecretory cells produce these hormones, which are released when these cells receive nerve impulses. As another example of the overlap between the nervous and endocrine systems, consider that endorphins on occasion travel in the bloodstream, but they act on nerve cells to alter their membrane potential. Also, norepinephrine is secreted by the adrenal medulla but is also a neurotransmitter in the sympathetic nervous system.

Environmental signals that act locally between adjacent cells. Neurotransmitter substances belong in this category, as do substances that are sometimes called local hormones. For example, when the skin is cut, histamine, released by mast cells, promotes the inflammatory response.

Today, hormones are categorized as one type of environmental signal. There are environmental signals that work at a distance between individuals (e.g., pheromones), at a distance between body parts (e.g., hormones), and locally between adjacent cells (e.g., neurotransmitters). Do 19.2 Critical Thinking, found at the end of the chapter.

SUMMARY

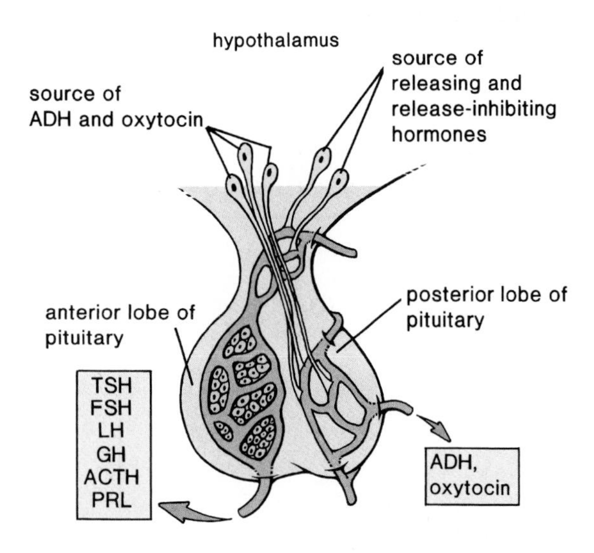

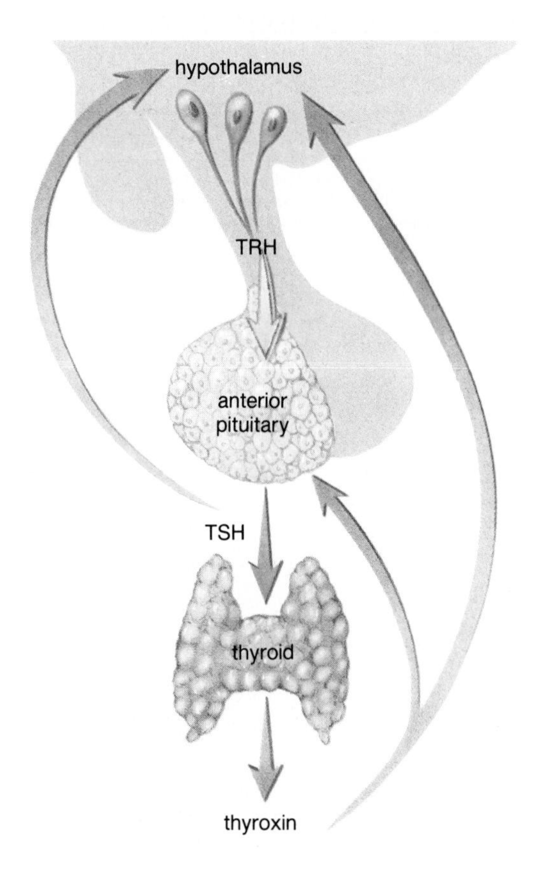

Hormones are chemical messengers having a metabolic effect on cells. The hypothalamus produces the hormones ADH and oxytocin, which are released by the posterior pituitary. The hypothalamus also produces releasing hormones and release-inhibiting hormones, which control the production of hormones by the anterior pituitary. In addition to GH and PRL, which affect the body directly, the anterior pituitary secretes hormones that control other endocrine glands: ACTH stimulates the adrenal cortex to release glucocorticoids; gonadotropic hormones (FSH and LH) stimulate the gonads to release the sex hormones; and TSH stimulates the thyroid to release thyroxin. The secretion of hormones is controlled by a negative feedback mechanism; for example, the blood level of thyroxin controls the secretion of TRH from the hypothalamus and TSH from the anterior pituitary.

Other hormones are PTH, the mineralocorticoids, norepinephrine, and insulin (table 19.1).

The most common illness due to hormonal imbalance is diabetes mellitus. This condition occurs when the islets of Langerhans within the pancreas fail to produce insulin. Insulin promotes the conversion of glucose to glycogen, thereby lowering blood glucose levels. Without the production of insulin, the blood glucose level rises, and some glucose spills over into the urine. The real problem in diabetes mellitus, however, is acidosis, which may cause the death of the diabetic if therapy is not begun.

There are 3 categories of chemical messengers: those that act at a distance between individuals (pheromones); those that act at a distance within the individual (traditional endocrine hormones and secretions of neurosecretory cells); and local messengers (such as prostaglandins and neurotransmitters). Since there is great overlap between these categories, perhaps the definition of a hormone now should be expanded to include all of them.

STUDY QUESTIONS

In order to practice **writing across the curriculum,** students should write out the answers to any or all of the study questions. The study questions are sequenced in the same order as the text.

1. Give the location in the human body of all the major endocrine glands. Name the hormones secreted by each gland, and describe their chief function. (pp. 347–349)
2. Give a definition of endocrine hormones that includes their most likely source, how they are transported in the body, and how they are received. What does "target" organ mean? (p. 344)
3. Categorize endocrine hormones according to their chemical makeup. (p. 344)
4. Tell how the 2 major types of hormones influence the metabolism of the cell. (p. 346)
5. Explain the relationship of the hypothalamus to the posterior pituitary and to the anterior pituitary. (pp. 347–348)
6. Explain a negative feedback mechanism, and give an example involving ADH. (pp. 346–347)
7. Explain why the anterior pituitary can be called the master gland. (p. 351) Give an example of the three-tiered relationship among the hypothalamus, the anterior pituitary, and other endocrine glands. (p. 351)
8. Diagram the contrary actions of insulin and glucagon. Use your diagram to explain the symptoms of type I diabetes mellitus. (p. 358)
9. Categorize environmental signals into 3 groups, and give examples of each group. (pp. 362–363)

OBJECTIVE QUESTIONS

1. The hypothalamus _______ the hormones _______ and _______, which are released by the posterior pituitary.
2. The _______ secreted by the hypothalamus control the anterior pituitary.
3. Generally, hormone production is self-regulated by a _______ mechanism.
4. Growth hormone is produced by the _______ pituitary.
5. Simple goiter occurs when the thyroid is producing _______ (too much, too little) _______.
6. ACTH, produced by the anterior pituitary, stimulates the adrenal _______.
7. An overproductive adrenal cortex results in the condition called _______.
8. PTH increases the level of _______ in blood.
9. Type I diabetes mellitus is due to a malfunctioning _______, while type II diabetes is due to limited uptake of insulin by _______.
10. Prostaglandins are not carried in _______ as are hormones secreted by the endocrine glands.
11. Complete the following table:

Acronym	Name of Hormone	Secreted by	Function
TSH			
ACTH			
PRL			
GH			
ADH			
PTH			

CRITICAL THINKING

In order to practice **writing across the curriculum,** students should write out the answers to any or all of the critical thinking questions. Suggested answers to the critical thinking questions are in appendix E.

19.1

1. If the pancreas is removed from an animal, what substance would you expect to find in the urine? Why?
2. Would your findings support the contention that the pancreas is the source of insulin? What do they prove?
3. How would you prove that insulin lowers blood sugar?
4. Once you have shown that insulin lowers blood sugar, do your findings in question 1 support the belief that the pancreas is the source of insulin? Why?
5. What could you do to prove that the pancreas is the source of insulin?

19.2

Environmental signals alter the behavior of target cells.

1. How would you modify this definition to be more comprehensive? (Hint: See figure 19.19.)
2. When is a cell, organ, or organism sensitive to a particular environmental signal?
3. How does the environmental signal glucose affect a pancreatic cell, insulin affect the liver, and pheromone affect a male moth?
4. The production of environmental signals is sometimes under the control of other signals. What causes insulin secretion? neurotransmitter release?

SELECTED KEY TERMS

acromegaly (ak″ ro-meg′ ah le) a condition resulting from an increase in GH production after adult height has been achieved.

anterior pituitary the portion of the pituitary gland that produces 6 types of hormones and is controlled by hypothalamic-releasing and release-inhibiting hormones.

cretinism (kre′ tin-izm) a condition resulting from improper development of the thyroid in an infant.

diabetes mellitus (di″ ah-be′tēz mĕ-li′ tus) condition characterized by a high blood glucose level and the appearance of glucose in the urine due to a deficiency of insulin production or uptake by cells.

endocrine gland a gland that secretes hormones directly into blood or body fluids.

exophthalmic goiter (ek″ sof-thal′mik goi′ ter) an enlargement of the thyroid gland accompanied by an abnormal protrusion of the eyes.

islet of Langerhans distinctive group of cells within the pancreas that secretes insulin and glucagon.

myxedema (mik″ să-de′ mah) a condition resulting from a deficiency of thyroid hormone in an adult.

pheromone (fer′o-mōn) a chemical substance secreted by one organism that influences the behavior of another.

posterior pituitary back lobe of the pituitary gland that stores and secretes ADH and oxytocin produced by the hypothalamus.

simple goiter condition in which an enlarged thyroid produces low levels of thyroxin.

FURTHER READINGS FOR PART FOUR

Alkon, D. L. July 1989. Memory storage and neural systems. *Scientific American.*

Aoki, C., and P. Siekevitz. December 1988. Plasticity in brain development. *Scientific American.*

Atkinson, M. A., and N. K. Maclaren. July 1980. What causes diabetes? *Scientific American.*

Barlow, R. B., Jr. April 1990. What the brain tells the eye. *Scientific American.*

Berns, M. B. June 1991. Laser surgery. *Scientific American.*

Boon, Thierry. March 1993. Teaching the immune system to fight cancer. *Scientific American.*

Cohen, I. R. April 1988. The self, the world and autoimmunity. *Scientific American.*

Cooper, G. M. 1992. *Elements of human cancer.* Boston: Jones and Bartlett.

Creager, J. G. 1992. *Human anatomy and physiology.* 2d ed. Dubuque, Ia.: Wm. C. Brown Publishers.

Fox, S. I. 1990. *Human physiology.* 3d ed. Dubuque, Ia.: Wm. C. Brown Publishers.

Freeman, W. J. February 1991. The physiology of perception. *Scientific American.*

Golde, D. W. December 1991. The stem cell. *Scientific American.*

Golub, E. S., and D. R. Green. 1992. *Immunology: A synthesis.* 2d ed. Sunderland, Mass.: Sinauer Associates.

Green, H. November 1991. Cultured cells for the treatment of disease. *Scientific American.*

Hales, D. 1992. *An invitation to health.* 5th ed. Redwood City, Calif.: Benjamin/Cummings Publishing Co.

Hegarty, V. 1988. *Decisions in nutrition.* St. Louis, Mo.: Times Mirror/Mosby College Publishing.

Hirshhorn, N., and W. B. Greenough III. May 1991. Progress in oral rehydration therapy. *Scientific American.*

Hole, J. W., Jr. 1992. *Essentials of human anatomy and physiology.* 4th ed. Dubuque, Ia.: Wm. C. Brown Publishers.

Holloway, M. March 1991. Rx for addiction. *Scientific American.*

Konishi, M. April 1993. Listening with two ears. *Scientific American.*

Koretz, J. F., and G. H. Handelman. July 1988. How the human eye focuses. *Scientific American.*

Little, R., and W. Little. 1989. *Physiology of the heart and circulation.* 4th ed. Chicago: Year Book Medical Publishers.

Mahowald, M. A., and C. Mead. May 1991. The silicon retina. *Scientific American.*

Marieb, E. N. 1991. *Essentials of human anatomy and physiology.* 3d ed. Redwood City, Calif.: Benjamin/Cummings Publishing Co.

Melzack, R. April 1992. Phantom limbs. *Scientific American.*

Moberg, C. L., and Z. A. Cohn. May 1991. Rene Jules Dubos. *Scientific American.*

Nathans, J. February 1989. The genes for color vision. *Scientific American.*

Orci, L., et al. September 1988. The insulin factory. *Scientific American.*

Powell, C. P. June 1991. Peering inward. *Scientific American.*

Ramachandran, V. S. May 1992. Blind spots. *Scientific American.*

Rasmussen, H. October 1989. The cycling of calcium as an intracellular messenger. *Scientific American.*

Rennie, J. December 1990. The body against itself. *Scientific American.*

Sataloff, R. T. December 1992. The human voice. *Scientific American.*

Schultz, J. S. August 1991. Biosensors. *Scientific American.*

Scientific American. September 1992. Mind and brain. Special Issue.

Scrimshaw, N. S. October 1991. Iron deficiency. *Scientific American.*

Selkoe, D. J. November 1991. Amyloid protein and Alzheimer's disease. *Scientific American.*

Smith, K. A. March 1990. Interleukin-2. *Scientific American.*

Spence, A. P. 1990. *Basic human anatomy.* 3d ed. Redwood City, Calif.: Benjamin/Cummings Publishing Co.

Tuomanen E. February 1993. Breaching the blood-brain barrier. *Scientific American.*

vonBoehmer, H., and P. Kisielow. October 1991. How the immune system learns about self. *Scientific American.*

Weissman, G. January 1991. Aspirin. *Scientific American.*

Young, J., and Z. Cohn. January 1988. How killer cells kill. *Scientific American.*

Zivin, J. A., and D. W. Choi. July 1991. Stroke therapy. *Scientific American.*

Part Five

HUMAN REPRODUCTION, DEVELOPMENT, AND INHERITANCE

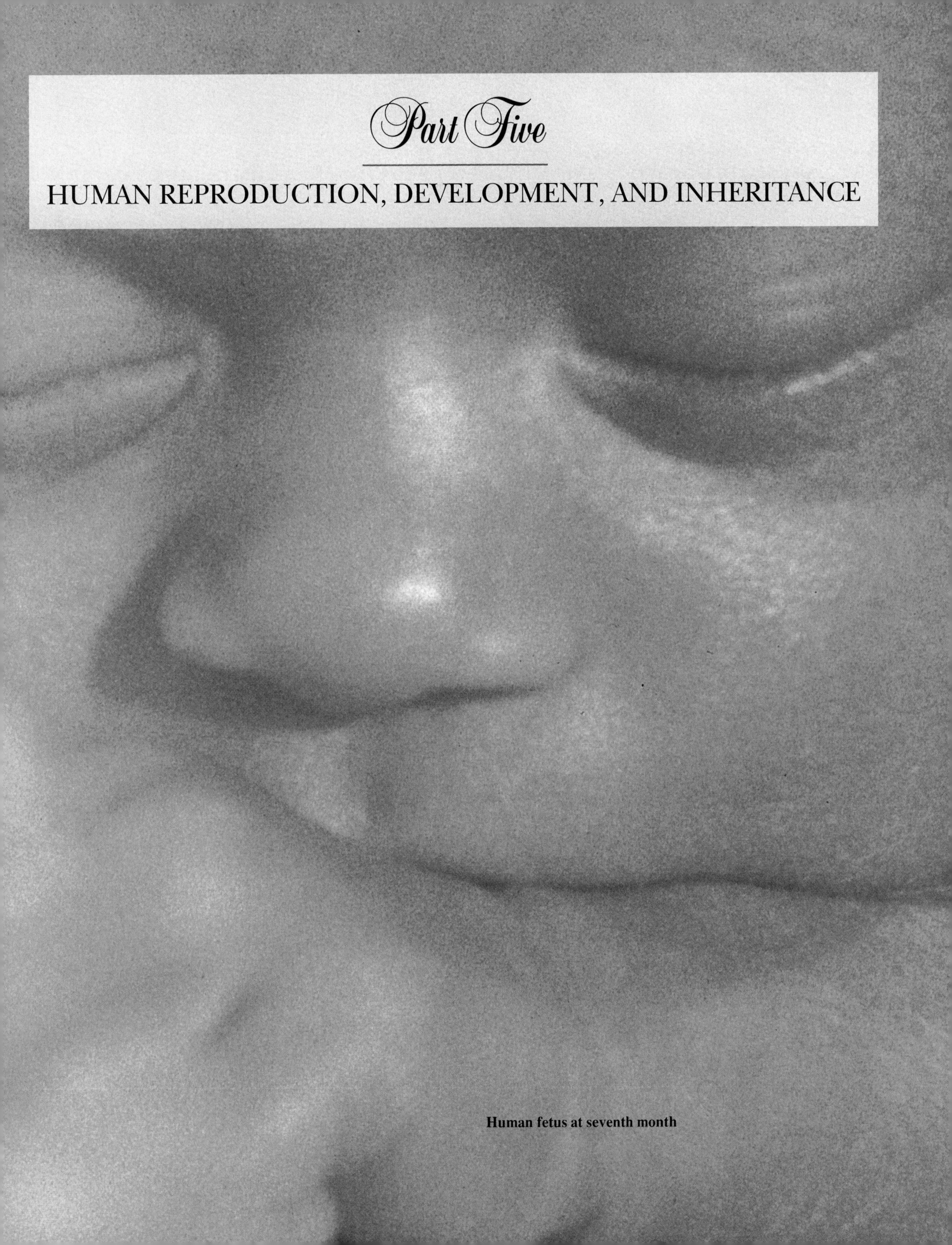

Human fetus at seventh month

HUMAN REPRODUCTION, DEVELOPMENT, AND INHERITANCE

The anatomy of the human male and female serves to bring the sperm to the egg, resulting in fertilization. Fertilization is followed by the gradual steps of development. Sexual reproduction results in a recombination of genes and therefore produces offspring that are at the same time both similar to and different from the parents. Knowledge of the genes carried by the parents sometimes makes it possible to predict certain features of the offspring. However, this simple relationship can be influenced by the interaction of genes during development and by the prenatal environment.

Biochemical knowledge of the makeup and the operation of the hereditary material has forged a biological revolution. It is now possible to manipulate the genes, an advance that may make it possible to cure genetic diseases and cancer.

20

THE REPRODUCTIVE SYSTEM

Chapter Concepts

1.
The male reproductive system is designed for the continuous production of a large number of sperm within a fluid medium. 370, 390

2.
The female reproductive system is designed for the monthly production of an egg and preparation of the uterus for possible implantation of the fertilized egg. 375, 390

3.
Hormones control the reproductive process and the sex characteristics of the individual. 378, 390

4.
Birth-control measures vary in effectiveness from those that are very effective to those that are minimally effective. 382, 390

5.
There are alternative methods of reproduction today, including in vitro fertilization followed by artificial implantation. 385, 390

6.
There are several serious and prevalent sexually transmitted diseases. 386, 390

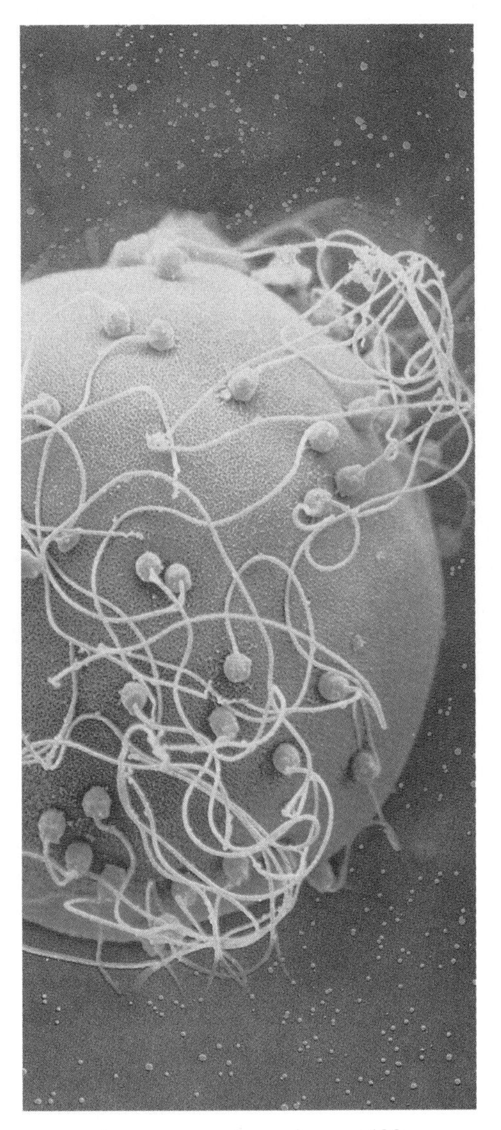

Sperm on surface of egg, ×480

Chapter Outline

Figure 20.1

Side view of the male reproductive system. **a.** The testes produce sperm, whose path is sketched in (**b**). The seminal vesicles, the prostate gland, and the Cowper's gland provide a fluid medium. Notice that the penis in this drawing is not circumcised—the foreskin is present. **b.** The path of sperm in the male genital tract.

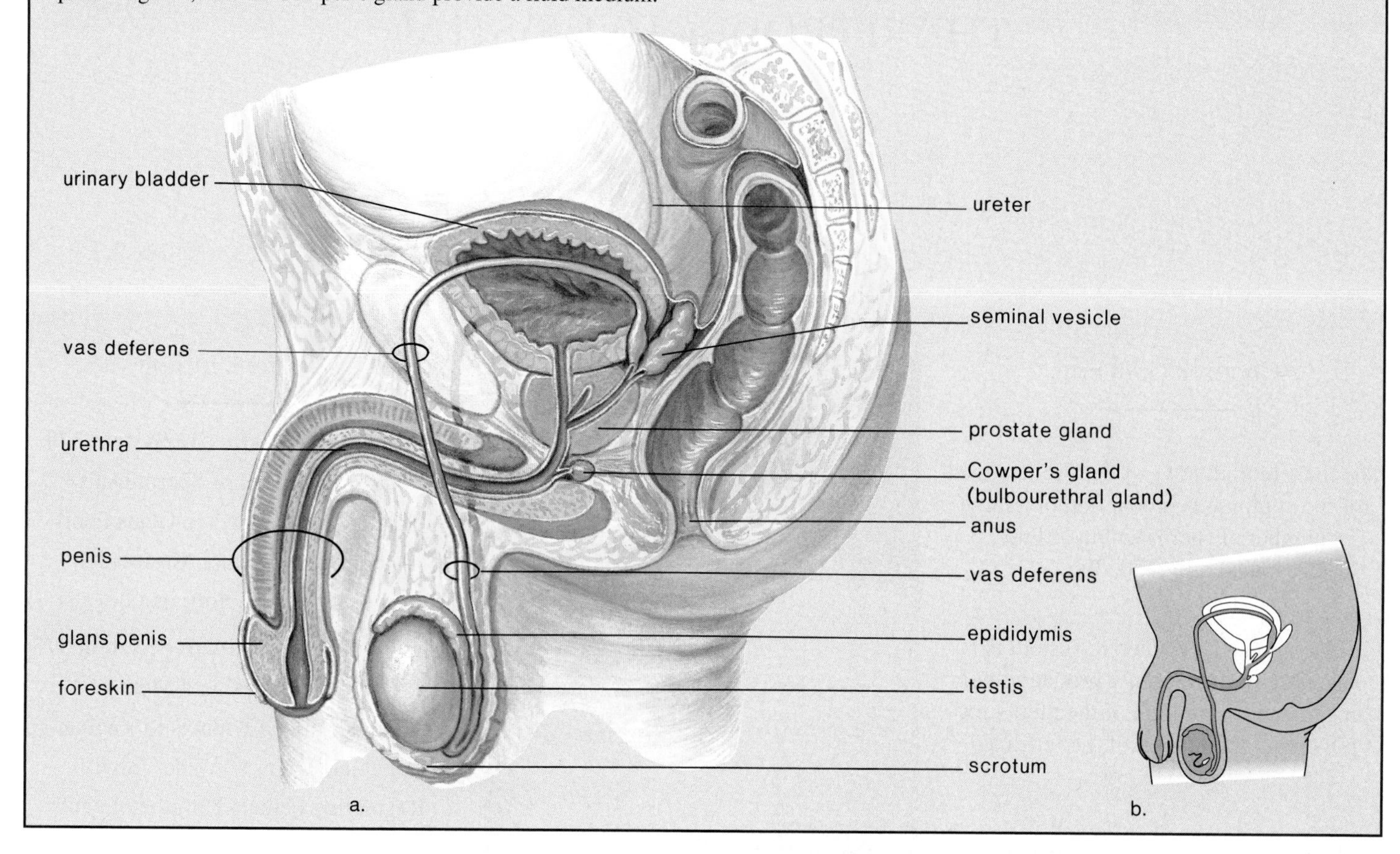

SEXUAL reproduction usually involves 2 types of gametes (sex cells), both of which contribute the same number of chromosomes to the new individual (see fig. 7.2). The sperm swim to the stationary egg, a much larger cell that contributes cytoplasm and organelles to the zygote. It seems reasonable that there are a large number of sperm to ensure that a few find the egg. After puberty, the human male continually produces sperm, which are temporarily stored before being released.

The Male Reproductive System

Figure 20.1 shows the reproductive system of the male, and table 20.1 lists the anatomical parts of this system.

Testes: Millions of Sperm Daily

The testes (sing., **testis**) lie outside the abdominal cavity of the male within the **scrotum.** The testes first begin to develop inside the abdominal cavity but descend into the scrotal sacs during the last 2 months of fetal development. If by chance the testes do not descend and the male is not treated or operated on to place the testes in the scrotum, *sterility*—the inability to produce offspring—usually results. This is because the internal temperature of the body (37°C) is too high to produce viable sperm; the temperature in the scrotum is about 34°C. Wearing tight clothing can increase this temperature and reduce sperm production. When the body is cold, the testes are normally held closer to the body, and this maintains an optimum temperature.

Table 20.1
Male Reproductive System

Organ	Function
Testis	Produces sperm and sex hormones
Epididymis	Stores sperm as they mature
Vas deferens	Conducts and stores sperm
Seminal vesicle	Contributes to seminal fluid
Prostate gland	Contributes to seminal fluid
Urethra	Conducts sperm
Cowper's gland (bulbourethral gland)	Contributes to seminal fluid
Penis	Serves as organ of copulation

Figure 20.2
Testis and sperm. The lobules of a testis contain seminiferous tubules, where spermatogenesis occurs. A sperm has a head, a middle piece, and a tail. The nucleus is in the head, capped by the enzyme-containing acrosome.

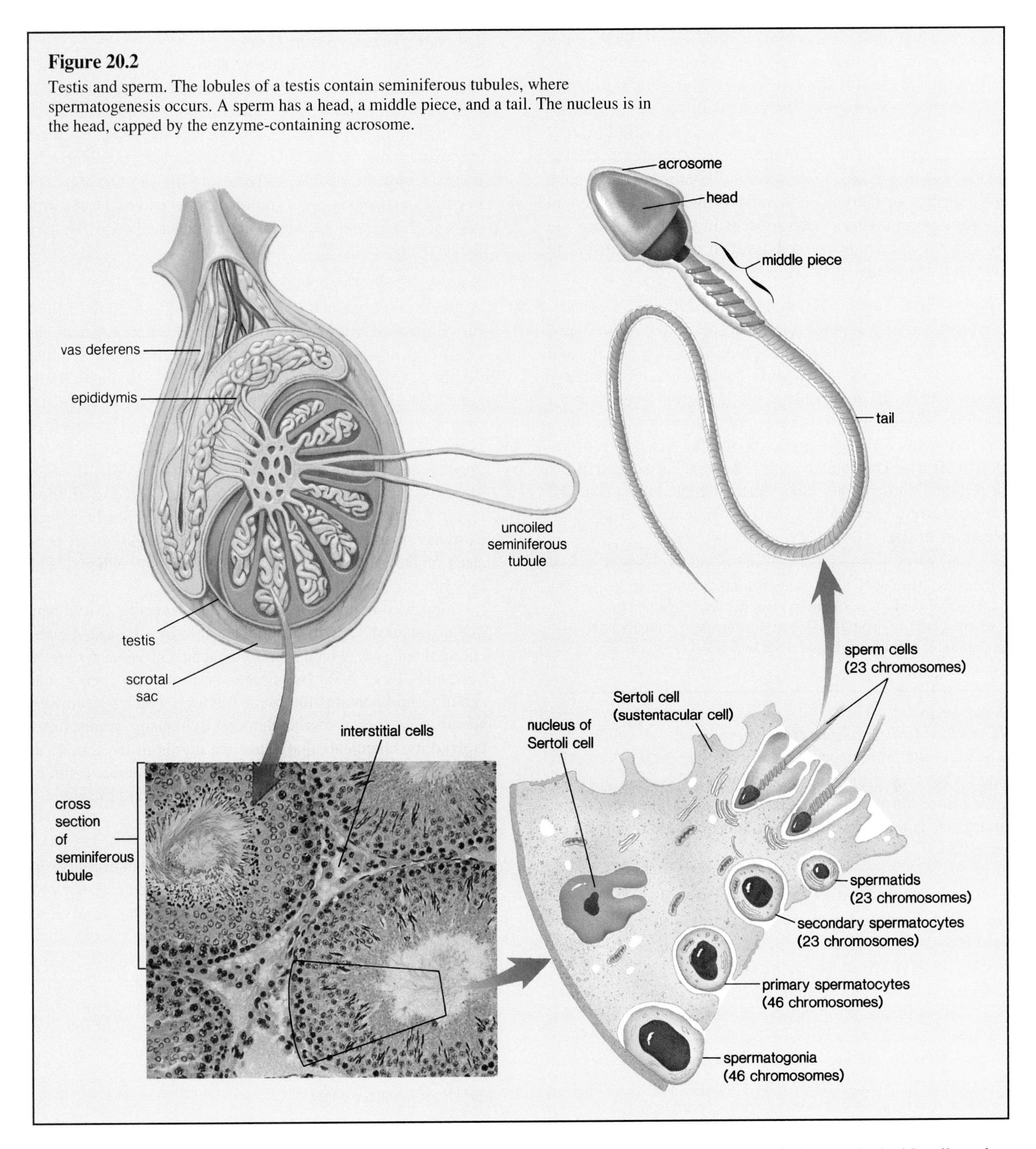

Seminiferous Tubules: Produce Sperm

Fibrous connective tissue forms the wall of a testis and divides it into lobules (fig. 20.2). Each lobule contains 1–3 tightly coiled **seminiferous tubules,** each of which is approximately 70 cm long when uncoiled. A microscopic cross section through a tubule shows that a tubule is packed with cells undergoing spermatogenesis (fig. 20.2). These cells are derived from undifferentiated cells called spermatogonia (sing., spermatogonium), which lie just inside the outer wall of a tubule and divide mitotically, always producing new spermatogonia.

Some newly formed spermatogonia move away from the outer wall to increase in size and become primary spermatocytes, which undergo meiosis, a type of cell division described in chapter 7. Primary spermatocytes, with 46 chromosomes, divide to give 2 secondary spermatocytes, each with 23 duplicated chromosomes. Secondary spermatocytes divide to produce 4 spermatids, also with 23 chromosomes, but each having only one chromatid. Spermatids then differentiate into sperm (spermatozoa). Also present in the tubules are the *Sertoli cells* (sustentacular cells), which support, nourish, and regulate the spermatogenic cells.

Sperm: Three Parts. The mature sperm (fig. 20.2) has 3 distinct parts: a head, a middle piece, and a tail. The *tail* contains the 9 + 2 pattern of microtubules typical of cilia and flagella (see fig. 3.11), and the *middle piece* contains energy-producing mitochondria. The *head* contains the 23 chromosomes within a nucleus. Adhering to the nucleus is a specialized lysosome called the **acrosome,** which contains enzymes that facilitate penetration of the egg. The human egg is surrounded by several layers of cells and a mucoprotein substance called the zona pellucida; the acrosomal enzymes help a sperm to digest its way into an egg.

In males, spermatogenesis occurs within the seminiferous tubules of the testes. Sperm have an acrosome-capped head, where 23 chromosomes reside in the nucleus, a mitochondria-containing middle piece, and a tail with a 9 + 2 pattern of microtubules.

Interstitial Cells: Produce Hormones

The male sex hormones, the androgens, are secreted by cells that lie between the seminiferous tubules. Therefore, they are called **interstitial cells** (fig. 20.2). The most important of the androgens is **testosterone,** whose functions are discussed on page 374.

Genital Tract: Testes to Glans Penis

Sperm are produced in the testes, but they mature and are stored in an **epididymis** (fig. 20.2), a tightly coiled tubule about 5–6 m (17 ft) in length. An epididymis lies just outside each testis. Each epididymis joins with a **vas** (ductus) **deferens,** which ascends through the *inguinal canal* and enters the abdomen, where it curves around the urinary bladder and empties into the urethra (fig. 20.1). Sperm are also stored in the first part of a vas deferens. They pass from each vas deferens into the urethra only when ejaculation (p. 373) is imminent.

Spermatic Cords: Hernia Risk

The testes are suspended in the scrotum by the *spermatic cords,* each of which consists of fibrous connective tissue and muscle fibers that enclose a vas deferens, the blood vessels, and the nerves. The region of the inguinal canal, where the spermatic cord passes into the abdomen, remains a weak point in the abdominal wall. As such, it is frequently the site of hernias. A **hernia** is an opening or separation of some part of the abdominal wall through which a portion of an internal organ, usually the intestine, protrudes.

Seminal Fluid: Three Sources

At the time of ejaculation, sperm leave the penis in a fluid called **seminal fluid** (also called semen). Three types of accessory glands add secretions to seminal fluid—the seminal vesicles, the prostate gland, and the Cowper's (bulbourethral) glands. The **seminal vesicles** lie at the base of the bladder, and each has a duct that joins with a vas deferens. The **prostate gland** is a single doughnut-shaped gland that surrounds the upper portion of the urethra just below the bladder. In older men, the prostate can enlarge and squeeze off the urethra, making urination painful and difficult. This condition can be treated medically or surgically. **Cowper's glands** are pea-sized organs that lie posterior to the prostate on either side of the urethra.

Each component of seminal fluid seems to have a particular function. Sperm are more viable in a basic solution, and seminal fluid, which is milky in appearance, has a slightly basic pH (about 7.5). Swimming sperm require energy, and seminal fluid contains the sugar fructose, which presumably serves as an energy source. Seminal fluid also contains prostaglandins, chemicals that cause the uterus in the female to contract. Some investigators now believe that uterine contraction is necessary to help propel the sperm toward the egg.

Male Orgasm: Upon Ejaculation

The **penis** (fig. 20.3) is the organ of sexual intercourse in males. The penis is a long shaft with an enlarged cone-shaped tip called the glans penis. At birth, the glans penis is covered by a layer of skin called the **foreskin,** or prepuce. Sometime near puberty, small glands located in the foreskin and the glans begin to produce an oily secretion. This secretion, along with dead skin cells, forms a cheesy substance known as smegma. In the child, no special cleansing method is needed to wash away smegma, but in the adult, the foreskin can be retracted to do so. **Circumcision** is the surgical removal of the foreskin, usually soon after birth.

Figure 20.3

Penis anatomy. **a.** Beneath the skin and the connective tissue lies the urethra, surrounded by erectile tissue. This tissue expands to form the glans penis, which in uncircumcised males is partially covered by the foreskin. **b.** Two other columns of erectile tissue in the penis are located dorsally.

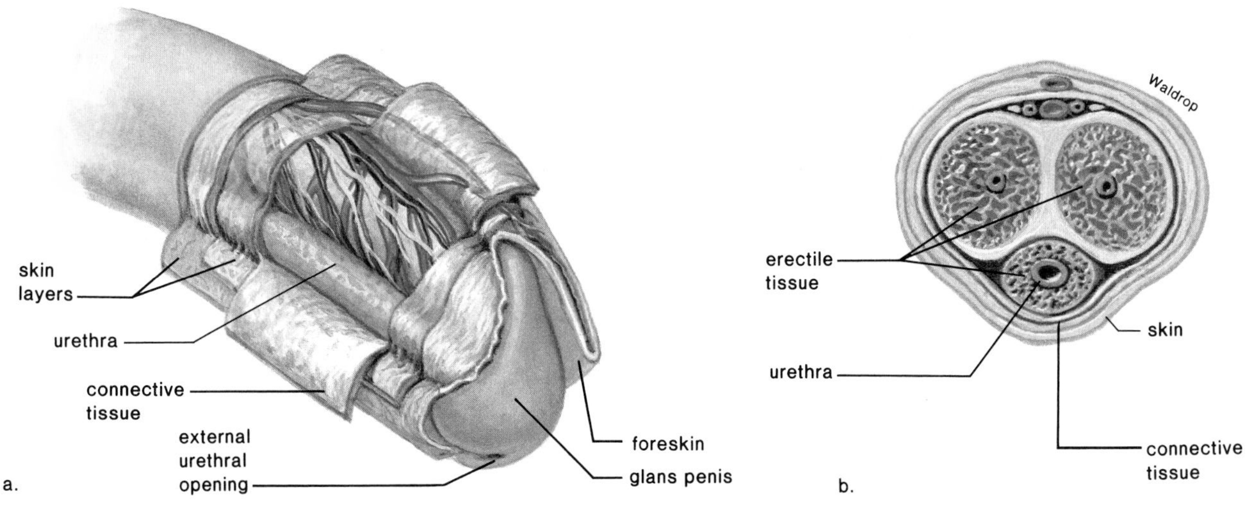

When the male is sexually aroused, the penis becomes erect and ready for intercourse. **Erection** is achieved because blood sinuses within the erectile tissue of the penis fill with blood. Parasympathetic impulses dilate the arteries of the penis, while the veins are passively compressed so that blood flows into the erectile tissue under pressure. **Impotency** is the condition in which erection cannot be achieved. There are medical and surgical remedies for impotency.

Ejaculation: Over 400 Million Sperm

As sexual stimulation intensifies, sperm enter the urethra from each vas deferens, and the accessory glands add their fluids. Once seminal fluid is in the urethra, rhythmic muscle contractions expel it from the penis in spurts. During **ejaculation**, a sphincter closes off the bladder so that no urine enters the urethra. (Notice that at different times the urethra carries either urine or seminal fluid.)

The contractions that expel seminal fluid from the penis are a part of male **orgasm,** the physiological and psychological sensations that occur at the climax of sexual stimulation. The psychological sensation of pleasure is centered in the brain, but the physiological reactions involve the genital (reproductive) organs and associated muscles, as well as the entire body. Marked muscular tension is followed by contraction and relaxation.

Following ejaculation and/or loss of sexual arousal, the penis returns to its normal flaccid state. The male typically experiences a refractory period, during which stimulation does not bring about an erection. The length of the refractory period increases with age.

There may be in excess of 400 million sperm in the 3.5 ml of seminal fluid expelled during ejaculation. The sperm count can be much lower than this, however, and fertilization (see fig. 20.7) still can occur.

> Sperm mature in the epididymides and are stored in a vas deferens before entering the urethra just prior to ejaculation. The accessory glands (seminal vesicles, prostate gland, and Cowper's glands) add secretions to seminal fluid, which leaves the penis during ejaculation.

Regulating Male Hormone Levels

The hypothalamus has ultimate control of the testes' sexual functions because it secretes gonadotropic-releasing hormone (GnRH), which stimulates the anterior pituitary to release the gonadotropic hormones. Two gonadotropic hormones, **follicle-stimulating hormone (FSH)** and **luteinizing hormone (LH),** are named for their function in females but exist in both sexes, stimulating the appropriate gonads in each. FSH promotes spermatogenesis in the seminiferous tubules, and LH promotes the production of testosterone in the interstitial cells. Sometimes, LH in males is given the name interstitial cell-stimulating hormone (ICSH).

Figure 20.4

The hypothalamus-pituitary-testes control relationship. GnRH (gonadotropic-releasing hormone) stimulates the anterior pituitary to secrete the gonadotropic hormones FSH and LH. FSH stimulates the testes to produce sperm, and LH stimulates the testes to produce testosterone. Testosterone and inhibin exert negative feedback control over the hypothalamus and the anterior pituitary, and this ultimately regulates the level of testosterone in blood.

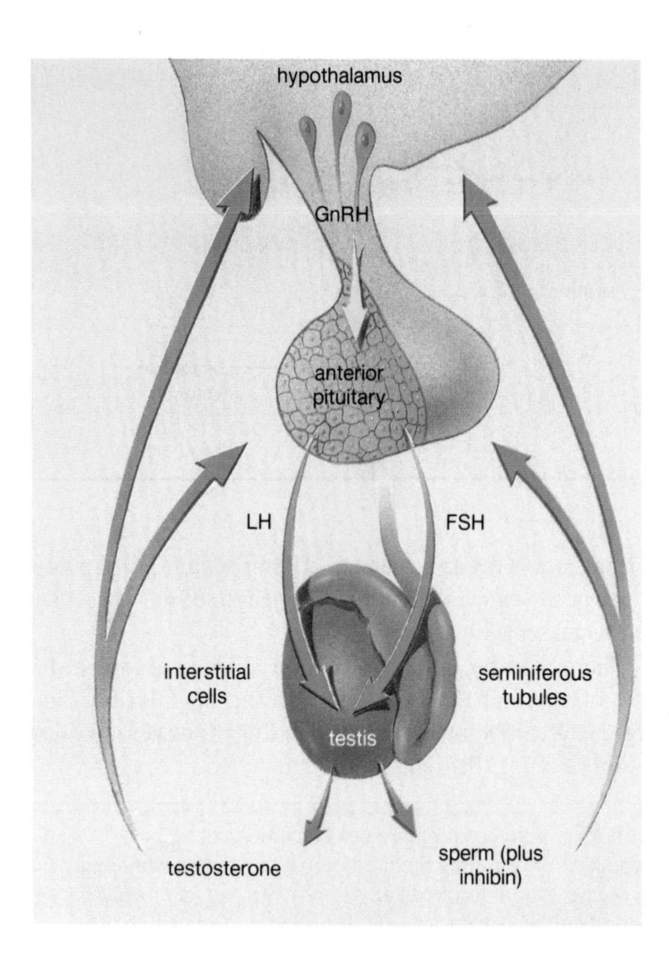

The hormones mentioned are involved in a negative feedback mechanism (fig. 20.4) that maintains the production of testosterone at a fairly constant level. For example, when the amount of testosterone in blood rises to a certain level, it causes the anterior pituitary to decrease its secretion of LH. As the level of testosterone begins to fall, the anterior pituitary increases its secretion of LH, and stimulation of the interstitial cells reoccurs. It should be emphasized that only minor fluctuations of testosterone level occur in the male and that the feedback mechanism in this case acts to maintain testosterone at a normal level. It had long been suspected that the seminiferous tubules produce a hormone that blocks FSH secretion. This substance, termed *inhibin,* has recently been isolated.

Testosterone: The Main Male Sex Hormone

The male sex hormone, testosterone, has many functions. It is essential for the normal development and function of the primary sex organs, those structures we have just discussed. It is also necessary for the production of sperm. FSH causes spermatogenic cells to take up testosterone, and it is testosterone that promotes their activity.

Greatly increased testosterone secretion at the time of puberty stimulates maturation of the penis and the testes. Testosterone also brings about and maintains the secondary sex characteristics in males, which develop at the time of puberty. Testosterone causes growth of a beard, axillary (underarm) hair, and pubic hair. It prompts the larynx and the vocal cords to enlarge, causing the voice to change. It is responsible for the greater muscular strength of males, and this is the reason some athletes take a supplemental *anabolic steroid,* which is either testosterone or a related chemical. The contraindications of anabolic steroid use are discussed in the reading on pages 360–361. Testosterone also causes oil and sweat glands in the skin to secrete; therefore, it contributes to acne and body odor. A side effect of testosterone activity is baldness. Genes for baldness are probably inherited by both sexes, but baldness is seen more often in males because of the higher level of testosterone.

Testosterone is believed to be largely responsible for the sex drive. It may even contribute to the supposed aggressiveness of males.

In males, FSH promotes spermatogenesis in the seminiferous tubules and LH promotes testosterone production by the interstitial cells. Testosterone stimulates growth of the male genitals during puberty and is necessary for maturation of sperm and development of the secondary sex characteristics. Do 20.1 Critical Thinking, found at the end of the chapter.

Figure 20.5
Side view of the female reproductive system. The ovaries produce one egg a month; fertilization occurs in the oviduct, and development occurs in the uterus. The vagina is the birth canal and the organ of sexual intercourse.

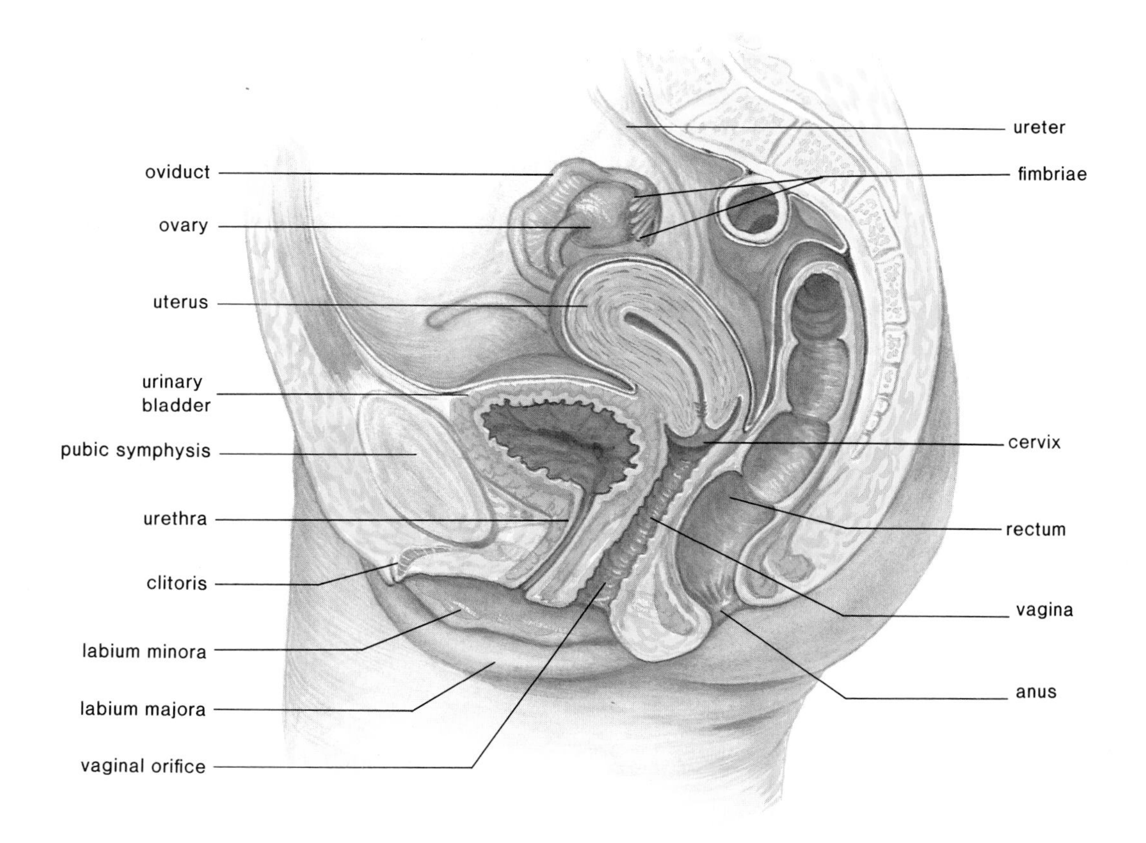

The Female Reproductive System

Figure 20.5 illustrates the female reproductive system, and table 20.2 lists the anatomical parts of this system.

Ovaries: An Egg a Month

The **ovaries** lie in shallow depressions, one on each side of the upper pelvic cavity. A longitudinal section through an ovary shows that it is made up of an outer cortex and an inner medulla. There are many saclike structures called **follicles** in the cortex, and each one contains an immature egg called an oocyte. A female is born with as many as 2 million follicles, but

Table 20.2
Female Reproductive System

Organ	Function
Ovary	Produces eggs and sex hormones
Oviduct (fallopian or uterine tube)	Conducts eggs toward uterus
Uterus (womb)	Houses developing fetus
Vagina	Receives penis during sexual intercourse and serves as birth canal

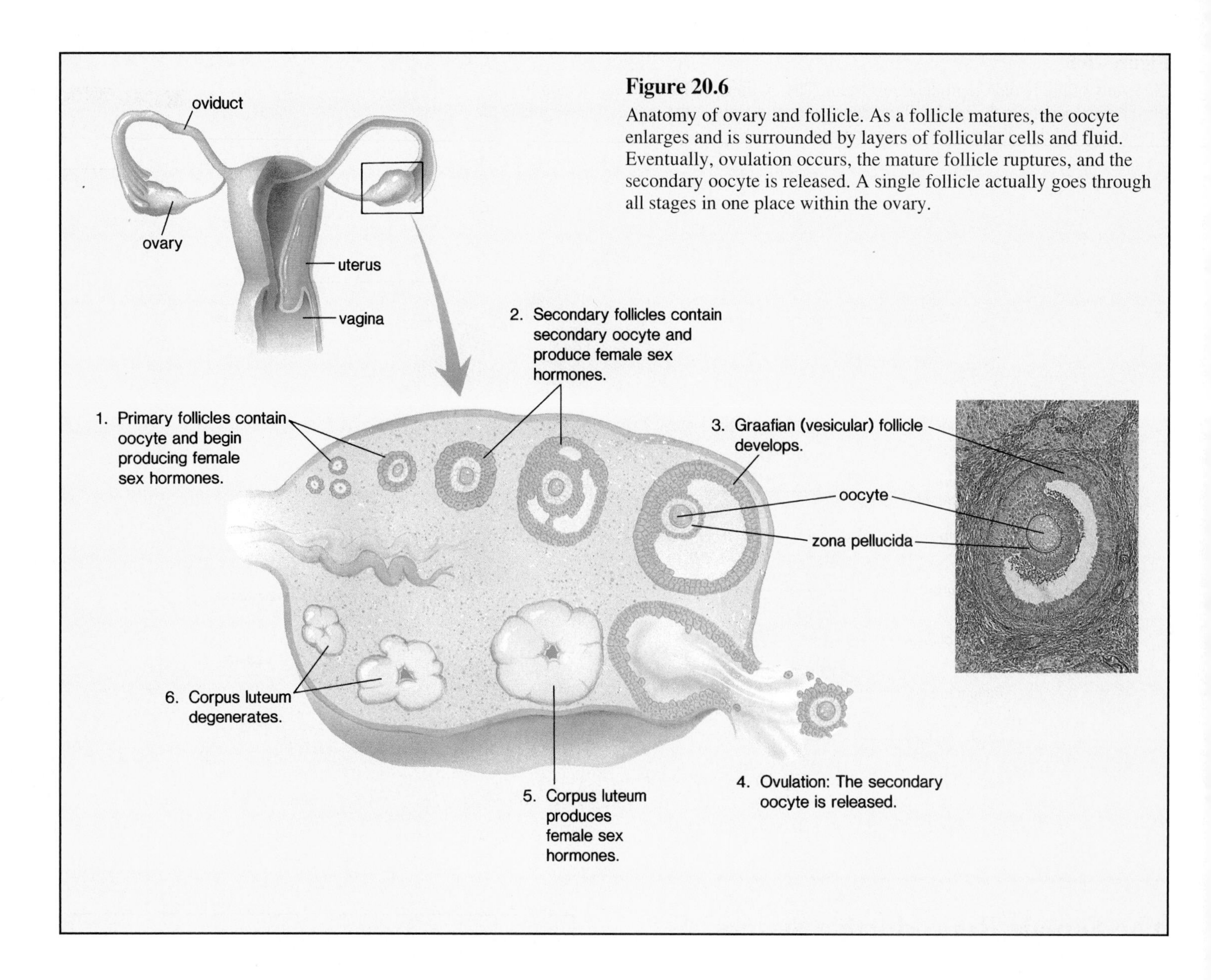

Figure 20.6

Anatomy of ovary and follicle. As a follicle matures, the oocyte enlarges and is surrounded by layers of follicular cells and fluid. Eventually, ovulation occurs, the mature follicle ruptures, and the secondary oocyte is released. A single follicle actually goes through all stages in one place within the ovary.

the number is reduced to 300,000–400,000 by the time of puberty. Only a small number of follicles (about 400) ever mature because a female usually produces only one egg per month during her reproductive years.

As the follicle undergoes maturation, it develops from a primary follicle to a secondary follicle to a **Graafian** (or vesicular) **follicle** (fig. 20.6). Oogenesis takes place in follicles. In a primary follicle, the primary oocyte divides meiotically into 2 cells, each having 23 chromosomes (see fig. 7.13). One of these cells, termed the *secondary oocyte,* receives almost all the cytoplasm. The other is a polar body, which disintegrates. A secondary follicle contains the secondary oocyte pushed to one side of a fluid-filled cavity. In a Graafian follicle, pressure within the fluid-filled cavity increases to the point that the follicle wall balloons out on the surface of the ovary and bursts, releasing the secondary oocyte (often called an egg for convenience) surrounded by a mucoprotein substance, the *zona pellucida,* and follicular cells. This is referred to as **ovulation.** Once a follicle has lost its egg, it develops into a **corpus luteum,** a hormone-secreting structure. If pregnancy does not occur, the corpus luteum begins to degenerate after about 10 days. If pregnancy does occur, the corpus luteum persists for 3–6 months. The follicle and the corpus luteum secrete the female sex hormones estrogen and progesterone, as discussed on page 378.

In females, oogenesis occurs within the ovaries, where one follicle reaches maturity each month. This follicle balloons out of the ovary and bursts to release the egg. The ruptured follicle develops into a corpus luteum. The follicle and the corpus luteum produce the female sex hormones estrogen and progesterone.

Do Well on Tests and Exams
Get Your Own Tutor!

Wm. C. Brown Publishers has now developed a take-home tutor that will help you achieve your goals in the classroom and prepare for careers in the nursing/allied health sciences. The ***Biology Learning Support Package*** is a complete set of additional study aids designed to facilitate learning and achievement. The package includes:

STUDENT STUDY GUIDE

ISBN 0-697-13702-3

For each text chapter there is a corresponding study guide chapter that includes suggestions for review, objective questions, and essay questions that will review the major chapter concepts.

•••

HOW TO STUDY SCIENCE

ISBN 0-697-14474-7

This excellent workbook offers you helpful suggestions for meeting the considerable challenges of a college science course. It offers tips on how to take notes, how to get the most out of laboratories, as well as how to overcome science anxiety. The book's unique design helps you develop critical thinking skills, while facilitating careful note taking.

•••

THE LIFE SCIENCE LEXICON

ISBN 0-697-12133-X

This portable, inexpensive reference will help you quickly master the vocabulary of the life sciences. Not a dictionary, it carefully explains the rules of word construction and derivation, in addition to giving complete definitions of all important terms.

•••

IS YOUR MATH READY FOR BIOLOGY?

ISBN 0-697-22677-8

This unique booklet provides a diagnostic test that measures your math ability. Part II of the booklet provides helpful hints on the necessary math skills needed to successfully complete a biological science course.

•••

ANIMATED ILLUSTRATIONS OF PHYSIOLOGICAL PROCESSES VIDEOTAPE

ISBN 0-697-21512-0

This videotape animates 13 illustrations of complex physiological processes. Through moving visual animations, these hard-to-understand concepts are made palatable.

To get your own take-home tutor, just ask your bookstore manager, or if unavailable through your bookstore, call toll-free 1-800-338-5578 and order using your credit card.

Genital Tract: Oviducts to Vagina

The female genital tract includes the oviducts, the uterus, and the vagina.

Oviducts: Tubes to the Uterus

The oviducts, also called uterine or fallopian tubes, extend from the uterus to the ovaries. The oviducts are not attached to the ovaries; instead, they have fingerlike projections called **fimbriae,** which sweep over the ovary at the time of ovulation. When the egg bursts (fig. 20.6) from the ovary during ovulation, it is usually swept up into an oviduct by the combined action of the fimbriae and the beating of cilia that line the oviducts.

Because the egg must cross a small space before entering an oviduct, it is possible for the egg to get lost and instead enter the abdominal cavity. Such eggs usually disintegrate, but in some rare cases, they have been fertilized in the abdominal cavity and have implanted themselves in the wall of an abdominal organ. Very rarely, such embryos have come to term, the child being delivered by surgery.

Once in the oviduct, muscular contractions and the cilia of epithelial cells propel the egg slowly toward the uterus. Fertilization (fig. 20.7), the completion of oogenesis, and zygote formation normally occur in an oviduct. The developing embryo usually arrives at the uterus after several days and then embeds, or implants, itself in the uterine lining, which has been prepared to receive it. Occasionally, the embryo becomes embedded in the wall of an oviduct, where it begins to develop. These "tubular" pregnancies cannot succeed because the oviducts are not anatomically capable of allowing full development to occur. An *ectopic pregnancy* is one that begins anywhere outside the uterus.

The Uterus: Upside-Down Pear

The **uterus** is a thick-walled, muscular organ about the size and the shape of an inverted pear. Normally, it lies above and is tipped over the urinary bladder. The oviducts join the uterus anteriorly, while posteriorly, the **cervix** enters into the vagina at nearly a right angle. A small opening in the cervix leads to the lumen of the vagina.

Figure 20.7

Fertilization. A single sperm enters the egg, and a new life begins. The reproductive systems of males and females are designed to bring about this union of the gametes.

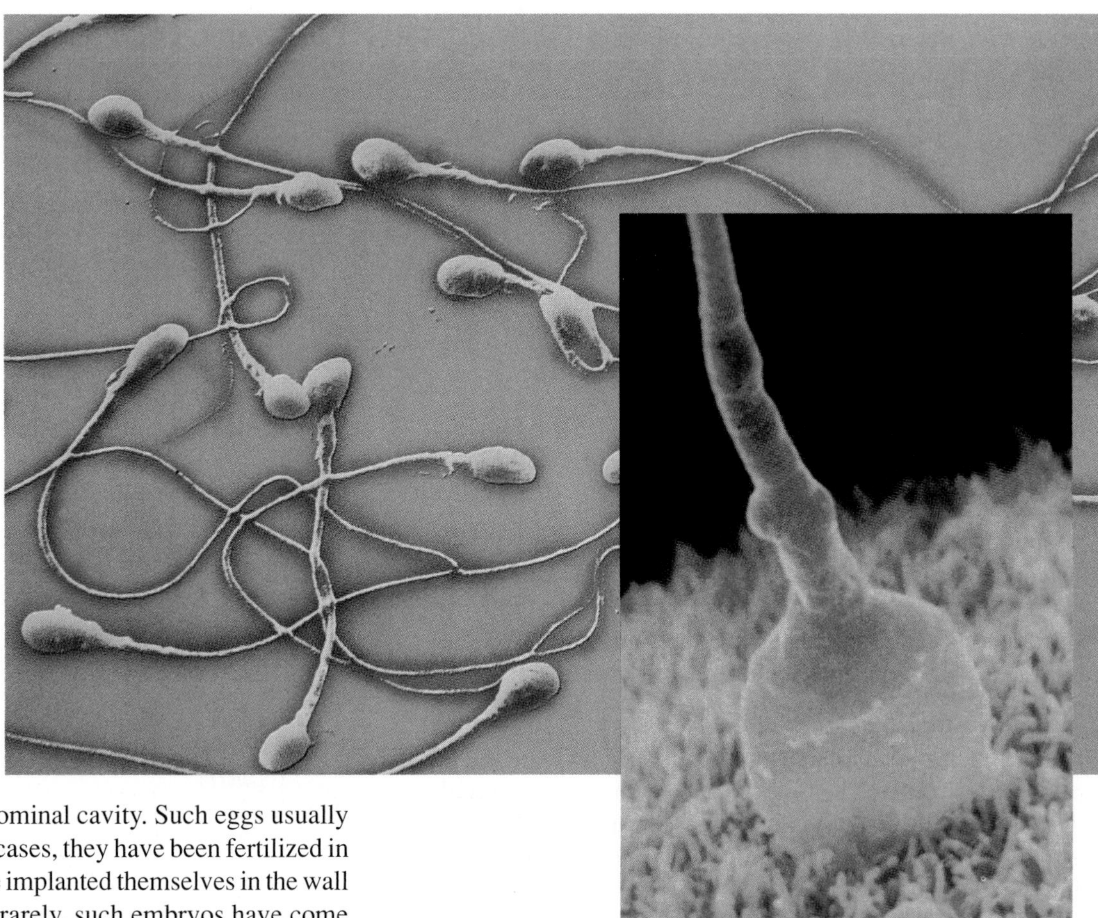

Development of the fetus normally takes place in the uterus. This organ, sometimes called the womb, is approximately 5 cm wide in its usual state but is capable of stretching to over 30 cm to accommodate the growing baby. The lining of the uterus, called the **endometrium,** participates in the formation of the placenta (p. 381), which supplies nutrients needed for fetal development. The endometrium has 2 layers: a basal layer and an inner functional layer. In the nonpregnant female, the functional layer of the endometrium varies in thickness according to a monthly reproductive cycle, called the uterine cycle (p. 379).

Cancer of the cervix is a common form of cancer in women. Early detection is possible by means of a **Pap smear,** which requires the removal of a few cells from the region of the cervix for microscopic examination. If the cells are cancerous, a *hysterectomy* may be recommended. A hysterectomy is the removal of the uterus. Removal of the ovaries in addition to the uterus is termed an *ovariohysterectomy.* Because the vagina remains, the woman can still engage in sexual intercourse.

Figure 20.8
External genitals of female. At birth, the opening of the vagina is partially blocked by a membrane called the hymen. Physical activities and sexual intercourse disrupt the hymen.

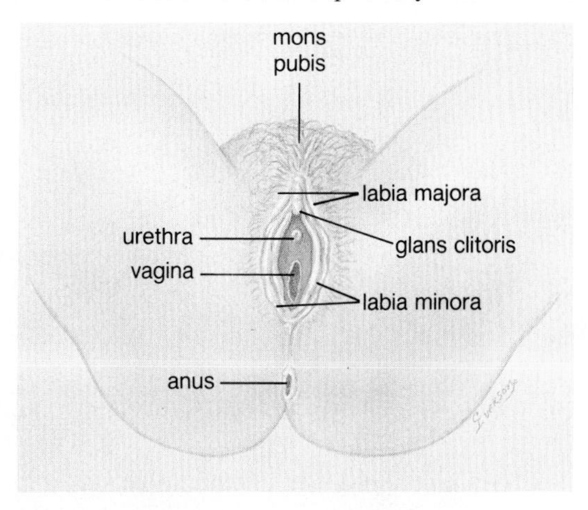

Vagina: Stretching Tube

The **vagina** is a tube that makes a 45-degree angle with the small of the back. The mucous membrane lining the vagina lies in folds, which extend as the wall stretches. This capacity to extend is especially important when the vagina serves as the birth canal, and it can also facilitate sexual intercourse.

External Genitals: Vulva

The external genitals of the female (fig. 20.8) are known collectively as the **vulva.** The vulva includes 2 large, hair-covered folds of skin called the **labia majora.** They extend backward from the *mons pubis,* a fatty prominence underlying the pubic hair. The **labia minora** are 2 small folds lying just inside the labia majora. They extend forward from the vaginal orifice (opening) to encircle and form a foreskin for the *clitoris,* an organ that is homologous to the penis. Although quite small, the clitoris has a shaft of erectile tissue and is capped by a pea-shaped glans. The glans clitoris also has sense receptors that allow it to function as a sexually sensitive organ.

The *vestibule,* a cleft between the labia minora, contains the orifices of the urethra and the vagina. The vagina may be partially closed by a ring of tissue called the hymen. The hymen is ordinarily ruptured by initial sexual intercourse; however, it can also be disrupted by other types of physical activities. If the hymen persists after sexual intercourse, it can be surgically ruptured.

Notice that the urinary and reproductive systems in the female are entirely separate. For example, the urethra carries only urine, and the vagina serves only as the birth canal and the organ for sexual intercourse.

> The egg enters the oviducts, which lead to the uterus, followed by the vagina. The vagina opens into the vestibule, the location of female external genitals.

Female Orgasm: Wide Variety

Sexual response in the female is not as distinct as in the male, but there are certain similarities. The clitoris is believed to be an especially sensitive organ for initiating sexual sensations. It is possible for the clitoris to become slightly erect as its erectile tissues engorge with blood, but vasocongestion is more obvious in the labia minora, which expand and deepen in color. Erectile tissue within the vaginal wall also expands with blood, and the added pressure in these blood vessels causes small droplets of fluid to squeeze through the vessel walls and to lubricate the vagina.

Release from muscular tension occurs in females during orgasm, especially in the region of the vulva and vagina but also throughout the entire body. Increased uterine motility may assist the transport of sperm toward the oviducts. Since female orgasm is not signaled by ejaculation, there is a wide range in normalcy of sexual response.

Regulating Female Hormone Levels

Hormone regulation in the female is quite complex, so we begin with a simplified presentation and follow with a more in-depth presentation for those who wish to study the topic in greater detail.

The following glands and hormones are involved in hormonal regulation.

Hypothalamus: secretes gonadotropic-releasing hormone (*GnRH*)
Anterior pituitary: secretes follicle-stimulating hormone (*FSH*) and luteinizing hormone (*LH*), the gonadotropic hormones
Ovaries: secrete **estrogen** and **progesterone,** the female sex hormones

A Short Look at Female Cycles

Ovarian Cycle: FSH and LH. The gonadotropic and sex hormones are not present in constant amounts in the female and instead are secreted at different rates during a monthly **ovarian cycle,** which lasts an average of 28 days but may vary widely in individuals. For simplicity's sake, it is convenient to emphasize that during the follicular phase of a cycle (days 1–13, table 20.3), FSH secreted by the anterior pituitary promotes the development of a follicle in the ovary and this follicle secretes estrogen. As the blood estrogen level rises, the hormone exerts negative feedback control over the anterior pituitary secretion of FSH so that this follicular phase comes to an end (fig. 20.9). The end of the follicular phase is marked by ovulation on the fourteenth day of the cycle lasting 28 days. Similarly, it can be emphasized that during the last half of the ovarian cycle (days 15–28, table 20.3), anterior pituitary secretion of LH promotes the development of a corpus luteum, which secretes progesterone. As the blood progesterone level rises, it exerts negative feedback control over anterior pituitary secretion of LH so that

Table 20.3
Ovarian and Uterine Cycles (Simplified)

Ovarian Cycle	Events	Uterine Cycle	Events
Follicular phase (Days 1–13)	FSH stimulates ovary Follicle matures Estrogen is released	Menstruation (Days 1–5) Proliferative phase (Days 6–13)	Endometrium breaks down Endometrium rebuilds
Ovulation (Day 14[*])			
Luteal phase (Days 15–28)	LH stimulates ovary Corpus luteum develops Progesterone is released	Secretory phase (Days 15–28)	Endometrium thickens and glands are secretory

[*]Assuming a cycle lasting 28 days

The hypothalamus produces GnRH (gonadotropic-releasing hormone)

GnRH stimulates the anterior pituitary to produce FSH (follicle-stimulating hormone) and LH (luteinizing hormone)

FSH stimulates the follicle to produce estrogen and LH stimulates the corpus luteum to produce progesterone

Estrogen and progesterone affect the sex organs (e.g., uterus) and the secondary sex characteristics, and exert feedback control over the hypothalamus and the anterior pituitary

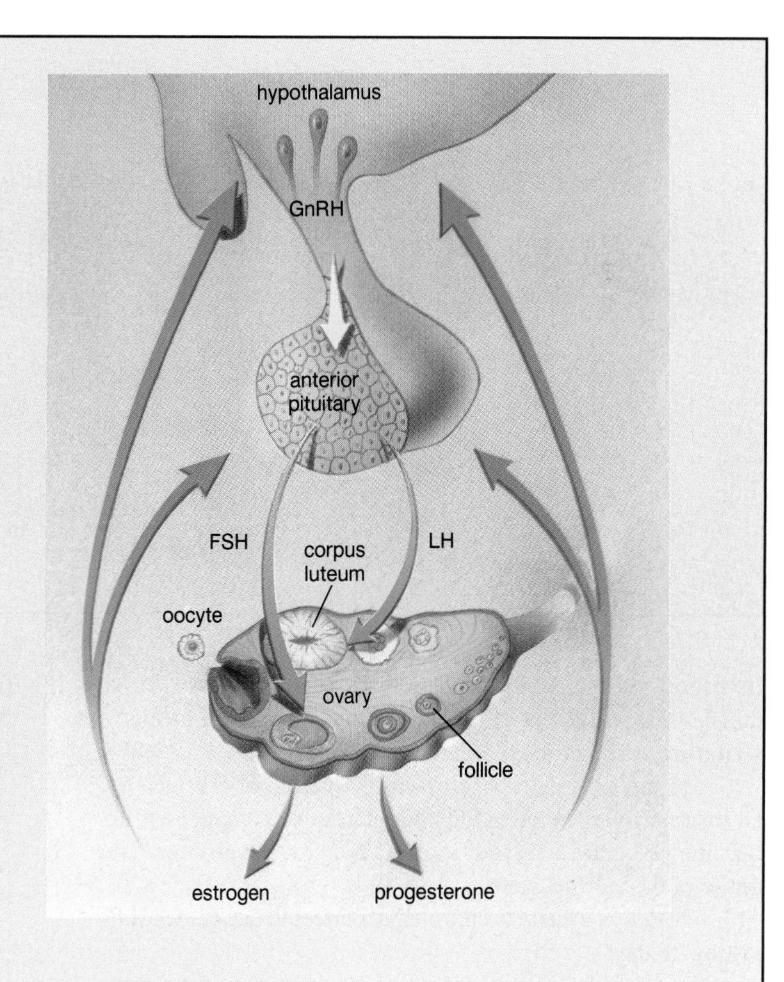

Figure 20.9
The hypothalamus-pituitary-ovary control relationship.

the corpus luteum begins to degenerate. As the luteal phase comes to an end, menstruation occurs.

Uterine Cycle: Estrogen and Progesterone. The female sex hormones estrogen and progesterone have numerous functions, one of which is discussed here. The effect these hormones have on the endometrium of the uterus causes the uterus to undergo a cyclical series of events known as the **uterine cycle** (table 20.3). Cycles that last 28 days are divided as follows.

During *days 1–5,* there is a low level of female sex hormones in the body, causing the uterine lining to disintegrate and

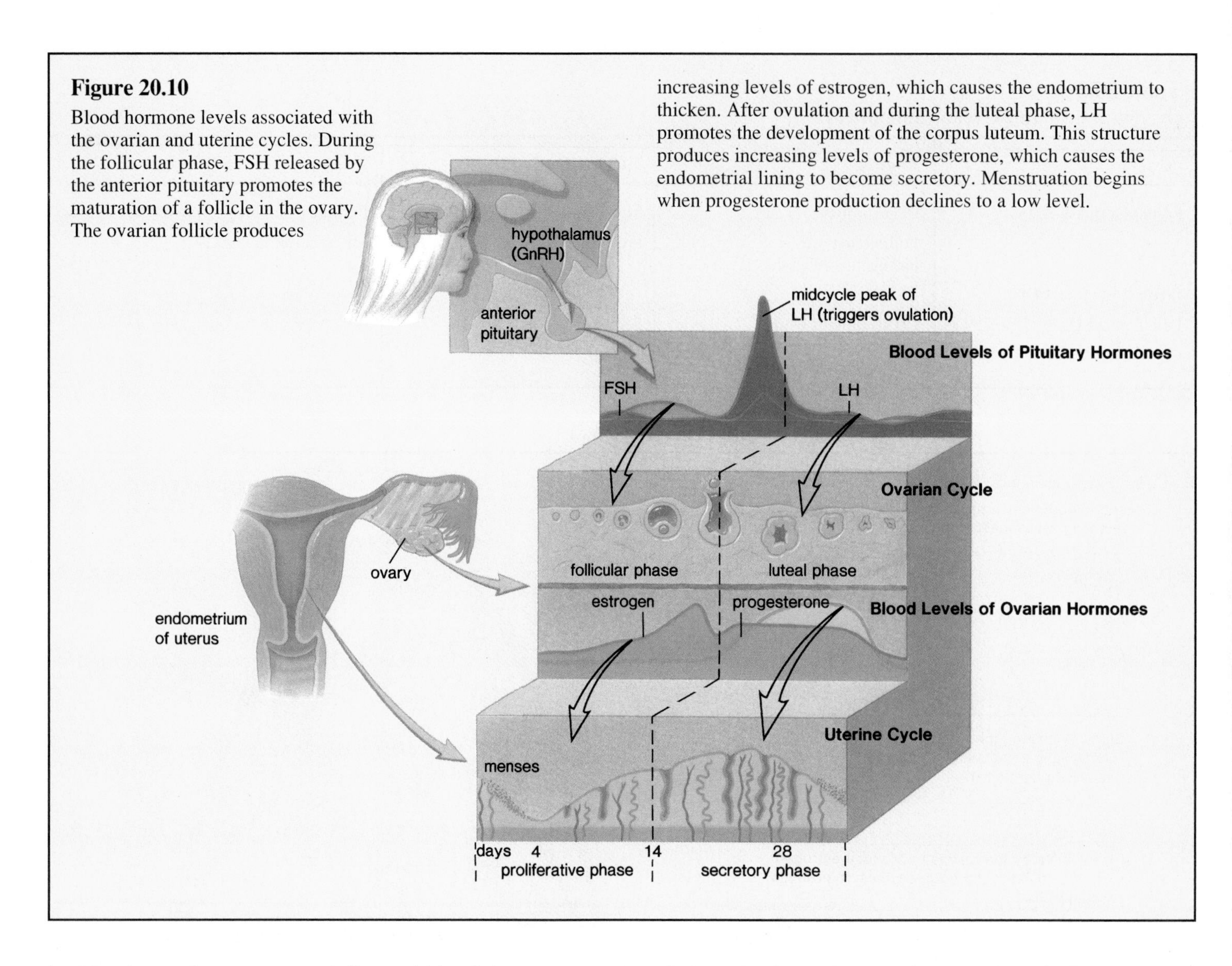

Figure 20.10

Blood hormone levels associated with the ovarian and uterine cycles. During the follicular phase, FSH released by the anterior pituitary promotes the maturation of a follicle in the ovary. The ovarian follicle produces increasing levels of estrogen, which causes the endometrium to thicken. After ovulation and during the luteal phase, LH promotes the development of the corpus luteum. This structure produces increasing levels of progesterone, which causes the endometrial lining to become secretory. Menstruation begins when progesterone production declines to a low level.

its blood vessels to rupture. A flow of blood, known as the *menses,* passes out of the vagina during a period of **menstruation,** also known as the menstrual period.

During *days 6–13,* increased production of estrogen by an ovarian follicle causes the endometrium to thicken and to become vascular and glandular. This is called the proliferative phase of the uterine cycle.

Ovulation usually occurs on the fourteenth day of the cycle lasting 28 days.

During *days 15–28,* increased production of progesterone by the corpus luteum causes the endometrium to double in thickness and the uterine glands to mature, producing a thick mucous secretion. This is called the secretory phase of the uterine cycle. The endometrium is now prepared to receive the developing embryo, but if pregnancy does not occur, the corpus luteum degenerates and the low level of sex hormones in the female body causes the uterine lining to break down. This is evident, due to the menstrual discharge that begins at this time. Even while menstruation is occurring, the anterior pituitary begins to increase its secretion of FSH and a new follicle begins to mature. Table 20.3 indicates how the ovarian cycle controls the uterine cycle.

A Long Look at Female Cycles

Figure 20.10 shows the changes in blood concentration of all 4 hormones participating in the ovarian and uterine cycles. Notice that all 4 of these hormones (FSH, LH, estrogen, and progesterone) are present during the entire 28 days of the cycle. Therefore, in actuality both FSH and LH *are* present during the follicular phase and both are needed for follicle development and egg maturation. The follicle secretes primarily estrogen and a very minimal amount of progesterone. Similarly, both LH and FSH are present in decreased amounts during the luteal phase. LH may be primarily responsible for corpus luteum formation, but the corpus luteum secretes both progesterone and estrogen. The effect that these hormones have on the endometrium has already been stated: Estrogen stimulates growth of the endometrium and readies it for reception of progesterone, which causes it to thicken and to become secretory.

Feedback Control: Negative and Positive. As the estrogen level increases during the first part of the follicular phase, FSH secretion begins to decrease due to negative feedback. However, the high level of estrogen is believed to exert *positive feedback on the hypothalamus,* causing it to secrete GnRH, after which the anterior pituitary momentarily releases an unusually large amount of FSH and LH. It is the surge of LH that is believed to promote ovulation. During the luteal phase, estrogen and progesterone bring about feedback inhibition as expected, and the levels of both LH and FSH decline steadily. In this way, all 4 hormones eventually reach their lowest levels, resulting in menstruation. Therefore, the corpus luteum degenerates unless pregnancy occurs. In some mammals, evidence suggests that prostaglandins (p. 362) are involved in degeneration, but this is not believed to be the case in humans.

During the first half of the ovarian cycle, FSH released by the anterior pituitary causes maturation of a follicle, which secretes estrogen. After ovulation and during the second half of the cycle, LH from the anterior pituitary converts the follicle into the corpus luteum, which produces progesterone. Estrogen and progesterone regulate the uterine cycle, in which the endometrium builds up and is then shed during menstruation.

Pregnancy: Hormones Change

Pregnancy occurs when the developing embryo embeds itself in the endometrium several days after fertilization. During **implantation**, a membrane surrounding the embryo produces a gonadotropic hormone called **human chorionic gonadotropic hormone (HCG)**, which prevents degeneration of the corpus luteum and instead causes it to secrete even larger quantities of progesterone. Progesterone (*pro* means *for*; *gestation* means *pregnancy*) also inhibits the motility of the uterus and together with estrogen prepares the breasts for lactation. The corpus luteum may be maintained for as long as 6 months, even after the placenta is fully developed.

The **placenta** (see fig. 21.17) originates from both maternal and fetal tissue and is the region of exchange of molecules between fetal and maternal blood, although there is no mixing of the 2 types of blood. After its formation, the placenta continues to produce HCG. It also begins to produce progesterone and estrogen, which have 2 effects: they shut down the anterior pituitary so that no new follicles mature, and they maintain the lining of the uterus so that the corpus luteum is not needed. There is no menstruation during the 9 months of pregnancy.

Pregnancy Tests: Looking for HCG. Pregnancy tests, which are readily available in hospitals, clinics, and now even drug- and grocery stores, detect the HCG present in the urine and blood of a pregnant woman.

Before the advent of monoclonal antibodies, only a hospital blood test using radioactive material was available to detect pregnancy before the first missed menstrual period.

Figure 20.11

Female breast anatomy. The female breast contains lobules consisting of ducts and alveoli. The alveoli are lined by milk-producing cells in the lactating (milk-producing) breast.

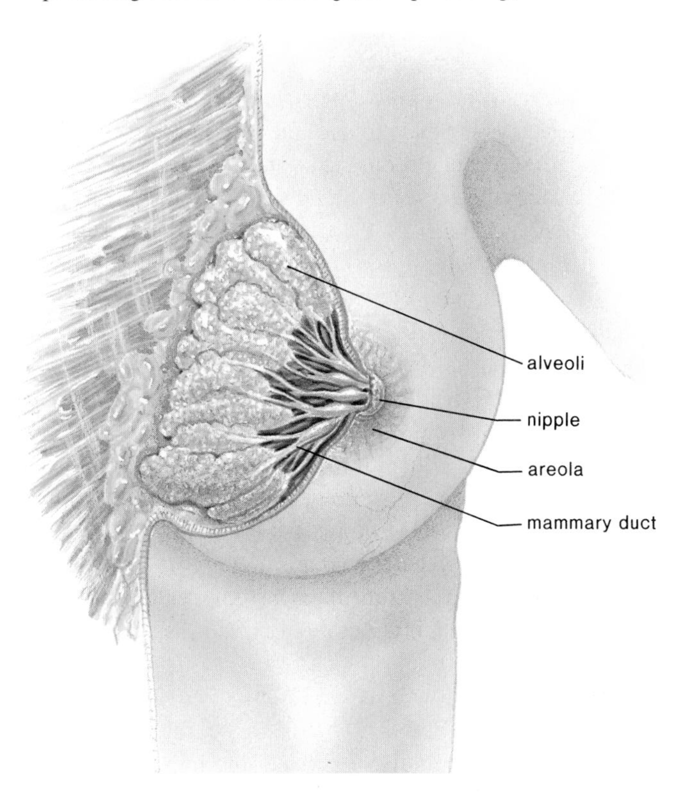

Now there is a monoclonal antibody (p. 231) test for the detection of pregnancy as early as 10 days after conception. This test can be done on a urine sample in a doctor's office, and the results are available within the hour.

The physical signs that oftentimes prompt a woman to have a pregnancy test are cessation of menstruation, increased frequency of urination, morning sickness, and increase in the size and the fullness of the breasts, as well as darkening of the *areolae,* the ring of pigmented skin that surrounds each nipple (fig. 20.11).

Female Sex Hormones: Estrogen and Progesterone

The female sex hormones estrogen and progesterone have many effects on the body. In particular, estrogen secreted at the time of puberty stimulates the growth of the uterus and the vagina. Estrogen is necessary for egg maturation and is largely responsible for the secondary sex characteristics in females. For example, it is responsible for the onset of the uterine cycle, as well as female body hair and fat distribution. In general, females have a more rounded appearance than males because of a greater accumulation of fat beneath the skin. Also, the pelvic girdle enlarges in females so that the pelvic cavity has a larger relative size compared to males; this means that females have wider hips. Both estrogen and progesterone are also required for breast development.

Female Breasts: Hormones Control Milk Production. A female breast contains 15–25 lobules (fig. 20.11), each with its own mammary duct, which opens at the nipple. The nipple is surrounded by a pigmented area called the **areola** (pl., areolae). Hair and sweat glands are absent from the nipples and areolae, but glands are present that secrete a saliva-resisting lubricant to protect the nipples, particularly during nursing. Smooth muscle fibers in the region of the areola may cause the nipple to become erect in response to sexual stimulation or cold. Within each lobe, the *mammary duct* divides into numerous other ducts that end in blind sacs called *alveoli*. In a lactating breast, the cells of the alveoli produce milk.

Milk is not produced during pregnancy. Prolactin (PRL) is needed for lactation (milk production) to begin, and the production of this hormone is suppressed because of the feedback inhibition estrogen and progesterone have on the anterior pituitary during pregnancy. It is not until a couple of days after delivery that milk production begins, and in the meantime, the breasts produce a watery, yellowish white fluid called **colostrum.** Colostrum, a source of passive immunity for the infant, differs from milk in that it contains more protein and less fat. The continued production of milk requires continued production of PRL, which occurs as long as the woman is breast-feeding.

The hormone oxytocin is necessary to milk letdown. When a breast is suckled, the nerve endings in the areola are stimulated, and nerve impulses travel to the hypothalamus, which causes oxytocin to be released by the posterior pituitary. When this hormone arrives at the breasts, it causes the lobules to contract so that milk flows into the ducts.

Menopause: Ovaries Don't Respond

Menopause, the period in a woman's life during which the ovarian and uterine cycles cease, is likely to occur between ages 45 and 55. The ovaries are no longer responsive to the gonadotropic hormones produced by the anterior pituitary, and the ovaries no longer secrete estrogen or progesterone. At the onset of menopause, the uterine cycle becomes irregular, but as long as menstruation occurs, it is still possible for a woman to conceive. Therefore, a woman usually is not considered to have completed menopause until there has been no menstruation for a year.

The hormone changes during menopause often produce physical symptoms such as "hot flashes," which are caused by circulatory irregularities, dizziness, headaches, insomnia, sleepiness, and depression. Again, there is great variation among women, and any of these symptoms may be absent altogether.

Women sometimes report an increased sex drive following menopause. It has been suggested that this may be due to androgen production by the adrenal cortex.

Estrogen and to some extent progesterone affect the female genitals, promote development of the egg, and maintain the secondary sex characteristics. PRL causes the breasts to secrete milk after the birth of a child, while another hormone, oxytocin, is responsible for milk letdown. When menopause occurs, FSH and LH are still produced by the anterior pituitary, but the ovaries are no longer able to respond to them. Do 20.2 Critical Thinking, found at the end of the chapter.

The Control of Reproduction

Several means are available to dampen or enhance our reproductive potential.

Birth Control: Variety of Methods

The most reliable method of birth control is abstinence, that is, the absence of sexual intercourse. This form of birth control has the added advantage of preventing sexually transmitted disease. Other, perhaps more common, means of birth control used in this country are given in table 20.4. The table gives the effectiveness for the various birth-control methods listed. For example, with the least effective method given in the table, we expect that within a year, 70 out of 100 women, or 70%, of sexually active women will not get pregnant, while 30 women will get pregnant. A few of the birth-control devices listed in the table are shown in figure 20.12.

Searching for Other Means of Birth Control

There has been a revival of interest in barrier methods of birth control, including the male condom, because these methods offer some protection against sexually transmitted diseases. A female condom is expected to be on the market soon. The closed end of a large plastic tube has a flexible ring that fits onto the cervix. The open end of the tube has a ring that covers the external genitals.

Investigators have long searched for a "male pill." Analogues of gonadotropic-releasing hormone have been used to prevent the hypothalamus from stimulating the anterior pituitary. Inhibin has also been used to prevent the anterior pituitary (fig. 20.4) from producing FSH. Testosterone and/or related chemicals have been used to inhibit spermatogenesis in males, but there are usually feminizing side effects because an excess of testosterone is changed to estrogen by the body.

Table 20.4
Common Birth-Control Methods

Name	Procedure	Methodology	Effectiveness	Risk
Vasectomy	Vas deferentia are cut and tied	No sperm in seminal fluid	Almost 100%	Irreversible sterility
Tubal ligation	Oviducts are cut and tied	No eggs in oviduct	Almost 100%	Irreversible sterility
Oral contraception (Pill)	Hormone medication taken daily	Anterior pituitary does not release FSH and LH	Almost 100%	Thromboembolism, especially in smokers
Depo-Provera	Four injections of progesterone-like steroid a year	Anterior pituitary does not release FSH and LH	About 99%	Breast cancer? Osteoporosis?
Norplant	Tubes of progestin (form of progesterone) are implanted under skin	Anterior pituitary does not release FSH and LH	More than 90%	Presently unknown
IUD	Plastic coil is inserted into uterus by physician	Prevents implantation	More than 90%	Infection (PID)
Vaginal sponge	Sponge permeated with spermicide is inserted in vagina	Kills sperm on contact	About 90%	Presently unknown
Diaphragm	Plastic cup is inserted into vagina to cover cervix before intercourse	Blocks entrance of sperm to uterus	With jelly, about 90%	Presently unknown
Cervical cap	Rubber cup is held by suction over cervix	Delivers spermicide near cervix	Almost 85%	Cancer of cervix?
Male condom	Latex sheath is fitted over erect penis at time of intercourse	Traps sperm and prevents STDs	About 85%	Presently unknown
Female condom	Plastic tubing is fitted inside vagina	Blocks entrance of sperm to uterus and prevents STDs	About 85%	Presently unknown
Coitus interruptus (withdrawal)	Male withdraws penis before ejaculation	Prevents sperm from entering vagina	75%	Presently unknown
Jellies, creams, foams	Contain spermicidal chemicals that are inserted before intercourse	Kill a large number of sperm	About 75%	Presently unknown
Rhythm method	Day of ovulation is determined by record keeping; various methods of testing	Intercourse avoided on certain days of the month	About 70%	Presently unknown
Douche	Vagina and uterus are cleansed after intercourse	Washes out sperm	Less than 70%	Presently unknown

Figure 20.12
Various birth-control devices. **a.** IUD. **b.** Vaginal sponge. **c.** Diaphragm. **d.** Birth-control pills. **e.** Vaginal spermicide. **f.** Male condom.

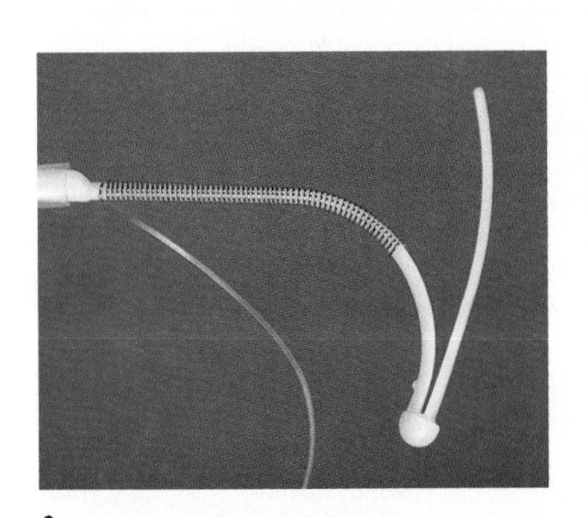
a.

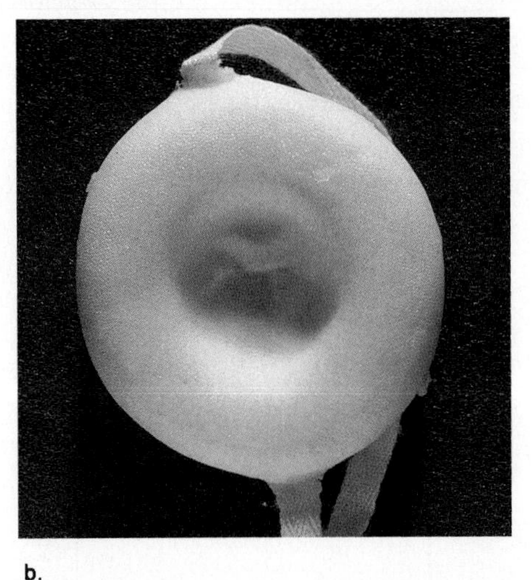
b.

c.

d.

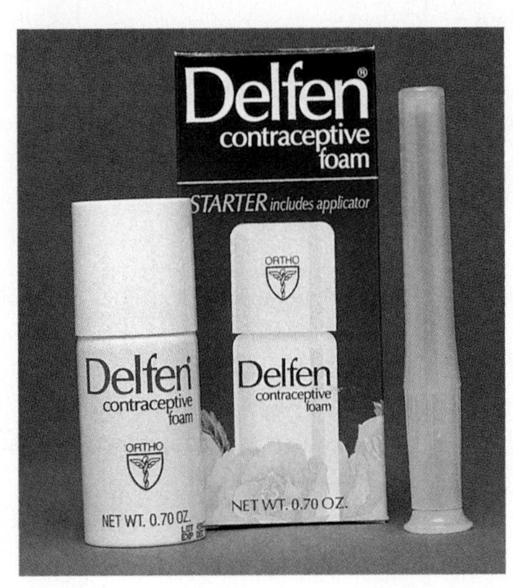

e.

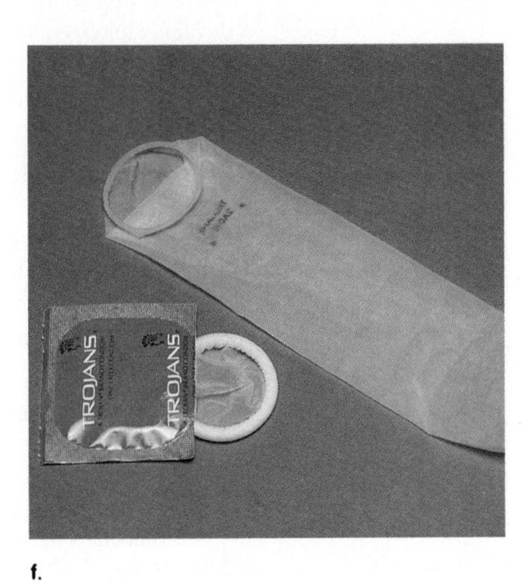

f.

Morning-after Pills. There are morning-after regimens available that, depending on when the woman begins medication, either prevent fertilization altogether or stop the fertilized egg from ever implanting. These regimens involve taking pills containing synthetic progesterone and/or estrogen in a manner prescribed by a physician. Many women do not realize that this method of birth control is available, and yet it is estimated that use of these regimens could greatly reduce the number of unintended pregnancies. Effective treatment sometimes causes nausea and vomiting, which can be severe.

RU-486 is a pill presently used in France and Britain that is now being considered for use in this country. RU-486 causes the loss of an implanted embryo by blocking the progesterone receptors of the cells in the uterine lining. Without functioning receptors for progesterone, the uterine lining sloughs off, carrying the embryo with it. When taken in conjunction with a prostaglandin to induce uterine contractions, RU-486 is 95% effective. It is possible that some day the medication will be used by women who are experiencing delayed menstruation without knowing if they are actually pregnant.

> There are numerous well-known birth-control methods and devices available to those who wish to prevent pregnancy. Their effectiveness varies. In addition, new methods are expected to be developed.

Figure 20.13
Mother and child. Sometimes couples utilize alternative methods of reproduction in order to experience the joys of parenthood.

Infertility: One Out of 4 Couples

Sometimes, couples do not need to prevent pregnancy; conception or fertilization does not occur despite frequent intercourse. The American Medical Association estimates that 15% of all couples in this country are unable to have any children and therefore are properly termed *sterile*; another 10% have fewer children than they wish and therefore are termed *infertile*. The latter assumes that the couple has been unsuccessfully trying to become pregnant for at least one year.

What Causes Infertility

The 2 major causes of infertility in females are blocked oviducts, possibly due to pelvic inflammatory disease (PID), discussed in a following section, and failure to ovulate due to low body weight. *Endometriosis,* the presence of uterine tissue outside the uterus, particularly in the oviducts and on the abdominal organs, can also contribute to infertility. Endometriosis occurs when the menstrual discharge flows up into the oviducts and out into the abdominal cavity. This backward flow allows living uterine cells to establish themselves in the abdominal cavity where they go through the usual uterine cycle, causing pain and structural abnormalities that make it more difficult for a woman to conceive.

Sometimes the causes of infertility can be corrected surgically and/or medically so that couples can experience the joys of parenthood (fig. 20.13). If no obstruction is apparent and body weight is normal, it is possible to give females HCG extracted from the urine of postmenopausal women. This treatment causes multiple ovulations and sometimes multiple pregnancies.

The most frequent cause of infertility in males is low sperm count and/or a large proportion of abnormal sperm. Disease, radiation, chemical mutagens, high testes temperature, and the use of psychoactive drugs can contribute to this condition.

When reproduction does not occur in the usual manner, many couples adopt a child. Others sometimes first try one of the alternative reproductive methods discussed in the following paragraphs. If all the alternative methods discussed are considered, it is possible to imagine that a baby has 5 parents: (1) sperm donor, (2) egg donor, (3) surrogate mother, and (4) and (5) adoptive mother and father.

Alternative Reproduction: How to Have 5 Parents

Artificial Insemination by Donor (AID). During artificial insemination, sperm are placed in the vagina by a physician. Sometimes a woman is artificially inseminated by her husband's sperm. This is especially helpful if the husband has a low sperm count—the sperm can be collected over a period of time and concentrated so that the sperm count is sufficient to result in fertilization. Often, however, a woman is inseminated by sperm acquired from a donor who is a complete stranger to her. At times, a mixture of husband and donor sperm are used.

A variation of AID is intrauterine insemination (IUI). IUI involves hormonal stimulation of the ovaries, followed by placement of the donor's sperm in the uterus rather than in the vagina.

In Vitro Fertilization (IVF). In the case of in vitro fertilization (IVF), hormonal stimulation of the ovaries is followed by laparoscopy. In this procedure, an aspiratory tube is used to retrieve preovulatory eggs (fig. 20.6). Alternately, it is possible to pierce the vaginal wall with a needle and to guide it, using ultrasound, to the ovaries, where the needle is used to retrieve the eggs. This method is called transvaginal retrieval.

Concentrated sperm from the male are placed in a solution that approximates the conditions of the female genital tract. When the eggs are introduced, fertilization occurs. The resultant zygotes (fertilized eggs) begin development, and after about 2 to 4 days, the embryos are inserted into the uterus of the woman, who is now in the secretory phase of her uterine cycle. If implantation is successful, development is normal and continues to term.

Gamete Intrafallopian Transfer (GIFT). Gamete intrafallopian transfer (GIFT) was devised as a means to overcome the low success rate (15–20%) of in vitro fertilization. The method is exactly the same as in vitro fertilization, except the eggs and the sperm are placed in the oviducts immediately after they have been brought together. This procedure is helpful to couples whose eggs and sperm never make it to the oviducts; sometimes the egg gets lost between the ovary and the oviducts, and sometimes the sperm never reach the oviducts. GIFT has an advantage in that it is a one-step procedure for the woman—the eggs are removed and are reintroduced all in the same time period. For this reason, it is less expensive—approximately $1,500 compared to $3,000 and up for in vitro fertilization.

Surrogate Mothers. In some instances, women are paid to have babies. These women are called surrogate mothers. Other individuals contribute sperm (or eggs) to the fertilization process in such cases.

> Some couples are infertile due to various physical abnormalities. When corrective medical procedures fail, today it is possible to consider an alternative method of reproduction in order to be a parent.

Sexually Transmitted Diseases

There are many diseases that are transmitted by sexual contact and are therefore called sexually transmitted diseases (STDs). Our discussion centers on those that are most prevalent. AIDS, genital herpes, and genital warts are viral diseases; therefore, they are difficult to treat because viral infections do not respond to traditional antibiotics. Other types of drugs have been developed to treat these, however. And although gonorrhea and chlamydia are treatable with appropriate antibiotic therapy, they are not always promptly diagnosed. Unfortunately, the human body does not become immune to STDs, and as yet there are no vaccines available for any of them.

AIDS: By 2000, 40 Million to Be Infected

Acquired immunodeficiency syndrome (AIDS) is caused by retroviruses called human immunodeficiency virus type 1 (HIV-1) and type 2 (HIV-2). The vast majority of infections in the United States are caused by HIV-1. These viruses infect helper T lymphocytes, and as the disease progresses, the number of these vital white cells declines. The World Health Organization estimates that between 8 million and 10 million adults and one million children are infected now and that by the year 2000, there will be 40 million persons infected worldwide.

Catching AIDS

In the United States, about 23% of persons infected with HIV are intravenous drug abusers who contracted the disease after using a contaminated needle. About 3% of AIDS cases involve persons who received infected blood during a blood transfusion. Because blood is now routinely tested for the presence of HIV, however, the chance of contracting AIDS in this manner is presently considered minimal.

AIDS is primarily a sexually transmitted disease. In Africa, where heterosexual transmission is the norm, as many women are infected as men. In the United States, about 64% of total cases are homosexual men who practice anal intercourse. (About 6% of these are also intravenous drug abusers.) Unlike the vagina, the epithelial lining of the rectum is a thin, single-celled layer, which is easily torn during intercourse. Heterosexual transmission is occurring in the United States, however, and the female population aged 15–44 and their babies are the most rapidly growing group of HIV-infected persons. Apparently, homosexual men are beginning to adopt behaviors that prevent the transmission of AIDS, and women must learn to do so also. The same may be said for college students, whose overall rate of infection is about 10 times that of the general population. One unhappy side effect of female infection is the fact that viruses and infected lymphocytes can pass to a fetus via the placenta or to an infant via the mother's milk.

Three Stages of AIDS

An HIV infection can be divided into 3 stages (fig. 20.14).

Carrier with No Symptoms. Only 1–2% of those newly infected with HIV have mononucleosis-like symptoms, which may include fever, chills, aches, swollen lymph glands, and an itchy rash. These symptoms disappear, and there are no other symptoms for 9 months or longer. Although the individual exhibits no symptoms during this stage, he or she is highly infectious. The results of standard HIV blood tests for the presence of antibody are positive during this stage because the concentration of HIV in the body is high.

AIDS-Related Complex (ARC). The most common symptom of AIDS-related complex (ARC), the second stage of infection, is swollen lymph glands in the neck, the armpits, or the groin persisting for 3 months or more. There is severe fatigue unrelated to exercise or drug use; unexplained persistent or recurrent fevers, often with night sweats; persistent cough not associated with smoking, a cold, or the flu; and persistent diarrhea. Also possible are signs of nervous system impairment, including loss of memory, inability to think clearly, loss of judgment, and/or depression.

The development of non-life-threatening and recurrent infections, such as thrush or herpes simplex, signals that full-blown AIDS will occur soon.

Full-Blown AIDS. The final stage of AIDS occurs when the helper T cell count dips to $200/mm^3$ or less. The patient experiences severe weight loss and weakness due to persistent diarrhea, and usually one of several opportunistic infections is also present. These infections are called opportunistic because the body can usually prevent them—only an impaired immune system gives them the opportunity to get started. These infections include the following:

***Pneumocystis carinii* pneumonia.** An infection of the lungs that can lead to death. There is not a single documented case of this pneumonia due to this organism in persons with normal immunity.

Toxoplasmic encephalitis. An infection of the brain that leads to loss of brain cells, seizures, and weakness in AIDS patients.

Figure 20.14

Balance of power between HIV and the immune system during the course of an HIV infection. During the time that a person is an asymptomatic carrier, the concentration of HIV in the body is high. Then the immune system becomes active, and antibodies are produced. While an infected person has ARC, the immune system begins to lose the battle, and eventually the battle is lost: the person develops full-blown AIDS. Source: Data from R. R. Redfield and D. S. Burke, "HIV Infection: The Clinical Picture" in *Scientific American*, October 1988.

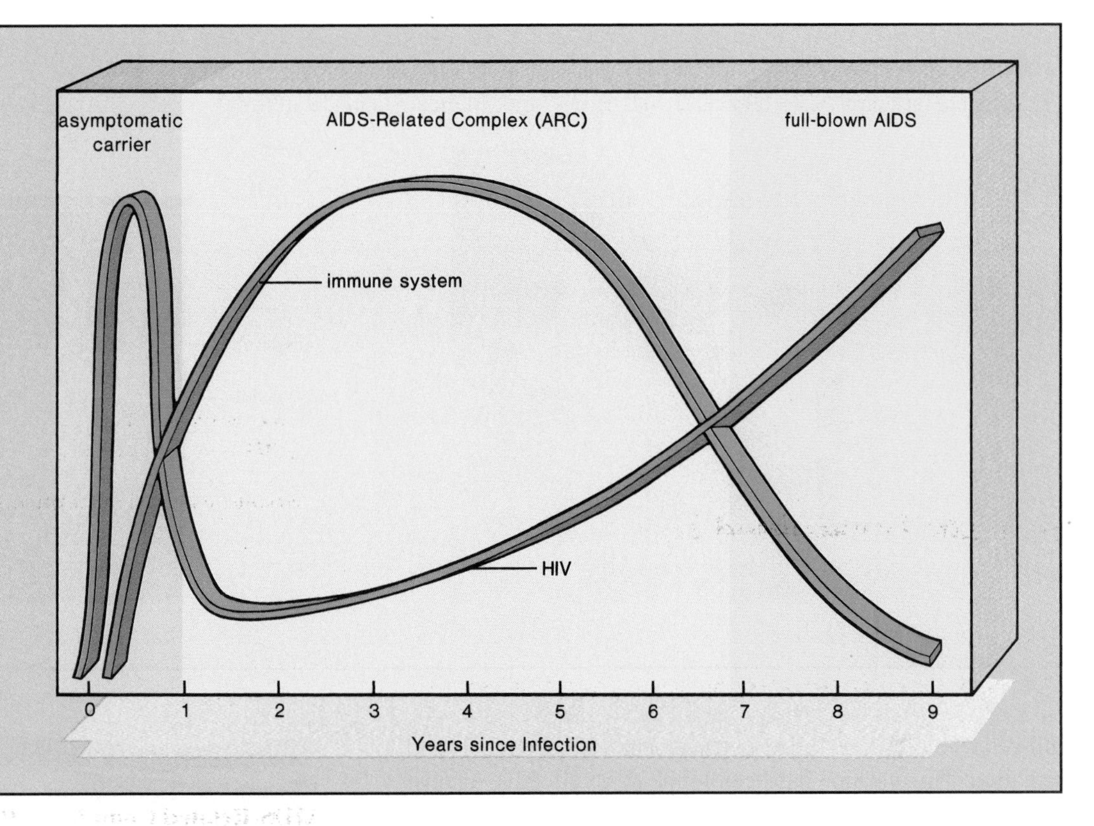

Kaposi's sarcoma. A cancer of the blood vessels that causes reddish purple, coin-sized spots and lesions on the skin in AIDS patients.

***Mycobacterium* tuberculosis.** Bacterial infection, usually of the lungs, that is seen more often as an infection of lymph nodes and other organs in patients with AIDS. Of special concern, tuberculosis is spreading into the general populace and is multidrug resistant.

Invasive cervical cancer. Cancer of the cervix, which spreads to nearby tissues. This condition has been added to the list because the incidence of AIDS has now increased in women.

Treating AIDS

Several drugs are available for treatment of those infected with HIV. Zidovudine (also known as AZT), didanosine (ddI), and now nevirapine are all inhibitors of the enzyme reverse transcriptase. It is hoped that a combination therapy will prove to be more effective than using each of the drugs separately. It appears that HIV mutates and thereby becomes resistant to each of these drugs when they are taken separately. A drug that inhibits proteases, enzymes HIV cells need to bud from the host cell, may also be available soon.

A number of different types of vaccines are in, or are expected to be in, the human trial stage of development. Several of these are subunit vaccines utilizing genetically engineered proteins that resemble those found in HIV. For example, HIV-1, the cause of most AIDS cases in the United States, has an outer envelope protein called GP120 (fig. 20.15).

Figure 20.15

HIV-1 has an envelope protein called GP120, which allows it to attach to CD4 receptors that project from a helper T cell. Infection of the helper T cell follows, and viruses eventually bud from the infected helper T cell. If the immune system can be trained by the use of a vaccine to attack and destroy all cells that bear GP120, a person could not be infected with HIV-1.

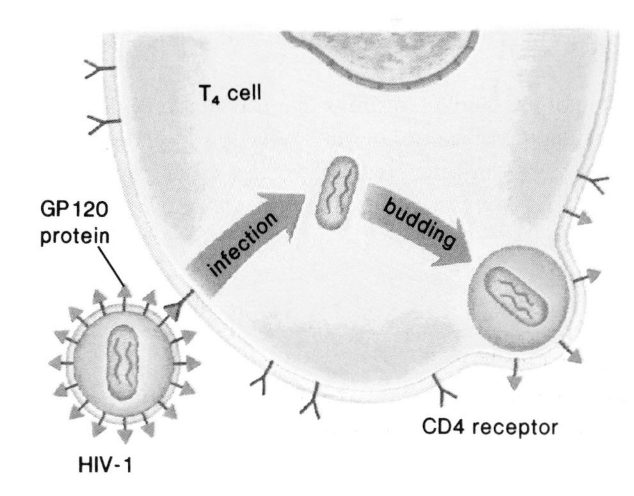

When GP120 combines with a CD4 receptor, which projects from a helper T cell, the virus enters the cell. There are subunit vaccines that use only the GP120 protein molecules to stimulate the production of antibodies.

An entirely different approach is being taken by Jonas Salk, who developed the polio vaccine. His AIDS vaccine

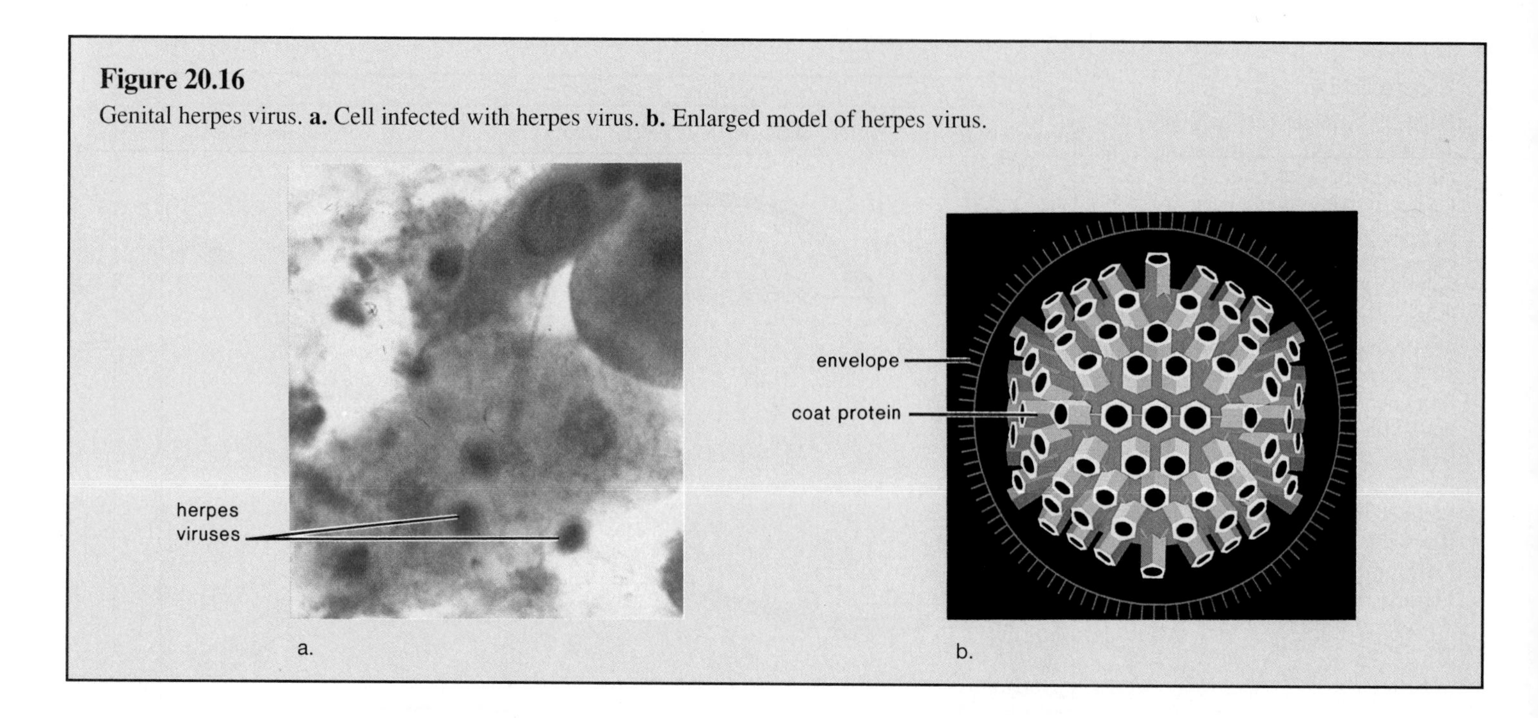

Figure 20.16
Genital herpes virus. **a.** Cell infected with herpes virus. **b.** Enlarged model of herpes virus.

utilizes whole HIV-1, killed by treatment with chemicals and radiation. This vaccine has been found to be effective against experimental HIV-1 infection in chimpanzees, and clinical trials in humans will begin soon.

How to Prevent AIDS

Shaking hands, hugging, social kissing, coughing or sneezing, and swimming in the same pool do not transmit the AIDS virus. You cannot get AIDS from inanimate objects such as toilets, doorknobs, telephones, office machines, or household furniture.

The following behaviors will help to prevent the spread of AIDS:

1. Do not use alcohol or drugs in a way that prevents you from being in control of your behavior. In particular, do not inject drugs into veins. If you are an intravenous drug user and cannot stop this behavior, however, always use a sterile needle or one cleansed by bleach.
2. Refrain from multiple sex partners, especially with homosexual or bisexual men or intravenous drug users of either sex. Either abstain from sexual intercourse or develop a long-term monogamous (always the same partner) sexual relationship with a partner who is free of HIV and who is not an intravenous drug user.
3. If you are uncertain about your partner, always use a male or female latex condom. Follow the directions, and also use a spermicide containing nonoxynol-9, which kills viruses and virus-infected lymphocytes. The risk of contracting AIDS is greater in persons who already have a sexually transmitted disease.

Genital Herpes: Estimated 500,000 New Cases per Year

Genital herpes is caused by the herpes simplex virus (fig. 20.16), of which there are 2 types: type 1, which usually causes cold sores and fever blisters, and type 2, which more often causes genital herpes.

Genital herpes is one of the more prevalent sexually transmitted diseases today. An estimated 40 million persons in the United States have it, with an estimated 500,000 new cases appearing each year. Immediately after infection, there are no symptoms, but the individual may experience a tingling or itching sensation before blisters appear at the infected site, usually within 2–20 days. Once the blisters rupture, they leave painful ulcers, which may take as long as 3 weeks or as little as 5 days to heal. These symptoms may be accompanied by fever, pain upon urination, and swollen lymph nodes.

After the ulcers heal, the disease is only dormant. Blisters can reoccur repeatedly at variable intervals. Sunlight, sexual intercourse, menstruation, and stress seem to cause the symptoms of genital herpes to reoccur. While the virus is dormant, it resides in nerve cells near the brain and the spinal cord. Herpes occasionally infects the eye, causing an eye infection that can lead to blindness. Type 2 was once thought to cause a form of cervical cancer, but this is no longer believed to be the case.

Infection of the newborn can occur if the child comes in contact with a lesion in the birth canal. In 1–3 weeks, the infant is gravely ill and can become blind, have neurological disorders including brain damage, or die. Birth by cesarean section prevents these adverse developments.

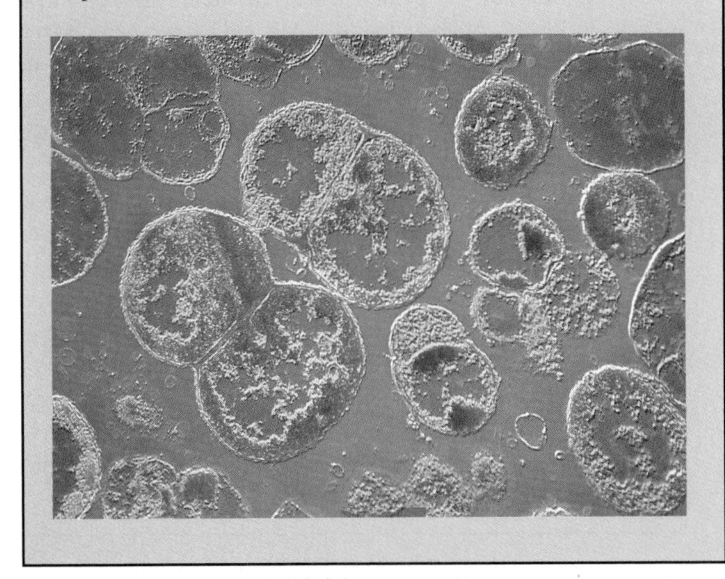

Figure 20.17
Gonorrheal bacteria (*Neisseria gonorrheae*). Notice that these round bacteria occur in pairs; for this reason, they are called diplococci.

Genital Warts: Linked to Cancer

Genital warts are caused by the human papillomaviruses (HPVs), which are sexually transmitted. Sometimes carriers do not have any sign of warts, although flat lesions may be present. When present, the warts are commonly seen on the penis and the foreskin of males and the vaginal orifice in females. If the warts are removed, they may reoccur.

HPVs, rather than genital herpes, are now associated with cancer of the cervix, as well as tumors of the vulva, the vagina, the anus, and the penis. Some researchers believe that the viruses are involved in 90–95% of all cases of cancer of the cervix. Teenagers who have or have had multiple sex partners seem to be particularly susceptible to HPV infections. More and more cases of cancer of the cervix are being seen in this age group.

Presently, there is no cure for an HPV infection, but warts can be effectively treated by surgery, freezing, application of an acid, or laser burning. A suitable medication to treat genital warts before cancer occurs is being sought. Efforts are also underway to develop a vaccine.

Gonorrhea: Early Detection Difficult in Women

Gonorrhea is caused by the bacterium *Neisseria gonorrheae.* This bacterium is a diplococcus, meaning that 2 cells generally stay together (fig. 20.17).

The diagnosis of gonorrhea in the male is not difficult as long as he displays typical symptoms (as many as 20% of males may be asymptomatic). The patient complains of pain on urination and has a thick, greenish yellow urethral discharge 3–5 days after contact with an infected partner. In the female, the bacteria may first settle within the urethra or near the cervix, from which they may spread to the oviducts, causing **pelvic inflammatory disease (PID).** As the inflamed tubes heal, they may become partially or completely blocked by scar tissue. As a result, the female is sterile or at best subject to ectopic pregnancy. Similarly, there may be inflammation in untreated males followed by scarring of each vas deferens. Unfortunately, 60–80% of females are asymptomatic until they develop severe PID-induced pains in the abdominal region. PID affects about one million women a year in the United States.

Homosexual males develop gonorrhea proctitis, or infection of the anus, with symptoms including pain in the anus and blood or pus in the feces. Oral sex can cause infection of the throat and the tonsils. Gonorrhea can also spread to other parts of the body, causing heart damage or arthritis. If, by chance, the person touches infected genitals and then his or her eyes, a severe eye infection can result.

Eye infection leading to blindness can occur as a baby passes through the birth canal. Because of this, all newborn infants receive eye drops containing antibacterial agents such as silver nitrate, tetracycline, or penicillin as a protective measure.

Chlamydia: Risking PID and Sterility

Chlamydia is named for the tiny bacterium that causes it, *Chlamydia trachomatis.* For years, chlamydiae were considered to be more closely related to viruses than to bacteria, but today it is known that they are prokaryotic cells. Even so, they are obligate parasites due to their inability to produce ATP molecules. After a cell phagocytizes them, they develop inside the phagocytic vacuole, which eventually bursts and liberates many new infective chlamydiae.

New chlamydial infections of the genitals occur at an even faster rate than gonorrheal infections. They are the most common cause of **nongonococcal urethritis (NGU).** About 8–21 days after infection, men experience a mild burning sensation on urination and a mucous discharge. Women may have a vaginal discharge, along with the symptoms of a urinary tract infection. Unfortunately, a physician may mistakenly diagnose a gonorrheal or urinary infection and prescribe the wrong type of antibiotic, or the person may never seek medical help. In either case, the infection can eventually cause PID, sterility, or ectopic pregnancy.

If a newborn is exposed to chlamydia during delivery, inflammation of the eyes or pneumonia can result. Some believe that chlamydial infections increase the possibility of premature births and stillbirths.

Detecting and Treating Chlamydia

New and faster clinical tests, including one utilizing a DNA probe, are now available for detection of chlamydia. Their expense sometimes prevents public clinics from using them, however. It's been suggested that these criteria could help physicians decide which women should be tested: no more than 24 years old; a new sex partner within the preceding 2 months; cervical discharge; bleeding during parts of the vaginal exam; and use of a nonbarrier method of contraception. Some doctors, however, are routinely prescribing additional antibiotics appropriate to treating chlamydia for anyone who has gonorrhea because 40% of females and 20% of males with gonorrhea also have chlamydia.

As with AIDS, condoms protect against both gonorrheal and chlamydial infections. The concomitant use of a spermicide containing nonoxynol-9 enhances protection.

> PID and sterility are common effects of a chlamydial infection in the female. This condition often accompanies a gonorrheal infection.

Syphilis: Three Stages

Syphilis is caused by a type of bacterium called *Treponema pallidum.* As with many other bacterial diseases, penicillin has been used as an effective antibiotic. Syphilis has 3 stages, which can be separated by latent stages in which the bacteria are resting before multiplying again. During the *primary stage,* a hard chancre (ulcerated sore with hard edges) indicates the site of infection. The chancre can go unnoticed, especially since it usually heals spontaneously, leaving little scarring. During the *secondary stage,* proof that bacteria have invaded and spread throughout the body is evident when the victim breaks out in a rash. Curiously, the rash does not itch and is seen even on the palms of the hands and the soles of the feet. There can be hair loss and infectious gray patches on the mucous membranes, including the mouth. These symptoms disappear of their own accord.

During the *tertiary stage,* which lasts until the patient dies, syphilis may affect the cardiovascular system: weakened arterial walls (aneurysms) are seen, particularly in the aorta. In other instances, the disease may affect the nervous system: an infected person may show psychological disturbances, for example. Gummas, large destructive ulcers, may develop on the skin or within the internal organs in another variety of the tertiary stage.

Congenital syphilis is caused by syphilitic bacteria crossing the placenta. The child is born blind and/or with numerous anatomical malformations.

> The sexually transmitted diseases—AIDS, genital herpes, genital warts, gonorrhea, chlamydia, and syphilis—are prevalent in the United States at this time. AIDS often results in death within a few years; herpes attacks reoccur throughout life; genital warts are associated with cervical cancer; and gonorrhea and chlamydia often lead to sterility. The best preventive measure against these diseases is abstinence or monogamy with a partner who is free of them.

SUMMARY

In males, spermatogenesis occurs within the seminiferous tubules of the testes, which also produce testosterone within the interstitial cells. Sperm mature in the epididymis and are stored in a vas deferens before entering the urethra, along with seminal fluid, prior to ejaculation. Hormonal regulation, involving secretions from the hypothalamus, the anterior pituitary, and the testes, maintains testosterone at a fairly constant level in the male.

In females, oogenesis occurs within the ovaries, where one follicle produces an egg each month. Fertilization, if it occurs, takes place in the oviducts, and the resulting embryo travels to the uterus, where it embeds itself in the uterine lining. In the nonpregnant female, hormonal regulation involves the ovarian and uterine cycles, which are dependent upon the hypothalamus, the anterior pituitary, and the female sex hormones estrogen and progesterone.

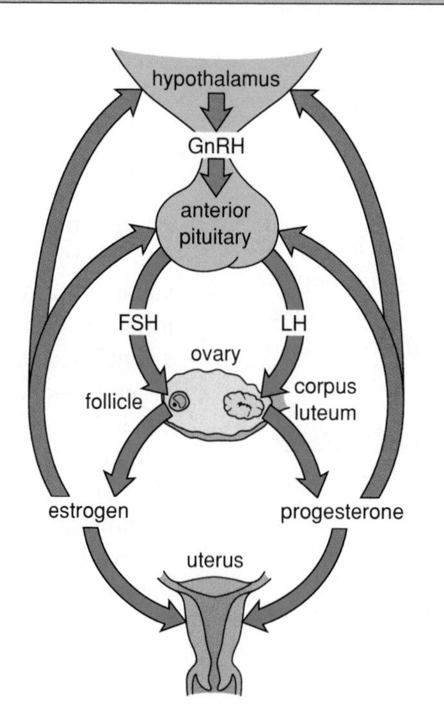

Numerous birth-control methods and devices are available for those who wish to prevent pregnancy. Infertile couples are increasingly resorting to alternative methods of reproduction.

Sexually transmitted diseases are of concern to all. AIDS, genital herpes, genital warts, and chlamydia are presently of the greatest concern, but still prevalent are gonorrhea and syphilis.

STUDY QUESTIONS

In order to practice **writing across the curriculum,** students should write out the answers to any or all of the study questions. The study questions are sequenced in the same order as the text.

1. Discuss the anatomy and the physiology of the testes. (p. 370) Describe the structure of sperm. (p. 372)
2. Give the path of sperm. (p. 372)
3. Which glands produce seminal fluid? (p. 372)
4. Discuss the anatomy and the physiology of the penis. (p. 372) Describe ejaculation. (p. 373)
5. Discuss hormonal regulation in the male. Name 3 functions for testosterone. (pp. 373–374)
6. Discuss the anatomy and the physiology of the ovaries. (pp. 375–376) Describe ovulation. (p. 376)
7. Give the path of the egg. Where do fertilization and implantation occur? Name 2 functions of the vagina. (p. 377)
8. Describe the external genitals in females. (p. 378)
9. Compare male and female orgasm. (p. 373, p. 378)
10. Discuss hormonal regulation in the female, either simplified or detailed. (p. 378) Give the events of the uterine cycle, and relate them to the ovarian cycle. (pp. 379–380) In what way is menstruation prevented if pregnancy occurs? (p. 381)
11. Name 4 functions of the female sex hormones. (p. 381) Describe the anatomy and the physiology of the breast. (p. 382)
12. Discuss the various means of birth control and their relative effectiveness. (p. 383)
13. Describe the most common types of sexually transmitted diseases. (pp. 386–390)

OBJECTIVE QUESTIONS

1. In tracing the path of sperm, the structure that follows the epididymis is the _______.
2. The prostate gland, the Cowper's glands, and the _______ all contribute to seminal fluid.
3. The main male sex hormone is _______.
4. An erection is caused by the entrance of _______ into sinuses within the penis.
5. In the female reproductive system, the uterus lies between the oviducts and the _______.
6. In the ovarian cycle, once each month a(n) _______ produces an egg. In the uterine cycle, the _______ lining of the uterus is prepared to receive the zygote.
7. The female sex hormones are _______ and _______.
8. Pregnancy in the female is detected by the presence of _______ in blood or urine.
9. In vitro fertilization occurs in _______.
10. Although a sexually transmitted disease, the AIDS virus mainly infects _______ cells.
11. Herpes simplex virus type 1 causes _______, and type 2 causes _______.
12. The most prevalent sexually transmitted disease today is _______.
13. Label this diagram of the male reproductive system, and trace the path of sperm.

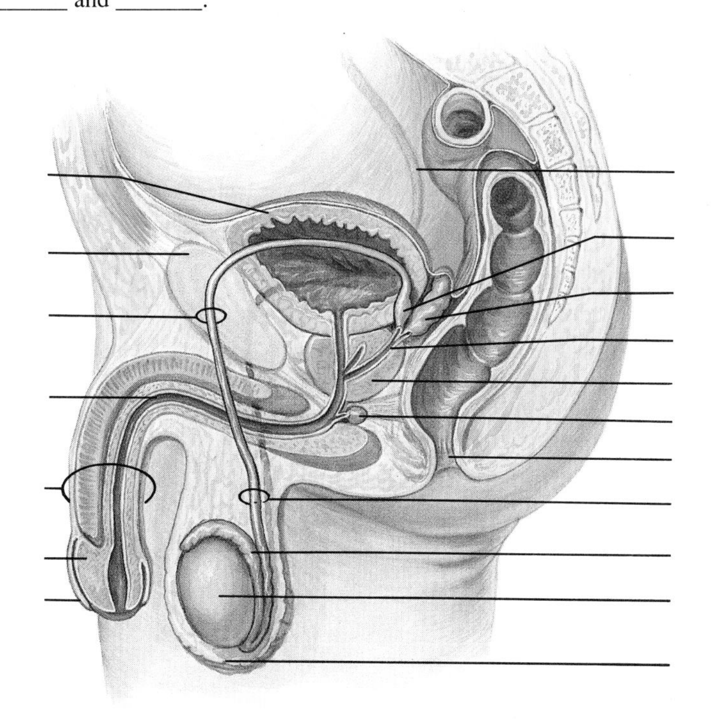

CRITICAL THINKING

In order to practice **writing across the curriculum,** students should write out the answers to any or all of the critical thinking questions. Suggested answers to the critical thinking questions are in appendix E.

20.1

1. Using figure 20.4 as a guide, hypothesize why anabolic steroids would shrink the size of the testes.
2. How might you test your hypothesis, for example in laboratory mice?
3. Would you predict that anabolic steroids raise or lower LDL blood levels (p. 360), leading to increased risk of heart disease?
4. How might you test your hypothesis, for example, in laboratory mice?

20.2

1. Using figure 20.9 as a guide, hypothesize why a pill that contains estrogen and progesterone could be used as a birth-control pill.
2. How might you test your hypothesis, for example, in laboratory mice?
3. In postmenopausal women (p. 382), there are usually increased levels of FSH and LH, but because the ovaries are unable to respond, there are decreased levels of estrogen and progesterone. How would you expect these levels to change in postmenopausal women taking birth-control pills?
4. How might you test your prediction, for example in postmenopausal women?

SELECTED KEY TERMS

endometrium (en″ do-me′ tre-um) the lining of the uterus, which becomes thickened and vascular during the uterine cycle.
erection the penis when it is turgid and erect, instead of being flaccid or lacking turgidity.
estrogen female sex hormone that, along with progesterone, maintains the primary sex organs and stimulates development of the female secondary sex characteristics.
Graafian follicle (graf′ e-an fol′ ĭ-k′l) mature follicle within the ovaries that contains a developing egg.
implantation the attachment and penetration of the embryo into the lining of the uterus (endometrium).
interstitial cell hormone-secreting cell located between the seminiferous tubules of the testes.
menopause termination of the ovarian and uterine cycles in older women.
menstruation loss of blood and tissue from the uterus at the end of a uterine cycle.
ovarian cycle monthly occurring changes in the ovary that determine the level of sex hormones in blood.
ovary the female gonad, the organ that produces eggs, estrogen, and progesterone.
penis external organ in males through which the urethra passes and which serves as the organ of sexual intercourse.
prostate gland gland located around the male urethra below urinary bladder; adds secretions to seminal fluid.
seminal fluid the sperm-containing secretion of males; also called semen.
seminiferous tubule (sem″ ĭ-nif″ er-us tu-būl) highly coiled duct within the male testis that produces and transports sperm.
testis the male gonad, the organ that produces sperm and testosterone.
testosterone the main male sex hormone responsible for development of primary and secondary sex characteristics in males.
uterine cycle monthly occurring changes in the characteristics of the uterine lining (endometrium).
uterus the womb, the organ located in the female pelvis where the fetus develops.
vagina organ that leads from the uterus to the vestibule and serves as the birth canal and the organ of sexual intercourse in females.
vas deferens tube that leads from the epididymis to the urethra in males.

21

DEVELOPMENT

Chapter Concepts

1.
The first stages of embryonic development in animals lead to the establishment of the embryonic germ layers. 395, 418

2.
Differentiation and morphogenesis are two processes that occur when specialized organs develop. 399, 418

3.
It is possible to precisely outline the steps in human embryonic and fetal development up until birth. 402, 418

4.
Investigation into aging shows hope of identifying underlying causes of degeneration and prolonging the health span of individuals. 414, 419

Development of a frog

Chapter Outline

THE study of development concerns the events and the processes that occur as a single cell becomes a complex organism. These same processes are also seen as the newly born or hatched organism matures, as a lost part regenerates, as a wound heals, and even during aging. Therefore, today it is customary to stress that the study of development encompasses not only embryology (development of the embryo), but these other events as well.

Development requires growth, differentiation, and morphogenesis. When an organism increases in size, we say that it has grown. During **growth,** cells divide, enlarge, and divide once again. **Differentiation** occurs when cells become specialized in structure and in function. A muscle cell looks and acts quite differently than a nerve cell, for example. **Morphogenesis** goes one step beyond growth and differentiation. It occurs when body parts are shaped and patterned into a certain form. There is a great deal of difference between your arm and your leg, for example, even though they contain the same types of tissues.

These processes are discussed as they apply to development of the embryo, but keep in mind that they also occur whenever an organism goes through any developmental change.

Growth, differentiation, and morphogenesis are 3 processes that are seen whenever a developmental change occurs.

Figure 21.1

Stages in early animal development. All animals go through the stages noted. The appearance of these stages in lancelets and humans is compared (l.s. = longitudinal section; c.s. = cross section).

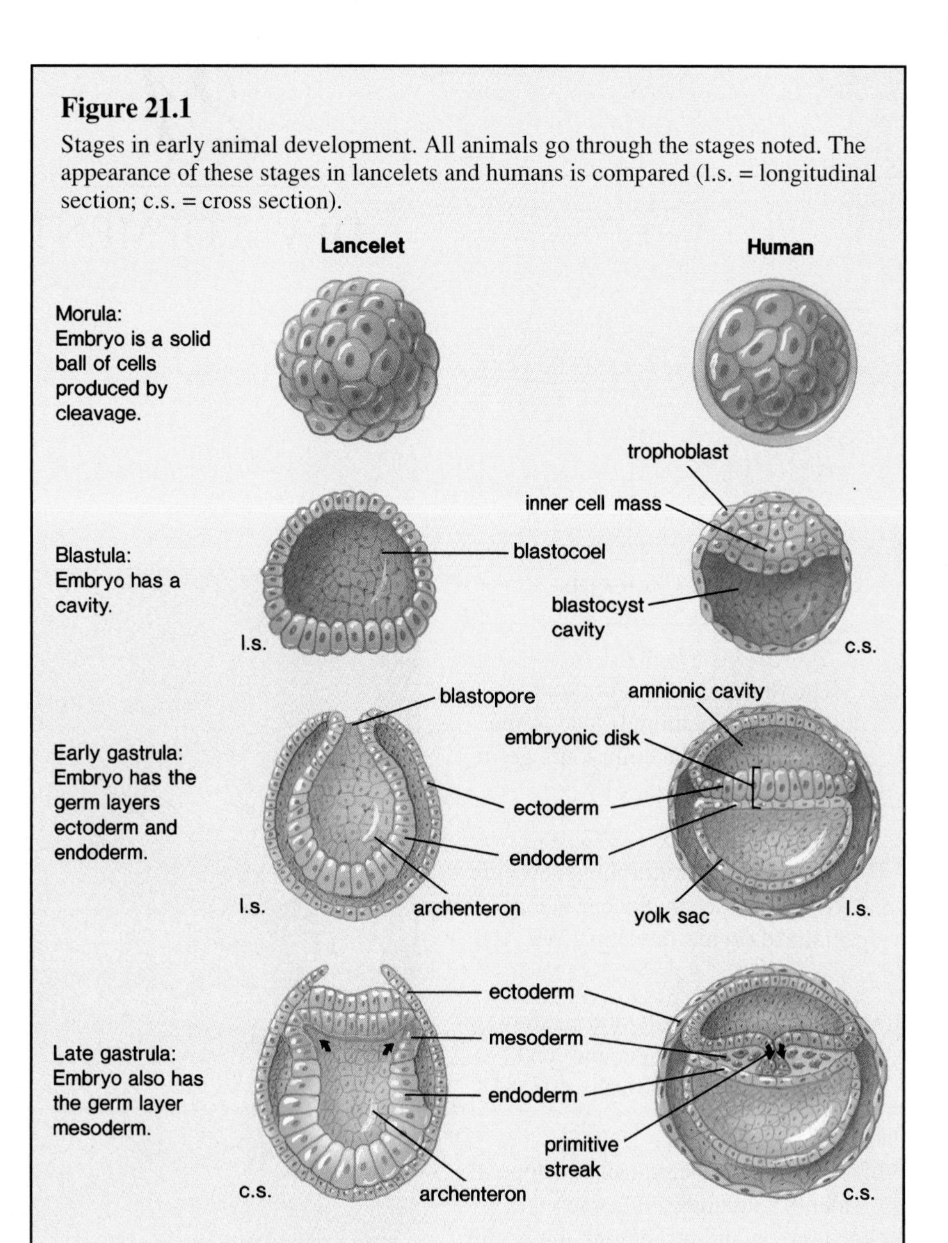

Early Developmental Stages

All chordate embryos go through the same early stages of development (fig. 21.1). Chordates are animals that at some time in their life history have an elastic supporting rod known as a notochord. In vertebrates, this rod is replaced by the vertebral column. All the animals discussed in this chapter are vertebrates except the lancelet, a small animal that superficially resembles a fish (p. 602).

Morula: Solid Ball

Cleavage, the first event of development, is cell division without growth. It is best observed in an embryo such as the lancelet, which has little **yolk,** a rich nutrient material. (The yellow portion of a chick egg is the yolk.) Because a lancelet egg has little yolk, cell division is about equal, and the cells are fairly uniform in size. Cleavage continues until there is a solid ball of cells called the *morula.*

Blastula: Hollow Ball

As cleavage continues, the cells of the morula more or less position themselves to create a cavity. In the lancelet, a completely hollow ball called the *blastula* results, and the cavity within the ball is called the *blastocoel*. The human blastula is called the *blastocyst*, and therefore the cavity is called the blastocyst cavity. The blastocyst has a mass of cells—the *inner cell mass*—at one end. Figure 21.1 compares the appearance of a human embryo to that of a lancelet during the first stages of development.

In lancelets, cleavage results in a morula, which becomes the blastula when the blastocoel develops. In humans, the morula becomes the blastocyst, which contains an inner cell mass at one end of the blastocyst cavity.

Gastrula: Forming 3 Layers

During gastrulation in the lancelet, certain cells begin to push, or *invaginate*, into the blastocoel, creating a double layer of cells. The outer layer is called **ectoderm,** and the inner layer is called **endoderm.** The space created by invagination becomes the gut and is called either the primitive gut or the **archenteron.** The pore, or hole, created by invagination is called the blastopore, and in a lancelet, as well as in vertebrates, this first pore becomes the anus. Since a second pore becomes the mouth, vertebrates are called deuterostomes (*second mouth*).

Gastrulation in the lancelet is not complete until a third, middle layer of cells, **mesoderm,** has formed. In the lancelet, this layer begins as outpocketings of the archenteron; these outpocketings grow in size until they meet and fuse. In effect, then, 2 layers of mesoderm are formed, and the space between them is called the coelom. A *coelom* is defined as a body cavity lined by mesoderm within which the internal organs form.

Figure 21.1 compares human gastrulation to that of the lancelet. In humans, a space called the amniotic cavity appears within the inner cell mass. The portion of the mass below this cavity is the embryonic disk, which elongates to form the primitive streak, the midline region of the embryo where invagination occurs (fig. 21.2). Some of the upper cells within the primitive streak invaginate and spread out between the remaining cells of the upper layer, now called ectoderm, and the cells of the lower layer, now called endoderm. The invaginating cells are the mesoderm.

Figure 21.2

Human embryo at 16 days (amnion removed). The primitive node marks the extent of the primitive streak, where invagination occurs to establish the germ layer.

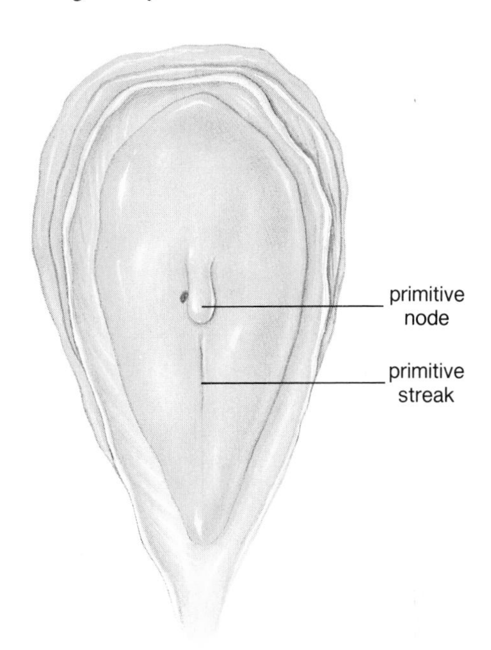

Because of the availability of chicken eggs, embryologists also have made a detailed study of this animal's development. In the chick, there is a primitive streak rather than a spherical gastrula because the yolk does not participate in the early stages of development. Human development resembles chick development, despite the fact that the human egg lacks yolk. The evolutionary history of these 2 animals can provide an answer to the amazing resemblance of their early developmental stages. Both birds (e.g., chicks) and mammals (e.g., humans) are related to reptiles, and this evolutionary relationship manifests itself in the manner in which development proceeds.

Germ Layers: From 3 Comes All

Ectoderm, mesoderm, and endoderm are called the *germ layers* because they give rise to all other tissues of the body. Early development in most animals involves gastrulation and the formation of 3 germ layers. It is possible to relate the development of future organs to these germ layers, as is done for humans in figure 21.3.

During gastrulation, the 3 embryonic germ layers (ectoderm, mesoderm, and endoderm) arise. Certain organs develop from each layer.

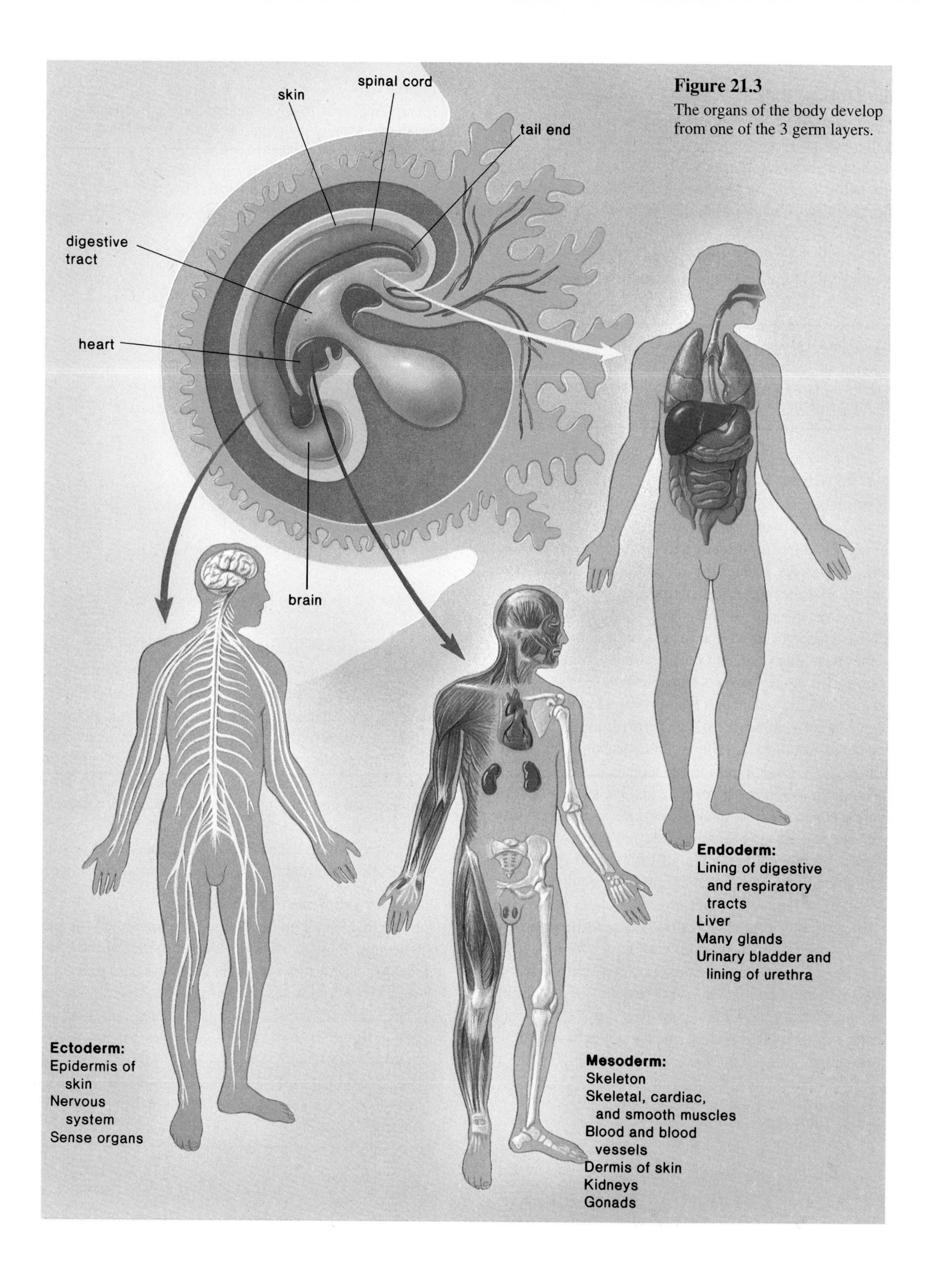

Figure 21.3
The organs of the body develop from one of the 3 germ layers.

Neurula: Neural Tube Forms

In chordate embryos, newly formed mesoderm cells that lie along the main axis are presumptive notochord—they will become a dorsal supporting rod called the *notochord*. (The notochord persists in lancelets, but in humans, it is later replaced by the vertebral column.) The nervous system develops from ectoderm located just above the notochord. At first, there is a thickening of cells called the neural plate; then, when *neural folds* fuse, a neural tube is formed (fig. 21.4). The neural tube develops into the spinal cord and the brain. In chordates, the neural tube (spinal cord) is a dorsal hollow nerve cord.

During neurulation, the neural tube develops just above the notochord. It becomes the spinal cord (dorsal hollow nerve cord) and the brain.

Midline mesoderm not contributing to the formation of the notochord becomes 2 longitudinal masses of tissue. From these, blocklike portions of mesoderm called *somites* develop. Somites become the muscles of the body and also the vertebrae of the spine. The coelom, an embryonic body cavity that forms at this time, is completely lined by mesoderm. In mammals, the coelom becomes the thoracic and abdominal cavities.

Figure 21.5*a* shows an intact human neurula and a generalized cross section of all chordate embryos. Another characteristic of all embryonic chordates is the presence of paired pharyngeal pouches (fig. 21.6). The pouches become functioning gills only in fishes and amphibian larvae. The fact that all vertebrates go through this embryonic stage indicates an evolutionary relationship among them. However, animal embryos proceed only through the stages that are consistent with their later development. For example, in humans, the first pair of pharyngeal pouches becomes the eustachian tubes. The second pair of pouches becomes the tonsils, while the third and fourth pairs become the thymus gland and the parathyroids. Therefore, pharyngeal pouches develop because of a phylogenetic relationship, but their eventual structure and function are suited to the particular animal.

Figure 21.4

Development of nervous system and coelom in vertebrates. The drawings show development of the neural tube in the frog.

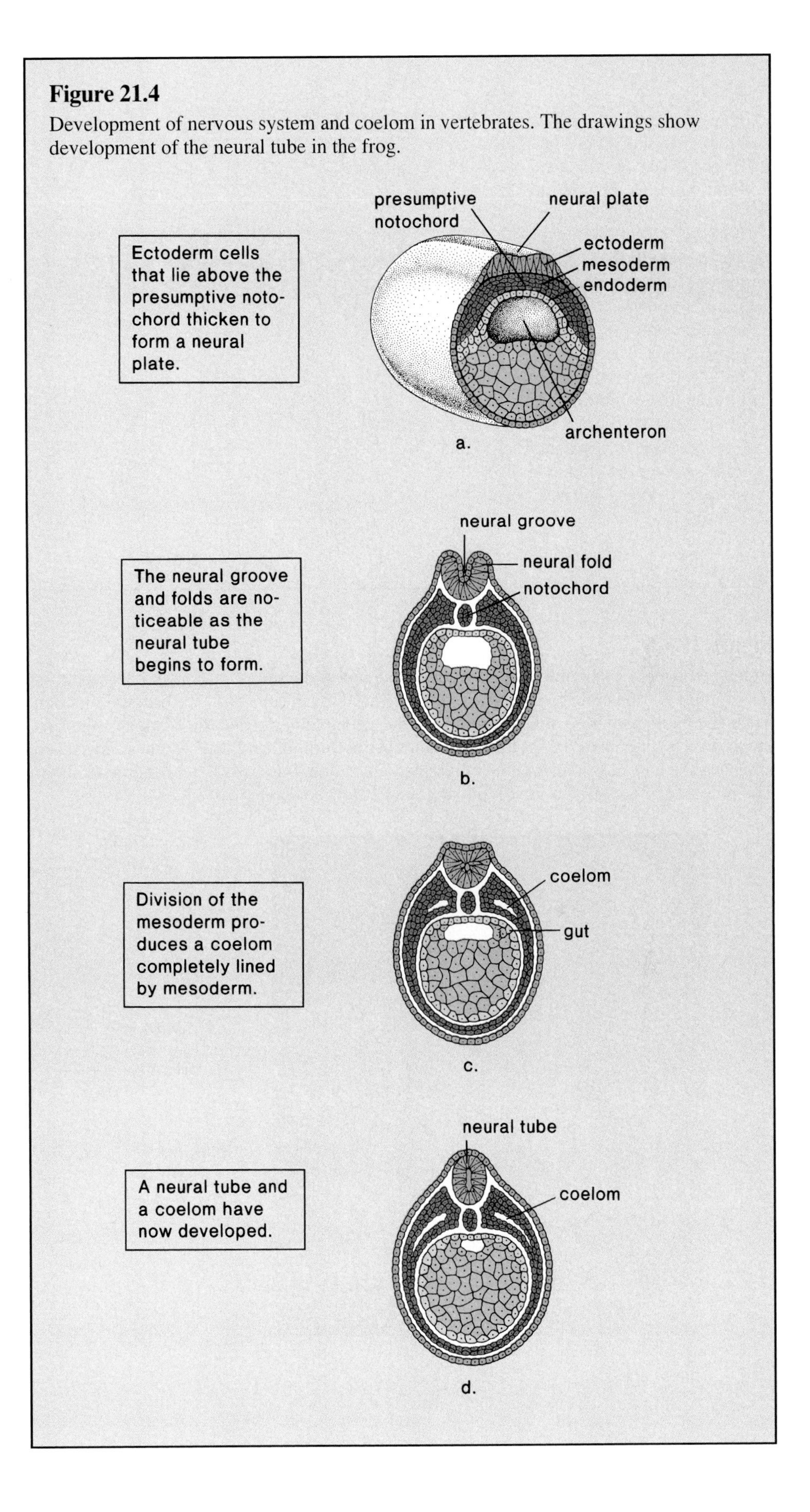

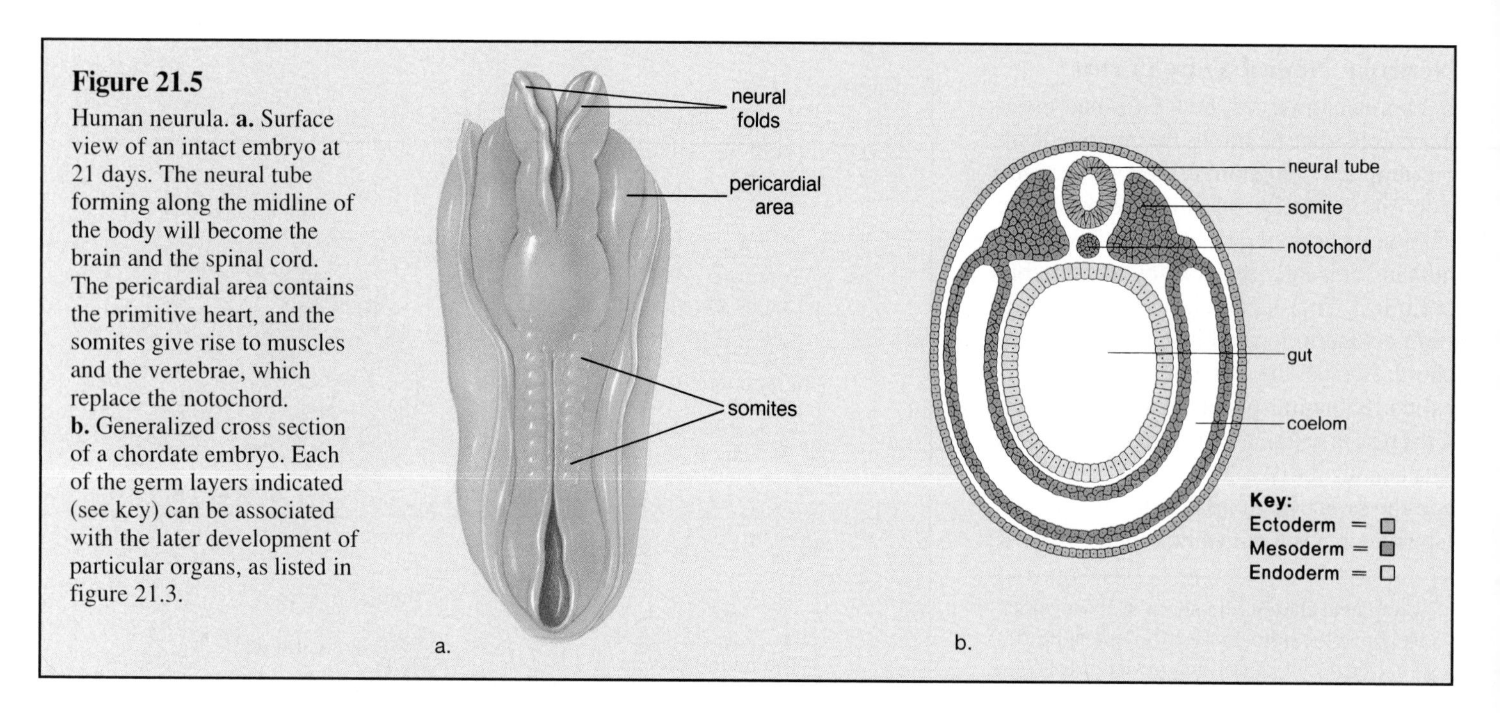

Figure 21.5

Human neurula. **a.** Surface view of an intact embryo at 21 days. The neural tube forming along the midline of the body will become the brain and the spinal cord. The pericardial area contains the primitive heart, and the somites give rise to muscles and the vertebrae, which replace the notochord. **b.** Generalized cross section of a chordate embryo. Each of the germ layers indicated (see key) can be associated with the later development of particular organs, as listed in figure 21.3.

Figure 21.6

Human embryo at beginning of fifth week. **a.** Scanning electron micrograph. **b.** The embryo is curled so that the head touches the heart, the 2 organs whose development is further along than the rest of the body. The organs of the gastrointestinal tract are forming, and the arms and the legs develop from the bulges that are called limb buds. The presence of the tail is an evolutionary remnant; its bones regress and become those of the coccyx (tailbone). The pharyngeal pouches become functioning gills only in fishes and amphibian larvae; in humans, the first pair of pharyngeal pouches becomes the eustachian tubes. The second pair becomes the tonsils, while the third and fourth become the thymus gland and the parathyroids.

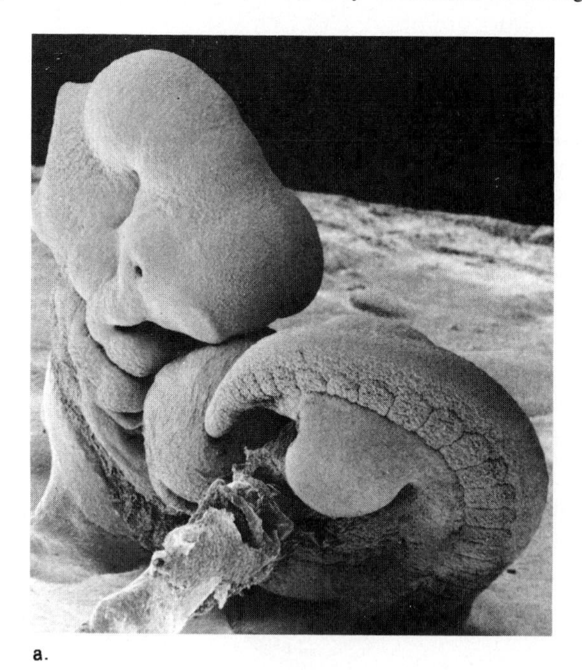

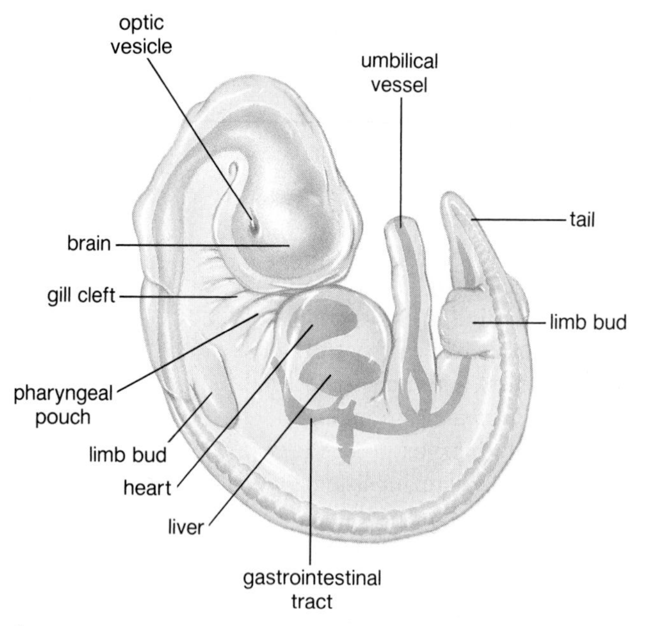

All chordates at some time in their development have a cross section with typical embryonic characteristics: a dorsal hollow nerve cord (neural tube), a notochord, and a coelom completely lined by mesoderm. Also, at some time in their embryonic history, chordates have pharyngeal pouches and gill clefts.

Differentiation and Morphogenesis

Differentiation and morphogenesis are 2 developmental processes. They account for the specialization of tissues and the formation of organs and determine the overall shape and form of an animal.

Differentiation: Unspecialized to Specialized

Differentiation has occurred when cells become specialized in structure and in function. The process of differentiation starts long before we can recognize different types of cells. Ectoderm, endoderm, and mesoderm cells in the gastrula look quite similar, but yet they must be different because they develop into different organs. What causes differentiation to occur, and when does it begin?

Figure 21.7 describes an experiment in which a nucleus from an intestinal cell of a tadpole is transplanted into an egg, the nucleus of which has been destroyed. Development proceeds normally, showing that embryonic nuclei are totipotent—they contain all the genetic information required to bring about complete development of the organism. Therefore, differentiation cannot be due to the parceling of genes into the various embryonic cells. Instead, it is hypothesized that gene-regulating substances of the zygote are distributed unequally in the cells of the morula:

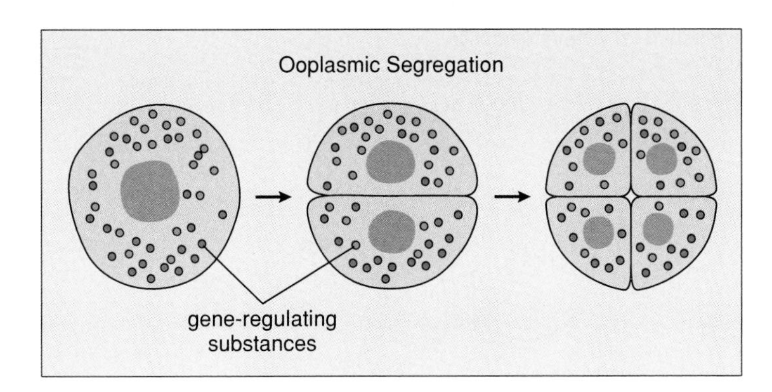

These gene-regulating substances play a role in determining which genes are activated in which cells. In support of this hypothesis, we can note that the cytoplasm of a frog's egg is not uniform. It is polar and has both an anterior/posterior axis and a dorsal/ventral axis, which can be correlated with the gray crescent, a gray area that appears after the sperm fertilizes the egg (fig. 21.8). Normally, the first cleavage gives each daughter cell half of the gray crescent. In this case, each of the daughter cells has the potential to become a tadpole (fig. 21.8*b*). However, if the researcher causes the egg to divide so that only one daughter cell receives the gray crescent, only that cell develops into a tadpole. We can therefore speculate that particular chemical signals within the gray crescent are needed to turn on the genes that control development.

Figure 21.7

Totipotency experiment. The haploid nucleus of a frog's egg is destroyed by ultraviolet irradiation. Now it can receive a diploid nucleus taken from an intestinal cell of a tadpole. In some cases, the reconstituted cell develops into a mature frog.

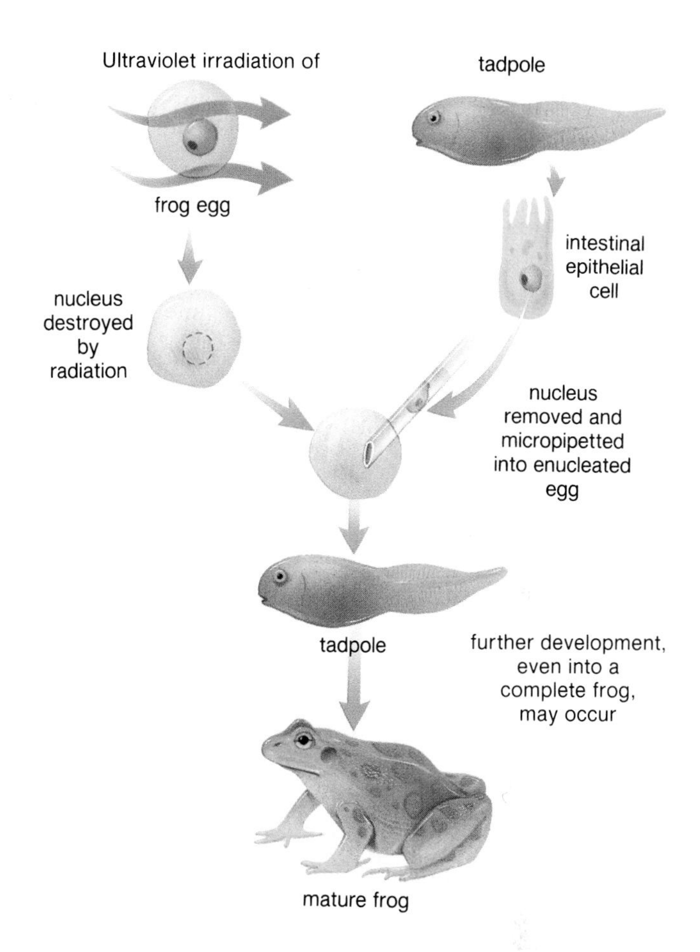

Cytoplasmic substances unequally distributed in the egg are parceled out during cleavage. These substances initially influence which genes are activated and how a cell differentiates.

Figure 21.8

Cytoplasmic influence on development. **a.** A frog's egg has an anterior/posterior and dorsal/ventral axis which correlates with the position of the gray crescent. **b.** The first cleavage normally divides the gray crescent in half, and each daughter cell is capable of developing into a complete tadpole. **c.** But if only one daughter cell receives the gray crescent, then only that cell can become a complete embryo. This shows that chemical messengers are not uniformly distributed in the egg's cytoplasm.

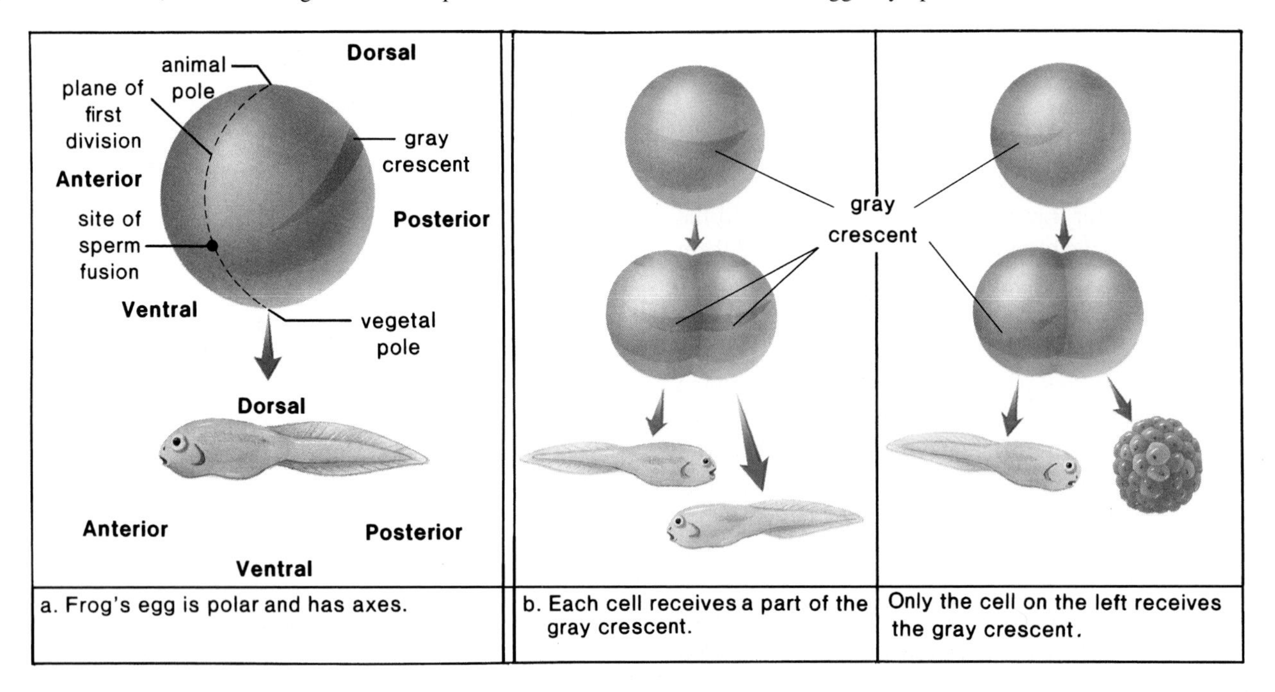

Morphogenesis: Taking Shape

As development proceeds, a cell's differentiation is influenced not only by its cytoplasmic contents, but also by signals given off by neighboring cells. Migration of cells occurs during gastrulation, and there is evidence that one set of cells can influence the migratory path taken by another set of cells. Some cells produce an extracellular matrix that contains fibrils, and in the laboratory, it can be shown that the orientation of these fibrils influences migratory cells. The cytoskeletons of the migrating cells are oriented in the same direction as the fibrils. Although this may not be an exact mechanism at work during gastrulation, it suggests that formation of the germ layers is probably influenced by environmental factors.

More specific information is known about neurulation, the process by which the nervous system develops. Neurulation involves **induction,** the ability of one tissue to influence the development of another tissue. Experiments have shown that the presumptive (potential) notochord induces formation of the neural plate, the first sign of the nervous system (fig. 21.9). If the presumptive nervous system, located just above the

Figure 21.9

Experiments showing importance of presumptive notochord. In experiment **a,** the presumptive nervous system (blue) does not develop into the neural plate if moved from its normal location. In experiment **b,** the presumptive notochord (red) can cause the belly ectoderm to develop into the neural plate (blue). This shows that the notochord induces ectoderm to become a neural plate, most likely by sending out chemical signals.

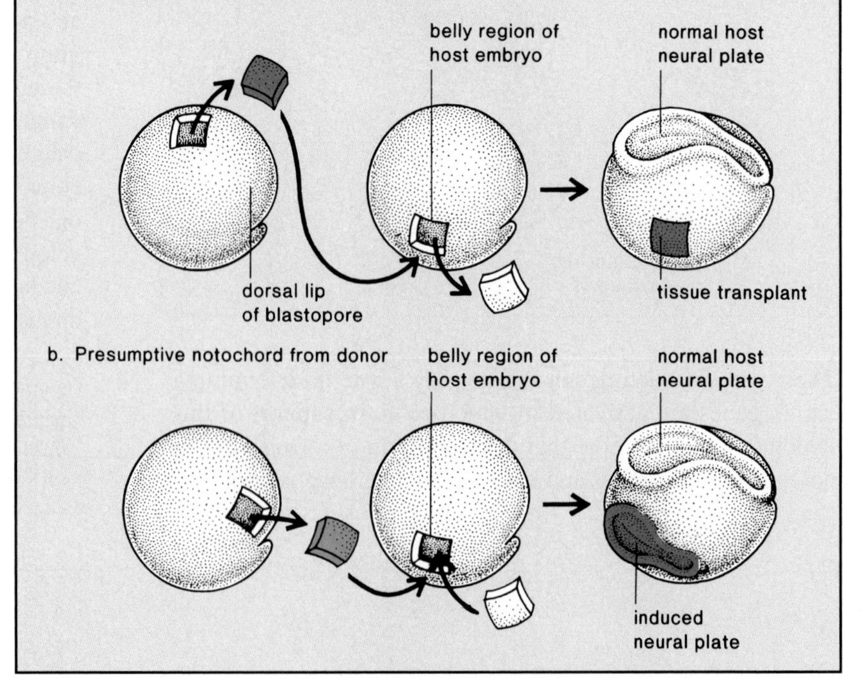

Figure 21.10

Development of the vertebrate eye. An optic vesicle induces the overlying ectoderm to thicken and become a lens vesicle. The lens vesicle induces formation of an optic cup, where the retina develops. The optic cup in turn induces formation of the lens and the cornea.

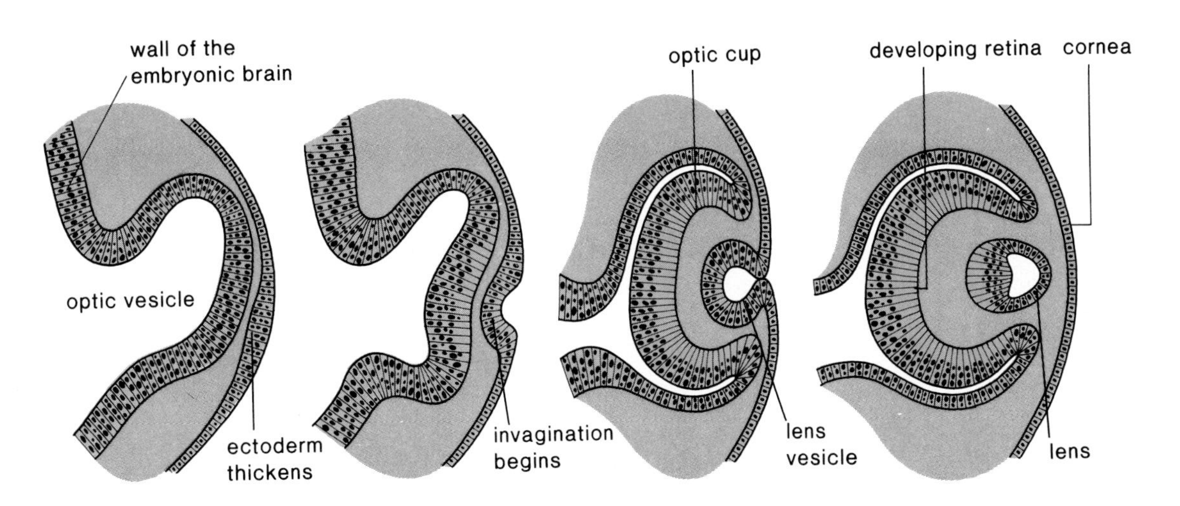

notochord, is cut out and transplanted to the belly region of the embryo, it will not form a neural plate. On the other hand, if presumptive notochord tissue is cut out and transplanted beneath what would be belly ectoderm, this ectoderm differentiates into a neural plate. A well-known series of inductions accounts for the development of the vertebrate eye (fig. 21.10). An optic vesicle, which is a lateral outgrowth of the embryonic brain, induces the overlying ectoderm to thicken and become a lens vesicle. The lens vesicle induces the optic vesicle to become an optic cup, where the retina develops. The optic cup in turn induces formation of the lens and the cornea.

Today, investigators believe the process of induction goes on continuously—neighboring cells are always influencing one another. Either direct contact or the production of a chemical acts as a signal that activates certain genes and brings about protein synthesis. This diagram shows how morphogenesis can be a sequential process:

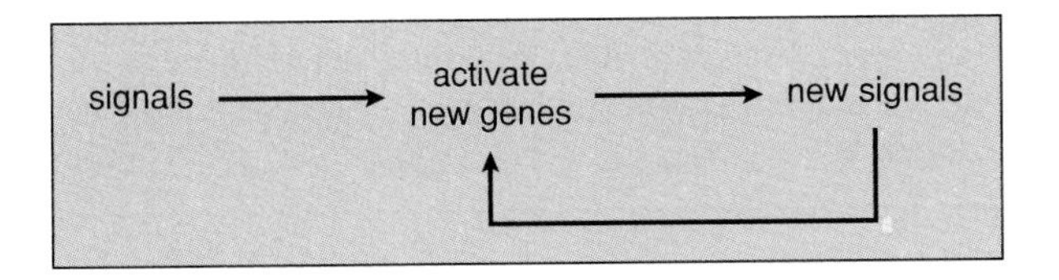

The diagram points out that there are both genetic and chemical aspects to morphogenesis. When the notochord induces the ectoderm to form the neural tube, a chemical signal is involved. The neural tube forms even if the 2 tissues are separated by a filter that allows only molecules to pass through. Presumably, this chemical signal activates particular genes in ectoderm, and this is why ectoderm forms the neural tube.

Investigators studying morphogenesis in *Drosophila* (fruit fly) have discovered that there are some genes that determine the animal's anterior/posterior and dorsal/ventral axes, others that determine the number and polarity of its segments, and still others, called *homeotic genes,* that determine how these segments develop. Homeotic genes have now been found in many other animals, and surprisingly, they all contain the same *homeobox,* a particular sequence of DNA. The homeobox codes for a sequence of amino acids in proteins that stay in the nucleus and regulate transcription of other genes during development. In keeping with the diagram, investigators envision that a protein produced by one homeotic gene turns on the next homeotic gene, and this orderly procession determines the overall organization (pattern formation) of the embryo.

Morphogenesis is dependent upon signals (either contact or chemical) from neighboring cells. These signals are believed to activate particular genes. Do 21.1 Critical Thinking, found at the end of the chapter.

Figure 21.11

Human embryo and fetus. Human development is divided into the embryonic period (first 2 months) and fetal development (months 3–9). **a.** Embryo is not recognizably human. **b.** Fetus is recognizably human.

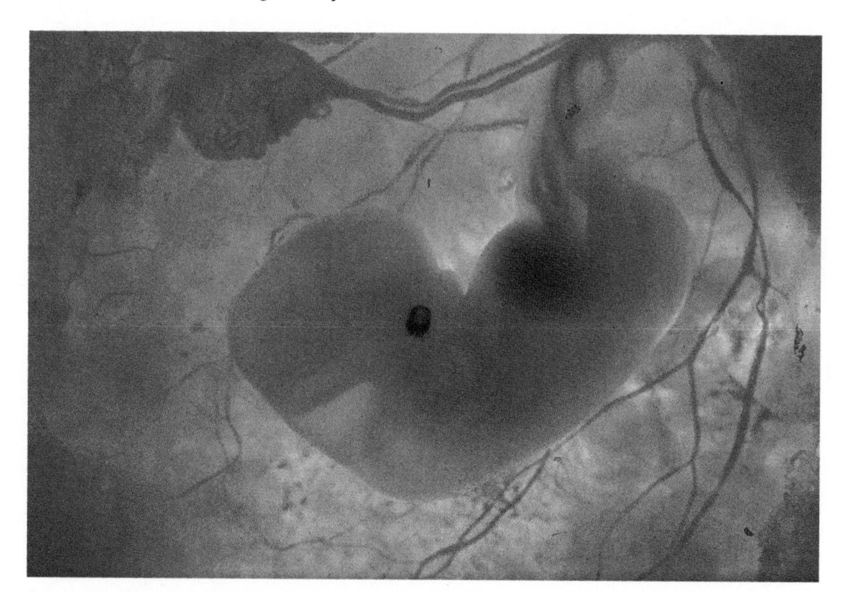

a.

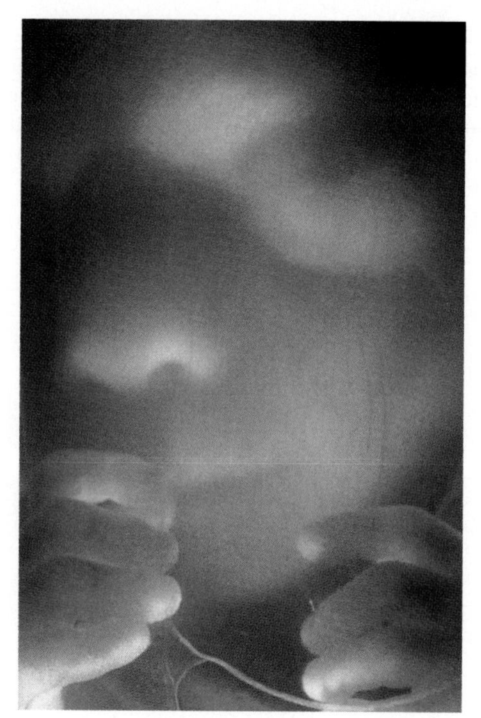

b.

Human Embryonic and Fetal Development

Human development is often divided into embryonic development (first 2 months) and fetal development (months 3–9) (fig. 21.11). The embryonic period consists of early formation of the major organs, and fetal development is the refinement of these structures.

Before we consider human development chronologically, we must understand the placement of **extraembryonic membranes.** Extraembryonic membranes are best understood by considering their function in reptiles and birds. In reptiles, these membranes made development on land first possible. If an embryo develops in the water, the water supplies oxygen for the embryo and takes away waste products. The surrounding water prevents desiccation, or drying out, and provides a protective cushion. For an embryo that develops on land, all these functions are performed by the extraembryonic membranes.

In the chick, the extraembryonic membranes develop from extensions of the germ layers, which spread out over the yolk. Each extraembryonic membrane consists of 2 germ layers (table 21.1). Figure 21.12 shows the chick within its hard shell surrounded by the membranes. The **chorion** lies next to the shell and carries on gas exchange. The **amnion** contains the protective amniotic fluid, which bathes the developing embryo. The **allantois** collects nitrogenous wastes, and the **yolk sac** surrounds the remaining yolk, which provides nourishment.

As figure 21.12 indicates, humans (and other mammals as well) also have these extraembryonic membranes. The chorion develops into the fetal half of the placenta; the yolk sac, which lacks yolk, is the first site of blood cell formation; the allantoic blood vessels become the umbilical blood vessels; and the amnion contains fluid to cushion and protect the fetus. Therefore, the function of the membranes in humans has been modified to suit internal development, but their very presence indicates our relationship to birds and to reptiles. It is interesting to note that all animals develop in water, either directly or within amniotic fluid.

The presence of extraembryonic membranes in reptiles made development on land possible. Humans also have these membranes, but their function has been modified for internal development.

Table 21.1
Extraembryonic Membranes in Chick and Human

Name	Germ Layers	Location in Chick	Function in Chick	Location in Human	Function in Human
Chorion	Outer layer of ectoderm and inner layer of mesoderm	Lies next to shell	Gas exchange	Fetal half of placenta	Exchange with mother's blood
Amnion	Outer layer of mesoderm and inner layer of ectoderm	Surrounds embryo	Protection; prevention of drying out and temperature changes	Same as chick	Protection and prevention of temperature changes
Allantois	Outer layer of mesoderm and inner layer of endoderm	Outgrowth of hindgut	Collection of nitrogenous wastes	Same as chick	Blood vessels become umbilical blood vessels
Yolk sac	Outer layer of mesoderm and inner layer of endoderm	Outgrowth of midgut; surrounds yolk	Provision of nourishment	Same; but contains no yolk	First site of blood cell formation

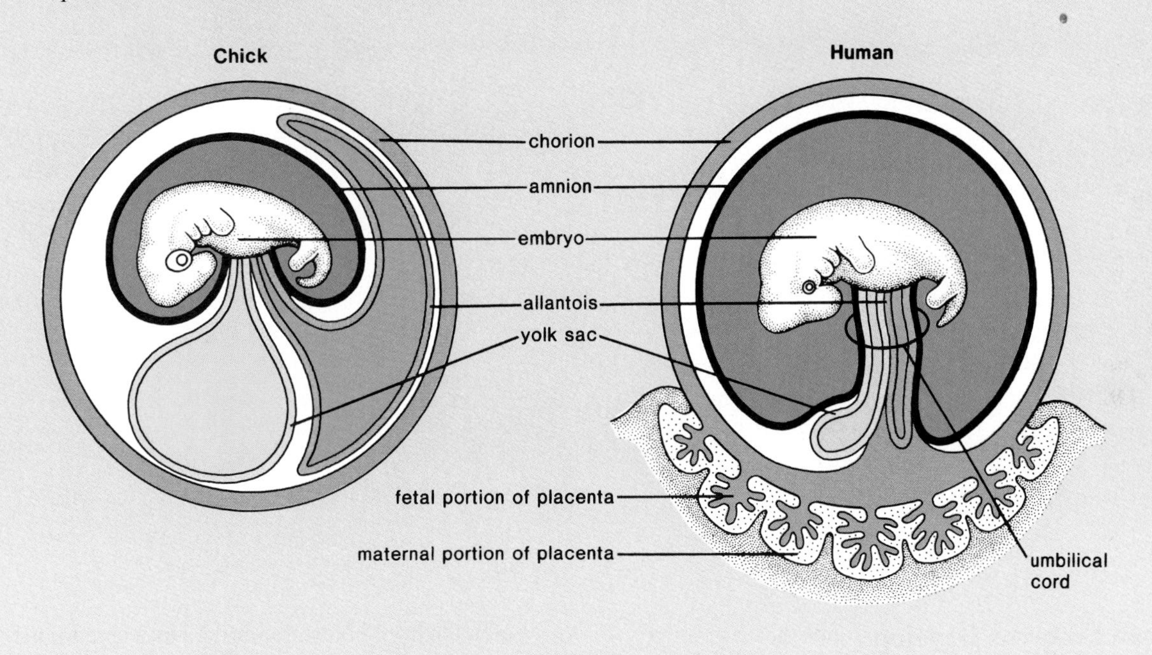

Figure 21.12
Extraembryonic membranes. The membranes, which are not part of the embryo, are found during the development of chicks and humans, where each has a specific function (table 21.1).

Figure 21.13

Human development before implantation. Structures and events proceed counterclockwise. At ovulation, the secondary oocyte leaves the ovary. A single sperm penetrates the zona pellucida, and fertilization occurs in the oviduct. As the zygote moves along the oviduct, it undergoes cleavage to produce a morula. The blastocyst forms and implants itself in the uterine lining.

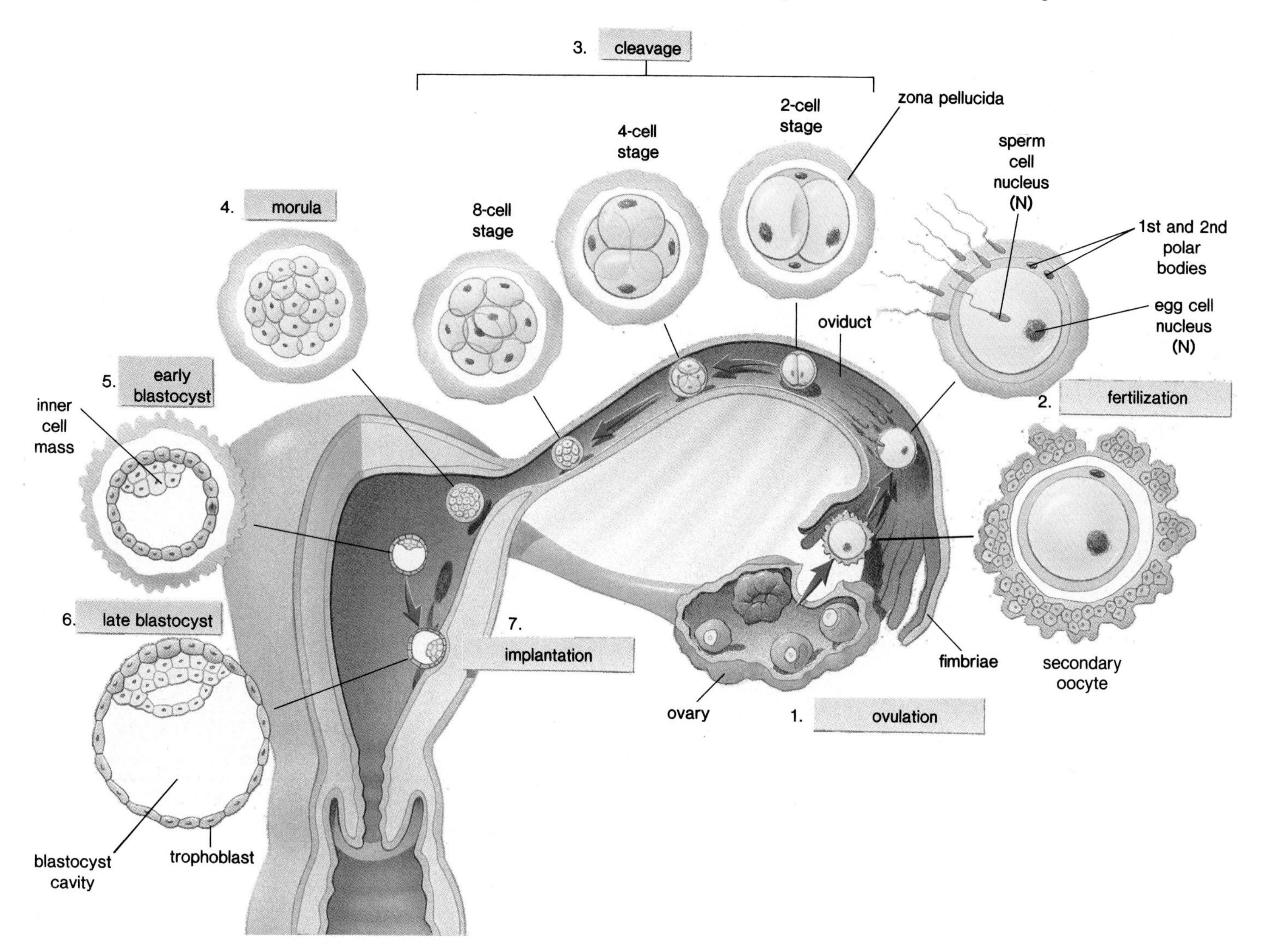

How Human Embryos Develop

Embryonic development includes the first 2 months of development.

First Week: Blastocyst Forms

Fertilization occurs in the upper third of an oviduct (fig. 21.13), and cleavage begins even as the embryo passes down this tube to the uterus. By the time the embryo reaches the uterus on the third day, it is a *morula.* The morula is not much larger than the zygote because even though multiple cell divisions have occurred, there has been no growth of these newly formed cells. By about the fifth day, the morula is transformed into the *blastocyst.* The blastocyst has a fluid-filled cavity, a single layer of outer cells called the **trophoblast,** and an inner cell mass. Later, the trophoblast, reinforced by a layer of mesoderm, gives rise to the *chorion,* one of the extraembryonic membranes (fig. 21.12). The *inner cell mass* eventually becomes the fetus. Each cell within the inner cell mass has the genetic capability of becoming a complete individual. Sometimes during human development, the inner cell mass splits, and 2 embryos start developing rather than one. These 2 embryos, which share the same placenta (p. 407), are identical twins because they have inherited exactly the same chromosomes. Fraternal twins, who arise when 2 different eggs are fertilized by 2 different sperm, do not have identical chromosomes. There are even cases of such "twins" having different fathers. Each fraternal twin has a separate placenta during development.

Figure 21.14

Stages showing the early appearance of the extraembryonic membranes and the formation of the umbilical cord in the human embryo. **a.** At 14 days, the amniotic cavity appears. **b.** At 21 days, the chorion and the yolk sac are apparent. **c.** At 28 days, the body stalk and the allantois form. **d.** At 35 days, the embryo begins to take shape as the umbilical cord starts forming. **e.** Eventually, at 42^{+} days, the umbilical cord is fully formed.

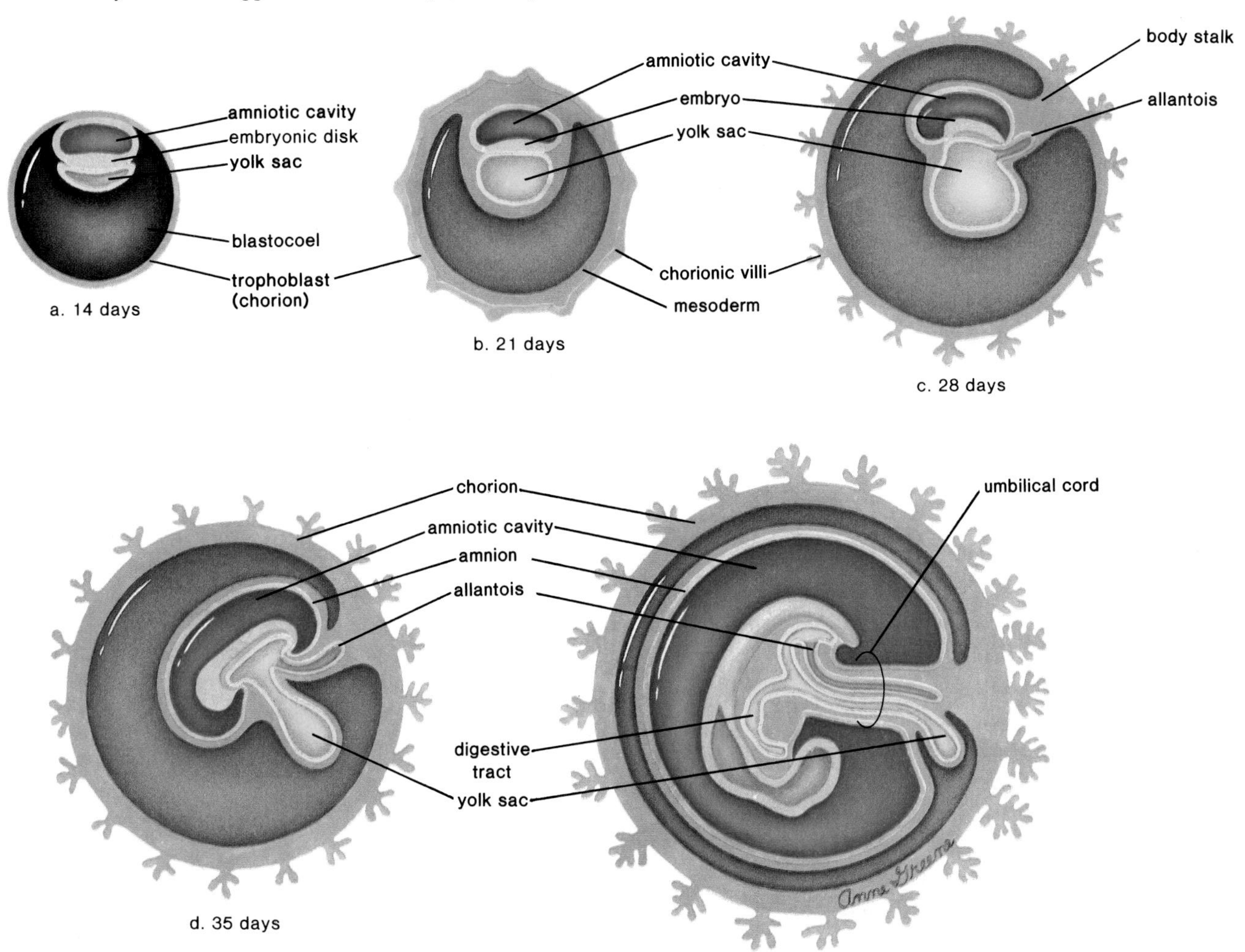

During the first week, the human embryo undergoes cleavage. The morula develops into the blastocyst, which has 2 main parts, the outer trophoblast (becomes the chorion) and the inner cell mass (becomes the fetus).

Second Week: Germ Layers Form

At the end of the first week, the embryo begins the process of *implanting* in the wall of the uterus. The trophoblast secretes enzymes to digest away some of the tissue and blood vessels of the uterine wall (fig. 21.13). The embryo is now about the size of the period at the end of this sentence. The trophoblast begins to secrete **human chorionic gonadotropic hormone (HCG),** the hormone that is the basis for the pregnancy test and that serves to maintain the corpus luteum past the time it normally disintegrates. Because of this, the endometrium is maintained and menstruation does not occur.

As the week progresses, the inner cell mass detaches itself from the trophoblast, and 2 more extraembryonic membranes form (fig. 21.14). The *yolk sac,* which forms below the embryonic disk, has no nutritive function as in chicks, but it is the first site of blood cell formation. However, the *amnion* and its cavity are where the embryo (and then the fetus) develops. In humans, amniotic fluid acts as an insulator against cold and heat and also absorbs any shock, such as a blow to the mother's abdomen.

Gastrulation occurs during the second week. The inner cell mass now has flattened into the *embryonic disk,* composed of 2 layers of cells: *ectoderm* above and *endoderm* below. Once the embryonic disk elongates to form the *primitive streak* (fig. 21.2), the third germ layer, mesoderm, forms by invagination of cells along the streak. The trophoblast is reinforced by mesoderm and becomes the chorion.

It is possible to relate the development of future organs to these germ layers (fig. 21.3).

> The germ layers (ectoderm, endoderm, and mesoderm) are laid down during the second week of development. Development of the organs can be related to these germ layers.

Third Week: First Major Organs

Two important organ systems make their appearance during the third week. The nervous system is the first organ system to be visually evident. At first, a thickening appears along the entire dorsal length of the embryo, and then invagination occurs as neural folds appear. When the neural folds meet at the midline, the neural tube, which later develops into the brain and the nerve cord, is formed (fig. 21.5). After the notochord is replaced by the vertebral column, the nerve cord is called the spinal cord.

Development of the heart begins in the third week and continues into the fourth week. At first, there are right and left heart tubes; when these fuse, the heart begins pumping blood, even though the chambers of the heart are not fully formed. The veins enter posteriorly and the arteries exit anteriorly from this largely tubular heart, but later the heart twists so that all major blood vessels are located anteriorly.

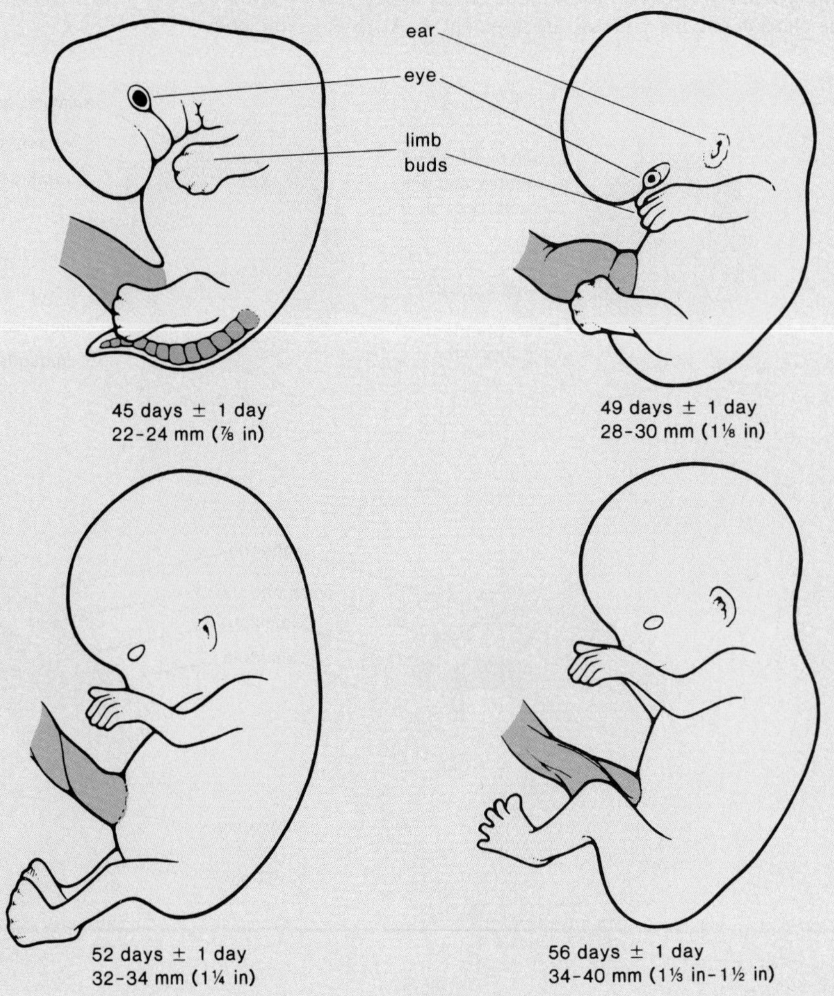

> During the third week, major organs, like the nerve cord and the heart, first make their appearance.

Figure 21.15
Human embryo at the days indicated and the sizes noted.

Fourth and Fifth Weeks: Head, Arms, Legs

At 4 weeks, the embryo is barely larger than the height of this print. A bridge of mesoderm called the body stalk connects the caudal (tail) end of the embryo with the chorion, which has projections called chorionic villi (fig. 21.14*c*). The fourth extraembryonic membrane, the *allantois,* is contained within this stalk, and its blood vessels become the umbilical blood vessels. The head and the tail then lift up, and the body stalk moves anteriorly by constriction (fig. 21.14*d*). Once this process is complete, the **umbilical cord,** which connects the developing embryo to the placenta, is fully formed (fig. 21.14*e*).

Little flippers called limb buds appear (fig. 21.15); later, the arms and the legs develop from the limb buds, and even the hands and the feet become apparent. At the same time—during the fifth week—the head enlarges and the sense organs become more prominent. It is possible to make out the developing eyes, ears, and even nose.

> During the fourth and fifth weeks, human features in regard to the arms and the legs begin to make their appearance.

Sixth through Eighth Weeks: Only 1.5 Inches Long

There is a remarkable change in external appearance during the sixth through eighth weeks of development (fig. 21.15), from a form that is difficult to recognize as human to one that is easily recognized as human. Concurrent with brain development, the

Figure 21.16

Anatomy of the placenta in a fetus at 6–7 months. The placenta is composed of both fetal and maternal tissues. Chorionic villi penetrate the uterine lining and are surrounded by maternal blood. Exchange of molecules between fetal and maternal blood takes place across the walls of the chorionic villi.

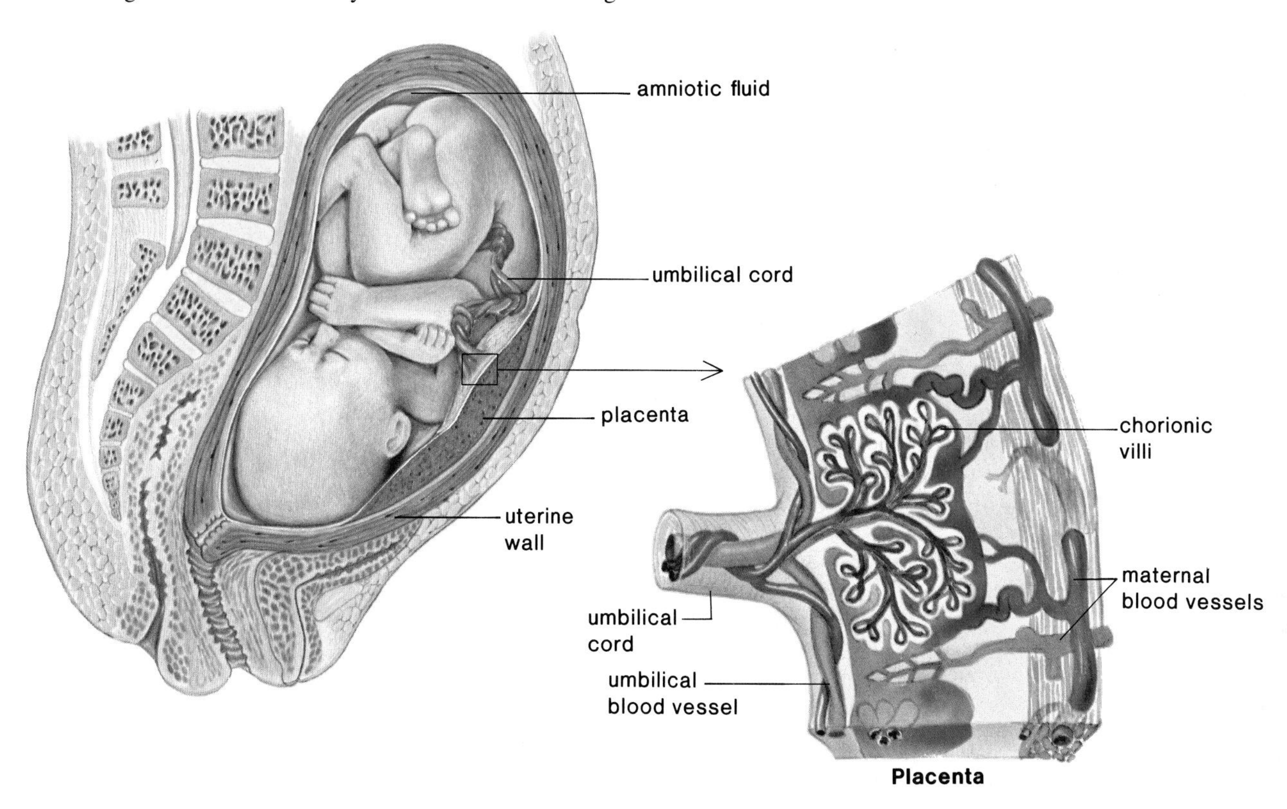

head achieves its normal relationship with the body as a neck region develops. The nervous system is developed well enough to permit reflex actions, such as a startle response to touch. At the end of this period, the embryo is about 38 mm (1 ½ in) long and weighs no more than an aspirin tablet, even though all organ systems are established.

Placenta: Forms by Tenth Week. The *placenta* begins formation once the embryo is fully implanted. Treelike extensions of the chorion called **chorionic villi** project into the maternal tissues. Later, these disappear in all areas except where the placenta develops. By the tenth week, the placenta (fig. 21.16) is fully formed and begins to produce progesterone and estrogen (fig. 21.17). These hormones have 2 effects: due to their negative feedback control of the hypothalamus and the anterior pituitary, they prevent any new follicles from maturing, and they maintain the lining of the uterus—now the corpus luteum is not needed. There is no menstruation during pregnancy.

Figure 21.17

Hormones during pregnancy. Human chorionic gonadotropin (gonadotropic hormone) is secreted by the trophoblast during the first 3 months of pregnancy. This maintains the corpus luteum, which continues to secrete estrogen and progesterone. At about 5 weeks of pregnancy, the placenta begins to secrete estrogen and progesterone in increasing amounts as the corpus luteum degenerates.

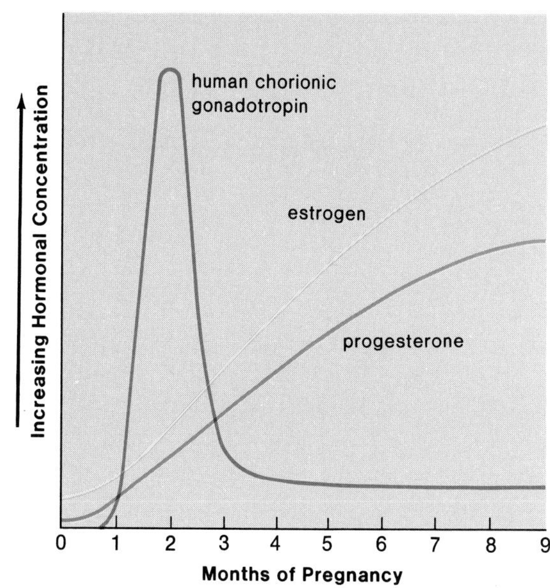

BIRTH DEFECTS CAN BE PREVENTED

It is believed that at least 1 in 16 newborns has a birth defect, either minor or serious, and the actual percentage may be even higher. Most likely, only 20% of all birth defects are due to heredity. Those that are hereditary can sometimes be detected before birth. Amniocentesis allows the fetus to be tested for abnormalities of development; chorionic villi sampling allows the embryo to be tested; and just recently, a method has been developed for screening eggs to be used for in vitro fertilization (fig. 21.A).

Treatment of the fetus in the womb is a rapidly developing area of medical expertise. Biochemical defects can sometimes be treated by giving the mother appropriate medicines. For example, if a baby is unable to use biotin efficiently, the mother can take these substances in doses large enough to prevent any untoward effects in the child. Structural defects can sometimes be corrected by intrauterine surgery. For example, if the fetus has water on the brain or is unable to pass urine, tubes that temporarily allow the fluid to pass out into the amniotic fluid can be inserted while the fetus is still in the womb. Physicians are hopeful that eventually all sorts of structural defects can be corrected by lifting the fetus from the womb long enough for corrective surgery to be performed.

It is recommended that all females take everyday precautions to protect any future and/or presently developing embryos and fetuses from defects that are not due to heredity. X-ray diagnostic therapy should be avoided during pregnancy because X rays cause mutations in the developing embryo or fetus. Children born to women who have received X-ray treatment are apt to have birth defects and/or to develop leukemia later on. Toxic chemicals, such as pesticides and many organic industrial chemicals, are also mutagenic. Cigarette smoke not only contains carbon monoxide but also some of these very same fetotoxic chemicals. Babies born to smokers are often underweight and subject to convulsions.

Pregnant Rh^- women should receive an Rh immunoglobulin injection to prevent the production of Rh antibodies. These antibodies can cause nervous system and heart defects.

Sometimes, birth defects are caused by microbes. Females can be immunized before the childbearing years for rubella (German measles), which in particular causes such birth defects as deafness. Unfortunately, immunization for sexually transmitted diseases is not possible. The AIDS virus can cross the placenta, and over 1,500 babies who contracted AIDS while in their mother's womb are now mentally retarded. When a mother has herpes, gonorrhea, or chlamydia, newborns can become infected as they

Figure 21.A

Three methods for genetic defect testing before birth. **a.** Amniocentesis cannot be done until the sixteenth week of pregnancy. A long needle is passed through the abdominal wall to withdraw a small amount of amniotic fluid, along with fetal cells. Since there are only a few cells in the amniotic fluid, testing may be delayed as long as 4 weeks until cell culture produces enough cells for testing purposes. About 40 tests are available for different defects.

b. Chorionic villi sampling can be done as early as the fifth week of pregnancy. The doctor inserts a long, thin tube through the vagina into the uterus. With the help of ultrasound, which gives a picture of the uterine contents, the tube is placed between the lining of the uterus and the chorion. Then a sampling of the chorionic villi cells is suctioned. Chromosome analysis and biochemical tests for several different genetic defects can be done immediately on these cells.

c. Screening eggs for genetic defects is a new technique. Preovulatory eggs are removed by aspiration after a telescope with a fiberoptic illuminator, called a laparoscope, is inserted into the abdominal cavity through a small incision in the region of the navel. The prior administration of FSH ensures that several eggs are available for retrieval and screening. Only the chromosomes within the first polar body are tested because if the woman is heterozygous for a genetic defect and it is found in the polar body, then the egg must be normal. Normal eggs undergo in vitro fertilization and are placed in the prepared uterus. At present, only 1 in 10 attempts results in a birth, but it is known ahead of time that the child will be normal for the genetic traits tested.

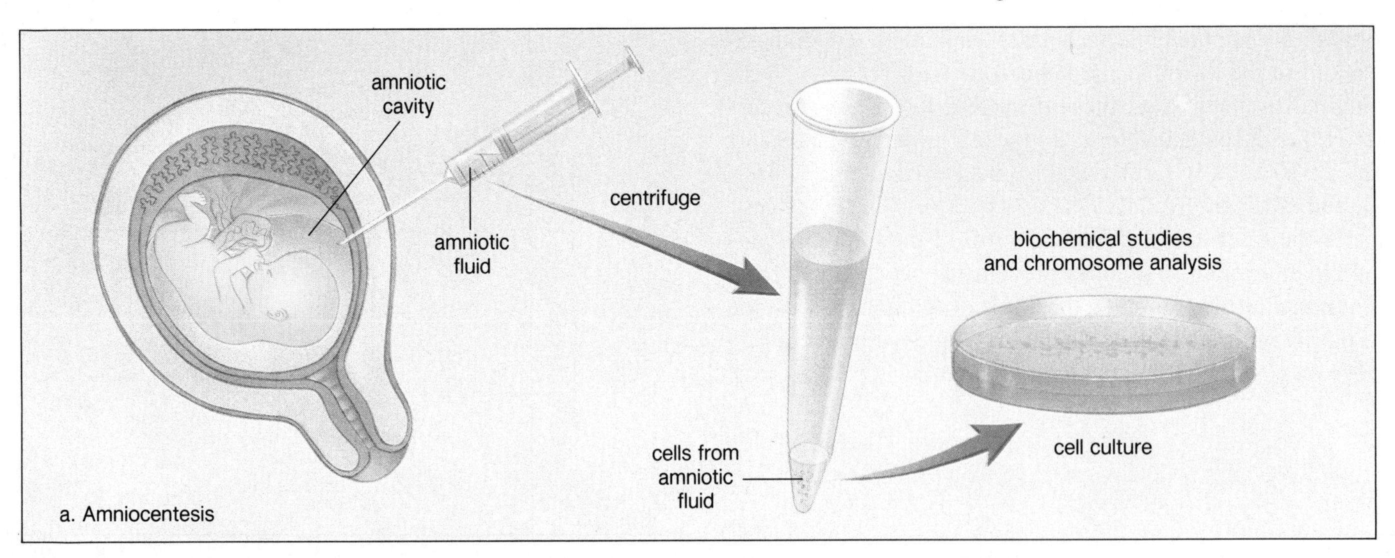

pass through the birth canal. Blindness and other physical or mental defects may develop. Birth by cesarean section could prevent these occurrences.

Pregnant women should not take any type of drug. Certainly illegal drugs, like marijuana, cocaine, and heroin, should be completely avoided. "Cocaine babies" now make up 60% of drug-affected babies. Severe fluctuations in blood pressure accompany the use of cocaine; these temporarily deprive the developing brain of oxygen. Cocaine babies have visual problems, lack coordination, and are mentally retarded. The drugs aspirin, caffeine (present in coffee, tea, and cola), and alcohol should be severely limited. It is not unusual for babies of drug addicts and alcoholics to display withdrawal symptoms and to have various abnormalities. Babies born to women who have about 45 drinks a month and as many as 5 drinks on one occasion are apt to have fetal alcohol syndrome (FAS). These babies have decreased weight, height, and head size, with malformation of the head and the face. Mental retardation is common in FAS infants.

Medications can also cause problems. When the synthetic hormone DES was given to pregnant women to prevent miscarriage, their daughters showed various abnormalities of the reproductive organs and an increased tendency toward cervical cancer. Other sex hormones, including birth-control pills, can possibly cause abnormal fetal development, including abnormalities of the sex organs. The tranquilizer thalidomide is well known for having caused deformities of the arms and legs in children born to women who took the drug. Therefore, a woman has to be very careful about taking medications while pregnant.

Now that physicians and laypeople are aware of the various ways in which birth defects can be prevented, it is hoped that the incidence of birth defects will decrease in the future.

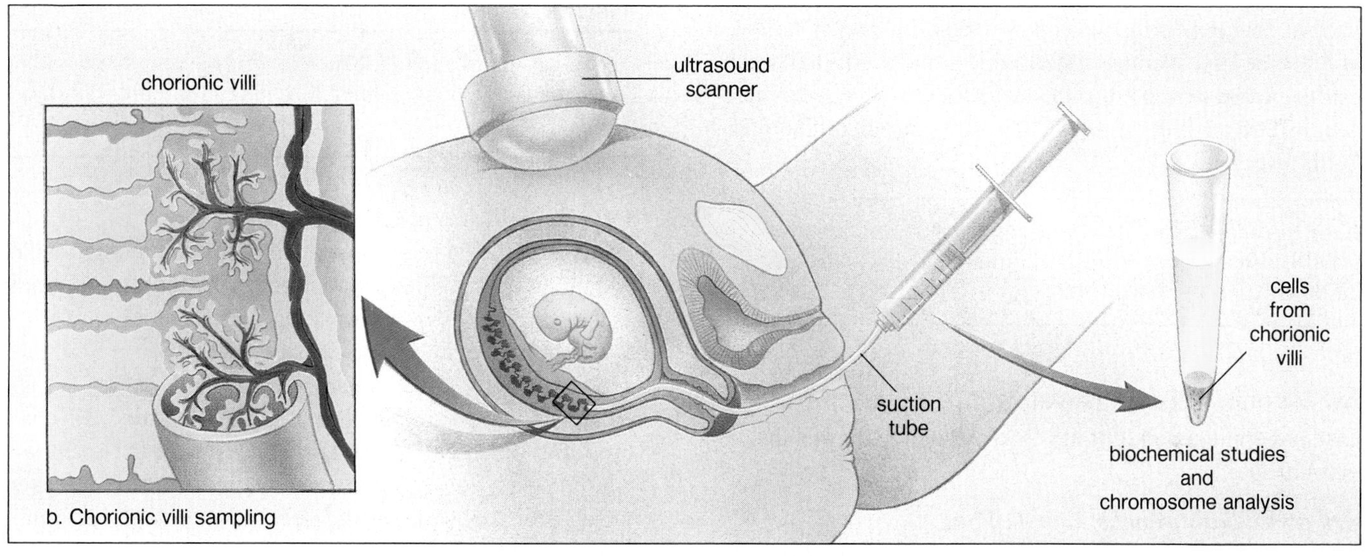

b. Chorionic villi sampling

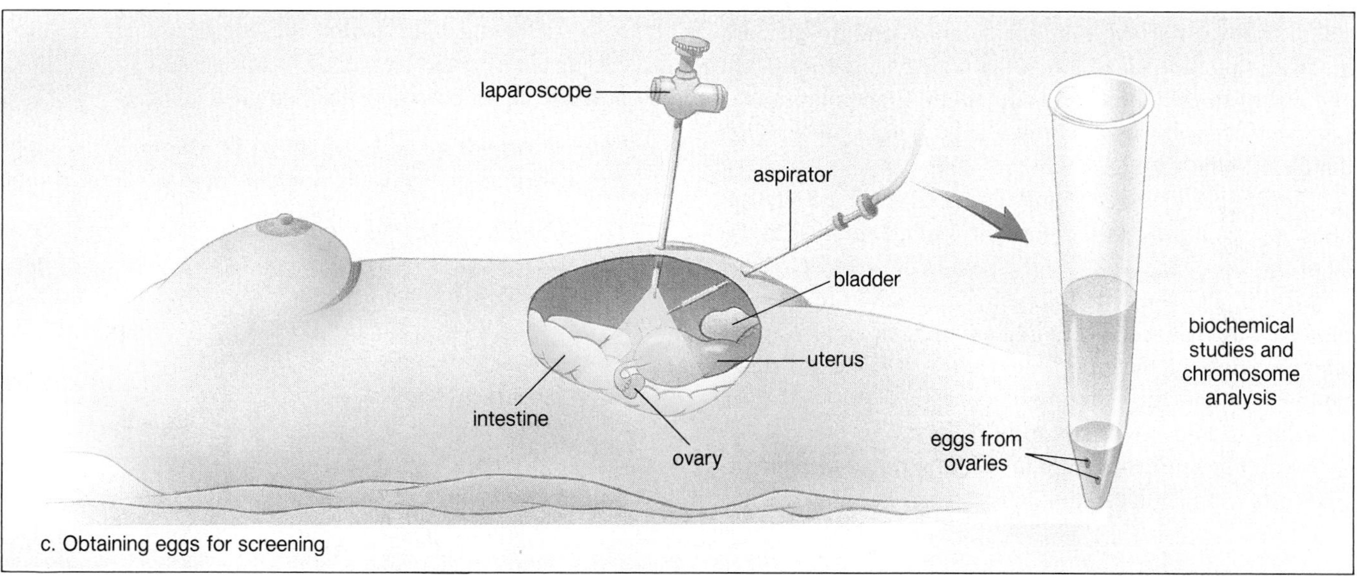

c. Obtaining eggs for screening

The placenta has a fetal side contributed by the chorion and a maternal side consisting of uterine tissues. Notice in figure 21.16 how the chorionic villi are surrounded by maternal blood sinuses; yet, maternal and fetal blood never mix since exchange always takes place across cell membranes. Carbon dioxide and other wastes move from the fetal side to the maternal side, and nutrients and oxygen move from the maternal side to the fetal side of the placenta. The umbilical cord stretches between the placenta and the fetus. Although it may seem that the umbilical cord travels from the placenta to the intestine, actually the umbilical cord is simply taking fetal blood to and from the placenta. The umbilical cord is the lifeline of the fetus because it contains the umbilical arteries and vein, which transport waste molecules (carbon dioxide and urea) to the placenta for disposal and take oxygen and nutrient molecules from the placenta to the rest of the fetal circulatory system.

Harmful chemicals can also cross the placenta. This is of particular concern during the embryonic period, when various structures are first forming. Each organ or part seems to have a sensitive period during which a substance can alter its normal function. The reading on pages 408–409 concerns the origin of birth defects.

> At the end of the embryonic period, all organ systems are established and the placenta is mature and fully functioning. The embryo is only about 38 mm ($1^1/_2$ in) long.

How Human Fetuses Develop

Fetal development includes the third through ninth months of development.

Third and Fourth Months: Can Tell Sex

At the beginning of the third month, the fetal head is still very large, the nose is flat, the eyes are far apart, and the ears are distinctively present. Head growth now begins to slow down as the rest of the body increases in length. Epidermal refinements, such as eyelashes, eyebrows, hair on head, fingernails, and nipples, appear.

Cartilage begins to be replaced by *bone* as ossification centers appear in most of the bones. Cartilage remains at the ends of the long bones, and ossification is not complete until age 18 or 20. The skull has 6 large membranous areas called *fontanels,* which permit a certain amount of flexibility as the head passes through the birth canal and allow rapid growth of the brain during infancy. Progressive fusion of the skull bones causes the fontanels to disappear by 16 months of age.

Sometime during the third month, it is possible to distinguish males from females. Apparently, the Y chromosome has a testis-determining factor gene (TDF), which triggers the differentiation of gonads into testes. Once the testes differentiate, they produce androgens, the male sex hormones. The androgens, especially testosterone, stimulate the growth of the male external genitals. In the absence of androgens, female genitals form. The ovaries do not produce estrogen because there is plenty of it circulating in the mother's bloodstream.

At this time, both testes and ovaries are located within the abdominal cavity, but later, in the last trimester of fetal development, the testes descend into the scrotal sacs (scrotum). Sometimes the testes fail to descend, and in that case, an operation may be done later to place them in their proper location, as discussed in the previous chapter.

During the fourth month, the fetal heartbeat is loud enough to be heard when a physician applies a stethoscope to the mother's abdomen. By the end of this month, the fetus is about 152 mm (6 in) in length and weighs about 171 g (6 oz).

> During the third and fourth months, it is obvious that the skeleton is becoming ossified. The sex of the individual can now be distinguished.

Fifth through Seventh Months: Fetus Moves

During the fifth through seventh months, the mother begins to feel movement. At first, there is only a fluttering sensation, but as the fetal legs grow and develop, kicks and jabs are felt. The fetus, though, is in the fetal position with the head bent down and in contact with the flexed knees.

The wrinkled, translucent, pink-colored skin is covered by a fine down called **lanugo.** This in turn is coated with a white, greasy, cheeselike substance called **vernix caseosa,** which probably protects the delicate skin from the amniotic fluid. The eyelids are now fully open, however.

At the end of this period, the length has increased to about 300 mm (12 in) and the weight is now about 1,380 g (3 lb). It is possible that if born now, the baby will survive.

Fetal Circulation. As figure 21.18 shows, the fetus has 4 circulatory features that are not present in adult circulation.

1. ***Oval opening,*** or *foramen ovale,* an opening between the 2 atria. This opening is covered by a flap of tissue that acts as a valve.
2. ***Arterial duct,*** or *ductus arteriosus,* a connection between the pulmonary artery and the aorta
3. ***Umbilical arteries*** and ***vein,*** vessels that travel to and from the placenta, leaving waste and receiving nutrients
4. ***Venous duct,*** or *ductus venosus,* a connection between the umbilical vein and the inferior vena cava

Figure 21.18

Fetal circulation. The umbilical arteries carry deoxygenated blood to the placenta, where gas exchange occurs. The umbilical vein carries oxygenated blood to the inferior vena cava via the venous duct. The inferior and superior venae cavae enter the right atrium. From there, mixed blood enters the left atrium by way of the oval window or it continues to the right ventricle. From the right ventricle, blood enters the pulmonary trunk and then the aorta via the arterial duct. In this way, blood is diverted from the lung. Notice, too, that the blood in the aorta is mixed blood with less oxygen.

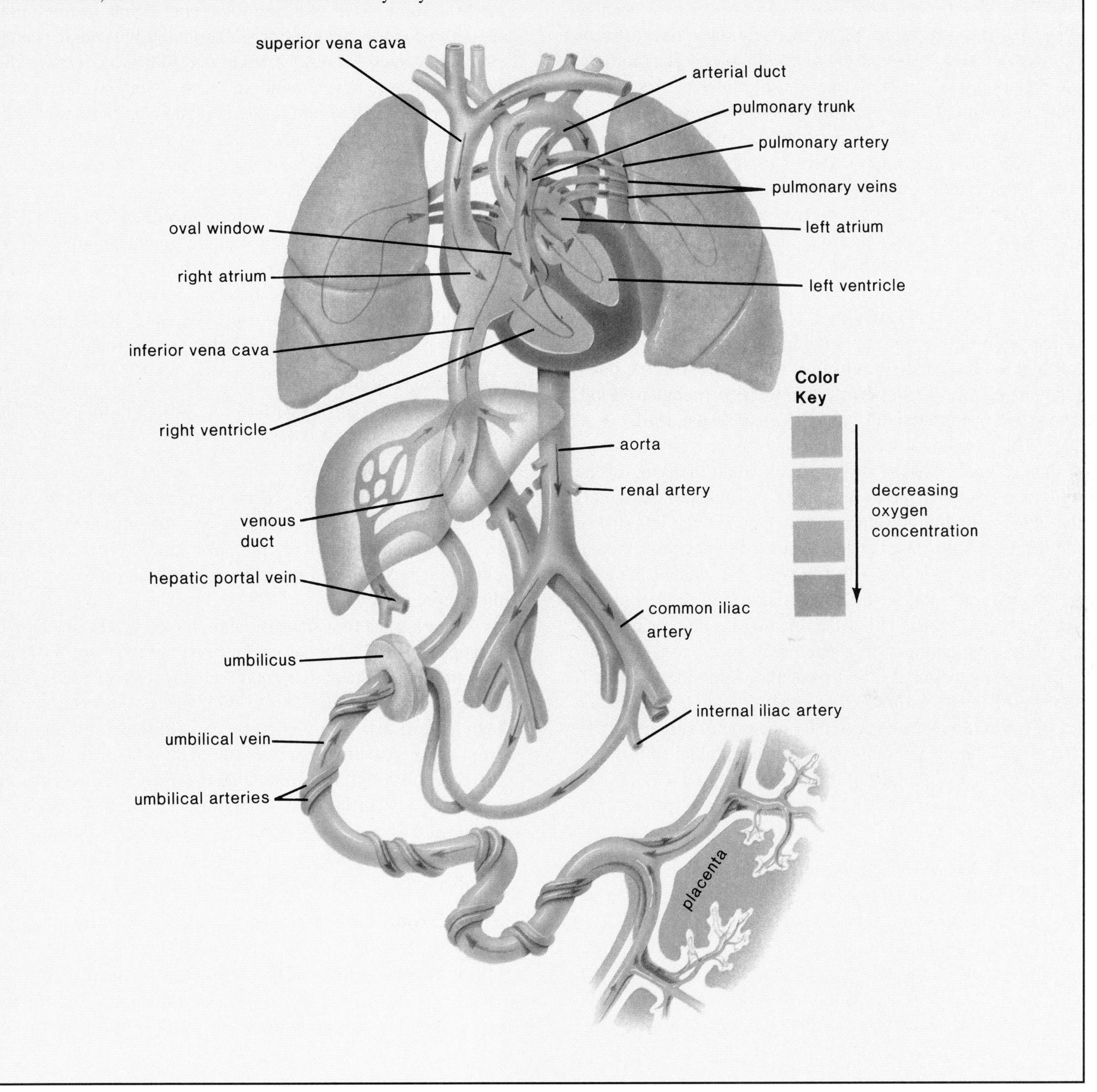

All of these features can be related to the fact that the fetus does not use its lungs for gas exchange since it receives oxygen and nutrients from the mother's blood by way of the placenta.

To trace the path of blood in the fetus, begin with the right atrium (fig. 21.18). From the right atrium, blood may pass directly into the left atrium by way of the oval opening or it may pass through the atrioventricular valve into the right ventricle. From the right ventricle, blood goes into the pulmonary artery, but because of the arterial duct, most blood then passes into the aorta. Therefore, by whatever route blood takes, most of it reaches the aorta instead of the lungs.

Blood within the aorta travels to the various branches, including the iliac arteries, which connect to the umbilical arteries leading to the placenta. Exchange between maternal and fetal blood takes place at the placenta. It is interesting to note that blood in the umbilical arteries, which travels to the placenta, is low in oxygen, but blood in the umbilical vein, which travels from the placenta, is high in oxygen. The umbilical vein enters the venous duct, which passes directly through the liver. The venous duct then joins with the inferior vena cava, a vessel that contains deoxygenated blood. The vena cava returns this "mixed blood" to the heart.

The most common of all cardiac defects in the newborn is the persistence of the oval opening. With the tying of the umbilical cord and the expansion of the lungs, blood enters the lungs in quantity. Return of this blood to the left side of the heart usually causes a flap to cover the opening. Incomplete closure occurs in nearly 1 out of 4 individuals, but even so, passage of blood from the right atrium to the left atrium rarely occurs because either the opening is small or it closes when the atria contract. In a small number of cases, the passage of deoxygenated blood from the right side to the left side of the heart is sufficient to cause a "blue baby." Such a condition can now be corrected by open-heart surgery.

The arterial duct closes because endothelial cells divide and block off the duct. Remains of the arterial duct and parts of the umbilical arteries and vein are later transformed into connective tissue.

Eighth and Ninth Months: More Growth

As the time of birth approaches, the fetus rotates so that the head is pointed toward the cervix (fig. 21.19*a*). If the fetus does not turn, then the likelihood of a breech birth (rump first) may call for a cesarean section. It is very difficult for the cervix to expand enough to accommodate this form of birth, and asphyxiation of the baby is more likely.

At the end of this time period, the fetus is about 525 mm (20 1/2 in) long and weighs about 3,380 g (7 1/2 lb). Weight gain is largely due to the accumulation of fat beneath the skin.

Birth: Occurs in Stages

The uterus characteristically contracts throughout pregnancy. At first, light, often indiscernible contractions lasting about 20–30 seconds occur every 15–20 minutes, but near the end of pregnancy, they become stronger and more frequent, so that the woman may falsely think that she is in labor. The onset of true labor is marked by uterine contractions that occur regularly every 15–20 minutes and last for 40 seconds or more. **Parturition,** which includes labor and expulsion of the fetus, usually is considered to have 3 stages (fig. 21.19*b–d*).

The events that cause parturition still are not known entirely, but there is now evidence suggesting the involvement of prostaglandins. It may be, too, that the prostaglandins cause the release of oxytocin from the mother's posterior pituitary. Both prostaglandins and oxytocin cause the uterus to contract, and either hormone can be given to induce parturition.

Three Stages of Birth

During the *first stage* of parturition, the cervix dilates; during the *second,* the baby is born; and during the *third,* the afterbirth is expelled.

Stage 1: Cervix Dilates. Prior to or at the same time as the first stage of parturition, there may be a "bloody show," caused by the expulsion of a mucus plug from the cervical canal. This plug prevents bacteria and sperm from entering the uterus during pregnancy.

During the first stage of labor, the cervical canal slowly disappears (fig. 21.19*b*) as the lower part of the uterus is pulled upward toward the baby's head. This process is called *effacement,* or *taking up the cervix.* With further contractions, the baby's head acts as a wedge to assist cervical dilation. The baby's head usually has a diameter of about 10 cm; therefore, the cervix has to dilate to this diameter in order to allow the head to pass through. If it has not occurred already, the amniotic membrane is apt to rupture now, releasing the amniotic fluid, which escapes out the vagina. The first stage of labor ends once the cervix is completely dilated.

Stage 2: Baby Emerges. During the second stage of parturition, the uterine contractions occur every 1–2 minutes and last about one minute each. They are accompanied by a desire to push, or bear down. As the baby's head gradually descends into the vagina, the desire to push becomes greater.

Figure 21.19

Three stages of parturition. **a.** Position of fetus just before birth begins. **b.** Dilation of cervix. **c.** Birth of baby. **d.** Expulsion of afterbirth.

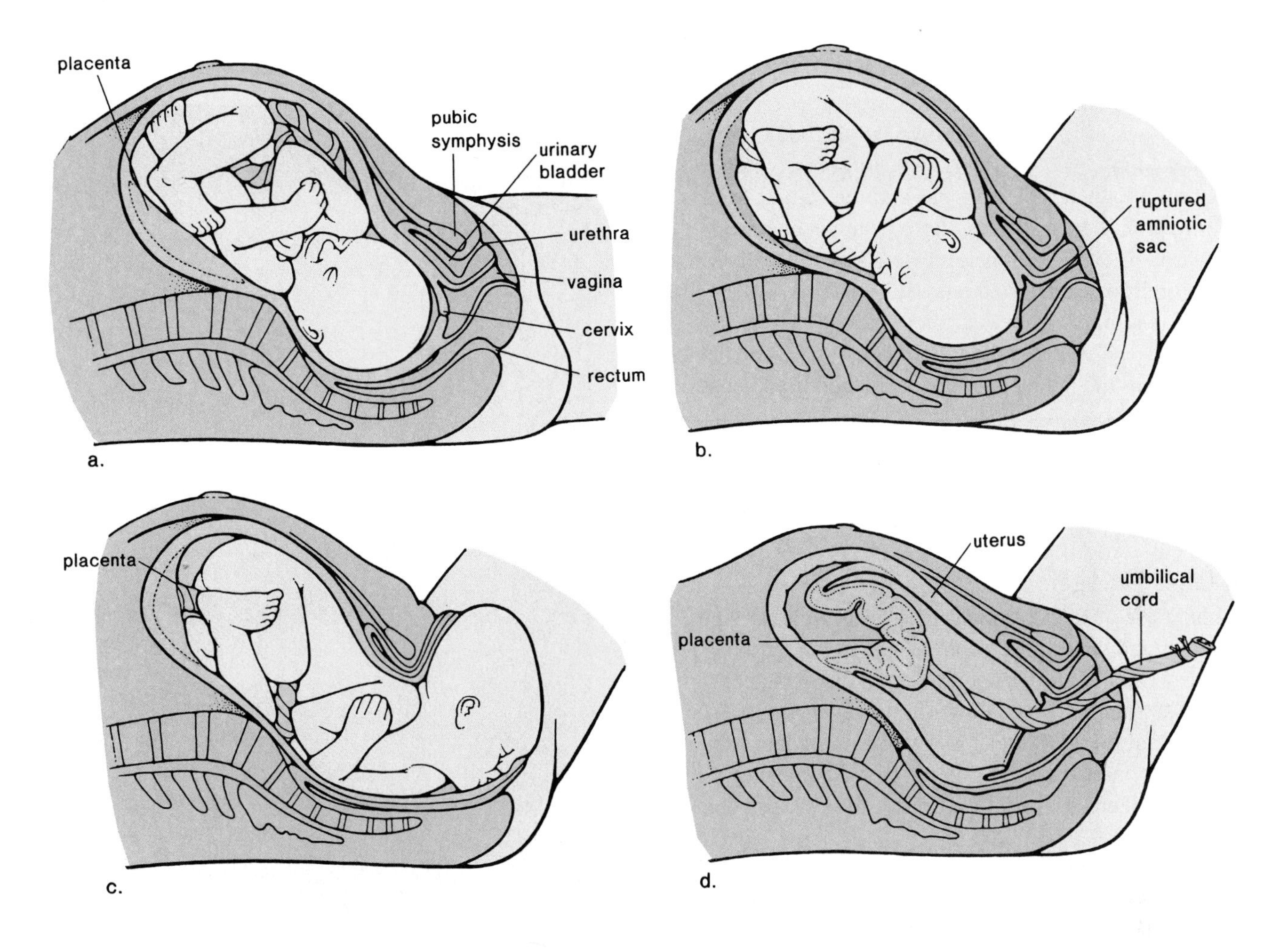

When the baby's head reaches the exterior, it turns so that the back of the head is uppermost (fig. 21.19*c*). Since the vagina may not expand enough to allow passage of the head without tearing, an *episiotomy* is often performed. This incision, which enlarges the vaginal opening, is stitched later and heals more perfectly than a tear. As soon as the head is delivered, the baby's shoulders rotate so that the baby faces either to the right or the left. The physician at this time may hold the head and guide it downward, while one shoulder and then the other emerges. The rest of the baby follows easily.

Once the baby is breathing normally, the umbilical cord is cut and tied, severing the child from the placenta. The stump of the cord shrivels and leaves a scar, which is the navel.

Stage 3: Expelling Afterbirth. The placenta, or *afterbirth*, is delivered during the third stage of labor (fig. 21.19*d*). About 15 minutes after delivery of the baby, uterine muscle contractions shrink the uterus and dislodge the placenta. The placenta is then expelled into the vagina. As soon as the placenta and its membranes are delivered, the third stage of labor is complete.

During the first stage of birth, the cervix dilates; during the second, the child is born; and during the third, the afterbirth is expelled.

Human Development after Birth

Development does not cease once birth has occurred but continues throughout the stages of life: infancy, childhood, adolescence, and adulthood.

Infancy lasts until about 2 years of age. It is characterized by tremendous growth and sensorimotor development. During *childhood,* the individual grows and the body proportions change. *Adolescence* begins with **puberty,** when the secondary sex characteristics appear and the sexual organs become functional. At this time, there is an acceleration of growth leading to changes in height, weight, fat distribution, and body proportions. Males commonly experience a growth spurt later than females; therefore, they grow for a longer period of time. Males are generally taller than females and have broader shoulders and longer legs relative to their trunk length.

Developmental changes keep occurring throughout infancy, childhood, adolescence, and adulthood.

Figure 21.20

Aging is a slow process during which the body undergoes changes that eventually bring about death even if no marked disease or disorder is present. Medical science is trying to extend the human life span and the health span, the length of time the body functions normally.

What Makes Us Age?

Young adults are at their physical peak in muscle strength, reaction time, and sensory perception. The organ systems at this time are best able to respond to altered circumstances in a homeostatic manner. From now on, however, there is an almost imperceptible, gradual loss in certain of the body's abilities. **Aging** encompasses these progressive changes that contribute to an increased risk of infirmity, disease, and death (fig. 21.20).

Today, there is great interest in **gerontology,** the study of aging, because there are now more older individuals in our society than ever before and the number is expected to rise dramatically. In the next half-century, those over age 75 will rise from the present 13 million to 35–45 million, and those over age 80 will rise from 3 million to 6 million individuals. The human life span is judged to be a maximum of 110–115 years. The present goal of gerontology is not to necessarily increase the life span but to increase the health span, the number of years that an individual enjoys the full functions of all body parts and processes.

Three Theories of Aging

There are many theories about what causes aging. Three of these are considered here.

Genetic in Origin. Several lines of evidence indicate that aging has a genetic basis. (1) The number of times a cell divides is species-specific. The maximum number of times human cells divide is around 50. Perhaps as we grow older, more and more cells are unable to divide any longer, and instead they undergo degenerative changes and die. (2) Some cell lines may become nonfunctional long before the maximum number of divisions has occurred. Whenever DNA replicates, mutations can occur, and this can lead to the production of nonfunctional proteins. Eventually, the number of inadequately functioning cells can build up, which contributes to the aging process. (3) The children of long-lived parents tend to live longer than those of short-lived parents. Recent work in lower animals suggests that when an animal produces fewer free radicals, it lives longer. *Free radicals* are molecules that are reduced, because they have taken electrons away from the macromolecules that make up a cell. In this way, production of free radicals leads to the destruction of the cell. There are genes that code for antioxidant enzymes that detoxify free radicals. This research suggests that animals with particular forms of these genes—and therefore more efficient antioxidant enzymes—live longer.

Whole-Body Processes. A decline in the hormonal system can affect many different organs of the body. For example, type II diabetes is common in older individuals. The pancreas makes insulin, but the cells lack the receptors that enable them to respond. Menopause in women occurs for a similar reason. There is plenty of FSH in the bloodstream, but the ovaries do not respond. Perhaps aging results from the loss of hormonal activities and a decline in the functions they control.

The immune system, too, no longer performs as it once did, and this can affect the body as a whole. The thymus gland gradually decreases in size, and eventually most of it is replaced by fat and connective tissue. The incidence of cancer increases among the elderly, which may signify that the immune system is no longer functioning as it should. This idea is substantiated, too, by the increased incidence of autoimmune diseases in older individuals.

It is possible, though, that aging is not due to the failure of a particular system that can affect the body as a whole, but to a specific type of tissue change that affects all organs and even the genes. It has been noticed for some time that proteins—such as collagen, which makes up the white fibers (p. 156) and is present in many support tissues—become increasingly cross-linked as people age. Undoubtedly, this cross-linking contributes to the stiffening and the loss of elasticity characteristic of aging tendons and ligaments. It may also account for the inability of such organs as the blood vessels, the heart, and the lungs to function as they once did. Some researchers have now found that glucose has the tendency to attach to any type of protein, which is the first step in a cross-linking process that ends with the formation of advanced glycosylation end products (AGEs).

AGEs not only explain why cataracts develop, they also may contribute to the development of atherosclerosis and to the inefficiency of the kidneys in diabetics and older individuals. Even DNA-associated proteins seem capable of forming glucose-derived cross-links, and perhaps this increases the rate of mutations as we age. These researchers are presently experimenting with the drug aminoguanidine, which can prevent the development of AGEs.

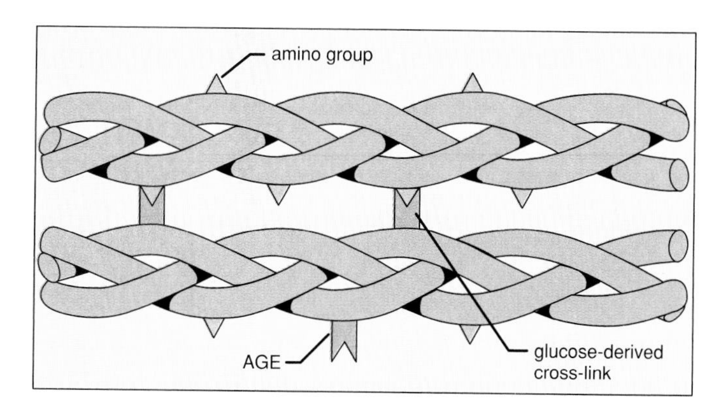

Extrinsic Factors. The current data about the effects of aging are often based on comparisons of the characteristics of the elderly to younger age groups, but perhaps today's elderly were not as aware when they were younger of the importance of, for example, diet and exercise to general health. It is possible, then, that much of what we attribute to aging is instead due to years of poor health habits.

Consider, for example, osteoporosis. This condition is associated with a progressive decline in bone density in both males and females so that fractures are more likely to occur after only minimal trauma. Osteoporosis is common in the elderly—by age 65, one-third of women will have vertebral fractures, and by age 81, one-third of women and one-sixth of men will have suffered a hip fracture. While there is no denying that there is a decline in bone mass as a result of aging, certain extrinsic factors are also important. The occurrence of osteoporosis itself is associated with cigarette smoking, heavy alcohol intake, and perhaps inadequate calcium intake. Not only is it possible to eliminate these negative factors by personal choice, it also is possible to add a positive factor. A moderate exercise program has been found to slow down the progressive loss of bone mass.

Even more important, an exercise program helps eliminate cardiovascular disease, the leading cause of death today. Experts no longer believe that the cardiovascular system necessarily suffers a large decrease in functioning ability with age. Persons 65 years of age and older can have well-functioning hearts and open coronary arteries if their health habits are good and they continue to exercise regularly.

Rather than collecting data on the average changes observed between different age groups, it might be more useful to note the differences within any particular age group. If this type of comparison is done, extrinsic factors that contribute to a decline as well as those that promote the health of an organ can be identified.

How Aging Affects Body Systems

Keeping in mind that we want to accept such data with reservations, we will still discuss in general the effects of aging on the various systems of the body. Figure 21.21 compares the percentage of function of various organs in a person 75–80 years old to that of a person 20 years old whose organs are assumed to function at 100% capacity. When making this comparison, we should keep in mind that the body has a vast functional reserve; it can still perform well even when not at 100% capacity.

Skin. As aging occurs, skin becomes thinner and less elastic because the number of elastic fibers decreases and the collagen fibers undergo cross-linking as discussed previously. Also, there is less adipose tissue in the subcutaneous layer; therefore, older people are more likely to feel cold. The loss of thickness accounts for skin sagging and wrinkling.

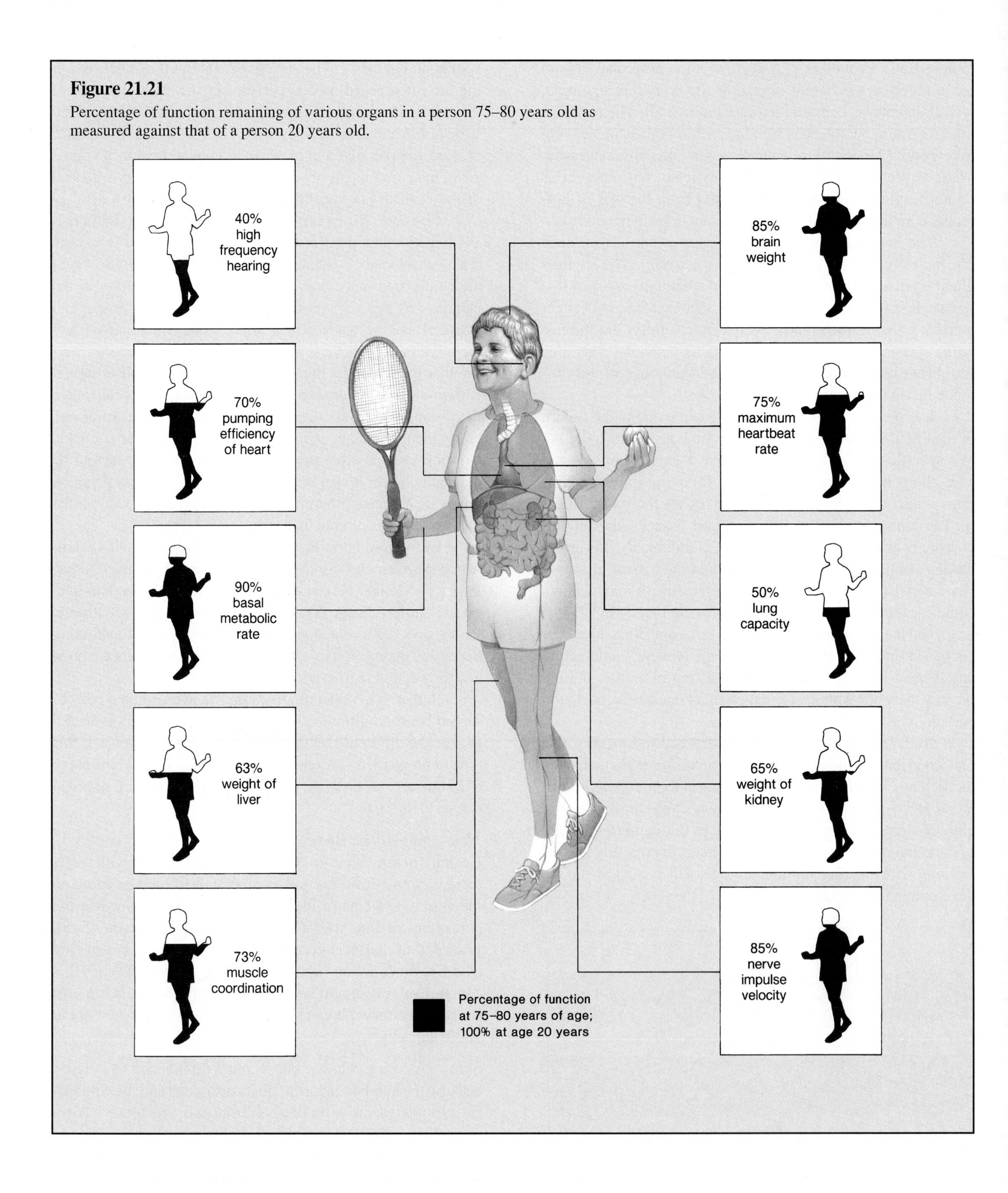

Figure 21.21
Percentage of function remaining of various organs in a person 75–80 years old as measured against that of a person 20 years old.

Homeostatic adjustment to heat is also limited because there are fewer sweat glands for sweating to occur. There are fewer hair follicles, so the hair on the scalp and the extremities thins out. The number of sebaceous glands is reduced, and the skin tends to crack. Older people also experience a decrease in the number of melanocytes, making hair gray and skin pale. In contrast, some of the remaining pigment cells are larger, and pigmented blotches appear in skin.

Processing and Transporting. Cardiovascular disorders are the leading cause of death among the elderly. The heart shrinks because there is a reduction in cardiac muscle cell size. This leads to loss of cardiac muscle strength and reduced cardiac output. Still, it is observed that the heart, in the absence of disease, is able to meet the demands of increased activity. It can increase its rate to double or triple the amount of blood pumped each minute even though the maximum possible output declines.

Because the middle coat of arteries contains elastic fibers, which most likely are subject to cross-linking, the arteries become more rigid with time, and their size is further reduced by plaque (p. 214). Therefore, blood pressure readings gradually rise. Such changes are common in individuals living in western industrialized countries but not in agricultural societies. As mentioned earlier, diet has been suggested as a way to control degenerative changes in the cardiovascular system (p. 188).

There is reduced blood flow to the liver, and this organ does not metabolize drugs as efficiently as before. This means that as a person gets older, less medication is needed to maintain the same level in the bloodstream.

Circulatory problems often are accompanied by respiratory disorders and vice versa. Growing inelasticity of lung tissue means that ventilation is reduced. Because we rarely use the entire vital capacity, these effects are not noticed unless there is increased demand for oxygen.

There is also reduced blood supply to the kidneys. The kidneys become smaller and less efficient at filtering wastes. Salt and water balance are difficult to maintain, and the elderly dehydrate faster than younger people. Difficulties involving urination include incontinence (lack of bladder control) and the inability to urinate. In men, the prostate gland may enlarge and reduce the diameter of the urethra, making urination so difficult that surgery is often needed.

The loss of teeth, which is frequently seen in elderly people, is more apt to be the result of long-term neglect than a result of aging. The digestive tract loses tone and secretion of saliva and gastric juice is reduced, but there is no indication of reduced absorption. Therefore, an adequate diet, rather than vitamin and mineral supplements, is recommended. There are common complaints of constipation, increased amount of gas, and heartburn, but gastritis, ulcers, and cancer can also occur.

Integration and Coordination. It is often mentioned that while most tissues of the body regularly replace their cells, some at a faster rate than others, the brain and the muscles do not. No new nerve or skeletal muscle cells are formed in the adult. However, contrary to previous opinion, recent studies show that few neural cells of the cerebral cortex are lost during the normal aging process. This means that cognitive skills remain unchanged even though there is characteristically a loss in short-term memory. Although the elderly learn more slowly than the young, they can acquire and remember new material as well. It is noted that when more time is given for the subject to respond, age differences in learning decrease.

Neurons are extremely sensitive to oxygen deficiency, and if neuron death does occur, it may not be due to aging itself but to reduced blood flow in narrowed blood vessels. Specific disorders, such as depression, Parkinson disease, and Alzheimer disease (p. 295), are sometimes seen, but they are not common. Reaction time, however, does slow, and more stimulation is needed for hearing, taste, and smell receptors to function as before. After age 50, there is a gradual reduction in the ability to hear tones at higher frequencies, and this can make it difficult to identify individual voices and to understand conversation in a group. The lens of the eye does not accommodate as well and also may develop a cataract. Glaucoma is more likely to develop because of a reduction in the size of the anterior cavity of the eye.

Loss of skeletal muscle mass is not uncommon, but it can be controlled by a regular exercise program. There is a reduced capacity to do heavy labor, but routine physical work should be no problem. A decrease in the strength of the respiratory muscles and inflexibility of the rib cage contribute to the inability of the lungs to expand as before, and reduced muscularity of the urinary bladder contributes to difficulties with urination.

As noted before, aging is accompanied by a decline in bone density. Osteoporosis, characterized by a loss of calcium and minerals from bone, is not uncommon, but there is evidence that proper health habits can prevent its occurrence. Arthritis, which causes pain upon movement of the joint, is also seen.

Weight gain occurs because the basal metabolism (p. 192) decreases and inactivity increases. Muscle mass is replaced by stored fat and retained water.

The Reproductive System. Females undergo menopause, and thereafter the level of female sex hormones in blood falls markedly. The uterus and the cervix are reduced in size, and there is a thinning of the walls of the oviducts and the vagina. The external genitals become less pronounced. In males, the level of androgens falls gradually over the age span 50–90, but sperm production continues until death.

It is of interest that as a group, females live longer than males. Although their health habits may be poorer, it is also possible that the female sex hormone estrogen offers to women some protection against circulatory disorders when they are younger. Males suffer a marked increase in heart disease in their forties, but an increase is not noted in females until after menopause. Then women lead men in the incidence of stroke. Men are still more likely than women to have a heart attack, however.

Conclusion: To Age Well, Start Health Habits Young

We have listed many adverse effects due to aging, but it is important to emphasize that while such effects are seen, they are not a necessary occurrence (fig. 21.22). We must discover any extrinsic factors that precipitate these adverse effects and guard against them. Just as it is wise to make the proper preparations to remain financially independent when older, it is also wise to realize that biologically successful old age begins with the health habits developed when we are younger.

Figure 21.22

The aim of gerontology is to allow the elderly to enjoy living. This requires studying the debilities that can occur with aging and then making recommendations as how best to forestall or prevent their occurrence.

SUMMARY

Embryonic development has certain stages. During cleavage, there is division but there is no overall increase in the size of the embryo. The result is a morula, which becomes the blastula when an internal cavity appears. During the gastrula stage, the germ layers (ectoderm, endoderm, and mesoderm) develop. At the neurula stage, the nervous system develops from midline ectoderm just above the notochord. It is possible at this point to draw a typical cross section of a vertebrate embryo.

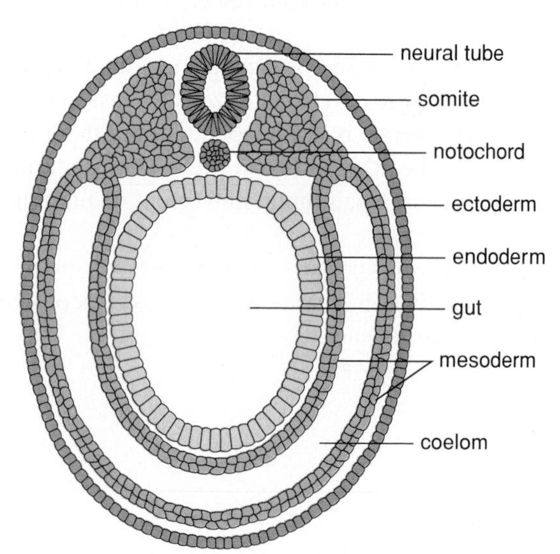

Differentiation begins with cleavage as the egg's cytoplasm is partitioned among the numerous cells. The cytoplasm is not uniform in content, and presumably each of the first few cells differ as to their cytoplasmic contents. Some probably contain substances that can influence gene activity—turning some genes on and others off. Morphogenesis involves the process of induction. Experiments show that the notochord induces formation of the nervous system, and the optic vesicles induce formation of the lens. Today, we envision induction as always present because cells are believed to constantly give off signals that influence the genetic activity of neighboring cells. This is the cause of morphogenesis.

Human development is divided into embryonic development and fetal development. The extraembryonic membranes appear early in human development. There are 4 membranes, as in the chick, but their function has been modified for internal development (table 21.1). During embryonic development, the body's various organs appear, and during fetal development, refinement of features occurs. The fetus is dependent upon the placenta for gas exchange and as a source of nutrient molecules. Birth has 3 stages: during the first stage, the cervix dilates; during the second, the child is born; and during the third, the afterbirth is expelled.

Development after birth consists of infancy, childhood, adolescence, and adulthood. Young adults are at their prime, and then the aging process begins. Aging encompasses progressive changes from about age 20 on that contribute to an increased risk of infirmity, disease, and death. Perhaps aging is genetic in origin, perhaps it is due to a change that affects the whole body, or perhaps it is due to extrinsic factors.

STUDY QUESTIONS

In order to practice **writing across the curriculum,** students should write out the answers to any or all of the study questions. The study questions are sequenced in the same order as the text.

1. State, define, and give examples of the 3 processes involved whenever a developmental change occurs. (p. 394)
2. Compare the process of cleavage and the formation of the blastula and the gastrula in the lancelet and the human. (p. 395)
3. Name the germ layers, and state organs derived from each. (p. 396)
4. Draw a cross section of a typical chordate embryo at the neurula stage, and label your drawing. (p. 398)
5. Give reasons for suggesting that differentiation begins with the embryonic stage of cleavage. (p. 399)
6. Describe an experiment that helped investigators to conclude that the notochord induces formation of the neural tube. (p. 400) Give another well-known example of induction between tissues. (p. 401)
7. Give reasons for suggesting that morphogenesis is dependent upon signals given off by neighboring cells. What do the signals bring about in the receiving cells? (p. 401)
8. List the human extraembryonic membranes, give a function for each, and compare this function to that in the chick. (p. 403)
9. Describe in general what happens during embryonic and fetal development of the human. (pp. 405–412)
10. Trace the path of blood in the fetus from the umbilical vein to the aorta, using 2 different routes. (p. 412)
11. Describe the 3 stages of parturition. (pp. 412–413)
12. Discuss 3 theories of aging. What are the major changes in body systems that have been observed as adults age? (pp. 414–415)

OBJECTIVE QUESTIONS

1. When cells take on a specific structure and function, _______ occurs.
2. The morula becomes the _______, a structure that contains the inner cell mass.
3. The _______ membranes include the chorion, the _______, the yolk sac, and the allantois.
4. The blastocyst _______ itself in the uterine lining.
5. Formation of germ layers occurs during _______.
6. The notochord _______ the formation of the nervous system.
7. During embryonic and fetal development, gas exchange occurs at the _______.
8. During development, there is a connection between the pulmonary artery and the aorta called the _______.
9. Fetal development begins with the _______ month.
10. If delivery is normal, the _______ appears before the rest of the body.
11. Label this diagram showing the extraembryonic membranes.

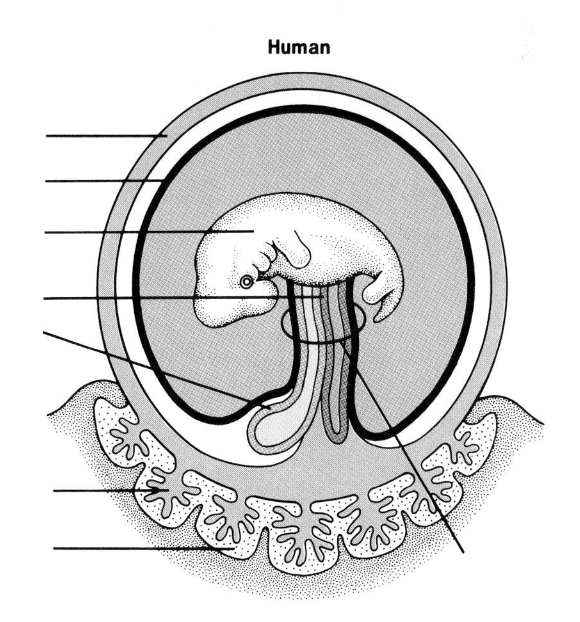

CRITICAL THINKING

In order to practice **writing across the curriculum,** students should write out the answers to any or all of the critical thinking questions. Suggested answers to the critical thinking questions are in appendix E.

21.1

1. With the help of these questions, develop a scenario to explain why a particular type of cell gives off *particular* signals.
   - **a.** What might be the effect on genes when a cell inherits a certain cytoplasmic composition?
   - **b.** What do activated genes do?
   - **c.** How might some of these proteins act?
2. With the help of these questions, develop a scenario to explain why embryonic development is so orderly.
   - **a.** Tissue A has just become differentiated. What does it give off to affect tissue B?
   - **b.** Having received a certain signal, what does tissue B do to affect tissue C?
   - **c.** Having received a certain signal, what does tissue C do?

SELECTED KEY TERMS

allantois (ah-lan´to-is) one of the extraembryonic membranes; in reptiles and birds, it is a pouch that collects nitrogenous waste; in mammals, it is a source of blood vessels to and from the placenta.

amnion (am´ne-on) an extraembryonic membrane; a fluid-containing sac around the embryo.

chorion (ko´re-on) an extraembryonic membrane; forms an outer covering around the embryo; in reptiles and birds, it functions in gas exchange; in mammals, it contributes to the formation of the placenta.

differentiation the process and the developmental stages by which a cell becomes specialized for a particular function.

ectoderm the outer germ layer of the embryonic gastrula; it gives rise to the nervous system and skin.

endoderm an inner layer of cells that lines the primitive gut of the gastrula; it becomes the lining of the digestive tract and associated organs.

extraembryonic membrane membrane that is not a part of the embryo but is necessary to the continued existence and health of the embryo.

induction a process by which one tissue gives off signals that control the development of another, as when the embryonic notochord induces the formation of the neural tube.

lanugo (lah-nu´go) downy hair on the body of a fetus; fetal hair.

mesoderm the middle germ layer of the embryonic gastrula; gives rise to the muscles, the connective tissue, and the circulatory system.

morphogenesis (mor˝fo-jen´i-sis) the movement of cells and tissues to establish the shape and the structure of an organism.

parturition the processes that lead to and include the birth of a human, and the expulsion of the afterbirth through the birth canal.

trophoblast (tro´fo-blast) the outer membrane surrounding the human embryo; when thickened by a layer of mesoderm, it becomes the chorion, an extraembryonic membrane.

umbilical cord cord connecting the fetus to the placenta, through which blood vessels pass.

vernix caseosa (ver´niks ka˝se-o´sah) cheeselike substance covering the skin of the fetus.

yolk sac one of the extraembryonic membranes that encloses yolk, except in most mammals.

22

PATTERNS OF GENE INHERITANCE

Chapter Concepts

1.
Alleles (alternative forms of a gene), located on chromosomes, are passed from one generation to the next. 422, 439

2.
Chromosomes and alleles separate and assort independently when the gametes form; this increases the variety of the offspring. 428, 439

3.
Many genetic disorders are inherited according to the laws first established by Gregor Mendel. 432, 440

4.
There are many exceptions to Mendel's laws, and these help to explain the wide variety in patterns of gene inheritance. 436, 440

African-American family inspecting tulips

Chapter Outline

Figure 22.1
Genes are passed from one generation to the next; therefore, a child resembles the parents and other family members.

WHEN a sperm fertilizes an egg, a new individual with the diploid number of chromosomes begins to develop. These chromosomes determine what the individual will be like; even if the zygote develops in a surrogate mother, the individual still will resemble the original parents (fig. 22.1).

The study of inheritance is an ongoing process that likely began as soon as people noticed the resemblance between parent and offspring. The latest findings about heredity are covered in chapter 24, but first we consider the results of Gregor Mendel's investigations into heredity. Mendel's studies form the basis for the particulate model of heredity, which assumes that genes are sections of chromosomes. For example, the letters on the homologous chromosomes in figure 22.2 stand for genes that control a **trait,** such as color of hair, type of fingers, or length of nose. According to the particulate model, genes, like the letters in the rectangles, are in definite sequence and remain in their spots, or loci, on the chromosomes. Alternate forms of a gene having the same position on a pair of homologous chromosomes and affecting the same trait are called **alleles.** In figure 22.2, *G* is an allele of *g*, and vice versa; also, *R* is an allele of *r*, and vice versa. *G* could never be an allele for *R* because *G* and *R* are at different loci.

The particulate model of heredity does not address the chemical nature of genes, that is, that they are composed of DNA, nor how DNA functions. It is a very useful model, though, because it provides an easy way to determine the probability of an offspring inheriting a particular characteristic.

Mendel's Laws

Gregor Mendel was a Catholic priest who in 1860 developed certain laws of heredity after doing crosses between garden

Figure 22.2
Diagrammatic representation of homologous chromosomes before and after replication. **a.** The letters represent alleles (alternate forms of a gene). Each allelic pair, such as *Gg* or *Zz*, is located on homologous chromosomes at a particular gene locus. **b.** Following replication, each sister chromatid carries the same alleles in the same order.

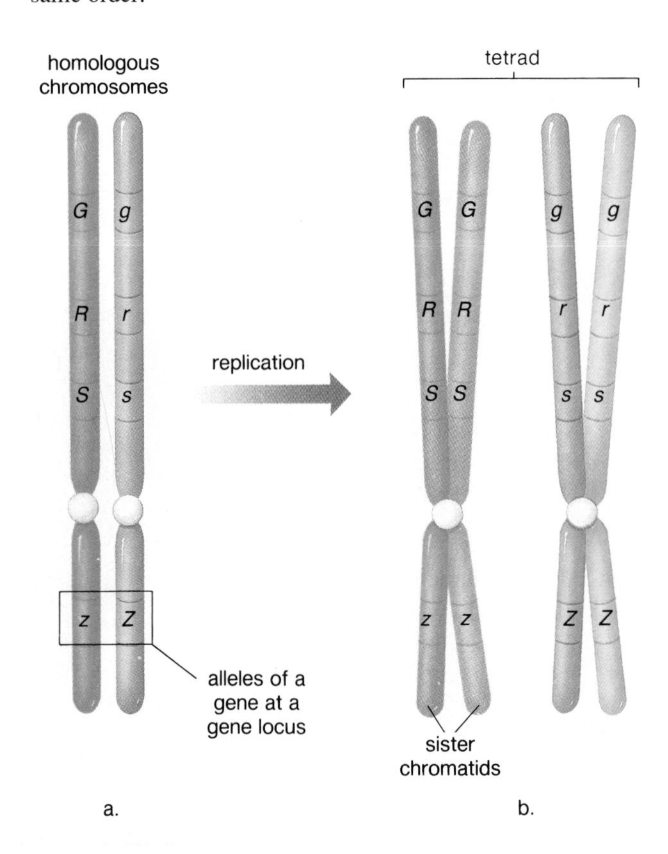

pea plants. His work is described in the reading on the next 2 pages. Mendel said that an individual has 2 factors for every trait, such as stem length. He observed that one of the factors controlling the same trait can be dominant over the other, which is recessive. For example, he found that a pea plant can show the dominant characteristic tallness, while the recessive factor for shortness, although present, is not expressed. Mendel's experiments led him to conclude that it is possible for such a tall pea plant to pass on a factor for shortness. Therefore, he reasoned that while the individual plant has 2 factors for each trait, the gametes (i.e., sperm and egg) contain only one factor for each trait. This is now known as Mendel's law of segregation.

The **law of segregation** states the following:

Each individual has 2 factors for each trait;
The factors segregate during the formation of the gametes;
Each gamete contains only one factor from each pair of factors;
Fertilization gives each new individual 2 factors for each trait.

MENDEL USHERS IN MODERN GENETICS

Mendel's use of pea plants as his experimental material was a good choice because pea plants are easy to cultivate, have a short generation time, and can be self-pollinated or cross-pollinated at will. Mendel selected certain traits for study, and before beginning his experiments, made sure his parental (P generation) plants bred true—he observed that when these plants self-pollinated, the offspring were like one another and like the parent plant. For example, a parent with yellow seeds always had offspring with yellow seeds; a plant with green seeds always had offspring with green seeds. Following that observation, Mendel cross-pollinated the plants by dusting the pollen of plants with yellow seeds on the stigma of plants with green seeds whose own anthers had been removed, and vice versa. Either way, the offspring (called F_1, or first filial generation) resembled the parents with yellow seeds. These results caused Mendel to allow F_1 plants to self-pollinate. Once he had obtained an F_2 generation, he observed the color of the peas produced. As table 22.A indicates, he counted over 8,000 plants and found an approximate 3 : 1 ratio (about 3 plants with yellow seeds for every plant with green seeds) in the F_2 generation.

Mendel realized that these results were explainable, assuming (a) there are 2 factors for every trait; (b) one of the factors can be dominant over the other, which is recessive; and (c) the factors separate when the gametes are formed. He assigned letters to these factors and displayed his results similar to this:

P	yellow *YY*	×	green *yy*
F_1	all yellow		
$F_1 \times F_1$	yellow *Yy*	×	yellow *Yy*
F_2	3 yellow : 1 green		

He believed that the F_2 plants with yellow seeds carried a dominant factor because his results could be related to the binomial equation $a^2 + 2ab + b^2$. He said if $a = Y$ and $b = y$, then the 4 F_2 plants were $YY + 2Yy + yy$.

Figure 22.A

Pea flower anatomy. Self-pollination normally occurs within the flowers of the pea plant. In order to cross the dominant by the recessive, Mendel removed the anthers of the plant with the dominant trait and used pollen from the plant with the recessive trait to bring about cross-fertilization. He also performed the reverse cross in the same manner.

Once seeds (peas) developed, they could be examined or planted in order to observe the results of a cross. The enlarged pod shows all possible shapes and colors of peas in a cross involving shape of the seed coat (smooth or wrinkled) and color of the seed coat (yellow or green).

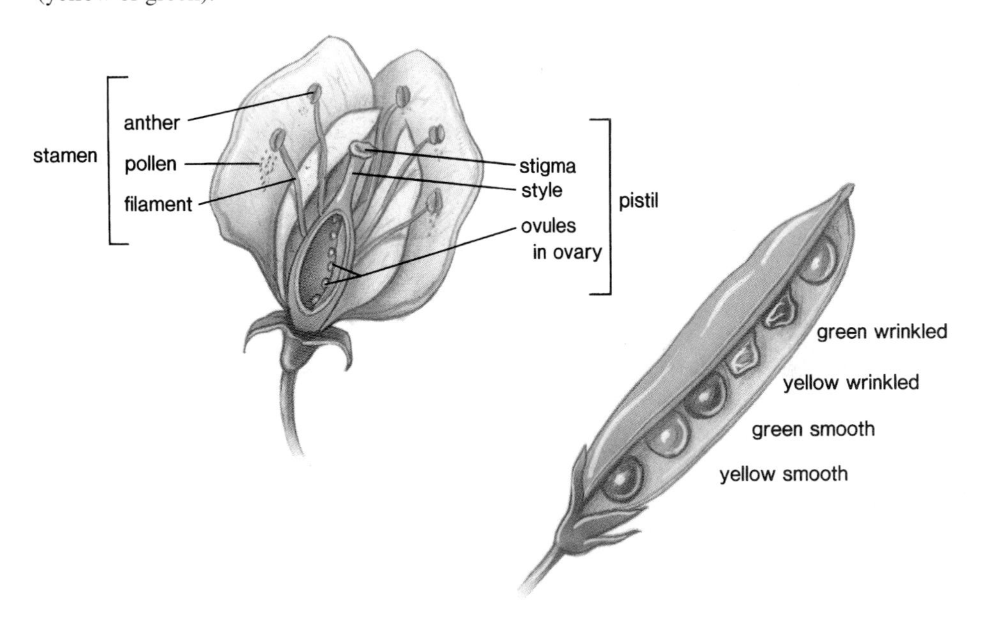

Table 22.A
Mendel's Results

Single-Trait Cross	F_1	F_2	Actual F_2 Ratio
Yellow × green	All yellow	6,022 yellow 2,001 green	3.0 : 1.0
Two-Trait Cross	**F_1**	**F_2**	**Actual F_2 Ratio**
Yellow, round × green, wrinkled	All round, yellow	315 yellow, round 101 yellow, wrinkled 108 green, round 32 green, wrinkled	9.8 : 2.9 : 3.1 : 1.0

—*Continued*

—*Continued from previous page*

And 3 plants with yellow seeds are expected for every plant with green seeds.

As a test to determine if the F_1 generation was indeed *Yy*, Mendel backcrossed it with the recessive parent, *yy*. His results of 1 : 1 indicated that he had reasoned correctly. Today, when a one-trait testcross is done, a suspected heterozygote is crossed with the recessive phenotype because this cross gives the best chance of producing the recessive phenotype.

Mendel performed a second series of experiments in which he crossed true-breeding plants that differed in 2 traits. For example, he crossed plants with yellow, round peas with plants with green, wrinkled peas. The F_1 generation always had both dominant characteristics; therefore, he allowed the F_1 plants to self-pollinate. Among the F_2 generation he achieved an almost perfect ratio of 9 : 3 : 3 : 1 (table 22.A). For example, for every plant that had green, wrinkled seeds, he had approximately 9 that had yellow, round seeds, and so forth. Mendel saw that these results were explainable if pairs of factors separate independently from one another when the gametes form, allowing all possible combinations of factors to occur in the gametes. This would mean that the probability of achieving any 2 factors together in the F_2 offspring is the product of their chance of occurring separately. Therefore, since the chance of yellow peas was 3/4 (in a one-trait cross) and the chance of round peas was 3/4 (in a one-trait cross), the chance of their occurring together was 9/16, and so forth.

Mendel achieved his success in genetics by studying large numbers of offspring, keeping careful records, and treating his data quantitatively. He showed that the application of mathematics to biology is extremely helpful in producing testable hypotheses.

Inheriting a Single Trait

Mendel suggested the use of letters to indicate factors, now called alleles. A capital letter indicates a **dominant allele**—a factor that is expressed when present in only single measure. A lowercase letter indicates a **recessive allele**—a factor that is expressed only in the absence of its dominant allele. Mendel's procedures and laws are applicable not only to peas but to all diploid individuals. Therefore, we now take as our example not peas, but human beings.

Figure 22.3 illustrates the difference between a widow's peak and a continuous hairline. In doing a problem concerning hairline, this *key* is suggested:

W = Widow's peak (dominant allele)
w = Continuous hairline (recessive allele)

The key tells us which letter of the alphabet to use for the gene in a particular problem. It also tells which allele is dominant, a capital letter signifying dominance.

Compare Genotype and Phenotype

When we indicate the genes of a particular individual, 2 letters must be used for each trait mentioned. This is called the **genotype** of the individual. The genotype can be expressed not only using letters but also with a short descriptive phrase, as table 22.1 shows. Therefore, the word **homozygous** means that the 2 members of the allelic pair in the zygote (*zygo*) are the same (*homo*); genotype *WW* is called *homozygous dominant* and *ww* is called *homozygous recessive*. The word **heterozygous** means that the members of the allelic pair are different (*hetero*); only *Ww* is heterozygous.

Figure 22.3
In humans, widow's peak (**a**) is dominant over continuous hairline (**b**).

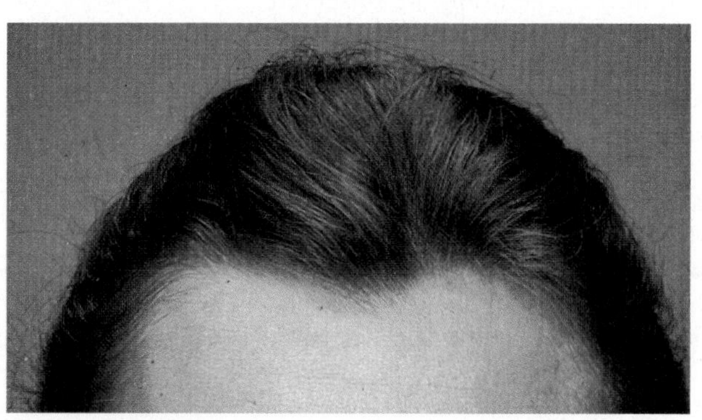
a.

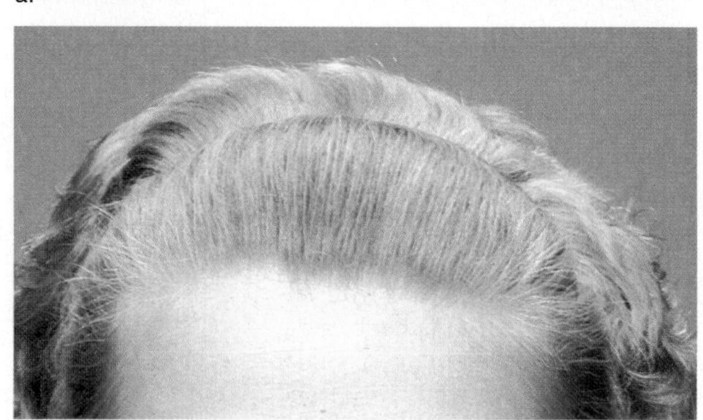
b.

Table 22.1
Genotype versus Phenotype

Genotype	Genotype	Phenotype
WW	Homozygous dominant	Widow's peak
Ww	Heterozygous	Widow's peak
ww	Homozygous recessive	Continuous hairline

As table 22.1 also indicates, the word **phenotype** refers to the physical characteristics of the individual—what the individual actually looks like. Also included in the phenotype are the microscopic and metabolic characteristics of the individual. Notice that both homozygous dominant (*WW*) and heterozygous (*Ww*) show the dominant phenotype.

Gametes: One Allele Per Trait

Whereas the genotype has 2 alleles for each trait, the gametes (i.e., sperm and egg) have only one allele for each trait in accordance with Mendel's law of segregation. This, of course, is related to the process of meiosis. The alleles are present on homologous chromosomes, and these chromosomes separate during meiosis. Therefore, the members of each allelic pair separate during meiosis, and there is only one allele for each trait in the gametes. When doing genetic problems, keep in mind that no 2 letters in a gamete should be the same. For this reason *Ww* represents a possible genotype, and the gametes for this individual could contain either a (*W*) or a (*w*). For easy recognition, we will circle the gametes.

Practice Problems 1*

1. For each of the following genotypes, give all possible gametes.
   a. *WW*
   b. *WWSs*
   c. *Tt*
   d. *Ttgg*
   e. *AaBb*
2. For each of the following, state whether a genotype or a gamete is represented.
   a. *D*
   b. *Ll*
   c. *Pw*
   d. *LlGg*

*Answers to problems are on page 699.

One-Trait Crosses in a Square

It is now possible for us to consider a particular cross. If a homozygous man with a widow's peak (fig. 22.3*a*) reproduces with a woman with a continuous hairline (fig. 22.3*b*), what kind of hairline will their children have?

In solving the problem, we use the key already established (p. 424) to indicate the genotype of each parent, we determine what the gametes are for each parent, we combine all possible gametes, and finally, we decide the genotypes and the phenotypes of all the offspring. In the format that follows, P stands for the *parental generation,* and the letters in the P row are the genotypes of the parents. The second row shows that each parent has only one type of gamete in regard to hairline, and therefore all the children (F = *filial generation*) have similar genotypes and phenotypes. The children are heterozygous (*Ww*) and show the dominant characteristic, a widow's peak.

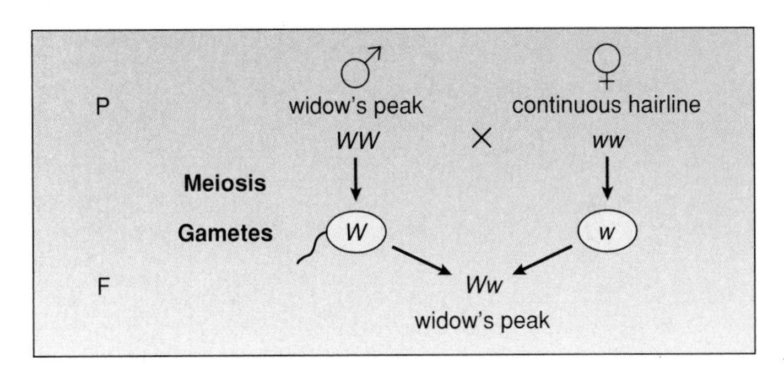

The children are **monohybrids;** that is, they are heterozygous for only one pair of alleles. If they reproduce with someone else of the same genotype, what type of hairline will their children have? In this problem (*Ww* × *Ww*), each parent has 2 possible types of gametes (*W* or *w*), and we must ensure that all types of sperm have equal chance to fertilize all possible types of eggs. One way to do this is to use a **Punnett square** (fig. 22.4), in which all possible types of sperm are lined up vertically and all possible types of eggs are lined up horizontally (or vice versa) and every possible combination of gametes occurs within the squares.

After we determine the genotypes and the phenotypes of the offspring, we see that 3 are expected to have a widow's peak and one is expected to have a continuous hairline. This 3 : 1 ratio is always expected for a monohybrid cross. The exact ratio is more likely to be observed if a large number of matings take place and if a large number of offspring result. Only then do all possible sperm have an equal chance to fertilize all possible eggs. Naturally, we do not routinely observe hundreds of offspring from a single type of cross in humans. The best

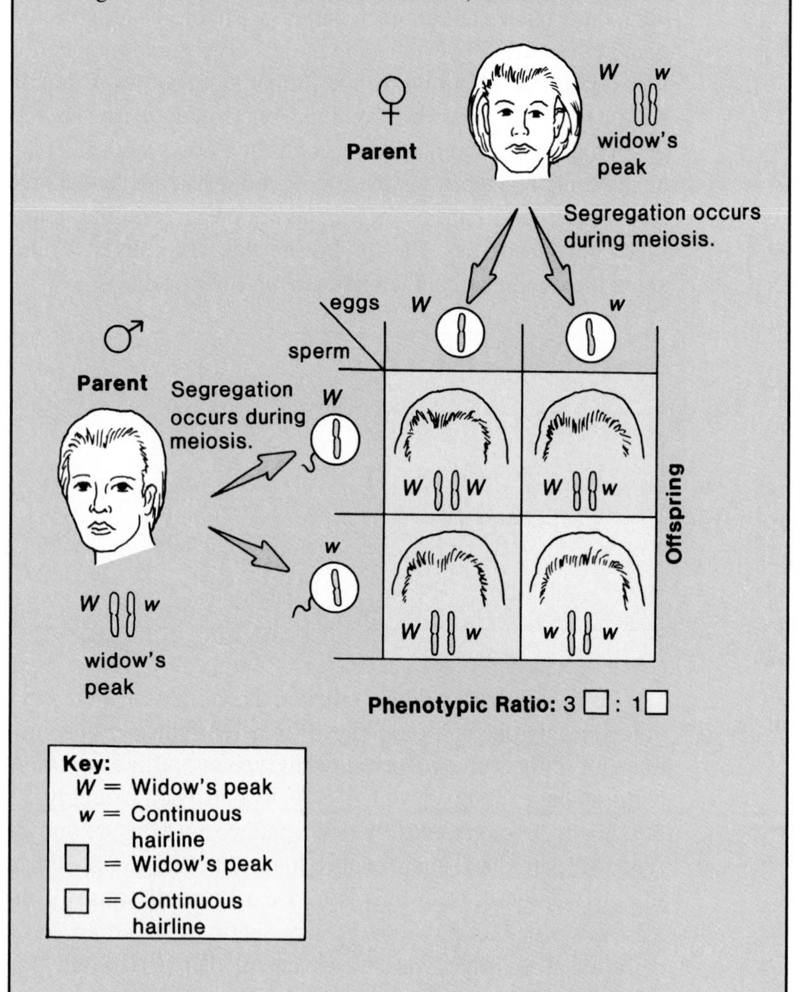

Figure 22.4
Monohybrid cross. The parents are heterozygous. Each parent can form 2 possible types of gametes because the alleles segregate (separate) during formation of the gametes. In order to calculate the expected ratio among the offspring, the Punnett square is a convenient way to make sure all possible sperm fertilize all possible eggs. Then we see that the chance of any child having a continuous hairline is one out of 4, or 25%.

interpretation of figure 22.4 in humans is to say that each child has 3 chances out of 4 to have a widow's peak or one chance out of 4 to have a continuous hairline. It is important to realize that *chance has no memory;* for example, if 2 heterozygous parents already have 3 children with a widow's peak and are expecting a fourth child, this child still has a 75% chance of a widow's peak and a 25% chance of a continuous hairline.

One-Trait Crosses and Probability. Another method of calculating the expected ratios uses the laws of probability. First, we must know that the probability (or chance) of 2 or more independent events occurring together is the product (multiplication) of their chance of occurring separately.

In the cross just considered ($Ww \times Ww$), what is the chance of obtaining either a W or a w from a parent?

The chance of $W = {^1/_2}$.
The chance of $w = {^1/_2}$.

Therefore, the probability of receiving these genotypes is as follows:

1. The chance of $WW = {^1/_2} \times {^1/_2} = {^1/_4}$
2. The chance of $Ww = {^1/_2} \times {^1/_2} = {^1/_4}$
3. The chance of $wW = {^1/_2} \times {^1/_2} = {^1/_4}$
4. The chance of $ww = {^1/_2} \times {^1/_2} = {^1/_4}$

Now we have to realize that the chance of an event that can occur in 2 or more independent ways is the sum (addition) of the individual chances. Therefore, the chance of offspring with a widow's peak (add chances of WW, Ww, or wW from the preceding) is $^3/_4$, or 75%. The chance of offspring with continuous hairline (only ww from the preceding) is $^1/_4$, or 25%.

When solving a genetics problem, it is assumed that all possible types of sperm fertilize all possible types of eggs. The results can be expressed as a probable phenotypic ratio; it is also possible to state the chances of an offspring showing a particular phenotype.

One-Trait Testcross: Who's Heterozygous?

A **testcross** occurs when an individual with the dominant phenotype is crossed with an individual having the recessive phenotype. It is not possible to tell by inspection if an individual expressing the dominant allele is homozygous dominant or heterozygous. However, the results of a testcross will most likely indicate which is the correct genotype.

Consider, for example, figure 22.5. It shows 2 possible results when a man with a widow's peak reproduces with a woman who has a continuous hairline. If the man is homozygous dominant, all his children will have a widow's peak. If

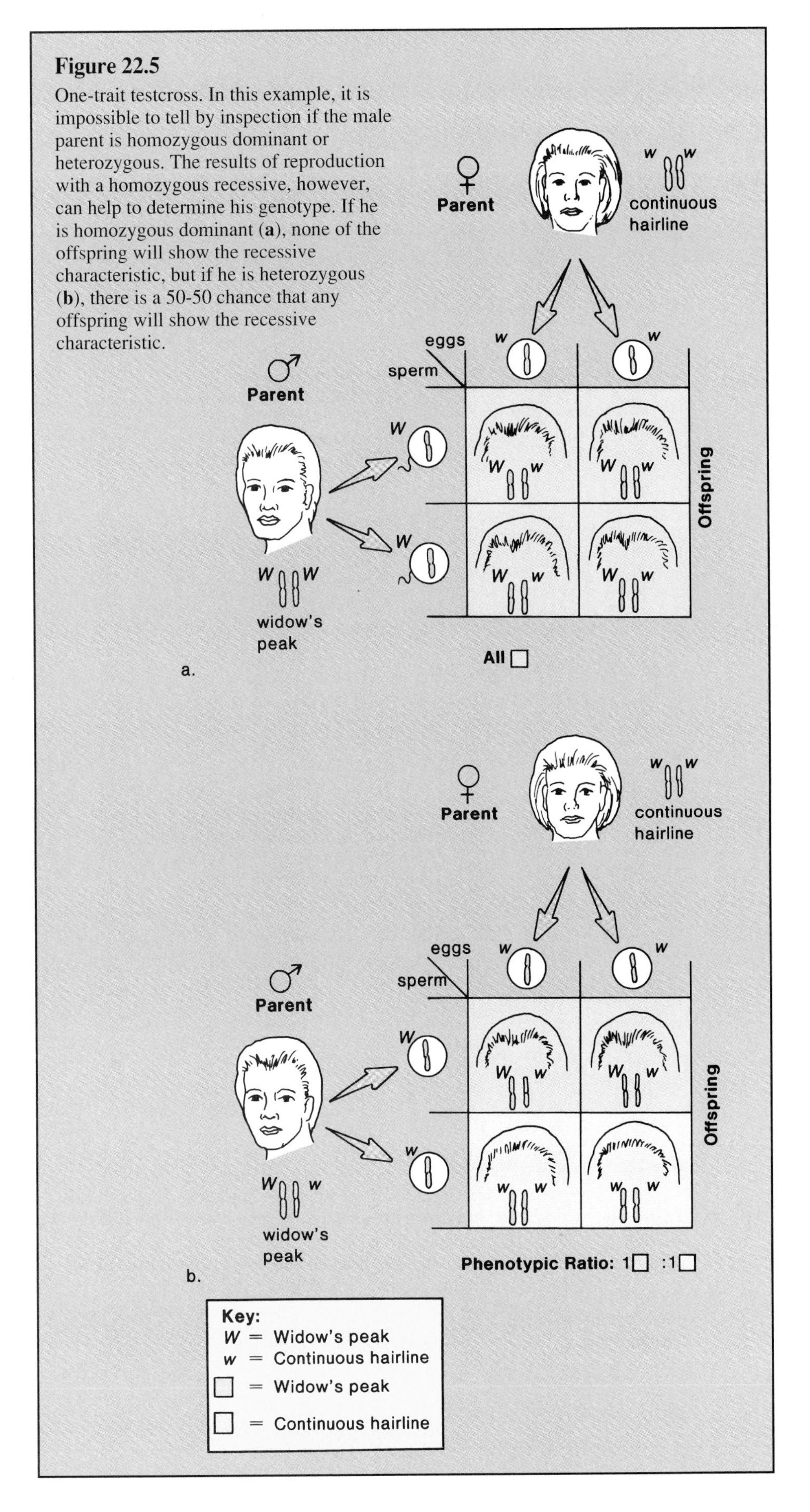

Figure 22.5

One-trait testcross. In this example, it is impossible to tell by inspection if the male parent is homozygous dominant or heterozygous. The results of reproduction with a homozygous recessive, however, can help to determine his genotype. If he is homozygous dominant (**a**), none of the offspring will show the recessive characteristic, but if he is heterozygous (**b**), there is a 50-50 chance that any offspring will show the recessive characteristic.

the man is heterozygous, each child has a 50% chance of a continuous hairline. The birth of just one child with a continuous hairline indicates that the man is heterozygous.

A testcross utilizes the homozygous recessive rather than the heterozygote. The cross *Aa* × *aa* has a better chance of producing the recessive phenotype than the cross *Aa* × *Aa* (50% chance compared to 25% chance).

The one-trait testcross determines whether an individual with the dominant phenotype is heterozygous or homozygous dominant. If any of the offspring of the testcross has the recessive phenotype, the parent with the dominant phenotype must be heterozygous.

Practice Problems 2*

1. Both a man and a woman are heterozygous for freckles. Freckles are dominant over no freckles. What are the chances that their child will have freckles?
2. A woman is homozygous dominant for short fingers. Short fingers are dominant over long fingers. Will any of her children have long fingers?
3. Both you and your sister or brother have attached earlobes, yet your parents have unattached earlobes. Unattached earlobes are dominant over attached earlobes. What are the genotypes of your parents?
4. A father has dimples, the mother does not have dimples, and all the children have dimples. Dimples are dominant over no dimples. Give the probable genotypes of all persons concerned.

*Answers to problems are on page 699.

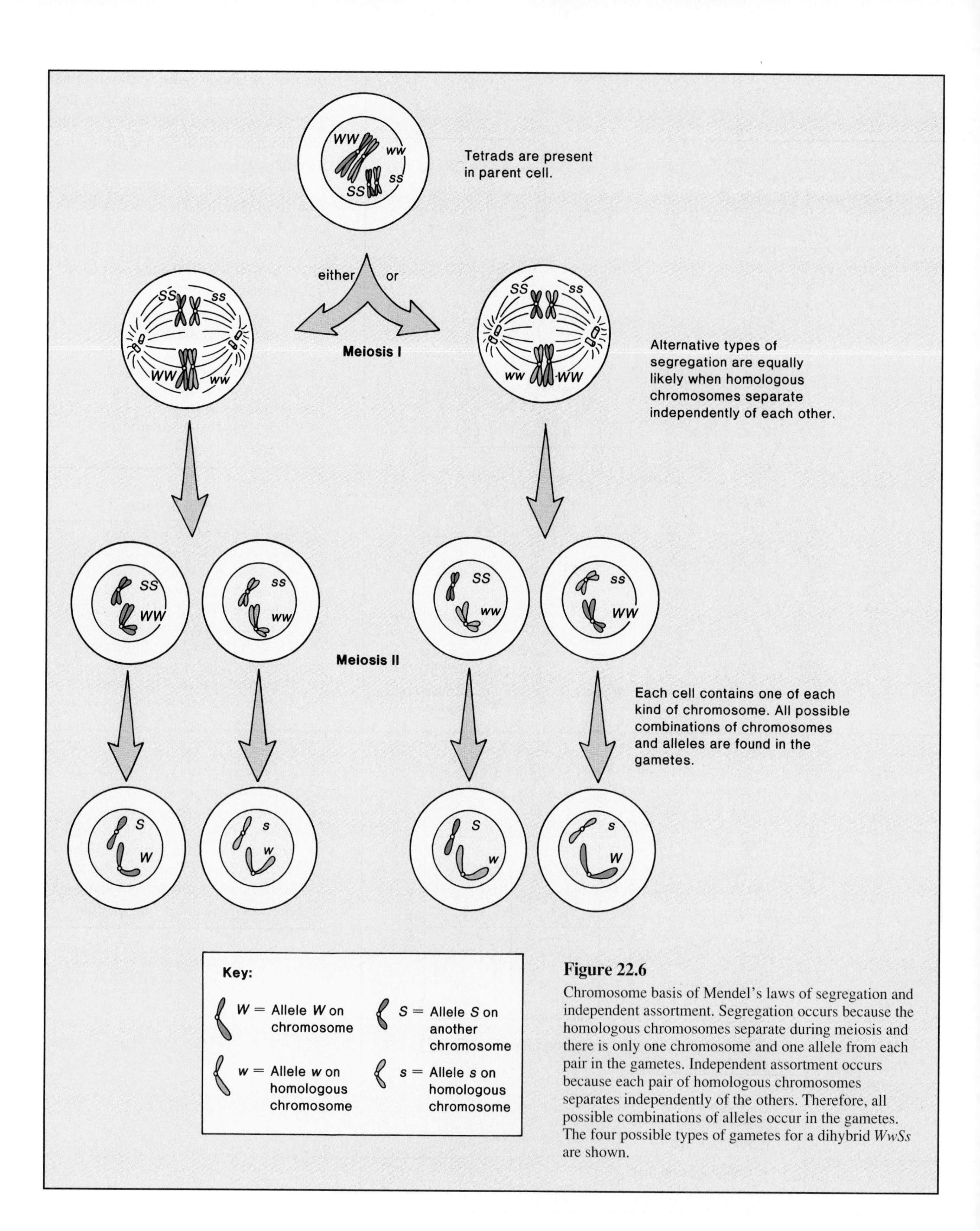

Figure 22.6

Chromosome basis of Mendel's laws of segregation and independent assortment. Segregation occurs because the homologous chromosomes separate during meiosis and there is only one chromosome and one allele from each pair in the gametes. Independent assortment occurs because each pair of homologous chromosomes separates independently of the others. Therefore, all possible combinations of alleles occur in the gametes. The four possible types of gametes for a dihybrid *WwSs* are shown.

Inheriting Many Traits

Although it is possible to consider the inheritance of just one trait, each individual actually passes on to his or her offspring an allele for each of many traits. In order to arrive at a general understanding of multitrait inheritance, the inheritance of 2 traits is considered. These genes are on different homologous chromosomes; therefore, the alleles are not linked. All the alleles on the same chromosome are said to form a linkage group.

Assorting Independently

When Mendel performed two-trait crosses, he noticed that his results were attainable only if sperm with every possible combination of factors fertilized every possible egg. This caused him to formulate his second law, the law of independent assortment.

> The **law of independent assortment** states the following:
>
> Each pair of factors segregates (assorts) independently of the other pairs;
>
> All possible combinations of factors can occur in the gametes (i.e., sperm and egg).

Figure 22.6 illustrates that the law of segregation and the law of independent assortment hold because of the manner in which meiosis occurs. The law of segregation is dependent on the separation of members of homologous pairs. The law of independent assortment is dependent on the random arrangement of homologous pairs at the equator of the spindle during metaphase I. Because of this, the homologous pairs separate independently of one another.

Two-Trait Crosses in a Square

When doing a two-trait cross, the genotypes of the parents require 4 letters because there is an allelic pair for each trait. Also, the gametes of the parents contain one letter of each kind in every possible combination, in accordance with Mendel's law of independent assortment. Finally, in order to produce the probable ratio of phenotypes among the offspring, all possible matings are presumed to occur.

To give an example (fig. 22.7), let us cross a person homozygous for widow's peak and short fingers (*WWSS*) with a person who has a continuous hairline and long fingers (*wwss*). Because each parent has only one type of gamete, the F_1 offspring will all have the genotype *WwSs* and the same phenotype (widow's peak with short fingers). This genotype is called a **dihybrid** because the individual is heterozygous in 2 regards: hairline and fingers.

When a dihybrid reproduces with a dihybrid, each F_1 parent has 4 possible types of gametes:

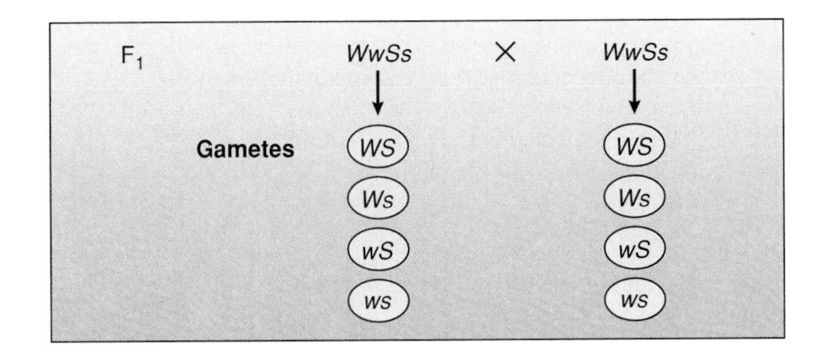

The Punnett square in figure 22.7 shows the expected genotypes among 16 F_2 offspring when all possible sperm fertilize all possible eggs. An inspection of the various genotypes in the square shows that the expected phenotypic ratio is 9 widow's peak and short fingers : 3 widow's peak and long fingers : 3 continuous hairline and short fingers : 1 continuous hairline and long fingers. This 9 : 3 : 3 : 1 phenotypic ratio is always expected for a dihybrid cross when simple dominance is present.

We can use this expected ratio to predict the chances of each child receiving a certain phenotype. For example, the chance of getting the 2 dominant phenotypes together is 9 out of 16, and the chance of getting the 2 recessive phenotypes together is one out of 16.

Two-Trait Crosses and Probability. Instead of using a Punnett square to arrive at the chances of the different types of phenotypes in the cross under discussion, it is possible to use the laws of probability we discussed before. For example, we already know the results for 2 separate monohybrid crosses are as follows:

1. Probability of widow's peak = $^3/_4$
   Probability of short fingers = $^3/_4$
2. Probability of continuous hairline = $^1/_4$
   Probability of long fingers = $^1/_4$

The probabilities for the dihybrid cross are, therefore, as follows:

Probability of widow's peak and short fingers = $^3/_4 \times {}^3/_4 = {}^9/_{16}$
Probability of widow's peak and long fingers = $^3/_4 \times {}^1/_4 = {}^3/_{16}$
Probability of continuous hairline and short fingers = $^1/_4 \times {}^3/_4 = {}^3/_{16}$
Probability of continuous hairline and long fingers = $^1/_4 \times {}^1/_4 = {}^1/_{16}$

The phenotypic ratio is 9 : 3 : 3 : 1.

Again, because all possible sperm must have an equal opportunity to fertilize all possible eggs to even approximate these results, a large number of offspring must be counted.

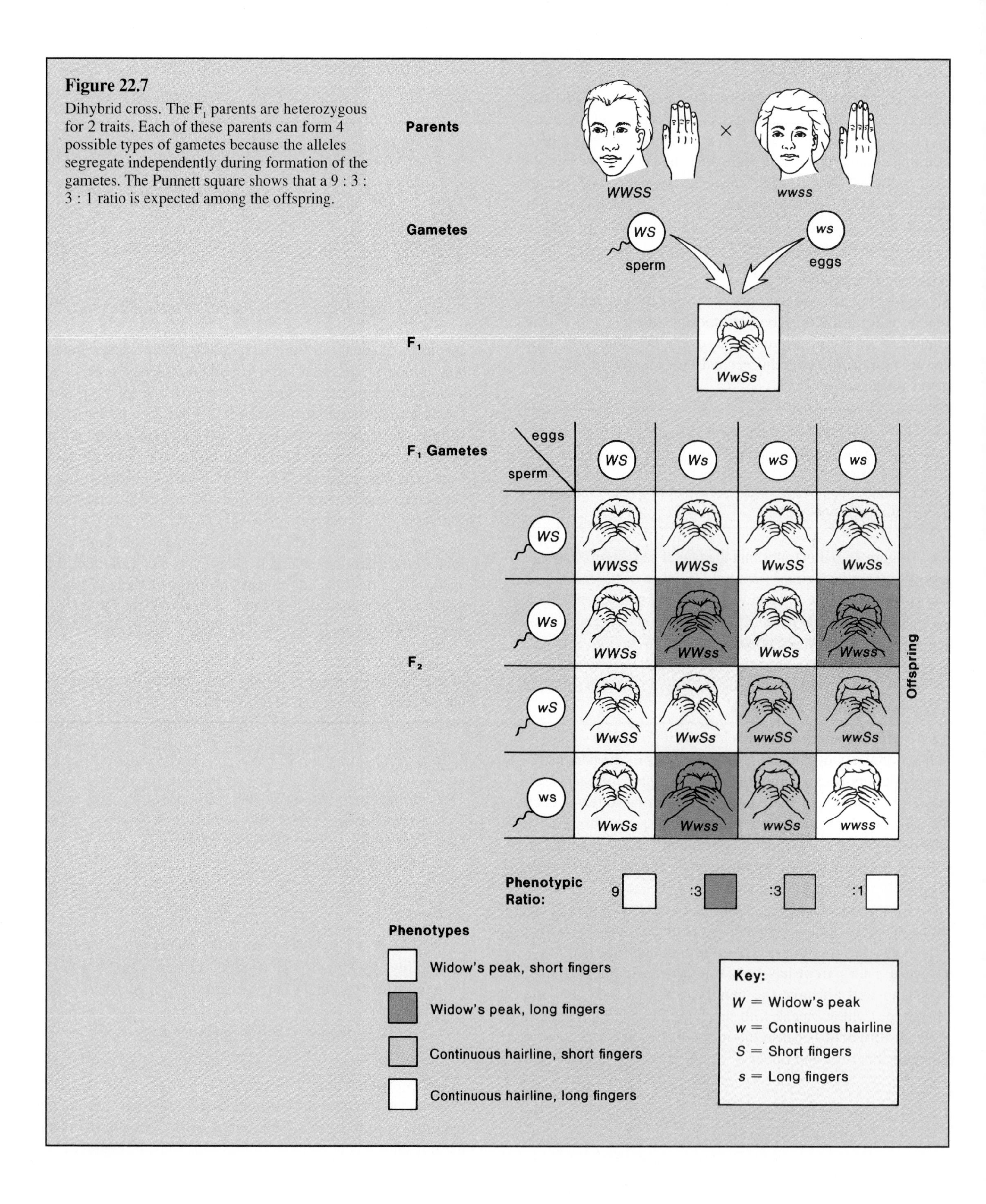

Figure 22.7

Dihybrid cross. The F_1 parents are heterozygous for 2 traits. Each of these parents can form 4 possible types of gametes because the alleles segregate independently during formation of the gametes. The Punnett square shows that a 9 : 3 : 3 : 1 ratio is expected among the offspring.

Two-Trait Testcross: Who's Heterozygous?

A two-trait testcross occurs when an individual with the dominant phenotype for 2 traits is crossed with a homozygous recessive for both traits. It is impossible to tell by inspection whether an individual expressing the dominant allele for 2 traits is homozygous dominant or heterozygous in regard to these traits. A cross with the homozygous recessive for both traits gives the best possible chance of producing an offspring with the recessive phenotype for both traits.

For example, if a man is homozygous dominant for widow's peak and short fingers, then all his children will have the dominant phenotypes, regardless with whom he reproduces. However, if a man is heterozygous for both traits, then each child has a 25% chance of showing one or both recessive traits. A Punnett square (fig. 22.8) shows that the expected ratio is 1 widow's peak with short fingers : 1 widow's peak with long fingers : 1 continuous hairline with short fingers : 1 continuous hairline with long fingers, or 1 : 1 : 1 : 1.

A two-trait testcross is used to determine whether an individual with the dominant phenotype is heterozygous or homozygous dominant for each trait. If any of the offspring of a two-trait testcross has a recessive phenotype, the parent with the dominant phenotype must be heterozygous for that trait. Do 22.1 Critical Thinking, found at the end of the chapter.

Figure 22.8

Two-trait testcross. In this example, it is impossible to tell by inspection if the male parent is homozygous dominant or if he is heterozygous for both traits. However, reproduction with a female who is recessive for both traits is likely to show which he is. If he is heterozygous, there is a 25% chance that the offspring will show both recessive characteristics and a 50% chance that they will show one or the other of the recessive characteristics.

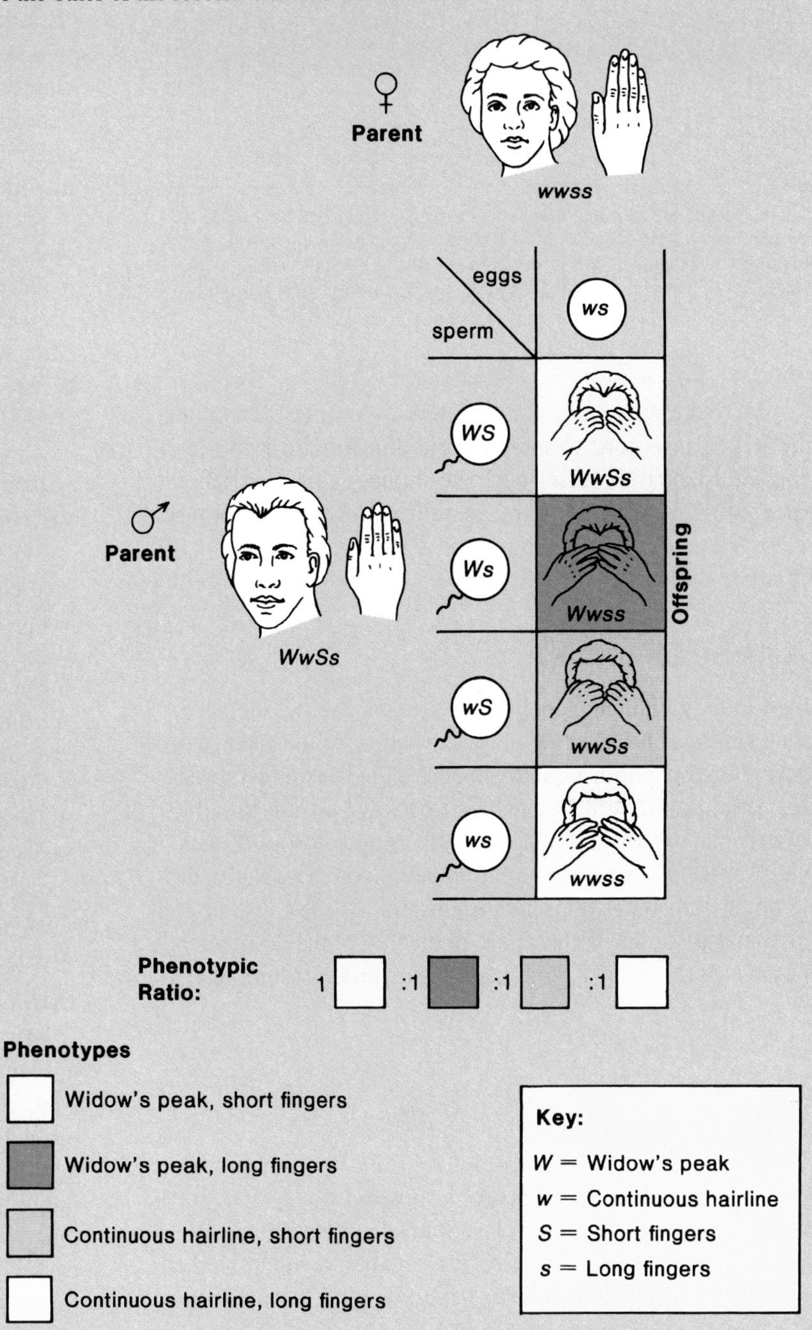

Practice Problems 3*

Using the information provided in Practice Problems 2, solve these problems.

1. What is the genotype of the offspring if a man homozygous recessive for type of earlobes and homozygous dominant for type of hairline reproduces with a woman who is homozygous dominant for earlobes and homozygous recessive for hairline?
2. If the offspring of this cross reproduces with someone of the same genotype, then what are the chances that this couple will have a child with a continuous hairline and attached earlobes?
3. A person who has dimples and freckles reproduces with someone who does not. This couple has a child who does not have dimples or freckles. What is the genotype of all persons concerned?

*Answers to problems are on page 699.

Table 22.2 Phenotypic Ratios of Common Crosses	
Genotypes	**Phenotypes**
Monohybrid × monohybrid	3 : 1 (dominant to recessive)
Monohybrid × recessive*	1 : 1 (dominant to recessive)
Dihybrid × dihybrid	9 : 3 : 3 : 1 (9 both dominant, 3 one dominant, 3 other dominant, 1 both recessive)
Dihybrid × recessive*	1 : 1 : 1 : 1 (all possible combinations in equal number)

*Called a backcross because it is as if the offspring were mated back to the recessive parent. Also called a testcross because it can be used to test if the individual showing the dominant gene is homozygous dominant or heterozygous. For a definition of all terms, see the end-of-chapter key terms.

We have now studied the pattern of simple Mendelian inheritance in regard to monohybrid and dihybrid crosses. Table 22.2 lists the crosses we have studied, along with their expected ratios. When doing genetics problems, it is not necessary to do a Punnett square if the expected ratio is already known.

Genetic Disorders

When studying human genetic disorders, biologists often construct pedigree charts, which show the pattern of inheritance of a characteristic within a group of people. Let us contrast 2 possible patterns of inheritance in order to show how it is possible to determine whether the characteristic is an autosomal recessive disorder or an autosomal dominant disorder. A faulty allele on an autosome (nonsex chromosome) is the cause of an autosomal disorder. If the allele is recessive, the disorder is recessive; if the allele is dominant, the disorder is dominant.

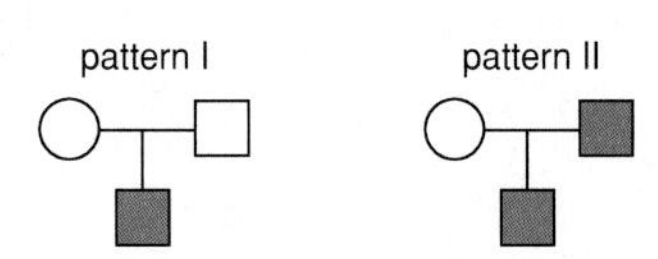

In both patterns, males are designated by squares and females are designated by circles. Shaded circles and squares indicate affected individuals. A line between a square and a circle represents a couple who have mated. A vertical line going downward leads, in these patterns, to a single child. (If there is more than one child, they are placed off a horizontal line.) Which pattern of inheritance do you suppose represents a dominant characteristic, and which represents a recessive characteristic?

In pattern I, the child is affected, but neither parent is; this can happen if the characteristic is recessively inherited. What are the chances that any offspring from this union will be affected? Because the parents are monohybrids, the chances are one in 4, or 25% (table 22.2). Notice that the parents also could be called **carriers** because they have a normal phenotype but are capable of having a child with a genetic disorder.

In pattern II, the child is affected, as is one of the parents. When a characteristic is dominant, an affected child usually has at least one affected parent. Of the 2 patterns, this one shows a dominant pattern of inheritance. What are the chances that any offspring from this union will be affected? Because this is a monohybrid by a recessive cross, the chances are 50% (table 22.2).

Autosomal Recessive Genetic Disorders

An autosomal recessive genetic disorder is caused by a recessive allele on an autosome. Figure 22.9 gives a sample pedigree chart and ways to recognize this pattern of inheritance. Many autosomal recessive disorders are known, but only 3—cystic fibrosis, Tay-Sachs disease, and phenylketonuria—are discussed here. Others that are well known are *albinism* (lack of pigment); *galactosemia* (accumulation of galactose in the liver and mental retardation); *thalassemia* (production of abnormal type of hemoglobin); and *xeroderma pigmentosum* (inability to repair ultraviolet-induced damage). The homozygous recessive phenotype is more likely to occur among a group of people who tend to marry each other, which may explain why autosomal recessive disorders are sometimes more prevalent among members of a particular ethnic group.

Cystic Fibrosis: Thick Mucus

Cystic fibrosis is the most common lethal genetic disease among Caucasians in the United States. About one in 5 Caucasians is a carrier, and about one in 2,000 children born to this group has the disorder. In these children, the mucus in the lungs and the digestive tract is particularly thick and viscous. In the lungs, the mucus interferes with gas exchange (fig. 22.10). Thick mucus also impedes the secretion of pancreatic juices, and food cannot be properly digested. This results in large, frequent, and foul-smelling stools.

In the past few years, much progress has been made in our understanding of cystic fibrosis, and new treatments have raised the average life expectancy to 17–20 years of age. Research has demonstrated that chloride ions (Cl^-) fail to pass through cell membrane channel proteins in these patients. Ordinarily, after chloride ions have passed through the membrane, water follows. It is believed that lack of water in the lungs is what causes the mucus to be abnormally thick.

The cystic fibrosis gene, which is located on chromosome 7, has been isolated and inserted into the lungs of living animals. The hope is that one day it will be possible to use an inhaler to carry copies of the normal gene into the lungs of cystic fibrosis patients.

Figure 22.9

Sample pedigree chart for an autosomal recessive genetic disorder. Only those affected are shaded. Which persons in the chart are carriers?

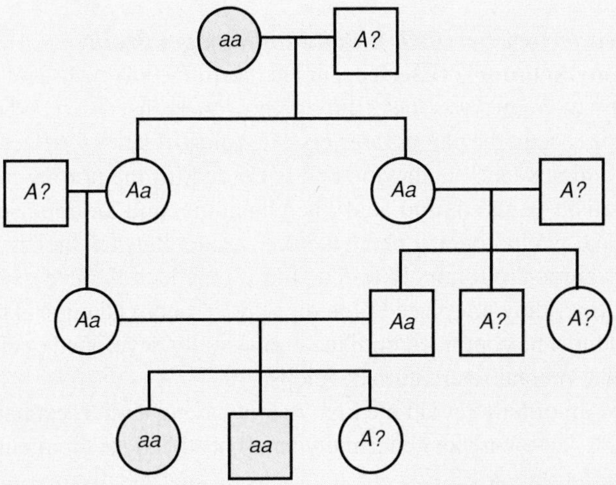

Autosomal Recessive Genetic Disorders

- Most affected children have normal parents.
- Heterozygotes have a normal phenotype.
- Two affected parents will always have affected children.
- Affected individuals who have noncarrier spouses will have normal children.
- Close relatives who marry are more likely to have affected children.
- Both males and females are affected with equal frequency.

Key:
aa = Affected
Aa = Carrier (appears normal)
AA = Normal

Figure 22.10

The mucus in the lungs of a child with cystic fibrosis should be periodically loosened by clapping the back. A new treatment destroys the cells that tend to build up in the lungs. The white cells are destroyed by one drug, and another drug does away with the DNA the cells leave behind.

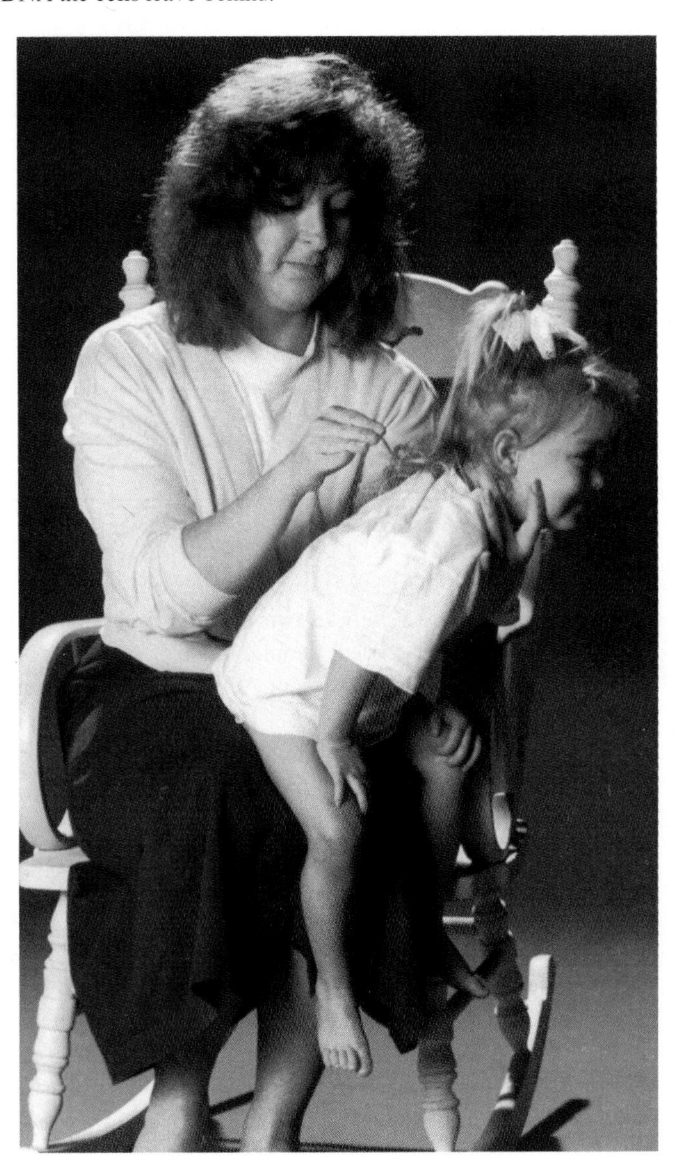

Figure 22.11

Tay-Sachs disease. A child with Tay-Sachs disease does not develop normally because of the lack of an enzyme called hexosaminidase A (Hex A). **a.** Normal cells have the enzyme and can break down glycosphingolipids. **b.** When Hex A is absent, these lipids accumulate in lysosomes and lysosomes accumulate in the cell.

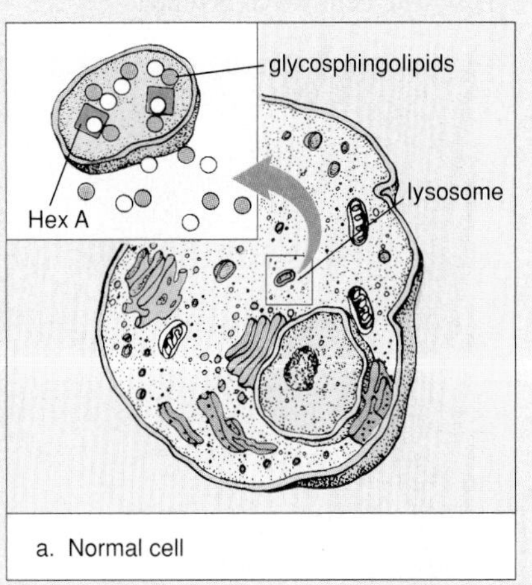

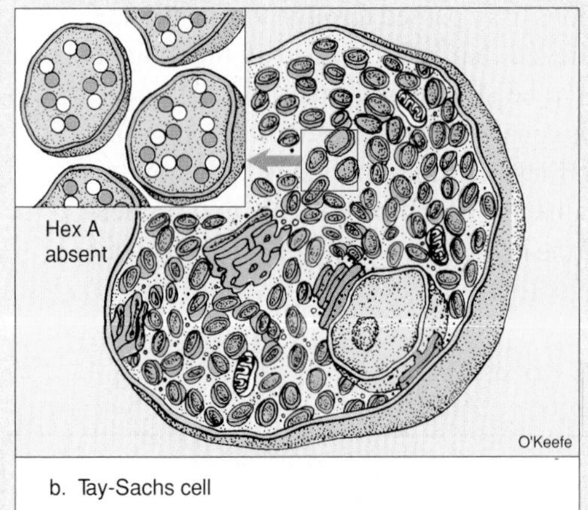

Tay-Sachs Disease: No Hex A

Tay-Sachs disease is a well-known genetic disease of high incidence among U.S. Jewish people, most of whom are of central and eastern European descent. At first, it is not apparent that a baby has Tay-Sachs disease. However, development begins to slow down between 4 months and 8 months of age, when neurological impairment and psychomotor difficulties become apparent. The child gradually becomes blind and helpless, develops uncontrollable seizures, and eventually becomes paralyzed. There is no treatment or cure for Tay-Sachs disease, and most affected individuals die by the age of 3 or 4.

So-called late-onset Tay-Sachs disease occurs in adults. The symptoms are progressive mental and motor deterioration, depression, schizophrenia, and premature death. The gene for late-onset Tay-Sachs disease has been sequenced, and this form of the disorder apparently is due to one changed pair of bases in the DNA of chromosome 1.

Tay-Sachs disease results from a lack of the enzyme hexosaminidase A (Hex A) and the subsequent storage of its substrate, a glycosphingolipid, in lysosomes. Although more and more lysosomes build up in many body cells (fig. 22.11), the primary sites of storage are the cells of the brain, which accounts for the onset of the progressive deterioration of psychomotor functions.

There is a test to detect carriers of Tay-Sachs disease. The test uses a sample of serum, white blood cells, or tears to determine if Hex A activity is present. Affected individuals have no detectable Hex A activity. Carriers have about half the level of Hex A activity found in normal individuals. Prenatal diagnosis of the disease also is possible following either amniocentesis or chorionic villi sampling.

Phenylketonuria (PKU): Buildup Damages Brain

Phenylketonuria (PKU) occurs in one in 10,000 births and so is not as frequent as the disorders previously discussed. When it does occur, the parents are very often close relatives. Affected individuals lack an enzyme that is needed for the normal metabolism of the amino acid phenylalanine, and an abnormal breakdown product, a phenylketone, accumulates in the urine. Newborns are routinely tested, and if they lack the necessary enzyme, they are placed on a diet low in phenylalanine. This diet must be continued until the brain is fully developed or else severe mental retardation develops.

In order to avoid the high risk of having a microcephalic child—one with an abnormally small head and severe mental retardation—a PKU woman should resume her limited diet several months before getting pregnant. The diet must not include the artificial sweetener NutraSweet™ because it contains phenylalanine.

Autosomal Dominant Genetic Disorders

An autosomal dominant genetic disorder is caused by a dominant allele on an autosome. Figure 22.12 gives a sample pedigree chart and ways to recognize this pattern of inheritance. Many autosomal dominant disorders are known, but only 2 are discussed here—neurofibromatosis and Huntington disease. Others that are well known are Marfan syndrome (connective tissue disorder); achondroplasia (dwarfism); brachydactyly (abnormally short fingers); porphyria (inability to metabolize porphyrins from hemoglobin breakdown); and hypercholesterolemia (elevated levels of cholesterol in blood).

Figure 22.12

Sample pedigree chart for an autosomal dominant genetic disorder. Are there any carriers?

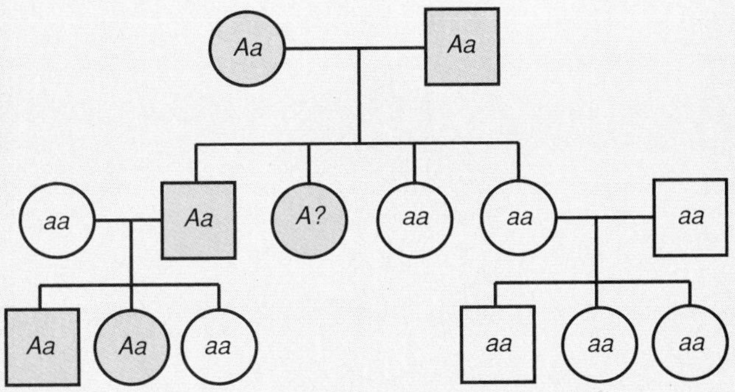

Aa — Aa

aa — Aa, A?, aa, aa — aa

Aa, Aa, aa; aa, aa, aa

Autosomal Dominant Genetic Disorders
- Affected children usually have an affected parent.
- Heterozygotes are affected.
- Two affected parents can produce an unaffected child.
- Two unaffected parents do not have affected children.
- Both males and females are affected with equal frequency.

Key:
AA = Affected
Aa = Affected
aa = Normal

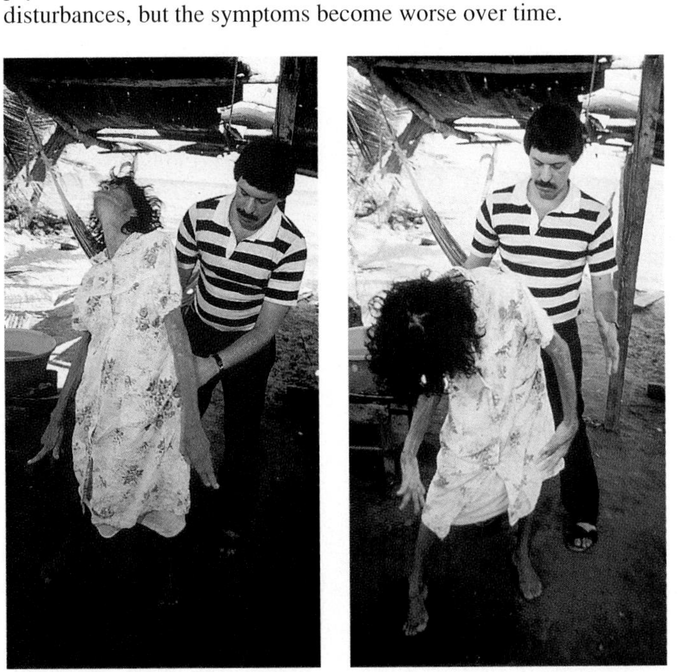

Figure 22.13

Huntington disease. Persons with this condition gradually lose psychomotor control of the body. At first there are only minor disturbances, but the symptoms become worse over time.

Neurofibromatosis: Common among Genetic Disorders

Neurofibromatosis (NF), sometimes called von Recklinghausen disease,[1] is one of the most common genetic disorders. It affects roughly one in 3,000 people, including an estimated 100,000 in the United States. It is seen equally in every racial and ethnic group throughout the world.

At birth or later, the affected individual may have 6 or more large tan spots on the skin. Such spots may increase in size and number and get darker. Small benign tumors (lumps) called neurofibromas may occur under the skin or in the muscles. Neurofibromas are made up of nerve cells and other cell types.

This genetic disorder shows *variable expressivity.* In most cases, symptoms are mild and patients live a normal life. In some cases, however, the effects are severe. Skeletal deformities, including a large head, are seen, and eye and ear tumors can lead to blindness and hearing loss. Many children with NF have learning disabilities and are hyperactive.

The NF gene is known to be on chromosome 17, and researchers have developed a test to diagnose the disorder. When the gene is functioning properly, it suppresses abnormal cell division. Cells that divide and produce tumors have a nonfunctioning gene.

1. Although neurofibromatosis is commonly associated with Joseph Merrick, the severely deformed nineteenth-century Londoner depicted in *The Elephant Man,* researchers today believe Merrick actually suffered from a much rarer disorder called Proteus syndrome.

Huntington Disease: Begins in Middle Age

One in 10,000 persons in the United States has **Huntington disease (HD),** a neurological disorder that affects specific regions of the brain. Most individuals who inherit the allele appear normal until middle age. Then, minor disturbances in balance and coordination lead to progressively worse neurological disturbances (fig. 22.13). The victim becomes insane before death occurs.

Much has been learned about Huntington disease. The gene for the disease is located on chromosome 4, and there is a test, of the type described on page 457, to determine if the dominant gene has been inherited. Because treatment is not available, however, few may want to have this information.

Research is being conducted, though, to determine the underlying cause of the disorder. It is known that the brain of a Huntington victim produces more than the usual amount of quinolinic acid, an excitotoxin that can overstimulate certain nerve cells. It is believed to lead to the death of these cells and to the subsequent symptoms of Huntington disease. Researchers are looking for chemicals that block quinolinic acid's action or inhibit quinolinic acid synthesis.

Some of the best-known genetic disorders in humans are either autosomal recessive or autosomal dominant disorders inherited in a simple Mendelian manner. Pedigree charts show the pattern of inheritance in a particular family.

Figure 22.14

Polygenic inheritance. **a.** When you record the heights of a large group of young men, (**b**) the values follow a bell-shaped curve. Such continuous distributions are seen when a trait is controlled by several sets of alleles.

a.

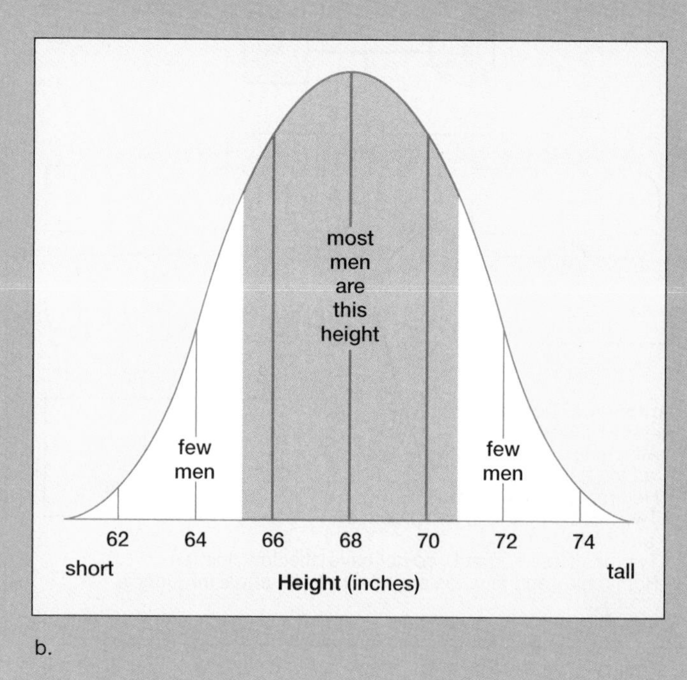

b.

Beyond Mendel's Laws

Certain traits, such as those just studied, follow the rules of simple Mendelian inheritance. There are, however, others that do not follow these rules.

Polygenic: Alleles Add Up

Polygenic inheritance occurs when 2 or more sets of alleles affect the same trait in an additive fashion. The result is a continuous variation of phenotypes between the extremes and a distribution of these phenotypes that resembles a bell-shaped curve. The more genes involved, the more continuous the variation and the distribution of the phenotypes. For example, we would expect height in humans to be controlled by polygenes because (1) height ranges from tall to short, (2) there are many variations in height between these extremes, and (3) few people are either short or tall and most humans have the mean height (fig. 22.14).

Inheriting Skin Color

Just how many pairs of alleles control skin color is not known, but a range in colors can be explained on the basis of 2 pairs. When a very dark person reproduces with a very light person, the children have medium brown skin, and when 2 people with medium brown skin reproduce with one another, the children range in skin color from very dark to very light. This can be explained by assuming that skin color is controlled by 2 pairs of alleles and that each capital letter contributes to the color of the skin:

Phenotype	Genotypes
Very dark	*AABB*
Dark	*AABb* or *AaBB*
Medium brown	*AaBb* or *AAbb* or *aaBB*
Light	*Aabb* or *aaBb*
Very light	*aabb*

Notice again that there is a range in phenotypes and that there are several possible phenotypes in between the 2 extremes (fig. 22.15). Therefore, the distribution of these phenotypes is expected to follow a bell-shaped curve—few people have the extreme phenotypes and most people have the phenotype that lies in the middle between the extremes.

Polygenic Disorders: Cleft Lip, Clubfoot

A number of serious genetic disorders, such as *cleft lip* or palate, *clubfoot, congenital dislocation of the hip,* and certain spinal conditions, are traditionally believed to be controlled by a combination of genes on autosomal chromosomes. This belief is being challenged by researchers who studied the inheritance of cleft palate in a large family in Iceland. These researchers reported the finding of a gene on the X chromosome that alone can cause cleft palate.

Figure 22.15
Inheritance of skin color. This white husband (*aabb*) and his intermediate wife (*AaBb*) had fraternal twins, one of whom is white and one of whom is intermediate.

A blood test is now available to couples concerned about the birth of a child with a neural tube defect. If the mother has a high serum level of *α-feto protein,* further testing is advised. An analysis of the amniotic fluid following amniocentesis (p. 408) can reveal if a neural tube substance has leaked into the fluid. If this has taken place, it is usually possible to diagnose the condition of the fetus.

When Multiple Alleles Control a Trait

ABO Blood Types

Three alleles for the same gene control the inheritance of ABO blood types. These alleles determine the presence or absence of antigens on the red blood cells.

A = A antigen on red blood cells
B = B antigen on red blood cells
O = No antigens on red blood cells

Each person has only 2 of the 3 possible alleles, and both *A* and *B* are dominant over *O*. Therefore, there are 2 possible genotypes for type A blood and 2 possible genotypes for type B blood. On the other hand, alleles *A* and *B* are fully expressed in the presence of the other. Therefore, if a person inherits one of each of these alleles, that person will have type AB blood. Type O blood can only result from the inheritance of 2 *O* alleles:

Phenotype	Possible Genotype
A	*AA, AO*
B	*BB, BO*
AB	*AB*
O	*OO*

An examination of possible matings between different blood types sometimes produces surprising results; for example,

Genotypes of parents: $AO \times BO$
Possible genotypes of children: *AB*, *OO*, *AO*, *BO*

Therefore, from this particular mating, every possible phenotype (types AB, O, A, B blood) is possible.

Blood typing can sometimes aid in paternity suits. However, a blood test of a supposed father can only suggest that he *might* be the father, not that he definitely *is* the father. For example, it is possible, but not definite, that a man with type A blood (having genotype *AO*) is the father of a child with type O blood. On the other hand, a blood test sometimes can definitely prove that a man is not the father. For example, a man with type AB blood cannot possibly be the father of a child with type O blood. Therefore, blood tests can be used legally only to exclude a man from possible paternity.

Rh Factor

The Rh factor is inherited separately from A, B, AB, or O blood types. In each instance, it is possible to be Rh positive (Rh^+) or Rh negative (Rh^-). When you are Rh positive, there is a particular antigen on the red blood cells, and when you are Rh negative, it is absent. It can be assumed that the inheritance of this antigen is controlled by a single allelic pair in which simple dominance prevails: the Rh-positive allele is dominant over the Rh-negative allele. Complications arise when an Rh-negative woman reproduces with an Rh-positive man and the child in the womb is Rh positive. Under certain circumstances, the woman may begin to produce antibodies that will attack the red blood cells of this baby or of a future Rh-positive baby. As discussed on page 234, the Rh problem can be eliminated by giving an Rh-negative woman an Rh immunoglobulin injection from midway through a pregnancy to no later than 72 hours after giving birth to any Rh-positive child.

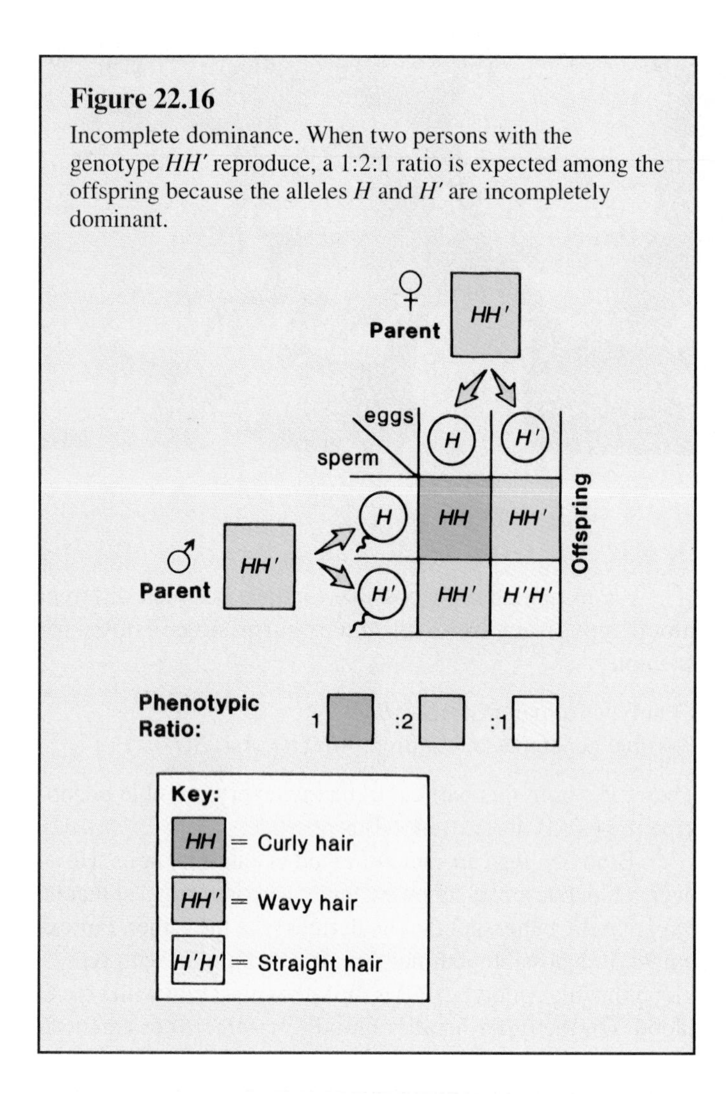

Figure 22.16
Incomplete dominance. When two persons with the genotype *HH'* reproduce, a 1:2:1 ratio is expected among the offspring because the alleles *H* and *H'* are incompletely dominant.

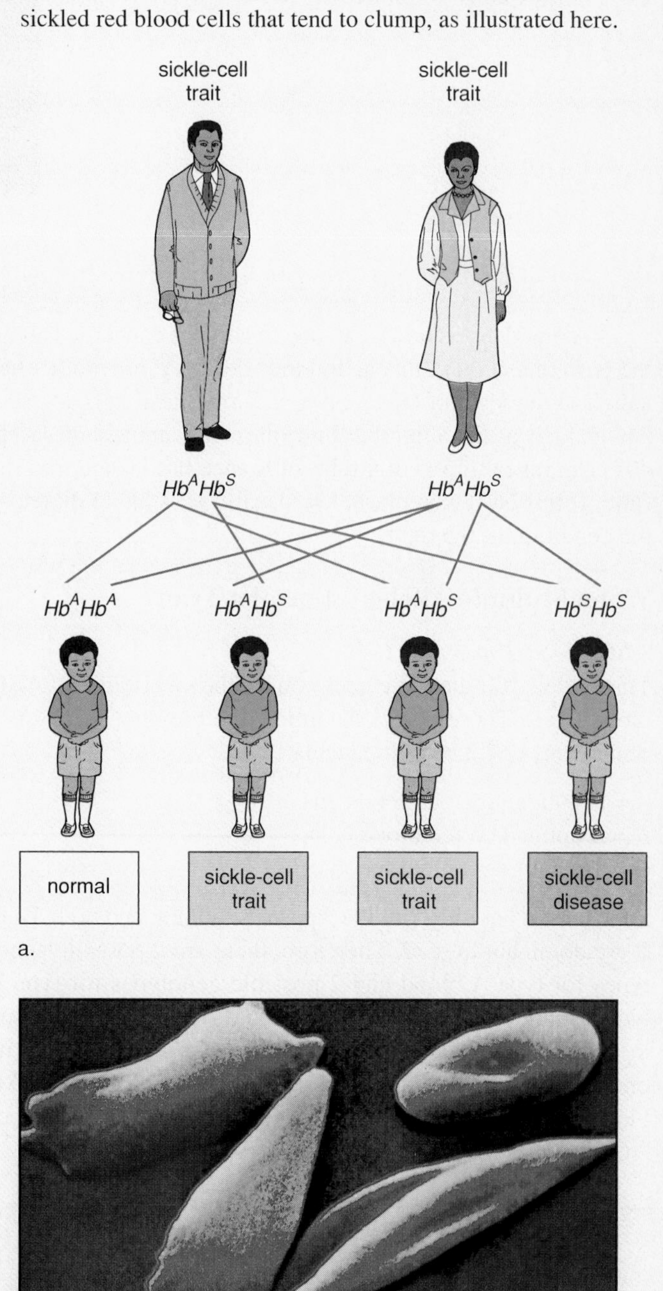

Figure 22.17
Inheritance of sickle-cell disease. **a.** In this example, both parents have the sickle-cell trait. Therefore, each child has a 25% chance of having sickle-cell disease or of being perfectly normal and a 50% chance of having the sickle-cell trait.
b. Sickled cells. Individuals with sickle-cell disease have sickled red blood cells that tend to clump, as illustrated here.

Degrees of Dominance

The field of human genetics also has examples of incomplete dominance and codominance. For example, when a curly-haired person reproduces with a straight-haired person, their children will have wavy hair. When two wavy-haired persons reproduce, the expected phenotypic ratio among the offspring is 1 : 2 : 1 (fig. 22.16). We have already mentioned that the multiple alleles controlling blood type are codominant. An individual with the genotype *AB* has type AB blood. Skin color, recall, is controlled by polygenes, and therefore it is possible to observe a range of skin colors from very dark to very light.

Sickle-Cell Disease

Sickle-cell disease is an example of a human disorder that is controlled by incompletely dominant alleles. Individuals with the genotype Hb^AHb^A are normal, those with the Hb^SHb^S genotype have sickle-cell disease, and those with the Hb^AHb^S genotype have *sickle-cell trait,* a condition in which the cells are sometimes sickle shaped. Two individuals with sickle-cell trait can produce a child with one of 3 possible phenotypes, as indicated in figure 22.17.

Among regions of malaria-infested Africa, infants with sickle-cell disease die, but infants with sickle-cell trait have a better chance of survival than the normal homozygote. Their sickled cells give protection against the malaria-causing parasite, which uses red blood cells during its life cycle. The parasite dies when potassium leaks out of the red blood cells as the cells become sickle shaped. The protection afforded by sickle-cell trait keeps the allele prevalent in populations exposed to malaria. As many as 60% of blacks in malaria-infected regions of Africa have the allele. In the United States, about 10% of the black population carries the allele.

The red blood cells in persons with sickle-cell disease cannot easily pass through small blood vessels. The sickle-shaped cells either break down or they clog blood vessels, and the individual suffers from poor circulation, anemia, and sometimes internal hemorrhaging. Jaundice, episodic pain in the abdomen and joints, poor resistance to infection, and damage to internal organs are all symptoms of sickle-cell disease.

Persons with sickle-cell trait do not usually have any difficulties unless they experience dehydration or mild oxygen deprivation. Although a recent study found that army recruits with sickle-cell trait are more likely to die when subjected to extreme exercise, previous studies of athletes do not substantiate these findings. At present, most investigators believe that no restrictions on physical activity are needed for persons with sickle-cell trait.

Innovative therapies are being attempted in persons with sickle-cell disease. For example, persons with sickle-cell disease produce normal fetal hemoglobin during development, and drugs that turn on the genes for fetal hemoglobin in adults are being developed. Mice have been genetically engineered to produce sickled red blood cells in order to test new antisickling drugs and various genetic therapies.

There are many exceptions to Mendel's laws. These include polygenic inheritance, multiple alleles, and degrees of dominance. Do 22.2 Critical Thinking, found at the end of the chapter.

Practice Problems 4*

1. What is the genotype of a person with straight hair? Could this individual ever have a child with curly hair?
2. What is the darkest child that could result from a mating between a light individual and a white individual?
3. What is the lightest child that could result from a mating between 2 intermediate individuals?
4. From the following blood types, determine which baby belongs to which parents:

Mrs. Doe	Type A
Mr. Doe	Type A
Mrs. Jones	Type A
Mr. Jones	Type AB
Baby 1	Type O
Baby 2	Type B

5. Prove that a child does not have to have the blood type of either parent by indicating what blood types *might* be possible when a person with type A blood reproduces with a person with type B blood.

*Answers to problems are on page 699.

SUMMARY

The genes are on the chromosomes; each gene has 2 alternative forms called alleles. Mendel's laws are consistent with the observation that each pair of alleles segregates independently of the other pairs during meiosis when the gametes form.

It is customary to use letters to represent the genotype of individuals. Homozygous dominant is indicated by 2 capital letters, and homozygous recessive is indicated by 2 lowercase letters. Heterozygous is indicated by a capital letter and a lowercase letter. In a monohybrid cross, each individual can form 2 types of gametes. In a dihybrid cross, each individual can form 4 types of gametes:

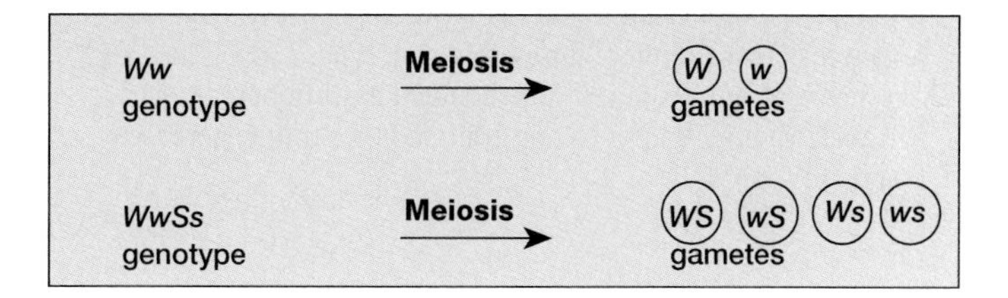

Use of the Punnett square allows us to make sure that all possible sperm have fertilized all possible eggs. The expected results of a monohybrid and a dihybrid cross can then be calculated as shown below. These results also tell us the chances of a child inheriting a particular phenotype. For example, for the monohybrid cross, there is a 25% chance of each child having the recessive phenotype and a 75% chance of each having the dominant phenotype.

Testcrosses are used to determine if an individual with the dominant phenotype is homozygous or heterozygous. If an individual expressing the dominant allele has offspring with the recessive phenotype, we know that the individual is heterozygous. The phenotypic ratios for testcrosses are given in table 22.2.

Studies of human genetics have shown that there are many autosomal genetic disorders that can be explained on the basis of simple Mendelian inheritance.

There are many exceptions to Mendel's laws. These include polygenic inheritance (skin color), multiple alleles (ABO blood type), and degrees of dominance (curly hair). There are genetic disorders associated with these patterns of inheritance also.

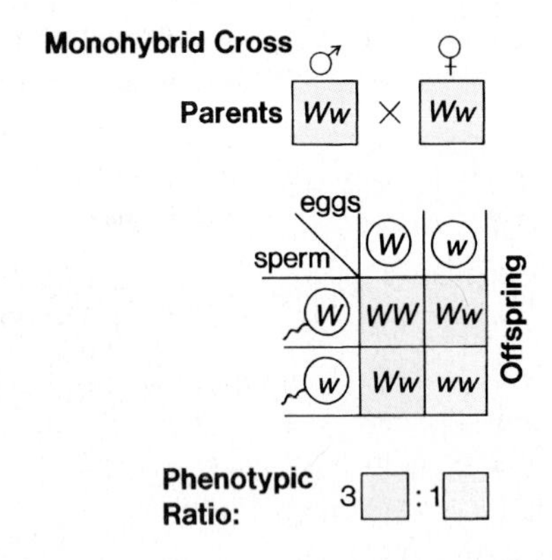

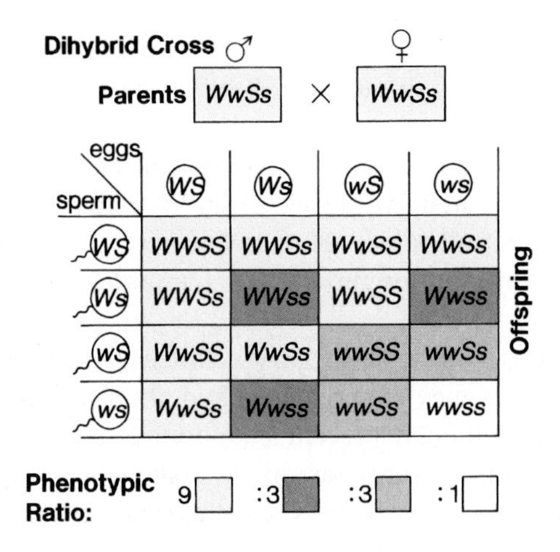

STUDY QUESTIONS

In order to practice **writing across the curriculum,** students should write out the answers to any or all of the study questions. The study questions are sequenced in the same order as the text.

1. What is Mendel's law of segregation? (p. 422) What do we call his factors today, and where are they located? (pp. 424–425)
2. What is the difference between the genotype and the phenotype of an individual? For which phenotype in a one-trait testcross are there 2 possible genotypes? (pp. 424–425)
3. What is Mendel's law of independent assortment? (p. 429) Relate Mendel's laws to one-trait and two-trait testcrosses. (pp. 429, 431)
4. What are the expected results of the following crosses? (p. 432)

   monohybrid × monohybrid
   monohybrid × recessive
   dihybrid × dihybrid
   dihybrid × recessive in both traits
5. Which of these crosses are called testcrosses? Why? (pp. 427, 431)
6. What are the chances of the dominant phenotype(s) for each of these crosses? (p. 432) What does the phrase "chance has no memory" mean? (p. 426)
7. List ways it is possible to tell an autosomal recessive genetic disease and an autosomal dominant genetic disease by examining a pedigree chart. (p. 435)
8. Give an example of these patterns of inheritance: polygenic inheritance, multiple alleles, and degrees of dominance. (pp. 436–438)

OBJECTIVE QUESTIONS

1. Whereas an individual has 2 alleles for every trait, the gametes have _______ allele(s) for every trait.
2. The recessive allele for the dominant gene *W* is _______.
3. Mary has a widow's peak, and John has a continuous hairline. This is a description of their _______.
4. *W* = widow's peak and *w* = continuous hairline; therefore, only the phenotype _______ could be heterozygous.
5. Two heterozygotes, each having a widow's peak, already have a child with a continuous hairline. The next child has what chance of having a continuous hairline? _______
6. In a testcross, an individual having the dominant phenotype is crossed with an individual having the _______ phenotype.
7. How many letters are required to designate the genotype of a dihybrid individual? _______
8. If a dihybrid is crossed with a dihybrid, how many of the 16 offspring are expected to have the dominant phenotype for both traits? _______
9. How many different phenotypes among the offspring are possible when a dihybrid is crossed with a dihybrid? _______
10. According to Mendel's law of independent assortment, a dihybrid can produce how many types of gametes having different combinations of genes? _______

ADDITIONAL GENETICS PROBLEMS

1. A woman heterozygous for polydactyly (dominant) reproduces with a homozygous normal man. What are the chances that their children will have 12 fingers and 12 toes? (p. 424)
2. John cannot curl his tongue (recessive), but both his parents can curl their tongue. Give the genotypes of all persons involved. (p. 424)
3. Parents who do not have Tay-Sachs disease (recessive) produce a child who has Tay-Sachs. What are the chances that each child born to this couple will have Tay-Sachs? (p. 424)
4. A man with a widow's peak (dominant) who cannot curl his tongue (recessive) reproduces with a woman who has a continuous hairline and who can curl her tongue. They have a child who has a continuous hairline and cannot curl the tongue. Give the genotypes of all persons involved. (p. 429)
5. Both Mr. and Mrs. Smith have freckles (dominant) and attached earlobes (recessive). Some of their children do not have freckles. What are the chances that their next child will have freckles and attached earlobes? (p. 429)
6. Is the characteristic represented by the darkened individuals inherited as an autosomal dominant or an autosomal recessive? (pp. 433, 435)

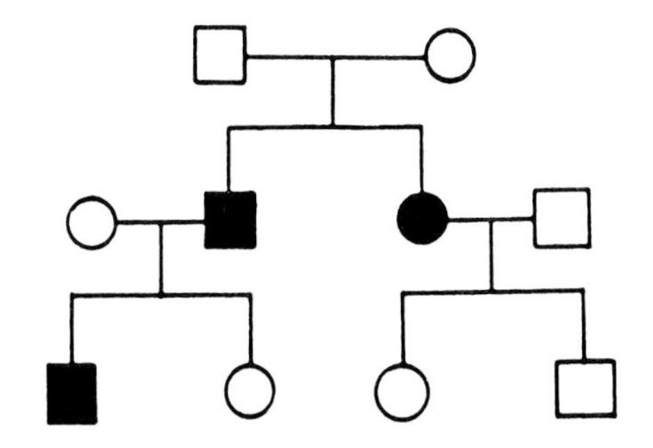

7. A woman with white skin has intermediate parents. If this woman marries a man with light skin, what is the darkest skin color possible for their children? the lightest? (p. 436)
8. A man has type AB blood. What is his genotype? Could this man be the father of a child with type B blood? If so, what blood types could the child's mother have? (p. 437)
9. Fill in this pedigree chart to give the probable genotypes of the twins pictured in figure 22.15. (p. 437)

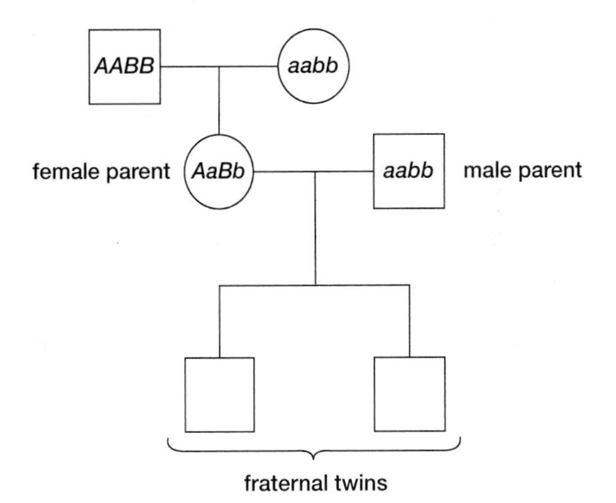

10. Mary has wavy hair (incomplete dominance) and reproduces with a man who has wavy hair. They have a child with straight hair. Give the genotypes of all persons involved. (p. 438)

CRITICAL THINKING

In order to practice **writing across the curriculum,** students should write out the answers to any or all of the critical thinking questions. Suggested answers to the critical thinking questions are in appendix E.

22.1

1. Before Mendel formulated his law of segregation, what 2 alternative hypotheses might he have formulated about the kinds of gametes for a parent that is *Yy* (*Y* = yellow peas; *y* = green peas)?
2. How did his results of 3 yellow : 1 green support one of these hypotheses and not the other?
3. Before Mendel formulated his law of independent assortment, what 2 alternative hypotheses might he have formulated about the kinds of gametes for a parent in figure 22.6?
4. How did his results of 9 : 3 : 3 : 1 support one of these hypotheses and not the other?

22.2

1. Some individuals are albinos—they have no melanin in any of their skin cells. Melanin is a molecule produced by a metabolic pathway (p. 75). What fault have albinos inherited?
2. Considering your answer to question 1, why would you have predicted that albinism is a recessive rather than a dominant disorder?
3. What possible crosses would produce an albino offspring? Are any of these individuals carriers?
4. Suppose you want to ensure that children born to an albino woman have normal pigmentation. Which of these—melanin, the enzyme to produce melanin, or a normal gene—would you inject into the egg?

SELECTED KEY TERMS

allele (ah-lēl´) an alternative form of a gene located at a particular chromosome site (locus).

dihybrid an individual that is heterozygous for 2 traits; shows the phenotype governed by the dominant alleles but carries the recessive alleles.

dominant allele the hereditary factor that expresses itself in the phenotype when the genotype is heterozygous.

genotype (je´ nə -tīp) the genes of any individual for (a) particular trait(s).

heterozygous having 2 different alleles (as *Aa*) for a given trait.

homozygous having identical alleles (as *AA* or *aa*) for a given trait; pure breeding.

monohybrid an individual that is heterozygous for one trait; shows the phenotype of the dominant allele but carries the recessive allele.

phenotype (fe´no-tīp) the outward appearance of an organism caused by the genotype and environmental influences.

Punnett square a gridlike device used to calculate the expected results of simple genetic crosses.

recessive allele a hereditary factor that expresses itself in the phenotype only when the genotype is homozygous.

testcross the backcross of a heterozygote with the recessive in order to determine the genotype.

trait specific term for a distinguishing phenotypic feature studied in heredity.

23

PATTERNS OF CHROMOSOME INHERITANCE

Chapter Concepts

1.
Normally, humans inherit 22 pairs of autosomes and 1 pair of sex chromosomes for a total of 46 chromosomes. 444, 458

2.
Abnormalities arise when humans inherit an extra autosome or an abnormal chromosome. 444, 458

3.
Normally, humans inherit 2 sex chromosomes. Males are XY and females are XX. 447, 458

4.
Abnormalities arise when humans inherit a broken X chromosome or an incorrect number of sex chromosomes. 449, 458

5.
Certain genes, called X-linked genes, occur on the X chromosome, and some control traits unrelated to the sex of the individual. 450, 458

6.
Males always express X-linked recessive disorders because they inherit only one X chromosome. 451, 458

7.
Various methods are being used to determine the order of the genes (mapping) of the human chromosomes. 454, 458

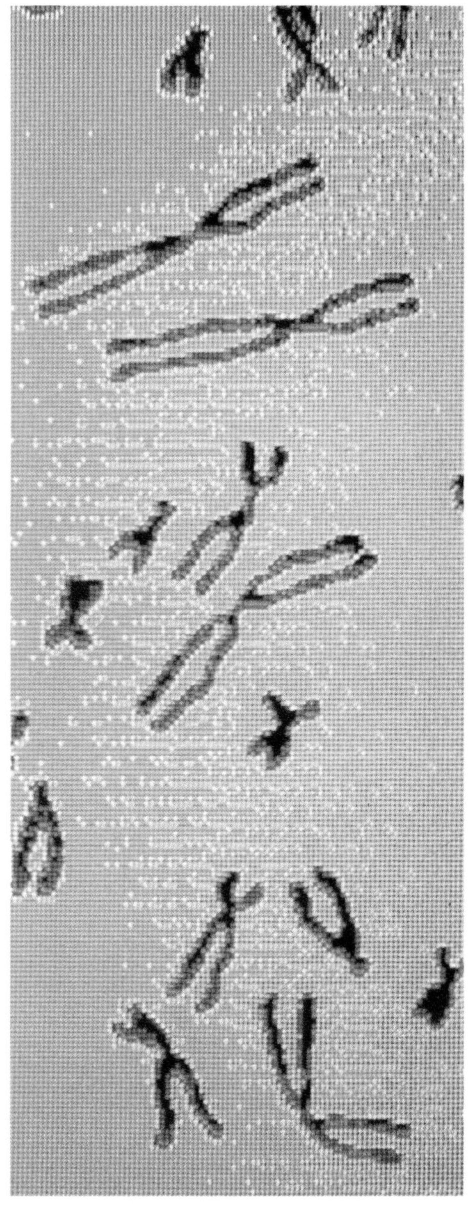

Human chromosomes, ×1,200

Chapter Outline

Figure 23.1
Nondisjunction of autosomes during oogenesis. Nondisjunction can occur during meiosis I if homologous chromosomes fail to separate and during meiosis II if the sister chromatids fail to separate completely. In either case, certain abnormal eggs carry an extra chromosome. Nondisjunction of chromosome 21 can lead to Down syndrome.

meiosis I nondisjunction
meiosis II nondisjunction
Meiosis I
Meiosis II
abnormal
normal

NDIVIDUALS inherit chromosomes from their parents. These determine what the individual will be like in regard to all the various traits that characterize the individual.

Inheritance of Autosomes

Normally, an individual receives 22 autosomes and one sex chromosome from each parent. Sometimes individuals are born with either too many or too few autosomes, most likely due to nondisjunction during the meiosis of either oogenesis or spermatogenesis. **Nondisjunction** can occur during meiosis I if the homologous chromosomes fail to separate or during meiosis II if the sister chromatids fail to separate completely (fig. 23.1).

Down Syndrome: Extra Chromosome

The most common autosomal abnormality is seen in individuals with **Down syndrome** (fig. 23.2). This syndrome is easily recognized by these characteristics: short stature; an oriental-like fold of the eyelids; stubby fingers; a wide gap between the first and second toes; a large, fissured tongue; a round head; a palm crease, the so-called simian line; and, unfortunately, mental retardation, which can sometimes be severe.

Persons with Down syndrome usually have 3 copies of chromosome 21 because the egg had 2 copies instead of one. (In 23% of the cases studied, however, the sperm had the extra chromosome 21.) A recent study has found that the incidence of nondisjunction is the same for the eggs of younger and older women. This was an unexpected finding because the chances of a woman having a Down syndrome child increase rapidly with age, starting at about age 35 (table 23.1). As a possible explanation, researchers suggest that older women may be more likely to carry a Down syndrome child to term because their bodies fail to recognize and abort abnormal embryos. It had been thought that older women had a greater tendency toward nondisjunction because as a woman ages, her eggs also age—a woman is born with all the eggs she will ever have, and they remain in an arrested condition until one matures each month.

Although an older woman is more likely to have a Down syndrome child, most babies with Down syndrome are born to women younger than age 35 because this is the age group having the most babies. As discussed in the reading on page 408, chorionic villi testing and amniocentesis followed by karyotyping can detect a Down syndrome child. However, young women are not encouraged to undergo such procedures because the risk of complications resulting from these tests is greater than the risk of having a Down syndrome child. Fortunately, there is now a test, based on substances in the blood, that can identify mothers who might be carrying a Down syndrome child, and only these individuals need to undergo further medical testing.

It is known that the genes that cause Down syndrome are located on the bottom third of chromosome 21 (fig. 23.2*b*), and extensive investigative work has been directed toward discovering the specific genes responsible for the characteristics of the syndrome. Thus far, investigators have discovered several genes that may account for various conditions seen in persons with Down syndrome. For example, they have located genes most likely responsible for the increased tendency toward

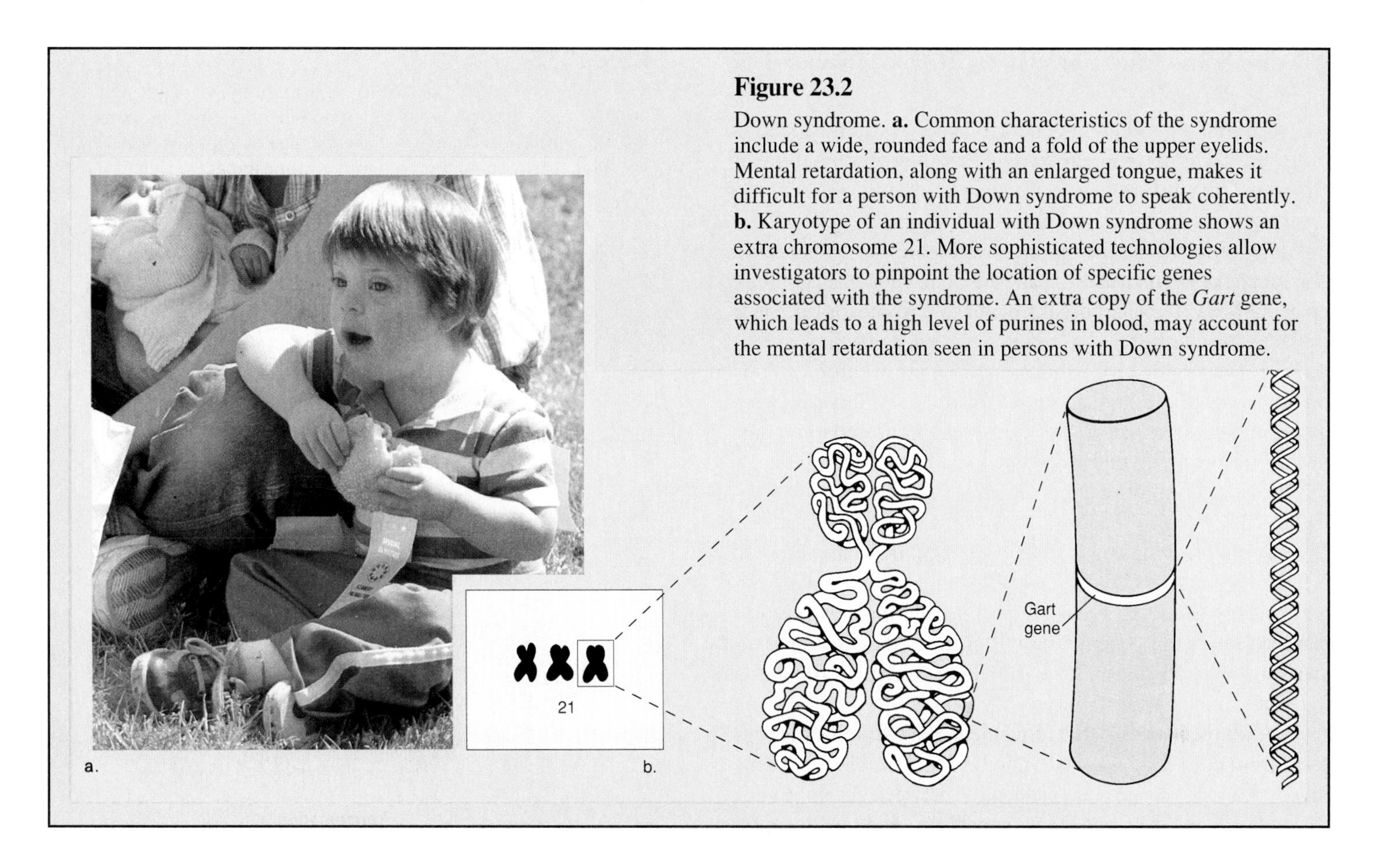

Figure 23.2

Down syndrome. **a.** Common characteristics of the syndrome include a wide, rounded face and a fold of the upper eyelids. Mental retardation, along with an enlarged tongue, makes it difficult for a person with Down syndrome to speak coherently. **b.** Karyotype of an individual with Down syndrome shows an extra chromosome 21. More sophisticated technologies allow investigators to pinpoint the location of specific genes associated with the syndrome. An extra copy of the *Gart* gene, which leads to a high level of purines in blood, may account for the mental retardation seen in persons with Down syndrome.

Table 23.1
Frequency and Effects of the Most Common Chromosome Abnormalities in Humans

Syndrome	Sex	Involved Chromosomes	Births	Fertility
Down				
Mothers under 40	M or F	Extra 21	1/800	Infertile
Mothers over 40			1/60	
Fragile X	M or F	Broken X	1/1,000 (M) 1/2,500 (F)	Infertile Varies
Turner	F	X	1/2,500–10,000	Sterile
Klinefelter	M	XXY	1/500–2,000	Sterile
Metafemale	F	XXX	1/1,000–2,000	Infertile
XYY	M	XYY	1/1,000	Normal

leukemia, cataracts, accelerated rate of aging, and mental retardation. The gene for mental retardation, dubbed the *Gart* gene, causes an increased level of purines in blood, a finding associated with mental retardation. It is hoped that someday it will be possible to control the expression of the *Gart* gene even before birth so that at least this symptom of Down syndrome does not appear.

Abnormal autosome inheritance can be due to the inheritance of extra chromosomes. Down syndrome is caused by the inheritance of an extra chromosome 21.

Chromosome Mutations: Parts Moved, Inverted, Lost, or Doubled

A mutation is a permanent genetic change. A change in chromosome structure that can be detected microscopically is a **chromosome mutation.** The process of crossing-over (p. 113) during meiosis can lead to a chromosome mutation, and various agents in the environment—radiation, certain organic chemicals, or even viruses—can cause chromosomes to break apart. Ordinarily, when breaks occur in chromosomes, the 2 broken ends reunite to give the same sequence of genes. Sometimes, however, the broken ends of one or more chromosomes do not rejoin in the same pattern as before, and this results in a chromosome mutation. Various types of chromosome mutations occur; an inversion, a translocation, a deletion, and a duplication of chromosome segments are all illustrated in figure 23.3.

An **inversion** occurs when a segment of a chromosome is turned around 180 degrees. You might think this is not a problem because the same genes are present, but the new position can lead to altered gene activity. It also can lead to abnormal crossing-over during meiosis, resulting in duplications and deletions in the chromosomes of the gametes.

A **translocation** is the movement of a chromosome segment from one chromosome to another, nonhomologous chromosome. In 5% of cases, a translocation between chromosomes 21 and 14 is the cause of Down syndrome. The translocation may have occurred in a parent or a relative who lived generations earlier. In the latter case, Down syndrome is not age related and instead tends to run in the family of either the father or the mother.

A **deletion** occurs when an end of a chromosome breaks off or when 2 simultaneous breaks lead to the loss of a segment. A **duplication,** the presence of a chromosome segment more than once in the same chromosome, can occur in 2 ways. A broken segment from one chromosome can simply attach to its homologue. The presence of a duplication in the middle of a chromosome is more likely due to unequal crossing-over. This can occur when homologous chromosomes are mispaired slightly; following crossing-over, there is both a gene duplication and a gene deletion.

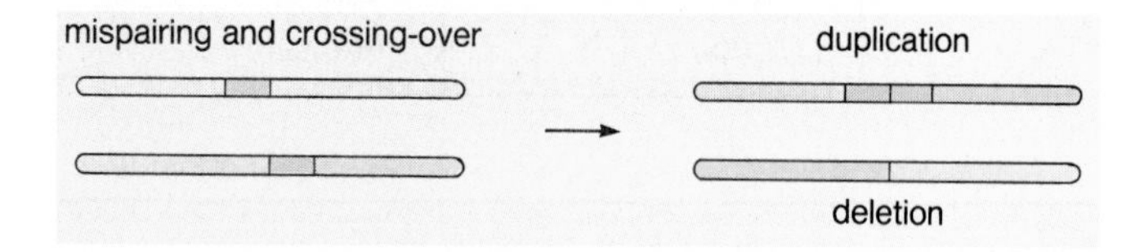

Instances of gene duplication occur in human cells. For example, human cells have multiple copies of the genes that code for the proteins found in hemoglobin. Each of these proteins is slightly different, and some occur only during fetal development.

Figure 23.3

Types of chromosome mutations. **a.** Inversion occurs when a chromosome segment breaks apart and then rejoins in reversed direction. **b.** Translocation is the exchange of chromosome segments between nonhomologous chromosomes. **c.** Deletion is the loss of a chromosome segment. **d.** Duplication occurs when the same segment is repeated within the chromosome.

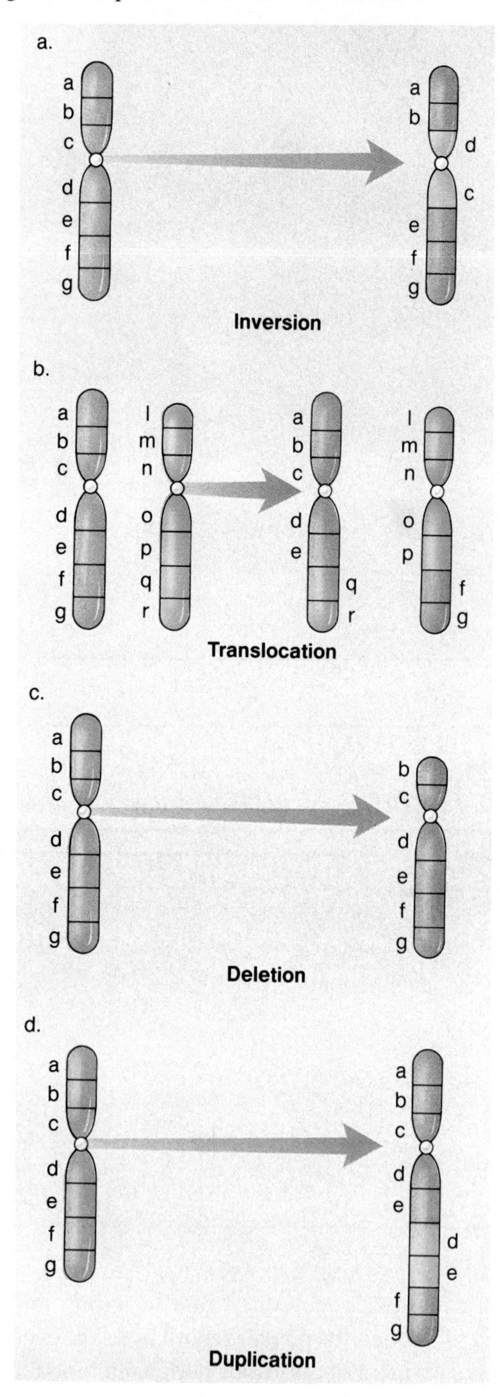

A chromosome deletion is responsible for **cri du chat** (cat's cry) **syndrome** (fig. 23.4). Affected individuals meow like a kitten when they cry, but more importantly, they tend to have a small head with malformations of the face and the

Figure 23.4

Cri du chat syndrome. An infant with this syndrome has a moon face, small head, and a cry that sounds like the meow of a cat. Normal persons have the chromosome-5 pair as shown on *bottom left*; persons with cri du chat syndrome have the pair as shown on *bottom right*.

body, and mental defectiveness usually causes retarded development. Chromosome analysis shows that a portion of chromosome 5 is missing (deleted), while the other chromosome 5 is normal.

Abnormal autosomal chromosome inheritance can be due to the inheritance of a mutated chromosome. Cri du chat syndrome occurs when a portion of chromosome 5 is deleted.

Sex Chromosome Inheritance

The sex chromosomes in humans are called X and Y. Since females are XX, an egg always bears an X, but since males are XY, a sperm can bear an X or a Y. The sex of the newborn child is determined by the father. If a Y-bearing sperm fertilizes the egg, then the XY combination results in a male. On the other hand, if an X-bearing sperm fertilizes the egg, the XX combination results in a female. All factors being equal, there is a

Figure 23.5

Inheritance of sex. In this Punnett square, the sperm and the eggs are shown as carrying only a sex chromosome. (Actually, they also carry 22 autosomes.) The offspring are either male or female, depending on whether an X or a Y chromosome is inherited from the male parent.

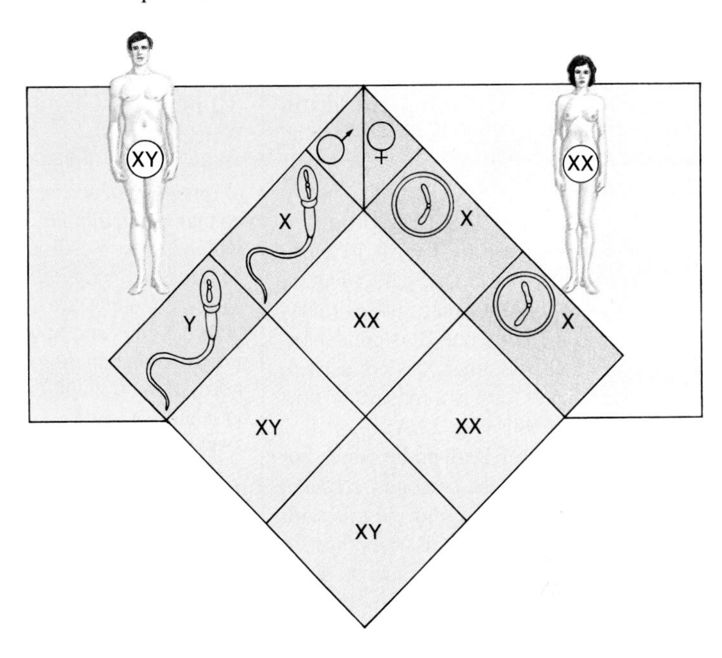

Table 23.2
Sex Ratios in the United States

Age	Males : Females
Birth	106 : 100
18 years	100 : 100
50 years	85 : 100
85 years	50 : 100
100 years	20 : 100

From John W. Hole, Jr., *Human Anatomy and Physiology*, 5th ed. Copyright © 1990 Wm. C. Brown Communications, Inc., Dubuque, Iowa. All Rights Reserved. Reprinted by permission.

50% chance of having a girl or a boy. It is possible to illustrate this probability with a *Punnett square* (fig. 23.5). In the square, all possible sperm are lined up on one side, all possible eggs are lined up on the other side (or vice versa), and every possible combination is determined. When this is done with regard to sex chromosomes, the results show one female to each male. However, for reasons that are not clear, more males than females are conceived, but from then on the death rate among males is higher. More males than females are spontaneously aborted, and this trend continues after birth until there is a dramatic reversal of the ratio of males to females (table 23.2).

WHAT MAKES A MALE AND A FEMALE

How do you tell a female from a male? A karyotype will tell you, but there are easier methods to determine if a person has XX chromosomes or XY chromosomes. It so happens that adults with XX chromosomes have small, darkly staining chromatin bodies, often called Barr bodies (after the person who first identified them), in their nuclei. A Barr body is a condensed, inactive X chromosome, as was first proposed by Mary Lyon. The validity of the *Lyon hypothesis* means that female cells function with a single X chromosome just as males do. Still, in some cells one X is condensed and in other cells the other X is condensed, so that the female body is a mosaic of genetically different cells (fig. 23.A).

Figure 23.A
Each female nucleus contains a Barr body, which is a condensed, inactive X chromosome. Females are mosaics for alleles carried on the X chromosome because chance alone dictates which X chromosome becomes a Barr body. These 2 females are heterozygous and have inherited one allele for normal sweat glands and another allele for inactive sweat glands. Where in the body the mutant allele expresses itself (purple color) varies from female to female.

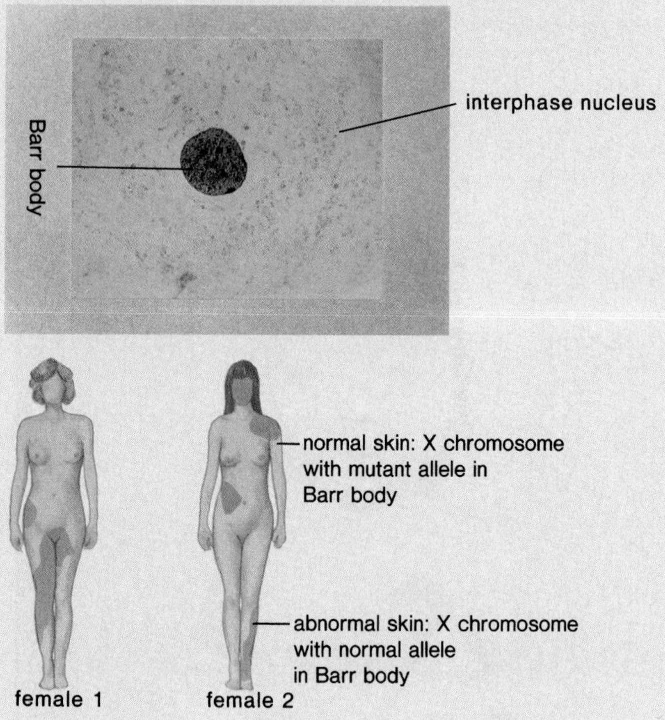

The presence of Barr bodies does not necessarily mean that the person is a female, however. After all, XXY individuals with Klinefelter syndrome have Barr bodies[1] in their cells. And then too, there are on occasion XX persons who have translocated genes for maleness on one of their X chromosomes. You should, however, be able to detect these persons by testing for the presence of the H-Y antigen in the cell membrane. The H-Y antigen is the product of a histocompatibility gene usually located on the Y chromosome. It is called an antigen because females produce antibodies against it. To test for maleness, it is possible to suspend a sample of white blood cells in a solution that contains some of these antibodies. If the cells carry the H-Y antigen, indicating that the person has male genes, the antibodies bind with them.

The absence of a Barr body and a positive test for the H-Y antigen signify that the person has an X and a Y chromosome. Most XY individuals are males, but there are on occasion XY females. For example, persons with androgen-insensitivity syndrome have testes located in the abdomen and the external genitals of a female. Their cells lack the receptors for testosterone, and therefore they have the secondary sex characteristics of females, even though the blood level of testosterone is typical of males.

1. How many Barr bodies does a person with Klinefelter syndrome have? a metafemale have?

The sex of a child is dependent on whether a Y-bearing or an X-bearing sperm fertilizes the X-bearing egg.

For some years, it had been proposed that a gene located on the Y chromosome brings about maleness. Embryos begin life with no evidence of a sex, but by about the third month of development, males can be distinguished from females (p. 410). Investigators have recently found a gene called the *Sry* gene (short for *sex-determining region Y* gene), which they believe is the best candidate for the gene that switches development to the male pathway. In other words, the embryo is basically female and will automatically develop into a female unless the *Sry* gene starts the development of testes instead of ovaries. When researchers inject XX mice embryos with a small fragment of Y chromosome containing the *Sry* gene, the mice develop into males, with testes and male behavior. Another gene, called the testes-determining factor, was previously thought to determine maleness, but it fails this test.

Once males and females develop, certain cellular differences can be detected. These are discussed in the reading on this page.

Figure 23.6
Fragile X chromosome. Fragile X syndrome is due to the inheritance of a fragile X chromosome. An arrow points out the fragile site on each chromatid.

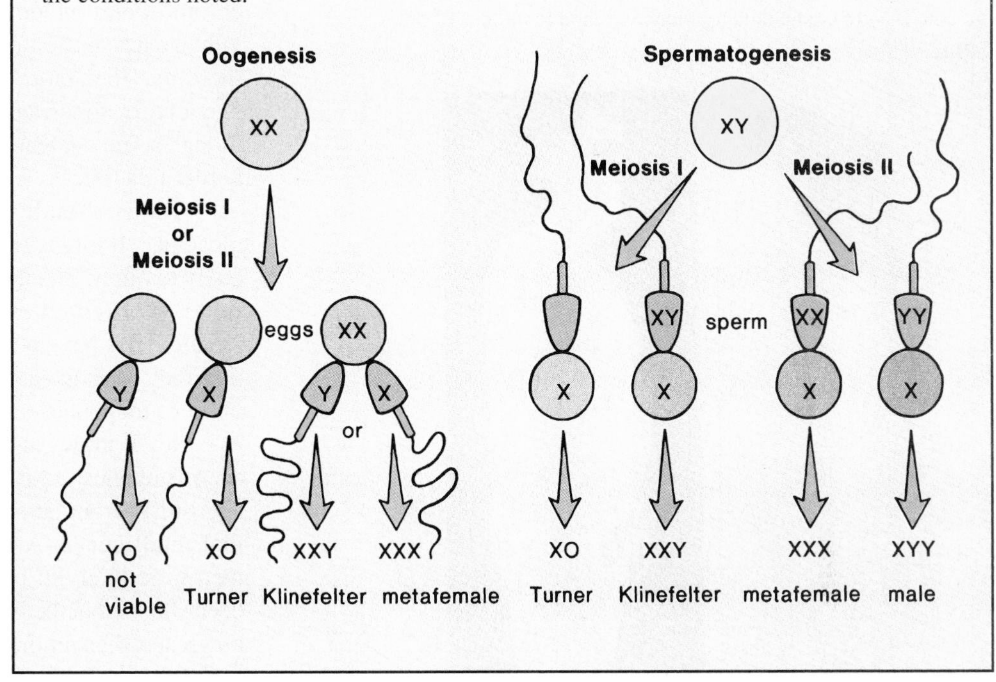

Figure 23.7
Nondisjunction of sex chromosomes. Nondisjunction of sex chromosomes during oogenesis followed by fertilization with normal sperm results in the conditions noted. Nondisjunction of sex chromosomes during spermatogenesis followed by fertilization of normal eggs results in the conditions noted.

Sex Chromosome Problems

Abnormal sex chromosome inheritance results from the inheritance of an abnormal X or an abnormal number of sex chromosomes.

Fragile X Syndrome: More Common in Males

Males outnumber females by about 25% in institutions for the mentally retarded. In some of these males, the X chromosome is nearly broken, leaving the tip hanging by a flimsy thread. These males are said to have fragile X syndrome (fig. 23.6).

As children, fragile X syndrome individuals may be hyperactive or autistic; their speech is delayed in development and often repetitive in nature. As adults, they have large testes and big, usually protruding ears. They are short in stature, but the jaw is prominent and the head is often large. Stubby hands, lax joints, and a heart defect may also occur.

Fragile X chromosomes occur in both males and females (table 23.1), but the syndrome is seen less often in females. When symptoms do appear in females, they tend to be less severe.

Too Many/Too Few Sex Chromosomes

Abnormal sex chromosome constituencies (table 23.1) are also due to nondisjunction. Nondisjunction of the sex chromosomes during oogenesis can lead to an egg with either 2 X chromosomes or to no X chromosome. Nondisjunction of the sex chromosomes during spermatogenesis can result in a sperm that has no sex chromosome, both an X and a Y chromosome, 2 X chromosomes, or 2 Y chromosomes. These abnormal gametes sometimes result in miscarriages, but other times, the zygote develops into an individual with one of the conditions listed in figure 23.7. The chromosome mutations noted in figure 23.3 for autosomal chromosomes are not seen in the sex chromosomes, most likely because the zygote is not viable.

> Sometimes a person inherits an abnormal combination of sex chromosomes due to nondisjunction of these chromosomes during meiosis.

Figure 23.8

Abnormal sex chromosome inheritance. **a.** Female with Turner (XO) syndrome, symptoms of which include a bull neck, short stature, and immature sexual features. **b.** A male with Klinefelter (XXY) syndrome, symptoms of which include male secondary sex characteristics that are often quite well-developed and testes that are small. Note breast development and female pattern of pubic hair growth.

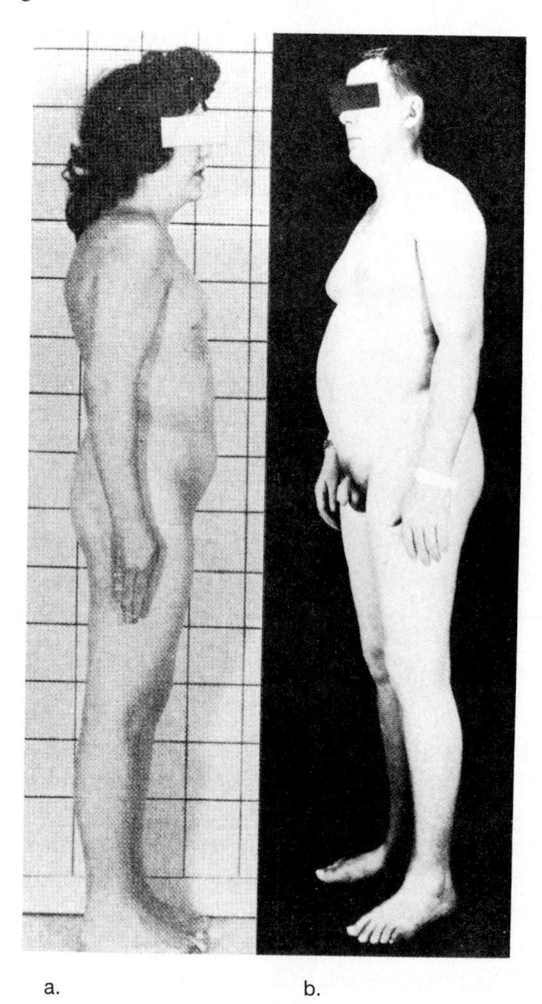

a. b.

From birth, an XO individual with **Turner syndrome** has only one sex chromosome, an X; the O signifies the absence of a second sex chromosome. Turner females are short, have a broad chest, and may have congenital heart defects. The ovaries never become functional, and in many they are simply white streaks. Turner females do not undergo puberty or menstruate, and there is a lack of breast development (fig. 23.8*a*). Although no overt mental retardation is reported, Turner females show reduced skills in interpreting spatial relationships.

A male with **Klinefelter syndrome** has 2 or more X chromosomes in addition to a Y chromosome. Affected individuals are sterile males; the testes are underdeveloped and there may be some breast development (fig. 23.8*b*). These phenotypic abnormalities are not apparent until puberty, although some evidence of subnormal intelligence may be seen before this time.

A **metafemale** is an individual with more than 2 X chromosomes. It might be supposed that the XXX female is especially feminine, but this is not the case. Although in some cases there is a tendency toward learning disabilities, most metafemales have no apparent physical abnormalities except that they may have menstrual irregularities, including early onset of menopause.

XYY males also can result from nondisjunction during spermatogenesis. Affected males usually are taller than average, suffer from persistent acne, and tend to have barely normal intelligence. At one time, it was suggested that these men were likely to be criminally aggressive, but it has since been shown that the incidence of such behavior among them is no greater than among XY males.

Individuals sometimes are born with the sex chromosomes XO (Turner syndrome), XXY (Klinefelter syndrome), XXX (metafemale), and XYY. No matter how many X chromosomes there are, an individual with a Y chromosome is usually a male.

Sex-Linked Inheritance

The genes that determine the development of the reproductive organs are on the sex chromosomes. Even so, most of the genes on the sex chromosomes have nothing to do with sexual development and instead are concerned with other body traits. **Sex-linked genes** are so called because they are on the sex chromosomesA few sex-linked alleles are presumed to be on the Y chromosome, but most of those discovered so far are only on the much larger X chromosome. These are called **X-linked genes.** The Y is blank for X-linked genes.

Solving X-Linked Genetics Problems

Recall that when solving autosomal genetics problems, we represent the genotypes of males and females similarly, as shown in the following example for humans.

Key	**Genotypes**
W = Widow's peak	WW, Ww, or ww
w = Continuous hairline	

In contrast, when we set up the key for a sex-linked gene, males and females must be indicated by sex.

Key

X^B = Normal vision
X^b = Color blindness

The possible genotypes in both males and females are as follows:

X^BX^B = Female who has normal color vision
X^BX^b = Carrier female who has normal color vision
X^bX^b = Female who is color blind
X^BY = Male who has normal vision
X^bY = Male who is color blind

Note that the second genotype is a *carrier* female because although a female with this genotype appears normal, she is capable of passing on an allele for color blindness. Color-blind females are rare because they must inherit the allele from both parents; color-blind males are more common since they need only one recessive allele to be color blind. The allele for color blindness has to be inherited from their mother because it is on the X chromosome; males inherit only the Y chromosome from their father.

Now, let us consider a particular cross. If a heterozygous woman reproduces with a man with normal vision, what are the chances of their having a color-blind daughter? a color-blind son?

$$\text{Parents: } X^BX^b \times X^BY$$

Inspection indicates that all daughters will have normal color vision because they all will receive an X^B from their father. The sons, however, have a 50% chance of being color blind, depending on whether they receive an X^B or an X^b from their mother. The inheritance of a Y chromosome from their father cannot offset the inheritance of an X^b from their mother.

Figure 23.9 illustrates the use of the Punnett square for solving X-linked problems. Notice that when indicating the results of a cross involving an X-linked gene, you give the phenotypic ratios for males and females separately.

Figure 23.10 gives a pedigree chart for an X-linked recessive gene. It also lists ways to recognize this pattern of inheritance.

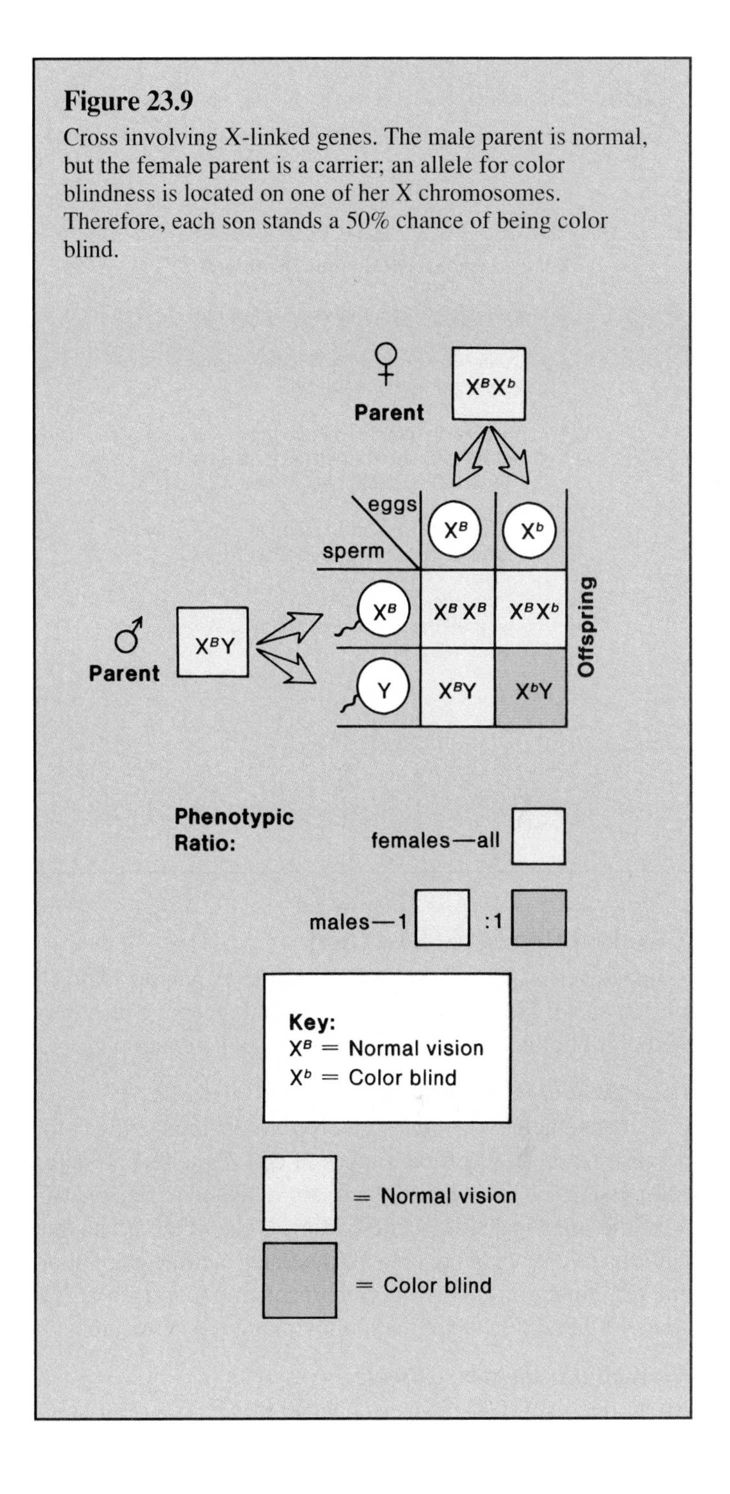

Figure 23.9
Cross involving X-linked genes. The male parent is normal, but the female parent is a carrier; an allele for color blindness is located on one of her X chromosomes. Therefore, each son stands a 50% chance of being color blind.

Figure 23.10

Sample pedigree chart for an X-linked recessive genetic disorder. Which females in the chart are carriers?

X-linked Recessive Genetic Disorders

- More males than females are affected.
- An affected son can have parents who have the normal phenotype.
- In order for a female to have the characteristic, her father must also have it. Her mother must have it or be a carrier.
- The characteristic often skips a generation from the grandfather to the grandson.
- If a woman has the characteristic, all of her sons will have it.

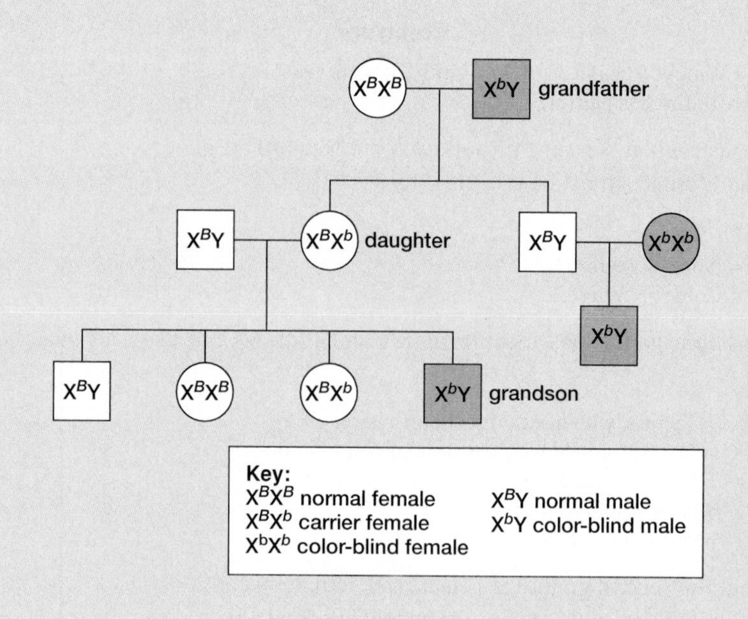

X-Linked Recessive Disorders

X-linked genes can be dominant or recessive, but most known are recessive. Three well-known X-linked recessive disorders are color blindness, hemophilia, and muscular dystrophy.

Color Blindness: Three Genes

In humans, there are 3 genes involved in distinguishing color because there are 3 different types of cones, the receptors for color vision (p. 332). Two of these are X-linked genes; one affects the green-sensitive cones, whereas the other affects the red-sensitive cones. About 5% of Caucasian men are color blind due to a mutation involving green perception, and about 2% are color blind due to a mutation involving red perception.

Hemophilia: Bleeder's Disease

About one in 10,000 males is a hemophiliac. The most common type of hemophilia is hemophilia A, due to the absence or minimal presence of a particular clotting factor called factor VIII. *Hemophilia* is called the bleeder's disease because the affected person's blood does not clot. Although hemophiliacs bleed externally after an injury, they also suffer from internal bleeding, particularly around joints. Hemorrhages can be checked with transfusions of fresh blood (or plasma) or concentrates of the clotting protein. Unfortunately, some hemophiliacs have contracted AIDS after using concentrated blood from untested donors, but this cannot occur if a purified form of the concentrate from donors who have been tested is used.

At the turn of the century, hemophilia was prevalent among the royal families of Europe, and all of the affected males could trace their ancestry to Queen Victoria of England (fig. 23.11). Because none of Queen Victoria's forebearers or relatives was affected, it seems that the gene she carried arose by mutation either in Victoria or in one of her parents. Her carrier daughters, Alice and Beatrice, introduced the gene into the ruling houses of Russia and Spain, respectively. Alexis, the last heir to the Russian throne before the Russian Revolution, was a hemophiliac. There are no hemophiliacs in the present British royal family because Victoria's eldest son, King Edward VII, did not receive the gene and therefore could not pass it on to any of his descendants.

Muscular Dystrophy: Muscles Waste Away

Muscular dystrophy, as the name implies, is characterized by a wasting away of the muscles. The most common form, *Duchenne muscular dystrophy,* is X linked and occurs in about one out of every 3,600 male births. Symptoms, such as waddling gait, toe walking, frequent falls, and difficulty in rising, may appear as soon as the child starts to walk. Muscle weakness intensifies until the individual is confined to a wheelchair. Death usually occurs by age 20; therefore, affected males are rarely fathers. The recessive allele remains in the population by passage from carrier mother to carrier daughter.

Recently, the gene for muscular dystrophy was isolated, and it was discovered that the absence of a protein now called dystrophin is the cause of the disorder. Much investigative

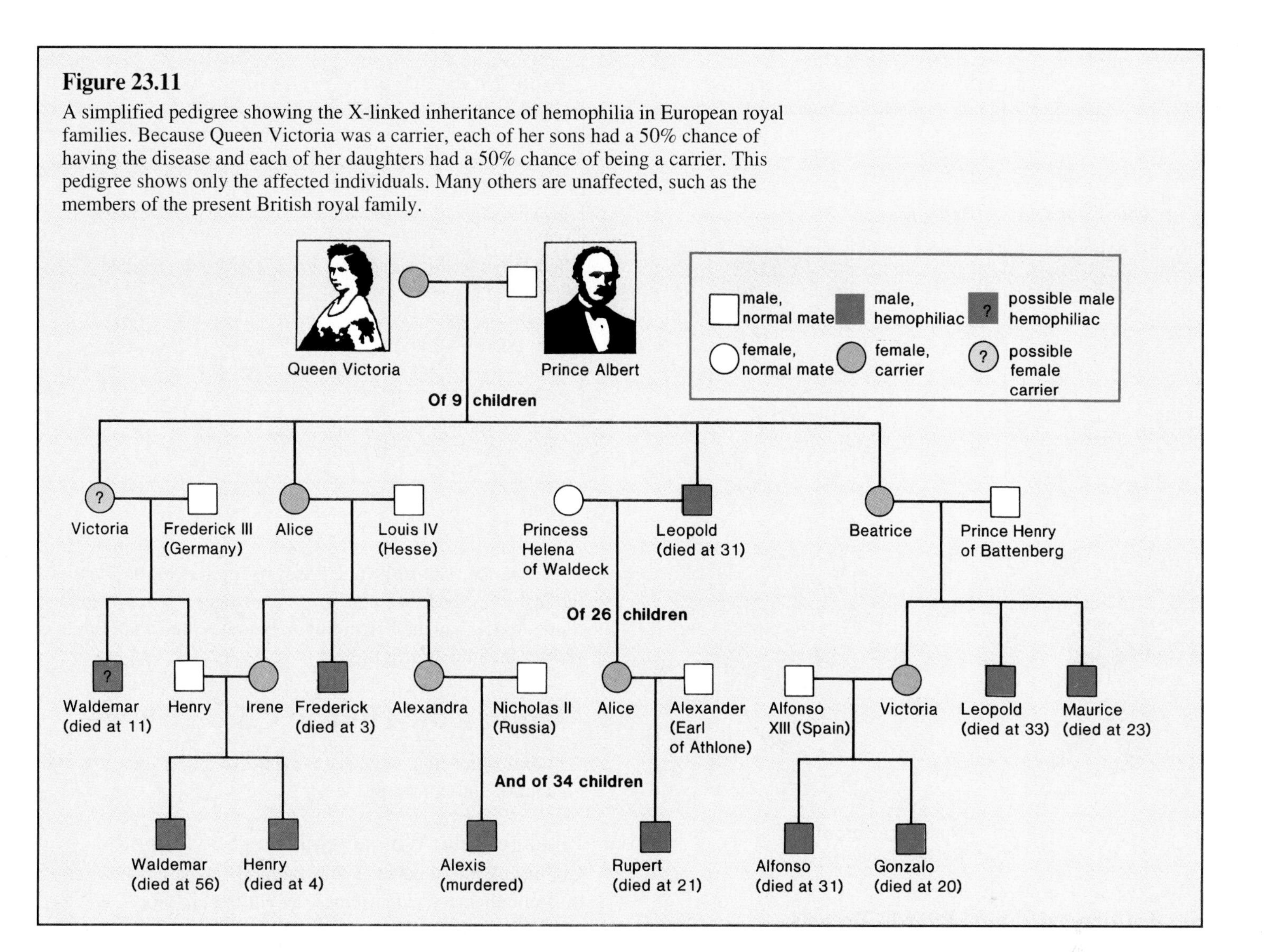

Figure 23.11
A simplified pedigree showing the X-linked inheritance of hemophilia in European royal families. Because Queen Victoria was a carrier, each of her sons had a 50% chance of having the disease and each of her daughters had a 50% chance of being a carrier. This pedigree shows only the affected individuals. Many others are unaffected, such as the members of the present British royal family.

work determined that dystrophin is involved in the release of calcium from the calcium-storage sacs (p. 314) in muscle fibers. The lack of dystrophin causes calcium to leak into the cell, which promotes the action of an enzyme that dissolves muscle fibers. When the body attempts to repair the tissue, fibrous tissue forms, and this cuts off the blood supply so that more and more cells die.

A test is now available to detect carriers for Duchenne muscular dystrophy. Also, various treatments are being attempted. Immature muscle cells can be injected into muscles, and for every 100,000 cells injected, dystrophin production occurs in 30–40% of muscle fibers. The gene for dystrophin has been inserted into the thigh muscle cells of mice, and about 1% of these cells then produced dystrophin.

There are sex-linked genes on the X chromosome that have nothing to do with sexual characteristics. Males have only one copy of these genes, and if they inherit a recessive allele, it is expressed. Do 23.1 Critical Thinking, found at the end of the chapter.

Practice Problems*

1. Both the mother and the father of a male hemophiliac appear to be normal. From whom did the son inherit the allele for hemophilia? What are the genotypes of the mother, the father, and the son?
2. A woman is color blind. What are the chances that her sons will be color blind? If she is married to a man with normal vision, what are the chances that her daughters will be color blind? will be carriers?
3. Both parents are right handed (R = right handed, r = left handed) and have normal vision. Their son is left handed and color blind. Give the genotypes of all persons involved.
4. Both the husband and the wife have normal vision. The wife gives birth to a color-blind daughter. What can you deduce about the girl's father?

*Answers to problems are on page 700.

Figure 23.12

Baldness, a sex-influenced characteristic. Due to hormonal influences, the presence of only one gene for baldness causes the condition in the male, whereas the condition does not occur in the female unless she possesses both genes for baldness.

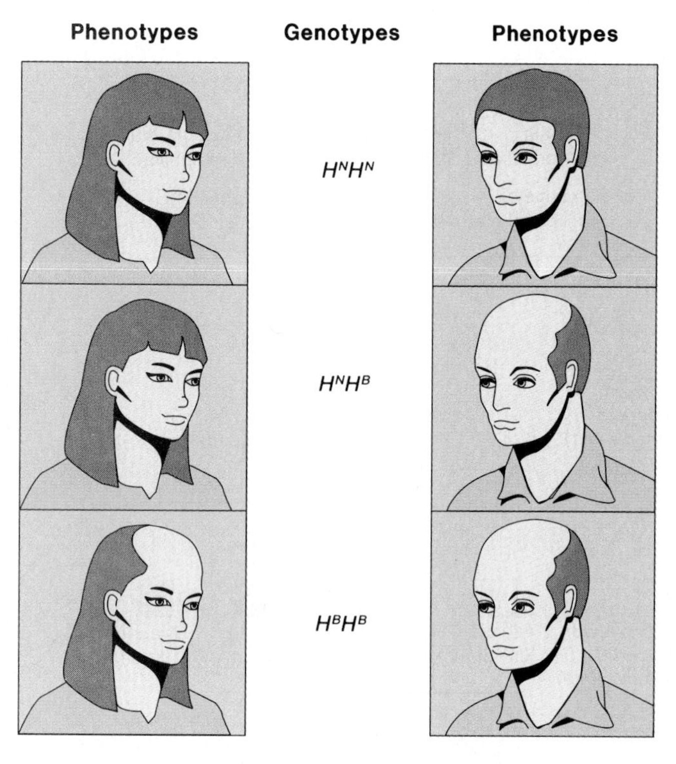

Sex-Influenced Traits: Beards, Breasts

Not all traits we associate with the sex of the individual are due to sex-linked genes. Some are simply sex-influenced traits. Sex-influenced traits are characteristics that often appear in one sex but only rarely appear in the other. It is believed that these traits are governed by genes that are turned on or off by hormones. For example, the secondary sexual characteristics, such as the beard of a male and the developed breasts of a female, probably are controlled by the balance of hormones.

Baldness (fig. 23.12) is believed to be caused by the male sex hormone testosterone because males who take the hormone to increase masculinity begin to lose their hair. A more detailed explanation has been suggested by some investigators. It has been reasoned that due to the effect of hormones, males require only one gene for the trait to appear, whereas females require 2 genes. In other words, the gene acts as a dominant in males but as a recessive in females. This means that males born to a bald father and a mother with hair *at best* have a 50% chance of going bald. Females born to a bald father and a mother with hair *at worst* have a 25% chance of going bald.

Table 23.3
Chromosome Theory of Inheritance

Chromosome Theory of Inheritance
The genes are on the chromosomes, and they are arranged in a definite sequence.
Both the chromosomes and the alleles occur in pairs in diploid cells.
Both the chromosomes and the alleles of each pair segregate independently during meiosis.
The gametes contain one of each type chromosome and one of each type allele in all possible combinations.
Fertilization restores the full number of chromosomes and alleles so that the zygote is diploid.
Chromosome mutations (abnormalities in number and type) affect the phenotype of the individual.

Another sex-influenced trait of interest is the length of the index finger. In women, the index finger is at least equal to if not longer than the fourth finger. In males, the index finger is shorter than the fourth finger.

Mapping the Human Chromosomes

A chromosome map indicates the order of the various gene loci on a particular chromosome.

Linkage Data: When Neighbors Stay Together

Our discussion of genetics thus far has been based on the chromosome theory of inheritance, as outlined in table 23.3. Figure 22.2 illustrates one aspect of this theory—that there are several alleles on each chromosome. All the alleles on one chromosome form a **linkage group** because they tend to be inherited together. Mendel's law of independent assortment does not hold for linked genes, and linked genes do consistently appear together in the same gamete. Therefore, traits controlled by linked genes tend to be inherited together. Figure 23.13*a* shows that if linkage is complete, a dihybrid produces only 2 types of gametes in equal proportion. However, figure 23.13*b* shows that when crossing-over occurs, linkage is incomplete and a dihybrid produces 4 types of gametes. (Crossing-over, you recall, is an exchange of segments between nonsister chromatids of a tetrad, as explained on page 113.)

To take an actual example, the genes for ABO blood types and the gene for a condition called nail-patella syndrome (NPS) are on the same chromosome. A person with NPS has fingernails and toenails that are reduced or absent and a kneecap (patella) that is small. NPS (*N*) is dominant, while the normal condition (*n*) is recessive. In one family, the spouses

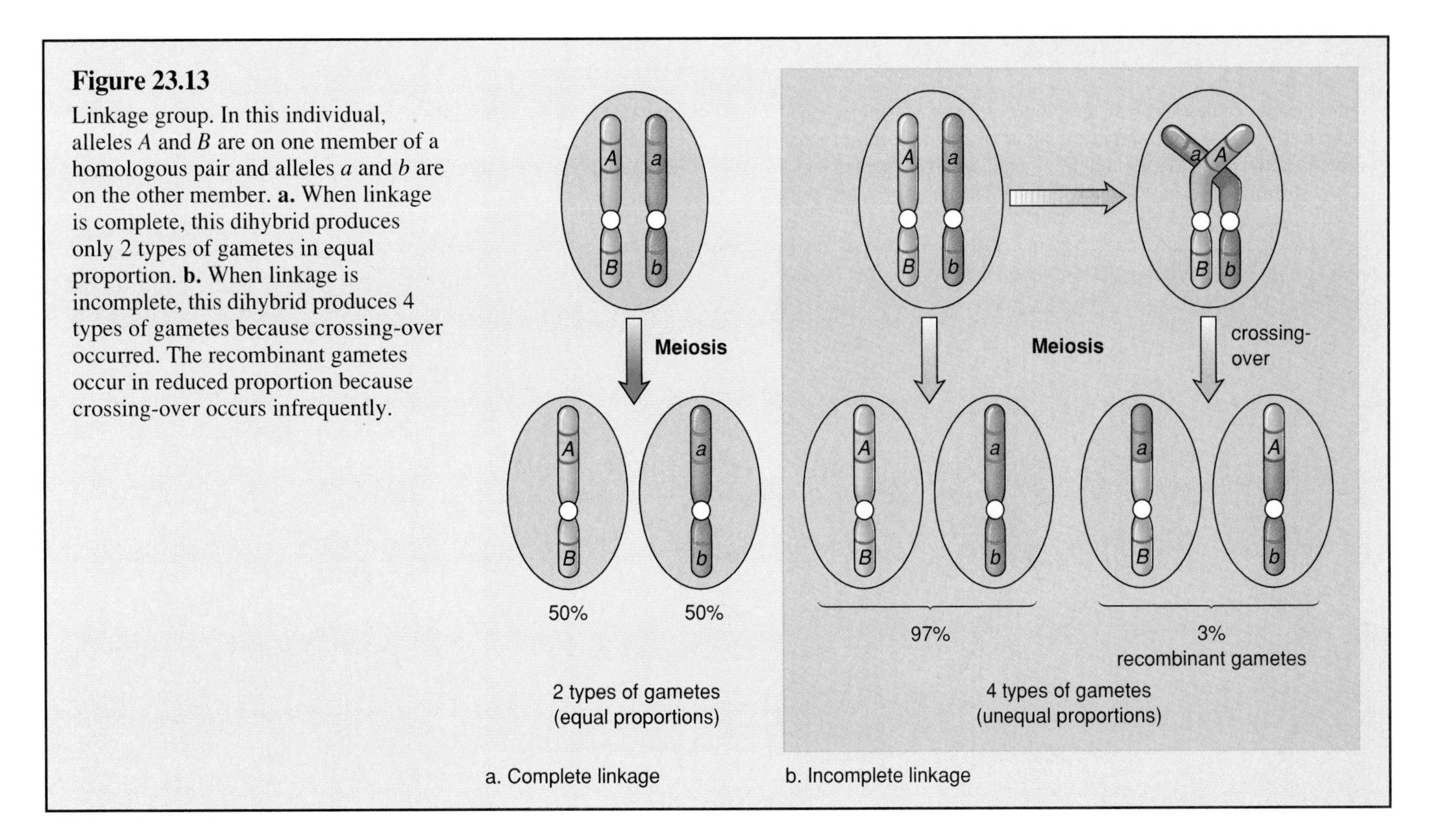

Figure 23.13
Linkage group. In this individual, alleles *A* and *B* are on one member of a homologous pair and alleles *a* and *b* are on the other member. **a.** When linkage is complete, this dihybrid produces only 2 types of gametes in equal proportion. **b.** When linkage is incomplete, this dihybrid produces 4 types of gametes because crossing-over occurred. The recombinant gametes occur in reduced proportion because crossing-over occurs infrequently.

had these chromosomes, and the results of their mating were predicted as follows:

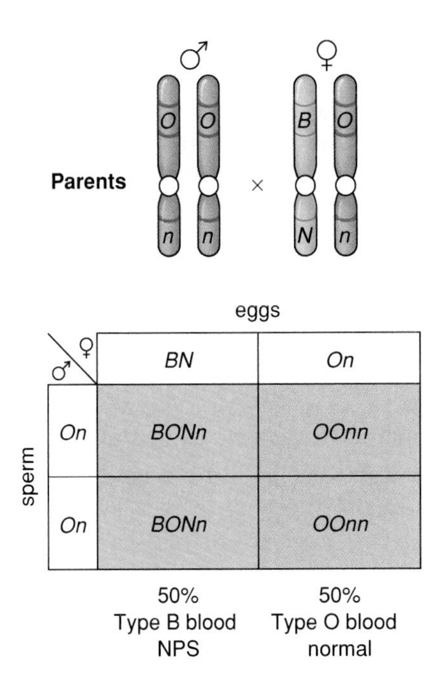

Linkage was not complete, however, and 10% of the children had recombinant phenotypes: 5% had type B blood and no NPS, and 5% had type O blood and NPS. To explain these results, we need only realize that these children received a recombinant chromosome from their mother:

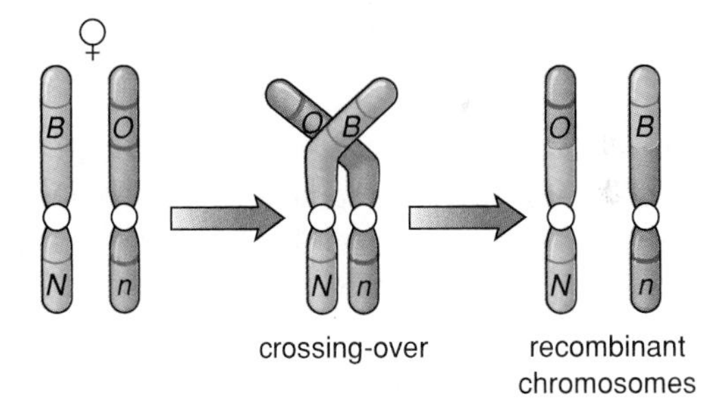

Crossing-over occurs more often between distant genes than between genes that are close together on a chromosome. For example, consider these homologous chromosomes:

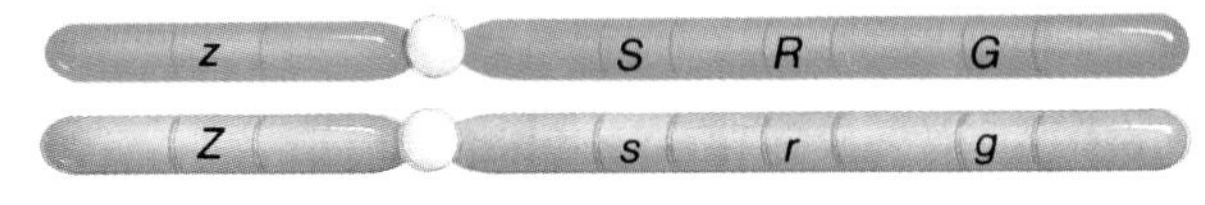

pair of homologous chromosomes

We expect recombinant gametes to include *G* and *z* more often than *R* and *s*. In fact, recombination frequencies are used to map the chromosomes. Each 1% of recombination frequency is taken to be equivalent to one map unit.

Figure 23.14

Human-mouse cell hybrids. In the presence of a fusing agent, human fibroblast cells sometimes join with mouse tumor cells to give hybrid cells having nuclei that contain both types of chromosomes. Subsequent cell division of the hybrid cell produces clones that have lost most of their human chromosomes, enabling the investigator to study these chromosomes separate from all other human chromosomes.

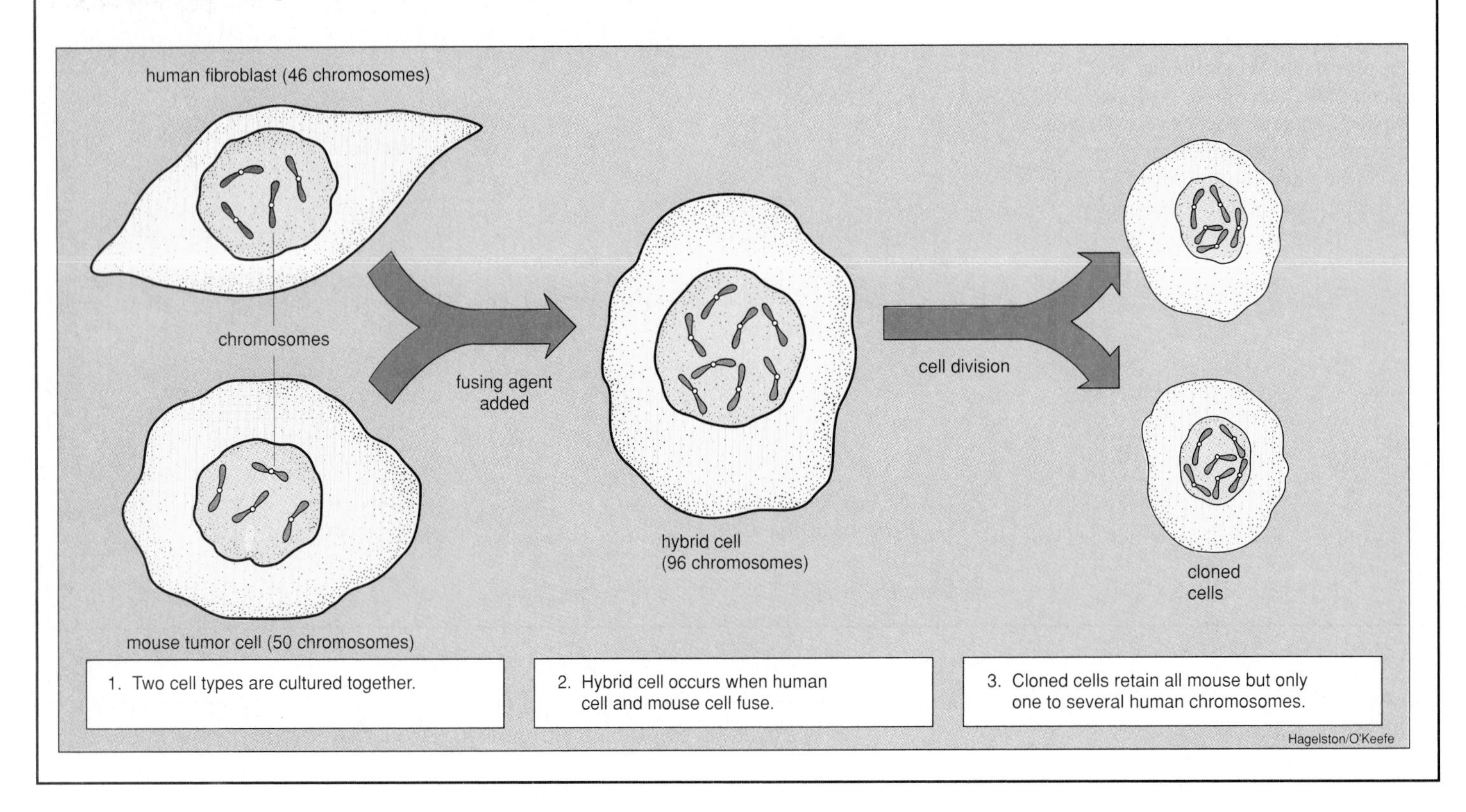

Linkage data have been used to map the chromosomes of the fruit fly *Drosophila,* but the possibility of using linkage data to map human chromosomes is limited because we can only work with matings that have occurred by chance. And this, coupled with the fact that humans tend to produce limited numbers of offspring, means that other methods have provided more extensive data for mapping human chromosomes.

> The presence of linkage groups changes the expected results of genetic crosses. The frequency of recombinant gametes that occur due to the process of crossing-over has been used to map chromosomes.

Human-Mouse Cell Data: Isolating Human Chromosomes

Human and mouse cells are mixed together in a laboratory dish, and in the presence of inactivated virus of a special type, they fuse (fig. 23.14). As the cells grow and divide, some of the human chromosomes are lost, and eventually the daughter cells contain only a few human chromosomes, each of which can be recognized by their distinctive banding pattern (see fig. 7.1). Analysis of the proteins made by the various human-mouse cells enables scientists to determine which genes to associate with which human chromosomes.

Sometimes it is possible to obtain a human-mouse cell that contains only one human chromosome or even just a portion of a chromosome. This technique has been very helpful to researchers who have been studying the genes located on chromosome 21 (p. 444).

Genetic Marker Data: Locating Particular Genes

The DNA that a person inherits from a parent has a unique sequence of base pairs. A genetic marker is a place on the chromosome where the sequence of base pairs differs from one person to another. These differences are usually seen in "filler DNA," sequences of bases between the genes.

Genetic markers are discovered by using special enzymes (called restriction enzymes, p. 489) that cleave DNA molecules into fragments. Because the sequence of bases differs among individuals, each person produces a particular pattern of different-length fragments. Scientists say that restric-

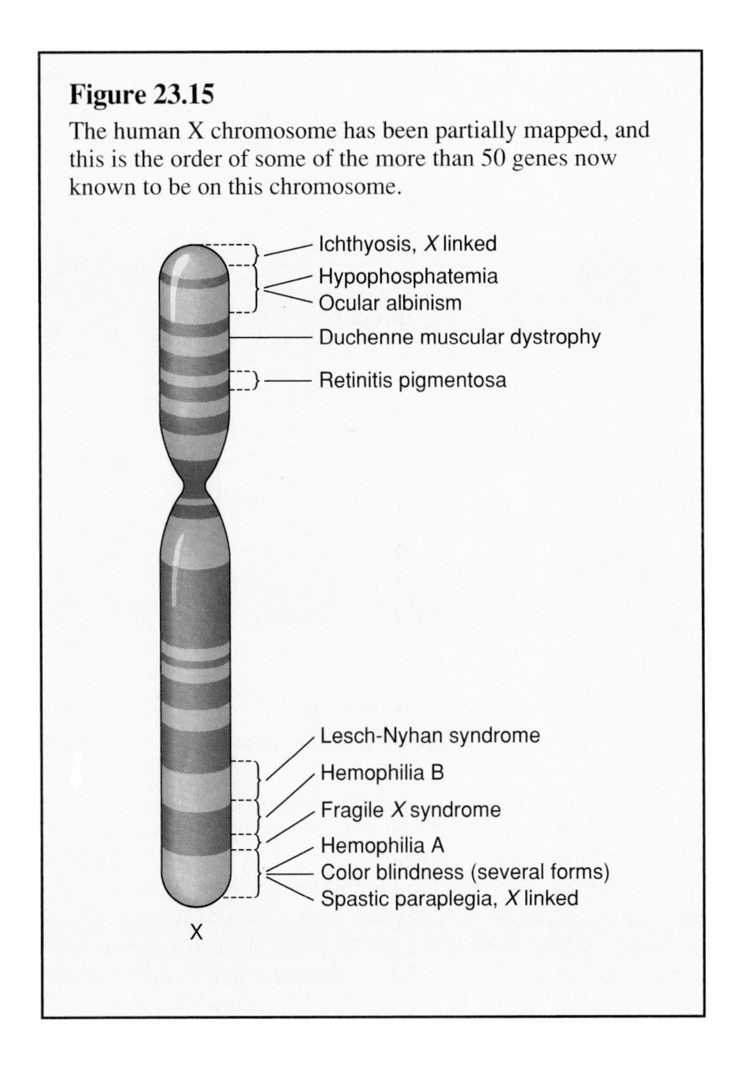

Figure 23.15
The human X chromosome has been partially mapped, and this is the order of some of the more than 50 genes now known to be on this chromosome.

tion fragment length polymorphisms, or *RFLPs* (pronounced "riflips"), are observed. *Polymorphism* means *many changes in structure*; in this case, polymorphisms exist in the structure of the chromosomes taken from different individuals.

Genes can be assigned a location on a chromosome according to their relative relationship to genetic markers. Therefore, the markers have helped scientists develop maps of the human chromosomes (fig. 23.15).

Genetic markers can also be used to test for a genetic disorder when a particular marker is always inherited with a particular genetic disorder. Individuals are tested for the presence of the marker instead of the specific faulty allele. For a marker to be dependable, it should be inherited with the faulty allele at least 98% of the time. The available tests for sickle-cell disease, Huntington disease, and Duchenne muscular dystrophy are all based on the presence of a marker.

Because each person has a different pattern of fragments, RFLP analysis is called the *DNA fingerprint* of a person. The DNA fingerprint can be used to match a child to his or her parents because RFLPs are inherited according to Mendel's laws.

Figure 23.16
Sequencing DNA using gel electrophoresis. The DNA is cleaved into fragments that vary in length by only one nucleotide. Each fragment ends with an *A, T, C,* or *G* marked by one of 4 fluorescent dyes. The fragments are subjected to gel electrophoresis, during which they migrate on a gel in an electric field. Shorter fragments migrate farther than longer fragments. Under proper lighting, the investigator (or a computer) can simply read off the sequence of the nucleotides.

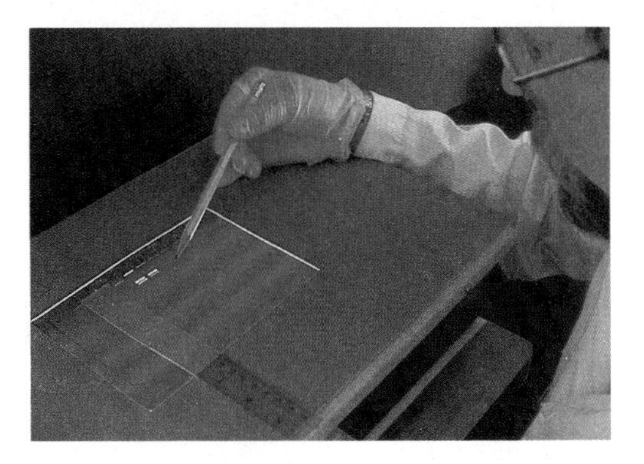

It can also be used to solve crimes because the fingerprint of a criminal matches that of any blood or tissue left at the scene of the crime (see figure 25.6).

Human Genome Project: Locating Genes and Bases

The goal of the *Human Genome Project* is to identify the location of the approximately 100,000 human genes on all the chromosomes. In order to create this genetic map, genetic markers will be used to index the chromosomes. Known and newly discovered genes will be assigned locations between the markers.

Eventually, those involved in the project also want to determine the sequence of the 3 billion bases in the human genome. In order to create this physical map, researchers will use laboratory procedures to determine the sequence of the DNA bases (fig. 23.16).

It has recently been discovered that genetic markers contain unique stretches of DNA called sequence-tagged sites, or STSs. Hopefully, these can be used to create links between the genetic map and the physical map.

The human genome project will require millions of dollars and many years to complete. Just how successful and useful it will be cannot be determined yet.

> Two new techniques—human-mouse cell preparations and genetic markers—now make it possible to map the human chromosomes at a faster rate than was formerly possible.

SUMMARY

Humans inherit 22 autosomes and one sex chromosome from each parent. Nondisjunction during meiosis can cause an abnormal number of autosomes to be inherited. Down syndrome results when an individual inherits 3 copies of chromosome 21. Chromosome mutations of the following types lead to phenotypic abnormalities: For example, in cri du chat syndrome, one copy of chromosome 5 has a deletion.

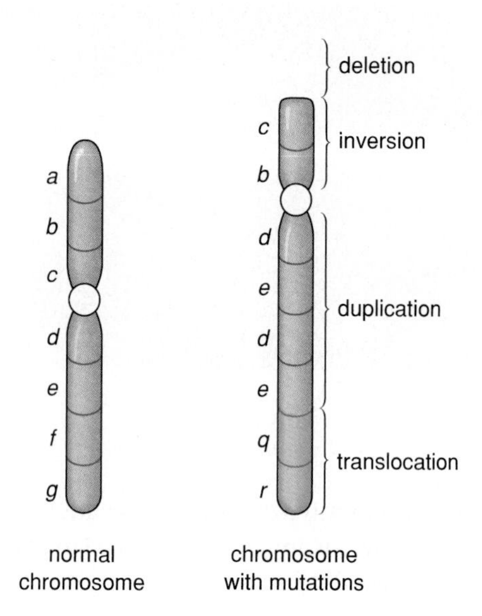

The father determines the sex of a child because the mother gives only an X chromosome while the father gives an X or a Y chromosome. Individuals with a broken X have fragile X syndrome, a chromosomal cause of mental retardation second only to Down syndrome. Nondisjunction of the sex chromosomes can also cause abnormal sex chromosome numbers in offspring. Females who are XO have Turner syndrome, and those who are XXX are metafemales. Males with Klinefelter syndrome are XXY. There are also XYY males.

Because males normally receive only one X chromosome, they are subject to disorders caused by the inheritance of a recessive allele on the X chromosome. For example, in a cross between a normal male and a carrier female, only the males have the X-linked disorder color blindness:

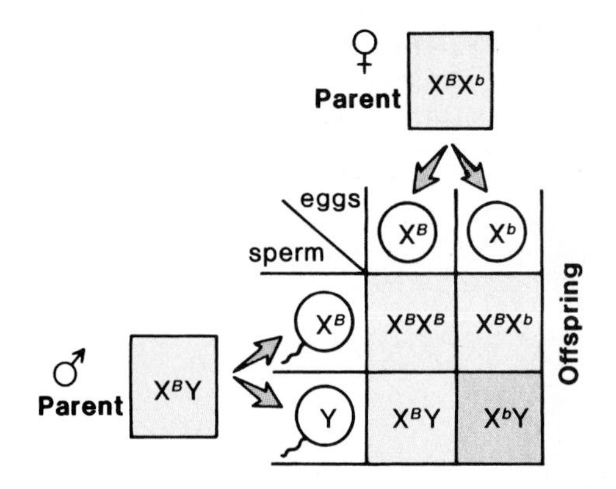

Phenotypic Ratio: females—all normal
males—1 normal:1 color blind

Other well-known X-linked disorders are hemophilia and Duchenne muscular dystrophy.

All the genes on one chromosome form a linkage group, which is broken only when crossing-over occurs. Genes that are linked do not follow Mendel's law of independent assortment because they tend to go together into the same gamete. If crossing-over occurs, a dihybrid cross gives all possible phenotypes among the offspring, but the expected ratio is greatly changed. Linkage data contribute to our knowledge of the location of genes on chromosomes. The frequency of recombinant phenotypes helps determine the order of genes since genes that are distant from one another tend to cross over more often than those that are close. Human-mouse cell data and genetic marker data are new ways to help map the human chromosomes.

STUDY QUESTIONS

In order to practice **writing across the curriculum,** students should write out the answers to any or all of the study questions. The study questions are sequenced in the same order as the text.

1. What is the normal chromosome inheritance of a human? (p. 444)
2. What is the most common autosomal abnormality seen in humans? What causes this abnormality? (p. 444)
3. Diagram how nondisjunction can occur during meiosis I and during meiosis II. (p. 444)
4. Name and describe 4 chromosome mutations. (p. 446)
5. Describe Turner syndrome, Klinefelter syndrome, metafemales, and XYY males. (p. 450)
6. Name 4 ways to recognize an X-linked recessive disorder. Why do males exhibit such disorders more often than females? (p. 452)
7. Explain the occurrence of sex-influenced traits. How do they differ from sex-linked traits? (p. 454)
8. What is a linkage group, and how can the occurrence of linkage groups help to map the human chromosomes? (pp. 454–456)
9. How do human-mouse cell data and genetic marker data help to map the human chromosomes? (p. 456)

OBJECTIVE QUESTIONS

1. An XXY individual has _______ syndrome.
2. A(n) _______ is the movement of a chromosome segment from one chromosome to another, nonhomologous chromosome.
3. Whereas females are XX, males are _______.
4. A female who is a carrier for color blindness has the genotype _______.
5. A male who is X^BY _______ (is, is not) color blind.
6. Genes *ABCDE* are all on the same chromosome. They are part of a(n) _______ group.
7. Among the genes listed in question 6, *AE* are _______ (more likely, less likely) to cross over than *CD*.
8. Genetic markers sometimes can be used to test an individual for a(n) _______.

ADDITIONAL GENETICS PROBLEMS

1. In fruit flies, X^R = red eye and X^r = white eye. a. If a white-eyed male reproduces with a homozygous red-eyed female, what phenotypic ratio is expected for males? for females? b. If a white-eyed female reproduces with a red-eyed male, what phenotypic ratio is expected for males? for females? (p. 451)
2. A woman who is homozygous dominant for widow's peak and is a carrier for color blindness reproduces with a man who is heterozygous for widow's peak and has normal vision. What are the genotypes of these parents? If a son is born, what are the chances he will be color blind with a widow's peak? (p.451)
3. In fruit flies, gray body (*G*) is dominant over black body (*g*). A female fly heterozygous for both gray body and red eyes reproduces with a red-eyed male heterozygous for gray body. What phenotypic ratio is expected for males? for females? (p. 451)
4. John is the only member of his family with hemophilia. What are the chances that a newborn brother will also be a hemophiliac? (p. 452)
5. Give 2 reasons for deciding that this is a pedigree chart for an X-linked trait. What is the genotype of the starred individual? (p. 452)

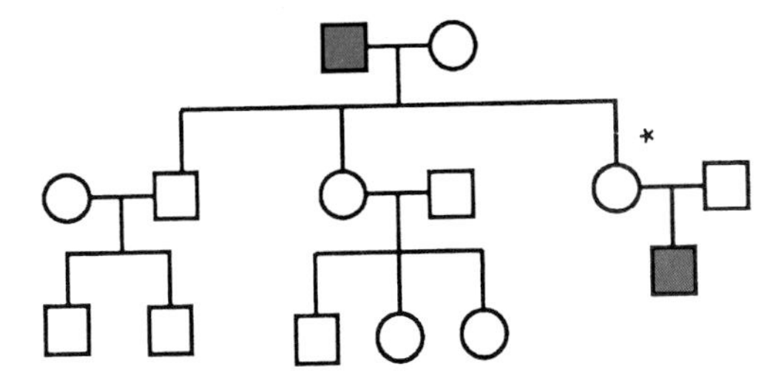

6. Imagine that the ability to curl the tongue is dominant and that this characteristic is linked to a rare form of mental retardation, which is also dominant. The parents are both dihybrids, with the 2 dominant alleles on one chromosome and the 2 recessive alleles on the other. What phenotypic ratio is expected among the offspring if crossing-over does not occur? (p. 454)

CRITICAL THINKING

In order to practice **writing across the curriculum,** students should write out the answers to any or all of the critical thinking questions. Suggested answers to the critical thinking questions are in appendix E.

23.1

1. Early in this century, geneticists performed this cross in *Drosophila* (fruit flies):

	♀		♂
P	Red-eyed	×	White-eyed
F_1	Red-eyed		Red-eyed

From these results, they knew which characteristic is dominant.

2. They went on to perform this cross:

	♀		♂
$F_1 \times F_1$	Red-eyed	×	Red-eyed
F_2	Red-eyed		1 : 1 red- to white-eyed

Are these results explainable if the allele for red/white eye color is on the Y chromosome but not on the X chromosome? on the X chromosome but not on the Y chromosome? Explain.

3. How do these results support the hypothesis that genes are on the chromosomes?

SELECTED KEY TERMS

chromosome mutation a variation in regard to the normal number of chromosomes inherited or in regard to the normal sequence of alleles on a chromosome; the sequence can be inverted, translocated from a nonhomologous chromosome, deleted, or duplicated.

Down syndrome human congenital disorder associated with an extra chromosome 21.

Klinefelter syndrome a condition caused by the inheritance of a chromosome abnormality in number; an XXY individual.

linkage group alleles on the same chromosome are linked in the sense that they tend to move together to the same gamete; crossing-over interferes with linkage.

metafemale a female who has 3 X chromosomes.

sex-linked gene gene located on the sex chromosomes.

Turner syndrome a condition caused by the inheritance of an abnormality in chromosome number; an X chromosome lacks a homologous counterpart—XO.

X-linked gene allele is located on the X chromosome.

XYY male a male who has an extra Y chromosome.

24

THE MOLECULAR BASIS OF INHERITANCE

Chapter Concepts

1.
DNA is the genetic material, and therefore its structure and functions constitute the molecular basis of inheritance. 463, 483

2.
DNA is able to replicate, to mutate, and to control the phenotype of the cell and the organism. 465, 483

3.
DNA directs protein synthesis, a process that also requires the participation of RNA. 468, 483

4.
Regulator genes control the activity (expression) of other genes. 475, 483

5.
Gene mutations range from those that have little effect to those that have an extreme effect. 478, 483

6.
Cancer develops when there is a loss of genetic control over genes involved in cell growth and/or cell division. 480, 483

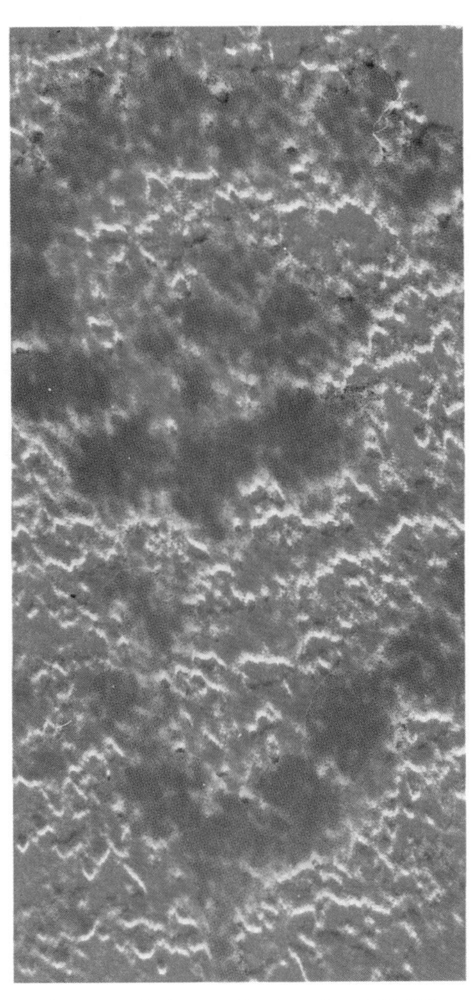

Polysome in human brain cell, ×240,000

Chapter Outline

Figure 24.1
DNA location and structure. DNA is highly compacted in chromosomes, but it is extended as chromatin during interphase. It is during interphase that DNA can be extracted from a cell and its structure and function studied.

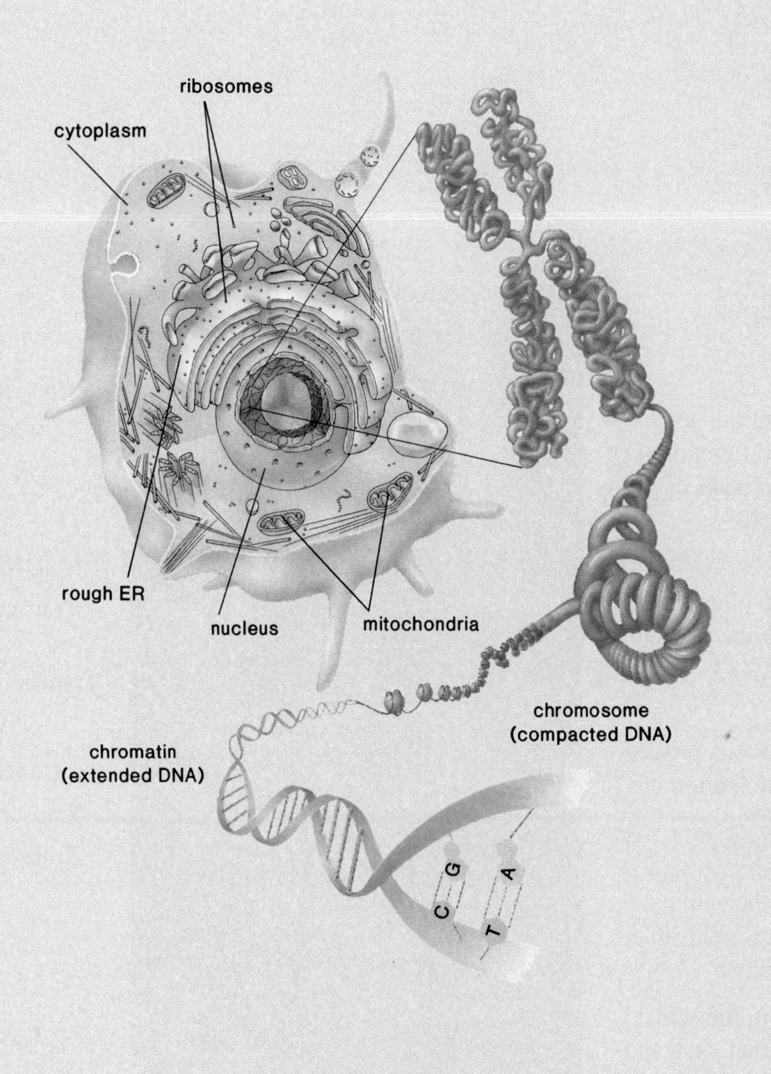

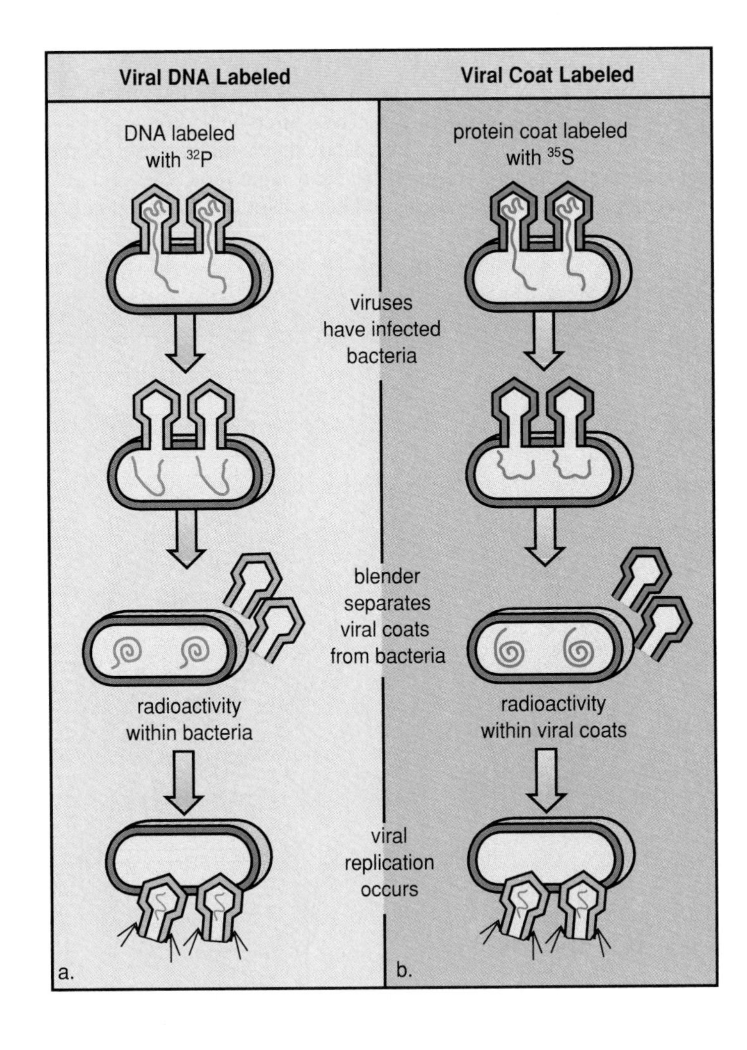

Figure 24.2
DNA, the genetic material. T_2 virus is composed of a DNA core and protein coat only. Whichever of these enters a bacterium and controls viral replication is the genetic material. **a.** In this experiment, viral DNA is labeled with ^{32}P. The radioactively labeled DNA enters the cell, and the coats are removed by agitation in a blender. Replication proceeds normally, and some of the new viruses are labeled. Therefore, DNA is the genetic material. **b.** In this experiment, the protein coat is labeled with ^{35}S. The radioactively labeled protein coats are removed when the cells are agitated in a blender. Replication proceeds normally, and the new viruses are not labeled. Therefore, protein is not the genetic material.

WHEN the sperm fertilizes the egg, a new individual comes into being. Each of the gametes contributes genes, which direct the function of the individual from conception to death. The genes are on the chromosomes in the nucleus of a cell (fig. 24.1), but of what are the genes composed?

DNA Structure and Functions

In the mid-1900s, scientists knew that the genes are on the chromosomes and that chromosomes contain both DNA and protein; they were uncertain, however, which of these was the genetic material. They turned to experiments with viruses to resolve this question because they knew that viruses are tiny particles having just 2 parts: an inner nucleic acid core and an outer protein coat.

They chose to work with a virus called a T_2 virus (the T_2 simply means *type 2*), which infects bacteria. They wanted to determine which part of a T_2 virus, the inner DNA core or the outer protein coat, enters a bacterium and takes over its machinery so that it produces more viruses. They prepared 2 batches of viruses. One batch had ^{32}P-labeled DNA, and the other had ^{35}S-labeled protein. Then they performed 2 experiments. In one experiment, bacteria were exposed to the viruses with labeled DNA, and in the other, bacteria were exposed to the viruses with labeled protein (fig. 24.2). They found that only labeled DNA enters a bacterium. Therefore, only DNA is needed for the reproduction of viruses, and only DNA is the genetic material.

Figure 24.3
Overview of DNA structure. **a.** The double-helix structure is a (**b**) twisted ladder. **c.** Unwound DNA shows that the sides of the ladder are composed of phosphate and sugar molecules and the rungs are complementary-paired bases. Notice that the strands of DNA run antiparallel to one another with the 5′ end (5-prime end) of one strand opposite the 3′ end (3-prime end) of the other strand (see fig. 24.4).

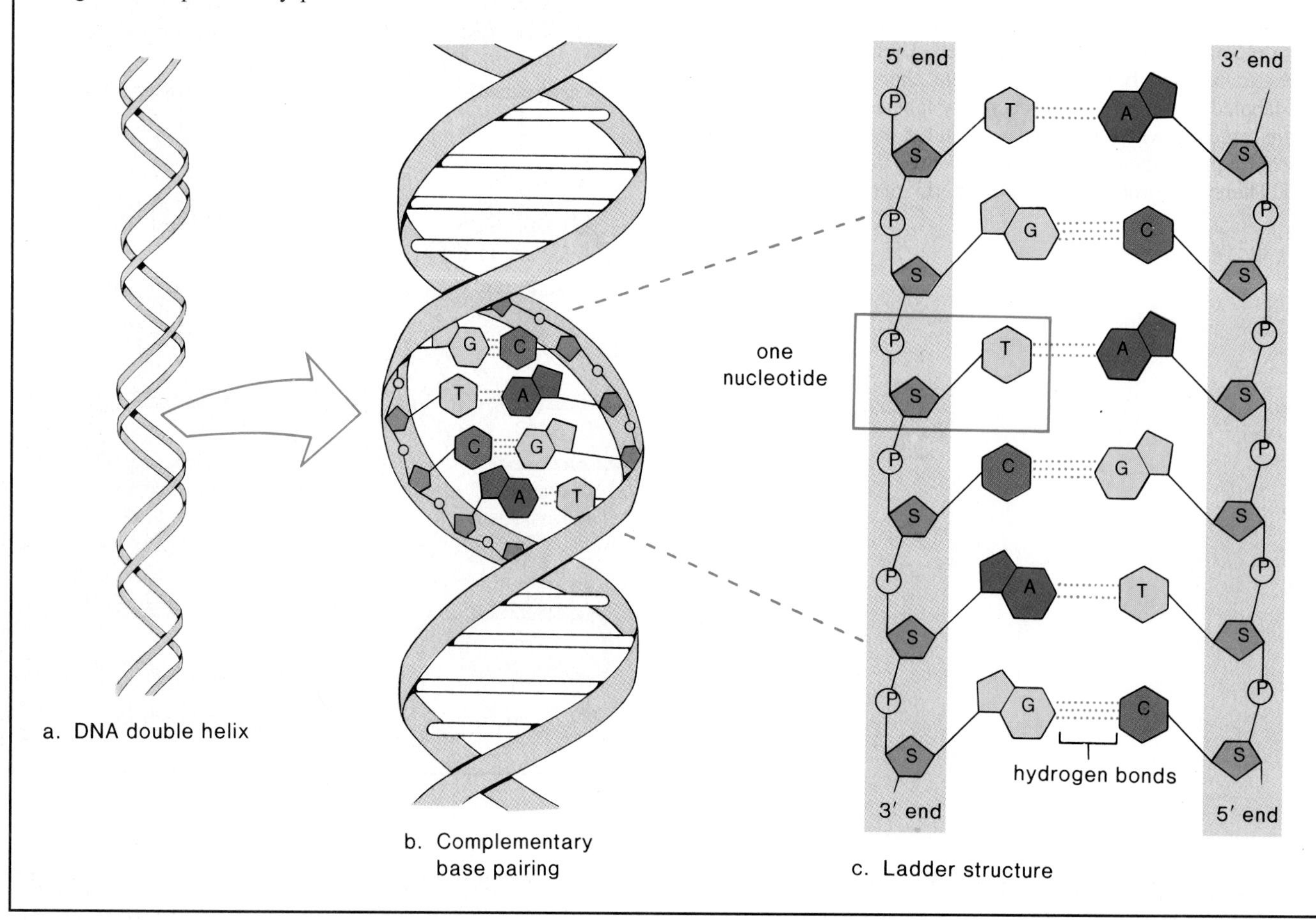

DNA Structure: Double Helix

DNA (deoxyribonucleic acid) is a polynucleotide; each nucleotide is a complex of 3 subunits—phosphoric acid (phosphate), a pentose sugar (deoxyribose), and a nitrogen-containing, organic base. There are 4 possible bases: 2 are **purines** with a double ring, and 2 are **pyrimidines** with a single ring (see fig. 2.29). The names of the bases are as follows:

Purines	**Pyrimidines**
Adenine (A)	Thymine (T)
Guanine (G)	Cytosine (C)

A polynucleotide *strand* has a backbone made up of alternating phosphate and sugar molecules. The bases are attached to the sugar but project to one side (see fig. 2.30). DNA has 2 such strands, and the 2 strands twist about one another in the form of a **double helix** (fig. 24.3*a* and *b*). The strands are held together by hydrogen bonding between the bases: A pairs with T, and G pairs with C, or vice versa. This is called **complementary base pairing.**

When the DNA helix unwinds, it resembles a ladder (fig. 24.3*c*). The sides of the ladder are the phosphate-sugar backbones, and the rungs of the ladder are the complementary-paired bases (fig. 24.4). The bases can be in any set order, and the number of any particular base pair can vary, but even so a purine is bonded to a pyrimidine. Therefore, the number of purines equals the number of pyrimidines.

> DNA is a double helix with phosphate-sugar backbones on the outside and paired bases on the inside. Complementary base pairing occurs: adenine (A) pairs with thymine (T) and guanine (G) pairs with cytosine (C).

The structure of DNA was determined by James Watson and Francis Crick in the early 1950s. The data they used and how they used the data to deduce DNA's structure are reviewed in the reading on page 466.

Figure 24.4
Complementary base pairing. Because a purine (A or G) pairs with a pyrimidine (T or C), the number of purines equals the number of pyrimidines in DNA. Also, the bases pair in such a way that the phosphate-sugar groups are oriented in different directions. This means that the strands of DNA end up running antiparallel to one another, with the 3′ end of one strand opposite the 5′ end of the other strand (see fig. 24.3*c*).

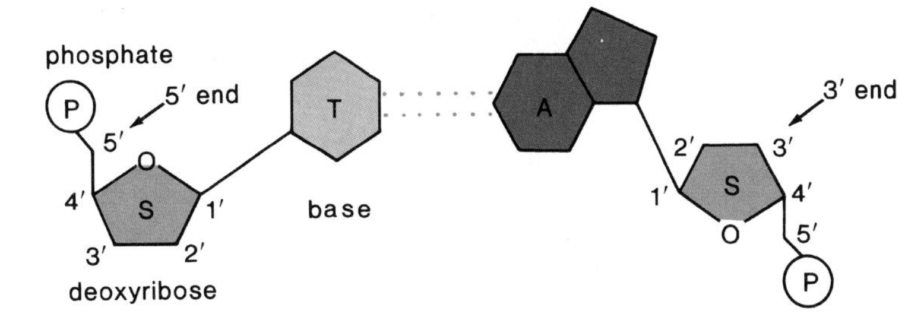

1. Thymine (T) is paired with adenine (A).

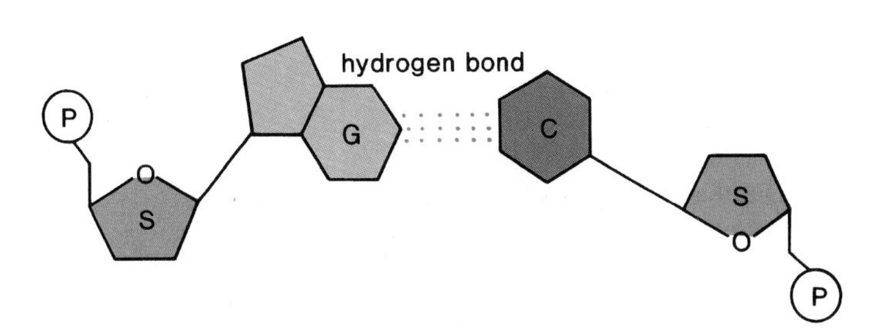

2. Guanine (G) is paired with cytosine (C).

Figure 24.5
Structure of RNA. Unlike DNA, RNA is single stranded, the backbone contains the sugar ribose instead of deoxyribose, and the bases are guanine (G), uracil (U), cytosine (C), and adenine (A).

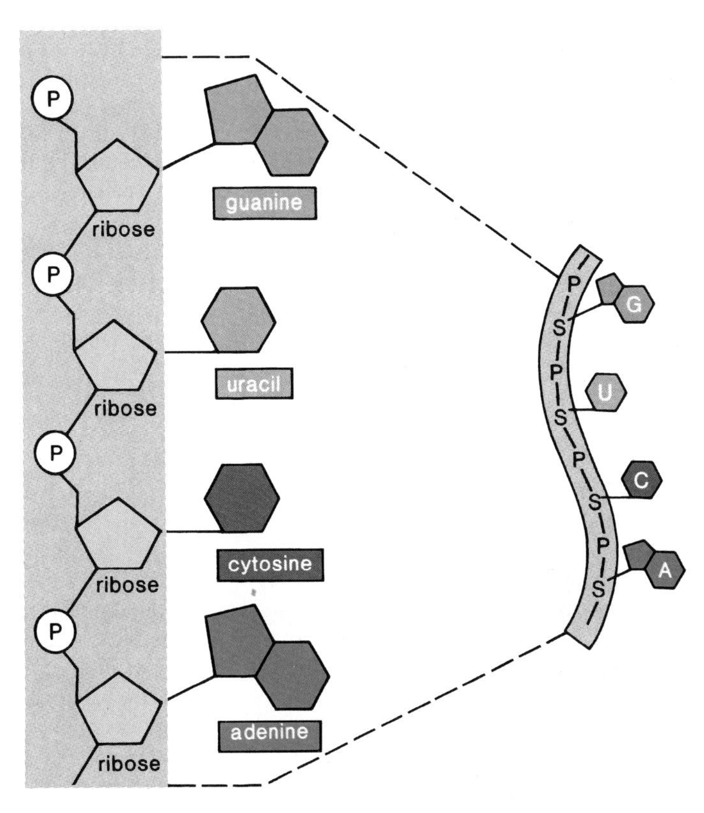

DNA versus RNA Structure

Like DNA, **RNA (ribonucleic acid)** is a polynucleotide (fig. 24.5). However, RNA contains the sugar ribose, not deoxyribose. Also, the pyrimidine thymine does not appear in RNA; it is replaced by the pyrimidine uracil. In RNA, adenine (A) pairs with uracil (U) and guanine (G) pairs with cytosine (C). Finally, RNA is single stranded and does not form a double helix like DNA does (table 24.1).

There are 3 different classes of RNA: messenger RNA (mRNA), ribosomal RNA (rRNA), and transfer RNA (tRNA). Each of these has specific functions during protein synthesis.

Table 24.1
DNA Structure Compared to RNA Structure

	DNA	RNA
Sugar	Deoxyribose	Ribose
Bases	Adenine, guanine, thymine, cytosine	Adenine, guanine, uracil, cytosine
Strands	Double stranded with base pairing	Single stranded
Helix	Yes	No

DNA Functions: At Least Three

DNA is the hereditary material. As such, it must

1. *replicate* with high accuracy prior to cell division when a copy is distributed to daughter cells. It must also be transmitted in the gametes from one generation to the next
2. *undergo rare changes* called **mutations.** Mutations provide genetic variability and are the raw materials for evolutionary change
3. *store information* that controls both the development and the metabolic activities of the cell and the organism.

DNA, the genetic material, must replicate with few errors, undergo mutations, and store information.

SOLVING THE DNA PUZZLE

In 1951, James Watson, an American biologist, began an internship at the University of Cambridge, England. There he met Francis Crick, an English physicist, who was interested in molecular structures. Together they set out to determine the structure of DNA and to build a model that would explain how DNA, the genetic material, varies from species to species and even from individual to individual. They also discovered the way DNA replicates (makes a copy of itself) so that daughter cells and offspring can receive a copy.

The bits and pieces of data available to Watson and Crick were like puzzle pieces they had to fit together. This is what they knew from the research of others:

1. DNA is a polymer of nucleotides, each one having a phosphate group, the sugar deoxyribose, and a nitrogen-containing, organic base. There are 4 types of nucleotides because there are 4 different bases: adenine (A) and guanine (G) are purines, while cytosine (C) and thymine (T) are pyrimidines.
2. A chemist, Erwin Chargaff, had determined in the late 1940s that regardless of the species under consideration, the number of purines in DNA always equals the number of pyrimidines. Further, the amount of adenine equals the amount of thymine (A = T), and the amount of guanine equals the amount of cytosine (G = C). These findings came to be known as *Chargaff's rules*.
3. Rosalind Franklin and Maurice Wilkins, working at King's College, London, had just prepared an X-ray diffraction photograph (fig. 24.A) of DNA. It showed that DNA is a double helix of constant diameter and that the bases are regularly stacked on top of one another.

Figure 24.A
X-ray diffraction photograph of DNA taken by Rosalind Franklin. The crossing pattern of dark spots in the center of the picture indicates that DNA is helical. The dark regions at the top and bottom of the photograph show that base pairs are stacked on top of one another.

Figure 24.B
A portion of the actual wire and tin model constructed by Watson and Crick.

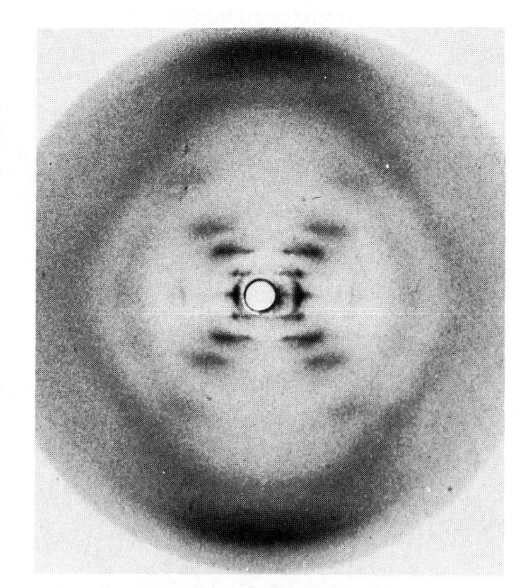

A.

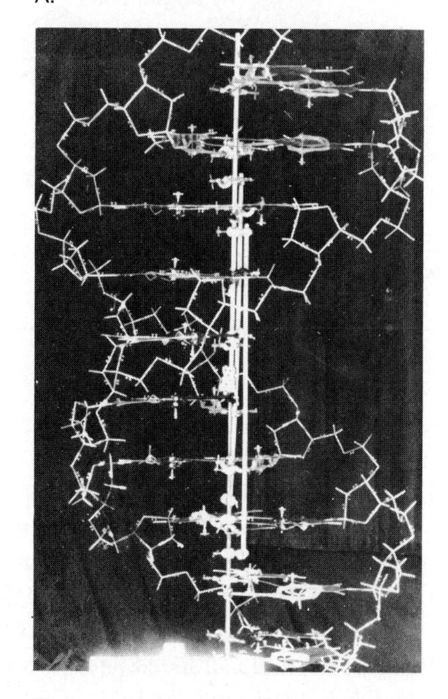

B.

Using these data, Watson and Crick deduced that DNA has a twisted, ladder-type structure; the sugar-phosphate molecules make up the sides of the ladder, and the bases make up the rungs. Further, they determined that if A is normally hydrogen bonded with T and G is normally hydrogen bonded with C (in keeping with Chargaff's rules), then the rungs always have a constant width (as required by the X-ray photograph).

Watson and Crick built an actual model of DNA out of wire and tin (fig. 24.B). This double-helix model does indeed allow for differences in DNA structure between species because the base pairs can be in any order. Also, the model suggests that complementary base pairing plays a role in the replication of DNA. As Watson and Crick pointed out in their original paper, "It has not escaped our notice that the specific pairing we have postulated immediately suggests a possible copying mechanism for the genetic material."

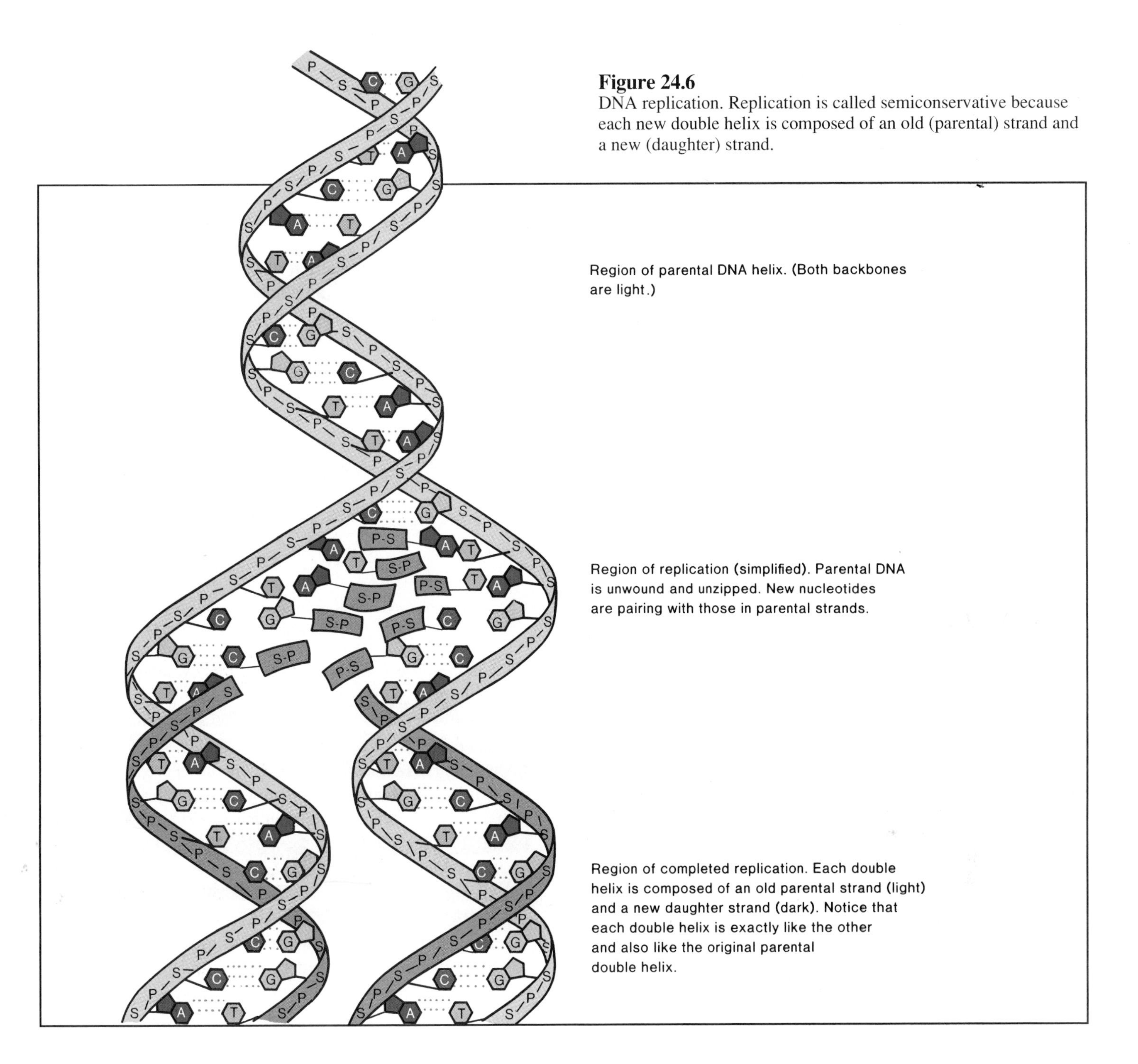

Figure 24.6
DNA replication. Replication is called semiconservative because each new double helix is composed of an old (parental) strand and a new (daughter) strand.

DNA Replication: Unzipping and Molding

The double-stranded structure of DNA aids replication because each strand can serve as a template for the formation of a complementary strand. A **template** is most often a mold used to produce a shape opposite to itself. In this case, each old (parental) strand is a template for each new (daughter) strand.

Replication has the following steps (fig. 24.6):

1. The 2 strands that make up DNA unwind and "unzip" (i.e., the weak hydrogen bonds between the paired bases break). A special enzyme called helicase causes the molecule to unwind.
2. New complementary nucleotides, always present in the nucleus, move into place by the process of complementary base pairing.
3. The complementary nucleotides are joined so that the new DNA molecule is again double stranded. This step is carried out by the enzyme **DNA polymerase.**

Because each old strand has produced a new strand through complementary base pairing, there are now 2 DNA helices identical to each other and to the original molecule.

This is an overview of DNA replication. DNA replication is actually an extremely complicated process involving many more steps and enzymes than those we have discussed here.

Replicating Semiconservatively. DNA replication is termed *semiconservative* because each new double helix has one old strand and one new strand. In other words, one of the parental strands is conserved, or present, in each new double helix.

During DNA replication, DNA unwinds and unzips, and a new strand that is complementary to each original strand forms. This is called semiconservative replication.

Replicating Accurately. The bases seldom pair incorrectly with one another as replication proceeds. The error rate is minimized because DNA polymerase has a "proofreading" function. It checks each pairing as soon as it occurs, and if it finds that a mistake has been made, it removes the incorrect nucleotide and replaces it with a correct one. If an error is not corrected, a gene mutation has occurred.

DNA Controls the Cell

The occurrence of inherited metabolic disorders first suggested that DNA controls the metabolism of the cell. In phenylketonuria (PKU) (p. 434), mental retardation is caused by the inability to convert phenylalanine to tyrosine. In albinism (p. 432), there is no natural pigment in the skin because tyrosine cannot be converted to melanin. Each condition is caused by the inheritance of a faulty enzyme:

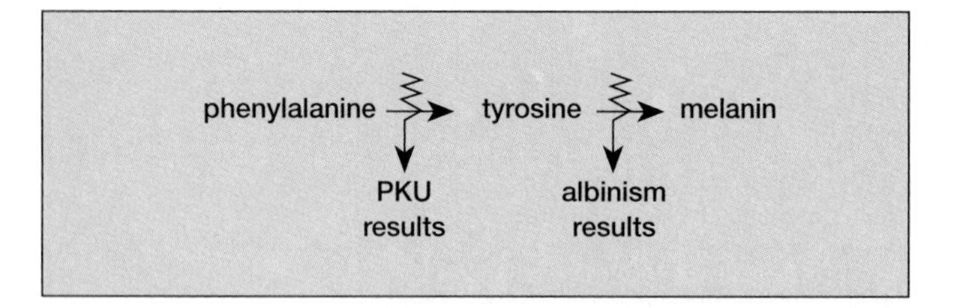

Even in the 1900s, these conditions were called *inborn errors* of metabolism.

Later laboratory investigations with the bread mold *Neurospora* led to the one gene–one enzyme hypothesis, which stated that each gene controls the production of a specific enzyme. This hypothesis was later broadened to the one gene–one protein hypothesis because not all proteins are enzymes—some are structural components of the cell. When it was pointed out that some proteins have more than one polypeptide, the hypothesis was finally modified to the one gene–one polypeptide hypothesis. A gene is a section of a DNA molecule that determines the sequence of amino acids in a polypeptide of a protein.

A gene (DNA) controls the primary structure (the sequence of amino acids) of a polypeptide. In this way, genes control the structure and the metabolism of the cell. Do 24.1 Critical Thinking, found at the end of the chapter.

Protein Synthesis

DNA controls the production of proteins even though in eukaryotes it is located in the nucleus and proteins are synthesized at the ribosomes in the cytoplasm (fig. 24.1). RNA, however, is not confined to the nucleus; it occurs in both the nucleus and the cytoplasm. The central dogma of modern genetics recognizes that DNA relies on RNA to control protein synthesis:

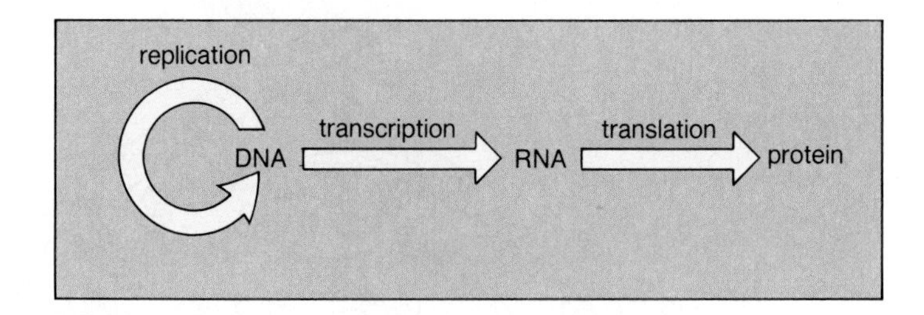

DNA not only serves as a template for its own replication, it is also a template for RNA formation.

Transcription is making an RNA molecule that is complementary to a portion of DNA. Following transcription, RNA moves into the cytoplasm. There are micrographs showing radioactively labeled RNA moving through a nuclear pore to the cytoplasm, where protein synthesis occurs. An RNA called **messenger RNA (mRNA)** carries the instructions for protein synthesis. During **translation,** the instructions—in the form of mRNA base sequences—are used to determine the order of amino acids in a polypeptide (fig. 24.7).

In ordinary speech, transcription means making a close copy of a document, and translation means putting the document in an entirely different language. In genetics, transcription is making a copy of RNA with the same base sequence as DNA; translation is going from a sequence of nucleotides (bases) to a sequence of amino acids.

DNA Base Sequence: Is Coded

Can 4 bases provide enough combinations to code for 20 amino acids? If the code were a doublet (any 2 bases stand for one amino acid), it would not be possible to code for 20 amino acids, but if the code were a triplet, then the 4 bases could supply 64 different triplets, far more than needed to code for 20 different amino acids. It should come as no surprise, then, to learn that the code is a triplet code.

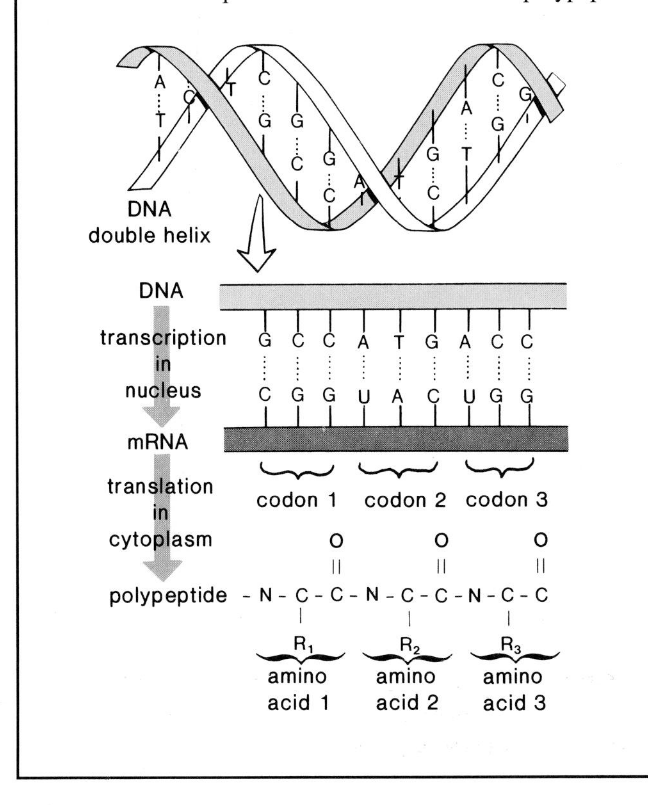

Figure 24.7
Overview of transcription and translation in protein synthesis. Transcription occurs in the nucleus when DNA acts as a template for mRNA synthesis. Translation occurs in the cytoplasm when the sequence of the mRNA codons determines the sequence of the amino acids in a polypeptide.

Figure 24.8
mRNA codons. Most of the 20 amino acids have more than one codon. In this chart, notice that each of the codons comprise a letter from the first, second, and third positions. For example, find the square where C (from the first position) and A (from the second position) come together, and then look across to the right. You will notice the letters for the third position of the codons for histidine and glutamine.

First Base	Second Base				Third Base
	U	C	A	G	
U	UUU phenylalanine	UCU serine	UAU tyrosine	UGU cysteine	U
U	UUC phenylalanine	UCC serine	UAC tyrosine	UGC cysteine	C
U	UUA leucine	UCA serine	UAA *stop*	UGA *stop*	A
U	UUG leucine	UCG serine	UAG *stop*	UGG tryptophan	G
C	CUU leucine	CCU proline	CAU histidine	CGU arginine	U
C	CUC leucine	CCC proline	CAC histidine	CGC arginine	C
C	CUA leucine	CCA proline	CAA glutamine	CGA arginine	A
C	CUG leucine	CCG proline	CAG glutamine	CGG arginine	G
A	AUU isoleucine	ACU threonine	AAU asparagine	AGU serine	U
A	AUC isoleucine	ACC threonine	AAC asparagine	AGC serine	C
A	AUA isoleucine	ACA threonine	AAA lysine	AGA arginine	A
A	AUG (*start*) methionine	ACG threonine	AAG lysine	AGG arginine	G
G	GUU valine	GCU alanine	GAU aspartate	GGU glycine	U
G	GUC valine	GCC alanine	GAC aspartate	GGC glycine	C
G	GUA valine	GCA alanine	GAA glutamate	GGA glycine	A
G	GUG valine	GCG alanine	GAG glutamate	GGG glycine	G

To crack the code, a cell-free experiment was done: artificial RNA was added to a medium containing bacterial ribosomes and a mixture of amino acids. Comparison of the bases in the RNA with the resulting polypeptide allowed investigators to decipher the code. Each three-letter unit of an mRNA molecule is called a **codon.** All 64 mRNA codons have been determined (fig. 24.8). Sixty-one triplets correspond to a particular amino acid; the remaining 3 are stop codons, which signal polypeptide termination. The one codon that stands for the amino acid methionine is also a start codon signaling polypeptide initiation.

DNA Code Is Universal

Research indicates that the genetic code is essentially universal. The same codons stand for the same amino acids in most bacteria, protists, plants, and animals. This illustrates the remarkable biochemical unity of living things and suggests that all living things have a common evolutionary ancestor.

DNA contains a code, and its message is passed to mRNA during transcription. Sixty-one of the 64 triplet codons stand for particular amino acids, and the other 3 codons are stop codons. During translation, the order of the codons in mRNA determines the order of the amino acids in a polypeptide.

Transcription: mRNA Base Sequence Has Codons

During transcription, DNA is a template for the production of mRNA. A segment of the DNA helix unwinds and unzips, and complementary RNA nucleotides pair with DNA nucleotides of one strand. The RNA nucleotides are joined by an enzyme called RNA polymerase, and an mRNA molecule results (fig. 24.9). Following transcription, mRNA has a sequence of bases complementary to DNA; wherever A, T, G, or C is present in the DNA template, U, A, C, or G is incorporated into the mRNA molecule. In this way, the code is transcribed, or copied. Now mRNA has a sequence of triplet codons with bases that are complementary to the DNA triplet code.

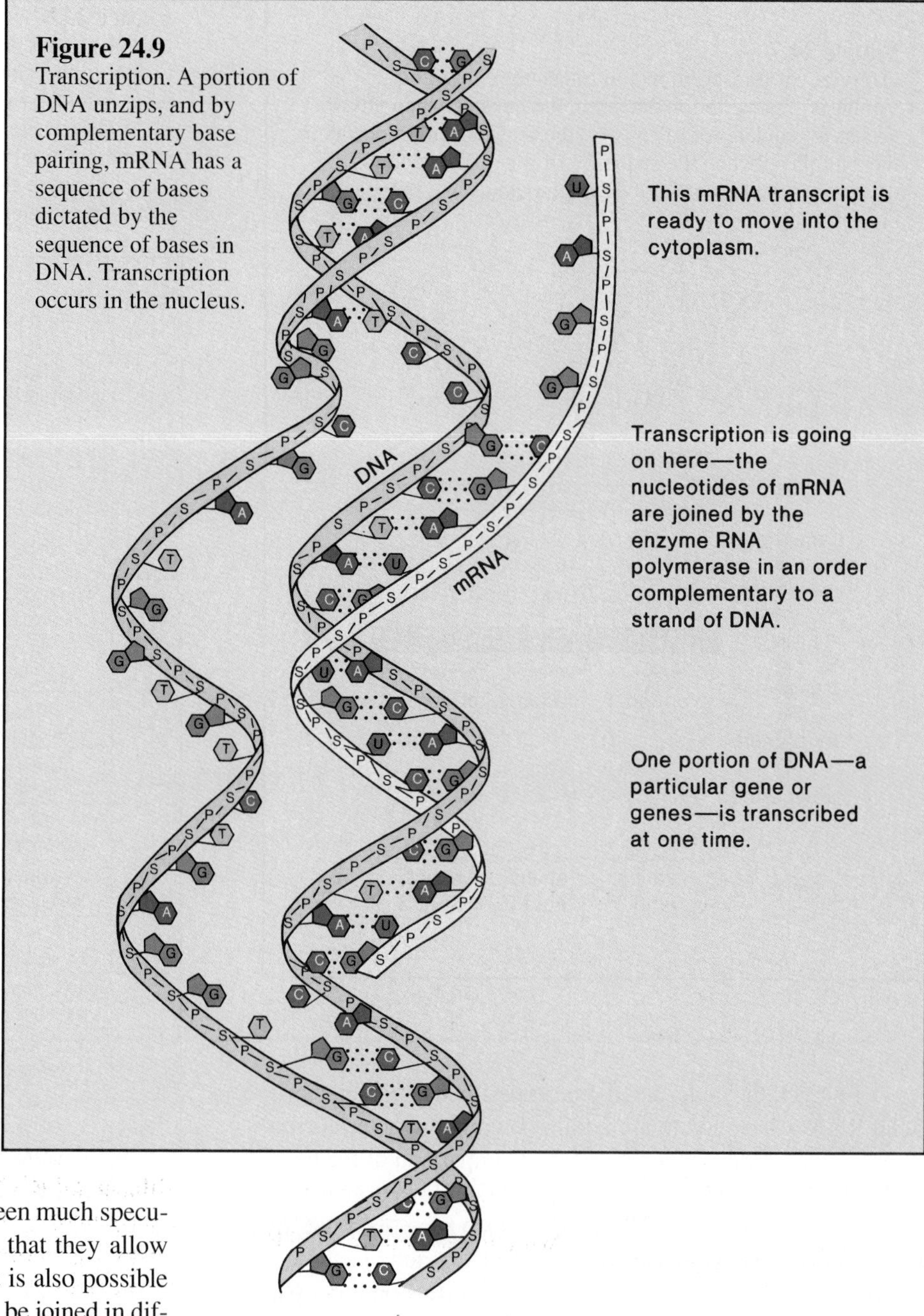

Figure 24.9
Transcription. A portion of DNA unzips, and by complementary base pairing, mRNA has a sequence of bases dictated by the sequence of bases in DNA. Transcription occurs in the nucleus.

Processing Strips Introns from mRNA

Most genes in humans are interrupted by segments of DNA that are not part of the gene. These portions are called introns because they are *intra*gene segments. The other portions of the gene are called exons because they are ultimately *ex*pressed. They result in a protein product.

When DNA is transcribed, the mRNA contains bases that are complementary to both exons and introns, but before the mRNA exits the nucleus, it is *processed*. During processing, the nucleotides complementary to the introns are enzymatically removed. There has been much speculation about the role of introns. It is possible that they allow crossing-over within a gene during meiosis. It is also possible that introns divide a gene into domains that can be joined in different combinations to give novel genes and protein products, facilitating the evolution of new species. The enzymes that remove introns are called **ribozymes** because they are RNA molecules. The discovery of ribozymes tells us that not all enzymes are proteins.

In eukaryotes, processing occurs in the nucleus. After the mRNA strand is processed, it passes from the cell nucleus into the cytoplasm. There it becomes associated with the ribosomes.

Following transcription, mRNA has a sequence of bases complementary to one of the DNA strands. It contains codons and moves into the cytoplasm, where it becomes associated with the ribosomes.

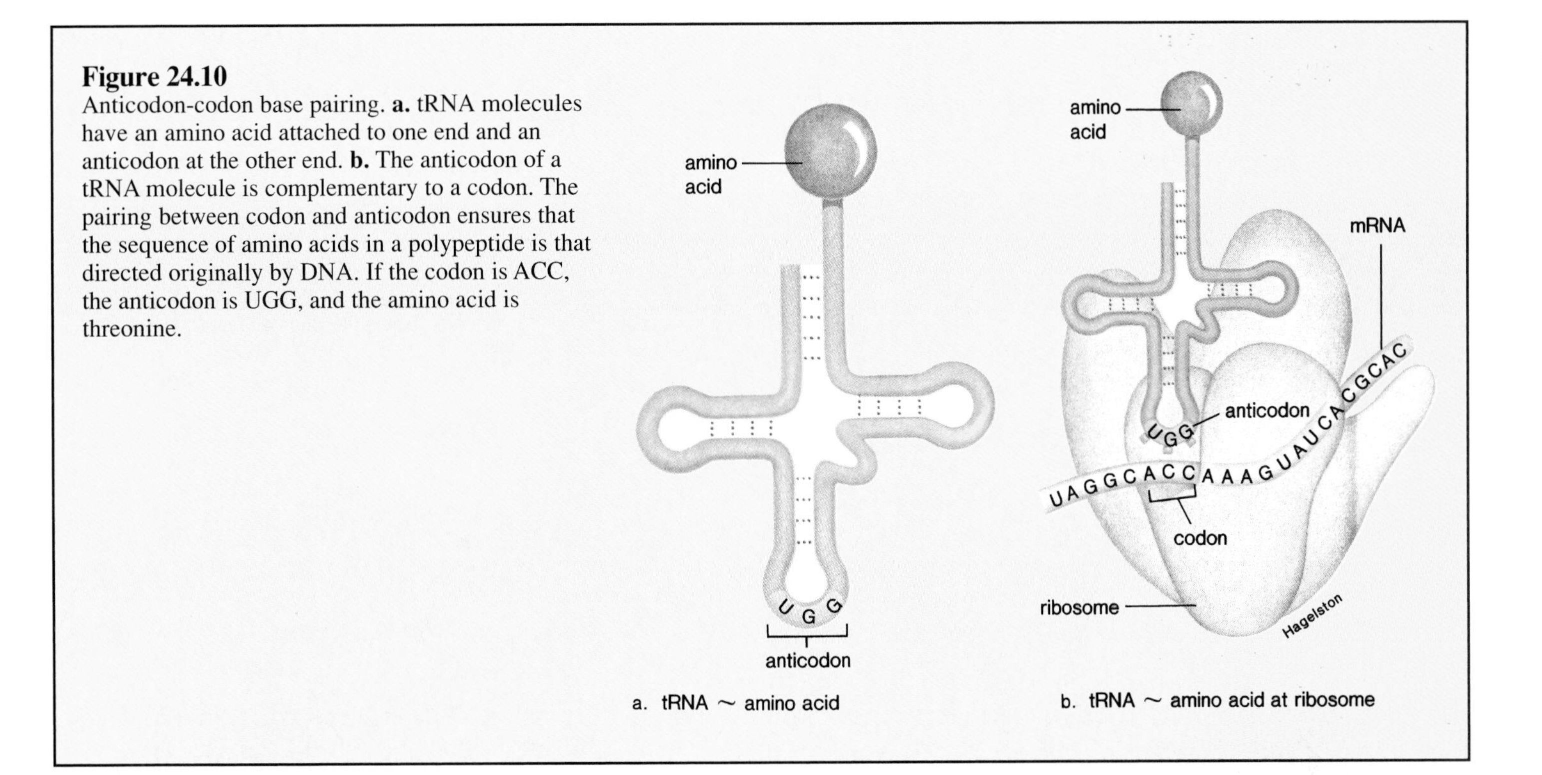

Figure 24.10
Anticodon-codon base pairing. **a.** tRNA molecules have an amino acid attached to one end and an anticodon at the other end. **b.** The anticodon of a tRNA molecule is complementary to a codon. The pairing between codon and anticodon ensures that the sequence of amino acids in a polypeptide is that directed originally by DNA. If the codon is ACC, the anticodon is UGG, and the amino acid is threonine.

Translation: From Codons to Polypeptide

During translation, the sequence of codons in mRNA dictates the order of amino acids in a polypeptide. This is called translation because the sequence of DNA and then RNA bases is translated into a sequence of amino acids. Translation requires several enzymes and 2 other types of RNA: transfer RNA and ribosomal RNA.

Transfer RNA Brings Amino Acids

Small molecules of **transfer RNA (tRNA)** located in the cytoplasm bring amino acids to the ribosomes, also located in the cytoplasm. A particular amino acid is attached to a tRNA molecule at one end (fig. 24.10*a*). Attachment requires ATP energy, and the resulting bond is a high-energy bond represented by a wavy line. Therefore, the entire complex is designated as tRNA ~ amino acid.

At the other end of each tRNA molecule, there is a specific **anticodon** complementary to an mRNA codon (fig. 24.10*b*). The tRNA molecules come to the ribosome, where each anticodon pairs with a codon. Let us consider an example: If the codon is ACC, what is the anticodon, and what amino acid will be attached to the tRNA molecule? Inspection of figure 24.8 allows us to determine this:

Codon	Anticodon	Amino Acid
ACC	UGG	Threonine

The order of the codons of the mRNA determines the order that tRNA ~ amino acids come to a ribosome and therefore the final sequence of amino acids in a polypeptide.

Ribosomal RNA: For Structure

Ribosomal RNA (rRNA) is called structural RNA because it makes up the ribosomes (see fig. 3.5) and is not involved in coding. Ribosomal RNA is produced in a nucleolus within the nucleus. There it joins with proteins manufactured in the cytoplasm. Ribosomal subunits then migrate to the cytoplasm, where they join just as protein synthesis begins.

Figure 24.11
Translation. During translation, a ribosome moves along an mRNA molecule. Each time it moves, one tRNA departs, making room for another tRNA anticodon to pair with its codon. The tRNA that departs gives its peptide to the newly arrived tRNA~ amino acid complex. In this way, the polypeptide elongates. Translation occurs in the cytoplasm.

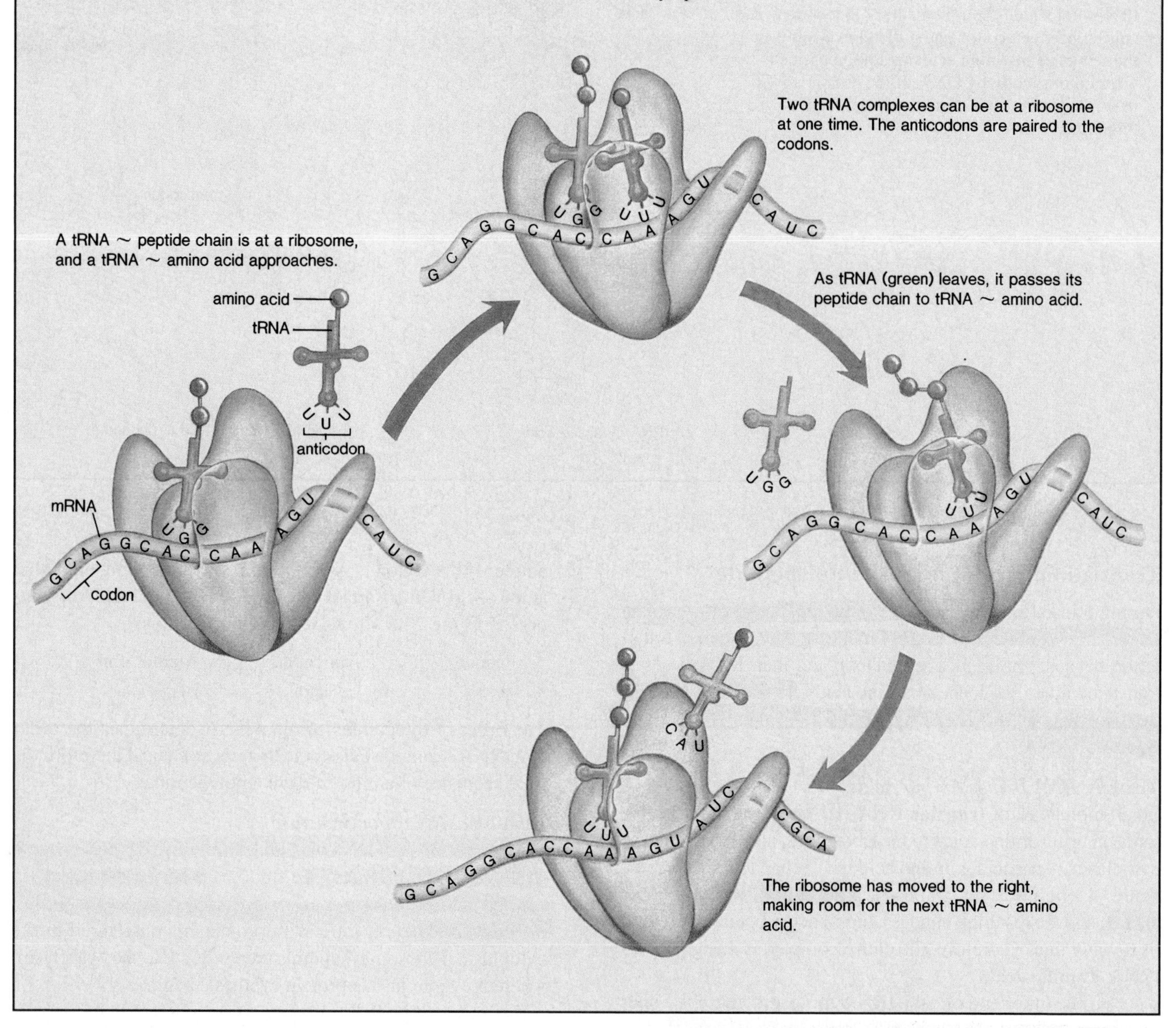

Translation in Three Steps

Polypeptide synthesis requires 3 steps: initiation, elongation, and termination. During *initiation,* mRNA binds to the smaller of the 2 ribosomal subunits; then the larger subunit joins the smaller one. During *elongation,* the polypeptide lengthens one amino acid at a time (fig. 24.11). A ribosome is large enough to accommodate 2 tRNA molecules: the incoming tRNA molecule and the outgoing tRNA molecule. The incoming tRNA ~ amino acid complex receives the peptide from the outgoing tRNA. The ribosome then moves laterally so that the next mRNA codon is available to receive an incoming tRNA ~ amino acid complex. In this manner,

Figure 24.12
Polysome structure. **a.** Several ribosomes, collectively called a polysome, move along an mRNA molecule at one time. They function independently of each other; therefore, several polypeptides can be made at the same time. **b.** Electron micrograph of a polysome.

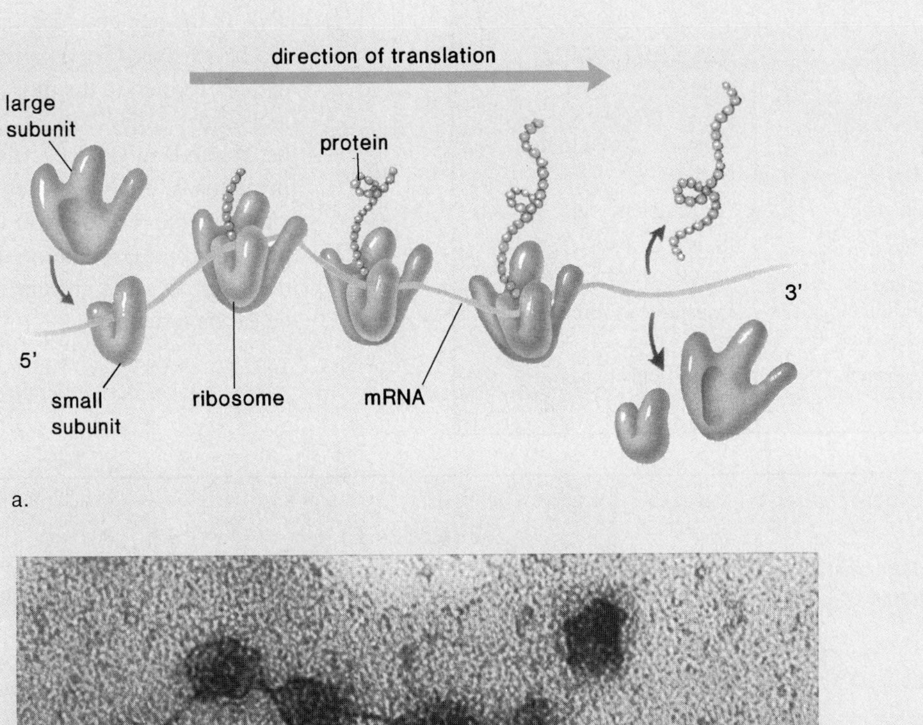

the peptide grows and the primary structure of a polypeptide comes about. (The secondary and tertiary structures of a polypeptide appear after termination, as the amino acids interact with one another.) *Termination* of synthesis occurs at a stop codon on the mRNA; there is no tRNA for this codon. The ribosome dissociates into its 2 subunits and falls off the mRNA molecule.

Several ribosomes, collectively called a **polysome,** can move along one mRNA at a time. Therefore, several polypeptides of the same type can be synthesized using one mRNA molecule (fig. 24.12).

During translation, tRNA molecules, each carrying a particular amino acid, travel to the mRNA. Through complementary base pairing between anticodon and codon, the tRNA molecules and therefore the amino acids in a polypeptide are sequenced in a particular order, the order determined by the DNA code.

Table 24.2 Participants in Protein Synthesis		
Name of Molecule	**Special Significance**	**Definition**
DNA	Code	Sequence of DNA bases in threes
mRNA	Codon	Sequence of RNA bases complementary to DNA code
rRNA	Ribosome	Site of polypeptide synthesis
tRNA	Anticodon	Sequence of 3 bases complementary to codon
Amino acid	Building block for polypeptide	Transported to ribosome by tRNA
Polypeptide	In a protein	Amino acids joined in a predetermined order

Protein Synthesis: Summary

The following list, along with table 24.2 and figure 24.13, provides a brief summary of the steps involved in protein synthesis within eukaryotes:

1. DNA in the nucleus contains a triplet *code*. Each group of 3 bases stands for a particular amino acid.
2. mRNA, formed in the nucleus, contains triplets of the nucleotides called *codons*. During transcription, one of the 2 strands of DNA serves as a template for the formation of mRNA, which then contains bases complementary to those in DNA.
3. mRNA is processed before leaving the nucleus. During mRNA processing, introns are spliced out so that only the exons remain.

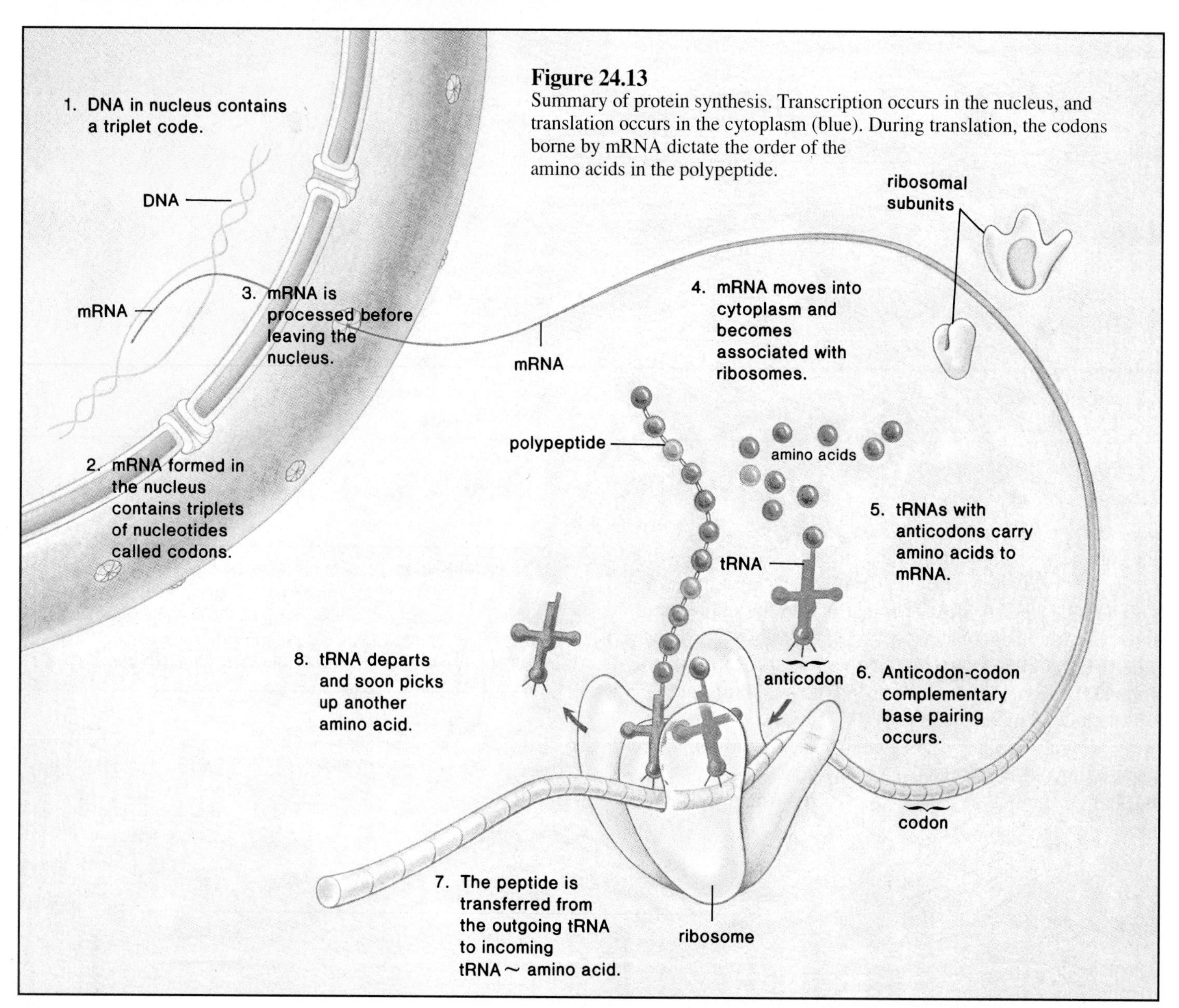

Figure 24.13
Summary of protein synthesis. Transcription occurs in the nucleus, and translation occurs in the cytoplasm (blue). During translation, the codons borne by mRNA dictate the order of the amino acids in the polypeptide.

4. mRNA moves into the cytoplasm and becomes associated with the ribosomes. *Ribosomes* are composed of rRNA and proteins.
5. tRNAs, with anticodons, carry amino acids to mRNA. The *anticodon* of each tRNA is complementary to a particular codon in mRNA.
6. Anticodon-codon complementary base pairing occurs. During translation, the order of the mRNA codons determines the order in which tRNA molecules and their attached amino acids come to a ribosome. Therefore, the sequence of codons determines the sequence of amino acids in a polypeptide.
7. The peptide is transferred from the outgoing tRNA to the incoming tRNA~ amino acid complex. A ribosome has 2 sites: one site is for the incoming tRNA~ amino acid complex, and the other is for the outgoing tRNA.
8. tRNA departs and soon picks up another amino acid. Each tRNA can be used over and over again to carry the same amino acid to the ribosome.

mRNA, transcribed in the nucleus, contains a sequence of triplet codons complementary to the triplet DNA code. Translation occurs at a ribosome in the cytoplasm, where tRNA anticodons bind to the mRNA codons. The sequence of the codons determines the sequence of tRNA binding and therefore the sequence of amino acids in the polypeptide. Do 24.2 Critical Thinking, found at the end of the chapter.

Control of Gene Expression

Gene expression results in a product, usually a functioning polypeptide. (There are exceptions because rRNA and tRNA are also gene products.) There are 4 levels at which regulation of gene expression can occur.

1. *Transcriptional control*—mechanisms that control which genes are transcribed and/or the rate at which transcription occurs.
2. *Posttranscriptional control*—differential processing of mRNA and/or the rate at which mRNA leaves the nucleus.
3. *Translational control*—how soon and how long the mRNA is active in the cytoplasm.
4. *Posttranslational control*—how soon the protein just translated becomes functional.

Although there are 4 levels of genetic control in eukaryotic cells (fig. 24.14), more is known about transcriptional control; therefore, we will limit our remarks to this level for both eukaryotes and prokaryotes.

Figure 24.14
Levels at which control of gene expression occurs in eukaryotic cells. Transcriptional and posttranscriptional control occur in the nucleus; translational and posttranslational control occur in the cytoplasm.

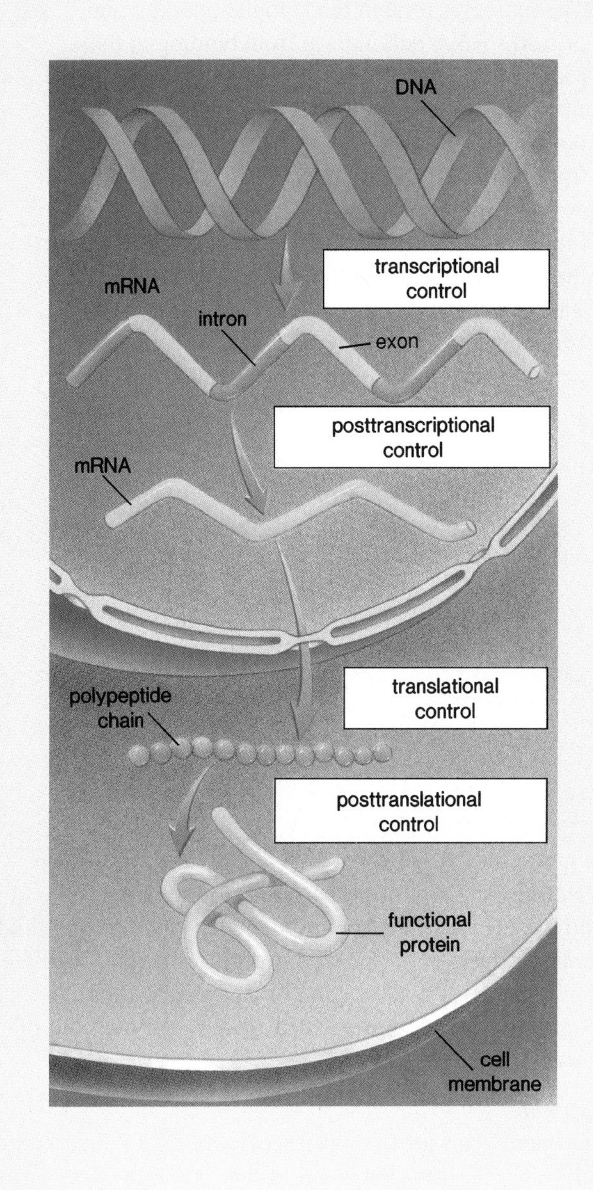

Transcriptional Control in Prokaryotes

The operon model is a well-known example of transcriptional control in prokaryotes. An **operon** includes the following elements:

Regulator gene—a gene that codes for a repressor protein. The repressor protein binds to the operator and prevents RNA polymerase from binding to the promoter.

Promoter—a short sequence of DNA where RNA polymerase first attaches when a gene is to be transcribed.

Operator—a short sequence of DNA where the repressor binds, preventing RNA polymerase from attaching to the promoter. This is often called the on/off switch of transcription.

Structural genes—one to several genes of a metabolic pathway that are transcribed as a unit.

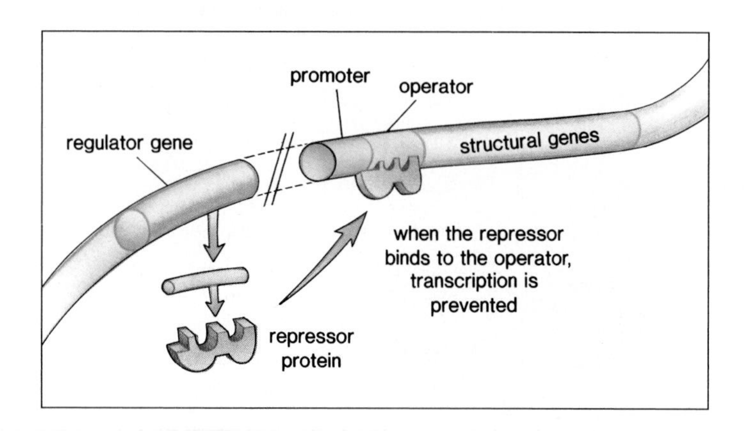

Notice the distinction between a regulator gene and a structural gene. A **structural gene** codes for protein, which functions in the cytoplasm, and a **regulator gene** regulates the activity of (a) structural gene(s). Therefore, although each cell contains a full complement of genes, only certain ones are active at any one time.

Lac Operon

The *lac* operon was the first operon discovered. Ordinarily, the bacterium *E. coli* uses glucose as its energy source; however, if it is denied glucose and is given the milk sugar lactose instead, it immediately begins to make 3 enzymes needed to metabolize lactose (fig. 24.15).

Notice that the structural genes in this operon are normally not transcribed because the regulator gene codes for an active repressor protein that automatically attaches to the operator, preventing transcription from occurring. The operon becomes active when the repressor joins with an inducer—lactose—to form an inactive repressor, which is unable to bind to the operator.

> The operon model explains one means of transcriptional control in prokaryotes—a way in which genes are turned on or off.

Transcriptional Control in Eukaryotes

It is hypothesized that transcriptional control in eukaryotes involves (1) the organization of chromatin and (2) regulatory proteins such as the one we have just observed in prokaryotes.

Chromosome Puffs

For a gene to be transcribed in eukaryotes, the chromosome in that region must first decondense (p. 118). During the development of an insect larva, such as a midge larva, first one and then another of the chromosome's bands bulge out, forming chromosome puffs (fig. 24.16). The use of radioactive uridine, a label specific for RNA, indicates that DNA is being actively transcribed at chromosome puffs. It could be that genes ordinarily are inactive in eukaryotes, and they must be turned on to be active.

Transcription Factors in Control

Although no operons like those of prokaryotic cells have been found in eukaryotic cells, investigations suggest that transcription is controlled by DNA-binding proteins called *transcription factors*. Every cell contains many different types of transcription factors, and a specific combination is believed to regulate the activity of any particular gene. After the right combination of transcription factors binds to DNA, an RNA polymerase attaches to DNA and begins the process of transcription.

As cells mature, they become specialized. Specialization is determined by which genes are active, and therefore perhaps by which transcription factors are present in that cell. Signals received from inside and outside the cell could turn on or off genes that code for certain transcription factors. For example, the gene for fetal hemoglobin ordinarily gets turned off as a newborn matures—one possible treatment for sickle-cell disease is to turn this gene on again.

> There is evidence of regulatory proteins in eukaryotic cells, even though no definite operons have been identified. Do 24.3 Critical Thinking, found at the end of the chapter.

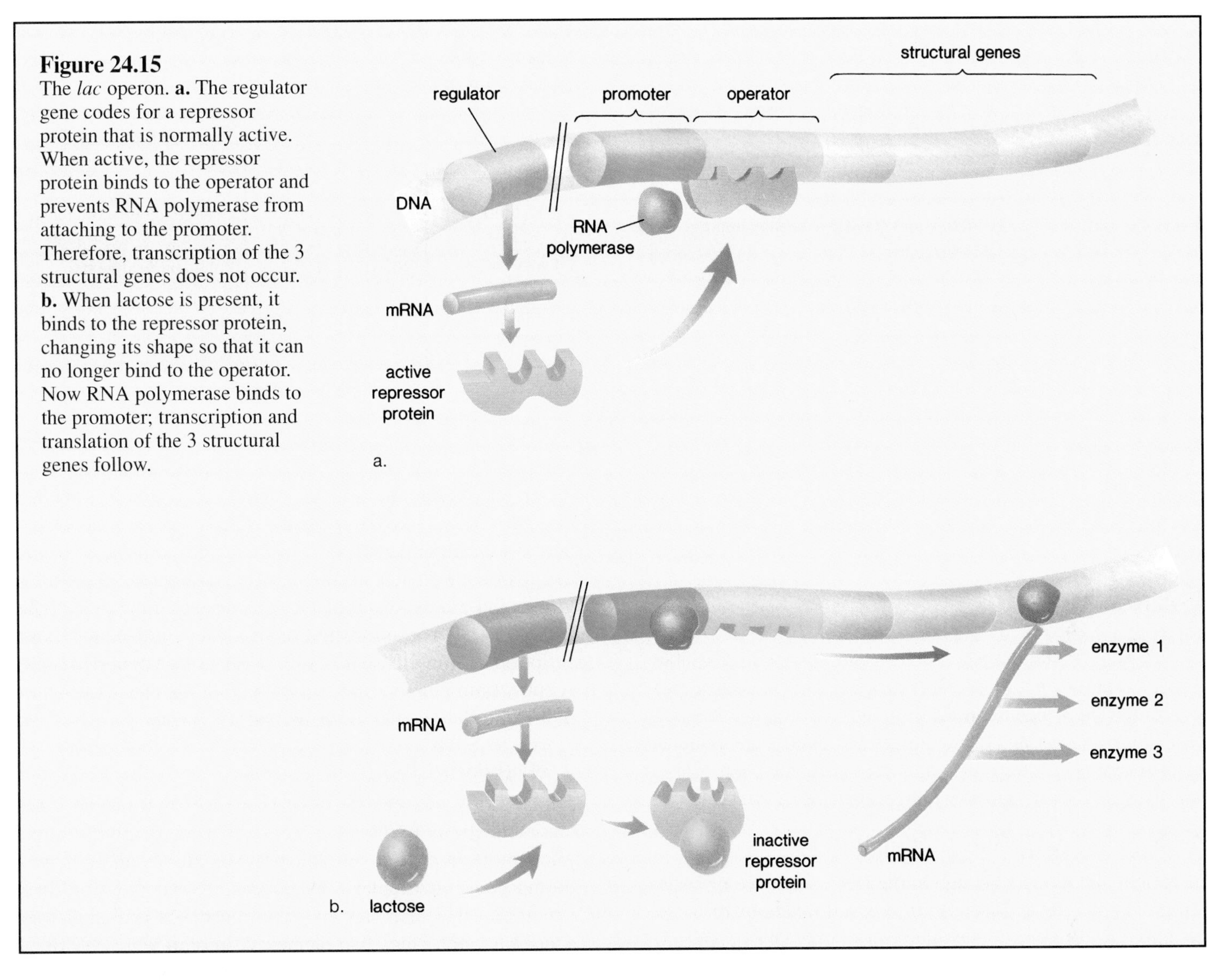

Figure 24.15
The *lac* operon. **a.** The regulator gene codes for a repressor protein that is normally active. When active, the repressor protein binds to the operator and prevents RNA polymerase from attaching to the promoter. Therefore, transcription of the 3 structural genes does not occur. **b.** When lactose is present, it binds to the repressor protein, changing its shape so that it can no longer bind to the operator. Now RNA polymerase binds to the promoter; transcription and translation of the 3 structural genes follow.

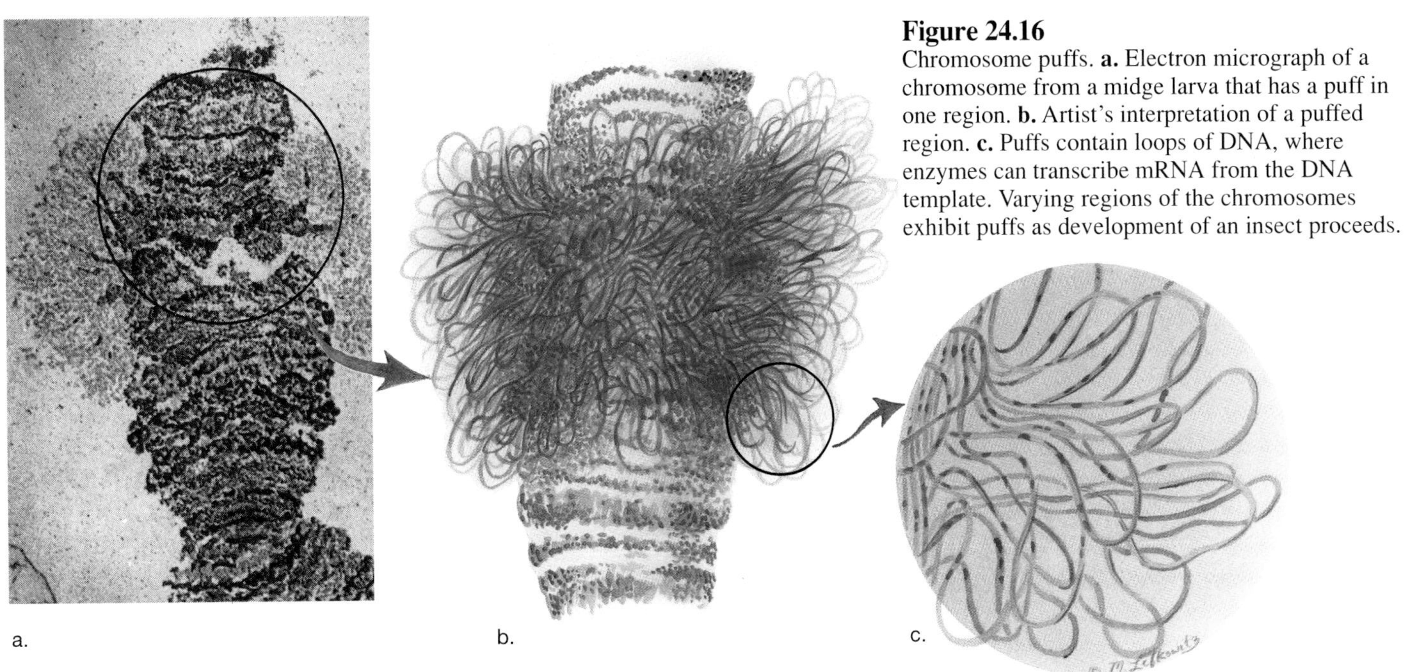

Figure 24.16
Chromosome puffs. **a.** Electron micrograph of a chromosome from a midge larva that has a puff in one region. **b.** Artist's interpretation of a puffed region. **c.** Puffs contain loops of DNA, where enzymes can transcribe mRNA from the DNA template. Varying regions of the chromosomes exhibit puffs as development of an insect proceeds.

Table 24.3
Types of Mutations

Type of Mutation	Definition
Chromosome mutation	A rearrangement of chromosome parts, as described in figure 23.3, which may or may not result in a change in the phenotype.
Gene mutation	A change in the genetic code for a gene or in the expression of the gene. Usually results in a change in the phenotype.
Germinal mutation	A mutation that manifests itself in the gametes so that it is passed on to offspring.
Somatic mutation	A mutation that occurs in the body cells and that very likely is not passed on to offspring.

Gene Mutations

Chromosome mutations were reviewed in chapter 23 (see table 23.3), and now we will consider gene mutations. As you can see in table 24.3, a **gene mutation** is any alteration in the code of a single gene or any change in its expression. Gene mutations do not necessarily have a deleterious effect; some have no effect at all, and some even have a beneficial effect.

Adding, Deleting, Substituting Bases

Mutations involving a change in the DNA sequence of bases are of 3 types (table 24.4). *Additions* and *deletions* of bases are apt to result in profound alterations in the DNA code and possibly an enzyme whose sequence of amino acids has been so greatly affected it can no longer function properly (table 24.4). When *substitutions* occur, there is a change in a single base. No effect at all is expected if the newly translated codon happens to stand for the same amino acid as the old codon

Table 24.4
Gene Mutations

Base Change		Worst Result
Normal	TAC´GGC´ATG	
Substitution	TAG´GGC´ATG	Change in one amino acid or change to stop signal
Deletion	ACG´GCA´TG	Polypeptide altered completely
Addition	*A*TA´CGG´CAT´G	Polypeptide altered completely

(fig. 24.8). On the other hand, a base substitution can sometimes lead to an amino acid substitution; for example, a change from mRNA codon GAG to mRNA codon GUG causes the amino acid glutamate to be replaced by the amino acid valine. This particular change in amino acids causes sickle-cell hemoglobin and a drastic and deleterious effect on the phenotype (fig. 24.17). If a base substitution results in a stop codon, transcription termination and an incomplete polypeptide result. A substitution having this effect may be the cause of hemophilia, the X-linked clotting disorder (p. 452).

Transposons: Jumping Genes

Transposons are specific DNA sequences that have the remarkable ability to move within and between chromosomes. Their movement to a new location sometimes alters neighboring genes, particularly by increasing or decreasing their expression. This can happen if the transposon is a regulatory gene. Although "movable elements" in corn were described 40 years ago, their significance was only realized recently. So-called jumping genes now have been discovered in bacteria, fruit flies, and humans, and it is likely that all organisms have such elements.

Figure 24.17

Sickle-cell disease in humans. **a.** Scanning electron micrograph of normal (*left*) and sickled (*right*) red blood cells. **b.** Portion of the polypeptide in normal hemoglobin (Hb^A) and in sickle-cell hemoglobin (Hb^S). Although the polypeptide is 146 amino acids long, the one change from glutamate to valine in the sixth position results in sickle-cell disease. **c.** Glutamate has a polar *R* group, while valine has a nonpolar *R* group, and this causes Hb^S to be less soluble and to precipitate out of solution. The Hb^S molecules stack up in long, semirigid rods, which push against the cell membrane and distort the red blood cell into a sickle shape.

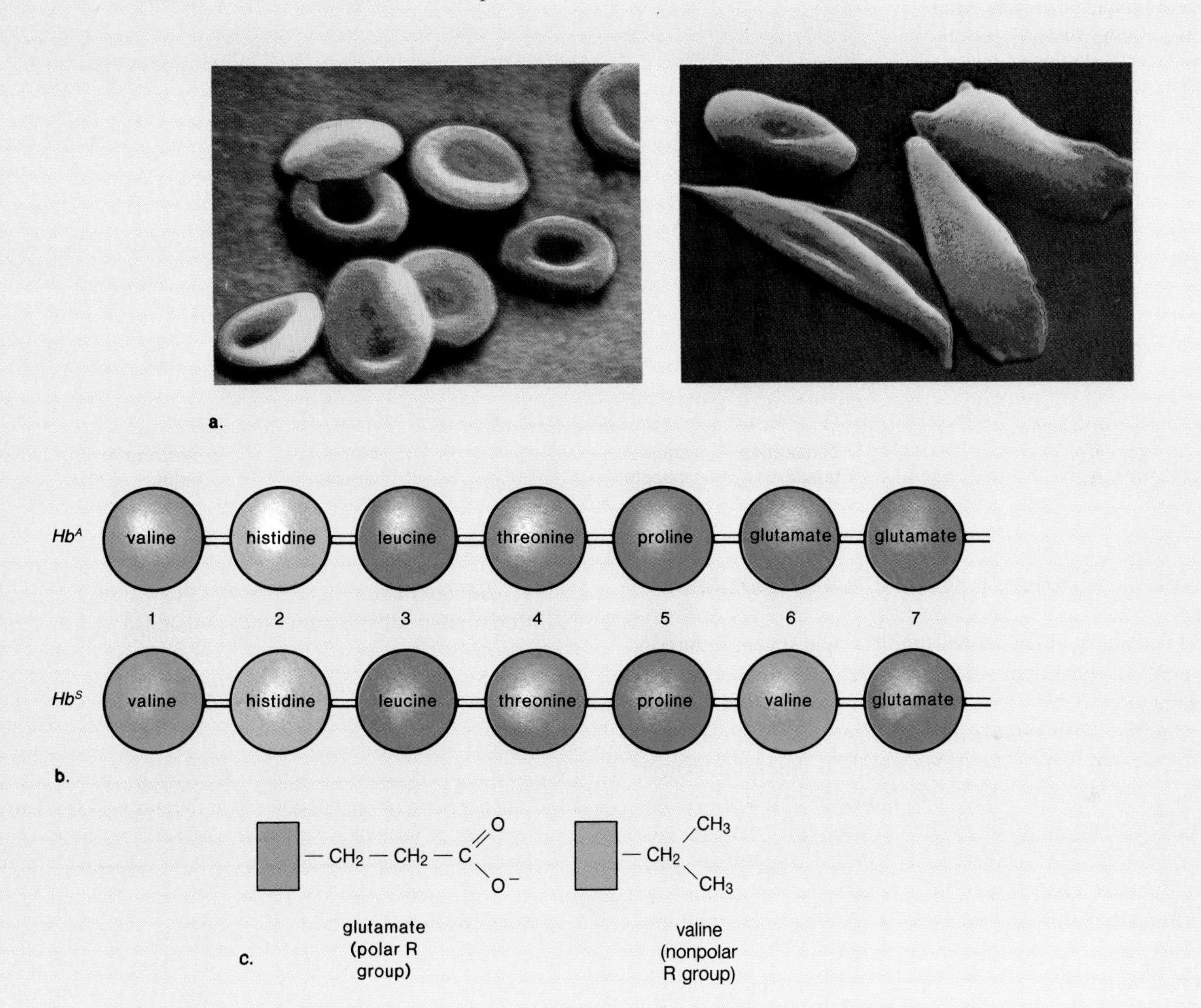

Figure 24.18
Metastasis. In the body, cancer cells form a tumor, a disorganized mass of cells undergoing uncontrolled growth. As a tumor grows, it may invade underlying tissues. Some of the cells leave this primary tumor and move through layers of tissue into blood vessels or lymphatic vessels. After traveling through these vessels, the metastatic cells start new tumors elsewhere in the body. A carcinoma is a cancer that begins in epithelial tissue; a sarcoma is one that begins in connective tissue.

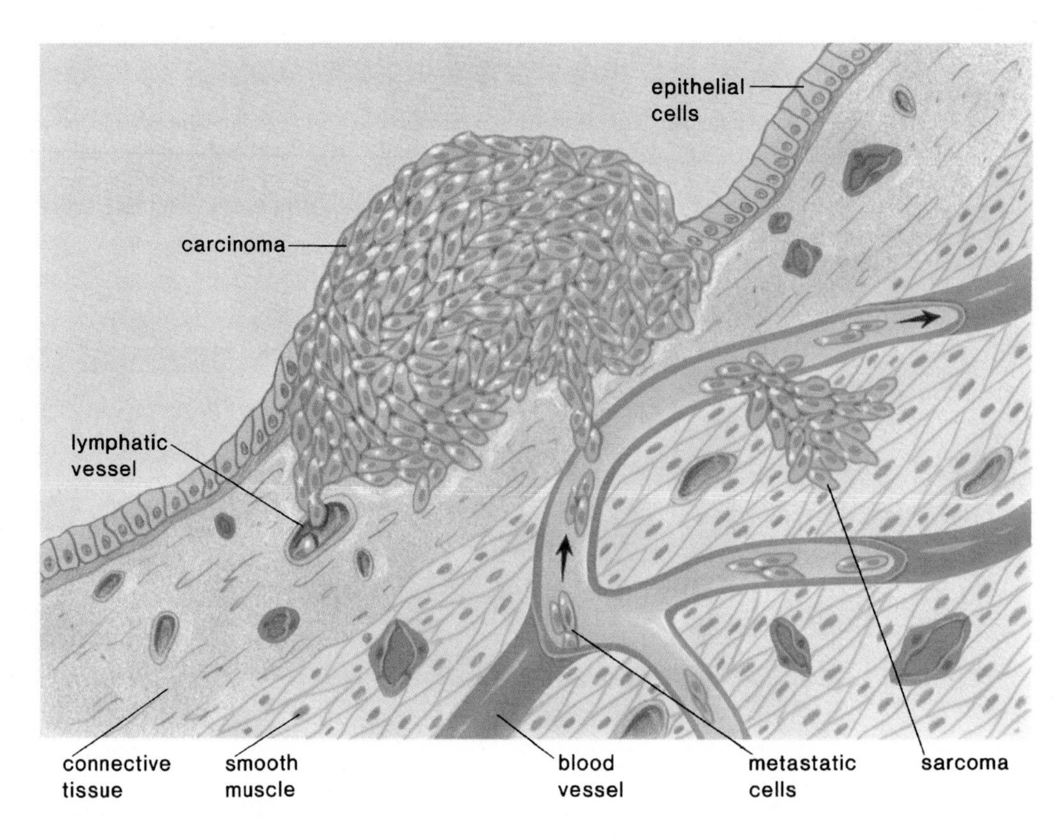

Cancer: A Failure of Genetic Control

Cancer cells have characteristics indicating a severe failure in the control of gene expression. These characteristics are described next.

Cancer cells exhibit uncontrolled and disorganized growth. Normal cells only divide about 50 times, but cancer cells enter the cell cycle (fig. 7.3) over and over again and never fully differentiate. In tissue culture, normal cells grow in only one layer because they adhere to the glass. They stop dividing once they make contact with their neighbors, a phenomenon called contact inhibition. Cancer cells have lost contact inhibition and grow in multiple layers, most likely because of cell-surface changes. In the body, cancer cells produce a tumor, which invades and destroys neighboring tissue (fig. 24.18). The nondifferentiated cells are disorganized and do not function as they should. To support their growth, cancer cells release a growth factor that causes neighboring blood vessels to branch into the cancerous tissue. Some modes of cancer treatment are aimed at preventing this phenomenon, termed vascularization.

Cancer cells detach from the tumor and spread around the body. Cancer cells produce hydrolytic enzymes, which enable them to invade underlying tissues. After traveling through the blood vessels or the lymphatic vessels, cancer cells start new tumors elsewhere in the body. This process is called metastasis. If a tumor is found before metastasis has occurred, the chances of a cure are greatly increased. This is the rationale for early detection of cancer.

Benign tumors have a slower growth rate, contain more differentiated cells, become encapsulated, and do not invade or metastasize. A wart is a benign tumor. Compared with benign tumors, *malignant tumors* are aggressive; they grow rapidly, contain many undifferentiated cells, and have a tendency to invade or metastasize.

Cancer cells grow and divide uncontrollably, and they often metastasize, forming new tumors wherever they relocate.

What Causes Cancer

One model for carcinogenesis (the development of cancer) suggests that it is a two-step process. This process involves (1) initiation and (2) promotion.

During initiation, a *carcinogen* brings about a gene or a chromosome mutation that helps bring on cancerous growth in the future. A papilloma viral infection, the cause of genital warts, may lead to cervical cancer. Radiation—ultraviolet (UV) radiation, radon (a radioactive gas produced by the decay of radium), and X rays—damages the normal bonding patterns between DNA nucleotides. Chemical carcinogens, such as pesticides, are known to cause mutations. Cigarette smoke contains chemical carcinogens and causes lung cancer.

A *promoter* of cancer is any influence that triggers uncontrolled growth of cells. A promoter can be a second gene mutation or cumulative mutations that result in cancerous

Figure 24.19
Summary of the causes of cancer. A virus can pass an oncogene to a cell. A normal gene, called a proto-oncogene, can become an oncogene because of a mutation caused by a chemical or radiation. The oncogene either expresses itself to a greater degree than normal or else expresses itself inappropriately. Thereafter, the cell is cancerous. Cancer cells are usually destroyed by the immune system, and the individual only develops cancer when the immune system fails to perform this function.

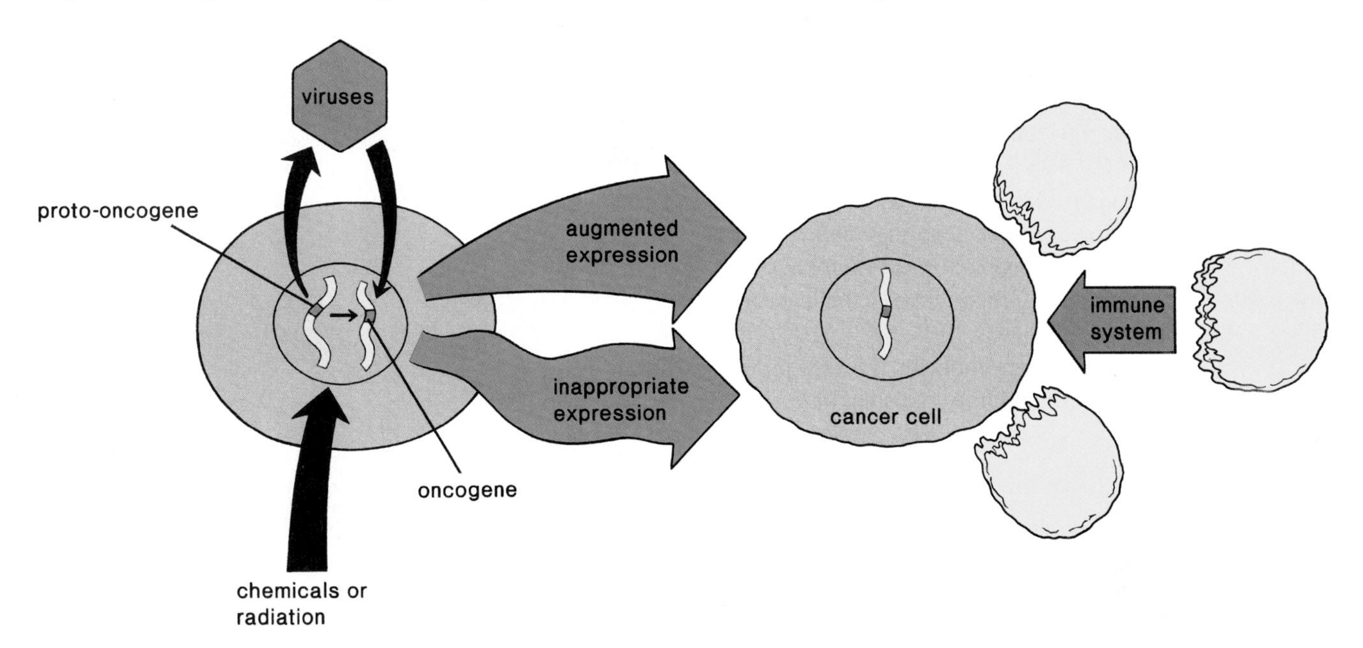

growth. It is also possible that a promoter only provides the environment that causes mutated cells to form a tumor. There is some evidence to suggest that a diet rich in saturated fats and cholesterol is a cancer promoter.

Considerable time may elapse between initiation and promotion. This is why cancer is more often seen in older rather than younger individuals.

> One model indicates that carcinogenesis involves a gene or a chromosome mutation (initiation) and a second influence that triggers cancerous growth (promotion).

Oncogenes: Normal Genes Mutate

It is now clear that cells contain *proto-oncogenes,* genes that can be transformed into **oncogenes,** or cancer-causing genes (fig. 24.19). These genes are not alien to the cell; they are normal, essential genes that have undergone a mutation that leads to augmented or inappropriate expression. An oncogene known to cause both lung and bladder cancer differs from a normal gene by a change in only one nucleotide. It is believed that almost any type of mutation can convert a proto-oncogene into an oncogene. An oncogene can also be introduced into a cell by a virus.

How Oncogenes Work. Several oncogenes have been located and studied. All of these are mutated forms of normal genes that regulate cell growth. They control the activity of growth factors (or growth factor inhibitors), receptors for growth factors, or nuclear proteins that regulate the transcription of a gene for a growth factor. When one of these elements malfunctions, the cell divides repeatedly and a tumor results.

It is of extreme interest to investigators that the oncogene known as the *src* gene codes for an enzyme called tyrosine kinase. This enzyme adds a phosphate group to tyrosine, an amino acid, and in some unknown way, this promotes cell division:

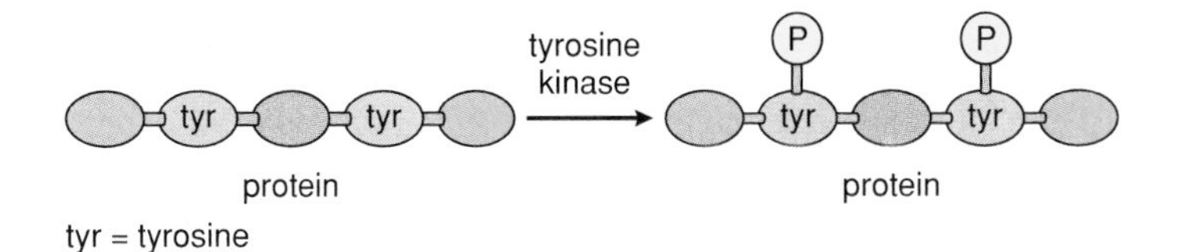

At least 40 different genes have been found to code for tyrosine kinases. Some of these enzymes are also cell membrane receptors for growth factors.

> Oncogenes are mutated copies of proto-oncogenes, normal genes coding for proteins that promote cell division. Some of these proteins are tyrosine kinases, enzymes that add phosphate to tyrosine, an amino acid.

Genes That Suppress Tumors

In addition to proto-oncogenes, there are also **tumor suppressor genes,** the mutation of which can lead to cancer. Ordinarily, tumor suppressor genes are *anti-oncogenes*—they oppose the action of oncogenes.

In patients with retinoblastoma, a cancer of the retina, a specific region of both copies of chromosome 13 is visibly missing. These patients do not have the gene *RB*. When *RB* is present, a cell does not become cancerous because *RB* is a tumor suppressor gene. *RB* has also been shown to reverse malignancy in both kidney and prostate cancer cells. Investigators have found another tumor suppressor gene they call *p53*. A mutated form of this gene is responsible for Li-Fraumeni syndrome, which is characterized by a high rate of lethal cancers among family members.

Some tumor suppressor genes code for tyrosine phosphatases, which remove a phosphate from tyrosine. Investigators are gathering evidence that suggests that tyrosine phosphatases are growth inhibitors.

Tumor suppressor genes are anti-oncogenes, which code for proteins that inhibit cell division. The possibility exists that some of these genes code for tyrosine phosphatases.

An examination of cancerous colon cells suggests that at least 5 mutations are required before cancer develops. One of these causes the oncogene called *RAS* to be activated. *RAS* codes for a protein that regulates the activity of a tyrosine kinase. The cancerous colon cells have also lost tumor suppressor genes from chromosomes 5, 17, and 18.

The development of cancer usually requires several mutations. Some of these mutations cause proto-oncogenes to become oncogenes, and some of these cause tumor suppressor genes to become ineffective.

Preventing Cancer

There is clear evidence that the risk of certain types of cancer can be reduced by adopting protective behaviors. In general, the avoidance of carcinogenic chemicals and excessive radiation is helpful. Avoiding excessive sunlight reduces the risk of skin cancer, and not smoking cigarettes and cigars reduces the risk of lung cancer and other types of cancer. Cigarette smoke contains many carcinogenic chemicals, agents that can cause mutations.

Figure 24.20
Some data suggest that diet can influence the development of cancer. Fresh fruits, especially those high in vitamins A and C, and vegetables, especially those in the cabbage family, are believed to reduce the risk of cancer.

A healthy diet also helps prevent cancer (fig. 24.20). To prevent cancer, it is believed helpful to modify the diet as follows:

Lower total fat intake. A high-fat intake has been linked to development of breast, colon, and prostate cancer.

Cut down on consumption of salt-cured, smoked, and nitrite-cured foods. Processed meats, such as luncheon meats and breakfast meats—for example, bacon and sausage—fall into this category. Some of the chemicals added to these foods may be cancer promoters.

Eat more high-fiber foods, such as those in whole-grain cereals, fruits, and vegetables. Studies have indicated that a high-fiber diet protects against colon cancer.

Increase consumption of foods rich in vitamins A and C. These vitamins are antioxidants, which prevent the formation of free radicals (organic ions that have a nonpaired electron) that can possibly damage DNA. Dark green, leafy vegetables, carrots, and various fruits contain beta-carotene, a precursor of vitamin A. Citrus fruits contain vitamin C.

Increase consumption of vegetables in the cabbage family. This includes cabbage, broccoli, and cauliflower. These foods seem to protect against the development of cancer.

Be moderate in the consumption of alcohol. People who drink and smoke are at an unusually high risk of cancer of the mouth, larynx, and esophagus.

SUMMARY

DNA is the genetic material; it can replicate, mutate, and store information. During replication, DNA "unzips," and then a complementary strand forms opposite to each original strand. DNA directs protein synthesis. During transcription, mRNA is made complementary to one of the DNA strands. mRNA, bearing codons, moves to the cytoplasm, where it becomes associated with the ribosomes. During translation, tRNA molecules, attached to their own particular amino acid, travel to a ribosome, and through complementary base pairing between anticodons and codons, the tRNAs and therefore the amino acids in a polypeptide are sequenced in a predetermined way.

The prokaryote operon model explains how one regulator gene controls the transcription of several structural genes, genes that code for proteins. In eukaryotes, the chromosome has to decompact before transcription can begin.

A gene mutation is an alteration in the normal sequence of bases within a gene. Cancer is characterized by a lack of control: the cells grow uncontrollably and metastasize. Cancer development is a multistep process involving the mutation of genes. Proto-oncogenes and tumor suppressor genes are normal genes that bring on cancer when they mutate because they code for factors involved in cell growth.

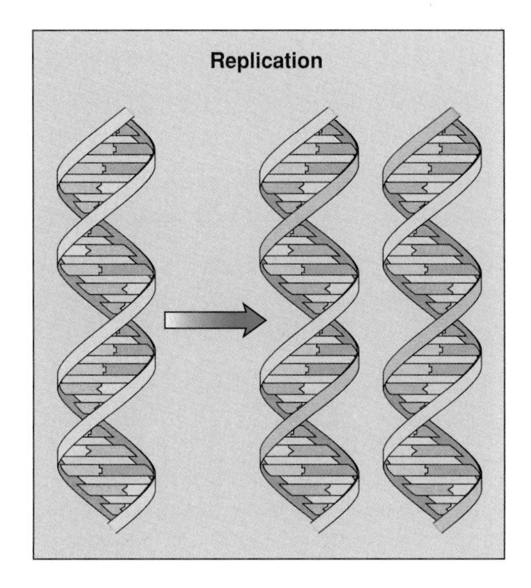

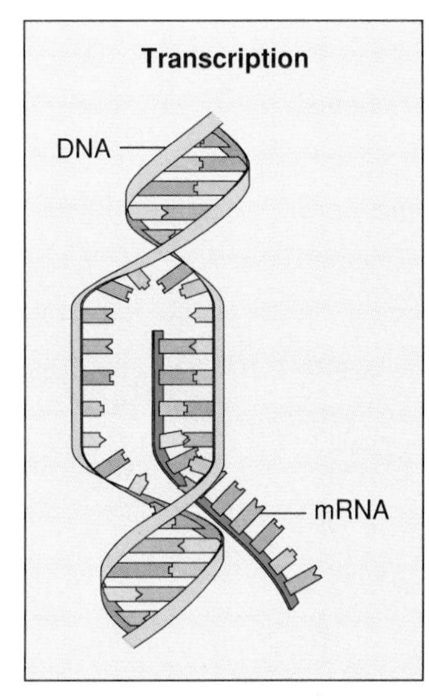

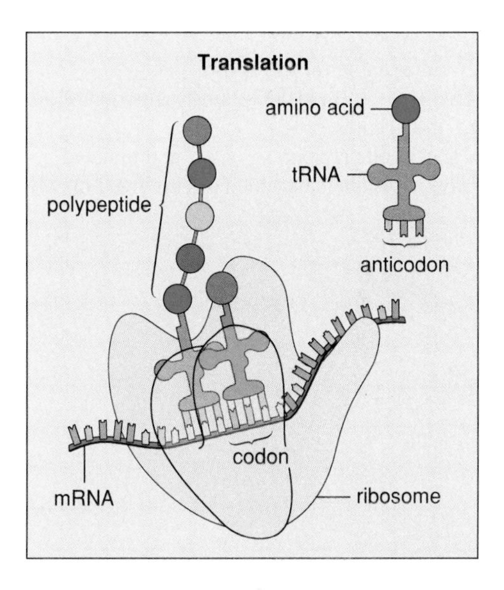

STUDY QUESTIONS

In order to practice **writing across the curriculum,** students should write out the answers to any or all of the study questions. The study questions are sequenced in the same order as the text.

1. Describe the experiment that designated DNA rather than protein as the genetic material. (p. 463)
2. Describe the structure of both DNA and RNA. (p. 465)
3. Explain how DNA replicates. (p. 467)
4. Name 2 genetic disorders that show DNA controls the formation of enzymes. (p. 468)
5. What happens during transcription? during translation? (pp. 470–472)
6. List the 8 steps involved in protein synthesis. (pp. 474–475)
7. What are the 4 levels of regulation of gene expression in eukaryotes? (p. 475)
8. What is the operon model of structural gene control? Describe the *lac* operon. (p. 476)
9. The substitution of one base for another base in DNA can have what effect on the phenotype? Why would you expect additions and deletions of bases to have a major effect on the phenotype? (p. 478)
10. What are 2 abnormal characteristics of cancer cells? (p. 480) Describe a two-step process that explains the development of cancer. (p. 480) What are a proto-oncogene and a tumor suppressor gene? (pp. 481–482)

OBJECTIVE QUESTIONS

1. The backbone of DNA is made up of _______ and _______ molecules.
2. Replication of DNA is semiconservative, meaning that each new helix is composed of a(n) _______ strand and a(n) _______ strand.
3. The base _______ in RNA replaces the base thymine in DNA.
4. The DNA code is a(n) _______ code, meaning that every 3 bases stands for a(n) _______.
5. The 3 types of RNA that are necessary to protein synthesis are _______, _______, and _______.
6. When mRNA is processed within the eukaryotic nucleus, the portions complementary to _______ in DNA are removed.
7. Which of the 3 types of RNA carries amino acids to the ribosomes? _______
8. Another name for transposons is _______.
9. The _______ model explains regulation of gene transcription in prokaryotes.
10. Cancer cells contain _______ and mutated _______ genes, the protein products of which are involved in cell growth.
11. This is a segment of a DNA molecule. (Remember that only the transcribed strand serves as the template.) What are (a) the RNA codons, (b) the tRNA anticodons, and (c) the sequence of amino acids in the polypeptide?

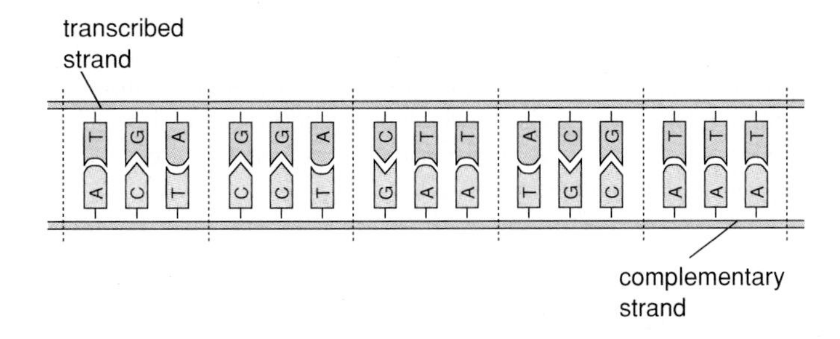

12. Label this diagram of an operon.

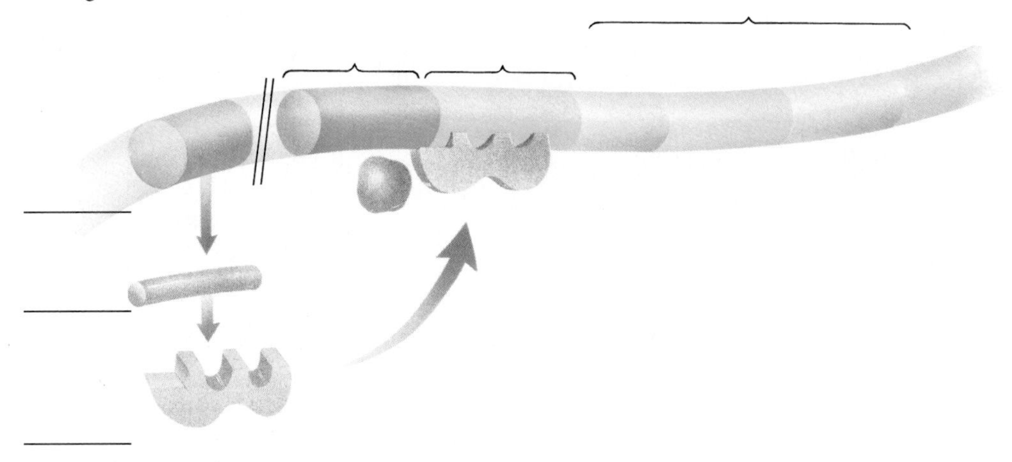

CRITICAL THINKING

In order to practice **writing across the curriculum,** students should write out the answers to any or all of the critical thinking questions. Suggested answers to the critical thinking questions are in appendix E.

24.1

1. DNA stores information. What is the information, and where is it stored? What is your evidence that DNA stores this information?
2. DNA replicates. What is there about DNA that allows it to replicate? What is your evidence that DNA replicates prior to cell division? What is your evidence that replication occurs in the nucleus?
3. DNA mutates. What part of DNA permits mutations to occur? What evidence in nature shows that DNA mutates?

24.2

1. Radioactive uridine is a label specific for RNA. If you supplied an actively metabolizing cell with radioactive uridine, where would you first and then later expect to find the label?
2. Suppose you have a sample of mRNA and rRNA and you know the sequence of bases for each. What cell-free experiment (p. 469) could you carry out to show that the mRNA and not the rRNA directs the synthesis of a polypeptide?
3. What physical evidence can you present that ribosomes (rRNA) move along the mRNA?
4. Assume you have a sample of tRNA and you know a biochemical test for amino acids as opposed to nucleotides. Would performing the test show that tRNA molecules carry amino acids? Why?

24.3

1. Multicellular organisms, like humans, have different phenotypes. Do the phenotypes of various cells in an organism differ? How do you know?
2. Do all the cells of an organism ordinarily possess a complete set of genes, including regulator genes and structural genes? How do you know?
3. Would you expect to find transcripts (mRNA) of all genes in all cells or only certain transcripts? Why?
4. How would you account for the fact that only certain structural genes are active in certain cells?

SELECTED KEY TERMS

anticodon a "triplet" of bases in tRNA that pairs with a complementary triplet (codon) in mRNA.

codon a "triplet" of bases in mRNA that directs the placement of a particular amino acid into a polypeptide.

complementary base pairing pairing of bases between nucleic acid strands; adenine pairs with either thymine (DNA) or uracil (RNA), and cytosine pairs with guanine.

double helix a double spiral; describes the three-dimensional shape of DNA.

messenger RNA (mRNA) ribonucleic acid complementary to DNA; has codons, which direct protein synthesis at the ribosome.

oncogene (ong´ko-jen) a gene that contributes to the transformation of a normal cell into a cancerous cell.

operon a group of structural and regulating genes that function as a single unit.

polysome a cluster of ribosomes attached to the same mRNA molecule; each ribosome is producing a copy of the same polypeptide.

purine (pu´ rēn) nitrogen-containing, organic base found in DNA and RNA that has 2 interlocking rings, as in adenine and guanine.

pyrimidine (pi-rĭm´ ə- dēn) nitrogen-containing, organic base found in DNA and RNA that has just one ring, as in cytosine, uracil, and thymine.

regulator gene gene that codes for a protein involved in regulating the activity of structural genes.

replication the duplication of DNA; occurs when the cell is not dividing.

ribosomal RNA (rRNA) RNA occurring in ribosomes, structures involved in protein synthesis.

structural gene gene that directs the synthesis of an enzyme or a structural protein in the cell.

template a pattern that serves as a mold for the production of an oppositely shaped structure; one strand of DNA is a template for a complementary strand.

transcription the process resulting in the production of a strand of mRNA that is complementary to a segment of DNA.

transfer RNA (tRNA) molecule of RNA that carries an amino acid to a ribosome engaged in the process of protein synthesis.

translation the process by which the sequence of codons in mRNA dictates the sequence of amino acids in a polypeptide.

tumor suppressor gene a gene that suppresses the development of a tumor; the mutated form contributes to the development of cancer.

25

RECOMBINANT DNA AND BIOTECHNOLOGY

Chapter Concepts

1.
Using recombinant DNA technology, bacteria can be bioengineered to mass-produce various products. 487, 498

2.
The polymerase chain reaction (PCR) makes multiple copies of any particular piece of DNA so that it can be analyzed. 492, 498

3.
There are several new types of biotechnology products and procedures. 494, 498

4.
Bacteria, agricultural plants, and farm animals have been bioengineered to improve the services they perform for humans. 494, 498

5.
Gene therapy replaces defective genes with healthy genes and otherwise uses genes to cure human ills. 497, 498

Analysis of bacterial DNA fragments

Chapter Outline

Figure 25.1
Biotechnology, an industrial endeavor. **a.** Laboratory procedures are adapted to mass-produce the product. **b.** Bacteria are grown in huge tanks. These tanks are called fermenters because they were first used for yeast fermentation in the production of wine. **c.** The product is purified and (**d**) packaged.

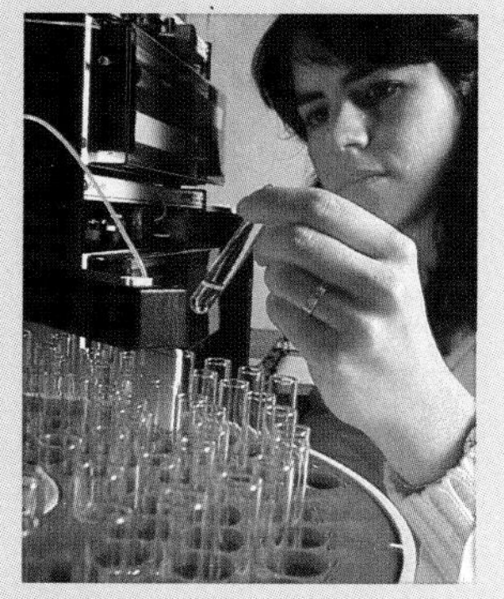

a.

b.

c.

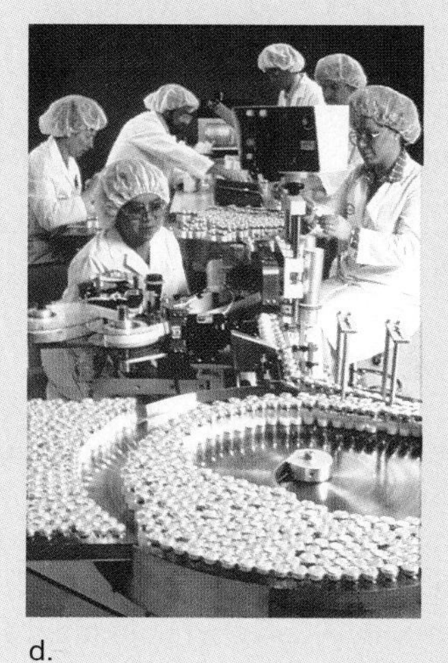
d.

Biotechnology originally meant the use of a natural biological system to produce a product or to achieve an end desired by human beings. Plants and animals have been bred to give a particular phenotype since the dawn of civilization. The biochemical capabilities of microorganisms have also been exploited for a very long time. For example, the production of both wine and bread is dependent on yeast cells to carry out fermentation reactions.

During the 1980s, biotechnology gave rise to an industry that provides products (fig. 25.1) made by **bioengineered** (genetically engineered) bacteria. These products include drugs that promote human health and proteins that are useful as vaccines or nucleic acids for laboratory research. Bioengineered bacteria are not necessarily confined to chemical plants or laboratories. They are also released into the environment to clean up pollutants, to increase the fertility of the soil, or to kill insect pests. Bioengineering extends beyond unicellular organisms; it is possible to alter the genotype and subsequently the phenotype of plants and animals. Gene therapy in humans is already undergoing clinical trials.

Scientists have found a way to alter naturally occurring proteins and indeed to make completely new ones. Someday it might even be possible to bypass bacteria in the production of human proteins. Antibodies can be altered to give them catalytic properties; since there are so many different types of antibodies, the variety of enzymes available might become enormous. Small molecules of RNA and DNA can be manufactured and used to shut down the activity of a gene or mRNA in an organism.

Recombinant DNA Techniques

The new era of biotechnology began in the 1970s when it became possible to make and use recombinant DNA.

Making Recombinant DNA

Recombinant DNA contains DNA from 2 or more different sources. A **vector** (carrier) is used to introduce recombinant DNA into cells.

Vectors: Carrying Foreign Genes

The most common vector is a plasmid (fig. 25.2*a*). **Plasmids** are small accessory rings of DNA found in bacteria. Plasmids used as vectors have been removed from bacteria and have had a foreign gene inserted into them. Treated bacteria take up a plasmid, and after it enters, the plasmid replicates. Whenever the host reproduces, each new cell contains one or more plasmids. Eventually, there are many copies of the plasmid and therefore many copies of the foreign gene. **Gene cloning** has occurred because there are now many exact copies of a foreign gene.

Figure 25.2
Gene cloning using bacteria and viruses. **a.** A plasmid is removed from a bacterium and is used to make recombinant DNA. After the recombined plasmid is taken up by a host cell, gene cloning is achieved when the plasmid reproduces. **b.** Viral DNA is removed from a virus and is used to make recombinant DNA. Virus containing the recombinant DNA infects a host bacterium. Cloning is achieved when the virus reproduces.

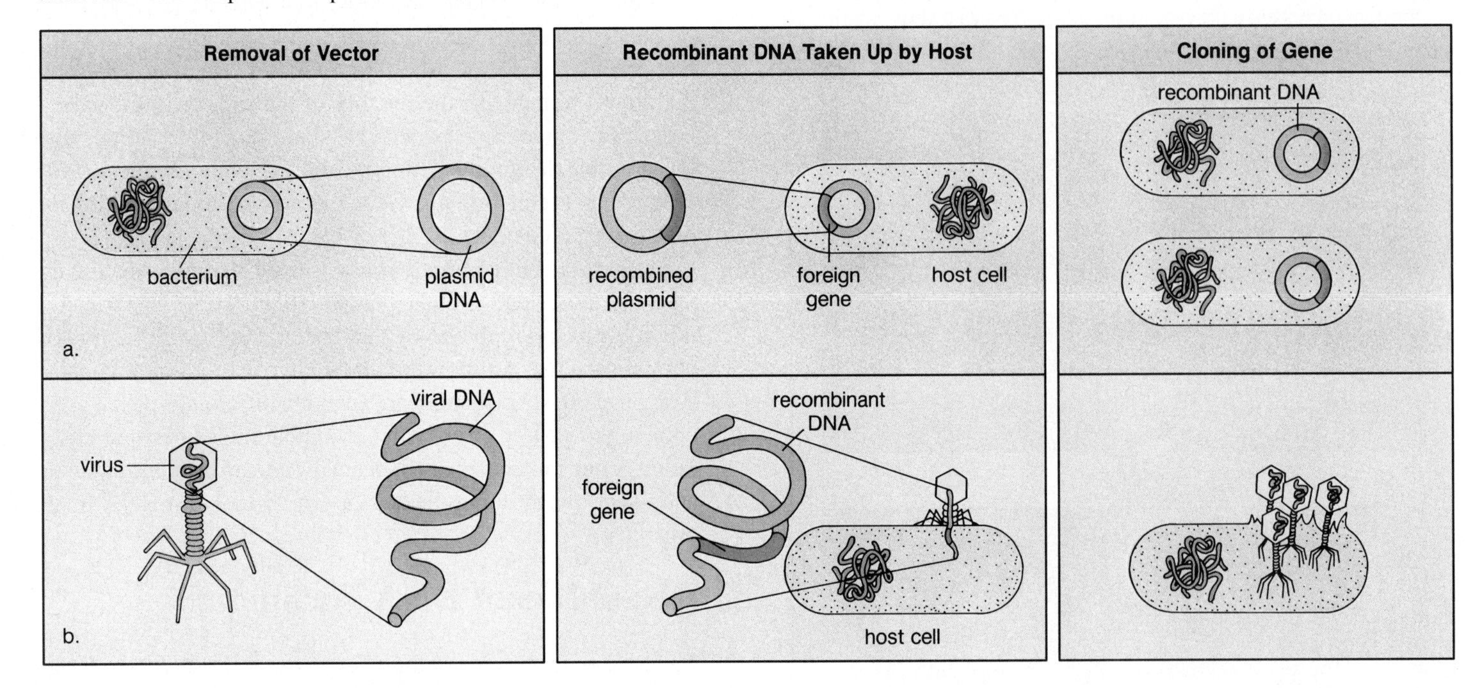

Viral DNA can also be used as a vector to carry recombinant DNA into a cell (fig. 25.2*b*). When a virus containing recombinant DNA infects a cell, the viral DNA enters. Here, it can direct the reproduction of many more viruses. Each virus derived from a viral vector contains a copy of the foreign gene. Therefore, viral vectors allow cloning of a particular gene, too. Viral vectors are also used to create genomic libraries. A **genomic library** is a collection of engineered viruses that together carry all the genes of a species. Since each virus carries only a short sequence of DNA, it takes about 10 million viruses to carry all the genes of a mouse.

Enzymes: To Cut or Seal

The introduction of foreign DNA into a plasmid or into viral DNA to give recombinant DNA is a two-step process (fig. 25.3). First, the plasmid (or viral) DNA is cut open, and then the foreign DNA is inserted into this opening. Both of these steps require a specific type of enzyme.

One type of enzyme has the ability to cut a DNA molecule into discrete pieces. Such enzymes occur naturally in some bacteria, where they stop viral reproduction by cutting up viral DNA. These are called **restriction enzymes** because they *restrict* the growth of viruses. Each type of restriction enzyme—and over a hundred are now known—cleaves DNA at a specific location called a *restriction site*. For example, one restriction enzyme always cleaves double-stranded DNA in this manner when the DNA has this sequence of bases:

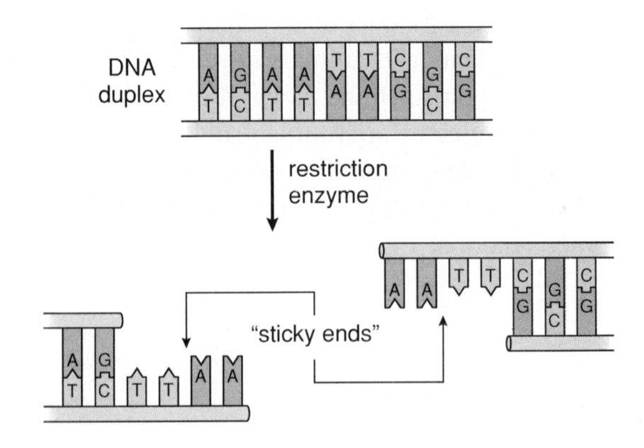

Notice that there is now a gap into which a piece of foreign DNA can be placed if it ends in bases complementary to those exposed by the restriction enzyme. To ensure this, it is only necessary to cleave the foreign DNA with the same type of restriction enzyme. The single-stranded but complementary ends of the 2 DNA molecules are called "sticky ends" because they adhere by complementary base pairing. These ends facilitate insertion of foreign DNA into vector DNA.

The second enzyme needed for preparation of a vector is **DNA ligase,** a bacterial enzyme that seals any breaks in a DNA molecule. Bioengineers use this enzyme to seal the foreign piece of DNA into the vector. Gene splicing is then complete, and a recombinant DNA molecule has been prepared.

Producing the Foreign Protein

Once the recombined plasmid has been taken up by a bacterium and the gene functions normally, the bacterium has been **transformed**—it can produce a protein it never produced before. The investigator can recover either the cloned gene (e.g., insulin gene) or the protein product (e.g., insulin) (fig. 25.3).

In order for mammalian gene expression to occur in a bacterium, the gene has to be accompanied by the proper regulatory regions. Also, the gene should not contain introns because bacterial cells do not have the necessary enzymes to process mRNA. It is possible to make a mammalian gene that lacks introns, however. The enzyme called reverse transcriptase (p. 540) can be used to make a DNA copy of processed mRNA. This DNA molecule, called *complementary DNA* (*cDNA*), does not contain introns. Alternatively, it is possible to manufacture small genes in the laboratory. A machine called a DNA synthesizer joins together the correct sequence of nucleotides, and this resulting gene also lacks introns.

Figure 25.3
Cloning of the human insulin gene. First, restriction enzyme is used to cleave human DNA in order to remove the insulin gene and to open a plasmid. Second, the human gene and the plasmid are sealed together by DNA ligase. Gene cloning is achieved when a host cell takes up the recombined plasmid and the plasmid reproduces. Multiple copies of the gene are now available to an investigator. If the insulin gene functions normally as expected, the product (insulin) may also be retrieved.

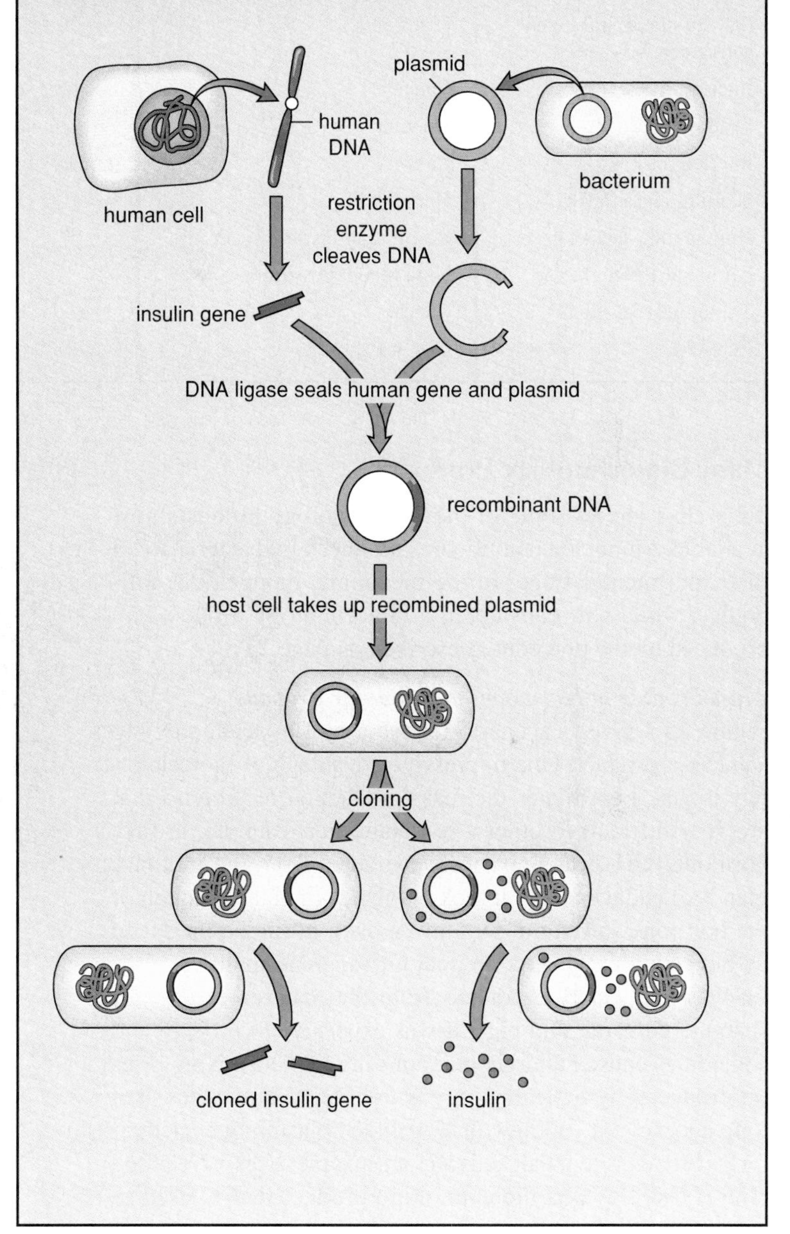

Table 25.1
Representative Biotechnology Products

Hormones and Similar Types of Proteins		Vaccines
Treatment of Humans	*For*	*Use in Humans*
Insulin	Diabetes	AIDS
Growth hormone	Pituitary dwarfism	Herpes (oral and genital)
tPA (tissue plasminogen activator)	Heart attack	Hepatitis A, B, and C
Interferons	Cancer	Lyme disease
Erythropoietin	Anemia	Whooping cough
Interleukin-2	Cancer	Chlamydia
Clotting factor VIII	Hemophilia	
Human lung surfactant	Respiratory distress syndrome	
Atrial natriuretic factor	High blood pressure	
Tumor necrosis factor	Cancer	
Ceredase	Gaucher disease	

Many Biotechnology Products

Table 25.1 shows some of the biotechnology products now available. Monoclonal antibodies produced by bacteria are in the experimental stage. In the meantime, monoclonal antibodies, while still considered a biotechnology product, are produced by the procedure described on page 231.

Mass-Producing Hormones and Similar Proteins

Figure 25.4 gives a scenario for the use of biotechnology to achieve a product. One impressive advantage of biotechnology is that it facilitates the mass production of proteins that are very difficult to otherwise obtain. For example, growth hormone (GH) was previously extracted from the pituitary glands of cadavers, and it took 50 glands to obtain enough of the hormone for one dose. Now growth hormone produced by biotechnology is used to treat growth abnormalities. Insulin was previously extracted from the pancreas glands of slaughtered cattle and pigs; it was expensive to procure and sometimes caused allergic reactions in recipients. Now insulin produced by biotechnology is used to treat diabetes. Not long ago, few of us knew of tPA (tissue plasminogen activator), a blood protein that activates an enzyme to dissolve blood clots. Now tPA is a biotechnology product that is used to treat heart attacks.

Table 25.1 lists other troublesome and serious afflictions in humans that are now treatable by biotechnology products. Clotting factor VIII is for the treatment of hemophilia; human lung surfactant treats respiratory distress syndrome in premature infants; and atrial natriuretic factor will help control hypertension. This list will grow because bacteria (or other cells) can be engineered to produce virtually any protein.

Several hormones are products of biotechnology used in animals. It is no longer necessary to feed steroids to farm animals; they can be given growth hormone, which produces a leaner meat that is more healthful for humans. Cows given bovine growth hormone (bGH) produce 25% more milk than usual, which should make it possible for dairy farmers to maintain fewer cows and to cut down on overhead expenses.

Making Safer Vaccines

Vaccines are used to make people immune to an infection so they do not become ill when exposed to the infectious organism (p. 229). In the past, vaccines were made from treated bacteria or viruses, and on occasion they caused the illness they were supposed to prevent.

Vaccines produced through biotechnology do not cause illness. Bacteria and viruses have surface proteins, and a gene for just one of these can be used to bioengineer bacteria. The copies of the surface protein that result can be used as a vaccine. Since only a portion of the microbe is present in the vaccine, no illness can result. A vaccine for hepatitis B is now available, and those for chlamydia, malaria, and AIDS are in experimental stages.

Vaccines are available through biotechnology for the inoculation of farm animals, too. There are vaccines for such illnesses as hoof-and-mouth disease and scours. These animal ailments were once a severe drain on the time, energy, and resources of farmers.

Biotechnology products include hormones and similar types of proteins, and vaccines. These products are of enormous importance to the fields of medicine and animal husbandry.

Figure 25.4
Possible biotechnology scenario. **a.** A polypeptide can be removed from a cell, and the amino acid sequence can be determined. From this, the sequence of nucleotides in DNA can be deduced. **b.** The DNA synthesizer can be used to make DNA probes for the gene. **c.** The probe can be used to test fetal cells for an inborn error of metabolism.

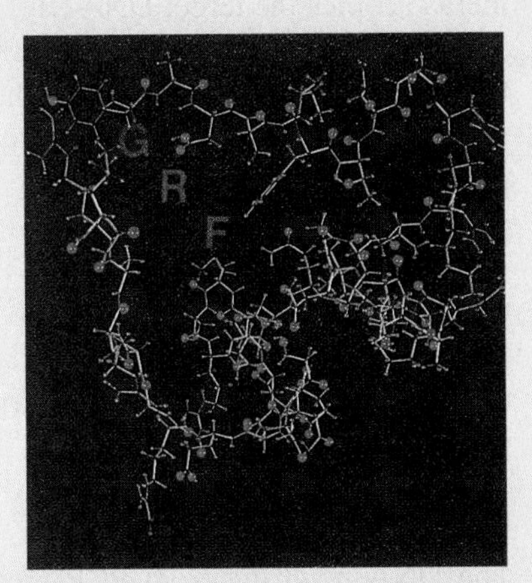

a.

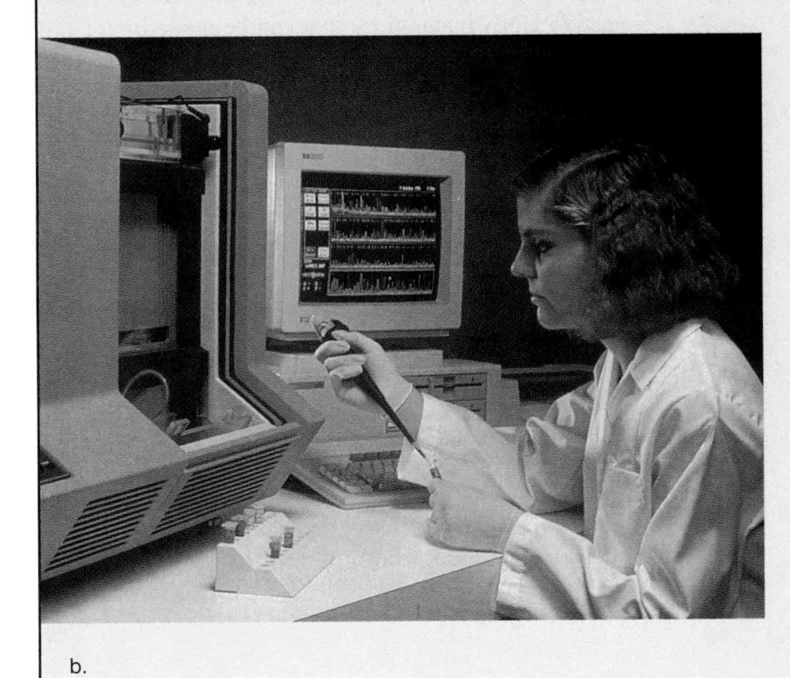

b.

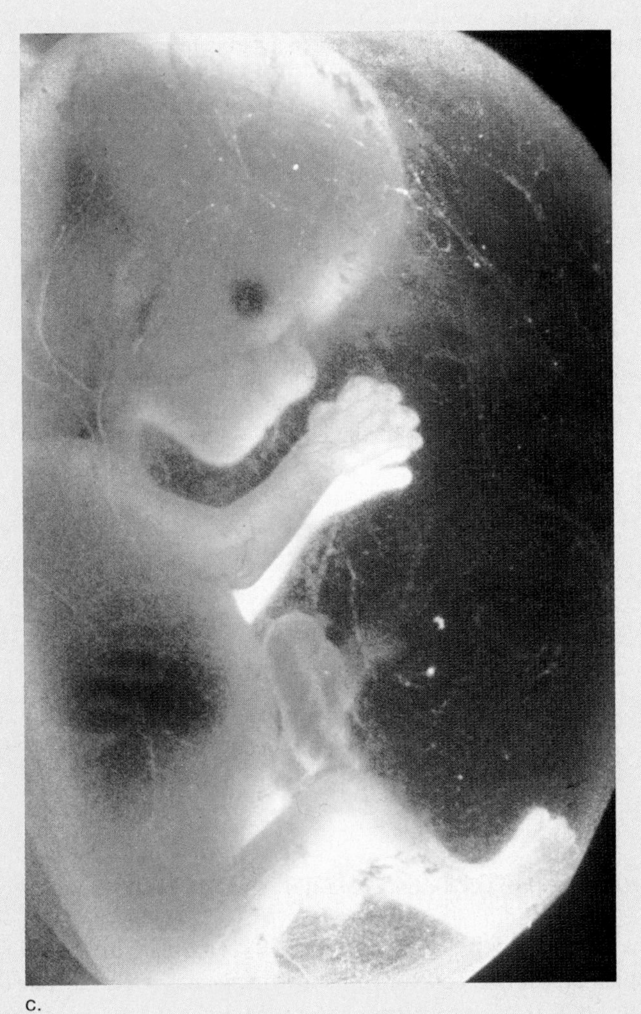

c.

Figure 25.5
PCR amplification and analysis. **a.** DNA is removed from a cell and is placed in a test tube along with appropriate primers, DNA polymerase, and a supply of nucleotides. **b.** Following PCR amplification, many copies of target DNA (red) are present. **c.** Binding of a labeled DNA probe (blue) enables the scientist to determine that a particular DNA segment was indeed present in the original sample.

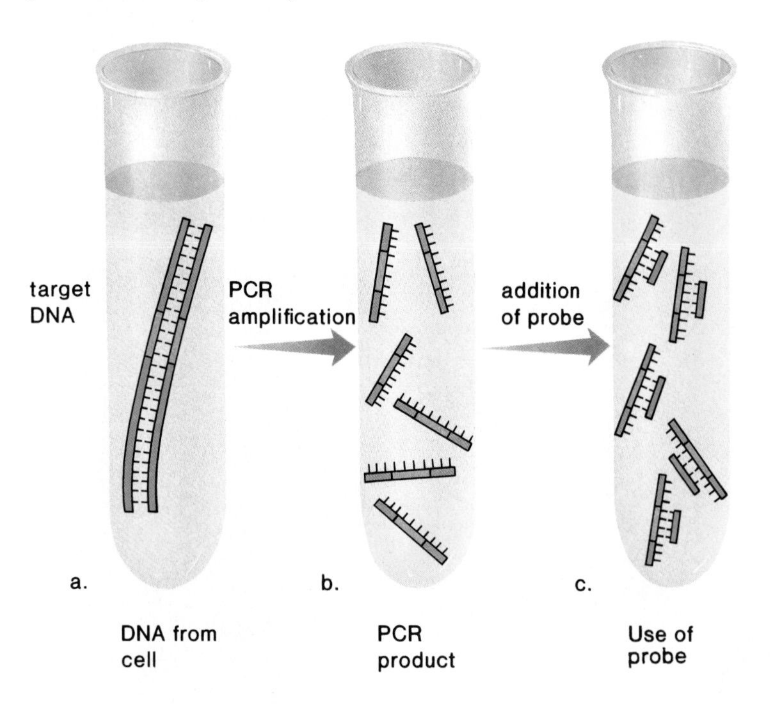

The Polymerase Chain Reaction

The *polymerase chain reaction (PCR)* can create millions of copies of a single gene or any specific piece of DNA in a test tube. PCR is very specific—the *targeted DNA sequence* can be less than one part in a million of the total DNA sample! This means that a single gene among all the human genes can be amplified (copied) using PCR.

PCR takes its name from DNA polymerase, the enzyme that carries out DNA replication in a cell. It is considered a chain reaction because DNA polymerase will carry out replication over and over again, until there are millions of copies of the targeted DNA. PCR does not replace gene cloning; cloning provides many more copies of a gene, and it still is used whenever a large quantity of a gene or a protein product is needed.

Before carrying out PCR, *primers*—sequences of about 20 bases that are complementary to the bases on either side of the "target DNA"—must be available. The primers are needed because DNA polymerase does not start the replication process—it only continues or extends the process. After the primers bind by complementary base pairing to the DNA strand, DNA polymerase copies the target DNA (fig. 25.5*a* and *b*).

PCR has been in use for several years, but the introduction of automated PCR machines is a recent advance. Now almost any laboratory can carry out the procedure. Automation became possible after a temperature-insensitive (thermostable) DNA polymerase was extracted from a bacterium. The availability of this enzyme means that there is no need to add more DNA polymerase each time a high temperature is used to separate double-stranded DNA so that replication can reoccur at the target area.

Analyzing DNA Strands

The DNA resulting from PCR amplification can be analyzed using the following procedures involving DNA probes. A **DNA probe** is a single strand of radioactive DNA nucleotides that can be made by a DNA synthesizer. Because a DNA probe is single stranded, it seeks out and binds to a complementary DNA strand (fig. 25.4*c*). A DNA probe that has a sequence of nucleotides complementary to the gene of a microbe can be used to diagnose an infection caused by the microbe. Specific DNA probes can also be used to diagnose tuberculosis or an HIV infection. In addition, a DNA probe can tell us whether a gene coding for a hereditary defect or causing a cell to be cancerous is present.

DNA probes are used during **DNA fingerprinting,** a process described in figure 25.6. The DNA is first treated with restriction enzymes, which cut it into fragments. Each organism's DNA results in a collection of different-sized fragments. Therefore, restriction fragment length polymorphisms (RFLPs) exist. During a process called gel electrophoresis, the fragments are separated according to their lengths, and the result is a pattern of bands that is different for each organism. The use of radioactive probes allows specific sequences to be separated from all the other fragments, and the resulting pattern to be recorded on X-ray film.

Figure 25.6
How to do a DNA fingerprint. DNA samples I and II are from the same individual. DNA sample III is from a different individual. Notice, therefore, that the restriction enzyme cuts are different. Gel electrophoresis separates the DNA fragments according to their length because shorter fragments migrate further in an electrical field than do longer fragments. The fragments are separated (denatured) and transferred to a membrane where a radioactive probe can be applied. The resulting pattern (the DNA fingerprint) can then be detected by autoradiography. Notice that samples I and II were from the same individual. In a rape case, sample I could be from the suspect's white blood cells, and sample II could be from sperm that were in the victim's vagina. Sample III could be from the victim's white blood cells.

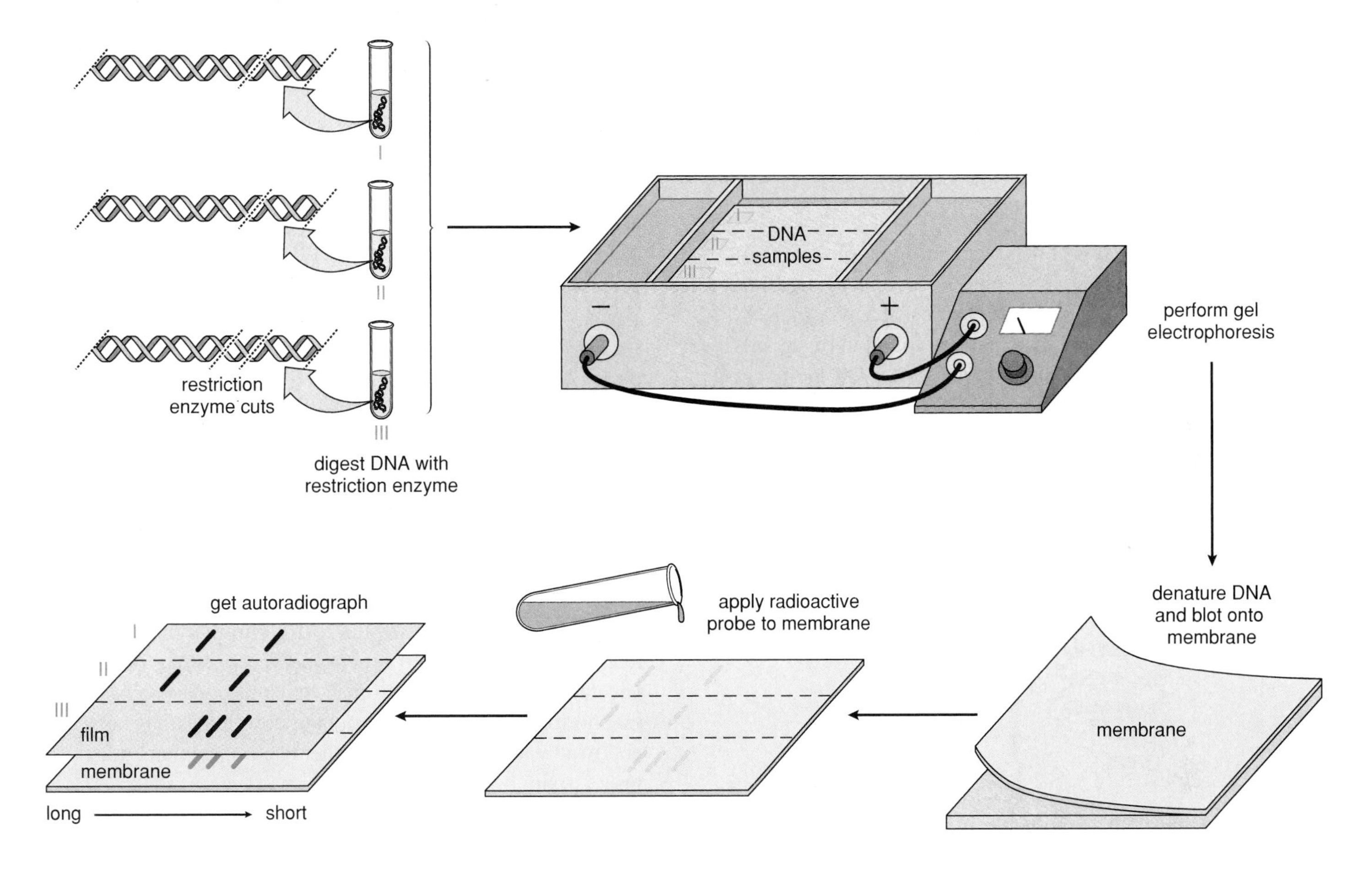

DNA fingerprinting can be done on any piece of DNA (fig. 25.6). It was used to successfully identify a teenage murder victim from remains 8 years old. Skeletal DNA was compared to that obtained from blood samples donated by the victim's parents. A DNA fingerprint resembles that of the parents because it is inherited. DNA from a single sperm is enough to identify a suspected rapist when PCR amplification precedes DNA fingerprinting.

DNA fingerprinting is also helpful to evolutionists. For example, it was used to determine that the quagga, an extinct zebralike animal, was a zebra rather than a horse. The only remains of the quagga consisted of dried skin. In other studies, DNA sequences from a 7,000-year-old mummified human brain and from a 17–20-million-year-old plant fossil were analyzed using PCR amplification followed by DNA fingerprinting.

Other Biotechnology Procedures

Ribozymes and abzymes are 2 new types of enzymes now being utilized by scientists. Ribozymes are RNA enzymes that splice out introns during mRNA processing, before the molecule leaves the nucleus of a eukaryotic cell. Ribozymes isolated from cells are being considered for medical applications. For example, they can perhaps be used to destroy the HIV viruses, which have RNA instead of DNA genes.

Abzymes are antibodies that have been modified to have catalytic properties. Because there are so many different types of antibodies, many different types of enzymes for various purposes may soon be available.

Antisense: Turning Off Genes

Antisense technology is a new biotechnology tool that promises to be helpful in research, medicine, and agriculture. *Antisense molecules* are short sequences of DNA or RNA nucleotides complementary to sequences found in organisms. An antisense DNA molecule is complementary to the nucleotide sequence of a gene. When an antisense DNA molecule binds to a gene, it turns the gene off. It is possible that someday, antisense DNA molecules could be used to turn off the genes of a microbe or an oncogene.

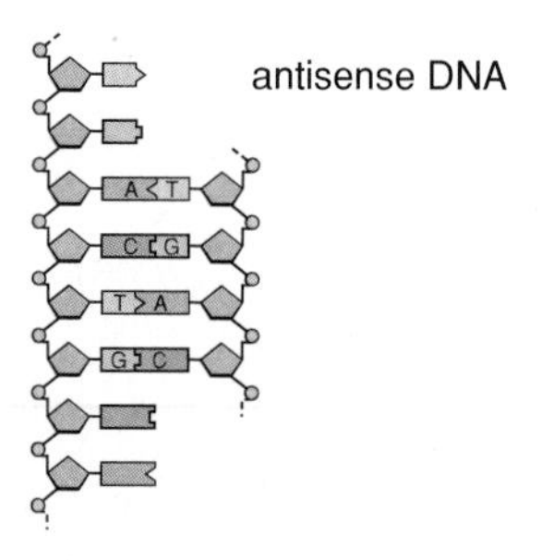

An antisense RNA molecule is complementary to the nucleotide sequence of an RNA. When an antisense RNA molecule binds to an mRNA molecule, the mRNA is inactivated. An antisense RNA might be useful to turn off the replication of HIV viruses because they have RNA genes.

Transgenic Organisms

Transgenic organisms have had a foreign gene inserted into them.

Transgenic Bacteria: Corn Fields to Gold Mines

As you know, bacteria are used to clone a gene or to mass-produce a product (fig. 25.3). They are also bioengineered to perform other services.

Protecting and Enhancing Plants

Bioengineered bacteria can be used to promote the health of plants. For example, bacteria that normally live on plants and normally encourage the formation of ice crystals have been changed from frost-plus to frost-minus bacteria. Field tests showed that these bioengineered bacteria protect the vegetative parts of plants from frost damage. Also, a bacterium that normally colonizes the roots of corn plants has now been endowed with genes (from another bacterium) that code for an insect toxin.

Many other recombinant DNA applications in agriculture are thought to be possible. For example, *Rhizobium* is a bacterium that lives in nodules on the roots of leguminous plants, such as bean plants. Here, the bacteria fix atmospheric nitrogen into a form that can be used by the plant (see fig. 33.10). It might be possible to transfer the necessary genes to other bacteria, which can then infect nonleguminous plants, such as corn, rice, and wheat. This would reduce the amount of fertilizer needed on agricultural fields.

Figure 25.7
Oil-eating bacteria, engineered and patented by investigator Dr. Chakrabarty. In the inset, the flask toward the front contains oil and no bacteria; the flask toward the rear contains the bacteria and is almost clear of oil. Now that engineered organisms (e.g., bacteria and plants) can be patented, there is even greater impetus to create them.

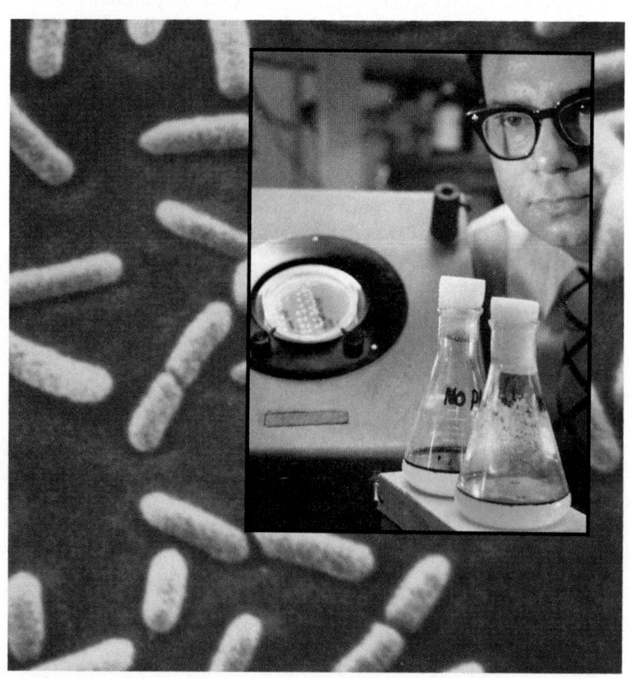

Bioremediation

There are naturally occurring bacteria that can degrade almost any type of chemical or material. Bacteria can be selected for their ability to degrade a particular substance, and then this ability can be enhanced by bioengineering. For example, naturally occurring bacteria that eat oil can be bioengineered to do an even better job of cleaning up beaches after oil spills (fig. 25.7).

Producing Chemicals

Organic chemicals are often synthesized by having catalysts act on precursor molecules or by using bacteria to carry out the synthesis. Today, it is possible to go one step further and to manipulate the genes that code for these enzymes. For example, biochemists discovered a strain of bacteria that is especially good at producing phenylalanine, an organic chemical needed to make aspartame, the dipeptide sweetener better known as NutraSweet™. They isolated, altered, and formed a vector for the appropriate genes so that various bacteria could be bioengineered to produce phenylalanine.

Processing Minerals

Many major mining companies already use bacteria to obtain various metals. Bioengineering may enhance the ability of bacteria to extract copper, uranium, and gold from low-grade sources. At least 2 mining companies plan to test bioengineered organisms having enhanced bioleaching capabilities.

Ecological Considerations

There are those who are very concerned about the deliberate release of genetically engineered microbes (GEMs) into the environment. Ecologists point out that these bacteria might displace those that normally reside in an ecosystem, and the effects could be deleterious. Others rely on past experience with GEMs, primarily in the laboratory, to suggest that these fears are unfounded. Tools are now available to detect, measure, and even disable cell activity in the natural environment. It is hoped that these will eventually pave the way for GEMs to play a significant role in agriculture and in environmental protection.

Transgenic Plants: Tougher Tomatoes

Plants, in particular, lend themselves to genetic manipulation because it is possible to grow plant cells in tissue culture, where each cell can be stimulated to produce an entire plant (fig. 25.8).

Figure 25.8
Cloning of entire plants from tissue cells. **a.** Sections of carrot root are cored, and thin slices are placed in a nutrient medium. **b.** After a few days, the cells form a callus, a lump of undifferentiated cells. **c.** After several weeks, the callus begins sprouting cloned carrot plants. **d.** Eventually, the carrot plants can be moved from culture medium to potting soil.

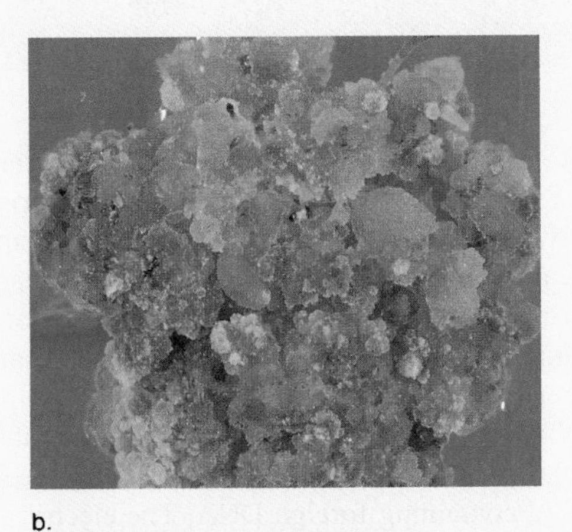

b.

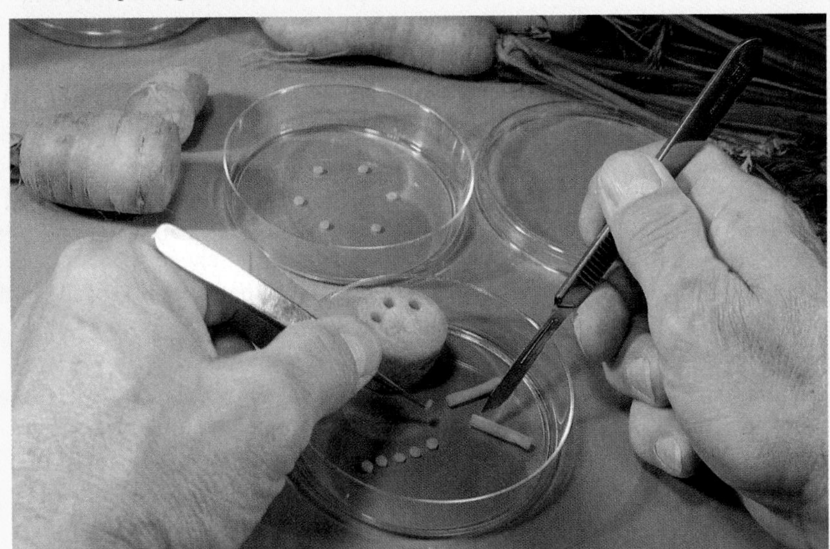

a.

c.

d.

Figure 25.9
Genetically engineered plants. The cotton boll on the *right* is from a plant that has not been bioengineered to resist cotton bollworm larvae and is heavily infested. The boll on the *left* is from a plant that was bioengineered to resist cotton bollworm larvae and will go on to give a normal yield of cotton.

Before plants will take up a plasmid, their cell walls must be removed to give "naked" cells called **protoplasts.** The only possible plasmid for bioengineering plant cells, the *Ti* plasmid, is transferred by its host, *Agrobacterium,* to many but not all plants. Unfortunately, the plants of greatest agricultural significance do not take up this plasmid. Therefore, other techniques have been developed to introduce foreign DNA into plant cells. For example, it is possible to treat protoplasts with an electric current while they are suspended in a liquid containing foreign DNA. The electric current makes tiny self-sealing holes in the cell membrane through which genetic material can enter. Presently, about 50 types of bioengineered plants that resist insects, viruses, or herbicides have now entered small-scale field trials (fig. 25.9). The major crops that can be improved in this way are soybean, cotton, alfalfa, and rice; however, even bioengineered corn may reach the marketplace by the year 2000.

It is hoped that one day bioengineered plants will include the following:

Plants that have an increased ability to grow under unfavorable environmental conditions, such as freezing temperatures. Heat-, cold-, drought-, and salt-tolerant plants are expected in the near future.

Plants that are more nutritious because the seed contains all the amino acids required by humans. Protein-enhanced beans, corn, soybeans, and wheat are now being developed. The food processing industry is also interested in plants that can be stored and transported without fear of damage. Already, bruise-resistant tomatoes in which fruit ripening is delayed have been produced using antisense RNA therapy (fig. 25.9).

Plants that require less fertilizer and are able to make use of nitrogen from the atmosphere. (Plants ordinarily take nitrogen from the soil only.) Certain bacteria can make use of atmospheric nitrogen. If the required genes were cloned, perhaps they could be transferred to plants.

Plants that produce chemicals and drugs are of interest to humans. Potatoes have been engineered to produce human albumin, and one day it is believed that plants will produce human proteins, such as hormones, in their seeds.

There are ecological considerations involving bioengineered plants, also. Some worry, for example, that if crops are resistant to herbicides, farmers will not hesitate to use massive amounts of herbicide to kill weeds, and the environment will be degraded. If plants are resistant to insects because they produce a toxin, perhaps resistant insects will evolve.

Transgenic Animals: Bigger Fishes and tPA Sheep

Animals, too, are being bioengineered. Because animal cells will not take up bacterial plasmids, other methods are used to insert genes into their eggs. Previously, the most common method was to microinject foreign genes into eggs before they were fertilized. During vortex mixing, a time-saving procedure, eggs are placed in an agitator with DNA and silicon-carbide needles. The needles make tiny holes through which the DNA can enter the eggs. Using this technique, many types of animal eggs have been injected with bovine growth hormone (bGH). The procedure has been used to produce larger fishes, cows, pigs, rabbits, and sheep. Bioengineered fishes are now being kept in ponds that offer no escape to the wild because there is much concern that they will upset or destroy natural ecosystems.

Transgenic farm animals are being developed to produce biotechnology products. For example, the milk of a transgenic cow now contains lactoferin, a protein that is involved in iron transport and has antibacterial activity. Similarly, the milk of a transgenic sheep contains tPA (see table 25.1). And there is now a pig that can produce human hemoglobin—it is hoped that one day artificial blood for humans will be a reality.

Transgenic bacteria, plants, and animals are now a reality. These organisms have been bioengineered to serve all sorts of purposes. Do 25.1 Critical Thinking, found at the end of the chapter.

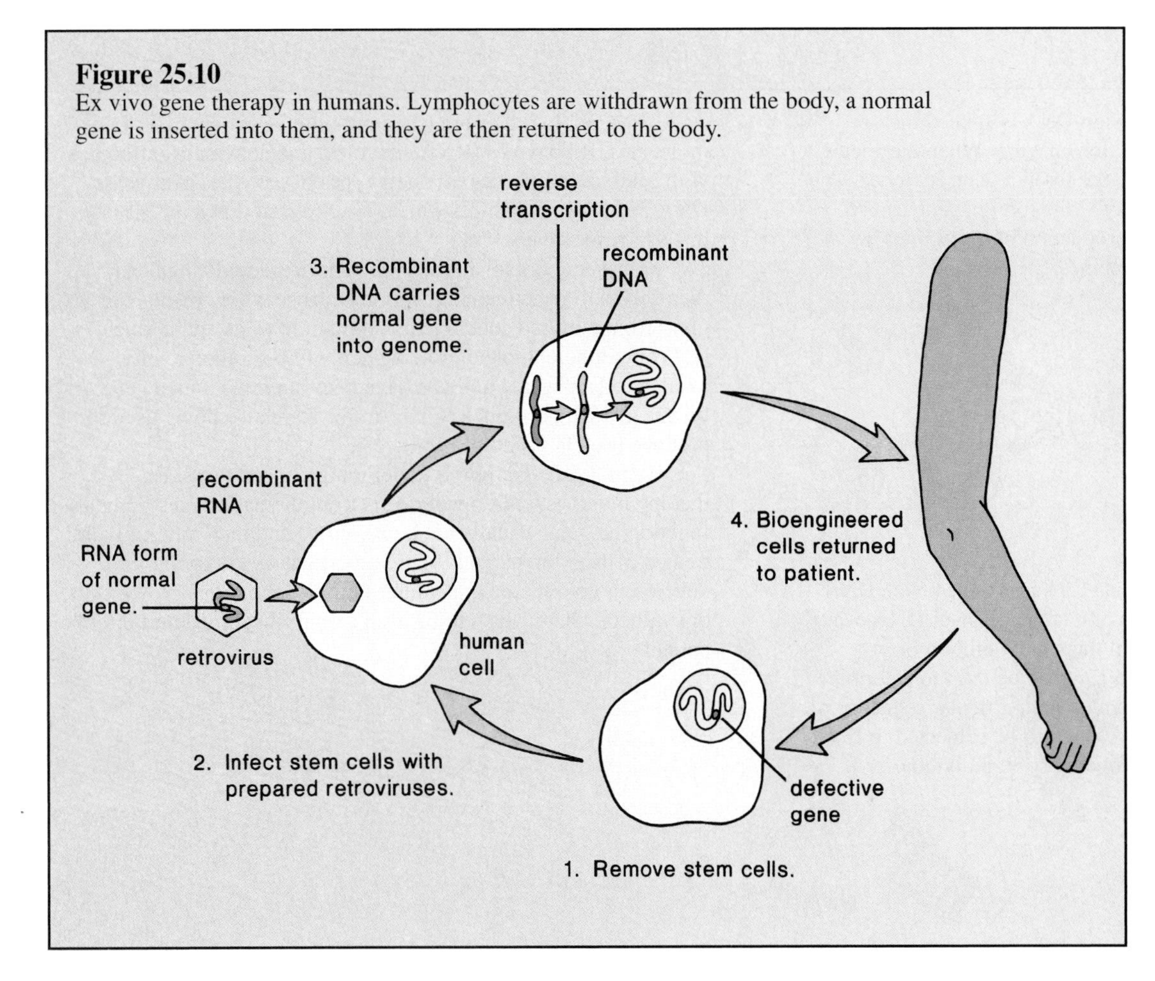

Figure 25.10
Ex vivo gene therapy in humans. Lymphocytes are withdrawn from the body, a normal gene is inserted into them, and they are then returned to the body.

Gene Therapy in Humans

Gene therapy replaces defective genes with healthy genes. Gene therapy also includes the use of genes to treat such human ills as diabetes and AIDS.

Ex Vivo: Cells Removed from Patient

During ex vivo (outside living organism) therapy, cells are removed from a patient, treated, and returned to the patient. A retrovirus, which has RNA genes instead of DNA genes, is used as a vector to carry normal genes into the cells of the patient, where they become incorporated into the genome. When a retrovirus is used for gene therapy, it has been equipped with recombinant RNA—viral RNA plus an RNA copy of the normal gene. After recombinant RNA enters a human cell, such as a bone marrow stem cell, reverse transcription occurs, and then recombinant DNA carrying the normal gene enters a human chromosome (fig. 25.10).

Ex vivo therapy clinical trials are underway. In one such trial, the gene needed by two girls with severe combined immune deficiency syndrome (SCID),[1] was introduced into their lymphocytes. These girls lack an enzyme that is involved in the maturation of T and B cells; as a precaution, they also received the enzyme (isolated from cows) as a medication. Recently, one of these girls received bioengineered bone marrow stem cells; this is preferred because stem cells are long-lived and their use may result in a permanent cure.

In Vivo: Genes Directly to Patient

Other gene therapy procedures use viruses, laboratory-grown cells, or even synthetic carriers to introduce genes directly into the patient. If in vivo (inside living organism) therapy is used, no cells are removed from the patient. For example, an adenovirus that contains a gene to treat cystic fibrosis patients has been placed in an aerosol spray. When laboratory animals inhale the spray, the virus enters the cells lining the lungs.

Perhaps it will be possible to use in vivo therapy to cure hemophilia, diabetes, Parkinson disease, or AIDS. To cure hemophilia, patients would get regular doses of cells containing normal clotting-factor genes. Or cells could be placed in organoids, artificial organs that can be implanted in the abdominal cavity. To cure Parkinson disease, dopamine-producing cells could be grafted directly into the brain. These procedures will use laboratory-grown cells that have been stripped of antigens (to decrease the possibility of an immune system attack). To cure AIDS, decoys that display the HIV CD4 receptor could be injected into the body. The HIV viruses would combine with these cells instead of with helper T lymphocytes.

Some investigators are pursuing gene therapy by direct injection of a protein-DNA complex, which is expected to enter liver cells for the purpose of treating diseases associated with the liver, including cancer.

1. SCID is often called the "bubble-baby" disease after David, a young person who lived under a plastic dome to prevent infection.

SUMMARY

To produce recombinant DNA, foreign DNA is spliced into a vector that has been cut by a restriction enzyme. When the vector (plasmids or viral DNA) replicates, the foreign gene is cloned. The cloned gene or the protein produced by the foreign gene can be collected. This is the basis for the production of biotechnology products, such as hormones and vaccines.

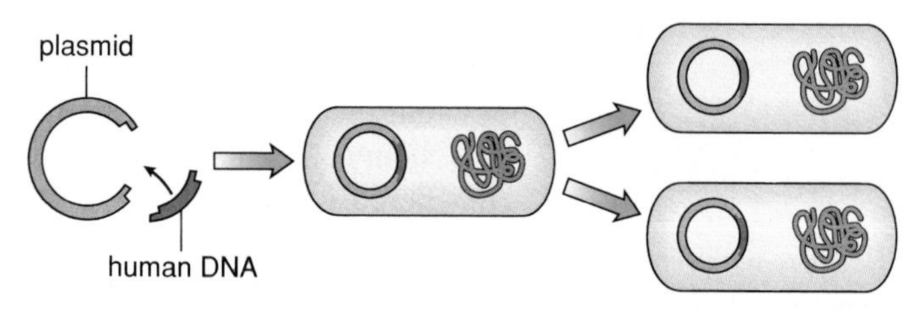

The polymerase chain reaction (PCR) uses the enzyme DNA polymerase to carry out replication of a target piece of DNA over and over again (amplification), until there are a million or so copies. A suitable radioactive probe can then be used to determine if the DNA of an infectious organism or any particular sequence of DNA is present. The copies of DNA can also be subjected to DNA fingerprinting, which can be used to identify an individual.

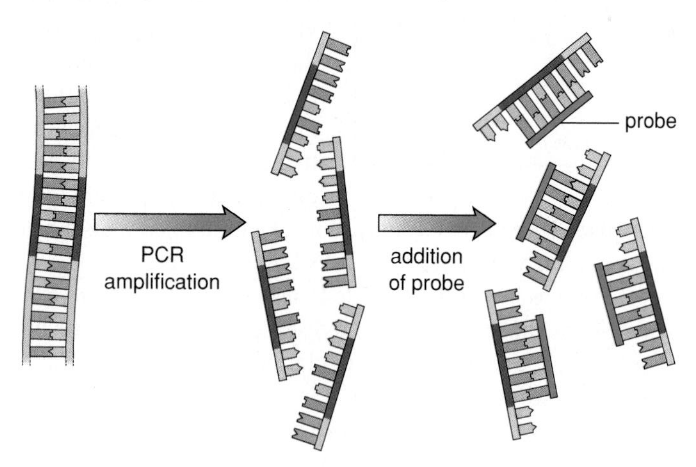

Biotechnology has branched out into the production of novel molecules. Ribozymes (RNA enzymes) and abzymes (antibodies with catalytic properties) are new types of enzymes. Antisense DNA and RNA sequences can be used to shut down a gene or mRNA, respectively.

Transgenic organisms also have been made. Plants lend themselves to genetic manipulation because whole plants will grow from cultured cells. Plants resistant to pests and herbicides and containing complete protein soon will be commercially available. Transgenic animals have been supplied with a gene for bovine growth hormone (bGH). Animals are also being used to produce protein products of interest.

Human gene therapy is undergoing clinical trials. Ex vivo therapy involves withdrawing cells from the patient, inserting a functioning gene, usually via a retrovirus, and then returning the treated cells to the patient. Many investigators are trying to develop in vivo therapy, in which viruses, laboratory-grown cells, or synthetic chemicals will be used to carry healthy genes into the patient.

STUDY QUESTIONS

In order to practice **writing across the curriculum,** students should write out the answers to any or all of the study questions. The study questions are sequenced in the same order as the text.

1. Explain the recombinant DNA technology for cloning a gene. (p. 488)
2. Give examples of the types of biotechnology products available today. (p. 490)
3. What is the polymerase chain reaction (PCR), and how is it carried out? (p. 492)
4. How and why are the strands resulting from PCR analyzed? (p. 492)
5. What new types of enzymes are being used in biotechnology procedures? What are antisense DNA and antisense RNA? (p. 494)
6. Bacteria have been bioengineered to perform what services? What are the ecological concerns regarding their release into the environment? (pp. 494–495)
7. Why are plants good candidates for bioengineering? In what ways have plants been bioengineered, and what type of bioengineering is expected in the future? (pp. 495–496)
8. In what ways have animals been bioengineered? Discuss any ecological concerns. (p. 496)
9. Define and give an example of ex vivo gene therapy. (p. 497) Define and give an example of in vivo gene therapy. (p. 497)

OBJECTIVE QUESTIONS

1. Plasmids and viruses can be used as _______ to carry foreign DNA into host cells.
2. When vectors replicate, the gene has been _______.
3. The _______ can produce a million copies of target DNA in a test tube.
4. A DNA probe is a sequence of bases that binds by _______ to a particular piece of DNA.
5. When _______ RNA binds to mRNA, the mRNA is shut down.
6. Plants can be bioengineered to be resistant to _______ and _______.
7. Many types of transgenic animals have now received a(n) _______ gene, which makes them grow larger.
8. The current ex vivo gene therapy clinical trials use a(n) _______ as a vector to insert healthy genes in the patient's cells.
9. The foreign gene to be inserted into this plasmid DNA, cleaved by a restriction enzyme, ends in what bases (a) from the left and (b) from the right?

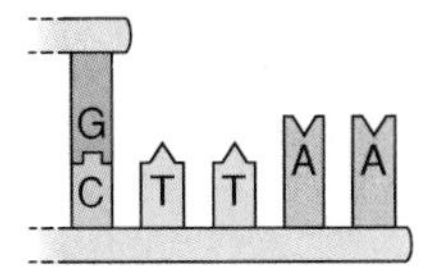

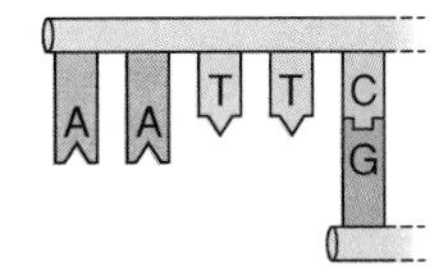

10. Label these drawings, using these terms: retrovirus, recombinant RNA (twice), human gene, recombinant DNA, reverse transcription, human genome.

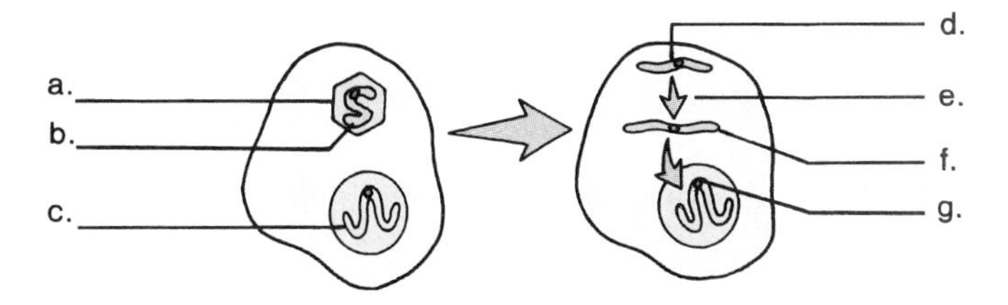

CRITICAL THINKING

In order to practice **writing across the curriculum,** students should write out the answers to any or all of the critical thinking questions. Suggested answers to the critical thinking questions are in appendix E.

25.1

1. What experiment would you suggest to support the hypothesis that insulin produced by biotechnology will have fewer side effects than insulin taken from the organs of livestock? Why do you expect your results to support the hypothesis?
2. What experiment would you suggest to support the hypothesis that meat from cattle bioengineered to contain extra growth hormone genes is safe for human consumption? Why do you expect your results to support the hypothesis? Your answer should consider the fact that growth hormone is a protein.
3. What experiment would you suggest to support the hypothesis that plants engineered to contain genes for an insect toxin are safe for human consumption? Why might your results not support the hypothesis?
4. What experiment would you suggest to support the hypothesis that bacteria bioengineered to clean up a pollutant disappear when the pollutant is gone? What would make them disappear?

SELECTED KEY TERMS

bioengineered alteration of the genome of an organism by technological processes; genetic engineering.

biotechnology the use of a natural biological system to produce a commercial product.

DNA fingerprinting using highly variable sequences of DNA to identify particular individuals.

DNA ligase an enzyme that links DNA fragments; used in bioengineering to join foreign DNA to vector DNA.

DNA probe single strand of radioactive DNA that can be used to find complementary DNA strands; can be used diagnostically to determine the presence of particular genes.

gene cloning to use recombinant DNA technology for the production of many copies of a gene.

gene therapy the use of biotechnology to treat genetic disorders and illnesses.

genomic library a collection of engineered viruses that together carry all of the genes of the species.

plasmid a circular DNA segment that is present in bacterial cells but is not part of the bacterial chromosome.

recombinant DNA DNA having genes from 2 different organisms, often produced in the laboratory by introducing foreign genes into a bacterial plasmid.

restriction enzyme enzyme that stops viral reproduction by cutting viral DNA; used in bioengineering to cut DNA at specific points.

transformed cell cell that has been altered by bioengineering and is capable of producing new protein.

transgenic organism an organism that has a foreign gene inserted into it.

vector a carrier, such as a plasmid or a virus, for recombinant DNA that introduces a foreign gene into a host cell.

FURTHER READINGS FOR PART FIVE

Anderson, R. M., and R. M. May. May 1992. Understanding the AIDS pandemic. *Scientific American.*

Aral, S. L., and K. K. Holmes. February 1991. Sexually transmitted diseases in the AIDS era. *Scientific American.*

Barton, J. H. March 1991. Patenting life. *Scientific American.*

Beardsley, T. August 1991. Smart genes. *Scientific American.*

Cummings, M. R. 1991. *Human heredity.* 2d ed. St. Paul, Minn.: West Publishing Co.

Cunningham, P. May 1991. The genetics of thoroughbred horses. *Scientific American.*

Drlica, K. 1992. *Understanding DNA and gene cloning.* 2d ed. New York: John Wiley and Sons.

Edelman, G. M. 1988. *Topobiology: An introduction to molecular embryology.* New York: Basic Books.

Edlin, G. 1990. *Human genetics: A modern synthesis.* Boston: Jones and Bartlett Publishers.

Feldman, M., and L. Eisenback. November 1988. What makes a tumor cell metastatic? *Scientific American.*

Gardner, E. J., M. J. Simmons, and D. P. Snostad. 1991. *Principles of genetics.* 8th ed. New York: John Wiley and Sons.

Gasser, C. S., and R. T. Fraley. June 1992. Transgenic crops. *Scientific American.*

Gilbert, S. 1991. *Developmental biology.* 3d ed. Sunderland, Mass.: Sinauer Associates.

Hampton, J. K., Jr. 1991. *Biology of human aging.* Dubuque, Ia.: Wm. C. Brown Publishers.

Hartl, D. L. 1988. *A primer of population genetics.* 2d ed. Sunderland, Mass.: Sinauer Associates.

Henig, R. M. 1985. *How a woman ages.* New York: Ballantine Books.

Holliday, R. June 1989. A different kind of inheritance. *Scientific American.*

Kalil, R. E. December 1989. Synapse formation in the developing brain. *Scientific American.*

Kart, C. S., E. K. Metress, and S. P. Metress. 1992. *Human aging and chronic disease.* Boston: Jones and Bartlett.

Liotta, L. A. February 1992. Cancer cell invasion and metastasis. *Scientific American.*

McNight, S. L. April 1991. Molecular zippers in gene regulation. *Scientific American.*

Mange, A. P., and E. J. Mange. 1990. *Genetics: Human aspects.* 2d ed. Sunderland, Mass.: Sinauer Associates.

Mills, J., and H. Masur. August 1990. AIDS-related infections. *Scientific American.*

Moore, K. L. 1988. *Essentials of human embryology.* Toronto: B. C. Decker, or St. Louis: Mosby Year Book.

Moyzis, R. K. August 1991. The human telomere. *Scientific American.*

Mullis, K. B. April 1990. The unusual origin of the polymerase chain reaction. *Scientific American.*

Murray, A. W., and M. W. Kirschner. March 1991. What controls the life cycle. *Scientific American.*

National Research Council, Institute of Medicine. 1990. *Developing new contraceptives: Obstacles and opportunities.* Washington, D.C.: National Academy Press.

Neher, E., and B. Sakmann. March 1992. The patch clamp technique. *Scientific American.*

Neufeld, P. J., and N. Colman. May 1990. When science takes the witness stand. *Scientific American.*

Nilsson, L. 1977. *A child is born.* Rev. ed. New York: Delacorte Press.

Pesmen, C. 1984. *How a man ages.* New York: Ballantine Books.

Ptashne, M. January 1989. How gene activators work. *Scientific American.*

Radman, M., and R. Wagner. August 1989. The high fidelity of DNA duplication. *Scientific American.*

Rennie, J. March 1993. DNA's new twists. *Scientific American.*

Rhodes, D., and A. Klug. February 1993. Zinc fingers. *Scientific American.*

Ross, J. April 1989. The turnover of messenger RNA. *Scientific American.*

Rusting, R. L. December 1992. Why do we age? *Scientific American.*

Sapienza, C. October 1990. Parental imprinting of genes. *Scientific American.*

Scientific American. October 1988. Entire issue is devoted to AIDS.

Steitz, J. A. June 1988. "Snurps." *Scientific American.*

Stine, G. J. 1992. *Biology of sexually transmitted diseases.* Dubuque, Ia.: Wm. C. Brown Publishers.

Tamarin, R. 1991. *Principles of genetics.* 3d ed. Dubuque, Ia.: Wm. C. Brown Publishers.

Ulmann, T., and D. Philibert. June 1990. RU 486. *Scientific American.*

Uvnas-Moberg, K. July 1989. The gastrointestinal tract in growth and reproduction. *Scientific American.*

Verma, I. M. November 1990. Gene therapy. *Scientific American.*

Weinberg, P. A. September 1988. Finding the anti-oncogene. *Scientific American.*

Weintraub, H. M. January 1990. Antisense RNA and DNA. *Scientific American.*

White, R., and J. Lalouel. February 1988. Chromosome mapping with DNA markers. *Scientific American.*

Wistreich, G. A. 1992. *Sexually transmitted diseases: A current approach.* Dubuque, Ia.: Wm. C. Brown Publishers.

Wolpert, L. 1991. *The triumph of the embryo.* New York: Oxford University Press.

Part Six

EVOLUTION AND DIVERSITY

Mandrill baboons with infant

Part Six

EVOLUTION AND DIVERSITY

A gradual increase in chemical complexity produced the first cell(s), which evolved into all the forms of life we see about us. Despite their diversity, living things are classified into just 5 kingdoms: monera, protists, fungi, plants, and animals. Evolution depends on the retention of phenotypic changes that have been tested by the environment. This process, termed natural selection, results in the great variety of living things. Life has a history that is traceable to the first cell(s). Since taxonomists try to classify living things according to their evolutionary relationship, when we study taxonomy, we are also studying evolutionary history.

26

THE ORIGIN OF LIFE

Chapter Concepts

1.
A chemical evolution proceeded from atmospheric gases to small organic molecules to macromolecules to protocell(s). 504, 512

2.
The first cell was bounded by a membrane and contained a replicating system, that is, DNA, RNA, and proteins. 506, 512

3.
The first cell was an anaerobic heterotroph. 507, 512

4.
Biological evolution proceeded from prokaryotic cell to eukaryotic cell to multicellularity to complex forms that diversified. 507, 512

5.
The diverse forms of life today are classified into 5 kingdoms. 510, 512

Volcanic region with geysers

Chapter Outline

Figure 26.1
A model for the origin of life.

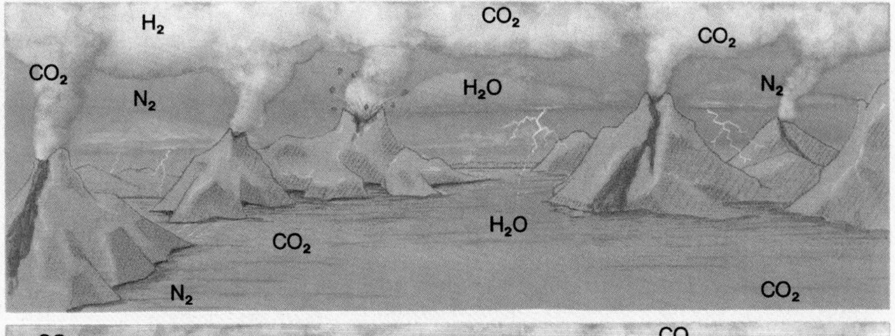

The primitive atmosphere contained gases, including water vapor, that escaped from volcanoes; as the water vapor cooled, some gases were washed into the oceans by rain.

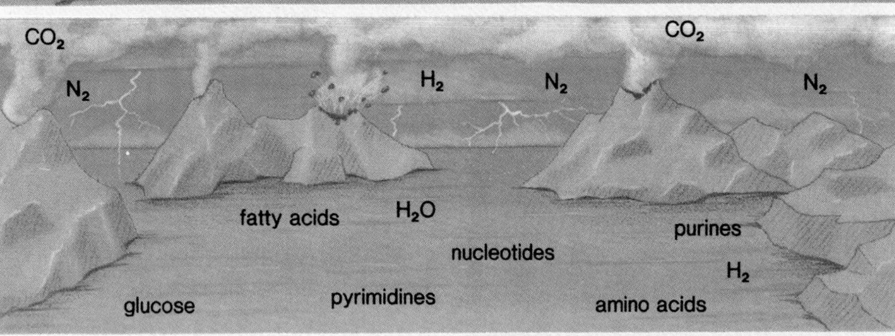

The availability of energy from volcanic eruption and lightning allowed gases to form simple organic molecules.

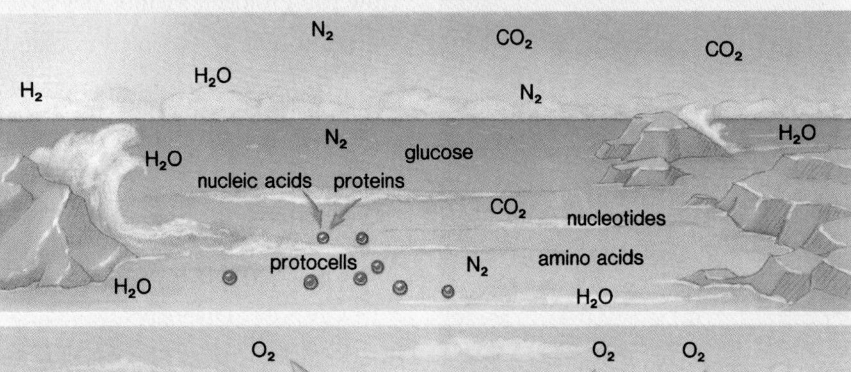

Simple organic compounds could have joined to form proteins and nucleic acids, which became incorporated into membrane-bounded spheres. The spheres became the first cells, called protocells.

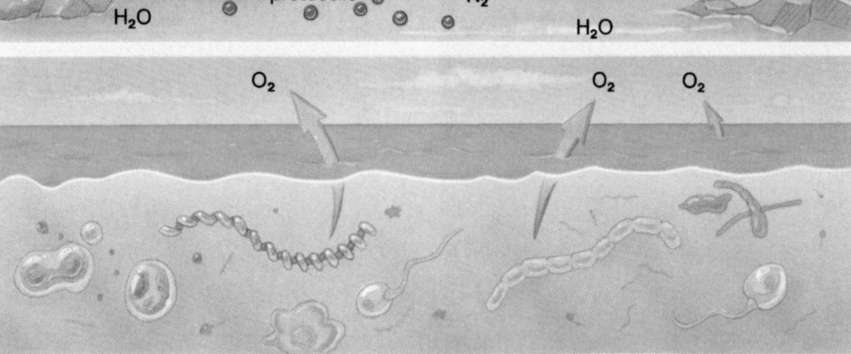

Eventually, various types of prokaryotes and then eukaryotes evolved. Some of the prokaryotes were oxygen-producing photosynthesizers. The presence of oxygen in the atmosphere was necessary for aerobic cellular respiration to evolve.

TODAY we do not believe that life arises spontaneously from nonlife, and we say that "life comes only from life." But then, how did the very first cell or cells come about? Since this cell was the very first living thing, it had to come from nonliving chemicals. Could there have been an increase in the complexity of the first chemicals—could a **chemical evolution** have produced the first cell(s) (fig. 26.1)?

Chemical Evolution

The sun and the planets probably formed from aggregates of dust particles and debris about 4.6 billion years ago. At first, the earth was such an inferno no living thing could have existed. The oldest fossils are about 3.5 billion years old; therefore, it seems that the first cell(s) must have arisen during the earth's first billion years or so, when the earth condensed and

Figure 26.2
A chemical evolution produced the protocell. There was an increase in the complexity of macromolecules, leading to a self-replicating system (DNA→RNA→protein) enclosed by a cell membrane. The protocell, a heterotrophic fermenter, underwent biological evolution, becoming a true cell, which then diversified.

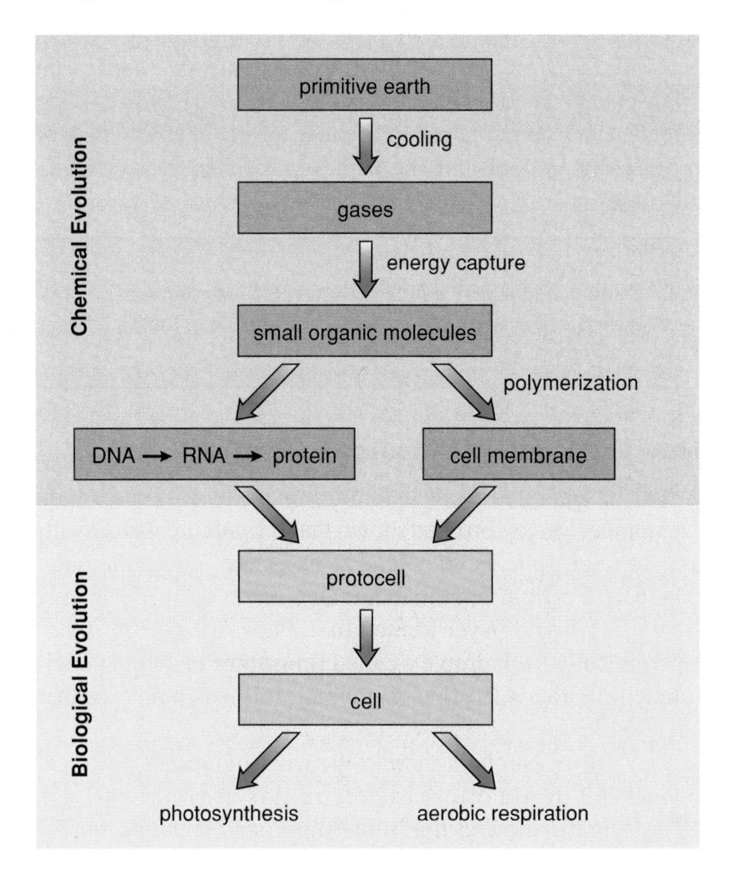

Figure 26.3
In Miller's experiment, gases (CH_4 [methane], NH_3 [ammonia], H_2 [hydrogen], H_2O [water]) were admitted to the apparatus, circulated past an energy source (electric spark), and cooled to produce a liquid, which could be withdrawn. Upon chemical analysis, the liquid was found to contain various simple organic molecules.

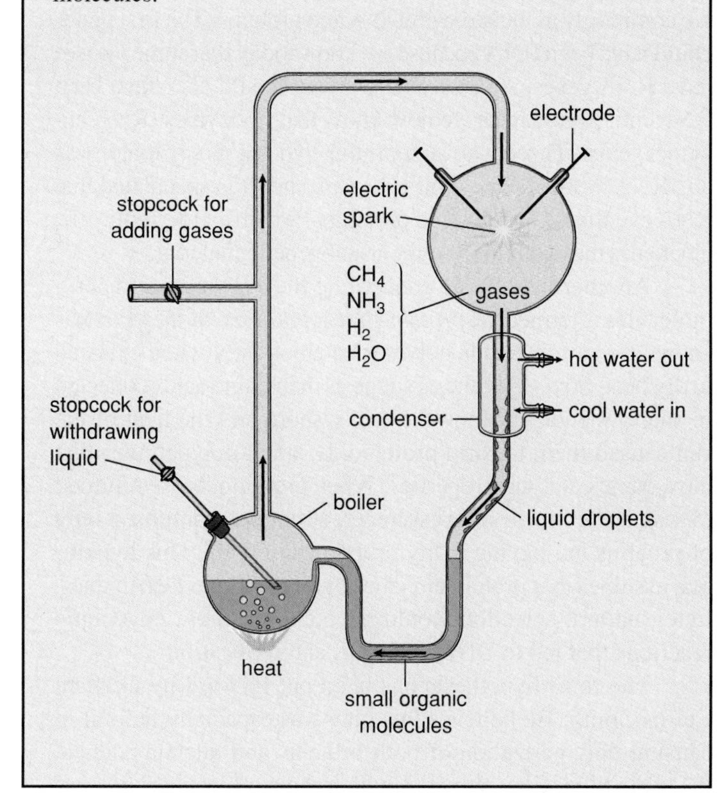

cooled. The steps outlined in figure 26.2 have been proposed as leading to the first cell(s).

Small Organic Molecules Evolve

The primitive atmosphere was not the same as the atmosphere today. It contained various gases (N_2, H_2, CO_2, water vapor) but no oxygen (O_2). Therefore, it was a **reducing atmosphere,** not the **oxidizing atmosphere** of today. This was fortuitous because oxygen attaches to organic molecules, preventing them from joining to form larger molecules. As the earth cooled, water vapor condensed to liquid water, and rain began to fall. It rained in such quantity that the oceans of the world were produced. The atmospheric gases, dissolved in rain, were carried down into newly forming oceans, where life may have originated.

As early as the 1920s, Soviet biochemist A. I. Oparin proposed that organic molecules could have been produced from the gases of the primitive atmosphere in the presence of a strong outside **energy source,** such as heat from volcanoes, powerful electric discharges in lightning, or ultraviolet radiation. In 1953, Stanley Miller provided support for Oparin's ideas with an ingenious experiment (fig. 26.3). Miller placed a mixture resembling a strongly reducing primitive atmosphere in a closed system, heated the mixture, and then circulated it past an electric spark (simulating lightning). After a week's run, Miller discovered that a variety of amino acids and organic acids had been produced.

This experiment and others similar to it support the hypothesis that the primitive gases could have reacted with one another to produce small organic molecules that accumulated in the ancient ocean, which became a thick, hot organic soup.

> Cooling caused water vapor to turn to rain, which formed the oceans. It is possible that here atmospheric gases reacted with one another under the influence of an outside energy source to produce simple organic molecules.

Macromolecules Evolve and Interact

The newly formed organic molecules could have polymerized to form macromolecules. There are 3 primary hypotheses concerning this stage in the origin of life. One of these, the *RNA-first hypothesis,* suggests that only the macromolecule RNA was needed to progress toward formation of the first cell or cells. In other words, RNA could have carried out the processes of life we commonly associate with DNA and proteins. The first genes could have been RNA because we know today that some viruses have RNA genes. And the first enzymes could also have been RNA molecules, since we now know that ribozymes (RNA enzymes) exist. Those who support this hypothesis say that it was an "RNA world" some 4 billion years ago. It is speculated that RNA eventually synthesized proteins—which can be more efficient enzymes—and DNA, the usual genetic material.

Another hypothesis concerning the evolution of macromolecules is termed the *protein-first hypothesis.* Sidney Fox has shown that amino acids polymerize abiotically when exposed to dry heat. Fox's hypothesis suggests that amino acids collected in shallow puddles along the rocky shore and the heat of the sun caused them to form proteinoids, small polypeptides that have some catalytic properties. When proteinoids are returned to water, they form **microspheres,** structures composed only of proteins but having many properties of cells. This hypothesis assumes that protein enzymes evolved before DNA genes. Later, natural selection would have chosen those enzymatic reactions that led to DNA synthesis and replication.

The third hypothesis has been put forward by Graham Cairns-Smith. He believes that clay was especially helpful in causing polymerization of both proteins and nucleic acids at the same time. Clay attracts small organic molecules and contains iron and zinc, which may have served as inorganic catalysts for polypeptide formation. In addition, clay has a tendency to collect energy from radioactive decay and to discharge it when the temperature and/or humidity changes. This source of energy could have helped polymerization take place. Cairns-Smith's hypothesis suggests that RNA nucleotides and amino acids became associated in such a way that polypeptides were ordered by and helped synthesize RNA. Later, DNA formed.

All 3 hypotheses assume that through molecular interaction, the DNA → RNA → protein self-replicating system evolved. Some of the proteins were enzymes that carried on crude energy-yielding metabolism. It is believed that all living things have a common origin because all use the same type of replicating system and the same type of energy-yielding metabolism.

> Small organic molecules polymerized to produce macromolecules, which interacted to arrive at the DNA→RNA→protein self-replicating system. Do 26.1 Critical Thinking, found at the end of the chapter.

How Did Protocell Membranes Evolve?

A cell is separated from its environment by a lipid-protein membrane. Sidney Fox has shown that if lipids are made available to microspheres, they tend to associate so that the outer boundary is a lipid-protein membrane. Even so, it is also possible that a lipid bilayer formed first. Phospholipid molecules automatically form droplets called **liposomes** in a liquid environment. Perhaps the first membrane formed in this manner (fig. 26.4).

Some researchers support the work of Oparin, who was one of the original researchers in this area. As early as 1938, Oparin showed that under appropriate conditions of temperature, ionic composition, and pH, concentrated mixtures of macromolecules tend to give rise to complex units called **coacervate droplets.** Coacervate droplets have a tendency to absorb and incorporate various substances from the

Figure 26.4
The cell membrane could have come about when (**a**) microspheres associated with lipids in the environment or when (**b**) phospholipids automatically formed liposomes in water.

a.

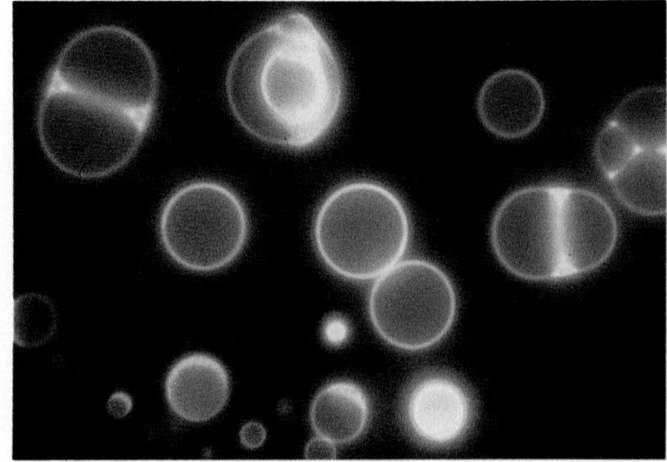
b.

surrounding solution. Eventually, a semipermeable type of boundary may form about the droplet.

Once the replicating system discussed previously was located within a cell membrane, a protocell had come into being. A **protocell** was the very first structure to function as a cell functions.

The first cell (protocell) had a self-replicating system and a cell membrane.

Protocells Were Heterotrophic

Nutrition would have been no problem for the protocell because it presumably existed in the ocean, which at that time was an **organic soup** containing small organic molecules. These molecules could have served as sources of building blocks and energy. Therefore, the protocell was likely a **heterotroph,** an organism that takes in preformed food. Notice that this hypothesis suggests that heterotrophs are believed to have preceded autotrophs, organisms that make their own food.

At first, the protocell may have used preformed ATP, but as this supply dwindled, natural selection favored any cells that could extract energy from carbohydrates in order to transform ADP to ATP. Glycolysis is a common metabolic pathway in living things, and this testifies to its early evolution in the history of life. Since there was no free oxygen, we can assume that the protocell carried on a form of fermentation.

It seems logical that the protocell at first had limited ability to break down organic molecules and that it took millions of years for glycolysis to evolve completely. It is of interest that Fox showed that microspheres have some catalytic ability and that Oparin found that coacervates do incorporate enzymes if they are available in the medium.

The protocell is hypothesized to have been a heterotrophic fermenter with some enzymatic ability.

Biological Evolution

Once a prokaryote arose, biological evolution would have begun. A protocell became a prokaryote once all cellular metabolic pathways were in place and self-replication was possible. **Biological evolution** depends on the presence of genes that can replicate, mutate, and control the characteristics of the cell.

Prokaryotic Cells: Photosynthesizing First

The first fossils, dated from about 3.5 billion years ago, are presumed to have been prokaryotic cells. Some of these fossils are found in *stromatolites* (*stroma* means *bed; lithos* means *stone*), which are pillarlike structures containing sedimentary layers and communities of autotrophic microorganisms (fig. 26.5). Living stromatolites exist even today in shallow waters off the west coast of Australia.

Figure 26.5
The oldest prokaryotic fossils date from 3.5 billion years ago. **a.** This filamentous prokaryotic fossil resembles the cells of cyanobacteria today. **b.** A fossilized stromatolite has layers that contain this type of fossil. **c.** Living stromatolites are located in shallow waters off the shores of western Australia.

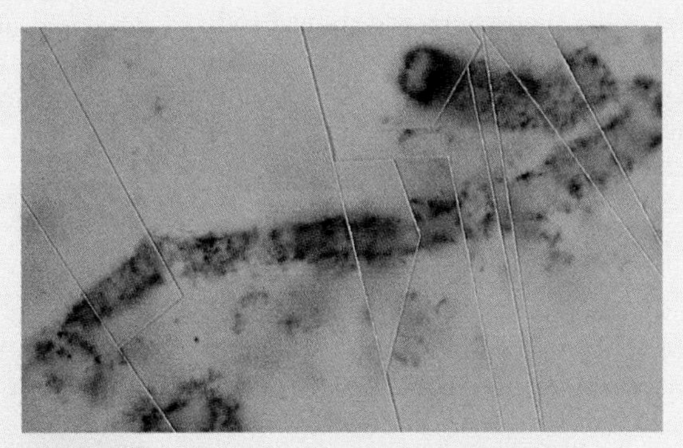
a.

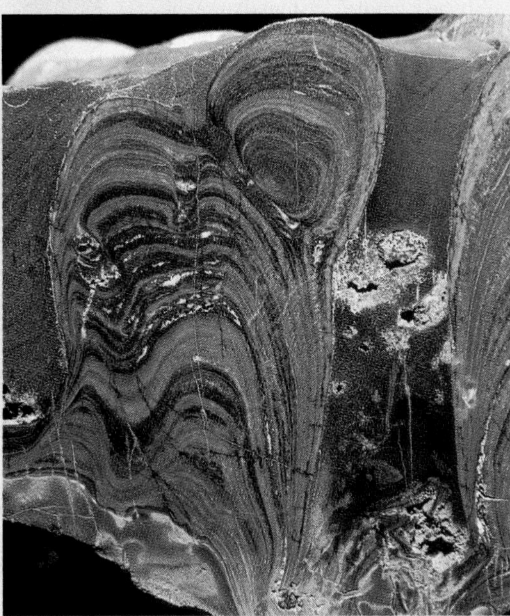
b.

c.

MOTHER EARTH: THE GAIA HYPOTHESIS

The Gaia hypothesis (after the classical Greek word for *our mother earth*) maintains that life regulates the climate and the chemical composition of the atmosphere, oceans, and soil so that they are suitable to life. For example, photosynthesizing organisms put oxygen (O_2) into the atmosphere, and its level has been no greater than 20% for the last half-billion years. Since oxygen tends to destroy macromolecules, what keeps atmospheric oxygen at a level suitable for all other living things? Those who support the Gaia hypothesis point to the presence of anaerobic bacteria that produce methane (CH_4). Methane combines with oxygen to form carbon dioxide (CO_2). Could there possibly be a control system that regulates how much methane should be produced in order to keep the level of oxygen constant?

Consider also this example. The sun has been heating up since its formation, yet the temperature of the earth has always been about the same. About 4 billion years ago, when the sun was perhaps 25% less luminous, the primitive atmosphere contained more ammonia (NH_3) and methane, which acted like the glass of a greenhouse to trap the heat of the sun near the earth. Today, carbon dioxide put into the atmosphere by all living things plays a more prominent role in creating a greenhouse effect. As the sun heats up, less and less of a greenhouse effect is needed to maintain the same temperature. Those who support the Gaia hypothesis believe that oceanic microorganisms with calcium carbonate ($CaCO_3$) shells are an essential part of a possible regulatory system. When carbon dioxide enters an ocean, it becomes bicarbonate and is utilized by these microorganisms to form their shells (fig. 26.A). When the organisms die, their shells sink to the bottom of the ocean. In effect, carbon dioxide is removed from the atmosphere in this way.

While many biologists agree that living things help to maintain suitable conditions for life on earth, they doubt there is any purposeful regulation. Quite often, those who support the Gaia hypothesis see the earth as a giant geophysiological organism that is purposefully maintaining itself. In this way, the Gaia hypothesis seems more like a religion than a science, and therefore it would be difficult to devise ways to test the hypothesis.

Figure 26.A
Coccolithophore. This microorganism has calcium carbonate ($CaCO_3$) plates embedded in its cell wall and exemplifies marine organisms that take calcium carbonate out of the sea. Calcium carbonate forms after carbon dioxide (CO_2) enters the sea from the air; therefore, this action helps to decrease the amount of carbon dioxide in the atmosphere. The sea is full of these tiny organisms, which are only 60 μm in diameter (a human hair is 100–200 μm thick).

Those who wish to refute the Gaia hypothesis show that inorganic processes also play a role in maintaining the chemical composition and the temperature of the earth. For example, when oxygen was first put into the atmosphere, it failed to build up quickly because much of it was taken up as iron became oxidized (there is much iron in the earth's crust). Also, carbon dioxide is washed out of the atmosphere by rain and in the process becomes carbonic acid. Carbonic acid combines with calcium silicate in rocks to give sediments containing carbon compounds. In other words, inorganic and organic processes have contributed to keeping the earth suitable for life.

Another argument against the Gaia hypothesis is that while some life forms have benefited from changed conditions, other forms have not. For example, when oxygen first entered the atmosphere, it may have helped aerobic organisms to evolve, but the populations of anaerobic organisms could only decline. The history of life contains many species of organisms that became extinct due to changing conditions. New ones more suited to the current conditions then evolve.

The Gaia hypothesis challenges us to see the earth at a geophysiological level of organization. Life is not passively acted upon by physical processes; it has contributed to changing and shaping the world in which we now live.

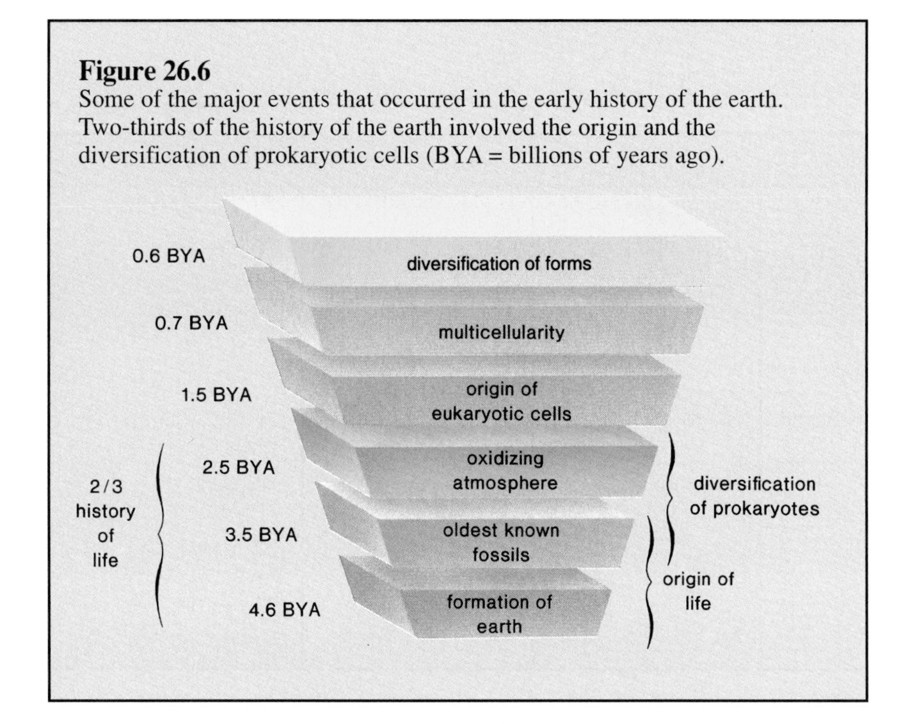

Figure 26.6
Some of the major events that occurred in the early history of the earth. Two-thirds of the history of the earth involved the origin and the diversification of prokaryotic cells (BYA = billions of years ago).

The autotrophs found in ancient stromatolites are believed to have been photosynthesizers. They most likely arose because once the supply of organic molecules in the organic soup began to decline, natural selection favored cells that could use solar energy to produce carbohydrates.

> The first prokaryotes are believed to have been photosynthesizers.

Life Created an Oxidizing Atmosphere

Once oxygen-producing photosynthesizers (cyanobacteria) evolved, the oxygen in the atmosphere would have gradually increased. By 2.5 billion years ago, the atmosphere had changed to an oxidizing atmosphere. This is an example of the profound effect that life has on the physical condition of the earth. As discussed in the reading on the previous page, there are those who have been so struck with this realization they have formulated a new hypothesis. The Gaia hypothesis states that living things function in such a way as to preserve the earth for life.

In contrast to a reducing atmosphere, which promotes the buildup of organic molecules, an oxidizing atmosphere tends to break down organic molecules. Also, oxygen in the upper atmosphere forms ozone (O_3), called an **ozone shield,** which filters out the ultraviolet rays of the sun. Before an ozone shield formed, ultraviolet radiation could have been a source of energy for the formation of new organic molecules. Now this is not possible. However, the ozone shield performs an important function in that it allows life to exist on land. Without it, ultraviolet radiation reaching the earth would be sufficient to destroy land-dwelling organisms. This is why there is such concern today about pollutants that act to break down the ozone shield (p. 679).

New organic molecules (and new life) do not come into being today. As mentioned, there is no adequate energy source (such as sufficient ultraviolet radiation) to bring about synthesis of macromolecules, and oxygen tends to break down organic molecules once formed. Today, an important biological concept states that life comes only from life.

> The evolution of photosynthesizing organisms caused oxygen to enter the atmosphere. Oxygen breaks down organic molecules and forms an ozone shield, which protects the earth from ultraviolet radiation.

Eukaryotic Cells: Evolved 1.5 Billion Years Ago

The presence of oxygen in the atmosphere meant that most environments were no longer suitable for anaerobic prokaryotes, and they began to decline in number and therefore importance. The photosynthetic cyanobacteria proliferated, and eventually the atmosphere contained a stable amount of oxygen. Still, it was not until 1.5 billion years ago that the eukaryotic cell arose. As figure 26.6 shows, a large portion of the history of life was devoted to prokaryotic evolution, during which time various metabolic pathways, including aerobic respiration, came into existence.

The eukaryotic cell contains a nucleus and all the other organelles we studied in chapter 3. It is believed the eukaryotic cell acquired its organelles gradually. It may be that the nucleus and at least a primitive form of mitosis arose once there were several chromosomes, the DNA being too extensive to be contained within a single chromosome. According to the *endosymbiotic theory,* the mitochondria of the eukaryotic cell probably were once free-living aerobic prokaryotes and the chloroplasts probably were free-living photosynthetic prokaryotes. A nucleated cell engulfed these prokaryotes, which then became organelles (see fig. 3.13).

It has been suggested that flagella (and cilia) also arose by endosymbiosis. First, slender undulating prokaryotes could have attached themselves to a host cell in order to share some of its food. Eventually, these prokaryotes could have been drawn inside the host cell to become the flagella and cilia we know today.

> Prokaryotes were alone on the earth for 2 billion years, during which time many biochemical pathways evolved. The first eukaryotes evolved about 1.5 billion years ago.

Table 26.1
Classification Criteria for 5 Kingdoms

Kingdoms	*Monera*	*Protista*	*Fungi*	*Plantae*	*Animalia*
Type of Cell	Prokaryotic	Eukaryotic	Eukaryotic	Eukaryotic	Eukaryotic
Complexity	Unicellular	Mostly unicellular	Multicellular	Multicellular	Multicellular
Type of Nutrition	Autotrophic by various means Heterotrophic by various means	Photosynthetic Heterotrophic by various means	Heterotrophic by absorption	Autotrophic by photosynthesis	Heterotrophic by ingestion
Motility	Sometimes by flagella	Sometimes by flagella (or cilia)	Nonmotile	Nonmotile	Motile by contractile fibers
Life cycle	Asexual usual	Various	Haplontic	Alternation of generations	Diplontic
Internal protection of zygotes	No	No	No	Yes	No in most aquatic forms, yes in many land forms
Nerve fibers	None	Usually no	No	No	Usually yes

Multicellularity: Sex and Sudden Diversity

Multicellularity probably evolved about 700 million years ago, and the first multicellular forms were most likely microscopic. It is also possible that the first multicellular organisms practiced sexual reproduction. Among protists today, we find colonial forms in which some cells are specialized to produce gametes needed for sexual reproduction. Separation of germ cells, which produce gametes from somatic cells, may have been an important first step toward the complex macroscopic animals that suddenly appeared 600 million years ago. It is also possible that the increase in complexity correlates with an increase in atmospheric oxygen at this time. Macroscopic animals with muscular tissue require a significantly higher concentration of oxygen than simpler forms. In any case, 600 million years ago, there was a burst of diversity, particularly in animals.

Diverse complex organisms appear in the fossil record about 600 million years ago, most likely a result of the evolution of multicellularity and sexual reproduction.

Five Kingdoms from Monera to Animals

The many diverse forms of life can be classified in various ways. This text recognizes 5 kingdoms, which differ according to the criteria given in table 26.1. The prokaryotic bacteria are placed in the kingdom Monera (chap. 28). Their mode of nutrition is diverse, and all forms of nutrition are seen except ingestion. Remember that prokaryotes were evolving about 2 billion years before eukaryotes came on the scene. During this time, they developed many and various metabolic pathways. Some are **autotrophs,** organisms that make their own food from inorganic chemicals. They are either *chemosynthetic* (using the oxidation of inorganic chemicals as a source of energy) or *photosynthetic* (using the sun as a source of energy). Some are heterotrophs, being either saprophytic or parasitic. A saprophyte breaks down dead organic matter and then absorbs the nutrient molecules.

The other kingdoms contain the eukaryotes. Single-celled eukaryotes and the multicellular algae are placed in the kingdom Protista (chap. 28). Among the single-celled eukaryotes are the protozoans, which are heterotrophic by ingestion or else parasitic. Water molds, which resemble fungi, are saprophytic. The algae, of course, are photosynthetic. Kingdom Fungi, kingdom Plantae, and kingdom Animalia contain only multicellular forms. Fungi (chap. 28), such as molds and mushrooms, are primarily saprophytic, although some are parasitic. Only some bacteria but all fungi are saprophytic. They are the decomposers, which release nutrients when they break down dead organic matter. Plants (chap. 29), of course, are photosynthetic. They are distinguishable from algae because they are multicellular and adapted to a land environment. Notably, they protect the zygote from drying out by retaining it within an internal organ. Animals (chap. 30) are the only organisms that have a nervous system. Typically, they move about in search of food and are heterotrophic by ingestion. Figure 26.7 diagrams the five-kingdom system of classification, indicating in a general way how the organisms might be related.

Figure 26.7
The five-kingdom system of classification. Kingdom Monera contains the monerans; kingdom Protista contains the protists; kingdom Fungi contains the fungi, which are heterotrophic by absorption; kingdom Plantae contains the plants, which are autotrophic by photosynthesis; and kingdom Animalia contains the animals, which are heterotrophic by ingestion.

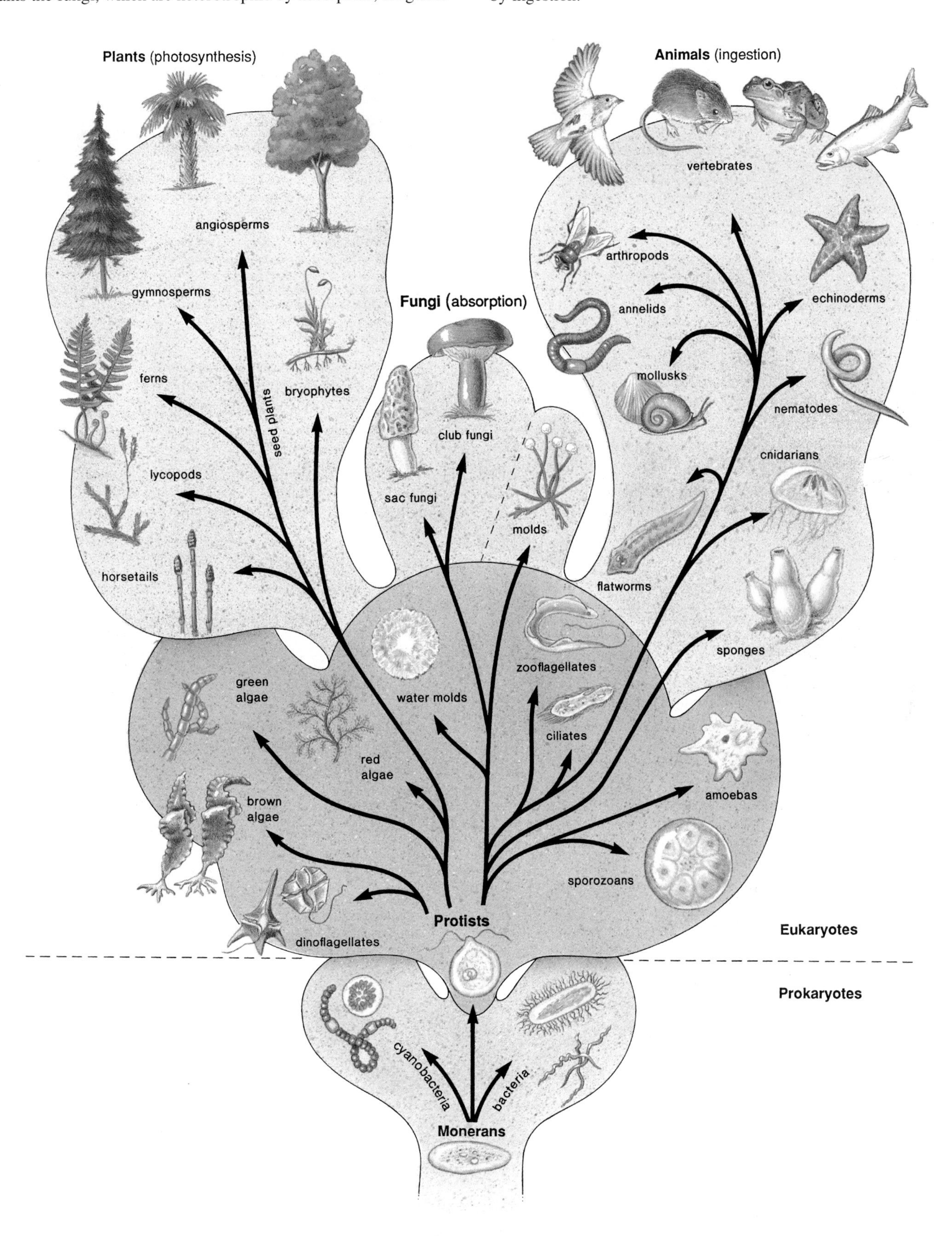

SUMMARY

The primitive atmosphere was a reducing atmosphere; it had no free oxygen gas. In the presence of an outside energy source, such as ultraviolet radiation, the primitive atmospheric gases reacted with one another to produce small organic molecules. This hypothesis is supported by experiments performed by Stanley Miller and others.

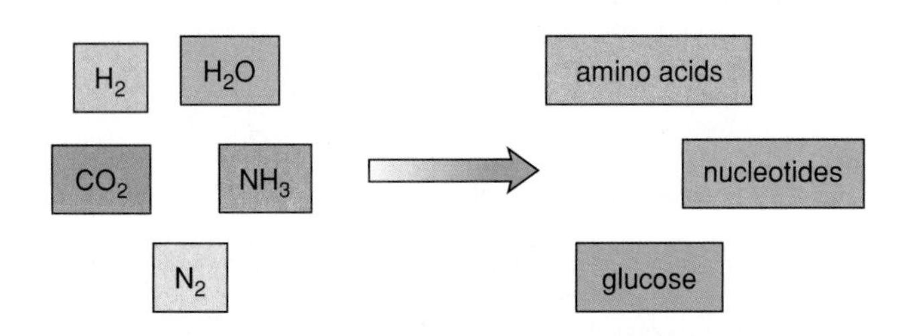

Next, macromolecules evolved and interacted. The RNA-first hypothesis is supported by the discovery of ribozymes, RNA enzymes. The protein-first hypothesis is supported by the observance that amino acids polymerize abiotically when exposed to dry heat. The Cairns-Smith hypothesis suggests that macromolecules could have originated in clay. All 3 hypotheses assume that eventually the DNA→RNA→protein self-replicating system evolved. Phospholipids readily form a spherical liposome, and perhaps this was the origin of the cell membrane.

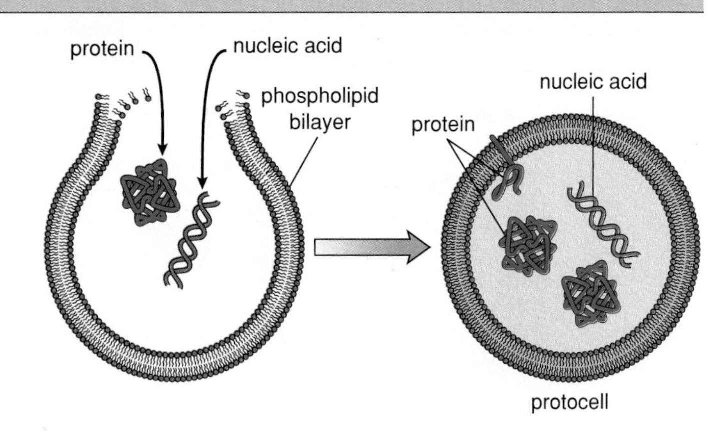

The protocell must have been a heterotrophic fermenter living on the preformed organic molecules in the organic soup. Once the protocell arose, biological evolution was possible, and the true cell evolved to give prokaryotes and eukaryotes. Prokaryotes existed alone on the surface of the earth for at least 2 billion years. Multicellularity and sexual reproduction arose about 700 million years ago, and by 600 million years ago, there was a rich supply of complex life forms.

The five-kingdom system of classification utilizes various criteria to place organisms in the kingdoms Monera, Protista, Fungi, Plantae, and Animalia (see table 26.1 and fig. 26.7).

STUDY QUESTIONS

In order to practice **writing across the curriculum**, students should write out the answers to any or all of the study questions. The study questions are sequenced in the same order as the text.

1. Trace the steps by which a chemical evolution may have produced a protocell. (p. 505)
2. Describe Stanley Miller's experiment, and discuss its significance. (p. 505)
3. Contrast the RNA-first hypothesis with the protein-first hypothesis. If polymerization occurred in clay, what macromolecules would have resulted at the same time? (p. 506)
4. How might a cell membrane have evolved? What is a protocell? (p. 506)
5. Why is it likely the protocell was a heterotrophic fermenter? (p. 507)
6. The earliest types of fossils are found in stromatolites. What kind of fossils are they? How old are they? What are stromatolites? (pp. 507–509)
7. Contrast the primitive atmosphere with today's atmosphere. Why is it unlikely that life could evolve today? (p. 509)
8. Relate the diversity of prokaryotes to the 2 billion years they existed before eukaryotes. (p. 509)
9. What is the importance of multicellularity and the evolution of sexual reproduction? (p. 510)
10. Describe the five-kingdom system of classification, and list the criteria by which organisms are placed in a particular kingdom. (p. 510)

OBJECTIVE QUESTIONS

1. A(n) _______ evolution is believed to have preceded an organic evolution.
2. Miller's experiment showed that the first gases could have reacted together to form _______.
3. The discovery of ribozymes lends support to the hypothesis that the first genes were made of _______.
4. For about two-thirds of the history of life, only _______ cells were present on earth.
5. The protocell carried on _______ nutrition and fed on the organic material in the oceans.
6. The protocell was a(n) _______ organism because there was no free oxygen in the atmosphere.
7. The earliest fossils found so far have been _______, organisms capable of making their own food.
8. The presence of oxygen in the atmosphere permitted _______ respiration to evolve.
9. Today, we believe that life comes only from _______.
10. Degree of complexity and type of _______ are especially helpful in classifying organisms.
11. Put these in the correct order: formation of earth, diversification of form, origin of eukaryotic cells, oxidizing atmosphere, multicellularity, oldest known fossils.
12. What theory is represented by this diagram? Explain the diagram.

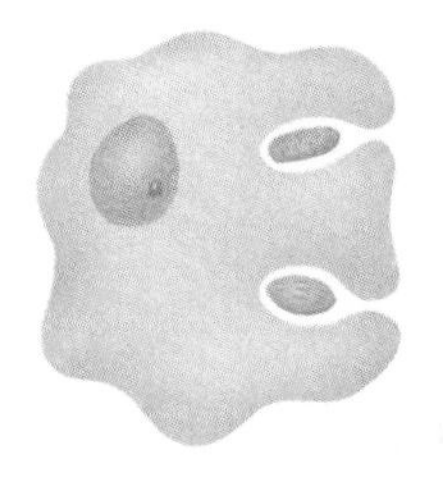

CRITICAL THINKING

In order to practice **writing across the curriculum,** students should write out the answers to any or all of the critical thinking questions. Suggested answers to the critical thinking questions are in appendix E.

26.1

1. Why did Fox show that polymerization of amino acids can occur without enzymes?
2. Why do investigators believe that proteins alone cannot form a cell? What properties do they lack?
3. By what mechanism would it be possible for RNA to replicate in the first cell?
4. Why was it necessary for this RNA to have enzymatic properties?

SELECTED KEY TERMS

biological evolution changes that have occurred in life forms from the origination of the first cell or cells to the many diverse forms in existence today.

chemical evolution a gradual increase in the complexity of chemical compounds that is believed to have brought about the origination of the first cell or cells.

coacervate droplet (ko-as´er-vāt drŏp-let) a mixture of polymers that may have preceded the origination of the first cell or cells.

energy source way by which energy from the environment can be made available to organisms.

heterotroph (het´er-o-trof˝) an organism that takes in preformed food.

liposome (li´pə-sōm) lipid bilayer sphere that forms when phospholipids are placed in a liquid environment.

microsphere structure composed only of protein that looks like a cell and carries on many cellular functions; a possible early step in cell evolution.

organic soup an expression used to refer to the oceans before the origin of life when they contained newly formed organic compounds.

oxidizing atmosphere an atmosphere that contains oxidizing molecules, such as O_2, rather than reducing molecules, such as H_2.

ozone shield a layer of O_3 present in the upper atmosphere that protects the earth from damaging ultraviolet radiation.

protocell the structure that preceded the true cell in the history of life.

reducing atmosphere an atmosphere that contains reducing molecules, such as H_2, rather than oxidizing molecules, such as O_2.

27

EVOLUTION

Chapter Concepts

1.
Many fields of biology provide evidence that common descent has occurred. 515, 533

2.
Evolution is defined in terms of population genetic changes. 520, 533

3.
There are several agents of evolutionary change, one of which is natural selection. 522, 533

4.
Only natural selection results in adaptation to the environment. 526, 533

5.
New species come about when populations are reproductively isolated from other, similar populations. 529, 533

Banyan pillar roots

Chapter Outline

Figure 27.1
The fossil record contains transitional fossils. **a.** This is *Archaeopteryx,* a link between reptiles and birds. It had feathers and wing claws. Most likely, it was a poor flier. Perhaps it ran over the ground on strong legs and climbed up into trees with the assistance of these claws. It also had a feather-covered reptilian type of tail, which shows up well in this artist's representation of the animal (**b**).

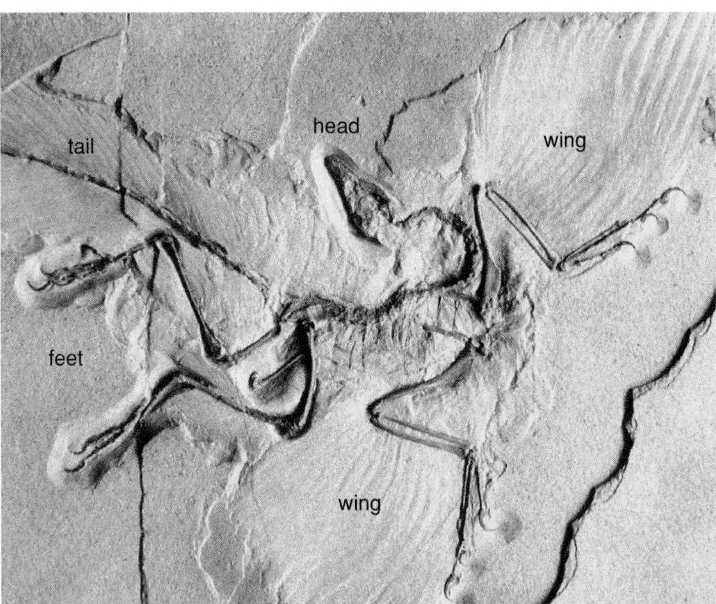

a.

b.

EVOLUTION is all the changes that have occurred in living things since the beginning of life. **Evolution** explains the unity and the diversity of life. All living things share the same fundamental characteristics because they are descended from a common ancestor. Life is diverse because the various living things are adapted to different ways of life.

Evidence of Common Descent

Many fields of biology provide evidence that supports the hypothesis of common descent. This is significant because the more varied the evidence supporting a hypothesis, the more certain it becomes. Darwin cited much of the evidence we will discuss, except he had no knowledge of biochemical evidence, which, of course, became available after his time.

Fossils Show the History of Life

The fossil record is the history of life as recorded by relics from the past (fig. 27.1). **Fossils** include such items as mineralized pieces of bone, impressions of plants pressed into shale, and even insects trapped in tree resin (which we know as amber). Over the last 2 centuries or so, paleontologists have studied fossils in the earth's strata (layers) all over the world and have pieced together the story of past life. Boundaries between the strata, where one mix of fossils gives way to another, provide the basis for dividing geological time as shown in table 27.1. We now know that many of these transitions were due to mass extinctions, which are indicated in the table. Notice that the names of the largest geological time divisions, the eras, pertain to life forms:

Eras	Cenozoic	Modern life
	Mesozoic	Middle life
	Paleozoic	Ancient life

The strata can tell us the sequence of organisms over time and provide relative dates for fossils. Absolute dates can be determined from a study of the radioactive isotopes in the rocks of each stratum. This method of dating fossils is explained in the reading on page 517.

Table 27.1
The Geological Time Scale: Major Divisions of Geological Time with Some of the Major Evolutionary Events of Each Geological Period

Era	Period	Epoch	Millions of Years Ago	Plant Life	Animal Life
Cenozoic (from the present to 65 million years ago)	Quaternary	Holocene	(recent)	Destruction of tropical rain forests by humanity accelerates extinctions	**Age of Human Civilization**
		Pleistocene	0.01–2	Herbaceous plants spread and diversify	Modern humans appear
	Extinction of Many Large Animals and Birds				
	Tertiary	Pliocene	2–5	Herbaceous angiosperms flourish	First hominids appear
		Miocene	5–25	Grasslands spread as forests contract	Apelike mammals and grazing mammals flourish; insects flourish
		Oligocene	25–38	Many modern families of flowering plants evolve	Browsing mammals and monkeylike primates appear
		Eocene	38–55	Subtropical forests with heavy rainfall thrive	First horses appear; all modern orders of mammals are represented
		Paleocene	55–65	Angiosperms diversify	Primitive primates, carnivores, and insectivores appear
Mesozoic (from 65 million to 248 million years ago)	***Extinction of Dinosaurs***				
	Cretaceous		65–144	Flowering plants spread; coniferous trees decline	Placental mammals appear; modern insect groups appear
	Jurassic		144–213	Gymnosperms flourish	**Age of Dinosaurs** Birds appear
	Extinction of 35% of Existing Animal Families				
	Triassic		213–48	Forests of gymnosperms and ferns dominate	First mammals appear; first dinosaurs appear
Paleozoic (from 248 million to 590 million years ago)	***Extinction of 50% of Existing Animal Families***				
	Permian		248–86	Conifers, cycads, and ginkgos appear	Reptiles diversify; amphibians decline
	Carboniferous		286–360	Age of great coal-forming forests: club mosses, horsetails, and ferns flourish	**Age of Amphibians**
	Extinction of 30% of Existing Animal Families				
	Devonian		360–408	First seed ferns appear	**Age of Fishes** First insects appear; first amphibians appear
	Silurian		408–38	Low-lying primitive vascular plants appear on land	First jawed fishes appear
	Ordovician		438–505	Marine algae flourish	Invertebrates spread and diversify; first vertebrates appear
	Extinction of 50% of Existing Animal Families				
	Cambrian		505–90	Marine algae flourish	**Age of Invertebrates** First fishes (jawless) appear
Precambrian (from 590 million to 4,500 million years ago)			700	Multicellular organisms appear	
			1,400	First complex (eukaryotic) cells appear	
			2,500	Oxygen-forming cyanobacteria appear	
			3,500	First prokaryotic cells—anaerobic bacteria—appear	
			4,500	Earth forms	

DATING FOSSILS: HOW OLD IS A FOSSIL?

Table 27.A Principal Decay Series Used for Mineral and Total-Rock Dating			
Parent Isotope	**Half-Life**	**Ultimate Stable Product**	**Effective Age Range**
Rubidium-87 (Rb)	47 billion years (b.y.)	Strontium-87 (Sr)	>100 million years (m.y.)
Thorium-232 (Th)	13.9 b.y.	Lead-208 (Pb)	>200 m.y.
Uranium-238 (u)	4.5 b.y.	Lead-206 (Pb)	>100 m.y.
Potassium-40 (K)	1.3 b.y.	Argon-40 (Ar)	>100,000 years
Uranium-235 (U)	0.71 b.y.	Lead-207 (Pb)	>100 m.y.
Carbon-14 (C)	5,730 years	Nitrogen-14 (N)	0–50,000 years

There are 2 ways to date fossils. The relative-dating method determines the relative order of fossils but does not determine the actual date they were formed. The relative-dating method is possible because fossil-containing sedimentary rocks occur in layers. Because the top layers are younger than the lower layers, the fossils in each successive layer are older than the fossils in the layer above.

The absolute-dating method assigns an actual rather than a relative date to a fossil. All radioactive isotopes have a particular half-life, the length of time it takes for half of the isotope in a specimen to decay and become nonradioactive (table 27.A). If the fossil has organic matter, half of the carbon-14 in it will be gone in 56,000 years. This is not very helpful unless we know how much carbon-14 was in the fossil to begin with. To make this determination, it is reasoned that organic matter always begins with the same amount of carbon-14. Now it is only necessary to use a radiation counter to compare the carbon-14 radioactivity of the fossil to that of a modern sample of organic matter. The older the fossil, the lower the counts per minute of radiation. In fact, after 50,000 years, there is not enough carbon-14 radioactivity left in the fossil to measure its age accurately.

Carbon-14 is the only isotope listed in table 27.A that organic matter contains, but it is possible to use the others listed in the table to measure the age of a rock containing a fossil. This also indicates the age of the fossil because the 2 most likely formed together. It is possible to measure potassium-40, a common constituent of rocks, and its decay product, the gas argon (Ar), which is usually trapped inside the rock. The ratio of potassium-40 to argon-40 in the rock can then be used to determine its age. For example, if the ratio is 1:1, then half of the potassium-40 has decayed and the rock is 1.3 billion years old. The isotopes rubidium-87, thorium-232, and uranium-238 are of no use for dating minerals less than about 100 million years old. They have such a long half-life that no perceptible decay will have occurred in a length of time shorter than this.

Particularly interesting are the fossils that serve as links between groups. For example, the famous fossil *Archaeopteryx* is intermediate between reptiles and birds (fig. 27.1). The dinosaur-like skeleton of this fossil has reptilian features, including jaws with teeth, and a long, jointed tail, but *Archaeopteryx* also had feathers and wings. Other intermediate fossils among the fossil vertebrates include the amphibious fish *Eustheopteron,* the reptilelike amphibian *Seymouria,* and the mammallike reptiles called synapsids. These fossils allow us to deduce this line of descent for the vertebrates:

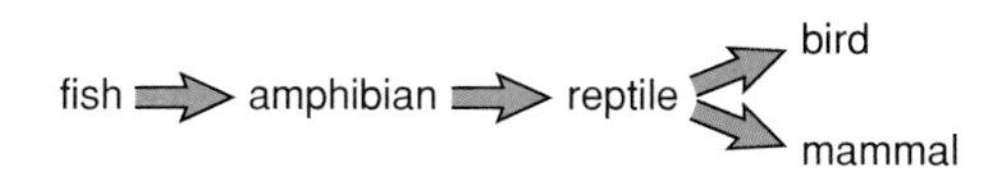

Sometimes, we can use the fossil record to trace the history of one particular organism, such as the modern-day horse, *Equus. Equus* evolved from *Hyracotherium,* which was about the size of a dog. *Hyracotherium* was adapted to the forestlike environment of the Eocene, an epoch of the Tertiary period. This small animal could have hidden among the trees for protection, and its low-crowned teeth were appropriate for browsing on leaves. In the Miocene and Pliocene epochs, however, grasslands began to replace the forests. Then the ancestors of *Equus* were subject to selective pressure for the development of strength, intelligence, speed, and durable grinding teeth. A larger size provided the strength needed for combat, a larger skull made room for a larger brain, elongated legs ending in hooves gave speed to escape enemies, and the durable grinding teeth enabled the animals to feed efficiently on grasses. In all cases, like this one, living organisms closely resemble the most recent fossils in their line of descent. Underlying similarities, however, allow us to trace a line of descent over vast amounts of time.

The fossil evidence supports common descent. Fossils can be linked over time because they show a similarity in form despite observed changes.

> The fossil record traces in broad terms the history of life and more specifically allows us to study the history of particular groups.

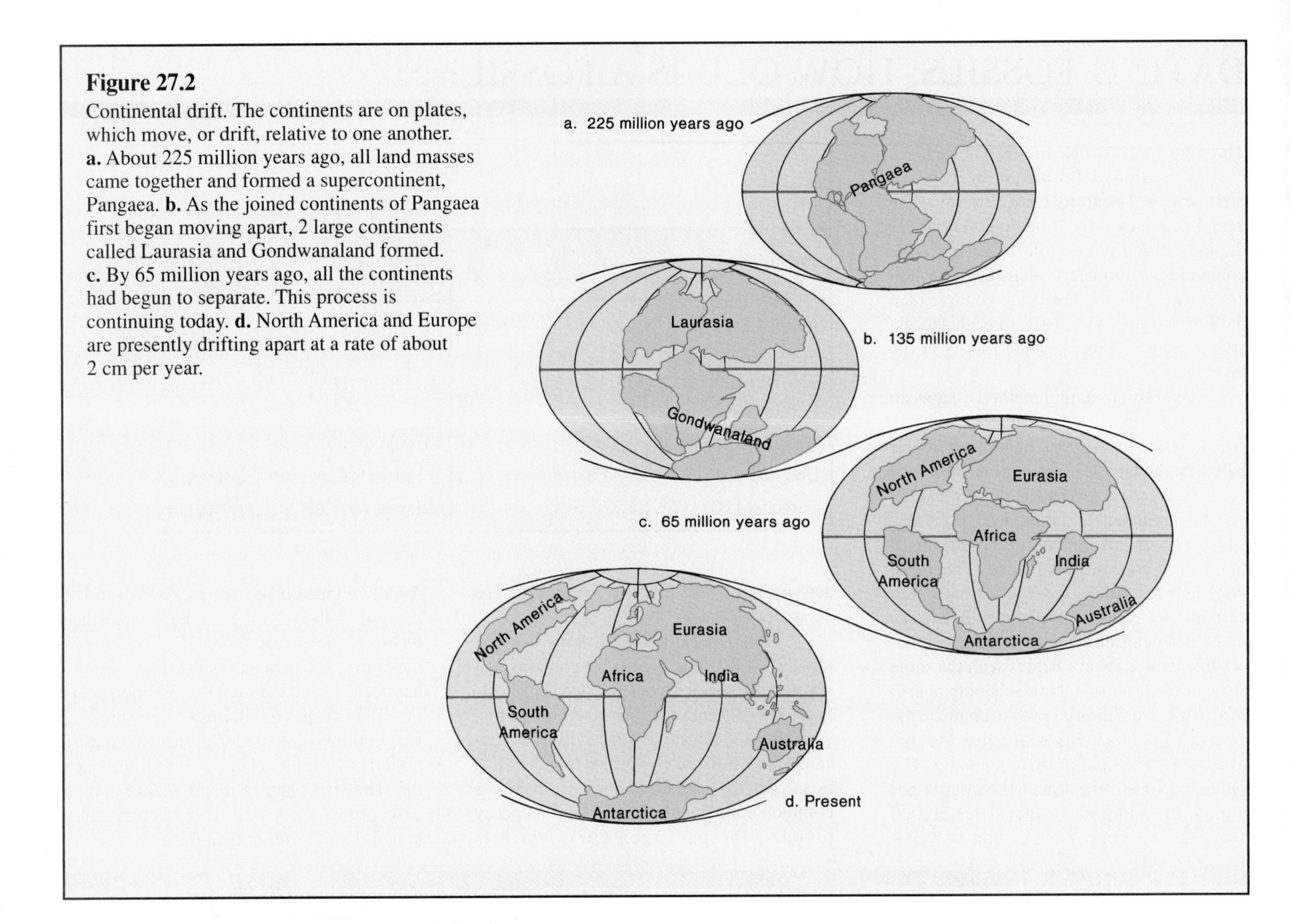

Figure 27.2
Continental drift. The continents are on plates, which move, or drift, relative to one another. **a.** About 225 million years ago, all land masses came together and formed a supercontinent, Pangaea. **b.** As the joined continents of Pangaea first began moving apart, 2 large continents called Laurasia and Gondwanaland formed. **c.** By 65 million years ago, all the continents had begun to separate. This process is continuing today. **d.** North America and Europe are presently drifting apart at a rate of about 2 cm per year.

Geography and Evolutionary Relationships

Biogeography is the study of the distribution of plants and animals throughout the world. Such distributions are consistent with the hypothesis that related forms evolve in one locale and then spread out into other regions. For example, there are no rabbits in South America because rabbits originated somewhere else and they had no means to reach South America.

Physical factors, such as the location of continents, often determine where a population can spread. For example, at one time in the history of the earth, South America, Antarctica, and Australia were all connected (fig. 27.2). Marsupials (pouched mammals) arose at this time and today are found in both South America and Australia. When Australia separated and drifted away, the marsupials diversified into many different forms suited to specific environments. These observations support common descent. It is quite plausible that a particular line of descent in a particular geographical region could give rise to many forms, each adapted to a different environment. We recognize that these many forms are related because they are anatomically similar. In this example, all the animals are marsupials and none are placental mammals (animals in which development occurs in the uterus).

> The distribution of organisms on the earth is various but explainable by assuming that related forms evolved in one locale, where they then diversified and/or spread out into other accessible areas. Do 27.1 Critical Thinking, found at the end of the chapter.

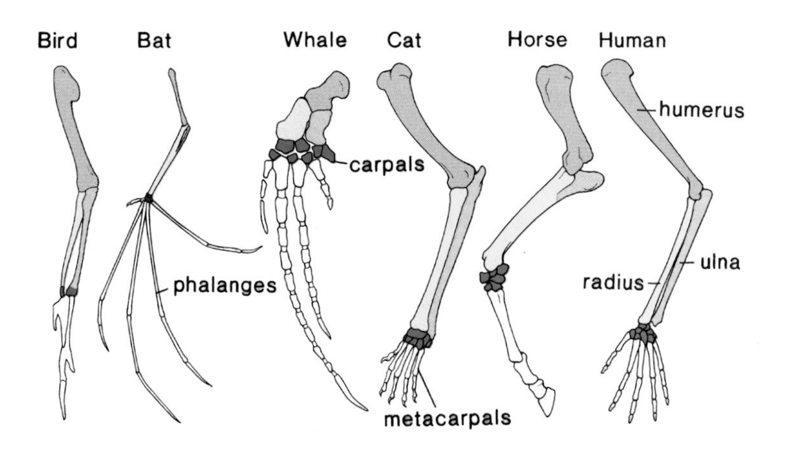

Figure 27.3
Bones in vertebrate forelimbs. The same bones are present (they are color coded) but designed for different functions. The unity of plan is evidence of a common ancestor.

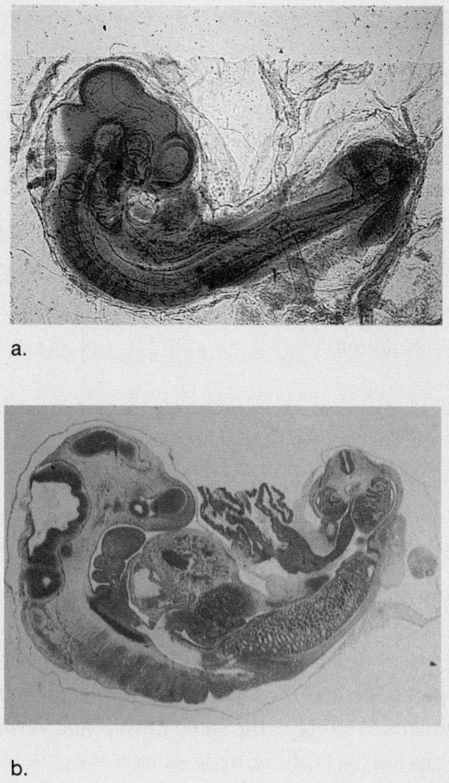

Figure 27.4
Chick (**a**) and pig (**b**) embryos at comparable stages of early development have many features in common, although they eventually are completely different animals. This is evidence that they evolved from a common ancestor.

Related Species Share Anatomy

Related species share a *unity of plan.* The reproductive organs of all flowering plants are basically similar, and all vertebrate animals have essentially the same type of skeleton. For example, vertebrate forelimbs are used for flight (birds and bats), orientation during swimming (whales and seals), running (horses), climbing (arboreal lizards), or swinging from tree branches (monkeys). Yet, all vertebrate forelimbs contain the same sets of bones organized in similar ways despite their dissimilar functions (fig. 27.3). The best explanation for this unity of plan is that the basic forelimb plan that originated with a common ancestor was modified in the succeeding groups as each continued along its own evolutionary pathway. Such structures are called **homologous structures.** In contrast to homologous structures, **analogous structures,** such as an insect wing and a bird wing, have similar functions but differ in anatomy.

Vestigial structures are anatomical features that are fully developed and functional in one group of organisms but are reduced and functionless in similar groups. Most birds, for example, have well-developed wings used for flight. Some bird species, however, have wings that are greatly reduced, prohibiting flight. Similarly, snakes have no use for hind limbs, yet some have remnants of a pelvic girdle and legs. Humans have a tailbone (coccyx) but no tail. Common descent explains the presence of vestigial structures. Vestigial structures occur because organisms inherit their anatomy from their ancestors; they are traces of an organism's evolutionary history.

Related Species Share Development

The unity of plan shared by vertebrates extends to their embryological development (fig. 27.4). At some time during development, all vertebrates have a dorsal supporting rod, called a notochord, and exhibit paired pharyngeal pouches. In fishes and amphibian larvae, these pouches become functioning gills. In humans, the first pair of pouches become the eustachian tubes. The second pair become the tonsils, while the third and fourth pairs become the thymus gland and the parathyroids.

Patterns of development are consistent with common descent. Organisms that share a unity of plan also share similar stages of development. Why should terrestrial vertebrates develop and then modify structures, like pharyngeal pouches, that have lost their original function? It is because animals tend to repeat the developmental stages of their ancestors.

Organisms share a unity of plan when they are closely related because of common descent. This is substantiated by comparative anatomy and embryological development.

Table 27.2
DNA Differences between Selected Pairs of Animals

Pairs of Animals	Percent Difference in Nucleotide Sequences
Human/chimpanzee	2.5
Human/gibbon	5.1
Human/green (old-world) monkey	9.0
Human/capuchin (new-world) monkey	15.8
Human/lemur	42.0
Mouse/rat	30.0

Source: Data from G. L. Stebbins, *Darwin to DNA, Molecules to Humanity,* 1982, page 320.

Related Organisms Have Similar DNA

Almost all living organisms use the same basic biochemical molecules, including DNA, ATP, and many identical or nearly identical enzymes. Further, organisms utilize the same DNA triplet code and the same 20 amino acids in their proteins. Organisms even share the same introns and variable regions. There is obviously no functional reason why these elements need to be so similar, but their similarity can be explained by descent from a common ancestor.

When the degree of similarity in DNA base sequences or the degree of similarity in amino acid sequences of proteins is examined, the data are as expected, assuming common descent. For example, investigators have determined the DNA base sequence differences between a number of organisms, some of which are given in table 27.2. The table shows that humans and chimpanzees have only a 2.5% difference in DNA base sequences and that humans and lemurs have a 42% difference in DNA base sequences. This is understandable by assuming that humans and chimpanzees share a more recent common ancestor than do humans and lemurs. These data are consistent with other available data regarding the anatomical similarity of these animals and therefore their relatedness.

Organisms have remarkable similarities in nonfunctional DNA elements and have DNA sequence differences that are based on the degree of their anatomical relatedness. The pattern of DNA base sequence differences (and the amino acid sequence differences) among organisms supports common descent.

All organisms have certain biochemical molecules in common. The degree of similarity between DNA base sequences and amino acid sequences indicates their degree of relatedness.

The Evolutionary Process

Evolution is one of the great unifying theories of biology. In science, the word *theory* is reserved for conceptual schemes that are supported by a large number of observations and have not yet been found lacking.

Modern evolutionists emphasize that individuals are members of a population. A study of population genetics allows us to see when and if evolution has occurred.

How to Detect Evolution

A **population** is all members of a single species occupying a particular area at the same time. A population could be all the green frogs in a frog pond, all the field mice in a barn, or all the English daisies on a hill. The members of a population reproduce with one another to produce the next generation.

Each member of a population is assumed to be free to reproduce with any other member, and when reproduction occurs, the genes of one generation are passed on in the manner described by Mendel's laws. Therefore, in this so-called Mendelian population of sexually reproducing individuals, the various alleles of all the gene loci in all the members make up a **gene pool** for the population. It is customary to describe the gene pool of a population in terms of allele frequencies for the various genes. Using this methodology, 2 investigators, G. H. Hardy, an English mathematician, and W. Weinberg, a German physician, discovered a law that bears their names.

Hardy-Weinberg law: As long as certain conditions are met, allele frequencies in a sexually reproducing population come to an equilibrium that is maintained generation after generation. The conditions are no mutations, random mating, no gene flow, no genetic drift, and no natural selection.

The Hardy-Weinberg law predicts that sexual reproduction alone cannot alter the allele frequencies in a population. For example, suppose it is known that one-fourth of all flies in a *Drosophila* population are homozygous dominant for long wings, one-half are heterozygous, and one-fourth are homozygous recessive for short wings. Therefore, in a population of 100 individuals, we have

$$25\ LL,\ 50\ Ll,\ \text{and}\ 25\ ll$$

What is the number of the allele L and the allele l in the population?

Number of *L* Alleles		Number of *l* Alleles	
LL $(2\,L \times 25)$	$= 50$	LL $(0\,l)$	$= 0$
Ll $(1\,L \times 50)$	$= 50$	Ll $(1\,l \times 50)$	$= 50$
ll $(0\,L)$	$= 0$	ll $(2\,l \times 25)$	$= 50$
	$100\,L$		$100\,l$

To determine the frequency of each allele, calculate its percentage of the total number of alleles in the population; in each case, 100/200 = 50% = 0.50. The sperm and the eggs produced by this population also contain these alleles in these frequencies. Assuming random mating (all possible gametes have an equal chance to combine with any other), we can calculate the ratio of genotypes in the next generation using a Punnett square.

eggs (from all females)

♂ \ ♀	0.50***L***	0.50***l***
0.50***L***	0.25***LL***	0.25***Ll***
0.50***l***	0.25***Ll***	0.25***ll***

sperm (from all males)

Frequency of **L** = $0.25\mathbf{L} + \frac{1}{2}\,0.50\mathbf{Ll} = 0.50$

Frequency of ***l*** = $0.25\mathbf{l} + \frac{1}{2}\,0.50\mathbf{Ll} = 0.50$

Figure 27.5 Calculating gene pool frequencies.

$$p^2 + 2pq + q^2$$

p^2 = % homozygous dominant individuals
p = frequency of dominant allele
q^2 = % homozygous recessive individuals
q = frequency of recessive allele
$2pq$ = % heterozygous individuals

Realize that $p + q = 1$ (there are only 2 alleles)
$p^2 + 2pq + q^2 = 1$ (these are the only genotypes)

Example

An investigator has determined by inspection that 16% of a human population has a continuous hairline (recessive trait). Using this information, we can complete all the genotypic and allele frequencies for the population, provided the conditions for Hardy-Weinberg equilibrium are met.

Given: $q^2 = 0.16$ = 16% are homozygous recessive individuals

Therefore, $q = \sqrt{0.16} = 0.4$ = frequency of recessive allele
$p = 1.0 - 0.4 = 0.6$ = frequency of dominant allele
$p^2 = (0.6)(0.6) = 0.36$ = 36% are homozygous dominant individuals
$2pq = 2(0.6)(0.4) = 0.48$ = 48% are heterozygous individuals
or
$= 1.00 - 0.52 = 0.48$

84% have the dominant phenotype

There is an important difference between this Punnett square and one used for a cross between individuals: here the sperm and the eggs are those produced by the members of a population—not those produced by individuals. As you can see, the results of the Punnett square indicate that the frequency for each allele in the next generation is still 0.50.

Practice Problems*

1. In a certain population, 21% are homozygous dominant, 49% are heterozygous, and 30% are homozygous recessive. What percentage of the next generation is predicted to be homozygous recessive, assuming a Hardy-Weinberg equilibrium?
2. Of the members of a population of pea plants, 1% are short. What are the frequencies of the recessive allele *t* and the dominant allele *T*? What are the genotypic frequencies in this population?
3. A student places 600 fruit flies with the genotype *Ll* and 400 with the genotype *ll* in a culture bottle. What will the genotypic frequencies be in the next generation and each generation thereafter, assuming a Hardy-Weinberg equilibrium?

*Answers to problems are on page 701.

G. H. Hardy and W. Weinberg were mathematicians who used the binomial expression ($p^2 + 2pq + q^2$) to calculate the genotypic and allele frequencies of a population. Figure 27.5 shows you how this is done. However, it is not necessary to do the mathematics in order to realize that sexual reproduction in and of itself cannot bring about a change in allele frequencies. Also, the dominant allele does not need to increase from one generation to the next. Dominance does not cause an allele to become a common allele.

In real life, the Hardy-Weinberg law does not hold because the conditions (i.e., no mutations, random mating, no gene flow, no genetic drift, and no natural selection) are rarely if ever met. The allele frequencies in the gene pool of a population *do* change from one generation to the next. Therefore, evolution has occurred. The significance of the Hardy-Weinberg law is that it tells us what factors cause evolution—those that violate the conditions mentioned. Evolution can be detected by any deviation from the Hardy-Weinberg equilibrium of allele frequencies in the gene pool of a population.

Figure 27.6
Microevolution. Microevolution has occurred when there is a change in gene pool frequencies—in this case, due to natural selection. The percentage of the dark-colored phenotype has increased because predatory birds can see light-colored moths against sooty tree trunks.

The accumulation of small changes in the gene pool over a relatively short period of 2 or more generations is called microevolution (fig. 27.6).

> The Hardy-Weinberg equilibrium provides a baseline by which to judge whether or not evolution has occurred. Any change from the initial allele frequencies in the gene pool of a population signifies that evolution has occurred.

Five Agents of Evolutionary Change

The list of conditions for genetic equilibrium stated previously implies that the opposite conditions can cause evolutionary change. These conditions are mutations, gene flow, nonrandom mating, genetic drift, and natural selection.

Mutations: Makes Raw Material

Mutations provide new alleles and therefore underlie all other mechanisms that produce variation, the raw material for evolutionary change. Investigations have demonstrated that high levels of molecular variation are the rule in natural populations. When various *Drosophila* enzymes were extracted and subjected to electrophoresis, it was found that a fly population is polymorphic at no less than 30% of all its gene loci and that an individual fly is likely to be heterozygous at about 12% of its loci.

Many of these mutations do not affect the phenotype and are not ordinarily detected when a population is already well adapted. In a changing environment, however, even a seemingly harmful mutation can be a source of variation that can help a population adapt to the new surroundings. For example, the water flea *Daphnia* ordinarily thrives at temperatures around 20°C and cannot survive temperatures of 27° or more. There is, however, a mutation that allows *Daphnia* to live at temperatures between 25°C and 30°C. The adaptive value of this mutation is entirely dependent on environmental conditions.

Gene Flow: Prevents Local Adaptation

Gene flow is the movement of alleles between populations by, for example, the migration of breeding individuals. Adult plants are not able to migrate, but their gametes are often either windblown or carried by insects. The wind, in particular, can carry pollen for long distances and can therefore be a factor in gene flow between plant populations.

Gene flow between 2 populations keeps their gene pools similar. It also prevents close adaptation to a local environment.

Figure 27.7
Genetic drift occurs when by chance some members of a population do not reproduce and do not pass on their genes to the next generation. If these individuals carry rare alleles, the allele frequencies of the next generation may markedly change.

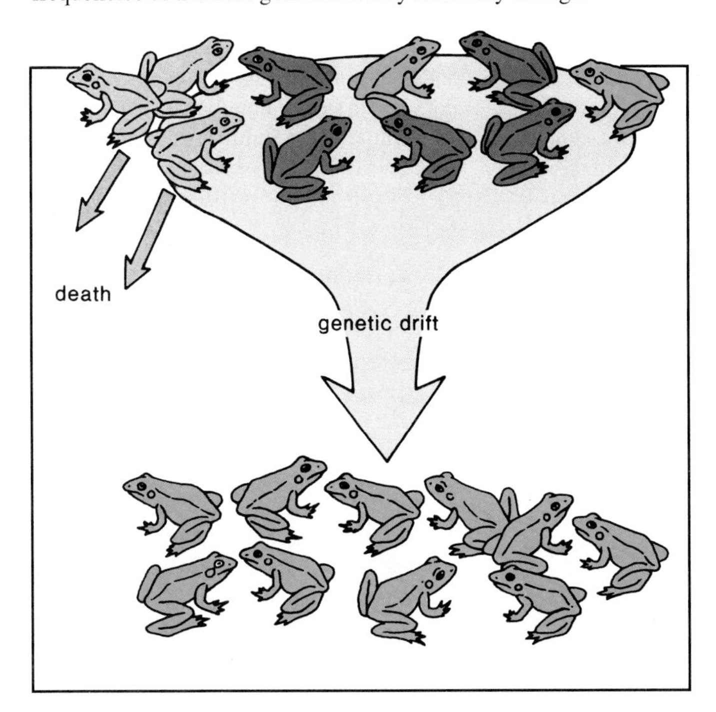

Nonrandom Mating: When Partners Choose

Nonrandom mating occurs when individuals pair up not by chance but according to their genotypes or phenotypes. Inbreeding, or mating between relatives to a greater extent than by chance, is an example of nonrandom mating. Inbreeding decreases the proportion of heterozygotes and increases the proportions of both homozygotes at all gene loci. In a human population, inbreeding increases the frequency of recessive abnormalities.

Genetic Drift: Founders and Bottlenecks

Genetic drift refers to changes in allele frequencies of a gene pool due to chance. When a population is small, there is a greater chance that some rare genotype will not participate at all in the production of the next generation (fig. 27.7). In nature, 2 situations, called founder effect and bottleneck effect, lead to small populations in which genetic drift drastically affects gene pool frequencies.

When a few individuals found a colony, only a fraction of the total genetic diversity of the original gene pool is represented. Which particular alleles are carried by the founders is dictated by chance alone. The Amish of Lancaster, Pennsylvania, are an isolated religious sect descended from German founders. Today, as many as one in 14 individuals in this group carries a recessive allele that causes an unusual form of dwarfism (it affects only lower arms and legs) and polydactylism (extra fingers) (fig. 27.8). In the population at large, only one in 1,000 individuals has this allele.

Figure 27.8
Founder effect. A member of the founding population of Amish in Pennsylvania had a recessive allele for a rare kind of dwarfism. The percentage of the Amish population now carrying this allele is much higher compared to that of the general population.

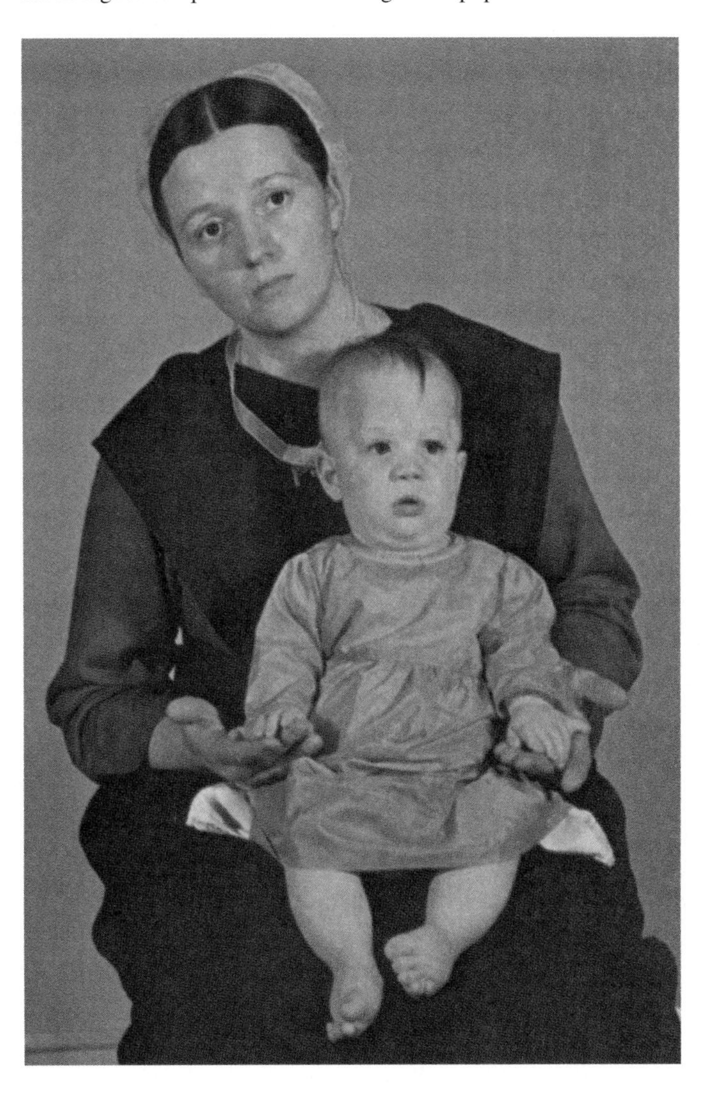

Sometimes a population is subjected to near extinction because of a natural disaster (e.g., earthquake or fire) or because of human aggression. Again, chance alone may determine which individuals survive these unfavorable times, which act as a bottleneck, preventing the majority of types of genotypes from participating in the production of the next generation. For example, the large genetic similarity found in cheetahs is believed to be due to a bottleneck. In a study of 47 different enzymes, each of which can come in several different forms, all the cheetahs had exactly the same form. This demonstrates that genetic drift can cause certain alleles to be lost from a population.

CHARLES DARWIN'S THEORY OF NATURAL SELECTION

At the age of 22, Charles Darwin signed on as a naturalist with the HMS Beagle, a ship that took a five-year trip around the world in the first half of the nineteenth century. Because the ship sailed in the Southern Hemisphere, where life is more abundant and varied, Darwin encountered forms of life very different from those in his native England.

Even though it was not his original intent, Darwin began to realize and to gather evidence that life forms change over time and from place to place. He read a book by Charles Lyell, a geologist, who suggested the world is very old and has been undergoing gradual changes for many, many years. Darwin found the remains of a giant ground sloth and an armadillo on the east coast of South America and wondered if these extinct forms were related to the living forms of these animals. When he compared the animals of Africa to those of South America, he noted that the African ostrich and the South American rhea, although similar in appearance, were actually different animals. He reasoned that they had a different line of descent because they were on different continents. When Darwin arrived at the Galápagos Islands, he began to study the 13 species of finches (fig. 27.13), whose adaptations could best be explained by assuming they had diverged from a common ancestor. With this type of evidence, Darwin concluded that species evolve (change) with time.

When Darwin returned home, he spent the next 20 years gathering data to support the principle of organic evolution. His most significant contribution to this principle was his theory of natural selection, which explains how a species becomes adapted to its environment. Before formulating the theory, he read an essay on human population growth written by Thomas Malthus. Malthus observed that although the reproductive potential of humans is great, there are many environmental factors, such as availability of food and living space, that tend to keep the human population within bounds. Darwin applied these ideas to all populations of organisms. For example, he calculated that a single pair of elephants could have 19 million descendants in 750 years. He realized that other organisms have even greater reproductive potential than this pair of elephants; yet, usually the number of each type of organism remains about the same. Darwin decided there is a constant struggle for existence, and only a few members of a population survive to reproduce. The ones that survive and contribute to the evolutionary future of the species are by and large the better adapted individuals. This so-called survival of the fittest causes the next generation to be better adapted than the previous generation.

Darwin's theory of natural selection was nonteleological. Organisms do not strive to adapt themselves to the environment, rather, the environment acts on them to select those individuals that are best adapted. These are the ones that have been "naturally selected" to pass on their characteristics to the next generation. In order to emphasize the nonteleological nature of Darwin's theory, it is often contrasted with the theory of Jean-Baptiste Lamarck, another nineteeth-century naturalist (fig. 27.A). The Lamarckian explanation for the long neck of the giraffe was based on the assumption that the ancestors of the modern giraffe were trying to reach into the trees to browse on high-growing vegetation. Continual stretching of the neck caused it to become longer, and this acquired characteristic was passed on to the next generation. Lamarck's theory is teleological because, according to him, the outcome is known ahead of time. This type of explanation has not stood the test of time, but Darwin's theory of evolution by natural selection has been fully substantiated by later investigations.

These are the critical elements of Darwin's theory.

- **Variations.** Individual members of a species vary in physical characteristics. Physical variations can be passed from generation to generation. (Darwin was never aware of genes, but we know today that the inheritance of the genotype determines the phenotype.)
- **Struggle for existence.** The members of all species compete with each other for limited resources. Certain members are able to capture these resources better than others.
- **Survival of the fittest.** Just as humans have artificial breeding programs to select which plants and animals reproduce, so there is a natural selection by the environment of which organisms survive and reproduce. While Darwin emphasized the importance of survival, modern evolutionists emphasize the importance of unequal reproduction. In any case, however, the selection process is nonteleological. Certain members of the population are selected to produce more offspring simply because they happen to have a variation that makes them better suited to the environment.

Figure 27.A
Mechanism of evolution. This diagram contrasts (**a**) Jean-Baptiste Lamarck's theory as to how evolution occurs, called the theory of acquired characteristics, with that of (**b**) Charles Darwin, called the theory of natural selection. Only Darwin's theory is supported by data.

- **Adaptation.** Natural selection causes a population of organisms and ultimately a species to become adapted to the environment. The process is slow, but each subsequent generation includes more individuals that are better adapted to the environment.

Can natural selection account for the origin of new species and for the great diversity of life? Yes, if we are aware that life has been evolving for a very long time and that variously adapted populations can arise from a common ancestor.

Darwin was prompted to publish his findings only after he received a letter from another naturalist, Alfred Russel Wallace, who had come to the exact same conclusions about evolution. Although both scientists subsequently presented their ideas at the same meeting of the famed Royal Society in London in 1858, only Darwin later gathered together detailed evidence in support of his ideas. He described his experiments and reasonings at great length in *The Origin of Species by Means of Natural Selection,* a book still studied by most biologists today.

Natural Selection: Adapting to the Environment

Natural selection is the process by which populations become adapted to their environment. The reading on pages 524–525 outlines how Charles Darwin, the father of evolution, explained evolution by natural selection. Here, we restate these steps in the context of modern evolutionary theory. In evolution by natural selection, the **fitness** of an individual is measured by how reproductively successful its offspring are in the next generation.

Evolution by natural selection requires

1. variation. The members of a population differ from one another.
2. inheritance. Many of these differences are heritable genetic differences.
3. differential adaptedness. Some of these differences affect how well an organism is adapted to its environment.
4. differential reproduction. Individuals that are better adapted to their environment are more likely to reproduce, and their fertile offspring will make up a greater proportion of the next generation.

Random gene mutations are the ultimate source of variation because they provide new alleles. However, in sexually reproducing organisms, recombination of alleles and chromosomes due to crossing-over during meiosis, independent assortment of chromosomes, and fertilization contribute greatly to variation. Recombination may at some time bring a more favorable combination of alleles together. After all, it is the combined phenotype that is subjected to natural selection. In fact, most of the traits on which natural selection acts are polygenic and controlled by more than one pair of alleles. Such traits have a range of phenotypes, the frequency distribution of which usually resembles a bell-shaped curve (p. 436).

Three types of natural selection have been described for any particular trait. They are stabilizing selection, disruptive selection, and directional selection.

Stabilizing Selection. **Stabilizing selection** occurs when an intermediate phenotype is favored (fig. 27.9). It can improve adaptation of the population to those aspects of the environment that remain constant. With stabilizing selection, extreme phenotypes are selected against, and individuals near the average are favored. As an example, consider the birth weight of human infants, which ranges from 0.89 kg–4.9 kg (2 lb–10.8 lb). The death rate is higher for infants who are at these extremes and lowest for babies who have a birth weight between 3.1 kg and 3.5 kg. Most babies have a birth weight within this range, which gives the best chance of survival. Similar results have been found in other animals, also.

Figure 27.9

Stabilizing selection. Natural selection favors the intermediate phenotype (see arrows) over the extremes. Today, it is observed that most human babies are of intermediate weight (about 3.2 kg [7 lb]), and very few babies are lightweight or heavy.

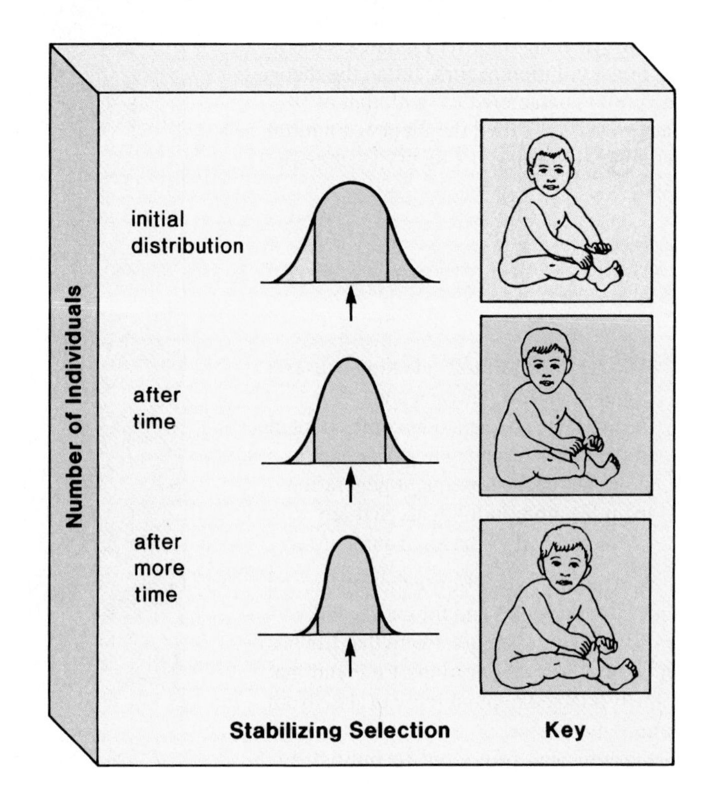

Disruptive Selection. In **disruptive selection,** 2 or more extreme phenotypes are favored over any intermediate phenotype (fig. 27.10). For example, British land snails (*Cepaea nemoralis*) have a wide habitat range that includes low-vegetation areas (grass fields and hedgerows) and forest areas. In low-vegetation areas, thrushes feed mainly on snails with dark shells that lack light bands, and in forest areas, they feed mainly on snails with light-banded shells. Therefore, these 2 different phenotypes are found in the population.

Directional Selection. **Directional selection** occurs when an extreme phenotype is favored and the distribution curve shifts in that direction (fig. 27.11). Such a shift can occur when a population is adapting to a changing environment. For example, the gradual increase in the size of the modern horse, *Equus,* can be correlated with a change in the environment from forestlike conditions to grassland conditions, as discussed earlier (p. 517). Nevertheless, the evolution of the horse should not be viewed as a straight line of descent; there were many side branches that became extinct.

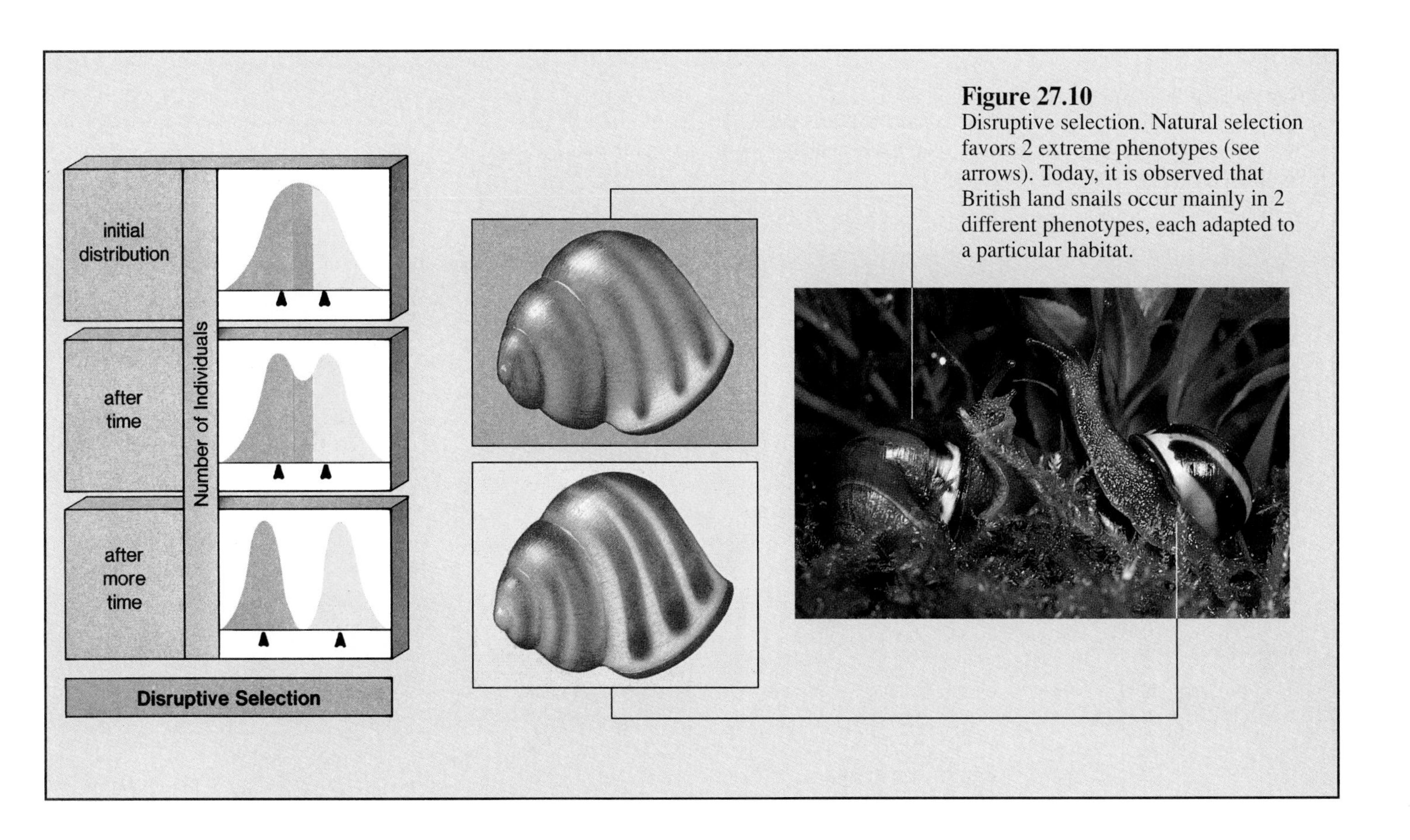

Figure 27.10
Disruptive selection. Natural selection favors 2 extreme phenotypes (see arrows). Today, it is observed that British land snails occur mainly in 2 different phenotypes, each adapted to a particular habitat.

Soot-Colored Moths and Other Natural Selection Cases
Industrial melanism is another good example of directional selection in which the selective agent is known. Moths rest on the trunks of trees during the day; if they are seen by predatory birds, they are eaten. As long as the tree trunks in the environment are light in color, the light-colored moths live to reproduce. But if the tree trunks turn black from industrial pollution, the dark-colored moths survive and reproduce to a greater extent than the light-colored moths. The dark-colored phenotype then becomes the more frequent one in the population (fig. 27.6). However, if pollution is reduced and the trunks of the trees regain their normal color, the light-colored moths again increase in number.

Pesticides and antibiotics are selective agents for insects and bacteria, respectively. The forms that survive exposure to these agents give rise to future generations that are resistant to these toxic substances.

A study of sickle-cell disease shows how natural selection can affect gene pool frequencies. Persons with sickle-cell disease have sickle-shaped red blood cells, leading to hemorrhaging and organ destruction. In parts of Africa, there is a high incidence of malaria caused by a parasite that lives in and destroys red blood cells (see fig. 28.13). Sickle-cell disease tends to be more common in such areas. A study of the 3 genotypes and phenotypes involved explains why.

Genotype	Phenotype	Result
Hb^AHb^A	Normal	Dies due to malarial infection
Hb^AHb^S	Sickle-cell trait	Lives due to protection from both
Hb^SHb^S	Sickle-cell disease	Dies due to sickle-cell disease

Persons with sickle-cell trait are more likely to survive to reproduce for 2 reasons. Most of the time they do not have circulatory problems because their red blood cells have a normal

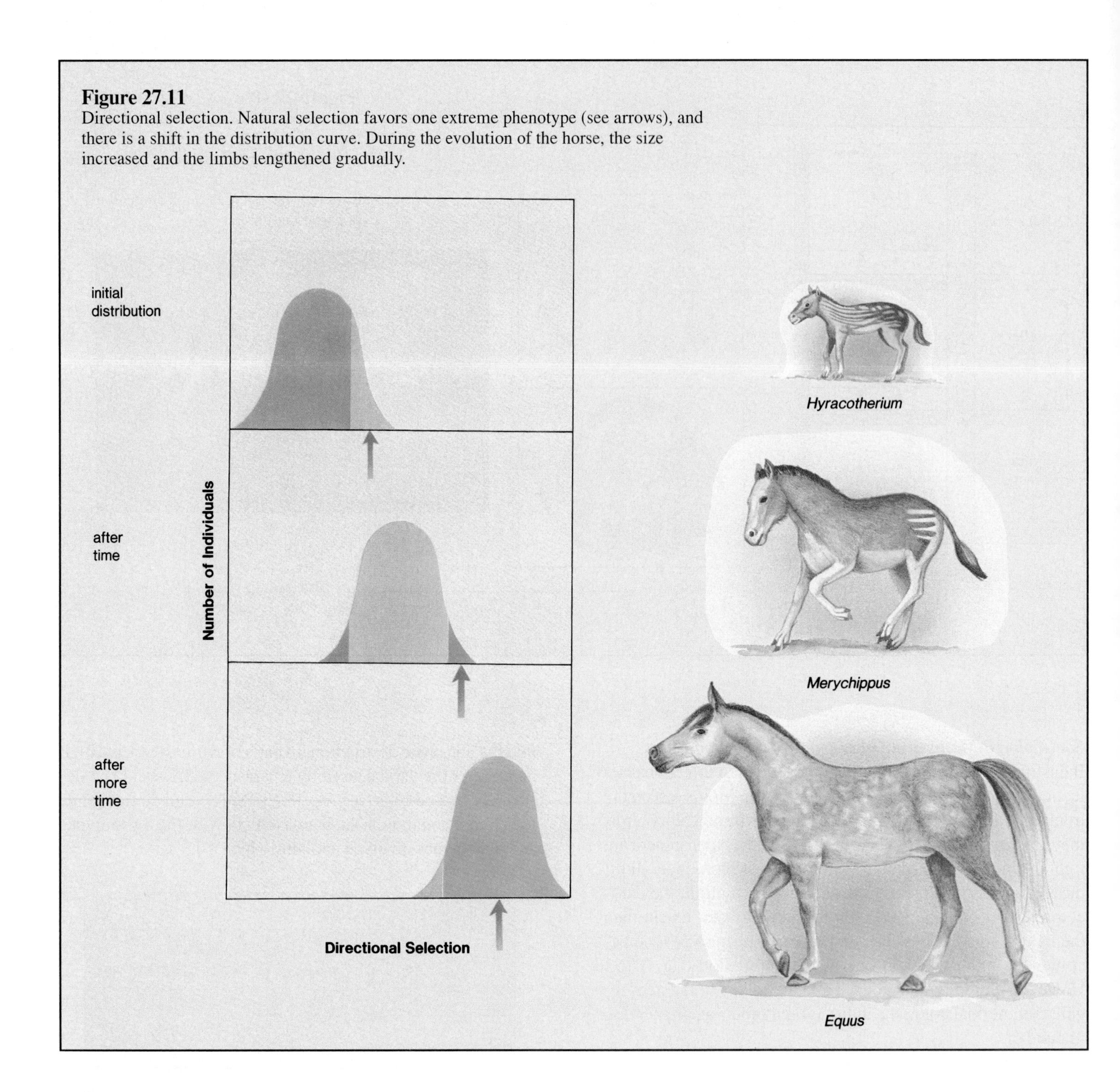

Figure 27.11
Directional selection. Natural selection favors one extreme phenotype (see arrows), and there is a shift in the distribution curve. During the evolution of the horse, the size increased and the limbs lengthened gradually.

shape. Even so, the malarial parasite cannot survive in their red blood cells. When the cells are sickled, they lose potassium and the parasite dies.

The frequency of the sickle-cell allele in some parts of Africa is 0.40, while among African-Americans it is only 0.05. The ability of the heterozygote to survive accounts for the greater frequency of the sickle-cell allele in Africa.

The agents of evolutionary change are mutation, gene flow, nonrandom reproduction, genetic drift, and natural selection. These processes cause changes in the gene pool frequencies of a population. Only natural selection results in adaptation to the environment. Do 27.2 Critical Thinking, found at the end of the chapter.

Table 27.3
Reproductive Isolating Mechanisms of Coexisting Species

Isolating Mechanisms	Example
Premating	
Habitat	Species at same locale occupy different habitats
Temporal	Species mate at different seasons or different times of day
Behavioral	In animals, courtship behavior differs or they respond to different songs, calls, pheromones, or other signals
Mechanical	Genitalia unsuitable to one another
Postmating	
Gamete mortality	Sperm cannot reach or fertilize egg
Zygote mortality	Hybrid dies before maturity
Hybrid sterility	Hybrid survives but is sterile, it cannot reproduce

Figure 27.12
First stage of speciation is geographic isolation (**a** and **b**). Second stage of speciation is reproductive isolation (**c** and **d**).

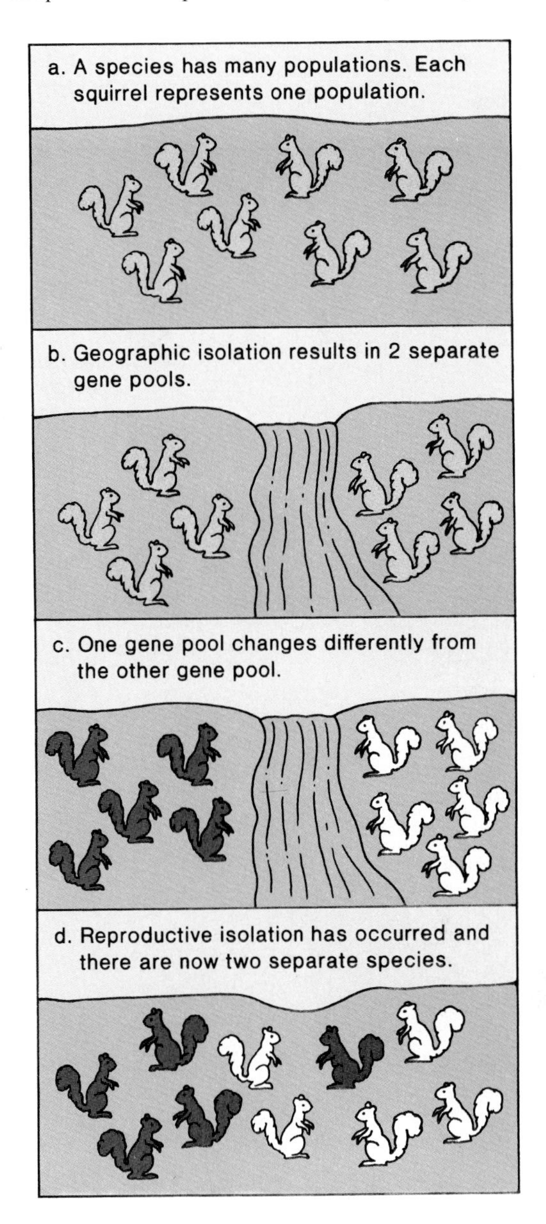

How Species Form

Usually, a species occupies a certain geographical range, within which there are several subpopulations. For our present discussion, **species** is defined as a group of interbreeding subpopulations that share a gene pool and that are isolated reproductively from other species. The subpopulations of the same species exchange genes, but different species do not exchange genes. Reproductive isolation of the gene pools of similar species is accomplished by such mechanisms as those listed in table 27.3. If **premating isolating mechanisms** are in place, reproduction is never attempted. If **postmating isolating mechanisms** are in place, reproduction may take place, but it does not produce fertile offspring.

Two Steps to New Species

Speciation has occurred when one species gives rise to 2 species. As an example, consider a species of squirrels that has several populations. In figure 27.12*a*, each population is represented by a single squirrel. All the populations share a common gene pool, and therefore all squirrels are the same color.

It is generally accepted that speciation is a two-stage process. Stage I is geographic isolation. Suppose that a canal dug to divert water from a nearby source separates the populations of squirrels into 2 groups. One group is now geographically isolated from the other (fig. 27.12*b*). Stage II is reproductive isolation. In figure 27.12*c*, we see that the 2 groups are colored differently. This symbolizes that each group is now evolving separately. There are several reasons for this: (1) Certain variations may be present in only one group—because the gene pools are smaller, these variations have a better chance of

Figure 27.13
Each of Darwin's finches is adapted to gathering and eating a different type of food; tree finches have beaks largely adapted to eating insects and, at times, plants; ground finches have beaks adapted to eating off the prickly-pear cactus or different size seeds. The woodpecker-type finch, a tool-user, uses a cactus spine or twig to probe in the bark of a tree for insects.

being passed on; (2) each gene pool now experiences different mutations and recombination of alleles; and (3) the environment is different for each group, and each one is subject to different selective pressures.

Given enough time, we expect that even if the physical barrier is removed, the 2 groups will not be able to reproduce with one another (fig. 27.12*d*). The second stage of speciation is complete—what was formerly one species has become 2 species because they are isolated reproductively by the mechanisms listed in table 27.3.

Speciation is the origin of species. This usually requires geographic isolation followed by reproductive isolation. Do 27.3 Critical Thinking, found at the end of the chapter.

Adaptive Radiation and Finches

One of the best examples of speciation is provided by the finches on the Galápagos Islands, which are often called Darwin's finches because Darwin first realized their significance as an example of how evolution works. The Galápagos Islands, located 600 miles west of Ecuador, South America, are volcanic but do have forest regions at higher elevations. The 13 species of finches (fig. 27.13), placed in 3 genera, are believed to be descended from mainland finches that migrated to one of the islands some years ago. Therefore, Darwin's finches are an example of *adaptive radiation,* or the proliferation of a species by adaptation to different ways of life. We can imagine that after the original population of a single island increased, some individuals dispersed to other islands. The islands are ecologically different enough to have promoted divergent feeding habits. This is apparent because although the birds physically resemble each other in many respects, they have different beaks, each of which is adapted to gathering and eating a different type of food. There are seed-eating ground finches, with beaks appropriate to cracking small-, medium-, or large-sized seeds; cactus-eating finches with beaks appropriate to feeding off of prickly-pear cacti; insect-eating tree finches, also with different-sized beaks; and a warbler-type finch, with a beak adapted to insect eating and nectar gathering. Among the tree finches, there is a woodpecker type, which lacks the long tongue of a true woodpecker but makes up for this by using a cactus spine or a twig to ferret out insects.

One frequently cited example of speciation is the evolution of several species of finches on the Galápagos Islands. This is also an example of adaptive radiation because the various species have different ways of life.

Table 27.4
Hierarchy of Classification

Category	Description
Kingdom	Contains related phyla
Phylum (Animals) Division (Plants)	Contains related classes
Class	Contains related orders
Order	Contains related families
Family	Contains related genera
Genus	Contains related species
Species	Reproductively isolated populations with phenotypic similarities

Studying and Classifying Big Changes

Macroevolution concerns major phenotypic changes and is the study of relationships between groups of organisms above the species level (table 27.4). Therefore, study of macroevolution is related to the science of taxonomy.

Taxonomy: The Science of Classifying

Taxonomy, the science of classifying organisms, utilizes the fossil record and comparative anatomy, including embryological data, to classify organisms. Homologous structures are especially helpful in deciphering relationships between species because they were inherited from a common ancestor. Figure 27.3 shows that the forelimbs of vertebrates are *homologous* because they contain the same sets of bones organized in the same general way. Modern taxonomists rely heavily on anatomical features; however, they maintain that members of a species have phenotypic traits in common because they belong to a group of organisms that is reproductively isolated from other groups in nature.

Taxonomists group species into ever larger categories. All the species in the same genus have many characteristics in common. For example, we would expect all species of oak trees in the genus *Quercus* to be very similar. All the species in the same kingdom, however, can be quite different from one another. For example, both roses and pine trees are in the plant kingdom. The 5 kingdoms recognized by this text are discussed in more detail on page 510.

Figure 27.14
Evolutionary trees are based on the way organisms are classified. The classification and the tree tell the phylogenic history of the organism. A species is most closely related to other species in the same genus than to those species in other genera in the same family and so forth, from order to class to phylum to kingdom.

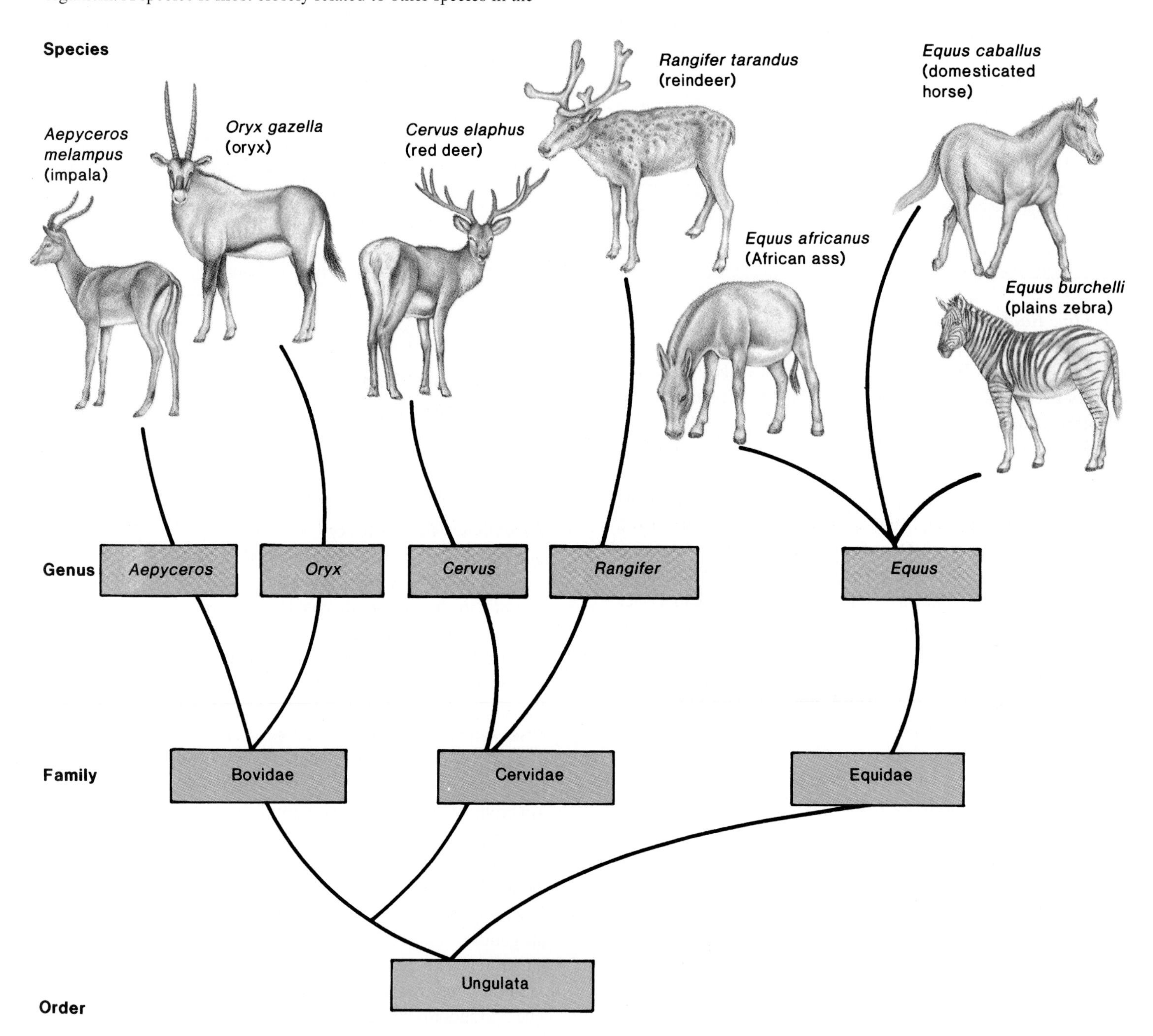

Drawing Evolutionary Trees

Many taxonomists attempt to classify organisms in a way that reflects phylogeny, or the evolutionary history of a group of organisms. The results are frequently portrayed in diagrams called **evolutionary trees,** in which each branching point shows where a common ancestor gave rise to other lineages (fig. 27.14). Notice how the most closely related species have a more recent common ancestor.

It is also possible to construct evolutionary trees based on molecular data. Amino acid sequences in certain proteins like hemoglobin and cytochrome c have been determined for various animals, as have DNA nucleotide differences between animals. When 2 lineages first diverge from a common ancestor, the genes and the proteins of the lineages are nearly identical. But as time goes by, each lineage accumulates its own changes in gene and protein structure. Therefore, the degree of differences in these molecules provides some indication of how long ago the 2 lineages diverged from a common ancestor. In other words, DNA and amino acid differences can be used as a "molecular clock" to indicate evolutionary time. Biochemical data and the fossil record can be used as independent ways to measure the length of time 2 lineages have been diverging.

> Evolutionary trees are a way to show the evolutionary history of a group of organisms and how they are related by way of common ancestors.

SUMMARY

The fossil record, biogeography, and comparative anatomy, embryology, and biochemistry all give evidence of evolution. The fossil record gives us the history of life in general and allows us to trace the descent of a particular group. Biogeography shows that the distribution of organisms on earth is explainable by assuming organisms evolved in one locale. Comparing the anatomy and the development of organisms reveals a unity of plan among those that are closely related. All organisms have certain biochemical molecules in common, and any differences indicate the degree of relatedness.

Evolution is described as a process that involves a change in gene frequencies within the gene pool of a sexually reproducing population. The Hardy-Weinberg law states that the gene pool frequencies arrive at an equilibrium that is maintained generation after generation unless disrupted by mutations, nonrandom mating, gene flow, genetic drift, or natural selection. Any change from the initial allele frequencies in the gene pool of a population signifies that evolution has occurred.

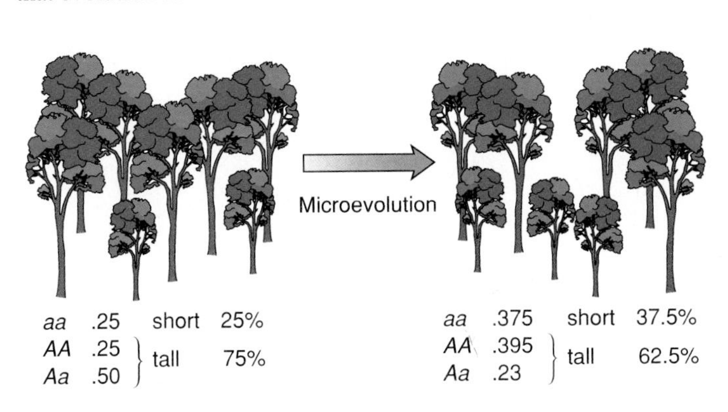

Speciation is the origin of species. This usually requires geographic isolation, followed by reproductive isolation.

The evolution of several species of finches on the Galápagos Islands is an example of adaptive radiation because each species has a different way of life, but all species came from one common ancestor.

Macroevolution pertains to evolution above the species level. Evolutionary trees are a way to show the evolutionary history of a group of organisms and how they are related by way of common ancestors.

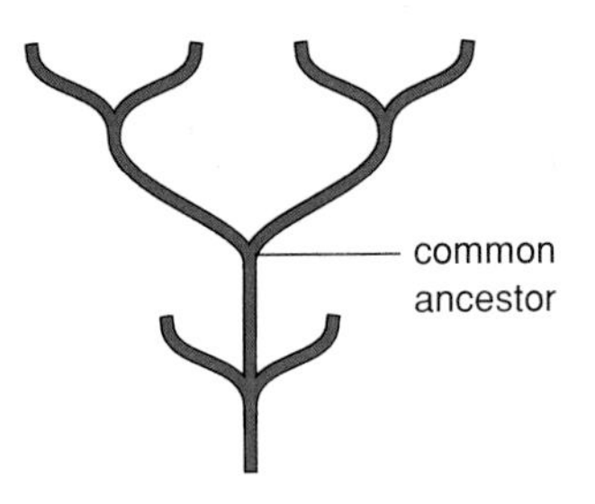

STUDY QUESTIONS

In order to practice **writing across the curriculum,** students should write out the answers to any or all of the study questions. The study questions are sequenced in the same order as the text.

1. Show that the fossil record, biogeography, and comparative anatomy, embryology, and biochemistry all give evidence that common descent has occurred. (pp. 515–520)
2. What are a population and a gene pool? (p. 520)
3. What is the Hardy-Weinberg law? What is its significance? (p. 520)
4. Name and define the agents of evolutionary change. (pp. 522–526)
5. What is the founder effect, and what is a bottleneck? (p. 523)
6. What are the 4 requirements for evolution by natural selection? (p. 526)
7. Name and give an example for each type of selection. (p. 526)
8. Define a species. How do new species originate? (p. 529)
9. When is adaptive radiation apt to take place? (p. 531)
10. What is macroevolution, and how does macroevolution relate to evolutionary trees? (p. 531)

OBJECTIVE QUESTIONS

1. If a population is in Hardy-Weinberg equilibrium, evolution _______ (always, does not) occur(s).
2. Twenty-one percent of a population is homozygous recessive. Assuming a Hardy-Weinberg equilibrium, what percentage is expected to be homozygous recessive in the next generation? _______
3. _______ is the ultimate source of genetic variation in a population.
4. A change in gene pool frequencies due to chance death of some members is an example of _______.
5. A gradual increase in the size of the human brain since humans evolved is an example of _______.
6. Gene flow between subpopulations of a species keeps their gene pools _______ (similar, dissimilar).
7. During the first stage of speciation, populations become _______.
8. During the second stage of speciation, populations become _______.
9. Two species of butterflies have different courtship behavior patterns. This is an example of a(n) _______ isolating mechanism of the _______ type.
10. When classifying organisms, similar families are assigned to the same _______. (Hint: Look back to chapter 1.)
11. Offer explanations for the following observations.
    a. Wings of bats and wings of insects perform the same functions but do not have similar embryological origins. _______
    b. Amphibians, reptiles, birds, and mammals all have pharyngeal pouches sometime during development. _______
    c. Cacti and euphorbias exist on different continents but both have spiny, water-storing, leafless stems. _______
    d. The tomato, potato, and jimsonweed are members of the nightshade family and have similar flowers and fruits. The plants have other characteristics that are quite diverse. _______
12. This diagram represents one species. Each circle is a population. What changes would you make to use the diagram to symbolize 2 species?

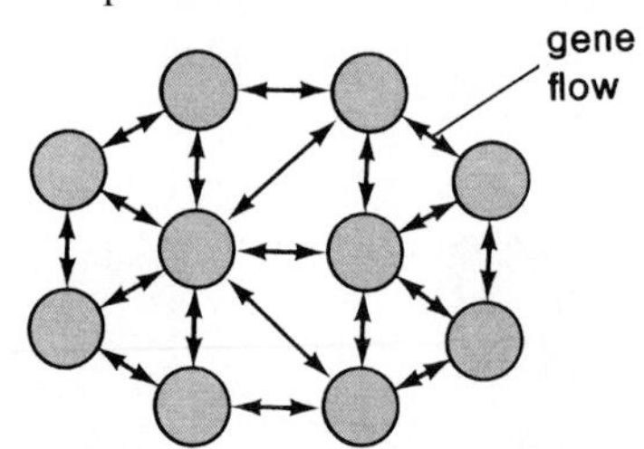

13. Using these values and the binomial expression $p^2 + 2pq + q^2$, state the genotypic frequencies in a population.

    $p = 0.2$ $\quad p^2 = 0.04$ $\quad 2pq = 0.32$

    $q = 0.8$ $\quad q^2 = 0.64$

    a. Homozygous recessive genotype = _______%
    b. Homozygous dominant genotype = _______%
    c. Heterozygous genotype = _______%

CRITICAL THINKING

In order to practice **writing across the curriculum,** students should write out the answers to any or all of the critical thinking questions. Suggested answers to the critical thinking questions are in appendix E.

27.1

1. Why would you expect to find coyotes in both North American and African grasslands?
2. Explain, on the basis of evolution, why there are coyotes in North American grasslands but jackals in African grasslands.
3. What evidence would you gather to support your explanation?

27.2

1. A scientist observes that members of a particular plant species at the top of a mountain are shorter than members at the bottom. Give an explanation based on natural selection.
2. The scientist gathers seeds from the plants at the top of the mountain and plants them at the base of the mountain. If your explanation is correct, what will the plants look like?
3. Explain these experimental results in terms of a particular type of selection.

27.3

1. Organisms are generally adapted to conserve energy. Which ones—premating or postmating isolating mechanisms (see table 27.3)—represent a waste of energy? Why?
2. During the process of speciation, which types of mechanisms most likely evolve first? Why?
3. Which of the premating isolating mechanisms would you expect to be operating among Darwin's finches? What other type of premating mechanism might there be that is not listed in table 27.3?

SELECTED KEY TERMS

analogous structure (ah-nal´ o-gus) structure similar to another in function but not in anatomy; particularly in reference to similar adaptations.
biogeography the study of the geographical distribution of organisms.
evolution changes that occur in populations of organisms with the passage of time, often resulting in increased adaptation of organisms to the environment.
evolutionary tree diagram describing the evolutionary relationship of groups of organisms.
fossil any remains of an organism that have been preserved in the earth's crust.
gene flow the movement of genes from one population to another via sexual reproduction between members of the populations.
gene pool the total of all the genes of all the individuals in a population.
genetic drift evolution by chance processes alone.
homologous structure (ho-mol´o-gus) structure similar to another in anatomy but not necessarily function; homologous structures in animals share a common ancestry.
natural selection the process by which populations become adapted to their environment.
population all the organisms of the same species in one place.
species a group of similarly constructed organisms capable of interbreeding and producing fertile offspring; organisms that share a common gene pool.
vestigial structure the remains of a structure that was functional in some ancestor but is no longer functional in the organism in question.

28

VIRUSES, MONERA, PROTISTS, AND FUNGI

Chapter Concepts

1.
Viruses are acellular. Whether they should be considered living organisms is questionable. 537, 558

2.
The monera are prokaryotes, while the protists and the fungi are eukaryotes. 542, 558

3.
Kingdom Monera includes bacteria, which are important organisms despite their small size. 542, 558

4.
Kingdom Protista contains algae, protozoa, and 2 types of molds (slime molds and water molds). Algae are the plant-like protists; protozoa are the animal-like protists; slime molds and water molds are the fungus-like protists. 546, 558

5.
Kingdom Fungi contains the most complex organisms to rely on saprophytic nutrition. 555, 558

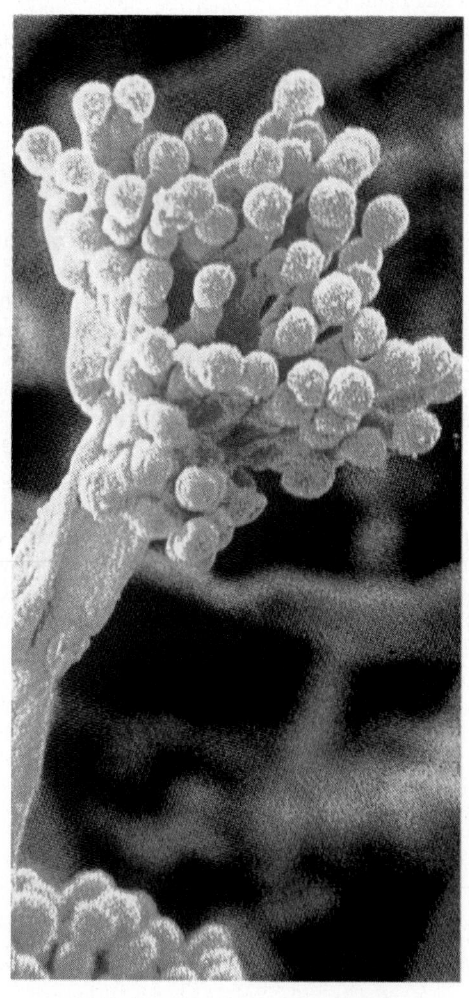

Conidia of *Penicillium,* ×11,500

Chapter Outline

IN this and the next 2 chapters, we will discuss living organisms, from the simple to the complex and from the primitive (earliest evolved) to the most advanced (most recently evolved). We can suggest how living organisms may be related only in the broadest terms, since detailed information is often lacking. It is also important to remember that no living group of organisms is the direct ancestor of another living group of organisms, although it is possible that 2 living groups once had a common ancestor.

We begin our discussion with viruses because they are on the borderline between living and nonliving things. Notice that viruses are not included in the classification of organisms given in appendix D.

Viruses

Viruses are acellular—they do not have a cellular type of organization. Regardless of size (25–200 nm), a virus is composed of only 2 parts: a protein coat and a nucleic acid core. The protein coat, specifically called a *capsid,* is sometimes surrounded by an outer envelope derived from the membrane of the host cell.

In general, viruses are classified according to whether DNA or RNA serves as the genome and whether the nucleic acid is single stranded or double stranded. The type of capsid is also important because it determines the overall shape of the virus. Commonly, the capsid has a helical, polyhedral, or complex organization (fig. 28.1). If the virus has an envelope, the capsid organization is not obvious.

Figure 28.1
Virus diversity. On the basis of structure, viruses can be classified into 4 main groups. In a helical virus, the capsid units join in a helical manner and the nucleic acid core is helical. In a polyhedral virus, the capsid is a polyhedron with 20 triangular faces and 12 corners. A complex virus is a composite. In the one shown here, the head is polyhedral and the tail is helical. In an enveloped virus, the capsid is covered by an envelope.

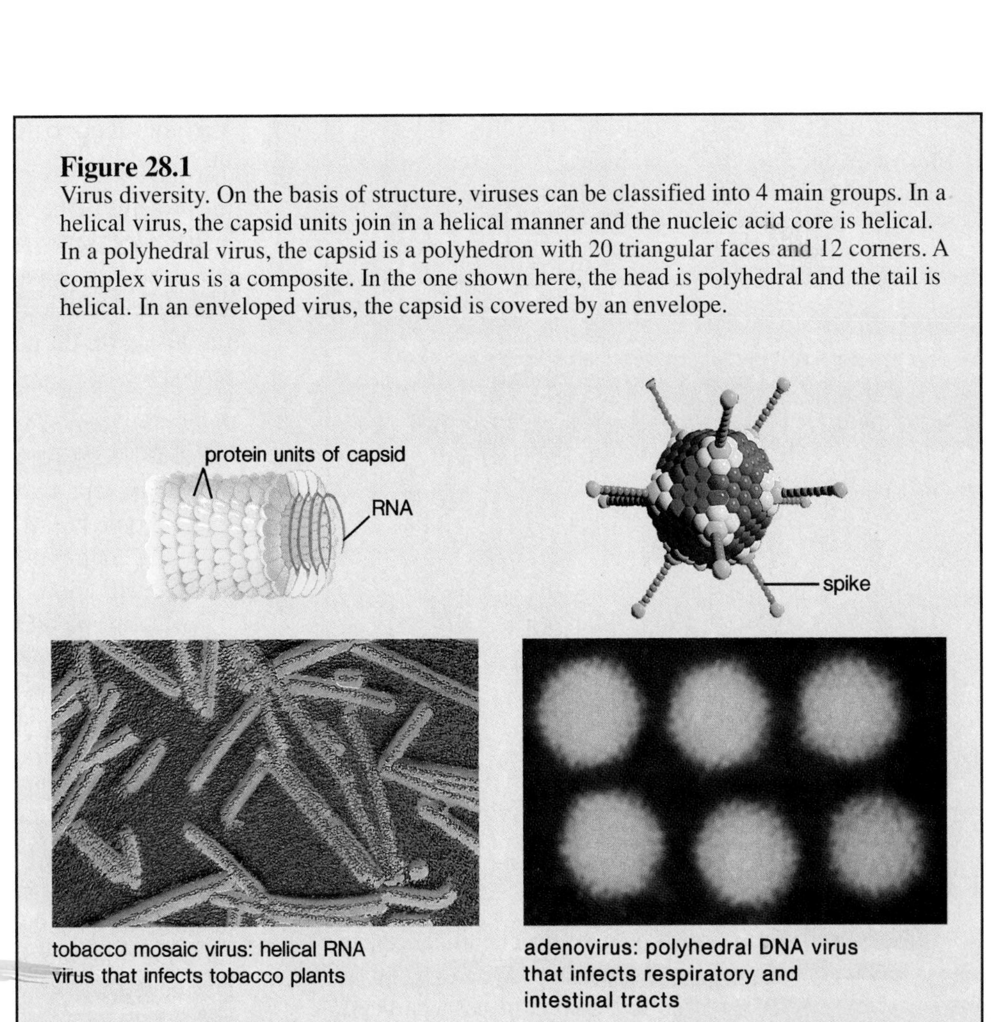

tobacco mosaic virus: helical RNA virus that infects tobacco plants

adenovirus: polyhedral DNA virus that infects respiratory and intestinal tracts

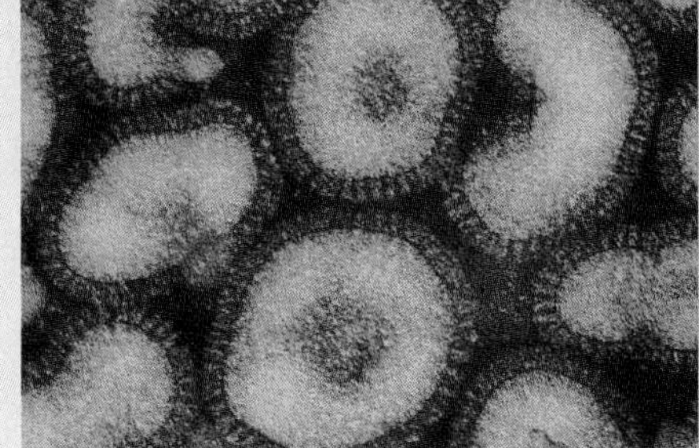

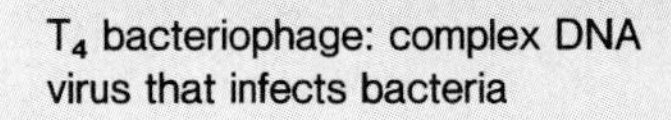

T_4 bacteriophage: complex DNA virus that infects bacteria

influenza virus: enveloped helical RNA virus that infects respiratory tract

Figure 28.2
Inoculation of live chick eggs with a virus. A virus only reproduces inside a living cell, not because it uses the cell for nutrients but rather because it takes over the metabolic machinery of the cell.

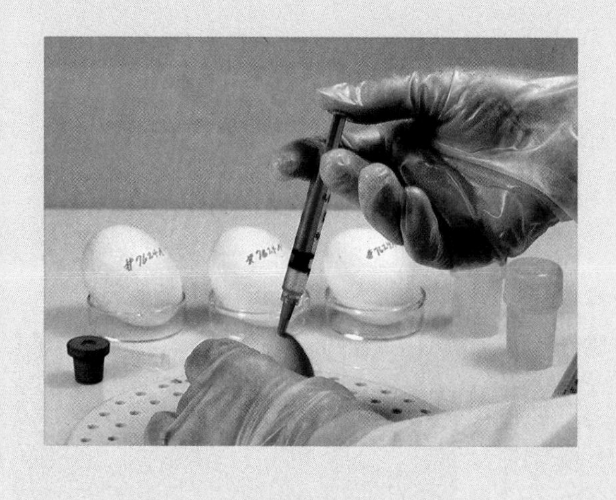

Host specificity can also be used for classification purposes. Viruses that attack bacteria, plants, fungi, invertebrates, and vertebrates are known. Further, each virus has a particular host; for example, plant viruses attack only specific plants, and animal viruses attack only specific animals. The human disease-causing viruses even reproduce in just certain types of cells. Because viruses reproduce only inside a living cell, they are called *obligate parasites.* In the laboratory, animal viruses are replicated by injecting them into living cells, such as live chick embryos (fig. 28.2). Outside living cells, viruses are nonliving and are stored in liquid nitrogen.

> Viruses are acellular obligate parasites that always have a protein coat and a nucleic acid core.

Viruses Reproduce Inside Living Cells

The nucleic acid within a virus contains the genes that code for the proteins in the capsid. In addition, it may contain genes for a few enzymes needed for the virus to reproduce. In large measure, however, a virus relies on the host's enzymes, ribosomes, tRNA, and ATP for its own reproduction. In other words, a virus takes over the metabolic machinery of the host cell when it reproduces.

Bacteriophages: Reproduce Inside Bacteria

Bacteriophages are viruses that reproduce inside bacteria (fig. 28.3). The reproductive cycle of the T-even viruses (meaning that the type [T] is designated by an even rather than an odd number) requires these stages: attachment, penetration, biosynthesis, maturation, and release.

During *attachment,* a T-even virus collides with an *E.coli* cell, and attachment sites on the tail fibers of the virus combine with receptors on the bacterial cell. (Attachment sites and receptor sites fit together as a key fits a lock, which accounts for the specificity of a virus for a particular host cell.) During *penetration,* a viral enzyme digests away part of the bacterial cell membrane, and viral DNA enters the bacterial cell by way of the tail. *Biosynthesis* of viral components begins after the virus brings about disintegration of host DNA. The virus takes over the machinery of the cell in order to carry out viral DNA replication and production of multiple copies of the coat protein. During *maturation,* viral DNA and capsids are assembled to produce several hundred viral particles. Lysozyme, an enzyme coded for by a viral gene, is produced; this disrupts the cell membrane, and *release* of phage particles then occurs. The bacterial cell then dies.

Animal Viruses: Uncoat and Enter Cells

Animal viruses follow the same reproductive process as bacteriophages, with slight modifications. One difference is that the entire animal virus (not just the nucleic acid) penetrates a host cell by endocytosis. During *uncoating,* the envelope, if present, and the capsid are removed. The viral genome, either DNA or RNA, is now free of its coverings and replication can proceed. Another difference is that if the virus has an envelope, release occurs by *budding.* A viral envelope consists of lipids, proteins, and carbohydrates that are present in the cell membrane of the host cell. Some of these proteins are specific to the virus and have been coded for by viral genes. Budding does not result in the death of the host cell.

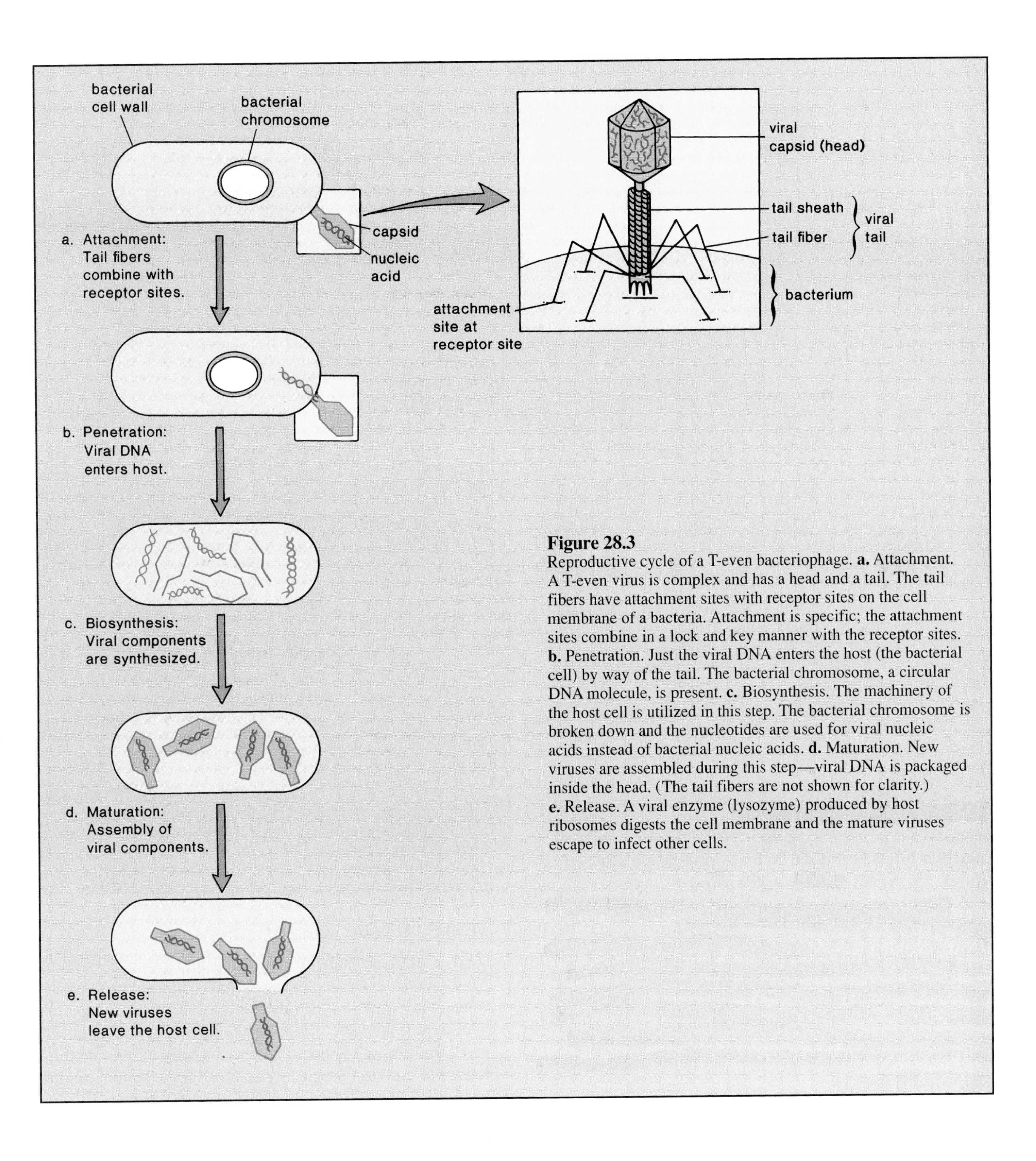

Figure 28.3
Reproductive cycle of a T-even bacteriophage. **a.** Attachment. A T-even virus is complex and has a head and a tail. The tail fibers have attachment sites with receptor sites on the cell membrane of a bacteria. Attachment is specific; the attachment sites combine in a lock and key manner with the receptor sites. **b.** Penetration. Just the viral DNA enters the host (the bacterial cell) by way of the tail. The bacterial chromosome, a circular DNA molecule, is present. **c.** Biosynthesis. The machinery of the host cell is utilized in this step. The bacterial chromosome is broken down and the nucleotides are used for viral nucleic acids instead of bacterial nucleic acids. **d.** Maturation. New viruses are assembled during this step—viral DNA is packaged inside the head. (The tail fibers are not shown for clarity.) **e.** Release. A viral enzyme (lysozyme) produced by host ribosomes digests the cell membrane and the mature viruses escape to infect other cells.

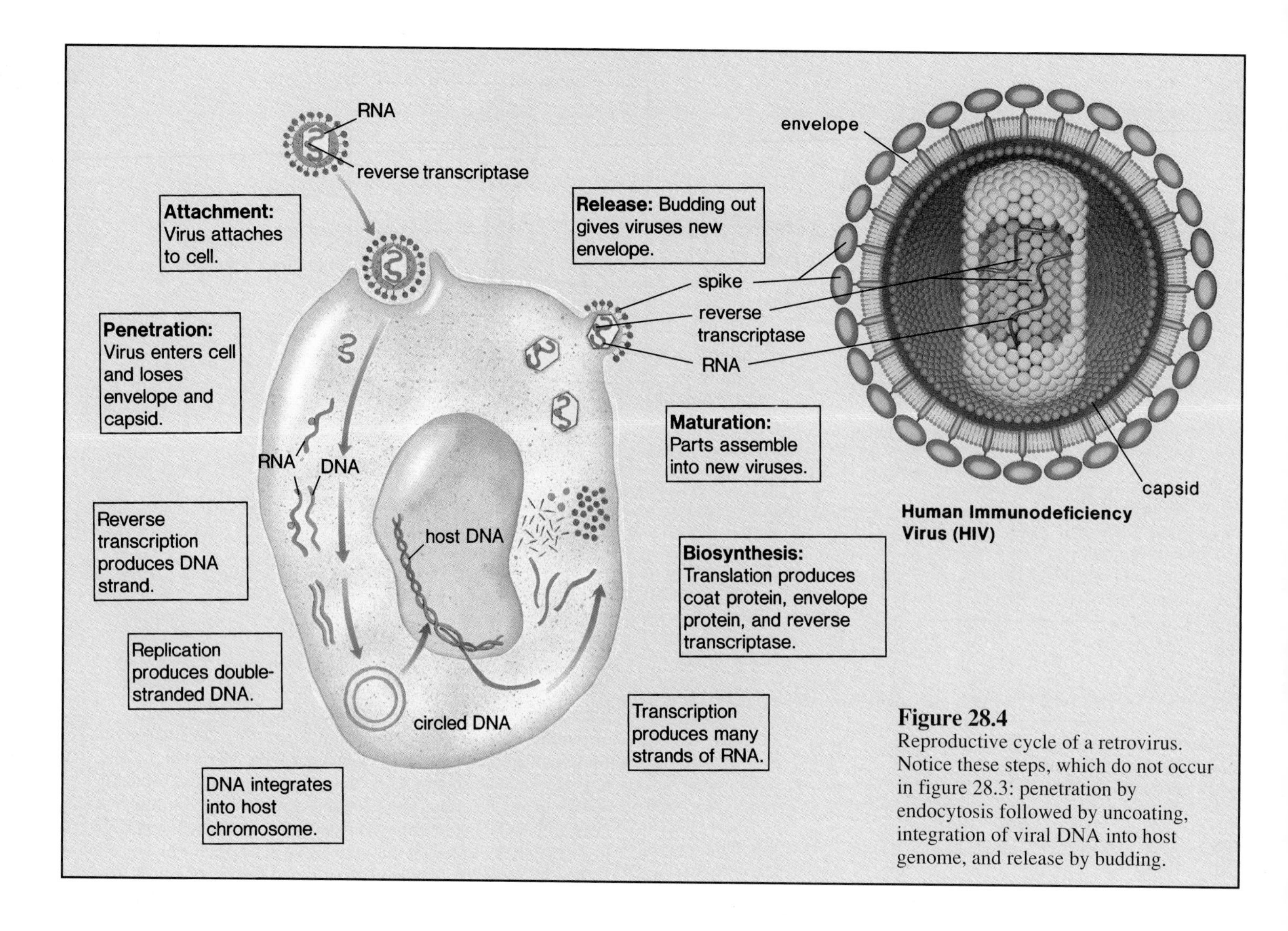

Figure 28.4
Reproductive cycle of a retrovirus. Notice these steps, which do not occur in figure 28.3: penetration by endocytosis followed by uncoating, integration of viral DNA into host genome, and release by budding.

Retroviruses: Have RNA Genes

Figure 28.4 illustrates the reproduction of a retrovirus. Much interest is currently focused on retroviruses because the HIV viruses, which cause AIDS, are retroviruses.

A **retrovirus** has an RNA genome and contains a special enzyme called *reverse transcriptase,* which carries out RNA→DNA transcription. Following replication, the resulting double-stranded DNA is integrated into the host genome. The viral DNA remains in the host genome and is replicated when host DNA is replicated. When and if this DNA is transcribed, new viruses are produced by the steps we have already cited: biosynthesis, maturation, and release not by destruction of the cell but by budding.

> Viral reproduction requires attachment, penetration, biosynthesis, maturation, and release. If an animal virus has an envelope, the virus is released by budding.

Viral Infections: From AIDS to Warts

Viruses are best known for causing infectious diseases in plants and animals. In plants, infectious diseases can be controlled only by destroying the plants that show the symptoms of disease. In animals, especially humans, viruses are controlled by administering vaccines and only recently by the administration of antiviral drugs, as discussed in the reading on the next page. Table 28.1 lists some well-known human diseases caused by viruses.

KILLING MICROORGANISMS

Viruses and bacteria are microbes that cause diseases in humans. The development of drugs to kill viruses has lagged far behind the development of those to kill bacteria. Viruses lack most enzymes and instead utilize the metabolic machinery of the host cell. Rarely has it been possible to find a drug that successfully interferes with viral reproduction without also interfering with host metabolism. One such drug, however, called vidarabine, was approved in 1978 for treatment of viral encephalitis, an infection of the nervous system. Acyclovir (ACV) seems to be helpful in treating genital herpes, and the drugs AZT (zidovudine), DDI (dideoxyinosine), and nevirapine are being used to treat AIDS patients.

An antibiotic is a chemical that selectively kills bacteria when it is taken into the body as a medicine. Since the introduction of the first antibiotics in the 1940s, there has been a dramatic decline in deaths due to pneumonia, tuberculosis, and other infectious diseases.

Most antibiotics are produced naturally by soil microorganisms. Penicillin is made by the fungus *Penicillium* (fig 28.A), and streptomycin, tetracycline, and erythromycin are all produced by the bacterium *Streptomyces*. Sulfa, a chemotherapeutic agent rather than an antibiotic, is an analogue of a bacterial growth factor. Sulfa is produced in the laboratory.

These antibiotics are metabolic inhibitors, specific for bacterial enzymes. This means that they poison bacterial enzymes without harming host enzymes. Penicillin blocks the synthesis of the bacterial cell wall; streptomycin, tetracycline, and erythromycin block protein synthesis; and sulfa prevents the production of a coenzyme.

There are problems associated with antibiotic therapy. Some patients are allergic to antibiotics, and the reaction can be fatal. Antibiotics not only kill off disease-causing bacteria, they also reduce the number of beneficial bacteria in the intestinal tract and other locations. These beneficial bacteria hold in check the growth of certain microbes that now begin to flourish. Diarrhea can result as can a vaginal yeast infection. The use of antibiotics can also prevent natural immunity from occurring, leading to recurring antibiotic therapy. Most important, perhaps, is the growing resistance of certain strains of bacteria to antibiotics. While penicillin used to be 100% effective against hospital strains of *Staphylococcus aureus,* today it is far less effective. Tetracycline and penicillin, long used to cure gonorrhea, now have a failure rate of more than 20% against certain strains of gonococcus.

Most physicians believe that antibiotics should be administered only when absolutely necessary. Some believe that if antibiotic use is not strictly limited, resistant strains of bacteria will completely replace present strains and antibiotic therapy will no longer be effective at all. They are much opposed to the current practice of adding antibiotics to livestock feed in order to make animals grow fatter. Bacteria that become resistant are easily transferred from animals to humans. Antibiotics have been a boon to humans, but they should be used with care.

Figure 28.A
Penicillium chrysogenum. The antibiotic penicillin is prepared from this fungus.

Table 28.1
Some Significant Viral Diseases in Humans

Category	Disease (Type of Virus)
Sexually transmitted diseases	AIDS (HIV) Genital warts (papilloma group) Genital herpes (herpes simplex type 2 and occasionally type 1)
Childhood diseases	Mumps (mumps virus) Measles (measles virus) Chicken pox (herpes zoster) German measles (rubella)
Respiratory diseases	Common cold (rhino group, corona group) Influenza (influenza group) Acute respiratory infection (adeno group)
Skin diseases	Warts (papilloma group) Fever blisters (herpes simplex 1) Shingles (herpes zoster)
Digestive tract diseases	Gastroenteritis (parvo group) Diarrhea (entero group, reo group)
Nervous system diseases	Poliomyelitis (polio) Rabies (rabies) Encephalitis (encephalitis)
Other diseases	Cancer (Epstein-Barr, HIV) Hepatitis (hepatitis group)

Kingdom Monera

In the classification system used for this text, the kingdom Monera contains all the different types of **bacteria.**

Structure of Bacteria: No Nucleus

Prokaryotic cells (see fig. 3.12) are very small (1–10 μm in length and 0.2–0.3 μm in width), and except for ribosomes (see table 3.2), they do not have the cytoplasmic organelles found in eukaryotic cells. They do have DNA, but it is not contained within a nuclear envelope; therefore, they are said to lack a nucleus. They have respiratory enzymes, but no mitochondria, and if they possess chlorophyll, it may be found within thylakoids, but there are no chloroplasts.

The eubacteria (true bacteria) have a cell wall containing unique amino sugars cross-linked by peptide chains. The cell wall may be surrounded by a capsule. Some bacteria move by means of flagella, and some adhere to surfaces by means of short, fine, hairlike appendages called pili (sing., pilus).

Bacteria occur in 3 basic shapes (fig. 28.5): *rod* (bacillus), *spherical* or round (coccus), and *spiral* (e.g., helical shape called a spirillum). The bacilli and the cocci may form chains of a length typical of the particular bacterium.

Prokaryotic cells lack a nucleus and most of the other organelles found in eukaryotic cells. Bacteria occur in 3 basic shapes: rod, spherical, and spiral.

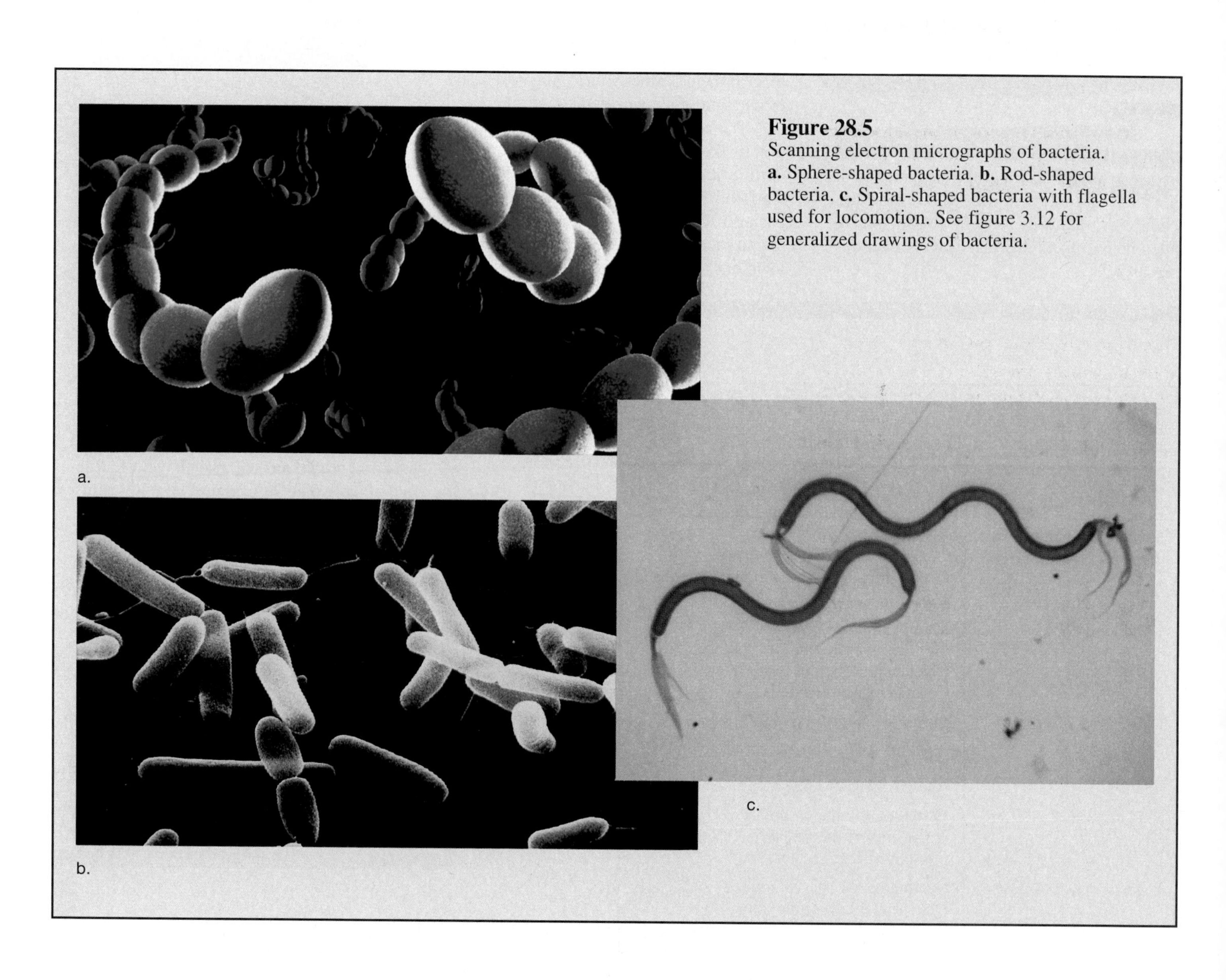

Figure 28.5
Scanning electron micrographs of bacteria. **a.** Sphere-shaped bacteria. **b.** Rod-shaped bacteria. **c.** Spiral-shaped bacteria with flagella used for locomotion. See figure 3.12 for generalized drawings of bacteria.

Reproduction of Bacteria: Asexual and Sexual

Bacteria reproduce asexually by means of **binary fission.** First, the single circular chromosome duplicates; then there are 2 chromosomes attached to the inside of the cell membrane. The chromosomes are separated by an elongation of the cell, which pushes the chromosomes apart. Then the cell membrane grows inward and the cell wall forms, dividing the cell into 2 daughter cells, each of which now has its own chromosome (fig. 28.6).

Sexual exchange of DNA occurs among bacteria in 3 ways. Conjugation takes place when the so-called male cell passes DNA to the female cell by way of a sex pilus. *Transformation* occurs when a bacterium binds to and then takes up DNA released into the medium by dead bacteria. During *transduction,* bacteriophages carry portions of DNA from one bacterium to another.

When faced with unfavorable environmental conditions, some bacteria form *endospores* (fig. 28.7). A portion of the cytoplasm and a copy of the chromosome dehydrate and are then encased by 3 heavy, protective spore coats. The rest of the bacterial cell deteriorates and the endospore is released. When environmental conditions are again suitable for growth, the endospore absorbs water and grows out of the spore coats. In time, it becomes a typical bacterial cell, capable of reproducing once again by binary fission.

Figure 28.6
Binary fission. In electron micrographs, it is possible to observe a bacterium dividing to become 2 bacteria. The diagrams depict chromosome duplication and distribution. First DNA replicates, and as the cell membrane lengthens, the 2 chromosomes separate. Upon fission, each bacterium has its own chromosome.

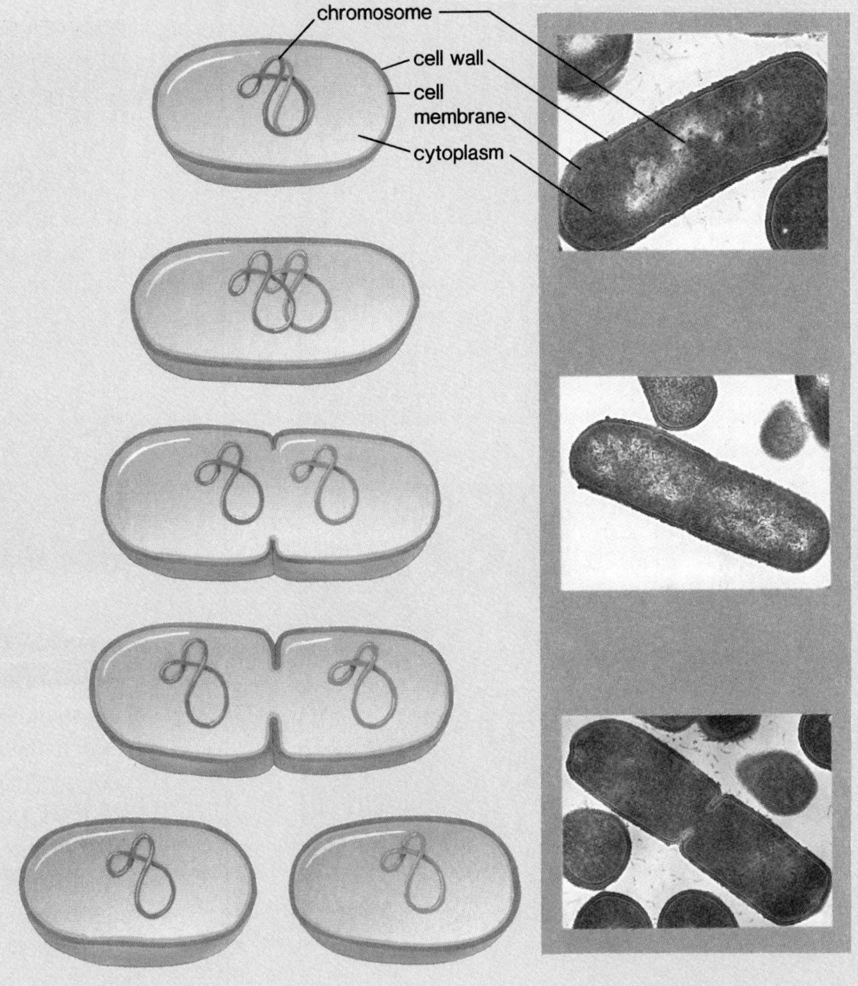

chromosome
cell wall
cell membrane
cytoplasm

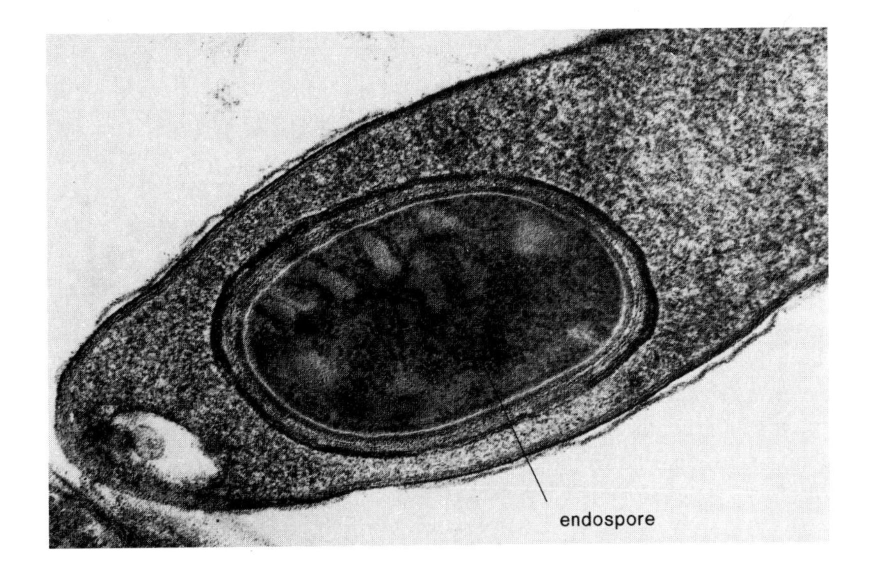

Figure 28.7
Endospore formation. This bacterium (*Clostridium botulinum*) contains an endospore, the dark oval at the lower end of the cell. An endospore normally protects the organism's DNA from exposure to environmental conditions that could destroy it. However, sterilization in an autoclave, a container that maintains steam under pressure, can kill endospores.

Figure 28.8
Diversity among the cyanobacteria. **a.** In *Gleocapsa,* single cells are grouped together in a common gelatin-like sheath. **b.** Filaments of cells occur in *Oscillatoria.*

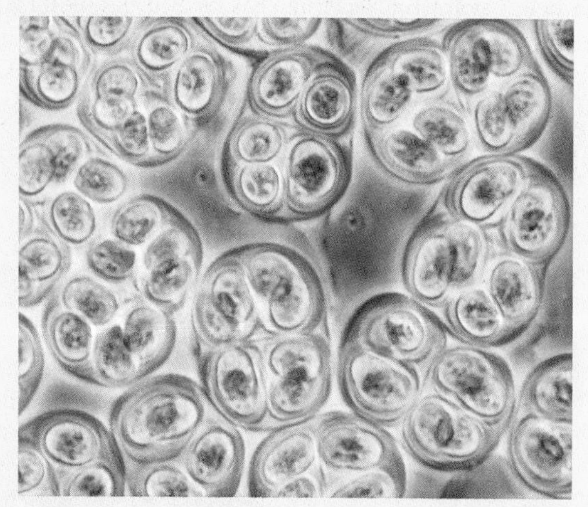

a.

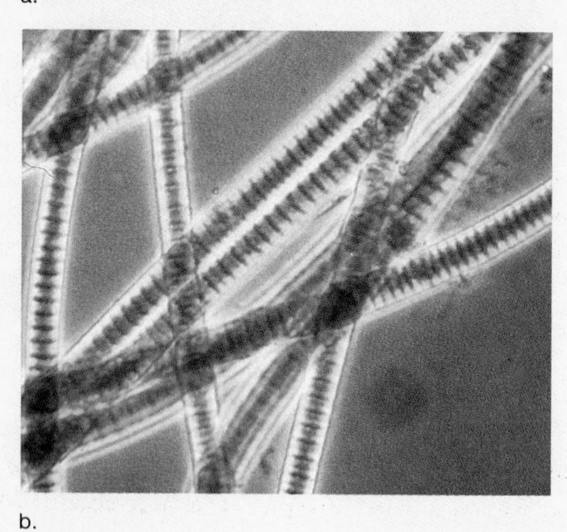

b.

Figure 28.9
Deep-sea ecosystem. At midoceanic ridges, there are dense clusters of tubeworms and clams plus other types of animals. Both tubeworms and clams are home to chemosynthetic bacteria that derive energy by oxidizing hydrogen sulfide (H_2S). The bacteria then produce organic molecules that are utilized by their hosts.

Metabolism in Bacteria: Almost Every Type of Nutrition

Some bacteria are *obligate anaerobes,* unable to grow in the presence of oxygen. A few serious illnesses, such as botulism, gas gangrene, and tetanus, are caused by anaerobic bacteria. Some other bacteria, called *facultative anaerobes,* are able to grow in either the presence or the absence of oxygen. Most bacteria, however, are *aerobic* and, like animals, require a constant supply of oxygen to carry out cellular respiration.

Every type of nutrition is found among bacteria except heterotropism by ingestion. Some bacteria are autotrophic by photosynthesis; they use light as a source of energy to produce their own food. Photosynthetic bacteria do not always give off oxygen. The cyanobacteria (fig. 28.8) do, however; they carry on photosynthesis in the same manner as plants. Some bacteria are autotrophic by **chemosynthesis.** They oxidize inorganic compounds to obtain the necessary energy to produce their own food. In the nitrogen cycle, which we will discuss further in the ecology chapter (p. 660), the nitrifying bacteria oxidize ammonia (NH_3) to nitrites (NO_2^-) and nitrites to nitrates (NO_3^-). Also of interest, chemosynthetic bacteria supply nutrients to organisms that live 2.5 km below sea level, where light never penetrates. Here, hot minerals, including hydrogen sulfide (H_2S), spew up out of the midoceanic ridge. The bacteria reside within the bodies of giant tubeworms and clams, where they oxidize hydrogen sulfide and supply organic molecules to their host (fig. 28.9).

Most types of bacteria are heterotrophic by absorption. They are **saprophytes,** organisms that carry on external digestion of organic matter and absorb the resulting nutrients across the cell membrane. Bacteria are called saprophytic decomposers when they are involved in recycling matter in ecosystems, thereby making inorganic molecules available to photosynthesizers. This "recycling" is necessary for the continued functioning of all ecosystems.

The metabolic capabilities of various heterotrophic bacteria are exploited by human beings who use them to perform services ranging from the digestion of sewage and oil to the production of such products as alcohol, vitamins, and even antibiotics. By means of gene splicing (see fig. 25.3), bacteria are now used to produce useful substances, such as human insulin and growth hormone.

Table 28.2
Some Significant Bacterial Diseases in Humans

Category	Disease (Type of Bacteria)
Sexually transmitted diseases	Syphilis (*Treponema pallidum*) Gonorrhea (*Neisseria gonorrhoeae*) Chlamydia (*Chlamydia trachomatis*)
Respiratory diseases	Strep throat (*Streptococcus pyogenes*) Scarlet fever (*Streptococcus pyogenes*) Tuberculosis (*Mycobacterium tuberculosis*) Pneumonia (*Streptococcus pneumoniae*) Legionnaire's disease (*Legionella pneumonophila*) Whooping cough (*Bordetella pertussis*)
Skin diseases	Erysipelas (*Streptococcus pyogenes*) Boils, carbuncles, impetigo, infections of surgical or accidental wounds and burns (*Staphylococcus aureus*) Acne (*Propionibacterium acnes*)
Digestive tract diseases	Gastroenteritis (*Salmonella*) Food poisoning (*Staphylococcus aureus*) Dysentery (*Shigella*) Cholera (*Vibrio cholerae*)
Nervous system diseases	Botulism (*Clostridium botulinum*) Tetanus (*Clostridium tetani*) Spinal meningitis (*Neisseria meningitidis*) Leprosy (*Myobacterium leprae*)
Systemic diseases	Plague (*Yersinia pestis*) Typhoid fever (*Salmonella typhi*) Diphtheria (*Corynebacterium diphtheriae*)
Other diseases	Gas gangrene (*Clostridium perfringens*) Puerperal fever (*Streptococcus pyogenes*) Toxic shock syndrome (*Staphylococcus aureus*) Lyme disease (*Borrelia burgdorferi*)

Bacteria are often symbiotic; they live in association with other organisms. The nitrogen-fixing bacteria in the nodules of legumes are mutualistic, as are the bacteria that live within our own intestinal tract. We provide a home for the bacteria, and they provide us with certain vitamins. Commensalistic bacteria reside on our skin, where they usually cause no problems. Parasitic bacteria are responsible for a wide variety of plant and animal diseases. Common human infections caused by parasitic bacteria include strep throat, diphtheria, typhoid fever, and gonorrhea (table 28.2).

The majority of bacteria are heterotrophic by absorption (saprophytic decomposers) and contribute significantly to recycling matter through ecosystems. Many are also symbiotic heterotrophs, including those that cause disease.

How to Classify Bacteria

There are many different types of bacteria. For example, *Bergey's Manual*, a standard for classification of bacteria since the 1920s, has 4 separate volumes that list the different types of bacteria. We will divide the 33 major groups of bacteria into the archaebacteria and the **eubacteria** (true bacteria).

Archaebacteria

Most likely, archaebacteria were the earliest prokaryotes. Their cell wall, cell membrane, and ribosomes do not have the same composition as those of the eubacteria.

The archaebacteria are able to live in the most extreme environments, perhaps representing the kinds of habitats that were available when the earth first formed. The methanogens are anaerobic and live in swamps and marshes, producing methane, also known as marsh gas. They also live in the guts of organisms, including humans. The halophiles live where it is salty, such as the Great Salt Lake in Utah. Curiously, a type of rhodopsin pigment (related to the one found in our own eyes) allows them to carry on a primitive form of photophosphorylation for ATP production. The thermoacidophiles live where it is both hot and acidic. Those that live in the hot sulfur springs of Yellowstone National Park obtain energy by oxidizing sulfur.

Eubacteria

Most bacteria are eubacteria, which can be divided in many different ways. Particularly, they can be differentiated on the basis of their cell wall construction.

Most eubacteria are heterotrophic, but some are autotrophic, as discussed on page 544. Among the photosynthetic bacteria, the cyanobacteria are of special importance; therefore, these are discussed further here.

Cyanobacteria. **Cyanobacteria** (fig. 28.8), formerly called blue-green algae, are the most prevalent of the photosynthetic bacteria. The cyanobacteria carry on photosynthesis in a manner similar to that of plants. They possess chlorophyll *a* and evolve oxygen. They also have other pigments that can mask the color of chlorophyll, giving them, for example, not only blue-green but also red, yellow, brown, or black colors. We mentioned in chapter 26 that the cyanobacteria are believed to be responsible for first introducing oxygen into the primitive atmosphere.

Cyanobacteria can be unicellular, filamentous, or colonial. The filaments and colonies are not considered multicellular because each cell is independent of the others. Cyanobacteria lack any visible means of locomotion, although some glide when in contact with a solid surface and others oscillate (sway back and forth) (fig. 28.8*b*). Some cyanobacteria have a special advantage because they possess heterocysts, thick-walled cells without nuclei where nitrogen fixation occurs. The ability to photosynthesize and also make use of atmospheric nitrogen (N_2) means that their nutritional require ments are minimal.

Cyanobacteria are common in fresh water, in soil, and on moist surfaces but are also found in inhospitable habitats, such as hot springs. In fresh water, cyanobacteria sometimes are responsible for the algal bloom associated with cultural eutrophication (overenrichment due to wastes such as treated sewage). They also form symbiotic relationships with a number of organisms, such as protozoa and even at times invertebrates, like sponges and corals. In association with fungi, they form lichens (p. 556), which can grow on rocks. Therefore, cyanobacteria may have been among the first organisms to colonize land.

Cyanobacteria are photosynthesizers that sometimes also fix atmospheric nitrogen. They first introduced oxygen into the atmosphere and probably were among the first organisms to colonize land.

Classification

Kingdom Protista
Eukaryotic, unicellular organisms (and the most closely related multicellular forms). Nutrition: heterotrophic by ingestion (protozoa), heterotrophic by absorption (fungi), or photosynthetic (algae).

Phylum Sarcodina: amoeboid protozoa
Phylum Ciliophora: ciliated protozoa
Phylum Zoomastigina: flagellated protozoa
Phylum Sporozoa: parasitic protozoa
Phylum Chlorophyta: green algae
Phylum Dinoflagellata: dinoflagellates
Phylum Euglenophyta: *Euglena* and relatives
Phylum Chrysophyta: diatoms
Phylum Rhodophyta: red algae
Phylum Phaeophyta: brown algae
Phylum Myxomycota: slime molds
Phylum Oomycota: water molds

Kingdom Protista

The protists are eukaryotes; their cells have all the organelles we studied in chapter 3. Unicellular organisms are predominant in kingdom Protista, and even the multicellular forms lack the tissue differentiation that is seen in more complex organisms. The protists are grouped according to their mode of nutrition and other characteristics into the following categories: (1) *protozoa,* which are animal-like because they are heterotrophic by ingestion and are motile; (2) *algae,* which are plant-like because they are photosynthetic; and (3) *slime molds* and *water molds,* which are fungus-like. Slime molds produce windblown spores like fungi, but unlike fungi, they are heterotrophic by ingestion. Water molds are heterotrophic by absorption like fungi. Unlike fungi, water molds produce swimming zoospores.

Table 28.3
Types of Protozoa

Common Name	Locomotion	Example
Amoebas	Pseudopods	*Amoeba proteus*
Ciliates	Cilia	*Paramecium caudatum*
Zooflagellates	Flagella	*Trichonympha collaris*
Sporozoa	Nonmotile	*Plasmodium vivax*

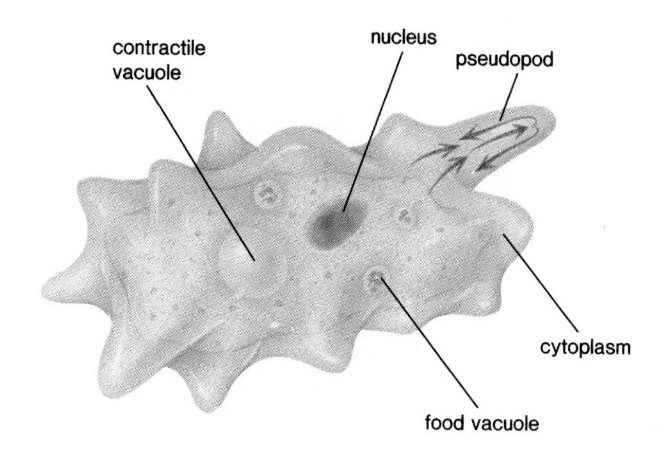

Figure 28.10
Amoeba proteus, a protozoan that moves by formation of pseudopods. Note also the unique organelles, including the food vacuoles and the contractile vacuole. Arrows indicate the flow of cytoplasm as a newly formed pseudopod takes shape.

Protozoa: Ingest Food and Move

Protozoa are small (2–1,000 μm), usually colorless unicellular organisms that lack a cell wall. Like animals, they tend to have special structures for food gathering and locomotion. Excretion and respiration are carried out by diffusion across the cell membrane. Sexual exchange does occur, but reproduction is by simple cell division. Protozoa are classified according to mode of locomotion (table 28.3).

Amoebas: Pseudopod Formers

An **amoeba,** such as *Amoeba proteus* (fig. 28.10), is a small mass of cytoplasm without any definite shape. It moves about and feeds by means of cytoplasmic extensions called pseudopods, or false feet. A pseudopod forms when the cytoplasm streams forward in a particular direction.

Figure 28.11
Paramecium caudatum. Despite its complexity, a paramecium is a unicellular organism. Trichocysts are poisonous, threadlike darts used for defense and for capturing prey.

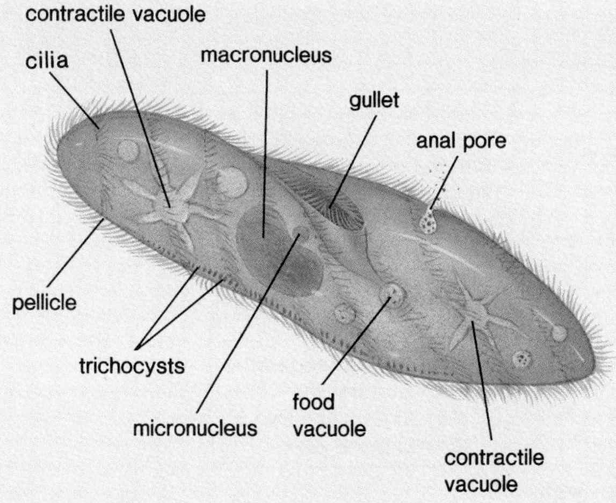

Figure 28.12
Trypanosome infection. A stained blood smear from a patient suffering from African sleeping sickness shows trypanosomes among the blood cells.

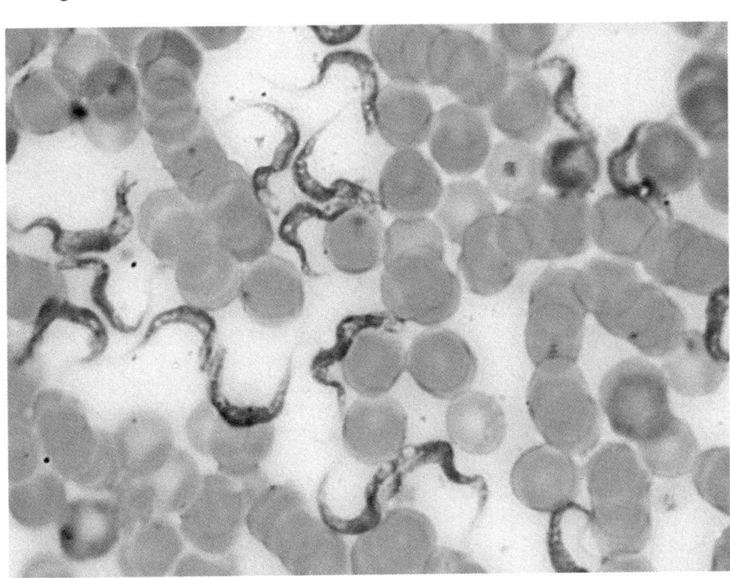

The organelles within an amoeba include food, or digestive, vacuoles and contractile vacuoles. *Food vacuoles* are formed by phagocytosis. A pseudopod surrounds a morsel of food and a vacuole results. Some white blood cells in humans are said to be amoeboid because they are phagocytic in the same maner as an amoeba. *Contractile vacuoles* first collect excess water from the cytoplasm and then appear to "contract," releasing the water through a temporary opening in the cell membrane. Contractile vacuoles are most often seen in freshwater protozoa.

Ciliates: Covered with Cilia

The **ciliates,** such as *Paramecium caudatum* (fig. 28.11), are the most complex protozoa. They move by cilia—hundreds of cilia project through tiny holes in a semirigid outer covering, or *pellicle*. Numerous oval capsules lying in the cytoplasm just beneath the pellicle contain **trichocysts.** When discharged, trichocysts are long, barbed threads, useful for capturing prey and for defense. When a paramecium feeds, food moves down a gullet, at the end of which food vacuoles form. Digestion occurs, and soluble nutrients enter the cytoplasm; nondigestible residue is eliminated at an anal pore.

Ciliates have 2 types of nuclei: a large macronucleus and one or more small micronuclei. The macronucleus controls the normal metabolism of the cell, while the micronuclei are concerned with reproduction. The micronuclei undergo meiosis, and then 2 ciliates exchange a haploid micronucleus during conjugation. **Conjugation** is the temporary union of 2 individuals during which there is an exchange of genetic material.

Zooflagellates: Move by Flagella

Protozoa that move by means of flagella are called **zooflagellates** to distinguish them from unicellular algae that have flagella.

Many zooflagellates are symbiotic. *Trichonympha collaris,* which lives in the gut of termites, enzymatically converts wood to soluble carbohydrates that can be used by its host. The trypanosomes (fig. 28.12) are the cause of African sleeping sickness in humans. Humans are infected by tsetse flies, which acquire the parasite when taking a blood meal from a diseased animal. In infected persons, white blood cells congregate around blood vessels leading to the brain, eventually cutting off circulation. The lethargy characteristic of the disease is caused by an inadequate supply of oxygen to the brain.

Figure 28.13
Life cycle of *Plasmodium vivax*. Asexual reproduction occurs in humans, while sexual reproduction takes place within the *Anopheles* mosquito.

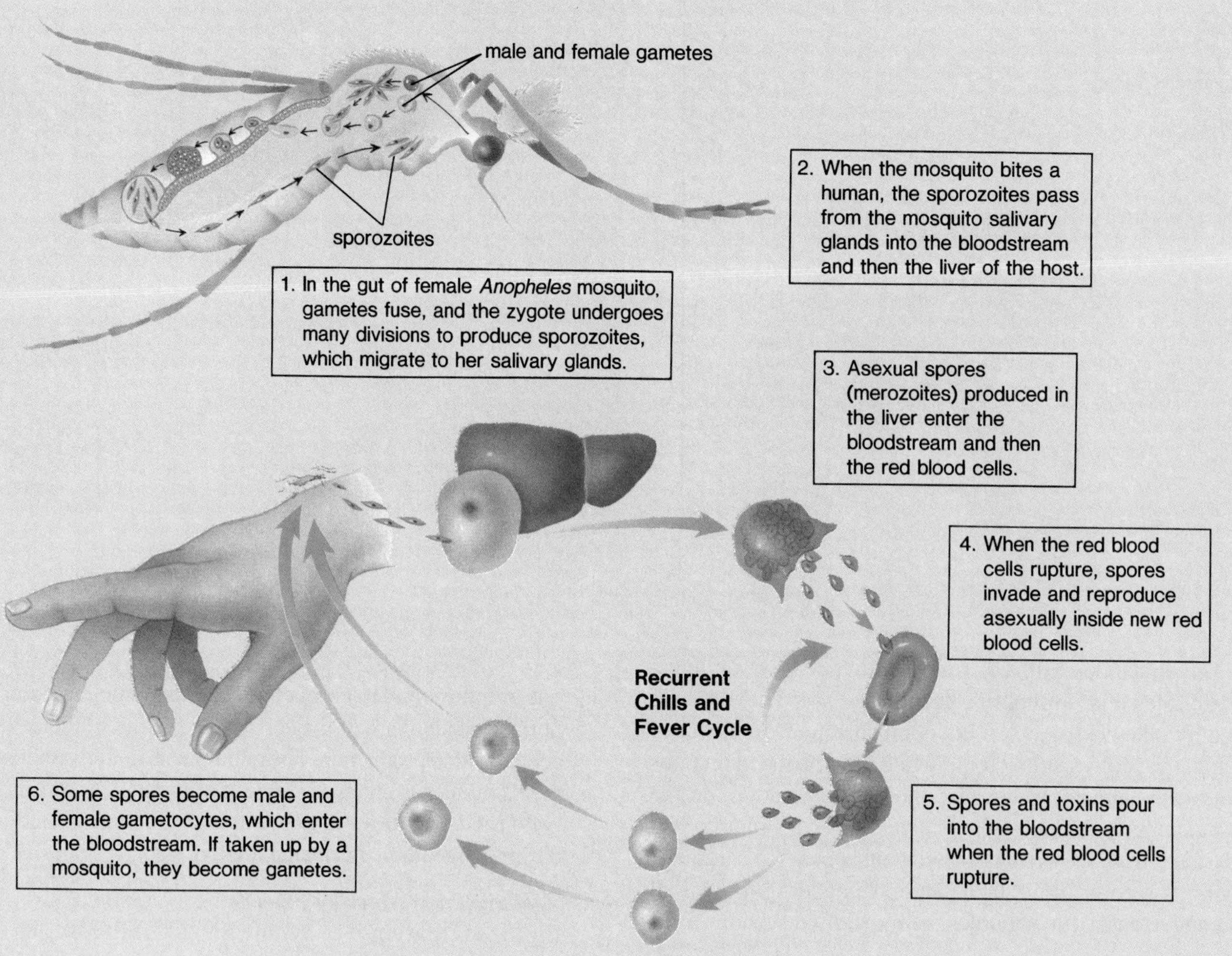

Sporozoa: Form Infecting Spores

The **sporozoa** are nonmotile parasites with a complicated life cycle that always involves the formation of asexual spores. The most important human parasite among the sporozoa is *Plasmodium vivax* (fig. 28.13), a causative agent of malaria. When humans are bitten by an infected female *Anopheles* mosquito, spores invade red blood cells, where they reproduce. The chills and fever of malaria are caused by the release of new spores and toxins into the bloodstream when infected cells burst.

The eradication of malaria has centered on the destruction of the mosquito, since without this host the disease cannot be transmitted from one human being to another. However, the use of pesticides has resulted in the development of resistant strains of mosquitoes. It is hoped that genetic engineering techniques will soon result in the production of a vaccine.

The protozoa are heterotrophic by ingestion and motile-like animals. They are classified according to the type of locomotor organelle employed (table 28.3).

Figure 28.14
Three types of life cycles in organisms. **a.** *Haplontic* life cycle is typical of algae and fungi. Notice that the adult is haploid, the gametes are sometimes isogametes (look alike), and the only diploid part of the cycle is the zygote, which undergoes meiosis to produce spores. **b.** *Alternation of generations* life cycle is typical of plants. Notice that there are 2 generations. The sporophyte (2N) produces the spores by meiosis, and the gametophyte (N) produces gametes. **c.** *Diplontic* life cycle is typical of animals. Notice that the adult is always diploid and meiosis produces heterogametes (egg and sperm). The earliest cycle may have been the haplontic, which could have led to both the alternation of generations and diplontic cycles.

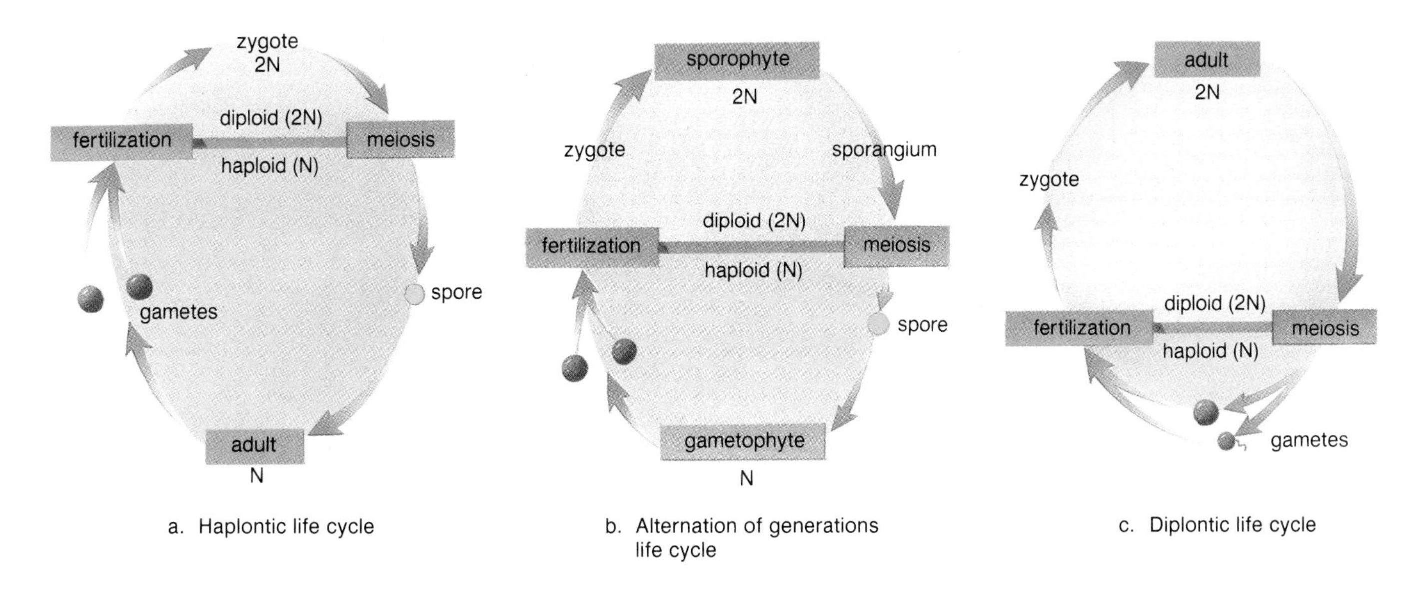

Algae: Live in Water and Photosynthesize

Algae are autotrophic by photosynthesis like plants. However, algae are aquatic and so they do not need to protect the zygote and embryo during sexual reproduction. Algae produce the food that maintains communities of organisms in both the oceans and bodies of fresh water. They are commonly named for the type of pigment they contain; therefore, there are green, golden brown, brown, and red algae. All algae contain chlorophyll, but they may also contain other pigments that mask the color of the chlorophyll. Algae are grouped according to their pigmentation and biochemical differences, such as the chemistry of the cell wall and the chemical compound used to store excess food.

Three Life Cycles

Three types of life cycles typical of very different organisms are found among the algae. In the **haplontic cycle,** the adult is haploid; in the **alternation of generations cycle,** a haploid form alternates with a diploid form; and in the **diplontic cycle,** the adult is always diploid. These cycles are diagrammed in figure 28.14 and are further contrasted in table 28.4.

Table 28.4
Life Cycles

Name	Chromosome Number in Adult(s)	Spores
Haplontic	Haploid only	Usually
Alternation of generations	Haploid ⇌ Diploid	Usually
Diplontic	Diploid only	No

In the haplontic and alternation of generations life cycles, meiosis usually produces **spores,** structures that develop into a haploid generation. In the diplontic life cycle, meiosis produces gametes, the only haploid structures found within the cycle.

Basically, there are 3 life cycles (haplontic, alternation of generations, and diplontic). All 3 of these are seen among the algae. Do 28.1 Critical Thinking, found at the end of the chapter.

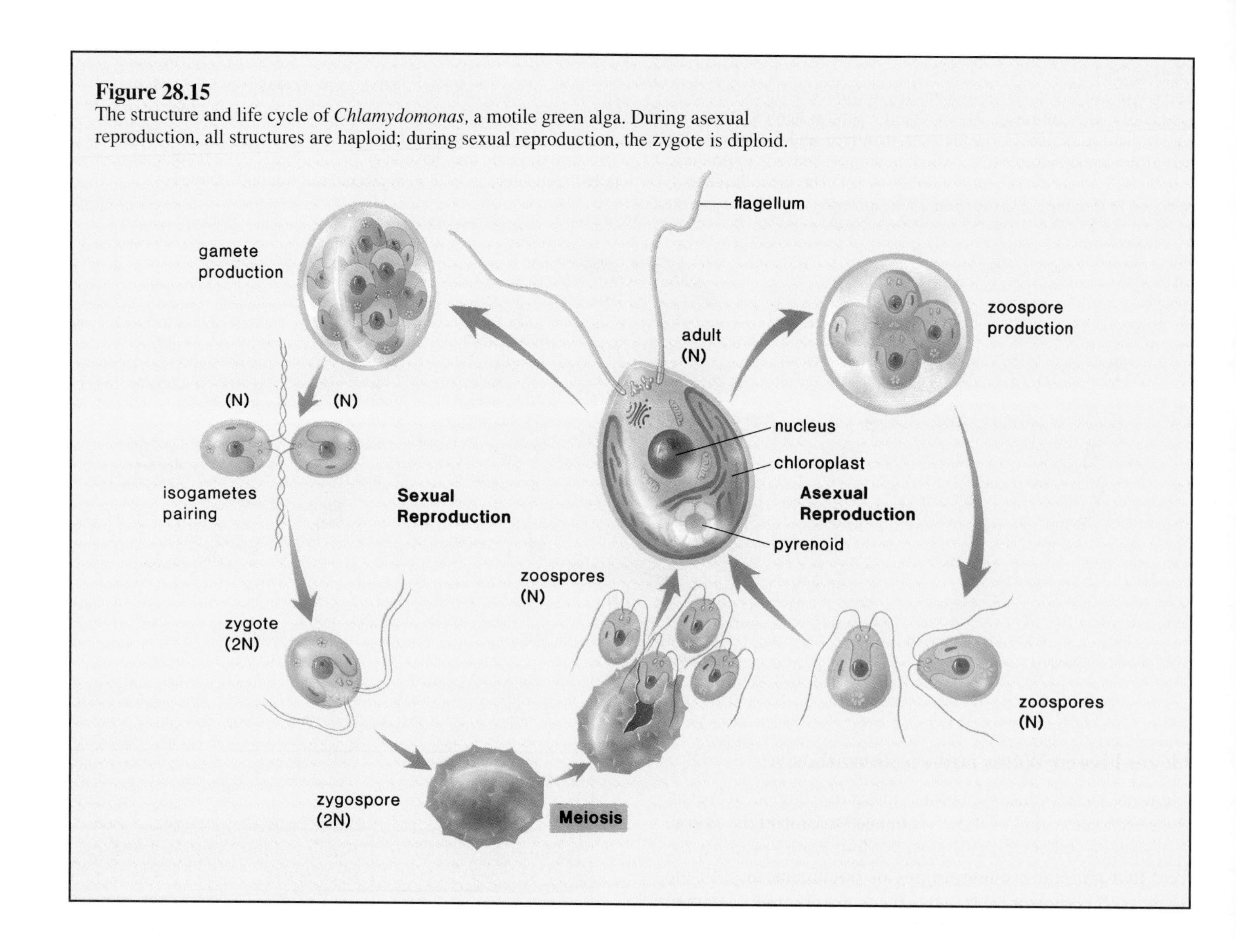

Figure 28.15
The structure and life cycle of *Chlamydomonas,* a motile green alga. During asexual reproduction, all structures are haploid; during sexual reproduction, the zygote is diploid.

Green Algae: Somewhat Like Plants

There are unicellular, colonial, filamentous, and multicellular green algae. A **filament** is a string of cells, and a **colony** is a cluster of cells that cooperate to a degree. It is sometimes suggested that the green algae are ancestral to the first plants because both of these groups possess chlorophylls *a* and *b,* both store reserve food as starch, and both have cell walls that contain cellulose.

Flagellated Green Algae. *Chlamydomonas* (fig. 28.15) is a unicellular green alga that has been studied in detail using the electron microscope. It has a definite cell wall and a single, large, cup-shaped chloroplast, which contains a pyrenoid, a dense body where starch is stored. A red-pigmented eyespot is sensitive to light, and 2 flagella, which project from the anterior end, move the cell freely toward the light, where photosynthesis can occur. *Chlamydomonas* has both animal-like and plant-like characteristics in that it is motile and yet makes its own food.

Chlamydomonas has a haplontic life cycle. Usually, this protist practices asexual reproduction, and the adult divides to give *zoospores* (flagellated spores) that resemble the parent cell. During sexual reproduction, gametes of 2 different strains join to form a zygote. A heavy wall forms around the zygote, and it becomes a zygospore. The zygospore is able to survive until conditions are favorable for germination and subsequent production of 4 zoospores by meiosis. The gametes shown in figure 28.15 are *isogametes;* that is, they look exactly alike.

The life cycle of *Chlamydomonas* illustrates that sexual reproduction is reproduction that involves gametes. Distinct and separate sexes are not required, nor are heterogametes (dissimilar gametes, such as egg and sperm).

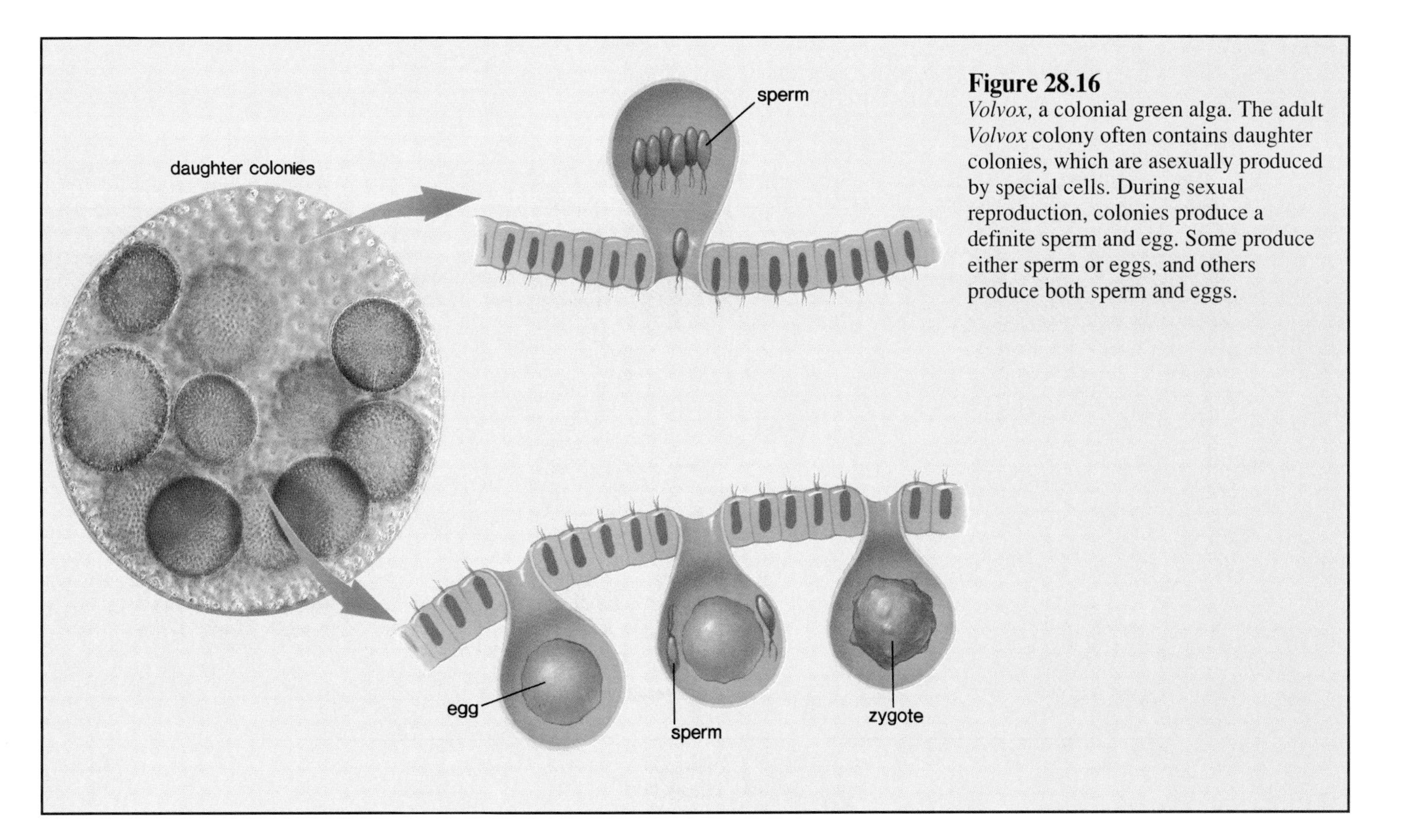

Figure 28.16
Volvox, a colonial green alga. The adult *Volvox* colony often contains daughter colonies, which are asexually produced by special cells. During sexual reproduction, colonies produce a definite sperm and egg. Some produce either sperm or eggs, and others produce both sperm and eggs.

Colonial Green Algae. There are a number of colonial forms among the flagellated green algae. A *Volvox* colony is a hollow sphere with hundreds of cells arranged in a single layer around a watery interior. The cells of a *Volvox* colony, each one of which resembles a *Chlamydomonas* cell, cooperate in that the flagella beat in a coordinated fashion. Some cells are specialized for reproduction, and each of these can divide asexually to form a new daughter colony (fig. 28.16). Daughter colonies reside for a time within the parent colony. A daughter colony leaves the parent colony by releasing an enzyme that dissolves a portion of the matrix of the parent colony, allowing it to escape. During sexual reproduction, there are **heterogametes**—large nonmotile eggs and small flagellated sperm (fig. 28.16).

Filamentous Green Algae. *Spirogyra* (fig. 28.17) is a filament found in green masses on the surface of ponds and streams. A ribbonlike chloroplast is arranged in a spiral within each cell. Sexual exchange occurs during *conjugation,* when the cell contents of one filament move into the cells of the other filament, forming 2N zygotes. These zygotes survive the winter, and in the spring, they undergo meiosis to produce new haploid filaments.

Multicellular Sheets. Multicellular *Ulva* is commonly called sea lettuce because of its leafy appearance (fig. 28.18). *Ulva* has an alternation of generations life cycle in which (1) the spores are flagellated, (2) there are isogametes, and (3) both generations look exactly alike and have equal dominance. In plants, which also have this life cycle, the spores are nonflagellated, there are heterogametes, and one generation is typically dominant over (longer lasting than) the other.

Green algae are a diverse group that have some characteristics in common with plants. *Ulva* has an alternation of generations life cycle.

Figure 28.17
Spirogyra, a filamentous green alga, in which each cell has a ribbonlike chloroplast. During conjugation, the cell contents of one filament enter the cells of another filament.

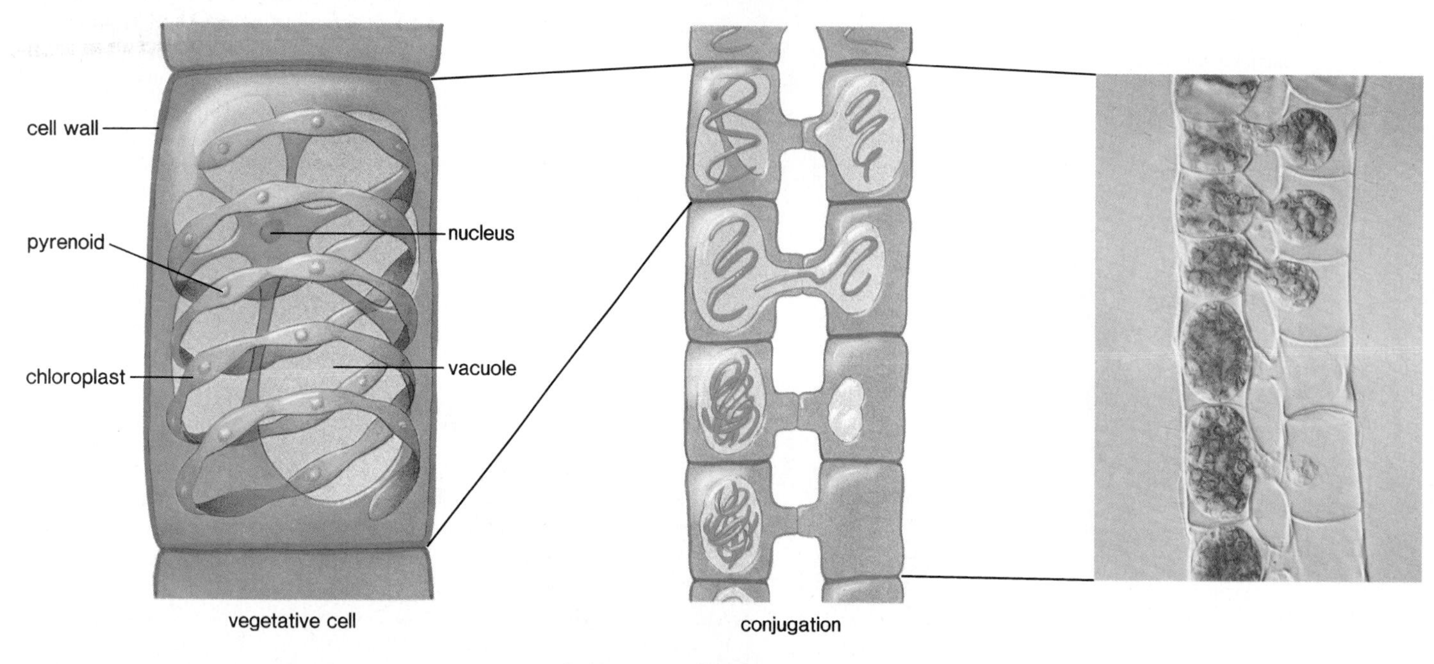

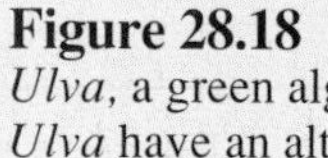

Figure 28.18
Ulva, a green alga. Members of the genus *Ulva* have an alternation of generations life cycle in which the sporophyte and the gametophyte have the same appearance. The sporophyte produces spores by meiosis, and the gametophyte produces isogametes, which form a zygote upon fertilization. The plus (+) and minus (–) signs represent opposite mating strains.

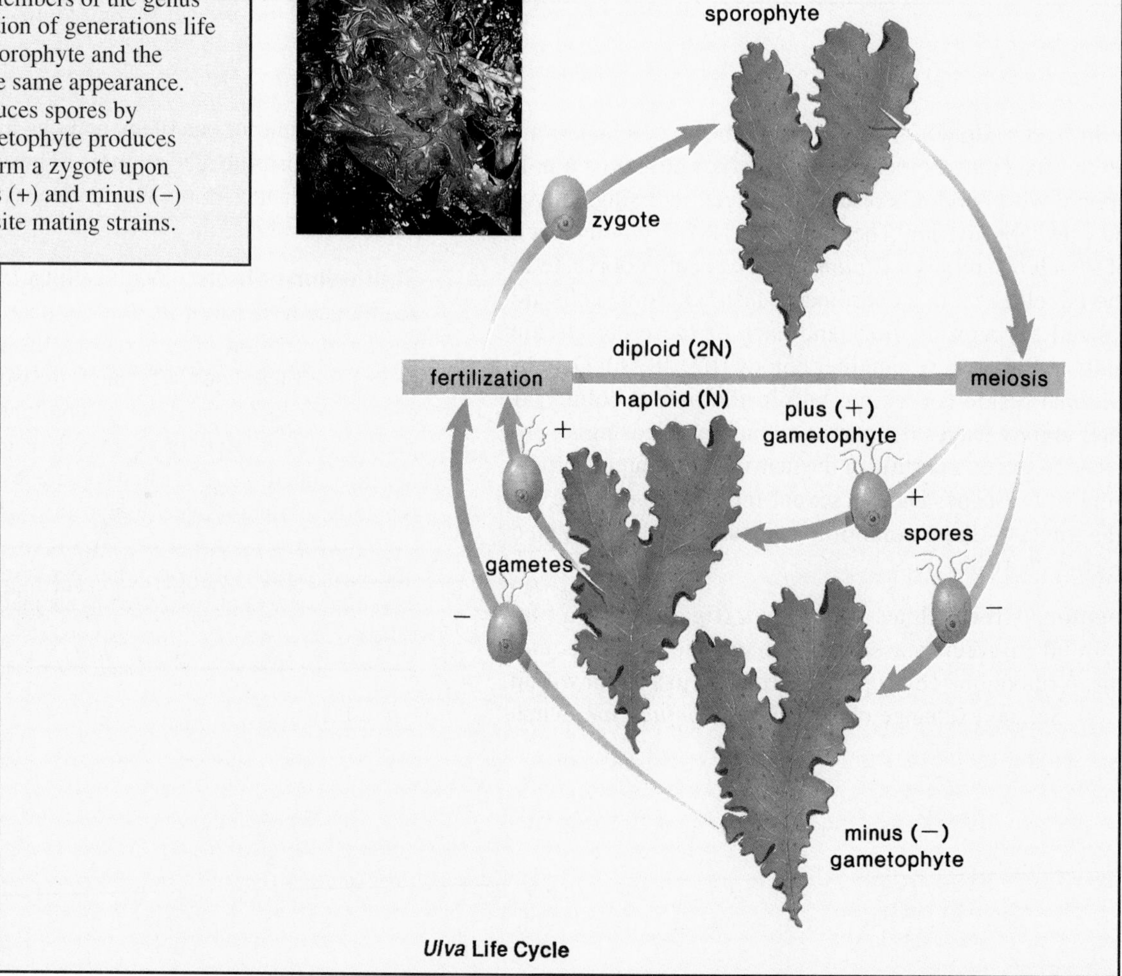

Figure 28.19
Diversification among the brown algae. Both *Laminaria,* a type of kelp, and *Fucus,* known as rockweed, are examples of brown algae that grow along the shoreline. Individuals of *Sargassum* sometimes break off from their holdfasts and form floating masses, where life forms congregate in the ocean. Brown algae provide food and habitat for marine organisms and in several parts of the world have even been harvested for human food and for fertilizer. They are also a source of algin, a pectinlike material that is added to ice cream, sherbet, cream cheese, and other products to give them a stable, smooth consistency.

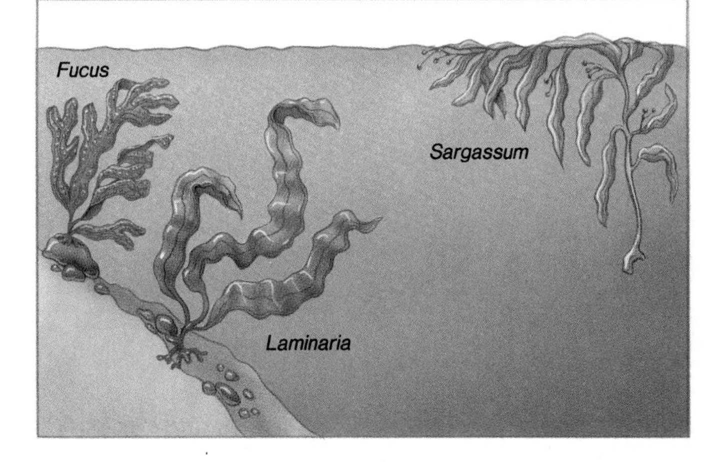

Figure 28.20
Euglena anatomy. *Euglena* is typical of those protists that have both animal-like and plant-like characteristics. A very long flagellum propels the body, which is enveloped by a flexible pellicle. A photoreceptor shaded by an eyespot allows *Euglena* to find light, after which photosynthesis can occur in the numerous chloroplasts. In addition to the pyrenoids, which store starch, there are starch granules in the cytoplasm.

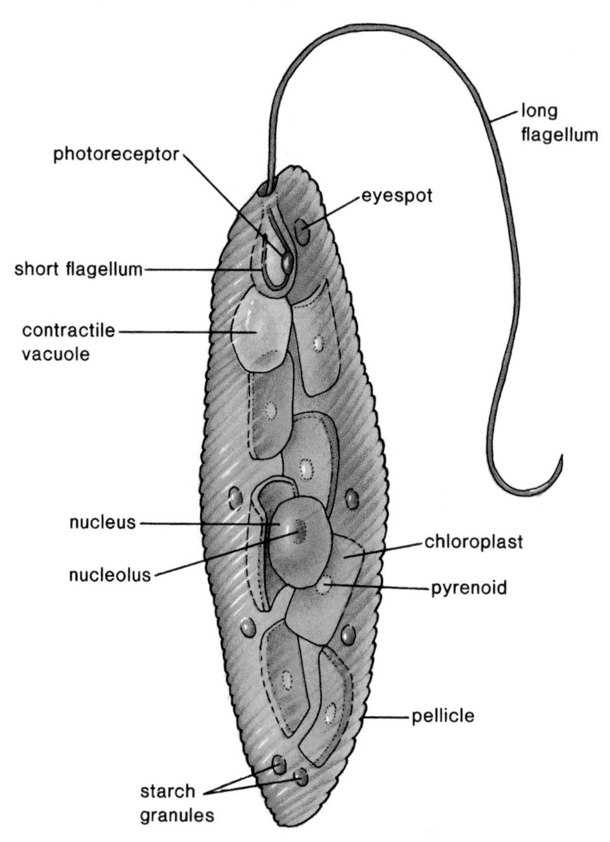

Seaweeds: Multicellular Marine Algae

Multicellular green algae (such as *Ulva*), red algae, and brown algae are all seaweeds. Although all seaweeds have chlorophyll, this green pigment sometimes is masked by red or brown pigments.

Brown algae range from small forms with simple filaments to large forms between 50 m and 100 m in length (fig. 28.19). Large brown algae are often observed along the rocky shoreline in the north temperate zone, where they are pounded by waves as the tide comes in and exposed to dry conditions when the tide goes out. These plants are firmly anchored by holdfasts, and when the tide is in, their broad flattened blades are buoyed by air bladders. When the tide is out, they do not dry up because their cell walls contain a mucilaginous, water-retaining material. Most brown algae have an alternation of generations life cycle, but some species of *Fucus* are unique in that they have the diplontic life cycle (fig. 28.14), in which meiosis produces gametes and the adult is always diploid, as in animals.

Like the brown algae, the red algae are multicellular, but they occur chiefly in warmer seawaters, growing both in shallow waters and as deep as light penetrates. Some forms of red algae are filamentous, but more often they are complexly branched. The branches having a feathery, flat, and expanded or ribbonlike appearance. Coralline algae are red algae that have cell walls impregnated with calcium carbonate ($CaCO_3$). In some instances, they contribute as much to the growth of coral reefs as coral animals do.

The seaweeds are multicellular protists that are grouped according to the type of accessory pigment they contain.

Other Algae: Mix Plant and Animal Features

There are several groups of exclusively unicellular algae in kingdom Protista. These include the euglenoid flagellates, the dinoflagellates, and the diatoms.

The freshwater *euglenoid flagellates* have both animal-like and plant-like characteristics (fig. 28.20). They are motile, moving by means of a long, anteriorly placed flagellum. They can assume different shapes when the underlying cytoplasm undulates or contracts because they have a flexible pellicle instead of a rigid cell wall. However, euglenoid flagellates also have chloroplasts. A photoreceptor shaded by an eyespot at the base of one flagellum enables them to detect light. After they move toward light, photosynthesis takes place, and carbohydrate is stored as starch in pyrenoids and starch granules.

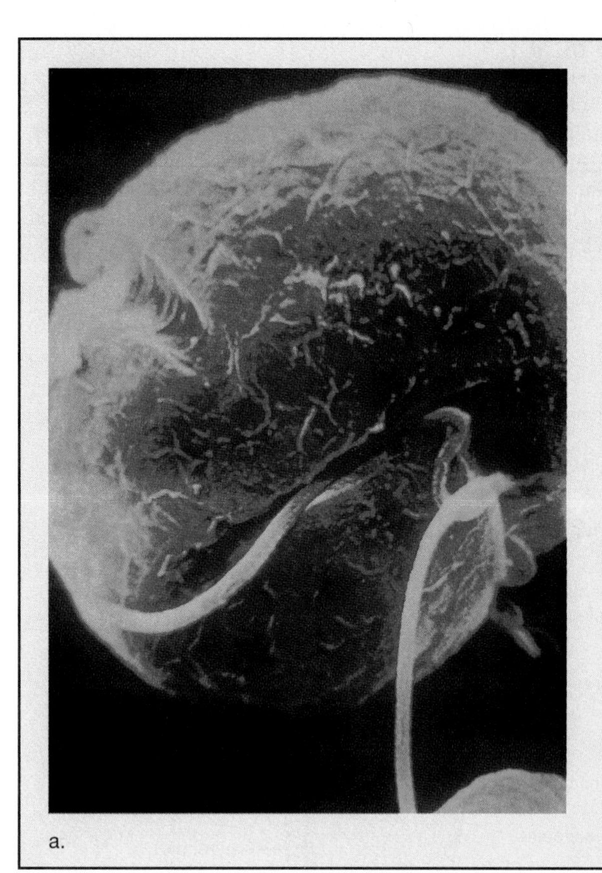

a.

Figure 28.21
Two types of unicellular algae. **a.** Scanning electron micrograph of a dinoflagellate pellicle. The pellicle contains cellulose and when thick is divided into plates like this. There are 2 flagella: one circles the cell within a transverse groove, while the other is free and extends posteriorly.
b. Diatoms may be variously colored, but they have a unique golden brown pigment in addition to chlorophyll within the chloroplasts. The beautiful pattern results from markings on the silica-embedded wall.

b.

Dinoflagellates have 2 flagella; one is free, but the other is located in a transverse groove that encircles the animal. The beating of these flagella causes the organism to spin like a top. The pellicle contains deposits of cellulose and, when thick, is divided into polygonal plates. At times, there are so many of these organisms in the ocean they cause a condition called "red tide." Toxins in red tides cause widespread fish kills and can paralyze humans who eat shellfish that have fed on the dinoflagellates.

Dinoflagellates are an important source of food for small animals in the ocean. They also live as symbiotes within the bodies of some invertebrates. For example, because corals usually contain large numbers of dinoflagellates, they grow much faster than otherwise possible.

Diatoms have a golden brown accessory pigment in their chloroplasts, which masks the color of chlorophyll (fig. 28.21). The structure of a diatom is often compared to a box because the cell wall has 2 halves, or valves, with the larger valve acting as a "lid" for the smaller valve. When diatoms reproduce, each receives only one old valve. The new valve fits inside the old one.

The cell wall of a diatom has an outer layer of silica, a common ingredient of glass. The valves are covered with a great variety of striations and markings, which form beautiful patterns when observed under the microscope. Diatoms are among the most numerous of all unicellular algae in the oceans. As such, they serve as an important source of food for other organisms. In addition, they produce a major portion of earth's oxygen supply. Their remains, called diatomaceous earth, accumulate on the ocean floor and are mined for use as filtering agents, soundproofing materials, and scouring powders.

The unicellular algae have unique characteristics. Some have characteristics of both plants and animals. Do 28.2 Critical Thinking, found at the end of the chapter.

Slime Molds and Water Molds: Like Fungi

Slime molds and water molds have 2 fungus-like characteristics. These molds live on dead organic matter, and they produce spores during one phase of their life cycle.

Slime Molds: Move and Produce Spores

There are 2 types of **slime molds:** cellular and acellular. They both have an amoeboid stage that lives on rotting logs or dead agricultural crops. In cellular slime molds, the total mass is composed of individual amoeboid cells, but in acellular slime molds, the mass is multinucleated and is called a **plasmodium.** In another part of their life cycle, slime molds form fruiting bodies, stalks that produce windblown spores. The spores germinate to give cells that join to form a zygote. This zygote eventually begins the cycle again.

Water Molds: In Both Water and Land

Some **water molds** live in the water, where they parasitize fishes, forming furry growths on the gills. In spite of their common name, others live on land and parasitize insects and plants. A water mold was responsible for the potato famine in the 1840s that caused many Irish to come to the United States. Most water molds are saprophytic and live off dead organic matter, however.

Water molds have a threadlike body like that of a fungus, but the cells are diploid, not haploid, and their life cycle is diplontic (fig. 28.14). Also, the cell walls are largely composed of cellulose, quite unlike fungi.

Classification

Kingdom Fungi
Eukaryotic organisms, usually having haploid or multinucleated hyphal filaments. Spore formation during both asexual and sexual reproduction. Nutrition: heterotrophic principally by absorption.

- Division Zygomycota: black bread molds
- Division Ascomycota: sac fungi
- Division Basidiomycota: club fungi
- Division Deuteromycota: imperfect fungi (i.e., means of sexual reproduction not known)

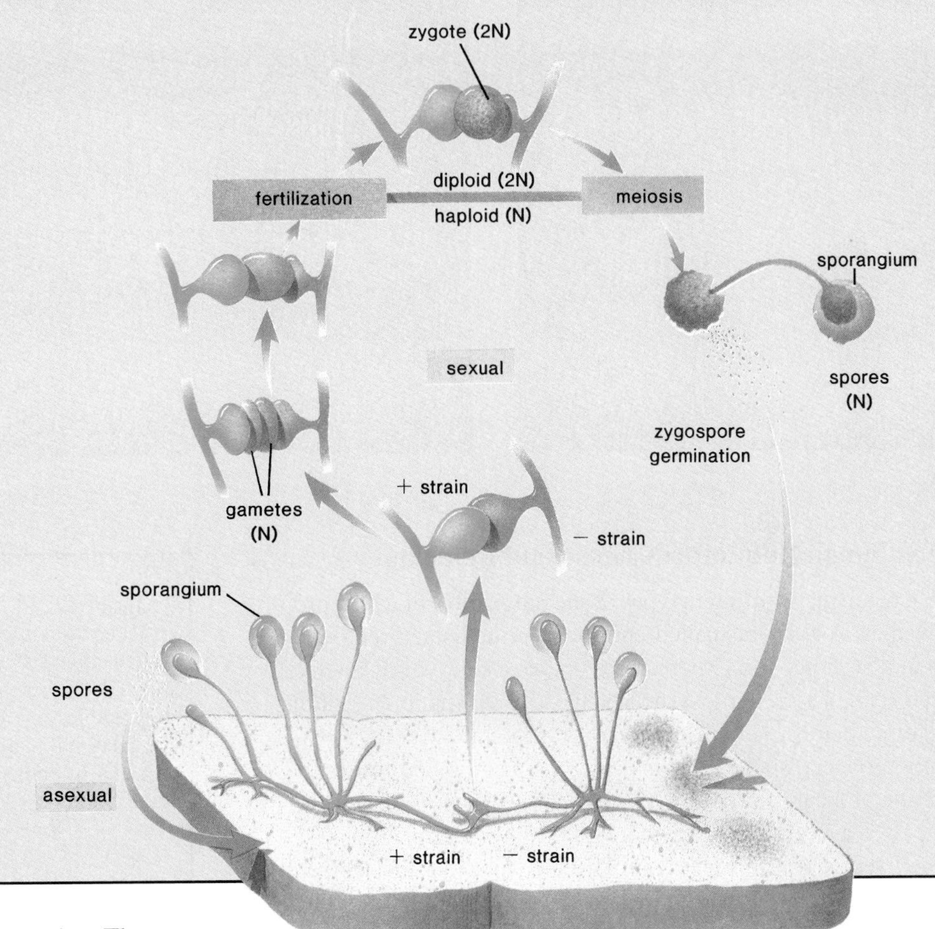

Figure 28.22
Life cycle of the black bread mold *Rhizopus*. (*lower left*) Asexually, a mycelium gives rise to spore-producing sporangia. (*above*) Sexually, the tip ends of hyphae from opposite mating strains fuse, forming a zygote, which thickens and becomes a zygospore. After a period of dormancy, meiosis is followed by zygospore germination and production of a sporangium. Windblown spores, an adaptation to land, produce mycelia.

Kingdom Fungi

Fungi (sing., **fungus**) are heterotrophic by absorption. They are multicellular and eukaryotic. Yet, like bacteria, fungi are usually saprophytic decomposers and assist in recycling nutrients in ecosystems. Some fungi are parasitic, causing serious diseases in plants and animals. The bodies of all fungi, except unicellular yeast, are made up of filaments called hyphae. A **hypha** is an elongated cylinder containing a mass of cytoplasm and many haploid nuclei, which may or may not be separated by cross-walls. A collection of hyphae is called a **mycelium.**

Fungi have a haplontic life cycle (fig. 28.14). Most are adapted to life on land—they produce windblown spores when they reproduce asexually and sexually. Classification is largely based on type of sexual spore and fruiting body.

> Fungi are heterotrophic by absorption. They are composed of hyphae (a collection is a mycelium). Fungi produce spores during both sexual and asexual reproduction, and the major groups of fungi are distinguished by type of sexual spore and fruiting body.

Black Bread Molds: On Bread and Fruit

Black bread molds belonging to the genus *Rhizopus* are often used as an example of fungi. These molds exist as a whitish or grayish haploid mycelium on bread or fruit (fig. 28.22). During asexual reproduction, some hyphae grow upright and bear a spherical **sporangium,** within which thousands of spores are formed. During sexual reproduction, the ends of hyphae from 2 mating strains (usually called plus and minus) act as gametes—they join to form a diploid zygote. The zygote darkens as it enlarges into a *zygospore*. After remaining dormant for several months, meiosis occurs and the zygospore germinates, producing a short haploid hypha and a sporangium. The sporangium releases windblown spores.

Figure 28.23
Sac fungi and club fungi. **a.** Colorful cup fungi (a sac fungus). **b.** Bracket fungi (a club fungus) growing on a dead tree limb. **c.** Mushrooms (a club fungus) of the genus *Mycena* have bell-shaped caps.

a.

b.

c.

Sac Fungi: Roquefort Cheese and Penicillin

There are many different types of sac fungi, most of which produce asexual spores called **conidia.** During sexual reproduction, sac fungi form spores called *ascospores* within saclike cells called asci. In most species, the asci are supported within fruiting bodies. A **fruiting body** is a structure that bears sexually produced spores. In cup fungi, the fruiting body takes the shape of a cup (fig. 28.23*a*).

Yeasts are sac fungi that do not form fruiting bodies. In fact, yeasts are different from all other fungi in that they are unicellular and most often reproduce asexually by budding. Yeasts, as you know, carry out fermentation as follows:

glucose → carbon dioxide + ethyl alcohol

The production of wines and beers is dependent on this reaction.

Blue-green molds, notably *Penicillium,* are also sac fungi. These molds grow on many different organic substances, such as oranges, bread, fabric, leather, and wood. They are used by humans to provide the characteristic flavor of Camembert and Roquefort cheeses; more important, they produce the antibiotic penicillin. Another mold, the red bread mold *Neurospora,* was used in the experiments that helped to decipher the function of genes.

Unfortunately, sac fungi are also the cause of chestnut tree blight and Dutch elm disease, 2 diseases that have killed most of these trees in the United States. Powdery mildew, apple scab, and ergot, a disease of cultivated cereals, are also caused by sac fungi.

A **lichen** is a symbiotic relationship between an alga and a fungus (fig. 28.24). The fungal portion of lichens is usually a

Figure 28.24
Lichen structure. Lichens come in many shapes and sizes, but each is a symbiotic relationship between a fungus and an alga. Whether the fungus contributes to the relationship is debatable, and it may even be parasitic on the alga. **a.** Diagrammatic longitudinal section of a crustose (flat) lichen. **b.** Photo of a crustose lichen. Lichens are important soil formers.

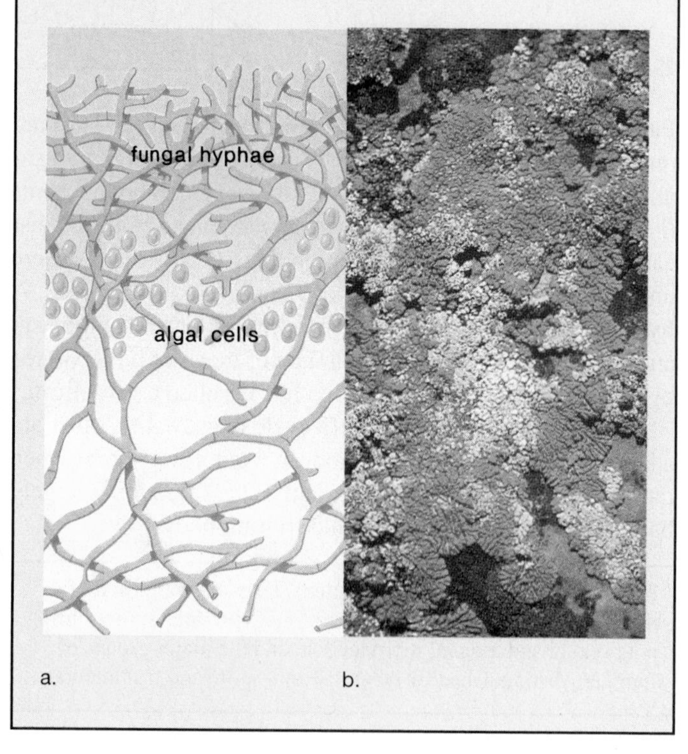

a. b.

Figure 28.25
Life cycle of a mushroom. (*beginning lower right*) Hyphae from opposite mating strains fuse and produce a mycelium, in which there are haploid nuclei from both strains. The mycelium gives rise to a mushroom, consisting of a stalk and a cap. The gills on the underside of the cap are lined with basidia (club-shaped structures). After nuclei fuse to form a diploid nucleus, meiosis in each basidium produces basidiospores, which are windblown. If conditions are favorable, each basidiospore germinates into a hypha. Scanning electron micrograph of a basidium and basidiospores.

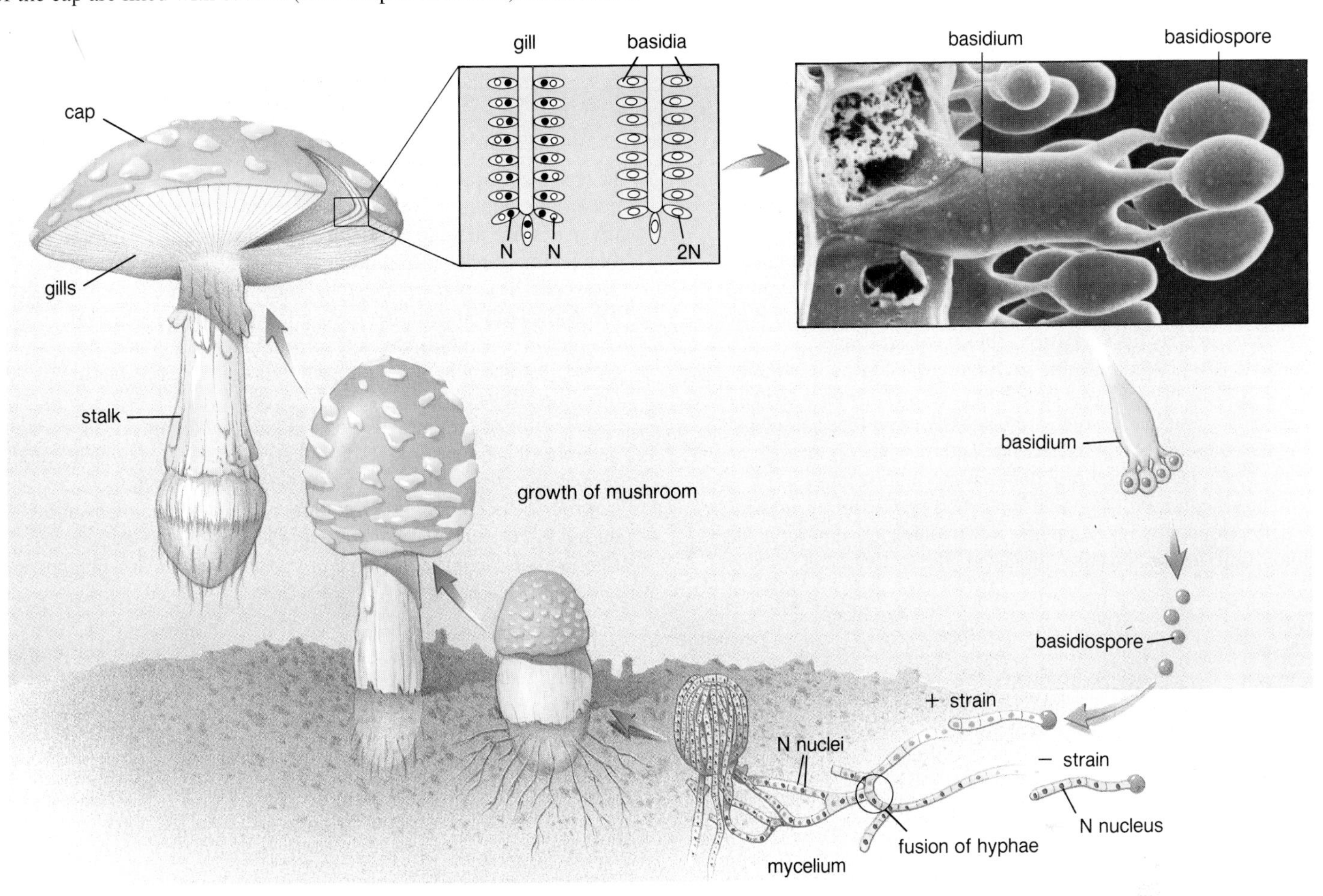

sac fungus, while the algal part of this symbiotic relationship is usually a green alga. Lichens can live on bare rock or in poor soil and are able to survive great extremes in temperature and moisture in all regions of the world. Reindeer moss is a lichen that is an important food source for arctic animals.

Club Fungi: Many Mushrooms

Among the club fungi, asexual reproduction is accomplished by formation of *conidia.* As a result of sexual reproduction, members of this group form club-shaped structures called *basidia,* often within fruiting bodies. The mushroom, the puffball, and the bracket (shelf) fungi are club fungi (fig. 28.23*b* and *c*). The visible portions of these are actually fruiting bodies, and the mycelia lie underground or within a tree or rotting log. On the underside of a mushroom cap, the basidia project from the gills. Within each basidium, a diploid nucleus undergoes meiosis to produce *basidiospores,* which are windblown (fig. 28.25).

Club fungi are economically important. Mushrooms are raised and sold commercially as a delicacy. Rusts and smuts are parasitic club fungi that attack grains, resulting in great economic loss and necessitating expensive control measures. They do not have conspicuous fruiting bodies, and generally occur as vegetative hyphae that produce spores of various kinds. On the other hand, the mycelia of club fungi that lie beneath the soil often form beneficial symbiotic relationships with plants, notably pine trees. These *mycorrhizae* (fungal roots) help the trees to garner nutrients from the soil and to grow at a faster rate. Therefore, foresters make sure the tree roots are exposed to club fungi before they are planted.

Other Fungi: No Observed Sex

Some fungi cannot be assigned to a definite group because the reproductive portion of the life cycle has not been observed. For this reason, these fungi are sometimes called "imperfect fungi." The fungi that cause ringworm and athlete's foot belong to this group, as does the yeast *Candida albicans,* which causes thrush, a mouth infection, and moniliasis, a fairly common vaginal infection in females who take the birth-control pill, diabetics, or those who wear tight clothing, including panty hose.

During sexual reproduction, the black bread molds produce spores in sporangia, the sac fungi produce spores in saclike cells, and the club fungi produce spores on club-shaped structures. The sac fungi and the club fungi typically have fruiting bodies. Do 28.3 Critical Thinking, found at the end of the chapter.

SUMMARY

Viruses are acellular obligate parasites that have a protein coat and a nucleic acid core. Viral DNA must enter a host cell before reproduction is possible. The kingdom Monera includes prokaryotic, usually unicellular, organisms, the archaebacteria and the eubacteria. The archaebacteria are adapted to living in extreme habitats. Most information relates to the eubacteria (true bacteria). Reproduction is by binary fission, but sexual exchange occasionally takes place. Some bacteria form endospores, which can survive the harshest of treatment except sterilization. Usually bacteria are aerobic, but some are facultative anaerobes or even obligate anaerobes. Most bacteria are heterotrophic by absorption, but the cyanobacteria are important photosynthesizers.

The kingdom Protista includes the eukaryotic protozoa, which are heterotrophic by ingestion, the algae, which are autotrophic by photosynthesis, and the slime molds and the water molds, which have some characteristics of fungi. Protozoa are classified according to the type of locomotor organelle. Green algae are diverse: some are unicellular or colonial flagellates, some are filamentous, and some are multicellular sheets. The latter are seaweeds, as are brown algae and red algae. Three types of algae are exclusively unicellular: euglenoids, dinoflagellates, and diatoms. Slime molds have an amoeboid stage and then form fruiting bodies, which produce windblown spores. Water molds have threadlike bodies.

The kingdom Fungi contains eukaryotic organisms that are heterotrophic by absorption. Fungi are composed of hyphae, which form a mycelium. Along with heterotrophic bacteria, they are saprophytic decomposers. The fungi produce windblown spores during both sexual and asexual reproduction. The major groups of fungi are distinguished by type of sexual spore and fruiting body.

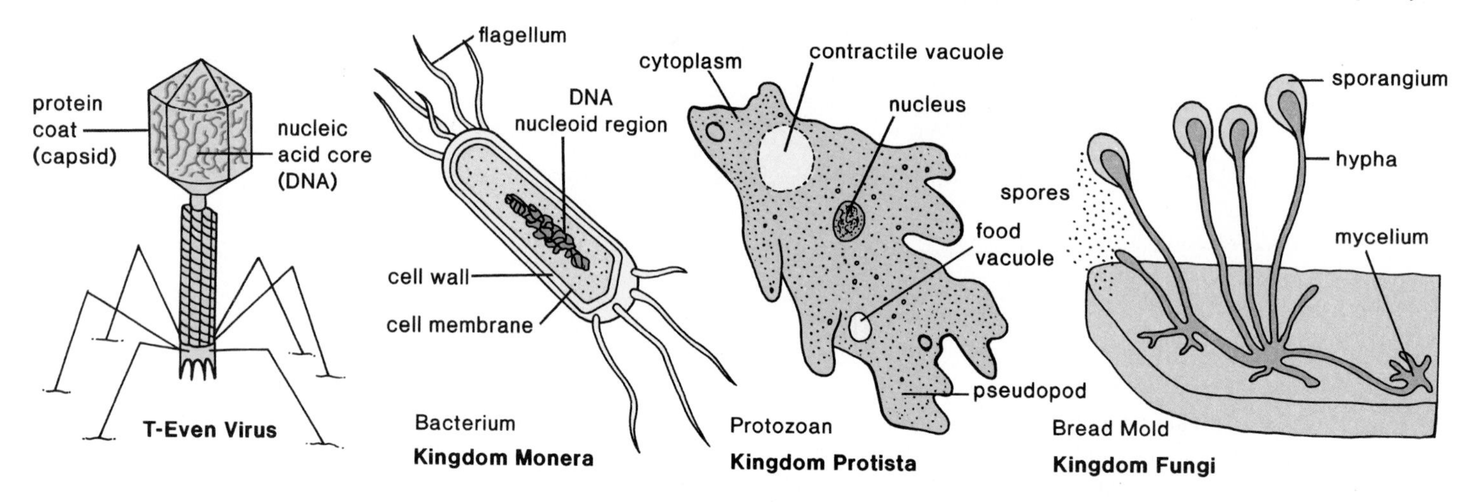

STUDY QUESTIONS

In order to practice **writing across the curriculum,** students should write out the answers to any or all of the study questions. The study questions are sequenced in the same order as the text.

1. Describe the reproductive cycle of a T-even virus. (p. 538)
2. Describe the life cycle of an RNA retrovirus. (p. 540)
3. In general, describe the structure of prokaryotic cells. (p. 542)
4. What are the 3 shapes of bacteria? (p. 542) How do bacteria reproduce? (p. 543) What are endospores? (p. 543)
5. Discuss the importance of bacteria, including cyanobacteria. (pp. 544–545)
6. Give an example of each type of protozoan studied. Describe the anatomy of those that are free living and the life cycle of a sporozoan that is parasitic. (pp. 546–548)
7. What are the 3 types of life cycles found among living things? (p. 549) Show that *Chlamydomonas* and black bread mold have the haplontic life cycle and that *Ulva* has an alternation of generations life cycle. (pp. 550–555) What type of life cycle does *Fucus* have? (p. 553)

8. List the distinguishing features of the euglenoid flagellates, the dinoflagellates, and the diatoms, mentioning any anatomical and ecological aspects that are of importance. (pp. 553–554)
9. Why are the slime molds and the water molds sometimes called the fungus-like protists? (p. 554)
10. Describe the anatomical features of the fungi, and tell how fungi are classified. (p. 555)
11. Describe the structure and the life cycle of black bread mold. (p. 555)
12. Define a fruiting body, and name 2 groups of fungi that typically have fruiting bodies. (p. 556)

OBJECTIVE QUESTIONS

1. Viruses always have a(n) _______ core and a(n) _______ coat called a capsid.
2. The _______ are RNA viruses that carry an enzyme for RNA → DNA transcription.
3. All different types of _______ are classified as monera.
4. Most bacteria are saprophytic decomposers, meaning that they _______.
5. In contrast, cyanobacteria are _______, using the energy of the sun to make their own _______.
6. Amoeba move by means of _______, and ciliates move by means of _______.
7. In the haplontic life cycle, the only diploid stage is the _______.
8. In *Spirogyra*, zygotes form following the process of _______.
9. *Ulva* has the _______ life cycle, as do land plants.
10. The body of a fungus is a(n) _______ that contains filamentous _______.
11. The _______ fungi and the _______ fungi both have fruiting bodies.
12. A lichen is a symbiotic relationship between a(n) _______ and a(n) _______.
13. Label this diagram of the *Chlamydomonas* life cycle. Give 2 reasons why each portion of the cycle is either asexual or sexual.

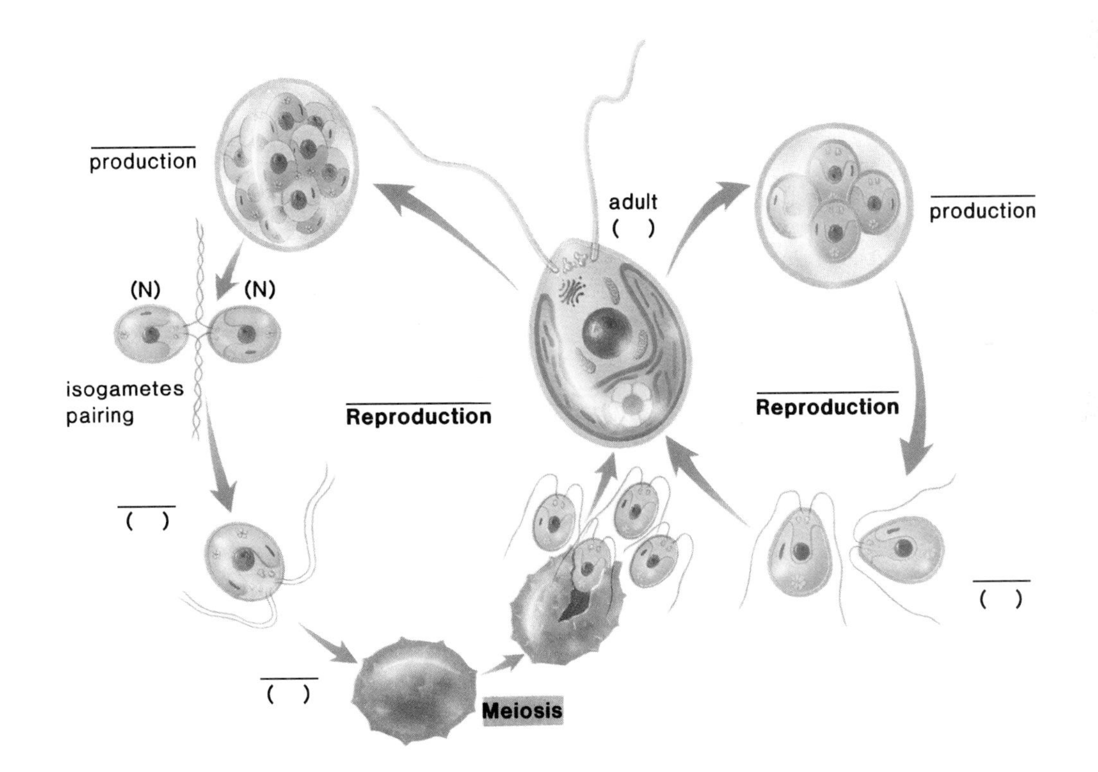

CRITICAL THINKING

In order to practice **writing across the curriculum,** students should write out the answers to any or all of the critical thinking questions. Suggested answers to the critical thinking questions are in appendix E.

28.1

1. In reference to the 3 life cycles (fig. 28.14), the timing of what one element determines whether or not a diploid adult results?
2. Hypothesize how the haplontic life cycle may have given rise to the alternation of generations life cycle.
3. Hypothesize how the alternation of generations life cycle may have given rise to the diplontic life cycle.

28.2

1. For reasons given in the text (p. 550), it is believed that green algae are ancestral to plants. Give 2 reasons why a member of the genus *Ulva* is a good candidate for this ancestor.
2. Some believe that euglenoid flagellates are ancestral to both plants and animals. How would a euglenoid flagellate have to change in order to become more plantlike? How would it have to change in order to become more animal-like?
3. Suppose you used cellular structure to determine which protist is ancestral to animals. Which organelle would be of greatest interest to you? Why?

28.3

1. Members of the kingdom Fungi are adapted to living on land. What characteristic of fungi would make you think that other organisms (such as plants and animals) must have been present on land before fungi?
2. Both *Chlamydomonas* and fungi have a haplontic life cycle. What elements in the cycle show that the former is adapted to living in the water and the latter is adapted to living on land?
3. The mycelia of sac fungi and club fungi are found in the ground, and the fruiting bodies usually appear after a rain. What does this tell you about the adaptation of fungi to living on land?

SELECTED KEY TERMS

bacteriophage (bak-te´re-o-fāj˝) a virus that infects a bacterial cell.

bacterium a unicellular organism that is prokaryotic—its single cell lacks the complexity of a eukaryotic cell; archaebacteria and eubacteria.

binary fission reproduction by division into 2 equal parts by a process that does not involve a mitotic spindle.

colony a cluster of specialized cells that cooperate to a degree.

conjugation sexual union between organisms in which the genetic material of one cell enters another.

cyanobacterium (si˝ah-no-bak-te´re-um) photosynthetic prokaryote that contains chlorophyll and releases O_2; formerly called a blue-green alga.

diatom (di´ah-tom) one of a large group of freshwater and marine unicellular algae having a cell wall consisting of 2 silica-impregnated valves that fit together as in a pillbox.

fungus a eukaryote, usually composed of strands called hyphae, that is usually saprophytic; e.g., mushroom and mold.

hypha (hi´fah) one filament of a mycelium, which constitutes the body of a fungus.

lichen (li´ken) fungi and algae coexisting in a symbiotic relationship.

mycelium (mi-se´le-um) a mass of hyphae that makes up the body of a fungus.

protozoan animal-like protist that is classified according to means of locomotion: amoeba, flagellate, ciliate.

retrovirus virus that contains only RNA and carries out RNA → DNA transcription, called reverse transcription.

saprophyte (sap´ro-fīt) a heterotroph such as a bacterium or a fungus that externally digests dead organic matter before absorbing the products.

sporangium (spo-ran´je-um) a structure within which spores are produced.

29

THE PLANT KINGDOM

Chapter Concepts

1.
Plants, unlike algae, protect the embryo. This is an adaptation that facilitates a land existence. 563, 578

2.
The nonvascular plants, which are low-growing, lack a means of water transport and internal support. 564, 578

3.
Vascular plants have a system that not only transports water but also provides internal support. 566, 578

4.
In nonseed plants, spores disperse the species, and in seed plants, seeds disperse the species. 567, 578

5.
In seed plants, a germinating pollen grain transports sperm to the egg. 570, 578

6.
The presence of vascular tissue and reproductive strategy are used to compare plants. 577, 578

Orchid (*Epidendrum*)

Chapter Outline

Figure 29.1

Representatives of the dominant types of land plants on earth today. **a.** Most plants today are flowering plants. This is a tiger lily. **b.** Mosses are low-growing and thrive in moist locations. This is hairy-cap moss. **c.** Ferns were much more prominent in ancient times. This is a bracken fern. **d.** The gymnosperms are seed plants, as are flowering plants. This is a conifer, in which the seeds are borne on cones.

a.

b.

c.

d.

Classification

Kingdom Plantae

Eukaryotic, terrestrial, multicellular organisms with rigid cellulose cell walls and chlorophyll *a* and chlorophyll *b*. Nutrition principally by photosynthesis. Starch is the reserve food.

- Division Bryophyta: mosses and liverworts
- Division Psilophyta: whisk ferns
- Division Lycophyta: club mosses
- Division Sphenophyta: horsetails
- Division Pterophyta: ferns
- Division Cycadophyta: cycads
- Division Ginkgophyta: ginkgos
- Division Coniferophyta: conifers
- Division Anthophyta: flowering plants
  - Class Dicotyledonae: dicots
  - Class Monocotyledonae: monocots

Table 29.1
Comparison of Water Environment to Land Environment

Water	Land
1. The surrounding water prevents the organism from drying out (desiccation).	1. To prevent drying out, the organism obtains water, transports it to all body parts, and possesses a covering that prevents evaporation.
2. The surrounding water buoys the organism.	2. An internal skeleton helps a large body oppose the pull of gravity.
3. The water facilitates transport of reproductive units.	3. In plants, the reproductive units may be adapted to transport by wind currents or by motile animals.
4. The surrounding water prevents the embryo from drying out.	4. The developing embryo is protected by parent from possible drying out.
5. The water maintains a relatively constant environment in regard to temperature, pressure, and moisture.	5. The organism may be capable of withstanding extreme external fluctuations in temperature, humidity, and wind.

LANTS can be distinguished from algae even by a very cursory examination (fig. 29.1). Some plant characteristics are shared by algae, but some are not.

Characteristics of Plants

Plants are photosynthetic organisms with the following characteristics.

1. Plants are photosynthetic; they contain chlorophyll *a* and chlorophyll *b* and store food as starch.
2. Plants are not motile and do not have nerve cells or muscle cells.
3. Plants are multicellular eukaryotes whose cells have cell walls.
4. Plants have an alternation of generations life cycle in which a diploid adult alternates with a haploid adult.
5. Plants have reproductive structures that protect the gametes and the diploid embryo from drying out.

The last characteristic listed is particularly useful to organisms that are adapted—as plants are—to living on the land. A land existence offers some advantages to plants. One advantage is the greater availability of light for photosynthesis since water, even if clear, filters out light. Another advantage is that carbon dioxide (CO_2) and oxygen (O_2) are present in higher concentrations and diffuse more readily in air than in water. Water, however, provides many services that must be fulfilled in other ways when an organism lives on land (table 29.1).

Figure 29.2
Alternation of generations life cycle. In this life cycle, the zygote develops into the sporophyte. The sporophyte produces haploid spores by meiosis. These develop into the gametophyte, and the gametophyte produces gametes. Fertilization results in a diploid zygote.

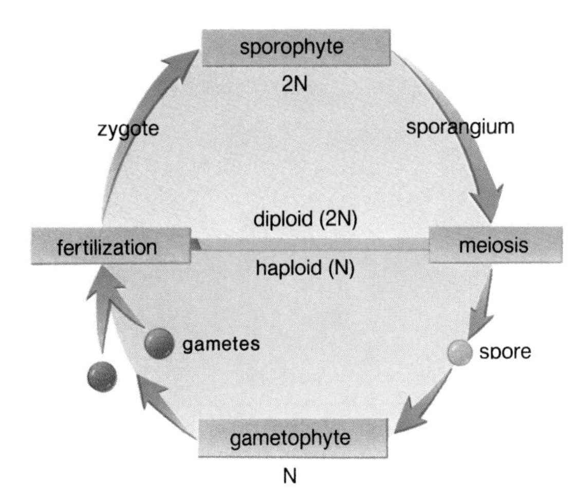

Generations Alternate: 2N, N, 2N, . . .

All plants have a two-generation life cycle known as **alternation of generations.** This means that a plant exists in 2 forms: the haploid generation, the **gametophyte,** which produces gametes, and the diploid generation, the **sporophyte,** which produces spores by meiosis. **Spores** are haploid structures that develop or mature into a gametophyte.

In plants, one generation, either the gametophyte or the sporophyte, is *dominant* over the other generation: the dominant generation lasts longer, is larger and more conspicuous, and is actually the one that laypeople refer to as the plant because they do not realize that the other generation exists. The alternation of generations life cycle is diagrammed in figure 29.2. Notice the following in this life cycle:

Meiosis produces spores, haploid (N) structures that develop into a gametophyte (haploid generation). The *gametophyte produces the gametes* (egg and sperm).
Fertilization results in a zygote (2N), which develops into a sporophyte (diploid generation). The *sporophyte produces haploid spores* by meiosis in structures called sporangia.

> Plants are multicellular photosynthesizers that protect the diploid embryo from drying out. They have an alternation of generations life cycle. Some plants have a dominant sporophyte and some have a dominant gametophyte.

Figure 29.3
Marchantia, a liverwort. *Marchantia* can reproduce asexually by means of gemmae—minute bodies that give rise to new plants. As shown here, gemmae are located in a cuplike structure called a gemma cup.

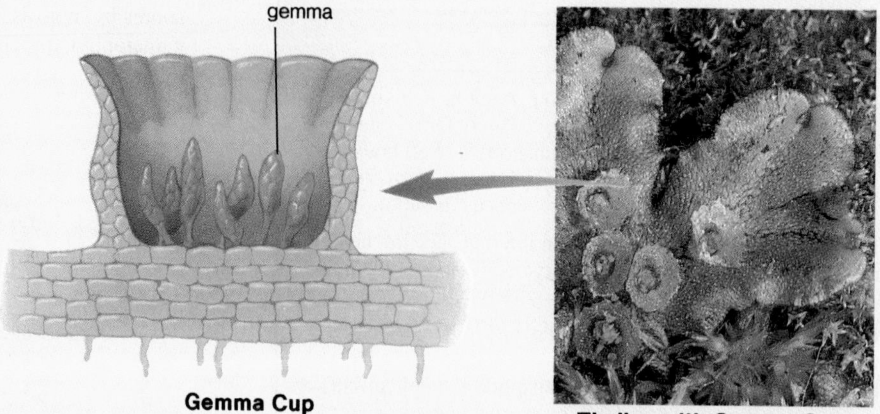

Nonvascular Plants

The only nonvascular plants are the bryophytes. All other plants have vascular tissue at some time during their life cycle.

Bryophytes: Liverworts and Mosses

The **bryophytes** include liverworts and mosses. The well-known liverwort *Marchantia* has a flattened, lobed body known as a thallus (fig. 29.3). The thallus has a smooth upper surface, and the lower surface bears numerous rhizoids (root-like hairs) projecting into the soil. *Marchantia* reproduces both asexually and sexually. Gemmae cups on the upper surface of the thallus contain gemmae, groups of cells that detach from the thallus and can start a new plant. Sexual reproduction involves male and female umbrella-like structures called gametophores, so called because they produce the gametes.

Mosses have a leafy appearance. The central stemlike structure bears spirally arranged leaflike structures (fig. 29.1*b*). Rhizoids anchor the plant and absorb minerals and water from the soil. None of these parts have vascular tissue; therefore, bryophytes lack true roots, stems, and leaves. Instead, they have rhizoids, stemlike structures, and leaflike structures.

The Gametophyte Dominates the Moss Life Cycle

In mosses, the gametophyte is dominant—it is longer lasting. In some mosses, there are separate male and female gametophytes (fig. 29.4). At the tip of a male gametophyte are antheridia (sing., **antheridium**), in which flagellated sperm are produced. After a rain or a heavy dew, the sperm swim to the tip of a female gametophyte, where eggs have been produced within the archegonia (sing., **archegonium**). Antheridia and archegonia are both multicellular structures, and each has an outer layer of cells that protects the enclosed gametes from desiccation, or drying out. After an egg is fertilized, the developing sporophyte is retained within the archegonium. The sporophyte, which is dependent—indeed parasitic—on the gametophyte, consists of a *foot,* which grows down into the gametophyte tissue, a *stalk,* and an upper *capsule,* or sporangium, where meiosis occurs and where haploid spores are produced. In some species of mosses, a hoodlike covering is carried upward by the growing sporophyte. When this covering and the capsule lid fall off, the spores are mature and ready to escape.

When a spore lands on an appropriate site, it germinates. The single row of cells that first appears branches, producing an algalike structure called a *protonema.* After about 3 days of favorable growing conditions, new moss plants are seen at intervals along the protonema. Each of these consists of the root-like rhizoids and the upright shoots of a moss gametophyte. The gametophytes produce gametes, and the moss life cycle begins again.

Bryophytes Need Water, Few Minerals

Bryophytes are capable of living where limited mineral availability prohibits invasion by other types of plants. There is no vascular tissue for water transport in these generally low-growing plants. There must be external water or mist at least periodically because sperm swim from the antheridia to the eggs in the archegonia. The embryo is protected from drying out, however, by remaining within the archegonium. Also, the organism is dispersed to new locations as windblown spores.

Importance: Rock Colonizers and Peat

Bryophytes help colonize bare rock and in this way help to convert the rocks to soil that can be used for the growth of other organisms. *Sphagnum,* also called bog or peat moss, has commercial importance. This moss has special nonliving cells that

Figure 29.4
Moss life cycle. The male and female gametophytes are leafy shoots. Flagellated sperm swim to the egg, and the zygote develops into the dependent sporophyte. The sporophyte consists of a foot, a stalk, and a sporangium, where haploid spores are produced.

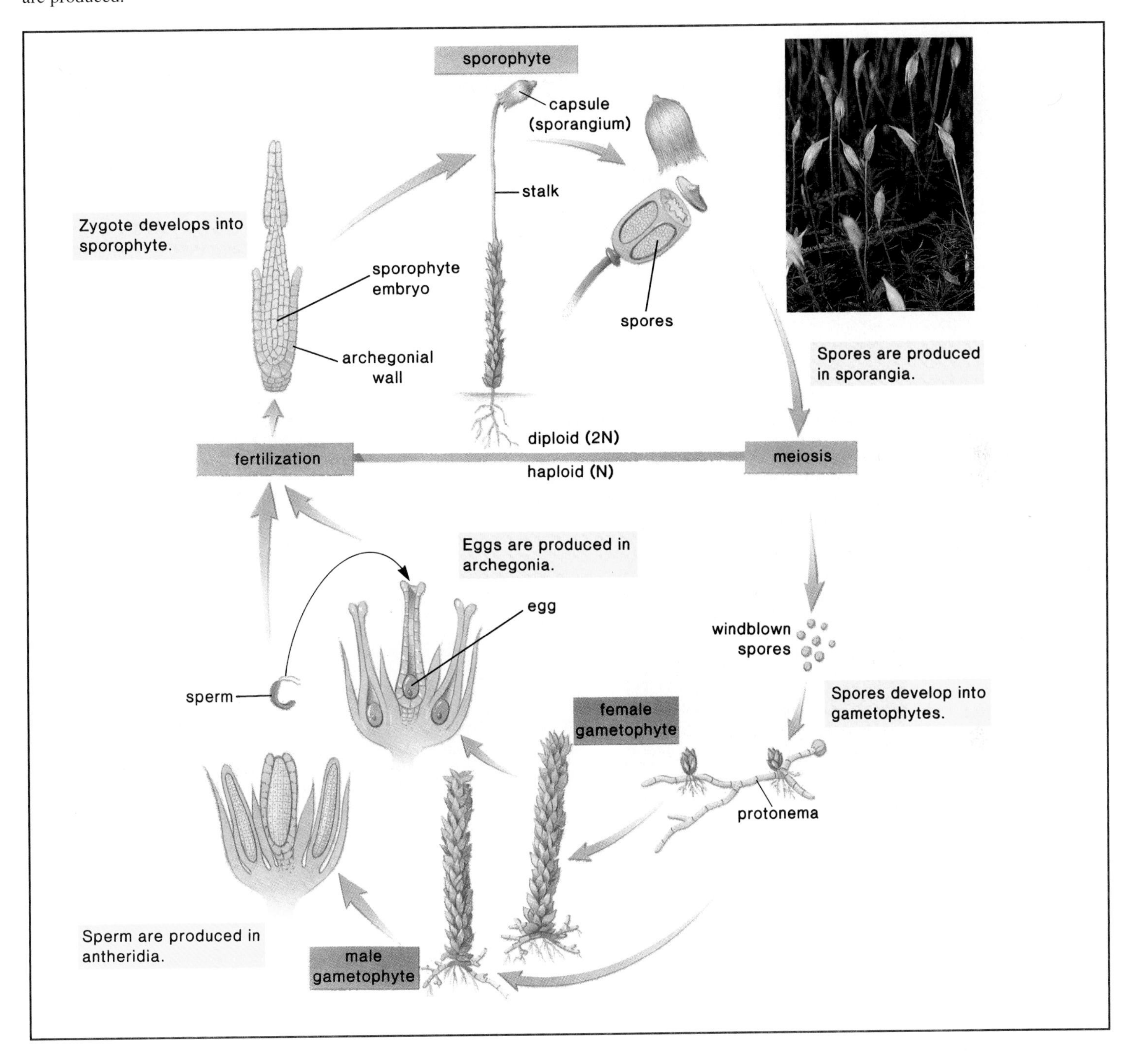

can absorb moisture, which is why peat moss is often used in gardens to improve the water-holding capacity of the soil. In some areas, like bogs, where the ground is wet and acidic, dead mosses, especially sphagnum, do not decay. The accumulated moss, called peat, can be used as fuel.

The bryophytes include the inconspicuous liverworts and the mosses, plants that have a dominant gametophyte. Bryophytes lack vascular tissue, and fertilization requires an outside source of moisture. Windblown spores disperse the species. Do 29.1 Critical Thinking, found at the end of the chapter.

Figure 29.5

Primitive vascular plants. **a.** Drawing of a Carboniferous swamp, showing primitive vascular plants and ferns when they were much larger than today. **b.** Whisk fern. **c.** Club moss. **d.** Horsetail.

Vascular Plants without Seeds

All the other plants we will study are **vascular plants,** so called because they have vascular tissue: xylem conducts water and minerals up from the soil, and phloem transports organic nutrients from one part of the plant body to another. The diploid sporophyte, the generation that has vascular tissue, is dominant. Xylem, with its strong-walled cells, supports the body of the plant against the pull of gravity. The tallest organisms in the world are vascular plants—the giant redwood trees of California, for example. Another advantage of having a dominant sporophyte relates to its being diploid. If a faulty allele is present, it can be masked by a functional allele.

During the Carboniferous period (see table 27.1), the nonseed vascular plants were abundant and treelike (fig. 29.5*a*). For some unknown reason, a large number of these plants died but did not decompose completely. Instead, they were compressed to form the coal that we still mine and burn today. (Oil has a similar origin but most likely formed in marine sedimentary rocks and included animal remains.)

Primitives: Whisk Ferns, Club Mosses, Horsetails

The primitive vascular plants include whisk ferns, club mosses, and horsetails (fig. 29.5 *b–d*). Among these, the whisk fern most closely resembles a rhyniophyte, an extinct group of plants known from the fossil record of the Silurian period (see table 27.1).

The rhyniophytes fulfill the qualifications for an ancestral prototype for all vascular plants. The dominant sporophyte had a horizontal underground stem called a rhizome, with erect branches above and rhizoids below (fig. 29.6). The stem and branches had vascular tissue, but there were no true roots or leaves. Sporangia were at the ends of some of the branches.

In general, whisk ferns have the same anatomy. The branches, which are green and carry on photosynthesis, bear sporangia where windblown spores are produced. The gametophyte is separate from the sporophyte. In fact, the whisk fern life cycle is very close to that of the fern.

> The nonseed vascular plants (whisk ferns, club mosses, horsetails) have vascular tissue and disperse the species by producing windblown spores. They have a life cycle that resembles that of the fern.

Ferns: Flagellated Sperm

Ferns vary in appearance. Many are low-growing, but there are also tall tree ferns in the tropics. Figure 29.7 shows the life cycle of a common fern of the temperate zone, with a horizontal underground stem (rhizome), from which hairlike roots project downward and **fronds** (large leaves) project upward. Young fronds grow in a curled-up form called a fiddlehead, which unrolls as it grows. A frond is often subdivided into a large number of leaflets.

The sporophyte is the dominant generation in a fern plant. Sporangia develop on the lower surface of a frond in clusters called sori (sing., **sorus**). A sorus is protected by a covering called the indusium. As a band of thickened cells on the rim of a sporangium dries out, it moves backward, and the spores are released.

Figure 29.6

Rhynia major, the simplest and one of the earliest known vascular plants. The leafless stem of this plant, which thrived about 350–400 million years ago, was green and carried on photosynthesis. The terminal sporangia apparently released their spores by splitting longitudinally.

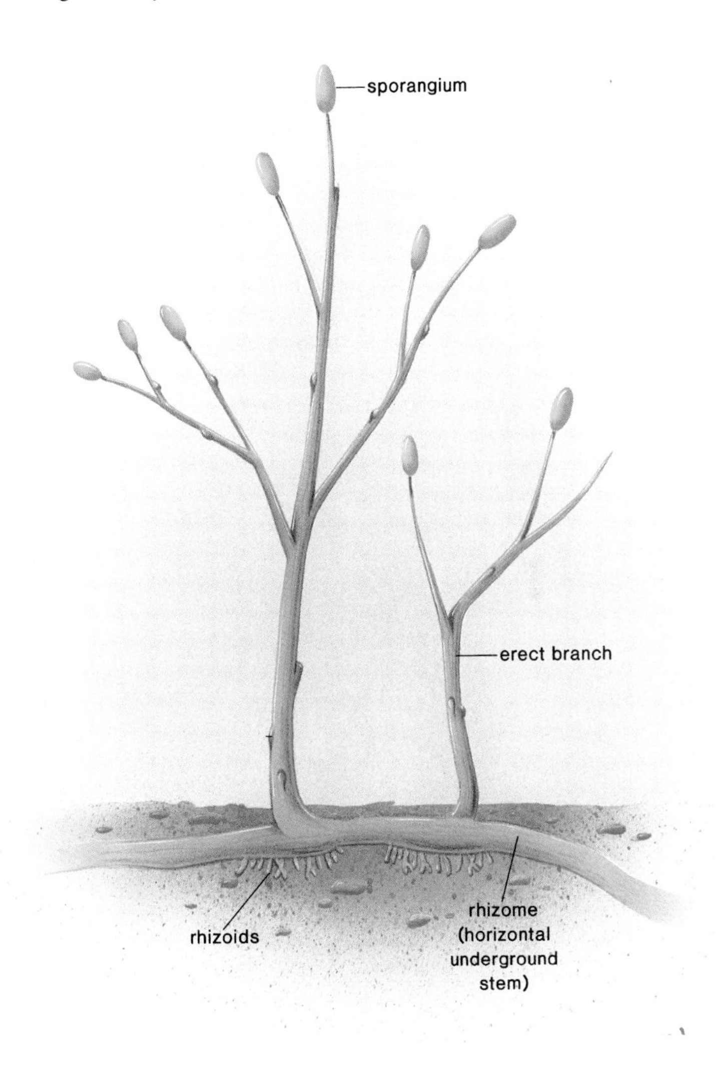

The gametophyte is a tiny (5–6 mm) heart-shaped structure called a **prothallus.** The antheridia and the archegonia develop on the lower surface of a prothallus. Typically, the archegonia are at the notch of a prothallus, and the antheridia are toward the tip between the rhizoids. Fertilization takes place when moisture is present because the sperm must swim in external water from the antheridia to the eggs within the

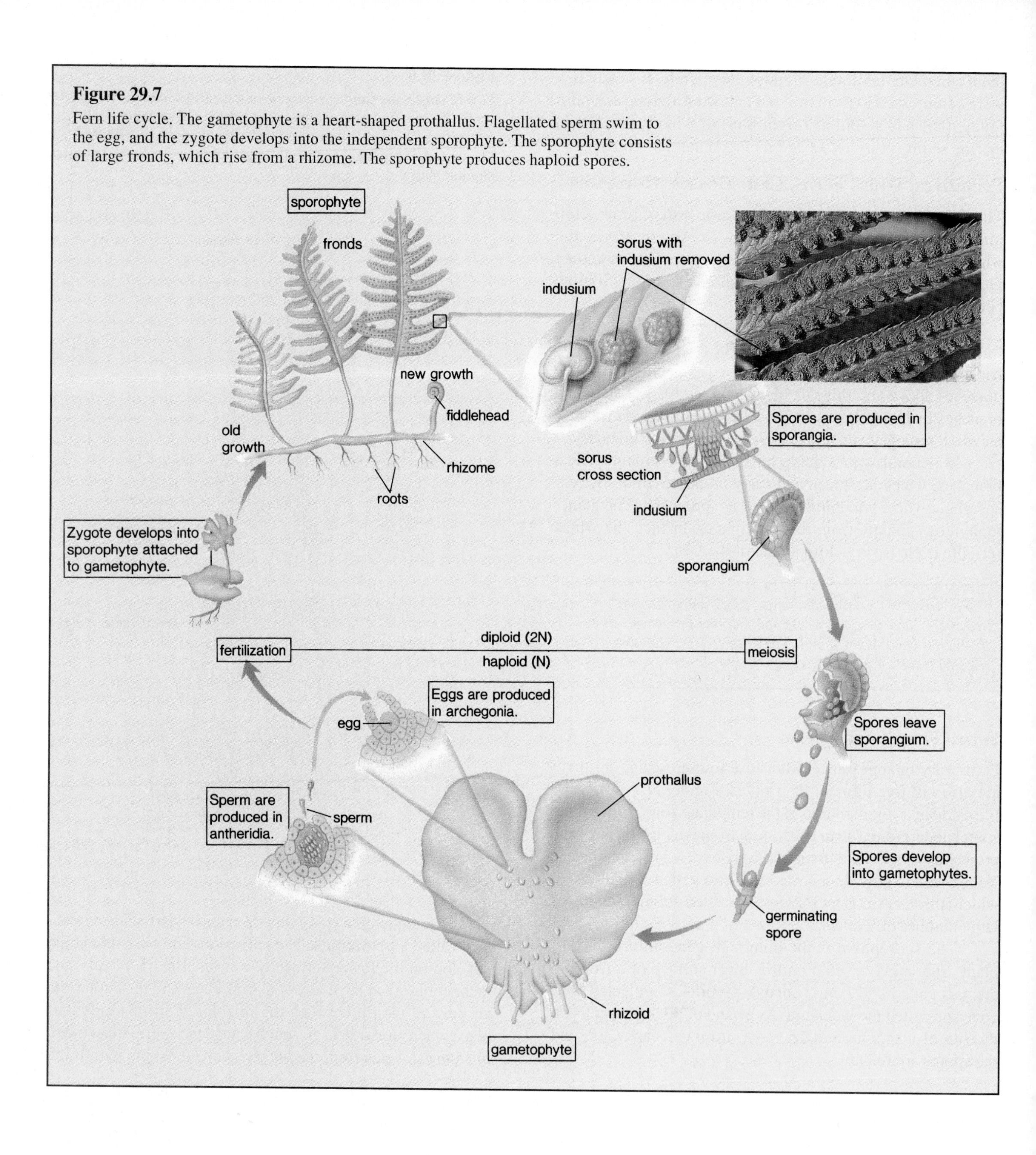

Figure 29.7

Fern life cycle. The gametophyte is a heart-shaped prothallus. Flagellated sperm swim to the egg, and the zygote develops into the independent sporophyte. The sporophyte consists of large fronds, which rise from a rhizome. The sporophyte produces haploid spores.

Figure 29.8

Fern diversity. **a.** Temperate-zone ferns are low-growing. **b.** Tropical-zone ferns are sometimes small trees.

a.

b.

archegonia. The resulting zygote begins its development inside an archegonium, but the embryo soon outgrows the available space. As a distinctive first leaf appears above the prothallus and as the roots develop below it, the sporophyte becomes visible. Often the sporophyte tissues are a distinctively different shade of green than the gametophyte tissues. The young sporophyte grows and develops into the familiar fern plant (fig. 29.8).

Adaptations of Ferns

Fern sporophytes have roots and leaves in addition to a stem. The well-developed leaves fan out, capture solar energy, and photosynthesize. The water-dependent gametophyte lacks vascular tissue and is separate from the sporophyte. Flagellated sperm require an outside source of water in order to swim to the eggs in the archegonia (fig. 29.7). Once established, some ferns, like the bracken fern (fig. 29.1*c*), can spread into drier areas by means of vegetative reproduction. As the rhizomes spread out, the fiddleheads grow up as new fronds.

> In nonseed vascular plants, such as ferns, there is a dominant sporophyte with vascular tissue. These plants usually are found in moist environments because the independent, nonvascular gametophyte produces flagellated sperm.

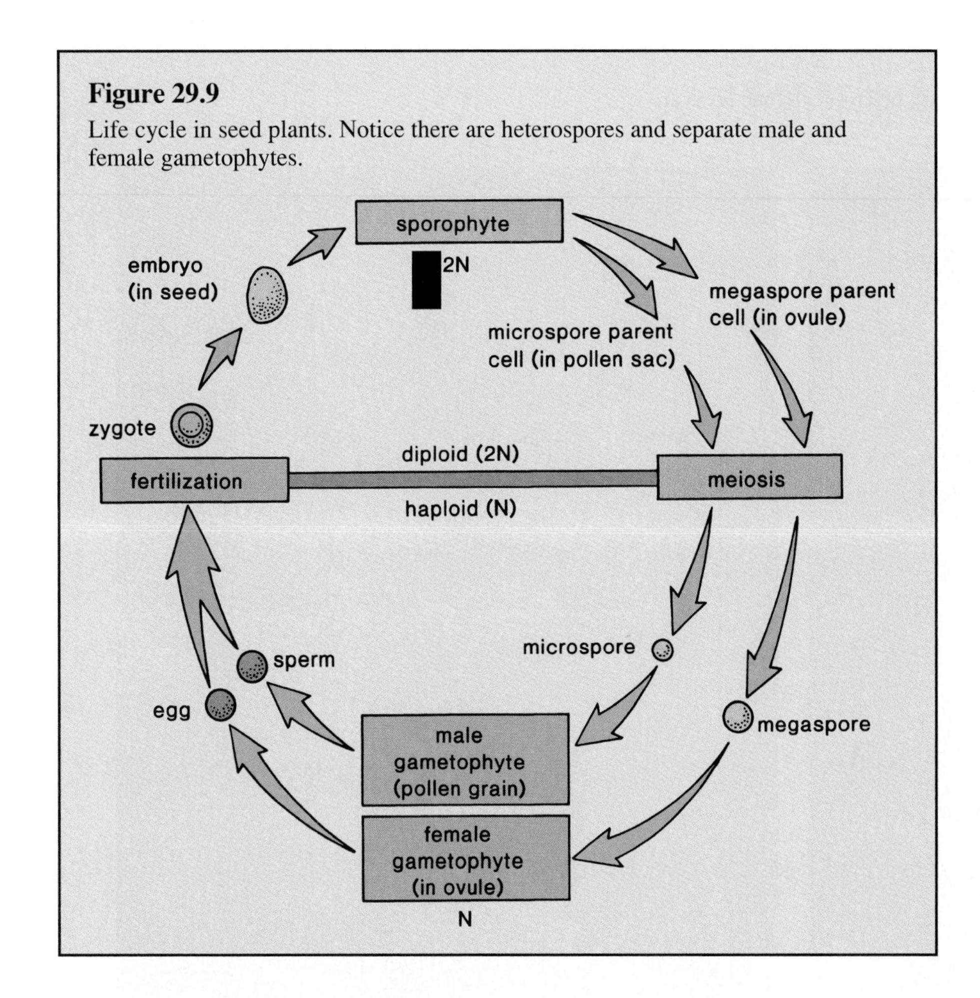

Figure 29.9
Life cycle in seed plants. Notice there are heterospores and separate male and female gametophytes.

Figure 29.10
Representatives of the lesser known gymnosperm divisions. **a.** Cycads resemble palm trees but are gymnosperms that produce cones. Male plants have pollen cones, and female plants have seed cones, like the one shown here. **b.** Ginkgos exist only as a single species—the maidenhair tree. This photograph of a female plant features the leaves and seeds. Male plants have pollen cones.

a.

b.

Seed Plants

There are 2 groups of plants that produce seeds: the gymnosperms and the angiosperms. The **gymnosperms** produce naked seeds; that is, they are not enclosed by fruits. A good example is the familiar winged seed of pine trees. **Angiosperms** produce covered seeds; that is, they are enclosed by fruits.

Seed plants have an alternation of generations life cycle; however, the cycle has been modified and they do not generally require external moisture to reproduce (fig. 29.9). Seed plants produce **heterospores,** called microspores and megaspores, instead of homospores, or identical spores. A **microspore** develops into a pollen grain, which later becomes a mature, sperm-bearing *male gametophyte*. Pollen is carried by wind or animals to the vicinity of the female gametophyte.

A **megaspore** develops into an egg-bearing *female gametophyte* while still retained within the body of the plant. After fertilization, the zygote becomes an embryo enclosed by a seed. While nonseed plants are dispersed by means of spores, seed plants are dispersed by seeds. A **seed** contains an embryonic sporophyte plus stored food, enclosed by a protective seed coat. Seeds are resistant to such adverse conditions as dryness and temperature extremes.

Gymnosperms: Naked Seeds

There are 4 divisions of gymnosperms, but only 3 of these are considered here. Cycads are cone-bearing, palmlike plants found mainly in tropical and subtropical regions (fig. 29.10*a*). There is only one species of ginkgo—the maidenhair tree (fig. 29.10*b*). The maidenhair tree has always been planted in Chinese and Japanese ornamental gardens, but now it is valued for its ability to do well in polluted areas. Because female trees produce rather smelly seeds, it is customary to use only male trees, propagated vegetatively, in city parks.

The largest group of gymnosperms is the cone-bearing **conifers** (fig. 29.1*d*), which include pine, cedar, spruce, fir, and

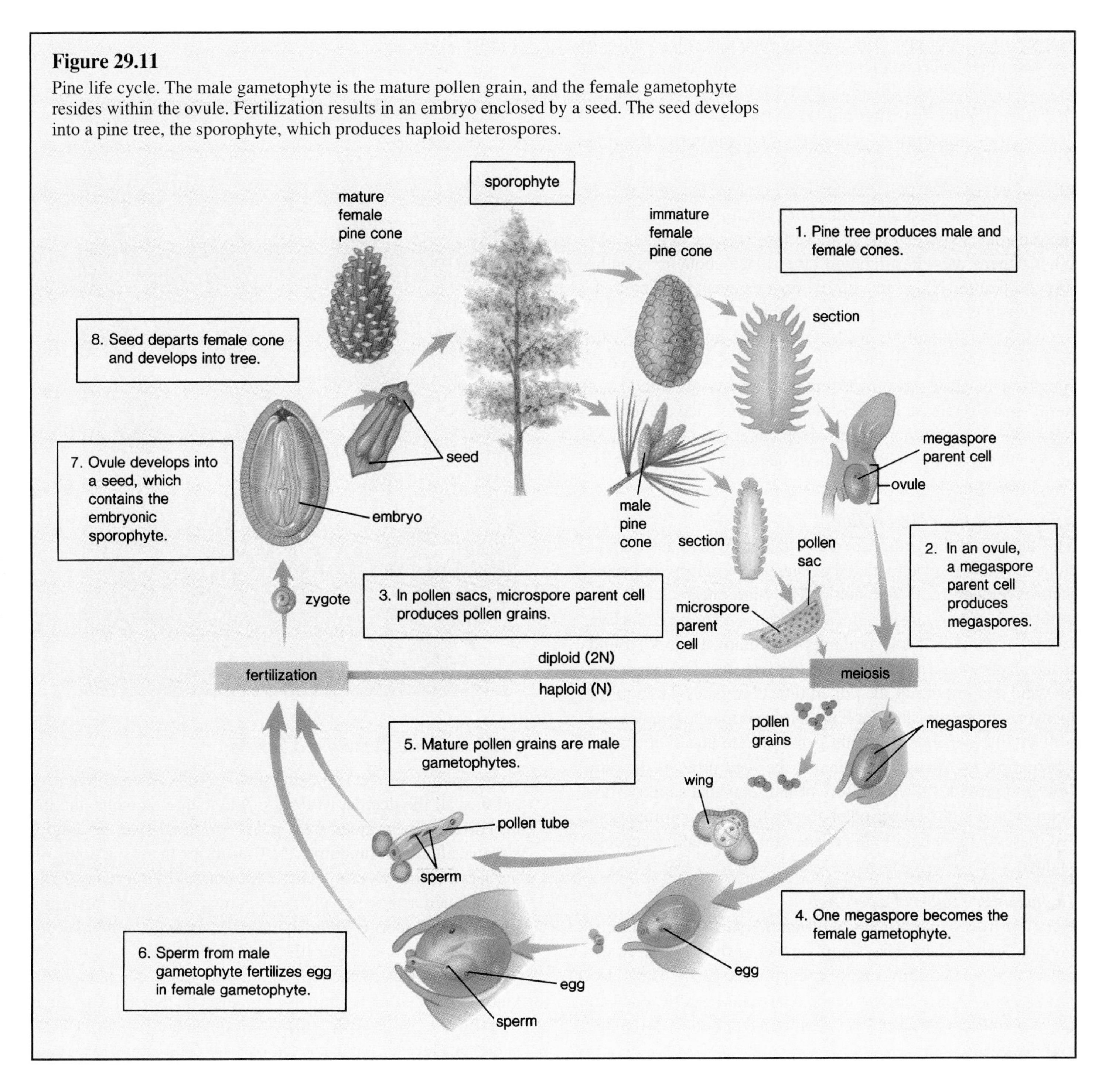

Figure 29.11

Pine life cycle. The male gametophyte is the mature pollen grain, and the female gametophyte resides within the ovule. Fertilization results in an embryo enclosed by a seed. The seed develops into a pine tree, the sporophyte, which produces haploid heterospores.

redwood trees. These trees have needlelike leaves, which are well adapted to not only hot summers but also cold winters and high winds. Most gymnosperms are evergreen trees—they keep their leaves all year long.

The Pine Life Cycle: Wind and Pollen Tubes

The pine life cycle is illustrated in figure 29.11. A pine tree produces 2 kinds of cones—male and female. Typically, the male pine cones are smaller than the female cones. Each scale of the male cone has 2 or more **pollen sacs** (microsporangia) on the lower surface. Here microspore parent cells undergo meiosis and produce haploid microspores, which become pollen grains. A mature **pollen grain** is the male gametophyte. The pollen grain of pine trees has 2 lobular wings and is carried by the wind. Pine trees release so many pollen grains during pollen season that everything in the area may be covered by a dusting of yellow, powdery pine pollen.

Each scale of the female cone has 2 **ovules** (containing megasporangia) on its upper surface. Inside each ovule, a megaspore parent cell undergoes meiosis, producing 4 haploid megaspores. Only one of these megaspores develops into an egg-bearing female gametophyte. The scales of the female cone open at the time of pollination, permitting entry of pollen, but they close tightly thereafter until seed release.

During **pollination,** pollen grains are transferred by wind from the male cones to the female cones. Once enclosed within the female cone, the pollen grain develops a *pollen tube,* which slowly grows toward the ovule. The pollen tube discharges 2 nonflagellated sperm. One of these fertilizes the egg, and the other degenerates. Fertilization takes place about 15 months after pollination and is an entirely separate event from pollination, which is simply the transfer of pollen.

After fertilization, the ovule matures and becomes the seed, composed of the embryo, its stored food, and a seed coat. Finally, in the third season, the female cone, by now woody and hard, opens to release its seeds, the wings of which are formed from a thin, membranous layer of the cone scale. When a seed germinates, the sporophyte embryo develops into a new pine tree, and the cycle is complete.

Adaptations for a Way of Life

The success of the gymnosperms is largely due to the adaptations they have made to a land existence. Gymnosperms have well-developed roots and stems. Many are tall trees that can withstand temperature extremes and dryness. The reproductive pattern of conifers has several important innovations not found in the plants we have considered so far. Transfer of pollen grains by wind and growth of the pollen tube eliminates the requirement of external moisture for flagellated sperm. The female cone protects the dependent female gametophyte and shelters the developing zygote as well. Finally, the seed protects the embryo and provides it with a store of nutrients that supports development for the first period of growth following germination. All these factors increase the chance for reproductive success on land.

Importance: Timber, Paper, Age

Conifers grow on large areas of the earth's surface and are economically important. They supply much of the wood used for construction of buildings and production of paper. They also produce many valuable chemicals, such as those extracted from resin, a substance that protects conifers from attack by fungi and insects.

Perhaps the oldest and largest trees in the world are conifers. Bristlecone pines in the Nevada mountains are known to be more than 4,500 years old (fig. 29.12), and a number of redwood trees in California are 2,000 years old and more than 90 m tall.

> A conifer is the most typical example of a gymnosperm. In its life cycle, windblown pollen grains replace flagellated sperm. Following fertilization, the seed develops from the ovule, a structure that has been protected within the body of the sporophyte. The naked seeds are dispersed by the wind.

Figure 29.12

Bristlecone pines, perhaps the oldest living plants in the world.

Angiosperms: Enclosed Seeds

Angiosperms are the flowering plants. All hardwood trees, including all the deciduous trees of the temperate zone and the broad-leaved evergreen trees of the tropical zone, are angiosperms, although sometimes the flowers are inconspicuous. All herbaceous (nonwoody) plants common to our everyday experience, such as grasses and most garden plants, are flowering plants. Angiosperms are adapted to every type of habitat, including water (e.g., water lilies and duckweed).

The flower of an angiosperm provides 3 distinct advantages: it often attracts animals (e.g., insects) that aid in pollination (fig. 29.13), it protects the developing female gametophyte, and it produces seeds enclosed by fruit. The angiosperms are divided into 2 classes: the monocots and the dicots. They differ by a number of features, which are described in figure 8.4, page 129.

Flower Anatomy

In angiosperms, the reproductive structures are located in the **flower** (fig. 29.14). The **sepals,** most often green, form a whorl about the **petals,** the color of which accounts for the attractiveness of many flowers. In the center of the flower is a small vaselike structure, the **pistil,** which usually has 3 parts: the **stigma,** an enlarged sticky knob; the **style,** a slender stalk; and

Figure 29.13

Pollinators. While gathering pollen and nectar, pollinators transfer pollen from the anther of one flower to the stigma of another flower. **a.** Butterflies can be pollinators. Composites, such as this flower, offer a broad surface of many individual flowers. **b.** Bees are well-known pollinators. In this enlargement, you can make out the loaded pollen basket on the bee's leg. The tongue is extended as if drinking nectar. **c.** Hummingbirds can be pollinators. As the bird inserts its beak to reach the rich supply of nectar, its head touches the reproductive structures.

a.

b.

c.

Figure 29.14

Anatomy of a flower. Flowers have an outer whorl of green leaflike sepals and an inner whorl of usually colorful petals. At the very center of a flower is the pistil, composed of the stigma, the style, and the ovary. Grouped about the pistil are the stamens, each of which has an anther and a filament. Sometimes the pistil is called the female part of the flower and the stamens are called the male part of the flower. This is not strictly correct because they do not produce gametes; they produce megaspores and microspores, respectively. A microspore goes on to become a male gametophyte, which contains a nucleus that will produce sperm, and a megaspore goes on to become a female gametophyte, which produces an egg.

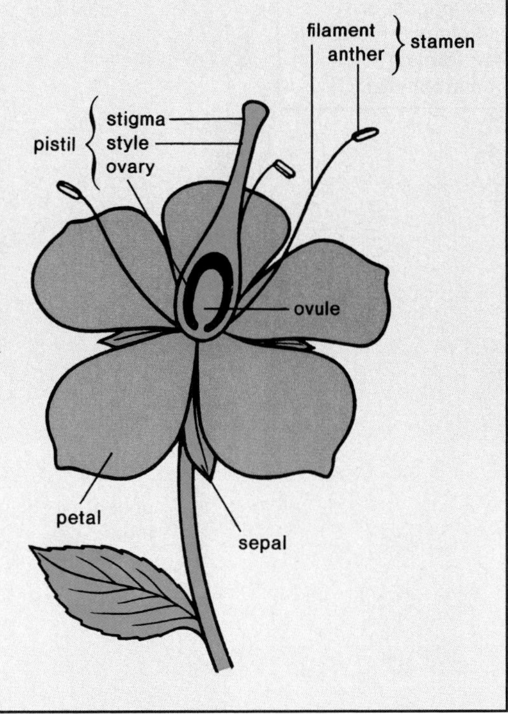

the **ovary,** an enlarged base. The ovary contains a number of ovules, which play a significant role in reproduction. Grouped about the pistil are a number of stamens, each of which has 2 parts: the **anther,** a saclike container, and the **filament,** a slender stalk.

Life Cycle: Animal Pollinators and Enclosed Seeds

In general, the flowering plant life cycle is like that of other seed plants (fig. 29.15). Specifically, the pollen sacs are in the anthers. Here microspore parent cells undergo meiosis, producing microspores. Each microspore becomes a pollen grain, which is the male gametophyte and contains sperm when mature. The ovules are in the ovary at the base of the pistil. In each ovule, a megaspore parent cell produces 4 megaspores by meiosis, but only one undergoes mitosis and develops into the mature **embryo sac,** which contains an egg. If the pollen grain is transported to the stigma of the pistil, it forms a pollen tube, which grows down through the pistil to an embryo sac in an ovule. Usually, 2 sperm pass down the pollen tube into the embryo sac, and *double fertilization* occurs. One sperm fertilizes the egg so that a zygote results. The other sperm unites with 2 central polar nuclei to form a 3N endosperm nucleus. Following cell division, the zygote becomes the embryo and the endosperm nucleus becomes the **endosperm,** food for the developing embryo and future seedling.

Figure 29.15

Flowering plant life cycle. The male gametophyte is the mature pollen grain, and the female gametophyte is the embryo sac. Fertilization results in an embryo enclosed by a fruit-covered seed. The seed develops into a flowering plant, the sporophyte, which produces haploid heterospores.

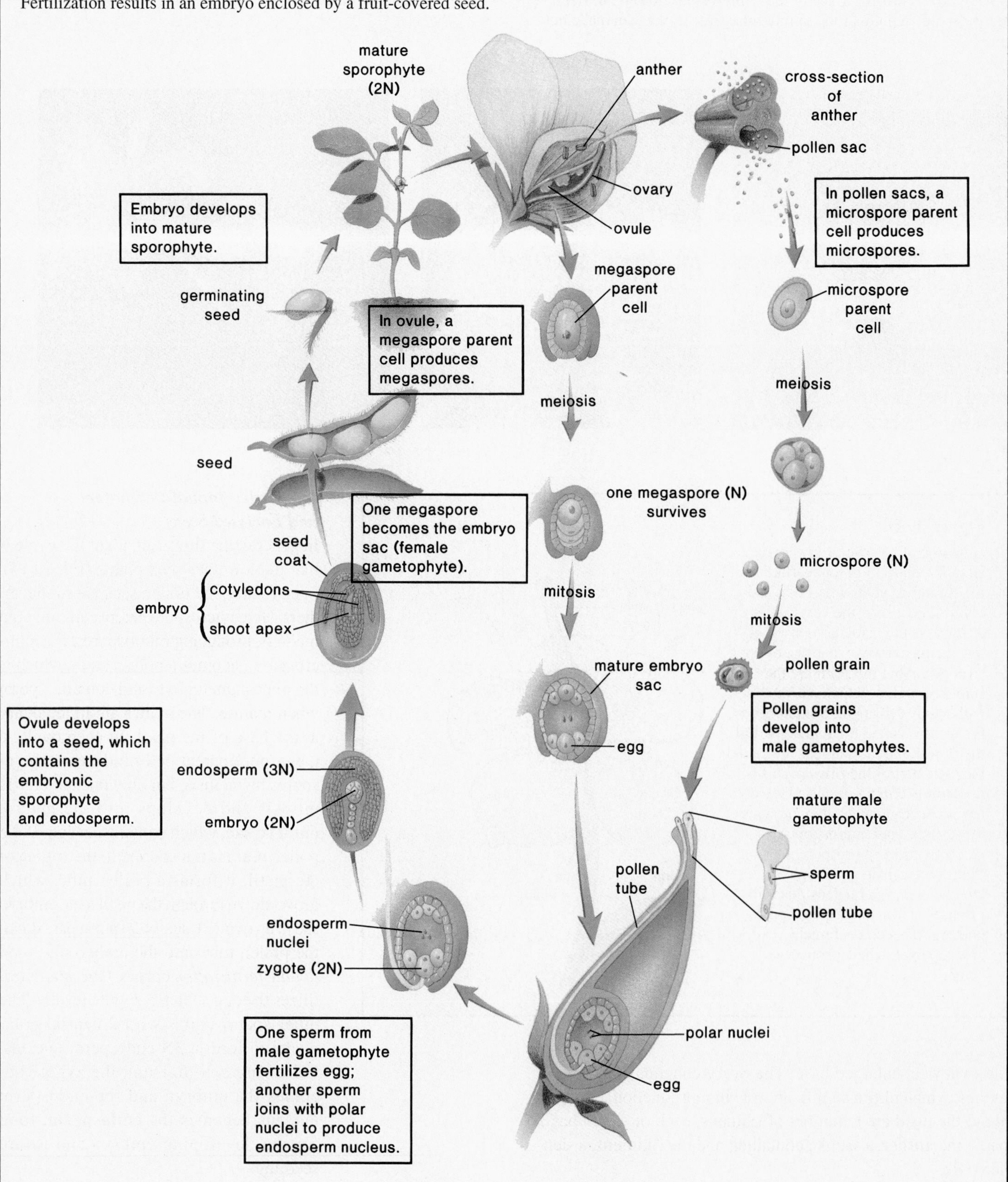

The ovule now develops into a seed. Its outer tissue layers harden and become the seed coat. The ovary and sometimes other floral parts develop into a fruit. A **fruit** is a mature ovary that contains the seeds. Therefore, angiosperms are said to have enclosed seeds.

Fruits: From Cocklebурs to Watermelon

The function of fruits is largely to disperse the seeds. Fruits can be simple or compound (table 29.2), but most are simple fruits; they are formed from a single ovary. Simple fruits can be dry or fleshy. Milkweed pods, legumes, like peas and beans, pecans and other nuts, and grains, such as rice and wheat, are all dry. Dry fruits are dispersed in various ways. Milkweed uses the wind, a cocklebur attaches to an animal coat, and nuts may float on the water. Tomatoes, oranges, and watermelons are all fleshy. When fleshy fruits are eaten by animals, the seeds pass through the digestive tract and are deposited in droppings.

Some fruits are compound fruits; they are formed from several ovaries. The ovaries can be from the same flower (blackberry and strawberry) or from different flowers (pineapple).

Table 29.2
Kinds of Fruits

Simple fruits	***Dry fruits***	
(formed from a single ovary)	Splits open	Milkweed, pea, lentil
	Remains intact	Sunflower, acorn, rice, wheat
	Fleshy fruits	
	Ovary wall becomes fleshy	Peach, plum, tomato
Compound fruits	***Aggregate fruits***	
(formed from many ovaries)	Ovaries from single flower	Blackberry, raspberry
	Multiple fruits	
	Ovaries from several flowers	Pineapple

Attractive Flowers and Other Adaptations

Angiosperms have well-developed vascular tissue and generally broad leaves. The reproductive organs are in the flowers, which often attract animal pollinators. Their ovules are located in ovaries, which develop into fruits. These angiosperm features, not seen in gymnosperms, have contributed to their success. Angiosperms are found in all sorts of habitats. Some have even returned to the water.

Importance: Food, Drugs, Even Cosmetics

As discussed in the reading on page 576, angiosperms are our basic food sources. Grains supply protein and more than half of our calories. Legumes and grasses are forage for livestock. Grasses for forage include fescue, Kentucky bluegrass, and timothy; legumes for forage include alfalfa and clover (fig. 29.16).

Humans use spices to add flavor, fragrance, and variety to foods. Peppercorns are small, berrylike structures from a vine, cinnamon comes from the bark of a tree, and cloves are dried flower buds. Coffee beans and tea leaves are used to prepare beverages that are consumed worldwide. The seeds of the cacao tree are processed into chocolate. Most vegetable oils are derived from seeds, except olive oil and palm oil, which come from high-fat fruit pulps.

Today, plant products are used medicinally in more than one-fifth of prescriptions. Digitalis, a plant steroid, is used to treat heart disease. Saponins, steroids from yam species, are used as molecular backbones for the synthesis of cortisone, hydrocortisone, and human sex hormones. Taxol, from the Pacific yew, has recently become a very promising active agent against advanced ovarian cancer, advanced breast cancer, and other types of tumors.

Psychoactive drugs occur in over 100 flowering plants. The leaves and flower of *Cannabis sativa* contain marijuana. Heroin is one of the opiates derived from the opium poppy. Alcohol, caffeine, and nicotine are all plant-derived drugs. Tobacco cultivation is an important industry even today.

Plant materials are used for clothing, shelter, and fuel. Cotton comes from the fibers of the plant's seed coat. Flax is a source of natural fibers for linen. Pulpwood chips provide the cellulose for rayon and acetate production. Jute is used for sacking and twine, and hemp is made into rope and canvas. Bamboo is a traditional construction material in tropical areas, although wood is more widely used for construction of family dwellings and for fuel.

Other valuable angiosperm products include vegetable dyes, tannins for leather processing, rubber, gums, pectins, starches, and waxes. Plant gums, such as gum arabic, are used to thicken and emulsify processed foods, medicines, and cosmetics. Pectins are used to thicken jellies, and starches are thickeners for paper and cardboard manufacture. A hard wax used in polishes, carnauba wax, is made from the leaves of a wild palm.

a.

Figure 29.16

Importance of angiosperms. Angiosperms are used by humans for many and varied purposes. **a.** Alfalfa is used as food for animals. **b.** Cooking oil is extracted from sunflower seeds. **c.** Morphine is in a latex obtained from the fruit of the poppy plant. **d.** Taxol, a medicine that shrinks tumors in cancer patients, is derived from the bark of the Pacific yew.

c.

d.

b.

TWELVE PLANTS THAT KEEP HUMANS ALIVE

Virtually all food plants are angiosperms, or flowering plants. This is not surprising when you recall that only the angiosperms have seeds enclosed by fruits and it is fruits that are used by humans for food. However, it may come as a surprise to learn that relatively few species of plants are involved. Some of these are nuts (e.g., walnuts), berries (e.g., blueberries), and fleshy fruits (e.g., apples and oranges). Of the 150 species that have been cultivated extensively, only 12 species are really important—indeed, it can be said that these 12 plants stand between humanity and starvation. If all 12 or even a few of these cultivated plants were eliminated from the earth, millions of people would starve.

Three of these all-important species are cereals—*wheat, corn,* and *rice*; the last alone supplies the energy required by 50% of the people of the world. It is a remarkable fact that each of these cereals, or grains, is associated with a different major culture or civilization—wheat with Europe and the Middle East, corn or maize with the Americas, and rice with the Far East. Three of the 12 food plants are so-called root crops—*white potato* (not a root but a tuber, an enlarged tip of a rhizome, or horizontal underground stem); *sweet potato*; and *cassava,* or *manioc* or *tapioca,* from which millions of people in the tropics of both hemispheres derive their basic food. Two of the 12 are sugar-producing plants—*sugarcane* and *sugar beet.* Another pair of species are legumes—the *common bean* and the *soybean,* both important sources of vegetable protein and hence sometimes referred to as the "poor person's meat." The final 2 plants of this august company are tropical tree crops—*coconut* and *banana* ....

From Tippo/Stern, *Humanistic Botany.*

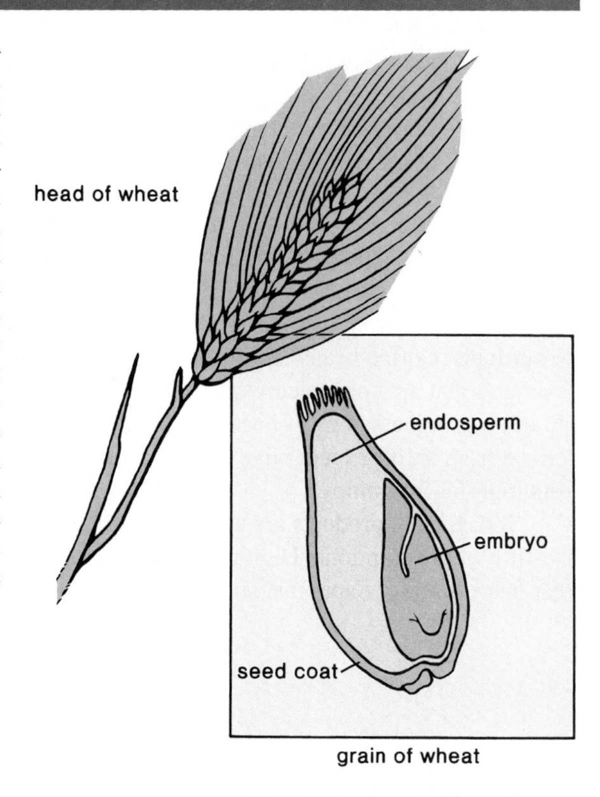

Table 29.3
Adaptation Summary

Plant	Generations	Reproduction	Representative
Nonseed Plants: Windblown spores disperse the species			
Bryophytes	Both generations lack vascular tissue. Sporophyte is dependent on gametophyte.	Flagellated sperm require a source of outside moisture.	Moss
Primitive vascular plants	Sporophyte has vascular tissue; gametophyte lacks vascular tissue and is separate and independent of the sporophyte.	Flagellated sperm require a source of outside moisture.	Fern
Seed Plants: Seeds disperse the species			
Gymnosperms (naked seeds)	Gametophyte is retained and protected from drying out by sporophyte, which has vascular tissue.	Pollen grains replace flagellated sperm. Windblown seeds.	Pine
Angiosperms (enclosed seeds)	Adapted in the same manner as gymnosperms.	Insect pollination. Fruits aid dispersal.	Flower

Figure 29.17
The relative significance of the sporophyte (diploid generation) and the gametophyte (haploid generation) among plants.

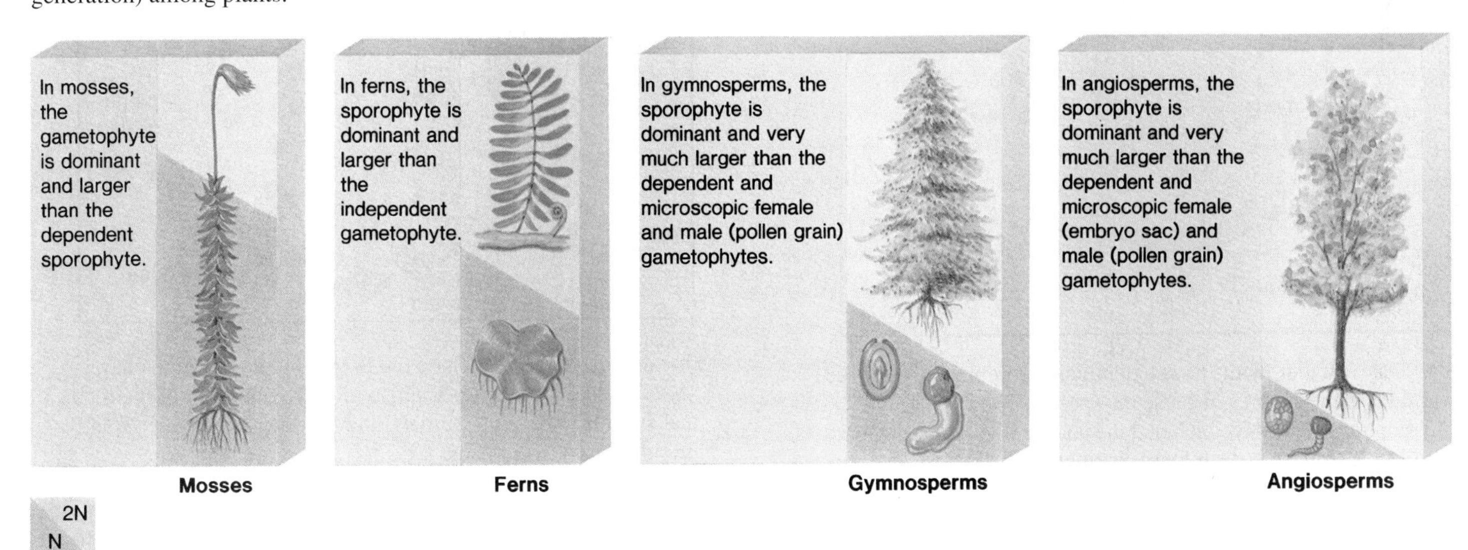

Comparisons among Plants

The presence or absence of vascular tissue distinguishes the plants we have studied. The bryophytes lack vascular tissue and are unable to transport water any distance; therefore, they are low-growing. The primitive vascular plants have vascular tissue but their roots and leaves generally are not well developed. Gymnosperms and angiosperms have well-developed stems, roots, and leaves. These groups include many types of tall trees, which transport water some distance from the roots to the leaves.

Table 29.3 compares the reproductive strategy of the nonseed plants to the seed plants, and figure 29.17 shows the relative importance of the diploid (2N) generation and the haploid (N) generation in plants. The gametophyte is dominant and larger than the sporophyte in bryophytes. In primitive vascular plants, including ferns, the small gametophyte is separate from the dominant sporophyte. The gametophyte lacks vascular tissue and fares better in moist habitats. In gymnosperms and angiosperms, the gametophyte is microscopic and dependent on the dominant sporophyte. Similarly, the seed plants are not dependent on moisture for reproduction. The nonseed plants have flagellated sperm, which require external moisture to travel from the antheridia to the eggs in the archegonia. In seed plants, pollen is generally windblown or carried by animals to the female gametophyte, and then a sperm is delivered to the egg by a pollen tube. External moisture is not necessary for fertilization.

The gymnosperms and angiosperms are widely distributed on land. The delicate heterospores, gametophytes, gametes, zygotes, and embryos (seeds) are enclosed by coverings that protect them from drying out.

Vascular tissue is lacking in the bryophytes but is present in all other plants. In bryophytes, the gametophyte is dominant and the sporophyte is dependent. In vascular plants, the sporophyte is dominant. The gametophyte is dependent and microscopic in the pine and the flower, which use seeds to disperse the species. Do 29.2 Critical Thinking, found at the end of the chapter.

SUMMARY

Plants are multicellular, photosynthetic organisms adapted to a land existence. Among the various adaptations, all plants protect the developing embryo from drying out. Plants have an alternation of generations life cycle, but some have a dominant gametophyte (haploid generation) and others have a dominant sporophyte (diploid generation). The bryophytes, which include the liverworts and the mosses, are nonvascular plants and therefore lack true roots, stems, and leaves. In the moss life cycle, spores disperse the species, the gametophyte is dominant, and external moisture is required for sperm to swim to the egg.

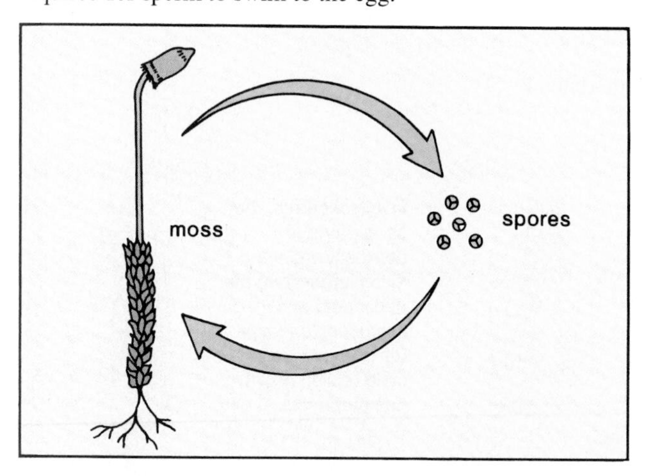

The vascular plants have a dominant sporophyte. In nonseed vascular plants, spores disperse the species. The extinct rhyniophytes may have included the ancestral vascular plant. Today, the primitive vascular plants include the whisk ferns, the club mosses, and the horsetails. In these plants, as well as in ferns, there is a separate and moisture-dependent gametophyte that produces flagellated sperm.

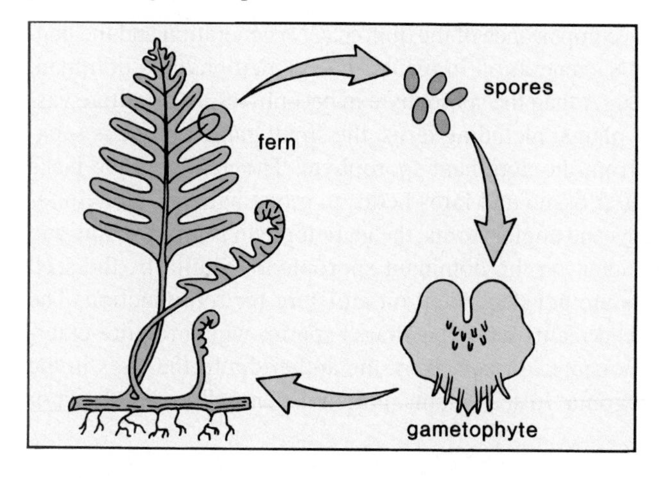

Gymnosperms and angiosperms are vascular plants that produce seeds. They have a life cycle that includes heterospores and male and female gametophytes. The male gametophyte is the mature pollen grain, and this structure replaces the flagellated sperm of nonseed vascular plants. The female gametophyte is microscopic and retained within the ovule, a sporophyte structure. Gymnosperms (meaning naked seeds) are exemplified by the conifers, in which the sporangia are located on cones. Angiosperms (meaning covered seeds) are the flowering plants, which have seeds enclosed by fruits. Angiosperms provide most of the food that sustains terrestrial animals, and they are the source of many products used by humans.

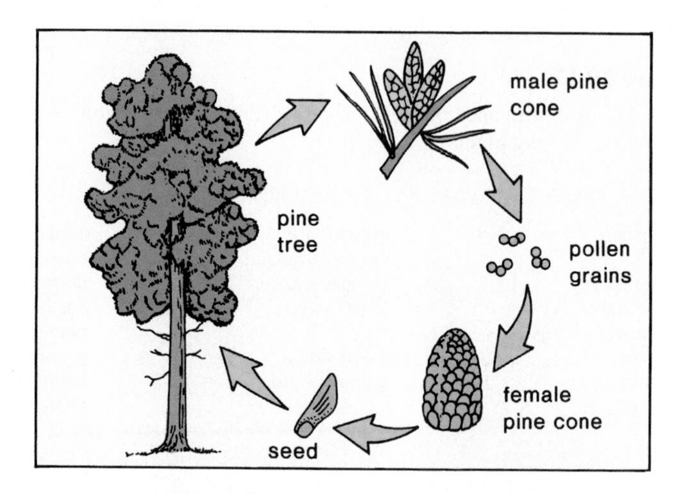

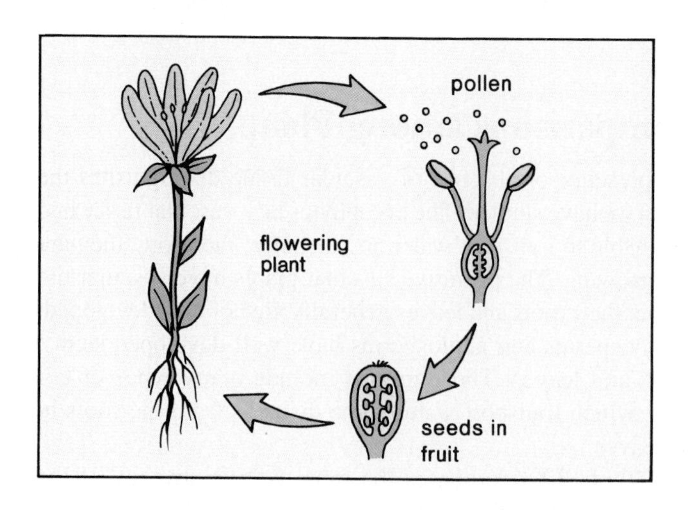

STUDY QUESTIONS

In order to practice **writing across the curriculum,** students should write out the answers to any or all of the study questions. The study questions are sequenced in the same order as the text.

1. What characteristics define plants? (p. 563)
2. Describe the moss life cycle, and point out significant features of this cycle. (p. 564)
3. The primitive vascular plants comprise which plants? At what time in the history of the earth were they larger and more abundant than today? (p. 566)
4. Describe the fern life cycle, and point out significant features of this cycle. (pp. 567–569)
5. Describe the pine life cycle, and point out significant features of this cycle. (pp. 571–572)
6. List 3 innovations that are observed in the gymnosperm plant life cycle that are not seen in the life cycle of ferns. (p. 572)
7. List all the ways that angiosperms are adapted to a land existence. (pp. 563, 572)
8. Describe the life cycle of flowering plants, pointing out any features not observed in gymnosperms. (pp. 573, 575)
9. List five ways angiosperms are important to humans, with an example for each. (p. 575)
10. Draw a diagram that shows the increasing dominance of the sporophyte generation among the 4 groups of plants studied. (p. 577)

OBJECTIVE QUESTIONS

1. All plants protect the _______ from drying out.
2. In the alternation of generations life cycle, the sporophyte (2N) produces _______.
3. In the moss life cycle, the dominant generation is the _______.
4. Mosses and ferns are apt to be found in moist locations because the sperm must _______ in external moisture to the egg.
5. In the fern life cycle, the gametophyte is independent and _______ from the sporophyte.
6. In the seed plant life cycle, there are _______ and therefore separate male and female gametophytes.
7. In the seed plant life cycle, _______ replace flagellated sperm.
8. In the life cycle of the pine tree, ovules are found on _______ pine cones.
9. Angiosperms practice double fertilization. One sperm unites with the _______, and the other unites with the _______ nuclei.
10. In angiosperms, the ovule becomes the _______ and the ovary becomes the _______.
11. Label this diagram of the alternation of generations life cycle.

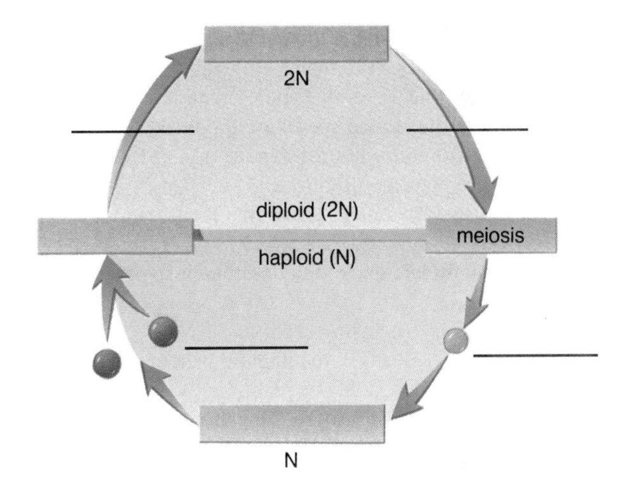

12. Complete this table to compare a moss to a fern.

	Moss	Fern
Description of gametophyte	a.	c.
Description of sporophyte	b.	d.

e. In which plant is the sporophyte dominant?

13. Complete this table to compare the pine life cycle to the life cycle of a flowering plant.

	Pine	Flower
Location of pollen sacs	a.	c.
Location of ovules	b.	d.

e. Which plant produces fruit?

CRITICAL THINKING

In order to practice **writing across the curriculum,** students should write out the answers to any or all of the critical thinking questions. Suggested answers to the critical thinking questions are in appendix E.

29.1

1. A student has learned that a plant sporophyte has vascular tissue. Do you expect the student to find vascular tissue in the moss sporophyte?
2. How could you test that bryophytes are adapted to limited mineral availability?
3. Why would you expect any mutations in bryophytes to be immediately tested by the environment?
4. Would diversification be promoted if the rudiments of vascular tissue evolved in bryophytes? Explain.

29.2

1. Most terrestrial organisms protect the gametes and the embryo from drying out. Compare the reproductive adaptations of humans to that of trees.
2. Most terrestrial organisms have a transport system. Compare the transport adaptations of a tree to that of humans.
3. Many terrestrial organisms have an internal skeleton to oppose the force of gravity. Compare the skeletal adaptations of a tree to that of humans.
4. In the temperate zone, humans remain active in the winter; deciduous trees do not. Explain.

SELECTED KEY TERMS

alternation of generations a life cycle typical of plants in which a diploid sporophyte alternates with a haploid gametophyte.

antheridium (an″ ther-id′ e-um) male organ in certain nonseed plants where flagellated sperm are produced.

archegonium (ar″ kĕ-go′ ne-um) female organ in certain nonseed plants where eggs are produced.

bryophyte (bri′ o-fīt) a plant group that includes mosses and liverworts.

embryo sac the female gametophyte of angiosperms that contains an egg.

fruit a mature ovary enclosing seed(s).

gametophyte (gam′ ĕ-to-fīt) the haploid generation that produces gametes in the life cycle of a plant.

heterospore a nonidentical spore, such as a microspore and a megaspore, produced by the same plant.

megaspore in seed plants, spore that develops into the female gametophyte.

microspore in seed plants, spore that develops into a pollen grain.

ovule (o′ vūl) in seed plants, a structure that contains the megasporangium, where meiosis occurs and the female gametophyte is produced; develops into the seed.

pollen grain mature male gametophyte of seed plants that contains sperm when mature.

prothallus (pro-thal′ us) a small, heart-shaped structure that is the gametophyte of the fern.

seed a mature ovule that contains an embryo with stored food enclosed by a protective coat.

sorus a cluster of sporangia found on the lower surface of fern leaves.

spore a haploid reproductive structure produced by the diploid sporophyte of a plant; asexually gives rise to the haploid gametophyte.

sporophyte the diploid generation that produces spores in the life cycle of a plant.

vascular plant a plant that has vascular tissue (xylem and phloem); includes ferns and seed plants.

30

THE ANIMAL KINGDOM

Chapter Concepts

1.
Animals are classified according to certain criteria, including symmetry, type of coelom, body plan, and presence of segmentation. 582, 617

2.
The animals that evolved first are less complex than those that evolved later. 585, 617

3.
Embryological observations are used to determine how groups of animals might be related. 592, 617

4.
A jointed skeleton and segmentation occur in both the arthropods and the vertebrates. Both groups also have land representatives. 598, 617

5.
Among mammals, primates are adapted to living in trees. 611, 617

6.
Human evolution diverged from ape evolution about 4 million years ago. 614, 617

Collared lizard and grasshopper

Chapter Outline

ANIMALS come in a great variety of sizes and shapes, but they all have certain characteristics that separate them from other organisms. These characteristics are as follows:

1. Animals are heterotrophic by ingestion; they take in preformed organic food.
2. Animals are motile during all or some part of their life cycle.
3. Animals are multicellular eukaryotes whose cells lack a cell wall.
4. Animals have a life cycle in which the adult is always diploid; they can all re-produce sexually.
5. Animals are able to respond quickly to outside stimuli because they have nerve and muscle cells.

Figure 30.1

Evolutionary tree of animals. Animals are believed to be descended from protozoa; however, the sponges may have evolved separately from protozoa.

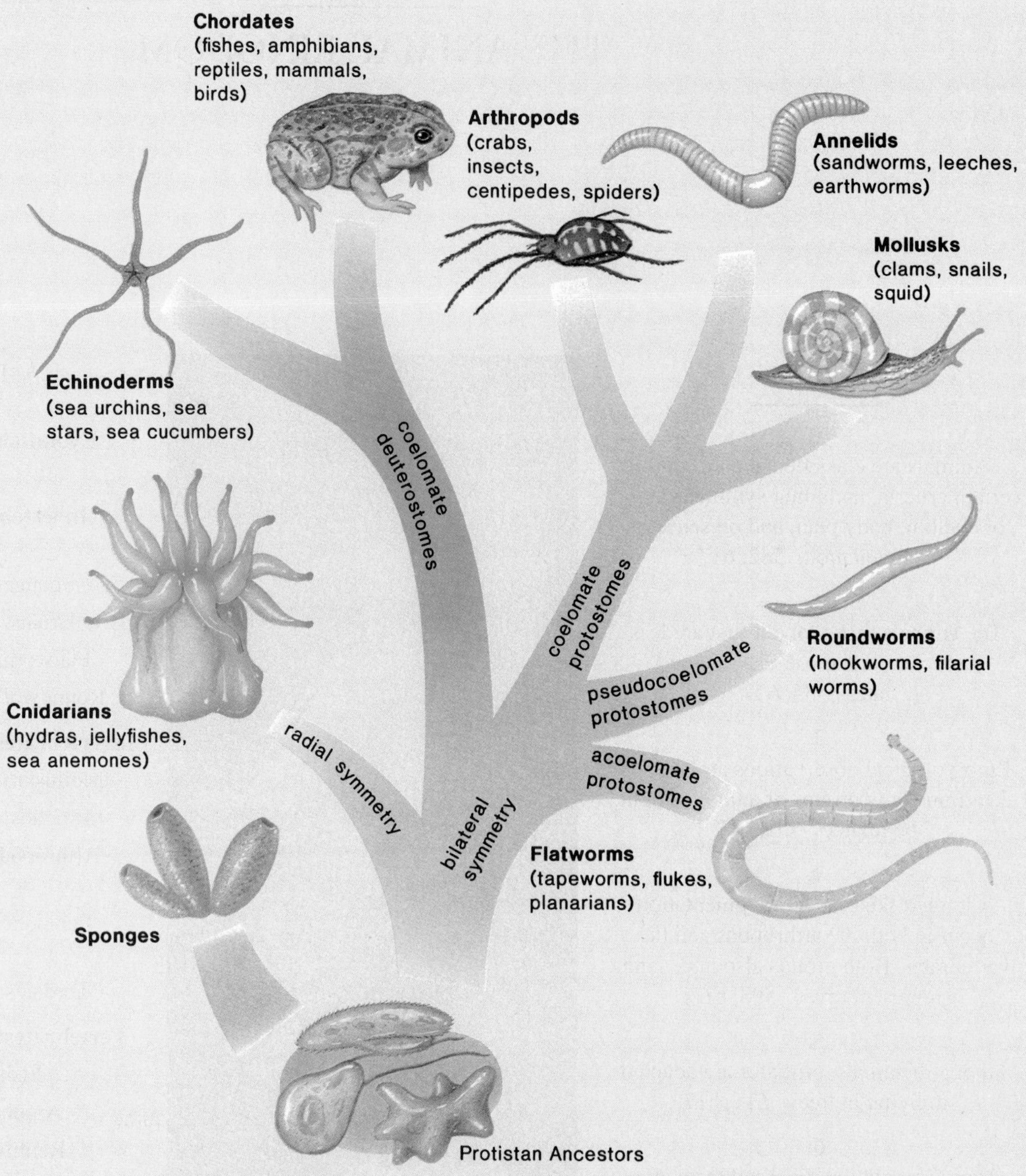

Evolution and Classification of Animals

The evolutionary tree of animals given in figure 30.1 shows that animals may have evolved from protozoa—perhaps a colonial form whose cells differentiated into various tissues. A number of features are used to classify animals and to determine how they might be related. For example, most animals have 3 germ layers (ectoderm, endoderm, mesoderm, p. 395) during development, but in a few animals there are only 2 germ layers (ectoderm and endoderm). Such animals have the *tissue level of organization.* Animals with 3 germ layers have the *organ level of organization.*

All the animal phyla placed in the main trunk of the evolutionary tree in figure 30.1 lack a true coelom, an internal body cavity lined by mesoderm. Some are acoelomates—they do not have a coelom at all (fig. 30.2). Some are pseudocoelomates—they have a cavity, but it is incompletely lined with mesoderm. There is a layer of mesoderm beneath the body wall

Figure 30.2

Comparison of mesoderm organization. **a.** In an acoelomate, such as a flatworm, mesoderm fills the space between ectoderm and endoderm. **b.** In a pseudocoelomate, such as a roundworm, the cavity is incompletely lined with mesoderm—there is mesoderm inside the ectoderm, but not adjacent to the gut endoderm. **c.** In a true coelomate, such as a human, the cavity, called a coelom, is completely lined by mesoderm—there is mesoderm both inside the ectoderm and adjacent to the gut endoderm. Mesenteries hold organs in place within the body cavity.

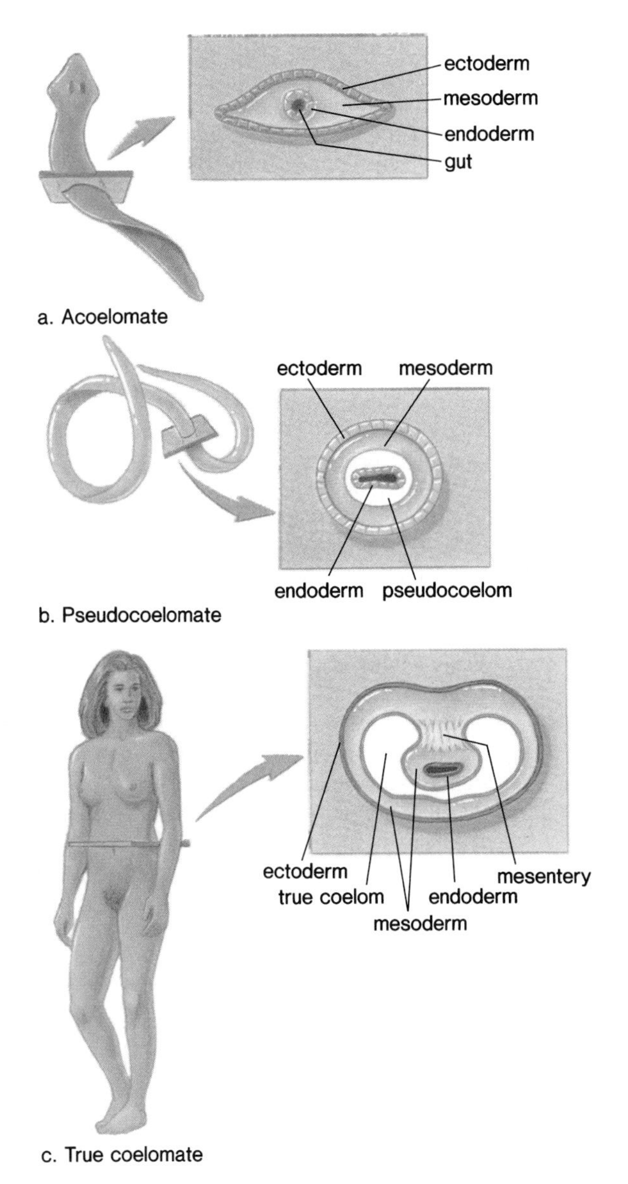

Figure 30.3

Types of symmetry. If an animal is radially symmetrical, any longitudinal cut, such as those shown, gives 2 identical halves. If an animal is bilaterally symmetrical, only the longitudinal cut shown gives 2 identical halves.

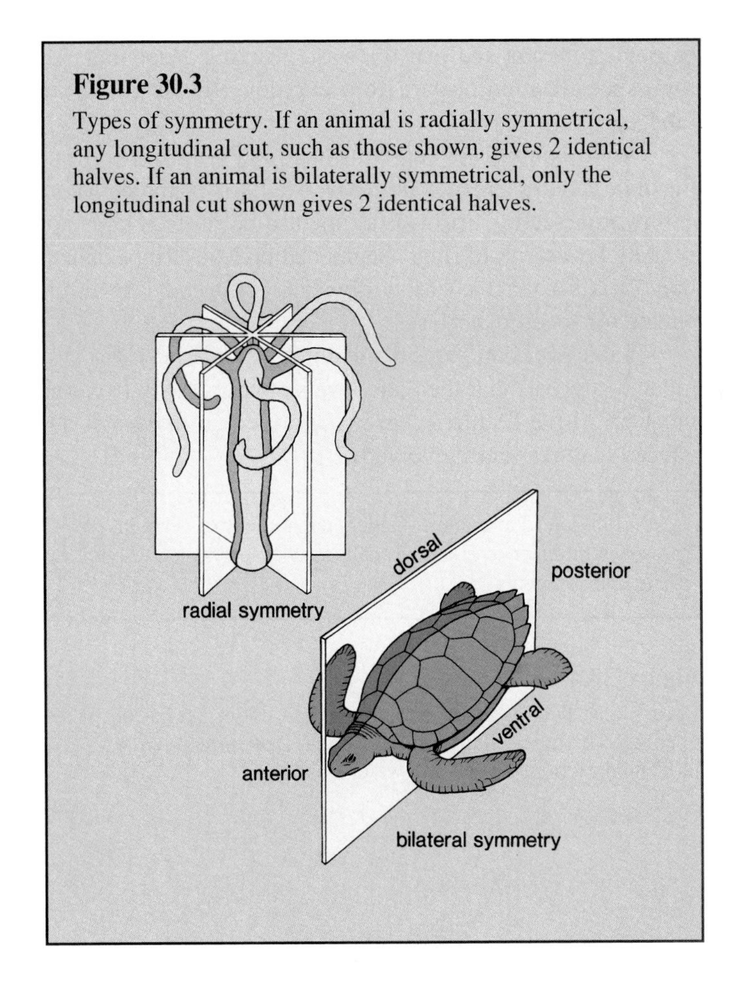

but not around the gut. All the rest of the animal phyla are true coelomates. They have a coelom that is completely lined with mesoderm. Besides serving as a location for the internal organs, a coelom can have other functions. In animals that lack blood vessels, coelomic fluids aid the movement of materials like nutrients or metabolic wastes. In animals that lack a skeleton, a fluid-filled coelom acts as a *hydrostatic skeleton*. It offers resistance to the contraction of muscles and permits flexibility. The animal can then change shape and perform a variety of movements.

Another characteristic that varies among animals is symmetry. **Asymmetry** means that the animal has no particular symmetry. **Radial symmetry** means that the animal is organized circularly, and just as with a wheel, any longitudinal cut yields 2 identical halves. **Bilateral symmetry** means that the animal has definite left and right halves; only one longitudinal cut down the center of the animal produces 2 equal halves (fig. 30.3). Radially symmetrical animals tend to be relatively inactive. Radial symmetry under these circumstances allows them to reach out in all directions. Bilaterally symmetrical animals tend to be active and to move forward with the anterior end. This end undergoes *cephalization,* the development of a head, which contains a brain and bears sense organs, so that the animal is acutely aware of its environment as it moves forward.

Two body plans (fig. 30.4) are observed in the animal kingdom: the **sac plan** and the **tube-within-a-tube plan.** Animals with the sac plan have an incomplete digestive system. It has only one opening, which is used both as an entrance for food and an exit for waste. Animals with the tube-within-a-tube plan have a complete digestive system, with a separate entrance for food and exit for waste. Having 2 openings allows specialization of parts to occur along the length of the tube.

Some animals are nonsegmented, and some have repeating units called segments. It is easy to tell, for example, that an earthworm (see fig. 30.14) is segmented because its body appears to be a series of rings. Segmentation leads to specialization of parts in that the various divisions of the body can differentiate for specific purposes.

Table 30.1 compares the features we have been discussing and suggests that there are levels of complexity from the most primitive features (earliest evolved) to the most advanced features (latest evolved).

> Classification of animals considers the number of germ layers, presence or absence of a true coelom, symmetry, type of body plan, and segmentation.

Figure 30.4
Body plans. The sac plan (incomplete digestive tract) has only one opening. The tube-within-a-tube plan has 2 openings. Arrows indicate direction of food and water flow.

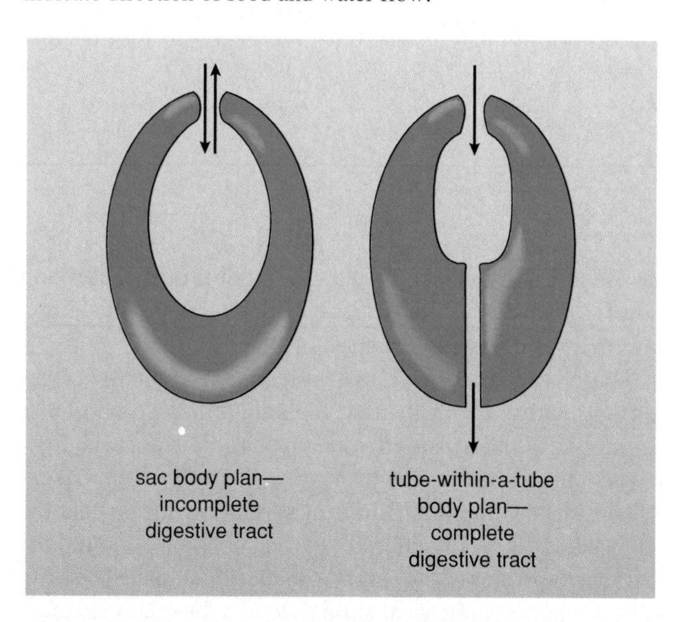

Classification

Kingdom Animalia
Eukaryotic, usually motile, multicellular organisms without cell walls or chlorophyll. Nutrition principally ingestive, with digestion in an internal cavity.

Invertebrates without a True Coelom

Phylum Porifera: sponges
Phylum Cnidaria: radially symmetrical marine animals (Portuguese man-of-war, jellyfishes, sea anemones, corals)
Phylum Platyhelminthes: flatworms
- Class Turbellaria: free-living flatworms
- Class Trematoda: parasitic flukes
- Class Cestoda: parasitic tapeworms

Phylum Nematoda: roundworms

Invertebrates with a True Coelom

Phylum Mollusca: soft-bodied, unsegmented animals
- Class Gastropoda: snails and slugs
- Class Cephalopoda: squid and octopuses
- Class Bivalvia: clams and mussels

Phylum Annelida: segmented worms
- Class Polychaeta: sandworms
- Class Oligochaeta: earthworms
- Class Hirudinea: leeches

Phylum Arthropoda: animals with chitinous exoskeleton and jointed appendages
- Class Crustacea: lobsters, crabs, barnacles
- Class Arachnida: spiders, scorpions, ticks
- Class Chilopoda: centipedes
- Class Diplopoda: millipedes
- Class Insecta: grasshoppers, termites, beetles

Phylum Echinodermata: marine, spiny, radially symmetrical animals (sea lilies, starfishes, brittle stars, sea urchins, sand dollars, sea cucumbers)
Phylum Chordata: supporting rod (notochord) at some stage; dorsal hollow nerve cord; pharyngeal pouches or slits
- Subphylum Urochordata: tunicates
- Subphylum Cephalochordata: lancelets
- Subphylum Vertebrata: vertebrates

Table 30.1
Classification Features

	Most Primitive	Primitive	Advanced	Most Advanced
Germ Layers	None	2	3	3
Level of Organization	None	Tissue	Organ	Organ system
Body Cavity	Acoelomate	Acoelomate	Pseudocoelom	True coelom
Symmetry	None	Radial	Bilateral	Bilateral with cephalization
Body Plan	None	Sac plan	Tube-within-tube plan	Tube-within-tube plan with specialization of parts
Segmentation	Nonsegmented	Nonsegmented	Segmented	Segmented with specialization of parts

Invertebrates without a True Coelom

Animals without a true coelom include sponges, cnidarians, flatworms, and roundworms. (These animals are all **invertebrates;** that is, they lack a backbone composed of vertebrae.) By studying the animals in this order, an increase in complexity that may reflect the order in which these animals evolved is observed. Nevertheless, it is difficult to determine their exact evolutionary relationships.

Sponges: Stay Put and Filter

Sponges branch early from the evolutionary tree (fig. 30.1), and it is possible they evolved separately from protozoa. Sponges are usually marine and abundant in warm ocean water near the coast. Some grow on rocks and are brightly colored, appearing almost lichenlike when seen from a distance. Sponges are often shaped like a vase (fig. 30.5). They all have a central cavity with an opening called an **osculum.** The body wall is perforated by numerous *pores* surrounded by contractile cells capable of regulating pore size. The wall also contains an outer layer of flattened *epidermal cells* and an inner layer of *flagellated collar cells.* The constant movement of the flagella produces water currents that flow through the pores into the central cavity and out through the osculum. Unlike any other animal, then, the main opening is used as an exit, not an entrance.

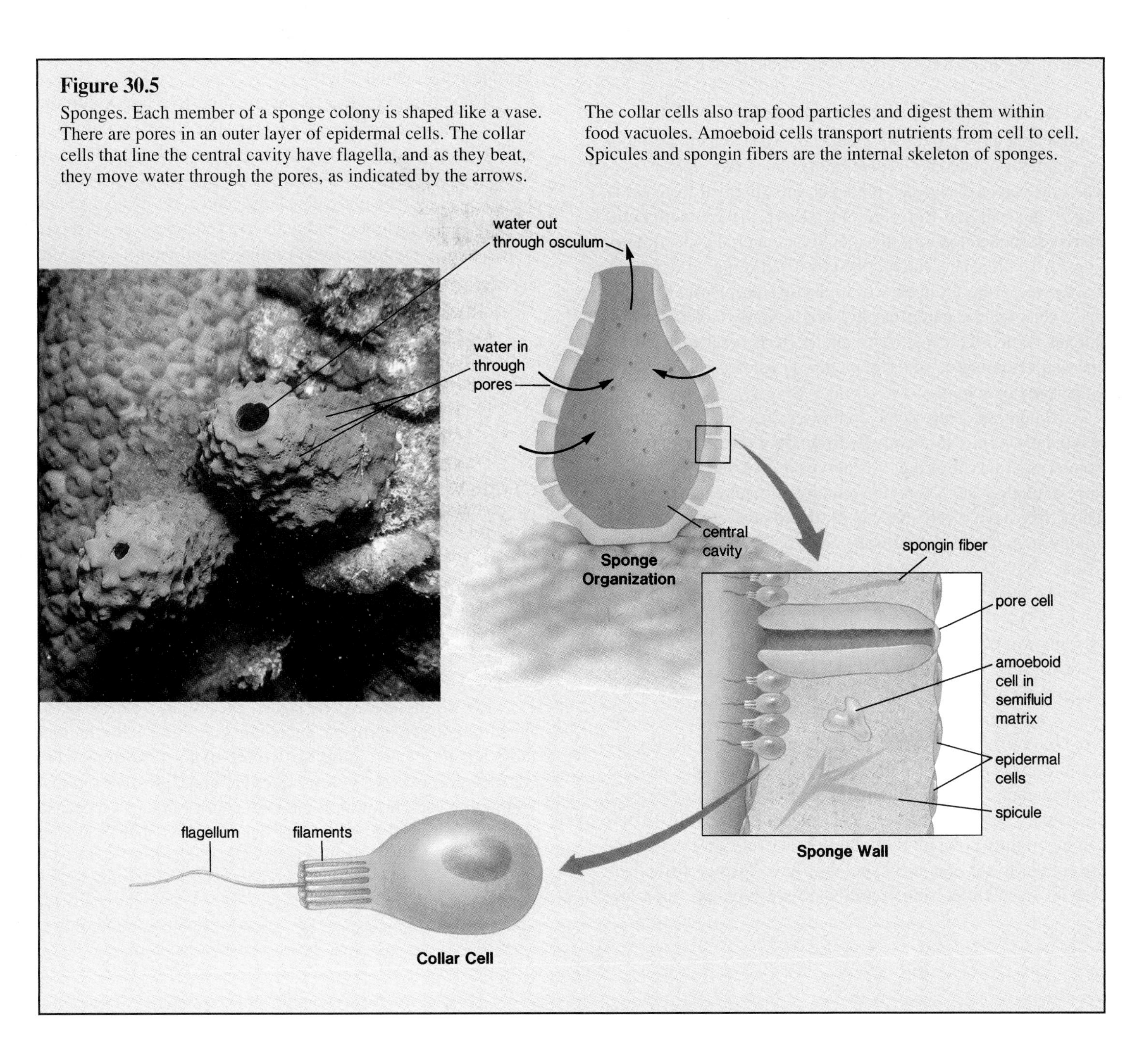

Figure 30.5
Sponges. Each member of a sponge colony is shaped like a vase. There are pores in an outer layer of epidermal cells. The collar cells that line the central cavity have flagella, and as they beat, they move water through the pores, as indicated by the arrows. The collar cells also trap food particles and digest them within food vacuoles. Amoeboid cells transport nutrients from cell to cell. Spicules and spongin fibers are the internal skeleton of sponges.

Amoeboid cells within the wall of a sponge act as a circulatory device for transporting nutrients from cell to cell. They also produce **spicules,** the needle-shaped structures that serve as the internal skeleton of sponges. Sponges are classified according to type of spicule; some have spicules of calcium carbonate, others of silicon dioxide, and still others also have spongin fibers and spicules. Spongin is a type of protein.

Sponges are **sessile filter feeders.** Adults stay in one place, and the food they acquire enters through the pores. Microscopic food particles carried in the water are either engulfed by the collar cells—which digest them in food vacuoles—or passed to the amoeboid cells for digestion.

Sponges reproduce sexually as well as asexually by budding. Budding results in whole colonies, which can be quite large. Like many less-specialized organisms, sponges are capable of **regeneration,** or growth of a whole from a small part.

Cnidarians: Jellyfish, Sea Anemones

Cnidarians have a sac body plan (fig. 30.4), which accounts for their former name—*coelenterate,* meaning *hollow gut.* The outer layer of the sac, the epidermis (derived from ectoderm), is separated from the inner layer, the **gastrodermis** (derived from endoderm), by a jellylike material called **mesoglea.** All cnidarians have specialized stinging cells called *cnidocytes,* from a Greek word meaning *sea nettles.* Within these cells are the **nematocysts,** long, spirally coiled, hollow threads. When the trigger of a cnidocyte is touched, the discharged nematocyst, which sometimes contains poison, stuns either prey or enemy.

Cnidarians have radial symmetry (fig. 30.3), and there is typically a ring of *tentacles* surrounding the mouth region. Some cnidarians, referred to as **polyps** or **hydroids** (fig. 30.6), have a tubular shape, with the mouth region directed upward. Those that have a bell shape, with the mouth region directed downward, are called **medusae** or jellyfish. A polyp is adapted to a sessile life, while a medusa is adapted to a floating existence. At one time, both body forms may have been a part of the life cycle of all cnidarians, because today in the life cycle of some we see an alternation of these forms. Polyps produce medusae, and the medusae, which produce eggs and sperm, disperse the species.

Cnidarians are quite diverse. Jellyfish often have an alternation of form life cycle, but the prominence of the 2 generations is not equal—the medusa is the primary stage and the polyp remains quite small and inconspicuous. *Sea anemones* are solitary polyps, often very large and with thick walls. They can be brightly colored, resembling beautiful flowers. **Corals** are similar to sea anemones, but they have calcium carbonate ($CaCO_3$) skeletons. Some corals are solitary, but most are colonial, with either flat and rounded or upright and branching colonies. The slow accumulation of coral skeletons has formed reefs in the Caribbean Sea and South Pacific. An ancient **coral reef** that now lies beneath Texas is the source of petroleum for that state.

Hydras: Typical Cnidarian

Hydras (fig. 30.6) are likely to be found attached to underwater plants or rocks in most lakes and ponds. The hydra body is a small, tubular polyp about 7.5 mm in length. Although hydras usually remain in one place, they may glide along on their base, or foot, or even move rapidly by somersaulting. Nerve cells form a connecting network throughout the mesoglea known as the **nerve net.** The nerve net makes contact with the outer epidermal cells and the inner gastrodermal cells, which are capable of contracting. The tentacles seize prey and stuff it down into the central cavity.

The gastrodermal cells secrete digestive juices into the central cavity, where digestion begins. Digestion is completed within food vacuoles after gastrodermal cells engulf small pieces of prey. Nutrient molecules are passed by diffusion to the other cells of the body. The large central cavity also makes it possible for gastrodermal cells to exchange gases directly with a watery medium. Because the central cavity carries on digestion and acts as a vascular system by distributing food and gases, it is called a **gastrovascular cavity.**

Hydras only exist as polyps, and there are no medusae. Still, they can reproduce sexually or asexually. During sexual reproduction, an ovary or a testis develops in the body wall. Like the sponge, a whole cnidarian can regenerate from a small piece. When conditions are favorable, hydras produce small outgrowths, or *buds,* that pinch off and begin to live independently.

Flatworms: Three Germ Layers

Flatworms have a combination of primitive and advanced features (table 30.1). They are nonsegmented acoelomates (fig. 30.2*a*) that have the sac body plan (fig. 30.4). Free-living flatworms are bilaterally symmetrical (fig. 30.3), have a well-developed nervous system, and have undergone cephalization. This combination makes them efficient predators.

Flatworms have 3 germ layers. The presence of mesoderm in addition to ectoderm and endoderm gives bulk to the animal and leads to greater complexity. Free-living flatworms have muscles and excretory, reproductive, and digestive organs. The worms lack respiratory and circulatory organs—because the body is flat, diffusion alone is adequate for the passage of oxygen and other substances from cell to cell.

There are 3 classes of flatworms: one is free living and 2 are parasitic.

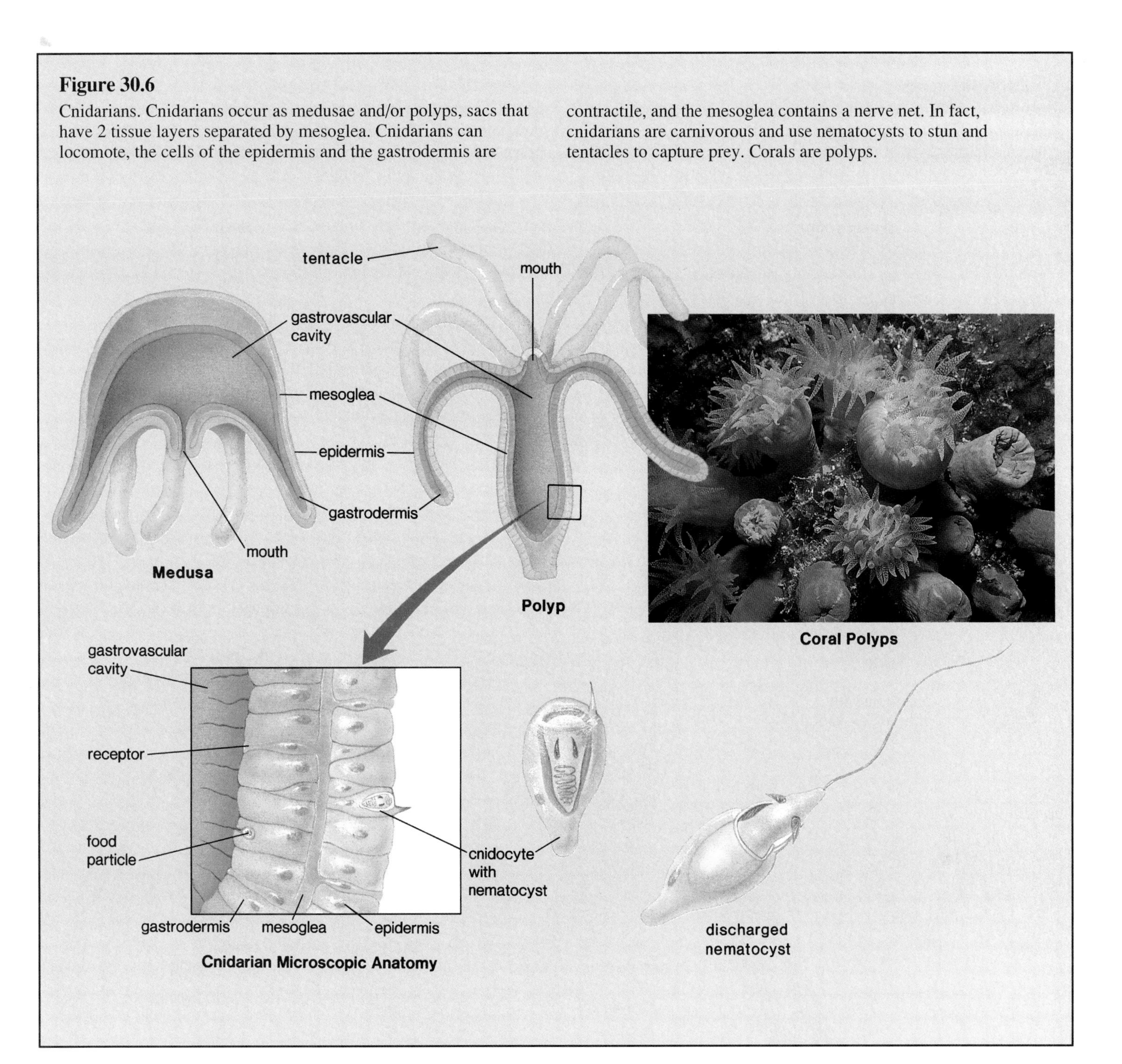

Figure 30.6

Cnidarians. Cnidarians occur as medusae and/or polyps, sacs that have 2 tissue layers separated by mesoglea. Cnidarians can locomote, the cells of the epidermis and the gastrodermis are contractile, and the mesoglea contains a nerve net. In fact, cnidarians are carnivorous and use nematocysts to stun and tentacles to capture prey. Corals are polyps.

Figure 30.7

Flatworm anatomy as exemplified by a planarian. Flatworms are hermaphrodites; they have both male and female reproductive organs. The pharynx leads to the digestive tract, which has 3 main branches. Excretory canals have flame cells (enlarged drawing), whose beating cilia draw in fluid that is excreted by way of excretory pores. The nervous system has a ladder appearance because cross-branches stretch between longitudinal fibers that extend the length of the animal.

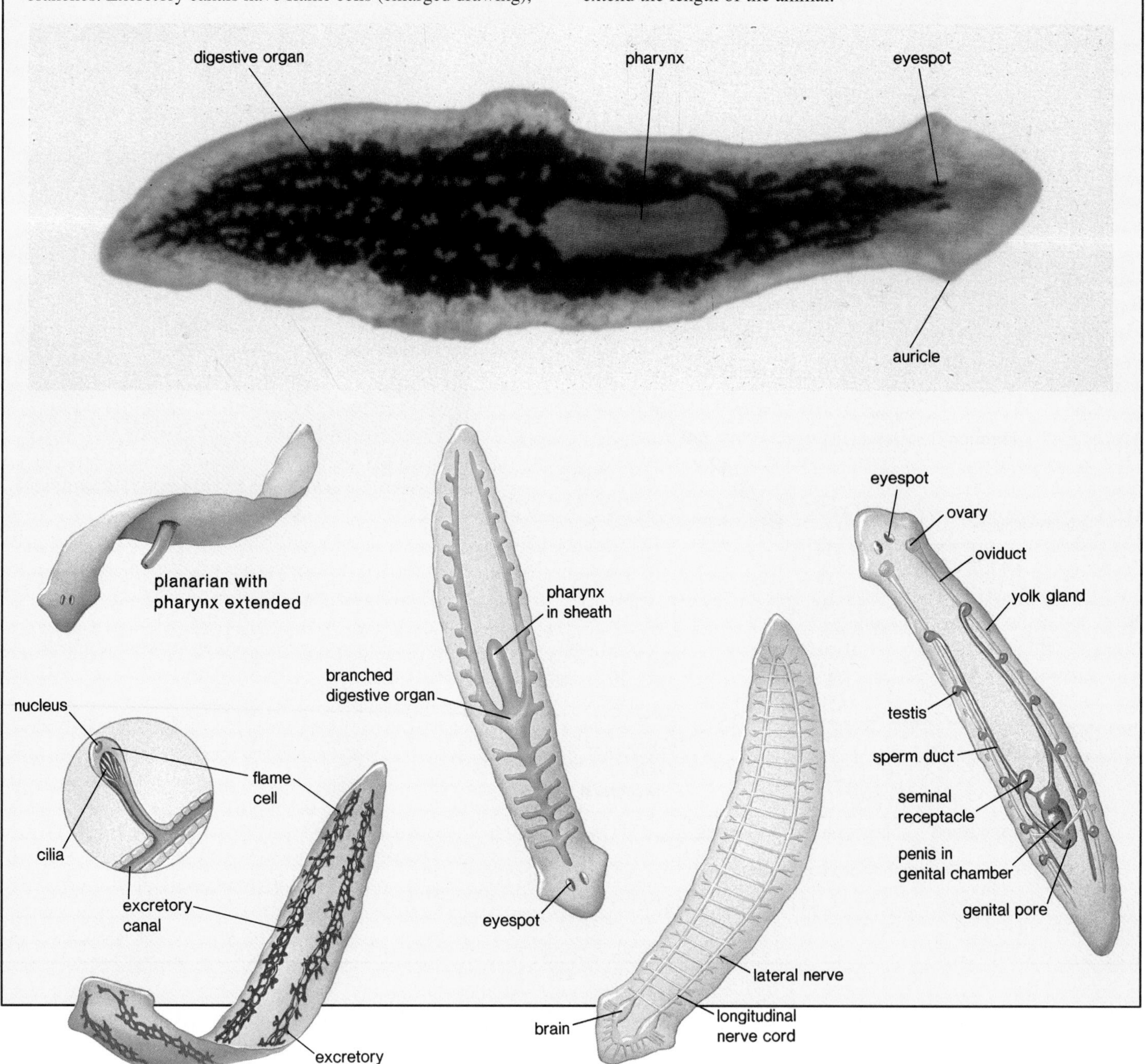

Planarians: Grow New Heads and Tails

Freshwater planarians (fig. 30.7) are small (several mm–several cm), literally flat worms. Some tend to be colorless; others have brown or black pigmentation. Planarians live in lakes, ponds, streams, and springs, where they feed on small living or dead organisms, such as worms and crustacea.

Because planarians live in fresh water, water tends to enter the body by osmosis. They have an excretory organ that serves primarily to rid the body of excess water. A network of interconnecting canals extends through much of the body. The beating of cilia in the **flame cells** (so named because the beating of the cilia reminded some early investigator of the flickering of a flame) keeps the water moving toward the excretory pores.

Table 30.2
Free-Living Worms versus Parasitic Worms

	Free-Living	Parasitic	
	Planarians	***Tapeworms***	***Flukes***
Body Wall	Ciliated epidermis	Glycocalyx covers tegument	Glycocalyx covers tegument
Cephalization	Yes—Eyespots and auricles	No—Scolex with hooks and suckers	No—Oral sucker (and hooks)
Nervous Connections	Nerves and brain	Reduced	Reduced
Digestive Organ	Branched	Absent	Reduced
Reproductive Organs	Hermaphroditic	Greatly increased in volume	Increased in volume
Larva	Absent	Present	Present

The digestive tract is incomplete because it has only one opening. The pharynx, which extends through the mouth, sucks food particles into the gastrovascular cavity, which has 3 branches.

Planarians have a **ladder-type nervous organ.** There is a small anterior brain and 2 lateral nerve cords joined by cross-branches. Planarians have undergone cephalization—aside from a brain, there are light-sensitive organs (the eyespots) and chemosensitive organs (located on the auricles). They have well-developed muscles, and their ciliated epidermis allows them to glide along a film of mucus.

Planarians are **hermaphrodites;** they possess both male and female sex organs. The worms practice cross-fertilization: the penis of one is inserted into the genital pore of the other, and there is a reciprocal transfer of sperm. The fertilized eggs hatch in 2–3 weeks as tiny worms. Like sponges and cnidarians, planarians can regenerate. If a worm is cut crosswise, each piece grows a new head or a new tail, as appropriate.

Parasitic Flatworms: Lost Features

There are 2 classes of parasitic flatworms: **flukes** and **tapeworms.** The structure of both these worms illustrates the modifications that occur in parasitic animals (table 30.2). Cephalization does not occur in these nonpredatory animals; instead, the anterior end carries hooks and suckers for attachment to the host. The parasite absorbs nutrients from the digestive tract of the host, and its own digestive system is essentially absent. The tegument, a specialized body wall resistant to host digestive juices, is covered by the glycocalyx, a mucopolysaccharide coating. The extensive development of the reproductive system, with the production of millions of eggs, may be associated with difficulties in dispersing the species. Both parasites utilize a *secondary,* or intermediate, *host* to transport the species from primary host to primary host. The *primary host* is infected with the sexually mature adult; the secondary host contains the larval stage or stages.

Both flukes and tapeworms cause serious illnesses in humans. Different fluke species infect the digestive tract, the bile duct, blood, and the lungs. Because the intermediate host can be a fish or a snail, humans can acquire the fluke in food. Active penetration can also result in infection. *Schistosomiasis,* a serious infection caused by blood flukes that enter the body by active penetration, is seen primarily in the Far East and Africa. Tapeworms can be acquired by eating poorly cooked beef or pork, as depicted in figure 30.8. A long series of proglottids, or segments, is found behind the **scolex** of the adult worm. Newly formed proglottids contain a full set of both male and female sex organs, but as proglottids mature, they become a sac filled with developing eggs. When these eggs are taken up by a pig (one tapeworm species) or a cow (another species), they develop into a bladder worm, which encysts in muscle. Here a *cyst* means a small, hard-walled structure that contains an immature worm.

Roundworms: Tube within a Tube

Roundworms, such as nematodes, are nonsegmented—they have a smooth, outside body wall. These worms, which are generally colorless and less than 5 cm in length, occur almost anywhere—in the sea, in fresh water, and in the soil—in such numbers that thousands of them can be found in a small area.

Roundworms possess 2 anatomical features not seen before: a tube-within-a-tube body plan (fig. 30.4) and a body cavity. With a tube-within-a-tube body plan, the digestive tract is complete; there is both a mouth and an anus. The body cavity is a **pseudocoelom** (fig. 30.2*b*), or a cavity incompletely lined with mesoderm. The fluid-filled pseudocoelom provides space for the development of organs, substitutes for a circulatory system by allowing easy passage of molecules, and provides a type of skeleton. Worms in general do not have an internal or external skeleton, but they do have a hydrostatic skeleton, a fluid-filled interior that supports muscle contraction and enhances flexibility.

When roundworms are analyzed according to table 30.1, they have features associated with advanced animals. Roundworms are thought to be a side branch to the main evolution of animals; they may have arisen from an ancestor that also produced true coelomates.

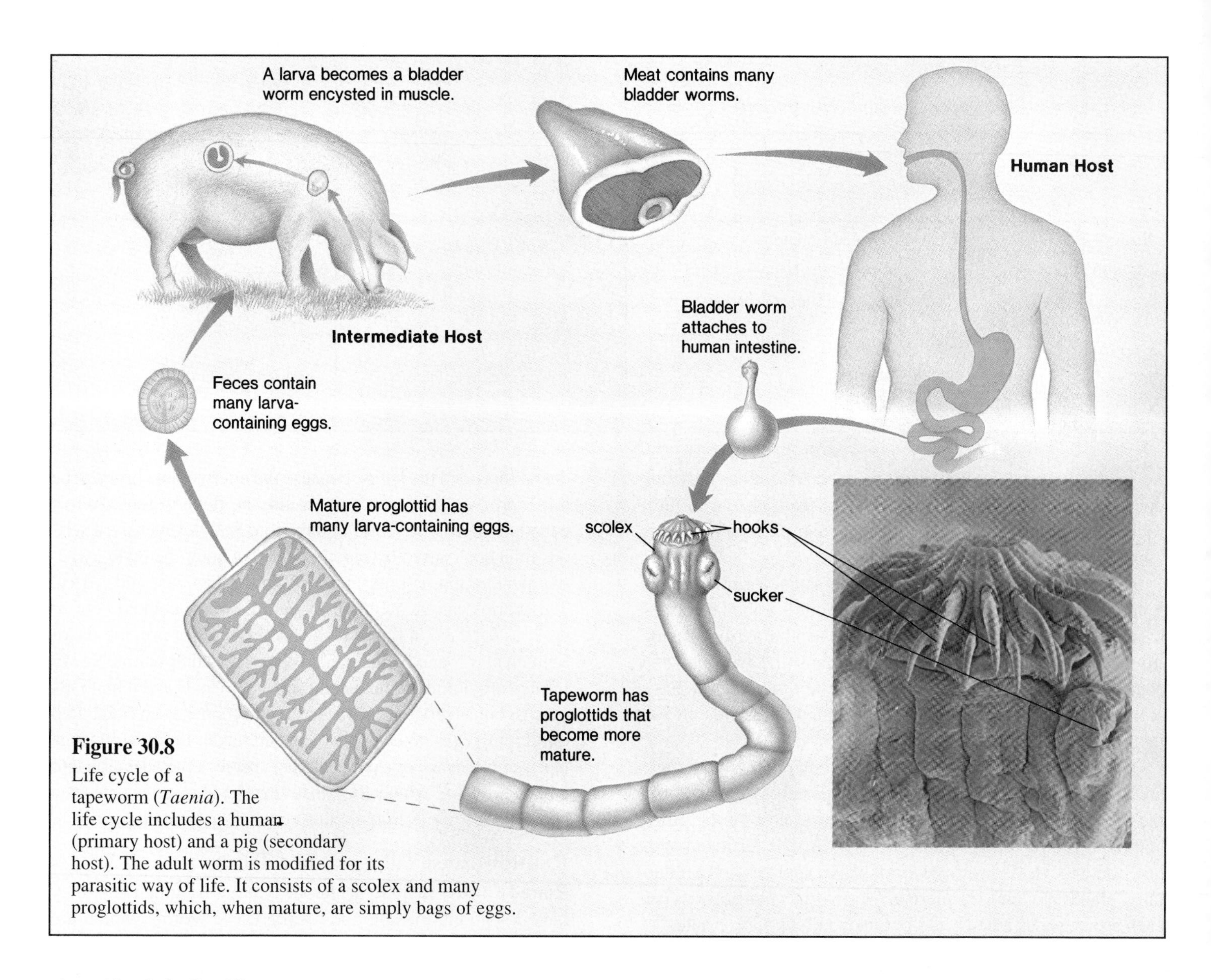

Figure 30.8
Life cycle of a tapeworm (*Taenia*). The life cycle includes a human (primary host) and a pig (secondary host). The adult worm is modified for its parasitic way of life. It consists of a scolex and many proglottids, which, when mature, are simply bags of eggs.

Ascarids: Only One Host

Most roundworms are free living, but a few are parasitic. *Ascaris,* a large parasitic roundworm, is often studied as an example of this phylum.

Ascaris (fig. 30.9) females tend to be larger (20–35 cm in length) than males. Both sexes move by means of a characteristic whiplike motion because only longitudinal muscles lie next to the body wall.

The *Ascaris* life cycle begins when larvae within a protective covering are swallowed. The worms then escape from the covering and burrow through the host's intestinal wall. Making their way through the organs of the host, they move from the intestine to the liver, the heart, and then the lungs. Growth takes place in the lungs, and after about 10 days, the larvae migrate up the windpipe to the throat, where they are swallowed, allowing them to once again reach the intestine. Then the mature worms mate, and the female produces larva-containing eggs, which pass out with the feces.

Other Parasites: Serious Business

Trichinosis is a serious infection of humans caused by *Trichinella* roundworms. Humans contract the disease when they eat rare pork containing encysted larvae. After maturation, the female adult burrows into the wall of the host's small intestine and produces live offspring, which are carried by the bloodstream to the skeletal muscles, where they encyst (fig. 30.9). The symptoms include muscular pain, weakness, fever, and anemia.

Elephantiasis is caused by a roundworm called the filarial worm, which utilizes the mosquito as a secondary host. Because the adult worms reside in lymphatic vessels, fluid collection is impeded, and the limbs of an infected human can swell to an enormous size. When a mosquito bites an infected person, it transports larvae to new hosts.

Other roundworm infections are more common in the United States. Children frequently acquire a pinworm infection, and hookworm is seen in the southern states. A hookworm

Figure 30.9
Nematodes. Roundworms, such as *Ascaris,* lack segmentation—the outer wall is smooth. They have a pseudocoelom and a complete digestive tract, with both a mouth and an anus. The sexes are separate—this is a male roundworm. Some roundworms, such as *Trichinella,* are parasites that encyst in muscle. Such an infection in humans is called trichinosis.

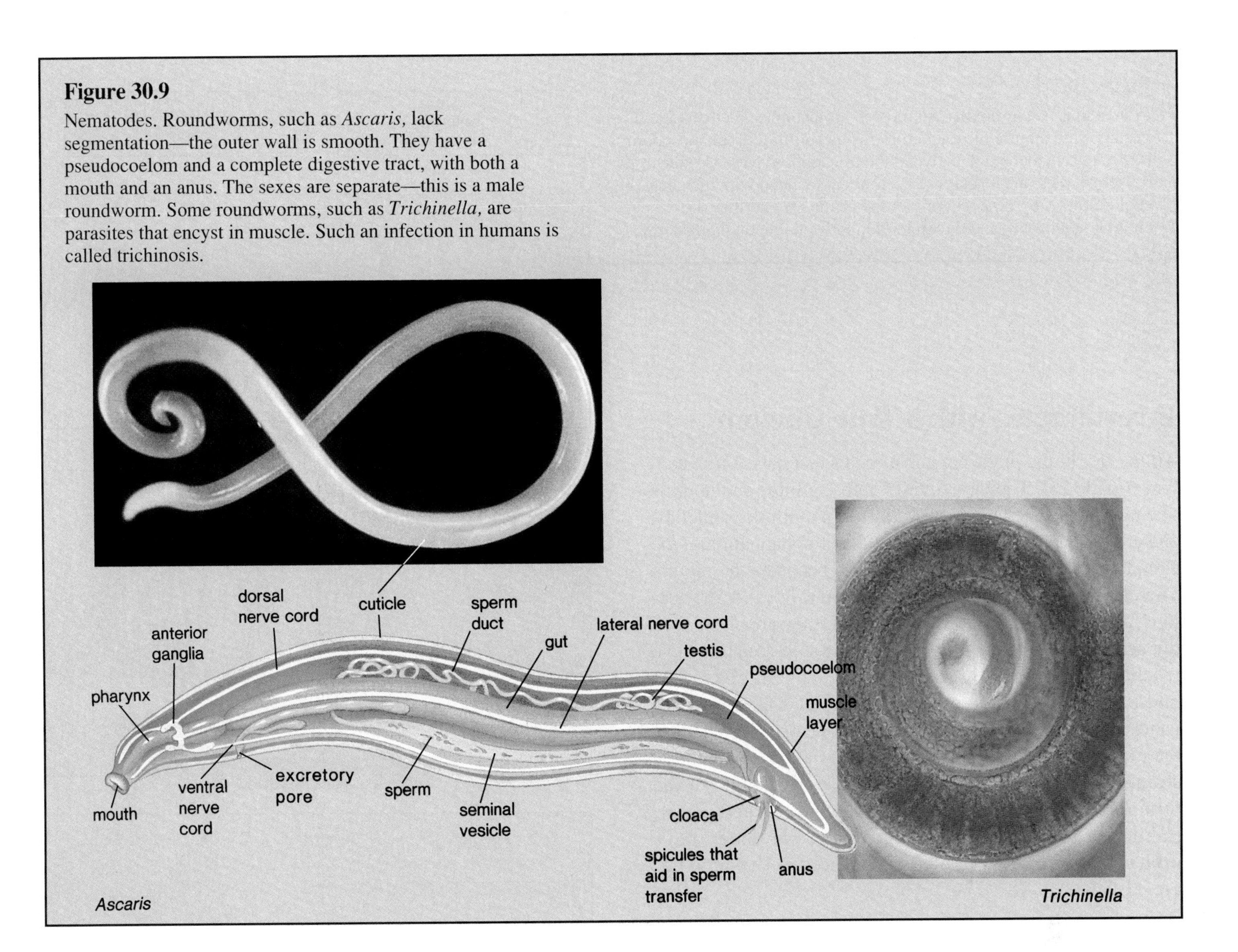

Table 30.3
Comparison of Animals without a True Coelom

	Sponges	Cnidarians	Flatworms	Roundworms
Germ Layers	None	2	3	3
Level of Organization	Cell	Tissue	Organ	Organ
Body Cavity	None	None	None	Pseudocoelom
Symmetry	Radial or none	Radial	Bilateral	Bilateral
Body Plan	None	Sac	Sac	Complete

infection can be very debilitating because the worms attach to the intestinal wall and feed on blood. Good hygiene, proper disposal of sewage, and cooking meat thoroughly usually protect a populace from parasitic roundworms.

Notice that we have completed a survey of the animal phyla that make up the main trunk of the evolutionary tree of animals (fig. 30.1). Note that complexity has increased and that roundworms have advanced features (table 30.3). Do 30.1 Critical Thinking, found at the end of the chapter.

Figure 30.10

Protostomes versus deuterostomes. In protostomes, the blastopore becomes the mouth, the coelom forms by splitting of the mesoderm (they are schizocoelomates), and the trochophore larva is typical. In deuterostomes, the blastopore becomes associated with the anus, the coelom forms by outpocketing of the primitive gut (they are enterocoelomates), and the dipleurula larva is found among some.

Protostomes
mollusks
annelids
arthropods

Deuterostomes
echinoderms
chordates

blastopore
mouth
primitive gut
anus
blastopore
anus
primitive gut
mouth
fate of blastopore

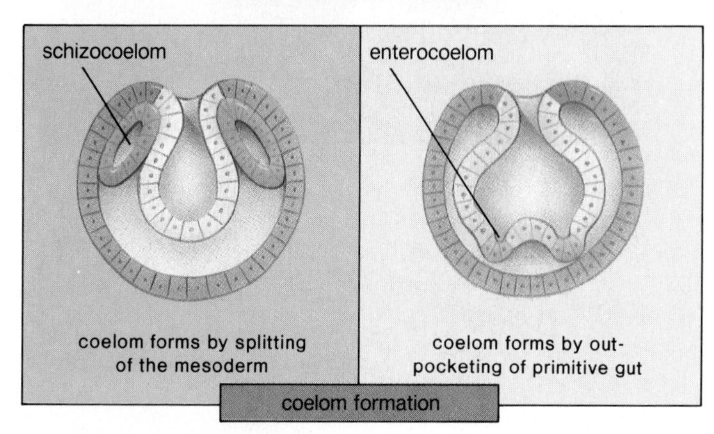

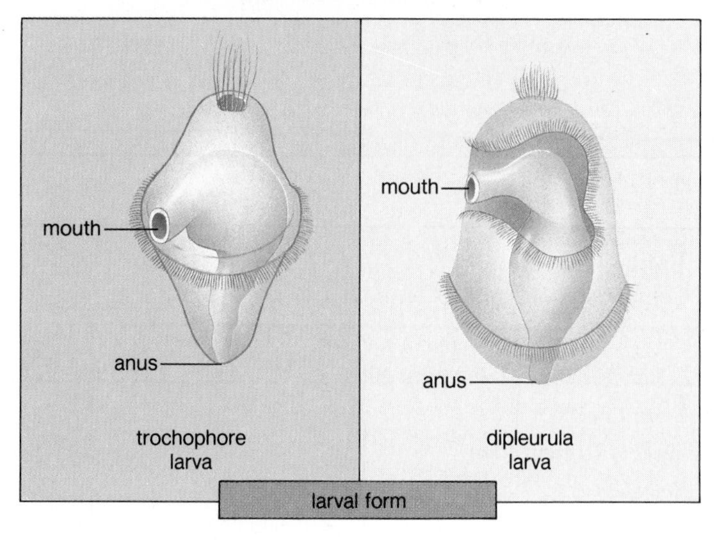

Invertebrates with a True Coelom

All the rest of the phyla contain animals that have a true coelom (fig. 30.2*c*). They are divided into 2 groups on the basis of embryological evidence (fig. 30.10). In mollusks, annelids, and arthropods, the blastopore (the site of invagination of the endoderm during development, p. 406) becomes the mouth. Therefore, they are called the **protostomes.** Because the coelom forms by splitting of the mesoderm, they are also called the **schizocoelomates.** Marine mollusks and annelids have **trochophore** larvae (top-shaped with a band of cilia at the midsection). In echinoderms and chordates, the blastopore becomes associated with the anus and a second opening becomes the mouth. Therefore, these animals are called the **deuterostomes.** Because the coelom forms by outpocketing of the primitive gut, they are also called the **enterocoelomates.** Echinoderms and certain invertebrate chordates have **dipleurula** larvae (bands of cilia are placed as shown in figure 30.10).

Mollusks: Mass, Foot, and Mantle

Mollusks are a very large and diversified group containing many thousands of living and extinct forms. However, all forms of mollusks have a body composed of at least 3 distinct parts.

1. **Visceral mass:** the soft-bodied portion that contains internal organs.
2. **Foot:** the strong, muscular portion used for locomotion.
3. **Mantle:** the membranous or sometimes muscular covering that envelops but does not completely enclose the visceral mass. The *mantle cavity* is the space between the 2 folds of the mantle. The mantle may secrete a shell.

In addition to these 3 parts, many mollusks show cephalization: they have a head region with eyes and other sense organs.

The division of the body into distinct areas seems to have facilitated diversification because there are many different types of mollusks adapted to various ways of life (fig. 30.11). Molluscan groups can be distinguished by a modification of the foot. In the *gastropods* (*belly-footed*), including nudibranchs, conchs, and snails, the foot is ventrally flattened, and the animal moves by muscle contractions that pass along the foot. While nudibranchs, also called sea slugs, lack a shell, conchs and snails have a coiled shell in which the visceral mass spirals. Some types of snails are adapted to life on land.

Figure 30.11
Molluscan diversity. **a.** A scallop, with sensory tentacles extended between the valves. Humans eat only the single, large muscle that holds the 2 halves of the shell together. **b.** A chiton has a flattened foot and a shell that consists of 8 articulating valves. It normally creeps along feeding on algae that it scrapes from the substratum, but it will roll up into a ball when dislodged. **c.** While most mollusks are marine, some types of snails are adapted to living on land, and their mantle tissue is capable of gas exchange with air. During copulation, each snail inserts a penis into the mantle cavity of its partner.

a.

b.

c.

For example, their mantle is richly supplied with blood vessels and functions as a lung when air is moved in and out through respiratory pores.

In *cephalopods* (*head-footed*), including octopuses and squid, the foot has evolved into tentacles about the head. Aside from the tentacles, which seize prey, cephalopods have powerful beaks and a radula (toothy tongue) to tear prey apart. Cephalization aids these animals in recognizing prey and in escaping enemies. The eyes are superficially similar to those of vertebrates—they have a lens and a retina with photoreceptors. However, its construction is so different from the vertebrate eye that we believe the eye actually evolved twice—once in the mollusks and once in the vertebrates. The brain is formed from a fusion of ganglia, and nerves leaving the brain supply various parts of the body. An especially large pair of nerves controls the rapid contraction of the mantle, allowing these animals to move quickly by a jet propulsion of water. Rapid movement and the secretion of a brown or black pigment from an ink gland help cephalopods to escape their enemies. Octopuses have no shell, and squid have only a remnant of one concealed beneath the skin.

Table 30.4
Comparison of Clam to Squid

	Clam	Squid
Food Gathering	Filter feeder	Active predator
Skeleton	Heavy shell for protection	No external skeleton
Circulation	Open	Closed
Cephalization	None	Marked
Locomotion	"Hatchet" foot	Jet propulsion
Nervous System	3 separate ganglia	Brain and nerves

In *bivalves*, such as clams, oysters, and scallops, the foot is laterally compressed. They are called bivalves because there are 2 parts to the shell. Notice in table 30.4 that a clam is adapted to a less active life and a squid is adapted to a more active life.

Figure 30.12

Clam anatomy. The shell and the mantle have been removed from one side. Trace the path of food from the incurrent siphon, past the gills, to the mouth, the stomach, the intestine, the anus, and the excurrent siphon. Locate the 3 ganglia: anterior, foot, and posterior. The heart lies in the reduced coelom. Do clams have an open or closed circulatory system?

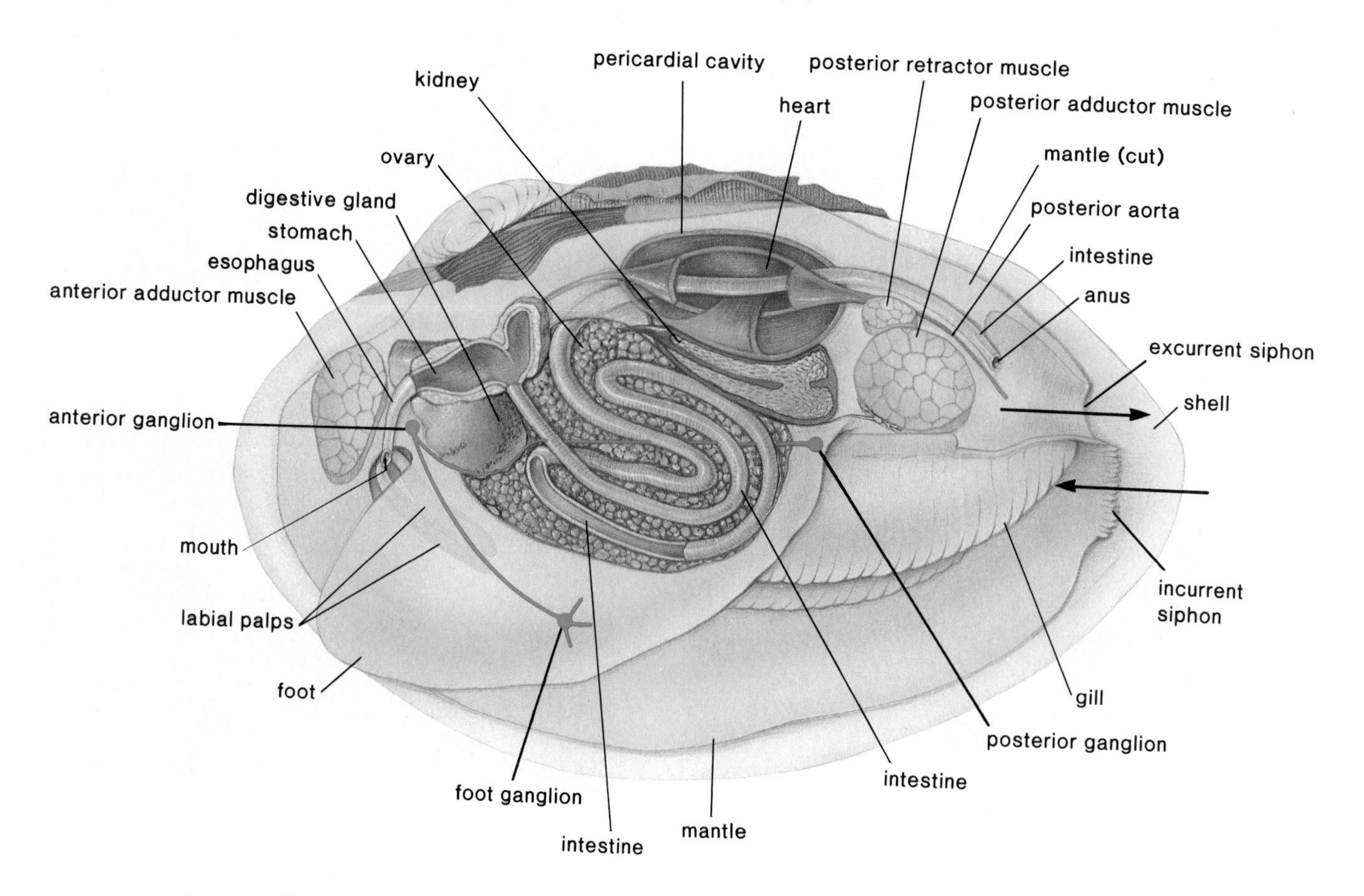

Clams: Filter Feeders

In a clam, such as the freshwater clam *Anodonta,* the shell, secreted by the mantle, is composed of protein and calcium carbonate with an inner layer of *mother-of-pearl.* If a foreign body is placed between the mantle and the shell, pearls form as concentric layers of shell are deposited about the particle.

Within the mantle cavity, the gills (fig. 30.12) hang down on either side of the visceral mass, which lies above the foot. **Gills** are composed of vascularized, highly convoluted, thin-walled tissue specialized for gas exchange.

The heart of a clam lies just below the hump of the shell within the pericardial cavity, the only remains of the coelom. Therefore, the *coelom is reduced.* The heart pumps blood into vessels that lead to the various organs of the body. Within the organs, however, blood flows through spaces, or sinuses, rather than through vessels. This is an *open circulatory system* because the blood is not contained within blood vessels all the time. This type of circulatory system can be associated with a relatively inactive animal because it is an inefficient means of transporting blood throughout the body. An active animal needs to have oxygen and nutrients transported quickly to rapidly working muscles, while an inactive animal is able to survive with a sluggish system of transport.

The nervous system of a clam (fig. 30.12) is composed of *3 pairs of ganglia* (anterior, foot, and posterior), which are all connected by nerves. Clams lack cephalization. The foot projects anteriorly from the shell, and by expanding the tip of the foot and pulling the body after it, the clam moves forward.

The clam is a filter feeder, meaning that it feeds on small particles that have been filtered from the water environment. Food particles and water enter the mantle cavity by way of the *incurrent siphon,* a posterior opening between the 2 valves. Mucous secretions cause smaller particles to adhere to the gills, and ciliary action sweeps them toward the mouth. Many inactive animals are filter feeders because this method of feeding does not require rapid movement.

The digestive system of the clam includes a mouth, a stomach, and an intestine, which coils about in the visceral mass and then goes right through the heart before ending in an anus. The anus empties at an *excurrent siphon,* which lies just

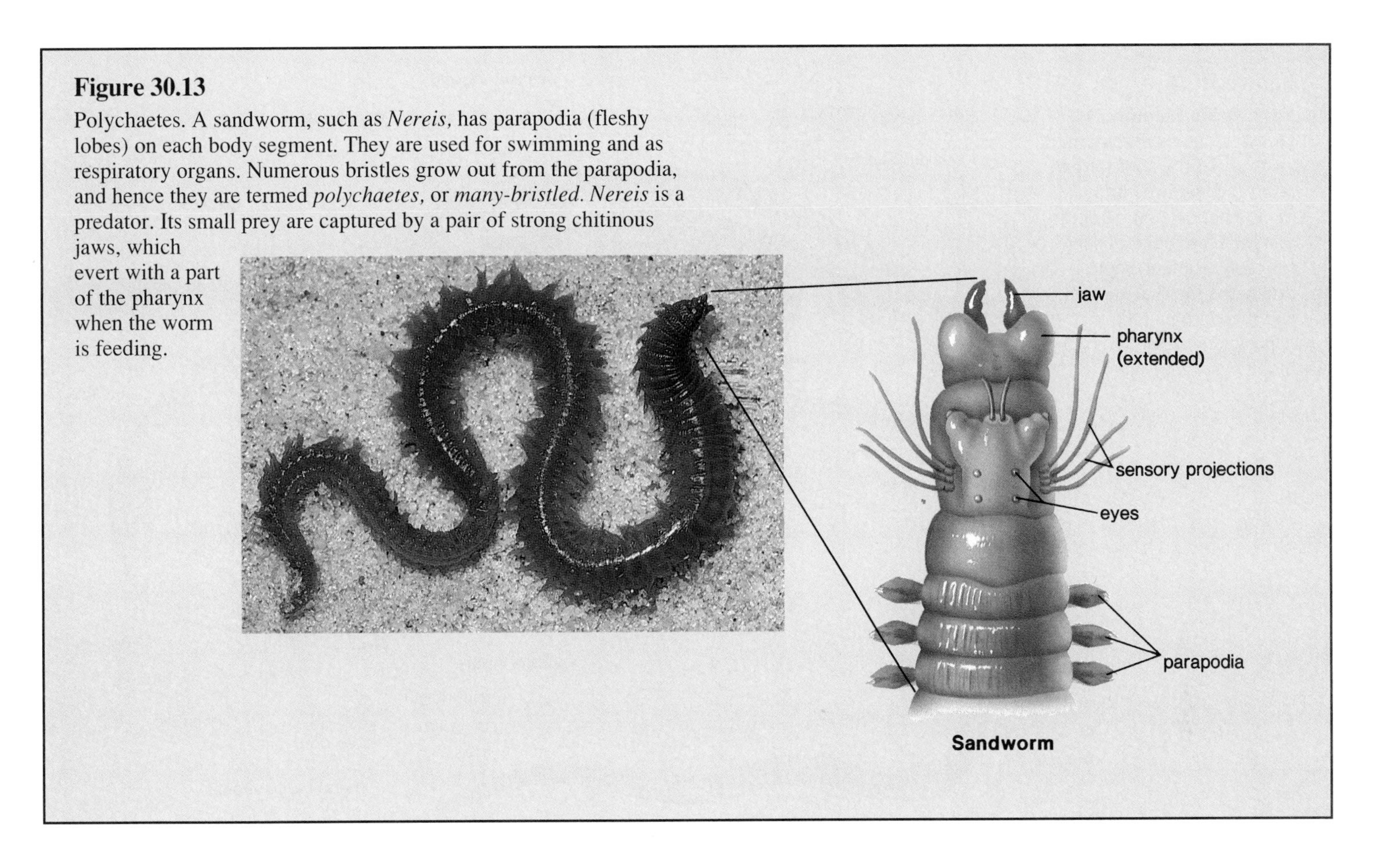

Figure 30.13
Polychaetes. A sandworm, such as *Nereis,* has parapodia (fleshy lobes) on each body segment. They are used for swimming and as respiratory organs. Numerous bristles grow out from the parapodia, and hence they are termed *polychaetes,* or *many-bristled. Nereis* is a predator. Its small prey are captured by a pair of strong chitinous jaws, which evert with a part of the pharynx when the worm is feeding.

above the incurrent siphon. There is also an accessory organ of digestion called a digestive gland. The 2 excretory kidneys in the clam (fig. 30.12), which lie just below the heart, remove waste from the pericardial cavity for excretion into the mantle cavity.

The sexes are separate. The gonad can be found about the coils of the intestine. While all clams have some type of larval stage, only marine clams have a trochophore larva. The presence of the trochophore larva (fig. 30.10) among some mollusks indicates a relationship to the annelids, some members of which also have this type of larval stage.

Annelids: Segmentation Appears

Annelids are the segmented worms. The body has obvious rings, and even the well-developed coelom is partitioned by membranous septa. While the earthworm is usually studied as an example of annelids, the marine sandworm *Nereis* may be more representative. A sandworm has a pair of fleshy lobes, the **parapodia,** on each body segment. These are used for swimming and as respiratory organs because their expanded surface area allows for exchange of gases. Numerous chitinous bristles grow from the parapodia; therefore, these worms are called *polychaetes,* or *many bristled.* A sandworm preys on crustacea and other small animals captured by a pair of strong, chitinous *jaws,* which evert when *Nereis* is feeding. Cephalization has occurred in the sandworm, as evidenced by a head region bearing sense organs, including eyes and sensory projections (fig. 30.13).

Earthworms: Inactive, Too

Earthworms (fig. 30.14) live in damp soil, where a moist body wall can readily be used for gas exchange. Cephalization has not occurred in the earthworm, and it is not a predator. It feeds on leaves or any other organic matter, living or dead, that can be conveniently taken into its mouth along with soil. Food drawn into the mouth by the action of the muscular pharynx is stored in a crop and is ground up in a thick, muscular gizzard. Digestion and absorption occur in a long intestine, the dorsal surface of which is expanded by a **typhlosole.** The typhlosole provides additional surface area for absorption into the blood. Notice that the tube-within-a-tube plan has allowed specialization of the digestive system to occur.

Each body segment has 4 pairs of **setae,** which are slender bristles. After setae are inserted into the dirt, the body is pulled forward. The worm can move and change its shape because there are both circular and longitudinal muscles in the body wall. Locomotion is aided by the fluid-filled coelomic compartments, which act as a hydrostatic skeleton. The nervous

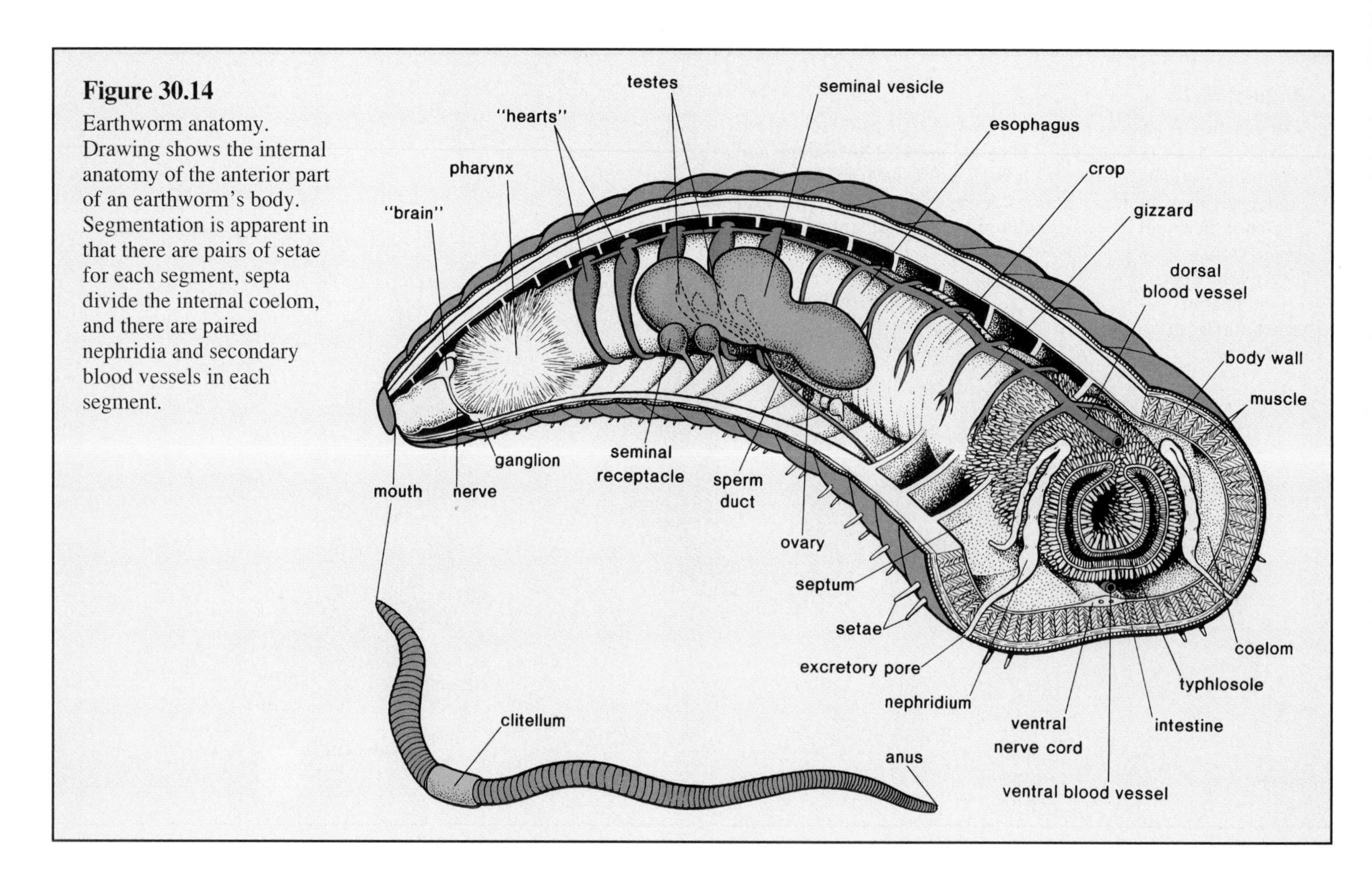

Figure 30.14
Earthworm anatomy. Drawing shows the internal anatomy of the anterior part of an earthworm's body. Segmentation is apparent in that there are pairs of setae for each segment, septa divide the internal coelom, and there are paired nephridia and secondary blood vessels in each segment.

system (fig. 30.14) consists of an anterior dorsal ganglionic mass, or a brain, and a long *ventral nerve cord with ganglia* and lateral nerves in each segment.

The excretory system consists of paired nephridia (sing., **nephridium**) (fig. 30.14), or coiled tubules, in each segment. A nephridium begins with a ciliated funnel that collects coelomic fluid, and ends at an excretory pore in the body wall. Between the 2 openings is a convoluted region where waste material is removed from the blood vessels about the nephridium.

The earthworm has an extensive *closed circulatory system.* Hemoglobin-containing blood moves anteriorly in a dorsal blood vessel and then is pumped by 5 pairs of hearts into a ventral blood vessel. As the ventral blood vessel takes blood toward the posterior regions of the worm's body, it gives off branches in every segment.

The worms are *hermaphrodites.* The clitellum, a smooth girdle about each worm, secretes mucus, and then there is a reciprocal transfer of sperm between 2 worms. After separation, the mucus becomes a slime tube, in which the eggs are fertilized and miniature worms develop. There is no larval stage.

The annelids have the most obvious segmentation of any animal phylum. Table 30.5 shows the many ways an earthworm's anatomy displays *segmentation.*

Table 30.5
Segmentation in the Earthworm

1. Body rings
2. Coelom divided by septa
3. Setae on each segment
4. Ganglia and lateral nerves in each segment
5. Nephridia in each segment
6. Branch blood vessels in each segment

Arthropods: Distinct Head, Outer Skeleton

The **arthropods** have more species (approximately 900,000) than any other group of animals and therefore are often said to be the most successful of all the animals. The phylum includes animals adapted for living in water, such as crayfish, lobsters, and shrimps, and animals adapted for living on land, such as spiders, insects, centipedes, and millipedes (fig. 30.15).

Arthropods have an external skeleton containing **chitin,** a strong, flexible polysaccharide. The skeleton serves many functions—protection, attachment for muscles, and

Figure 30.15

Arthropod diversity. There are 5 major classes of arthropods. **a.** Class Crustacea is represented by this crab, with a large carapace and 5 pairs of legs. The first pair are pinching claws. **b.** Class Arachnida is represented by this land-dwelling scorpion, with 4 pairs of legs, poisonous claws, and stinging tail. **c.** Class Chilopoda is represented by this centipede, a carnivorous animal, with a pair of appendages on every segment. **d.** Class Diplopoda is represented by this millipede, a scavenger that seems to have 2 pairs of appendages on each segment because every 2 segments are fused. Class Insecta is represented by **(e)** a swallowtail butterfly, with wings, and **(f)** an ant, with clearly defined head, thorax, and abdomen. Insects have 3 pairs of legs.

a.

b.

d.

c.

f.

e.

prevention of desiccation if the animal lives on land. The phylum name, *arthropod,* means *jointed appendage.* Such appendages are adaptive for locomotion on land. Because the skeleton does not grow, arthropods **molt,** or shed the skeleton periodically.

Specialization of parts is readily seen in the arthropod because the body does not have a series of like segments; rather, the body usually has 3 parts—head, thorax, and abdomen. The occurrence of cephalization is apparent. The sense organs include **antennae** (or feelers) and eyes. The eyes are of 2 types: **compound eye** and **simple eye.** The compound eye is not seen in any other phylum. It is composed of many complete visual units grouped in a composite structure. Each visual unit contains a separate lens and a light-sensitive cell. In the simple eye, a single lens covers many light-sensitive cells. Compound eyes are small and lightweight. Vision is not acute, but this type of eye sees details of movement that we cannot make out. Flies, as we know, are very difficult to catch!

The coelom, so well developed in the annelids, is highly *reduced* in adult arthropods. Arthropods, like most mollusks, have an open circulatory system. Instead of a coelomic cavity, there is a **hemocoel,** or blood cavity, consisting of vessels and sinuses (open spaces), where the blood flows about the organs. The dorsal heart keeps the blood moving through the hemocoel.

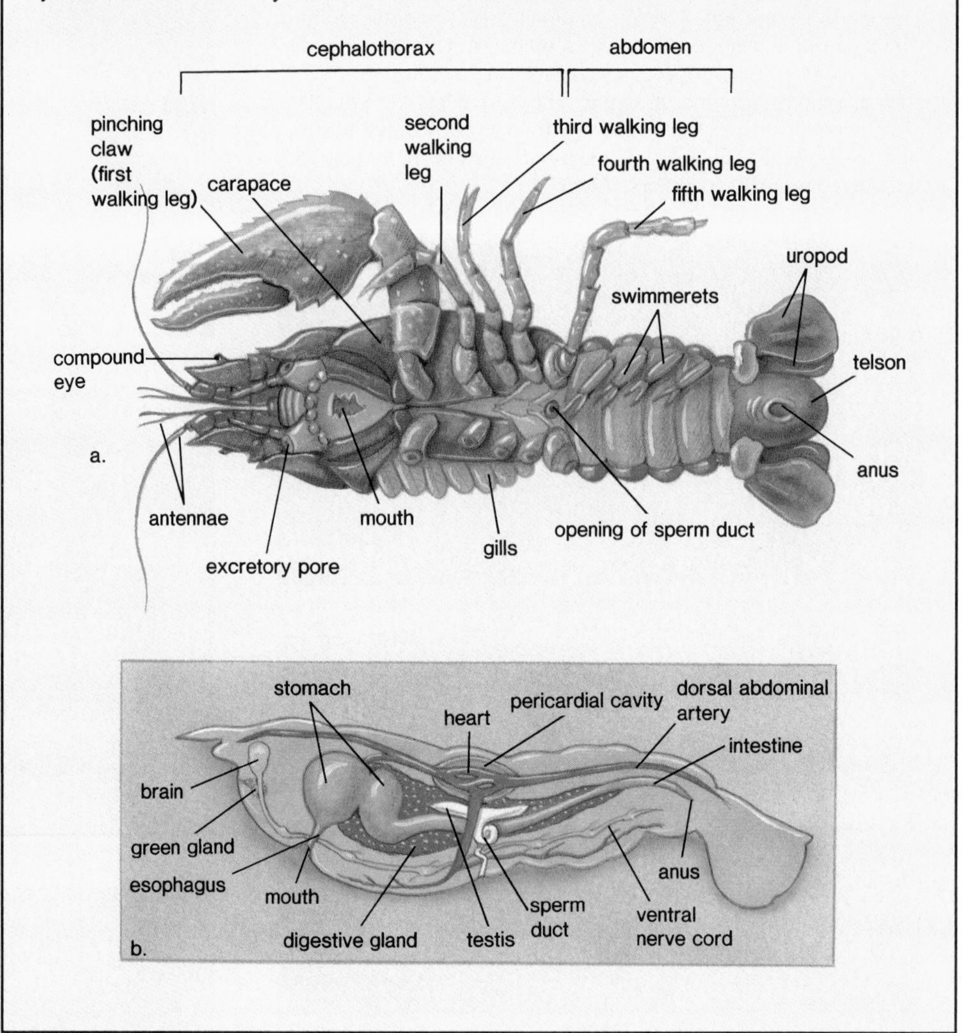

Figure 30.16

Anatomy of a crayfish. **a.** Externally, it is possible to observe the jointed appendages, including the swimmerets, the walking legs, and the claws. These appendages, plus a portion of the carapace, have been removed from the right side so that the gills are visible. **b.** Internally, the parts of the digestive system are particularly visible. The circulatory system can also be clearly seen. Note the ventral solid nerve cord.

Crayfish: Adapted to Water

Crayfish, as well as lobsters, shrimps, copepods, and crabs, are crustacea. Figure 30.16*a* gives a view of the external anatomy of a crayfish. The head and the thorax are fused into a **cephalothorax,** which is covered on the top and sides by a nonsegmented **carapace.** The abdominal segments are marked off clearly. The head region bears a pair of stalked compound eyes and 2 pairs of antennae. Chitinous jaws and mouthparts are also present. In the thorax, the appendages include accessory mouthparts and 5 pairs of *walking legs* (the first pair are chelipeds or *pinching claws*). The abdominal segments are equipped with *swimmerets,* small paddlelike structures. The last 2 segments bear the *uropods* and the *telson,* which make up a fan-shaped tail.

Ordinarily, the crayfish lies in wait for prey. It faces out from an enclosed spot with the claws extended and the antennae moving about. The claws seize any small animal, dead or alive, that happens by and carry it to the mouth. When a crayfish moves about, it generally crawls slowly but may swim rapidly backwards by using the heavy abdominal muscles and the tail.

Gills, for gas exchange (fig. 30.16*a*), lie above the walking legs, protected by the carapace. The crayfish has *blue blood* containing the pigment hemocyanin, which aids in the transport of oxygen. The digestive system (fig. 30.16*b*) includes a stomach, the anterior portion of which has a *gastric* *mill*—chitinous teeth that grind coarse food—and the posterior portion of which acts as a filter to prevent coarse particles from entering the digestive glands, where absorption takes place. *Green glands,* lying in the head region anterior to the esophagus, excrete metabolic wastes through a duct that opens ventrally at the base of the antennae. The nervous system is quite similar to that of the earthworm. There is a brain, and a *ventral nerve cord* passes posteriorly. Along the length of the nerve cord, fused instead of segmental ganglia give off lateral nerves.

The sexes are separate in the crayfish, and the gonads are located just ventral to the pericardial cavity. In the male, a coiled sperm duct opens to the outside at the base of the fifth walking leg. Sperm transfer is accomplished by the modified first 2 swimmerets of the abdomen. In the female, an oviduct opens at the base of the third walking leg. A stiff fold between the bases of the fourth and fifth pair serves as a seminal receptacle. Following fertilization, the eggs are attached to the swimmerets of the female.

Grasshoppers: Adapted to Land

Grasshoppers are insects. Insects compose the largest animal group—both in number of species and in number of individuals. Most insects have wings, which help them escape enemies, find food, and disperse the species.

Every system of a grasshopper (fig. 30.17) is adapted to life on land. There are *3 pairs* of legs, and the hindmost of these pairs is suited to jumping. There are 2 pairs of wings; the forewings are tough and leathery, and when folded back at rest, they protect the broad, thin hindwings. On the lateral surface, the first abdominal segment bears a large **tympanum** on each side for the reception of sound waves. The posterior region of the exoskeleton in the female has 2 pairs of projections that form an ovipositor, which is used to dig a hole in which eggs are laid.

Figure 30.17

Anatomy of a female grasshopper. **a.** Externally, the hard skeleton prevents loss of water. There are spiracles, openings in the skeleton that admit air into tracheae (air tubes for respiration). The tympanum uses air waves for sound reception, and the hopping legs and the wings are for locomotion. The ovipositor deposits eggs in soil. **b.** Internally, the digestive system is adapted to digesting grass. The Malpighian tubules excrete a solid nitrogenous waste (uric acid). A seminal receptacle receives sperm from the male, which has a penis. Internal fertilization prevents the gametes from drying out.

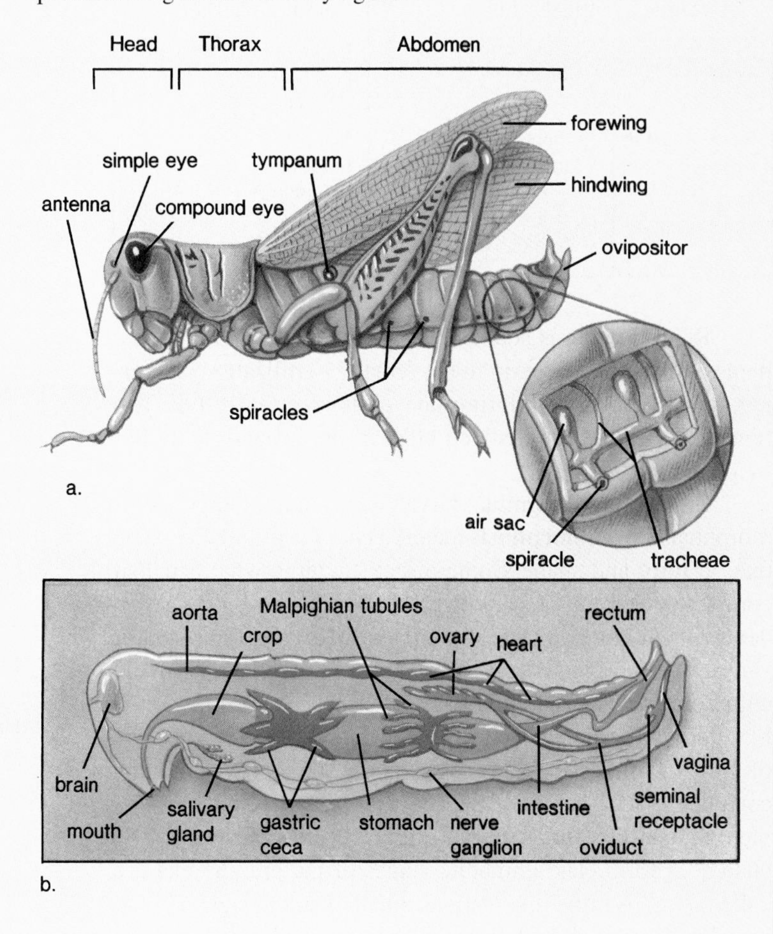

The digestive system (fig. 30.17*b*) is suitable for a plant diet. In the mouth, food is broken down mechanically by mouthparts and enzymatically by salivary secretions. Food is temporarily stored in the crop before passing into the gizzard, where it is finely ground. Digestion is completed in the stomach, and nutrients are absorbed into the hemocoel from outpockets called gastric ceca. The excretory system consists of **Malpighian tubules,** which extend into the hemocoel and collect nitrogenous wastes that are concentrated and excreted into the digestive tract. The formation of a solid nitrogenous waste, namely uric acid, conserves water.

The respiratory system begins with openings in the exoskeleton called **spiracles.** From here, the air then enters small tubules called **tracheae.** The tracheae branch and rebranch, finally ending in moist areas, where the actual exchange of gases takes place. The movement of air through this complex of tubules is not a passive process; air is pumped through by a series of bladderlike structures (air sacs), which are attached to the tracheae near the spiracles. Air enters the anterior 4 spiracles and exits by the posterior 6 spiracles.

The circulatory system contains a slender, tubular heart that lies against the dorsal wall of the abdominal exoskeleton and passes blood into the hemocoel, where it circulates before finally returning to the heart. The *blood is colorless*—it lacks a respiratory pigment because the tracheal system transports gases.

Table 30.6
Comparison of Crayfish to Grasshopper

	Crayfish	Grasshopper
Locomotion	Legs and uropods	Hopping legs and wings
Respiration	Gills	Tracheae
Excretion	Liquid waste by way of green glands	Solid waste by way of Malpighian tubules
Circulation	Blue blood	Colorless blood
Nervous System	Cephalization with antennae	Cephalization with one pair of antennae and a tympanum
Reproduction	In male, modified swimmerets pass sperm to female; in female, eggs are attached to swimmerets	Penis in male, ovipositor in female; metamorphosis

Reproduction is adapted to life on land. The male has a penis, which passes sperm to the female. Fertilization is internal, usually occurring during late summer or early fall. The female deposits the fertilized eggs in the ground using her ovipositor.

Insects have wormlike larval stages and undergo metamorphosis. **Metamorphosis** means a *change,* usually a drastic one, in form and shape. Some insects undergo what is called complete metamorphosis, in which case they have 4 stages of development: *egg, larval stage, pupal stage,* and finally *adult stage.* Metamorphosis occurs during the pupal stage, when the animal is enclosed within a hard covering. The animal that is best known for metamorphosis is the butterfly, the larval stage of which is the caterpillar and the pupal stage of which is the cocoon; the adult is the butterfly. Grasshoppers undergo incomplete metamorphosis, which is a gradual rather than a drastic change in form. The immature stages of the grasshopper are called nymphs rather than larvae, and they are recognizable as grasshoppers even though they differ somewhat in shape and form. Metamorphosis is controlled by hormones.

Table 30.6 compares the crayfish to the grasshopper to illustrate how one is adapted to the water and the other is adapted to the land.

This completes our survey of the protostomes, animals in which the mouth is the first opening observed during embryological development.

The segmentation first observed in annelids has led to specialization of body parts in arthropods. Like annelids, arthropods have a ventral nerve cord, but only arthropods have undergone marked cephalization. Besides being segmented, arthropods have a rigid and jointed skeleton. These features facilitate complex body movements and are adaptive to the land environment. We will see that this winning combination is seen not only in the arthropods but also in the chordates (vertebrates).

Echinoderms: Starfish, Urchins, Sand Dollars

The echinoderms include only marine animals—starfish, sea urchins, sea cucumbers (fig. 30.18), feather stars, sea lilies, and sand dollars. An echinoderm begins life as the bilateral dipleurula larva and then becomes a *radially symmetrical* adult, with a body plan based on *5 parts.* Their other unique feature is a **water vascular system,** which is used as a means of locomotion. They also have a calcium carbonate **endoskeleton,** the projecting spines of which give the phylum its name, which means *spiny skin.*

Starfish: Five Arms

Starfish, sometimes called sea stars, are commonly found along rocky coasts. A starfish has a five-rayed body plan with an *oral* (mouth) side and an *aboral* (anus) side (fig. 30.18). The oral side is actually the underside, and the aboral side is the upper side. Various structures project through the body wall on the aboral side: (1) spines that project from the endoskeletal plates; (2) pincerlike structures called *pedicellarie,* which keep the surface free of small particles; and (3) skin gills, which serve for respiratory exchange. The mouth is located on the oral surface, where each of the 5 arms has a groove lined by little **tube feet.**

Starfish feed on mollusks. When a starfish attacks a clam, it arches its body over the shell, and by the concerted action of the tube feet, forces the clam open. Then it everts the cardiac portion of its stomach through the mouth to digest the contents of the clam. The narrower pyloric portion of the stomach is connected to a short intestine. The anus opens on the aboral, or upper, side of the animal.

The 5 arms of a starfish project from a central disk. Each arm contains a well-developed coelom, a pair of large digestive glands, which secrete powerful enzymes into the pyloric portion of the stomach, and gonads (either male or female), which open on the aboral surface by very small pores. The nervous system consists of a central nerve ring, which supplies radial nerves to each arm. A light-sensitive eyespot is at the tip of each arm.

Coelomic fluid, circulated by ciliary action, performs many of the normal functions of a circulatory system; the water vascular system is purely for locomotion. Water enters this system through a structure on the aboral side called the **sieve plate,** or madreporite. From there, it passes through a short canal, called the *stone canal,* to a *ring canal,* which surrounds the mouth. From the ring canal, *5 radial canals* extend into the arms. From the radial canals, many lateral canals extend into the tube feet. One lateral canal goes to each tube foot, where it ends in the *ampulla.* When the ampulla contracts, the water is forced into the tube foot, expanding it. When the foot touches a surface, the center is withdrawn, giving it suction so that it can adhere to the surface. By alternating the expansion and contraction of the tube feet, a starfish moves slowly along.

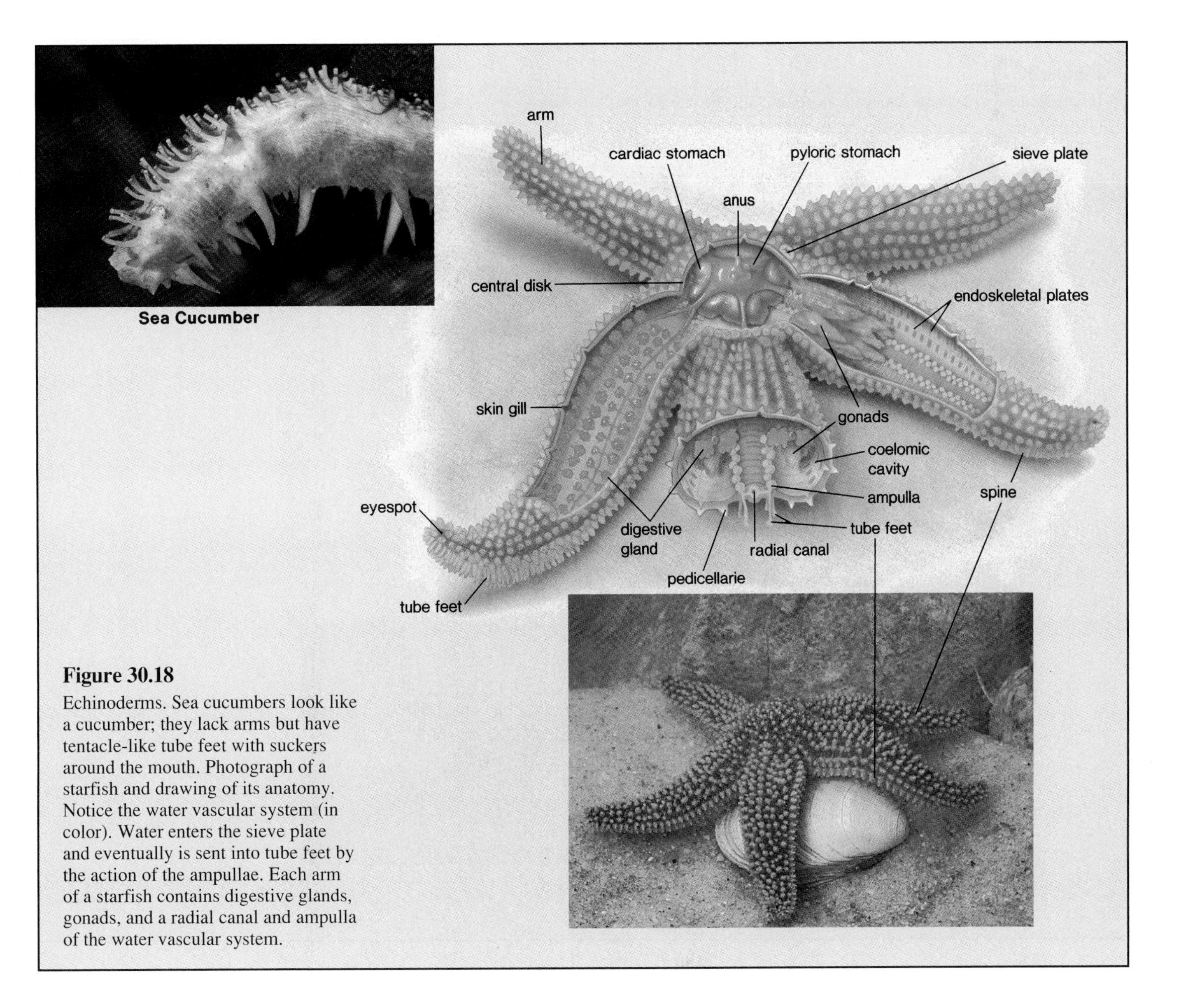

Figure 30.18

Echinoderms. Sea cucumbers look like a cucumber; they lack arms but have tentacle-like tube feet with suckers around the mouth. Photograph of a starfish and drawing of its anatomy. Notice the water vascular system (in color). Water enters the sieve plate and eventually is sent into tube feet by the action of the ampullae. Each arm of a starfish contains digestive glands, gonads, and a radial canal and ampulla of the water vascular system.

Relationship of Echinoderms to Chordates

Based on embryological evidence, echinoderms are believed to be closely related to the invertebrate chordates. (The invertebrate chordates lack the vertebral column of vertebrates.) Most likely, these 2 groups share a common bilateral ancestor, but the echinoderms became radially symmetrical as an adaptation to a fairly inactive way of life. The invertebrate chordates, on the other hand, gave rise to the **vertebrates,** which are not only segmented but also have a jointed skeleton.

Chordates: 3 Basic Characteristics

Among the chordates are those animals with which we are most familiar, including human beings. All members of this phylum are observed to have the following 3 basic characteristics at some time in their life history.

1. A dorsal supporting rod called a **notochord,** which is replaced by the vertebral column in adult vertebrates.
2. A *dorsal hollow nerve cord,* which lies above the notochord. By hollow, it is meant that the cord contains a canal that is filled with fluid. In vertebrates, the nerve cord, more often called the spinal cord, is protected by the vertebrae.
3. *Pharyngeal pouches,* which are seen only during embryological development in most vertebrate groups. In the invertebrate chordates, fishes, and amphibian larvae, the pharyngeal pouches become functioning gills. Water passing into the mouth and the pharynx goes through the gill slits, which are supported by gill bars.

Figure 30.19

Invertebrate chordates. **a.** Tunicate anatomy. Gill slits are the only chordate feature retained by the adult. **b.** Lancelet anatomy. This animal retains all 3 chordate characteristics as an adult.

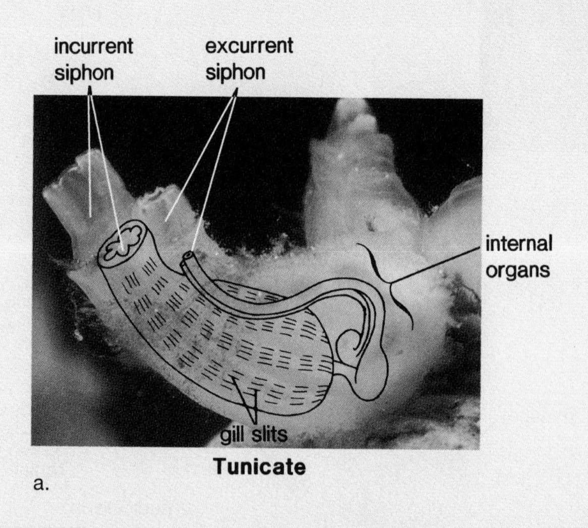

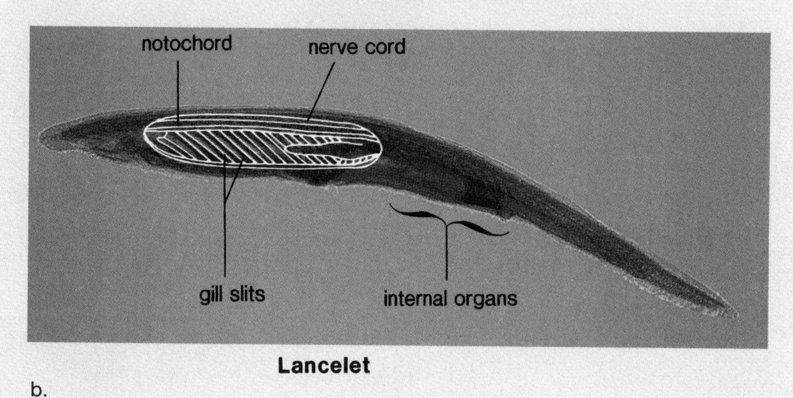

Invertebrate Chordates: Three Basics, but No Backbone

The tunicates and the lancelets are the *invertebrate chordates.* A **tunicate,** or sea squirt (fig. 30.19*a*), appears to be a thick-walled, squat sac with 2 openings, an incurrent siphon and an excurrent siphon. The oversized pharynx has numerous gill slits, which function in filter feeding. Gill slits are the only chordate feature retained by the adult. The larva of the tunicate, however, has a tadpole shape and possesses the 3 chordate characteristics. It has been suggested that such a larva may have become sexually mature without developing the other adult tunicate characteristics. If so, it may have evolved into a fishlike vertebrate.

A **lancelet** (fig. 30.19*b*) is a filter-feeding chordate that has the *3 chordate characteristics as an adult.* In addition, segmentation is present, as witnessed by the fact that the muscles are segmentally arranged and the nerve cord gives off periodic branches.

Classification

Subphylum Vertebrata: vertebrates

Class Agnatha: jawless fishes (lampreys, hagfishes)
Class Chondrichthyes: cartilaginous fishes (sharks, rays)
Class Osteichthyes: bony fishes
Class Amphibia: frogs, toads, salamanders
Class Reptilia: snakes, lizards, turtles
Class Aves: birds
Class Mammalia: mammals
Subclass Prototheria: egg-laying mammals
Order Monotremata: duckbilled platypuses, spiny anteaters
Subclass Metatheria: marsupial mammals
Order Marsupialia: opossums, kangaroos
Subclass Eutheria: placental mammals
Order Insectivora: shrews, moles
Order Chiroptera: bats
Order Rodentia: rats, mice, squirrels
Order Lagomorpha: rabbits and hares
Order Cetacea: whales, dolphins, porpoises
Order Carnivora: dogs, bears, weasels, cats, skunks
Order Perissodactyla: horses, hippopotamuses, zebras
Order Artiodactyla: pigs, deer, cattle
Order Primates: lemurs, monkeys, apes, humans
Suborder Prosimii (the prosimians): lemurs, tree shrews, tarsiers, lorises, pottos
Suborder Anthropoidea (the anthropoids): monkeys, apes, humans
Superfamily Ceboidea: new-world monkeys
Superfamily Cercopithecoidea: old-world monkeys
Superfamily Hominoidea (the hominoids): apes and humans
Family Hylobatidae: gibbons
Family Pongidae: chimpanzees, gorillas, orangutans
Family Hominidae (the hominids): *Australopithecus,* Homo habilis,* Homo erectus,* Homo sapiens neanderthalensis,* Homo sapiens sapiens*

*extinct

Vertebrates: Adapted for Action

In vertebrates (fig. 30.20), the embryonic notochord is replaced by a *vertebral column* composed of individual **vertebrae.** The vertebrae protect the spinal cord (nerve cord). The main axis of the internal and jointed skeleton consists not only of a vertebral column, but also a skull, which encloses and protects the brain. Vertebrates have undergone an extreme degree of cephalization. The eyes begin as outgrowths of the brain. The ears are primarily equilibrium devices in most aquatic vertebrates, and also function as sound wave receivers in land vertebrates.

The vertebrates are extremely motile and have well-developed muscles and paired limbs. They have *bilateral symmetry* and are *segmented,* as witnessed by the vertebral column. There is a large coelom and a complete gut. The *circulatory system is closed;* the blood is contained within blood vessels. Vertebrates have an efficient means of extracting oxygen from water or air, as appropriate. The kidneys are important excretory and water-regulating organs; they conserve or rid the body of water as necessary. The sexes are generally separate, and reproduction is usually sexual.

Vertebrates are distinguished in particular by these features:

3 chordate characteristics	paired limbs
jointed internal skeleton	closed circulatory system
extreme degree of cephalization	efficient respiration

In short, vertebrates are adapted to an active life-style.

Figure 30.20

Evolutionary tree of vertebrates. The animals on the main trunk are ancestral to the modern-day animals on the branches. Reptiles, birds, and mammals evolved on land.

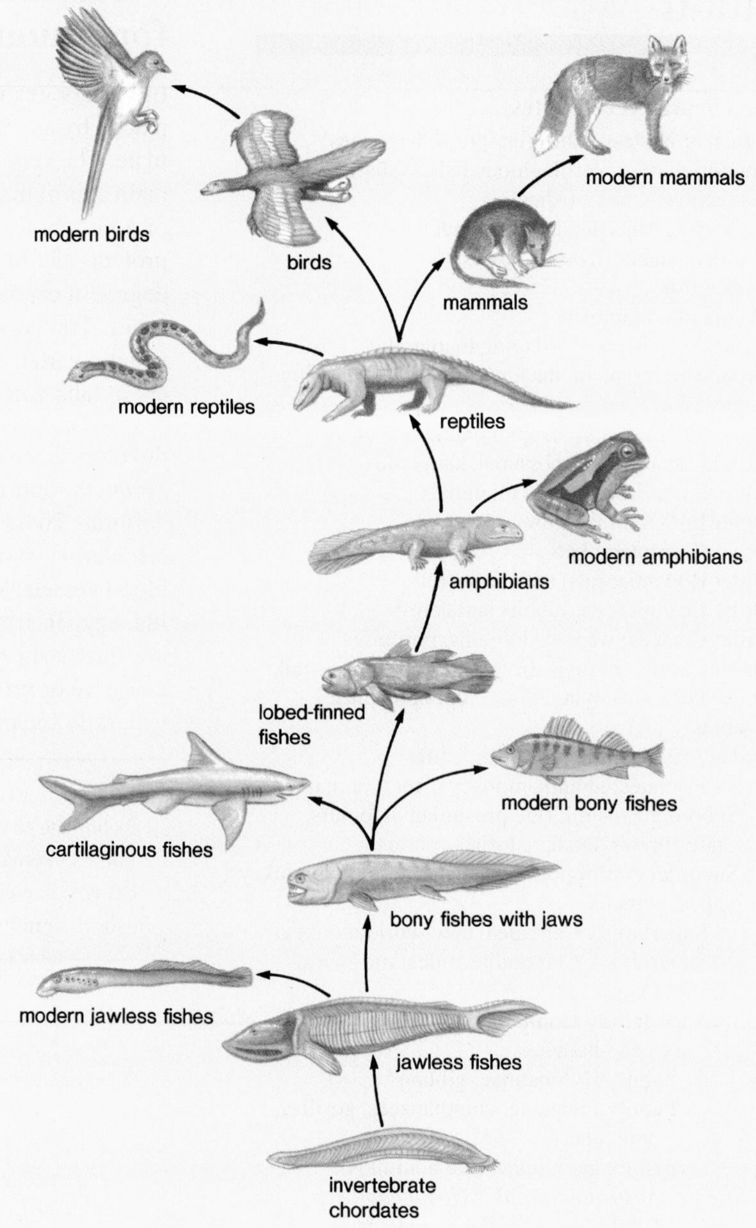

Fishes: Live in Water

Three living classes of vertebrates are commonly called fishes: the jawless fishes, the cartilaginous fishes, and the bony fishes. Living representatives of the *jawless fishes* are cylindrical and up to a meter long. They have smooth, scaleless skin and no jaws or paired fins. There are 2 families of jawless fishes: *hagfishes* and *lampreys*. The hagfishes are scavengers, feeding mainly on dead fishes, while some lampreys are parasitic. When parasitic, the round mouth of the lamprey serves as a sucker. The lamprey attaches itself to another fish, tapping into its circulatory system.

Cartilaginous fishes are the sharks, the rays, and the skates, which have skeletons of cartilage instead of bone. The small dogfish shark is often dissected in biology laboratories. One of the most dangerous sharks inhabiting both tropical and temperate waters is the hammerhead shark. The largest sharks, the whale sharks, feed on small fishes and marine

a.

b.

Figure 30.21
Fishes are adapted to living in water, and adult frogs are adapted to living on land. **a.** Fishes breathe by means of gills and locomote by using their fins and the muscles of the body wall, which produce an undulating motion. **b.** Frogs breathe by means of lungs supplemented by gas exchange through the skin. They locomote by means of paired limbs.

invertebrates and do not attack humans. Skates and rays are rather flat fishes that live partly buried in the sand and feed on mussels and clams.

Bony fishes (fig. 30.21*a*) are by far the most numerous and varied of the fishes. Most of the fish we eat, such as perch, trout, salmon, and haddock, are a type of bony fish called *ray-finned fishes.* These fishes have a *swim bladder,* which usually serves as a buoyancy organ. By secreting gases into the bladder or by absorbing gases from it, these fishes can change their density and thus go up or down in the water. *Ray-finned* refers to the fact that the fins are thin and supported by bony rays. Ancestors of another type of bony fish, called the *lobe-finned fishes,* evolved into the amphibians (fig. 30.20). These fishes not only had fleshy appendages that could be adapted to land locomotion, they also had a lung, which was used for respiration. A type of lobe-finned fish called the coelacanth, which exists today, is the only "living fossil" among the fishes. The coelacanth, however, does not have a lung.

Fishes are adapted to life in the water. Their streamlined shape, fins, and muscle action are all suited to locomotion in the water. Their skin is covered by *scales,* which protect the body but do not prevent water loss. Fishes breathe by means of *gills,* respiratory organs that are kept continuously moist by the passage of water through the mouth and out the gill slits. As the water passes over the gills, oxygen is absorbed by blood and carbon dioxide is given off. The heart of a fish is a simple pump, and the blood flows through the chambers, including a nondivided atrium and ventricle, to the gills. Oxygenated blood leaves the gills and goes to the body proper, eventually returning to the heart for recirculation.

Generally speaking, reproduction in fishes requires external water; sperm and eggs are usually shed into the water, where fertilization occurs. The zygote develops into a swimming larva, which can fend for itself until it develops into the adult form.[1]

Most fishes today are ray-finned fishes. They have the following characteristics:

bony skeleton	gills	two-chambered heart
skin with scales	swim bladder	jaws

1. Some fishes, such as sharks, practice internal fertilization and retain their eggs during development. Their young are born alive.

Amphibians: Double Life

The living amphibians include frogs, toads, newts, and salamanders (fig. 30.21*b*). The adult has distinct walking legs, each with 5 or fewer toes. Respiration is accomplished by the use of small, relatively *inefficient lungs,* supplemented by gas exchange through the smooth, moist, and glandular skin. All amphibians possess 2 nostrils, which are directly connected to the mouth cavity. Air enters the mouth by way of the nostrils, and when the floor of the mouth is raised, air is forced into the lungs.

With the development of lungs, there is a change in the circulatory system. The amphibian heart has a divided atrium but a single ventricle. The right atrium receives impure blood with little oxygen from the body proper, and the left atrium receives purified, oxygenated blood from the lungs. These 2 types of blood are partially mixed in the single ventricle. Mixed blood is then sent to all parts of the body, some to the skin, where it is further oxygenated.

Nearly all the members of this class lead an amphibious life—that is, the larval stage lives in the water and the adult stage lives on the land. The adults must return to the water, however, to reproduce. The sperm and the eggs are discharged into the water, and fertilization results in a zygote that develops into the familiar tadpole. The tadpole undergoes metamorphosis into the adult before taking up life on the land.

These features in particular distinguish adult amphibians:

usually 4 limbs	lungs	metamorphosis
smooth, moist, and glandular skin	three-chambered heart	

Figure 30.22

Reptilian egg, a major advancement because it enabled reptiles to reproduce on land. **a.** Baby crocodile hatching out of its shell. Note that the shell is leathery and flexible, not brittle like birds' eggs. **b.** Inside the egg, the embryo is surrounded by membranes. The chorion aids gas exchange, the yolk sac provides nutrients, the allantois stores waste, and the amnion encloses a fluid that prevents drying out and provides protection.

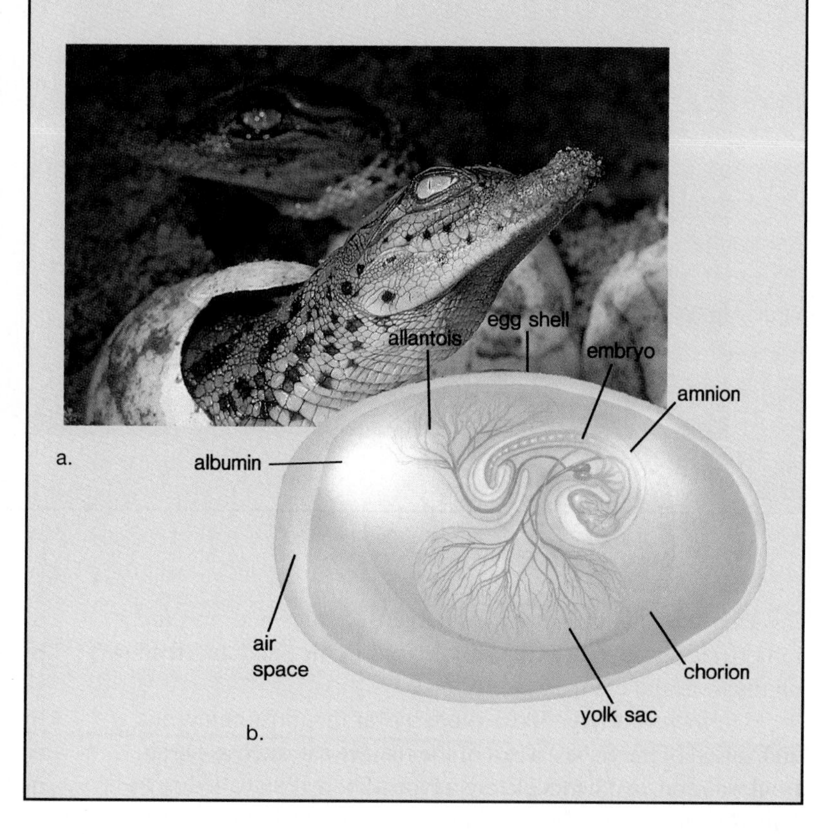

Reptiles: Adapted to Land

The reptiles living today are turtles, alligators, snakes, and lizards. Reptiles with limbs, such as lizards, are able to lift their body off the ground, and the body is covered with hard, *horny scales,* which protect the animal from desiccation and from predators. Both of these features are adaptations to life on land.

Reptiles have well-developed lungs enclosed by a protective rib cage. When the rib cage expands, the lungs expand and air rushes in. The creation of a partial vacuum establishes a negative pressure, which causes air to rush into the lungs. The atrium of the heart is always separated into right and left chambers, but division of the ventricle varies. An interventricular septum is incomplete in certain species; therefore, there is some exchange of oxygenated and deoxygenated blood between the ventricles.

Perhaps the most outstanding adaptation of the reptiles is that they have a means of reproduction suitable to a land existence. The penis of the male passes sperm directly to the female. *Fertilization* is internal, and the female lays leathery, flexible, shelled eggs. The *shelled egg* made development on land possible and eliminated the need for a swimming-larval stage during development. It provides the developing embryo with oxygen, food, and water; it removes nitrogenous wastes; and it protects the embryo from drying out and from mechanical injury. This is accomplished by the presence of *extraembryonic membranes* (fig. 30.22).

Figure 30.23
Respiratory system in birds. Because of the presence of air sacs, there is a one-way flow of air through the lungs. Upon inhalation (blue arrows), air moves into the air sacs; upon exhalation, air moves through the lungs (purple arrows).

Sometimes, animals that cannot maintain a constant temperature—that is, fishes, amphibians, and reptiles—are called *cold blooded.* Actually, however, they take on the temperature of the external environment. If it is cold externally, they are cold internally; if it is hot externally, they are hot internally. Reptiles try to regulate body temperatures by exposing themselves to the sun if they need warmth or by hiding in the shadows if they need cooling off. This works reasonably well in most areas of the world.

These features in particular distinguish reptiles:

usually 4 limbs	lungs with expandable rib cage	shelled egg
dry, scaly skin		

Birds: Usually Can Fly

Birds (fig. 30.23) are characterized by the presence of feathers, which are actually modified reptilian scales. There are many orders of birds, including birds that are flightless (ostriches), web footed (penguins), divers (loons), fish eaters (pelicans), waders (flamingos), broad billed (ducks), birds of prey (hawks), vegetarians (fowl), shorebirds (sandpipers), nocturnal (owls), small (hummingbirds), and songbirds, the most familiar of the birds.

Nearly every anatomical feature of a bird can be related to its *ability to fly.* The anterior pair of appendages (wings) are adapted for flight; the posterior are variously modified, depending on the type of bird. Some are adapted for swimming, some for running, and some for perching on limbs. The breastbone is enormous and has a ridge, the keel, to which the flight muscles are attached. Respiration is efficient since the lobular lungs form *air sacs* throughout the body, including the bones. The presence of these sacs means that the air circulates one way through the lungs; "used" air is not trapped in the lungs (fig. 30.23). Another benefit of air sacs is that the air-filled bones lighten the body and aid flying.

Birds have a four-chambered heart (see p. 201), which completely separates oxygenated blood from deoxygenated blood. Oxygenated blood is sent under pressure to the muscles. Birds have no bladder and excrete uric acid in a semidry state.

Birds have well-developed brains, but they rely on instinctive behavior. Therefore, birds follow very definite patterns of migration and nesting.

Birds are *warm blooded.* Like mammals, their internal temperature is constant because they generate and maintain metabolic heat. This may be associated with their efficient nervous, respiratory, and circulatory systems. Also, their feathers provide insulation.

These features, in particular, distinguish birds:

feathers	usually wings for flying	hard-shelled egg
air sacs	four-chambered heart	constant internal temperature

Do 30.2 Critical Thinking, found at the end of the chapter.

Figure 30.24

Three types of mammals. **a.** The duck-billed platypus is a monotreme, which lays shelled eggs. **b.** The koala is a marsupial, whose young are born immature and complete their development within the mother's pouch. **c.** The white-tailed deer is a placental mammal, whose young develop within a uterus.

a.

b.

Mammals: Mammary Glands

The chief characteristics of mammals are *hair* and *mammary glands.* In humans, mammary glands, which produce milk to nourish the young, are called breasts.

Mammals are classified according to their means of reproduction. There are *egg-laying* mammals, mammals with *pouches* for immature embryos, and **placental mammals.**

Monotremes: Lay Eggs

Monotremes are egg-laying mammals and are represented by the spiny anteater and the duck-billed platypus (fig. 30.24*a*). The female monotreme incubates her eggs in the same manner as birds, but after hatching, the young are dependent upon the milk that seeps from glands on the abdomen of the female. Therefore, monotremes retained the reptilian mode of reproduction while evolving hair and mammary glands. The young are blind, helpless, and completely dependent on the parent for some months. The mouth is variously modified among the monotremes. The platypus has a leathery, bill-like structure somewhat resembling that of a duck, while the anteater has an elongated, cylindrical snout.

c.

Marsupials: Have Pouches

Another primitive group of mammals is the *marsupials,* such as opossums, kangaroos, and koalas (fig. 30.24*b*). In marsupials, the young are born in a very immature state and finish their development in the mother's abdominal pouch, called the marsupium. When a marsupial such as an opossum is born—after only 12–16 days of gestation—it is blind, naked, grublike, and no larger than a honeybee. Using clawed forelimbs, the newborn crawls toward the mother's fur-lined pouch. Once there, it attaches itself to a nipple. After 4 or 5 weeks in the pouch, an opossum spends an additional 8 or 9 weeks clinging to the mother's back.

Placental Mammals: Placenta Supplies Needs

The vast majority of living mammals are placental mammals (fig. 30.24*c*). In these mammals, the extraembryonic membranes (fig. 30.22) have been modified for internal development within the uterus of the female. The chorion contributes to the fetal portion of the placenta, while a part of the uterine wall contributes to the maternal portion. Here, nutrients, oxygen, and waste are exchanged between fetal and maternal blood.

Mammals are adapted to life on land and have limbs that allow them to move rapidly. In fact, an evaluation of mammalian features leads us to the obvious conclusion that they lead active lives. The brain is well developed; the lungs are expanded not only by the action of the rib cage but also by the contraction of the *diaphragm,* a horizontal muscle that divides the thoracic cavity from the abdominal cavity; and the heart has *4 chambers.* The internal temperature is constant, and hair, when abundant, helps to insulate the body.

The mammalian brain is enlarged due to the expansion of the foremost part—the cerebral hemispheres. These have become convoluted and have expanded to such a degree that they hide many other parts of the brain from view. The brain is not fully developed for some time after birth, and there is a long period of dependency on the parents, during which the young learn to take care of themselves.

Classification of placental mammals is based on mode of locomotion and methods of obtaining food. For example, bats (order Chiroptera) have membranous wings supported by digits; horses (order Perissodactyla) have long, hoofed legs; and whales (order Cetacea) have paddlelike forelimbs. The specific shape and size of the teeth may be associated with whether the mammal is an herbivore (eats vegetation), a carnivore (eats meat), or an omnivore (eats both meat and vegetation). For example, mice (order Rodentia) have continuously growing incisors; horses (order Perissodactyla) have large, grinding molars; and dogs (order Carnivora) have long canine teeth.

These features in particular distinguish placental mammals:

body hair	muscular diaphragm	differentiated teeth
mammary glands	constant internal temperature	well-developed brain
internal development		infant dependency

Figure 30.25

Primate evolution. There are at least 4 lines of evolution among the primates, which arose from mammalian insectivores: prosimians, including lemurs and tarsiers; monkeys, including new-world and old-world monkeys; apes, including gibbons and great apes (orangutans, gorillas, and chimpanzees); and humans.

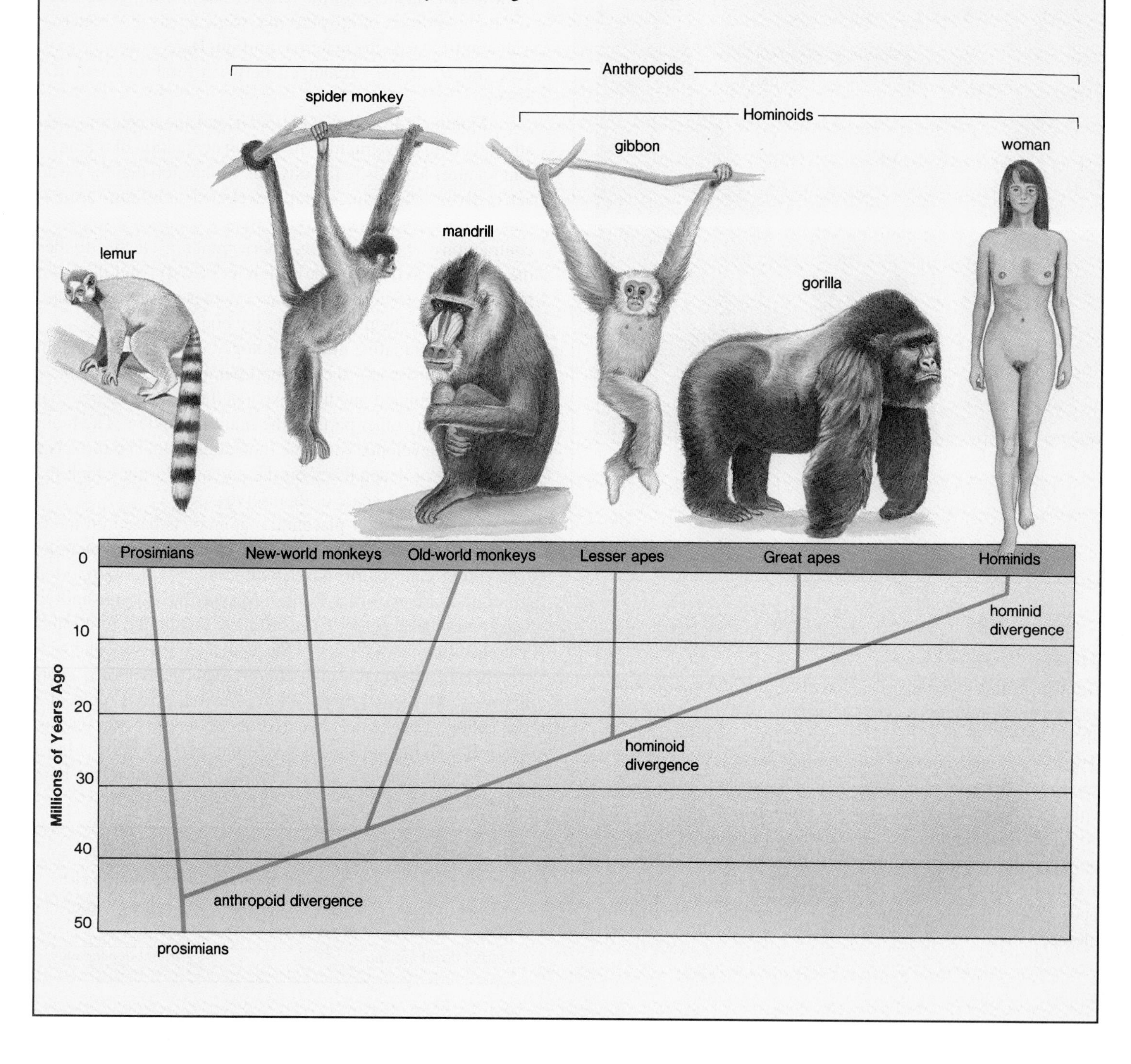

Human Evolution

All of the animals depicted in figure 30.25 and listed in the classification box are primates.

Primates: Adapted to Life in Trees

Primates are adapted for living in trees. Their limbs are mobile, as are the hands because the thumb (and in nonhuman primates, the big toe as well) is opposable; that is, the thumb can touch each of the other fingers. Therefore, a primate can easily reach out and bring food, such as fruits, to the mouth. A tree limb can be grasped and released freely because nails have replaced claws.

A snout is common in animals in which a sense of smell is of primary importance. In primates, the sense of sight is more important, and the snout is shortened considerably, allowing the eyes to move to the front of the head. This results in stereoscopic vision (or depth perception), permitting primates to make accurate judgments about the distance and position of adjoining tree limbs.

One birth at a time is the norm in primates; it would have been difficult to care for several offspring as large as primates while moving from limb to limb. The forebrain is especially large, and there is an emphasis on learned behavior. The juvenile period of dependency is extended, and most primates tend to form social units that include individuals of all ages and sexes.

These characteristics especially distinguish primates from other mammals:

opposable thumb (and in some cases big toe)	expanded forebrain
nails (not claws)	emphasis on learned behavior
single birth	extended period of parental care

Prosimians and Anthropoids: On the Road to Humans

The first primates were **prosimians,** a term meaning *premonkey.* The prosimians are represented today by several types of animals. Lemurs have a squirrel-like appearance, and tarsiers are curious mouse-sized creatures with enormous eyes suitable for their nocturnal way of life. The prosimians are believed to have evolved from an insectivore-type mammal.

Monkeys, apes, and humans are all **anthropoids.** There are 2 types of monkeys: new-world monkeys, which have long prehensile (grasping) tails and flat noses, and old-world monkeys, which lack such tails and have protruding noses. Two of the well-known new-world monkeys are the spider monkey and the capuchin, the "organ grinder's monkey." Some of the better known old-world monkeys are now ground dwellers, such as the baboon and the rhesus monkey.

Humans are more closely related to apes than to monkeys; only apes and humans are **hominoids**. There are 4 types of apes: gibbons, orangutans, gorillas, and chimpanzees (fig. 30.26). The gibbon is the smallest of the apes, with a body weight ranging from 5 kg to 10 kg (12–25 lb). Gibbons have extremely long arms, which are specialized for swinging between tree limbs. The **orangutan** is large (75 kg/165 lb) but nevertheless spends a great deal of time in trees. In contrast, the gorilla, the largest of the apes (185 kg/400 lb), spends most of its time on the ground. Chimpanzees, which are at home both in the trees and on the ground, are the most humanlike of the apes in appearance.

Molecular data tell us that humans are genetically very similar to chimpanzees and gorillas. Nucleotide sequences in genes, amino acid sequences in proteins, and immunological properties of various molecules all indicate that we are even more closely related to these apes than orangutans are related to them. Humans can be distinguished from apes, however, by locomotion and posture, dental features, and other characteristics.

Figure 30.27 compares the locomotion of a monkey, a gorilla, and a human. When moving in a tree, a monkey is quadrupedal (walks on all 4 limbs) and leaps from branch to branch. Notice that in keeping with this method of locomotion, the vertebral column is arched and the shoulder joint faces downward. In contrast to monkeys, apes swing from branch to branch. Consistent with this type of locomotion, the vertebral column is straight and the shoulder joint faces outward, rather than downward. Also, the arms are elongated. Humans are not as specialized for moving from limb to limb as the apes, but some tendency in this direction may have facilitated bipedalism, or walking on 2 feet. Humans have an S-shaped vertebral column, which provides a way to transmit the upper body's weight to the pelvis. The pelvis is stronger because it is shorter and the sacrum fits like a keystone between the 2 pelvic bones. The broad pelvic bones serve as attachments for muscles that maintain stability as first one leg and then the other leaves the ground while walking.

Figure 30.26

Ape diversity. **a.** Of the apes, gibbons are the most distantly related to humans. They dislike coming down from trees, even at watering holes. They extend a long arm into the water and then drink collected moisture from the back of the hand. **b.** Orangutans are solitary except when they reproduce. Their name means *forest man*; early Malayans believed that they were intelligent and could speak but did not because they were afraid of being put to work. **c.** Gorillas are terrestrial and live in groups in which a silver-backed male, such as this one, is always dominant. **d.** Of the apes, chimpanzees sometimes seem the most humanlike.

a.

b.

c.

d.

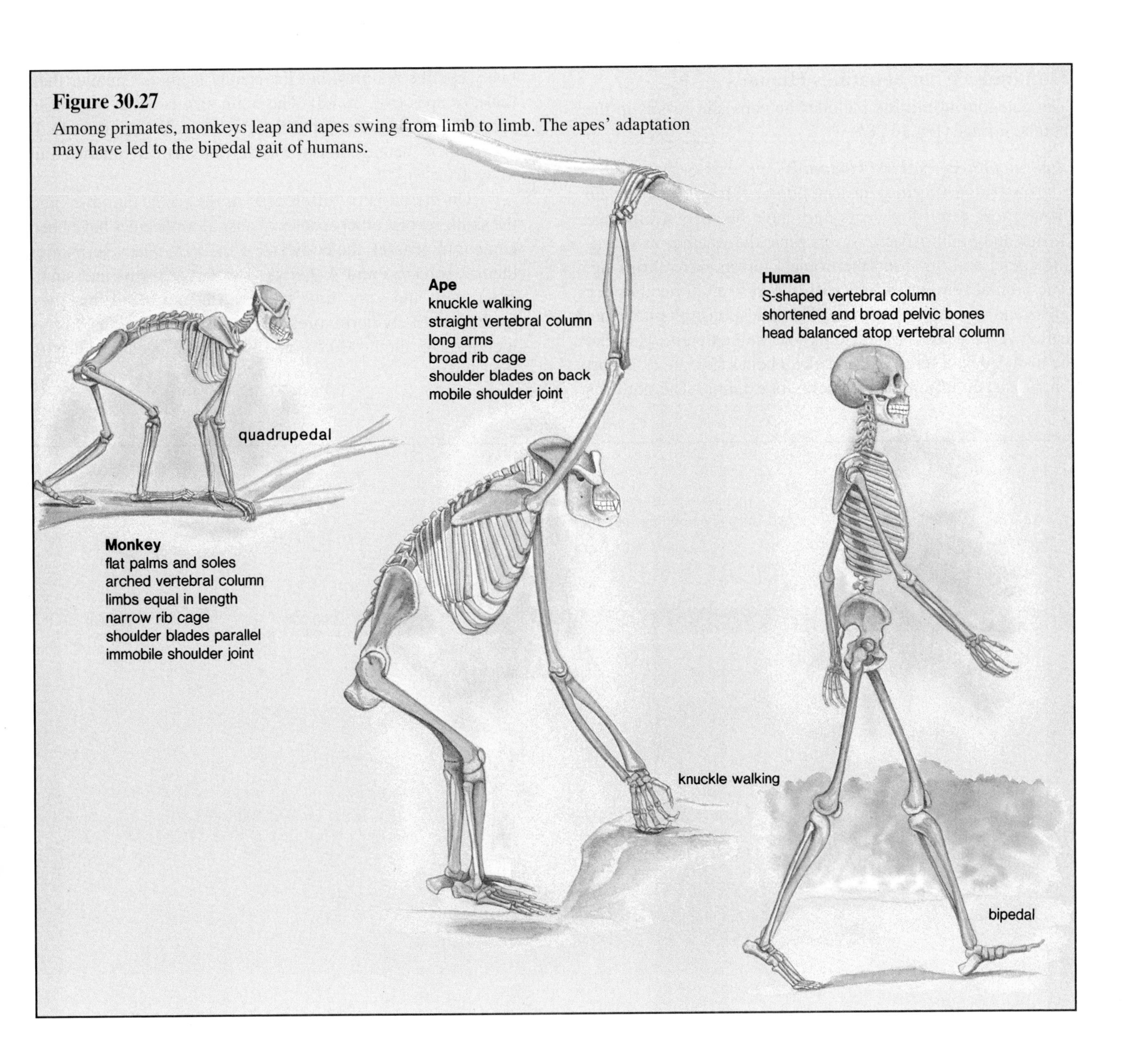

Figure 30.27
Among primates, monkeys leap and apes swing from limb to limb. The apes' adaptation may have led to the bipedal gait of humans.

Hominids: What Separates Humans

The category **hominids** includes humans and fossils in the human lineage (fig. 30.28).

Australopithecines: First Hominids

The **australopithecine** (*Australopithecus*) fossils, which date from about 4 million years ago, have been found in East Africa and South Africa. At this time, the weather was turning cooler and drier and the tropical forests were shrinking. The animal remains found with the australopithecines are grassland, not forest, animals. The oldest australopithecine, called *A. afarensis,* was bipedal and walked erect. This can be deduced by a study of the skeletal bones (fig. 30.29); short but broad pelvic bones are evident, for example. The skull has many apelike features, but the canine teeth are smaller than those of apes (fig. 30.30). The brain size is about that of an ape (400 cc), however, and there is no evidence of tools. It seems, then, that bipedalism was the first distinctly human trait to evolve.

On the basis of differences in the teeth, the jaws, and the skull, several other species of australopithecines have been named. In general, the body size of *A. africanus* was smaller than *A. robustus* and *A. boisei*. The larger forms had small front teeth and very large back teeth, indicating that they were most likely herbivores. Brain size in these later fossils increases to about 600 cc, and tools have been found with their bones.

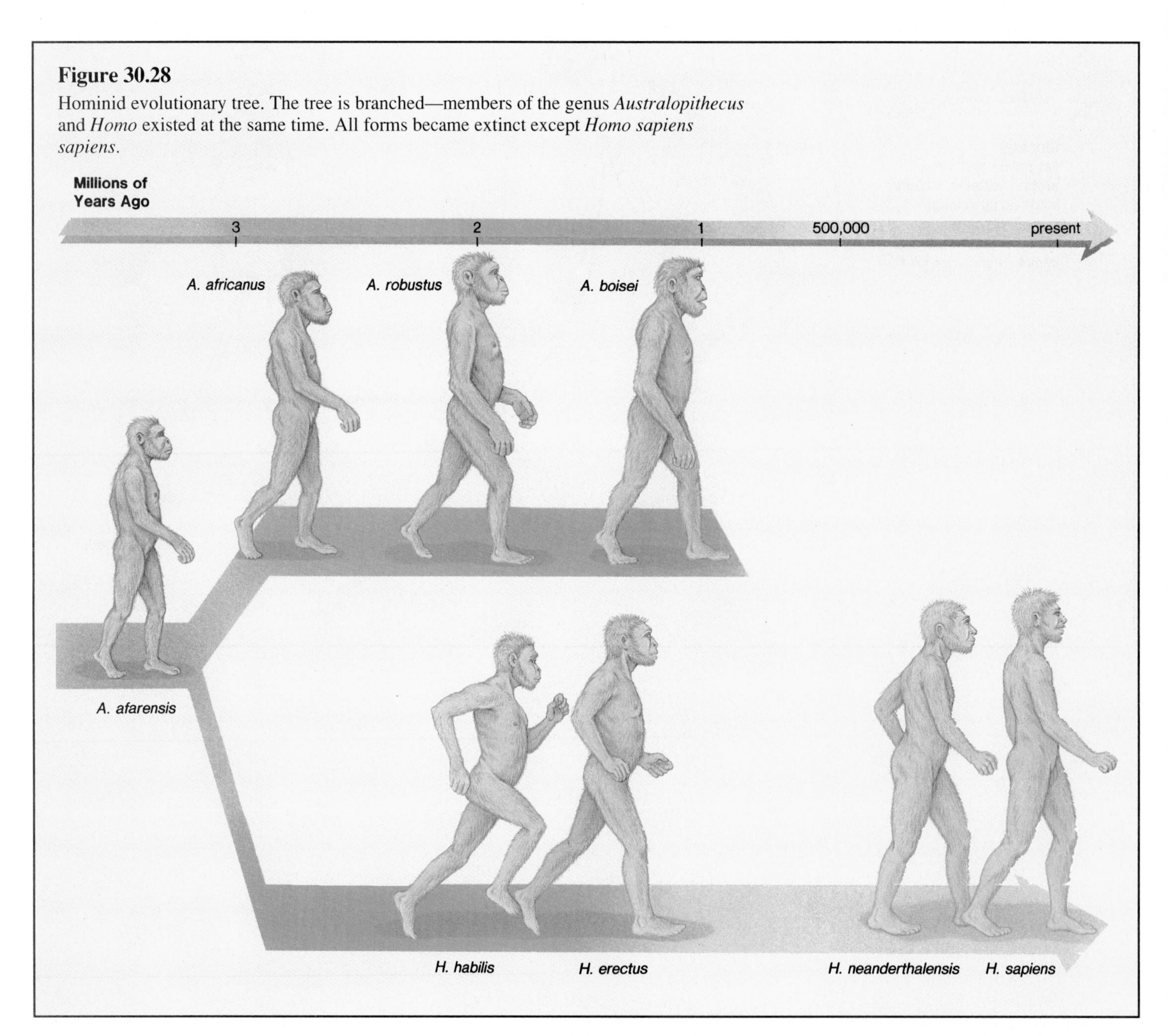

Figure 30.28

Hominid evolutionary tree. The tree is branched—members of the genus *Australopithecus* and *Homo* existed at the same time. All forms became extinct except *Homo sapiens sapiens*.

Homo habilis: Tool Maker

Fossils now classified as *Homo habilis* date between 1.6 million and 2 million years ago and, like *Australopithecus,* are found in East Africa and South Africa. Although the face still projects, this species had a larger brain (700 cc) and smaller teeth than those of many australopithecines. Some believe that these people were the first to fashion crude stone tools; in fact, the species' name means *handy man.*

Other experts suggest that the fossils designated as *Homo habilis* are actually australopithecine fossils. They also say that tool production began with the australopithecines.

Homo erectus: Fire User

The fossils of ***Homo erectus,*** which date to about 1.5 million years ago, indicate that at least some members of this species were about the size of modern humans and had our brain size (1,300 cc). The *Homo erectus* skull had a mixture of more-advanced and primitive features (fig. 30.30). Skeletal features suggest greater endurance for walking and running than we have. The pelvis was narrower and the femur had a longer neck (portion between ball and shaft). A wider pelvis is not as adaptive for walking, but it accommodates the birth of offspring with larger brains.

Figure 30.29

a. Reassembled skeleton of *A. afarensis,* named Lucy by her discoverer. Lucy walked erect, as indicated, for example, by the short pelvis with prominent blades. **b.** Lucy's (or some other australopithecine's) footprints, discovered at Latoli in Tanzania, confirm that this hominid walked erect.

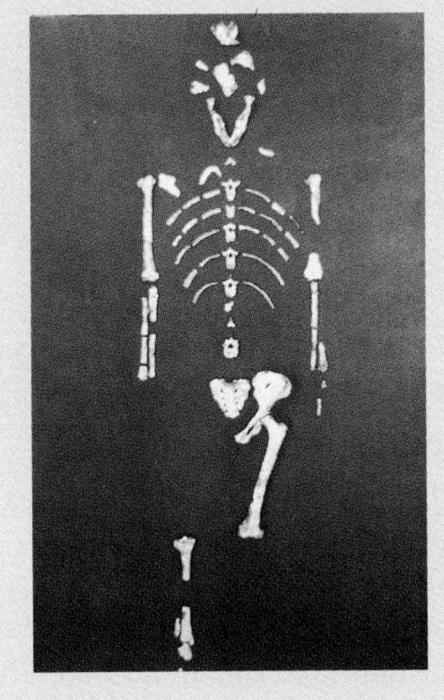

a.

b.

Figure 30.30

Comparative hominid skull anatomy and tools. Oldowan tools are crude, Acheulean tools are better made and more varied, and Aurignacian tools are well designed for specific purposes.

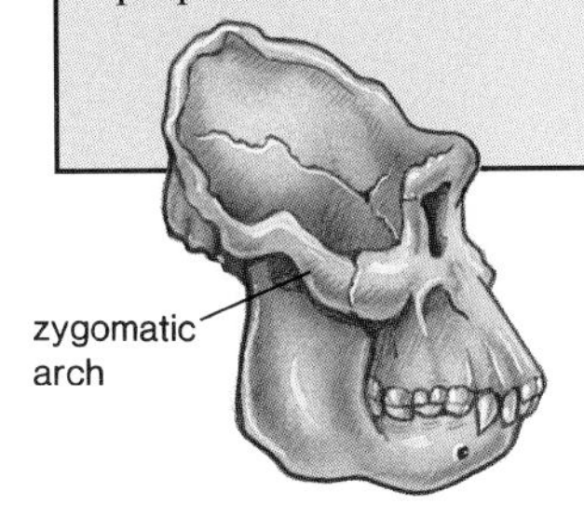

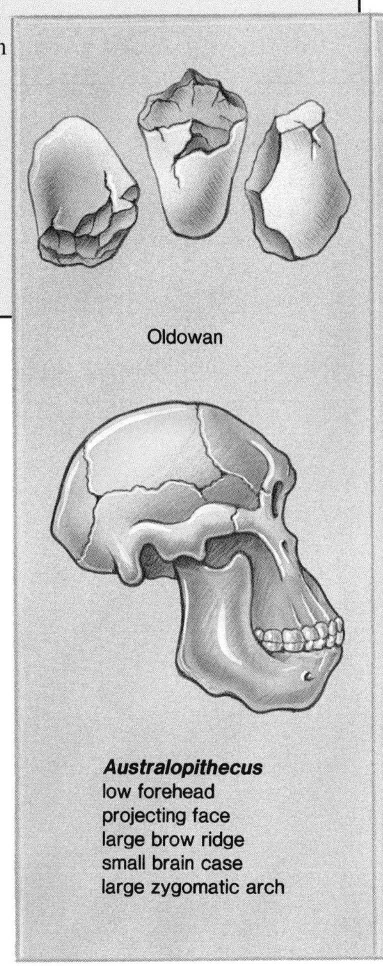

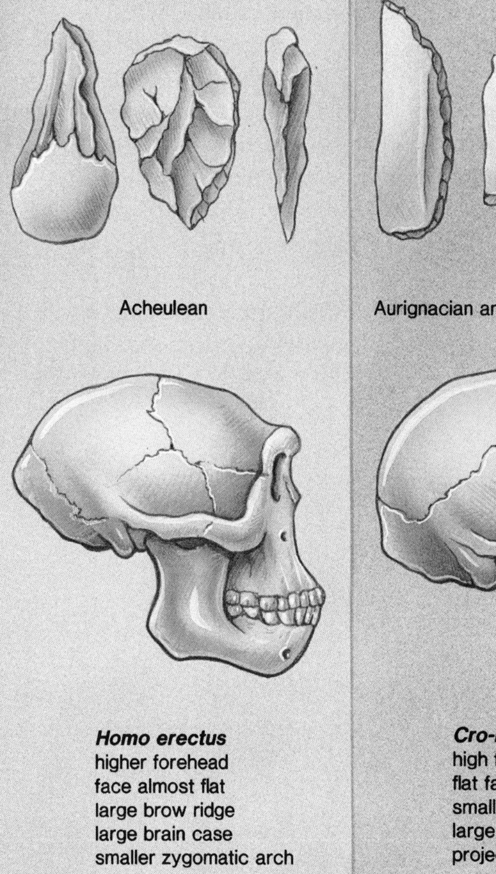

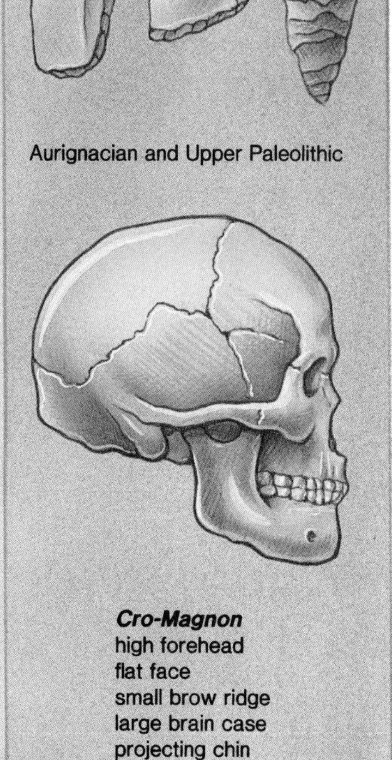

Homo erectus ranged widely. Fossils are not only found in Africa but also Europe, China (called Peking man) and Java (called Java man). These people made more advanced tools, including hand axes, and it is possible they were big-game hunters. They had knowledge of fire and could have cooked their meat before eating it.

Homo sapiens: Big Brains

Fossils of the species *Homo sapiens neanderthalensis* are dated from about 100,000 years ago to about 35,000 years ago. The term **Neanderthal** is derived from the Neander Valley in Germany, where the species was first found. The Neanderthal skull resembles that of *Homo erectus,* except the brain was quite large—slightly larger than that of modern humans (1,330 cc). The body was shorter and more massive than that of modern humans; the larger brain may correlate with the need for control over larger muscles. It is now believed that the Neanderthals stood fully erect, but they possibly moved more slowly than we do. The Neanderthals died out about 50,000–40,000 years ago for reasons unknown.

Fossils identified as *Homo sapiens sapiens* date back to about this time period and are found in Africa, Asia, and Europe. Sometimes these fossils are collectively referred to as **Cro-Magnon,** after a cave site in southern France where remains were found. The facial features and brain size of Cro-Magnon distinctly resemble modern humans (fig. 30.30). Studies based on mitochondrial DNA, which are now in dispute, suggest that modern humans originated in Africa and then spread out into Europe, giving rise to the various human races. Others believe that the human races originated in several geographical regions, but they became one species because of gene flow.

Both Neanderthals and Cro-Magnon peoples were excellent hunters. Big-game hunting requires cooperation, which may have fostered societal living and the beginning of language. When the Neanderthals buried their dead, they seemed to have prepared them for a future life by supplying flint, tools, and food. Evidence of culture, however, is particularly evident with Cro-Magnon. For example, Cro-Magnon people painted beautiful drawings of animals on cave walls in Spain and France (fig. 30.31) and sculpted many small figurines.

Figure 30.31

Cro-Magnon people painting on cave walls. Some of these paintings can still be observed today. Of all the animals, only humans developed a culture that includes technology and the arts.

We have seen that these features in particular distinguish humans from apes:

bipedalism
small canine teeth and other facial features
brain size (400 cc for apes compared to 1,330 cc for humans)
toolmaking
language
culture

Do 30.3 Critical Thinking, found at the end of the chapter.

SUMMARY

Classification of animals considers number of germ layers, level of organization, symmetry, type of coelom, body plan, and presence of segmentation. Among the animals without a true coelom, there is an increase in complexity from the sponges to the roundworms. The roundworms have a tube-within-a-tube body plan, bilateral symmetry, 3 germ layers, and a pseudocoelom.

Among animals with a true coelom, the protostomes are so named because the blastopore becomes the mouth, and the deuterostomes are so named because the blastopore becomes related to the anus and a second opening becomes the mouth. Annelids are the first of the animals to have all the features mentioned plus segmentation. Segmentation and a jointed skeleton are seen in arthropods and vertebrates that have groups adapted to a land environment.

Chordates at some time in their life history have a notochord, dorsal hollow nerve cord, and pharyngeal pouches. In vertebrates, the notochord is replaced by the vertebral column. Adaptation to land among vertebrates begins in amphibians and continues in reptiles, which are able to reproduce on land due to a shelled egg having extraembryonic membranes. Only birds and mammals are able to maintain a constant internal temperature.

Among mammals, primates—prosimians, monkeys, apes, and humans—are adapted for living in trees. Monkeys leap but apes swing from limb to limb. This may have been an adaptation that led to bipedalism in humans. The first hominid (humans and immediate ancestors) was *A. afarensis,* which could walk erect but had only a small brain. Later-appearing australopithecines may have manufactured stone tools, but *H. habilis* certainly did. *H. erectus* was the first fossil to have a brain size of more than 1,000 cc. The fossils of this species appear throughout Africa, Europe, and Asia. They used fire and may have been big-game hunters. *H. sapiens neanderthalensis* fossils have been found in Europe and Asia. The Neanderthals did not have the physical traits of modern humans, but they did have culture. There is some dispute over the manner in which *H. sapiens sapiens* (Cro-Magnon) arose, but it is clear that modern humans appear in the fossil record at the time the Neanderthals disappear from it.

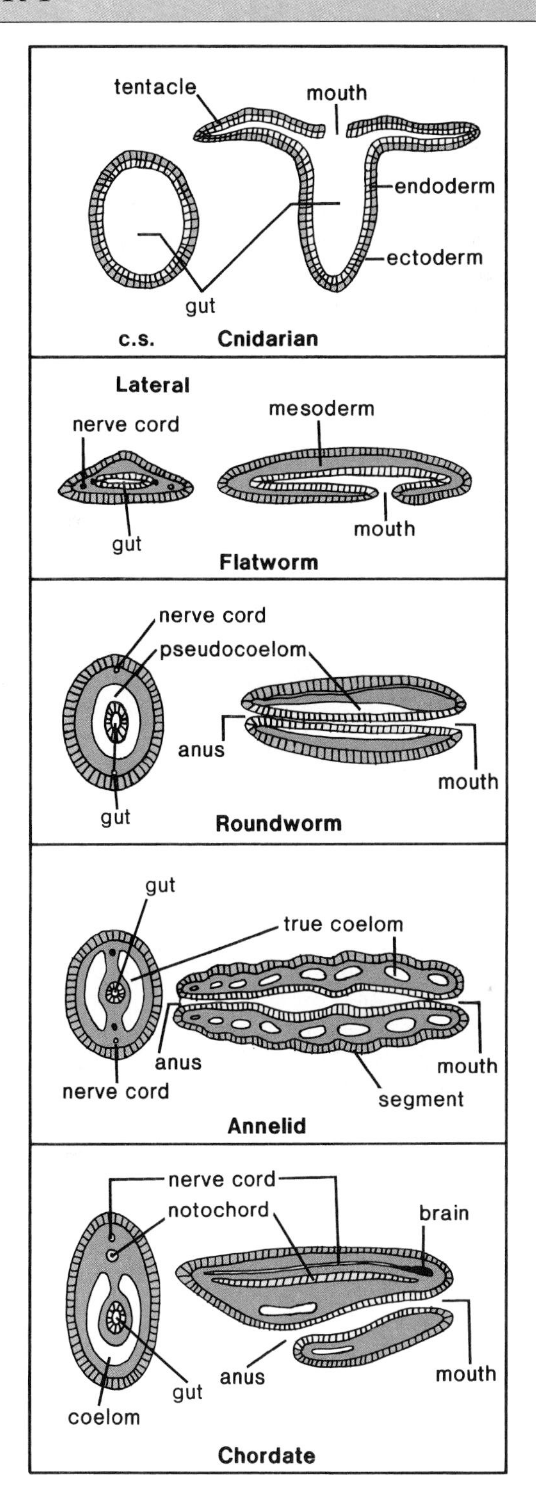

STUDY QUESTIONS

In order to practice **writing across the curriculum,** students should write out the answers to any or all of the study questions. The study questions are sequenced in the same order as the text.

1. The animals without a true coelom include which phyla of animals? (p. 584) Compare the representatives of these animals in regard to germ layers, level of organization, presence or absence of a coelom, symmetry, body plan, and segmentation. (p. 584)
2. Compare free-living forms of the first 3 phyla of animals in regard to nerve organization, digestive cavity, and means of reproduction. (pp. 585–589)
3. Describe the life cycle and the structure of a tapeworm. (p. 590) Compare the anatomy of the free-living flatworm to that of the fluke and the tapeworm. (p. 589)
4. What biological data are used to divide animals with a true coelom into 2 groups? (p. 592)
5. Compare the adaptations of the clam to those of the squid to show that the clam is adapted to an inactive life and the squid is adapted to an active life. (p. 593)
6. Compare the clam, the earthworm, the crayfish, the grasshopper, and the starfish with respect to nervous, digestive, skeletal, excretory, circulatory, and respiratory systems and means of reproduction and locomotion. (pp. 594–601)
7. Compare the adaptations of the crayfish to those of the grasshopper to show that the crayfish is adapted to water while the grasshopper is adapted to land. (p. 600)
8. Name and describe unique features of echinoderm anatomy and physiology. (p. 600)
9. Compare the adaptations to a land existence among vertebrates. (pp. 606–607)
10. Name several primate characteristics still retained by humans. (p. 611)
11. Draw an evolutionary tree that includes all primates. Discuss each member of the tree. (pp. 610–613)
12. Draw and discuss an evolutionary tree for hominids. (pp. 614–616)

OBJECTIVE QUESTIONS

1. The function of collar cells in a sponge is _______.
2. Cnidarians have the _______ body plan and are _______ symmetrical.
3. Planarians have the _______ type of nervous system and a(n) _______ excretory system.
4. The intermediate hosts for a tapeworm are either _______ or _______.
5. Pinworm, trichinosis, hookworm, and elephantiasis are all infections caused by a(n) _______.
6. In protostomes, the first embryonic opening becomes the _______.
7. In today's mollusks, the coelom is much _______ and limited to the region around the _______.
8. Earthworms have external rings, signifying that they are _______ animals.
9. The water vascular system of echinoderms consists of canals and _______ feet.
10. The 3 chordate characteristics are a(n) _______, _______, and _______.
11. The _______ and the _______ are primitive chordates.
12. The 3 classes of fishes, _______, _______, and _______, indicate in general their order of evolution.
13. Amphibians evolved from _______ fishes, which had primitive lungs.
14. Whereas amphibians must return to the _______ to reproduce, reptiles lay _______ that contain _______ membranes.
15. Both _______ and mammals maintain a constant internal _______.
16. There are 3 types of mammals: _______, _______, and _______.
17. Dogs, cats, horses, mice, rabbits, bats, whales, and humans are all _______ mammals.
18. Place each animal in the highest category possible.

a. Neanderthal	1. prosimian
b. lemur	2. anthropoid
c. australopithecine	3. hominid
d. chimpanzee	4. <u>Homo</u>

19. Primates are adapted to life in the _______.
20. *A. afarensis* could probably walk _______ but had a(n) _______ brain.
21. _______ are fossils classified as *Homo sapiens sapiens.*
22. Label this diagram of the animal evolutionary tree.

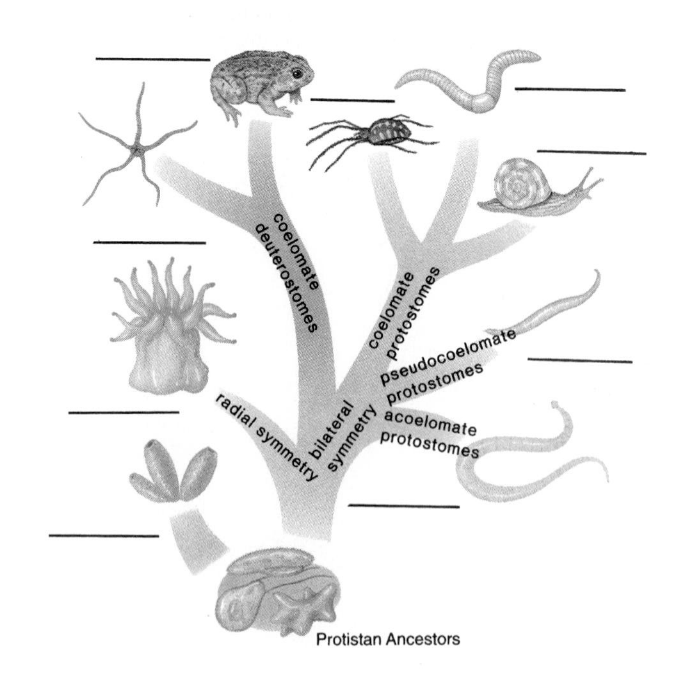

CRITICAL THINKING

In order to practice **writing across the curriculum,** students should write out the answers to any or all of the critical thinking questions. Suggested answers to the critical thinking questions are in appendix E.

30.1

1. Using the listing on page 563 for the characteristics of plants, list, if possible, a corresponding characteristic for animals.
2. What one animal characteristic can be associated with an animal's need to acquire food?
3. Why is it reasonable to assume that animals arose from protozoans?
4. Some animals can regenerate from a small part of the whole. Why would you expect only simple animals and not complex ones to be able to do this?

30.2

Both insects and birds are adapted to life on land.

1. How does each carry on respiration?
2. How does each prevent desiccation of the animal?
3. How does each prevent desiccation of the gametes and the embryo?
4. What is the reproductive strategy of each animal? How do they differ?

30.3

How can you tell that humans are

1. animals?
2. chordates and vertebrates?
3. mammals?
4. primates?

SELECTED KEY TERMS

anthropoid (an´ thro-poid) a group of primates that includes only monkeys, apes, and humans.

australopithecine (aw˝ strah-lo-pith´ e-sīn) referring to one of the 3 species of *Australopithecus,* the first generally recognized hominids.

bilateral symmetry having a right half and a left half so that only one vertical cut gives 2 equal halves.

Cro-Magnon the common name for the first fossils to be accepted as representative of modern humans.

deuterostome (du´ ter-o-stōm˝) member of a group of animal phyla in which the anus develops in relation to the blastopore and a second opening becomes the mouth.

hermaphrodite (her´ maf-ro-dīt) an animal having both male and female sex organs.

hominid (hom ĭ-nid) member of a family of upright, bipedal primates that includes australopithecines and modern humans.

Homo erectus the earliest nondisputed species of humans, named for their erect posture, which allowed them to have a bipedal gait.

Homo habilis an extinct species that may include the earliest humans, having a small brain but quality tools.

invertebrate an animal that lacks a vertebral column.

mesoglea (mes˝ o-gle´ ah) a jellylike packing material between the ectoderm and the endoderm in cnidarians.

Neanderthal the common name for an extinct subspecies of humans whose remains are found in Europe, Asia, and Africa.

nematocyst (nem´ ah-to-sist) a threadlike structure in stinging cells of cnidarians that can be expelled to numb and to capture prey.

notochord (no´ to-kord) dorsal supporting rod that exists in all chordates sometime in their life history; replaced by the vertebral column in vertebrates.

primate animal that belongs to the order Primates, the order of mammals that includes prosimians, monkeys, apes, and humans.

prosimian (pro-sim´ e-an) primitive primate, such as the lemur, tarsier, and tree shrew.

protostome member of a group of animal phyla in which the mouth develops from the blastopore.

radial symmetry regardless of the angle of a vertical cut made at the midline of an organism, 2 equal halves result.

sessile filter feeder an animal that stays in one place and obtains its food, usually in small particles, by filtering it from water.

trachea (tra´ ke-ah) an air tube of insects.

vertebrate animal possessing a backbone composed of vertebrae.

FURTHER READINGS FOR PART SIX

Alexander, R. M. April 1991. How dinosaurs ran. *Scientific American.*

Avers, C. J. 1989. *Process and pattern in evolution.* New York: Oxford University Press.

Bar-Yosef, O., and B. Vandermeersch. April 1993. Modern humans in the Levant. *Scientific American.*

Beehler, B. M. December 1989. The birds of paradise. *Scientific American.*

Blumenschine, R. J., and J. A. Cavallo. October 1992. Scavenging and human evolution. *Scientific American.*

Bonatti, E. March 1987. The rifting of continents. *Scientific American.*

Brock, T. D., and M. T. Madigan. 1988. *Biology of microorganisms.* 5th ed. Englewood Cliffs, N.J.: Prentice-Hall.

Brusca, R. C., and G. J. Brusca. 1990. *Invertebrates.* Sunderland, Mass.: Sinauer Associates.

Cavalli-Sforza, L. L. November 1991. Genes, people and language. *Scientific American.*

Colbert, E. H., and M. Morales. 1991. *Evolution of the vertebrates.* 4th ed. New York: John Wiley and Sons.

Davies, N. B., and M. Brooke. January 1991. Coevolution of the cuckoo and its hosts. *Scientific American.*

DeDuve, C. 1991. *Blueprint for a cell: The nature and origin of life.* Burlington, N.C.: Neil Patterson Publishers.

del Pino, E. M. May 1989. Marsupial frogs. *Scientific American.*

DeRobertis, E. M., G. Oliver, and C. V. E. Wright. July 1990. Homeobox genes and the vertebrate body plan. *Scientific American.*

Eckert, R., and D. Randall. 1983. *Animal physiology.* 2d ed. San Francisco: W. H. Freeman.

Evans, H. E. 1984. *Insect biology: A textbook of entomology.* Redwood City, Calif.: Addison-Wesley Publishing Co.

Feder, M. E., and W. W. Burggren. November 1985. Skin breathing in vertebrates. *Scientific American.*

Fischetti, V. F. June 1991. Streptococcal M protein. *Scientific American.*

Futuyma, E. J. 1986. *Evolutionary biology.* 2d ed. Sunderland, Mass.: Sinauer Associates.

Gallo, R. C. December 1986. The first human retrovirus. *Scientific American.*

Gamlin, L., and G. Vines, eds. 1991. *The evolution of life.* New York: Oxford University Press.

Gould, S. G. 1989. *Wonderful life: The Burgess shale and the nature of history.* New York: W.W. Norton and Co.

Grant, P. R. October 1991. Natural selection and Darwin's finches. *Scientific American.*

Hadley, N. F. July 1986. The arthropod cuticle. *Scientific American.*

Hickman, Z. P., and L. S. Roberts. 1988. *Integrated principles of zoology.* 8th ed. St. Louis: C.V. Mosby Co.

Hirsch, M. S., and J. C. Kaplan. April 1987. Antiviral therapy. *Scientific American.*

Hogle, J. M. March 1987. The structure of poliovirus. *Scientific American.*

Horgan, J. February 1991. In the beginning. *Scientific American.*

Knoll, A. H. October 1991. End of the proterozoic eon. *Scientific American.*

Levinton, J. S. November 1992. The big bang of animal evolution. *Scientific American.*

Li, W. H., and D. Graver. 1991. *Fundamentals of molecular evolution.* Sunderland, Mass.: Sinauer Associates.

Lovelock, J. E. 1987. *Gaia: A new look at life on earth.* New York: Oxford University Press.

Margulis, L. 1984. *Early life.* Boston, Mass.: Jones and Bartlett Publishers.

Pechenik, J. A. 1991. *Biology of invertebrates.* 2d ed. Dubuque, Iowa: Wm. C. Brown Publishers.

Raven, P. H., et al. 1986. *Biology of plants.* 4th ed. New York: Worth Publishers, Inc.

Rismiller, P. D., and R. S. Seymour. February 1991. The echidna. *Scientific American.*

Rose, M. R. 1991. *Evolutionary biology of aging.* New York: Oxford University Press.

Ross, P. E. May 1992. Eloquent remains. *Scientific American.*

Schopf, J. W. 1992. *Major events in the history of life.* Boston, Mass.: Jones and Bartlett Publishers.

Storch, G. February 1992. The mammals of island Europe. *Scientific American.*

Strickberger, M. W. 1990. *Evolution.* Boston, Mass.: Jones and Bartlett Publishers.

Tattersall, I. August 1992. Evolution comes to life. *Scientific American.*

Tortora, G. J., B. R. Funke, and C. L. Case. 1992. *Microbiology: An introduction.* 4th ed. Redwood City, Calif.: Benjamin/Cummings Publishing Co.

Toth, N., D. Clark, and G. Ligabue. July 1992. The latest stone ax makers. *Scientific American.*

Vollrath, F. March 1992. Spider webs and silks. *Scientific American.*

Wenke, R. J. 1990. *Patterns in prehistory.* 3d ed. New York: Oxford University Press.

Wilson, A. C., and R. L. Cann. April 1992. The recent African genesis of humans. *Scientific American.*

York, D. January 1993. The earliest history of the earth. *Scientific American.*

Part Seven

BEHAVIOR AND ECOLOGY

A Bengal tiger approaches

BEHAVIOR AND ECOLOGY

The behavior of organisms increases their chances of survival and allows them to interact with both their own kind and other species. Species interact within ecosystems, units of the biosphere in which energy flows and chemicals cycle. Mature natural ecosystems contain populations that remain relatively constant in size and require about the same amount of energy and chemicals each year.

In contrast, the worldwide human population continues to increase in size and uses more energy and raw materials each year. Because energy is used inefficiently and raw materials are not cycled properly, the human population is dependent on natural ecosystems to absorb pollutants. The natural ecosystems are no longer able to support humanity in this manner, and we must find ways to use energy more efficiently and to recycle materials so that sustainable growth is possible. Furthermore, the preservation of the natural communities, called biomes, is beneficial to all ecosystems. Preserving the biomes helps to ensure the continuance of the biosphere upon which *all* species depend.

The countries of the world are divided into 2 groups. The more-developed countries enjoy a comparatively high standard of living, consume more resources, and create much of the pollution in the world. The less-developed countries have a comparatively low standard of living and consume fewer resources, but they are responsible for most of the world's population increase.

31

ANIMAL BEHAVIOR

Chapter Concepts

1.
The nervous and endocrine systems have immediate control over behavior. 624, 633

2.
Behavior has a genetic basis, but interaction with the environment is also involved. 624, 633

3.
Natural selection influences such behaviors as methods of feeding, selecting a home, and reproducing. 628, 633

4.
Animals living in societies have various means of communicating with one another. 630, 633

5.
Apparently, altruistic behavior only occurs if it actually benefits the animal. 632, 633

Gentoo penguin feeding 2 chicks in Antarctica

Chapter Outline

ANIMALS carry on many activities that help them get food, avoid predators, find shelter, seek mates, and reproduce. **Behavior** encompasses all those activities that enable an animal to meet its needs.

Mechanisms of Behavior

Mechanisms of behavior pertain to the immediate cause of behavior. The ultimate (evolutionary) cause of behavior is not considered.

Physiology's Role in Behavior: Ringdoves

The role of the nervous and endocrine (hormones) systems in animal behavior is exemplified very well by a study of ringdove reproductive behavior. When male and female ringdoves are put together in a cage, the male begins courting by repeatedly bowing and cooing (fig. 31.1). Because castrated males do not do this, it can be reasoned that the hormone testosterone readies the male for this behavior. The sight of the male courting causes the pituitary gland in the female to release FSH. This, in turn, causes her ovaries to produce eggs and to release estrogen into the bloodstream. Now both male and female birds construct a nest, during which time copulation takes place. The hormone progesterone causes birds of either sex to incubate the eggs, and the hormone prolactin causes the crop (a portion of the lower esophagus) to grow so that both parents are capable of feeding their young a secretion called crop milk.

Figure 31.1

Ringdove mating behavior. **a.** Male and female respond to each other. **b.** Copulation takes place. **c.** Male and female create a nest. **d.** Chicks are fed crop milk. The reproductive behavior of ringdoves is controlled by the nervous and endocrine systems.

a.

b.

c.

d.

To determine whether the incubation behavior of the female is actually controlled by this sequence of events, investigators studied 3 experimental groups: females housed alone, females housed with males only, and females housed with a male and nesting material. Each group of females was presented with a nest containing eggs. All females housed with a male and nesting material incubated eggs, but this was not so for the other groups.

> Complex interactions between external cues and the nervous and endocrine systems control the reproductive cycle of ringdoves.

Innate and Learned Behaviors: The Sequence

Behavior has a genetic basis, but development is also required for the final pattern to appear. There is an interaction with the environment during development. Two early ethologists (biologists who study behavior), Konrad Lorenz and Niko Tinbergen, were particularly interested in innate versus learned behaviors. They discovered that the degree of programming and the flexibility of responses reflect an evolutionary sequence from invertebrates to vertebrates (fig. 31.2).

Figure 31.2

Innate types of behavior (taxes and reflexes) are more frequently utilized by invertebrates, and learned types of behavior are more frequently utilized by vertebrates.

Source: Data from V. G. Dethier and Eliot Stellar, *Animal Behavior*, 3d ed., page 91, 1970.

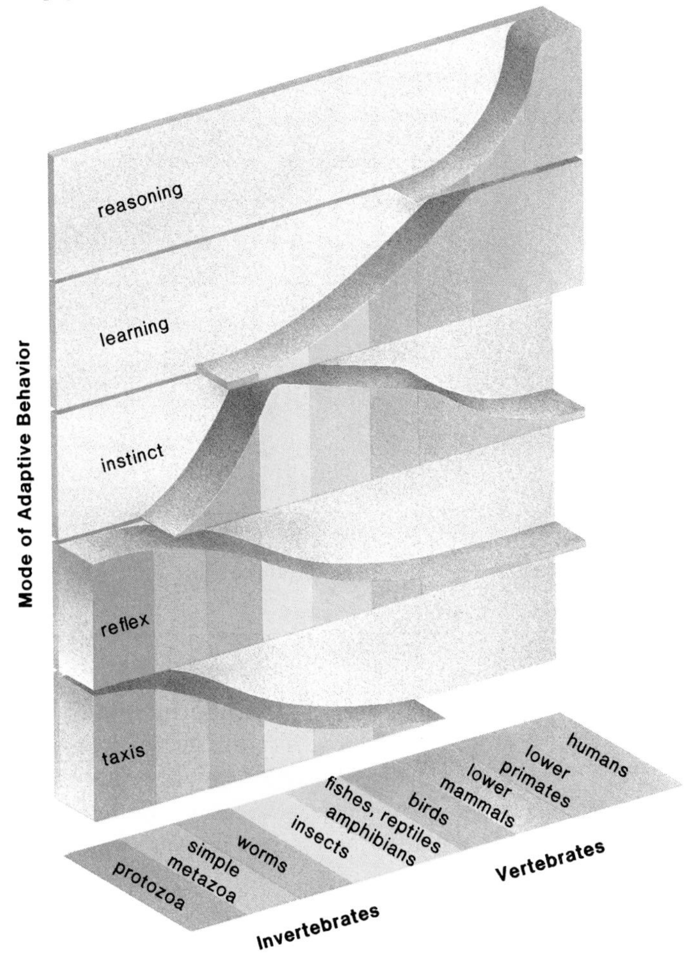

Innate Behaviors: Same Sign, Same Action

Taxes (sing., **taxis**) are directed movements in reaction to a stimulus. For example, phototaxis is movement toward or away from light, and geotaxis is movement in the same or opposite direction as gravity. Some **innate behaviors** seem like automatic reflexes. The members of a species all perform the exact same sequence of actions under the same environmental conditions. In humans, when a knee is hit by a mallet, the lower leg jerks in a characteristic manner. Tinbergen used the term **fixed-action pattern (FAP)** for a behavior that occurs automatically, as if it were a composite of reflex actions. The behavior pattern is stereotyped—it always occurs the same way. A stimulus that initiates a fixed-action pattern is called a **sign stimulus.** For example, male robins attack a tuft of red feathers rather than an exact replica of a male robin without a red breast (fig. 31.3). The color red, not the entire bird, is the sign stimulus that releases the aggressive behavior. Presumably, sign stimuli have been chosen through the process of natural selection.

Animals performing FAPs seem to be acting in a purposeful manner, but actually FAPs always proceed from start to finish in a prescribed manner, regardless of the circumstance. For example, certain solitary female digger wasps dig a hole, seek out a caterpillar, and then paralyze it with a series of stings along the undersurface. The wasp carries the prey to the hole and pulls it in. After laying her egg on the side of the caterpillar, she begins to close the hole. If at this point the researcher removes the caterpillar and puts it on the ground nearby, the wasp continues to cover the hole, even though the caterpillar is in full view.

Fixed-action patterns are innate. They are largely controlled by genes.

Figure 31.3

Aggressive behavior in male robins. **a.** A male robin approaches a tuft of red feathers and a model without a red breast. **b.** The male robin attacks the tuft of red feathers rather than the model. The color red is a sign stimulus for the aggressive behavior, a fixed-action pattern.

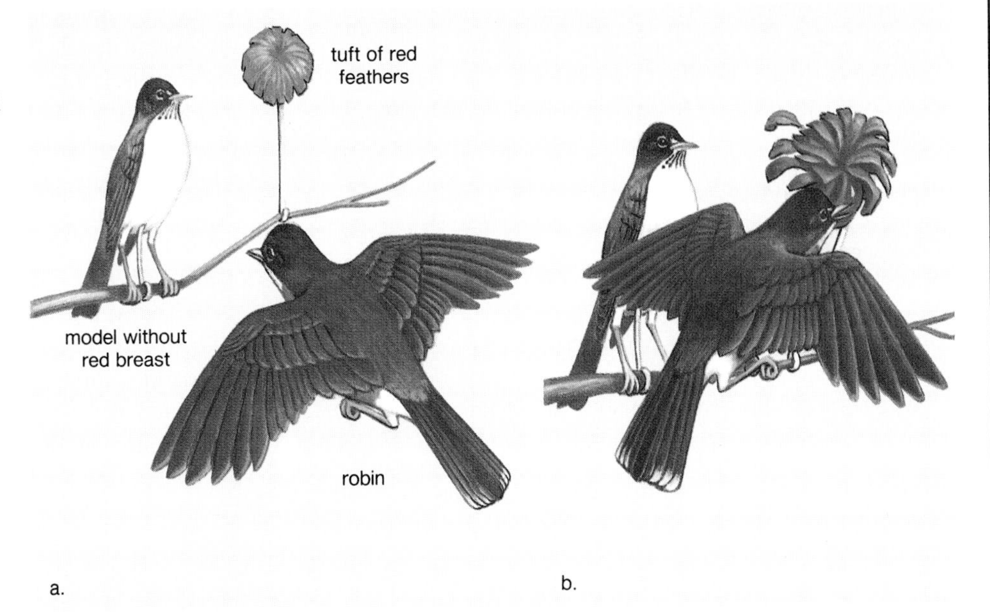

Figure 31.4
Modification of behavior. **a.** Chicks peck at the red spot on a parent's bill when they want to be fed. **b.** Over time, chicks peck at a more exact replica of a parent's bill (first model). In each example, the top bar represents the pecking frequency of newly hatched chicks and the bottom bar represents the pecking frequency of chicks 2–5 days old.

It is not unusual for innate behaviors to be modified by experience. For example, baby gulls peck at a parent's beak to induce the parent to feed them. Research with various models has shown that the chicks are not very discriminatory at first and choose a model that has a large red spot (the sign stimulus), even if the model does not resemble the parent. Over time, however, chicks become progressively more selective and choose a model that more nearly resembles the parent (fig. 31.4).

Most behavior probably contains all of these components:

Preprogramming. The genetic makeup of an organism predisposes it to respond to certain stimuli in a certain way.

Practice. Experience affects the development of the behavior.

Potentiation. The physiological state of the animal readies the animal to perform the behavior (sometimes called motivation or drive).

Performance. The resulting series of acts and internal changes that accompany the behavior.

Learned Behaviors: Result of Experience

Learning is a change in behavior as a result of experience. The capacity to learn is inherited and enables an organism to change its behavior to suit the environment.

Imprinting and Critical Periods. Konrad Lorenz observed that hatchlings become attached to and follow the first moving object they see. He termed this behavior **imprinting** and suggested that it is a means by which organisms learn to recognize their own species. Ordinarily, the object followed is the mother; however, in Lorenz's experiments, goslings became imprinted on him (fig. 31.5) and later chose him over their own mother when given the opportunity.

Lorenz found that if goslings are totally isolated for the first 2 days after hatching, they fail to undergo the process of imprinting. Apparently, there is a *critical period* of time during which imprinting is possible. Since Lorenz's time, it has been discovered that there are critical periods for other types of learning in birds. For example, a male white-crowned sparrow reared in isolation sings a song, but the song is less complicated than the normal song. If the bird is permitted to hear another adult white-crowned sparrow sing during a critical period of 10–50 days following hatching, however, it later learns to sing the normal song. White-crowned sparrows never learn to sing the song of another bird species.

Today, the term *imprinting* is used in a larger context to refer to any type of learning that has a critical period, even when older animals are involved. For example, adult birds are imprinted on their young after hatching and accept as their own any chick, even one of another species, if the chick is put in the nest during the critical period.

Figure 31.5

a. Normally, goslings are imprinted on their mother. **b.** Goslings follow Konrad Lorenz, who, like Niko Tinbergen, performed experiments in the field. Lorenz found that goslings learn to follow the first moving object they see and therefore can be imprinted on an improper object.

a.

b.

Other Modes of Learning. Habituation. When an organism is presented with the same stimuli time and time again, the original response ceases. For example, at first chicks crouch in fear when even a leaf flutters overhead, but then they learn to disregard this movement and even that of certain birds. Habituation is useful because it prevents animals from wasting energy on unnecessary responses.

Operant conditioning (trial-and-error learning). An animal that is rewarded (positively reinforced) for making a proper choice learns to make this response without hesitation. This type of learning, employed by animal trainers, can be used successfully to teach animals all sorts of tricks. Characteristically, however, conditioned learning is maintained only as long as a reward is given for the behavior. If the reward is withdrawn, the behavior also disappears, a result called extinction. Extinction is also a type of learning.

Classical conditioning (associative learning). An animal can learn to respond to an irrelevant stimulus. Pavlov's dogs expected food and began to salivate when a bell rang because they had learned to associate the ringing of the bell with food. Associative learning can explain behavior that seems out of place—advertisers teach us to associate power and prestige with their particular product, for example.

Insight learning. Some animals are able to solve a problem using previous experiences to think through to a solution. To the observer, it seems as if the animal employing insight needs no practice to successfully reach a goal. For example, apes can devise the means to get to bananas placed out of their arms' reach: they pile up boxes or use a pole in order to reach the food (fig. 31.6).

Figure 31.6

Apes are capable of insight learning—reasoning out a solution to a problem without trial and error. In (**a**), a chimpanzee is unable to reach a banana. In (**b**), it stands on boxes to reach the banana. If an animal can reason, does it have a cognitive awareness of itself and its surroundings? Researchers are now wrestling with this question.

a.

b.

Table 31.1
Review of Terms Related to Behavior

Term	Definition
Fitness	The ability of an organism to survive and reproduce in its local (immediate) environment.
Natural selection	The environment selects certain members of a population for survival and reproduction. In the end, adaptive traits become more prevalent in the population (p. 520).
Altruistic behavior	Unselfish acts that help others in a group so that they, not the individual, have a better chance of surviving and reproducing.
Selfish behavior	Acts that increase the likelihood that only the individual will survive and reproduce offspring.
Inclusive fitness	The ability of kin (relatives who share common genes) to survive and reproduce.

Natural Selection and Behavior

Behavior patterns enhance **fitness**—the ability of an organism to survive and reproduce (table 31.1). This premise applies to such behaviors as feeding, finding and defending a territory, and mating.

Feeding Behavior and Energy Costs

Regardless of how an animal feeds, the energy benefit of the food itself should be greater than the energy cost of getting the food. Using this as a principle, we can see why great blue herons have a more varied diet in less-productive northern lakes than in the more-productive waters of Florida. Still, there is a genetic component to food preference. Isolation-reared garter snakes from the Florida Everglades readily accept fish as food, but not so for garter snakes from Massachusetts.

Natural selection not only promotes efficient feeding techniques, it also reduces the likelihood of being eaten. Animals have antipredator defenses, including concealment, startle display, warning coloration, and vigilance (fig. 31.7).

Home Territory for Better Reproduction

In general, animals tend to select habitats that increase their reproductive success. For example, among plant parasites called aphids, the first females to hatch in the spring settle on large leaves, leaving the small ones for latecomers. The females on the large leaves have greater reproductive success than the females on small leaves.

Figure 31.7
Antipredator defenses. **a.** Startle display. The South American lantern fly has a large false head, which resembles an alligator's head. This may frighten a predator into thinking it is facing a dangerous animal. **b.** Warning coloration. Poison-arrow frogs are so poisonous they were used by natives to make their arrows instant lethal weapons. The coloration of these frogs warns others to beware.

a.

b.

Territoriality means that an animal defends a certain area, preventing certain other members of the same species from utilizing it. Even the aphid females defend their leaves against other females and, as mentioned, those utilizing the largest leaves reproduce the most. Among great tits (a type of bird), John Krebs found that 92% of woodland nests but only 22% of hedgerow nests had offspring. He removed 6 pairs of birds from the woodlands and observed that within a few days the hedgerow birds had moved in. Territoriality in birds improves reproductive success because the better territories provide more food for offspring.

Figure 31.8
Male elephant seals fight to establish dominance. In their polygamous system, males invest little in the offspring, which are cared for by females.

Reproductive Behavior: Monogamy and Polygamy

The reproductive behavior of an organism is expected to increase individual fitness. At least for the season, songbirds are often monogamous—they have only one mate. In the case of birds, incubating the eggs or gathering enough food requires the effort of both parents. If the parents did not cooperate, neither would have any offspring. On the other hand, mammals tend to be polygamous—they have more than one mate. A female produces only a few hundred eggs during her lifetime, but a male produces millions of sperm. A female is certain in a way no male can be that the offspring is her own. Under these circumstances, it is more adaptive for a female to invest a great deal of energy in the rearing of offspring, and it is adaptive for males to impregnate as many females as possible.

In polygamous systems, males often compete intensely for the right to mate with females. In elephant seals, males fight with one another to establish a dominance hierarchy (fig. 31.8). A **dominance hierarchy** exists when animals form a relationship in which a higher ranking animal has access to resources before a lower ranking animal. One study found that the top 5 elephant seal males do at least 50% of the copulating. Since pups are born a year after copulation, males have no way of knowing if an offspring is theirs. They have been observed to trample a pup, possibly their own, while striving to inseminate a female.

Females participate in the selection process. They are observed to preferentially mate with a genetically superior male. For example, female elephant seals scream loudly during copulation, alerting dominant males, who chase away less-dominant males. At times, female selection even seems to be responsible for the evolution of male attributes, such as the multicolored face of male mandril baboons and the splendid tail of male peacocks. Courtship rituals also probably evolved as a way for females to make a choice among males. Only males performing the correct ritual are selected as mates.

> Animal behaviors, as exemplified by the way in which an animal gets food, finds and occupies a home territory, and carries on reproduction, are expected to increase individual fitness. Do 31.1 Critical Thinking, found at the end of the chapter.

Societies

A *society* is a group of individuals belonging to the same species that are organized in a cooperative manner extending beyond sexual and parental behavior. **Sociobiology** applies the principles of evolutionary biology to the study of social behavior in animals.

There are benefits and disadvantages to living in a group. Possible benefits include the following:

Protection against predators. Members of the society give alarm signals and group together for defense.

Finding and procuring food. Members of a society signal each other as to the location of food and sometimes work together to procure food.

Division of labor among specialists. In insect societies, members have specialized roles to perform.

Possible disadvantages of living in a group include the following:

Increased competition for resources. Members compete with one another for shelter, food, and mates.

Increased chance of disease and parasites. Members give diseases and spread parasites from one to the other.

Interference with reproduction. Members of a society are sometimes aggressive toward the young of other members.

> Animals socialize when the benefits of the behavior outweigh the disadvantages. Benefit is judged by increased fitness of the individual member.

Four Kinds of Communication

In particular, members of a society must communicate with one another. *Communication* is defined as the transmission of a signal from one animal to another such that the sender usually benefits from the recipient's response.

Many animals (even humans) practice *chemical communication.* The term *pheromone* is used to designate chemical signals that are passed between members of the same species. For example, female moths secrete chemicals from special abdominal glands. These chemicals are detected downwind by receptors on male antennae. This signaling method is extremely efficient: it has been estimated that only 40 of the 40,000 receptors on the male antennae need to be activated in order for the male to respond.

Visual communication includes many sign stimuli. For example, male birds and fishes sometimes undergo a color change that indicates they are ready to mate. When female baboons are willing to mate, there is a reddening of the flesh on the buttocks. Defense and courtship displays are ritualized behaviors having exaggerated movements that are always performed in the same way so that the sign stimuli are clear.

Tactile communication occurs when one animal touches another. For example, a gull chick pecks at a parent's beak in order to induce the parent to feed it. In primates, grooming—when one animal cleans the coat of another—helps cement social bonds within a group. The communication of honeybees is remarkable because the so-called language of the bees uses a variety of stimuli to impart information about the environment. Karl von Frisch, another famous ethologist, carried out many detailed bee experiments in the 1940s and was able to determine that when a foraging bee returns to the hive, it performs a *waggle dance* (fig. 31.9). The dance, which indicates the distance and the direction of a food source, has a figure-eight pattern. As the bee moves between the 2 loops of the figure 8, it buzzes noisily and shakes its entire body in so-called waggles. Distance to the food source is believed to be indicated by the number of waggles and/or the amount of time taken to complete the straight run. The straight run also indicates the location of the food. When the dance is performed outside the hive, the straightaway indicates the exact direction of the food, but when it is done inside the hive, the angle of the straightaway to that of the direction of gravity is the same as the angle of the food source to the sun. In other words, a 40-degree angle to the left of vertical means that food is 40 degrees to the left of the sun. As mentioned earlier, honeybees can use the sun as a compass because their biological clock (see reading on the next page) allows them to compensate for the movement of the sun in the sky. In the dark hive, bees use a combination of tactile and auditory communication. Through touch, bees can determine the direction and waggles of the dance. Both the waggles and the buzzing noises of the dance tell the distance to the food source.

Figure 31.9

Tactile and auditory communication among bees relates information not about the bee but about the environment. Honeybees do a waggle dance to indicate the direction of food. **a.** If the dance is done outside the hive on a horizontal surface, the straight run of the dance points to the food source. **b.** If the dance is done inside the hive on a vertical surface, the angle of the straightaway to that of the direction of gravity is the same as the angle of the food source to the sun.

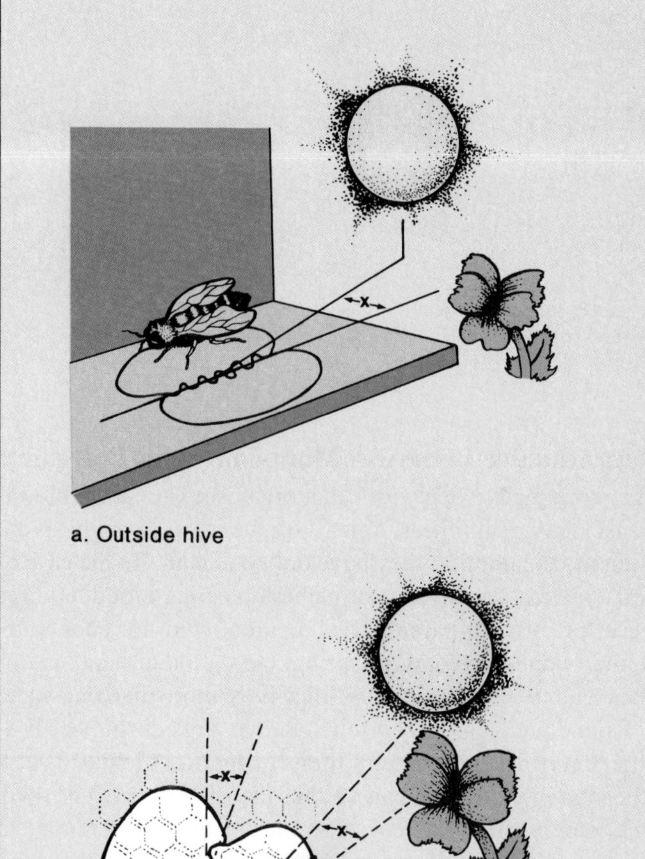

Auditory (sound) communication has some advantages over other kinds of communication. It is faster than chemical communication and, unlike visual communication, it can be sent and received even in the dark. Auditory communication can be modified not only by loudness but also by pattern,

HOW NATURE KEEPS TIME

Certain behaviors in animals reoccur at regular intervals. Behaviors that occur on a daily basis are said to have a *circadian* (about a day) *rhythm.* For example, some animals, like humans, are usually active during the day and sleep at night. Others, such as bats, sleep during the day and hunt at night. There are also behaviors that occur on a yearly basis. For example, in the Northern Hemisphere, birds migrate south in the fall, and the young of many animals are born in the spring. Such behaviors have a circannual rhythm.

Originally, it was assumed that environmental changes such as the coming of night or the length of the day directly controlled cyclical behaviors in animals. But it is now known that cyclical behaviors occur even when the associated stimulus (daylight or darkness) is lacking. For example, fiddler crabs are dark in color during the day and light in color at night, even when kept in a constant environment. But if the crabs are kept in the constant environment indefinitely, the timing of the daily change tends to drift out of synchronization with the natural cycle. For this reason, it has been suggested that rhythmic behavior is under the control of an innate, internal biological clock that runs on its own but is reset by external stimuli.

Aside from keeping time, a biological clock must also be able to bring about the change in behavior. In the fiddler crab, for example, the behavior must be able to stimulate the processes that cause the shell to change color. Therefore, a biological clock system must have the following components (fig. 31.A):

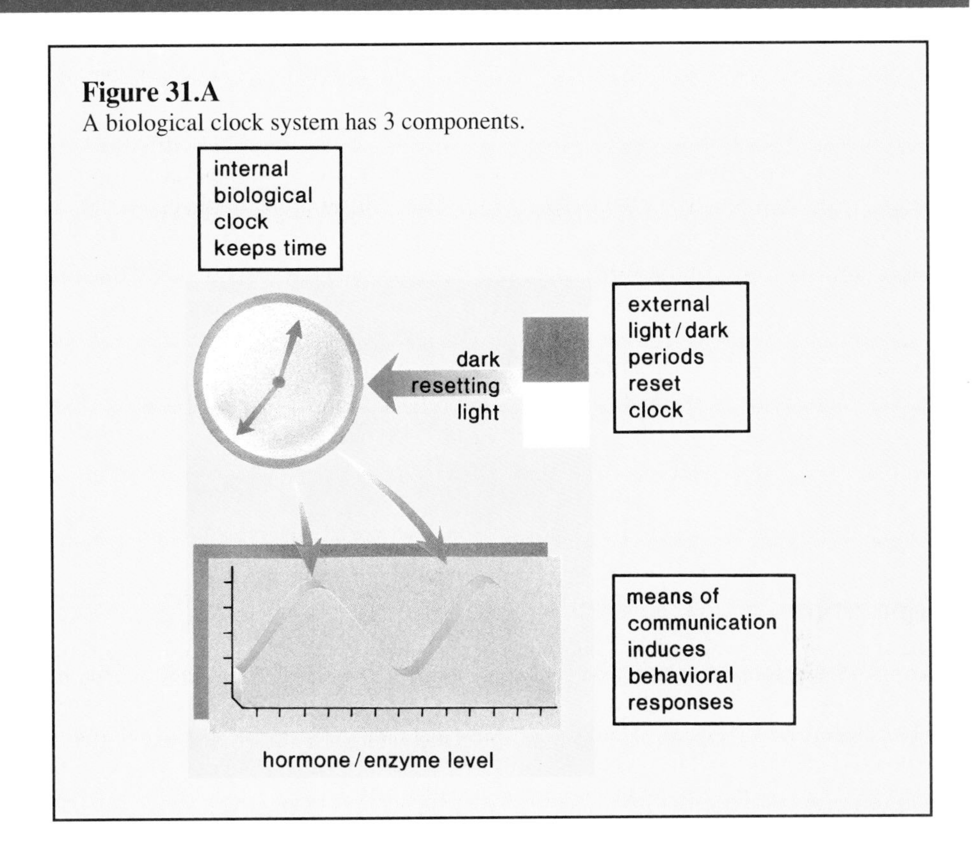

Figure 31.A
A biological clock system has 3 components.

A biological clock (timekeeping mechanism) that keeps time independent of external stimuli (i.e., a minute is always a minute).

A receptor that is sensitive to light/dark periods and can reset the clock for circadian rhythms or can indicate a change in the length of the day/night for circannual rhythms.

A means of communication by which the clock induces the behavioral response.

A review of the discussion concerning flowering in plants shows that the last 2 components of a biological clock system have been tentatively identified. Phytochrome (p. 144) is believed to be the receptor sensitive to light and dark periods, and plant hormones are believed to be the means by which flowering is induced. The timekeeping mechanism has been not identified, however.

In animals, we know that melatonin is produced at night by the pineal gland; therefore, the amount produced begins to increase in the fall and decrease in the spring. Melatonin is believed to inhibit development of the reproductive organs, which explains why some animals reproduce in the spring. For animals in which the pineal gland is the "third eye," both the clock and the receptor are presumed to be in the pineal gland.

For some behaviors, it appears that only the timekeeping mechanism is required. For example, bees and some birds use the sun and the stars as compasses, even though these compasses change position throughout the day and the night.

duration, and repetition. Male crickets have calls and male birds have songs for a number of different occasions. Sailors have long heard the songs of humpback whales because they are transmitted through the hull of a ship. One purpose of the song could be sexual by serving to advertise the availability of the singer.

Language is the ultimate auditory communication, but only humans have the biological ability to produce a large number of different sounds and to put them together in many different ways. Although chimpanzees can be taught to use an artificial language, they never progress beyond the capability level of a child 2 years old (fig. 31.10).

Altruism and Hidden "Selfishness"

Altruism is behavior performed for the good of others rather than for self-interest. For example, in ants and bees, only the queen lays eggs but most of the work is done by female workers, which do not have any offspring at all. How can such a society increase the fitness of the workers? The answer may lie in the fact that the male parent is haploid; therefore, siblings have on average three-fourths of their genes in common (fig. 31.11). Because offspring have only one-half of their genes in common with their parents, it may actually increase the fitness of the worker ants to raise siblings, some of which may become queens rather than workers.

Altruism, exemplified by the willingness of daughters to help the queen rather than to have their own offspring, can be explained by noting that *kin selection* increases the animal's **inclusive fitness.** In other words, survival of the close relatives (kin) increases the frequency of an animal's genes in the next generation. Therefore, the animal's overall (inclusive) fitness is also increased.

Figure 31.10

A chimpanzee with a researcher. Chimpanzees are unable to speak but can learn to use a visual language consisting of symbols. Some researchers believe that chimps are capable of creating their own sentences, but others believe that they only mimic their teachers and never understand the cognitive use of language. Here the experimenter shows Nim the sign for "drink." Nim copies.

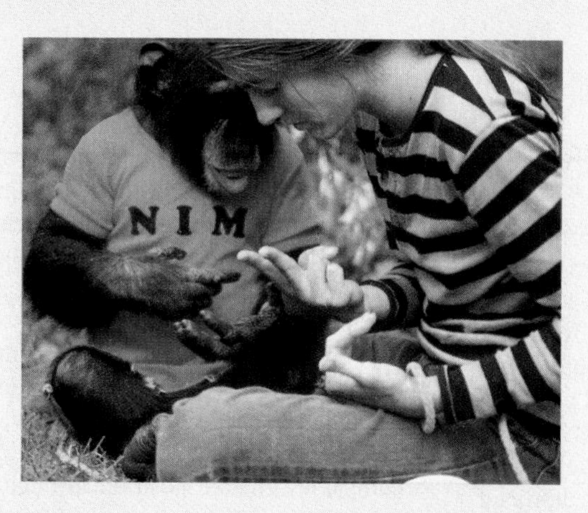

Figure 31.11

Sociobiology and the behavior of ants. It is genetically more advantageous for female ants and bees to assist in caring for siblings rather than to produce their own offspring. Offspring possess one-half of the queen's genes, but because the male parent is haploid, siblings share, on the average, three-fourths of their genes.

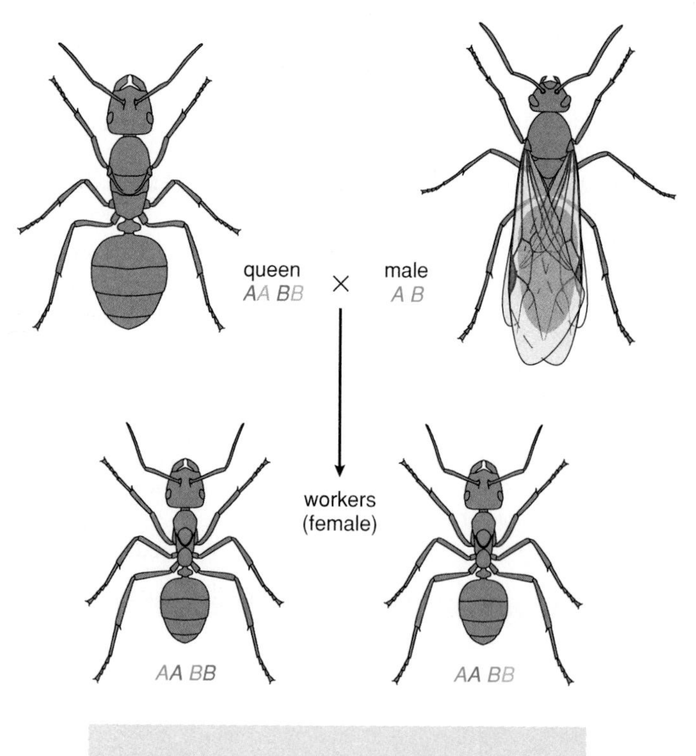

Another example of altruism is alarm calling. Evolutionary theory suggests that an animal always acts to ensure its own individual safety. However, it is observed that among animals that live in groups, members give alarm calls that put themselves in danger. This behavior, too, can be explained on the basis of kin selection and inclusive fitness. If alarm calling means the survival of relatives, the increase in inclusive fitness may outweigh any loss of individual fitness.

Sociobiologists interpret human behavior according to this same principle. For example, parental love is clearly selfish in that it promotes the likelihood that an individual's genes will be present in the next generation's gene pool. People tend to be less aggressive toward blood relatives than nonrelatives. We would expect small towns to have less crime than large cities because small communities have a greater percentage of related residents.

Human reproductive behavior can be similarly examined. Human infants are born helpless and have a much better chance of developing properly if both parents contribute to the effort. The human female, unlike other mammals, is continuously receptive to sexual intercourse. (In other mammals, females are receptive only during estrus, a physiological stage that recurs periodically, sometimes only once a year.) The fact that sex is continuously available may help assure that the human male will remain with and help the female raise the young.

While it may seem as if members of a group are altruistic, it can be shown that their motives are most likely selfish and that their behavior serves to increase their own fitness. Sometimes it is necessary to utilize the concept of inclusive fitness when interpreting an animal's behavior.

SUMMARY

The nervous and endocrine systems have immediate control over behavior. Ultimately, all behavior has a genetic basis, but even so, an interaction between inheritance and the environment determines behavior. The genetic component of behavior is more noticeable in innate (instinctive) behaviors. Fixed-action patterns, which appear to be a series of reflexes, occur as a response to a sign stimulus. FAPs are largely inherited, although they sometimes increase in efficiency with practice. In learned behaviors such as imprinting, habituation, operant conditioning, classical conditioning, and insight learning, the environmental component is believed to be more influential.

The role of natural selection in behavior is obvious in an animal's method of feeding, selecting a home, and reproducing. Animals try to increase their individual fitness by first securing resources and then by reproducing with many mates (males) or the best mate (females).

Animals live in a society when the benefits outweigh the drawbacks. In a society, animals have various means of communicating with one another. Living in a society is expected to increase the chance of reproduction; altruistic behavior to increase inclusive fitness is observed in such situations.

STUDY QUESTIONS

In order to practice **writing across the curriculum,** students should write out the answers to any or all of the study questions. The study questions are sequenced in the same order as the text.

1. Describe the role of the nervous and endocrine systems in the reproductive behavior of ringdoves. (p. 624)
2. Give an example to show that a fixed-action pattern can be modified by learning. (p. 626)
3. What are the 4 components of most behavior patterns? Explain each one. (p. 626)
4. Explain the concept of critical period in relation to imprinting. (p. 626)
5. Name 5 types of learning, and explain the significance of each. (pp. 626–627)
6. What premises underlie the various ways animals get food, establish a territory, and carry on reproductive behavior? (p. 628)
7. List and describe the ways that animals communicate with one another. (pp. 630–632)
8. What are the benefits and the disadvantages of living in a group? (p. 630)
9. What is a biological clock system, and what are the functions of its components? (p. 631)
10. Explain altruism from the sociobiologist's point of view. (p. 632)

OBJECTIVE QUESTIONS

1. A(n) _______ pattern is observed when male robins in breeding plumage attack a tuft of red cotton.
2. _______ has occurred when chicks no longer cringe and hide when a long-necked bird flies overhead.
3. Feeding behavior, like all behaviors, is determined by the genotype and the _______.
4. The function of the nervous system is a(n) _______ cause of behavior.
5. To recognize that it is adaptive for male robins to attack an object with a red breast is to consider the role of _______ in behavior.
6. Animals acquire resources by staking out _______, where they reproduce.
7. Males impregnate as many females as possible because they are never sure offspring _______.
8. Altruism increases the _______ fitness of the individual.
9. Helpers are usually _______ to the breeders they assist in reproducing.
10. For each item that follows, indicate whether it is a form of learning or a form of communication and then tell what specific form it is. The first is done for you as an example.

	Communication	**Chemical**
a. pheromone		
b. critical period	_______	_______
c. language	_______	_______
d. positive reward	_______	_______
e. waggle dance	_______	_______
f. courtship display	_______	_______

CRITICAL THINKING

In order to practice **writing across the curriculum,** students should write out the answers to any or all of the critical thinking questions. Suggested answers to the critical thinking questions are in appendix E.

31.1

In an experiment to determine the control of behavior, young guinea pigs are placed in a cage and are fed a type of food they do not ordinarily eat in the wild. When released, the guinea pigs are offered various types of food, including the one type they were fed while caged.

1. How might you attempt to show that the nervous system is involved in food gathering and eating?
2. What will you conclude if the guinea pigs choose only the type of food they were recently fed?
3. What will you conclude if the guinea pigs choose only the type of food they ordinarily eat in the wild?

SELECTED KEY TERMS

altruism (al´ troo-izm) behavior performed for the benefit of others without regard to its possible detrimental effect on the performer.

dominance hierarchy a system in which animals arrange themselves in a pecking order; the animal above takes precedence over the one below.

fitness the ability of an organism to survive and reproduce in its local (immediate) environment.

fixed-action pattern (FAP) a sequence of reflexes that always occurs in the same order under the same environmental conditions.

imprinting the tendency of a newborn animal to become attached to and follow the first moving object it sees.

inclusive fitness the fitness of closely related group members is part of the fitness of the individual.

innate behavior activities that are instinctive, inborn, and not having to be learned.

learning a change in behavior as a result of experience.

sign stimulus stimulus that releases a fixed-action pattern.

sociobiology an analysis of behavior, particularly social behavior, according to the tenets of evolutionary theory.

taxis (tak´ sis) a movement in relation to a stimulus; phototaxis is movement toward or away from light.

32

THE BIOSPHERE

Chapter Concepts

1.
Communities of organisms on land, called biomes, are adapted to climate. 636, 648

2.
Only forests have adequate rainfall to support large populations of trees. 641, 648

3.
The trees of a temperate deciduous forest lose their leaves in the fall because cold weather sets in; the trees of a tropical rain forest keep their leaves year-round because of adequate rainfall and temperature. 641, 648

4.
Aquatic communities are divided into 2 types: freshwater communities and saltwater communities. 646, 648

5.
Life in coastal communities, where nutrients are adequate, is much more abundant than in the open seas, where nutrients are scarce. 647, 648

Squirrel monkeys in the Amazon

Chapter Outline

THE earth is enveloped by the **biosphere**—a thin realm composed of water, land, and air where organisms are found. The biosphere contains large communities of populations, which on land are called **biomes** (fig. 32.1). There is no equivalent term for large aquatic communities. This is unfortunate because large communities, whether on land or in water, have unique, defined characteristics.

Biomes

The earth's surface can be divided into various zones, each of which contains a number of biomes (fig. 32.2*a*). *Climate,* which can be described largely in terms of temperature and rainfall, determines the geographic location of a biome. The sun's rays fall perpendicular to the equator, but solar energy also creates air and ocean currents, which distribute this heat. Gigantic ocean currents bring the equator's warmth to certain continents while cooling others. In addition, solar energy drives the *water cycle.* Fresh water evaporates from seawater and then rises into the atmosphere. After cooling, it falls as rain over the oceans and the land. Rain is not distributed evenly; it is heaviest along the equator and tapers off toward the poles. Other features—like mountain ranges and ocean and wind currents—determine how much rain falls on the various parts of a continent. Tropical rain forests occur where rainfall is most plentiful, and deserts occur where it is least plentiful (fig. 32.2*c*).

Climate influences where the different biomes are found on the surface of the earth.

Treeless Biomes: From Hot to Freezing

The treeless biomes include deserts, tundra, grasslands, and scrubland (fig. 32.2*b*). These biomes contain few, if any, trees.

Deserts: Less than 25 cm of Rain

Deserts (fig. 32.1*c*) occur in regions where annual rainfall is less than 25 cm (10 in). The rain that does fall is subject to rapid runoff and evaporation. The days are hot because there is no cloud cover to block the sun's rays, but the nights are cold because heat escapes easily into the atmosphere.

The Sahara in Africa, and a few other deserts, have little or no vegetation. In contrast, most deserts have a variety of plants. The best-known desert perennials in this country are the succulent (water-storing), leafless cacti, which have stems that store water and also carry on photosynthesis. All cacti have extensive root systems that can absorb great quantities of water during brief periods of rainfall. Their spines provide a means of defense against desert herbivores. Desert vegetation also includes small-leaved, nonsucculent shrubs, such as sagebrush, a densely branched evergreen, and creosote, with leaves that turn brown and drop off. The mesquite tree, which has deep roots, is also common. Desert annuals exist most of the year as seeds; they burst into flower during the limited period of time when moisture and temperature are favorable.

Certain animals are adapted to the desert environment. A desert has numerous insects, some of which, like the annual plants, have a compressed life cycle. They pass through the stages of development from the parental pupa to the offspring pupa within a very short time, and these pupa remain inactive until it rains again. Reptiles, especially lizards and snakes, are perhaps the most characteristic group of vertebrates found in deserts, but running birds (e.g., the roadrunner) and rodents (e.g., the kangaroo rat) are also well known. Large mammals, like the coyote, prey on the rodents, as do hawks.

Most deserts are located in high-temperature areas that receive less than 25 cm (10 in) of rainfall a year. Plants and animals living in deserts have adaptations that protect them from the sun's rays and the shortage of water.

Figure 32.1

Three major biomes in the United States, each containing its own mix of plants and animals. Temperature and rainfall largely determine the type of biome. **a.** A temperate deciduous forest is typical of the eastern United States. **b.** A tall-grass prairie biome is found in the Midwest. **c.** A desert is located in the Southwest.

b.

a.

c.

Figure 32.2

Biome distribution. **a.** Geographic zones of earth. **b.** Temperature and rainfall determine the nature of the biome to a large extent. For example, rain forests are found in tropical zones, where temperatures are always warm. Deserts, on the other hand, are found in both tropical and temperate zones where rainfall is minimal. **c.** A map showing the location of the major biomes of the world.

Geographic Zones of Earth

Name	Latitude	Climate
Tropical zone	North and south of equator (from equator to Tropic of Cancer and from equator to Tropic of Capricorn)	Seasons determined by changes in rainfall; temperature is always warm.
Subtropical zones	Portions of the temperate zones bordering on the tropical zone	Somewhat like the tropics.
Temperate zones	North and south of tropical zone (Tropic of Cancer to Arctic Circle and Tropic of Capricorn to Antarctic Circle)	Seasons determined by changes in temperature.
Arctic zones	Around the North Pole and the South Pole (from Arctic Circle to North Pole and from Antarctic Circle to South Pole)	Temperature is always cold.

a.

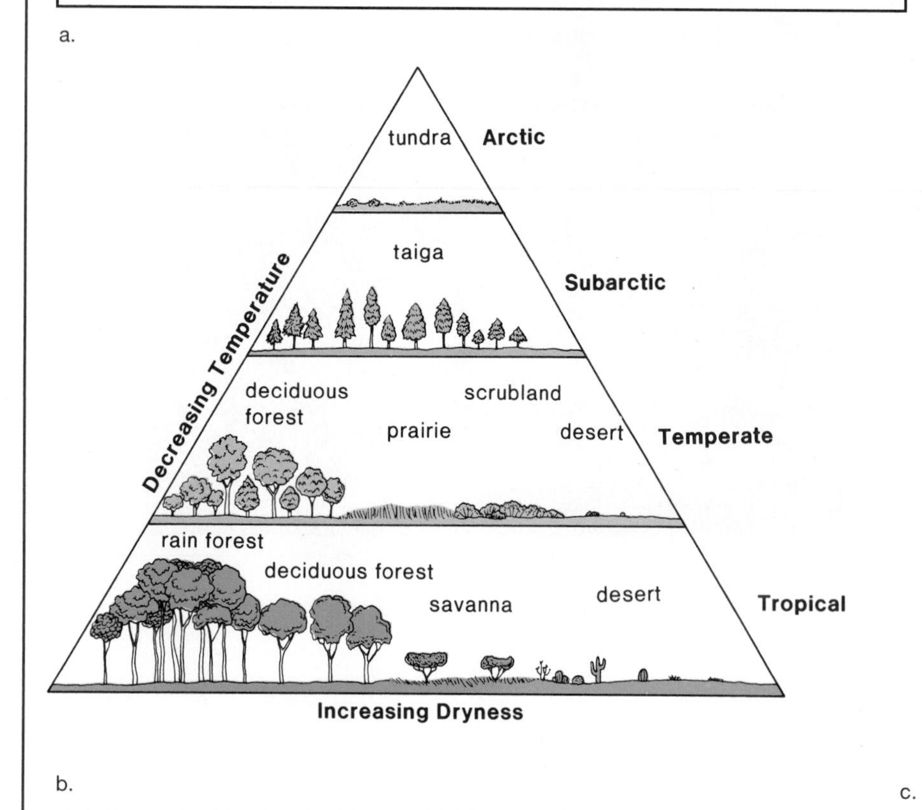

b.

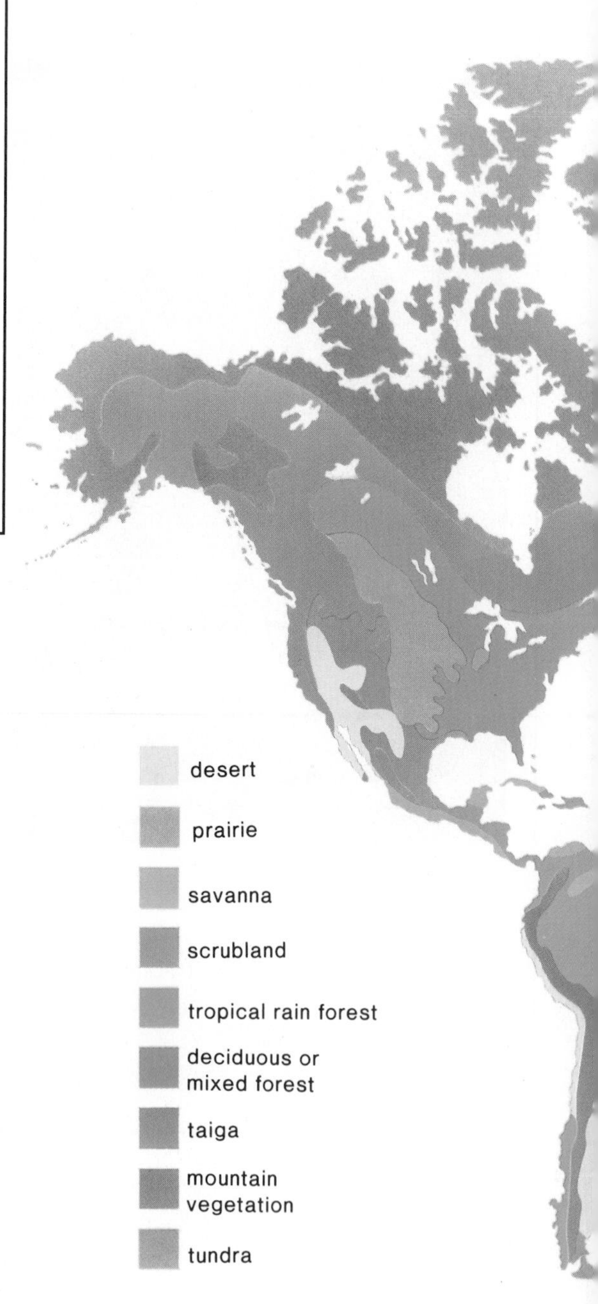

c.

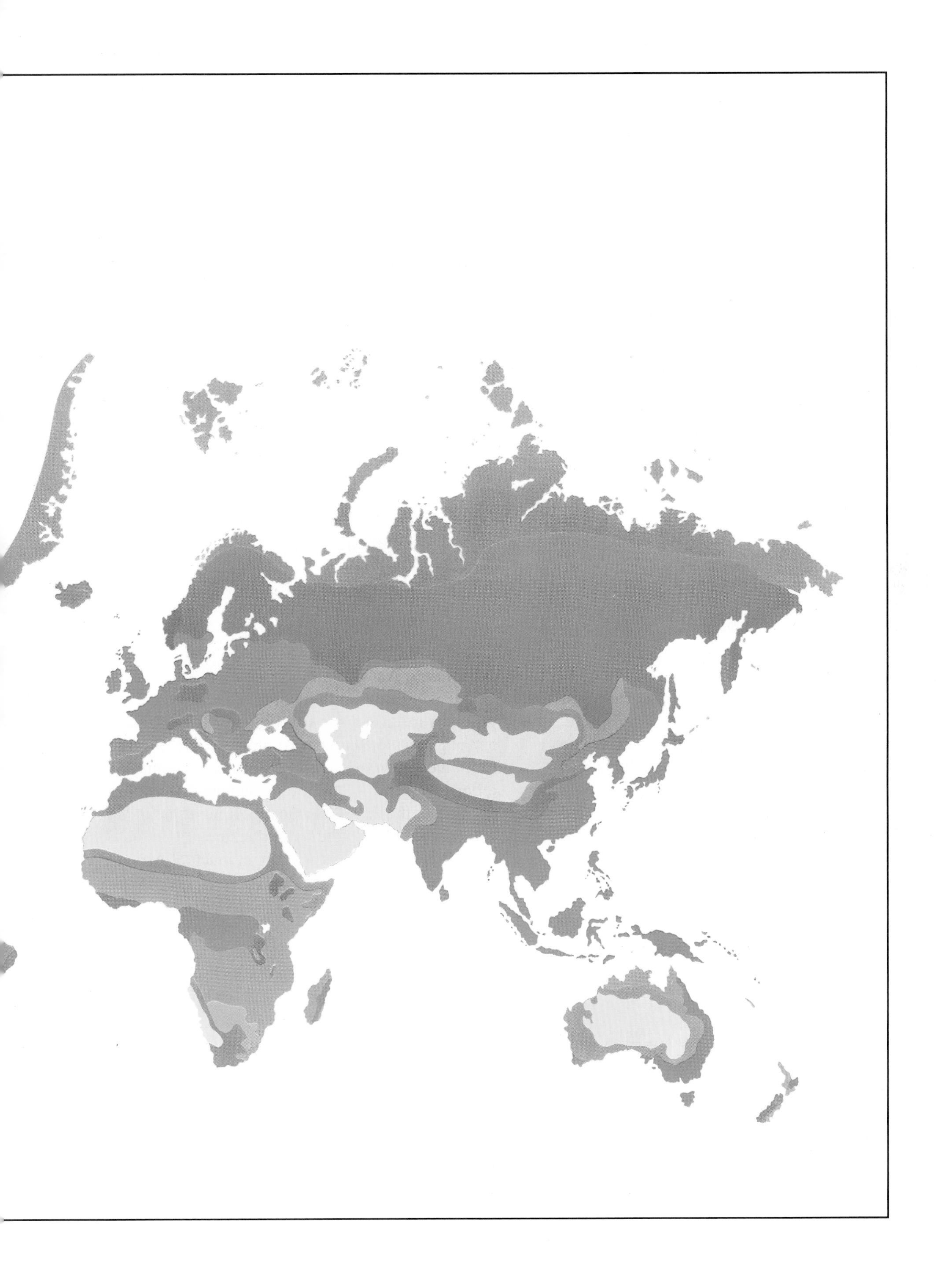

Figure 32.3
Tundra biome in the autumn. Caribou bull grazes on the low-lying vegetation.

Figure 32.4
Savanna biome. **a.** The grasses provide food for many types of herbivores, such as these zebras. **b.** Male lions are carnivores that feed on zebras.

a.

Tundra: Only the Top Layer Thaws

The arctic **tundra** (fig. 32.3) biome encircles the earth just south of ice-covered polar seas in the Northern Hemisphere. (A similar community called the alpine tundra occurs above the timberline on mountain ranges.) The arctic tundra is cold and dark much of the year. Because rainfall amounts to only about 20 cm a year, the tundra could possibly be considered a desert, but melting snow makes water plentiful in the summer, especially because so little evaporates. Only the topmost layer of earth thaws; the **permafrost** beneath this layer is always frozen.

Trees are not found in the tundra because the growing season is too short, their roots cannot penetrate the permafrost, and they cannot become anchored in the boggy soil of summer. In the summer, the ground is nearly covered with sedges and shortgrasses, but there are also numerous patches of lichens and mosses. Dwarf woody shrubs flower and seed quickly while there is plentiful sun for photosynthesis.

Animals that stay in the tundra for the winter are adapted to survive the cold and snowy weather. For example, the ratlike lemming stays beneath the snow; the ptarmigan, a grouse, burrows in the snow during storms; and the musk-ox conserves heat with its thick coat and short, squat body. In the summer, the tundra is alive with numerous insects and birds, particularly shorebirds and waterfowl that migrate inland to feed and nest. Caribou and reindeer also come, along with the wolves that prey upon them. Polar bears are common near the coast.

Grasslands: Not Wet Enough for Trees

Grasslands occur where rainfall is greater than 25 cm but is generally insufficient to support trees. The extensive root system of grasses allows them to quickly recover from drought and fire, which occur frequently in this biome. The matted roots also efficiently absorb surface water and prevent invasion by most trees. Over the years, organic matter builds up in the rich soil, which is often exploited for agriculture.

In the tropics, particularly in much of Africa, there is a grassland called the **savanna** (fig. 32.4). The savanna has few trees because of a severe dry season. One tree that is found here, however, is the flat-topped acacia, which sheds its leaves during a drought, thus relieving it of most of its need for water.

The African savanna supports the greatest variety and number of large herbivores of all the biomes. Elephants and giraffes are browsers that feed on tree vegetation. Antelopes, zebras, wildebeests, water buffalo, and rhinoceroses are grazers that feed on grasses. Any plant litter that is not consumed by grazers is attacked by a variety of small organisms, among them termites. Termites also build towering nests, in which they tend fungal gardens, a source of food. The herbivores support a large population of carnivores. Lions and hyenas hunt in packs, cheetahs hunt singly by day, and leopards hunt singly by night.

A **prairie** is a temperate grassland. Much of the U.S. Midwest was a prairie until it was converted to farmland. A tall-grass prairie (fig. 32.1*b*) gradually gave way to the short-grass prairie. Although grasses dominate both types of prairie, they

b.

are interspersed by forbs (herbs other than grasses). Forbs often catch the eye because they have colorful flowers, whereas grasses do not.

Because there are so few trees, grazers are very common in the prairie. Small mammals, such as mice, prairie dogs, and rabbits, typically live in burrows in the ground, although they usually feed above ground. Hawks, snakes, badgers, coyotes, and foxes feed on these mammals. Large herds of buffalo—estimated at hundreds of thousands—once roamed the prairies, as did herds of pronghorn antelope.

Grasslands occur where rainfall is greater than 25 cm but is insufficient to support many trees. The savanna supports the greatest number of different types of herbivores, which serve as food for carnivores. Most of the U.S. Midwest was part of a prairie biome until converted to farmland.

Scrubland: Shrubs and Very Dry Summers

In parts of South Africa, western Australia, and central Chile, around the Mediterranean Sea, and in California, most of the rain falls in winter and the summers are very dry. Here, there is a dense scrubland called **chaparral** in this country. The scrubby shrubs have small but thick evergreen leaves, which are often coated with a waxy material that prevents loss of moisture. Their thick underground stems can survive dry summers and frequent fires and can also sprout new growth. Rodents and reptiles abound in this biome.

Figure 32.5

Taiga biome in the winter. This biome stretches around the globe in the subarctic zone and contains narrow-leaved coniferous trees that are adapted to harsh conditions.

A Variety of Forests

Taiga: Cold and Evergreen

The **taiga** is a coniferous forest extending in a broad belt across northern Eurasia and North America. The climate in this belt is characterized by cold winters and cool summers, with a growing season of about 130 days. Rainfall ranges between 40 cm and 100 cm per year, with much of it in the form of heavy snows. The great stands of evergreen narrow-leaved trees, such as spruce, fir, and pine (fig. 32.5), are interrupted by many lakes and swamps. (*Taiga* is Russian for *swampland.*) Little light penetrates the dense tree canopy (upper layer of leaves that is the first to receive sunlight), but a ground cover of lichens, mosses, and ferns is usually found.

Compared to other forests, the taiga has relatively few consumer species. In the summer, insects attack the trees and are themselves eaten by warblers and flycatchers. Other birds, such as crossbills and grosbeaks, extract seeds from the cones of the trees. Birds of prey and other carnivores—weasels, lynx, and wolves—feed on small animals, such as rodents. The large herbivores—moose and bears—are apt to be found in clearings or near the water's edge, where small trees and shrubs are found.

Temperate Forests: Wet, Moderate Climate

Temperate forests (fig. 32.1*a*) are found south of the taiga in eastern North America, eastern Asia, and much of Europe. The climate in these areas is moderate, with relatively high rainfall (75–150 cm per year). The seasons are well defined, and the growing season ranges between 140 days and 300 days. The trees, such as oak, beech, and maple, have broad leaves and are termed **deciduous trees;** they lose their leaves in the fall and grow them in the spring. A forest dominated by these trees is called a temperate **deciduous forest.**

TROPICAL RAIN FORESTS: CAN WE LIVE WITHOUT THEM?

So far only about 1.7 million species of organisms have been discovered and named; yet, scientists estimate that the total number of species is more like 5 million–30 million. Most of the unknown species live in the tropical rain forests, a green belt spanning the planet on both sides of the equator. A square kilometer of South American forest provides a home for hundreds of bird and thousands of insect species. Forty-three ant species were counted on one tree in the Peruvian Amazon—more kinds of ants than are found in the entire British Isles. Some 1,200 different plant species were observed in less than one square kilometer of forest at the biological station in Ecuador. The Amazon forest may be home to 30,000 plant species, twice the number found in the United States. Altogether, the tropics contain two-thirds of the world's known plants, 90% of the nonhuman primates, 40% of birds of prey, and 80% of the world's insect species.

Tropical forests cover only 6–7% of the total land surface of the earth—an area roughly equivalent to our lower 48 states. Every year humans destroy an area of forest equivalent to the size of Oklahoma. At this rate, these forests and the species they contain (fig. 32.A) will disappear completely in just a few more decades. Such a loss of biodiversity has not been seen since the end of the Cretaceous period, when the dinosaurs and many other types of organisms became extinct. If the forest areas now legally protected survive, 56–72% of all tropical forest species will still be lost. Aside from other considerations, don't these organisms have a "right to life" as much as we do?

The loss of tropical forests results from an interplay of social, economic, and political pressures. Many people already live in the forest, and as their numbers increase, more of the land is being cleared for farming. Other people are moving to the forests because of internationally financed projects that build roads and open the forests up for exploitation. Small-scale farming is the largest single cause of tropical deforestation, accounting for about 60%, followed by commercial logging, cattle ranching, and mining. International demand for timber promotes destructive logging of rain forests in Southeast Asia, and the fast-food market for low-grade beef encourages the conversion of forest to pastures for cattle in South America, the so-called hamburger connection. The lure of gold draws miners to rain forests in Costa Rica and Brazil.

Figure 32.A

Biodiversity in tropical rain forests. Destroying the tropical rain forests will greatly reduce the variety of life on earth. Already vulnerable are (**a**) the green iguana, (**b**) the South American ocelot, and (**c**) the blue-yellow macaw.

a.

b.

In a temperate deciduous forest, enough sunlight penetrates the canopy for the growth of a well-developed understory—a layer of shrubs followed by herbaceous plants and then often a ground cover of mosses and ferns. Animal life is plentiful. Birds and rodents provide food for bobcats, wolves, and foxes. The white-tailed deer has increased in number of late, while the black bear, an omnivore, has decreased in number, although it is still commonly found in some areas.

Tropical Forests: Wet and Warm

In tropical regions, some forests are deciduous—the broad-leaved trees lose their leaves because of a dry season. We are more interested in the **tropical rain forests,** which occur in South America, Africa, and the Indo-Malayan region near the equator. Here, the weather is always warm (between 20°C and

The destruction of tropical rain forests gives only short-term gain, but there are many long-term reasons they should be saved. The forests act like a giant sponge, soaking up rainfall during the wet season and releasing it during the dry season. Without them, a regional yearly regime of flooding followed by drought is likely; this would destroy property and reduce agricultural harvests. Worldwide, there could be changes in climate that would affect the entire human race. One-fourth of the medicines we currently use come from tropical plants. For example, the rosy periwinkle from Madagascar has produced 2 potent drugs for use against Hodgkin disease, leukemia, and other blood cancers. It is hoped that many of the still-unknown plants will provide medicines for other human ills.

Studies show that if the forests were used as a sustainable source of nonwood products, such as nuts, fruits, and latex rubber, they would generate as much or more revenue while continuing to perform their various ecological functions. And their rich biodiversity would continue to exist for pharmacological discoveries, scientific study, and the "genetic bank" we need to be able to draw upon to improve our crops.

All nations must be concerned about and accept responsibility for preserving the tropical rain forests. Biodiversity preserves should be established in areas of critical importance, and ways to use tropical rain forests in a sustainable manner must be devised. Humans must be able to pursue healthy and productive lives in tropical rain forests without destroying them. Brazil is exploring the concept of "extractive reserves," in which plant and animal products are harvested, but the forest itself is not cleared. Ecologists have also proposed "forest farming" systems, which mimic the natural forest as much as possible while providing abundant yields. But for such plans to work maximally, the human population size and the resource consumption per person must be stabilized.

Preserving tropical rain forests is a wise investment. Such action will promote the survival of most of the world's species—indeed, the human species, too.

c.

25°C) and rainfall is plentiful (with a minimum of 190 cm per year). This is the richest biome, both in the different kinds of species found and the total biomass, that is, the amount of living matter. The need to preserve tropical rain forests is discussed in the reading on this and the previous page.

The tropical rain forest has a complex structure, with many levels of life (fig. 32.6). Some of the broad-leaved evergreen trees grow to 50 m or more, some to 35 m, and some to 15 m. These tall trees often have trunks buttressed at ground level to prevent their toppling over. *Lianas,* or woody vines, which encircle the tree as it grows, also help to strengthen the trunk. The rain forest understory becomes dense in open areas or clearings, where light contributes to the development of a thick jungle.

Although there is animal life on the ground (pacas, agoutis, peccaries, armadillos, and coatis), most animals live in the trees. Insect life is so abundant that the majority of species have not been identified yet. The various birds, such as hummingbirds, parakeets, parrots, and toucans, are often beautifully colored. Amphibians and reptiles are well represented by many types of frogs, snakes, and lizards. Monkeys are well-known primates that feed on the fruits of the trees. The largest carnivores are the big cats—the jaguars in South America and the leopards in the Old World.

Many animals spend their entire life in the canopy, as do some plants. **Epiphytes** are plants that grow on other plants but do not parasitize them. Instead, some have roots that absorb moisture and minerals leached from the canopy and others catch rain and debris in hollows produced by overlapping leaf bases. The most common epiphytes are related to pineapples, orchids, or ferns.

Whereas the soil of a temperate deciduous forest is rich enough for agricultural purposes, the soil of a tropical rain forest is not. Numerous organisms quickly break down any litter, and nutrients are recycled immediately to the plants. Of the minerals, aluminum and iron sometimes remain near the surface, producing a red-colored soil known as laterite. When the trees are cleared, laterite bakes in the hot sun to a bricklike consistency that will not support crops. Slash-and-burn agriculture is the only type that so far has been successful in the tropics. Trees are felled and burned, and the ashes produced provide enough nutrients for several harvests. Thereafter, the forest is allowed to regrow, and a new section is utilized for agriculture. These matters are discussed further on page 672.

Forests require adequate rainfall. The taiga has the least amount of rainfall. The temperate deciduous forest has trees that grow and shed their leaves with the seasons. The tropical rain forest is the least studied and the most complex of all the biomes. Do 32.1 Critical Thinking, found at the end of the chapter.

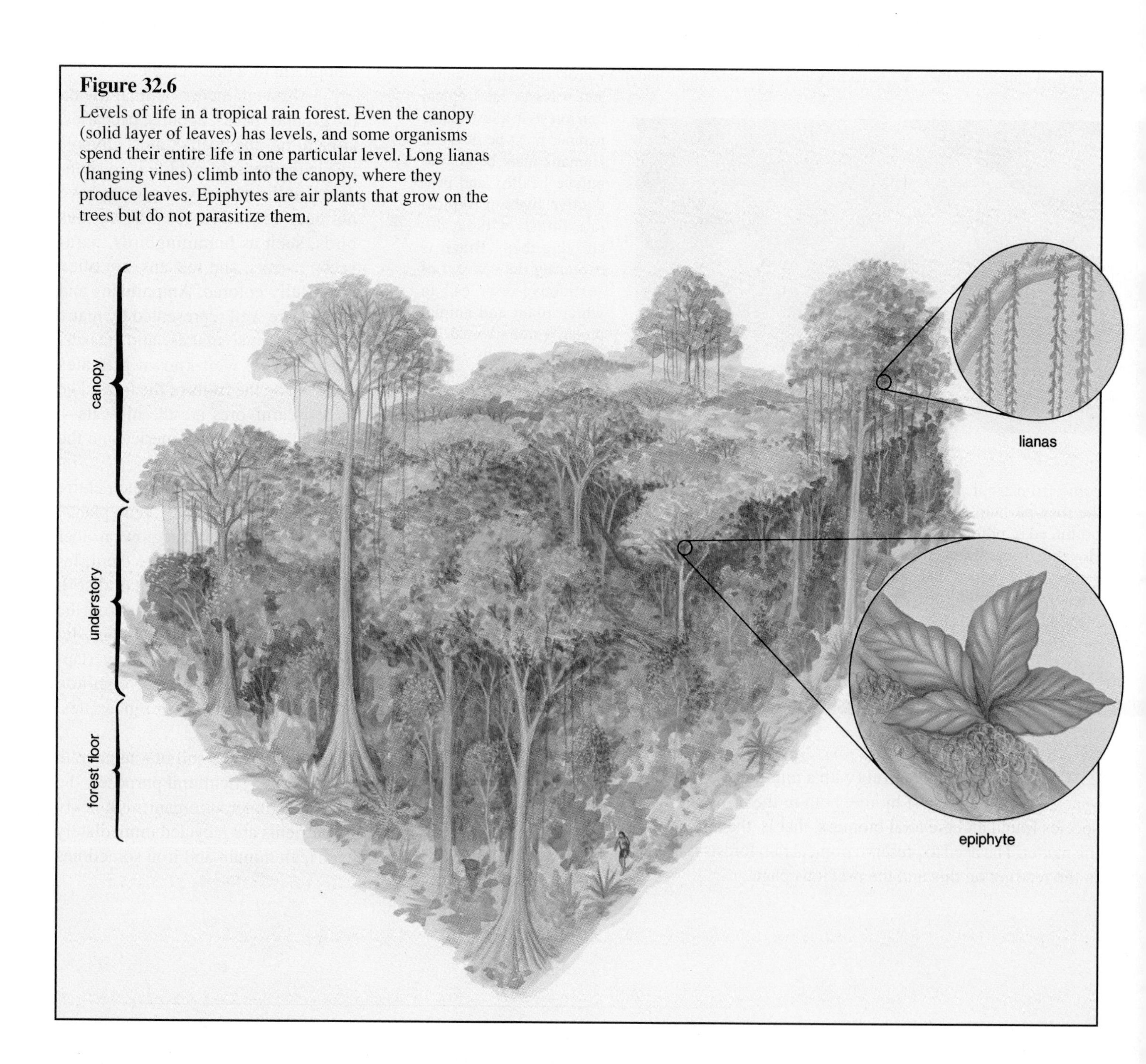

Figure 32.6
Levels of life in a tropical rain forest. Even the canopy (solid layer of leaves) has levels, and some organisms spend their entire life in one particular level. Long lianas (hanging vines) climb into the canopy, where they produce leaves. Epiphytes are air plants that grow on the trees but do not parasitize them.

Latitude and Altitude: Same Series of Biomes

Latitude (distance north and south of equator) determines the biome, but so does altitude (distance above the earth's surface).

Latitude

If you travel from the equator to the North Pole, it is possible to observe first a tropical rain forest, followed by a temperate deciduous forest, and then the taiga and the tundra, in that order. This shows that the location of the biomes is influenced by latitude because latitude determines temperature.

Altitude

It is also possible to observe a similar sequence of biomes by traveling from the bottom to the top of a mountain (fig. 32.7). These transitions are largely due to decreasing temperature as the altitude increases, but soil conditions and rainfall are also important.

> In general, the same sequence of biomes is observed when latitude increases as when altitude increases.

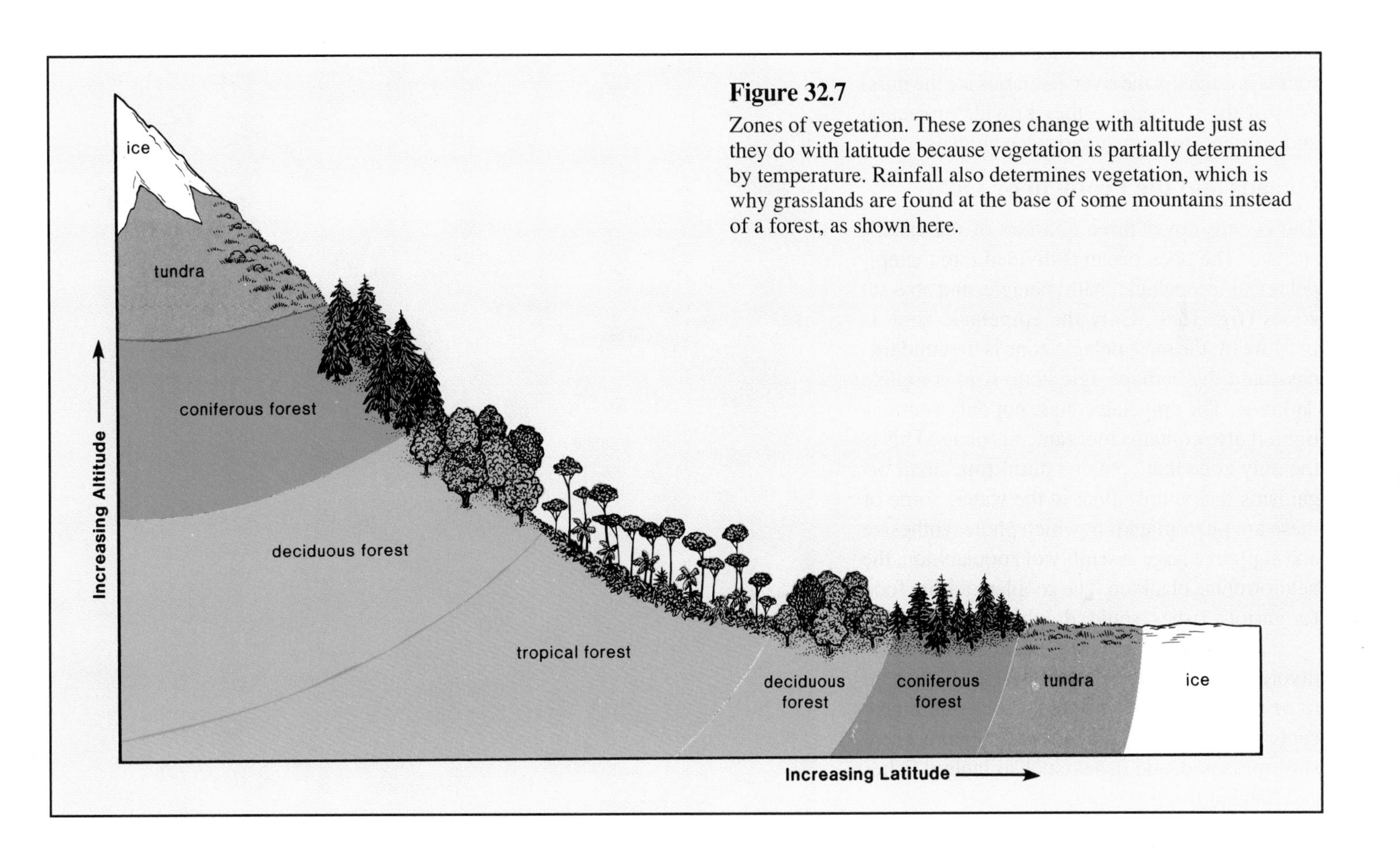

Figure 32.7
Zones of vegetation. These zones change with altitude just as they do with latitude because vegetation is partially determined by temperature. Rainfall also determines vegetation, which is why grasslands are found at the base of some mountains instead of a forest, as shown here.

Aquatic Communities

Aquatic communities can be divided into 2 types: freshwater (inland) communities and saltwater (oceanic or marine) communities. Freshwater communities are lakes, ponds, rivers, and streams. Figure 33.5 in the next chapter describes the food web of a freshwater pond. Saltwater communities occur along the coast and in the ocean itself. They include sandy beaches, rocky shores, coral reefs, the open ocean, and estuaries. An **estuary,** a place where a river empties into the sea, is a nutrient trap; the tides bring nutrients from the sea and at the same time prevent the seaward escape of nutrients brought by the river. Estuaries are the nurseries of the sea because they provide protection and nutrients to immature marine animals.

Oceans and the Problem of Light

The oceans cover three-quarters of the earth's surface. The open ocean is divided into the epipelagic, mesopelagic, bathypelagic, and abyssal zones (fig. 32.8). Only the epipelagic zone is brightly lit; the mesopelagic zone is in semidarkness, and the bathypelagic zone is in complete darkness. The epipelagic zone not only contains light, it also contains inorganic nutrients. This is the only zone that contains **plankton,** small organisms that simply float in the water. Some of these are phytoplankton, which photosynthesize and support a large assembly of zooplankton, the heterotrophic plankton. The zooplankton are food for various fishes, squid, dolphins, and whales.

Animals in the mesopelagic zone are carnivores adapted to the absence of light. They tend to be translucent, red colored, or even luminescent—they give off light. There are luminescent shrimps, squid, and fishes, such as lantern fishes and hatchet fishes.

The bathypelagic zone, the largest zone, is in complete darkness except for an occasional flash of bioluminescent light. Strange-looking fishes with distensible mouths and abdomens and small, tubular eyes feed on infrequent prey. Because of the cold temperature (averaging 2°C) and intense pressure (300–500 atmospheres) in the abyssal (bottom) zone of the ocean, it was once thought that only a few specialized animals live at the ocean bottom. Yet, a diverse assemblage of organisms has been found. Debris that falls from the other zones is taken in by filter feeders, such as the sea lilies, which rise above the seafloor, and the clams and tubeworms that burrow in the mud. Other animals, such as sea cucum-

Figure 32.8

The zones of the ocean. The epipelagic zone contains the most organisms because this is where you find the phytoplankton and the zooplankton. Only carnivores are found in the mesopelagic zone. Both carnivores and scavengers are found in the bathypelagic zone. The abyssal zone is characterized especially by certain echinoderms (sea cucumbers and sea lilies) and tubeworms. A few kinds of fishes, such as the tripod fish, are also found here.

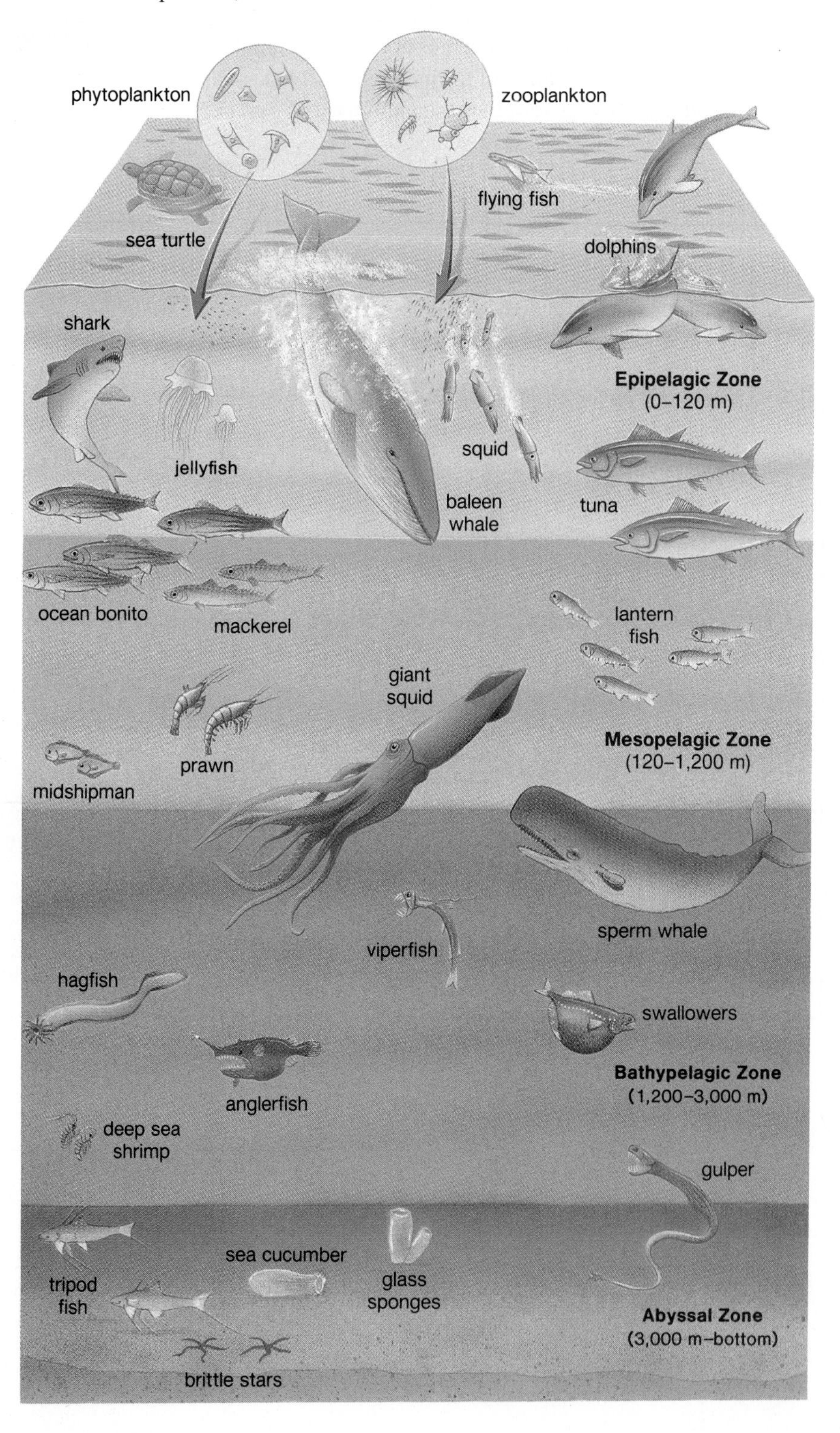

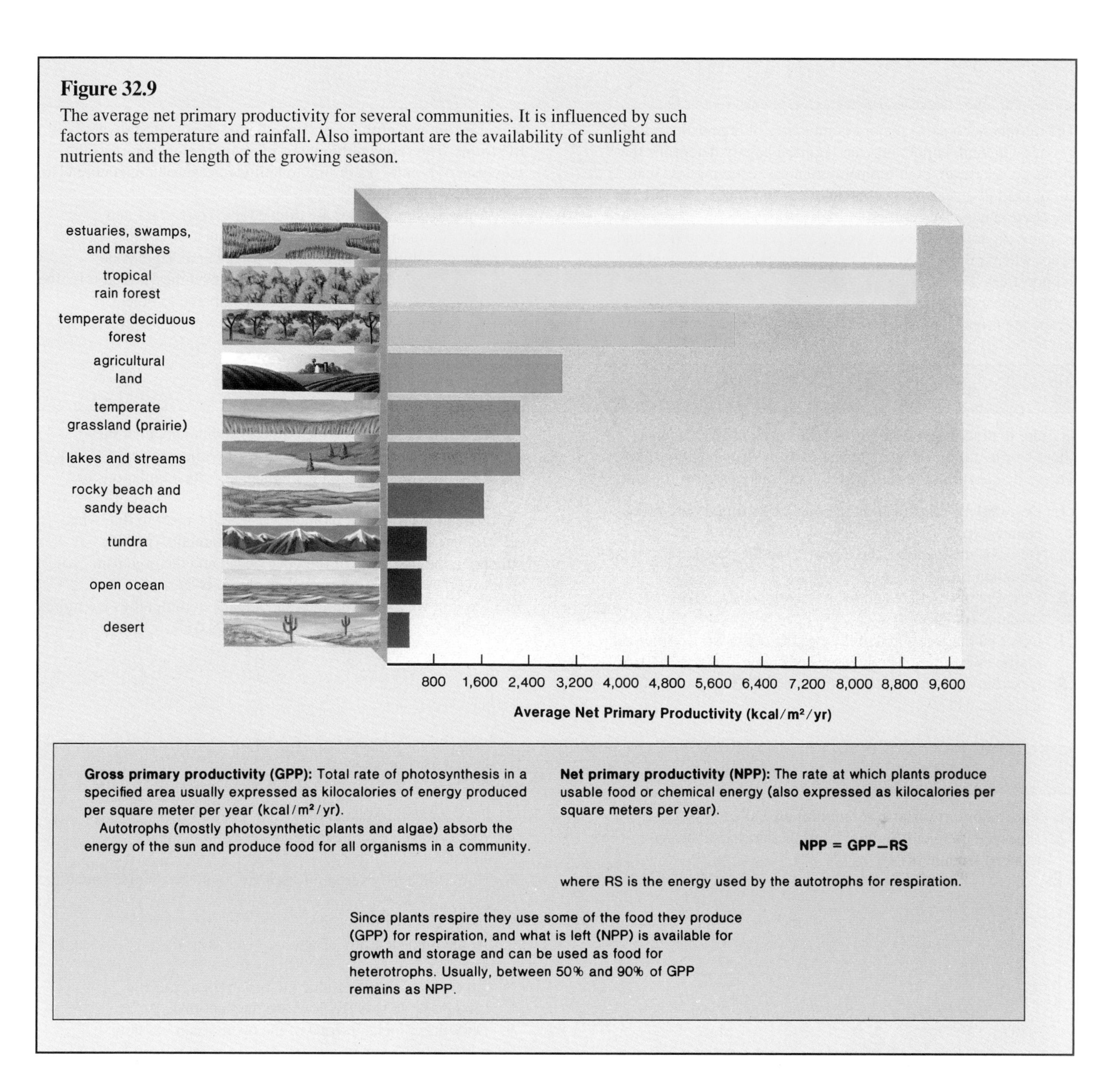

Figure 32.9
The average net primary productivity for several communities. It is influenced by such factors as temperature and rainfall. Also important are the availability of sunlight and nutrients and the length of the growing season.

bers and sea urchins, crawl around on the sea bottom, living on detritus, which is organic matter produced by bacteria of decay. They, in turn, are food for predaceous brittle stars and other larger animals found in the abyssal zone.

Of the pelagic zones of the ocean, only the epipelagic zone receives sunlight and contains the organisms with which we are most familiar. The organisms of the abyssal zone are dependent upon debris that floats down from above. Do 32.1 Critical Thinking, found at the end of the chapter.

Productivity: How an Ocean Is like a Desert

The primary productivity (organic energy produced) of the communities discussed is given in figure 32.9. Productivity is influenced by such factors as temperature and rainfall. The availability of sunlight and nutrients and the length of the growing season are also important. The more productive a community, the more life it can support per unit area. Notice that the productivity of the open ocean is not much greater than a desert.

SUMMARY

The biosphere contains major communities of organisms called biomes. On land, temperature and rainfall largely determine the biomes. Deserts are high-temperature areas receiving less than 25 cm of rainfall a year. They contain organisms that are adapted to a dry environment. The arctic tundra is the northernmost biome and contains organisms that are adapted to a cold climate. Grasslands occur where rainfall is greater than 25 cm but is insufficient to support trees. Forests require adequate rainfall. The taiga, the northernmost of the forests, contains coniferous trees, and a temperate forest contains deciduous trees. Tropical rain forests, with a warm, wet climate all year, contain the greatest diversity of life forms. The series of biomes on mountain slopes mirrors the sequence of biomes according to latitude because temperature also decreases with altitude.

Aquatic communities are divided into freshwater and saltwater communities. Estuaries are the nurseries of the sea. The open ocean is divided into zones. Only the epipelagic zone receives adequate sunlight to support photosynthesis; therefore, the productivity of the open ocean is low compared to other communities.

STUDY QUESTIONS

In order to practice **writing across the curriculum,** students should write out the answers to any or all of the study questions. The study questions are sequenced in the same order as the text.

1. Describe the climate and the populations of a desert in this country. (p. 636)
2. Describe the location, the climate, and the populations of the arctic tundra. (p. 640)
3. Describe the climate and the populations of the African savanna. (p. 640)
4. Describe the location and the climate of the North American prairie. What are its populations of organisms? (pp. 640–641)
5. Describe the location, the climate, and the populations of the taiga. (p. 641)
6. Describe the location, the climate, and the populations of temperate deciduous forests in North America. (pp. 641–642)
7. Describe the location, the climate, and the populations of tropical rain forests. (pp. 642–644)
8. Name the terrestrial biomes you would expect to find when going from the base to the top of a mountain. (p. 645)
9. Describe the zones of the open ocean and the organisms you would expect to find in each zone. (p. 646)
10. Explain why some biomes and aquatic communities are more productive than other communities. (p. 647)

OBJECTIVE QUESTIONS

1. The major terrestrial communities are called _______.
2. Trees are not plentiful in a desert or a grassland because of the reduced amount of _______.
3. The tropical grassland of Africa is called a(n) _______.
4. Broad-leaved evergreen trees are found in a(n) _______ forest.
5. Narrow-leaved evergreen trees are found in a biome called the _______.
6. In the U.S. Midwest, the _______ biome is found.
7. At the highest altitudes and latitudes, a(n) _______ biome is found.
8. An estuary is very productive because it acquires _______ brought by both river flow and tidal action.
9. Only the epipelagic zone of the open ocean has both _______ and _______.
10. Label this diagram showing the vegetation zones of a mountain.

CRITICAL THINKING

In order to practice **writing across the curriculum,** students should write out the answers to any or all of the critical thinking questions. Suggested answers to the critical thinking questions are in appendix E.

32.1

1. If a deciduous forest is destroyed by fire, what less complex biome might take its place? If a grassland biome is overgrazed, what less complex biome might take its place?
2. The greater amount of life in a tropical rain forest is related to a plentiful supply of sunlight. Explain why this might be so.
3. How might the great diversity of a tropical rain forest be related to a plentiful supply of varied foods?

32.2

1. Give 2 reasons, based on the principle of adaptation to the environment, why you would expect to find a squid in the ocean and not in a tropical rain forest.
2. Give 2 reasons why you would expect to find a monkey in a tropical rain forest and not in a grassland.
3. Give 2 reasons why you would expect to find a zebra in a grassland and not in a tropical rain forest.
4. Give 2 reasons why you would expect to find a polar bear along the coast in the arctic tundra and not in a desert.
5. What general conclusions can you draw from these examples?

SELECTED KEY TERMS

biome (bi´ ōm) one of the major land communities in the biosphere, characterized by a particular mix of plants and animals.

chaparral (shap-ə-ral´) a biome of broad-leaved evergreen shrubs forming dense thickets.

deciduous tree tree that sheds leaves at certain seasons.

desert an arid biome characterized especially by plants such as cacti, which are adapted for receiving less than 25 cm of rain per year.

epiphyte (ep´ ĭ-fīt) nonparasitic plant, such as the arboreal orchid and Spanish moss, that grows on the surface of other plants.

estuary (es´ tu-a-re) an area where fresh water meets the sea; therefore, an area with salinity intermediate between fresh water and salt water.

permafrost earth that remains permanently frozen beneath the surface in the tundra.

plankton floating microscopic organisms found in most bodies of water.

prairie a grassland biome of the temperate zone that occurs where rainfall is greater than 25 cm but less than 40 cm.

savanna a grassland biome that has occasional trees and is commonly associated with Africa.

taiga (ti´ gah) a biome that forms a worldwide northern belt of coniferous trees.

tropical rain forest a biome of equatorial forests that remains warm year-round and receives abundant rain.

tundra a biome characterized by lack of trees due to cold temperatures and the presence of permafrost year-round.

33

ECOSYSTEMS

Chapter Concepts

1.
Ecosystems are units of the biosphere in which populations interact with each other and with the physical environment. 651, 662

2.
Energy is dissipated and does not cycle in an ecosystem; therefore, there is a need for a continual supply of solar energy. 652, 662

3.
Chemical elements do cycle through an ecosystem; therefore, there is no need for a continual outside supply. 657, 662

4.
In contrast to mature natural ecosystems, the human ecosystem is characterized by an ever-greater use of energy and materials each year. 661, 662

White-footed mouse with young

Chapter Outline

Figure 33.1
Primary succession. These photos show a possible sequence of events by which bare rock becomes a climax community. **a.** Lichens growing on bare rock. **b.** Individual plants taking hold. **c.** Perennial plants spreading out over the area.

a.

b.

c.

WHEN life first arose, it was confined to the oceans, but about 450 million years ago, organisms began to colonize the bare land. Eventually, the land supported many complex communities of living things. Similarly, today we can observe a sequence of events by which bare rock becomes capable of sustaining many organisms. We call this primary succession (fig. 33.1). During **succession,** a number of communities replace one another, until finally there is a climax community, a mix of plants and animals that is typical of that area. Secondary succession is also observed when a climax community that has been disturbed returns to its former state. Abandoned farmland goes through a series of stages as it becomes like the surrounding area.

When we study a community, we are considering only the populations of organisms that make up that community, but when we study an **ecosystem,** we are concerned with the community plus its physical environment. **Ecology** is the study of the interactions of organisms with each other and with the physical environment. A review of ecological terms is given in table 33.1.

Table 33.1
Ecological Terms

Term	Definition
Ecology	Study of the interactions of organisms with each other and with the physical environment.
Population	All the members of the same species that inhabit a particular area.
Community	All the populations that are found in a particular area.
Ecosystem	A community and its physical environment; has nonliving (abiotic) and living (biotic) components.
Biosphere	The portion of the surface of the earth (air, water, and land) where living things exist.

Ecosystem Composition

An ecosystem possesses both nonliving (abiotic) and living (biotic) components. The abiotic components include soil, water, light, inorganic nutrients, and weather. The biotic components of the ecosystem have a habitat and a niche. The **habitat** of an organism is its place of residence, that is, where it can be found, such as under a log or at the bottom of a pond. The **niche** of an organism is its profession or total role in the community.

The diversity of organisms in an ecosystem is explained by the *competitive exclusion principle,* which states that no 2 species can occupy the same niche at the same time. It might seem as if all the different types of monkeys in a tropical rain forest, for example, are competing directly with one another. However, if all the aspects of their niches (table 33.2) are carefully examined, they are expected to prefer living at a different height above ground, feeding on slightly different foods, or feeding at different times. How organisms acquire food is an important aspect of their niche.

Table 33.2
Aspects of Niche

Plants	Animals
Season of year for growth and reproduction	Time of day for feeding and season of year for reproduction
Sunlight, water, and soil requirements	Habitat and food requirements
Relationships with other organisms	Relationships with other organisms
Effect on abiotic environment	Effect on abiotic environment

Figure 33.2
Ecosystem composition. A diagram illustrating energy flow and chemical cycling through an ecosystem. Energy does not cycle because all that is derived from the sun eventually dissipates as heat.

Producers are autotrophic organisms with the ability to carry on photosynthesis and to make food for themselves (and indirectly for the other populations as well). In terrestrial ecosystems, the producers are predominantly green plants, while in freshwater and saltwater ecosystems, the dominant producers are various species of algae.

Consumers are heterotrophic organisms that eat preformed food. It is possible to distinguish 4 types of consumers based on their food source. **Herbivores** feed directly on green plants; they are termed *primary consumers.* **Carnivores** feed only on other animals and are therefore *secondary* or *tertiary consumers.* **Omnivores** feed on both plants and animals. Therefore, a caterpillar feeding on a leaf is an herbivore; a green heron feeding on a fish is a carnivore; and a human eating both leafy green vegetables and beef is an omnivore. **Decomposers** are organisms of decay, such as bacteria and fungi. Decomposers break down **detritus,** nonliving organic matter, to inorganic matter that can be used again by producers. In this way, the same chemical elements can be used over and over again in an ecosystem. Other ways by which organisms relate are discussed in the reading on page 653.

When we diagram the components of an ecosystem, as in figure 33.2, it is possible to illustrate that every ecosystem is characterized by 2 fundamental phenomena: energy flow and chemical cycling. *Energy flow* begins when producers absorb solar energy, and *chemical cycling* begins when producers take in inorganic nutrients from the physical environment. Thereafter, producers make food for themselves and indirectly for

ORGANISMS LIVING TOGETHER

Predation is not the only way that organisms acquire energy from an ecosystem. *Symbiotic relationships* are close relationships between members of 2 different species. In *parasitism,* one organism, called the *parasite*, acquires food from another organism, called the host. Therefore, the parasitic species is benefited and the host species is harmed. An efficient parasite does not kill the host, at least until its own life cycle is complete. Viruses are always parasites, as are a number of bacteria, protists, plants, and animals. Smaller parasites tend to be endoparasites; they live in the body of the host. Larger parasites tend to be ectoparasites; they remain attached to the exterior of the host by means of specialized organs and appendages. A tapeworm has hooks and suckers for attachment to the intestinal lining of its host (see fig. 30.8). It lacks the cephalization that is seen in its free-living relatives and has reduced organ systems except for the reproductive system. Like many other parasites, the tapeworm life cycle is complex and includes a larval stage in an intermediate host that precedes an adult stage in a new host.

In *commensalism,* one organism is benefited and the other organism is neither benefited nor harmed. Often the benefited species acquires a home and/or transportation from the host species. Barnacles attach themselves to the backs of whales and the shells of horseshoe crabs. Remoras are fishes that attach themselves to the bellies of sharks by means of a modified dorsal fin acting as a suction cup. The remoras obtain a free ride and also feed on the remains of the shark's meals. Epiphytes grow in the branches of trees, where they receive light, but they take no nourishment from the trees. Instead, their roots obtain nutrients and water from the air. Clownfishes live safely among the tentacles of sea anemones, apparently immune to the poisonous tentacles that other fishes avoid (fig. 33.A).

Figure 33.A
Example of commensalism. Clownfishes live among a sea anemone's tentacles and yet are not seized and eaten as prey. The reason this relationship is maintained is not known.

In *mutualism,* both species benefit from the relationship. Often, mutualistic relationships allow organisms to obtain food or avoid predation. Bacteria that reside in the human intestinal tract are provided with food, but they also provide humans with vitamins, which are molecules we are unable to synthesize for ourselves. Termites would not be able to digest wood if not for the protozoa that inhabit their intestinal tract. The bacteria in the protozoa digest cellulose, which termites cannot. Mycorrhizae, also called fungal roots, are symbiotic associations between the roots of plants and fungal hyphae. Mycorrhizal hyphae improve the uptake of nutrients by the plant, protect the plant's roots from microbes, and provide plant growth hormones. In return, the fungus obtains carbohydrates from the plant. As we discussed on page 573, flowers and their pollinators have coevolved and are dependent upon one another. The flower is benefited when the pollinator carries pollen to another flower, assuring cross-fertilization, and the flower provides food for the pollinator.

the other populations of the ecosystem. Energy flow occurs because all the energy content of organic food is eventually lost to the environment as heat. Therefore, most ecosystems cannot exist without a continual supply of solar energy. The original inorganic elements are cycled back to the producers, however, and no new input is required.

> Within an ecosystem, energy flows and chemicals cycle.

Energy flows through an ecosystem because when one form of energy is transformed into another form, there is always a loss of some usable energy as heat. For example, the conversion of energy in one molecule of glucose to 38 molecules of ATP represents less than 50% of the available energy in a glucose molecule. The rest is lost as heat. This means that as one population feeds on another and as decomposers work on detritus, all of the captured solar energy that was converted to chemical-bond energy by algae and plants is returned to the atmosphere as heat. Therefore, energy flows through an ecosystem and does not cycle.

Figure 33.3
Examples of food chains. **a.** Terrestrial. **b.** Aquatic.

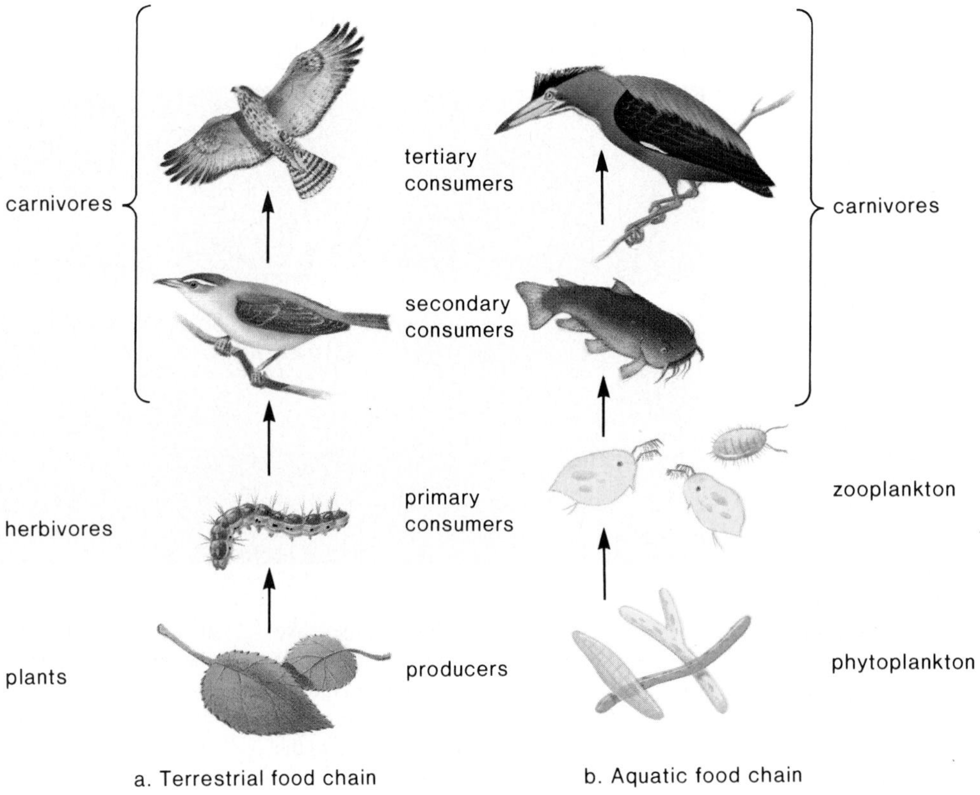

Food Chains Join and Overlap

Energy flows through an ecosystem as the individuals of one population feed on those of another. A **food chain** indicates who eats whom in an ecosystem. Figure 33.3 depicts examples of a terrestrial food chain and an aquatic food chain. It is important to realize that each represents just one path of energy flow through an ecosystem. Natural ecosystems have numerous food chains, each linked to others to form a complex **food web.** For example, figure 33.4 shows a deciduous forest ecosystem in which plants are eaten by a variety of insects, and in turn, these are eaten by several different birds, while any one of the latter may be eaten by a larger bird, such as a hawk. Therefore, energy flow is better described in terms of **trophic** (feeding) **levels,** each one further removed from the producer population, the first (photosynthetic) trophic level. All animals acting as primary consumers are part of a second trophic level, and all animals acting as secondary consumers are part of the third level, and so on.

The populations in an ecosystem form food chains, in which the producers make food for the other populations, which are consumers. While it is convenient to study food chains, the populations in an ecosystem actually form a food web, in which food chains join with and overlap one another.

Figure 33.4
A deciduous forest ecosystem. The arrows indicate the flow of energy in a food web.

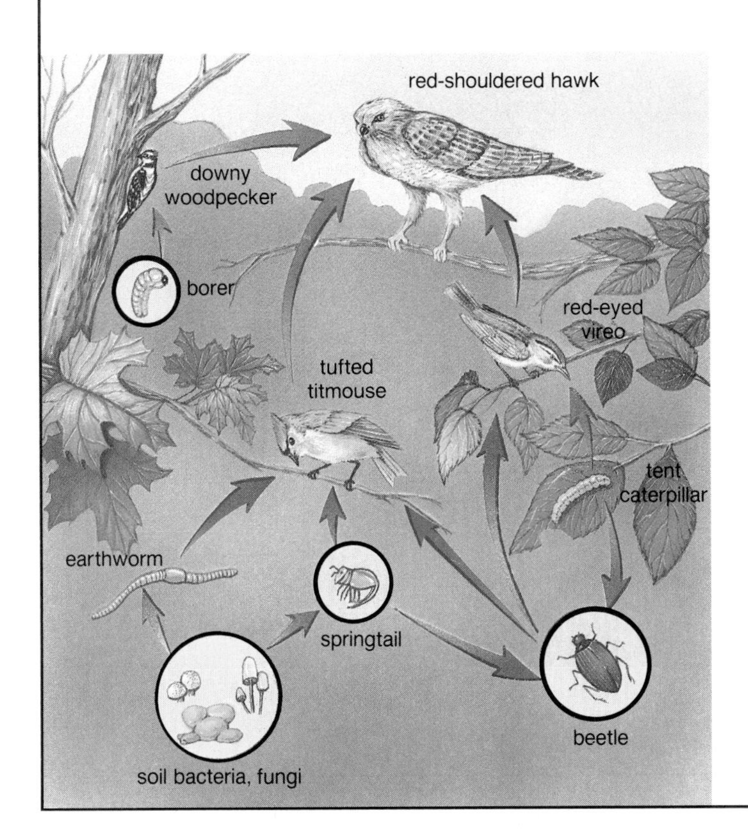

Figure 33.5
A freshwater pond ecosystem. The arrows indicate the flow of energy in a food web.

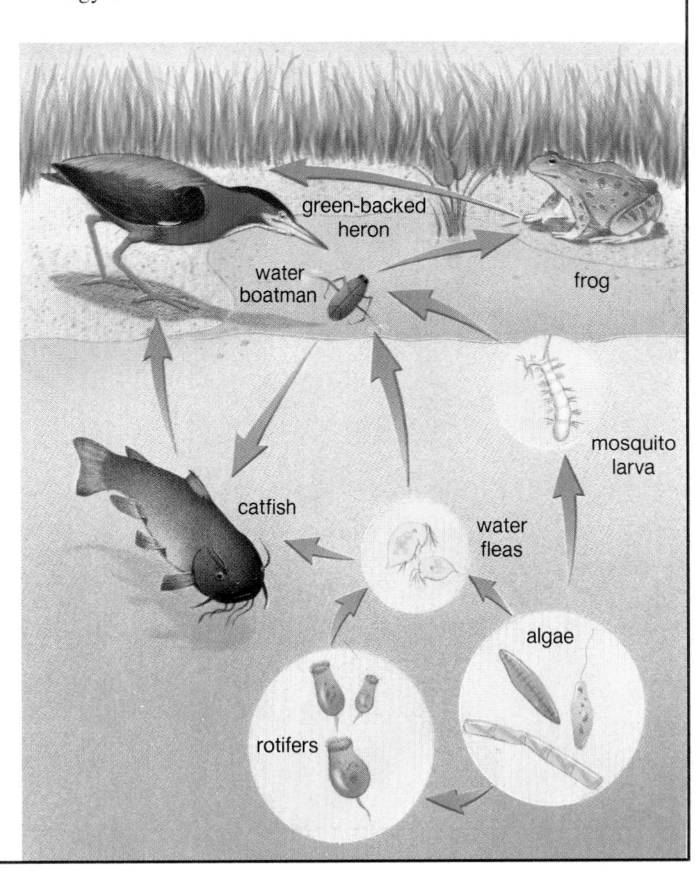

One of the food chains depicted in figure 33.3 is part of the forest food web shown in figure 33.4 and the other food chain is part of the aquatic food web from the freshwater pond ecosystem shown in figure 33.5. Both of these are called grazing food chains because the primary consumer feeds on a photosynthesizer. In some ecosystems (forests, rivers, and marshes), the primary consumer feeds mostly on detritus. The *detritus food chain* accounts for more energy flow than the grazing food chain whenever most organisms die before they are eaten. In the forest, an example of a detritus food chain is

detritus → soil bacteria → earthworms

A detritus food chain is often connected to a grazing food chain, as when earthworms are eaten by a robin. However, as dead organisms decompose, all the solar energy that was taken up by the producer populations eventually dissipates as heat. Therefore, energy does not cycle.

Figure 33.6
Pyramid of energy. At each step in the pyramid, an appreciable portion of energy originally trapped by the producer dissipates as heat. Accordingly, organisms in each trophic level pass on less energy than they receive.

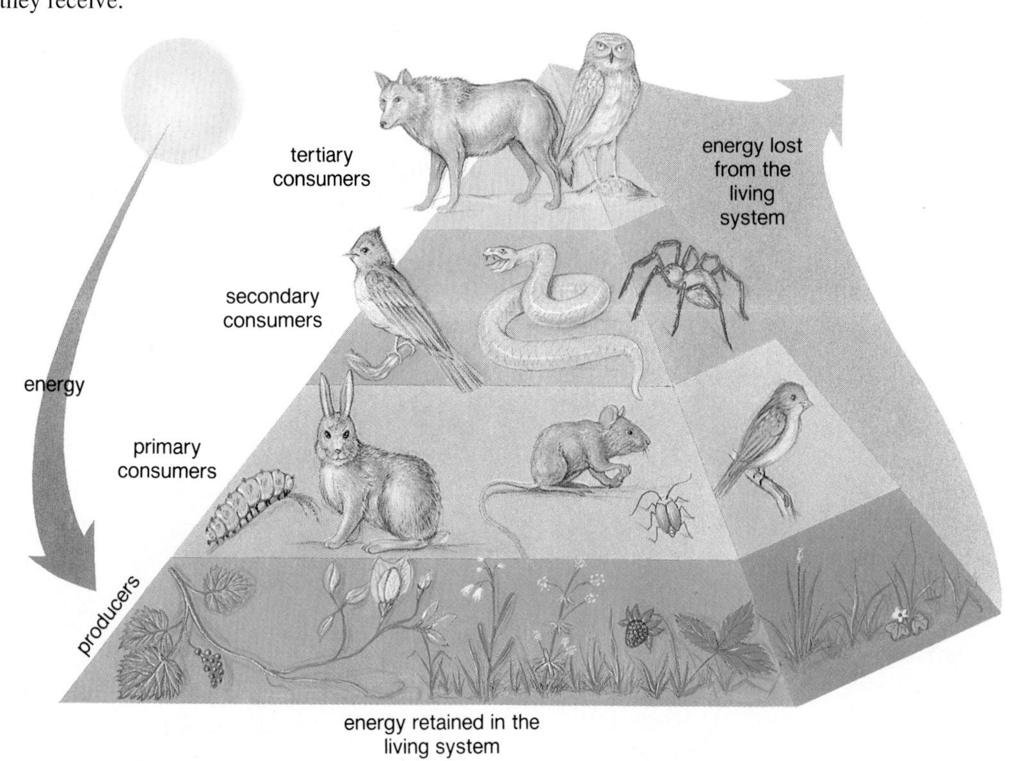

Ecological Pyramids: Because Energy Is Lost

The trophic structure of an ecosystem can be summarized in the form of an **ecological pyramid.** The base of the pyramid represents the producer trophic level, and the apex is the highest level consumer, called the top predator. The other consumer trophic levels are in between the producer and the top-predator levels.

There are 3 kinds of ecological pyramids. One is a *pyramid of numbers,* based on the number of organisms at each trophic level. A second is the *pyramid of biomass. Biomass* is the weight of living material at some particular time. To calculate the biomass for each trophic level, an average weight for the organisms at each level is determined and then the number of organisms at each level is estimated. Multiplying the average weight by the estimated number gives the approximate biomass for each trophic level. A third pyramid, the *pyramid of energy* (fig. 33.6), shows that there is a decreasing amount of energy available at each successive trophic level. Less energy is available at each step for the following reasons:

1. Only a certain amount of food is captured and eaten by the next trophic level.
2. Some of the food that is eaten cannot be digested and exits the digestive tract as waste.
3. Only a portion of the food that is digested becomes part of the organism's body. The rest is used as a source of energy.

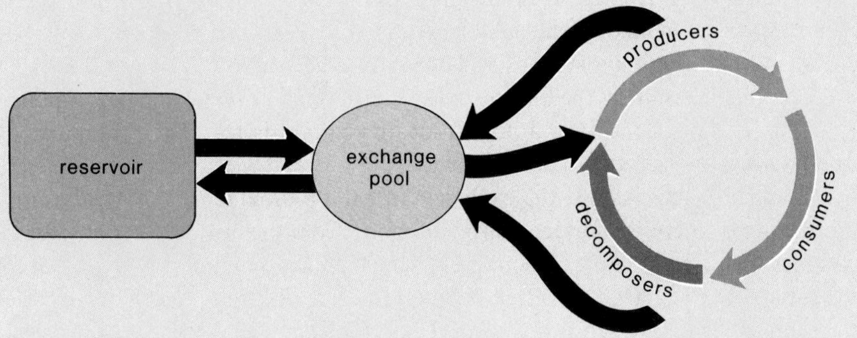

Figure 33.7
Components of a chemical cycle. The reservoir stores the chemical, and the exchange pool makes it available to producers. The chemical then cycles through food chains, which are the biotic community. Decomposition returns the chemical to the exchange pool once again if it has not already returned by another process (see figs. 33.8 and 33.9).

In regard to the last point, we have to realize that a significant portion of food molecules is used as an energy source for ATP buildup in mitochondria. This ATP is needed to build the proteins, carbohydrates, and lipids that compose the body. ATP is also needed for such activities as muscle contraction, nerve conduction, and active transport.

The energy considerations associated with ecological pyramids have implications for the human population. It is generally stated that only about 10% of the energy available at a particular trophic level is incorporated into the tissues of animals at the next level. This being the case, it can be estimated that 100 kg of grain could, if consumed directly, result in 10 human kg; however, if fed to cattle, the 100 kg of grain would result in only 1 human kg. Therefore, a larger human population can be sustained by eating grain than by eating grain-fed animals. Humans generally need some meat in their diet, however, because this is the most common source of the essential amino acids, as discussed in chapter 11.

In a food web, each successive trophic level has less total energy content. This is because some energy is lost and so not transferred from one trophic level to the next. Do 33.1 Critical Thinking, found at the end of the chapter.

Chemical Cycling

In contrast to energy, inorganic nutrients (chemicals) do cycle through large natural ecosystems. Because there is minimal input from the outside, the various chemical elements essential for life are used over and over. For each element, the cycling process (fig. 33.7) involves (1) a reservoir—that portion of the earth that acts as a storehouse for the element; (2) an exchange pool—that portion of the environment from which the producers take their nutrients; and (3) the biotic community, through which elements move along food chains to and from the exchange pool.

Carbon Cycle: Photosynthesizing and Respiring

The relationship between photosynthesis and aerobic cellular respiration should be kept in mind when discussing the carbon cycle. Recall that for simplicity's sake, this equation in the forward direction represents aerobic cellular respiration; in the other direction, it is used to represent photosynthesis:

$$C_6H_{12}O_6 + 6\ O_2 \rightleftharpoons 6\ CO_2 + 6\ H_2O$$

The equation tells us that respiration releases carbon dioxide (CO_2), the molecule needed for photosynthesis. However, photosynthesis releases oxygen (O_2), the molecule needed for respiration. Animals are dependent on green organisms, not only to produce organic food and energy but also to supply the biosphere with oxygen. However, since producers both photosynthesize *and* respire, they can function independently of the animal world.

In the carbon cycle, organisms in both terrestrial and aquatic ecosystems (fig. 33.8) exchange carbon dioxide with the atmosphere. On land, plants take up carbon dioxide from the air, and through photosynthesis, they incorporate carbon into food that is used by themselves and heterotrophs alike. When any organism respires, a portion of this carbon is returned to the atmosphere as carbon dioxide.

In aquatic ecosystems, the exchange of carbon dioxide with the atmosphere is indirect. Carbon dioxide from the air combines with water to give carbonic acid, which breaks down to bicarbonate ions (HCO_3^-). Bicarbonate ions are a source of carbon for algae, which produce food for themselves and for heterotrophs. Similarly, when aquatic organisms respire, the carbon dioxide they give off becomes bicarbonate. The amount of bicarbonate in the water is in equilibrium with the amount of carbon dioxide in the air.

Carbon Reservoirs: The World's Trees, for Example

Living and dead organisms contain organic carbon and serve as one of the reservoirs for the carbon cycle. The world's biota (all living things), particularly trees, contain 800 billion tons of organic carbon, and an additional 1,000 billion–3,000 billion tons are estimated to be held in the remains of plants and animals in the soil. Before decomposition could occur, some of these remains were subjected to physical processes that transformed them into coal, oil, and natural gas. We call these materials the fossil fuels. Most of the fossil fuels were formed during the Carboniferous period, 280 million–350 million years ago, when an exceptionally large amount of organic matter was buried before decomposing. Another reservoir is the calcium carbonate that accumulates in limestone and in calcium carbonate shells. The oceans abound with organisms,

Figure 33.8
Carbon cycle. Photosynthesizers take up carbon dioxide (CO_2) from the air or the bicarbonate ion (HCO_3^-) from the water. They and all other organisms return carbon dioxide to the environment. The carbon dioxide level is also increased when volcanoes erupt and fossil fuels are burned. Presently, the oceans are a primary reservoir for carbon in the form of limestone and calcium carbonate shells.

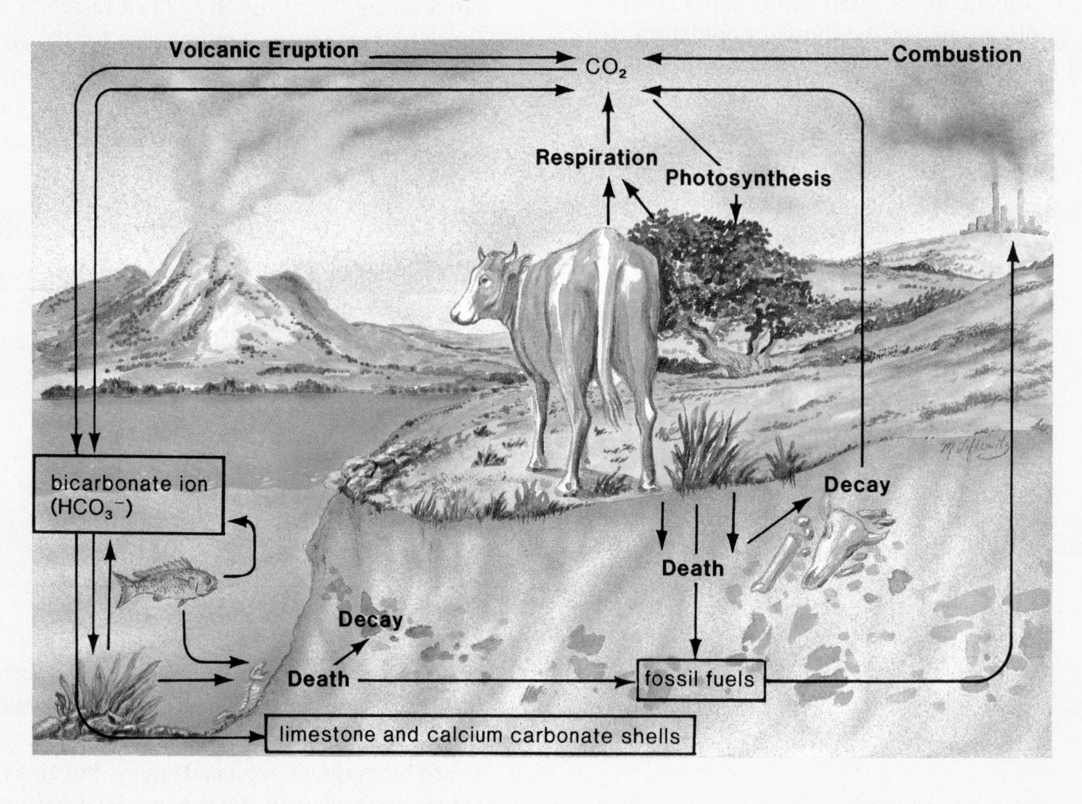

some microscopic, that have calcium carbonate shells. After these organisms die, their shells accumulate in ocean bottom sediments. Limestone is formed from these sediments by geological transformation.

How People Influence the Carbon Cycle

The activities of human beings have increased the amount of carbon dioxide (CO_2) and other gases in the atmosphere. Data from monitoring stations recorded an increase of 20 ppm (parts per million) in carbon dioxide in only 22 years. (This is equivalent to 42 billion tons of carbon.) This buildup is attributed primarily to the burning of fossil fuels and the destruction of the world's tropical rain forests (fig. 33.9). When forests are destroyed, a reservoir that takes up excess carbon dioxide is reduced. At this time, the oceans are believed to be taking up most of the excess carbon dioxide; the burning of fossil fuels in the last 22 years has probably released 78 billion tons of carbon, yet the atmosphere registers an increase of "only" 42 billion tons.

As discussed on page 678, there is much concern that an increased amount of carbon dioxide (and other gases) in the atmosphere is causing global warming. These gases allow the sun's rays to pass through to the earth, but they also absorb and reradiate heat to the earth, a phenomenon called the *greenhouse effect*.

In the carbon cycle, carbon dioxide is removed from the atmosphere by photosynthesis but is returned by aerobic cellular respiration. Living things and dead matter are carbon reservoirs. The oceans, because they abound with calcium carbonate shells and limestone, are also major carbon reservoirs.

Figure 33.9
Burning of trees in tropical rain forest. When trees are burned, carbon dioxide (CO_2) is released from one of the reservoirs in the carbon cycle.

Figure 33.10
Nitrogen cycle. Several processes and types of bacteria are at work, as described in the table accompanying this figure.

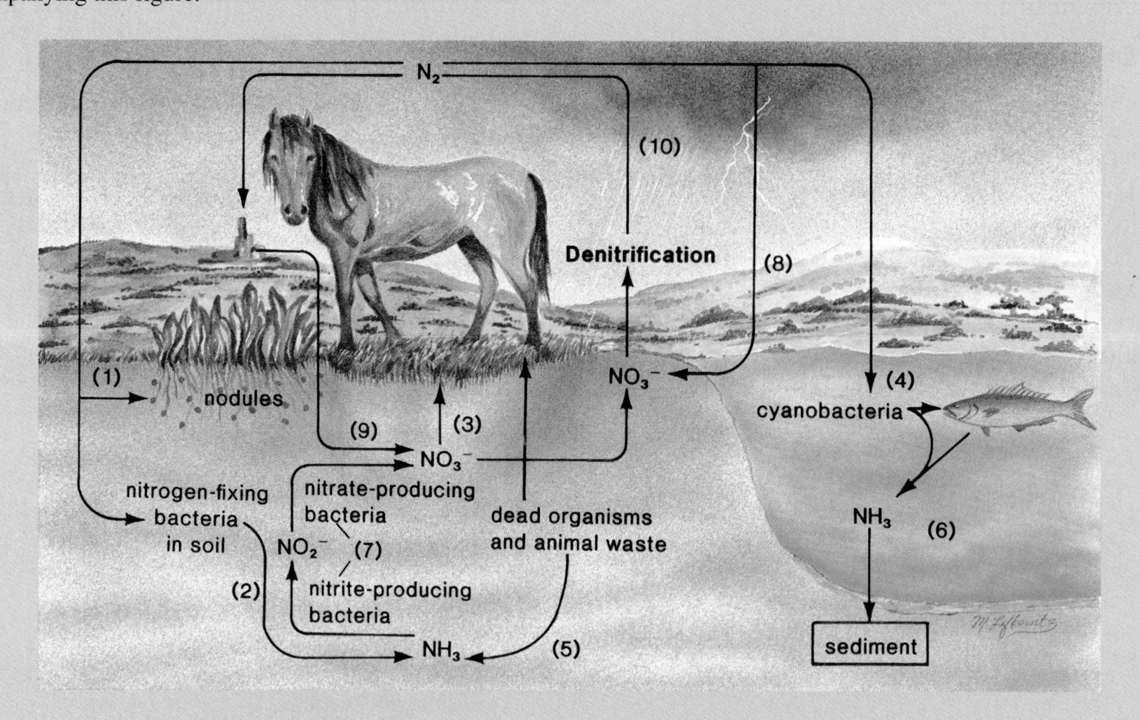

Nitrogen fixation: Reduction and incorporation of nitrogen into organic compounds.

*Nitrogen-fixing bacteria in nodules of legumes reduce nitrogen gas and produce organic compounds (1).

*Nitrogen-fixing bacteria in soil reduce nitrogen gas to ammonia (2).

*Plants reduce nitrates in soil and produce organic compounds (3).

*Cyanobacteria reduce nitrogen gas and produce organic compounds (4).

Decomposition: Decomposing bacteria break down dead organic remains and give off ammonia (5), (6).

Nitrification: Production of nitrates that can be used by plants.

*Nitrite-producing and nitrate-producing bacteria convert ammonia to nitrate (7).

*Lightning converts nitrogen gas to nitrate (8).

*Humans convert nitrogen gas to nitrate for use in fertilizers (9).

Denitrification: Denitrifying bacteria convert nitrate to nitrogen gas (10).

Nitrogen Cycle: Important Bacteria

Nitrogen is an abundant element in the atmosphere. Nitrogen (N_2) makes up about 78% of the atmosphere by volume, yet nitrogen deficiency commonly limits plant growth. Plants cannot incorporate nitrogen into organic compounds and therefore depend on various types of bacteria to make nitrogen available to them (fig. 33.10).

Nitrogen Fixation: Bacteria Reduce Gas

Nitrogen fixation occurs when nitrogen (N_2) is reduced and added to organic compounds. Some cyanobacteria in aquatic ecosystems and some free-living bacteria in soil are able to reduce nitrogen gas to ammonia (NH_3). Other *nitrogen-fixing bacteria* infect and live in nodules on the roots of legumes (fig. 33.10). They make reduced nitrogen and organic compounds available to the host plant.

Nitrogen fixation also occurs after plants take up nitrates (NO_3^-) from the soil and produce amino acids and nucleic acids.

Nitrification: Almost Balances Denitrification

Nitrification is the production of nitrates. Nitrogen gas (N_2) is converted to nitrate (NO_3^-) in the atmosphere when cosmic radiation, meteor trails, and lightning provide the high energy needed for nitrogen to react with oxygen. Also, humans make a most significant contribution to the nitrogen cycle when they convert nitrogen gas to nitrate for use in fertilizers.

Ammonia (NH_3) in the soil is converted to nitrate by certain soil bacteria in a two-step process. First, nitrite-producing bacteria convert ammonia to nitrite (NO_2^-), and then nitrate-producing bacteria convert nitrite to nitrate. These 2 groups of bacteria are called the *nitrifying bacteria.* Notice the subcycle in the nitrogen cycle that involves only ammonia, nitrites, and nitrates. This subcycle does not depend on the presence of nitrogen gas at all (fig. 33.10).

Denitrification is the conversion of nitrate to nitrogen gas. There are *denitrifying bacteria* in both aquatic and terrestrial ecosystems. Denitrification counterbalances nitrogen fixation but not completely. More nitrogen fixation occurs, especially due to fertilizer production.

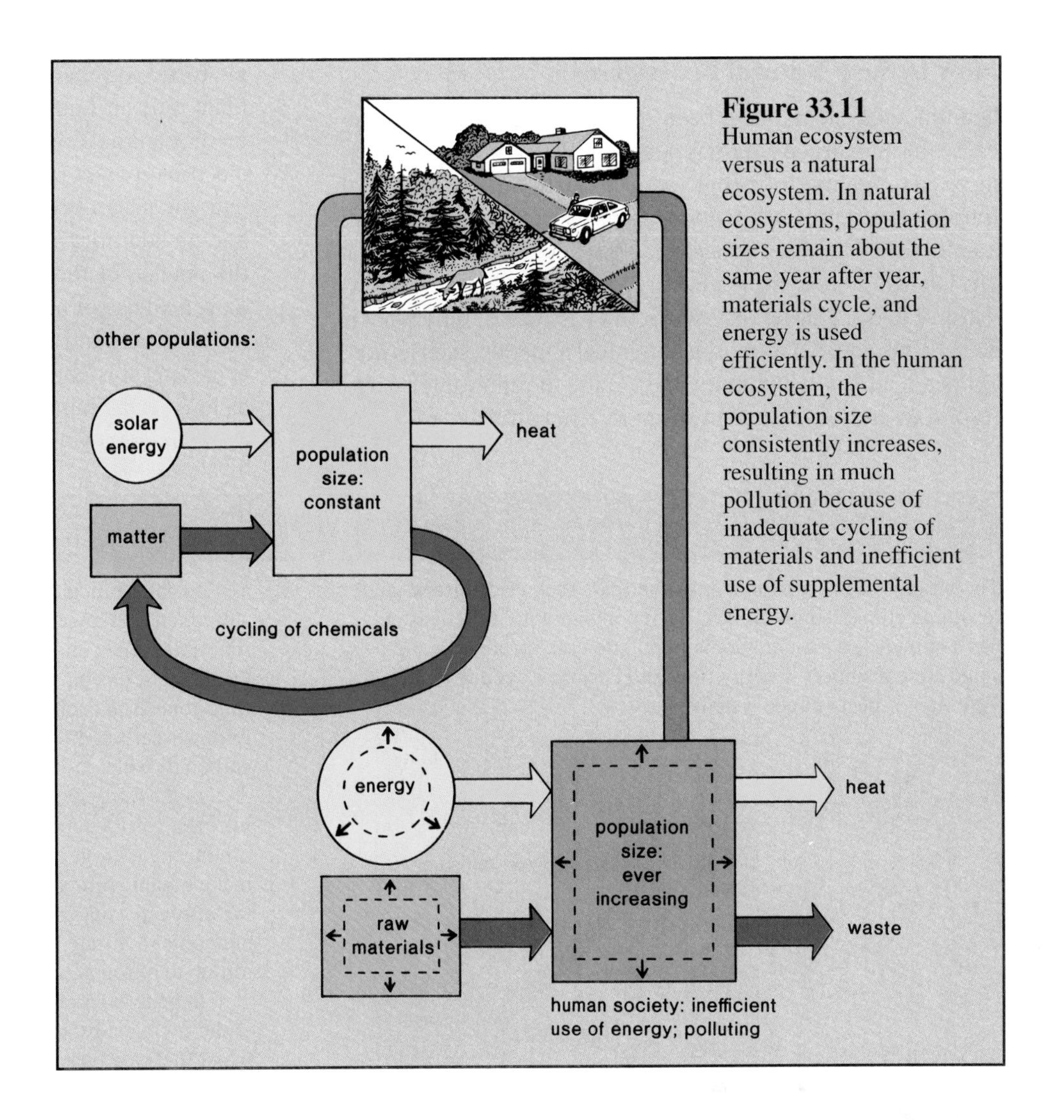

Figure 33.11
Human ecosystem versus a natural ecosystem. In natural ecosystems, population sizes remain about the same year after year, materials cycle, and energy is used efficiently. In the human ecosystem, the population size consistently increases, resulting in much pollution because of inadequate cycling of materials and inefficient use of supplemental energy.

In the nitrogen cycle, nitrogen-fixing bacteria (in nodules and in the soil) reduce nitrogen gas, and thereafter nitrogen can be incorporated into organic compounds; nitrifying bacteria convert ammonia to nitrate; and denitrifying bacteria convert nitrate back to nitrogen gas.

The Human Ecosystem

Mature natural ecosystems tend to be stable. The sizes of the many and varied populations are held in check by the interactions between species, such as predation and parasitism; the energy that enters and the amount of matter that cycles is appropriate to support these populations. **Pollution,** defined as any undesirable change in the environment that can be harmful to humans and other life, does not normally occur.

Human beings have replaced natural ecosystems with one of their own making, as depicted in figure 33.11. This ecosystem essentially has 2 parts: the *country,* where agriculture and animal husbandry are found, and the *city,* where most people live and where industry is located. This representation of the human ecosystem, although simplified, allows us to see that the system requires 2 major inputs: *fuel energy* and *raw materials* (e.g., metals, wood, synthetic materials). The use of these necessarily results in *pollution* and *waste.*

Just as the city is not self-sufficient and depends on the country to supply it with food, so the whole human ecosystem is dependent on the natural ecosystems to provide resources and to absorb wastes. Fuel combustion by-products, sewage, fertilizers, pesticides, and solid wastes are all added to natural ecosystems in the hope that these systems will cleanse the biosphere of these pollutants. But we have replaced natural ecosystems with our human ecosystem and have exploited natural ecosystems for resources, adding even more pollutants, to the extent that the remaining natural ecosystems have become overloaded.

How to Save Natural Ecosystems

Natural ecosystems have been destroyed and overtaxed because the human ecosystem is noncyclical and because an ever-increasing number of people want to maintain a standard of living that requires many goods and services. But we can call a halt to this spiraling process if we achieve zero population growth and if we conserve energy and raw materials. Conservation can be achieved in 3 ways: (1) wise use of only what is actually needed; (2) recycling of nonfuel minerals, such as iron, copper, lead, and aluminum; and (3) use of renewable energy resources and development of more efficient ways to utilize all forms of energy. As a practical example, consider the plant built in Lamar, Colorado, that produces methane from feedlot animals' wastes. The methane is burned in the city's electrical power plant, and the heat given off is used to incubate the anaerobic digestion process that produces the methane. In addition, a protein feed supplement is produced from the residue of the digestion process. This system represents a cyclical use of material and an efficient use of energy, similar to that found in nature. Many other such processes for achieving this end have been and will be devised. However, as long as the human ecosystem on the whole remains inefficient and noncyclical, it will continue to cause pollution.

SUMMARY

The process of succession from either bare rock or disturbed land results in climax communities. Each population in an ecosystem has a habitat and a niche. Some populations are producers and some are consumers. Energy flow and chemical cycling are important aspects of ecosystems.

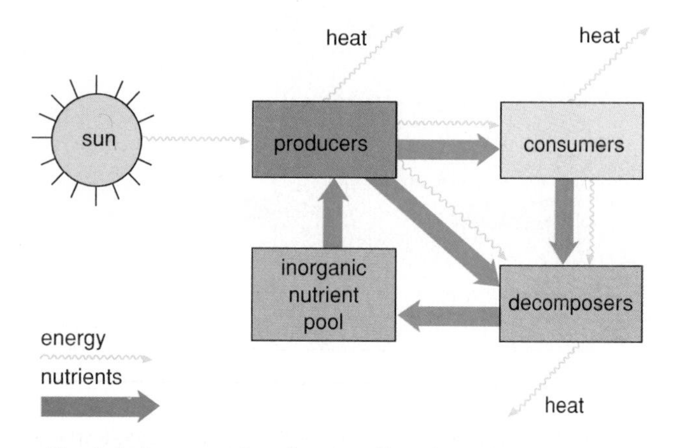

Food chains are paths of energy flow through an ecosystem. Both grazing food chains and detritus food chains exist. Eventually, the very same chemicals are made available to producer populations, but the energy dissipates as heat. The various food chains form an intricate food web, in which there are various trophic (feeding) levels. A pyramid of energy illustrates that each succeeding trophic level has less energy than the preceding level.

Each chemical cycle involves a reservoir, where the element is stored; an exchange pool, from which the populations take and return nutrients; and the populations themselves. In the carbon cycle, the reservoir is organic matter, calcium carbonate shells, and limestone. The exchange pool is the atmosphere: photosynthesis removes carbon dioxide (CO_2), and respiration and combustion add carbon dioxide.

In the nitrogen cycle, the reservoir is the atmosphere, but nitrogen gas (N_2) must be converted to nitrate (NO_3^-) for use by producers. Nitrogen-fixing bacteria, particularly in root nodules, make organic nitrogen available to plants. Other bacteria active in the nitrogen cycle are the nitrifying bacteria, which convert ammonia to nitrate, and the denitrifying bacteria, which convert nitrate to nitrogen.

In general, humans affect the carbon and nitrogen cycles by withdrawing substances from the reservoirs. For example, when fossil fuels are burned, organic carbon is removed from a reservoir. Nitrogen gas (N_2) is also converted to ammonia (NH_3), and thereby nitrogen gas is removed from the air.

In mature natural ecosystems, the populations usually remain the same size and need the same amount of energy each year. Additional material inputs are minimal because matter cycles. In the human ecosystem, the population size constantly increases, more energy is needed each year, and additional material inputs are necessary. Therefore, there is much pollution. It would be beneficial for us and for future generations to find ways to use excess heat and to recycle materials.

STUDY QUESTIONS

In order to practice **writing across the curriculum,** students should write out the answers to any or all of the study questions. The study questions are sequenced in the same order as the text.

1. What is succession, and how does it result in a climax community? (p. 651)
2. Define habitat and niche. (p. 651)
3. Name 4 different types of consumers found in natural ecosystems. (p. 652)
4. What is the difference between a food chain and a food web? Define a trophic level. (p. 654)
5. Give an example of a grazing food chain and a detritus food chain for a terrestrial ecosystem and an aquatic ecosystem. (p. 655)
6. Draw a pyramid of energy, and explain why such a pyramid can be used to verify that energy does not cycle. (p. 656)
7. What are the reservoir and the exchange pool of a chemical cycle? (p. 657)
8. Describe the carbon cycle. How do humans contribute to this cycle? (pp. 658–659)
9. Describe the nitrogen cycle. How do humans contribute to this cycle? (p. 660)
10. Contrast the characteristics of mature natural ecosystems with those of the human ecosystem. (p. 661)

OBJECTIVE QUESTIONS

1. Chemicals cycle through the populations of an ecosystem, but energy is said to _______ because all of it eventually dissipates as heat.
2. When organisms die and decay, chemical elements are made available to _______ populations once again.
3. Organisms that feed on plants are called _______.
4. A pyramid of energy illustrates that there is a loss of energy from one _______ level to the next.
5. There is a loss of energy because one form of energy can never be completely _______ into another form.
6. Forests are a(n) _______ for carbon in the carbon cycle.
7. In the carbon cycle, when organisms _______, carbon dioxide (CO_2) is returned to the exchange pool.
8. Humans make a significant contribution to the nitrogen cycle when they convert nitrogen gas (N_2) to _______ for use in fertilizers.
9. During the process of denitrification, nitrate is converted to _______.
10. Natural ecosystems utilize the same amount of energy per year, but the human ecosystem utilizes a(n) _______.
11. In reference to figure 33.5, which organisms would you place at *a–e* in this diagram?

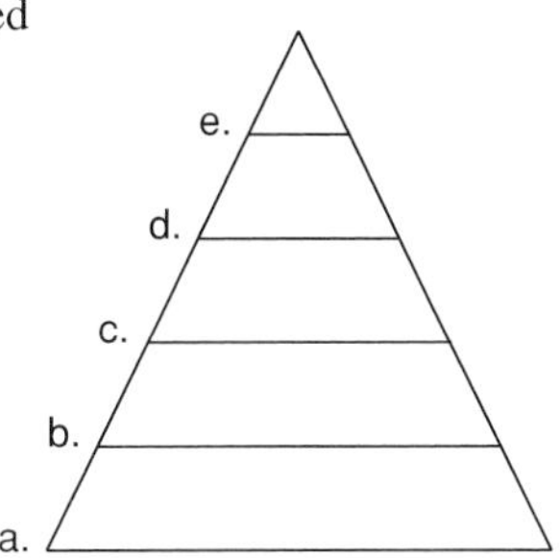

12. In this simplified diagram of the carbon cycle, label each lettered arrow as combustion, photosynthesis, or respiration.

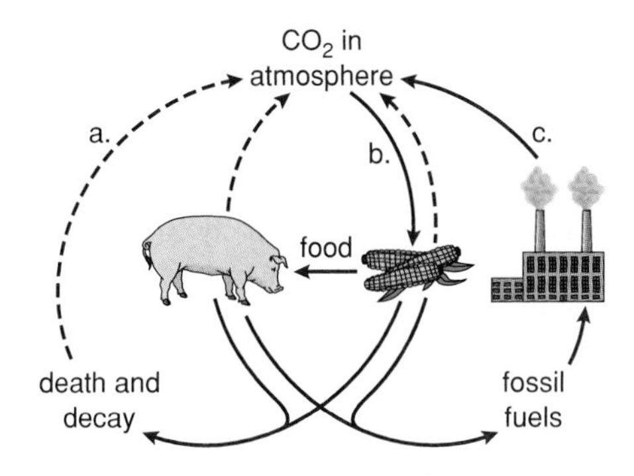

CRITICAL THINKING

In order to practice **writing across the curriculum,** students should write out the answers to any or all of the critical thinking questions. Suggested answers to the critical thinking questions are in appendix E.

33.1

1. Using the data given in figure 33.6, explain why you would expect mice (herbivores) to be more common than weasels, foxes, or hawks (carnivores) in the environment.
2. Explain why you would expect food chains to be short—4 or 5 links at most.
3. The size of a top-predator population is not held in check by another predator population. Again with reference to figure 33.6, why doesn't a top-predator population constantly increase in size?
4. What would you expect to happen to an ecosystem if one of the secondary consumer populations were to suffer a collapse?

SELECTED KEY TERMS

consumer a member of a population that feeds on members of other populations in an ecosystem.

decomposer organism of decay (fungus and bacterium) in an ecosystem.

denitrification (de-ni˝ trĭ-fĭ-ka´ shun) the process of converting nitrate to nitrogen; part of the nitrogen cycle.

detritus (di-tri´ tus) nonliving organic matter.

ecological pyramid pictorial graph representing biomass, organism number, or energy content of each trophic level in a food web, from the producer to the final consumer populations.

ecology the study of the interactions of organisms with each other and with the physical environment.

ecosystem a setting in which populations interact with each other and with the physical environment.

food chain a succession of organisms in an ecosystem that are linked by an energy flow and the order of who eats whom.

food web the complete set of food links between populations in a community.

habitat the natural abode of an animal or plant species.

niche (nich) total description of an organism's functional role in an ecosystem, from activities to reproduction.

nitrogen fixation a process whereby nitrogen is reduced prior to the incorporation of nitrogen into organic compounds.

pollution detrimental alteration of the normal constituents of air, land, and water due to human activities.

producer organism that produces food and is capable of synthesizing organic compounds from inorganic constituents of the environment; usually the green plants and the algae in an ecosystem.

succession a series of ecological stages by which the community in a particular area gradually changes until there is a climax community that can maintain itself.

trophic level (tro´ fik lev´ el) a categorization of species in a food web according to their feeding relationships from the first-level autotrophs through succeeding levels of herbivores and carnivores.

34

HUMAN POPULATION CONCERNS

Chapter Concepts

1.
A population undergoing exponential growth has an ever-greater increase in numbers and a shorter doubling time, and it may outstrip the carrying capacity of the environment. 666, 680

2.
The world is divided into the more-developed countries and the less-developed countries; mainly the less-developed countries are presently undergoing exponential population growth. 669, 680

3.
Human activities cause land, water, and air pollution and threaten the integrity of the biosphere. 671, 680

4.
A sustainable world is possible if economic growth is accompanied by ecological preservation. 680, 681

Hundred-mile-long traffic jam

Chapter Outline

Figure 34.1

More-developed countries (MDCs) versus less-developed countries (LDCs). **a.** In the MDCs, most people enjoy a high standard of living. **b.** In the LDCs, the majority of people are poor and have few amenities.

a.

b.

THE countries of the world today are divided into 2 groups (fig. 34.1). The more-developed countries (MDCs), typified by countries in North America and Europe, are those in which population growth is under control and the people enjoy a good standard of living. The less-developed countries (LDCs), typified by countries in Latin America, Africa, and Asia, are those in which population growth is out of control and the majority of people live in poverty. (Sometimes the term *third-world countries* is used to mean the less-developed countries. This term was introduced by those who thought of the United States and Europe as the first world and the former USSR as the second world.)

Before we explore why the world is now divided into MDCs and LDCs, it is necessary to study exponential population growth in general.

Exponential Population Growth

The human growth curve is J-shaped (fig. 34.2). In the beginning, growth of the human population was relatively slow, but as more reproducing individuals were added, growth increased until the curve began to slope steeply upward. It is apparent from the position of 1995 on the growth curve in figure 34.2 that growth is quite rapid now. The world population increases at least the equivalent of a medium-sized city every day (200,000) and the equivalent of the combined populations of the United Kingdom, Norway, Ireland, Iceland, Finland, and Denmark every year. These startling figures are a reflection of the fact that a very large world population is undergoing exponential growth.

Mathematically speaking, **exponential growth,** or geometric increase, occurs in the same manner as compound interest; that is, the percentage increase is added to the principal before the next increase is calculated. Referring specifically to populations, consider the hypothetical population sizes in table 34.1. This table illustrates the circumstances of world population growth at the moment: the percentage increase has decreased, yet the size of the population grows by a greater amount each year. The increase in size is dramatically large because the world population is very large.

In our hypothetical examples (table 34.1), an initial increase of 2% added to the original population size followed by

Figure 34.2

Growth curve for human population. The human population is now undergoing rapid exponential growth. Since the growth rate is declining, it is predicted that the population size will level off at 8 billion, 10.5 billion, or 14.2 billion, depending upon the speed with which the growth rate declines.

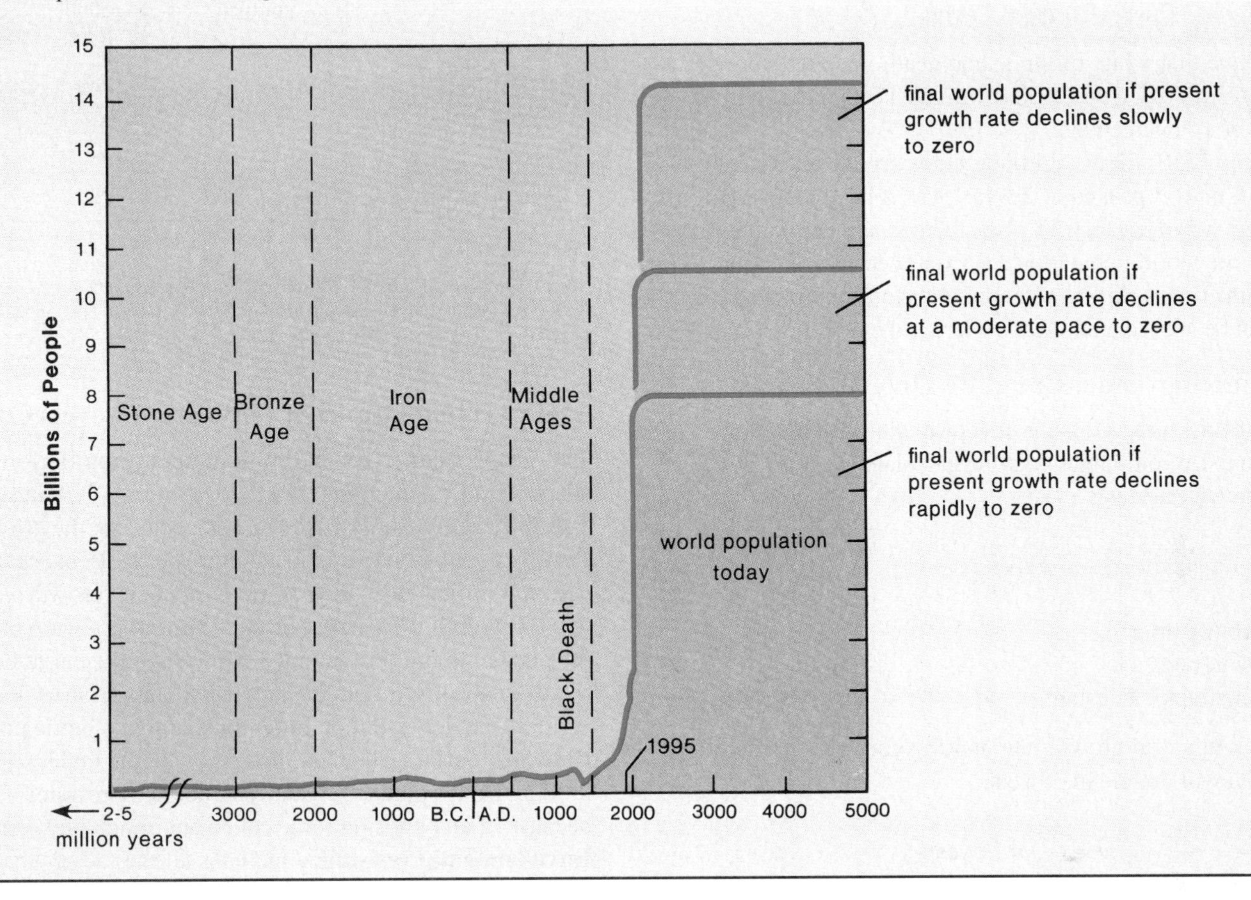

Table 34.1

Exponential Growth of Hypothetical Populations

Population Size	Increase (%)	Actual Increase in Numbers	Population Size	Increase (%)	Actual Increase in Numbers	Population Size
500,000,000	2.00	10,000,000	510,000,000	1.99	10,149,000	520,149,000
3,000,000,000	2.00	60,000,000	3,060,000,000	1.99	60,894,000	3,120,894,000
5,000,000,000	2.00	100,000,000	5,100,000,000	1.99	101,490,000	5,201,490,000

a 1.99% increase results in the third-generation size listed in the last column. Notice the following:

1. In each instance, the second generation has a larger increase than the first generation even though the growth rate decreased from 2% to 1.99%.
2. The larger the population, the larger the increase for each generation.

The percentage increase is termed the growth rate, which is calculated per year.

The Growth Rate: Comparing Births to Deaths

The **growth rate** of a population is determined by considering the difference between the number of persons born (birthrate, or natality) and the number of persons who die per year (death rate, or mortality). It is customary to record these rates per 1,000 persons. For example, Canada at the present time has a birthrate of 15 per 1,000 per year, but it has a death rate of 7 per 1,000 per year. This means that Canada's population growth, or simply its growth rate, is

$$\frac{15-7}{1,000} = \frac{8}{1,000} = \frac{0.8}{100} = 0.8\%$$

Notice that while birthrate and death rate are expressed in terms of 1,000 persons, the growth rate is expressed per 100 persons, or as a percentage.

After 1750, the world population growth rate steadily increased, until it peaked at 2% in 1965. It has fallen slightly since then, however, to 1.7%. Yet, there is an ever-greater increase in the world population each year because of exponential growth. The explosive potential of the present world population can be appreciated by considering the doubling time.

The Doubling Time: Could Be Only 39 Years

The **doubling time** (*d*)—the length of time it takes for the population size to double—can be calculated by dividing 70 by the growth rate (*gr*):

$$d = \frac{70}{gr}$$

d = Doubling time
gr = Growth rate
70 = Demographic constant

If the present world growth rate of 1.8% continues, the world population will double in 39 years.

$$d = \frac{70}{1.8} = 39 \text{ years}$$

This means that in 39 years, the world will need double the amount of food, jobs, water, energy, and so on to maintain the same standard of living.

It is of grave concern to many that the amount of time needed to add each additional billion persons to the world population has taken less and less time (table 34.2). However, if the growth rate continues to decline, this trend will reverse itself, and eventually there will be *zero population growth* when births = deaths. Then population size will remain steady. Therefore, figure 34.2 shows 3 possible logistic curves: the population may level off at 8 billion, 10.5 billion, or 14.2 billion, depending on the speed with which the growth rate declines.

Table 34.2
World Population Increase

Billions of People	Time Needed*	Year of Increase
First	2–5 million	1800
Second	130	1930
Third	30	1960
Fourth	15	1975
Fifth	12	1987
Sixth (projected)	11	1998

*Measured in years

Source: Data from Elaine M. Murphy, *World Population: Toward the Next Century*. Washington, DC: Population Reference Bureau, November 1981, page 3.

The Carrying Capacity: Population Levels Off

The growth curve for many nonhuman populations is S-shaped—the population tends to level off at a certain size. For example, figure 34.3 is based on actual data for the growth of a fruit fly population reared in a culture bottle. Because the fruit flies were adjusting to their new environment, growth was slow in the beginning. Then, because food and space were plentiful, they began to multiply rapidly. Notice that the curve began to rise dramatically, just as the human population curve does now. At this time, a population is demonstrating its biotic potential. **Biotic potential** is the maximum growth rate under ideal conditions. Biotic potential usually is not demonstrated for long because of an opposing force called environmental resistance. **Environmental resistance** includes all the factors that cause early death of organisms and therefore prevents the population from producing as many offspring as it might otherwise do. As far as the fruit flies are concerned, we can speculate that environmental resistance included the limiting factors of food and space. The waste given off by the fruit flies also may have limited the population size. When environmental resistance sets in, biotic potential is overcome, and the slope of the growth curve begins to decline. This is the inflection point of the curve.

The eventual size of any population represents a compromise between biotic potential and environmental resistance. This compromise occurs at the carrying capacity of the environment. The **carrying capacity** is the maximum population that the environment can support for an indefinite period. The carrying capacity of the earth for humans has not

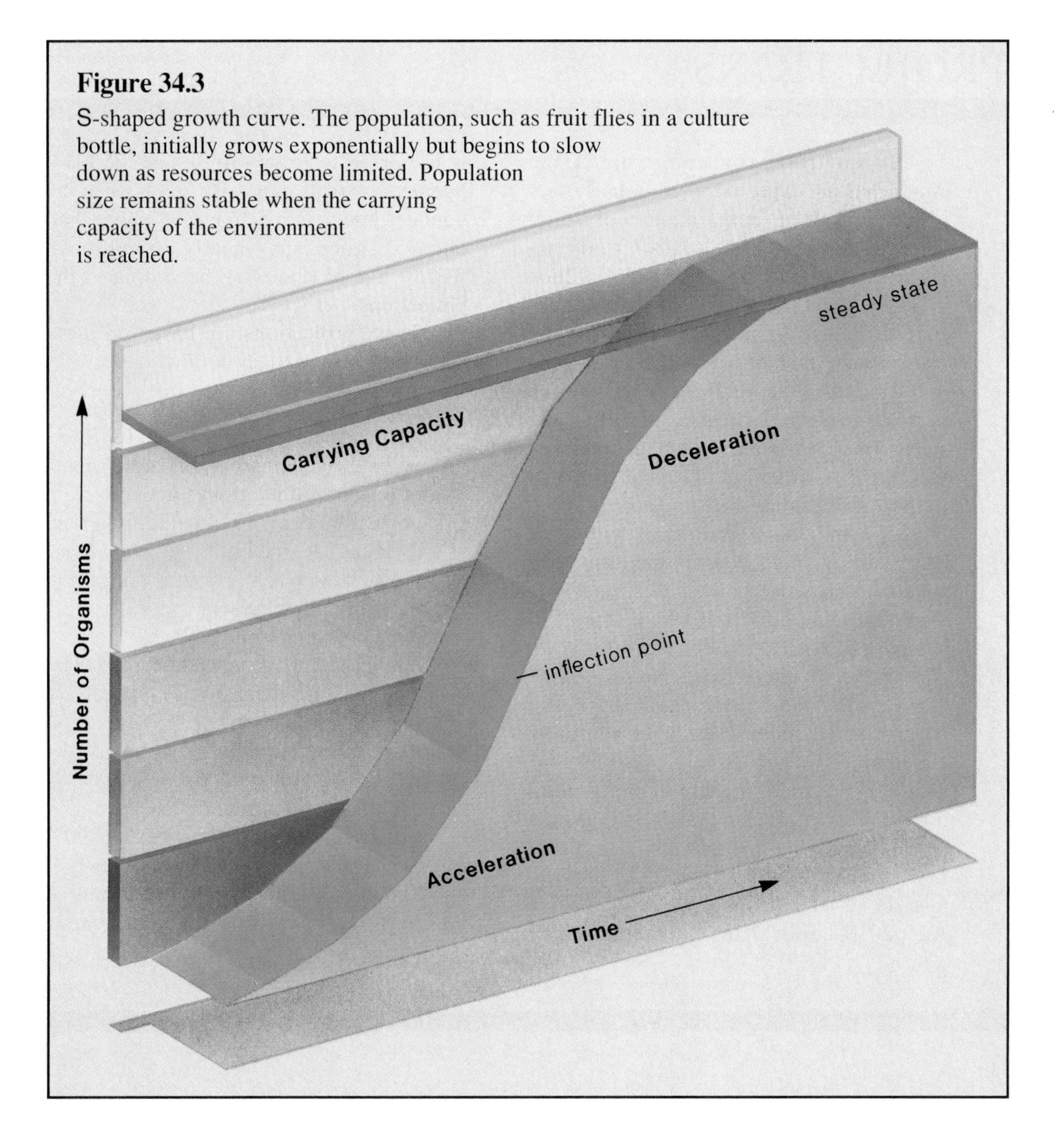

Figure 34.3
S-shaped growth curve. The population, such as fruit flies in a culture bottle, initially grows exponentially but begins to slow down as resources become limited. Population size remains stable when the carrying capacity of the environment is reached.

been determined. Some authorities think the earth is potentially capable of supporting 50 billion–100 billion people. Others think we already have more humans than the earth can adequately support, especially if each person is to be provided the opportunity to develop all his/her potential to learn, create, and enjoy life.

Populations have a biotic potential for increase. Biotic potential is normally held in check by environmental resistance so that a population's growth curve is S-shaped and levels off at the carrying capacity of the environment. Do 34.1 Critical Thinking, found at the end of the chapter.

Human Population Growth

The human population has undergone 3 periods of exponential growth. *Toolmaking* may have been the first technological advance that enabled the human population to enter a period of exponential growth. Farming may have resulted in a second phase of growth, and the *Industrial Revolution,* which occurred about 1850, promoted the third phase.

More-Developed Countries versus Less-Developed Countries

Russia and Japan and the countries of Europe and North America were the first to become industrialized. These nations, often referred to collectively as the *more-developed countries (MDCs),* doubled their populations between 1850 and 1950. This was largely due to a decline in the death rate, the result of modern medicine and improved socioeconomic conditions. The decline in the death rate was followed shortly thereafter by a decline in the birthrate, so that populations in the MDCs experienced only modest growth between 1950 and 1975. This sequence of events (i.e., decreased death rate followed by decreased birthrate) is termed a **demographic transition.**

The growth rate for the MDCs as a whole has now stabilized at about 0.5%. The populations of a few of the MDCs—Italy, Denmark, Hungary, Sweden—are not growing or are actually decreasing in size. The United States has a higher growth rate (0.8%) than average because of the factors discussed in the reading on the next page.

Most countries in Africa, Asia, and Latin America are known collectively as the *less-developed countries (LDCs)* because they are not fully industrialized. Although the death

U.S. POPULATION PROJECTIONS

Based on the 1990 census, the Census Bureau has made certain projections concerning the U.S. population. It is projected that the U.S. population will reach 274.8 million by the year 2000 and ultimately 382.7 million by 2050. The projected population for 2050 is more than 80 million higher than expected. In the new projections, there is no leveling off and no end in sight to population growth.

The U.S. racial/ethnic mix will grow increasingly diverse, due to a combination of factors: slow growth among non-Hispanic whites, steady growth among African-Americans and Native Americans, and rapid growth among Hispanics and Asian Americans. Over the next 60 years, the share of the population that is non-Hispanic white should decline steadily—from 76% in 1990 to 68% in 2010 and 60% in 2030. By 2050, a bare majority of Americans (53%) will be non-Hispanic whites.

The nonwhite population itself is expected to become more diverse as well. While the numbers of non-Hispanic blacks are expected to grow steadily, reaching 38.2 million in 2010 and 57.3 million in 2050, their share of the population is expected to grow more slowly, reaching 13% in 2010 and 15% in 2050.

By contrast, Hispanics and Asian Americans are expected to grow fairly rapidly, the result of immigration and natural increase. Between 1990 and 2010, Hispanics are expected to grow from 22.4 million to 39.3 million. About that time, Hispanics should replace African-Americans as the nation's largest minority group. By 2050, one in 5 Americans—a total of over 80 million persons—will be Hispanic. Non-Hispanic Asian Americans are expected to grow even more rapidly—from 7 million in 1990 to 16.5 million in 2010 and 38.8 million in 2050. By 2050, one in 10 Americans will be of Asian descent. As for Native Americans, they should continue to grow steadily, reaching over 4 million by 2050. However, they will compose barely 1% of all Americans.

The report also projects the pace of the graying of America, as the baby-boom generation continues to age. (The unusually large number of babies born between 1947 and 1964 is called the baby-boom generation.) For example, the population ages 65 and older will continue growing—steadily at first, from 31 million persons (12.5% of the total) in 1990 to 39.7 million (13%) in 2010. After 2010, however, the elderly population is projected to grow rapidly, as the baby boomers enter its ranks. By 2030, when the younger boomers reach age 65, more than one in 5 Americans—nearly 70 million—will be age 65 and over, according to the projections.

These projections are based on these assumptions:

- Future fertility is assumed to remain near current levels. However, fertility will slowly increase as the proportion of higher fertility groups in the population rises. The total fertility rate (average lifetime births per woman) is projected to reach 2.12 in 2050.
- Future immigration is assumed to remain near current levels. Net immigration is assumed to be 880,000 annually (including 200,000 illegal immigrants per year).
- Fertility and mortality differentials by race/ethnic group are assumed to continue their current trends.
- Life expectancy based on 1980–1990 trends is projected to improve slowly. The middle projection is increasing from 75.8 years in 1992 to 82.1 years by 2050.

rate began to decline steeply in these countries with the importation of modern medicine from the MDCs following World War II, the birthrate remained high. The LDCs are unable to adequately cope with such rapid population expansion, and many people in these countries are underfed, poorly housed, unschooled, and living in abject poverty.

The growth rate of the LDCs peaked at 2.4% between 1960 and 1965. Since that time, the death rate decline has slowed and the birthrate has fallen. The growth rate is expected to be 1.8% by the end of the century. At that time, however, more than two-thirds of the human population will live in the LDCs. The reason for a decline in the growth rate is not clear, but the greatest decline is seen in countries with good family-planning programs supported by community leaders.

> The history of the world population shows that the more-developed countries underwent a demographic transition between 1950 and 1975; the less-developed countries are just now undergoing demographic transition.

Comparing Age Structure

Populations have 3 age groups: dependency, reproductive, and postreproductive. One way of characterizing population is by these age groups. This is best visualized when the proportion of individuals in each group is plotted on a bar graph, thereby producing an age-structure diagram (fig. 34.4).

Laypeople are sometimes under the impression that if each couple has 2 children, zero population growth will take place immediately. However, **replacement reproduction,** as it is called, will still cause most countries today to continue growing due to the age structure of the population. If there are more young women entering the reproductive years than there are older women leaving them behind, then replacement reproduction will give a positive growth rate.

Many MDCs have a stabilized age-structure diagram (fig. 34.4), but most LDCs have a youthful profile—a large proportion of the population is younger than the age of 15. Since there

Figure 34.4

Age-structure diagram for MDCs and LDCs, 1989. The diagrams illustrate that the MDCs are approaching stabilization, whereas the LDCs will expand rapidly due to the shape of their age-structure diagram.

Source: Data from *World Population Profile:* 1989, WP-89.

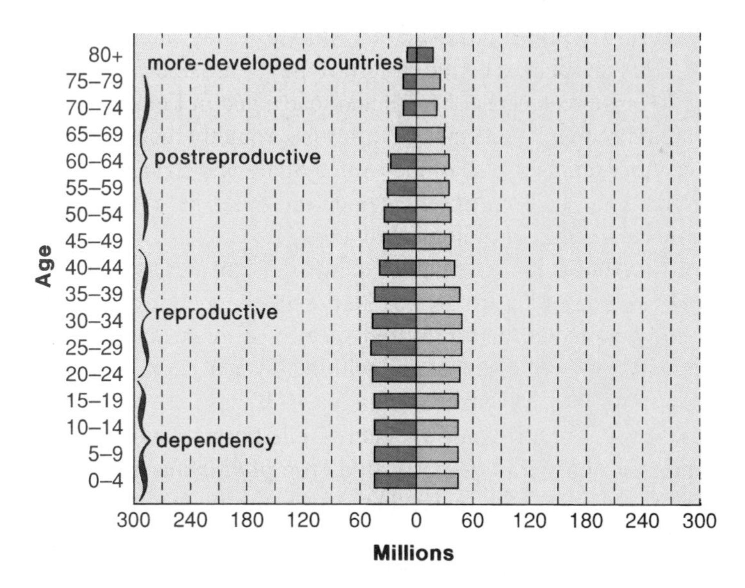

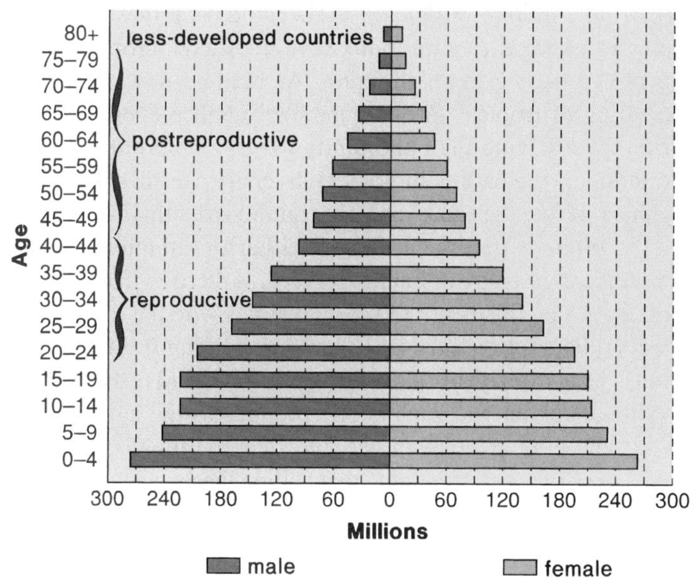

are so many young women entering the reproductive years, the population will still expand greatly, even after replacement reproduction is attained. The more quickly replacement reproduction is achieved, however, the sooner zero population growth will result.

Figure 34.5

Contour farming. Crops are planted according to the lay of the land to reduce soil erosion. This farmer has planted alfalfa in between the strips of corn to replenish the nitrogen content of the soil. Alfalfa, a legume, has root nodules that contain nitrogen-fixing bacteria.

The Human Population and Pollution

As the human population increases in size, more energy and materials are consumed. Because the human population does not use energy efficiently and does not recycle materials (see fig. 33.11), pollutants are added to all components of the biosphere—land, water, and air.

How People Degrade Land

The land has been degraded in many ways. Here, we will discuss only those of the greatest concern.

Soil Erodes and Deserts Grow

Soil erosion causes the productivity of agricultural lands to decline. It occurs when wind and rain carry away the topsoil, leaving the land exposed and without adequate cover. The U.S. Department of Agriculture estimates that erosion is causing a steady drop in the productivity of farmland equivalent to the loss of 0.5 million hectares per year. To maintain the productivity of eroding land, more fertilizers, more pesticides, and more energy must be applied.

One answer to the problem of erosion is to adopt soil conservation measures. For example, farmers could use strip-cropping and contour farming (fig. 34.5).

Desertification is the transformation of marginal lands to desert conditions because of overgrazing and overfarming. Desertification has been particularly evident along the southern edge of the Sahara Desert in Africa, where it is estimated that 350,000 mi^2 of once-productive grazing land has become desert in the last 50 years. However, desertification also occurs in this country. The U.S. Bureau of Land Management, which opens up federal lands for grazing, reports that much of the rangeland it manages is in poor or bad condition, with much of its topsoil gone and with greatly reduced ability to support forage plants.

What's Causing Tropical Rain Forest Destruction

Virgin rain forests (fig. 34.6) exist in Southeast Asia and Oceania, Central and South America, and Africa. These forests are severely threatened by human exploitation. The MDCs' demand for wood products has created a market for beautiful and costly tropical woods. Indonesia is cutting heavily to feed its new wood-exporting business, and Brazil is racing to catch up. Much of this wood goes to Japan, the United States, and Europe.

Tropical rain forests are also undergoing destruction as a result of the needs of the people who live there. For example, in Brazil, a large sector of the population has no means of support. To ease social unrest, the government allows citizens to own any land they clear in the Amazon forest (occurs along the Amazon River). In tropical rain forests, it is customary to practice **slash-and-burn agriculture,** in which trees are cut down and burned to provide space to raise crops. Unfortunately, the fertility of the land is sufficient to sustain agriculture for only a few years. Once the cleared land is incapable of sustaining crops, the farmer moves on to another part of the rain forest to slash and burn again.

Cattle ranchers are the greatest beneficiaries of deforestation, and increased ranching is therefore another reason for tropical rain forest destruction. The ranchers in Central and South America are so ruthless, they actually force colonists to sell them newly cleared land at gunpoint. Some of the cattle provide beef for export, including to the U.S. A newly begun pig-iron industry also indirectly results in further exploitation of the rain forest. The pig iron must be processed before it is exported, and smelting it requires the use of charcoal. The largest pig-iron company acknowledges having paid for construction of 1,500 small makeshift ovens used by peasants who burn trees from the rain forest to produce the charcoal.

> There are currently 3 primary reasons for tropical rain forest destruction: (1) logging to provide hardwoods for export, (2) slash-and-burn agriculture, and (3) cattle ranching. Industrialization in countries having extensive tropical rain forests no doubt will become another major reason.

Losing Biological Diversity. As is discussed in the reading on pages 642–643, biological diversity is much greater in tropical rain forests than in temperate forests. For example, temperate forests across the entire United States contain about 400 tree species. In the rain forest, a typical 10-ha^2 (one hectare = 2,471 acres) area holds as many as 750 types of trees. The fresh waters of South America are inhabited by an estimated 5,000 fish species; on the eastern slopes of the Andes, there are 980 or more species of frogs and toads; and in Ecuador, there are more than 1,200 species of birds—roughly twice as many as those inhabiting all of the United States and Canada. Therefore, a very serious side effect of deforestation in tropical countries is the loss of biological diversity.

Altogether, over half of the world's species are believed to live in tropical forests. A National Academy of Sciences study estimated that a million species of plants and animals are in danger of disappearing within 20 years as a result of deforestation in tropical countries. Many of these life forms have never been studied, and yet many could possibly be useful. At present, our entire domesticated crop production around the world relies on the fewer than 30 species of plants and animals domesticated during the last 10,000 years! It is quite possible that many additional species of wild plants and animals now living in tropical forests could be domesticated, or at least their genes used to improve traditional crops through genetic engineering techniques. As matters now stand, the clearing of tropical forests very likely will prevent humans from ever having the opportunity to utilize more than a tiny fraction of the earth's biological diversity. Besides, these organisms have the same inherent right to exist that we do.

While it may seem like an either/or situation—either biological diversity or human survival—there is growing recognition that this is not the case. Natives who harvest rubber from rubber trees (fig. 34.6*b*) can earn a living from the same trees year after year. A recent study calculated the market value of rubber and such exotic produce as the aguaje palm fruit and nuts, which can be harvested continually from the Amazon forest. It concluded that selling these products would yield more than twice the income of either lumbering or cattle ranching.

> There is much worldwide concern about the loss of biological diversity due to the destruction of tropical rain forests. The myriad of plants and animals that live there could possibly benefit human beings.

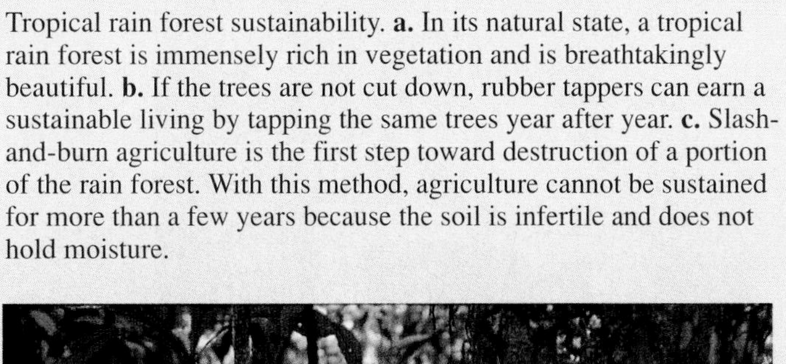

Figure 34.6

Tropical rain forest sustainability. **a.** In its natural state, a tropical rain forest is immensely rich in vegetation and is breathtakingly beautiful. **b.** If the trees are not cut down, rubber tappers can earn a sustainable living by tapping the same trees year after year. **c.** Slash-and-burn agriculture is the first step toward destruction of a portion of the rain forest. With this method, agriculture cannot be sustained for more than a few years because the soil is infertile and does not hold moisture.

a.

b.

c.

Figure 34.7
Dump sites. These not only cause land and air pollution, they also allow chemicals to enter groundwater; later these chemicals may become part of human drinking water.

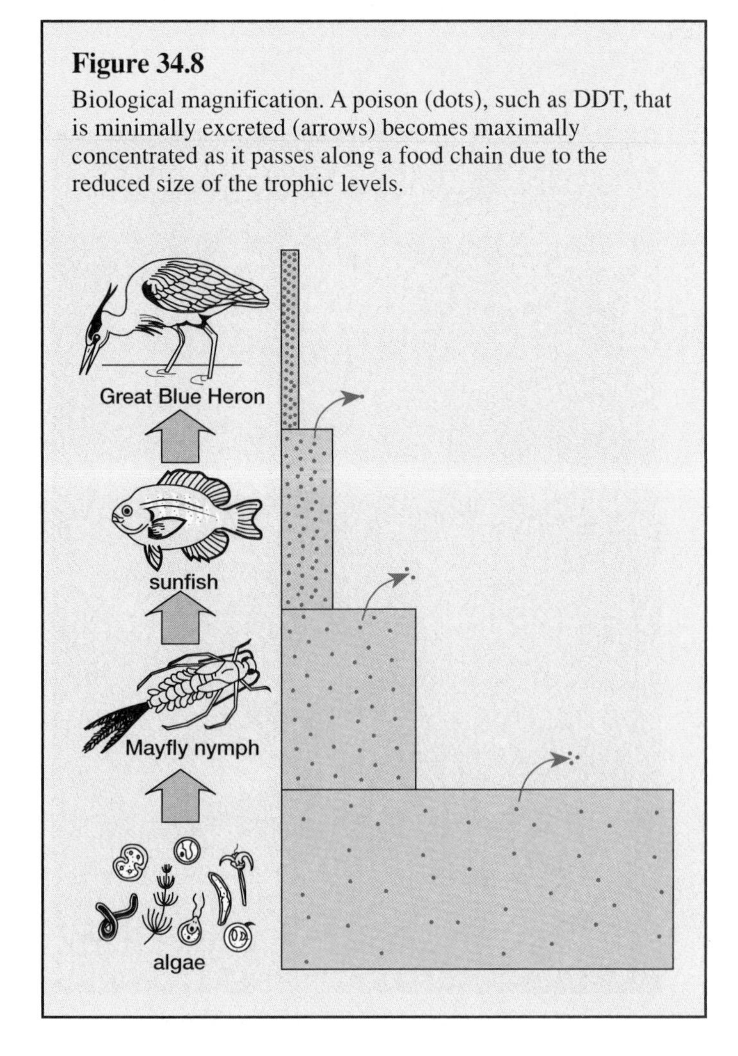

Figure 34.8
Biological magnification. A poison (dots), such as DDT, that is minimally excreted (arrows) becomes maximally concentrated as it passes along a food chain due to the reduced size of the trophic levels.

Waste Disposal and Dangerous Trash

Every year, the U.S. population discards billions of metric tons of solid wastes, much of it on land. Solid wastes include not only household trash (fig. 34.7) but also sewage sludge, agricultural residues, mining refuse, and industrial wastes. Some of these solid wastes contain substances that cause illness and sometimes even death; they are called **hazardous wastes.**

Hazardous wastes, such as heavy metals, chlorinated hydrocarbons (organochlorines), and nuclear wastes, enter bodies of water and are subject to biological magnification (fig. 34.8). Decomposers are unable to break down these wastes. They enter and remain in the body because they are not excreted. Therefore, they become more concentrated as they pass along a food chain. Notice in figure 34.8 that the dots representing DDT become more concentrated as they pass from producer to tertiary consumer. Biological magnification is most apt to occur in aquatic food chains—there are more links in aquatic food chains than there are in terrestrial food chains. Humans are the final consumers in both types of food chains, and in some areas, human milk contains detectable amounts of DDT and PCBs, which are organochlorines.

The dumping of hazardous wastes directly endangers public health. Chemical wastes buried over a quarter of a century ago in Love Canal, near Niagara Falls, have seriously damaged the health of some residents there. Similarly, the town of Times Beach, Missouri, was abandoned because workers spread an organochlorine (dioxin)-laced oil on the city streets; this resulted in a myriad of illnesses among its citizens. In other places, such as Holbrook, Massachusetts, manufacturers have left thousands of waste-filled metal drums in abandoned or uncontrolled sites. Toxic chemicals are oozing out of the drums and into the ground and are contaminating the water supply. Illnesses, especially forms of cancer, are quite common not only in Holbrook but also in adjoining towns.

Water Pollution from Rivers to Oceans

Pollution of surface water, groundwater, and the oceans is of major concern today.

Many Sources Pollute Surface Water

All sorts of pollutants from various sources enter surface waters, as depicted in figure 34.9. Sewage treatment plants help degrade organic wastes, which can otherwise cause oxygen depletion in lakes and rivers. As the oxygen level decreases, the diversity of life is greatly reduced. Also, human feces may contain the microbes that cause cholera, typhoid fever, and

Figure 34.9

Sources of water pollution. Many bodies of water are dying due to the introduction of sediments and surplus nutrients.

Source: Adapted from U.S. Environmental Protection Agency, Office of Water Supply and Solid Waste Management Programs, "Waste Disposal Practices and Their Effects on Ground Water," Executive Summary. Washington, DC: U.S. Government Printing Office, 1977.

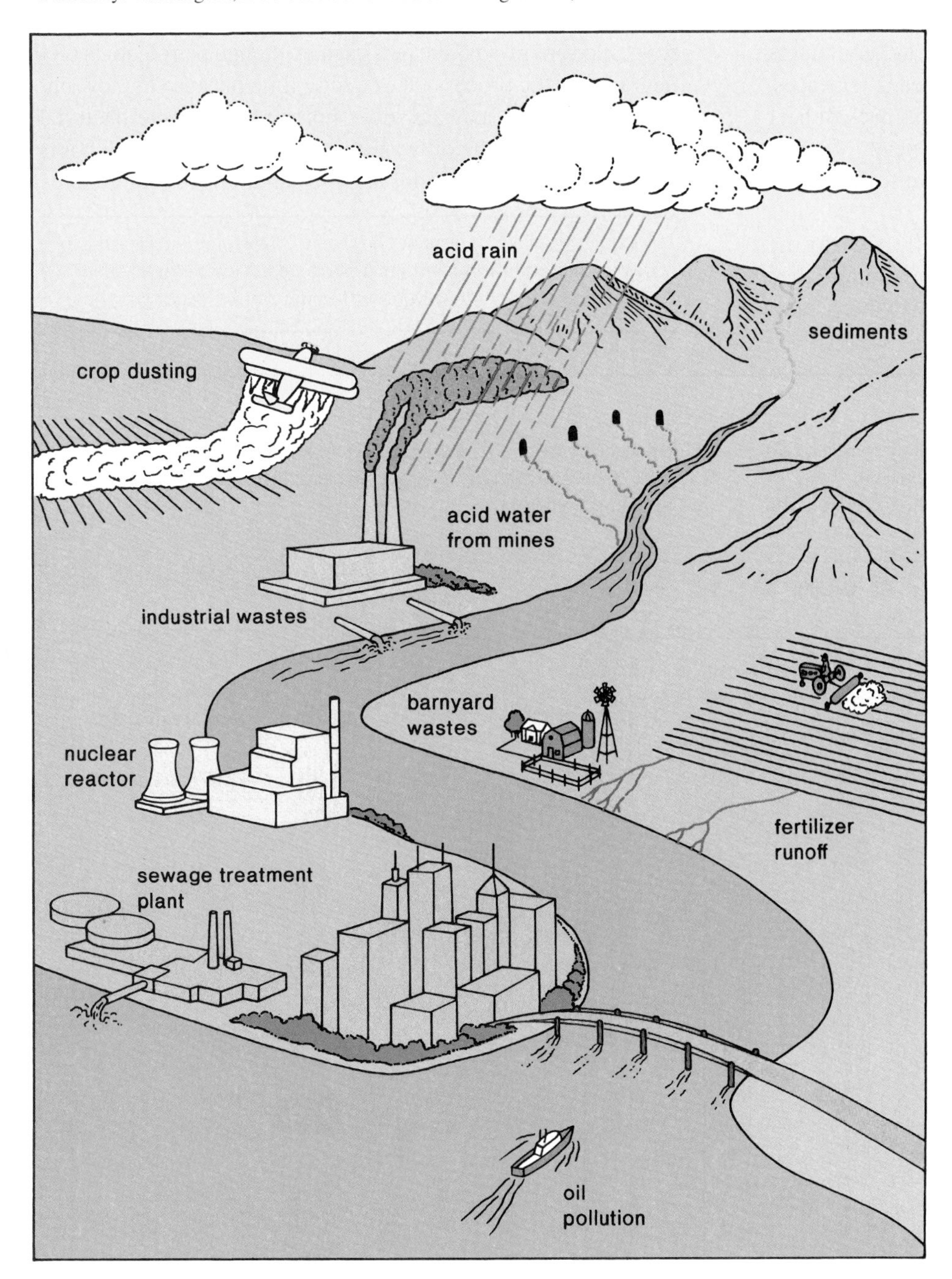

dysentery. In LDCs, where the population is growing and where waste treatment is practically nonexistent, many children die each year from these water-borne diseases.

Typically, sewage treatment plants use bacteria to break down organic matter to inorganic nutrients, like nitrates and phosphates, which then enter surface waters. These types of nutrients, which can also enter waters by fertilizer runoff and soil erosion, lead to **cultural eutrophication,** the enrichment of bodies of water due to human activities. First, the nutrients cause overgrowth of algae. Then, when the algae die, oxygen is used up by the decomposers, and the water's capacity to support life is reduced. Massive fish kills are sometimes the result of cultural eutrophication.

Industrial wastes include heavy metals and organochlorines, such as pesticides. These materials are not readily degraded under natural conditions nor in conventional sewage treatment plants. Sometimes they accumulate in the mud at the bottom of deltas, and environmental problems result if estuaries of very polluted rivers are disturbed. Industrial pollution is being addressed in many MDCs but usually has low priority in LDCs.

Groundwater Pollution: Farms and Industry

In areas of intensive animal farming or where there are many septic tanks, ammonia (NH_3) released from animal and human waste is converted by soil bacteria to soluble nitrate, which moves down through the soil (percolates) into underground water supplies. Between 5% and 10% of all wells examined in the United States have nitrate levels higher than the recommended maximum. Nitrates are converted to nitrites in the digestive tract. Nitrites are poisons because they combine with hemoglobin, forming methoglobin, which has a reduced oxygen-carrying capacity.

Industry also pollutes aquifers (underground rivers). Previously, industry would run wastewater into a pit from which the pollutants would seep into the ground. Wastewater and chemical wastes were also injected into deep wells, from which the pollutants constantly discharged. Both of these customs have been or are in the process of being phased out. It is very difficult for industry to find other ways to dispose of wastes, especially since citizens do not wish to live near waste treatment plants. The emphasis today, therefore, is on prevention of wastes in the first place. Industry is trying to use processes that do not create wastes and/or to recycle the wastes they do generate.

Ocean Water Pollution: The Final Dumping Ground

Coastal regions are not only the immediate receptors for local pollutants, they are also the final receptors for pollutants carried by rivers that empty at the coast. Waste dumping also occurs at sea, and ocean currents sometimes transport both trash and pollutants back to shore. Examples are nuclear wastes and the nonbiodegradable plastic bottles, pellets, and containers that now commonly litter beaches and the oceans' surfaces. Some of these, such as the plastic that holds a six-pack of beer and fishing lines, have been implicated in the death of birds, fishes, and marine mammals that mistook them for food and became entangled in them.

Offshore mining and shipping add pollutants to the oceans. Some 5 million metric tons of oil a year, or more than one g per 100 m^2 of the oceans' surfaces, end up in the oceans. Large oil spills kill plankton, fish larvae, and shellfishes, as well as birds and marine mammals. One of the largest spills occurred on March 24, 1989, when the tanker *Exxon Valdez* struck a reef in Alaska's Prince William Sound and leaked 44 million l of crude oil. During the war with Iraq, 120 million l were released from onshore storage tanks into the Persian Gulf, an event that was called environmental terrorism. Although petroleum is biodegradable, the process takes a long time because the low nutrient content of seawater does not support a large bacterial population. Once the oil washes up onto beaches, many hours of work and millions of dollars to clean it up are required.

> Adequate sewage treatment and waste disposal are necessary to prevent the pollution of rivers and oceans. New methods also are needed to prevent pollution of underground water supplies.

Figure 34.10

Air pollutants. These are the gases, along with their sources, that contribute to 4 environmental effects of major concern: photochemical smog, acid deposition, the greenhouse effect, and the destruction of the ozone shield. An examination of the sources of these gases shows that vehicle exhaust and fossil fuel burning are the chief contributors.

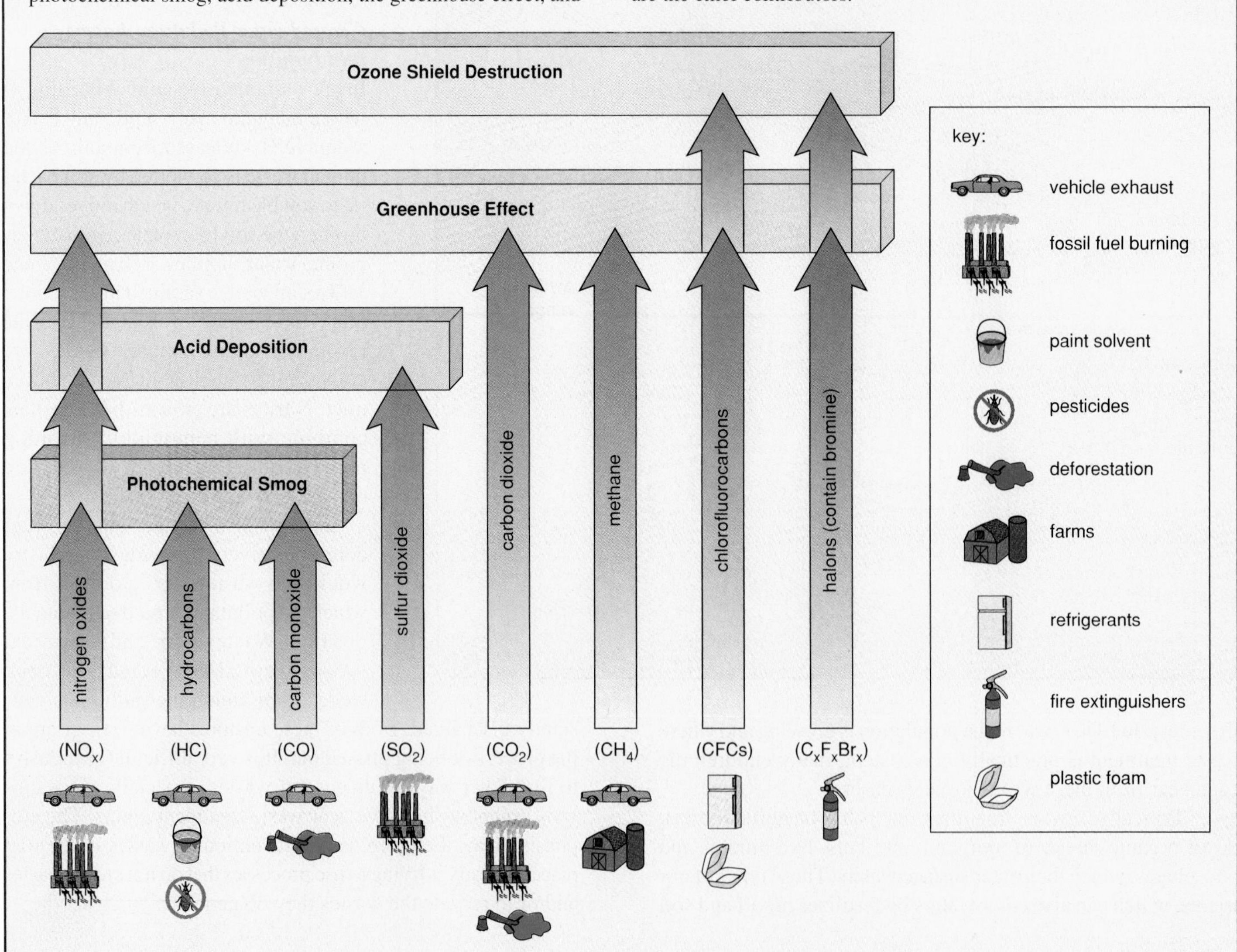

Air Pollution: Four Threats

The atmosphere has 2 layers, the stratosphere and the troposphere. The stratosphere is a layer that lies 15–50 km above the surface of the earth. Here, the energy of the sun splits oxygen molecules (O_2). These individual oxygen atoms (O) then combine with molecular oxygen to form ozone (O_3). This ozone layer acts as a shield because it absorbs the ultraviolet rays of the sun, preventing them from striking the earth. If these rays did penetrate the atmosphere, life on earth would not be possible because living things cannot tolerate heavy doses of ultraviolet radiation.

The troposphere is the atmospheric layer closest to the earth's surface. It ordinarily contains the gases nitrogen (N_2)—78%, oxygen (O_2)—21%, and carbon dioxide (CO_2)—0.3%.

Four major concerns—photochemical smog, acid deposition, global warming, and the destruction of the ozone shield—are associated with the air pollutants listed in figure 34.10. You can see that fossil fuel burning and vehicle exhaust are primary sources of gases associated with air pollution. These 2 sources are related because gasoline is derived from petroleum, a fossil fuel.

Photochemical Smog: Making Ozone and PAN

Photochemical smog contains 2 air pollutants—nitrogen oxides (NO_x) and hydrocarbons (HC)—that react with one another in the presence of sunlight to produce ozone (O_3) and PAN (peroxylacetyl nitrate). Both nitrogen oxides and hydrocarbons come from fossil fuel combustion, but additional hydrocarbons come from various other sources as well, including industrial solvents.

Ozone and PAN are commonly referred to as oxidants. Breathing ozone affects the respiratory and nervous systems, resulting in respiratory distress, headache, and exhaustion. These symptoms are particularly apt to appear in young people; therefore, in Los Angeles, where ozone levels are often high, schoolchildren must remain inside the school building whenever the ozone level reaches 0.35 ppm (parts per million by weight). Ozone is especially damaging to plants, resulting in leaf mottling and reduced growth (fig. 34.11).

Carbon monoxide (CO) is another gas that results from the burning of fossil fuels in the industrial Northern Hemisphere. High levels of carbon monoxide increase the formation of ozone. Carbon monoxide also combines preferentially with hemoglobin and thereby prevents hemoglobin from carrying oxygen. Breathing large quantities of automobile exhaust can even result in death because of this phenomenon. Of late, it has been discovered that the amount of carbon monoxide over the Southern Hemisphere is equal to the amount over the Northern Hemisphere. The source of the carbon monoxide here, however, is the burning of tropical rain forests.

Normally, warm air near the ground is able to escape into the atmosphere. Sometimes, however, air pollutants, including photochemical smog and soot, are trapped near the earth due to a long-lasting thermal inversion. During a **thermal inversion,** cold air is found at ground level beneath a layer of warm, stagnant air above. This often occurs at sunset, but turbulence usually mixes these layers during the day. Areas surrounded by hills are particularly susceptible to the effects of a thermal inversion because the air tends to stagnate.

Acid Deposition: From Burning Coal, Oil, Gasoline

Power plants that burn coal and oil have high sulfur dioxide (SO_2) emissions, and automobile exhaust contains nitrogen oxides (NO_x); both of these are converted to acids when they combine with water vapor in the atmosphere, a reaction that is promoted by ozone in photochemical smog. These acids return to earth as either wet deposition (acid rain or snow) or dry deposition (sulfate and nitrate salts).

As discussed earlier in the reading on page 26, **acid deposition** (the return to earth of acid particles in rain or snow) is now associated with dead or dying lakes and forests, particularly in North America and Europe. Acid deposition also corrodes marble, metal, and stonework, an effect that is noticeable in cities. It can also degrade water supplies by leaching heavy metals from the soil into drinking water sources. Similarly, acid water dissolves copper from pipes and from lead solder that is used to join pipes.

Figure 34.11
Effect of ozone on plants. The milkweed in (**a**) was exposed to ozone and appears unhealthy; the milkweed in (**b**) was grown in an enclosure with filtered air and appears healthy.

a.

b.

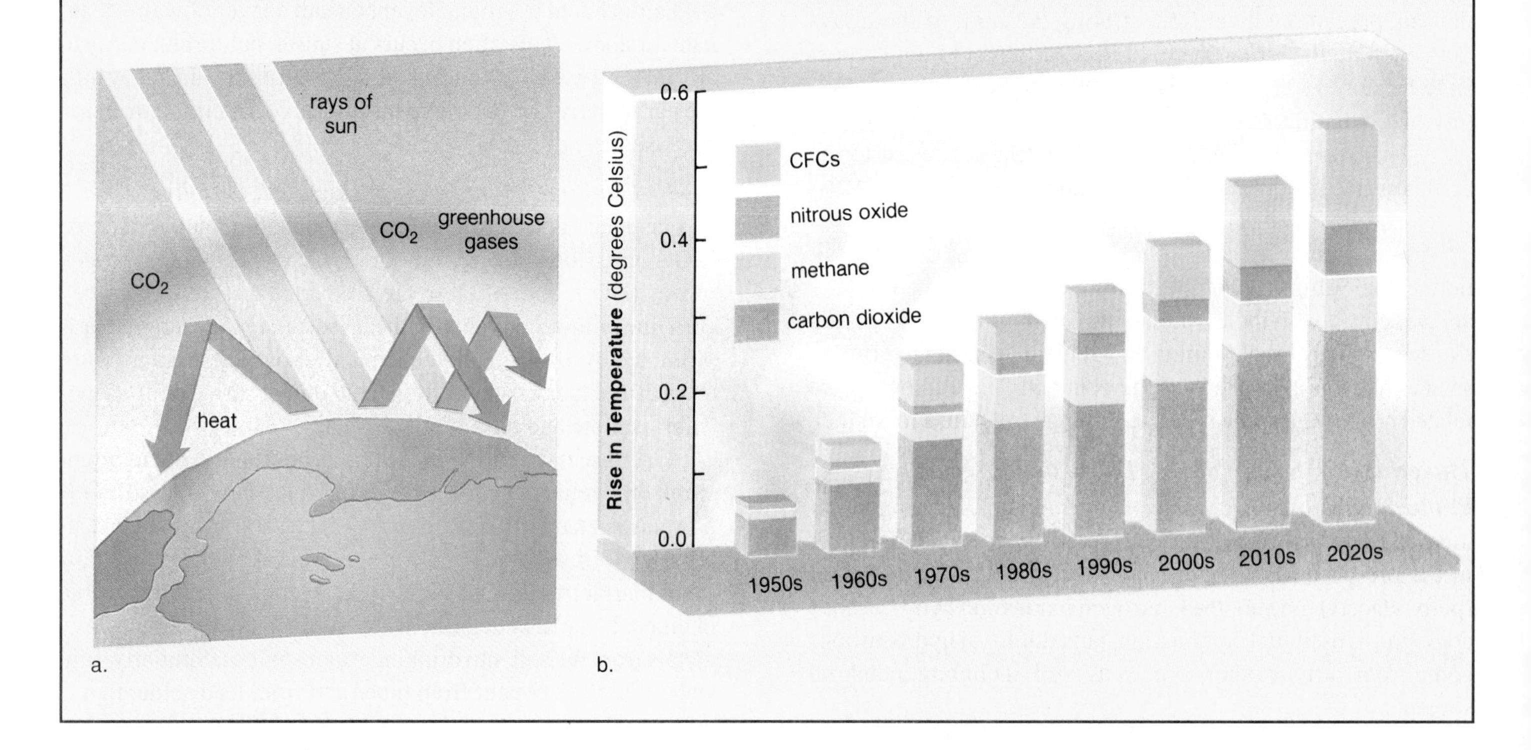

Figure 34.12
Global warming. **a.** The greenhouse effect is caused by the atmospheric accumulation of certain gases, such as carbon dioxide (CO_2), that allow the rays of the sun to pass through but absorb and reradiate heat to the earth. **b.** The greenhouse gases. This graph shows the fraction of warming caused by the gases carbon dioxide, methane, nitrous oxide, and CFCs for the decades from 1950 to 2020. There was no accumulation of CFCs in the 1950s because they were not being manufactured to any degree then. By 2020, carbon dioxide and all of the other gases taken together will each contribute about 50% to the projected global warming.

Global Warming: From Greenhouse Gases

Certain air pollutants allow the sun's rays to pass through but then absorb and reradiate the heat to the earth (fig. 34.12*a*). This is called the **greenhouse effect** because the glass of a greenhouse allows sunlight to pass through but then traps the resulting heat inside the structure. The air pollutants responsible for the greenhouse effect are known as the greenhouse gases. They are as follows:

Carbon dioxide (CO_2) from fossil fuel and wood burning
Nitrous oxide (NO_2) primarily from fertilizer use and animal wastes
Methane (CH_4) from biogas, bacterial decomposition (particularly in the guts of animals), sediments, and flooded rice paddies
Chlorofluorocarbons (CFCs) from Freon, a refrigerant
Halons (halocarbons, [$C_xF_xBr_x$]) from fire extinguishers

If nothing is done to control the level of greenhouse gases in the atmosphere, a rise in global temperature is expected. Figure 34.12 predicts a rise of over 0.5°C by the year 2020, but some authorities predict the rise in temperature could be as high as 5°C by 2050. It is possible that global warming has already begun: the 4 hottest years on record have occurred during the 1980s; there has been greater warming in the winters than in the summers; there has been greater warming at high latitudes than near the equator; and the stratosphere is cooler and the lower atmosphere is warmer than before. All of these effects had been predicted by computer models of the greenhouse effect.

The ecological effects of a 5°C rise in global temperature would be severe. The sea level would rise—melting of the polar ice caps would add more water to the sea, and in any case, water expands when it heats up. There would be coastal flooding and the possible loss of many cities, like New York, Boston, Miami, and Galveston in the United States. Coastal ecosystems, such as marshes, swamps, and bayous, would normally move inland to higher ground as the sea level rises, but many of these ecosystems are blocked by artificial structures and may be unable to move inland. If so, the loss of fertility would be immense.

There may also be food loss because of regional changes in climate. Because of greater heat and drought in the midwestern United States, the suitable climate for growing wheat and corn may shift as far north as Canada, where the soil is not as suitable.

It is clear from figure 34.12 that carbon dioxide accounts for at least 50% of the predicted rise in global temperature. Therefore, a sharp decrease in consumption of fossil fuels is recommended, and we must find more efficient ways to acquire energy from cleaner fuels, such as natural gas. We must also use alternative energy sources, such as solar and geothermal energy and perhaps even nuclear power, more aggressively. In addition to fossil fuel consumption, deforestation is a major contributor to the rise in carbon dioxide. Burning an acre of primary forest releases 200,000 kg of carbon dioxide into the air; moreover, the trees are no longer available to act as a sink to take up carbon dioxide during photosynthesis. Therefore, tropical rain forest deforestation should be halted and extensive reforesting all over the globe should take place.

The other greenhouse gases combined account for the other 50% predicted rise in global temperature. A complete phaseout of chlorofluorocarbon use would be most beneficial. Fortunately, in an effort to arrest ozone shield destruction, the United States and the European countries have agreed to reduce CFC production by 85% as soon as possible and to stop their production altogether by the end of the century.

Destroying Our Ozone Shield

The **ozone shield,** you will recall, is a layer of ozone (O_3) in the upper atmosphere that protects the earth from ultraviolet radiation. Ozone shield destruction is primarily caused by chlorofluorocarbons (CFCs) and halons (halocarbons [$C_xF_xBr_x$]). CFCs are heat-transfer agents used in refrigerators and air conditioners. They are also used as foaming agents in such products as styrofoam cups and egg cartons. In the past, they were used as propellants in spray cans, but this application is now banned in the United States. Halons are antifire agents used in fire extinguishers.

Scientists knew that CFCs would drift up into the stratosphere, but it was believed that they would be nonreactive there during their 150-year life span. It is now apparent, however, that solar energy causes CFCs to release chlorine, which attaches to ozone molecules. Thereafter, ozone breaks down to chlorine oxide and oxygen molecules (O_2), and the ozone shield is depleted. Over the South Pole, the entire scenario is accentuated because the dry, cold air of Antarctica is filled with ice crystals on which chlorine and ozone join and react. The result is a 60% loss of ozone each spring; such a severe loss is called an *ozone hole* (fig. 34.13).

Figure 34.13

Ozone depletion. NASA's Nimbus 7 satellite has been tracking the depletion of ozone at the South Pole of Antarctica since the late 1970s. The so-called ozone hole appears as a large white area at the center of this picture, which is typical of pictures taken each spring.

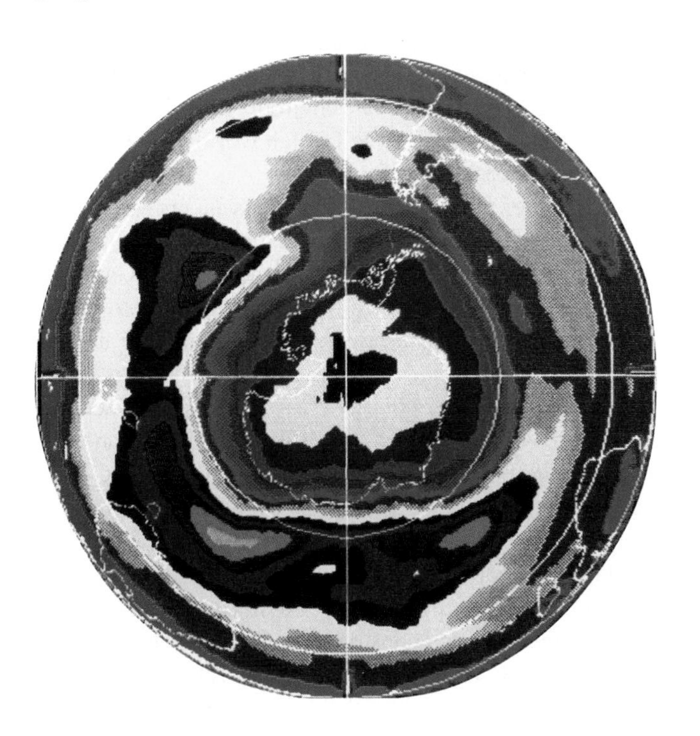

The Arctic and the Northern Hemisphere are not as vulnerable as Antarctica. Still, scientists find that at various points during the year, the Northern Hemisphere appears to lose as much as 6% of its ozone shield.

Depletion of the ozone shield will allow more ultraviolet rays to reach the earth. The incidence of human cancer, especially skin cancer, is expected to increase, and plants and animals living in the top microlayer of the oceans will begin to die. Increased ultraviolet radiation will also hasten the rate at which smog forms. Many believe that the current plans to reduce CFCs and halon emissions are insufficient and that much more stringent measures are necessary.

> Outdoor air pollutants are involved in causing 4 major detrimental environmental effects: photochemical smog, acid deposition, global warming, and ozone shield destruction. Each pollutant may be involved in more than one of these effects.

A Sustainable World

Economic growth is often accompanied by environmental degradation, and there is great concern that as the LDCs become more developed, environmental degradation will increase to the point that the human population will outstrip the carrying capacity of the planet. Without economic growth, however, the demographic transition may not occur, and the sheer number of people in the LDCs will provoke environmental degradation to such a degree that the effects will be felt worldwide, not just in the immediate area.

The answer to this dilemma is economic growth without the side effect of environmental degradation. This is called sustainable growth. Certain MDCs, such as the Scandinavian countries and to a degree the United States, are learning to protect the environment. Energy consumption is decreasing even as economic growth continues. Industries are beginning to recycle their wastes to prevent environmental pollution. Citizens are learning to recycle their trash. More should be done, and ecologically sound practices must be exported to the LDCs very quickly. Only sustainable economic development will ensure the continuance of the world's human population.

Once zero population growth has been achieved, we can begin to consider the possibility of a steady state, in which population and resource consumption remain constant. Environmental preservation is paramount in a steady state (fig. 34.14).

What would our culture be like if we had steady-state manufacturing and a steady-state population? Perhaps it would be greatly improved. Certainly, there are no limits to growth in knowledge, education, art, music, scientific research, human rights, justice, and cooperative human interactions. In a steady-state world, the general sense of fearful competition among peoples might diminish, allowing human compassion and creativity to prosper as never before.

Figure 34.14
In the steady state, environmental preservation is an important consideration.

It is hoped that a sustainable world will result in a steady-state society, where there is no yearly increase in population or resource consumption. Do 34.2 Critical Thinking, found at the end of the chapter.

SUMMARY

Populations have a biotic potential for increase in size. Biotic potential is normally held in check by environmental resistance, thereby producing an S-shaped growth curve and leveling off at the carrying capacity of the environment.

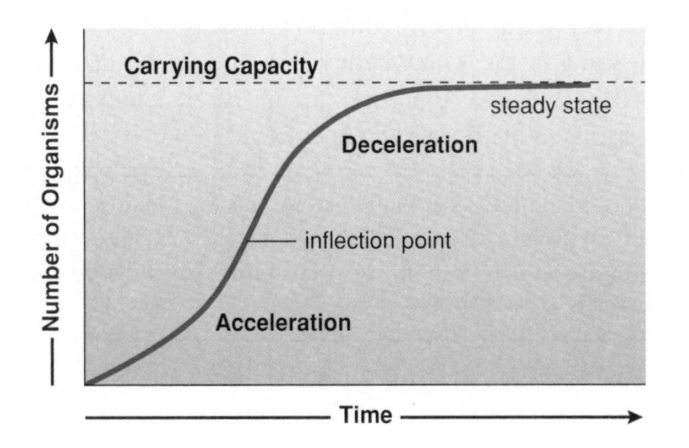

The human population is expanding exponentially, and it is unknown when growth will level off. Presently, each year exhibits a large increase, and the doubling time is now about 39 years. The MDCs underwent a demographic transition between 1950 and 1975, but the LDCs are just now undergoing demographic transition. In these countries, where the average age is less than 15, it will be many years before reproduction replacement equals zero population growth.

On land, soil erosion sometimes reduces soil quality and leads to desertification. The tropical rain forests are being reduced in size, with a potentially immense loss of biological diversity. Solid wastes, including hazardous wastes, are deposited on land and at sea. These are not biodegradable and are subject to biological magnification.

Surface waters, groundwater, and the oceans are all being polluted. Organic materials can be broken down in sewage treatment plants, but the nutrients made available to algae from this process can lead to cultural eutrophication. It is difficult to rid surface waters and particularly groundwater of hazardous wastes.

The oceans are the final recipients of all the pollutants that enter water. In addition, some materials are dumped directly into the oceans, either purposefully or accidentally.

In the air, hydrocarbons and nitrogen oxides (NO_x) react to form photochemical smog, which includes ozone and PAN. Sulfur dioxide (SO_2) and nitrogen oxides react with water vapor to form acids that contribute to acid deposition. Several gases—carbon dioxide (CO_2), nitrous oxide (NO_2), methane (CH_4), and CFCs—are called the greenhouse gases because they trap heat and lead to global warming. Destruction of the ozone shield is associated particularly with CFCs, which rise into the stratosphere and react with frozen particles within clouds to release chlorine. Chlorine causes ozone breakdown.

Economic growth is required particularly in the LDCs. All countries of the world should strive for sustainable growth, which necessitates preservation of the environment. Eventually, it may be possible to have a steady state in which neither population nor resource consumption increases.

STUDY QUESTIONS

In order to practice **writing across the curriculum,** students should write out the answers to any or all of the study questions. The study questions are sequenced in the same order as the text.

1. Define exponential growth. (p. 666) Draw a growth curve to represent exponential growth, and explain why a curve representing population growth usually levels off. (pp. 667–668)
2. Calculate the growth rate and the doubling time for a population in which the birthrate is 20 per 1,000 and the death rate is 2 per 1,000. (pp. 667–668)
3. Define demographic transition. When did the MDCs undergo demographic transition? When did the LDCs undergo demographic transition? (pp. 669–670)
4. Distinguish between the MDCs and the LDCs. (p. 669)
5. Give 2 reasons why the quality of the land is being degraded today. What is desertification? (pp. 671–672)
6. Give 3 reasons why tropical rain forests are being destroyed. What is another potential reason? (p. 672)
7. What is the primary ecological concern associated with the destruction of tropical rain forests? (p. 672)
8. What are the 3 types of hazardous wastes that contribute to pollution on land? (p. 674)
9. What are several ways in which underground water supplies can be polluted? (pp. 674–675)
10. What substances contribute to air pollution? What are their sources? Which are associated with each of the following phenomena: photochemical smog, acid deposition, global warming, and destruction of the ozone shield? (pp. 677–679)

OBJECTIVE QUESTIONS

1. After a country has undergone the demographic transition, the death rate and the birthrate are both _______ (high, low).
2. If a country has a pyramid-shaped age-structure diagram, most individuals are _______ (prereproductive, reproductive, postreproductive).
3. LDCs are not as _______ as MDCs.
4. When a population is undergoing exponential growth, the increase in number of people each year is _______ (higher, lower) than the year before.
5. The gas most appropriately associated with global warming is _______ .
6. The chemicals most appropriately associated with ozone depletion are _______.
7. Sewage is biodegradable, but the nutrients released in the process can lead to _______ of surface waters.
8. Pesticides and radioactive wastes are both subject to biological _______.
9. Associate each of these air pollutants with one of the environmental problems listed below: hydrocarbons (HC), sulfur dioxide (SO_2), carbon dioxide (CO_2), chlorofluorocarbons (CFCs).

a. photochemical smog
b. ozone shield destruction
c. global warming
d. acid deposition

CRITICAL THINKING

In order to practice **writing across the curriculum,** students should write out the answers to any or all of the critical thinking questions. Suggested answers to the critical thinking questions are in appendix E.

34.1

Imagine a watering hole that can support 100 rabbits.

1. If at first there are only 2 rabbits, and each new pair of rabbits produces only 4 rabbits, how many future generations can be added before the carrying capacity of the watering hole is overcome? Why does this rate of increase have to cease long before there are 100 rabbits?
2. How many rabbits are there after the addition of the next generation? Is there an ever-greater increase each year? What kind of growth is this?
3. Plot the growth curve for this rabbit population using number of rabbits versus time. How would you describe the shape of your curve? If the rabbit population happens to outstrip the carrying capacity of the environment, what will happen to the curve?

34.2

Consider this definition of overpopulation:

"...where there are more people than can live on the earth in comfort, happiness, and health and still leave the world a fit place for future generations."[1]

1. Do comfort and happiness mean the typical standard of living seen in the MDCs or in the LDCs? Should everyone in the world have the same standard of living? Why or why not?
2. What standard of health is acceptable for the MDCs? for the LDCs? Whose responsibility is it to achieve this end?
3. Should citizens and private industry work to find ways to make the world an ecologically fit place for future generations? Why?
4. When discussing overpopulation, should we think in terms of the world, the country, or the area?

1. From George Morris, *Overpopulation: Everyone's Baby*. London: Priory Press Limited, 1973, p. 24.

SELECTED KEY TERMS

acid deposition (as´ id dep˝ o-zish´ un) the return to earth as rain or snow of the sulfate salts or nitrate salts of acids produced by commercial and industrial activities on earth.

biotic potential the maximum population growth rate under ideal conditions.

carrying capacity the largest number of organisms of a particular species that can be maintained indefinitely in an ecosystem.

cultural eutrophication (kul´ tu-ral u˝ tro-fĭ-ka´ shun) enrichment of a body of water due to human activities, causing excessive growth of producers and then death of these and other inhabitants.

demographic transition the change from a high birthrate to a low birthrate so that the growth rate is lowered.

desertification (dez-ert˝ ĭ-fĭ-ka´ shun) desert conditions caused by human misuse of land.

doubling time the number of years it takes for a population to double in size.

environmental resistance sum total of factors in the environment that limit the numerical increase of a population in a particular region.

exponential growth growth, particularly of a population, in which the increase occurs in the same manner as compound interest.

greenhouse effect carbon dioxide (CO_2) buildup in the atmosphere as a result of fossil fuel combustion; retains and reradiates heat, causing an abnormal rise in the earth's average temperature.

hazardous waste waste containing chemicals that endanger life.

ozone shield a layer of ozone (O_3) present in the upper atmosphere that protects the earth from damaging ultraviolet light. Nearer the earth, ozone is a pollutant.

photochemical smog air pollution that contains nitrogen oxides and hydrocarbons, which react to produce ozone and peroxylacetyl nitrate (PAN).

replacement reproduction a population in which each person is replaced by only one child.

FURTHER READINGS FOR PART SEVEN

Alcock, J. 1988. *Animal behavior, an evolutionary approach.* 4th ed. Sunderland, Mass.: Sinauer Associates.

Berner, R. A., and A. C. Lasaga. March 1989. Modeling the geochemical carbon cycle. *Scientific American.*

Brown, B. E., and J. Ogden. January 1993. Coral bleaching. *Scientific American.*

Brown, L. R., et al. 1990. *State of the world: 1990.* New York: W. W. Norton and Co.

Colinvaux, P. A. May 1989. The past and future Amazon. *Scientific American.*

Corson, W. M., ed. 1990. *The global ecology handbook: What you can do about the environmental crisis.* Boston: Beacon Press.

Cunningham, W. P., and B. W. Saigo. 1992. *Environmental science: A global concern.* 2d ed. Dubuque, Iowa: Wm. C. Brown Publishers.

Enger, E. D., and B. F. Smith. 1991. *Environmental science: The study of interrelationships.* Dubuque, Iowa: Wm. C. Brown Publishers.

Evans, H. E., and K. M. O'Neill. August 1991. Beewolves. *Scientific American.*

FitzGerald, G. J. April 1993. The reproductive behavior of the stickleback. *Scientific American.*

Francis, C. A., C. B. Flora, and L. D. King, eds. 1990. *Sustainable agriculture in temperate zones.* New York: John Wiley and Sons.

Goudie, A. 1990. *The human impact on the natural environment.* 3d ed. Cambridge: MIT Press.

Goulding, M. March 1993. Flooded forests of the Amazon. *Scientific American.*

Holloway, M., and J. Horgan. October 1991. Soiled shores. *Scientific American.*

Homer-Dixon, T. F., J. H. Boutwell, and G. W. Rathjens. February 1993. Environmental change and violent conflict. *Scientific American.*

Horn, M. H., and R. N. Gibson. January 1988. Intertidal fishes. *Scientific American.*

Houghton, R. A., and G. M. Woodwell. April 1989. Global climatic change. *Scientific American.*

Jones, P. D., and T. M. L. Wigley. August 1990. Global warming trends. *Scientific American.*

Lohmann, K. J. January 1992. How sea turtles navigate. *Scientific American.*

Miller, J. T. 1992. *Living in the environment.* 7th ed. Belmont, Calif.: Wadsworth.

Mohnen, V. A. August 1988. The challenge of acid rain. *Scientific American.*

Newell, R. E., H. G. Reichle, Jr., and W. Seiler. October 1989. Carbon monoxide and the burning earth. *Scientific American.*

Nichol, J., and C. Newton. 1990. *The mighty rain forest.* London: Newton Abbott.

Odum, E. P. 1989. *Ecology and our endangered life-support systems.* Sunderland, Mass.: Sinauer Associates.

O'Leary, P. R., P. W. Walsh, and R. H. Ham. December 1988. Managing solid waste. *Scientific American.*

Regenold, J. P., R. I. Papendik, and J. F. Parr. June 1990. Sustainable agriculture. *Scientific American.*

Rennie, J. January 1992. Living together. *Scientific American.*

Repetto, R. April 1990. Deforestation in the tropics. *Scientific American.*

Repetto, R. June 1992. Accounting for environmental assets. *Scientific American.*

Ricklefs, R. E. 1989. *Ecology.* 4th ed. New York: Chiron Press.

Ruddiman, W. F., and J. E. Kutzbach. March 1991. Plateau uplift and climatic change. *Scientific American.*

Scientific American. September 1989. Planet Earth. Special issue.

Scientific American. September 1990. Energy for Planet Earth. Special issue.

Seymour, R. S. December 1991. The brush turkey. *Scientific American.*

Smith, R. E. 1990. *Ecology and field biology.* 4th ed. New York: Harper and Row Publishers.

Strobel, G. July 1991. Biological control of weeds. *Scientific American.*

Sumich, J. L. 1992. *An introduction to the biology of marine life.* 5th ed. Dubuque, Iowa: Wm. C. Brown Publishers.

Trivers, R. 1985. *Social evolution.* Redwood City, Calif.: Benjamin/Cummings Publishing Co.

Tumlinson, J. H., W. J. Lewis, and L. E. M. Vet. March 1993. How parasitic wasps find their hosts. *Scientific American.*

White, R. M. July 1990. The great climate debate. *Scientific American.*

Whitmore, T. C. 1991. *An introduction to tropical rain forests.* New York: Oxford University Press.

Wursig, B. April 1988. The behavior of baleen whales. *Scientific American.*

Young, J. A. March 1991. Tumbleweed. *Scientific American.*

APPENDIX A

Periodic Table of the Elements

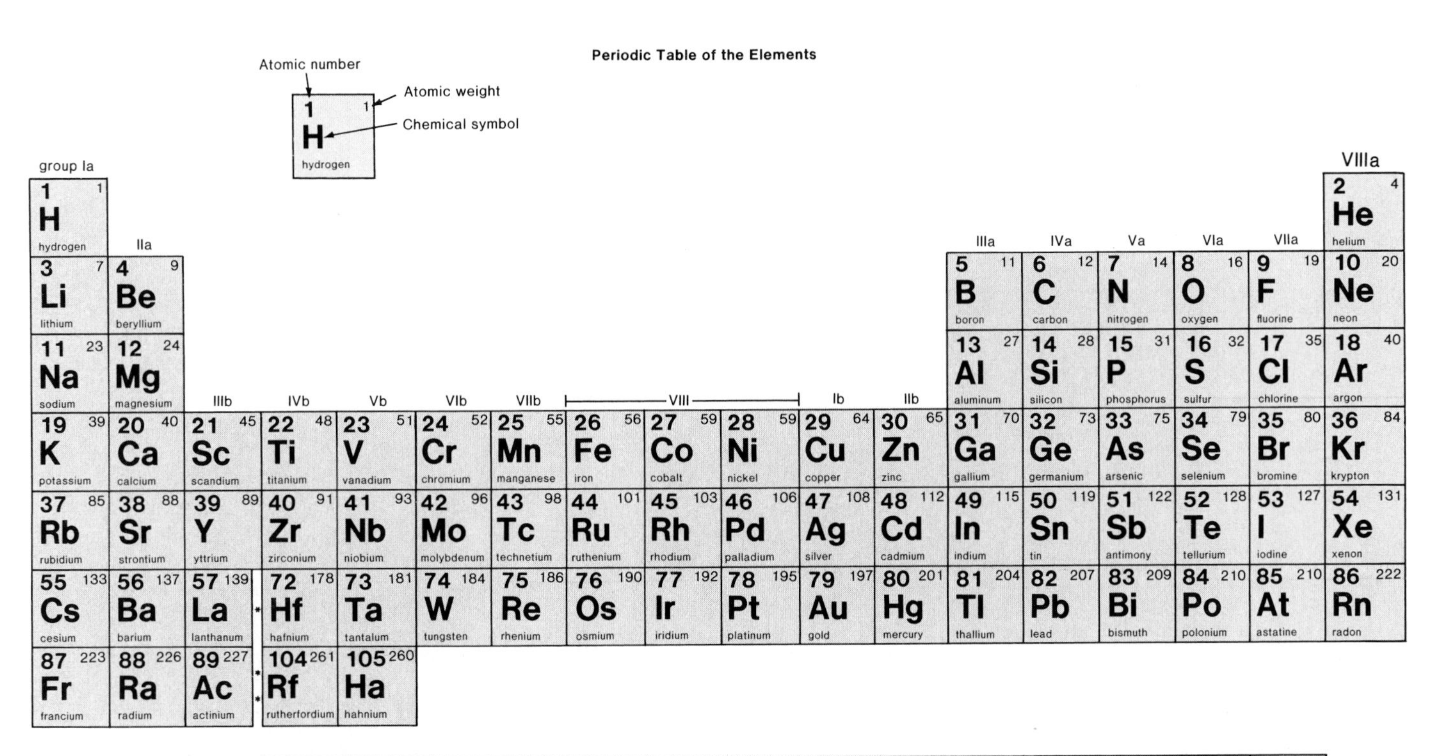

APPENDIX B

Drugs of Abuse

	Drugs	Often Prescribed Brand Names	Medical Uses	Potential Physical Dependence	Potential Psychological Dependence	Tolerance
Narcotics	Opium	Dover's Powder, Paregoric	Analgesic, antidiarrheal	High	High	Yes
	Morphine	Morphine	Analgesic	High	High	Yes
	Codeine	Codeine	Analgesic, antitussive	Moderate	Moderate	Yes
	Heroin	None	None	High	High	Yes
	Meperidine (Pethidine)	Demerol, Pethadol	Analgesic	High	High	Yes
	Methadone	Dolophine, Methadone, Methadose	Analgesic, heroin substitute	High	High	Yes
	Other Narcotics	Dilaudid, Leritine, Numorphan, Percodan	Analgesic, antidiarrheal, antitussive	High	High	Yes
	Chloral Hydrate	Noctec, Somnos	Hypnotic	Moderate	Moderate	Probable
Depressants	Barbiturates	Amytal, Butisol, Nembutal, Phenobarbital, Seconal, Tuinal	Anesthetic, anticonvulsant, sedation, sleep	High	High	Yes
	Glutethimide	Doriden	Sedation, sleep	High	High	Yes
	Methaqualone	Optimil, Parest, Quaalude, Somnafac, Sopor	Sedation, sleep	High	High	Yes
	Tranquilizers	Equanil, Librium, Miltown Serax, Tranxene, Valium	Antianxiety, muscle relaxant, sedation	Moderate	Moderate	Yes
	Other Depressants	Clonopin, Dalmane, Dormate, Noludar, Placydil, Valmid	Antianxiety, sedation, sleep	Possible	Possible	Yes
Stimulants	Cocaine*	Cocaine	Local anesthetic	Possible	High	Yes
	Amphetamines	Benzedrine, Biphetamine, Desoxyn, Dexedrine	Hyperkinesis, narcolepsy, weight control	Possible	High	Yes
	Phenmetrazine	Preludin	Weight control	Possible	High	Yes
	Methylphenidate	Ritalin	Hyperkinesis	Possible	High	Yes
	Other Stimulants	Bacarate, Cylert, Didrex, Ionamin, Plegine, Pondimin, Pro-Sate, Sanorex, Voranil	Weight control	Possible	Possible	Yes
Hallucinogens	LSD	None	None	None	Degree unknown	Yes
	Mescaline	None	None	None	Degree unknown	Yes
	Psilocybin-Psilocyn	None	None	None	Degree unknown	Yes
	MDA	None	None	None	Degree unknown	Yes
	PCP†	Sernylan	Veterinary anesthetic	None	Degree unknown	Yes
	Other Hallucinogens	None	None	None	Degree unknown	Yes
Cannabis	Marijuana, Hashish, Hashish Oil	None	Glaucoma	Degree unknown	Moderate	Yes

Source: *Drugs of Abuse,* produced by the Affairs in Cooperation with the Office of Public Science and Technology.

*Designated a narcotic under the Controlled Substances Act.
†Designated a depressant under the Controlled Substances Act.

Duration of Effects (in hours)	Usual Methods of Administration	Possible Effects	Effects of Overdose	Withdrawal Syndrome
3–6	Oral, smoked	Euphoria, drowsiness, respiratory depression, constricted pupils, nausea	Slow and shallow breathing, clammy skin, convulsions, coma, possible death	Watery eyes, runny nose, yawning, loss of appetite, irritability, tremors, panic, chills and sweating, cramps, nausea
3–6	Injected, smoked			
3–6	Oral, injected			
3–6	Injected, sniffed			
3–6	Oral, injected			
12–24	Oral, injected			
3–6	Oral, injected			
5–8	Oral			
1–16	Oral, injected	Slurred speech, disorientation, drunken behavior without odor of alcohol	Shallow respiration, cold and clammy skin, dilated pupils, weak and rapid pulse, coma, possible death	Anxiety, insomnia, tremors, delirium, convulsions, possible death
4–8	Oral			
4–8	Oral			
4–8	Oral			
4–8	Oral			
2	Injected, sniffed	Increased alertness, excitation, euphoria, dilated pupils, increased pulse rate and blood pressure, insomnia, loss of appetite	Agitation, increased body temperature, hallucinations, convulsions, possible death	Apathy, long periods of sleep, irritability, depression, disorientation
2–4	Oral, injected			
2–4	Oral			
2–4	Oral			
2–4	Oral			
Variable	Oral	Illusions and hallucinations (with exception of MDA), poor perception of time and distance	Longer, more intense "trip" episodes, psychosis, possible death	Withdrawal syndrome not reported
Variable	Oral, injected			
Variable	Oral			
Variable	Oral, injected, sniffed			
Variable	Oral, injected, smoked			
Variable	Oral, injected, sniffed			
2–4	Oral, smoked	Euphoria, relaxed inhibitions, increased appetite, disoriented behavior	Fatigue, paranoia, possible psychosis	Insomnia, hyperactivity, and decreased appetite reported in a limited number of individuals

APPENDIX C

Metric System

Standard Metric Units	Abbreviations
Unit of length: meter	m
Unit of weight: gram	g
Unit of volume: liter	l

Length

Units and Abbreviations (by Increasing Order)	Metric Equivalent	Metric-to-English Conversion Factor	English-to-Metric Conversion Factor
nanometer (nm)	= 10^{-9} m (10^{-3} μm)	—	—
micrometer (μm)	= 10^{-6} m (10^{-3} mm)	—	—
millimeter (mm)	= 0.001 (10^{-3}) mm	mm→in: 0.039	—
centimeter (cm)	= 0.01 (10^{-2}) m	cm→in: 0.39	in→cm: 2.54 ft→cm: 30.5
meter (m)	= 100 (10^{2}) cm = 1000 mm	m→in: 39 m→ft: 3.28 m→yd: 1.09	— ft→m: 0.305 yd→m: 0.91
kilometer (km)	= 1000 (10^{3}) m	km→mi: 0.62	mi→km: 1.61

How to Use This Table

To convert English to metric (mi→km):

Multiply	English unit	×	English-to-metric factor	=	metric
	5 mi	×	1.61	=	8.05 km

To convert metric to English (km→mi):

Multiply	metric unit	×	metric-to-English factor	=	English
	8.05 km	×	0.62	=	5 mi

Think Metric*

Length

1. The speed of a car is 60 mph or _______ km per hour.
2. A man who is _______ feet tall is 180 cm.
3. A six-inch ruler is _______ cm.
4. _______ yard is almost a meter (0.9 m).

*See answers on page 690.

Weight

Units and Abbreviations (by Increasing Order)	Metric Equivalent	Metric-to-English Conversion Factor		English-to-Metric Conversion Factor	
nanogram (ng)	= 10^{-9} g	—		—	
microgram (μg)	= 10^{-6} g	—		—	
milligram (mg)	= 10^{-3} (0.001) g	mg→grains:	approx. 0.015		—
gram (g)	= 1000 mg	g→grains:	15.43	—	
		g→oz:	0.035	oz→g:	28.3
				lb→g:	453.6
kilogram (kg)	= 1000 (10^3)g	kg→lb:	2.2	lb→kg:	0.45
metric ton (t)	= 1000 kg	t→ton:	1.10	ton→t:	0.90

How to Use This Table

To convert English to metric (lb→kg):

Multiply	English unit	×	English-to-metric factor	=	metric
	5 lb	×	0.45	=	2.27 kg

To convert metric to English (kg→lb):

Multiply	metric unit	×	metric-to-English factor	=	English
	2.27 kg	×	2.2	=	5 lb

Think Metric*

Weight

1. One pound of hamburger is _______ grams.
2. The average human male brain weighs 1.4 kg (_______ lb _______ oz).
3. A person who weighs _______ lb weighs 70 kg.
4. Lucia Zarate weighed _______ kg (13 lb) at age 20.

*See answers on page 690.

Volume

Units and Abbreviations (by Increasing Order)	Metric Equivalent	Metric-to-English Conversion Factor	English-to-Metric Conversion Factor
microliter (μl)	= 10^{-6} l (10^{-3} ml)	—	—
milliliter (ml)	= 10^{-3} l = 1 cu cm (cc) = 1000 cu mm	ml→drops: approx. 15–16 ml→tsp: approx. 1/4 ml→fl oz: 0.03 — —	— tsp→ml: approx. 5 fl oz→ml: 30 pt→ml: 47 qt→ml: 95
liter (l)	= 1000 ml	l→qt: 1.06 l→pt: 2.1 l→gal: 0.26	qt→l: 0.95 pt→l: 0.47 gal→l: 3.79
kiloliter (kl)	= 1000 l	kl→gal: 264.17	—

How to Use This Table

To convert English to metric (fl oz→ml):

Multiply	English unit	×	English-to-metric factor	=	metric
	5 oz	×	30	=	150 ml

To convert metric to English (ml→fl oz):

Multiply	metric unit	×	metric-to-English factor	=	English
	150 ml	×	.03	=	5 oz

Think Metric

Volume

1. One can of soda (12 oz) contains _______ ml.
2. The average human body contains between 10 pt and 12 pt of blood or between _______ l and _______ l.
3. One cubic foot of water (7.48 gal) is _______ l.
4. If a gallon of unleaded gasoline costs $1.00, a l costs _______ .

Answers

Length: (1) 97; (2) 6; (3) 15; (4) One
Weight: (1) 453.6; (2) 3, 1.3; (3) 154; (4) 5.85
Volume: (1) 360; (2) 4.7, 5.6; (3) 28.4; (4) 26 cents

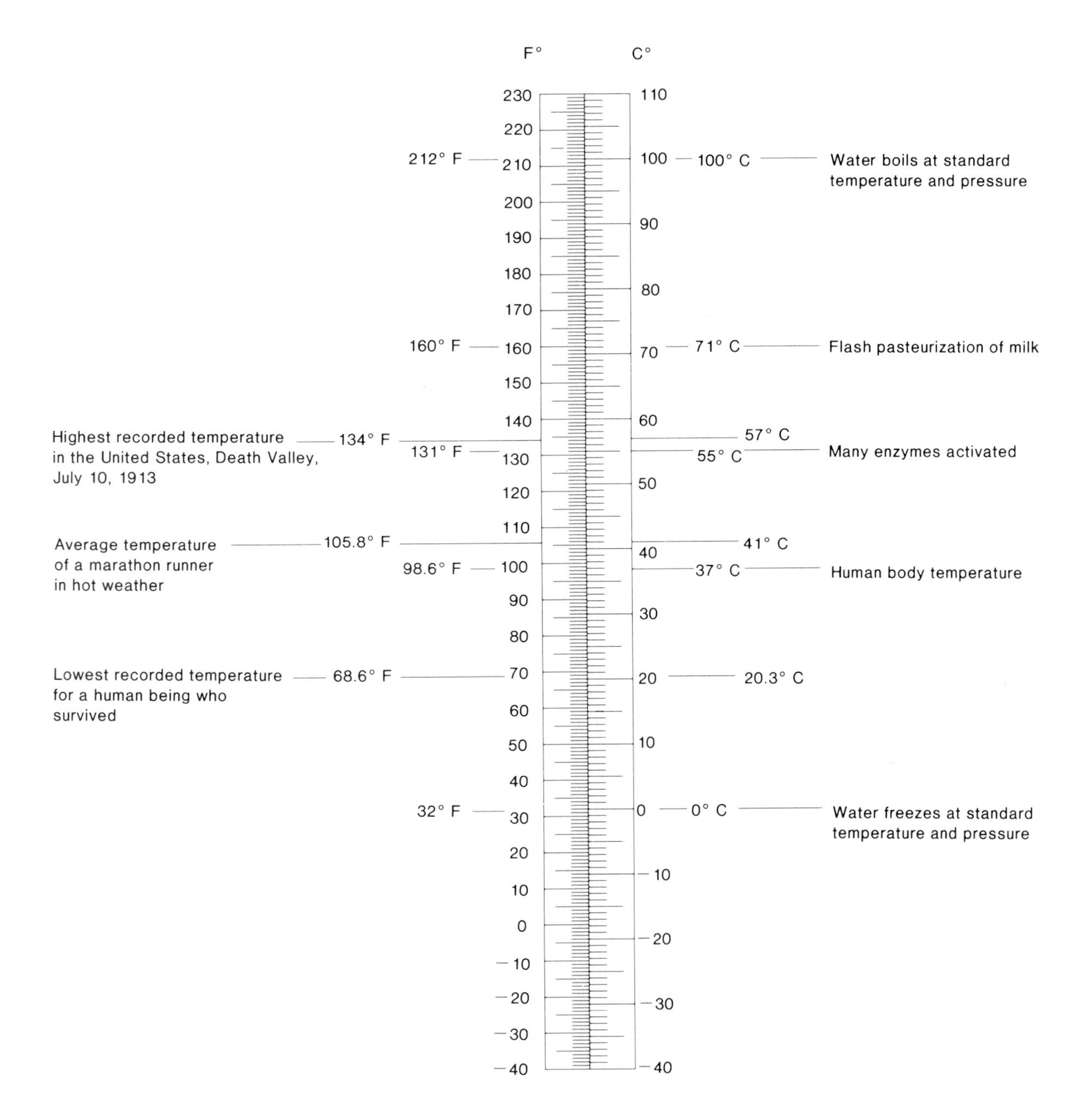
F°
C°
230
220
210
200
190
180
170
160
150
140
130
120
110
100
90
80
70
60
50
40
30
20
10
0
−10
−20
−30
−40
110
100
90
80
70
60
50
40
30
20
10
0
−10
−20
−30
−40
212° F
100° C
Water boils at standard temperature and pressure
160° F
71° C
Flash pasteurization of milk
Highest recorded temperature in the United States, Death Valley, July 10, 1913
134° F
131° F
57° C
55° C
Many enzymes activated
Average temperature of a marathon runner in hot weather
105.8° F
98.6° F
41° C
37° C
Human body temperature
Lowest recorded temperature for a human being who survived
68.6° F
20.3° C
32° F
0° C
Water freezes at standard temperature and pressure

APPENDIX D

Classification of Organisms

The classification system given here is a simplified one, containing all the major kingdoms, as well as the major divisions (called phyla in the kingdom Protista and the kingdom Animalia). The text does not discuss all the divisions and phyla listed here.

Kingdom Monera

Prokaryotic, unicellular organisms. Nutrition principally by absorption, but some are photosynthetic or chemosynthetic.

- Division ARCHAEBACTERIA: methanogens, halophiles, and thermoacidophiles
- Division EUBACTERIA: all other bacteria, including cyanobacteria (formerly called blue-green algae)

Kingdom Protista

Eukaryotic, unicellular organisms (and the most closely related multicellular forms). Nutrition heterotrophic by ingestion (protozoa), absorption (fungi), or photosynthesis (algae).

- Phylum SARCODINA: amoeboid protozoa
- Phylum CILIOPHORA: ciliated protozoa
- Phylum ZOOMASTIGINA: flagellated protozoa
- Phylum SPOROZOA: parasitic protozoa
- Phylum CHLOROPHYTA: green algae
- Phylum DINOFLAGELLATA: dinoflagellates
- Phylum EUGLENOPHYTA: *Euglena* and relatives
- Phylum CHRYSOPHYTA: diatoms
- Phylum RHODOPHYTA: red algae
- Phylum PHAEOPHYTA: brown algae
- Phylum MYXOMYCOTA: slime molds
- Phylum OOMYCOTA: water molds

Kingdom Fungi

Eukaryotic organisms, usually having haploid or multinucleated hyphal filaments. Spore formation during both asexual and sexual reproduction. Nutrition heterotrophic principally by absorption.

- Division ZYGOMYCOTA: black bread molds
- Division ASCOMYCOTA: sac fungi
- Division BASIDIOMYCOTA: club fungi
- Division DEUTEROMYCOTA: imperfect fungi, that is, means of sexual reproduction not known

Kingdom Plantae

Eukaryotic, terrestrial, multicellular organisms with rigid cellulose cell walls and chlorophylls *a* and *b*. Nutrition principally by photosynthesis. Starch is the reserve food.

- Division BRYOPHYTA: mosses and liverworts
- Division PSILOPHYTA: whisk ferns
- Division LYCOPHYTA: club mosses
- Division SPHENOPHYTA: horsetails
- Division PTEROPHYTA: ferns
- Division CYCADOPHYTA: cycads
- Division GINKGOPHYTA: ginkgos
- Division GNETOPHYTA: gnetae
- Division CONIFEROPHYTA: conifers
- Division ANTHOPHYTA: flowering plants
    - Class DICOTYLEDONAE: dicots
    - Class MONOCOTYLEDONAE: monocots

Kingdom Animalia

Eukaryotic, usually motile multicellular organisms without cell walls or chlorophyll. Nutrition principally ingestive, with digestion in an internal cavity.

- Phylum PORIFERA: sponges
- Phylum CNIDARIA: radially symmetrical marine animals
    - Class HYDROZOA: hydras, Portuguese man-of-war
    - Class SCYPHOZOA: jellyfish
    - Class ANTHOZOA: sea anemones and corals
- Phylum PLATYHELMINTHES: flatworms
    - Class TURBELLARIA: free-living flatworms
    - Class TREMATODA: parasitic flukes
    - Class CESTODA: parasitic tapeworms
- Phylum NEMATODA: roundworms
- Phylum ROTIFERA: rotifers
- Phylum MOLLUSCA: soft-bodied, unsegmented animals
    - Class POLYPLACOPHORA: chitons
    - Class MONOPLACOPHORA: *Neopilina*
    - Class GASTROPODA: snails and slugs
    - Class CEPHALOPODA: squid and octopuses
    - Class BIVALVIA: clams and mussels

Phylum ANNELIDA: segmented worms
- Class POLYCHAETA: sandworms
- Class OLIGOCHAETA: earthworms
- Class HIRUDINEA: leeches

Phylum ARTHROPODA: animals with chitinous exoskeleton and jointed appendages
- Class CRUSTACEA: lobsters, crabs, barnacles
- Class ARACHNIDA: spiders, scorpions, ticks
- Class CHILOPODA: centipedes
- Class DIPLOPODA: millipedes
- Class INSECTA: grasshoppers, termites, beetles

Phylum ECHINODERMATA: spiny, radially symmetrical marine animals
- Class CRINOIDEA: sea lilies and feather stars
- Class ASTEROIDEA: starfishes
- Class OPHIUROIDEA: brittle stars
- Class ECHINOIDEA: sea urchins and sand dollars
- Class HOLOTHUROIDEA: sea cucumbers

Phylum CHORDATA: dorsal supporting rod (notochord) at some stage; dorsal hollow nerve cord; pharyngeal pouches or slits
- Subphylum UROCHORDATA: tunicates
- Subphylum CEPHALOCHORDATA: lancelets
- Subphylum VERTEBRATA: vertebrates
  - Class AGNATHA: jawless fishes (lampreys, hagfishes)
  - Class CHONDRICHTHYES: cartilaginous fishes (sharks, rays)
  - Class OSTEICHTHYES: bony fishes
    - Subclass SARCOPTERYGII: lobe-finned fishes
    - Subclass ACTINOPTERYGII: ray-finned fishes
  - Class AMPHIBIA: frogs, toads, salamanders
  - Class REPTILIA: snakes, lizards, turtles
  - Class AVES: birds
  - Class MAMMALIA: mammals
    - Subclass PROTOTHERIA: egg-laying mammals
      - Order MONOTREMATA: duckbilled platypuses, spiny anteaters
    - Subclass METATHERIA: marsupial mammals
      - Order MARSUPIALIA: opossums, kangaroos
    - Subclass EUTHERIA: placental mammals
      - Order INSECTIVORA: shrews, moles
      - Order CHIROPTERA: bats
      - Order EDENTATA: anteaters, armadillos
      - Order RODENTIA: rats, mice, squirrels
      - Order LAGOMORPHA: rabbits, hares
      - Order CETACEA: whales, dolphins, porpoises
      - Order CARNIVORA: dogs, bears, weasels, cats, skunks
      - Order PROBOSCIDEA: elephants
      - Order SIRENIA: manatees
      - Order PERISSODACTYLA: horses, hippopotamuses, zebras
      - Order ARTIODACTYLA: pigs, deer, cattle
      - Order PRIMATES: lemurs, monkeys, apes, humans
        - Suborder PROSIMII: lemurs, tree shrews, tarsiers, lorises, pottos
        - Suborder ANTHROPOIDEA: monkeys, apes, humans
          - Superfamily CEBOIDEA: new-world monkeys
          - Superfamily CERCOPITHECOIDEA: old-world monkeys
          - Superfamily HOMINOIDEA: apes and humans
            - Family HYLOBATIDAE: gibbons
            - Family PONGIDAE: chimpanzees, gorillas, orangutans
            - Family HOMINIDAE: *Australopithecus*,* *Homo habilis*,* *Homo erectus*,* *Homo sapiens sapiens*

*extinct

APPENDIX E

Answers to the Critical Thinking and Objective Questions

Chapter 1

Answers to Objective Questions

1. cells **2.** materials, energy. **3.** responding **4.** reproduce **5.** adapted **6.** homeostasis **7.** inductive **8.** experimentation **9.** mathematical **10.** *theory* **11.** and **12.** See chapter 1 summary.

Answers to 1.1 Critical Thinking

1. There is no observation or experiment that can be performed to prove the hypothesis false.
2. You can perform a controlled experiment. The experimental group is fed biotin-free food. The control group is expected to remain healthy, while the experimental group is expected to get sick.
3. Religious beliefs are not subjected to the process described in figure 1.8; scientific beliefs arise from this process.

Answers to 1.2 Critical Thinking

1. Sweetener S is being varied in order to determine its effects on the body.
2. The control mice are subjected to all conditions (e.g., living in a cage kept at a certain temperature and eating the same food) except they are not receiving sweetener S.
3. Chance of bladder cancer development is dependent upon quantity of sweetener S consumed.
4. Yes, the conditions mentioned in answer 2 are constant.
5. A control group increases confidence in results. For example, if there is no bladder cancer in control mice, it is more likely that development of bladder cancer in test mice is caused by sweetener S.

Chapter 2

Answers to Objective Questions

1. protons **2.** electrons, protons **3.** share **4.** Buffers **5.** increases, decreases **6.** Enzymes **7.** peptide **8.** primary **9.** glucose **10.** cellulose **11.** saturated **12.** glycerol **13.** nucleotides **14.** double **15.** See figure 2.15 (p. 29) in text.

Answers to 2.1 Critical Thinking

1. Water does not contain carbon.
2. Water contains a small number of atoms.
3. Water has covalent bonding.
4. Water (a) absorbs and gives off heat—this helps to keep the body warm; (b) is cohesive—this helps in fluid transport; and (c) is a solvent in bodies—this helps in chemical reactions.

Answers to 2.2 Critical Thinking

1. The sequence of their amino acids makes actin and myosin differ.
2. Starch is not as branched as glycogen.
3. The length of the chain and the placement of unsaturated bonds in the 2 acids may be different.

Chapter 3

Answers to Objective Questions

1. resolving **2.** protein, phospholipid **3.** nucleus **4.** ribosomes **5.** hydrolytic **6.** Golgi apparatus **7.** mitochondria **8.** chloroplasts **9.** cytoskeleton **10.** centrioles
11. *a.* nucleus—DNA directs
b. nucleolus—RNA helps
c. rough ER produces
d. smooth ER transports
e. Golgi apparatus packages and secretes
12. *b.* Mitochondria and *chloroplasts* are both membranous structures involved in energy conversion.
c. Centrioles and *flagella* both contain microtubules; centrioles give rise to the basal bodies of flagella.
d. ER and *ribosomes* are rough ER, which produces proteins.

Answers to 3.1 Critical Thinking

1. 1/10,000,000
2. 1,000
3. 100
4. Yes, it is much smaller.

Answers to 3.2 Critical Thinking

1. Carbon is found in carbohydrates and fats, while sulfur is unique to the amino acids cysteine and methionine and therefore proteins.
2. Radiation will appear first at the ribosomes (polysomes), then at the region of the nuclear pore, and finally in the nucleus.
3. Radiation will appear first at rough ER, then at the Golgi apparatus, and finally at the cell membrane.

Answers to 3.3 Critical Thinking

1. There are no centrioles in plant cells.
2. Centrioles are present in the microtubule organizing region.
3. Centrioles give rise to basal bodies, which organize the microtubules in cilia and flagella. Cilia and flagella are associated with animal cells.

Chapter 4

Answers to Objective Questions

1. wall **2.** concentration gradient **3.** lose, shrink **4.** turgor pressure, press **5.** facilitated **6.** against **7.** sodium-potassium pump **8.** vesicle **9.** See figure 4.1 (p. 61) in text. **10.** *a.* isotonic *b.* hypotonic *c.* hypertonic

Answers to 4.1 Critical Thinking

1. Proteins are made at the ribosomes. Carbohydrate chains might be added in the Golgi apparatus.
2. Proteins travel to the cell membrane via vesicles formed by the Golgi apparatus.
3. Because the CD4 receptor is missing, it is not likely that this person will be infected with the HIV virus.
4. The proteins of the cell membrane are structural proteins.

Answers to 4.2 Critical Thinking

1. Alcohol can cross the lipid bilayer by diffusion. Water diffuses by way of a channel protein.
2. Na^+ is pumped across by the sodium-potassium pump. Cl^- diffuses out by way of a channel protein.
3. Amino acids enter by facilitated diffusion. Proteins enter by endocytosis.
4. The proteins would be digested after the endocytic vesicle fuses with a lysosome containing hydrolytic enzymes.

Chapter 5

Answers to Objective Questions

1. sun **2.** ATP **3.** heat **4.** energy **5.** substrate **6.** energy (heat) **7.** pH **8.** coenzymes **9.** oxidizes **10.** reduces **11.** See figure 5.4 (p. 77) in text. The shape of an enzyme is important to its activity because it allows an enzyme-substrate complex to form. The substrates are specific to the enzyme because their shapes fit together as a key fits a lock.

Answers to 5.1 Critical Thinking

1. According to the second law of thermodynamics, when nutrients derived from food are converted into the energy of ATP, there is a loss of energy.
2. A food chain always begins with a photosynthesizer.
3. Without glucose breakdown, there is no ATP buildup.

Answers to 5.2 Critical Thinking

1. Correct pH and a warm temperature are the recommended conditions.
2. The yield could be increased by adding more pepsin or more egg white, whichever is in short supply.
3. With irreversible inhibition, the reaction stops; with reversible inhibition, the reaction continues at a reduced rate.

Chapter 6

Answers to Objective Questions

1. the sun **2.** ATP, $NADH_2$ **3.** water **4.** fixes **5.** PGAL **6.** carbon dioxide and water **7.** PYR, 2, 2 **8.** transition, Krebs **9.** the electron transport system (or chemiosmotic ATP synthesis) **10.** intermembrane **11.** *a.* granum (or thylakoids)—light-dependent reactions and chemiosmotic ATP synthesis *b.* stroma—light-independent reactions, Calvin cycle, reduction of carbon dioxide *c.* cristae—electron transport system *d.* matrix—Krebs cycle, transition reaction *e.* intermembrane space—chemiosmotic ATP synthesis

Answers to 6.1 Critical Thinking

1. A plant takes CO_2 and H_2O from the environment and gives O_2 to the environment.
2. O_2 comes from the thylakoids in grana after the breakdown of water. CO_2 goes into the stroma and is reduced to carbohydrate.
3. O_2 is used for cellular respiration. CO_2 comes from cellular respiration.

Answers to 6.2 Critical Thinking

1. O_2 is the final acceptor for hydrogen atoms at the end of the electron transport system. The system can continue to produce ATP only if O_2 is present.
2. CO_2 is produced as molecules are broken down by the transition reaction and the Krebs cycle.
3. Carbohydrates are a source of glucose. The body realizes 38 ATP molecules for each molecule of glucose that is broken down completely to carbon dioxide and water.

Answers to 6.3 Critical Thinking

1. The matrix of a mitochondrion compares to the stroma of a chloroplast.
2. The processes are opposite. In the matrix of a mitochondrion, glucose products are oxidized (CO_2 is given off); in the stroma of a chloroplast, CO_2 is reduced to carbohydrate.
3. The crista of the mitochondrion compares to the thylakoid membrane of the chloroplast. The intermembrane space of the mitochondrion compares to the thylakoid space of the chloroplast.
4. The processes are the same. Both cristae and the thylakoid membrane contain a cytochrome system for ATP production. Both the intermembrane space and the thylakoid space are hydrogen ion reservoirs.

Chapter 7

Answers to Objective Questions

1. chromatin **2.** 6 **3.** sister chromatids **4.** 24 **5.** centrioles **6.** Anaphase **7.** cleavage furrow, cell plate **8.** 4 **9.** homologous chromosomes **10.** different **11.** *a.* chromosome *b.* centriole *c.* spindle fiber *d.* nuclear envelope fragments *e.* aster **12.** The right cell represents metaphase I because tetrads are present.

Answers to 7.1 Critical Thinking

1. You can tell that chromatin is actively directing protein synthesis because it is (a) the extended form of the genetic material, and (b) it is seen in metabolically active cells, such as those shown in figure 3.4 (p. 48).
2. Yes, they might differ. It is not necessary to inherit exactly the same form of a gene from each parent.
3. Various disorders result because there are too many proteins (enzymes) of the same kind.
4. If a particular gene is not needed for the maturation of the sperm and the egg, a defective form cannot have an effect.

Answers to 7.2 Critical Thinking

1. This is useful because nutrients enter and wastes exit cytoplasm at the cell membrane.
2. This is useful because newly formed cells must receive organelles from the dividing cell.

3. Examples of such specialized cells are nerve and muscle cells.

Answers to 7.3 Critical Thinking

1. Colchicine disrupts the spindle apparatus. Specifically, it prevents microtubule assembly.
2. Asexual reproduction, requiring only mitosis, produces cells (offspring) that have the same kinds of chromosomes as the parent cell.
3. Meiosis is a part of sexual reproduction. Because of meiosis, the daughter cells can have any combination of the haploid number chromosomes, and the zygote has a different combination of the haploid number of chromosomes than either parent.
4. Yes, the production of variation allows new types of organisms to evolve.

Chapter 8

Answers to Objective Questions

1. Meristem **2.** root hairs, stomata **3.** vessel elements, tracheids **4.** maturation **5.** endodermis **6.** scattered **7.** annual rings **8.** palisade **9.** *a.* epidermis *b.* cortex *c.* endodermis *d.* phloem *e.* xylem **10.** *a.* cork *b.* phloem, *c.* vascular cambium *d.* xylem *e.* pith **11.** *a.* upper epidermis *b.* palisade mesophyll *c.* leaf vein *d.* spongy mesophyll *e.* lower epidermis

Answers to 8.1 Critical Thinking

1. Leaf epidermis prevents drying out; it is covered by a waxy cuticle. Leaf epidermis allows gas exchange; it contains stomata.
2. The spongy layer carries on gas exchange; there are air spaces next to these cells. The spongy layer carries on photosynthesis; the cells contain chloroplasts.
3. Leaf veins transport water and minerals and organic substances; they contain xylem and phloem.

Chapter 9

Answers to Objective Questions

1. transpiration **2.** potassium ions, water **3.** pressure-flow **4.** will not **5.** Phytochrome **6.** alternation of generations **7.** ovary, anther **8.** cotyledon **9.** See figure 9.10 (p. 148) in text.

Answers to 9.1 Critical Thinking

1. Atmospheric pressure is the pressure of the air. This shows that atmospheric pressure cannot raise water to the height of a tall tree.
2. Transpiration occurs and creates a pull on water.
3. This suggests that transpiration could raise water to the top of trees.

Chapter 10

Answers to Objective Questions

1. tissues **2.** cuboidal **3.** layered (stratified), cilia, columnar (elongated) **4.** connective **5.** striated **6.** neurons **7.** dermis **8.** keratin **9.** epithelial, loose connective **10.** constancy, tissue **11.** *a.* columnar epithelium, lining of intestine (digestive tract), protection and absorption *b.* compact bone, skeleton, support and protection *c.* cardiac muscle, wall of heart, pump blood

Answers to 10.1 Critical Thinking

1. Epithelial cells have flat surfaces; therefore, they can be easily placed next to one another. This makes them suitable for covering a surface.
2. The equivalent feature is the Casparian strip. Both tight junctions and the Casparian strip prevent materials from moving between cells.
3. The long, tubular cells contain actin filaments and myosin filaments, and these account for their ability to contract. When the cells contract, the muscle contracts. The object would move left.
4. Nerve cells conduct nerve impulses, sometimes over long distances. The long, skinny process, or fiber, of a nerve cell makes this possible.

Answers to 10.2 Critical Thinking

1. It fluctuates.
2. Internal conditions do not stay exactly the same—they fluctuate above and below normal.
3. Hot and cold body temperatures activate the receptor and the regulator center.
4. Yes, this accounts for fluctuation because conditions have to become either hot or cold in order to stimulate the receptor and the regulator center. Then there is an adaptive response in the opposite direction.

Chapter 11

Answers to Objective Questions

1. amylase, maltose **2.** epiglottis **3.** esophagus, protein **4.** bile, emulsifies **5.** duodenum **6.** trypsin, pancreatic amylase, lipase **7.** strongly acidic, slightly basic **8.** villi **9.** glycogen **10.** essential amino acids **11.** See figure 11.1 (p. 172) in text. **12.** *a.* no digestion, enzyme missing *b.* no digestion, wrong enzyme *c.* digestion, both bile salts and pancreatic lipase are present

Answers to 11.1 Critical Thinking

1. The epithelial portion of mucous membrane produces the enzymes.
2. The digestive system provides nutrients needed by cells.
3. The liver regulates the output of glucose to keep the amount in blood fairly constant.
4. Glucose is used for energy; amino acids are used for protein synthesis.

Answers to 11.2 Critical Thinking

1. No. Experiment is as follows: tube 1—pepsin, water, and egg white so pH is neutral; tube 2—pepsin, water, egg white, and $NaHCO_3$ so pH is basic; tube 3—same as tube 4 in figure 11.13, which is expected to show the best digestion.
2. No. Experiment is as follows: tube 1—pepsin, HCl, water (control); tube 2—pepsin, HCl, water, starch; tube 3—same as tube 4 in figure 11.14, which is expected to be the only tube that shows digestion.
3. No. Experiment is as follows: 3 tubes, all having the contents of test tube 4 in figure 11.13. Place one in the cold, one at room temperature, and one in the incubator at body temperature. The last tube is expected to show the best digestion.
4. The contents are bile, fat, water, lipase, and $NaHCO_3$.

Answers to 11.3 Critical Thinking

1. The cell is lacking a particular enzyme.
2. A mutation must have occurred.
3. The plant would die. Plants take in only inorganic nutrients and must make all their organic nutrients.
4. Animals take in preformed food, so they can depend on their diet to supply amino acids they cannot make.

Chapter 12

Answers to Objective Questions

1. away from **2.** aorta **3.** high **4.** SA **5.** blood pressure, skeletal muscle contraction **6.** oxygen, fight infection **7.** fibrin **8.** neutrophil **9.** cholesterol, fat **10.** See figure 12.3 (p. 200) in text. **11.** *a.* arterial end *b.* venous end *c.* water, oxygen, nutrients *d.* water, carbon dioxide *e.* plasma proteins

Answers to 12.1 Critical Thinking

1. a. upper right side of heart; b. lower left side of heart; c. upper left side of heart; d. lower right side of heart.
2. The pulmonary circuit is not functioning. No, the aorta does not carry fully oxygenated blood because in the embryo and the fetus, blood is oxygenated at the placenta.

Answers to 12.2 Critical Thinking

1. It is expected to lodge in the lungs.
2. It is expected to lodge in the brain.
3. It is expected to lodge in the liver.
4. Coronary arteries are the first blood vessels off the aorta; most likely, any clots in the coronary arteries or the capillaries formed right there.

Chapter 13

Answers to Objective Questions

1. tissue fluid, subclavian **2.** purify **3.** neutrophil **4.** thymus **5.** plasma, memory B **6.** antibody **7.** APC **8.** lymphokines **9.** A, B, no **10.** Rh^-, Rh^+ **11.** histamine **12.** vaccines **13.** monoclonal **14.** helper (produces lymphokines); suppressor (shuts down response); memory (retains ability to kill same type of infected cell); cytotoxic (kills infected cells)

Answers to 13.1 Critical Thinking

1. When blood pressure rises, more fluid is pushed out of the capillary; when blood pressure decreases, less fluid is pushed out of the capillary.
2. When osmotic pressure decreases, less fluid is retrieved by blood; when osmotic pressure rises, more fluid is retrieved by blood.
3. Osmotic pressure will be much lower because it is largely dependent on a protein concentration gradient between blood and tissue fluid. Blood pressure will decrease because less fluid is being retrieved by blood. Tissue fluid will continue to form as long as blood pressure is higher than osmotic pressure.

Answers to 13.2 Critical Thinking

1. B cells produce antibodies when stimulated by helper T cells. T cell maturation occurs in the thymus.
2. No, all B cells do not bind because each B cell is specific for only one type of antigen.
3. T cells do not recognize an antigen unless it is presented by an APC.
4. They communicate by surface-to-surface interaction and by chemical signals (e.g., lymphokine molecules).

Answers to 13.3 Critical Thinking

1. Type O has neither antigen A nor antigen B on the red blood cells and theoretically would not clump in a recipient's blood. It now is known that matching blood is more involved than just this consideration.
2. Type AB has neither anti-A antibody nor anti-B antibody in the plasma and theoretically would not cause clumping of red blood cells. It now is known that matching blood is more involved than just this consideration.
3. It is inherited. Genes exist for blood type.
4. Most likely there are other antigens aside from A and B on the red blood cells.

Chapter 14

Answers to Objective Questions

1. larynx **2.** alveoli **3.** CO_2, H^+ **4.** expanded **5.** bicarbonate **6.** the globin portion of hemoglobin **7.** diffusion **8.** lungs **9.** cigarette smoking **10.** bronchi **11.** See figure 14.2 (p. 242) in text.

Answers to 14.1 Critical Thinking

1. With negative pressure, air is drawn in, and with positive pressure, air is pushed in. Frogs force air into the lungs by gulping it.
2. There is no mixing of used air with new air coming in; therefore, more O_2 integration occurs. Only mechanical and biochemical mechanisms are involved.
3. Frogs practice skin breathing. Reptiles have better developed lungs.
4. Movement of the diaphragm assists in creating negative pressure, which draws air into the lungs.

Answers to 14.2 Critical Thinking

1. CO_2 stimulates breathing; O_2 does not.
2. The buildup of CO_2 stimulates breathing.
3. CO_2 in blood raises the pH of blood by the formation of carbonic acid.
4. Sense receptors usually are stimulated by the presence of something.

Chapter 15

Answers to Objective Questions

1. urea **2.** bile pigments, hemoglobin **3.** urethra **4.** glomerulus **5.** Water **6.** Urea **7.** distal convoluted tubule **8.** ADH **9.** volume, pH **10.** dialysis **11.** See figure 15.8 (p. 267) in text.

Answers to 15.1 Critical Thinking

1. Urea is a single molecule (p. 262). Urine is a mixture of molecules and ions (table 15.2).
2. The force is osmotic pressure.
3. This increases blood pressure in the glomerulus.
4. The rate increases because osmotic pressure would decrease in blood and increase in Bowman's capsule.

Answers to 15.2 Critical Thinking

1. To say that urine is 95% water only indicates how much water per solutes there is.
2. Carriers can only work so fast. The fluid is moving in the proximal convoluted tubule, and in the meantime, glucose has gone by.
3. Refer to this equation:

   $$CO_2 + H_2O \leftrightarrow H_2CO_3 \leftrightarrow H^+ + HCO_3^-$$

   When lungs excrete CO_2, the equation is driven to the left and blood becomes more basic. When kidneys excrete HCO_3^-, the equation is driven to the right and blood becomes more acidic.

4. The pH affects enzymes, causing a change in shape so that they do not function as well.

Answers to 15.3 Critical Thinking

1. Filtration, to a degree, and excretion are part of hemodialysis. Reabsorption and tubular excretion are absent.
2. The membrane is semipermeable, and proteins are too large to pass through.
3. Reabsorption into blood does not occur during hemodialysis. The dialysate is always the area of lesser urea concentration.
4. Glucose in the same concentration as is normal for blood should be added.

Chapter 16

Answers to Objective Questions

1. axon **2.** sodium, inside **3.** synaptic cleft **4.** AChE **5.** muscles **6.** cranial, motor, or parasympathetic; internal organs **7.** interneuron **8.** meninges **9.** cerebrum **10.** cerebellum **11.** See figure 16.2 (p. 280) in text.

Answers to 16.1 Critical Thinking

1. The nerve impulse travels along a membrane and is dependent on the movement of Na^+ and K^+ across the membrane.
2. A reading lower than –65 mV is expected. The resting potential is –65 mV, and inhibitory neurotransmitters increase the polarity.
3. Synaptic vesicles occur only at one end of an axon.

Answers to 16.2 Critical Thinking

1. Interneurons can take nerve impulses across the spinal cord from one side to the other.
2. Neither leg would respond because nerve impulses would never reach the cord.
3. The right leg would still be able to respond.
4. Neither leg would respond because interneurons would be destroyed.

Chapter 17

Answers to Objective Questions

1. axial **2.** spinal cord **3.** radius, ulna **4.** synovial **5.** antagonistic **6.** tetanus **7.** myofibrils, sarcomeres **8.** creatine phosphate **9.** neuromuscular **10.** calcium **11.** *a.* sarcoplasm *b.* sarcolemma *c.* sarcoplasmic reticulum *d.* T tubules *e.* mitochondrion *f.* myofibril *g.* sarcomere

Answers to 17.1 Critical Thinking

1. Bone is living tissue: it grows and heals, it is supplied with blood and nerves, and it contains cells.
2. Bone strength has to equal muscle strength or movement of muscles can cause bones to crack or to break.
3. These are for attachment of muscles.
4. The wide pelvis may be associated with childbirth.

Answers to 17.2 Critical Thinking

1. The tendons of muscles extend across joints. When muscles contract, they pull on the bone to which a tendon is inserted.
2. When muscles contract, they can only shorten—they do not get longer.
3. The legs support the weight of the body, and therefore the muscles of the legs are larger.

Answers to 17.3 Critical Thinking

1. Yes, all myofibrils contract because a muscle fiber does not have degrees of contraction.
2. Yes, it moves closer to the center.
3. Myoglobin has the higher affinity or else it could never receive oxygen from hemoglobin.
4. Mitochondria in muscle fibers use the oxygen to receive hydrogen atoms (H) from the electron transport system.

Chapter 18

Answers to Objective Questions

1. Proprioception **2.** chemoreceptors **3.** rods, cones, retina **4.** color, bright (day) **5.** rounds up (accommodation) **6.** distant, concave **7.** hammer, anvil, stirrup **8.** dynamic equilibrium **9.** cochlear, cochlea **10.** brain **11.** See figure 18.5 (p. 327) in text.

Answers to 18.1 Critical Thinking

1. One possible categorization: focusing—lens, cornea, humors, ciliary body; vision—retina (rods, cones, fovea centralis), optic nerve; other—iris, pupil, choroid, sclera (except for cornea). Justification: Some parts of the eye are concerned with focusing the light, some with bringing about vision, and some do not have either of these functions. Glasses usually correct focusing.
2. Pigments are all colored molecules that usually are capable of absorbing energy. These 3 pigments absorb solar energy.
3. There must be a neural pathway between the eyes and the pineal gland.

Answers to 18.2 Critical Thinking

1. The evolution of the ear in a sequence of animals is needed to support the hypothesis. You would expect to find stages by which changes led to the mammalian ear.
2. Most likely, the inner ear evolved from the lateral line. Most likely, the outer ear and the middle ear evolved otherwise. The human inner ear has mechanoreceptors sensitive to fluid pressure waves. The outer ear receives sound waves in the air, and the middle ear transmits and amplifies these.

Chapter 19

Answers to Objective Questions

1. produces, ADH, oxytocin **2.** hormones **3.** negative feedback **4.** anterior **5.** too little, thyroxin **6.** cortex **7.** Cushing syndrome **8.** calcium **9.** pancreas, cells **10.** blood **11.** See table 19.1 (p. 349) in text.

Answers to 19.1 Critical Thinking

1. You would expect to find glucose in the urine because the body would contain no insulin; blood sugar would rise and spill over into the urine.
2. No, your findings only prove that blood sugar rises when the pancreas is missing.
3. To prove this, you have to get a supply of pure insulin, inject it in an animal, and show that the blood sugar lowers.
4. Yes, you now know that the presence of both the pancreas and insulin lowers blood sugar. The logical conclusion is that the pancreas is the source of insulin.
5. You could actually extract insulin from the pancreas.

Answers to 19.2 Critical Thinking

1. Change the definition to "Environmental signals alter the behavior of target cells, organs, and organisms."
2. The target cell is sensitive when the cell, organ, or organism has receptors for the environmental signal.

3. Glucose causes a pancreatic cell to secrete insulin; insulin causes the liver to store glucose as glycogen; a pheromone causes a male moth to fly toward a female.
4. Glucose causes insulin secretion; a nerve impulse causes neurotransmitter substance release.

Chapter 20

Answers to Objective Questions

1. vas deferens **2.** seminal vesicles **3.** testosterone **4.** blood **5.** vagina **6.** follicle, endometrial **7.** estrogen, progesterone **8.** HCG **9.** laboratory glassware **10.** helper T **11.** cold sores, genital herpes **12.** chlamydia **13.** See figure 20.1 (p. 370) in text.

Answers to 20.1 Critical Thinking

1. Due to negative feedback, the intake of anabolic steroids causes the anterior pituitary to stop producing gonadotropic hormones, leading to atrophy of the interstitial cells of the testes.
2. To test the hypothesis, administer anabolic steroids to mice and collect data on resulting blood levels of gonadotropic hormones. Remove the testes, and using a microscope, look for atrophy of tissues.
3. Anabolic steroids raise the level of LDL in blood. This could lead to increased risk of heart disease.
4. To test the hypothesis, administer anabolic steroids to mice and collect data on resulting blood levels of LDL. Remove coronary blood vessels and look for plaque.

Answers to 20.2 Critical Thinking

1. Due to negative feedback, the administration of estrogen and progesterone causes the anterior pituitary to stop secreting FSH and LH, and no follicles or oocytes mature. Without egg production, there can be no pregnancy.
2. To test, administer this birth-control pill to mice. Collect data on blood levels of FSH and LH. Remove the ovaries, and examine them for the presence of mature follicles.
3. Postmenopausal women who take birth-control pills should have minor or no levels of FSH and LH in their blood.
4. To test, administer birth-control pills to postmenopausal women, and collect data on blood levels of FSH and LH.

Chapter 21

Answers to Objective Questions

1. differentiation **2.** blastocyst **3.** extraembryonic, amnion **4.** implants **5.** gastrulation **6.** induces **7.** placenta **8.** arterial duct **9.** third **10.** head **11.** See figure 21.13 (p. 404) in text.

Answers to 21.1 Critical Thinking

1. a. When a cell inherits a certain cytoplasmic composition, only certain genes are activated. b. These activated genes begin to direct the synthesis of particular proteins. c. Some of these proteins may be secreted to act as specific signals for other cells.
2. a. Tissue A gives off certain signals that influence the morphogenesis of tissue B. b. Because of this, tissue B gives off certain signals that influence the morphogenesis of tissue C. c. Tissue C then gives off signals, and so forth.

Chapter 22

Answers to Objective Questions

1. one **2.** *w* **3.** phenotype **4.** widow's peak **5.** 25% **6.** recessive **7.** 4 **8.** 9 **9.** 4 **10.** 4

Answers to Additional Genetics Problems

1. 50%
2. John = *tt;* parents = *Tt*
3. 25%
4. Man = *Wwtt;* woman = *wwTt;* child = *wwtt*
5. 75%
6. Mary and husband = *HH´;* child = *H´H´*
7. *AB;* yes; A, B, O, AB
8. Light; white
9. Autosomal recessive
10. *AaBb* and *aabb*

Answers to Practice Problems

Practice Problems 1

1. a. (*W*) b. (*WS*), (*Ws*) c. (*T*), (*t*) d. (*Tg*), (*tg*) e. (*AB*), (*Ab*), (*aB*), (*ab*) **2.** a. gamete b. genotype c. gamete d. genotype

Practice Problems 2

1. 75% or 3:1
2. No
3. Heterozygous
4. *DD* × *dd; Dd*

Practice Problems 3

1. Dihybrid
2. 1/16
3. *DdFf* × *ddff; ddff*

Practice Problems 4

1. *H´H´;* no
2. Light
3. White
4. Baby 1 = Doe; baby 2 = Jones
5. AB, O, A, B

Answers to 22.1 Critical Thinking

1. Alternative hypotheses: (a) factors do not segregate—therefore, all parental gametes would be the same, that is, *Yy;* (b) factors do segregate—therefore, 2 parental gametes are possible, that is, *Y* and *y*.
2. If the gametes were always *Yy*, then the phenotype green would not have appeared. Since green does appear, then the second hypothesis is supported.
3. Alternative hypotheses for figure 22.6: (a) factors do not assort independently—therefore, the gametes always will be, for example, either *WS* or *ws;* (b) factors do assort independently—therefore, 4 gametes are possible, *WS, Ws, wS, ws.*
4. If the factors do not assort, then there would be fewer phenotypes among the offspring. Since there are all 4 possible phenotypes among the offspring, the factors have to assort independently of one another.

Answers to 22.2 Critical Thinking

1. The fault is an inability to produce melanin because of an enzyme defect.
2. An inability to produce a normal enzyme is most likely recessive.
3. The possible crosses are *aa* × *aa; Aa* × *aa; Aa* × *Aa.* Yes, *Aa* individuals are carriers.
4. A functioning gene would ensure that all skin cells are capable of producing melanin.

Chapter 23

Answers to Objective Questions

1. Klinefelter **2.** translocation **3.** XY **4.** X^BX^b **5.** is not **6.** linkage **7.** more likely **8.** genetic disorder

Answers to Additional Genetic Problems

1. 50% **2.** a. Males, all red eyes; females, all red eyes. b. Males, all white eyes; females, all red eyes **3.** Mother: WWX^BX^b, father: WwX^BY; 50% **4.** Males: 3 gray body with red eyes: 1 black body with red eyes: 3 gray body with white eyes: 1 black body with white eyes; females: 3 gray body with red eyes: 1 gray body with red eyes **5.** 3 mentally retarded children who can curl the tongue : 1 child with normal intelligence who cannot curl the tongue **6.** Only boys have the trait; it skips from grandfather to grandson; X^AX^a

Answers to Practice Problems

1. His mother; X^HX^h, X^HY, X^hY
2. 100%; none; 100%
3. $RrX^BX^b \times RrX^BY$; rrX^bY
4. The husband is not the father.

Answers to 23.1 Critical Thinking

1. Red eye is dominant.
2. No, the results are not explainable because females do not have a Y chromosome, and yet they have red eye color. Yes, because males have only one X chromosome; this explains why only males have white eyes in the F_2 generation.
3. The results are explainable only on the basis that the red/white eye color allele is on the X chromosome.

Chapter 24

Answers to Objective Questions

1. sugar (deoxyribose), phosphate **2.** old, new **3.** uracil **4.** triplet, amino acid **5.** messenger RNA (mRNA), ribosomal RNA (rRNA), transfer RNA (tRNA) **6.** introns **7.** tRNA **8.** jumping genes **9.** operon **10.** oncogenes, tumor suppressor **11.** *a.* ACU´CCU´GAA´UGC´AAA *b.* UGA´GGA´CUU´ACG´UUU *c.* threonine-proline-glutamate-cysteine-lysine **12.** See figure 24.15*a* (p. 477) in text.

Answers to 24.1 Critical Thinking

1. The information is the DNA's code for protein synthesis, and it is stored in the sequence of bases. The evidence is that the sequence of amino acids in a protein parallels the code in DNA.
2. DNA can replicate because complementary base pairing occurs between DNA strands. Your evidence is the existence of duplicated chromosomes. Chromosomes are found only in the nucleus.
3. The sequence of bases in DNA can change, and this permits mutation. Living things differ from one another. Also, inborn errors of metabolism occur.

Answers to 24.2 Critical Thinking

1. It would show up first in the nucleus and then in the cytoplasm.
2. Put your sample mRNA and rRNA in a test tube with the cellular elements needed for protein synthesis. Analyze the sequence of amino acids in the resulting polypeptide, and see which of these apparently is directing the sequence.
3. Figure 24.12*a* shows ribosomes moving along the mRNA.
4. Yes, this test shows that tRNA molecules carry amino acids because there are nucleotides, not amino acids, in RNA. If amino acids are present, they must have combined with tRNA.

Answers to 24.3 Critical Thinking

1. Yes, the phenotypes of cells differ. For example, muscle cells look quite different from nerve cells.
2. Yes, all cells contain all genes. The process of mitosis ensures that all cells receive a full complement of genes.
3. Only certain mRNA transcripts will be found in different cell types. Cells that look different must be constructed differently and must contain different proteins. For example, muscle cells contain the proteins actin and myosin.
4. Regulator genes control which structural genes are active.

Chapter 25

Answers to Objective Questions

1. vectors 2. cloned **3.** polymerase chain reaction **4.** complementary base pairing **5.** antisense **6.** insects, herbicides **7.** bovine growth hormone **8.** retrovirus **9.** *a.* AATT *b.* TTAA **10.** *a.* retrovirus *b.* recombinant RNA *c.* human genome *d.* recombinant RNA *e.* reverse transcription *f.* recombinant DNA *g.* human gene

Answers to 25.1 Critical Thinking

1. Inject a large number of diabetics with both types of insulin (at different times), and observe any effects. The bioengineered insulin is expected to show fewer side effects because it is human insulin, not cattle or pig insulin—the sequence of amino acids is expected to be closer to that of the individual receiving the insulin. Also, it might be purer—it does not contain any substances other than insulin.
2. Feed the meat to 2 groups of human volunteers, and observe any effects. Since growth hormone is a protein, any present in the meat is denatured upon cooking or digested upon eating.
3. Same experiment as described in the answer to question **2.** First, feed the plants to animals, and if no effects are observed, then feed the plants to humans. The toxin might be harmful to humans.
4. Keep testing for the presence of the pollutant and the bacteria. The bacteria disappear because they run out of food, that is, the pollutant.

Chapter 26

Answers to Objective Questions

1. chemical **2.** small organic molecules **3.** RNA **4.** prokaryotic **5.** heterotrophic **6.** anaerobic **7.** photosynthesizers **8.** aerobic **9.** life **10.** nutrition **11.** See figure 26.6 in text (p. 509). **12.** The endosymbiotic theory is represented. An amoeba like cell is engulfing a prokaryote capable of cellular respiration and a prokaryote capable of photosynthesis.

Answers to 26.1 Critical Thinking

1. There would have been no enzymes present before the first protein formed.
2. Proteins are unable to store genetic information or to replicate.

3. The mechanism is the same as in DNA—complementary base pairing.
4. Enzymes are needed for replication to occur and for other metabolic processes.

Chapter 27

Answers to Objective Questions

1. does not **2.** 21% **3.** Mutation **4.** genetic drift **5.** directional selection **6.** similar **7.** geographically isolated **8.** reproductively isolated **9.** premating, behavioral **10.** order **11.** *a.* Both bats and insects are adapted to flying; their wings are analogous structures. *b.* All vertebrates share a common ancestor, one which had pharyngeal pouches during development. *c.* Two different continents can have similar environments and therefore unrelated organisms that are similarly adapted. *d.* This demonstrates that diversification has occurred in the nightshade family. **12.** Do away with the 5 middle arrows so that 2 sets of circles remain. The circles of each set are still connected by arrows. **13.** *a.* 64% *b.* 4% *c.* 32%

Answers to Practice Problems

1. 30% **2.** $q = 0.1$, $p = 0.9$, homozygous recessive = 1%, homozygous dominant = 81%, heterozygous = 18% **3.** homozygous recessive = 49%, homozygous dominant = 9%, heterozygous = 42%

Answers to 27.1 Critical Thinking

1. Coyotes are adapted to a grassland environment.
2. Coyotes have their own line of descent in North America, and jackals have their own line of descent in Africa.
3. Fossils that show the different lines of descent are evidence of this explanation.

Answers to 27.2 Critical Thinking

1. The plant species contains variations that are inheritable. On the mountaintop, plants that were shorter tended to survive and to reproduce, until only shorter plants were observed there.
2. You expect the plants to still be short because only these genes are now present in the gene pool.
3. Directional selection (toward shorter plants) was followed by stabilizing selection (most plants, then, tend to have genes for shortness).

Answers to 27.3 Critical Thinking

1. Postmating is wasted energy because animals have put energy into mating, and the offspring are not viable/fertile.
2. Premating evolves first because it represents incomplete reproductive isolation, while postmating represents complete reproductive isolation.
3. Habitat and behavioral mechanisms are good candidates. The visual mechanism is not listed, and most likely the birds are able to visually recognize their own species.

Chapter 28

Answers to Objective Questions

1. nucleic acid, protein **2.** retroviruses **3.** bacteria **4.** break down dead organic matter **5.** photosynthetic, food **6.** pseudopods, cilia **7.** zygote **8.** conjugation **9.** alternation of generations **10.** mycelium, hyphae **11.** sac, club **12.** fungus, alga **13.** See figure 28.15 (p. 550) in text. Left side is asexual: one parent, no gametes. Right side is sexual: 2 parents, gametes.

Answers to 28.1 Critical Thinking

1. If the zygote undergoes meiosis, there is no diploid adult.
2. If meiosis is delayed until after there is a diploid adult, the alternation of generations life cycle occurs.
3. If haploid spores join (then they act like gametes), the haploid generation is eliminated and the diplontic life cycle occurs.

Answers to 28.2 Critical Thinking

1. Members of the *Ulva* genus are multicellular, and they have the alternation of generations life cycle.
2. The organism would have to lose the flagella and gain a cell wall to be more plantlike. To be more animal-like, the organism would have to lose the chloroplasts.
3. You would look for centrioles because they give rise to cilia and flagella, structures associated with animal cells.

Answers to 28.3 Critical Thinking

1. Fungi live on dead organic matter. Without other living things, there is no dead organic matter.
2. In the *Chlamydomonas* cycle, the adult, the spores, and the gametes are flagellated. In the fungus cycle, the adult, the spores, and the gametes are not flagellated.
3. The body of a fungus may not have the ability to withstand dry conditions.

Chapter 29

Answers to Objective Questions

1. embryo **2.** spores **3.** gametophyte **4.** swim **5.** separate **6.** heterospores **7.** pollen grains **8.** female **9.** egg, polar **10.** seed, fruit **11.** See figure 29.2 (p. 563) in text. **12.** *a.* leafy shoot *b.* stalk and capsule *c.* heart-shaped prothallus *d.* frond has large leaves *e.* fern **13.** *a.* on male cone *b.* on female cone *c.* in anther *d.* in ovary *e.* flower

Answers to 29.1 Critical Thinking

1. No, vascular tissue would not be found because bryophytes lack vascular tissue.
2. Try to grow them in areas of limited mineral availability. Observe results.
3. The gametophyte (N) is dominant, and it bears the burden of adaptation to the environment.
4. If vascular tissue evolved, the bryophytes might spread into more habitats on land, but only if they also lost their dependence on water for reproduction.

Answers to 29.2 Critical Thinking

1. In humans, the male passes sperm directly to the female, who retains the egg within her body. In trees, pollen, which can resist drying out, carries the sperm to the vicinity of the egg. Then the pollen germinates to give a pollen tube, through which the sperm passes.
2. Humans have a blood vascular system; trees have xylem and phloem.
3. Humans have an internal skeleton of bone; trees are supported by xylem.
4. Humans can maintain a warm internal temperature; deciduous trees lose their leaves and become inactive.

Chapter 30

Answers to Objective Questions

1. to keep water moving through the central cavity **2.** sac, radially **3.** ladder, flame-cell **4.** cattle or pigs **5.** roundworm **6.** mouth **7.** reduced, heart **8.** segmented **9.** tube **10.** notochord, dorsal hollow nerve cord, pharyngeal pouches **11.** tunicates, lancelets **12.** jawless, cartilaginous, bony **13.** lobe-finned **14.** water, eggs, extraembryonic **15.** birds, temperature **16.** monotremes, marsupials, placental mammals **17.** placental **18.** *a.* 4 *b.* 1 *c.* 3 *d.* 2 **19.** trees **20.** upright, small **21.** Cro-Magnons **22.** See figure 30.1 (p. 582) in text.

Answers to 30.1 Critical Thinking

1. Animals (a) are heterotrophic, (b) locomote by means of contractile fibers, (c) are multicellular with specialized tissues, (d) have a diplontic life cycle, (e) have sex organs, and (f) do not always protect the zygote and the embryo.
2. An animal locomotes by means of contractile fibers.
3. Protozoans locomote and are heterotrophic.
4. The highly differentiated cells of complex animals are unable to give rise to less specialized cells so that development can begin again.

Answers to 30.2 Critical Thinking

1. Insects have tracheae, and birds have lungs.
2. Insects have an external skeleton, and birds have a dry outer coat of dead cells.
3. The male passes sperm to the female in both. Insects deposit eggs in water or in soil or provide them with a hard covering. Birds lay hard-shelled eggs.
4. Insects typically lay thousands of eggs and do not tend the young. Birds lay a few eggs and care for each offspring for a few months.

Answers to 30.3 Critical Thinking

1. Humans take in preformed food and have the power of locomotion by means of contractile fibers.
2. As embryos, humans have the 3 chordate characteristics: notochord, pharyngeal pouches, and dorsal hollow nerve cord. In adults, the vertebral column replaces the notochord.
3. Humans have hair and mammary glands.
4. Humans have long arms and legs, 5 fingers on each hand and 5 toes on each foot, opposable thumbs, nails (not claws), no snout, and binocular vision with a poor sense of smell, and a large brain.

Chapter 31

Answers to Objective Questions

1. fixed-action **2.** Habituation **3.** environment **4.** immediate **5.** natural selection **6.** territory **7.** are their own **8.** inclusive **9.** related **10.** *a.* communication, chemical *b.* learning, imprinting *c.* communication, sound *d.* learning, operant conditioning *e.* communication, tactile *f.* communication, visual.

Answers to 31.1 Critical Thinking

1. Use means to prevent the guinea pigs from using their senses to detect that food is present.
2. Immediate environmental influence strongly affects what guinea pigs eat.
3. Long-term environmental influence, and possibly genetic inheritance, strongly influence what guinea pigs eat.

Chapter 32

Answers to Objective Questions

1. biomes **2.** rainfall **3.** savanna **4.** tropical rain **5.** taiga **6.** prairie or grassland **7.** tundra **8.** nutrients **9.** light and nutrients **10.** See figure 32.7 (p. 645) in text.

Answers to 32.1 Critical Thinking

1. A grassland might replace a deciduous forest, and a desert might replace a grassland.
2. The more sunlight, the more photosynthesis and the more life supported.
3. Varied and plentiful food sources provide different ways of getting food for various life forms.

Chapter 33

Answers to Objective Questions

1. flow **2.** producer **3.** herbivores **4.** trophic **5.** transformed **6.** reservoir **7.** respire **8.** nitrate **9.** nitrogen gas **10.** increasing amount **11.** *a.* algae *b.* rotifers, mosquito larva *c.* water fleas, water boatman *d.* catfish, frog *e.* green-backed heron **12.** *a.* respiration *b.* photosynthesis *c.* combustion

Answers to 33.1 Critical Thinking

1. Secondary consumers have more energy available to them (in the form of food) than tertiary consumers.
2. The pyramid indicates that there is less energy available at each trophic level. Eventually, there is not enough energy to support another population.
3. The size of a top-predator population is controlled by the amount of food energy available to it.
4. The other secondary consumer populations would increase in size due to less competition, and the ecosystem would remain about the same.

Chapter 34

Answers to Objective Questions

1. low **2.** prereproductive **3.** industrialized **4.** higher **5.** carbon dioxide (CO_2) **6.** CFCs **7.** cultural eutrophication **8.** magnification **9.** *a.* hydrocarbons (HC) *b.* chlorofluorocarbons (CFCs) *c.* carbon dioxide (CO_2) *d.* sulfur dioxide (SO_2)

Answers to 34.1 Critical Thinking

1. There can be 4 future generations. At that time, there will be 62 rabbits—32 newly born and 30 previous parents. If each new pair of rabbits then produces 4, there will be 126 rabbits, too many for the watering hole to support.
2. First generation: 2 rabbits; second generation: 6 rabbits; third generation: 14 rabbits; fourth generation: 30 rabbits; fifth generation: 62 rabbits. Yes, there is an ever greater increase. This is exponential growth.
3. It is a J-shaped curve. The curve will fall dramatically.

Answers to 34.2 Critical Thinking

1. Formerly, it was assumed that each country was responsible for its own standard of living. This attitude is changing because planners now think in global terms.
2. Increasingly, people in the MDCs believe that health care for all people should be the same.
3. Most people now believe that private citizens and industry should find ways to ensure an ecologically fit world for future generations. If we do not, the standard of living will decrease dramatically.
4. Increasingly, people are thinking in global terms.

GLOSSARY

A

accommodation Lens adjustment in order to see close objects. *360*

acetylcholine (ACh) A neurotransmitter substance secreted at the ends of many neurons; responsible for the transmission of a nerve impulse across a synaptic cleft. *285*

acetylcholinesterase (AChE) An enzyme that breaks down acetylcholine. *285*

acid A solution in which pH is less than 7; a substance that contributes or liberates hydrogen ions (protons) in a solution. *25*

acid deposition (as´id dep˝o-zish´un) The return to earth as rain or snow of the sulfate salts or nitrate salts of acids produced by commercial and industrial activities on earth. *677*

acromegaly (ak˝ro-meg´ah-le) A condition resulting from an increase in growth hormone production after adult height has been achieved. *350*

acrosome Covering on the tip of a sperm that contains enzymes necessary for fertilization. *372*

ACTH (adrenocorticotropic hormone) Hormone secreted by the anterior lobe of the pituitary gland that stimulates activity in the adrenal cortex. *351*

actin One of 2 major proteins of muscle; makes up thin filaments in myofibrils of muscle fibers. *See* myosin. *316*

actin filament An extremely thin fiber found within the cytoplasm that is composed of the protein actin; involved in the maintenance of cell shape and the movement of cell contents. *53*

action potential The change in potential propagated along the membrane of a neuron; the nerve impulse. *284*

active site The region on the surface of an enzyme where the substrate binds and where the reaction occurs. *77*

active transport Transfer of a substance into or out of a cell from a region of lower concentration to a region of higher concentration by a process that requires a carrier and an expenditure of energy. *68*

adaptation The fitness of an organism for its environment, including the process by which it becomes fit and is able to survive and to reproduce; also, a decrease in the excitability of receptors in response to continuous constant-intensity stimulation. *323*

adenosine triphosphate *See* ATP.

adrenocorticotropic hormone *See* ACTH.

aerobic Growing or metabolizing only in the presence of oxygen (O_2) as in aerobic respiration. *90*

aerobic cellular respiration The complete breakdown of glucose to carbon dioxide and water; requires glycolysis, transition reaction, Krebs cycle, and electron transport system. *52, 90*

aerobic respiration Respiration in the presence of oxygen. *90, 509*

agglutination (ag-gloo˝ tĭ-na-shun) Clumping of cells, particularly in reference to red blood cells involved in an antigen-antibody reaction. *233*

aging Progressive changes over time, leading to loss of physiological function and eventual death. *414*

albumin (al-bu´ min) Plasma protein of the blood having transport and osmotic functions. *209*

aldosterone A hormone secreted by the adrenal cortex that functions in regulating sodium and potassium concentrations of blood. *272, 355*

alga Aquatic organism that carries on photosynthesis. *549*

allantois (ah-lan-to-is) One of the extraembryonic membranes; in reptiles and birds, it is a pouch that collects nitrogenous waste; in mammals, it is a source of blood vessels to and from the placenta. *402*

allele (ah-lēl´) An alternative form of a gene located at a particular chromosome site (locus). *422*

all-or-none response Phenomenon in which a muscle fiber contracts completely when it is exposed to a stimulus of threshold strength. *311*

alternation of generations A life cycle, typical of plants, in which a diploid sporophyte alternates with a haploid gametophyte. *146, 549, 563*

altruism Behavior performed for the benefit of others without regard to its possible detrimental effect on the performer. *632*

alveolus Air sac of a lung. *243*

amino acid A monomer of a protein; takes its name from the fact that it contains an amino group ($-NH_2$) and an acid group ($-COOH$). *28*

amnion (am´ ne-on) An extraembryonic membrane; a fluid-containing sac around the embryo. *402*

amoeba Protozoan that moves by means of pseudopods. *546*

ampulla (am-pūl´ lah) Base of a semicircular canal in the inner ear. *336*

amylase A starch-digesting enzyme secreted by the salivary glands (salivary amylase) and the pancreas (pancreatic amylase). *182*

anabolic steroid A synthetic steroid that mimics the effect of testosterone. *359*

anaerobic Growing or metabolizing in the absence of oxygen (O_2). *90*

analogous structure (ah-nal´o-gus struk´tūr) Structure similar to another in function but not in anatomy; particularly in reference to similar adaptations. *519*

anaphase Stage in mitosis during which chromatids separate, forming chromosomes. *111*

anemia Inefficient oxygen-carrying ability of blood due to hemoglobin shortage. *210*

angiosperm A seed plant having seeds that develop within ovaries and that are eventually enclosed by fruits. *570*

antenna Sensory organ located on the arthropod head. *598*

anterior pituitary The portion of the pituitary gland that produces 6 types of hormones and is controlled by hypothalamic-releasing and release-inhibiting hormones. *346*

anther That portion of a stamen in which pollen is formed. *146, 573*

antheridium (an″ ther-id´ e-um) Male organ in certain nonseed plants where flagellated sperm are produced. *564*

anthropoid (an´thro-poid) A group of primates that includes only monkeys, apes, and humans. *603, 611*

antibody A protein produced in response to the presence of an antigen; an antibody combines with the antigen to produce a nonharmful complex. *225*

antibody-mediated immunity Body line of resistance with antibody-producing B cells. *226*

anticodon A "triplet" of bases in tRNA that pairs with a complementary triplet (codon) in mRNA. *471*

antidiuretic hormone (ADH) (an″tĭ-di″u-ret´ik hōr´mōn) Sometimes called vasopressin, a hormone secreted by the posterior pituitary that controls the degree to which water is reabsorbed by the kidneys. *272, 348*

antigen (ant´i-jen) A foreign substance, usually a protein, that stimulates the immune system to react, such as to produce antibodies. *225*

anus Inferior outlet of the digestive tube. *178*

anvil The middle bone of the 3 ossicles of the middle ear. *336*

aorta (ā-or´tah) Major systemic artery that receives blood from the left ventricle. *206*

appendicular skeleton Portion of the skeleton forming the upper extremities, the pectoral girdles, the lower extremities, and the pelvic girdle. *306*

appendix A small tubular appendage that extends outward from the cecum of the large intestine. *178*

aqueous humor Watery fluid that fills the anterior chamber of the eye. *327*

archaebacteria Monera that are able to live under adverse circumstances; represent an early branch of living organisms. *545*

archegonium (ar″kē-go´ne-um) Female organ of certain nonseed plants where eggs are produced. *564*

archenteron (ar-ken´ter-on) Central cavity or primitive gut in the animal embryo. *395*

areola Pigmented region surrounding the nipple of the breast. *382*

arterial duct Ductus arteriosus; fetal connection between the pulmonary artery and the aorta. *410*

arteriole (ar-te´ re-ōl) Vessel that takes blood from an artery to capillaries. *198*

artery Vessel that takes blood away from the heart to arterioles; characteristically possessing thick elastic and muscular walls. *198*

aster In animal cells, short microtubule that extends outward from a spindle pole during cell division. *110*

astigmatism A condition of blurred vision due to an irregular curvature of the cornea or the lens. *333*

asymmetry Lacking symmetry. *583*

atom Smallest unit of matter that cannot be divided by chemical means. *18*

atomic number The number of protons within the nucleus of an atom. *18*

atomic weight The number of protons plus the number of neutrons within the nucleus of an atom. *18*

ATP (adenosine triphosphate) A compound containing adenine, ribose, and 3 phosphates, 2 of which are high-energy phosphates; the "common currency" of energy for most cellular processes. *37, 75*

atrioventricular A structure in the heart that pertains to both the atria and the ventricles; for example, an atrioventricular valve is located between an atrium and a ventricle. *200*

atrioventricular node *See* AV node.

atrium (a´ tre-um) Chamber; particularly an upper chamber of the heart lying above the ventricles; either the left atrium or the right atrium. *200*

auditory canal A tube in the external ear that lies between the pinna and the tympanic membrane. *335*

auditory nerve A nerve sending the signal for sound from the inner ear to the temporal lobe of the brain. *338*

australopithecine (aw″strah-lo-pith´ə-sīn) Referring to one of 3 species of *Australopithecus,* the first generally recognized hominids. *614*

autosome Chromosome other than a sex chromosome. *104*

autotroph (aw-to-trōf) An organism that is capable of making its food (organic molecules) from inorganic molecules. *510*

AV (atrioventricular) node A small region of neuromuscular tissue that transmits impulses received from the SA node to the ventricular walls. *203*

axial skeleton Portion of the skeleton that supports and protects the organs of the head, the neck, and the trunk. *305*

axon Fiber of a neuron that conducts nerve impulses away from the cell body. *279*

B

bacteriophage (bak-te´re-o-fāj″) A virus that infects a bacterial cell. *538*

bacterium A unicellular organism that is prokaryotic—its single cell lacks the complexity of a eukaryotic cell; archaebacteria and eubacteria. *542*

bark All tissues outside the vascular cambium; includes phloem, cork cambium, and cork. *132*

basal body Short cylinder having a circular arrangement of 9 microtubule triplets (9 + 0 pattern) located within the cytoplasm at the base of cilia and flagella. *54*

base A solution in which pH is greater than 7; a substance that contributes or liberates hydroxide ions (OH^-) in a solution; alkaline; opposite of acidic. Also, a term commonly applied to one of the components of a nucleotide. *25*

behavior All responses made by an organism to changes in the environment. *7, 624*

bilateral symmetry Having a right half and a left half so that only one vertical cut gives 2 equal halves. *583*

bile A secretion of the liver that is temporarily stored in the gallbladder before being released into the small intestine, where it emulsifies fat. *180*

binary fission Reproduction by division into 2 equal parts by a process that does not involve a mitotic spindle. *543*

bioengineered Alteration of the genome of an organism by technological processes; genetic engineering. *487*

biogeography The study of the geographical distribution of organisms. *518*

biological evolution Changes that have occurred in life forms from the origin of the first cell or cells to the many diverse forms in existence today. *507*

biome (bi´ōm) One of the major land communities in the biosphere, characterized by a particular mix of plants and animals. *636*

biosphere That part of the earth's surface and atmosphere where living organisms exist. *636*

biotechnology Use of a natural biological system to produce a commercial product. *487*

biotic potential The maximum population growth rate under ideal conditions. *668*

black bread mold A fungus of the genus *Rhizopus* that forms a whitish or grayish mycelium on bread or fruit. *555*

blade The main portion of a leaf. *134*

blind spot Area of the eye containing no rods or cones and where the optic nerve passes through the retina. *328*

blood Connective tissue composed of cells separated by plasma that transports substances in the cardiovascular system. *160*

blood pressure The pressure of blood against the wall of a blood vessel. *204*

B lymphocyte A lymphocyte that matures in the bone marrow, and when stimulated by the presence of a specific antigen, gives rise to antibody-producing plasma cells. *225*

bone Connective tissue having a hard matrix of calcium salts deposited around protein fibers. *160*

Bowman's capsule A double-walled cup that surrounds the glomerulus at the beginning of the nephron. *265*

bradykinin (brad″e-ki´nin) A substance found in damaged tissue that initiates nerve impulses resulting in the sensation of pain. *223*

breathing Entrance and exit of air into and out of the lungs. *239*

bronchiole (brong-ke´ōl) One of the smaller air passages in the lungs that eventually terminate in alveoli. *243*

bronchus (brong-kus) One of 2 major divisions of the trachea leading to the lungs. *243*

bryophyte (bri´ o-fīt) A nonvascular plant, including liverworts and mosses. *564*

buffer A substance or compound that prevents large changes in the pH of a solution. *27*

C

calcitonin Hormone secreted by the thyroid gland that helps to regulate blood calcium level. *352*

Calvin cycle The primary (C_3) pathway of the light-independent reaction of photosynthesis; converts CO_2 to carbohydrate. *88*

capillary (kap´ĭ-lar″e) Microscopic vessel connecting arterioles to venules through the thin walls of which molecules either exit or enter blood. *198*

carapace Upper covering (shell) of some animals. *598*

carbaminohemoglobin Hemoglobin carrying carbon dioxide. *253*

carbohydrate One of a class of organic compounds characterized by the presence of CH_2O groups; includes monosaccharides, disaccharides, and polysaccharides. *30*

carbonic anhydrase An enzyme that catalyzes the formation and breakdown of carbonic acid, thereby allowing carbon dioxide to be carried as the bicarbonate ion in the blood. *253*

cardiac muscle Specialized type of muscle tissue found only in the heart. *161*

carnivore An animal that feeds only on other animals. *652*

carotenoid (ka-rot´en-oid″) An orange or yellow pigment that serves as an accessory to chlorophyll in photosynthesis. *85*

carrier A molecule that combines with a substance and transports it through the cell membrane. Also an individual that unknowingly transmits an infectious or genetic disease. *432*

carrier protein A protein molecule that combines with a substance and transports it through the cell membrane. *68*

carrying capacity The largest number of organisms of a particular species that can be maintained indefinitely in an ecosystem. *668*

cartilage A type of tissue characterized by cells separated by a matrix that often contains fibers. *159*

Casparian strip A waxy ring around endodermal cells of plants that prevents passage of water and minerals other than through the cells. *129*

cell The structural and functional unit of an organism; the smallest structure capable of performing all the functions necessary for life. *44*

cell body Portion of a neuron that contains the nucleus and from which the nerve fibers extend. *279*

cell cycle A repeating sequence of events in eukaryotic cells consisting of interphase, when growth and DNA synthesis occurs, and mitosis, when cell division occurs. *106*

cell-mediated immunity Body line of resistance in which T cells destroy antigen-bearing cells. *227*

cell membrane A membrane that surrounds the cytoplasm of cells and regulates the passage of molecules into and out of the cell. *61*

cell plate A double-layered membrane that precedes the formation of the cell wall as a part of cytokinesis in plant cells. *111*

cellular respiration The metabolic reactions that provide ATP energy to a cell. *239*

cellulose A polysaccharide composed of glucose molecules; the chief constituent of a plant's cell wall. *33*

cell wall A protective barrier outside the cell membrane of a bacterial, fungal, algal, or plant cell. *45*

central canal Tube within the spinal cord that is continuous with the ventricles of the brain and contains cerebrospinal fluid. *291*

central nervous system (CNS) The brain and the spinal cord in vertebrate animals. *279*

centriole (sen´trē-ōl) A short, cylindrical organelle in animal cells that contains microtubules in a 9 + 0 pattern; associated with the formation of basal bodies and the spindle during cell division. *54*

centromere (sen´tro-mēr) A region of attachment of a chromosome to spindle fibers that is generally seen as a constricted area. *104*

cephalothorax Fusion of head and thoracic regions displayed by some arthropods. *598*

cerebellum The part of the vertebrate brain that controls muscular coordination. *293*

cerebral hemisphere (ser´ĕ-bral hem´ĭ sfĕr) One of the large paired structures that together constitute the cerebrum of the brain. *293*

cerebrospinal fluid Fluid found in the ventricles of the brain, the central canal of spinal cord, and in association with the meninges. *291*

cerebrum The largest portion of the brain, consisting of the right and left cerebral hemispheres. *293*

cervix Narrow end of the uterus, which leads into the vagina. *377*

chaparral (shap-ə-ral´) A biome of broad-leaved evergreen shrubs forming dense thickets. *641*

chemical evolution A gradual increase in the complexity of chemical compounds that is believed to have brought about the origin of the first cell or cells. *504*

chemiosmotic ATP synthesis The production of ATP by coupling the transport of hydrogen ions across membranes with the phosphorylation of ADP. *87, 94*

chemosynthesis The process of making food by using energy derived from the oxidation of reduced molecules in the environment. *544*

chitin (ki´tin) Flexible, strong polysaccharide forming the exoskeleton of arthropods. *596*

chlorophyll The green pigment that converts solar energy to chemical energy during photosynthesis. *53, 85*

chloroplast (klo´ro-plast) A membranous organelle that contains chlorophyll and is the site of photosynthesis. *52*

chorion (ko´re-on) An extraembryonic membrane; forms an outer covering around the embryo; in reptiles and birds, it functions in gas exchange; in mammals, it contributes to the formation of the placenta. *402*

chorionic villi Treelike extensions of the chorion of the mammalian embryo projecting into the maternal tissues. *407*

choroid (ko´roid) The vascular, pigmented middle layer of the eyeball. *326*

chromatid (kro´ma-tid) One of the 2 identical parts of a chromosome following replication of DNA. *104*

chromatin Threadlike network in the nucleus that is made up of DNA and proteins. *46*

chromosome Rodlike structure in the nucleus seen during cell division; contains the hereditary units, or genes. *46*

chromosome mutation A variation in regard to the normal number of chromosomes inherited or in regard to the normal sequence of alleles on a chromosome; the sequence can be inverted, translocated from a nonhomologous chromosome, deleted, or duplicated. *446*

ciliary body (sil´e-er″e bod´e) Structure associated with the choroid layer that contains the ciliary muscle, which controls the shape of the lens of the eye. *326*

ciliate Protozoan that moves by means of cilia. *547*

cilium Hairlike projection used for locomotion by many unicellular organisms and having various purposes in higher organisms. *54*

circadian rhythm A regular physiological or behavioral event that occurs approximately every 24 hours. *359*

circumcision Removal of the foreskin of the penis. *372*

citric acid cycle *See* Krebs cycle.

cleavage furrow An indentation that begins the process of cleavage, by which animal cells undergo cytokinesis. *111*

clotting Process of blood coagulation, usually when injury occurs. *213*

cnidarian (ni-dah´re-an) Small aquatic animal having radial symmetry and bearing stinging cells with nematocysts. *586*

coacervate droplet A mixture of polymers that may have preceded the origin of the first cell or cells. *507*

cochlea (kok´le-ah) That portion of the inner ear that resembles a snail's shell and contains the organ of Corti, the sense organ for hearing. *336*

cochlear canal Canal within the cochlea that bears small hair cells that function as hearing receptors. *336*

codon A "triplet" of bases in mRNA that directs the placement of a particular amino acid into a polypeptide. *469*

coelom (se´lom) Body cavity of higher animals that contains internal organs, such as those of the digestive system. *165*

coenzyme A nonprotein molecule that aids the action of the enzyme to which it is loosely bound. *79, 91*

coenzyme A Coenzyme that participates in the transition reaction and carries the organic product to the Kreb's cycle. *91*

cohesion-tension theory Explanation for upward transportation of water in xylem based upon transpiration-created tension and the cohesive properties of water molecules. *140*

collecting duct A tube that receives urine from the distal convoluted tubules of several nephrons. *265*

colon The large intestine. *178*

colony A cluster of specialized cells that cooperate to a degree. *550*

colostrum Watery, yellowish white fluid produced by the breasts. *382*

columnar epithelium Pillar-shaped cells usually having the nucleus near the bottom of each cell and found lining the digestive tract, for example. *156*

compact bone Hard bone consisting of Haversian systems cemented together. *160, 308*

companion cell A small nucleated cell that lies adjacent to and helps with the activities of a sieve-tube cell. *142*

complementary base pairing Pairing of bases between nucleic acid strands; adenine pairs with either thymine (DNA) or uracil (RNA), and cytosine pairs with guanine. *464*

complement system A series of proteins in plasma that counteracts a microbe invasion in a variety of ways; complements the antigen-antibody reaction. *224*

compound Two or more atoms of different elements that are chemically combined. *19*

compound eye Arthropod eyes composed of multiple lenses. *598*

cone Bright-light receptor in the retina of the eye that detects color and provides visual acuity. Also, a specialized structure composed of scale-shaped leaves in conifers. *328*

conidia Spores produced by sac and club fungi during asexual reproduction. *556*

conifer A cone-bearing seed plant; mostly trees such as pines. *570*

conjugation Sexual union between organisms in which the genetic material of one cell enters another. *547*

connective tissue A type of tissue characterized by cells separated by a matrix that often contains fibers. *156*

consumer A member of a population that feeds on members of other populations in an ecosystem. *652*

control In experimentation, a sample that undergoes all the steps in the experiment except the one being tested. *13*

coral A cnidarian that has a calcium carbonate skeleton; remains accumulate to form reefs. *586*

coral reef A structure found in tropical waters that is formed by the buildup of coral skeletons and where many and various types of organisms reside. *586*

cork cambium Meristem that produces cork. *132*

coronary artery Artery that supplies blood to the wall of the heart. *207*

corpus callosum A mass of white matter within the brain that is composed of nerve fibers connecting the right and left cerebral hemispheres. *294*

corpus luteum A body, yellow in color, that forms in the ovary from a follicle that has discharged its egg. *376*

cortex In animals, the outer layer of an organ; in plants, a tissue composed mainly of parenchyma cells that is found between the vascular tissue and the epidermis in stems and roots. *129*

cortisol A glucocorticoid secreted by the adrenal cortex. *354*

cotyledon (kot″ĭ-lē´- don) The seed leaf of the embryonic plant. *126*

covalent bond A chemical bond between atoms that results from the sharing of a pair of electrons. *22*

covalent reaction A chemical change that involves the formation of a covalent bond. *22*

Cowper's gland Either of 2 small structures located below the prostate gland in males. *372*

cranial nerve Nerve that arises from the brain. *287*

creatine phosphate Compound unique to muscles that contains a high-energy phosphate bond. *316*

creatinine Excretion product from creatine phosphate breakdown. *262*

cretinism (kre´tin-izm) A condition resulting from improper development of the thyroid in an infant. *352*

cri du chat syndrome A group of body malfunctions caused by a deletion of chromosome 5. *446*

Cro-Magnon The common name for the first fossils to be accepted as representative of modern humans. *616*

crossing-over The exchange of corresponding segments of genetic material between nonsister chromatids of homologous chromosomes during synapsis of meiosis I. *113*

cuboidal epithelium Cube-shaped cells found lining the kidney tubules, for example. *156*

cultural eutrophication (kul´tu-ral u˝tro-fĭ-ka´shun) Enrichment of a body of water due to human activities, causing excessive growth of producers and then death of these and other inhabitants. *675*

cyanobacterium (si˝ah-no-bak-te´re-um) Photosynthetic prokaryote that contains chlorophyll and releases O_2; formerly called a blue-green alga. *545*

cystic fibrosis A lethal genetic disease involving problems with the functions of the mucous membranes in the respiratory and digestive tracts. *432*

cytokinesis (si˝to-ki-ne´sis) Division of the cytoplasm of a cell following telophase of mitosis and meiosis I and II. *111*

cytoplasm The contents of the cell; located between the nucleus and the cell membrane. *45*

cytoskeleton Filamentous protein structures found throughout the cytoplasm that help maintain the shape of the cell, anchor the organelles, and allow the cell and its organelles to move. *53*

cytotoxic T cell T lymphocyte that attacks and kills antigen-bearing cells; killer T cell. *227*

D

data Experimentally derived facts. *12*

deamination Removal of an amino group ($-NH_2$) from an amino acid or other organic compound. *97*

deciduous Plants that shed their leaves at certain seasons. *641*

decomposer Organism of decay (fungus and bacterium) in an ecosystem. *652*

deletion A chromosome mutation caused by the breakage and loss of a fragment of a chromosome. *446*

demographic transition The change from a high birthrate to a low birthrate so that the growth rate is lowered. *669*

dendrite Fiber of a neuron, typically branched, that conducts nerve impulses toward the cell body. *279*

denitrification (de-ni˝trĭ-fi-ka´shun) The process of converting nitrate to nitrogen; part of the nitrogen cycle. *661*

deoxyribonucleic acid *See* DNA.

dermis (der´mis) The layer of thick skin that lies beneath the epidermis. *162*

desert An arid biome characterized especially by plants such as cacti, which are adapted for receiving less than 25 cm of rain per year. *636*

desertification (dez-ert˝ ĭ-fi-ka´shun) Desert conditions caused by human misuse of land. *672*

detritus (di-tri´tus) Nonliving organic matter. *652*

deuterostome (du´ter-o-stōm˝) Member of a group of animal phyla in which the anus develops from the blastopore and a second opening becomes the mouth. *592*

development All the changes that take place during the life of an organism. *7*

diabetes insipidus Condition characterized by an abnormally large production of urine due to a deficiency of antidiuretic hormone. *348*

diabetes mellitus (di˝ah-bĕ´tēz me-li´tus) Condition characterized by a high blood glucose level and the appearance of glucose in the urine due to a deficiency of insulin production or uptake by cells. *358*

diaphragm A sheet of muscle that separates the thoracic cavity from the abdominal cavity in higher animals. Also, a birth-control device inserted in front of the cervix in females. *244*

diastole (di-as´to-le) Relaxation of a heart chamber. *201*

diatom (di´ah-tom) One of a large group of freshwater and marine unicellular algae having a cell wall consisting of 2 silica-impregnated valves that fit together as in a pillbox. *554*

dicot Dicotyledon; a type of flowering plant distinguished particularly by the presence of 2 cotyledons in the seed, such as beans and geraniums. *126*

differentially permeable Having degrees of permeability; the cell is impermeable to some substances and allows others to pass through at varying rates. *63*

differentiation The process and the developmental stages by which a cell becomes specialized for a particular function. *394*

diffusion The movement of molecules from a region of higher concentration to a region of lower concentration. *65*

dihybrid An individual that is heterozygous for 2 traits; shows the phenotype governed by the dominant alleles but carries the recessive alleles. *429*

dipeptide A molecule consisting of only 2 amino acids joined by a peptide bond. *29*

dipleura A larval form unique to the deuterostomes that indicates they are related. *592*

diploid (dĭp´loid) The 2N number of chromosomes; twice the number of chromosomes found in gametes. *105*

diplontic life cycle Life cycle typical of animals in which the adult is always diploid because meiosis occurs after maturity is reached. *549*

directional selection Natural selection that favors an atypical phenotype. *526*

disaccharide A sugar that contains 2 units of a monosaccharide; e.g., maltose. *32*

disruptive selection Selection that favors the extreme phenotypes for a particular characteristic rather than the intermediate phenotype. *526*

dissociate The breakdown of a compound into its ionic or elemental components. *25*

distal convoluted tubule Highly coiled region of a nephron that is distant from Bowman's capsule. *265*

DNA (deoxyribonucleic acid) A nucleic acid found in the cells; the genetic material that directs protein synthesis in cells. *36, 464*

DNA ligase (li´gās) An enzyme that links DNA fragments; used in bioengineering to join foreign DNA to vector DNA. *489*

DNA polymerase An enzyme catalyzing the union of complementary base pairs in the formation of a DNA strand. *467*

DNA probe Single strand of radioactive DNA that can be used to find a complementary DNA strand; can be used diagnostically to determine the presence of particular genes. *492*

dominance hierarchy A system in which animals arrange themselves in a pecking order; the animal above takes precedence over the one below. *629*

dominant allele Hereditary factor that expresses itself in the phenotype when the genotype is heterozygous. *424*

dorsal-root ganglion A mass of sensory neuron cell bodies located in the dorsal root of a spinal nerve. *288*

double helix A double spiral; describes the three-dimensional shape of DNA. *464*

doubling time The number of years it takes for a population to double in size. *668*

Down syndrome Human congenital disorder associated with an extra chromosome 21. *444*

duodenum (du″o-de´num) The first portion of the small intestine in vertebrates into which ducts from the gallbladder and pancreas enter. *177*

duplication Chromosome mutation in which the chromosome segment occurs more than once on the same chromosome. *446*

dyad A chromosome having 2 chromatids held together at a centromere. *113*

E

ecological pyramid Pictorial graph representing biomass, organism number, or energy content of each trophic level in a food web, from the producer to the final consumer populations. *656*

ecology The study of the interactions of species with each other and with the physical environment. *651*

ecosystem A setting in which populations interact with each other and with the physical environment. *9,* 651

ectoderm The outer germ layer of the embryonic gastrula; it gives rise to the nervous system and skin. *395*

edema Swelling due to tissue fluid accumulation in the intercellular spaces. *222*

effector A structure such as a muscle or a gland that allows an organism to respond to environmental stimuli. *279*

elastic cartilage Cartilage composed of elastic fibers, which allow greater flexibility. *160*

electrocardiogram (ECG or **EKG)** A graphic recording of the electrical activity associated with the heartbeat. *204*

electroencephalogram (EEG) A graphic recording of the brain's electrical activity. *294*

electron A subatomic particle that has almost no weight and carries a negative charge; orbits in a shell about the nucleus of an atom. *18*

electron transport system A chain of electron carriers in the thylakoid membranes of chloroplasts and the cristae of mitochondria that utilize released energy to produce ATP. *85, 90, 94*

element The simplest of substances, consisting of only one type of atom; e.g., carbon, hydrogen, oxygen. *18*

elephantiasis (el″ə-fan-ti´ah-sis) A disease caused by a parasitic nematode that blocks a lymphatic vessel; characterized by extreme swelling of a limb. *590*

embryo sac The female gametophyte of flowering plants that contains an egg cell. *573*

emulsification The act of dispersing one liquid in another. *35*

endocrine gland A gland that secretes hormones directly into the blood or body fluids. *344*

endocrine system An organ system consisting of ductless glands whose secretions, called hormones, are carried in blood. *344*

endocytosis (en″do-si-to´sis) A process in which a vesicle is formed at the cell membrane to bring a substance into the cell. *69*

endoderm An inner layer of cells that lines the primitive gut of the gastrula; it becomes the lining of the digestive tract and associated organs. *395*

endodermis Plant tissue consisting of a single layer of cells that surrounds and regulates the entrance of minerals, particularly into the vascular cylinder of roots. *129*

endometrium (en″do-me´tre-um) The lining of the uterus, which becomes thick and vascular during the uterine cycle. *377*

endoplasmic reticulum (ER) (en-do-plaz´mik rĕ-tik´u-lum) A membranous system of tubules, vesicles, and sacs in cells sometimes having attached ribosomes. Rough ER has ribosomes; smooth ER does not. *50*

endoskeleton Calcium carbonate supportive internal tissue of echinoderms and vertebrates. *600*

endosperm A nutrient material for the developing plant embryo. *573*

energy Capacity to do work and bring about change; occurs in a variety of forms. *5,* 74

energy source Way by which energy from the environment can be made available to organisms. *505*

enterocoelomate (ent´-ə-rō-sēl-o-māt) An animal in which the coelom forms as an outpocketing of the primitive gut. *592*

environmental resistance Sum total of factors in the environment that limit the numerical increase of a population in a particular region. *668*

enzyme An organic catalyst that speeds up a specific reaction or a specific type of reaction in cells. *28, 76*

epicotyl (ep″-ĭ-kot´il) The plant embryo portion above the cotyledons; contributes to stem development. *149*

epidermis (ep″ĭ-der´mis) The outer layer of cells of plants and other organisms; in humans, the outer layer of skin, composed of stratified squamous epithelium. *127, 129, 162*

epididymis Coiled tubule next to the testes where sperm mature and may be stored for a short time. *372*

epiglottis A structure that covers the glottis during the process of swallowing. *175, 241*

epiphyte (ep´ĭ-fīt) Nonparasitic plant, such as the arboreal orchid and Spanish moss, that grows on the surface of other plants. *643*

epithelial tissue (ep″ĭ-the´le-al tish´u) A type of tissue that covers the external surface of the body and lines its cavities. *156*

erection Referring to a structure such as the penis when it is turgid and erect as opposed to flaccid and lacking turgidity. *373*

erythrocyte (ə-rith´ro-sīt) A red blood cell that contains hemoglobin and carries oxygen from the lungs to the tissues in vertebrates. *160, 209*

essential amino acid One of 9 different amino acids required in the human diet because the body cannot make them. *184*

estrogen Female sex hormone that, along with progesterone, maintains the primary sex organs and stimulates development of the female secondary sex characteristics. *378*

estuary (es´tu-a-re) An area where fresh water meets the sea; therefore, an area with salinity intermediate between fresh water and seawater. *646*

eubacteria Group containing most species of bacteria except for the archaebacteria. *545*

eukaryotic cell A cell that possesses a nucleus and the other membranous organelles characteristic of complex cells. *45*

eustachian tube Extension from the middle ear to the nasopharynx for equalization of air pressure on the eardrum. *336*

evolution Changes that occur in the members of a species with the passage of time, often resulting in increased adaptation of organisms to the environment. *7, 515*

evolutionary tree Diagram describing the evolutionary relationship of groups of organisms. *532*

excretion Removal of metabolic wastes from the body. *260*

exocytosis (eks″o-si-to´sis) A process in which an intracellular vesicle fuses with the cell membrane so that the vesicle's contents are released outside the cell. *70*

exophthalmic goiter (ek″sof-thal´mik goi´ter) An enlargement of the thyroid gland accompanied by an abnormal protrusion of the eyes. *352*

expiration Process of expelling air from the lungs; exhalation. *239*

exponential growth Growth, particularly of a population, in which the total number increases in the same manner as compound interest. *666*

external respiration Exchange between blood and alveoli of carbon dioxide and oxygen. *239*

extraembryonic membrane Membrane that is not a part of the embryo but is necessary to the continued existence and health of the embryo. *402*

F

facilitated diffusion Passive transfer of a substance into or out of a cell along a concentration gradient by a process that requires a carrier. *68*

FAD (flavin adenine dinucleotide) A coenzyme of oxidation; a dehydrogenase that participates in hydrogen (electron) transport within the mitochondria. *93*

fatty acid An organic molecule having a long chain of carbon atoms and ending in an acid group. *34*

fermentation Aerobic breakdown of carbohydrates that results in organic end products such as alcohol and lactic acid. *90, 95*

fibrin Insoluble fibrous protein formed from fibrinogen during blood clotting. *213*

fibrinogen Plasma protein that is converted into fibrin threads during blood clotting. *213*

fibroblast Cell that forms fibers in connective tissues. *156*

fibrocartilage Cartilage with a matrix of strong collagenous fibers. *160*

fibrous connective tissue Tissue composed mainly of closely packed collagenous fibers that is found in tendons and ligaments. *159*

filament A threadlike structure such as the thick (myosin) and thin (actin) filaments found in myofibrils of muscle fibers. Also, in flowering plants, the stalk that supports the anther within a stamen. *146, 550, 573*

filter feeder An animal that obtains its food, usually in small particles, by filtering it from water. *586*

fimbria (fim´bre-ah) Fingerlike extension from the oviduct near the ovary. *377*

fitness The ability of an organism to survive and reproduce in its local (immediate) environment. *526, 628*

fixed-action pattern (FAP) A sequence of reflexes that always occurs in the same order under the same environmental conditions. *625*

flagellum (flah-jel´um) Slender, long extension used for locomotion by some protozoans, bacteria, and sperm. *54*

flame cell Excretory organ of flatworms. *588*

flower The blossom of a plant; contains the reproductive organs of angiosperms. *572*

fluid-mosaic model Proteins form a mosaic pattern within a bilayer of lipid molecules having a fluid consistency. *61*

fluke A parasitic flatworm; member of class Trematoda. *589*

focusing Manner by which light rays are bent by the cornea and lens, creating an image on the retina. *329*

follicle A structure in the ovary that produces the egg and, in particular, the female sex hormone, estrogen. *375*

follicle-stimulating hormone *See* FSH.

food chain A succession of organisms in an ecosystem that are linked by an energy flow and the order of who eats whom. *654*

food web The complete set of food links between populations in a community. *654*

foreskin Skin covering the glans penis in uncircumcised males. *372*

formed element A constituent of blood that is either cellular (red blood cells and white blood cells) or at least cellular in origin (platelets). *209*

fossil Any remains of an organism that have been preserved in the earth's crust. *515*

fovea centralis (fo´ve-ah sen-tral´is) Region of the retina, consisting of densely packed cones, that is responsible for the greatest visual acuity. *328*

frond The large leaf of a fern plant, containing many leaflets. *567*

frontal lobe Area of the cerebrum responsible for voluntary movements and higher intellectual processes. *293*

fruit A mature ovary enclosing seed(s). *575*

fruiting body A spore-bearing structure found in certain types of fungi, such as mushrooms. *556*

FSH (follicle-stimulating hormone) A hormone secreted by the anterior pituitary gland that stimulates the development of an ovarian follicle in a female or the production of sperm in a male. *373*

fungus A eukaryote, usually composed of strands called hyphae, that is usually saprophytic; e.g., mushroom and mold. *555*

G

gallbladder A saclike organ associated with the liver that stores and concentrates bile. *180*

gamete (gam´ēt) One of 2 types of reproductive cells that join in fertilization to form a zygote; most often an egg or a sperm. *105*

gametophyte (gam´ē-to-fīt) The haploid generation that produces gametes in the life cycle of a plant. *105, 563*

ganglion (gang´gle-on) A collection of neuron cell bodies within the peripheral nervous system. *286*

gastric gland Gland within the stomach wall that secretes gastric juice. *176*

gastrodermis Layer of cells found lining the body cavity in cnidarians. *586*

gastrovascular cavity A central cavity, having only one opening, of a lower animal in which digestion takes place and where nutrients are distributed to the cells lining the cavity. *586*

gene cloning To use recombinant DNA technology for production of many copies of a gene. *487*

gene flow The movement of genes from one population to another via sexual reproduction between members of the populations. *522*

gene mutation An alteration in the code of a single gene with a subsequent change in its expression. *478*

gene pool The total of all the genes of all the individuals in a population. *520*

gene therapy The use of biotechnology to treat genetic disorders and illnesses. *497*

genetic drift Evolution by chance processes alone. *523*

genomic library A collection of engineered viruses that together carry all of the genes of the species. *488*

genotype (ge´nə-tīp) The genes of any individual for (a) particular trait(s). *424*

gerontology The study of aging. *414*

gill Organ of gas exchange found in fishes and other types of marine and freshwater animals. *594*

gland A cell or group of epithelial cells specialized to secrete a substance. *156*

glomerular filtrate (glo-mer´u-lar fil´trāt) The filtered portion of blood contained within Bowman's capsule. *267*

glomerulus (glo-mer´u-lus) A cluster; for example, the cluster of capillaries surrounded by Bowman's capsule in a nephron. *267*

glottis Slitlike opening to the larynx between the vocal cords. *174, 241*

glucagon Hormone secreted by the islets of Langerhans in the pancreas that causes glycogen to release glucose. *357*

glycerol An organic compound that serves as a building block for fat molecules. *34*

glycogen The storage polysaccharide found in animals that is composed of glucose molecules joined in a linear fashion but having numerous branches. *32*

glycolysis (gli-kol´i-sis) A metabolic pathway found in the cytoplasm that participates in aerobic cellular respiration and fermentation; converts glucose to 2 molecules of pyruvate. *90, 91*

Golgi apparatus (gol´je) An organelle consisting of concentrically folded saccules that functions in the packaging, storage, and distribution of cellular products. *50*

gonadotropic hormone A type of hormone that regulates the activity of the ovaries and testes; principally FSH and LH (ICSH). *351*

Graafian follicle (graf´e-an fol´ĭ-k´l) A mature follicle within the ovaries that contains a developing egg. *376*

granum (pl., grana) A stack of thylakoids within chloroplasts. *52, 84*

greenhouse effect Carbon dioxide buildup in the atmosphere as a result of fossil fuel combustion; retains and reradiates heat, effecting an abnormal rise in the earth's average temperature. *678*

growth An increase in the number of cells and/or the size of these cells. *394*

growth hormone (GH) Hormone released by the anterior lobe of the pituitary gland that promotes the growth of the organism; also termed somatotropin. *348*

growth rate The yearly percentage of increase or decrease in the size of a population. *667*

guard cell A bean-shaped epidermal cell; one found on each side of a leaf stoma; their activity controls stoma size. *135*

gymnosperm A seed plant with uncovered seeds (not enclosed in an ovary); the conifers, for example. *570*

H

habitat The natural abode of an animal or plant species. *651*

hammer The middle ear ossicle adhering to the tympanic membrane. *336*

haploid The N number of chromosomes; half the diploid number; the number characteristic of gametes that contain only one set of chromosomes. *105*

haplontic life cycle Life cycle typical of protists in which the adult is always haploid because meiosis occurs after zygote formation and before maturity is reached. *549*

hard palate Bony anterior portion of the roof of the mouth. *174*

hazardous waste Waste containing chemicals hazardous to life. *674*

HCG (human chorionic gonadotropic) hormone A gonadotropic hormone produced by the chorion that functions to maintain the uterine lining. *381, 405*

heart Muscular organ located in the thoracic cavity that is responsible for maintenance of blood circulation. *199*

helper T cell T lymphocyte that releases lymphokines and stimulates certain other immune cells to perform their respective functions. *227*

hemocoel (he´mo-sēl) Residual coelom found in arthropods that is filled with blood. *598*

hemoglobin A red iron-containing pigment in blood that combines with and transports oxygen. *210, 248*

hepatic portal vein Vein leading to the liver that is formed by merging blood vessels from the villi of the small intestine. *180*

herbaceous Nonwoody. *130*

herbivore An animal that feeds directly on plants. *652*

hermaphrodite (her´maf-ro-dīt) An animal having both male and female sex organs. *589*

hernia An opening and separation of the abdominal wall, through which part of an organ protrudes. *372*

heterogamete A different kind of sex cell; a large and nonmotile egg or a small and flagellated sperm. *551*

heterospore A nonidentical spore, such as a microspore and a megaspore, produced by the same plant. *570*

heterotroph An organism that takes in preformed foods. *507*

heterozygous Having 2 different alleles (as *Aa*) for a given trait. *424*

hexose A 6-carbon monosaccharide. *32*

histamine A substance produced by basophil-derived mast cells in connective tissue that causes capillaries to dilate and release immune and other substances. *223*

homeostasis The maintenance of the internal environment, such as temperature, blood pressure, and other body conditions, within narrow limits. *7*

hominid (hom´ĭ-nid) Member of a family of upright, bipedal primates that includes australopithecines and modern humans. *603, 611*

hominoid A member of a superfamily containing humans and the great apes. *603, 611*

Homo erectus The earliest nondisputed species of humans, named for their erect posture, which allowed them to have a bipedal gait. *615*

Homo habilis An extinct species that may include the earliest humans, having a small brain but making quality tools. *615*

homologous Similarly constructed; homologous chromosomes have the same shape and contain genes for the same traits; homologous structures in animals share a common ancestry. *519*

homologous chromosome (ho-mol´o-gus kro´mo-sōm) Similarly constructed; homologous chromosomes have the same shape and contain genes for the same traits. *112*

homozygous Having identical alleles (as *AA* or *aa*) for a given trait; pure breeding. *424*

hormone A chemical messenger produced in small amounts in one region of the body that is transported to another region. *344*

Huntington disease A fatal genetic disease marked by neurological disturbances and failure of brain regions. *435*

hyaline cartilage (hi´ah-lĭn kar´tĭ-lij) Cartilage composed of very fine collagen fibers and a matrix having a milk-glass appearance. *159*

hydrogen bond A weak attraction between a hydrogen atom carrying a partial positive charge and an atom of another molecule carrying a partial negative charge. *24*

hydroid A tubular-shaped polyp displayed by some cnidarians. *586*

hydrolysis The splitting of a bond within a larger molecule by the addition of the components of water. *28*

hydrolytic enzyme An enzyme that catalyzes a reaction in which the substrate is broken down with the addition of water. *182*

hypertonic solution One that has a higher concentration of solute and a lower concentration of water than the cell. *67*

hypha (hi´fah) One filament of a mycelium, which constitutes the body of a fungus. *555*

hypocotyl (hi˝po-kot´il) The plant embryo portion below the cotyledons; contributes to stem development. *149*

hypothalamus A region of the brain—the floor of the third ventricle—involved with homeostasis. *292*

hypothesis A statement that is capable of explaining present data and is used to predict the outcome of future experimentation. *12*

hypotonic solution One that has a lower concentration of solute and a higher concentration of water than the cell. *67*

I

implantation The attachment and penetration of the embryo into the lining of the uterus (endometrium). *381*

impotency Failure of the penis to achieve erection. *373*

imprinting The tendency of a newborn animal to become attached to and follow the first moving object it sees. *626*

inclusive fitness The fitness of closely related group members is part of the fitness of the individual. *632*

induced-fit model A description of enzyme-substrate binding that includes changes in enzyme shape that allow the enzyme to fit more closely around the substrate. *77*

induction A process by which one tissue gives off signals that control the development of another, as when the embryonic notochord induces the formation of the neural tube. *400*

inflammatory reaction A tissue response to injury that is characterized by dilation of blood vessels and accumulation of fluid in the affected region. *223*

innate Instinctive, inborn, and not having to be learned. *625*

inner ear The portion of the ear consisting of a vestibule, semicircular canals, and the cochlea where balance is maintained and sound is transmitted. *336*

innervate (in´er-vāt) To activate an organ, muscle, or gland by motor neuron stimulation. *279*

insertion The end of a muscle that is attached to a movable bone. *310*

inspiration The act of breathing in. *239*

insulin A hormone produced by the pancreas that regulates carbohydrate storage. *357*

interferon (in″ter-fēr´on) A protein formed by a cell infected with a virus that can increase the resistance of other cells to the virus. *224*

internal respiration Exchange between blood and tissue fluid of oxygen and carbon dioxide. *239*

interneuron A neuron found within the central nervous system that takes nerve impulses from one portion of the system to another. *279*

interphase The interval between successive cell divisions; during this time, the chromosomes are in an extended state and are active in directing protein synthesis. *106*

interstitial cell Hormone-secreting cell located between the seminiferous tubules of the testes. *372*

inversion A chromosome mutation that occurs when a fragment of a chromosome is turned around 180 degrees. *446*

invertebrate An animal that lacks a vertebral column. *585*

ion An atom or group of atoms carrying a positive or negative charge. *20*

ionic bond A bond created by an attraction between oppositely charged ions. *21*

ionic reaction A chemical reaction in which atoms acquire or lose electrons. *20*

iris A muscular ring that surrounds the pupil and regulates the passage of light through this opening. *326*

islet of Langerhans (lahng´ər-hanz) Distinctive group of cells within the pancreas that secretes insulin and glucagon. *357*

isotonic solution One that contains the same concentration of solute and water as the cell. *66*

isotope One of 2 or more atoms with the same atomic number that differs in the number of neutrons and therefore in weight. *18*

K

karyotype (kar´e-o-tīp) The arrangement of all the chromosomes within a cell by pairs in a fixed order. *104*

kidney An organ in the urinary system that produces and excretes urine. *264*

Klinefelter syndrome A condition caused by the inheritance of a chromosome abnormality in number; an XXY individual. *450*

Krebs cycle A cyclical metabolic pathway found in the matrix of mitochondria that participates in aerobic cellular respiration; breaks down acetyl groups to carbon dioxide. Also called the citric acid cycle because the reactions begin and end with citric acid (citrate). *90*

L

labium (la´be-um) A fleshy border or liplike fold of skin, as in the labia majora and labia minora of the female genitals. *378*

lacteal (lak´te-al) A lymphatic vessel in a villus of the intestinal wall of mammals. *177*

lacuna (lah-ku´nah) A small pit or hollow cavity, as in bone or cartilage, where a cell or cells are located. *159*

ladder-type nervous organ Planarian nervous system consisting of a small brain and 2 lateral nerve cords joined by cross-bridges. *589*

lancelet A type of protochordate that has the 3 chordate characteristics as an adult; formerly called amphioxus. *602*

lanugo (lah-nu´go) Downy hair on the body of a fetus; fetal hair. *410*

large intestine The last major portion of the digestive tract, extending from the small intestine to the anus and consisting of the cecum, the colon, and the rectum. *178*

larynx (lar´ingks) Cartilaginous organ located between the pharynx and the trachea that contains the vocal cords; voice box. *241*

leaf vein The structure that contains vascular tissue in a leaf. *134, 241*

learning A change in behavior as a result of experience. *626*

lens A clear membranelike structure found in the eye behind the iris; brings objects into focus. *326*

lenticel A pocket of loosely arranged cells in cork that permits gas exchange. *132*

leukocyte White blood cell of which there are several types, each having a specific function in protecting the body from invasion by foreign substances and organisms. *160, 211*

lichen (li´ken) A fungus and an alga coexisting in a symbiotic relationship. *556*

ligament Dense fibrous connective tissue that joins bone to bone at a joint. *159, 307*

light-dependent reaction The first stage of photosynthesis, in which solar energy is stored temporarily as ATP and often $NADPH_2$. *85*

light-independent reaction The second stage of photosynthesis, in which energy produced by a light-dependent reaction converts CO_2 to a carbohydrate. *85*

limbic system A portion of the brain concerned with memory and emotions. *294*

linkage group Alleles on the same chromosome are linked in the sense that they tend to move together to the same gamete; crossing-over interferes with linkage. *454*

lipase A fat-digesting enzyme secreted by the pancreas. *182*

lipid One of a class of organic compounds that are insoluble in water; notably fats, oils, and steroids. *34*

liposome Lipid bilayer sphere that forms when phospholipids are placed in a liquid environment. *506*

loop of Henle U-shaped portion of the nephron. *265*

loose connective tissue Tissue composed mainly of fibroblasts that are widely separated by a matrix containing collagen and elastin fibers; found beneath epithelium. *156*

lumen The cavity inside any tubular structure, such as the lumen of the digestive tract. *175*

luteinizing hormone (LH) Hormone produced by the anterior pituitary gland that stimulates the development of the corpus luteum in females and the production of testosterone in males. *373*

lymph Fluid having the same composition as tissue fluid and carried in lymphatic vessels. *221*

lymphatic system A one-way vascular system that takes up excess fluid in the tissues and transports it to cardiovascular veins in the shoulders. *221*

lymphokine (lim´fo-kīn) Molecule secreted by T lymphocytes that has the ability to affect the activity of all types of immune cells. *231*

lysosome (li´so-sōm) An organelle in which digestion takes place due to the action of hydrolytic enzymes. *51*

M

macroevolution Evolution of taxa higher than the species level and commonly involving major morphological changes. *531*

macrophage A large phagocytic cell derived from a monocyte that ingests microbes and debris. *223*

Malpighian tubule An organ of excretion, notably in insects. *599*

maltose A disaccharide composed of 2 glucose units. *32, 183*

mantle Fleshy fold that envelops the visceral mass of mollusks. *592*

matrix The secreted basic material or medium of biological structures, such as the matrix of cartilage or bone. *156*

medulla oblongata The lowest portion of the brain, which is concerned with the control of internal organs. *292*

medusa A bell-shaped, free-swimming stage resembling a jellyfish that is capable of sexual reproduction in the life cycle of some sessile cnidarians. *586*

megaspore In seed plants, a spore that develops into the female gametophyte. *146, 570*

meiosis (mi-o´sis) A type of cell division occuring during the production of gametes in animals by means of which the 4 daughter cells have the haploid number of chromosomes. *112*

memory B cell One of a persistent population of B cells ready to produce antibodies specific to a particular antigen; accounts for the development of active immunity. *226*

memory T cell A T cell that is ready to recognize an antigen that previously invaded the body. *227*

meninges (mə-nin-jēz) Protective membranous coverings about the central nervous system. *291*

menopause Termination of the ovarian and uterine cycles in older women. *382*

menstruation Loss of blood and tissue from the uterus at the end of a uterine cycle. *380*

meristem Plant tissue that always remains undifferentiated and capable of dividing to produce new cells. *127*

mesoderm The middle germ layer of embryonic gastrula; gives rise to the muscles, the connective tissue, and the circulatory system. *395*

mesoglea (mes″o-gle´ah) A jellylike packing material between the ectoderm and the endoderm in cnidarians. *586*

mesophyll The middle portion of a leaf made up of parenchyma cells, which carries on photosynthesis and gas exchange. *135*

messenger RNA (mRNA) A nucleic acid (ribonucleic acid) complementary to genetic DNA; has codons, which direct cell protein synthesis at the ribosomes. *468*

metabolism All of the chemical reactions within a cell (or an organism), including breakdown reactions (catabolism) and synthetic reactions (anabolism). *7, 75*

metabolic pool Substrates in a cell that are used for biosynthesis. *97*

metafemale A female who has 3 X chromosomes. *450*

metal Class of elements that, in reactions, characteristically lose electrons and become positively charged ions. *21*

metamorphosis Change in form, as when a tadpole becomes an adult frog or when an insect larva develops into the adult. *600*

metaphase Stage in mitosis during which chromosomes are at the equator of the mitotic spindle. *110*

MHC (major histocompatibility complex) protein A membrane protein that serves to identify the cells of a particular individual. *228*

microbe Microscopic infectious agent, such as a bacterium or a virus. *208*

microsphere Structure composed only of protein that looks like a cell and carries on many cellular functions; a possible early step in cell evolution. *506*

microspore In seed plants, a spore that develops into a pollen grain maturing into the male gametophyte. *146, 570*

microtubule An organelle composed of 13 rows of globular proteins; found in multiple units within other organelles, such as the centriole, cilia, and flagella, as well as spindle fibers. *53*

middle ear A portion of the ear consisting of the tympanic membrane, the oval and round windows, and the ossicles; where sound is amplified. *335*

mineral An inorganic, homogeneous substance. *188*

mitochondrion (mi´to-kon´dre-on) A membranous organelle in which cellular respiration produces the energy molecule, ATP. *52*

mitosis (mi-to´sis) Type of cell division in which daughter cells receive the exact chromosome and genetic makeup of the parent cell; occurs during growth and repair. *107*

mixed nerve A nerve containing both sensory nerve fibers and motor nerve fibers that conducts impulses to and from the central nervous system. *286 – 287*

molecule A chemical consisting of 2 or more atoms bonded together; smallest unit of a compound that has the properties of the compound. *19*

molt To shed all or part of an outer covering; in arthropods, periodic shedding of parts of the exoskeleton allows an increase in size. *598*

monoclonal antibody An antibody of the same type produced by a hybridoma—a lymphocyte that has fused with a cancer cell. *231*

monocot (mon´o-kot) Monocotyledon; a type of flowering plant in which the seed has only one cotyledon, such as corn and lily. *126*

monohybrid An individual that is heterozygous for one trait; shows the phenotype of the dominant allele but carries the recessive allele. *425*

monosaccharide A simple sugar; a carbohydrate that cannot be decomposed by hydrolysis. *32*

morphogenesis (mor″fo-jen´ĭ-sis) The movement of cells and tissues to establish the shape and the structure of an organism. *394*

motor nerve A nerve that conducts an impulse from the central nervous system to a muscle or gland. *286*

motor neuron A neuron that takes nerve impulses from the central nervous system to an effector. *279*

muscle action potential An electrochemical change due to increased sarcolemma permeability that is propagated down the T system and results in muscle contraction. *317*

muscular (contractile) tissue A type of tissue that contains cells capable of contracting; skeletal muscles are attached to the skeleton, smooth muscle is found within the walls of internal organs, and cardiac muscle makes up the heart. *160*

mutation A change in the genetic material. *465*

mycelium (mi-se´le-um) A mass of hyphae that makes up the body of a fungus. *555*

myelin sheath (mi´ĕ-lin shēth) The Schwann cell membranes that cover long neuron fibers and give them a white, glistening appearance. *279*

myocardium Heart muscle. *199*

myofibril The contractile portion of a muscle fiber. *314*

myosin One of 2 major proteins of muscle; makes up thick filaments in myofibrils and is capable of breaking down ATP. *See* actin. *316*

myxedema (mik″să-de-mah) A condition resulting from a deficiency of thyroid hormone in an adult. *352*

N

NAD A coenzyme of oxidation; accepts hydrogen atoms (H) from a substrate and carries them to another acceptor. *79*

NADP A coenzyme of reduction; $NADPH_2$ donates hydrogen atoms (H) to substrate. *80*

natural selection The process by which populations become adapted to their environment. *526*

Neanderthal The common name for an extinct subspecies of humans whose remains are found in Europe, Asia, and Africa. *616*

negative feedback A self-regulatory mechanism that is activated by an imbalance and results in a fluctuation above and below a mean. *167*

nematocyst (nem´ah-to-sist) A threadlike structure in stinging cells of cnidarians that can be expelled to numb and to capture prey. *586*

nephridium (nə-frid-e-um) Excretory tubule found in invertebrates; notably, in segmented worms. *596*

nephron (nef´ron) The anatomical and functional unit of the vertebrate kidney; kidney tubule. *264*

nerve A bundle of long nerve fibers that run to and/or from the central nervous system. *162*

nerve impulse An electrochemical change due to increased membrane permeability that is propagated along a neuron from the dendrite to the axon following excitation. *281*

nerve net Neuron organization in which neurons are directly linked to receptors, muscles, and to each other; permits diffuse response to stimuli. *586*

neurilemma (nūr″ə-lem´ah) The outermost wrapping of a nerve fiber; promotes regeneration. *279*

neurofibromatosis (nūr″o-fi-bro-mah-to´sis) A genetic disease marked by development of neurofibromas under skin and muscles. *435*

neuroglial cell A nervous system cell that supports and protects neurons. *162, 279*

neuromuscular junction The point of contact between a nerve cell and a muscle fiber. *317*

neuron (nu´ron) Nerve cell that characteristically has 3 parts: dendrite, cell body, axon. *161, 279*

neurotransmitter substance A chemical found at the ends of axons that is responsible for transmission across a synapse. *284*

neutron A subatomic particle that has a weight of one atomic mass unit, carries no charge, and is found in the nucleus of an atom. *18*

niche (nich) Total description of an organism's functional role in an ecosystem, from activities to reproduction. *651*

nitrogen fixation A process whereby free atmospheric nitrogen is converted into compounds, such as ammonia and nitrates, usually by soil bacteria. *660*

node of Ranvier Gap in the myelin sheath around a nerve fiber. *284*

nondisjunction The failure of homologous chromosomes or sister chromatids to separate during the formation of gametes. *444*

nonrandom mating Mating among individuals on the basis of their phenotypic or genotypic similarities or differences rather than on a completely random basis. *523*

norepinephrine (NE) (nor″ep´ĭ-nef´ron) Excitatory neurotransmitter active in the peripheral and central nervous systems. *285, 354*

notochord (no´to-kord) Dorsal supporting rod that exists in all chordates sometime in their life history; replaced by the vertebral column in vertebrates. *601*

nuclear envelope The double membrane that surrounds the nucleus and is continuous with the endoplasmic reticulum. *46*

nucleic acid A large organic molecule made up of nucleotides joined together; for example, DNA and RNA. *36*

nucleolus (nu-kle´o-lus) An organelle found inside the nucleus; a special region of chromatin that produces rRNA for ribosome formation. *46*

nucleotide A monomer of a nucleic acid that forms when a nitrogen-containing organic base, a pentose sugar, and a phosphate join. *37*

nucleus A large organelle containing the chromosomes and acting as a control center for the cell. Also, the center of an atom. *18, 46*

nutrient A portion of food the body can use as a source of energy or building material. *184*

O

obesity Condition in which body weight is more than 20% of the ideal weight. *193*

occipital lobe (ok-sip´ĭ-tal lōb) Area of the cerebrum responsible for vision, visual images, and other sensory experiences. *293*

omnivore An animal that feeds on both plants and animals. *652*

oncogene (ong´-ko-jen) A gene that contributes to the transformation of a normal cell into a cancer cell. *481*

oogenesis (o″o-jen´ĕ-sis) Production of an egg in females by the process of meiosis and maturation. *116*

operator The sequence of DNA in an operon to which the repressor protein binds. *476*

operon A group of structural and regulating genes that function as a single unit. *476*

optic nerve A nerve that carries nerve impulses from the retina of the eye to the brain. *328*

orangutan One of the great apes; large with long red hair. *611*

organ A structure composed of 2 or more tissues functioning as a unit. *162*

organelle Specialized structures within cells (e.g., nucleus, mitochondria, and endoplasmic reticulum). *45*

organic soup An expression used to refer to the oceans before the origin of life when they contained newly formed organic compounds. *507*

organ of Corti A portion of the inner ear that contains the receptors for hearing. *336*

organ system A group of related organs working together. *162*

orgasm Physical and emotional climax during sexual intercourse; results in ejaculation in the male. *373*

origin End of a muscle that is attached to a relatively immovable bone. *310*

osculum (os´ku-lum) Opening to the exterior of the central cavity in the sponge. *585*

osmosis (oz-mo´sis) The movement of water from an area of higher concentration of water to an area of lower concentration of water across a differentially permeable membrane. *66*

osmotic pressure Pressure generated by and due to the osmotic flow of water; created by the solute in a solution. *66*

ossicle One of the small bones of the middle ear— hammer, anvil, stirrup. *336*

osteocyte A mature bone cell. *309*

otolith (o-to-lith) Calcium carbonate granule associated with ciliated cells in the utricle and the saccule. *336*

outer ear Portion of the ear consisting of the pinna and the auditory canal. *335*

oval opening An opening between the 2 atria in the fetal heart; also called the foramen ovale. *410*

oval window Opening between the stapes and the inner ear. *336*

ovarian cycle Monthly occurring changes in the ovary that determine the level of sex hormones in blood. *378*

ovary The female gonad, the organ that produces eggs, estrogen, and progesterone. Also, the base of the pistil in angiosperms. *146, 375, 573*

ovulation The discharge of a mature egg from the follicle within the ovary. *376*

ovule (o´vūl) In seed plants, a structure that contains megasporangium, where meiosis occurs and the female gametophyte is produced; develops into the seed. *146, 571*

oxidation The loss of electrons (usually inorganic); also, the removal of hydrogen atoms (H) (usually organic). *23, 79*

oxidizing atmosphere An atmosphere that contains oxidizing molecules, such as O_2, rather than reducing molecules, such as H_2. *505*

oxygen debt Oxygen that is needed to metabolize lactate, a compound that accumulates during vigorous exercise. *316*

oxytocin A hormone released by the posterior pituitary that causes contraction of uterus and milk letdown. *348*

ozone shield A layer of ozone (O_3) present in the upper atmosphere that protects the earth from damaging ultraviolet light. Nearer the earth, ozone is a pollutant. *509, 679*

P

pacemaker *See* SA node.

pancreas An elongate, flattened organ in the abdominal cavity that secretes enzymes into the small intestine (exocrine function) and hormones controlling blood sugar (endocrine function). *179, 357*

Pap smear An analysis done on cervical cells for detection of cancer. *377*

parapodium Footlike fleshy lobe found on the segments of marine annelids. *595*

parasite An organism that resides on or within another organism and does harm to this organism. *653*

parasitism Symbiotic relationship in which an organism derives nourishment from and does harm to a host. *653*

parasympathetic nervous system That part of the autonomic nervous system that usually promotes activities associated with a normal state. *289*

parathyroid hormone (PTH) A hormone secreted by the parathyroid glands that affects the level of calcium and phosphate in the blood. *353*

parenchyma (pah-reng´kĭ-mah) Relatively unspecialized cells that make up the fundamental tissue of plants. *127*

parietal lobe Area of the cerebrum responsible for sensations involving temperature, touch, pressure, and pain, as well as speech. *293*

parturition The processes that lead to and include the birth of a mammal, and the expulsion of the extraembryonic membranes through the terminal portion of the female reproductive tract. *412*

pelvic inflammatory disease (PID) A disease state of the reproductive organs caused by a sexually transmitted organism. *389*

pelvis A bony ring formed by the coxal bones. Also, a hollow chamber in the kidney that lies inside the medulla and receives freshly prepared urine from the collecting ducts. *306*

penis External organ in males through which the urethra passes and which serves as the organ of sexual intercourse. *372*

pentose A 5-carbon sugar; deoxyribose is the pentose sugar found in DNA; ribose is a pentose sugar found in RNA. *32*

pepsin A protein-digesting enzyme secreted by the gastric glands. *182*

peptidase An intestinal enzyme that breaks down short chains of amino acids to individual amino acids that are absorbed across the intestinal wall. *182*

peptide bond The covalent bond that joins 2 amino acids. *29*

pericycle (per″ĭ-si-kl) A single layer of tissue interior to the endodermis that produces secondary roots. *129*

periodontitis (per″e-o-don-ti´tis) Inflammation of the gums. *174*

peripheral nervous system (PNS) Nerves and ganglia that lie outside the central nervous system. *279*

peristalsis A rhythmic contraction that serves to move the contents along in tubular organs, such as the digestive tract. *175*

peritubular capillary Capillary that surrounds a nephron and functions in reabsorption during urine formation. *267*

permafrost Earth that remains permanently frozen beneath the surface in the tundra. *640*

petal The often colored leaf of a flower. *146, 572*

petiole A structure connecting a leaf to a stem. *134*

PG *See* prostaglandins.

PGAL (phosphoglyceraldehyde) A 3-carbon phosphorylated carbohydrate; an important molecule in both photosynthesis and glycolysis. *89*

pH A measure of the hydrogen ion concentration; any pH below 7 is acidic and any pH above 7 is basic. *25*

phagocytosis (fag″o-si-to´sis) The taking in of bacteria and/or debris by engulfing; cell eating. *69*

pharynx (far´ingks) A common passageway (throat) for both food intake and air movement. *174, 241*

phenotype (fe´no-tīp) The outward appearance of an organism caused by the genotype and environmental influences. *425*

phenylketonuria (PKU) A genetic disease stemming from the lack of an enzyme to metabolize the amino acid phenylalanine. *434*

pheromone (fer´o-mōn) A chemical substance secreted by one organism that influences the behavior of another. *362*

phloem (flo´em) The vascular tissue in plants that transports organic nutrients. *127*

photochemical smog Air pollution that contains nitrogen oxides (NO_2) and hydrocarbons, which react to produce ozone and peroxylacetyl nitrate (PAN). *677*

photoperiodism A response to light and dark; particularly in reference to flowering in plants. *143*

photosynthesis The process by which plants make their own food using the energy of the sun. *53*

photosystem A cluster of light-absorbing pigment molecules within thylakoid membranes. *85*

photosystems I and II Molecular units located within the membrane of a thylakoid that capture solar energy, making photophosphorylation possible. *85*

pH scale A measure of the hydrogen ion concentration [H^+]; any pH below 7 is acidic and any pH above 7 is basic. *25*

phytochrome (fi´to-krōm) A plant pigment that induces a photoperiodic response in plants. *144*

pineal gland A gland either at the skin surface (fishes, amphibians) or in the third ventricle of the brain, where it produces melatonin. *359*

pinna Outer, funnel-like structure of the ear that picks up sound waves. *335*

pinocytosis (pin″o-si-to′sis) The taking in of fluid along with dissolved solutes by engulfing; cell drinking. *70*

pistil Part of the flower that contains a stigma, a style, and an ovary. *146, 572*

pith Central tissue composed of parenchyma cells that occurs in dicot stems. *130*

pituitary gland A small gland lying just below the hypothalamus that is important for its hormone storage and production activities. *346*

placenta Structure in the uterine wall through which the embryo (later the fetus) is nourished. *381*

placental mammal A mammal having internal fetal development supported by the presence of a placenta. *608*

plankton Floating microscopic organisms found in most bodies of water. *646*

plasma The liquid portion of blood, consisting of all components except the formed elements. *209*

plasma cell A cell derived from a B-cell lymphocyte that is specialized to mass-produce antibodies. *225*

plasmid A circular DNA segment that is present in bacterial cells but is not part of the bacterial chromosome. *487*

plasmodium (plaz-mōd′ē-um) Multinucleated acellular mass in slime molds. *554*

plasmolysis (plas-mol′-ĭ-sis) Contraction of the cell contents due to the loss of water. *67*

platelet A component of blood that is necessary to blood clotting. *213*

pleural membrane A serous membrane that encloses the lungs. *244*

polar body Nonfunctioning daughter cell, formed during oogenesis, that has little cytoplasm. *116*

polar molecule A molecule that has partially positive and negative portions; the charges are caused by the ability of one portion to attract electrons more strongly than another portion. *24*

pollen grain Mature male gametophyte of seed plants that contains sperm when mature. *147, 571*

pollen sac Structure in which microspores are produced and develop into pollen grains; also microsporangium. *146, 571*

pollination The delivery of pollen by wind or animals to the stigma of a pistil in flowering plants. *147, 572*

pollution Detrimental alteration of the normal constituents of air, land, and water due to human activities. *661*

polygenic inheritance A pattern of inheritance in which many genes control a trait; each gene contributes in an additive and like manner. *436*

polyp The sedentary stage in the life cycle of cnidarians. Also, a benign growth. *586*

polypeptide A molecule composed of many amino acids linked together by peptide bonds. *29*

polysaccharide A macromolecule composed of many units of sugar. *32*

polysome A cluster of ribosomes attached to the same mRNA molecule; each ribosome is producing a copy of the same polypeptide. *473*

population All the organisms of the same species in one place. *520*

posterior pituitary Back lobe of the pituitary gland that stores and secretes ADH and oxytocin produced by the hypothalamus. *346*

postganglionic axon Axon that is located after an autonomic ganglion. *289*

postmating isolating mechanism An anatomical or physiological difference between 2 species that prevents successful reproduction after mating has taken place. *529*

postsynaptic membrane In a synapse, the membrane of the neuron opposite the presynaptic membrane. *284*

prairie A grassland biome of the temperate zone that occurs when rainfall is greater than 25 cm but less than 40 cm. *640*

preganglionic axon Axon that is located before an autonomic ganglion. *289*

premating isolating mechanism An anatomical or behavioral difference between 2 species that prevents the possibility of mating. *529*

pressure filtration The movement of small molecules from the glomerulus into Bowman's capsule due to the action of blood pressure. *267*

pressure-flow theory Explanation for phloem transport; osmotic pressure following active transport of sugar into phloem brings about a flow of sap from a source to a sink. *142*

presynaptic membrane In a synapse, the membrane of the neuron opposite the postsynaptic membrane. *284*

primate Animal that belongs to the order Primates, the order of mammals that includes prosimians, monkeys, apes, and humans. *611*

producer Organism that produces food and is capable of synthesizing organic compounds from inorganic constituents of the environment; usually the green plants and the algae in an ecosystem. *652*

progesterone Female sex hormone secreted by the corpus luteum of the ovary and by the placenta. *378*

prokaryotic cell A cell lacking a nucleus and the membranous organelles found in complex cells; bacteria, including cyanobacteria. *56*

prolactin (PRL) *See* lactogenic hormone.

promoter A sequence of DNA in an operon where DNA polymerase begins transcription. *476*

prophase Early stage in mitosis during which chromatin condenses so that chromosomes appear. *107*

proprioception (pro″pre-o-sep′shun) The sense of knowing the position of the limbs. *323*

prosimian (pro-sim′e-an) Primitive primate, such as the lemur, tarsier, and tree shrew. *611*

prostaglandin (PG) Hormone that has various and powerful local effects. *362*

prostate gland Gland located around the male urethra below the urinary bladder; adds secretions to seminal fluid. *372*

protein One of a class of organic compounds that is composed of either one or several polypeptides. *28*

prothallus (pro-thal′us) A small, heart-shaped structure that is the gametophyte of the fern. *567*

prothrombin Plasma protein that is converted to thrombin during the process of blood clotting. *213*

protocell The structure that preceded the true cell in the history of life. *507*

proton A subatomic particle found in the nucleus of an atom that has a weight of one atomic mass unit and carries a positive charge; a hydrogen ion. *18*

protoplast A plant cell from which the cell wall has been removed. *496*

protostome Member of a large group of animal phyla in which the mouth develops from the blastopore. *592*

protozoan Animal-like protist that is classified according to means of locomotion: amoeba, flagellate, or ciliate. *546*

proximal convoluted tubule Highly coiled region of a nephron near Bowman's capsule. *265*

pseudocoelom (su″do-se′lom) A coelom incompletely lined by mesoderm. *589*

puberty Developmental stage after birth during which the reproductive organs become functional. *414*

pulmonary circuit That part of the circulatory system that takes deoxygenated blood to and oxygenated blood away from the gas-exchanging surfaces in the lungs. *206*

pulse Vibration felt in arterial walls due to expansion of the aorta following contraction of the ventricle. *201*

Punnett square Gridlike device used to calculate the expected results of simple genetic crosses. *425*

pupil An opening in the center of the iris of the eye. *326*

pure *See* homozygous.

purine (pu´rēn) Nitrogen-containing, organic base found in DNA and RNA that has 2 interlocking rings, as in adenine and guanine. *464*

pus Thick, yellowish fluid composed of dead phagocytes, dead tissue, and bacteria. *223*

pyrimidine (pi-rim´ə-dēn) Nitrogen-containing, organic base found in DNA and RNA that has just one ring, as in cytosine, uracil, and thymine. *464*

pyruvate The end product of glycolysis; pyruvic acid. *91*

R

radial symmetry Regardless of the angle of a vertical cut made at the midline of an organism, 2 equal halves result. *583*

radicle The embryonic root of a plant. *149*

radioactive isotope An isotope that spontaneously emits radiation, allowing it to be detected in vivo and in vitro. *19*

receptor A structure specialized to receive information from the environment and to generate nerve impulses. Also, a structure found in the membrane of cells that combines with a specific chemical in a lock and key manner. *279*

recessive allele Hereditary factor that expresses itself in the phenotype only when the genotype is homozygous. *424*

recombinant DNA DNA having genes from 2 different organisms, often produced in the laboratory by introducing foreign genes into a bacterial plasmid. *487*

red blood cell *See* erythrocyte.

red bone marrow Tissue located in the cavity of bones that forms blood cells. *308*

reduced hemoglobin Hemoglobin that is not carrying oxygen. *253*

reducing atmosphere An atmosphere that contains reducing molecules, such as H_2, rather than oxidizing molecules, such as O_2. *505*

reduction The gain of electrons (inorganic); the addition of hydrogen atoms (H) (organic). *23, 80*

reflex An involuntary response to the stimulation of a receptor. *288*

regeneration Regrowth of tissue; formation of a complete organism from a small portion. *586*

regulator gene Gene that codes for a protein involved in regulating the activity of structural genes. *476*

REM sleep A stage in sleep that is characterized by eye movements and dreaming. *294*

renal cortex The outer portion of the kidney. *264*

renal medulla The inner portion of the kidney including the renal pyramids (loops of Henle and collecting ducts). *264*

renal pelvis A hollow chamber in the kidney that lies inside the renal medulla and receives freshly prepared urine from the collecting ducts. *264*

replacement reproduction A population in which each person is replaced by only one child. *670*

replication The duplication of DNA; occurs when the cell is not dividing. *467*

reproduce To make a copy similar to oneself; for example, bacteria dividing to produce more bacteria, or egg and sperm joining to produce offspring in more advanced organisms. *7*

residual volume The amount of air remaining in the lungs after a forceful expiration. *247*

resting potential The voltage recorded from inside a neuron when it is not conducting nerve impulses. *282*

restriction enzyme Enzyme that stops viral reproduction by cutting viral DNA; used in bioengineering to cut DNA at specific points. *488*

retina (ret´ĭ-nah) The innermost layer of the eyeball that contains the rods and the cones. *328*

retrovirus Virus that contains only RNA and carries out RNA $\rightarrow$ DNA transcription, called reverse transcription. *540*

rhodopsin (ro-dop´sin) Visual purple, a pigment found in the rods of a type of receptor in the retina of the eye. *331*

rib Bone hinged to the vertebral column and sternum that, with muscle, defines the top and sides of the thoracic cavity. *244, 306*

rib cage The top and sides of the thoracic cavity; contains ribs and intercostal muscles. *244*

ribonucleic acid *See* RNA.

ribosomal RNA (rRNA) RNA occurring in ribosomes, structures involved in protein synthesis. *471*

ribosome (ri´bō-sōm) Minute particle found attached to the endoplasmic reticulum or loose in the cytoplasm that is the site of protein synthesis. *50*

ribozyme Enzyme that carries out mRNA processing. *470*

ribulose bisphosphate (RuBP) (ri-bu-lōs bis-fos´fāt) The 5-carbon molecule that unites with CO_2 in the Calvin cycle. *88*

RNA (ribonucleic acid) A nucleic acid found in cells that assists DNA in controlling protein synthesis. *36, 465*

rod Dim-light receptor in the retina of the eye that detects motion but not color. *328*

root cap Thimble-shaped mass of parenchyma cells that protects the apical meristem of a root. *129*

rough endoplasmic reticulum (RER) Endoplasmic reticulum having attached ribosomes. *50*

round window A membrane-covered opening between the inner ear and the middle ear. *336*

RuBP *See* ribulose bisphosphate.

S

saccule (sak´ūl) A saclike cavity that makes up part of the membranous labyrinth of the inner ear; contains receptors for static equilibrium. *336*

sac plan Body plan possessed by animals having a single opening. *583*

salivary amylase An enzyme in the saliva that initiates the digestion of starch. *182*

salivary gland A gland associated with the mouth that secretes saliva. *174*

SA (sinoatrial) node Small region of neuromuscular tissue that initiates the heartbeat; also called the pacemaker. *203*

saprophyte (sap´ro-fīt) A heterotroph such as a bacterium or a fungus that externally digests dead organic matter before absorbing the products. *544*

sarcolemma The membrane that surrounds striated muscle cells. *314*

sarcomere Structural and functional unit of a myofibril; contains actin and myosin filaments. *314*

sarcoplasm The cytoplasm within a muscle fiber. *314*

sarcoplasmic reticulum Membranous network of channels and tubules within a muscle fiber; corresponds to the endoplasmic reticulum of other cells. *314*

savanna A grassland biome that has occasional trees and is commonly associated with Africa. *640*

schizocoelomate (skiz´o-se-lə-māt) An animal in which the coelom forms by the splitting of the mesoderm. *592*

scientific method Process by which scientists test their conclusions; consists of hypothesis generation and observation and experimentation, and results in theories. *12*

sclera (skle´rah) White, fibrous outer layer of the eyeball. *326*

sclerenchyma (skle-reng´ki-mah) A support tissue in plants made of hollow cells with thickened walls. *127*

scolex The head region of a tapeworm. *589*

scrotum A pouch of skin that encloses the testes. *370*

secretion A cell product; the act of releasing a cell product from the cell. *70*

seed A mature ovule that contains an embryo with stored food enclosed by a protective coat. *570*

selective reabsorption Movement of nutrient molecules, as opposed to waste molecules, from the contents of the nephron into blood at the proximal convoluted tubule. *269*

semicircular canal Tubular structure within the inner ear that contains the receptors responsible for the sense of dynamic equilibrium. *336*

seminal fluid The sperm-containing secretion of males; also called semen. *372*

seminal vesicle A convoluted, saclike structure attached to the vas deferens near the base of the urinary bladder in males. *372*

seminiferous tubule (sem″ĭ-nif´er-us tu-būl) Highly coiled duct within the male testis that produces and transports sperm. *371*

sensory nerve A nerve that conducts an impulse from a receptor to the central nervous system. *286*

sensory neuron A neuron that takes nerve impulses to the central nervous system: typically has a long dendrite and a short axon; afferent neuron. *279*

sepal The leafy division of the calyx found in a whorl at the base of petals. *146, 572*

septum A partition or wall, such as the septum in the heart, that divides the right half from the left half. *200*

serum Light yellow liquid left after clotting of blood. *213*

sessile Organisms that lack locomotion and remain stationary in one place, such as plants or sponges. *586*

seta (se´tah) Bristle, especially one of those found on segmented worms. *595*

sex chromosome Chromosome responsible for the development of characteristics associated with gender; an X or Y chromosome. *104*

sex-linked gene Gene located on sex chromosomes. *450*

sickle-cell disease A genetic disorder due to the homozygous genotype of the sickle-cell gene, producing sickle-shaped cells and loss of oxygen-carrying power in blood. *438*

sieve plate A large pore through which water enters an echinoderm; also termed madreporite. *142, 600*

sieve-tube cell A phloem cell that functions in transport of organic nutrients. During development, sieve-tube cells align vertically and form a continuous pathway for transport. *142*

sign stimulus Stimulus that releases a fixed-action pattern. *625*

simple eye A sensory organ in which a single lens covers light-sensitive cells. *598*

simple goiter Condition in which an enlarged thyroid produces low levels of thyroxin. *352*

sinoatrial node *See* SA node.

skeletal muscle The contractile tissue that makes up the muscles attached to the skeleton; also called striated muscle. *161*

slash-and-burn agriculture The cutting down and burning of trees to provide space to raise crops. *672*

sliding filament theory The movement of actin in relation to myosin; accounts for muscle contraction. *316*

slime mold Funguslike protist; cellular and acellular types exist. *554*

small intestine In the digestive tract, the long, tubelike chamber between the stomach and the large intestine. *177*

smooth endoplasmic reticulum (SER) Endoplasmic reticulum that does not have attached ribosomes. *50*

smooth muscle The contractile tissue that makes up the muscles found in the walls of internal organs. *161*

sociobiology An analysis of behavior, particularly social behavior, according to the tenets of evolutionary theory. *629*

soft palate Entirely muscular posterior portion of the roof of the mouth. *174*

solute A substance dissolved in a solvent to form a solution. *65*

solvent A fluid, such as water, that dissolves solutes. *65*

somatic cell In animals, any cell other than those that undergo meiosis and become a sperm or egg; body cell. *105*

somatic nervous system That part of the peripheral nervous system containing motor neurons that control skeletal muscles. *288*

sorus (pl., sori) A cluster of sporangia found on the lower surface of fern leaves. *567*

species A group of similarly constructed organisms capable of interbreeding and producing fertile offspring; organisms that share a common gene pool. *529*

spermatogenesis (sper″mah-to-jen´ĕ-sis) Production of sperm in males by the process of meiosis and maturation. *116*

sphincter A muscle that surrounds a tube and closes or opens the tube by contracting and relaxing. *176*

spicule (spik´ūl) Needle-shaped structure produced by some sponges that functions as a supportive inner skeleton. *586*

spinal nerve A nerve that arises from the spinal cord. *287*

spindle The structure that brings about the movement of chromosomes during cell division; microtubular fibers stretch between poles, which are surrounded by microtubular asters. *110*

spindle fiber Microtubule bundle in eukaryotic cells that is involved in the movement of chromosomes during mitosis and meiosis. *111*

spiracle Respiratory opening in arthropods. *599*

sponge A member of the phylum Porifera. *585*

spongy bone Porous bone found at the ends of long bones. *160*

sporangium (spo-ran´-je-um) A structure within which spores are produced. *555*

spore A haploid reproductive cell produced by the diploid sporophyte of a plant; asexually gives rise to the haploid gametophyte. *146, 549, 563*

sporophyte The diploid generation that produces spores in the life cycle of a plant. *563*

sporozoan Nonmotile parasitic protozoans. *548*

squamous epithelium Flat cells found lining the lungs and blood vessels, for example. *156*

stabilizing selection Effect of natural selection that eliminates atypical phenotypes. *526*

stamen Part of a flower, composed of filament and anther, where pollen grains are produced. *146*

starch The storage polysaccharide found in plants that is composed of glucose molecules joined in a linear fashion. *32*

stereoscopic vision The product of 2 eyes and both cerebral hemispheres functioning together to allow depth perception. *331*

sternum The breastbone, to which the ribs are ventrally attached. *199, 304*

steroid A lipid-soluble, biologically active molecule having 4 interlocking rings; examples are cholesterol, progesterone, and testosterone. *36*

stigma The uppermost part of a pistil. *146, 572*

stirrup Middle ear ossicle adhering to the oval window. *336*

stoma (pl., stomata) Opening in the leaves of plants through which gas exchange takes place. *134*

stretch receptor Muscle fiber that, upon stimulation, causes muscle spindles to increase the rate at which they fire. *323*

striated Having bands; cardiac and skeletal muscle are striated with bands of light and dark. *161*

stroma The fluid component of chloroplasts surrounding the grana. *85*

structural gene Gene that directs the synthesis of an enzyme or a structural protein in the cell. *476*

style The long slender part of the pistil. *146, 572*

subcutaneous layer (sub″ku-ta′ne-us la′er) A tissue layer found in vertebrate skin that lies just beneath the dermis and tends to contain adipose tissue. *164*

substrate A reactant in a reaction controlled by an enzyme. *76*

succession A series of ecological stages by which the community in a particular area gradually changes until there is a climax community that can maintain itself. *651*

sucrose A disaccharide composed of a glucose molecule and a fructose molecule bonded together; e.g., cane sugar. *32*

summation Ever-greater contraction of a muscle due to constant stimulation that does not allow complete relaxation to occur. *285*

suppressor T cell T lymphocyte that suppresses certain other T and B lymphocytes from continuing to divide and perform their respective functions. *227*

symbiosis (sim″bi-o′sis) An intimate association of 2 dissimilar species, including commensalism, mutualism, and parasitism. *653*

sympathetic nervous system That part of the autonomic nervous system that usually causes effects associated with emergency situations. *289*

synapse (sin′aps) The region between 2 nerve cells where the nerve impulse is transmitted from one to the other, usually from axon to dendrite. *284*

synapsis (sĭ-nap′sis) The attracting and pairing of homologous chromosomes during prophase I of meiosis. *113*

synaptic cleft Small gap between presynaptic and postsynaptic membranes. *284*

synovial joint A freely movable joint. *307*

synthesis To build up, such as the combining of 2 small molecules to form a larger molecule. *28*

systemic circuit That part of the circulatory system that serves body parts other than the gas-exchanging surfaces in the lungs. *206*

systole (sis′to-le) Contraction of a heart chamber. *201*

systolic blood pressure Arterial blood pressure during the systolic phase of the cardiac cycle. *204*

T

taiga (ti′gah) A biome that forms a worldwide northern belt of coniferous trees. *641*

tapeworm Parasitic flatworm; member of class Cestoda. *589*

tarsal A bone of the ankle in humans. *307*

taste bud Organ containing the receptors associated with the sense of taste. *325*

taxis (tak′sis) A movement in relation to a stimulus; phototaxis is movement toward or away from light. *625*

taxonomy The science of naming and classifying organisms. *9*

Tay-Sachs disease An inherited lysosomal storage disease that causes neurological impairment and death. *434*

tectorial membrane Membrane within the organ of Corti that transmits nerve impulses to the brain. *336*

telophase Stage of mitosis during which the diploid number of daughter chromosomes are located at each pole. *111*

template A pattern that serves as a mold for the production of an oppositely shaped structure; one strand of DNA is a template for a complementary strand. *467*

temporal lobe Area of the cerebrum responsible for hearing and smelling, as well as the interpretation of sensory experience and memory. *293*

tendon Dense fibrous connective tissue that joins muscle to bone. *159, 310*

testcross The backcross of a heterozygote with the recessive in order to determine the genotype. *426*

testis The male gonad, the organ that produces sperm and testosterone. *370*

testosterone The main male sex hormone responsible for development of primary and secondary sex characteristics in males. *372*

tetanus Sustained muscle contraction without relaxation. *311*

tetany Severe twitching caused by involuntary contraction of the skeletal muscles due to a lack of calcium. *353*

tetrad A set of 4 chromatids resulting from the pairing of homologous chromosomes during prophase I of meiosis. *353*

thalamus A mass of gray matter located at the base of the cerebrum in the wall of the third ventricle that receives sensory input. *292*

theory A concept supported by a large number of conclusions drawn by using the scientific method. *12*

thermal inversion Temperature inversion such that warm air traps cold air and its pollutants near the earth. *677*

thrombin An enzyme that converts fibrinogen to fibrin threads during blood clotting. *213*

thylakoid (thi′lah-koid) One of the flattened sacs within the grana of chloroplasts, the walls (thylakoid membranes) of which are sites of the light-dependent reactions of photosynthesis. *84*

thymus An organ that lies in the neck and thoracic area and is absolutely necessary to the development of immunity. *359*

thyroid-stimulating hormone *See* TSH.

thyroxin The hormone produced by the thyroid that speeds up the metabolic rate. *352*

tibia The shinbone found in the lower leg. *306*

tidal volume Amount of air normally moved in the human body during an inspiration or expiration. *247*

T lymphocyte A lymphocyte that matures in the thymus and occurs in 4 varieties, one of which kills antigen-bearing cells outright. *225*

tone The continuous partial contraction of muscle. Also, the quality of a sound. *312*

tonicity The degree to which the concentration of solute versus solvent causes fluids to move into or out of cells. *66*

trachea (tra´ke-ah) In vertebrates, a tube supported by C-shaped cartilaginous rings that lies between the larynx and the bronchi; also called the windpipe. Also, an air tube of insects. *243, 599*

tracheid (tra´ke-id) A component of xylem made of long, tapered nonliving cells. *139*

trait Specific term for a distinguishing phenotypic feature studied in heredity. *422*

transcription The process resulting in the production of a strand of mRNA that is complementary to a segment of DNA. *468*

transfer RNA (tRNA) Molecule of RNA that carries an amino acid to a ribosome engaged in the process of protein synthesis. *471*

transformed cell Cell that has been altered by bioengineering and is capable of producing new protein. *489*

transgenic organism Organism that has a foreign gene inserted into it. *494*

transition reaction A reaction within aerobic cellular respiration during which hydrogen atoms and carbon dioxide are removed from pyruvate; results in acetyl groups and connects glycolysis to the Krebs cycle. *90, 91*

translation The process by which the sequence of codons in mRNA dictates the sequence of amino acids in a polypeptide. *468*

translocation Chromosome mutation caused by the movement of a chromosome fragment from one chromosome to another, nonhomologous chromosome. *446*

transpiration The evaporation of water from a leaf; pulls water from the roots through a stem to leaves. *140*

trichinosis (trik″ĭ-no´sis) Disease caused by a roundworm in which the larvae are found in cysts within muscle cells. *590*

trichocyst Threadlike darts released by some ciliates that may help in defense against predators or in capturing prey. *547*

trochophore (tro´ko-fōr) A larval form unique to the protostomes that indicates they are related. *592*

trophic level (tro´fik lev´el) A categorization of species in a food web according to their feeding relationships from the first-level autotrophs through succeeding levels of herbivores and carnivores. *654*

trophoblast (tro´fə-blast) The outer membrane surrounding the human embryo; when thickened by a layer of mesoderm, it becomes the chorion, an extraembryonic membrane. *404*

tropical rain forest A biome of equatorial forests that remains warm year-round and receives abundant rain. *642*

trypsin A protein-digesting enzyme secreted by the pancreas. *182*

TSH (thyroid-stimulating hormone) Hormone that causes the thyroid to produce thyroxin. *351*

tube feet Rows of small tube-shaped appendages in echinoderms that are used in locomotion. *600*

tube-within-a-tube plan Body plan of animals that have 2 openings. *583*

tubular excretion The movement of certain molecules from blood into the distal convoluted tubule so that they are added to urine. *270*

tumor suppressor gene A gene that suppresses the development of a tumor; the mutated form contributes to the development of cancer. *482*

tundra A biome characterized by lack of trees due to cold temperatures and the presence of permafrost year-round. *640*

tunicate A type of invertebrate chordate in which only the larval stage has the 3 chordate characteristics; of these 3, only gills are found in the adult. *602*

turgor pressure Internal pressure that adds to the strength of the cell and builds up when water moves by osmosis into a cell. *67*

Turner syndrome A condition caused by the inheritance of an abnormality in chromosome number; an X chromosome lacks a homologous counterpart—XO. *450*

twitch A brief muscle contraction followed by relaxation. *311*

tympanic membrane (tim-pan´ik mem´brān) Membrane located between outer ear and the middle ear that receives sound waves; the eardrum. *335*

tympanum Sound receptor found in terrestrial animals, such as the grasshopper. *599*

typhlosole (tif´lo-sōl) A longitudinal fold in the intestine of annelids that enhances absorption capacity. *595*

U

ulcer An open sore in the wall of the digestive tract. *176*

ulna An elongated bone found within the lower arm. *306*

umbilical arteries and vein Fetal blood vessels that travel to and from the placenta. *410*

umbilical cord Cord connecting the fetus to the placenta through which blood vessels pass. *406*

urea (u-re´ah) Primary nitrogenous waste of mammals derived from amino acid breakdown. *262*

ureter (u´re´ter) One of 2 tubes that take urine from the kidneys to the urinary bladder. *264*

urethra (u´re´thrah) Tube that takes urine from the bladder to the outside. *264*

uric acid Waste product of nucleotide metabolism. *262*

urinalysis A medical procedure in which the composition of a patient's urine is determined. *274*

urinary bladder An organ where urine is stored before being discharged by way of the urethra. *264*

uterine cycle Monthly occurring changes in the characteristics of the uterine lining (endometrium). *379*

uterus The womb, the organ located in the female pelvis where the fetus develops. *377*

utricle (u´tre-k´l) Saclike cavity that makes up part of the membranous labyrinth of the inner ear; contains receptors for static equilibrium. *386*

V

vaccine Antigens prepared in such a way that they can promote active immunity without causing disease. *229*

vacuole A membranous cavity, usually filled with fluid. *51*

vagina Organ that leads from the uterus to the vestibule and serves as the birth canal and the organ of sexual intercourse in females. *378*

valve Membranous extension of a vessel or the heart wall that opens and closes, ensuring one-way flow; common to the systemic veins, the lymphatic veins, and the heart. *199*

vascular bundle Structure that includes xylem and phloem; typically found in herbaceous plant stems. *130*

vascular cambium A meristem that produces secondary phloem and secondary xylem, which add to the girth of a plant. *132*

vascular cylinder A central region of roots; contains vascular and other tissues. *129*

vascular plant A plant that has vascular tissue (xylem and phloem); includes ferns and seed plants. *566*

vas deferens Tube that leads from the epididymis to the urethra in males. *372*

vector A carrier, such as a plasmid or a virus, for recombinant DNA that introduces a foreign gene into a host cell. *487*

vein Vessel that takes blood to the heart from venules; characteristically having nonelastic walls. *198*

vena cava (ve´nah ka´vah) A large systemic vein that returns blood to the right atrium of the heart; either the superior or inferior vena cava. *201*

venous duct Fetal connection between the umbilical vein and the inferior vena cava; also called the ductus venosus. *410*

ventilation Breathing; the process of moving air into and out of the lungs. *244*

ventricle Cavity in an organ, such as a lower chamber of the heart (either the left ventricle or the right ventricle) or the ventricles of the brain. *200*

venule Vessel that takes blood from capillaries to a vein. *199*

vernix caseosa (ver´niks ka″se-o´sah) Cheeselike substance covering the skin of the fetus. *410*

vertebrae A bone of the vertebral column. *603*

vertebral column The backbone of vertebrates through which the spinal cord passes. *305*

vertebrate Animal possessing a backbone composed of vertebrae. *601*

vessel element An individual conducting cell in xylem. During development, vessel elements lose their contents and end walls so that they form a continuous vertical pipeline for transport of water and minerals. *139*

vestigial structure The remains of a structure that was functional in some ancestor but is no longer functional in the organism in question. *519*

villus (vil´us) Fingerlike projection from the wall of the small intestine that functions in absorption. *177*

visceral mass Soft-bodied portion of a mollusk that includes internal organs. *592*

vital capacity Maximum amount of air moved in or out of the human body with each breathing cycle. *247*

vitamin Essential requirement in the diet, needed in small amounts. They are often part of coenzymes. *79, 188*

vitreous humor (vit´re-us hu´ mor) The substance that fills the posterior chamber of the eye. *327*

vocal cord Fold of tissue within the larynx; creates vocal sounds when it vibrates. *243*

vulva The external genitals of the female that surround the opening of the vagina. *378*

W

water mold Protist, usually saprophytic, that has a threadlike body similar to fungi. *554*

water vascular system A series of canals that takes water to the tube feet of an echinoderm, allowing them to expand. *600*

white blood cell *See* leukocyte.

wood Secondary xylem of gymnosperms and angiosperms. *132*

X

X-linked gene Allele located on the X chromosome. *450*

xylem (zi´lem) The vascular tissue in plants that transports water and minerals. *127*

XYY male A male who has an extra Y chromosome. *450*

Y

yolk A rich nutrient material in the egg of certain vertebrate embryos. *394*

yolk sac The extraembryonic membrane that encloses yolk, except in most mammals. *402*

Z

zooflagellate (zo″o-flaj´ə-lāt) Protozoan that moves by means of flagella. *547*

zygote (zi´gōt) Diploid cell formed by the union of 2 gametes; the product of fertilization. *105*

CREDITS

Photographs

History of Biology

Leeuwenhoek: The Bettmann Archives; **Darwin:** The Bettmann Archives; **Pasteur:** The Bettmann Archives; **Koch:** The Bettmann Archives; **Pavlov:** The Bettmann Archives; **Lorenz:** The Bettmann Archives; **McClintock:** AP/Wide World Photos; **Franklin:** Cold Springs Harbor Laboratory; **Pauling:** The Bettmann Archives

Table of Contents and Part Openers

Part One: © Kjell B. Sanved; **Part Two:** © Cabisco/Visuals Unlimited; **Part Three:** © Leonard Lessin/Peter Arnold, Inc.; **Part Four:** © John Bavosi/Science Photo Library/Photo Researchers, Inc.; **Part Five:** © Petite Format/Nestle/Science Source/Photo Researchers, Inc.; **Part Six:** © David J. Cross/Peter Arnold, Inc.; **Part Seven:** © Patrick Martin-Vegue/Tom Stack & Associates

Chapter 1

Opener: © Kjell B. Sanved; **Figure 1.1** ***(top)***, ***(bottom right)*****:** © Mickey Gibson/Animals Animals; **1.1** ***(middle left*** **and** ***bottom left)*****:** © Michael Dick/OSF/Animals Animals; **1.1** ***(middle right)*****:** © Roland Seitre/Peter Arnold, Inc.; **1.1** ***(background)*****:** © Mickey Gibson/Earth Scenes; **1.3*a*:** © Joe McDonald/Visuals Unlimited; **1.3*b*:** © Dennis Schmidt/Valan Photos; **1.4:** © Runk/Schoenbuger/Grant Heilman, Inc.; **1.5*a*** ***(left to right)*****:** © Scott Camazinc/Photo Researchers, Inc., © John Cunningham/Visuals Unlimited, © Hermann Eisenbeiss/Photo Researchers, Inc., **1.5*b*** ***(all)*****:** © Dwight Kuhn; **1.7*a*:** © Carl Sams II/Peter Arnold, Inc.; **1.7*b*:** © Louis Goldman/Photo Researchers, Inc.; **1.A** ***(page 11)*****:** © Tom McHugh/Steinhurt Aquarium/Photo Researchers, Inc.; **1.9:** © Leonard Lessin/Peter Arnold, Inc.

Chapter 2

Opener: © Kjell B. Sanved; **2.9:** © Marty Cooper/Peter Arnold, Inc.; **2.18:** © Lawrence Migdale/Photo Researchers, Inc.; **2.21*b*:** © Jeremy Burgess/Science Photo Library/Photo Researchers, Inc.; **2.22*b*:** © Don Fawcett/Photo Researchers Inc.; **2.23*b*:** © Ulrike Welsch/Photo Researchers, Inc.

Chapter 3

Opener: © Dr. Gopal Murti/Science Photo Library/Photo Researchers, Inc.; **3.1*a*:** © BioPhoto Assoc./Photo Researchers, Inc.; **3.1*b*:** Courtesy of Stephen L. Wolfe; **3.1*c*:** © CNRI/SPL/Photo Researchers, Inc.; **3.1*d*:** © D. Wechsler/Earth Scenes; **3.3*a*:** © Don Fawcett/Photo Researchers, Inc.; **3.4*a*:** © Richard Rodewald/Biological Photo Service; **3.4*b*:** © W. P. Wergin, University of Wisconsin, Courtesy E. A. Newcomb/Biological Photo Service; **3.5:** © Warren Rosenberg, Iona College/Biological Photo Service; **3.6** ***(left)*****:** © David Phillips/Visuals Unlimited; **3.6** ***(right)*****:** © K. G. Nurti/Visuals Unlimited; **3.7:** Courtesy of Dr. Keith Porter; **3.8:** Courtesy of Herbert W. Israel, Cornell University; **3.10:** © Photo Researchers, Inc.; **3.11:** © BioPhoto Associates/Science Source/Photo Researchers, Inc.; **3.12*a*:** © David Phillips/Visuals Unlimited; **3.12*b*:** © BioPhoto Associates/Photo Researchers, Inc.

Chapter 4

Opener: © Tektoff-RM, CNRI/Science Photo Library/Photo Reserchers, Inc.; **4.A** ***(page 64)*****:** © Dr. Christine Jaeger/Cytotheraputics

Chapter 5

Opener: © Francis Leroy, Biocosmos/Science Photo Library/Photo Researchers, Inc.; **5.1** ***(left)*****:** © John Cunningham/Visuals Unlimited; **5.1** ***(right)*****:** © Sunset/Peter Arnold, Inc.

Chapter 6

Opener: © Bill Longcore/Photo Researchers, Inc.; **6.1*a*:** Courtesy of Herbert W. Israel, Cornell University; **6.6*a*:** Courtesy of Dr. Keith Porter

Chapter 7

Opener: © David M. Phillips/Visuals Unlimited; **7.1*b*:** © CNRI/Science Photo Library/Photo Researchers, Inc.; **7.5** ***(all)*****:** © Michael Abbey/Photo Researchers, Inc.; **7.6*a*:** © Francis Leroy, Biocosmos/Science Photo Library/Photo Researchers, Inc.; **7.7:** © David Phillips/Photo Researchers, Inc.; **7.8** ***(all)*****:** Courtesy of Dr. Andrew Bajer

Chapter 8

Opener: © Wolfgang Kaehler; **8.1** ***(top)*****:** © Philip Harris/Photo Researchers, Inc.; **8.1** ***(bottom)*****:** © John Cunningham/Visuals Unlimited; **8.3*b*:** Carolina Biological Supply Company; **8.5:** © John Cunningham/Visuals Unlimited; **8.6:** © J. R. Waaland, University of Washington/Biological Photo Service; **8.7*a*, 8.8*a*:** Carolina Biological Supply Company; **8.8*b*,*c*:** © Runk/Schoenberger/Grant Heilman; **8.10*b*:** Carolina Biological Supply Company; **8.10*c*:** © R. Gregory/Visuals Unlimited; **8.12:** © Dr. Jeremy Burgess/Science Photo Library/Photo Researchers, Inc.

Chapter 9

Opener: © Michael J. Johnson/Valan Photos; **9.1*b*:** © Biological Photo Service; **9.3** ***(both)*****:** © Dr. Jeremy Burgess/Science Photo Library/Photo Researchers, Inc.; **9.4:** © George Wilder/Visuals Unlimited; **9.A** ***(page 145)*****:** Courtesy of R. J. Weaver; **9.7:** Courtesy of Frank B. Salisbury; **9.10:** © Stanley Flegler/Visuals Unlimited; **9.11*a–c*:** © Michael Godfrey/Godfrey Bowes and Company.

Chapter 10

Opener: © Professor P. Motta/Dept. of Anatomy/University "La Sapienza," Rome/Science Photo Library/Photo Researchers, Inc.; **10.1** ***(all)*****:** © Ed Reschke/Peter Arnold, Inc.; **10.2** ***(left)*****:** © David Phillips/Visuals Unlimited; **10.2** ***(top right)*****:** © Don Fawcett/Visuals Unlimited; **10.2** ***(bottom right)*****:** © D. Allbertini/D. Fawcett/Visuals Unlimited; **10.3*a–d*, 10.5*a*,*b*:** © Edwin Reschke; **10.5*c*:** © Michael Abbey/Science Source/Photo Researchers, Inc.; **10.6:** © Edwin Reschke; **10.8** ***(left)*****:** © James Stevenson/Science Photo Library/Photo Researchers, Inc.; **10.8** ***(right)*****:** © Dr. P. Marazzi/Science Photo Library/Photo Researchers, Inc.; **10.8** ***(middle)*****:** © Ken Greer/Visuals Unlimited; **10.A** ***(page 165)*****:** © John Feingersh/Tom Stack & Associates

Chapter 11

Opener: © Secchi-Lecaque-Roussel-UCLAF/CNRI/Science Photo Library/Photo Researchers, Inc.; ***page 173*****:** © Ken Lucas/Biological Photo Service; **11.5*b*:** From R. G. Kessel & R. H. Kardon, *Tissues and Organs: A Text Atlas of Scanning Electron Microscopy*, © 1979 W.H. Freemann Company; **11.6** ***(right)*****:** © Ed Reschke/Peter Arnold, Inc.; **11.6** ***(bottom)*****:** © St. Bartholomew's Hospital/Science Photo Library/Photo Researchers, Inc.; **11.7:** © Manfred Kage/Peter Arnold, Inc.; **11.12:** Courtesy of Dr. Sheril D. Burton; **11.15:** © Robert Frerck/Odyssey Productions; **11.18*a*,*b*:** © BioPhoto Associates/Science Source/Photo Researchers, Inc.; **11.18*c*:** Courtesy of Centers for Disease Control, Alanta, Georgia; **11.18*d*:** © Ken Greer/Visuals Unlimited; **11.19:** © Michael P. Godomski/Photo Researchers, Inc.; **11.21:** © Alan Carey/The Image Works

Chapter 12

Opener: © Ed Reschke/Peter Arnold, Inc.; **12.12*a*:** © Lennart Nilsson: Behold Man; Little Brown and Company, Boston; **12.14:** © Boehringer International, GmbH, Courtesy Lennart Nilsson; **12.17*b*:** © NIBSC/Science Photo Library/Photo Researchers, Inc.; **12.A*a* (*page 215*):** © Lewis Lainey

Chapter 13

Opener: © David Scharf/Peter Arnold, Inc.; **13.1 (*top*):** © John Cunningham/Visuals Unlimited; **13.1 (*bottom*):** © Astrid & Hans-Frieder Michler/Science Photo Library/Photo Researchers, Inc.; **13.2*a–e*:** © John Cunningham/Visuals Unlimited; **13.6:** From J. Rini, et al., "Structural Evidence for Induced Fit as a Mechanism for Antibody Antigen Recognition," *Science,* 255: 959–965, Feb. 1992. © 1992 by the AAAS; **13.7 (*both*):** © Boehringer Ingelheim International/Lennart Nilsson; **13.8:** Courtesy of Dr. Kirk Ziegler; **13.9:** © Matt Meadows /Peter Arnold, Inc.; **13.11:** © Erika Stone/ Peter Arnold, Inc.; **13.12:** Courtesy of Schering-Plough Corporation; **13.13*a*:** Courtesy of Stuart I. Fox

Chapter 14

Opener: © CNRI/Science Photo Library/Photo Researchers, Inc.; **14.1:** © Hanson Carroll/Peter Arnold Inc.; **14.5:** © John Watney Photo Library; **14.A (*both, page 251*):** © Martin Rotker/Martin Rotker Photography

Chapter 15

Opener: © Manfred Kage/Peter Arnold, Inc.; **15.2:** © Dave Fleetham/Tom Stack & Associates; **15.3:** © Stephen J. Kraseman, The National Audubon Society Collection/Photo Researchers, Inc.; **15.7:** © CNRI/Science Photo Library/Photo Researchers, Inc.; **15.10*a*:** © J. Jennaro, Jr./Photo Researchers, Inc.; **15.A (*page 271*):** Courtesy of the Children's Hospital of Pittsburgh. Photo by Ed Eckman

Chapter 16

Opener: CNRI/Science Photo Library/Photo Researchers, Inc.; **16.1:** © David Madison/Bruce Coleman, Inc.; **16.3*b*:** © J. D. Robertson; **16.4:** Courtesy of Linda Bartlett; **16.15:** © Tim Davis/Photo Researchers, Inc.; **16.20:** © Ray Pfortner/Peter Arnold, Inc.;
16.22: © Scott Camazine/Photo Researchers, Inc.; **16.23*a*:** © Frerck/Odyssey Productions; **16.23*b*:** © Ogden Gigli/Photo Researchers, Inc.

Chapter 17

Opener: © CNRI/Science Photo Library/Photo Researchers, Inc; **p. 305:** © Joe McDonald/Bruce Coleman, Inc.; **17.A (*page 313*):** Courtesy of Alan D. Levenson; **17.6:** © Robert Brons/Biological Photo Service; **17.8*a*:** Courtesy of International Biomedical, Inc.; **17.10*b*:** Courtesy of H. E. Huxley; **17.12 (*top*):** © Eric Grave/Photo Researchers, Inc.; **17.12 (*bottom*):** © Victor Eichler

Chapter 18

Opener: © Omikron/Photo Researchers, Inc.; **18.1:** © Kaiser Porcelain, Ltd.; **18.9:** © Lennart Nilsson/ The Incredible Machine; **18.15:** © Motta/Dept of Anatomy/University "La Sapienza," Rome/Science/ Photo Library/Photo Researchers, Inc.; **18.A (*both, page 339*):** Courtesy of Professor J. E. Hawkins and Robert S. Preston, Kresage

Chapter 19

Opener: © Sidney Moulds/Science Photo Library/Photo Researchers, Inc.; **19.6:** © Bettina Cirone/Photo Researchers, Inc.; **19.7*a, b*:** © Dr. Charles A. Blake/The Pituitary Gland/Carolina Reader #118, (*page 11*); **19.9:** © BioPhoto Associates/Photo Researchers, Inc.; **19.10:** Courtesy of F. A. Davis Company, Philadelphia and Dr. R. H. Kampmeier; **19.11:** From Arthur Grollman, *Clinical Endocrinology and Its Physiological Basis,* © 1964 J. B. Lippincott Company; **19.12:** © Lester V. Bergman & Associates, Inc.; **19.15:** From Arthur Grollman, *Clinical Endocrinology and Its Physiological Basis,* © 1964 J.B. Lippincott Company; **19.16:** © Science VU/Visuals Unlimited; **19.17:** © Astrid & Hans-Frieder Michler/Science Photo Library/ Photo Researchers, Inc.

Chapter 20

Opener: © David M. Phillips/Photo Researchers, Inc.; **20.2:** © BioPhoto Associates/Photo Researchers, Inc.; **20.6:** © Ed Reschke/Peter Arnold, Inc.; **20.7 (*background*):** © David M. Phillips/Photo Researchers, Inc.; **20.7 (*inset*):** © David M. Phillips/ Visuals Unlimited; **20.12*a*:** © Ray Ellis/Photo Researchers, Inc.; **20.12*b*:** © Bob Coyle; **20.12*c, d*:** © Ray Ellis/Photo Researchers, Inc.; **20.12*e, f*:** © Bob Coyle; **20.13:** © M. Long/Visuals Unlimited; **20.16*a*:** © C. Lighdale, M.D./Photo Researchers, Inc.; **20.16*b*:** © Robert Settineri/Sierra Production; **20.17:** © CNRI/Science Photo Library/ Photo Researchers, Inc.

Chapter 21

Opener: © Dwight Kuhn; **21.6*a*:** © Lennart Nilsson; **21.11*a, b*:** © Petit Format/Nestle/Science Source/Photo Researchers, Inc.; **21.20:** © Joseph Nettis/Photo Researchers, Inc.; **21.22:** © Martin R. Jones/Unicorn Stock Photos

Chapter 22

Opener: © Erica Stone/Peter Arnold, Inc.; **22.1:** © Will & Deni McIntyre/Photo Researchers, Inc.; **22.3*a, b*:** © Bob Coyle; **22.10:** Courtesy The Cystic Fibrosis Foundation; **22.13 (*both*):** Courtesy of Steve Uzzell; **22.14*a*:** © Daemmrich/The Image Works; **22.15:** © London Express; **22.17*b*:** © Bill Longcore/ Photo Researchers, Inc.

Chapter 23

Opener: © Alfred Pasieka/Peter Arnold, Inc.; **23.2*a*:** © Jill Cannafax/EKM–Nepenthe; **23.4 (*both*):** Courtesy of Dr. Jerome Lejune, Universite de Paris, France; **23.6:** From C. J. Harrison, E. M. Jack, T. D. Allen, and R. Harris, "The Fragile X: A Scanning Electron Microscope Study," *Journal of Medical Genetics* 20:280–285, 1983.; **23.A (*page 448*):** © George Wilder/Visuals Unlimited; **23.8*a*:** © F. A. Davis Company, Philadelphia and R. H. Kampmeier; **23.8*b*:** From M. Bartalos & T. A. Baramski: MEDICAL CYTO-GENETICS, Copyright 1967 Williams and Wilkins Company

Chapter 24

Opener: © Secchi-Lecaque-Roussel-UCLAF/CNRI/ Science Photo Library/Photo Researchers, Inc.; **24.A (*page 466*):** Courtesy of Biophysics Department, Kings College, London; **24.B (*page 466*):** From "The Double Helix" by James D. Watson, Atheneum Press, New York, 1968. Courtesy of Cold Spring Harbor Laboratory Archives; **24.12*b*:** Courtesy of Alexander Rich; **24.16*a*:** © A. L. Olins, University of Tennessee/ Biological Photo Service; **24.17 (*both*):** © Bill Longcore/Photo Researchers, Inc.; **24.20:** © Bob Coyle

Chapter 25

Opener: © David M. Dennis/Tom Stack & Associates; **25.1*a*:** Courtesy of Genentech, Inc.; **25.1*b*:** © Hank Morgan/Photo Researchers, Inc.; **25.1*c, d*:** Courtesy of Genentech, Inc.; **25.4*a*:** © Sanofi Recherche; **25.4*b*:** © Applied Biosystems/Peter Arnold, Inc.; **25.4*c*:** © Petit Format/Nestle/Science Source/Photo Researchers, Inc. **25.7 (*both*):** General Electric Research and Development Center; **25.8*a-d*:** © Runk Schoenberger/Grant Heilman; **25.9:** Courtesy of Monsanto Company

Chapter 26

Opener: © Gerald & Buff Corsi/Tom Stack & Associates; **26.4*a*:** © Science VU/Sidney Fox/Visuals Unlimited; **26.4*b*:** Courtesy of David W. Deamer; **26.5*a*:** © S. M. Awramik/University of California/ Biological Photo Service; **26.5*b*:** © Sinclair Stammers/ Science Photo Library/Photo Researchers, Inc.; **26.5*c*:** © Francois Gohier/Photo Researchers, Inc.; **26.A (*page 508*):** Venrick/Reid, Scripps Institute of Oceanography

Chapter 27

Opener: © W. H. Hodge/Peter Arnold, Inc.; **27.1*a*:** © John Cunningham/Visuals Unlimited; **27.1*b*:** © American Museum of Natural History; **27.4*a*:** © BioPhoto Associates/Photo Researchers, Inc.; **27.4*b*:** © Eric Grave/Photo Researchers, Inc.; **27.6 (*left*):** © J. A. Bishop & L. M. Cook; **27.6 (*right*):** © Michael Tweedie/Photo Researchers, Inc.; **27.8:** Courtesy of Dr. McKusick; **27.10:** © Oxford Scientific Films/Animals Animals

Chapter 28

Opener: © CNRI/Science Photo Library/Photo Researchers, Inc.; **28.1*a*:** © Science Source/Photo Researchers, Inc.; **28.1*b* (*top*):** © NIH/R. Feldman/ Visuals Unlimited; **28.1*b* (*bottom*):** © Robert Caughey/ Visuals Unlimited; **28.1*c*:** © Michael Wurtz/ Biozentrum, University of Basel/Science Photo Library/ Photo Researchers, Inc.; **28.1*d*:** © K. G. Murti/Visuals Unlimited; **28.2:** © Dr. Ed Degginger/Color-Pic, Inc.; **28.4:** © George V. Kelvin/Scientific American, Cover, January, 1987; **28.A (*page 541*):** © Pfizer, Inc.; **28.5*a*:** © Francis Leroy; **28.5*b*:** © D. M. Phillips/Visuals Unlimited; **28.5*c*:** © John Cunningham/Visuals Unlimited; **28.6 (*all*):** © Biological Photo Service; **28.7:** © T. J. Beveridge/University of Guelph/Biological Photo Service; **28.8*a*:** © R. Knauft/Biology Media/ Photo Researchers, Inc.; **28.8*b*:** © Eric Grave/Science Source/Photo Researchers, Inc.; **28.9:** © R. R. Hessler/ Scripps Institution of Oceanography; **28.12:** © G. W. Willis/Biological Photo Service; **28.16:** © R. Knauft/ Photo Researchers, Inc.; **28.17:** © M. I. Walker/Science Source/Photo Researchers, Inc.; **28.18:** © John Cunningham/Visuals Unlimited; **28.21*a*:** © David M. Phillips/Visuals Unlimited; **28.21*b*:** © BioPhoto Associates/Photo Researchers, Inc.; **28.23*a*:** © Bob Coyle; **28.23*b*:** © Kitty Kohout/Root Resources; **28.23*c*:** © Bob Coyle; **28.24*b*:** © W. J. Weber/Visuals Unlimited; **28.25:** © Gordon Leedale/BioPhoto Associates

Chapter 29

Opener: © Kjell B. Sanved; **29.1*a*:** © Leonard Lee Rue III/Bruce Coleman, Inc.; **29.1*b*:** © John Shaw/Tom Stack & Associates; **29.1*c*:** © William F. Ferguson; **29.1*d*:** © Kent Dannen/Photo Researchers, Inc.; **29.3:** © Ed Reschke/Peter Arnold, Inc.; **29.4:** © David M. Dennis/Tom Stack & Associates; **29.5*a*:** Field Museum of Natural History, Chicago (Geo 75400C); **29.5*b*:** © David S. Addison/Visuals Unlimited; **29.6*c*:** © Barbara J. O'Donnell/Biological Photo Service; **29.6*d*:** © W. H. Hodge/Peter Arnold, Inc.; **29.7, 29.8*a,b*:** © Wolfgang Kaehler; **29.10*a*:** © Jack D. Swenson/Tom Stack & Associates; **29.10*b*:** © John D. Cunningham/Visuals Unlimited; **29.12:** USDA Forest Service; **29.13*a*:** © H. Eisenbeiss/Photo Researchers, Inc.; **29.13*b*:** © Jeremy Burgess/Science Photo Library/ Photo Researchers, Inc.; **29.13*c*:** © Anthony Mecieca/ Photo Researchers; **29.16*a*:** © W. H. Hodge/Peter Arnold, Inc.; **29.16*b,c*:** © Biological Photo Service; **29.16*d*:** © Tom & Pat Leeson/Photo Researchers, Inc.

Chapter 30

Opener: © Ken Highfill/Photo Researchers, Inc.; **30.5:** © M. Graybill/J. Hodder/Biological Photo Service; **30.6:** © Ron and Valerie Taylor/Bruce Coleman, Inc.; **30.7:** © Michael DiSpezio; **30.8:** Courtesy of Dr. Fred Whittaker; **30.9 *(top)*:** © Arthur Siegelman/Visuals Unlimited; **30.9 *(bottom)*:** © James Solliday/Biological Photo Service; **30.11*a*:** © BioPhoto Associates; **30.11*b*:** © Fred Bavendam/Peter Arnold, Inc.; **30.11*c*:** © Rick Poley; **30.13 *(left)*:** © Michael DiSpezio; **30.15*a*:** © Robert Evans/Peter Arnold, Inc.; **30.15*b*:** © Rob and Melissa Simpson/Valan Photos; **30.15*c*:** © Dwight Kuhn; **30.15*d*:** © John MacGregor/Peter Arnold, Inc.; **30.15*e and f*:** John Fowler/Valan Photos; **30.18 *(top left)*:** © Brian Parker/Tom Stack & Associates; **30.18 *(bottom right)*, 30.19*a*:** © Michael DiSpezio; **30.19*b*:** © Robert Brons/Biological Photo Service; **30.21*a*:** © Douglas Faulkner/Sally Faulkner Collection; **30.21*b*:** © Leonard Lee Rue III; **30.22*a*:** © Bruce Davidson/Animals Animals; **30.23:** © P. R. Ehrlich/Stanford University/Biological Photo Service; **30.24*a*:** © Tom McHugh/Photo Researchers, Inc.; **30.24*b*:** © Dallas Heaton/Uniphoto; **30.24*c*:** © Leonard Lee Rue; **30.26*a*:** © Martha Reeves/Photo Researchers, Inc.; **30.26*b*:** © Tom McHugh/Photo Researchers, Inc.; **30.26*c*:** © George Holton/Photo Researchers, Inc.; **30.26*d*:** © Tom McHugh/Photo Researchers, Inc.; **30.29*a*:** © Science VU/Visuals Unlimited; **30.29*b*:** © T. White; **30.31:** Courtesy Department of Library Service/American Museum of Natural History

Chapter 31

Opener: © Wolfgang Kaehler; **31.1*a–d*:** Courtesy of Dr. Rae Silver; **31.5*a*:** © Steve Kaufman/Peter Arnold, Inc.; **31.5*b*:** Nina Leen, Life Magazine © 1964 Time, Inc.; **31.7*a*:** © National Audubon Society/Photo Researchers, Inc.; **31.7*b*:** © Z. Leszczynski/Animals Animals;
31.8: © Fred Bruemmer/Peter Arnold, Inc.;
31.10: © Susan Kuklin/Photo Researchers, Inc.

Chapter 32

Opener: © Wolfgang Kaehler; **32.1*a*:** © Peter Kaplan/Photo Researchers, Inc.; **32.1*b*:** © Frank Miller/Photo Researchers, Inc.; **32.1*c*:** © Pat O'Hara; **32.3:** © John Shaw/Bruce Coleman, Inc.; **32.4*a*:** © Joe McDonald/Tom Stack & Associates; **32.4*b*:** © Jim & Julie Bruton/Photo Researchers, Inc.; **32.5:** © Bob Pool/Tom Stack & Associates; **32.A*a* *(page 642)*:** © Stephen Dalton/Photo Researchers, Inc.; **32.A*b*:** © Edwin and Peggy Bauer/Bruce Coleman, Inc.; **32.A*c*:** © Bruce Coleman/Bruce Coleman, Inc.

Chapter 33

Opener: © Dwight Kuhn; **33.1*a*:** © Stephen Kraseman/Peter Arnold, Inc.; **33.1*b*:** © Richard E. Ferguson; **33.1*c*:** © Mary Thacher/Photo Researchers, Inc.; **33.2 *(background)*:** © B. Milley/Biological Photo Service; **33.A *(page 653)*:** © Jeffrey L. Rotman/Peter Arnold, Inc.; **33.9:** © Jacques Jangoux/Peter Arnold, Inc.

Chapter 34

Opener: © John Cancalosi/Tom Stack & Associates; **34.1*a*:** © Helga Lade/Peter Arnold, Inc.; **34.1*b*:** © Wolfgang Kaehler; **34.5:** © Bob Coyle; **34.6*a*:** © Sydney Thomson/Animals Animals/Earth Scenes; **34.6*b*:** © Nichols/Magnum; **34.6*c*:** © G. Prance/Visuals Unlimited; **34.7:** © Thomas Kitchin/Tom Stack & Associates; **34.11*a,b*:** Courtesy of Dr. John Skelly; **34.13:** NASA; **34.14:** © Wolfgang Kaehler

Line Art

Chapter 1

1.10: Copyright © Mark Lefkowitz.

Chapter 11

11.3: From Kent M. Van De Graaff, *Human Anatomy*, 3d ed. Copyright © 1992 Wm. C. Brown Communications, Inc., Dubuque, Iowa. All Rights Reserved. Reprinted by permission.

Chapter 12

12.2, 12.3: Copyright © Mark Lefkowitz; **12.8:** From Stuart Ira Fox, *Human Physiology*, 4th ed. Copyright © 1993 Wm. C. Brown Communications, Inc., Dubuque, Iowa. All Rights Reserved. Reprinted by permission; **page 218 *(first column)*:** Copyright © Mark Lefkowitz.

Chapter 14

14.10: From John W. Hole, Jr., *Human Anatomy and Physiology*, 5th ed. Copyright © 1990 Wm. C. Brown Communications, Inc., Dubuque, Iowa. All Rights Reserved. Reprinted by permission; **reading, page 252:** Reproduced by permission of the American Cancer Society, Inc.

Chapter 15

15.4: From Kent M. Van De Graaff, *Human Anatomy*, 2d ed. Copyright © 1988 Wm. C. Brown Communications, Inc., Dubuque, Iowa. All Rights Reserved. Reprinted by permission; **15.5, 15.8 *(art)*:** From Kent M. Van De Graaff and Stuart Ira Fox, *Concepts of Human Anatomy and Physiology*, 3d ed. Copyright © 1992 Wm. C. Brown Communications, Inc., Dubuque, Iowa. All Rights Reserved. Reprinted by permission; **15.B *(page 271)*:** Courtesy of United Network for Organ Sharing (UNOS), Richmond, VA; ***page 276*:** From Kent M. Van De Graaff and Stuart Ira Fox, *Concepts of Human Anatomy and Physiology*, 3d ed. Copyright © 1992 Wm. C. Brown Communications, Inc., Dubuque, Iowa. All Rights Reserved. Reprinted by permission.

Chapter 16

16.1b: From Kent M. Van De Graaff, *Human Anatomy*, 3d ed. Copyright © 1992 Wm. C. Brown Communications, Inc., Dubuque, Iowa. All Rights Reserved. Reprinted by permission; **16.9; 16.12:** Copyright © Mark Lefkowitz; **16.16:** From John W. Hole, Jr., *Human Anatomy and Physiology*, 5th ed. Copyright © 1990 Wm. C. Brown Communications, Inc., Dubuque, Iowa. All Rights Reserved. Reprinted by permission.

Chapter 17

17.12c,d: From John W. Hole, Jr., *Human Anatomy and Physiology*, 5th ed. Copyright © 1990 Wm. C. Brown Communications, Inc., Dubuque, Iowa. All Rights Reserved. Reprinted by permission.

Chapter 18

18.8: From Joan G. Creager, *Human Anatomy and Physiology*, 2d ed. Copyright © 1992 Joan G. Creager. Reprinted by permission of Wm. C. Brown Communications, Inc., Dubuque, Iowa. All Rights Reserved. Reprinted by permission.

Chapter 19

19.17: From Kent M. Van De Graaff and Stuart Ira Fox, *Concepts of Human Anatomy and Physiology*, 3d ed. Copyright © 1992 Wm. C. Brown Communications, Inc., Dubuque, Iowa. All Rights Reserved. Reprinted by permission.

Chapter 20

20.2*c*, 20.3: From Kent M. Van De Graaff and Stuart Ira Fox, *Concepts of Human Anatomy and Physiology*, 3d ed. Copyright © 1992 Wm. C. Brown Communications, Inc., Dubuque, Iowa. All Rights Reserved. Reprinted by permission; **20.5:** From John W. Hole, Jr., *Human Anatomy and Physiology*, 5th ed. Copyright © 1990 Wm. C. Brown Communications, Inc., Dubuque, Iowa. All Rights Reserved. Reprinted by permission; **20.11:** From Kent M. Van De Graaff and Stuart Ira Fox, *Concepts of Human Anatomy and Physiology*, 3d ed. Copyright © 1992 Wm. C. Brown Communications, Inc., Dubuque, Iowa. All Rights Reserved. Reprinted by permission.

Chapter 21

21.4, 21.18: From John W. Hole, Jr., *Human Anatomy and Physiology*, 5th ed. Copyright © 1990 Wm. C. Brown Communications, Inc., Dubuque, Iowa. All Rights Reserved. Reprinted by permission; **21.20:** From Kent M. Van De Graaff and Stuart Ira Fox, *Concepts of Human Anatomy and Physiology*, 3d ed. Copyright © 1992 Wm. C. Brown Communications, Inc., Dubuque, Iowa. All Rights Reserved. Reprinted by permission.

Chapter 23

23.1: From Robert F. Weaver and Philip W. Hedrick, *Genetics*, 2d ed. Copyright © 1992 Wm. C. Brown Communications, Inc., Dubuque, Iowa. All Rights Reserved. Reprinted by permission; **23.12:** From E. Peter Volpe, *Biology and Human Concerns*, 3d ed. Copyright © 1983 Wm. C. Brown Communications, Inc., Dubuque, Iowa. All Rights Reserved. Reprinted by permission.

Chapter 24

24.16: Copyright © Mark Lefkowitz.

Chapter 28

28.16*b*: From *Kendall/Hunt Plate Series*, 1974. Copyright © 1974 Kendall/Hunt Publishing Company, Dubuque, Iowa. Reprinted by permission.

Chapter 31

31.4*b*: From J. P. Hailman, *Behaviour Supplement #15*. Copyright © E. J. Brill Publisher, Leiden, Holland.

Illustrators

Molly Babich

2.21*a*, 2.22*a*, 2.23*a*, 3.3*b*, 3.5*b*, 3.6*b*, 5.1*b*, 6.1*b,c*, 6.6*b*, 7.6*b*, 7.7*b*, 8.12*a*, 9.1*a,c*, 9.3*c,d*, 9.4*b*, 10.1*e*, 10.2*b*, 11.7*a,c*, 13.1*a*, 13.6*a,b*, 13.8*b*, 14.5*a*, 16.1*b*, 18.9*b*, 20.2*a,c*, 20.6, 21.4, 24.12*a*, 26.7, 27.6*c*, 27.10*a*, 27.11, 28.1 (*top left, middle*), 28.4*a*, 28.6*b*, 28.16*b*, 28.17*a*, 28.20, 29.3*a*, 29.4*a*, 29.7*a*, 30.5*b*, 30.6*a*, 30.7*b*, 30.8*b*, 30.9 (*middle*), 30.13*b*, 30.18*c*, 30.19 (*top right, bottom left*), 30.20, 30.22*b*, 30.23*b*, 30.25, 30.27; box art, page 173

Todd Buck

3.9, 6.4, 6.7, 6.10, 21.18; text art, page 99 (*bottom*)

Chris Creek

11.1, 14.2, 15.1, 15.5, 15.8*a*, 19.1; text art, pages 195, 255, 276

John Frieberg/Mary Albury-Noyes

18.8

Peg Gerrity

21.1, 22.A (page 423), 28.19, 30.16, 30.17, 30.30; box art, page 329

Anne Greene

18.5, 21.5*b*, 21.14; text art, page 341

Kathleen Hagelston
6.14, 7.A (page 118), 11.8, 16.3*a*, 20.14, 20.15, 24.10, 24.13, 24.15, 30.3, 30.12; text art, page 101; box art, page 240

Hagelston/Carlyn Iverson
12.11

Hagelston/Marjorie Leggitt
11.16, 15.11, 23.2*b*

Hagelston/Laurie O'Keefe
13.4, 13.7*c*, 23.14

Illustrious, Inc.
2.2, 2.25, 2.30, 2.31, 5.4, 11.11, 12.17*a*, 20.1*b*, 22.17*a*, 23.13, 23.15, 25.3, 26.2, 26.3, 27.5, 30.4, 31.11, 34.4, 34.10; text art, pages 28 (*right*), 72 (*right*), 79, 80 (*top left*), 81, 119 (*bottom left, right*), 137 (3), 150 (2), 168, 169, 194, 219, 235, 236, 244, 275, 300 (2), 318, 390, 399, 401, 419, 425, 428, 432, 455 (*bottom left, middle right*), 458 (*left*), 481, 483, 489, 494, 498 (2), 499 (*top*), 512 (2), 517, 521, 529, 533 (*bottom left, right*), 662, 663 (2), 680, 681

Carlyn Iverson
1.2, 2.14, 2.17, 2.27, 2.29, 3.2, 3.7*b*, 3.8*b*, 3.9, 3.11*a*, 4.4, 4.5, 4.9, 5.2, 5.5*b*, 6.2, 6.3, 6.5, 6.8, 6.9, 6.12, 6.13, 7.2, 7.3, 7.4, 7.5, 7.8, 7.9, 7.11, 7.12, 7.13, 8.2, 8.4, 8.10*a*, 9.6, 9.8, 9.9, 10.7, 10.10, 11.2, 11.5*a*, 11.9, 12.4, 12.13, 12.18, 13.5, 14.3, 14.4, 14.11, 14.12, 15.7*a*, 15.10*b*, 15.12, 15.13, 16.7, 16.14, 16.18, 16.19, 17.6, 18.A (*page 339, left*), 18.7, 18.11, 18.12, 18.13, 18.14, 18.15, 20.8, 20.9, 20.10, 21.2, 21.5*a*, 21.6*b*, 22.4, 22.5, 22.6, 22.7, 22.8, 22.16, 23.7, 24.1, 24.4, 24.5, 24.18, 25.5, 25.10, 26.1, 27.7, 27.9, 27.12, 27.13, 28.3, 28.10, 28.11, 28.13, 28.15, 29.9, 29.A (*page 575*), 29.11, 29.17, 30.1, 30.4, 30.28, 31.4*a,b*, 32.6, 33.6, 34.9; text art, pages 39, 57, 58, 63, 70, 120 (2), 151, 217, 254, 272 (*left*), 340, 364 (*left*), 440, 441, 458 (*right*), 499 (*bottom*), 533 (*top right*), 558, 559, 578, 617; box art, page 281

Iverson/Hagelston
5.6, 7.1*a*, 10.12, 11.4, 11.11, 12.1, 17.7, 18.2, 18.4

Mark Lefkowitz
1.10, 12.2, 16.10, 16.13, 24.16*b,c*, 33.8, 33.10; text art, page 618; box art, page 261

Lefkowitz/Hagelston
12.3

Robert Margulies
21.3

Steve Moon
20.11

Leonard Morgan/Illustrious, Inc.
7.14

Diane Nelson
12.A (*page 215, right*), 14.10

Laurie O'Keefe
10.4, 22.11*a,b*, 27.14

Precision Graphics
11.17, 11.20, 14.8, 16.21, 19.14, 22.14*b*, 23.10, 23.12

Rolin Graphics
1.6, 1.8, 1.11, 2.1, 2.3, 2.4, 2.5, 2.6, 2.7, 2.8, 2.A (*page 26*), 2.10, 2.11, 2.12, 2.13, 2.15, 2.16, 2.19, 2.20, 2.24, 2.26, 2.28, 3.4*a,b* (*top*), 3.13, 4.1, 4.7, 4.8, 4.10, 5.5*a*, 7.10, 7.15, 8.3*a*, 8.3*c*, 8.7*b*, 8.9, 9.2, 10.5*a,c* (*right*), 10.11, 11.13, 11.14, 12.10, 12.12*b*, 13.12*a*, 14.6, 14.7, 14.9, 16.4, 16.5, 17.1*a*, 18.3, 18.6, 19.2, 19.3, 19.4, 19.5, 19.8, 19.18, 19.19, 20.4, 21.A (*pages 408–409*), 21.7, 21.8, 21.14, 22.2, 23.3, 24.2, 24.17*b*, 25.2, 26.6, 27.2, 27.A (*page 525*), 28.14, 28.24*a*, 28.25*a*, 29.2, 29.5, 31.2, 31.6, 31.9, 31.A (*page 631*), 32.2*c*, 32.8, 32.9, 33.2*b*, 33.4, 33.5, 33.11, 34.3, 34.12*a,b*; text art, pages 29, 35, 72 (*left*), 75, 77 (2), 99, 151, 287, 364 (*right*), 415, 468 (*right*), 484, 526, 579

Rolin/Hagelston
3.10*a*, 3.12*a,b* (*top*), 4.2, 4.3, 4.6, 5.3, 5.7, 8.11, 9.5, 12.5, 12.9, 12.15, 13.14, 15.19, 17.2, 17.3, 17.4, 17.5, 17.9, 30.10; text art, page 476

Rolin/Iverson
8.1*c*, 12.16, 28.22, 29.15

Mike Schenk
11.3, 19.17*a*, 21.19

Tom Waldrop
16.8, 16.12, 17.12*c,d*, 19.A (*page 361*), 20.1*a*, 20.3, 20.5

Yvonne Walston
12.6, 13.3, 15.6, 16.11, 16.16

John Walters & Associates
12.8

Other artwork throughout the text was rendered by Brian Evans, Julie Janke, and Lisa Shoemaker

INDEX

How to Use This Index

1. A main entry has subentries listed below it. Page numbers next to the main entry give inclusive information about a general topic. Page numbers next to the subentry give specific information about a specific topic.
2. If you are looking for a topic with an acronym such as *AIDS,* try the acronym. Sometimes you will then be directed to the full name, in this case, *Acquired immune deficiency syndrome.*
3. If you are interested in a subtopic such as *function of cell membrane,* look under the entry *cell membrane* and follow the subentries until you come to *function of.*
4. If you are looking for a topic having a two-part name such as *chromosome number,* you can look under the word *chromosome* and follow down the column of subentries until you come to *number.* (If directed to do so, then look under the full name *chromosome number.)*
5. The systems (digestive, respiration, etc.) in the index pertain to humans. For systems in other animals, see the particular animal.

The entries in color are human issue topics. These are the topics that interest most students and show the relevancy of the text.

A

C

D

E

F

G

H

I

N

O

P

S

U

V

W

X

Y

Z

Robert Koch

Ivan Pavlov

Konrad Z. Lorenz

HISTORY OF BIOLOGY *CONTINUED*

Year	Name	Country	Contribution
1880	Walther Flemming	Germany	Studies the movement of chromosomes during mitosis.
1882	Robert Koch	Germany	Establishes the germ theory of disease and develops many techniques in bacteriology.
1888	Wilhelm Roux	Germany	Founds the science of embryology by performing experiments on embryos.
1892	August Weismann	Germany	Formulates the germ plasma theory, which states that only germ plasma is passed from generation to generation.
1897	Eduard Buchner	Germany	Extracts enzymes from yeast and uses them to bring about fermentation.
1900	Hugo De Vries Erich von Tschermak Karl Correns	Holland Austria Germany	Independently rediscover Mendel's laws.
1900	Walter Reed	United States	Discovers that yellow fever virus is transmitted by a mosquito.
1901	Hugo De Vries	Holland	States that mutations account for the presence of variations among members of a species.
1901	Santiago Ramón y Cajal	Spain	Suggests that neurons are separated by synapses.
1902	Walter S. Sutton Theodor Boveri	United States Germany	Suggests that genes are on the chromosomes, after noting the similar behavior of genes and chromosomes.
1902	William M. Bayliss E. H. Starling	Britain	Found the study of endocrinology by demonstrating the action of the hormone secretin.
1903	Karl Landsteiner	Austria	Discovers ABO blood types.
1904	Ivan Pavlov	Russia	Shows that conditioned reflexes affect behavior based on experiments with dogs.
1905	Paul Ehrlich	Germany	Discovers first antibiotic, named Salvarsan, to cure syphilis.
1910	Thomas H. Morgan	United States	States that each gene has a locus on a particular chromosome, based on experiments with *Drosophila*.
1914	Robert Feulgen	Germany	Devises a stain for DNA and shows that chromosomes contain DNA.
1922	Sir Frederick Banting Charles Best	Canada	Isolate insulin from the pancreas.
1924	Hans Spemann Hilde Mangold	Germany	Show that induction occurs during development, based on experiments with frog embryos.
1927	Hermann J. Muller	United States	Proves that X rays cause mutations.
1929	Sir Alexander Fleming	Britain	Discovers the toxic effect of a mold product he called penicillin on certain bacteria.
1931	Frederick Griffith	Britain	Discovers that nonvirulent bacteria can be transformed and become virulent if exposed to dead virulent bacteria within mice.
1932	Albert Szent-Györgyi von Nagyrapolt	Hungary	Isolates ascorbic acid from tissues and proves it is the anti-scurvy vitamin C.
1937	Konrad Z. Lorenz	Austria	Founds the study of ethology and shows the importance of imprinting as a form of early learning.
1937	Sir Hans A. Krebs	Britain	Discovers the reactions of a cycle that produces carbon dioxide during cellular respiration.
1940	George Beadle Edward Tatum	United States	Develop the one gene – one enzyme theory, based on red bread mold studies.